T0223992

Lecture Notes in Computer Science 9814

Commenced Publication in 1973
Founding and Former Series Editors:
Gerhard Goos, Juris Hartmanis, and Jan van Leeuwen

More information about this series at http://www.springer.com/series/7410

Matthew Robshaw · Jonathan Katz (Eds.)

Advances in Cryptology – CRYPTO 2016

36th Annual International Cryptology Conference
Santa Barbara, CA, USA, August 14–18, 2016
Proceedings, Part I

 Springer

Editors
Matthew Robshaw
Impinj, Inc.
Seattle, WA
USA

Jonathan Katz
University of Maryland
College Park, MD
USA

ISSN 0302-9743   ISSN 1611-3349   (electronic)
Lecture Notes in Computer Science
ISBN 978-3-662-53017-7   ISBN 978-3-662-53018-4   (eBook)
DOI 10.1007/978-3-662-53018-4

Library of Congress Control Number: 2016945783

LNCS Sublibrary: SL4 – Security and Cryptology

Printed on acid-free paper

This Springer imprint is published by Springer Nature
The registered company is Springer-Verlag GmbH Berlin Heidelberg

Preface

The 36th International Cryptology Conference (Crypto 2016) was held at UCSB, Santa Barbara, CA, USA, during August 14–18, 2016. The workshop was sponsored by the International Association for Cryptologic Research.

Crypto continues to grow. This year the Program Committee evaluated a record 274 submissions out of which 70 were chosen for inclusion in the program. Each paper was reviewed by at least three independent reviewers, with papers from Program Committee members receiving at least five reviews. Reviewers with potential conflicts of interest for specific papers were excluded from all discussions about those papers, and this policy was extended to the program chairs as well.

The 44 members of the Program Committee were aided in this complex and time-consuming task by many external reviewers. We would like to thank them all for their service, their expert opinions, and their spirited contributions to the review process. It was a tremendously difficult task to choose the program for this conference, as the quality of the submissions was very high. It was even harder to identify a single best paper, but our congratulations go to Elette Boyle, Niv Gilboa, and Yuval Ishai from IDC Herzliya, Ben Gurion University, and the Technion, respectively, whose paper "Breaking the Circuit Size Barrier for Secure Computation Under DDH" was awarded Best Paper. Our congratulations also go to Mark Zhandry of MIT and Princeton University who won the award for the Best Student Paper "The Magic of ELFs."

The invited speakers at Crypto 2016 were Brian Sniffen, Chief Security Architect at Akamai Technologies, Inc., and Paul Kocher, founder of Cryptography Research. Brian's presentation cast a fascinating light on the issues of real-world cryptographic deployment while Paul's presentation, a joint invitation from the program co-chairs of both Crypto 2016 and CHES 2016, marked 20 years since his publication of the first paper on side-channel attacks at Crypto 1996.

We are, of course, indebted to Brian LaMacchia, the general chair, as well as the local Organizing Committee, who together proved ideal liaisons for establishing the layout of the program and for supporting the speakers. Our job as program co-chairs was made much easier by the excellent tools developed by Shai Halevi; both Shai and Brian were always available at short notice to answer our queries. Finally, we would like to thank all the authors who submitted their work to Crypto 2016. Without you the conference would not exist.

August 2016

Matthew Robshaw
Jonathan Katz

Crypto 2016

The 36th IACR International Cryptology Conference

University of California, Santa Barbara, CA, USA
August 14–18, 2016

Sponsored by the *International Association for Cryptologic Research*

General Chair

Brian LaMacchia — Microsoft

Program Chairs

Matthew Robshaw — Impinj, USA
Jonathan Katz — University of Maryland, USA

Program Committee

Alex Biryukov — University of Luxembourg, Luxembourg
Anne Canteaut — Inria, France
Dario Catalano — Università di Catania, Italy
Nishanth Chandran — Microsoft Research, India
Melissa Chase — Microsoft Research, USA
Joan Daemen — STMicroelectronics, Belgium and Radboud University, The Netherlands
Martin Van Dijk — University of Connecticut, USA
Itai Dinur — Ben-Gurion University, Israel
Pierre-Alain Fouque — Université Rennes 1, France
Steven Galbraith — Auckland University, New Zealand
Sanjam Garg — University of California, Berkeley, USA
S. Dov Gordon — George Mason University, USA
Jens Groth — University College London, UK
Sorina Ionica — Université de Picardie, France
Tetsu Iwata — Nagoya University, Japan
Aggelos Kiayias — National and Kapodistrian University of Athens, Greece
Gregor Leander — Ruhr Universität Bochum, Germany
Shengli Liu — Shanghai Jiao Tong University, China
Alexander May — Ruhr Universität Bochum, Germany
Willi Meier — FHNW, Switzerland
Payman Mohassel — Visa Research, USA

Elke De Mulder	Cryptographic Research, France
Steven Myers	Indiana University, USA
Phong Nguyen	Inria, France and CNRS/JFLI and University of Tokyo, Japan
Kaisa Nyberg	Aalto University, Finland
Kenny Paterson	Royal Holloway University of London, UK
Thomas Peyrin	Nanyang Technological University, Singapore
Benny Pinkas	Bar-Ilan University, Israel
David Pointcheval	École Normale Supérieure, France
Manoj Prabhakaran	University of Illinois, USA
Bart Preneel	KU Leuven, Belgium
Mariana Raykova	Yale University, USA
Christian Rechberger	TU-Graz, Austria and DTU, Denmark
Mike Rosulek	Oregon State University, USA
Rei Safavi-Naini	University of Calgary, Canada
Alessandra Scafuro	Boston University and Northeastern University, USA
Patrick Schaumont	Virginia Tech, USA
Dominique Schröder	Saarland University, Germany
Jae Hong Seo	Myongji University, Korea
Yannick Seurin	ANSSI, France
Abhi Shelat	University of Virginia, USA
Nigel Smart	University of Bristol, UK
Ron Steinfeld	Monash University, Australia
Mehdi Tibouchi	NTT Secure Platform Laboratories, Japan

Additional Reviewers

Michel Abdalla
Masayuki Abe
Arash Afshar
Shashank Agrawal
Shweta Agrawal
Ayo Akinyele
Martin Albrecht
Gergely Alpar
Jacob Alperin-Sheriff
Elena Andreeva
Daniel Apon
Gilad Asharov
Gilles Van Assche
Nuttapong Attrapadung
Saikrishna
  Badrinarayanan
Josep Balasch

Foteini Baldimtsi
Paulo Barreto
Gilles Barthe
Lejla Batina
Christof Beierle
Mihir Bellare
Fabrice Benhamouda
Sanjay Bhattacherjee
Jean-Francois Biasse
Begul Bilgin
Gaetan Bisson
Nir Bitansky
Simon Blackburn
Olivier Blazy
Matthieu Bloch
Céline Blondeau
Andrej Bogdanov

Dan Boneh
Jonathan Bootle
Raphael Bost
Christina Boura
Florian Bourse
Cyril Bouvier
Elette Boyle
Zvika Brakerski
Lus Brandão
Anne Broadbent
Christina Brzuska
Christian Cachin
Ran Canetti
Angelo De Caro
Guilhem Castagnos
Andrea Cerulli
Pyrros Chaidos

André Chailloux
Jie Chen
Céline Chevalier
Chongwon Cho
Seung Geol Choi
Ashish Choudhury
Sherman Chow
Kai-Min Chung
Michele Ciampi
Michael Clear
Ran Cohen
Geoffroy Couteau
Dana Dachman-Soled
Deepesh Data
Jean Paul Degabriele
David Derler
Daniel Dinu
Christoph Dobraunig
Yevgeniy Dodis
Nico Döttling
Natnatee Dokmai
Leo Ducas
Tuyet Duong
Keita Emura
Frederic Ezerman
Pooya Farshim
Sebastian Faust
Dario Fiore
Marc Fischlin
Joe Fitzsimons
Nils Fleischhacker
Emmanuel Fouotsa
Georg Fuchsbauer
Eiichiro Fujisaki
Martin Gagne
François Le Gall
Chaya Ganesh
Juan Garay
Christina Garman
Romain Gay
Essam Ghadafi
Benedikt Gierlichs
Niv Gilboa
Vipul Goyal
Frédéric Grosshans
Aurore Guillevic

Divya Gupta
Felix Günther
Shai Halevi
Mike Hamburg
Shuai Han
Helena Handschuh
Christian Hanser
Carmit Hazay
Ethan Heilman
Ryan Henry
Gottfried Herold
Felix Heuer
Viet Tung Hoang
Dennis Hofheinz
Ziyuan Hu
Yan Huang
Michael Hutter
Malika Izabachene
Håkon Jacobsen
Mahavir Jhawar
Dingding Jia
Keting Jia
Thomas Johansson
Aaron Johnson
Kimmo Järvinen
Yael Tauman Kalai
Bhavana Kanukurthi
Petteri Kaski
Marcel Keller
Nathan Keller
Carmen Kempka
Iordanis Kerenidis
Dmitry Khovratovich
Dakshita Khurana
Eike Kiltz
Jinsu Kim
Taechan Kim
Paul Kirchner
Elena Kirshanova
Susumu Kiyoshima
Simon Knellwolf
Stefan Koelbl
Vlad Kolesnikov
Takeshi Koshiba
Luke Kowalczyk
Thorsten Kranz

Daniel Kraschewski
Anna Krasnova
Hugo Krawczyk
Fernando Krell
Stephan Krenn
Ranjit Kumaresan
Alptekin Kupcu
Fabien Laguillaumie
Virginie Lallemand
Enrique Larraia
Changmin Lee
Hyung Tae Lee
Kwangsu Lee
Nikos Leonardos
Tancrède Lepoint
Anthony Leverrier
Benoit Libert
Fuchun Lin
Rachel Lin
Yehuda Lindell
Feng-Hao Liu
Yi-Kai Liu
Patrick Longa
Steve Lu
Stefan Lucks
Atul Luykx
Anna Lysyanskaya
Lin Lyu
Vadim Lyubashevsky
Mohammad Mahmoody
Hemanta Maji
Giulio Malavolta
Tal Malkin
Alex Malozemoff
Mark Marson
Daniel Masny
Takahiro Matsuda
Florian Mendel
Bart Mennink
Thyla van der Merwe
Peihan Miao
Christof Michel
Ian Miers
Andrew Miller
Brice Minaud
Kazuhiko Minematsu

Ilya Mironov
Ameer Mohammad
Amir Moradi
Tal Moran
Nicky Mouha
Pratyay Mukherjee
Jörn Müller-Quade
Valérie Nachef
Michael Naehrig
Maria Naya-Plasencia
Soheil Nemati
Khoa Nguyen
Ivica Nikolic
Ventzi Nikov
Ryo Nishimaki
Anca Nitulescu
Adam O'Neill
Miyako Ohkubo
Go Ohtake
Tatsuaki Okamoto
Ozgur Oksuz
Cristina Onete
Claudio Orlandi
Elisabeth Oswald
Léo Paul Perrin
Jiaxin Pan
Giorgos Panagiotakos
Omkant Pandey
Kostas
    Pappagiannopoulos
Anat Paskin-Cherniavsky
Rafael Pass
Valerio Pastro
Arpita Patra
Souradyuti Paul
Christopher Peikert
Rene Peralta
Trevor Perrin
Giuseppe Persiano
Christophe Petit
Rafael Del Pino
Oxana Poburinnaya
Antigoni Polychroniadou
Orazio Puglisi
Baodong Qin
Max Rabkin

Carla Rafols
Srinivasan Raghuraman
Vanishree Rao
Manuel Reinert
Oscar Reparaz
Silas Richelson
Thomas Ristenpart
Damien Robert
Alon Rosen
Adeline Roux-Langlois
Arnab Roy
Tim Ruffing
Hansol Ryu
Sondre Rønjom
Akshayaram Srinivasan
Amin Sakzad
Katerina Samari
Ruediger Schack
Christian Schaffner
John Schanck
Thomas Schneider
Peter Scholl
Peter Schwabe
Sven Schäge
Adam Sealfon
Setareh Sharifian
Tom Shrimpton
Sandeep Shukla
Siang Meng Sim
Luisa Siniscalchi
Daniel Slamanig
Yongsoo Song
Kannan Srinathan
Akshayaram Srinivasan
Douglas Stebila
Damien Stehlé
John Steinberger
Marc Stevens
Valentin Suder
Willy Susilo
Björn Tackmann
Katsuyuki Takashima
Qiang Tang
Stefano Tessaro
Aishwarya
    Thiruvengadam

Jean-Pierre Tillich
Yosuke Todo
Yiannis Tselekounis
Michael Tunstall
Himanshu Tyagi
Aleksei Udovenko
Jon Ullman
Dominique Unruh
Prashant Vasudevan
Vesselin Velichkov
Muthu
    Venkitasubramaniam
Frederik Vercauteren
Damien Vergnaud
Jorge Villar
Dhinakaran
    Vinayagamurthy
Ivan Visconti
Michael Walter
Pengwei Wang
Qingju Wang
Xiao Wang
Hoeteck Wee
Mor Weiss
Yunhua Wen
Carolyn Whitnall
Daniel Wichs
Xiaodi Wu
Keita Xagawa
Sophia Yakoubov
Shota Yamada
Kan Yasuda
Arkady Yerukhimovich
Ouyang Yingkai
Thomas Zacharias
Mark Zhandry
Bingsheng Zhang
Liang Feng Zhang
Xiao Zhang
Yupeng Zhang
Hong-Sheng Zhou
Vassilis Zikas
Dionysis Zindros

Contents – Part I

Symmetric Primitives

Provable Security for Symmetric Cryptography

Key-Alternating Ciphers and Key-Length Extension: Exact Bounds and Multi-user Security

Viet Tung Hoang$^{(\boxtimes)}$ and Stefano Tessaro

Department of Computer Science, University of California Santa Barbara,
Santa Barbara, USA
tvhoang@engr.ucsb.edu, tessaro@cs.ucsb.edu

Abstract. The best existing bounds on the concrete security of key-alternating ciphers (Chen and Steinberger, EUROCRYPT '14) are only *asymptotically* tight, and the quantitative gap with the best existing attacks remains numerically substantial for concrete parameters. Here, we prove *exact* bounds on the security of key-alternating ciphers and extend them to XOR cascades, the most efficient construction for key-length extension. Our bounds essentially match, for any possible query regime, the advantage achieved by the best existing attack.

Our treatment also extends to the multi-user regime. We show that the multi-user security of key-alternating ciphers and XOR cascades is very close to the single-user case, i.e., given enough rounds, it does not substantially decrease as the number of users increases. On the way, we also provide the first explicit treatment of multi-user security for key-length extension, which is particularly relevant given the significant security loss of block ciphers (even if ideal) in the multi-user setting.

The common denominator behind our results are new techniques for information-theoretic indistinguishability proofs that both extend and refine existing proof techniques like the H-coefficient method.

Keywords: Symmetric cryptography · Block ciphers · Provable security · Tightness · Multi-user security

1 Introduction

Precise bounds on the security of symmetric constructions are essential in establishing when and whether these constructions are to be deployed. This paper revisits the question of proving *best-possible* security bounds for *key-alternating ciphers* and *key-length extension schemes*.

Our contribution is twofold. First, we prove *exact* bounds on the security of key-alternating ciphers and related methods for key-length extensions (i.e., XOR cascades) which essentially match what is achieved by the best-known attack. This is a substantial improvement over previous bounds, which are only *asymptotically* optimal. Second, we extend our treatment to the multi-user setting, where no non-trivial bounds are known to date for these constructions.

© International Association for Cryptologic Research 2016
M. Robshaw and J. Katz (Eds.): CRYPTO 2016, Part I, LNCS 9814, pp. 3–32, 2016.
DOI: 10.1007/978-3-662-53018-4_1

Our results are built on top of new *conceptual* insights in information-theoretic indistinguishability proofs, generalizing previous approaches such as the H-coefficient technique [9,24].

KEY-ALTERNATING CIPHERS. *Key-alternating ciphers* (KACs) generalize the Even-Mansour construction [13] over multiple rounds. They abstract the structure of AES, and this fact has made them the object of several recent analyses [1,7–9,11,25]. Given t permutations $\pi = (\pi_1, \ldots, \pi_t)$ on n-bit strings, as well as n-bit subkeys $L_0, L_1, \ldots, L_t$, the t-round KAC construction KAC$[\pi, t]$ outputs, on input M, the value

$$L_t \oplus \pi_t(L_{t-1} \oplus \pi_{t-1}(\cdots \pi_1(M \oplus L_0)\cdots)) . \tag{1}$$

Here, we are specifically interested in (strong) prp security of KAC$[\pi, t]$, i.e., its indistinguishability from a random permutation (under random secret sub-keys) for adversaries that can query both the construction and its inverse. Analyses here are in the random-permutation model: The permutations $\pi_1, \ldots, \pi_t$ are independent and random, and the distinguisher is given a budget of q on-line construction queries, and $p_1, \ldots, p_t$ queries to each of the permutations. The currently best bound is by Chen and Steinberger (CS) [9], who prove that the distinguishing advantage of any such distinguisher A satisfies (using $N = 2^n$ and $p_1 = \cdots = p_t = p$)

$$\mathsf{Adv}^{\pm\mathrm{prp}}_{\mathrm{KAC}[\pi,t]}(A) \leq (t+2)\left(\frac{q(6p)^t}{N^t} \cdot t^2(t+1)^{t+1}\right)^{1/(t+2)} . \tag{2}$$

Note that the best known distinguishing attack achieves advantage roughly qp^t/N^t. The bound from (2) is asymptotically "tight", i.e., the attacker needs to spend about $\Omega\left(N^{t/(t+1)}\right)$ queries for the bound to become constant, as in the attack. However, there is a substantial gap between the curve given by the bound and the advantage achieved by the best attack, and the constant hidden inside the Ω notation (which depends on t) is fairly significant.

EXACT BOUNDS FOR KACs. Our first contribution is a (near-)exact bound for KACs which matches the best-known attack (up to a small factor-four loss in the number of primitive queries necessary to achieve the same advantage). Concretely, we show that for A as above,

$$\mathsf{Adv}^{\pm\mathrm{prp}}_{\mathrm{KAC}[\pi,t]}(A) \leq \frac{q(4p)^t}{N^t} . \tag{3}$$

The core of our proof inherits some of the combinatorial tools from CS's proof. However, we use them in a different (and simpler) way to give a much sharper bound. We elaborate further at the end of this introduction. Clearly, our new bound substantially improves upon the CS bound from (2). For example, for realistic AES-like parameters ($n = 128$ and $t = 10$), and $q = p = 2^{110}$, the CS bound is already vacuous (indeed, the advantage starts becoming substantial at around 2^{100}), and in contrast, our new bound still gives us 2^{-50}. Another feature is that our bound does not make any assumptions on q and p — we can for example set $q = N$ and still infer security as long as p is sufficiently small. In contrast, the CS bound (and the technique behind it) assumes that $p, q \leq N/3$.

We note in passing that Lampe et al. [19] already proved a similar bound for the (simpler) case of a specific non-adaptive distinguisher. If one wants however to extend their bound to the adaptive case, a factor-two loss in the number of rounds becomes necessary.

MULTI-USER SECURITY. Similar to all prior works, the above results only consider a single user. Yet, block ciphers are typically deployed *en masse* and attackers are often satisfied with compromising *some* user among many. This can be substantially easier. For example, given multiple ciphertexts encrypted with a single k-bit key, a brute-force key-search attack takes effort roughly 2^k to succeed. However, if the ciphertexts are encrypted with u different keys, the effort is reduced to $2^k/u$. Overall we effectively lose $\log(u)$ bits of security, which can be substantial. Note that this loss is only inherent if exhaustive key-search *is* the best attack — it may be that a given design is subject to better degradation, and assessing what is true is crucial to fix concrete parameters.

The notion of multi-user (mu) security was introduced and formalized by Bellare et al. [2] in the context of public-key encryption. Unfortunately, until recently, research on *provable* mu security for block-cipher designs has been somewhat lacking, despite significant evidence of this being the right metric (cf. e.g. [6] for an overview). Recent notable exceptions are the works of Mouha and Luykx [22] and Tessaro [26]. The former, in particular, provided a tight analysis of the Even-Mansour cipher in the mu setting, and is a special case of our general analysis for $t = 1$.

MULTI-USER SECURITY FOR KACS. First recall that in the mu setting, the adversary makes q queries to multiple instances of $KAC[\pi, t]$ (and their inverses), each with an independent key (but all accessing the same π), and needs to distinguish these from the case where they are replaced by independent random permutations. The crucial point is that *we do not know* a per-instance upper bound on the number of the distinguisher queries, which are distributed adaptively across these instances. Thus, in the *worst-case*, at most q queries are made on some instance and by a naive hybrid argument,[1]

$$\mathsf{Adv}^{\pm\mathrm{mu\text{-}prp}}_{KAC[\pi,t]}(A) \leq \frac{u \cdot q(4(p+qt))^t}{N^t} \leq \frac{q^2(4(p+qt))^t}{N^t} \ , \tag{4}$$

where u is an upper bound on the number of different instances (or "users") for which A makes a query, which again can be at most q. Note that such additional multiplicative factor q is significant: e.g., for $t = 1$, it would enforce $q < N^{1/3}$.

---

[1] The increase from p to $p + qt$ is due to the fact that in the reduction to su prp security, the adversary needs to simulate queries to all but one of the instances with direct permutation queries.

As our second contribution, we show that this loss is not necessary, and that in fact essentially the same bound as in the single-user case holds, i.e.,

$$\mathsf{Adv}^{\pm\mathrm{mu\text{-}prp}}_{\mathrm{KAC}[\pi,t]}(A) \leq 2\frac{q(4(p+qt))^t}{N^t} \ . \tag{5}$$

To get a sense of why the statement holds true, note that we could prove this bound easily *if we knew* that the adversary makes at most q_i queries for the i-th user, and $q = \sum_i q_i$. In this case, the naive hybrid argument would yield the bound from (5), but we do not have such q_i's. Our proof relies on a "transcript-centric" hybrid argument, i.e., we use a hybrid argument to relate the real-world and ideal-world probabilities that the oracles of the security game behave according to a certain *a-priori fixed transcript*, for which the quantities q_i *are* defined. The fact that looking at these probabilities suffice will be at the core of our approach, discussed below.

KEY-LENGTH EXTENSION AND MULTI-USER SECURITY. A fundamental problem in symmetric cryptography, first considered in the design of "Triple-DES" (3DES), is that of building a cipher with a "long" key from one with a "short" key to mitigate the effects of exhaustive key search. Analyses of such schemes (in the ideal-cipher model) have received substantial attention [4,10,14–17,20], yet the practical relevance of these works is often put in question given existing designs have already sufficient security margins. However, *the question gains substantial relevance in the multi-user setting* – indeed, the mu PRP security of an ideal cipher with key length k is at most $2^{k/2}$, i.e., 64 bits for AES-128.

In this paper, we analyze XOR-cascades [14,20], which have been shown to give the best possible trade-off between number of rounds and achievable security. Given a block cipher E with k-bit keys and n-bit blocks, the t-round XOR cascade $\mathrm{XC}[E,t]$ uses sub-keys $J_1,\ldots,J_t,L_0,\ldots,L_t$, and on input M, outputs

$$L_t \oplus E_{J_t}(L_{t-1} \oplus E_{J_{t-1}}(\cdots E_{J_1}(M \oplus L_0)\cdots)) \ . \tag{6}$$

A connection between analyzing XC in the ideal-cipher model and KAC in the random permutation model was already noticed [14,15], but the resulting reduction is far from tight. Here, we give a tight reduction, and use our result on $\mathrm{KAC}[\pi,t]$ to show that for every adversary making q construction queries and at most p queries to an ideal cipher, *if the keys $J_1,\ldots,J_t$ are distinct,*

$$\mathsf{Adv}^{\pm\mathrm{prp}}_{\mathrm{XC}[E,t]}(A) \leq q\left(\frac{4p}{2^{k+n}}\right)^t \ . \tag{7}$$

Our bound does not make any assumption on q (which can be as high as 2^n) and p. If the keys are independent (and may collide), an additional term needs to be added to the bound — a naive analysis gives $t^2/2^k$, which is usually good enough, and this is what done in prior works. This becomes interesting when moving to the multi-user case. For the distinct-key case, we can apply our techniques to inherit the bound from (7) (replacing p with $p + q \cdot t$), noting that we are allowing keys to collide across multiple users, just same-user keys need to be distinct. An important feature of this bound (which is only possible thanks to the fact that we are not imposing any restrictions on query numbers in our original bound for $\mathrm{KAC}[\pi, t]$) is that it also gives guarantees when $q \gg 2^n$ and queries are necessarily spread across multiple users. This is particularly interesting when n is small (e.g., $n = 64$ for DES, or even smaller if E is a format-preserving encryption (FPE) scheme).

However, for the independent-key case, the naive analysis here gives us a term $ut^2/2^k$, where u is the number of users (and $u = q$ may hold). This term is unacceptably large – in particular, if $u = q \gg 2^n$. To this end, we significantly improve (in the single-user case already) the additive term $t^2/2^k$. In the multi-user setting, the resulting bound is going to be extremely close to the one for distinct keys, if $t \neq 3.^2$ We leave the question open of reducing the gap (or proving its necessity) for $t = 3$.

OUR TECHNIQUES. A substantial contribution of our work is conceptual. Section 3.1 below presents our tools in a general fashion, making them amenable to future re-use. We give an overview here.

All of our results rely on establishing a condition we call *point-wise proximity*: That is, we show that there exists an $\epsilon = \epsilon(q)$ such that for all possible transcripts τ describing a possible ideal- or real-world interaction (say with q queries), the probabilities $\mathsf{p}_0(\tau)$ and $\mathsf{p}_1(\tau)$ that the ideal and real systems, respectively, answer consistently with τ (when asked the queries in τ) satisfy

$$\mathsf{p}_0(\tau) - \mathsf{p}_1(\tau) \leq \epsilon \cdot \mathsf{p}_0(\tau) .$$

This directly implies that the distinguishing advantage of any q-query distinguisher is at most ϵ. This method was first used by Bernstein [5], and can be seen as a special case of Patarin's H-coefficient method [24] (recently revisited and re-popularized by Chen and Steinberger [9]) and Nandi's "interpolation method" [23], where we do not need to consider the possibility of some transcripts "being bad". It turns out that when we do not need such bad set, the notion becomes robust enough to easily allow for a number of arguments.

TRANSCRIPT-CENTRIC REDUCTIONS. Our first observation is that point-wise proximity makes a number of classical proof techniques *transcript-centric*, such as hybrid arguments and reductions. For example, assume that for a pair of systems with transcript probabilities p_0 and p_1, we have already established that $\mathsf{p}_0(\tau) - \mathsf{p}_1(\tau) \leq \epsilon \cdot \mathsf{p}_0(\tau)$. Now, to establish that for some other p_0' and p_1' we also have $\mathsf{p}_0'(\tau) - \mathsf{p}_1'(\tau) \leq \epsilon \cdot \mathsf{p}_0'(\tau)$, it is enough to exhibit a function φ, mapping transcripts into transcripts, such that

$$\frac{\mathsf{p}_1'(\tau)}{\mathsf{p}_0'(\tau)} = \frac{\mathsf{p}_1(\varphi(\tau))}{\mathsf{p}_0(\varphi(\tau))}$$

---

[2] We note that in practice, it is easy for a user to enforce that her t keys are distinct, making this part of the key sampling algorithm. Still, our bound shows that this is not really necessary for $t \neq 3$.

for every τ such that $\mathsf{p}_0'(\tau) > 0$. This is effectively a reduction, but the key point is that the reduction φ maps *executions* into *executions* (i.e., transcripts), and thus can exploit some global after-the-fact properties of this execution, such as the number of queries of a certain particular type. This technique will be central e.g. to transition (fairly generically) from single-user to multi-user security in a tight way. Indeed, while a hybrid argument does not give a tight reduction from single-user to multi-user security, such a reduction can be given when we have established the stronger property of single-user point-wise proximity.

THE EXPECTATION METHOD. Our main quantitative improvement over the CS bound is due to a generalization of the H-coefficient method that we call the *expectation method*.

To better understand what we do, we first note that through a fairly involved combinatorial analysis, the proof of the CS bound [9] gives (implicitly) an exact formula for the ratio $\epsilon(\tau) = 1 - \frac{\mathsf{p}_1(\tau)}{\mathsf{p}_0(\tau)}$ for every "good transcript" τ. The issue here is that $\epsilon(\tau)$ depends on the transcript τ, in particular, on numbers of paths of different types in a transcript-dependent graph $G = G(\tau)$. To obtain a sharp bound, CS enlarge the set of bad transcripts to include those where these path numbers excessively deviate from their expectations, and prove a unique bound $\epsilon^* \geq \epsilon(\tau)$ for all good transcripts. As these quantities do not admit overly strong concentration bounds, only Markov's inequality applies, and this results in excessive slackness. In particular, an additional parameter appears in the bound, allowing for a trade-off between the probability δ^* of τ being bad and the quality of the upper bound ϵ^*, and this parameter needs to be optimized to give the sharpest bound, which however still falls short of being exact.

The problem here is that the H-coefficient technique takes a worst-case approach, by unnecessarily requiring one *single* ϵ^* to give us an upper bound for *all* (good) transcripts. What we use here is that given a *transcript-dependent* $\epsilon = \epsilon(\tau)$ for which the above upper bound on the ratio holds, then one can simply replace ϵ^* in the final bound with the *expected value* of $\epsilon(\tau)$ for an *ideal-world* transcript τ. This expected value is typically fairly straightforward to compute, since the ideal-world distribution is very simple.

We in fact do even more than this, noticing that for KACs point-wise proximity can be established, and this will allow us to obtain many of the applications of this paper. In fact, once we do not need to enlarge the set of bad transcripts any more as in CS, we observe that every transcript is potentially good. Only in combination with the key (which is exposed as part of the transcript in CS) transcripts can be good or bad. We will actually apply the expectation method on every *fixed* transcript τ, the argument now being only over the choice of the random sub-keys $L_0, L_1, \ldots, L_t$ – this makes it even simpler.

A PERSPECTIVE. The above techniques are all fairly simple in retrospect, but they all indicate a conceptual departure from the standard "good versus bad" paradigm employed in information-theoretic indistinguishability proofs. CS already suggested that one can generalize their methods beyond a two-set partition, but in a way, what we are doing here is an extreme case of this, where every set in the partition is a singleton set.

It also seems that the issue of using Markov's inequality has seriously affected the issue of proving "exact bounds" (as opposed to asymptotically tight ones). Another example (which we also revisit) is the reduction of security of XOR cascades to that of KACs [14, 15].

2   Preliminaries

NOTATION. For a finite set S, we let $x \leftarrow\!\!{}_\$ S$ denote the uniform sampling from S and assigning the value to x. Let $|x|$ denote the length of the string x, and for $1 \leq i < j \leq |x|$, let $x[i, j]$ denote the substring from the ith bit to the jth bit (inclusive) of x. If A is an algorithm, we let $y \leftarrow A(x_1, \ldots; r)$ denote running A with randomness r on inputs $x_1, \ldots$ and assigning the output to y. We let $y \leftarrow\!\!{}_\$ A(x_1, \ldots)$ be the resulting of picking r at random and letting $y \leftarrow A(x_1, \ldots; r)$.

MULTI-USER PRP SECURITY OF BLOCKCIPHERS. Let $\Pi : \mathcal{K} \times \mathcal{M} \to \mathcal{M}$ be a blockcipher, which is built on a family of independent, random permutations $\pi : \text{Index} \times \text{Dom} \to \text{Dom}$. (Note that here Index could be a secret key, in this case π will model an ideal cipher, or just a small set of indices, in which case π models a (small) family of random permutations.) We associate with Π a key-sampling algorithm Sample. Let A be an adversary. Define

$$\mathsf{Adv}^{\pm\text{mu-prp}}_{\Pi[\pi], \text{Sample}}(A) = \Pr[\text{Real}^A_{\Pi[\pi], \text{Sample}} \Rightarrow 1] - \Pr[\text{Rand}^A_{\Pi[\pi], \text{Sample}} \Rightarrow 1]$$

where games Real and Rand are defined in Fig. 1. In these games, we first use Sample to sample keys $K_1, K_2, \ldots \in \mathcal{K}$ for Π, and independent, random permutations $f_1, f_2, \ldots$ on $\mathcal{M}$. The adversary is given four oracles PRIM, PRIMINV, ENC, and DEC. In both games, the oracles PRIM and PRIMINV always give access to the primitive π and its inverse respectively. The ENC and DEC oracles gives access to $f_1(\cdot), f_2(\cdot), \ldots$ and their inverses respectively in game Rand, and access to $\Pi[\pi](K_1, \cdot), \Pi[\pi](K_2, \cdot), \ldots$ and their inverses in game Real. The adversary finally needs to output a bit to tell which game it's interacting.

For the special case that and adversary A only queries $\text{PRIM}(\cdot), \text{ENC}(1, \cdot)$, and their inverses, we write $\mathsf{Adv}^{\pm\text{prp}}_{\Pi[\pi], \text{Sample}}(A)$ to denote the advantage of A.

If Sample is the uniform sampling of $\mathcal{K}$ then we only write $\mathsf{Adv}^{\pm\text{prp}}_{\Pi[\pi]}(A)$ and $\mathsf{Adv}^{\pm\text{mu-prp}}_{\Pi[\pi]}(A)$. If Π doesn't use π then $\mathsf{Adv}^{\pm\text{prp}}_{\Pi}(A)$ coincides with the conventional (strong) PRP advantage of A against Π.

3   Indistinguishability Proofs via Point-Wise Proximity

This section discusses techniques for information-theoretic indistinguishability proofs. A reader merely interested in our theorems can jump ahead to the next sections — the following tools are not needed to understand the actual statements, only their proofs.

proc INITIALIZE() $\text{Real}^A_{\Pi[\pi],\text{Sample}}$	proc INITIALIZE() $\text{Rand}^A_{\Pi[\pi],\text{Sample}}$
for $i = 1, 2, \ldots$ do $K_i \leftarrow_\$ \text{Sample}()$	for $i = 1, 2, \ldots$ do $f_i \leftarrow_\$ \text{Perm}(\mathcal{M})$
proc ENC(i, x) {ret $\Pi_{K_i}[\pi](x)$}	proc ENC(i, x) {ret $f_i(x)$}
proc DEC(i, y) {ret $\Pi^{-1}_{K_i}[\pi](y)$}	proc DEC(i, y) {ret $f_i^{-1}(y)$}
proc PRIM(J, u) {ret $\pi_J(u)$}	proc PRIM(J, u) {ret $\pi_J(u)$}
proc PRIMINV(J, v) {ret $\pi_J^{-1}(v)$ }	proc PRIMINV(J, v) {ret $\pi_J^{-1}(v)$}

Fig. 1. Games defining the multi-user security of a blockcipher $\Pi : \mathcal{K} \times \mathcal{M} \to \mathcal{M}$. This blockcipher is based on a family of independent, random permutations π : Index $\times$ Dom $\to$ Dom. The game is associated with a key-sampling algorithm Sample. Here Perm$(\mathcal{M})$ denotes the set of all permutations on $\mathcal{M}$.

3.1 The Indistinguishability Framework

Let us consider the setting of a distinguisher A (outputting a decision bit) interacting with one of two "systems" $\mathbf{S}_0$ and $\mathbf{S}_1$. These systems take inputs and produce outputs, and are randomized and possibly stateful. We dispense with a formalization of the concept of a system, as an intuitive understanding will be sufficient. Still, this can be done via games [4], random systems [21], ITMs, or whichever other language permits doing so. In this paper, these systems will provide a construction oracle ENC with a corresponding inversion oracle DEC, and a primitive oracle PRIM with a corresponding inversion oracle PRIMINV, but our treatment here is general, and thus does not assume this form.

The interaction between $\mathbf{S}_b$ and A (for $b \in \{0, 1\}$) defines a *transcript* $\tau = ((u_1, v_1), \ldots, (u_q, v_q))$ containing the ordered sequence of query-answer pairs describing this interaction. We denote by X_b the random variable representing this transcript. In the following, we consider the problem of upper bounding the statistical distance

$$\text{SD}(X_0, X_1) = \sum_\tau \max\{0, \Pr[X_1 = \tau] - \Pr[X_0 = \tau]\} , \tag{8}$$

of the transcripts, where the sum is over all possible transcripts. It is well known that $\text{SD}(X_0, X_1)$ is an upper bound on the distinguishing advantage of A, i.e., the difference between the probabilities of A outputting one when interacting with $\mathbf{S}_1$ and $\mathbf{S}_0$, respectively.

DESCRIBING SYSTEMS. Following [21], a useful way to formally describe the behavior of a system $\mathbf{S}$ is to associate with it a function $\mathsf{p_S}$ mapping a possible transcript $\tau = ((u_1, v_1), \ldots, (u_q, v_q))$ with a probability $\mathsf{p_S}(\tau) \in [0, 1]$. This is to be interpreted as the probability that if all queries $u_1, \ldots, u_q$ in τ are asked to $\mathbf{S}$ in this order, the answers are $v_1, \ldots, v_q$. Note that this is not a probability distribution (i.e., summing $\mathsf{p_S}(\tau)$ over all possible τ's does not give one). Moreover, $\mathsf{p_S}$ is independent of any possible distinguisher — it is a description of the system. (And in fact, this is precisely how [21] defines a system.)

Because our distinguishers are computationally unbounded, it is sufficient to assume them to be *deterministic* without loss of generality. A simple key observation is that for deterministic distinguisher A, given the transcript distribution X of the interaction with $\mathbf{S}$, we always have $\Pr[X = \tau] \in \{0, \mathsf{ps}(\tau)\}$. This is because, if $\tau = ((u_1, v_1), \ldots, (u_q, v_q))$, then either A is such that it asks queries $u_1, \ldots, u_q$ when fed answers $v_1, \ldots, v_q$ (in which case $\Pr[X = \tau] = \mathsf{ps}(\tau)$), or it is not, in which case clearly $\Pr[X = \tau] = 0$.

Let $\mathcal{T}$ denote the set of transcripts τ such that $\Pr[X_1 = \tau] > 0$. We call such transcripts *valid*. Also, note that if $\tau \in \mathcal{T}$, then we also have $\Pr[X_0 = \tau] = \mathsf{ps}_0(\tau)$. Therefore, we can rewrite (8) as

$$\mathsf{SD}(X_0, X_1) = \sum_{\tau \in \mathcal{T}} \max\{0, \mathsf{ps}_1(\tau) - \mathsf{ps}_0(\tau)\} . \tag{9}$$

Note that which transcripts are valid depends on A, as well as on the system $\mathbf{S}_1$.

THE H-COEFFICIENT METHOD. Let us revisit the well-known H-coefficient technique [9, 24] within this notational framework. (This is also very similar to alternative equivalent treatments, like the "interpolation method" presented in [5, 23].) The key step is to partition valid transcripts $\mathcal{T}$ into two sets, the *good* transcripts Γ_{good} and the *bad* transcripts Γ_{bad}. Then, if we can establish the existence of a value ϵ such that for all $\tau \in \Gamma_{\text{good}}$, we have $1 - \frac{\mathsf{ps}_0(\tau)}{\mathsf{ps}_1(\tau)} \leq \epsilon$, then we can conclude that

$$\mathsf{SD}(X_0, X_1) = \sum_{\tau} \max\{0, \mathsf{ps}_1(\tau) - \mathsf{ps}_0(\tau)\}$$

$$\leq \sum_{\tau \in \Gamma_{\text{good}}} \mathsf{ps}_1(\tau) \cdot \max\left\{0, 1 - \frac{\mathsf{ps}_0(\tau)}{\mathsf{ps}_1(\tau)}\right\} + \sum_{\tau \in \Gamma_{\text{bad}}} \mathsf{ps}_1(\tau) \cdot 1$$

$$\leq \epsilon + \Pr[X_1 \in \Gamma_{\text{bad}}] .$$

We note that in the typical treatment of this method, many authors don't notationally differentiate explicitly between e.g. $\Pr[X_0 = \tau]$ and $\mathsf{ps}_0(\tau)$ (and likewise for X_1 and $\mathbf{S}_1$), even though this connection is implicitly made. (For example, for typical cryptographic systems, the order of queries is re-arranged to compute $\Pr[X_0 = \tau]$ without affecting the probability, which is a property of ps_0, since queries may not appear in that order for the given A.) Treating these separately will however be very helpful in the following.

THE EXPECTATION METHOD. In the H-coefficient method, ϵ typically depends on some *global* properties of the distinguisher (e.g., the number of queries) and the system (key length, input length, etc.). However, this can be generalized: Assume that we can give a non-negative function $g : \mathcal{T} \to [0, \infty)$ such that

$$1 - \frac{\mathsf{ps}_0(\tau)}{\mathsf{ps}_1(\tau)} \leq g(\tau) \tag{10}$$

for all $\tau \in \Gamma_{\text{good}}$, then we can easily conclude, similar to the above, that

$$\mathsf{SD}(X_0, X_1) \leq \sum_{\tau \in \Gamma_{\text{good}}} \mathsf{ps}_1(\tau) \cdot g(\tau) + \Pr[X_1 \in \Gamma_{\text{bad}}]$$

$$\leq \mathbf{E}[g(X_1)] + \Pr[X_1 \in \Gamma_{\text{bad}}] .$$

Note that we have used the fact that the function g is non-negative for the first term to be upper bounded by the expectation $\mathbf{E}[g(X_1)]$. We refer to this method as the *expectation method*, and we will see below that this idea is very useful.

The H-coefficient technique corresponds to the special case where g is "constant", whereas here the value may depend on further specifics of the transcript at hand. Obviously, good choices of g, Γ_{good}, and Γ_{bad} are specific to the problem at hand. We also note that one can set $g(\tau) = 1$ for bad transcripts, and then dispense with the separate calculation of the probability. (The way we present it above, however, makes it more amenable to the typical application.) Note that Chen and Steinberger [9] explain that in the H-coefficient method one may go beyond the simple partitioning in good and bad transcripts. In a sense, what we are doing here is going to the extreme, partitioning Γ_{good} into singleton sets.

3.2   Point-Wise Proximity

A core observation is that for some pairs of systems $\mathbf{S}_0$ and $\mathbf{S}_1$ (and this will be the case for those we consider), we are able to establish a stronger "point-wise" proximity property.

Definition 1 (Point-wise proximity). *We say that two systems $\mathbf{S}_0$ and $\mathbf{S}_1$ satisfy ϵ-point-wise proximity if, for every possible transcript τ with q queries,*

$$\Delta(\tau) = \mathsf{ps}_1(\tau) - \mathsf{ps}_0(\tau) \leq \mathsf{ps}_1(\tau) \cdot \epsilon(q) . \tag{11}$$

Note that ϵ is a function of q, and often we will let it depend on more fine-grained partitions of the query complexity. (Also in some cases, the query complexity will be implicit.) In particular, for a certain q-query distinguisher A, by Eq. (9), it is clear that ϵ-point-wise proximity implies that $\mathsf{SD}(X_0, X_1) \leq \epsilon$, which is also a bound on A's advantage. Observe that point-wise proximity is a *property of a pair of systems $\mathbf{S}_0$ and $\mathbf{S}_1$*, independent of the adversaries interacting with them. Also, it is *equivalent* to the fact that $1 - \frac{\mathsf{ps}_0(\tau)}{\mathsf{ps}_1(\tau)} \leq \epsilon$ for all τ such that $\mathsf{ps}_1(\tau) > 0$.

In other words, establishing ϵ-proximity corresponds to applying the H-coefficient method without bad transcripts. This is exactly the special case considered by Bernstein [5]. Of course, this method is not always applicable, but when it is, it will bring numerous advantages.

THE EXPECTATION METHOD. We outline a general method to prove ϵ-point-wise proximity based on the above general expectation method.

As the starting point, we extend the system $\mathbf{S}_0$ to depend on some auxiliary random variable S (e.g., a secret key). In particular, we write $\mathsf{ps}_0(\tau, s)$ to be the probability that $\mathbf{S}_0$ answers queries according to τ and that $S = s$.

Further, we define $\mathsf{ps}_{\mathbf{S}_1}(\tau, s) = \mathsf{ps}_{\mathbf{S}_1}(\tau) \cdot \Pr[S = s]$, i.e., we think of $\mathbf{S}_1$ as also additionally sampling an auxiliary variable S with the same marginal distribution as in $\mathbf{S}_0$, except that the behavior of $\mathbf{S}_1$ remains independent of S. Then, for every transcript τ,

$$\Delta(\tau) = \sum_s \mathsf{ps}_{\mathbf{S}_1}(\tau, s) - \sum_s \mathsf{ps}_{\mathbf{S}_0}(\tau, s) = \sum_s \mathsf{ps}_{\mathbf{S}_1}(\tau, s) - \mathsf{ps}_{\mathbf{S}_0}(\tau, s) \ .$$

Now, we establish the following lemma, that is based on the above expectation method.

Lemma 1 (The expectation method). *Fix a transcript τ for which $\mathsf{ps}_{\mathbf{S}_1}(\tau) > 0$. Assume that there exists a partition Γ_{good} and Γ_{bad} of the range $\mathcal{U}$ of S, as well as a function $g : \mathcal{U} \to [0, \infty)$ such that $\Pr[S \in \Gamma_{\text{bad}}] \leq \delta$ and for all $s \in \Gamma_{\text{good}}$,*

$$1 - \frac{\mathsf{ps}_{\mathbf{S}_0}(\tau, s)}{\mathsf{ps}_{\mathbf{S}_1}(\tau, s)} \leq g(s) \ .$$

Then,

$$\Delta(\tau) \leq (\delta + \mathbf{E}(g(S))) \cdot \mathsf{ps}_{\mathbf{S}_1}(\tau) \ .$$

Proof. Note that $s \in \mathcal{U}$ implies $\Pr[S = s] > 0$, and thus $\mathsf{ps}_{\mathbf{S}_1}(\tau, s) > 0$. We can easily compute

$$\Delta(\tau) \leq \sum_{s \in \mathcal{U}} \mathsf{ps}_{\mathbf{S}_1}(\tau, s) - \mathsf{ps}_{\mathbf{S}_0}(\tau, s)$$

$$= \mathsf{ps}_{\mathbf{S}_1}(\tau) \cdot \sum_{s \in \mathcal{U}} \Pr[S = s] \cdot \left(1 - \frac{\mathsf{ps}_{\mathbf{S}_0}(\tau, s)}{\mathsf{ps}_{\mathbf{S}_1}(\tau, s)}\right)$$

$$\leq \mathsf{ps}_{\mathbf{S}_1}(\tau) \cdot \left(\sum_{s \in \Gamma_{\text{bad}}} \Pr[S = s] + \sum_{s \in \Gamma_{\text{good}}} \Pr[S = s] \cdot g(s)\right)$$

$$\leq (\delta + \mathbf{E}(g(S))) \cdot \mathsf{ps}_{\mathbf{S}_1}(\tau) \ . \quad \square$$

We stress that the partitioning into Γ_{good} and Γ_{bad}, as well as the function g and the random variable S, are all allowed to depend on τ (and in fact will depend on them in applications).

TRANSCRIPT REDUCTION. Lemma 1 gives us one possible approach to prove ϵ-point-wise proximity. Another technique we will use is to simply reduce this property to ϵ-point-wise proximity for another pair of systems.

Typically, we will assume that we are in the above extended setting, where we have enhanced the systems $\mathbf{S}_0$ and $\mathbf{S}_1$ with some auxiliary random variable S. Here, in contrast to the above, we assume that S is not necessarily independent of the behavior of the system $\mathbf{S}_1$. Further, assume that we are given two other systems $\mathbf{S}_0'$ and $\mathbf{S}_1'$ for which ϵ-point-wise proximity holds. To this end, we are simply going to provide an explicit reduction $\mathscr{R}$ which is going to map every (τ, s) for $\mathbf{S}_0$ and $\mathbf{S}_1$ into a transcript $\mathscr{R}(\tau, s)$ for $\mathbf{S}_0'$ and $\mathbf{S}_1'$ such that

$$\frac{\mathsf{ps}_{\mathbf{S}_0}(\tau, s)}{\mathsf{ps}_{\mathbf{S}_1}(\tau, s)} = \frac{\mathsf{ps}_{\mathbf{S}_0'}(\mathscr{R}(\tau, s))}{\mathsf{ps}_{\mathbf{S}_1'}(\mathscr{R}(\tau, s))} \ .$$

whenever $\mathsf{ps}_1(\tau, s) > 0$. This will be sufficient for our purposes, because (with $\mathcal{U}$ being the set of s such that $\mathsf{ps}_1(\tau, s) > 0$)

$$\Delta(\tau) \leq \sum_{s \in \mathcal{U}} \mathsf{ps}_1(\tau, s) \cdot \left(1 - \frac{\mathsf{ps}_0(\tau, s)}{\mathsf{ps}_1(\tau, s)} \right)$$

$$= \sum_{s \in \mathcal{U}} \mathsf{ps}_1(\tau, s) \cdot \left(1 - \frac{\mathsf{ps}'_0(\mathscr{R}(\tau, s))}{\mathsf{ps}'_1(\mathscr{R}(\tau, s))} \right) \leq \epsilon \cdot \mathsf{ps}_1(\tau) \ .$$

Note that here $\epsilon = \epsilon(q')$, where q' is the number of queries in $\mathscr{R}(\tau, s)$.

3.3   From Single-User to Multi-user Security

There is no generic way to derive a *tight* bound on the multi-user security of a construction given a bound on its single-user security — the naive approach uses a hybrid argument, but as we have no bounds on the per-user number of queries of the attacker (which may vary adaptively), this leads to a loss in the reduction. Here, we show how given point-wise proximity for the single-user case, a bound for multi-user security can generically be found via a hybrid argument.

We assume now we are in the above multi-user prp security setting presented in Sect. 2, and we let $\mathsf{p}_{\mathsf{real}}$ and $\mathsf{p}_{\mathsf{rand}}$ describe the oracles available in the real and random experiments (which we can see as systems in the framework above). Assume that we already established ϵ-point-wise proximity for the single-user case for transcripts with at most p primitive queries and q function queries (and we think of $\epsilon = \epsilon(p, q)$ as a function of p and q). That is, we have shown that for *every* transcript τ such that all function queries have form $\text{ENC}(i, x)$ and $\text{DEC}(i, y)$ for the same i (whereas $\text{PRIM}(J, u)/\text{PRIMINV}(J, v)$ are unrestricted),

$$\mathsf{p}_{\mathsf{rand}}(\tau) - \mathsf{p}_{\mathsf{real}}(\tau) \leq \mathsf{p}_{\mathsf{rand}}(\tau) \cdot \epsilon(p, q) \ . \tag{12}$$

Let m be the number of calls to π/π^{-1} that a single call to Π/Π^{-1} makes. Also assume now that ϵ satisfies the following properties: (i) $\epsilon(x, y) + \epsilon(x, z) \leq \epsilon(x, y + z)$, for every $x, y, z \in \mathbb{N}$, and (ii) $\epsilon(\cdot, z)$ is an increasing function on $\mathbb{N}$, for every $z \in \mathbb{N}$. Property (ii) usually holds, because asking more queries should only increase the adversary's advantage. Property (i) is also usually satisfied by typical functions we use to bound distinguishing advantages. Further, we assume that $\epsilon(p + qm, q) \leq 1/2$. Then, we show the following.

Lemma 2. (From su to mu point-wise proximity). *Assume all conditions above are met. Then for all transcripts τ with at most q function queries (for arbitrary users) and p primitive queries,*

$$\mathsf{p}_{\mathsf{rand}}(\tau) - \mathsf{p}_{\mathsf{real}}(\tau) \leq \mathsf{p}_{\mathsf{rand}}(\tau) \cdot 2\epsilon(p + q \cdot m, q) \tag{13}$$

Proof. Fix some transcript τ, and assume that in τ, function queries are made for r users $u_1, \ldots, u_r \in \mathbb{N}$. For each $i \in \{0, 1, \ldots, r\}$, consider the hybrid system

$\mathbf{S}_i$ which provides a compatible interface with the real and random games, and answers primitives queries in the same way, but queries for user u_j for $j > i$ are answered with the actual construction Π and Π^{-1}, whereas queries for u_j with $j \leq i$ are answered by i independent random permutations. Then clearly $p_{\mathbf{S}_0}(\tau) = p_{\text{real}}(\tau)$ and $p_{\mathbf{S}_r}(\tau) = p_{\text{rand}}(\tau)$. We can thus rewrite

$$p_{\text{rand}}(\tau) - p_{\text{real}}(\tau) = \sum_{i=1}^{r} p_{\mathbf{S}_i}(\tau) - p_{\mathbf{S}_{i-1}}(\tau) \ .$$

Suppose that τ contains q_i queries to $\text{ENC}(u_i, \cdot)/\text{DEC}(u_i, \cdot)$. We'll prove that for any $i \in \{1, \ldots, r\}$,

$$p_{\mathbf{S}_i}(\tau) - p_{\mathbf{S}_{i-1}}(\tau) \leq p_{\mathbf{S}_i}(\tau) \cdot \epsilon(p + qm, q_i) \ . \tag{14}$$

This claim will be justified later. Now Eq. (14) implies that

$$p_{\mathbf{S}_{i-1}}(\tau) \geq (1 - \epsilon(p + qm, q_i)) \cdot p_{\mathbf{S}_i}(\tau)$$

for every $i \in \{1, \ldots, r\}$. Thus for any $i \in \{1, \ldots, r\}$,

$$p_{\mathbf{S}_0}(\tau) \geq p_{\mathbf{S}_i}(\tau) \prod_{j=1}^{i}(1 - \epsilon(p + qm, q_j)) \geq p_{\mathbf{S}_i}(\tau)\left(1 - \sum_{j=1}^{i} \epsilon(p + qm, q_j)\right)$$

$$\geq p_{\mathbf{S}_i}(\tau)\left(1 - \sum_{j=1}^{r} \epsilon(p + qm, q_j)\right) \geq p_{\mathbf{S}_i}(\tau)\left(1 - \epsilon(p + qm, q)\right) \geq \frac{1}{2}p_{\mathbf{S}_i}(\tau) \ .$$

The first inequality is due to the fact that $(1 - x)(1 - y) \geq 1 - (x + y)$ for every $0 \leq x, y \leq 1$; the second last inequality is due to the property (i) of function ϵ; and the last one is due to the assumption that $\epsilon(p + qm, q) \leq 1/2$. Combining this with Eq. (14),

$$\sum_{i=1}^{r} p_{\mathbf{S}_i}(\tau) - p_{\mathbf{S}_{i-1}}(\tau) \leq \sum_{i=1}^{r} p_{\mathbf{S}_i}(\tau) \cdot \epsilon(p + qm, q_i)$$

$$\leq \sum_{i=1}^{r} 2p_{\mathbf{S}_0}(\tau) \cdot \epsilon(p + qm, q_i) \leq 2p_{\mathbf{S}_0}(\tau) \cdot \epsilon(p + qm, q) \ .$$

What's left is to prove Eq. (14). To this end, fix $i \in \{1, \ldots, r\}$, and we are going to use the transcript reduction technique presented above. First off, enhance $\mathbf{S}_{i-1}$ and $\mathbf{S}_i$ with an auxiliary variable S which contains (i) the transcript of all internal $\text{PRIM}/\text{PRIMINV}$ caused by querying $\text{ENC}(u_j, \cdot)/\text{DEC}(u_j, \cdot)$, and (ii) the keys K_j of users u_j, for $j > i$. Now, given (τ, s), note that if we start by removing all queries from τ for users u_j for $j < i$ (which are answered by random permutations in both $\mathbf{S}_{i-1}$ and $\mathbf{S}_i$), obtaining a transcript τ', then we necessarily have

$$\frac{p_{\mathbf{S}_{i-1}}(\tau, s)}{p_{\mathbf{S}_i}(\tau, s)} = \frac{p_{\mathbf{S}_{i-1}}(\tau', s)}{p_{\mathbf{S}_i}(\tau', s)} \ .$$

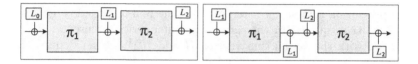

Fig. 2. Left: Illustration of $\mathrm{KAC}[\pi, 2]$. **Right:** Illustration of $\mathrm{KACX}[\pi, 2]$.

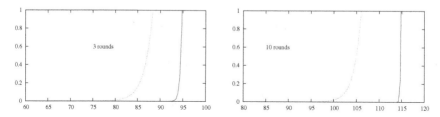

Fig. 3. Su PRP security of KAC **on 3 rounds (left) and 10 rounds (right) on 128-bit strings: our bounds versus CS's.** The solid lines depict our bounds, and the dashed ones depict CS's bounds. In both pictures, $p = q$, and the x-axis gives the log (base 2) of p, and the y-axis gives upper bounds on the PRP security of KAC.

This is because the distribution of these answers is independent of what is in τ', s in both $\mathbf{S}_{i-1}$ and $\mathbf{S}_i$, and in both cases the distribution is identical. Then, given τ' and a value s for S (in either of the system), we can easily construct a transcript $\mathscr{R}(\tau', s)$ where all function queries for users u_j for $j > i$ are removed, all primitive queries in s are made directly to the PRIM and PRIMINV oracles in τ', and all keys K_j of users u_j for $j > i$ are removed. It is easy to verify that

$$\frac{\mathsf{ps}_{\mathbf{S}_{i-1}}(\tau, s)}{\mathsf{ps}_{\mathbf{S}_i}(\tau, s)} = \frac{\mathsf{ps}_{\mathbf{S}_{i-1}}(\mathscr{R}(\tau', s))}{\mathsf{ps}_{\mathbf{S}_i}(\mathscr{R}(\tau', s))},$$

because (i) the function queries of users u_j can be derived from the primitive queries and K_j, and (ii) the keys K_j for $j > i$ are independent of what's used for user i. However, note $\mathscr{R}(\tau', s)$ contains q_i ENC/DEC queries, all for users u_i, and at most $p + q \cdot m$ queries to PRIM/PRIMINV. As for those transcripts we have already established ϵ-point-wise proximity, Eq. (14) follows by the transcript reduction method. □

4   Exact Bounds for Key-Alternating Ciphers

4.1   Results and Discussion

This section provides a comprehensive single- and multi-user security analysis of key-alternating ciphers. After reviewing the construction, and the concrete bound proved by Chen and Steinberger [9], we state and discuss our main results, starting with the single-user security case.

KEY-ALTERNATING CIPHERS. Let us review the key-alternating cipher construction. Let t and n be positive integers, and let $\pi : \mathbb{N} \times \{0,1\}^n \to \{0,1\}^n$ be a family of permutations on $\{0,1\}^n$. We write $\pi_i(\cdot)$ to denote $\pi(i, \cdot)$, and N for 2^n. The Key-Alternating Cipher (KAC) construction gives a blockcipher $\mathrm{KAC}[\pi, t] : (\{0,1\}^n)^{t+1} \times \{0,1\}^n \to \{0,1\}^n$ as follows. On input x and keys $K = (L_0, \ldots, L_t) \in (\{0,1\}^n)^{t+1}$, $\mathrm{KAC}[\pi, t](K, x)$ returns y_t, where $y_0 = x \oplus L_0$, and $y_i = \pi_i(y_{i-1}) \oplus L_i$ for every $i \in \{1, \ldots, t\}$. It is a direct generalization of the classic Even-Mansour construction [12]. See Fig. 2 for an illustration of $\mathrm{KAC}[\pi, 2]$.

THE CS BOUND. Chen and Steinberger (CS) [9] shows that if an adversary makes at most q queries to ENC/DEC, and at most $p \le N/3$ queries to PRIM$(i, \cdot)$ and PRIMINV$(i, \cdot)$ for every $i \in \{1, \ldots, t\}$, then

$$\mathsf{Adv}^{\pm\mathrm{prp}}_{\mathrm{KAC}[\pi,t]}(A) \le \frac{qp^t}{N^t} \cdot Ct^2(6C)^t + \frac{(t+1)^2}{C} \tag{15}$$

for any $C \ge 1$. Since Eq. (15) holds for *any* $C \ge 1$, to determine the best upper bound for $\mathsf{Adv}^{\pm\mathrm{prp}}_{\mathrm{KAC}[\pi,t]}(A)$ according to this inequality, one needs to find the *minimum* of the right-hand side of Eq. (15). For each fixed p, q and t, from the inequality of arithmetic and geometric means:

$$\frac{qp^t}{N^t} \cdot Ct^2(6C)^t + \frac{(t+1)^2}{C} = \frac{qp^t}{N^t} \cdot Ct^2(6C)^t + \frac{(t+1)}{C} + \cdots + \frac{(t+1)}{C}$$

$$\ge (t+2) \left(\frac{qp^t Ct^2(6C)^t}{N^t} \cdot \frac{(t+1)}{C} \cdots \frac{(t+1)}{C} \right)^{1/(t+2)}$$

$$= (t+2) \left(\frac{q(6p)^t}{N^t} \cdot t^2(t+1)^{t+1} \right)^{1/(t+2)} \quad .$$

The equality happens if $C = \left(\frac{N^t(t+1)}{qt^2(6p)^t} \right)^{(t+2)}$. Eq. (15) can be rewritten as

$$\mathsf{Adv}^{\pm\mathrm{prp}}_{\mathrm{KAC}[\pi,t]}(A) \le (t+2) \left(\frac{q(6p)^t}{N^t} \cdot t^2(t+1)^{t+1} \right)^{1/(t+2)} \quad .$$

(This bound is slightly smaller than the claimed result in [9, Corollary 1].) While this bound is asymptotically optimal, meaning that the adversary needs to spend about $N^{t/(t+1)}$ queries for the bound to become vacuous, it's concretely much weaker than the best possible bound, which is roughly qp^t/N^t [14].

SINGLE-USER SECURITY OF KACS. We establish the following theorem, which gives a near-exact bound on the PRP security of the $\mathrm{KAC}[\pi, t]$ construction in the ideal-permutation model. Following the theorem, we first give some comments. The proof is found in Sect. 4.2, where we also give a high-level overview.

Theorem 1 (Su PRP security of KACs). Let t and n be positive integers, and let $\pi : \mathbb{N} \times \{0,1\}^n \to \{0,1\}^n$ be a family of ideal permutations on $\{0,1\}^n$.

Let $\mathrm{KAC}[\pi,t]$ be as above. For an adversary A that makes at most q queries to $\mathrm{ENC}/\mathrm{DEC}$, and at most p_i queries to $\mathrm{PRIM}(i,\cdot)$ and $\mathrm{PRIMINV}(i,\cdot)$ for every $i \in \{1,\ldots,t\}$, it holds that

$$\mathsf{Adv}^{\pm\mathrm{prp}}_{\mathrm{KAC}[\pi,t]}(A) \le 4^t q p_1 \cdots p_t / N^t . \tag{16}$$

This bound constitutes a significant improvement over the CS bound. For example, consider $n = 128$ and $t = 3$. For $p = 2^{96}$ and $q = 2^{64}$, CS's result yields $\mathsf{Adv}^{\pm\mathrm{prp}}_{\mathrm{KAC}[\pi,3]}(A) \le 0.71$, whereas according to Theorem 1, $\mathsf{Adv}^{\pm\mathrm{prp}}_{\mathrm{KAC}[\pi,3]}(A) \le 2^{-26}$. See Fig. 3 for a graphical comparison of CS's bound and ours for the case $p = q$ and both $t = 3$ and $t = 10$ rounds. Note that the latter case is the one matching AES-128 the closest. In particular, here, we see that the advantage starts to become noticeable roughly at $q = p = 2^{100}$ for the CS bound, whereas this happens only at 2^{113} for our new bound. One of the issues in the CS bound is that the $1/(t+2)$ exponent smoothes the actual bound considerably, and thus gives a much less sharp transition from small advantage to large as t increases.

QUERY REGIMES. Let us point out two important remarks on the bound. First off, it is important that our bound does *not* require any bound on q and $p_1,\ldots,p_t$. Any of these values can equal N, and the construction remains secure as long as $4^t q p_1 \cdots p_t / N^t$ remains small enough. Dealing with such $q = N$ and $p_i = N$ case requires in fact a completely novel approach, which we introduce and explain below in Sect. 4.2. This will be important when using our bounds in the proof for the analysis of XOR cascades, which we want to hold true *even* if N is small (e.g., in the case of format-preserving encryption (FPE) [3]) and the attacker distributes $q \gg N$ queries across multiple users, possibly exhausting all possible queries for some of these users.

On the other hand, one might worry that an adversary may *adaptively* distribute the number of queries among the permutations $\pi_1,\ldots,\pi_t$, and want a bound in terms of p, the total number of queries to π. Naively, the bound in Theorem 1 is only $q(4p)^t/N^t$. However, we can exploit our point-wise proximity based approach to get a sharper bound: In each transcript τ, the number of queries $p_i[\tau]$ to π_i is completely determined, and thus Eq. (17) in the proof of Theorem 1 can be rewritten as

$$\mathsf{ps}_1(\tau) - \mathsf{ps}_0(\tau) \le \mathsf{ps}_1(\tau) \cdot \frac{4^t q p_1[\tau] \cdots p_t[\tau]}{N^t}$$

$$\le \mathsf{ps}_1(\tau) \cdot \frac{4^t q (p_1[\tau] + \cdots + p_t[\tau])^t}{N^t t^t} \le \mathsf{ps}_1(\tau) \cdot \frac{q(4p)^t}{N^t t^t} .$$

Then $\mathsf{Adv}^{\pm\mathrm{prp}}_{\mathrm{KAC}[\pi,t]}(A) \le q(4p)^t/(Nt)^t$.

VARIANTS. Consider the following natural variant $\mathrm{KACX}[\pi,t]$ of $\mathrm{KAC}[\pi,t]$. It uses only t subkeys $(L_1,\ldots,L_t) \in (\{0,1\}^n)^t$. On input x, it returns returns y_t, where $y_0 = x$, and $y_i = \pi_i(y_{i-1} \oplus L_i) \oplus L_i$ for every $i \in \{1,\ldots,t\}$. See Fig. 2 for an illustration of KACX. Note that KACX is KAC with effective key $(L_1, L_1 \oplus L_2, L_2 \oplus L_3, \ldots, L_{t-1} \oplus L_t, L_t)$, or in other words, we have chosen random keys *under the constraint that their checksum equals* 0^n.

While we do not give the concrete proof, we note that the same security bound and proof will continue to work: in the proof, whenever we need to use the independence of the subkeys, we consider only t subkeys at a time. We note that for $t = 1$ this is exactly the statement that the security of Even-Mansour is not affected when one sets both keys to be equal.

4.2   Proof of Theorem 1

This section is devoted to the proof of Theorem 1. We begin with a high-level overview of the proof structure. Following the notational framework of Sect. 3.1, let $\mathbf{S}_0$ and $\mathbf{S}_1$ be the systems associated by the real and ideal game in the prp security definition. In particular, transcripts τ for these systems contain two different types of entries:

- ENC/DEC *queries*. Queries to ENC$(1, x)$ returning y and DEC$(1, y)$ returning x are associated with an entry (enc, x, y).
- PRIM/PRIMINV *queries*. Queries to PRIM(j, x), returning y, and those to PRIMINV(j, y), returning x, are associated with an entry (prim, j, x, y).

Note that a further distinction between entries corresponding to forward and backward queries is not necessary, as this will not influence the probabilities $\mathsf{p}_{\mathbf{S}_0}(\tau)$ and $\mathsf{p}_{\mathbf{S}_1}(\tau)$ that a certain transcript occurs. Similarly, these probabilities are invariant under permuting the entries of τ. We also assume without loss of generality that no repeated entries exist in τ (this corresponds to the fact that an attacker asks no redundant queries).

OVERVIEW. Our goal is to establish the point-wise proximity for $\mathbf{S}_0$ and $\mathbf{S}_1$, i.e., for any transcript τ containing q entries $(\mathsf{enc}, \cdot, \cdot)$, and at most p_i entries of form $(\mathsf{prim}, i, \cdot, \cdot)$ for $i = 1, \ldots, t$, we show

$$\mathsf{p}_{\mathbf{S}_1}(\tau) - \mathsf{p}_{\mathbf{S}_0}(\tau) \le \mathsf{p}_{\mathbf{S}_1}(\tau) \cdot \frac{4^t q p_1 \cdots p_t}{N^t} \quad . \tag{17}$$

In particular, the proof of (17) is made by two parts:

- **Case 1.** $q, p_1, \ldots, p_t \le N/4$. Then, we give a direct proof of (17) using the expectation method from Lemma 1, where the auxiliary variable S will consist of the secret keys $L_0, L_1, \ldots, L_t$ (in $\mathbf{S}_0$). Our proof will resemble in some aspects that of Chen and Steinberger [9], but it will be much simpler due to the fact that the queries are fixed by τ, and we will only argue over the probability of S. We will still resort to the involved and elegant "path-counting" lemma of [9], but it will only be used to define a function g for which computing the expectation of $g(S)$ will be fairly easy.
- **Case 2.** At least one of $q, p_1, \ldots, p_t$ is bigger than $N/4$. We'll use the transcript reduction method, where the other two systems $\mathbf{S}_0'$ and $\mathbf{S}_1'$ on which we assume we have established point-wise proximity provide the real and ideal games for a $(t-1)$-round KAC.

Therefore, our proof for Eq. (17) uses induction on the number of rounds of the KAC. If all queries are smaller than $N/4$ then we can give a direct proof, otherwise the transcript reduction lands us back to the induction hypothesis. To this end, note that although KAC is defined for $t \geq 1$ rounds, we can also define $\mathrm{KAC}[\pi, 0](K, x) = x \oplus K$ for every $x \in \{0,1\}^n$, and the bound degenerates to 1. This is our base case in which Eq. (17) vacuously holds.

Now suppose that Eq. (17) holds for KAC of $0, \dots, t - 1$ rounds. We now prove that it also holds for KAC of t rounds as well. We'll consider the following two cases.[3]

Case 1: $q, p_1, \dots, p_t \leq N/4$. Fix a transcript τ. We use the expectation method. Let S be the random variable for the key of $\mathrm{KAC}[\pi, t]$ in $\mathbf{S}_0$, and let $\mathcal{K} = (\{0,1\}^n)^{t+1}$ be the key space. Then S is uniformly distributed over $\mathcal{K}$. For each key $s = (L_0, \dots, L_t) \in \mathcal{K}$, define the graph $G(s)$ as follows:

- Its set of vertices are partitioned into $t + 1$ sets $V_0, \dots, V_t$, each of 2^n elements. For each $j \in \{0, \dots, t\}$, use the elements of $\{j\} \times \{0,1\}^n$ to uniquely label the elements of V_j.
- For each entry $(\mathtt{prim}, j, x, y)$ in τ, connect the vertices $(j - 1, x \oplus L_{j-1})$ and (j, y).

For a path P in $G(s)$, let $|P|$ denote the number of edges in this path. (A vertex is a also a path that has no edge.) We define the following notion of good and bad keys.

Definition 2 (Bad and good keys). *We say that a key $s = (L_0, \dots, L_t)$ is bad if τ contains an entry (enc, x, y) such that in the graph $G(s)$, there's a path P_0 starting from $(0, x)$ and a path P_1 starting from $(t, y \oplus L_t)$ such that $|P_0| + |P_1| \geq t$. If a key is not bad then we'll say that it's good. Let Γ_{bad} be the set of bad keys, and let $\Gamma_{\mathrm{good}} = \mathcal{K} \backslash \Gamma_{\mathrm{bad}}$.*

Let $Z_s(i, j)$ be the number of paths from vertices in V_i to vertices in V_j of $G(s)$. For $0 \leq a < b \leq t$, let $\mathcal{B}(a, b)$ be the collection of sets $\sigma = \{(i_0, i_1), (i_1, i_2), \dots, (i_{\ell-1}, i_\ell)\}$, with $a = i_0 < \cdots < i_\ell = b$. Let the ENC entries of τ be $(\mathrm{enc}, x_1, y_1), \dots, (\mathrm{enc}, x_q, y_q)$. For $k \in \{1, \dots, q\}$, let $\alpha_k[s]$ be the length of the longest path starting from $(0, x_k)$, and $t - \beta_k[s]$ be the length of the longest path ending at (t, y_k). For $0 \leq a < b \leq t$, let $R_{a,b,k}[s] = 1$ if $\alpha_k[s] \geq a$ and $\beta_k[s] \leq b$, and let $R_{a,b,k}[s] = 0$ otherwise. Note that if s is good then $\alpha_k[s] < \beta_k[s]$ for every $k \in \{1, \dots, q\}$.

Recall that in the expectation method, one needs to find a non-negative function $g : \mathcal{K} \to [0, \infty)$ such that $g(s)$ bounds $1 - \mathsf{ps}_0(\tau, s)/\mathsf{ps}_1(\tau, s)$ for all $s \in \Gamma_{\mathrm{good}}$. The function g is directly given in the following technical lemma. The proof, which is based on the main combinatorial lemma of [9], is in Appendix A of the full version of this paper.

---

[3] Note that here the unusual thing is that Case 1 is handled via a direct proof.

Lemma 3. *For any $s \in \Gamma_{\mathrm{good}}$, it holds that*

$$1 - \frac{\mathsf{ps}_0(\tau, s)}{\mathsf{ps}_1(\tau, s)} \leq \sum_{k=1}^{q} \sum_{0 \leq a < b \leq t} R_{a,b,k}[s] \cdot \sum_{\sigma \in \mathcal{B}(a,b)} \prod_{(i,j) \in \sigma} \frac{Z_s(i,j)}{N - p_j - q} \ .$$

Before we continue the proof, a few remarks are needed. First, note that Lemma 3 only needs $q + p_i < N$ for every $i \in \{1, \dots, t\}$. Therefore, one in fact can consider Case 1 for $q, p_1, \dots, p_t \leq N/\lambda$, for an arbitrary constant $\lambda > 2$, and Case 2 for $\max\{q, p_1, \dots, p_t\} > N/\lambda$. This will lead to the bound around $q(cp/N)^t$, where $c = \max\{\lambda, 2(\lambda - 1)/(\lambda - 2)\}$. To minimize this, the best choice of λ is $2 + \sqrt{2}$, but we use $\lambda = 4$ for simplicity.

We finally have everything in place to apply the expectation method. Note that

$$\mathbf{E}[g(S)] = \mathbf{E}\left(\sum_{k=1}^{q} \sum_{0 \leq a < b \leq t} R_{a,b,k}[S] \cdot \sum_{\sigma \in \mathcal{B}(a,b)} \prod_{(i,j) \in \sigma} \frac{Z_S(i,j)}{N - p_j - q} \right)$$

$$\leq \sum_{k=1}^{q} \mathbf{E}\left(\sum_{0 \leq a < b \leq t} R_{a,b,k}[S] \cdot \sum_{\sigma \in \mathcal{B}(a,b)} \prod_{(i,j) \in \sigma} \frac{2Z_S(i,j)}{N} \right),$$

where the last inequality is due to the hypothesis that $p_1, \dots, p_t, q \leq N/4$. We will need the following technical lemma below; the proof is in Appendix B of the full version of this paper.

Lemma 4. *For $k \in \{1, \dots, q\}$,*

$$\mathbf{E}\left(\sum_{0 \leq a < b \leq t} R_{a,b,k}[S] \cdot \sum_{\sigma \in \mathcal{B}(a,b)} \prod_{(i,j) \in \sigma} \frac{2Z_S(i,j)}{N} \right) \leq \frac{(4^t - t - 1)p_1 \cdots p_t}{N^t} \ .$$

Note that expectation in Lemma 4 is over the *uniform* choices of the key vector $S = (S_0, S_1, \dots, S_t)$, and the proof of Lemma 4 can actually compute the *exact* value of this expectation. Hence, from Lemmas 1, 3, and 4, to get our bound for Case 1, it suffices to prove that

$$\Pr[S \in \Gamma_{\mathrm{bad}}] \leq (t+1)qp_1 \cdots p_t/N^t \ . \tag{18}$$

To justify Eq. (18), let $S = (S_0, \dots, S_t)$. If $S \in \Gamma_{\mathrm{bad}}$ then τ must contain entries $(\mathtt{enc}, x, y), (\mathtt{prim}, 1, u_1, v_1), (\mathtt{prim}, 2, u_2, v_2), \dots, (\mathtt{prim}, t, u_t, v_t)$ such that one of the following happens:

- $u_1 = x \oplus S_0$, and $u_i = v_{i-1} \oplus S_i$ for every $i \in \{2, \dots, t\}$, or
- $v_t = y \oplus S_t$, and $u_i = v_{i-1} \oplus S_i$ for every $i \in \{2, \dots, t\}$, or
- $u_1 = x \oplus S_0$, $v_t = y \oplus S_t$, and there is some $\ell \in \{2, \dots, t\}$ such that $u_i = v_{i-1} \oplus S_i$ for every $i \in \{2, \dots, t\} \setminus \{\ell\}$.

Since $S_0, \ldots, S_t$ are uniformly and independently random in $\{0,1\}^n$, the chance that S is bad is at most $(t+1)qp_1 \ldots p_t/N^t$.

Case 2: $N/4 < \max\{q, p_1, \ldots, p_t\} \le N$. Fix a transcript τ. We have three subcases below, each needs a different way to define S and uses a different transcript reduction.

We now give an intuition for the proof. We want to derive from (τ, s) a transcript $\mathscr{R}(\tau, s)$ for a system $\mathbf{S}_0'$ that implement the real game for a $(t-1)$-round KAC. In most cases (Cases 2.1 and 2.2), this KAC construction is KAC$[\pi, t-1]$, and S consists of the last subkey L_t and some additional query-answer pairs. In this case $\mathsf{ps}_{\mathbf{S}_1}(\tau, s)$ means the probability that $\mathbf{S}_1$ behaves according to the entries in (τ, s), and that $L_t \leftarrow\!\!\$\ \{0,1\}^n$ independent of $\mathbf{S}_1$ agrees with the subkey in s.

The target transcript $\mathscr{R}(\tau, s)$ consists of the PRIM entries to $\pi_1, \ldots, \pi_{t-1}$ in (τ, s), and the query-answer pairs to KAC$[\pi, t-1]$ that one can infer from the entries $(\mathtt{enc}, \cdot, \cdot)$, the entries $(\mathtt{prim}, t, \cdot, \cdot)$, and the last subkey as specified in (τ, s). The random variable S and the system $\mathbf{S}_1'$ that implements the ideal game for KAC$[\pi, t-1]$ will be constructed so that for every $b \in \{0,1\}$, the event that $\mathbf{S}_b$ behaves according to (τ, s) consists of two independent events: (i) $\mathbf{S}_b'$ behaves according to $\mathscr{R}(\tau, s)$, and (ii) π_t behaves according to the entries in (τ, s), and L_t agrees with what's specified in s. Since (ii) doesn't use ENC and DEC oracles, the reduction preserves the ratio $\mathsf{ps}_{\mathbf{S}_0}(\tau, s)/\mathsf{ps}_{\mathbf{S}_1}(\tau, s)$.

Case 2.1: $p_1, \ldots, p_t \le N/4$ but $N/4 < q \le N$. We'll in fact give an even stronger bound

$$\mathsf{ps}_{\mathbf{S}_1}(\tau) - \mathsf{ps}_{\mathbf{S}_0}(\tau) \le \mathsf{ps}_{\mathbf{S}_1}(\tau) \cdot \frac{4^{t-1}p_1 \ldots p_t}{N^{t-1}} \ .$$

Let S be the random variable for the last subkey L_t in $\mathbf{S}_0$ and the $(N-q)$ ENC queries/answers that τ lacks. (We stress that here S has only a *single* subkey, so a value s for S will have the form $\langle L_t, (\mathtt{enc}, x_1, y_1), \ldots, (\mathtt{enc}, x_{N-q}, y_{N-q})\rangle$.) It suffices to show that for any s such that $\mathsf{ps}_{\mathbf{S}_1}(\tau, s) > 0$,

$$\mathsf{ps}_{\mathbf{S}_1}(\tau, s) - \mathsf{ps}_{\mathbf{S}_0}(\tau, s) \le \mathsf{ps}_{\mathbf{S}_1}(\tau, s) \cdot \frac{4^{t-1}p_1 \ldots p_t}{N^{t-1}} \ . \tag{19}$$

Let $\mathbf{S}_0'$ be the system that implements the real game on KAC$[\pi, t-1]$. Let f be the ideal permutation that $\mathbf{S}_1$ uses for answering ENC/DEC queries. Let f' be the permutation such that $f'(x) = \pi_t^{-1}(f(x))$ for every $x \in \{0,1\}^n$, and thus f' is also an ideal permutation. The permutation f can be viewed as the cascade of f' and π_t (meaning that $f(x) = \pi_t(f'(x))$ for every $x \in \{0,1\}^n$). Let $\mathbf{S}_1'$ be a system that provides the ideal game on KAC$[\pi, t-1]$ and uses f' to answer ENC/DEC queries.

For any $b \in \{0,1\}$, although there are N ENC entries in (τ, s) for $\mathbf{S}_b$, since there are only p_t query-answer pairs to π_t, one can only "backtrack" p_t ENC query-answer pairs for $\mathbf{S}_b'$. Let $\mathscr{R}(\tau, s)$ be the transcript consisting of these p_t backtracked pairs and the query-answer pairs to $\pi_1, \ldots, \pi_{t-1}$. Formally, for any entry $(\mathtt{prim}, i, u, v)$ in (τ, s), add this to $\mathscr{R}(\tau, s)$ if $i \le t-1$. Next, for any entry $(\mathtt{prim}, t, u, v)$ in τ, there is exactly one entry $(\mathtt{enc}, x, y)$ in (τ, s) such that

$v \oplus L_t = y$, so add (enc, x, u) to $\mathscr{R}(\tau, s)$ as the corresponding backtracked query-answer pair. Then $\mathscr{R}(\tau, s)$ has p_t ENC entries and p_i query-answer pairs for π_i, for every $i \leq t-1$. Now, for $\mathbf{S}_b$ to behave according to (τ, s), it means that (i) $\mathbf{S}'_b$ must behave according to $\mathscr{R}(\tau, s)$, (ii) the subkey in S—recall that S contains only the last subkey L_t—must agree with what is specified in s, and (iii) π_t must be completely determined from $\mathbf{S}'_b$, the last subkey L_t, and the N ENC entries of (τ, s). Since π_t is independent of $\mathbf{S}'_b$ and L_t,

$$\mathsf{ps}_b(\tau, s) = \frac{1}{N \cdot N!} \cdot \mathsf{ps}'_b(\mathscr{R}(\tau, s)) \ .$$

Hence

$$\frac{\mathsf{ps}_0(\tau, s)}{\mathsf{ps}_1(\tau, s)} = \frac{\mathsf{ps}'_0(\mathscr{R}(\tau, s))}{\mathsf{ps}'_1(\mathscr{R}(\tau, s))} \ .$$

But from the induction hypothesis,

$$1 - \frac{\mathsf{ps}'_0(\mathscr{R}(\tau, s))}{\mathsf{ps}'_1(\mathscr{R}(\tau, s))} \leq \frac{4^{t-1} p_1 \ldots p_t}{N^{t-1}} \ .$$

Case 2.2: $p_1, \ldots, p_{t-1} \leq N/4$ but $p_t > N/4$. We'll in fact give an even stronger bound

$$\mathsf{ps}_1(\tau) - \mathsf{ps}_0(\tau) \leq \mathsf{ps}_1(\tau) \cdot \frac{4^{t-1} q p_1 \ldots p_{t-1}}{N^{t-1}} \ .$$

Let S be the random variable for the last subkey L_t in $\mathbf{S}_0$ and the $(N - p_t)$ queries/answers to π_t that τ lacks. From now on, this case is exactly the same as Case 2.1, except that since there are now N queries to π_t but only q ENC queries in (τ, s), we can only backtrack q ENC queries in $\mathbf{S}'_b$.

Case 2.3: There is some index $i \in \{1, \ldots, t-1\}$ such that $N/4 < p_i \leq N$. We'll give an even stronger bound

$$\mathsf{ps}_1(\tau) - \mathsf{ps}_0(\tau) \leq \mathsf{ps}_1(\tau) \cdot \frac{4^{t-1} q}{N^{t-1}} \prod_{j \in \{1,\ldots,t\}\backslash\{i\}} p_j \ .$$

Let S be the random variable for the subkey L_i in $\mathbf{S}_0$ and the other $(N - p_i)$ query-answer pairs to π_i that τ lacks. Fix s such that $\mathsf{ps}_1(\tau, s) > 0$. It suffices to prove that

$$\mathsf{ps}_1(\tau, s) - \mathsf{ps}_0(\tau, s) \leq \mathsf{ps}_1(\tau, s) \cdot \frac{4^{t-1} q}{N^{t-1}} \prod_{j \in \{1,\ldots,t\}\backslash\{i\}} p_j \ .$$

In this case, we'll need to build another $(t - 1)$-round KAC. Intuitively, we "collapse" the ith and $(i+1)$th round of KAC$[\pi, t]$ into a single round. Formally, construct $\pi' : \mathbb{N} \times \{0,1\}^n \to \{0,1\}^n$ from π and the subkey L_i in s as follows. For every $j < i$, we have $\pi'(j, \cdot) = \pi(j, \cdot)$. For every $j > i$, let $\pi'(j, \cdot) = \pi(j+1, \cdot)$. Finally, let $\pi'(i, x) = \pi(i + 1, \pi(i, x) \oplus L_i)$ for every $x \in \{0,1\}^n$. Thus π' is also a family of independent, ideal permutations on $\{0,1\}^n$. Let $\mathbf{S}'_0$ be a system that provides the real game on $\text{KAC}[\pi', t-1]$. Let f be the ideal permutation that $\mathbf{S}_1$ uses for answering ENC/DEC queries and let $\mathbf{S}'_1$ be a system that provides the ideal game on $\text{KAC}[\pi', t-1]$ and uses f to answer ENC/DEC queries.

Now, in (τ, s), we have N query-answer pairs for π_i and p_{i+1} query-answer pairs for π_{i+1}. One thus can "connect" those pairs to obtain p_{i+1} query-answer pairs for π'_i, which is the cascade of π_i and π_{i+1}. Formally, for any entry (prim, j, a, b) in (τ, s), if $j < i$ then add this entry to $\mathscr{R}(\tau, s)$ as a query for π'_j, and if $j > i+1$ then add $(\text{prim}, j-1, a, b)$ to $\mathscr{R}(\tau, s)$ as a query for π'_{j-1}. Next, for every entry $(\text{prim}, i+1, u, v)$ in τ, there is exactly one entry (prim, i, x, y) in (τ, s) such that $y \oplus L_i = u$, so add (prim, i, x, v) to $\mathscr{R}(\tau, s)$ as the corresponding connecting query. Hence $\mathscr{R}(\tau, s)$ has q ENC queries and p_j queries to π'_j if $j < i$, and p_{j+1} queries to π'_j if $j \geq i$.

For each $b \in \{0, 1\}$, for $\mathbf{S}_b$ to behave according to (τ, s), it means that (i) $\mathbf{S}'_b$ must behave according to $\mathscr{R}(\tau, s)$, (ii) the subkey in S must agree with what's specified in s, and (iii) π_t must behave according to the N entries specified by (τ, s). Note that π'_i is the cascade of π_i and π_{i+1}, and since π_{i+1} is independent of π_i, so is π'_i. Hence

$$\mathsf{ps}_{\mathbf{S}_b}(\tau, s) = \frac{1}{N \cdot N!} \cdot \mathsf{ps}_{\mathbf{S}'_b}(\mathscr{R}(\tau, s)) \ .$$

Hence

$$\frac{\mathsf{ps}_{\mathbf{S}_0}(\tau, s)}{\mathsf{ps}_{\mathbf{S}_1}(\tau, s)} = \frac{\mathsf{ps}_{\mathbf{S}'_0}(\mathscr{R}(\tau, s))}{\mathsf{ps}_{\mathbf{S}'_1}(\mathscr{R}(\tau, s))} \ .$$

But from the induction hypothesis,

$$1 - \frac{\mathsf{ps}_{\mathbf{S}'_0}(\mathscr{R}(\tau, s))}{\mathsf{ps}_{\mathbf{S}'_1}(\mathscr{R}(\tau, s))} \leq \frac{4^{t-1}q}{N^{t-1}} \prod_{j \in \{1, \dots, t\} \setminus \{i\}} p_j \ .$$

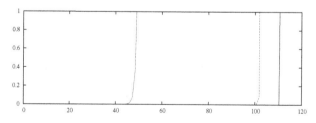

Fig. 4. Mu PRP security of 10-round KAC on 128-bit strings. From left to right: the naive bound by using the hybrid argument with CS's result, the naive bound by using the hybrid argument with the su PRP result in Theorem 1, and the bound in Theorem 2. We set $p = q = u$, where u is the number of users. The x-axis gives the log (base 2) of p, and the y-axis gives upper bounds on the mu PRP security of KAC.

4.3   Multi-user Security of KAC

In this section, we consider the multi-user security of KAC. The bounds are immediate, and rely on the fact that the actual *proof* of Theorem 1 established point-wise proximity. Indeed, from Eq. (17) in the proof of Theorem 1 and Lemma 2, we obtain Theorem 2. The analogous claims also hold for the variant KACX we discussed above.

Theorem 2 (Mu PRP security of KACs). Let t and n be positive integers, and let $\pi : \mathbb{N} \times \{0,1\}^n \to \{0,1\}^n$ be a family of ideal permutations on $\{0,1\}^n$. Let A be an adversary that makes at most q queries to ENC/DEC, and at most p_i queries to PRIM$(i,\cdot)$/PRIMINV$(i,\cdot)$ for every $i \in \{1,\ldots,t\}$. Then

$$\mathsf{Adv}^{\pm\text{mu-prp}}_{\text{KAC}[\pi,t]}(A) \leq \frac{2 \cdot 4^t q (p_1 + qt) \cdots (p_t + qt)}{N^t} \quad.$$

We note that this bound is essentially the same as the one from Theorem 1, with an additional factor two and the additive term qt. This additive term plays a significant role when t is small, but its role decreases as q grows. Concretely, for $t = 1$, we recover the Even-Mansour multi-user bound of Mouha and Luykx [22], i.e., $\mathsf{Adv}^{\pm\text{mu-prp}}_{\text{KAC}[\pi,1]}(A) \leq \frac{8(qp+q^2)}{N}$. The $O(q^2/N)$ term takes into account collisions on the keys across multiple users, which allows to easily distinguish and is therefore tight. Note that for $t = 1$, the distinction between single-key or two-key Even-Mansour is exactly the distinction between KAC and KACX, and our bounds are identical.

BEATING THE HYBRID ARGUMENT. We would like to stress once more the importance of giving direct bounds for mu security, as opposed to using a naive hybrid argument. Indeed, if we used the hybrid argument on our su PRP result in Theorem 1 then we would obtain an inferior bound with form

$$\mathsf{Adv}^{\pm\text{mu-prp}}_{\text{KAC}[\pi,t]}(A) \leq \frac{u \cdot 4^t q (p_1 + qt) \cdots (p_t + qt)}{N^t}$$

where u is the number of users. If one used the hybrid argument on CS's original bound, then the bound becomes

$$\mathsf{Adv}^{\pm\text{mu-prp}}_{\text{KAC}[\pi,t]}(A) \leq u(t+2) \left(\frac{q(6p + 6qt)^t}{N^t} \cdot t^2 (t+1)^{t+1} \right)^{1/(t+2)} \quad.$$

This makes one important point apparent: While the exponent $1/(t+2)$ in CS's bound is already undesirable in the su PRP setting, in the mu PRP case, it's much worse, as illustrated in Fig. 4. If one models AES as a 10-round KAC on 128-bit strings then our mu PRP result suggests that AES has about 110-bit security. Using the hybrid argument with our su PRP result decreases it to 100-bit security, whereas using the hybrid argument on CS's result plummets to 45-bit security.

5   XOR Cascades

In this section, we apply the above results to study XOR cascades for blockcipher key-length extension. Variants of XOR cascades have been studied in the literature [14,15,17,18,20] and the connection with KACs was already observed. However, we improve these results along two different axes: Tightness (we give a much better reduction to the security of KACs than the one of [15], using point-wise proximity), and multi-user security. In particular, to the best of our knowledge, this is the first work studying multi-user key-length extension, a problem we consider to be extremely important, given the considerable security loss in the multi-user regime.

THE XOR-CASCADE CONSTRUCTION. Let $E : \{0,1\}^k \times \{0,1\}^n \rightarrow \{0,1\}^n$ be a blockcipher. Let $t \geq 1$ be an integer, and let $\mathcal{K} = (\{0,1\}^k)^t \times (\{0,1\}^n)^{t+1}$. Let Sample be a sampling algorithm that samples $L_0, \ldots, L_t \leftarrow_\$ \{0,1\}^n$, and samples without replacement $J_1, \ldots, J_t$ from $\{0,1\}^k$, and outputs $(J_1, \ldots, J_1, L_0, \ldots, L_t)$. The XOR-Cascade construction $XC[E, t]$, on a key $K = (J_1, \ldots, J_t, L_0, \ldots, L_t) \in \mathcal{K}$, describes a permutation on $\{0,1\}^n$ as follows. On input x, $XC[E, t](x)$ returns y_t, where $y_0 = x \oplus L_0$, and $y_i = E_{J_i}(y_{i-1}) \oplus L_i$ for every $i \in \{1, \ldots, t\}$. See Fig. 5 for an illustration of $XC[E, 2]$.

We also define – in analogy with KACX above – a version of XC with t subkeys $L_1, \ldots, L_t$ (rather than $t + 1$), which xor's L_i to the input and the output of E_{J_i} in the i-th round. We refer to this as $XCX[E, t]$, and note that it is simply the t-fold sequential composition of DESX [18].

SINGLE-USER SECURITY OF $XC[E, t]$. The following theorem establishes the single-user security for $XC[E, t]$ in the ideal-cipher model, and, in contrast to previous analyses [14,15,20], the resulting bound is essentially exact. We require the keys $J_1, \ldots, J_t$ to be sampled by Sample as random yet distinct. This is no big loss – an additional $t^2/2^k$ term can added to take this into account, but this term is going to be large when moving to the multi-user case. Below, we'll develop a better bound for the independent-key case, and for now, stick with distinct keys.

Theorem 3 (Su PRP security of XC, distinct subkeys). Let t be a positive integer. Let $E : \{0,1\}^k \times \{0,1\}^n \rightarrow \{0,1\}^n$ be a blockcipher and let $XC[E, t]$ and Sample be as above. Then in the ideal-cipher model, for any adversary A that makes at most q ENC/DEC queries, and at most p PRIM/PRIMINV queries,

$$\mathsf{Adv}^{\pm\mathrm{prp}}_{XC[E,t],\mathrm{Sample}}(A) \leq \frac{4^t q p^t}{2^{t(k+n)}} \ . \tag{20}$$

The proof is in Appendix C of the full version of this paper. Here we point out a few remarks. First off, we note the bound above (and its proof) can easily adapted to analyze $XCX[E, t]$. Moreover, the proof itself is a direct application of point-wise proximity combined with the transcript reduction technique to

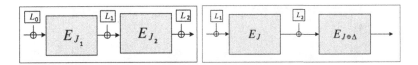

Fig. 5. Left: The XC$[E, 2]$ construction. **Right:** The 2XOR$[E]$ construction.

reduce XC case to the KAC case. This will give a tight relationship, substantially improving on the previous results by Gaži [14] and its generalization by Gaži et al. [15], which actually used an *adversarial* reduction, and needed to resort to Markov-like arguments which, once again, we avoid. Concretely, if we combine the reduction in [14,15] with our KAC result in Theorem 1, we'll obtain the following weak bound

$$\mathsf{Adv}^{\pm\mathrm{prp}}_{\mathrm{XC}[E,t],\mathrm{Sample}}(A) \leq 4^t \cdot (2t + 2) \left(\frac{qp^t}{2^{t(k+n)}} \right)^{1/(t+1)} .$$

As illustrated in Fig. 6, the gap between the bound above and ours is substantial.

MULTI-USER SECURITY OF XC. We now consider the multi-user security of XC. Since the *proof* of Theorem 3 actually establishes pointwise proximity, from Lemma 2, we obtain Theorem 4 below. If we instead use the hybrid argument on the su PRP security then we obtain an inferior bound

$$\mathsf{Adv}^{\pm\mathrm{mu\text{-}prp}}_{\mathrm{XC}[E,t],\mathrm{Sample}}(A) \leq u \cdot 4^t q(p + qt)^t / 2^{t(k+n)}$$

where u is the number of users. If we use the hybrid argument on the bound obtained by combining the reduction in [14,15] with our KAC result in Theorem 1, we'll obtain an even weaker bound

$$\mathsf{Adv}^{\pm\mathrm{prp}}_{\mathrm{XC}[E,t],\mathrm{Sample}}(A) \leq u \cdot 4^t(2t + 2) \left(\frac{q(p + qt)^t}{2^{t(k+n)}} \right)^{1/(t+1)} .$$

The three bounds are illustrated in Fig. 7.

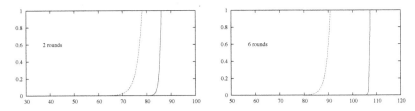

Fig. 6. Su PRP security (distinct subkeys) of XC on **2 iterations (left) and 6 iterations (right)** on $k = 56$ and $n = 64$: **our bound versus the results in** [14,15]. The solid lines depict the bound in Theorem 3, and the dashed ones depict the bound obtained by combining the reduction in [14,15] and our result in Theorem 1. In both pictures, $q = 2^n$, and the x-axis gives the log (base 2) of p, and the y-axis gives upper bounds on the su PRP security of XC.

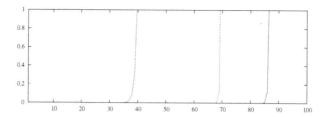

Fig. 7. Mu PRP security (distinct subkeys) of 3-round XC on $k = 56$ and $n = 64$: our bound versus naive ones from the hybrid argument. From left to right: the naive bound by using the hybrid argument with the bound obtained by combining the reduction in [14, 15] with our KAC result in Theorem 1, the naive bound by using the hybrid argument with the su PRP result in Theorem 3, and the bound in Theorem 4. We set $p = q = u$, where u is the number of users. The x-axis gives the log (base 2) of p, and the y-axis gives upper bounds on the mu PRP security of XC.

Theorem 4 (Mu PRP security of XC, distinct subkeys). Let t be a positive integer. Let $E : \{0,1\}^k \times \{0,1\}^n \to \{0,1\}^n$ be a blockcipher and let $XC[E, t]$ and Sample be as above. Then in the ideal-cipher model, for any adversary A that makes at most q ENC/DEC queries, and at most p PRIM/PRIMINV queries,

$$\mathsf{Adv}^{\pm\text{mu-prp}}_{XC[E,t],\text{Sample}}(A) \leq 2 \cdot 4^t q(p + qt)^t / 2^{t(k+n)} .$$

We stress here that q is allowed to be larger than $N = 2^n$ — nothing in the theorem limits this, and security is obtained as long $2 \cdot 4^t q(p + qt)^t / 2^{t(k+n)}$ is sufficiently small. This is conceptually very important. Indeed, we may want to apply our result even to ciphers for which N is very small (these arise in the setting of FPE [3], where one could have $N \approx 2^{30}$, or even less), and a multi-user attacker can exhaust the domain for multiple keys. In passing, we note that the reason such a strong result is possible is inherited directly from the fact that Theorem 1 does not make any restrictions on q.

There are some variants of XC in the literature. For example, Gaži and Tessaro (GT) [17] gave a variant of $XC[E, 2]$ that they call 2XOR. This construction, as illustrated in Fig. 5, uses a shorter key and saves one additional xor, compared to $XC[E, 2]$. While its su PRP security appears to be the same as $XC[E, 2]$, as GT's result suggests, in Appendix E of the full version, we show that it has much weaker mu PRP security by giving an attack.

ON UNIFORM SUBKEYS. So far we have considered security of the XC construction when each key $K = (J_1, \ldots, J_t, L_0, \ldots, L_t)$ is chosen so that the subkeys $J_1, \ldots, J_t$ are distinct. A natural question is to bound the degradation when $J_1, \ldots, J_t \leftarrow\!\!\!{}^\$ \{0,1\}^k$. First consider the su setting. A simple solution is to add a term $t^2 / 2^k$ to account for the probability that there are some $i \neq j$ such that $J_i = J_j$. This is fine for the su setting, but when one moves to the mu setting, this term blows up to $ut^2 / 2^k$, where u is the number of users. This happens even in the ideal case where the adversary distributes the queries evenly among users. To avoid this undesirable term, in Proposition 1 below, we take a different approach. Intuitively, even if there are only $\ell \leq t$ distinct subkeys, then at least our construction should achieve security level $\epsilon(\ell)$ similar to the bound in Theorem 3 for $\mathrm{XC}[E, \ell]$. Let L be the random variable for the number of distinct subkeys in $\mathrm{XC}[E, t]$, for example, $\Pr[L = t] \geq 1 - t^2/2^k$. Then our bound would be the expectation $\mathbf{E}(\epsilon(L))$. The gap between this bound and the naive one with the term $t^2/2^k$ may not be large on practical values of n and k, but it allows us to use Lemma 2 to obtain a good mu PRP bound.

Proposition 1 (Su PRP security of XC, uniform subkeys). Let $t \geq 2$ be an integer. Let $E : \{0,1\}^k \times \{0,1\}^n \to \{0,1\}^n$ be a blockcipher and let $\mathrm{XC}[E, t]$ be as above. Then in the ideal-cipher model, for any adversary A that makes at most q ENC/DEC queries, and at most p PRIM/PRIMINV queries,

(a) If $t \geq 3$ then $\mathsf{Adv}^{\pm\mathrm{prp}}_{\mathrm{XC}[E,t]}(A) \leq \frac{4^t qp^t}{2^{(n+k)t}} + \frac{qt^2}{2^k}\left(\frac{t}{2^k} + \frac{4p}{2^{k+n}}\right)^{t-2}$.

(b) If $t = 2$ then $\mathsf{Adv}^{\pm\mathrm{prp}}_{\mathrm{XC}[E,t]}(A) \leq \frac{q(4p)^2}{2^{2(n+k)}} + \frac{4qp}{2^{2k+n}} + \frac{2q}{2^{k+n/2}}$.

The proof of Proposition 1 is in Appendix D of the full version, and it also establishes pointwise proximity. From Lemma 2, we obtain Theorem 5 below. As illustrated in Fig. 8, this bound is much better than the naive one obtained via adding a term $ut^2/2^k$ to the bound in Theorem 4 (to account for the probability that there is a user whose subkeys are not distinct), where u is the number of users. When one increases the number of rounds then our bound shows that the security substantially improves (from 80-bit to 90-bit security), but the naive bound still stays at 50-bit security, since the bound $ut^2/2^k$ is the bottleneck, and it gets *worse* when t increases.

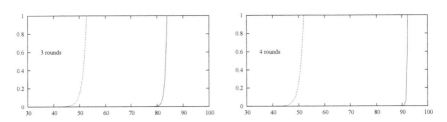

Fig. 8. Mu PRP security of XC (uniform subkeys) on 3 iterations (left) and 4 iterations (right) on $k = 56$ and $n = 64$: our bound versus naive one. The dashed lines depict the bound obtained by adding a term $ut^2/2^k$ to the bound in Theorem 4, and the solid ones depict the bound in Theorem 5, where u is the number of users. In both pictures, $p = q = u$, and the x-axis gives the log (base 2) of p, and the y-axis gives upper bounds on the mu PRP security of XC.

Theorem 5 (Mu PRP security of XC, uniform subkeys). Let $t \geq 2$ be an integer. Let $E : \{0,1\}^k \times \{0,1\}^n \to \{0,1\}^n$ be a blockcipher and let $\mathrm{XC}[E,t]$ be as above. Then in the ideal-cipher model, for any adversary A that makes at most q ENC/DEC queries, and at most p PRIM/PRIMINV queries,

(a) If $t \geq 3$ then $\mathsf{Adv}^{\pm\mathrm{mu\text{-}prp}}_{\mathrm{XC}[E,t]}(A) \leq \frac{2 \cdot 4^t q(p+qt)^t}{2^{(n+k)t}} + \frac{2qt^2}{2^k}\left(\frac{t}{2^k} + \frac{4p+4qt}{2^{k+n}}\right)^{t-2}$.

(b) If $t = 2$ then $\mathsf{Adv}^{\pm\mathrm{mu\text{-}prp}}_{\mathrm{XC}[E,t]}(A) \leq \frac{2q(4p+8q)^2}{2^{2(n+k)}} + \frac{8q(p+2q)}{2^{2k+n}} + \frac{4q}{2^{k+n/2}}$.

INTERPRETING THE BOUNDS IN THEOREM 5. For the case $t = 3$, there's a considerable gap compared to the matching attack. See Fig. 9 for an illustration of the degradation of the bound in Theorem 5 compared to that in Theorem 4. This gap is probably an artifact of the proof technique rather than reflecting a true security loss when using uniform subkeys: for example, in the su case, if $J_1 = \cdots = J_t$ then we give up, but of course even in this extreme case, the construction should still retain some reasonable security. For $t \geq 4$ and all practical choices of n and k, the bounds in Theorems 5 and 4 are close: the former is just about $t^2 + 1$ times worse than the latter. To justify this, note that we can assume that $4(p + qt)/2^n > 2^{k/2}$, otherwise both bounds are tiny. Then

$$\frac{qt^2}{2^k}\left(\frac{t}{2^k} + \frac{4p+4qt}{2^{k+n}}\right)^{t-2} \approx \frac{qt^2}{2^k}\left(\frac{4p+4qt}{2^{k+n}}\right)^{t-2} < t^2 \cdot \frac{4^t q(p+qt)^t}{2^{(n+k)t}} \quad .$$

Pictorially, as shown in Fig. 9, the two bounds are too close, and we have to choose very small n and k so that the gap between the two lines is still visible to the naked eye. Likewise, for $t = 2$ and and all practical choices of n and k, the bound in Theorem 5 is about twice worse than that of Theorem 4. (In Proposition 1, for $t = 2$, if $J_1 = J_2$ then we don't give up, but show that the construction still retains security bound up to $\frac{4qp}{2^{k+n}} + \frac{2q}{2^{n/2}}$. However, this method fails to work for $t = 3$. It's why the bound in Theorem 5 is still sharp for $t = 2$, but deteriorates for $t = 3$.)

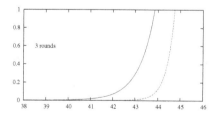 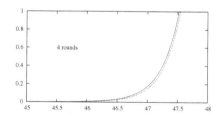

Fig. 9. Mu PRP security of XC on **3 iterations (left)** and **4 iterations (right)** on $k = n = 32$: **uniform versus distinct subkeys.** The dashed lines depict the bound in Theorem 4, and the solid ones depict the bound in Theorem 5. In both pictures, $p = q$, and the x-axis gives the log (base 2) of p, and the y-axis gives upper bounds on the mu PRP security of XC. The parameters n and k are chosen to be small so that in the right picture, the gap between the two lines is visible to the naked eye.

Acknowledgments. We thank Mihir Bellare for insightful feedback, and Daniel J. Bernstein for providing relevant pointers. We also wish to thank Atul Luykx and Bart Mennink for pointing out a glitch in a previous version of this write up. Finally, we thank the CRYPTO 2016 reviewers for many insightful comments.

This research was partially supported by NSF grants CNS-1423566 and CNS-1553758 (CAREER).

References

1. Andreeva, E., Bogdanov, A., Dodis, Y., Mennink, B., Steinberger, J.P.: On the indifferentiability of key-alternating ciphers. In: Canetti, R., Garay, J.A. (eds.) CRYPTO 2013, Part I. LNCS, vol. 8042, pp. 531–550. Springer, Heidelberg (2013)
2. Bellare, M., Boldyreva, A., Micali, S.: Public-key encryption in a multi-user setting: security proofs and improvements. In: Preneel, B. (ed.) EUROCRYPT 2000. LNCS, vol. 1807, pp. 259–274. Springer, Heidelberg (2000)
3. Bellare, M., Ristenpart, T., Rogaway, P., Stegers, T.: Format-preserving encryption. In: Jacobson Jr., M.J., Rijmen, V., Safavi-Naini, R. (eds.) SAC 2009. LNCS, vol. 5867, pp. 295–312. Springer, Heidelberg (2009)
4. Bellare, M., Rogaway, P.: The security of triple encryption and a framework for code-based game-playing proofs. In: Vaudenay, S. (ed.) EUROCRYPT 2006. LNCS, vol. 4004, pp. 409–426. Springer, Heidelberg (2006)
5. Bernstein, D.J.: How to stretch random functions: the security of protected counter sums. J. Cryptol. **12**(3), 185–192 (1999)
6. Bernstein, D.J.: Break a dozen secret keys, get a million more for free (2015). http://blog.cr.yp.to/20151120-batchattacks.html
7. Bogdanov, A., Knudsen, L.R., Leander, G., Standaert, F.-X., Steinberger, J., Tischhauser, E.: Key-alternating ciphers in a provable setting: encryption using a small number of public permutations. In: Pointcheval, D., Johansson, T. (eds.) EUROCRYPT 2012. LNCS, vol. 7237, pp. 45–62. Springer, Heidelberg (2012)
8. Chen, S., Lampe, R., Lee, J., Seurin, Y., Steinberger, J.: Minimizing the two-round even-mansour cipher. In: Garay, J.A., Gennaro, R. (eds.) CRYPTO 2014, Part I. LNCS, vol. 8616, pp. 39–56. Springer, Heidelberg (2014)
9. Chen, S., Steinberger, J.: Tight security bounds for key-alternating ciphers. In: Nguyen, P.Q., Oswald, E. (eds.) EUROCRYPT 2014. LNCS, vol. 8441, pp. 327–350. Springer, Heidelberg (2014)
10. Dai, Y., Lee, J., Mennink, B., Steinberger, J.: The security of multiple encryption in the ideal cipher model. In: Garay, J.A., Gennaro, R. (eds.) CRYPTO 2014, Part I. LNCS, vol. 8616, pp. 20–38. Springer, Heidelberg (2014)
11. Dunkelman, O., Keller, N., Shamir, A.: Minimalism in cryptography: the even-mansour scheme revisited. In: Pointcheval, D., Johansson, T. (eds.) EUROCRYPT 2012. LNCS, vol. 7237, pp. 336–354. Springer, Heidelberg (2012)
12. Even, S., Mansour, Y.: A construction of a cipher from a single pseudorandom permutation. In: Imai, H., Rivest, R.L., Matsumoto, T. (eds.) ASIACRYPT 1991. LNCS, vol. 739, pp. 210–224. Springer, Heidelberg (1993)
13. Even, S., Mansour, Y.: A construction of a cipher from a single pseudorandom permutation. J. Cryptol. **10**(3), 151–162 (1997)
14. Gaži, P.: Plain versus randomized cascading-based key-length extension for block ciphers. In: Canetti, R., Garay, J.A. (eds.) CRYPTO 2013, Part I. LNCS, vol. 8042, pp. 551–570. Springer, Heidelberg (2013)

15. Gaži, P., Lee, J., Seurin, Y., Steinberger, J., Tessaro, S.: Relaxing full-codebook security: a refined analysis of key-length extension schemes. In: Leander, G. (ed.) FSE 2015. LNCS, vol. 9054, pp. 319–341. Springer, Heidelberg (2015)
16. Gaži, P., Maurer, U.: Cascade encryption revisited. In: Matsui, M. (ed.) ASIACRYPT 2009. LNCS, vol. 5912, pp. 37–51. Springer, Heidelberg (2009)
17. Gaži, P., Tessaro, S.: Efficient and optimally secure key-length extension for block ciphers via randomized cascading. In: Pointcheval, D., Johansson, T. (eds.) EUROCRYPT 2012. LNCS, vol. 7237, pp. 63–80. Springer, Heidelberg (2012)
18. Kilian, J., Rogaway, P.: How to protect DES against exhaustive key search. In: Koblitz, N. (ed.) CRYPTO 1996. LNCS, vol. 1109, pp. 252–267. Springer, Heidelberg (1996)
19. Lampe, R., Patarin, J., Seurin, Y.: An asymptotically tight security analysis of the iterated even-mansour cipher. In: Wang, X., Sako, K. (eds.) ASIACRYPT 2012. LNCS, vol. 7658, pp. 278–295. Springer, Heidelberg (2012)
20. Lee, J.: Towards Key-length extension with optimal security: cascade encryption and xor-cascade encryption. In: Johansson, T., Nguyen, P.Q. (eds.) EUROCRYPT 2013. LNCS, vol. 7881, pp. 405–425. Springer, Heidelberg (2013)
21. Maurer, U.M.: Indistinguishability of random systems. In: Knudsen, L.R. (ed.) EUROCRYPT 2002. LNCS, vol. 2332, pp. 110–132. Springer, Heidelberg (2002)
22. Mouha, N., Luykx, A.: Multi-key security: the Even-Mansour construction revisited. In: Gennaro, R., Robshaw, M.J.B. (eds.) CRYPTO 2015, Part I. LNCS, vol. 9215, pp. 209–223. Springer, Heidelberg (2015)
23. Nandi, M.: A simple and unified method of proving indistinguishability. In: Barua, R., Lange, T. (eds.) INDOCRYPT 2006. LNCS, vol. 4329, pp. 317–334. Springer, Heidelberg (2006)
24. Patarin, J.: The "Coefficients H" technique. In: Avanzi, R.M., Keliher, L., Sica, F. (eds.) SAC 2008. LNCS, vol. 5381, pp. 328–345. Springer, Heidelberg (2009)
25. Steinberger, J.: Improved security bounds for key-alternating ciphers via hellingerdistance. Cryptology ePrint Archive, Report 2012/481 (2012). http://eprint.iacr.org/2012/481
26. Tessaro, S.: Optimally secure block ciphers from ideal primitives. In: Iwata, T., et al. (eds.) ASIACRYPT 2015. LNCS, vol. 9453, pp. 437–462. Springer, Heidelberg (2015). doi:10.1007/978-3-662-48800-3_18

Counter-in-Tweak: Authenticated Encryption Modes for Tweakable Block Ciphers

Thomas Peyrin[1] and Yannick Seurin[2(✉)]

[1] SPMS, NTU, Singapore, Singapore
thomas.peyrin@ntu.edu.sg
[2] ANSSI, Paris, France
yannick.seurin@m4x.org

Abstract. We propose the Synthetic Counter-in-Tweak (SCT) mode, which turns a tweakable block cipher into a nonce-based authenticated encryption scheme (with associated data). The SCT mode combines in a SIV-like manner a Wegman-Carter MAC inspired from PMAC for the authentication part and a new counter-like mode for the encryption part, with the unusual property that the counter is applied on the tweak input of the underlying tweakable block cipher rather than on the plaintext input. Unlike many previous authenticated encryption modes, SCT enjoys provable security beyond the birthday bound (and even up to roughly 2^n tweakable block cipher calls, where n is the block length, when the tweak length is sufficiently large) in the nonce-respecting scenario where nonces are never repeated. In addition, SCT ensures security up to the birthday bound even when nonces are reused, in the strong nonce-misuse resistance sense (MRAE) of Rogaway and Shrimpton (EUROCRYPT 2006). To the best of our knowledge, this is the first authenticated encryption mode that provides at the same time close-to-optimal security in the nonce-respecting scenario and birthday-bound security for the nonce-misuse scenario. While two passes are necessary to achieve MRAE-security, our mode enjoys a number of desirable features: it is simple, parallelizable, it requires the encryption direction only, it is particularly efficient for small messages compared to other nonce-misuse resistant schemes (no precomputation is required) and it allows incremental update of associated data.

Keywords: Authenticated encryption · Tweakable block cipher · Nonce-misuse resistance · Beyond-birthday-bound security · CAESAR competition

1 Introduction

BACKGROUND ON AUTHENTICATED ENCRYPTION. Confidentiality and authenticity of data are the two main security properties that one must ensure when communicating over an insecure channel. In the symmetric key setting, it has long been known how to ensure both of them independently, e.g., starting from

© International Association for Cryptologic Research 2016
M. Robshaw and J. Katz (Eds.): CRYPTO 2016, Part I, LNCS 9814, pp. 33–63, 2016.
DOI: 10.1007/978-3-662-53018-4_2

a secure block cipher, by using a suitable encryption mode for confidentiality [7] and a block cipher-based MAC for authenticity [9]. However, how exactly to combine both tools has long been left to the practitioners, leading to major security breaches [3,18,37]. Sometimes, protocol designers even overlooked that authenticity was a necessary requirement besides confidentiality, as exemplified by padding oracle attacks [59]. Even when the combination of the encryption and the MAC schemes is properly done, it might not be the most efficient solution, especially when the two parts rely on two different primitives. For these reasons, interest has shifted towards designing "integrated" Authenticated Encryption (AE) schemes ensuring jointly authenticity and confidentiality of data, which are more efficient and less likely to be incorrectly used. Besides, it has become standard for an AE scheme to have the ability to handle so-called associated data (AD), which are authenticated but not encrypted [51] (such a scheme was for a time called an AEAD scheme, but since this feature is so important in practice, virtually all modern AE schemes provide it; we will only talk of AE in this paper, implicitly meaning AEAD). Even though ad-hoc AE schemes were already used since a long time, the formal treatment of these constructions only started around 2000 [10,11,36]. At about the same time, provably secure AE designs started to appear, such as IAPM [34,35], XCBC [22], CCM [61], OCB [52,54], or GCM [43]. The CAESAR competition [1] for authenticated encryption, started in 2014, recently put this research topic in the limelight. Various AE schemes were proposed, from purely ad-hoc designs to (tweakable) block cipher operating modes.

NONCE-MISUSE RESISTANCE. Since most symmetric-key primitives (in particular block ciphers) from which AE schemes are built are deterministic, a random IV or a nonce (i.e., a value which must never be repeated for the same secret key) is a necessary ingredient for achieving strong security goals. Failing to ensure the corresponding requirement (high entropy for an IV, non-repetition for a nonce) can have dramatic consequences for security. For example, reusing a nonce just a single time for encrypting two messages in OCB completely breaks confidentiality: an attacker can immediately detect repeated message blocks since the corresponding ciphertext blocks will be equal. The non-repeating requirement on the nonce can be challenging to fulfill in some contexts, for example when encryption is implemented in a stateless device.[1] It is likely (and it has happened before) that some implementations will, e.g., simply generate nonces at random, "hoping" that no collision will occur. For that reason, a recent trend in AE has been to design schemes achieving *nonce-misuse resistance*, which informally means that the impact on security of a nonce repetition should be as limited as possible. This goal was first put forward by Rogaway and Shrimpton [55], who formalized the notion of *misuse-resistant AE* (MRAE). For a scheme enjoying this property, authenticity is not harmed by nonce repetitions, while confidentiality is only damaged insofar as the adversary can detect whether the same triple of nonce, AD and message values is repeated. Example of schemes achieving this

---

[1] Similarly, the high-entropy requirement on the IV is hard to meet when no good randomness source is available.

security notion are EAX [12], SIV [55], AEZ [26], or GCM-SIV [25]. Because the MRAE notion cannot be achieved for an online scheme (since each bit of the ciphertext must depend on every bit of the plaintext), Fleischmann *et al.* [21] proposed a relaxation of the MRAE notion called *online AE* (OAE), which can be achieved with a single pass on the input. Examples of schemes ensuring this security property are McOE [21], COPA [5], or POET [2]. However, the interest in the OAE notion has been recently reduced by some serious security concerns, notably the so-called chosen-prefix/secret-suffix (CPSS) generic attack [27], that shares some similarities with the BEAST attack [18].

BIRTHDAY AND BEYOND-BIRTHDAY SECURITY. Another important shortcoming of most AE operating modes is that they provide only birthday-bound security with respect to the block length of the underlying primitive. Since virtually all existing block ciphers have block length at most 128 bits (in particular the current block cipher standard AES), this means that security is lost at 2^{64} block cipher calls at best, which is low given modern security requirements (for 64-bit block ciphers, the situation is even more problematic). Moreover, this is rarely a problem with the tightness of the security proof: attacks matching the bound are often known. For example, Ferguson [19] described a simple collision-based attack on OCB that breaks authenticity with 2^{64} blocks of messages. Recently, some AE schemes providing security beyond the birthday bound (BBB) were proposed [28,29], but they usually come at an expensive performance price. One could argue that using a double-block-length block cipher would provide the expected security, but this solution comes with an important efficiency penalty (as can be seen in generic double-block-length block cipher constructions) and would be highly problematic for hardware implementations where internal state size is a major contribution to the total area cost.

AE FROM TWEAKABLE BLOCK CIPHERS. Compared with a conventional block cipher, a tweakable block cipher (TBC) $\widetilde{E}$ takes an additional input called a *tweak* bringing inherent variability to the primitive (equivalently, a TBC can be seen as a family of block ciphers indexed by the tweak). In the same paper that formalized the corresponding security notion [41], it was pointed out that a TBC was a very convenient starting point for building various schemes. In particular, for AE schemes, two prominent examples are the sibling modes TAE [41] and ΘCB [38] (the TBC-based generalization of OCB). They have "perfect" security in the sense that, when used with an ideal TBC, the advantage of any adversary is zero against confidentiality and close to $2^{-\tau}$ against authenticity, where τ is the tag length. However, as already pointed out, a weakness of both TAE and ΘCB (even when used with an ideal TBC) is that their security completely collapses as soon as a nonce is repeated. As a matter of fact, existing AE schemes built from an ideal TBC either ensure perfect security in the nonce-respecting scenario only (like TAE or ΘCB), or fulfill the weak OAE notion only (e.g. COPA, once recast to use an ideal TBC), or ensure MRAE-security but only up to the birthday bound, even if nonces are not repeated (like AEZ). The PIV construction by Shrimpton and Terashima [58] allows to construct a variable-input-length TBC with BBB-security, which in turn allows to construct (via the Encode-then-Encipher method) an AEAD scheme with BBB-security against nonce-respecting adversaries and birthday-bound security against nonce-misusing ones. However, PIV requires as a building block a fixed-input-length TBC with *variable* tweak length (comparable to the maximal input length of the PIV construction), which in turn requires to appeal to universal hash functions with key length comparable to the maximal tweak length. Hence, the resulting AEAD scheme must use very large keys to ensure BBB-security for large messages. As of today, there is no AEAD scheme based on a fixed-tweak-length TBC that ensures both BBB-security in the nonce-respecting scenario and (at least) birthday-bound security in the nonce-misuse scenario. Yet this seems a very desirable goal since such a scheme would at the same time yield very high (BBB) security guarantees in the nominal, nonce-respecting use case and retain acceptable (birthday-bound) security when inadvertently misused.

OUR CONTRIBUTIONS. In this paper, we propose the SCT (Synthetic Counter-in-Tweak) nonce-based AE mode for tweakable block ciphers and prove that it ensures BBB-security in the nonce-respecting scenario, and birthday-bound security in the nonce-misuse scenario (in the strong MRAE sense [55]). More precisely, for the nonce-respecting case, when using a ideal TBC with block length n and "effective" tweak length[2] w, SCT is secure up to roughly 2^n TBC calls when $w \geq n$, and up to roughly $2^{(n+w)/2}$ TBC calls when $w \leq n$, which is always larger than $2^{n/2}$. The SCT mode requires two passes (as is inevitable for MRAE-security), but it is simple, parallelizable, it requires the encryption direction only, it is particularly efficient for small messages compared to other nonce-misuse resistant schemes (no precomputation is required) and it allows incremental update of associated data.

With respect to how authentication and encryption are combined, our design draws inspiration from the SIV generic composition method [55]: the nonce N, the associated data A, and the message M are first input to a keyed function F_K, yielding an pseudorandom initial value IV, which will serve as authentication tag. The message is then encrypted, using the generated IV *and the nonce N* (see Fig. 4). This "recycling" of the nonce in the encryption part of the mode is what makes our high-level construction (called NSIV) crucially different from SIV and allows to reach BBB-security in the nonce-respecting case.[3]

It remains to instantiate the two components of the NSIV construction, the keyed function F_K and the encryption scheme. Since we aim at BBB-security in the nonce-respecting case, a natural starting point for F_K is the Wegman-Carter paradigm [14,56,60]. Hence, we propose a nonce-based MAC mode called PWC (*Parallel Wegman-Carter*) which combines a xor-universal hash function inspired from PMAC [13,52] applied to the AD and the message, and a simple

---

[2] The SCT mode uses 5 tweak prefixes to separate the different usages of the TBC. The "effective" tweak length is what remains once 3 bits have been used to encode the prefix.

[3] While SIV corresponds to generic composition method A4 in the nomenclature of Namprempre et al. [46], NSIV does not fit any of the NRS schemes.

pseudorandom function applied to the nonce. In order to achieve nonce-misuse resistance (which in general Wegman-Carter MACs do not provide), we add an additional encryption layer, which results in the EPWC (*Encrypted PWC*) mode.

The real challenge lies in designing an encryption scheme which is BBB-secure in the nonce-respecting case. Since on one hand it seems hard to leverage on the non-repeating property of the nonce without actually giving the nonce as input to the encryption mode, and on the other hand we need to make use in some way of the pseudorandom IV computed from F_K,[4] it appears that what we need to design is actually a *combined nonce- and IV-based encryption scheme (nivE scheme* for short). To the best of our knowledge, this notion has never appeared before, and we introduce it in this paper. The encryption mode that we propose, called CTRT (*CounTeR in Tweak*), is a counter-like mode with the unusual property that the counter is applied on the tweak input of the underlying TBC rather than on the plaintext input, where the nonce comes in. The combination of EPWC and CTRT through the NSIV construction (the IV generated by EPWC being used as initial counter in CTRT) yields the SCT mode, illustrated in Fig. 1.

For completeness, we also describe in the full version of this paper [50] the CTPWC (CTRT-*then*-PWC) mode, an *online* nonce-based AE scheme which combines in an "encrypt-then-MAC" manner a slight variant of the CTRT encryption mode and the PWC authentication mode. The security guarantees provided by CTPWC are similar to those of ΘCB, but it is roughly twice less efficient, so that we do not claim that it is of particularly high interest. One small advantage that we see for this mode compared with ΘCB is that the nonce length can be as large as the block length of the underlying TBC, whereas for ΘCB the sum of the nonce length and of the maximal length of encrypted messages must be less than the tweak length of the underlying tweakable block cipher, which might be restrictive in some settings (e.g., for KIASU-BC [32]). It might also escape the patent issues which hindered the adoption of OCB.

INSTANTIATING THE TBC. As just discussed, our new AE modes offer BBB-security (in the nonce-respecting case) when used with an ideal TBC. If one aims at leveraging this security level in the real world, one must instantiate the TBC with care. Most existing TBCs are built from conventional block ciphers in a generic way, the prominent example being the XE/XEX construction [52] which only ensures security up to the birthday bound. Hence, using XE/XEX in our schemes would in a sense waste their nice security promises.[5] To remedy this problem, one can use either generic TBC constructions with BBB-security [39,40,44,45] (but they are often inefficient or provably secure in the ideal cipher model only), or ad-hoc TBC designs without known weaknesses. The second option was chosen for a number of CAESAR candidates [24,30–32]. In fact, the SCT mode was explicitly designed as a replacement to the COPA mode used in versions 1.1 and 1.2 of CAESAR candidates Deoxys [30] (128-bit

---

[4] This excludes for example a simple OCB-like encryption mode since it is only nonce-based, not IV-based.

[5] Similarly, the only reason why OCB is secure up to the birthday bound whereas ΘCB is "perfectly" secure is because it relies on XE/XEX for instantiating the TBC.

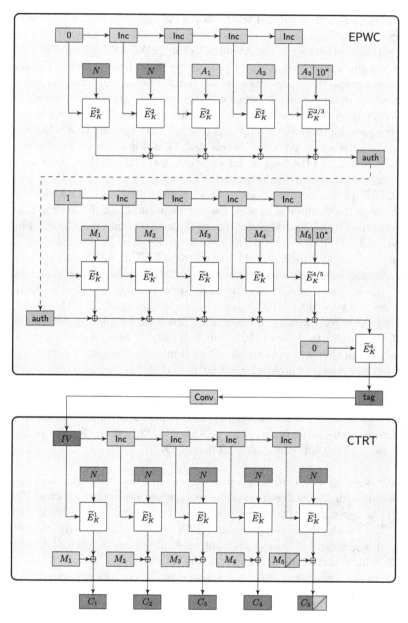

Fig. 1. The SCT mode, using a TBC $\widetilde{E}$ with tweak space $\{1, \ldots, 5\} \times \mathcal{T}$ and domain $\mathcal{X} = \{0,1\}^n$. For each call to $\widetilde{E}_K$, the tweak enters left while the plaintext enters on top. We denote $\widetilde{E}_K^i(T, X)$ for $\widetilde{E}_K((i, T), X)$ and $\widetilde{E}_K^{i/j}$ means that prefix i is used when the input block is complete and unpadded, whereas j is used when the input block is incomplete and padded. Function Inc is a cyclic permutation of $\mathcal{T}$, and Conv is a regular function from $\mathcal{X}$ to $\mathcal{T}$ (e.g., truncation when $\mathcal{X} = \{0,1\}^n$ and $\mathcal{T} = \{0,1\}^w$, $w \le n$).

blocks, 128-bit tweaks) and Joltik [31] (64-bit blocks, 64-bit tweaks).[6] We refer to the submission documents of these two candidates for a detailed report on implementation results, which are quite competitive. Other potential candidates for instantiating the SCT mode are Scream [24] and Threefish, the TBC on which the hash function Skein [20] is based. There is currently a shift towards designing dedicated TBCs achieving higher security and efficiency than generic BC-based constructions, and we hope to see more and more TBC proposals that could be used with SCT.

OPEN PROBLEMS AND FUTURE WORK. The CTRT encryption scheme has the notable feature that its security degrades gracefully with the maximal number of repetitions of nonces: when the nonce repetitions are limited, security remains close to the security bound in the nonce-respecting case. In contrast, the security of the EPWC authentication mode (and more generally of any encrypted Wegman-Carter MAC) falls back to birthday bound as soon as the adversary can repeat one single nonce twice (see Remark 2 in Sect. 5). It remains a pending question to modify EPWC so that it provides graceful security degradation with the maximal number of nonce repetitions as well. This would make the resulting AE scheme a good candidate for high security in both nonce-respecting and nonce-misuse models for most practical scenarios. Another challenging open problem would be to construct an AE scheme which remains BBB-secure even when nonces are arbitrarily repeated. The main difficulty is to build a deterministic, stateless, BBB-secure MAC, which is known to be notably hard [17,62]. Another possible direction for future work would be to design a mode similar to SCT using only one pass and achieving online nonce-misuse resistance in the OAE sense [21]. Such a feature would allow users to smoothly choose the best security achievable, depending on whether two passes can be tolerated or not by the application. Finally, it would be interesting to analyze how to strengthen SCT against other misuse scenarios such as release of unverified plaintext [4], and to study how its security is affected by tag truncation [26].

ORGANIZATION. In Sect. 2 we provide a high-level description of the various possibilities that we considered for constructing a BBB-secure nivE scheme. After introducing the notation and standard security notions in Sect. 3, we describe the CTRT encryption scheme and prove its security in Sect. 4, while we describe the PWC and EPWC nonce-based MAC schemes and prove their security in Sect. 5. Finally, we explain how to combine CTRT and EPWC using the NSIV construction to build the nonce-based AE mode SCT and prove its security in Sect. 6.

2   Counter-in-Tweak for Beyond-Birthday Security

As motivated in introduction, our goal is to design a simple TBC-based AE scheme that provides BBB-security in the nonce-respecting setting and (at least)

---

[6] The tweak prefixes used in this paper were chosen for ease of exposition and are slightly different from the ones used in Deoxys and Joltik v1.3, which were chosen mainly for efficiency reasons.

birthday-bound MRAE-security. As already mentioned, an encrypted Wegman-Carter MAC solves the problem for the authentication part, so that we focus here on encryption. Hence, our problem is as follows: given a nonce and a synthetic IV generated pseudorandomly from the nonce, the AD and the message, how do we use them to encrypt the message with BBB-security? We give a quick overview of the various constructions that we considered and why, except the CTRT encryption mode we propose, they fail or are unsatisfactory.

A natural direction to explore is to start from a scheme providing BBB-security for non-repeating nonces, such as TAE [41] or ⊖CB [38], which are similar with regard to encryption: it simply consists in a "tweakable" code-book mode, the tweak holding the nonce and a message block counter. This is obviously not nonce-misuse resistant: repeating the nonce a single time will lead to a complete break of confidentiality since a constant message block leads to a constant ciphertext block. One could incorporate the IV by simply concate-nating it to the nonce and the counter to form the tweak. However, this would require a TBC with larger tweak, which is usually very costly to achieve.[7] Rather than concatenating the nonce and the IV, one could try to combine them into a single shorter string S, but this would presumably result in birthday security even in the nonce-respecting scenario (since a collision on S would directly break confidentiality). Hence, a codebook encryption mode does not seem to be a very convenient starting point.

For this reason, we preferred to consider a counter mode (note that this was the encryption mode favored by Rogaway and Shrimpton to instantiate the SIV composition method [55]). The question now is: how do we feed the nonce, the IV, and the i-th counter to the TBC in order to create the mask that will be xored to the i-th message block? We considered several possibilities (we do not claim this to be exhaustive):

(a) One can put the nonce in the tweak input, and the sum of the IV and the counter in the plaintext input. The problem is that confidentiality caps at birthday bound even in the nonce-respecting scenario: the adversary can query the encryption of a single message with $2^{n/2}$ equal blocks, and observe that no collision occurs in the corresponding ciphertext blocks (since the nonce is fixed and all TBC calls use the same tweak), which will distin-guish the ciphertext from a random string (for which a collision would be expected).

(b) One can concatenate the nonce and the counter in the tweak input, and use the IV for the plaintext input. Since the tweak is different for each message block position, this solves the issue of the previous solution. We conjecture that this mode meets our security objectives, but an important drawback is that a larger tweak length is required and, as mentioned before, this is very costly.

---

[7] For example, for TBCs following the TWEAKEY approach [30,31,33], there is a large gap in the number of rounds needed to make the TBC secure as the tweak length increases.

(c) One can put the sum of the nonce and the counter in the tweak input (instead of concatenating them) and the IV in the plaintext input. This mode might meet our security objectives, however the adversary can very easily provoke collisions on tweak inputs even in the nonce-respecting scenario, which might complicate the proof of BBB-security. Another drawback is that in the nonce-misuse scenario, a collision on the IV immediately breaks confidentiality, which dashes any hope for BBB nonce-misuse resistance. Note that one could imagine variants where the nonce and the IV are first encrypted before being used, but it is not clear if this would prevent the issues just mentioned and this would presumably make the security proof quite complex.

(d) Finally, one can put the sum of the IV and the counter in the tweak input and the nonce in the plaintext input, which is exactly the CTRT mode. We will prove in Sect. 4 that it meets our security goal. The first idea is that the counter in the tweak input ensures that all the calls to the internal TBC will use different tweaks for one single message query, so that the ciphertext looks uniformly random in that case. Thus, the adversary has to query several messages with different nonce values and hope that many collisions will occur between tweak inputs in order to observe a divergence from uniformity in the ciphertexts. However, these collisions are hard to control since they depend on the pseudorandom IV (in contrast with other modes discussed above, where the tweak input can be easily controlled by the adversary). We will show in Sect. 4, using a "balls-into-bins" analysis, that the number of tweak collisions remains small, so that the distribution of the ciphertexts remains close to uniform. Moreover, the nonce-misuse scenario helps the adversary only if it can repeat the same nonce a very high number of times until a collision happens on the tweaks, so that the security of the CTRT mode degrades gracefully with the maximal number of nonce repetitions.

3   Preliminaries

NOTATION. Given a string $X \in \{0,1\}^*$, $|X|$ denotes its length. If X and Y are respectively n-bit and m-bit strings, $n < m$, then $X \oplus Y$ denotes the n-bit string obtained by xoring X with the n leftmost bits of Y. Given some implicit length n and a bit-string X of length $1 \le |X| < n$, we denote $X10^*$ the string obtained by appending a single 1 and $(n - |X| - 1)$ 0's to X. Given two sets $\mathcal{X}$ and $\mathcal{Y}$, the set of all functions from $\mathcal{X}$ to $\mathcal{Y}$ is denoted $\mathsf{Func}(\mathcal{X}, \mathcal{Y})$. A function $F \in \mathsf{Func}(\mathcal{X}, \mathcal{Y})$ is said *regular* if all $Y \in \mathcal{Y}$ have the same number of preimages by F (this obviously requires $|\mathcal{X}|$ to be a multiple of $|\mathcal{Y}|$).

TWEAKABLE BLOCK CIPHERS. A *tweakable block cipher* (TBC) with key space $\mathcal{K}$, tweak space $\mathcal{T}$, and domain $\mathcal{X}$ is a mapping $\widetilde{E} : \mathcal{K} \times \mathcal{T} \times \mathcal{X} \to \mathcal{X}$ such that for any key $K \in \mathcal{K}$ and any tweak $T \in \mathcal{T}$, $X \mapsto \widetilde{E}(K, T, X)$ is a permutation of $\mathcal{X}$. We often write $\widetilde{E}_K(T, X)$ or $\widetilde{E}_K^T(X)$ in place of $\widetilde{E}(K, T, X)$. We denote $\mathsf{TBC}(\mathcal{K}, \mathcal{T}, \mathcal{X})$ the set of all tweakable block ciphers with key space $\mathcal{K}$, tweak space $\mathcal{T}$, and domain $\mathcal{X}$. A *tweakable permutation* with tweak space $\mathcal{T}$ and domain $\mathcal{X}$ is a mapping $\widetilde{P} : \mathcal{T} \times \mathcal{X} \to \mathcal{X}$ such that for any tweak $T \in \mathcal{T}$, $X \mapsto \widetilde{P}(T, X)$ is a permutation of $\mathcal{X}$. We often write $\widetilde{P}^T(X)$ in place of $\widetilde{P}(T, X)$. We denote $\mathsf{TP}(\mathcal{T}, \mathcal{X})$ the set of all tweakable permutations with tweak space $\mathcal{T}$ and domain $\mathcal{X}$. The security of a TBC is defined as follows.

Definition 1 (TPRP Security). *Let $\widetilde{E} \in \mathsf{TBC}(\mathcal{K}, \mathcal{T}, \mathcal{X})$ and A be an adversary with oracle access to a tweakable permutation with tweak space $\mathcal{T}$ and domain $\mathcal{X}$. The advantage of A in breaking the TPRP-security of $\widetilde{E}$ is defined as*

$$\mathbf{Adv}_{\widetilde{E}}^{\mathrm{TPRP}}(\mathsf{A}) = \left| \Pr\left[K \leftarrow_\$ \mathcal{K} : \mathsf{A}^{\widetilde{E}_K} = 1 \right] - \Pr\left[\widetilde{P} \leftarrow_\$ \mathsf{TP}(\mathcal{T}, \mathcal{X}) : \mathsf{A}^{\widetilde{P}} = 1 \right] \right|.$$

Note that we do not need the strongest "two-sided" version of TPRP-security (where the adversary also has access to a decryption oracle) since all constructions considered in this paper only use the forward (encryption) direction of the underlying TBC.

TWEAK SEPARATION. Let $\widetilde{E}$ be a TBC with tweak space of the form $\mathcal{T}' = \mathcal{I} \times \mathcal{T}$ for some subset $\mathcal{I} \subset \mathbb{N}$ and some set $\mathcal{T}$. We call $\mathcal{T}$ the *effective* tweak space of $\widetilde{E}$. Then, for $i \in \mathcal{I}$, we denote $\widetilde{E}^i$ the tweakable block cipher with the same key and message spaces as $\widetilde{E}$ and tweak space $\mathcal{T}$ defined by

$$\widetilde{E}^i(K, T, X) = \widetilde{E}(K, (i, T), X).$$

By the same convention as before, we write $\widetilde{E}_K^i(T, X)$ or $\widetilde{E}_K^{i,T}(X)$ for $\widetilde{E}^i(K, T, X)$. Clearly, when $\widetilde{E}$ is an ideal TBC drawn uniformly at random from $\mathsf{TBC}(\mathcal{K}, \mathcal{T}', \mathcal{M})$, then each $\widetilde{E}^i$ is an independent ideal TBC drawn uniformly at random from $\mathsf{TBC}(\mathcal{K}, \mathcal{T}, \mathcal{M})$. Given a bit-string X of length $1 \leq |X| \leq n$, we compactly write

$$\widetilde{E}^{i/j}(K, T, X(10^*)), \quad \widetilde{E}_K^{i/j}(T, X(10^*)), \quad \text{or} \quad \widetilde{E}_K^{i/j,T}(X(10^*))$$

to mean $\widetilde{E}^i(K, T, X)$ when $|X| = n$ and $\widetilde{E}^j(K, T, X10^*)$ when $|X| < n$.

STANDARD SECURITY NOTIONS. We give the security definitions of a nonce-based PRF, a nonce-based MAC, and a nonce-based Authenticated Encryption scheme. All these are standard, except maybe the nonce-based PRF notion which is a straightforward adaptation of the classical definition of a PRF to the nonce-based setting. Our definition of the security of a MAC is indistinguishability-based (which will be more convenient later), but it is easy to see that it is equivalent to the more conventional unforgeability-based definition. In the following, a nonce-based keyed function is a function $F : \mathcal{K} \times \mathcal{N} \times \mathcal{D} \to \mathcal{Y}$, where $\mathcal{K}$ is the key space, $\mathcal{N}$ the nonce space, $\mathcal{D}$ the domain and $\mathcal{Y}$ the range.

Definition 2 (Nonce-Based PRF). *Let $F : \mathcal{K} \times \mathcal{N} \times \mathcal{D} \to \mathcal{Y}$ be a nonce-based keyed function, and let us write $F_K(N, D)$ for $F(K, N, D)$. Let A be an adversary* *with oracle access to a function from $\mathcal{N} \times \mathcal{D}$ to $\mathcal{Y}$. The advantage of* A *against the PRF-security of F is defined as*

$$\mathbf{Adv}_F^{\mathrm{PRF}}(\mathsf{A}) = \left| \Pr\left[K \leftarrow_\$ \mathcal{K} : \mathsf{A}^{F_K} = 1 \right] \right.$$
$$\left. - \Pr\left[R \leftarrow_\$ \mathsf{Func}(\mathcal{N} \times \mathcal{D}, \mathcal{Y}) : \mathsf{A}^R = 1 \right] \right|.$$

The adversary is said nonce-respecting if it never repeats a nonce $N \in \mathcal{N}$ in its oracle queries. In that case, we denote its advantage $\mathbf{Adv}_F^{\mathrm{nPRF}}(\mathsf{A})$.

Definition 3 (Nonce-Based MAC). *Let F be as in Definition 2. Let* B *be an adversary with oracle access to two oracles, the first oracle being a function from $\mathcal{N} \times \mathcal{D}$ to $\mathcal{Y}$, the second oracle with inputs in $\mathcal{N} \times \mathcal{D} \times \mathcal{Y}$ and outputs in $\{1, \bot\}$. The advantage of* B *against the MAC-security of F is defined as*

$$\mathbf{Adv}_F^{\mathrm{MAC}}(\mathsf{B}) = \left| \Pr\left[K \leftarrow_\$ \mathcal{K} : \mathsf{B}^{F_K, \mathsf{Ver}_K} = 1 \right] - \Pr\left[K \leftarrow_\$ \mathcal{K} : \mathsf{B}^{F_K, \mathsf{Rej}} = 1 \right] \right|,$$

where Ver_K is an oracle which takes as input a triple $(N, D, \mathsf{tag}) \in \mathcal{N} \times \mathcal{D} \times \mathcal{Y}$ and returns 1 if $F_K(N, D) = \mathsf{tag}$, and $\bot$ otherwise, and Rej is an oracle which always returns $\bot$. The adversary is not allowed to ask a verification query (N, D, tag) if a previous query (N, D) to F_K returned tag. The adversary is said nonce-respecting if it never repeats a nonce $N \in \mathcal{N}$ in its queries to the first oracle F_K. In that case, we denote its advantage $\mathbf{Adv}_F^{\mathrm{nMAC}}(\mathsf{B})$.

Note that in the general case where the adversary is allowed to repeat nonces, F can be seen as a standard (i.e., not nonce-based) keyed function with domain $\mathcal{N} \times \mathcal{D}$, in which case one recovers the standard definitions of a PRF and a MAC (hence our notation of the advantage when the adversary is unrestricted w.r.t. nonces). While it is a well-known fact that if F is a secure PRF, then it is a secure MAC [9,23], we stress that this is not true for the nonce-based variants of the two notions, which are in fact incomparable.[8]

We then give the definition of a nonce-based Authenticated Encryption (nAE) scheme (with associated data), for which we first recall the syntax. Let $\mathcal{K}$, $\mathcal{N}$, $\mathcal{A}$, and $\mathcal{M}$ be non-empty sets. A nAE scheme is a tuple $\Pi = (\mathcal{K}, \mathcal{N}, \mathcal{A}, \mathcal{M}, \mathsf{Enc}, \mathsf{Dec})$, where Enc and Dec are deterministic algorithms. The encryption algorithm Enc takes as input a key $K \in \mathcal{K}$, a nonce $N \in \mathcal{N}$, associated data $A \in \mathcal{A}$, and a message $M \in \mathcal{M}$, and outputs a binary string $C \in \{0, 1\}^*$ (we assume that Enc returns $\bot$ if one of the inputs is not in the intended set). The decryption algorithm Dec takes as input a key $K \in \mathcal{K}$, a nonce $N \in \mathcal{N}$, associated data $A \in \mathcal{A}$, and a binary string $C \in \{0, 1\}^*$, and outputs either a message $M \in \mathcal{M}$, or a special symbol $\bot$. We require that $\mathsf{Dec}(K, N, A, \mathsf{Enc}(K, N, A, M)) = M$ for all tuples $(K, N, A, M) \in \mathcal{K} \times \mathcal{N} \times \mathcal{A} \times \mathcal{M}$. We write $\mathsf{Enc}_K(N, A, M)$ for $\mathsf{Enc}(K, N, A, M)$ and $\mathsf{Dec}_K(N, A, C)$ for $\mathsf{Dec}(K, N, A, C)$.

---

[8] E.g., an nPRF-secure function F might depend only on the nonce, in which case it is trivial to forge and break nMAC-security, while an nMAC-secure function F might have all its outputs starting with a 0 bit, which allows to trivially break nPRF-security.

Definition 4 (Nonce-Based AE). *Let* $\Pi = (\mathcal{K}, \mathcal{N}, \mathcal{A}, \mathcal{M}, \mathsf{Enc}, \mathsf{Dec})$ *be a nAE scheme. The advantage of an adversary* A *in breaking* Π *is defined as*

$$\mathbf{Adv}_{\Pi}^{\mathrm{AE}}(\mathsf{A}) = \left| \Pr\left[K \leftarrow_{\$} \mathcal{K} : \mathsf{A}^{\Pi.\mathsf{Enc}_K(\cdot,\cdot,\cdot),\, \Pi.\mathsf{Dec}_K(\cdot,\cdot,\cdot)} = 1 \right] \right.$$
$$\left. - \Pr\left[\mathsf{A}^{\mathsf{Rand}(\cdot,\cdot,\cdot),\, \mathsf{Rej}(\cdot,\cdot,\cdot)} = 1 \right] \right|,$$

where Rand *is an oracle which on input* $(N, A, M) \in \mathcal{N} \times \mathcal{A} \times \mathcal{M}$ *outputs a random[9] string of length* $|\Pi.\mathsf{Enc}_K(N, A, M)|$ *and* Rej *is an oracle which always outputs* $\bot$. *The adversary is not allowed to make a decryption query* (N, A, C) *if a previous encryption query* (N, A, M) *returned* C. *The adversary is said nonce-respecting if it never repeats a nonce* $N \in \mathcal{N}$ *in its encryption queries, in which case we denote its advantage* $\mathbf{Adv}_{\Pi}^{\mathrm{nAE}}(\mathsf{A})$.

Note that in the general case where the adversary is allowed to repeat nonces, Π can be seen as a deterministic AE scheme [55] with header space (in the terms of [55]) $\mathcal{N} \times \mathcal{A}$, so that one exactly recovers the definition of the MRAE notion of Rogaway and Shrimpton [55] (which we simply abbreviate to AE here).

ADVERSARY CHARACTERISTICS. In all the paper, given some implicit parameter n, a (q, m, ℓ, σ, t)-adversary against a nonce-based scheme is an adversary:

- which makes at most q oracle queries; when the adversary has access to two oracles (i.e., when attacking the MAC-security of a keyed function or a nAE scheme), this means q queries in total to both oracles;
- which uses any nonce at most m times throughout its queries ($m = 1$ for a nonce-respecting adversary); when the adversary has access to two oracles, this only applies to queries to its first oracle (MAC or encryption oracle);
- such that the length of any of its queries (nonce excluded) is at most ℓ blocks of n bits; for a keyed function with domain $\mathcal{D} = \mathcal{A} \times \mathcal{M}$ or a nAE scheme, this means that *both* the AD length and the message length of any query is at most ℓ blocks of n bits;
- such that the total length of all its queries (nonce excluded) is at most σ blocks of n bits; for a keyed function with domain $\mathcal{D} = \mathcal{A} \times \mathcal{M}$ or a nAE scheme, this means the sum of the AD and the message length over all queries;
- which runs in time at most t.

4   The CTRT Encryption Mode

4.1   Syntax and Security of nivE Schemes

Most existing encryption schemes are either nonce-based [53] or IV-based [7], i.e., they employ an externally provided value which either should not repeat (nonce), or should be selected uniformly at random (IV). (See also [46]).

---

[9] We assume that Rand returns the same output if a query is repeated.

Here, we introduce the notion of combined nonce- and IV-based encryption scheme (*nivE* for short).

Syntactically, a nivE scheme is a tuple $\Pi = (\mathcal{K}, \mathcal{N}, \mathcal{IV}, \mathcal{M}, \mathsf{Enc}, \mathsf{Dec})$ where $\mathcal{K}$, $\mathcal{N}, \mathcal{IV}$ and $\mathcal{M}$ are non-empty sets and Enc and Dec are deterministic algorithms. The encryption algorithm Enc takes as input a key $K \in \mathcal{K}$, a nonce $N \in \mathcal{N}$, an initial value $IV \in \mathcal{IV}$, and a message $M \in \mathcal{M}$, and outputs a binary string $C \in \{0,1\}^*$ (we assume that Enc returns $\bot$ if one of the inputs is not in the intended set). The decryption algorithm Dec takes as input a key $K \in \mathcal{K}$, a nonce $N \in \mathcal{N}$, an initial value $IV \in \mathcal{IV}$, and a binary string $C \in \{0,1\}^*$, and outputs either a message $M \in \mathcal{M}$, or a special symbol $\bot$. We require that

$$\mathsf{Dec}(K, N, IV, \mathsf{Enc}(K, N, IV, M)) = M$$

for all tuples $(K, N, IV, M) \in \mathcal{K} \times \mathcal{N} \times \mathcal{IV} \times \mathcal{M}$.

We denote $\mathsf{Enc}^\$$ the probabilistic algorithm which takes as input $(K, N, M) \in \mathcal{K} \times \mathcal{N} \times \mathcal{M}$, internally generates a uniformly random $IV \leftarrow_\$ \mathcal{IV}$, computes $C = \mathsf{Enc}(K, N, IV, M)$, and outputs $(IV, C) \in \mathcal{IV} \times \{0,1\}^*$. We write $\mathsf{Enc}_K(N, IV, M)$ for $\mathsf{Enc}(K, N, IV, M)$ and $\mathsf{Enc}_K^\$(N, M)$ for $\mathsf{Enc}^\$(K, N, M)$. The security of a nivE scheme is defined as follows.

Definition 5 (Security of a nivE Scheme). *Let* $\Pi = (\mathcal{K}, \mathcal{N}, \mathcal{IV}, \mathcal{M}, \mathsf{Enc}, \mathsf{Dec})$ *be a nivE scheme. The advantage of an adversary* A *in breaking* Π *is defined as*

$$\mathbf{Adv}_\Pi^{\mathrm{ivE}}(\mathsf{A}) = \left| \Pr\left[K \leftarrow_\$ \mathcal{K} : \mathsf{A}^{\Pi.\mathsf{Enc}_K^\$(\cdot,\cdot)} = 1 \right] - \Pr\left[\mathsf{A}^{\mathsf{Rand}(\cdot,\cdot)} = 1 \right] \right|,$$

where Rand *is an oracle which on input* $(N, M) \in \mathcal{N} \times \mathcal{M}$ *outputs a random string of length* $|\Pi.\mathsf{Enc}_K^\$(N, M)|$. *The adversary is said* nonce-respecting *if it never repeats a nonce* $N \in \mathcal{N}$ *in its oracle queries, in which case we denote its advantage* $\mathbf{Adv}_\Pi^{\mathrm{nivE}}(\mathsf{A})$.

Note that when the adversary is allowed to repeat nonces, Π can be seen as a family of purely IV-based encryption (ivE) schemes [46] indexed by the nonce space $\mathcal{N}$, hence our notation of the advantage in that case.

4.2 Definition and Analysis of the CTRT Mode

We now define the CTRT (*CounTeR in Tweak*) mode, turning a tweakable block cipher into a nivE scheme. Let $\mathcal{K}$ and $\mathcal{T}$ be non-empty sets, and let $\widetilde{E} \in \mathsf{TBC}(\mathcal{K}, \mathcal{T}', \mathcal{X})$ be a tweakable block cipher with key space $\mathcal{K}$, tweak space[10] $\mathcal{T}' = \{1\} \times \mathcal{T}$, and domain $\mathcal{X} = \{0,1\}^n$. Let Inc be a cyclic permutation of $\mathcal{T}$. We construct from $\widetilde{E}$ a nivE scheme $\mathsf{CTRT}[\widetilde{E}]$ with key space $\mathcal{K}$, nonce space $\mathcal{N} = \mathcal{X} = \{0,1\}^n$, IV space $\mathcal{IV} = \mathcal{T}$, and message space $\mathcal{M} = \{0,1\}^*$ as defined in Fig. 2 and illustrated on bottom of Fig. 1.

---

[10] The CTRT mode does not need tweak separation per se. We use a single 1 prefix in order to conveniently combine CTRT with EPWC later.

---

1 **algorithm** $\mathsf{CTRT}[\widetilde{E}].\mathsf{Enc}_K(N, IV, M)$
2    $\ell := |M|/n$
3   parse M as $M_1\|\cdots\|M_\ell$, with $|M_1| = \ldots = |M_{\ell-1}| = n$ and $1 \le |M_\ell| \le n$
4   **for** $i = 1$ **to** ℓ **do**
5      $C_i := M_i \oplus \widetilde{E}_K^{1,IV}(N)$
6      $IV := \mathsf{Inc}(IV)$
7   **return** $C_1\|C_2\|\cdots\|C_\ell$

Fig. 2. Definition of the CTRT mode, using a TBC $\widetilde{E} \in \mathsf{TBC}(\mathcal{K}, \mathcal{T}', \mathcal{X})$ with $\mathcal{T}' = \{1\} \times \mathcal{T}$ and $\mathcal{X} = \{0,1\}^n$.

The security of CTRT is captured by Theorem 1 below. Logarithms are in base 2 and $t_{\mathsf{CTRT}}(\sigma)$ is an upper bound on the time needed for computing $\mathsf{CTRT}[\widetilde{E}].\mathsf{Enc}_K$ on inputs of total message length at most σ blocks of n bits when calls to $\widetilde{E}_K$ cost unit time.

Theorem 1 (Security of CTRT). *Let* $\widetilde{E} \in \mathsf{TBC}(\mathcal{K}, \mathcal{T}', \mathcal{X})$ *with* $\mathcal{X} = \{0,1\}^n$, $\mathcal{T}' = \{1\} \times \mathcal{T}$, *and* $|\mathcal{T}| \ge 8$. *Let* A *be a* (q, m, ℓ, σ, t)-*adversary against* $\mathsf{CTRT}[\widetilde{E}]$ *with* $\ell \le |\mathcal{T}|$. *Then there exists an adversary* A' *against the TPRP-security of* $\widetilde{E}$, *making at most* σ *oracle queries and running in time at most* $t + t_{\mathsf{CTRT}}(\sigma)$, *such that*

$$\mathbf{Adv}^{\mathrm{ivE}}_{\mathsf{CTRT}[\widetilde{E}]}(A) \le \mathbf{Adv}^{\mathrm{TPRP}}_{\widetilde{E}}(A') + \frac{2(m-1)\sigma + 1}{|\mathcal{T}|} + f(\sigma),$$

where

$$f(\sigma) = \frac{2\sigma \log^2 \sigma}{|\mathcal{X}|} \qquad \text{when } 8 \le \sigma \le |\mathcal{T}|,$$

$$= \frac{2\sigma^2 \log^2 |\mathcal{T}|}{|\mathcal{X}||\mathcal{T}|} \qquad \text{when } \sigma \ge |\mathcal{T}|.$$

In particular, if A *is nonce-respecting* $(m = 1)$, *one has*

$$\mathbf{Adv}^{\mathrm{nivE}}_{\mathsf{CTRT}[\widetilde{E}]}(A) \le \mathbf{Adv}^{\mathrm{TPRP}}_{\widetilde{E}}(A') + \frac{1}{|\mathcal{T}|} + f(\sigma).$$

Before proceeding to the proof, we comment the security bound of this theorem. Consider first the case of a nonce-respecting adversary A. Assuming $|\mathcal{T}| \ge |\mathcal{X}|$, then A must makes queries of total message length σ blocks of n bits with σ close to $|\mathcal{X}| = 2^n$ (neglecting logarithmic factors) before being able to distinguish the outputs of $\mathsf{CTRT}[\widetilde{E}]$ from random.[11] On the other hand, if

---

[11] Note that in that case the size of the tweak space $\mathcal{T}$ only impacts the maximal message length ℓ, not the security bound.

$|\mathcal{T}| = 2^w < |\mathcal{X}|$, then $\mathsf{CTRT}[\widetilde{E}]$ is secure up to roughly $2^{(n+w)/2}$ TBC calls (again, neglecting logarithmic factors), which is always larger than $2^{n/2}$. In particular, if $w = n/2$ (as e.g. for KIASU-BC [32]), then security is ensured up to roughly $2^{3n/4}$ TBC calls. In the nonce-misuse scenario, note that the additional term $2(m-1)\sigma/|\mathcal{T}|$ remains small as long as nonces are not repeated too many times (e.g. $m \le 100$) and $\sigma \ll |\mathcal{T}|$, and turns into a birthday-like term only in the extreme case where a few nonces are repeated close to σ times. This means that a few nonce repetitions will not hurt and that nonces must be "seriously" mishandled before security goes down to birthday bound.

PROOF OF THEOREM 1. Fix a $(q', m, |\mathcal{T}|, \sigma, t)$-adversary A against $\mathsf{CTRT}[\widetilde{E}]$ (we denote q' the maximal number of adversarial queries and will later use q for the actual number of queries in a specific attack). The first part of the proof is standard, and consists in introducing an intermediate game where all calls to $\widetilde{E}_K$ in the CTRT construction are replaced by calls to a random tweakable permutation $\widetilde{P}$. Consider the following adversary A' against the TPRP-security of $\widetilde{E}$. Let $G \in \{\widetilde{E}_K, \widetilde{P}\}$ be the oracle to which A' has access. Adversary A' runs A, answers its encryption queries (N, M) by drawing a random IV and executing the code in Fig. 2 on input (N, IV, M), replacing calls to $\widetilde{E}_K$ by oracle calls to G, and finally outputs the same bit as A. Clearly, A' makes at most σ oracle queries and runs in time at most $t + t_{\mathsf{CTRT}}(\sigma)$. Moreover, it is easy to see that

$$\mathbf{Adv}_{\mathsf{CTRT}[\widetilde{E}]}^{\mathrm{ivE}}(\mathsf{A}) \le \mathbf{Adv}_{\widetilde{E}}^{\mathrm{TPRP}}(\mathsf{A}') + \delta, \tag{1}$$

where

$$\delta = \left| \Pr\left[\widetilde{P} \leftarrow_{\$} \mathsf{TP}(\mathcal{T}, \mathcal{X}) : \mathsf{A}^{\mathsf{CTRT}[\widetilde{P}].\mathsf{Enc}^{\$}} = 1 \right] - \Pr\left[\mathsf{A}^{\mathsf{Rand}} = 1 \right] \right| \tag{2}$$

and $\mathsf{CTRT}[\widetilde{P}]$ is a slight abuse of notation for the CTRT construction based on an arbitrary tweakable permutation $\widetilde{P}$.

Upper bounding δ is now a purely information-theoretic problem, so that we allow A to be computationally unbounded, and hence, wlog, deterministic. The adversary is now trying to distinguish between $\mathsf{CTRT}[\widetilde{P}]$ for a random $\widetilde{P}$ (thereafter called the "real world") and Rand (thereafter called the "ideal world"). We assume wlog that A always makes queries of length a multiple of the block length n, and of total length σ blocks (if not, we pad all queries whose final block is incomplete with zeros for free, which can only increase the adversary's advantage).

Following the H-coefficients method [15,48], we summarize the interaction of A with its oracle in the so-called transcript of the attack

$$\tau = ((N_1, M_1, IV_1, C_1), \dots, (N_q, M_q, IV_q, C_q)),$$

where (N_i, M_i) denotes the i-th query of the attacker and (IV_i, C_i) the corresponding answer of the oracle. Furthermore, we denote $M_i = M_{i,1}\| \cdots \|M_{i,\ell_i}$, $C_i = C_{i,1}\| \cdots \|C_{i,\ell_i}$, where ℓ_i is the number of blocks of the i-th message, and

$IV_{i,j} = \mathsf{Inc}^{j-1}(IV_i)$ the j-th counter for the i-th message, $j = 1, \ldots, \ell_i$. Let Θ_{re}, resp. Θ_{id}, denote the distribution of the transcript in the real world, resp. ideal world. We say that a transcript τ is A-attainable (or simply attainable) if the probability to obtain τ in the ideal world is non-zero. Note that the number of queries q and the lengths of the queries $\ell_1, \ldots, \ell_q$ are themselves random variables (they can vary for distinct A-attainable transcripts), yet by the assumption that the attacker always asks the maximal number of allowed blocks throughout its queries, one always has $\sum_{i=1}^q \ell_i = \sigma$.

From τ we define for each possible tweak $T \in \mathcal{T}$ the "load" of the tweak as

$$L(T) = |\{(i,j) : IV_{i,j} = T\}|.$$

In words, $L(T)$ is the number of times the tweak T appears as a counter when encrypting the queries of the adversary. Clearly, one has

$$\sum_{T \in \mathcal{T}} L(T) = \sigma. \tag{3}$$

The proof relies on the fundamental lemma of the H-coefficients technique (see e.g. [15] for the proof).

Lemma 1. *Assume that the set of* A-*attainable transcripts is partitioned into two disjoint sets* GoodT *and* BadT, *and that there exists* ε_1 *and* ε_2 *such that for any* $\tau \in$ GoodT, *one has*

$$\frac{\Pr\left[\Theta_{\mathrm{re}} = \tau\right]}{\Pr\left[\Theta_{\mathrm{id}} = \tau\right]} \geq 1 - \varepsilon_1,$$

and $\Pr\left[\Theta_{\mathrm{id}} \in \mathsf{BadT}\right] \leq \varepsilon_2$. *Then* $\delta \leq \varepsilon_1 + \varepsilon_2$, *with* δ *as defined by* (2).

We say that an attainable transcript τ is bad if one of the two following conditions are met:

(C-1) there exists $(i,j) \neq (i',j')$ such that $IV_{i,j} = IV_{i',j'}$ and $N_i = N_{i'}$;
(C-2) there exists $(i,j) \neq (i',j')$ such that $IV_{i,j} = IV_{i',j'}$, $N_i \neq N_{i'}$, and $M_{i,j} \oplus C_{i,j} = M_{i',j'} \oplus C_{i',j'}$.

Note that condition (C-1) can only be satisfied for a nonce-misuse adversary, since (by the assumption that the length of each query is at most $|\mathcal{T}|$ blocks of n bits so that counters do not loop) $IV_{i,j} = IV_{i',j'}$ requires $i \neq i'$, which implies that $N_i \neq N_{i'}$ for a nonce-respecting adversary. Note also that the condition (C-2) can only be satisfied in the ideal world. Indeed, in the real world, $M_{i,j} \oplus C_{i,j} = \widetilde{P}(IV_{i,j}, N_i)$ and $M_{i',j'} \oplus C_{i',j'} = \widetilde{P}(IV_{i',j'}, N_{i'})$, so that if $IV_{i,j} = IV_{i',j'}$ and $N_i \neq N_{i'}$ one necessarily has $M_{i,j} \oplus C_{i,j} \neq M_{i',j'} \oplus C_{i',j'}$.

We let BadT be the set of bad transcripts, and GoodT be the set of attainable transcripts which are not bad, henceforth called good transcripts. We first consider good transcripts.

Lemma 2. *Let* $\tau \in \mathsf{GoodT}$ *be a good transcript. Then*

$$\frac{\Pr[\Theta_{\mathrm{re}} = \tau]}{\Pr[\Theta_{\mathrm{id}} = \tau]} \geq 1.$$

Proof. Note that a good transcript has the property that for each $(i, j) \neq (i', j')$ such that $IV_{i,j} = IV_{i',j'}$, one has $N_i \neq N_i'$ and $M_{i,j} \oplus C_{i,j} \neq M_{i',j'} \oplus C_{i',j'}$. In other words, the transcript encodes a partial tweakable permutation, where for each tweak $T \in \mathcal{T}$ there are exactly $L(T)$ distinct values N_i mapped to some value $M_{i,j} \oplus C_{i,j}$. The probability to obtain a good transcript τ in the ideal and real worlds can now be easily computed. In the ideal world, since the IV_i's and the C_i's are uniformly random, one has

$$\Pr[\Theta_{\mathrm{id}} = \tau] = \frac{1}{|\mathcal{IV}|^q \cdot |\mathcal{X}|^\sigma}.$$

In the real world, the IV_i's are random as well, but now the probability to obtain the C_i's can easily be seen to be the probability that the random tweakable permutation $\widetilde{P}$ is compatible with the partial tweakable permutation encoded by τ [15]. Hence, one has

$$\Pr[\Theta_{\mathrm{re}} = \tau] = \frac{1}{|\mathcal{IV}|^q} \cdot \prod_{T \in \mathcal{T}} \frac{1}{(|\mathcal{X}|)_{L(T)}},$$

where $(a)_b$ denotes the falling factorial $a(a-1)\cdots(a-b+1)$, with $(a)_0 = 1$ by convention. From this, we deduce that

$$\frac{\Pr[\Theta_{\mathrm{re}} = \tau]}{\Pr[\Theta_{\mathrm{id}} = \tau]} = \frac{|\mathcal{X}|^\sigma}{\prod_{T \in \mathcal{T}} (|\mathcal{X}|)_{L(T)}} \overset{(3)}{=} \prod_{T \in \mathcal{T}} \frac{|\mathcal{X}|^{L(T)}}{(|\mathcal{X}|)_{L(T)}} \geq 1.$$

$\square$

It remains to upper bound the probability to obtain a bad transcript in the ideal world. For $i \in \{1, 2\}$, let BadT_i be the set of attainable transcripts satisfying condition (C-i). We first consider condition (C-1).

Lemma 3. *One has*

$$\Pr[\Theta_{\mathrm{id}} \in \mathsf{BadT}_1] \leq \frac{2(m-1)\sigma}{|\mathcal{T}|}.$$

Proof. Consider two distinct queries (N_i, M_i) and $(N_{i'}, M_{i'})$. If the nonces are the same ($N_i = N_{i'}$), then the probability, over the random draw of IV_i and $IV_{i'}$ in $\mathcal{T}$, that there exists j and j' such that $IV_{i,j} = IV_{i',j'}$, is $(\ell_i + \ell_{i'} - 1)/|\mathcal{T}|$. If the nonces are distinct, then clearly condition (C-1) cannot be satisfied for i and i'. Hence, summing over all possible nonces, we have

$$\Pr[\Theta_{\mathrm{id}} \in \mathsf{BadT}_1] \leq \sum_{N \in \mathcal{N}} \sum_{\substack{1 \leq i < i' \leq q \\ N_i = N_{i'} = N}} \frac{\ell_i + \ell_{i'} - 1}{|\mathcal{T}|}.$$

Fix some nonce N, and assume for notational simplicity that the first q' queries use nonce N, with $q' \leq m$ by assumption. Then the probability that condition (C-1) is met for nonce N is at most

$$
\sum_{i=1}^{q'-1} \sum_{i'=i+1}^{q'} \frac{\ell_i + \ell_{i'} - 1}{|\mathcal{T}|} \leq \sum_{i=1}^{q'-1} \frac{(q'-1)\ell_i + \ell(N)}{|\mathcal{T}|}
$$
$$
\leq \frac{2(q'-1)\ell(N)}{|\mathcal{T}|}
$$
$$
\leq \frac{2(m-1)\ell(N)}{|\mathcal{T}|},
$$

where $\ell(N)$ is the total length of queries using nonce N. The result follows by summing over all possible nonces, using $\sum_{N \in \mathcal{N}} \ell(N) = \sigma$. $\square$

We handle condition (C-2) in the following lemma.

Lemma 4. *One has*

$$
\Pr\left[\Theta_{id} \in \mathsf{BadT}_2\right] \leq \frac{1}{|\mathcal{T}|} + \min\{\sigma, |\mathcal{T}|\} \cdot \frac{(L_{max})^2}{2|\mathcal{X}|},
$$

where $L_{max} = 2\log\sigma$ when $8 \leq \sigma \leq |\mathcal{T}|$ and $L_{max} = \frac{2\sigma \log |\mathcal{T}|}{|\mathcal{T}|}$ when $\sigma \geq |\mathcal{T}|$.

Proof. Let BadT_3 be the set of transcripts satisfying the following condition (L_{max} being defined as in the statement of the lemma):

(C-3) there exists $T \in \mathcal{T}$ such that $L(T) \geq L_{max}$.

Then

$$
\Pr[\Theta_{id} \in \mathsf{BadT}_2] \leq \Pr\left[\Theta_{id} \in \mathsf{BadT}_2 \mid \Theta_{id} \notin \mathsf{BadT}_3\right] + \Pr\left[\Theta_{id} \in \mathsf{BadT}_3\right].
$$

Note that in the ideal world, the values $L(T)$ only depend on the random draw of the IV_i's, and that once the $L(T)$'s are fixed, condition (C-2) only depends on the random draw of the $C_{i,j}$'s. In particular, since in the ideal world the $C_{i,j}$'s are uniformly random, one has

$$
\Pr\left[\Theta_{id} \in \mathsf{BadT}_2 \mid \Theta_{id} \notin \mathsf{BadT}_3\right]
$$
$$
\leq \sum_{T \in \mathcal{T}} \sum_{\substack{(i,j),(i',j') \\ IV_{i,j}=IV_{i',j'}=T}} \Pr\left[M_{i,j} \oplus C_{i,j} = M_{i',j'} \oplus C_{i',j'}\right]
$$
$$
\leq \sum_{T \in \mathcal{T}} \frac{L(T)(L(T)-1)}{2|\mathcal{X}|}
$$
$$
\leq \min\{\sigma, |\mathcal{T}|\} \cdot \frac{(L_{max})^2}{2|\mathcal{X}|},
$$

where for the third inequality we used that there are at most $\min\{\sigma, |\mathcal{T}|\}$ tweaks T such that $L(T) \geq 1$.

It remains to upper bound the probability that condition (C-3) is satisfied, which can be recast as a "balls-into-bins" problem. Thinking of each tweak T as a bin, each random IV_i determines a sequence of ℓ_i consecutive[12] bins where a ball is thrown. If the attacker only made queries of length one block, then this would be a standard "balls-into-bins" problem, where each ball is thrown independently in a bin chosen uniformly at random, and we could use classical results about the maximal occupancy of any bin directly. However, the attacker can choose the length of each message at will and we need to take this into account.[13] Intuitively, for some fixed total number σ of balls, using messages of length $\ell_i > 1$ should *lower* the maximal occupancy since balls thrown in consecutive bins cannot end in the same bin. We formalize this intuition in a separate Lemma below, which implies that

$$\Pr\left[\Theta_{\mathrm{id}} \in \mathsf{BadT}_3\right] \leq \frac{1}{|T|}.$$

The result follows.   □

The lemma below is a simple variant on the standard balls-into-bins problem. A similar result was proved in [6] (and potentially in many other papers).

Lemma 5. *Consider a set of $|T| \geq 8$ bins and $\sigma \geq 8$ balls. Fix an integer $q \leq \sigma$ and a sequence of integers $(\ell_1, \ldots, \ell_q)$ with $1 \leq \ell_i \leq |T|$ and $\sum_{i=1}^{q} \ell_i = \sigma$. Consider the following random process: for $i = 1, \ldots, q$, a chain of ℓ_i balls is thrown in consecutive bins, the initial bin being chosen independently and uniformly at random. Then the probability that, at the end of the process, any bin contains $L_{\max}$ balls or more, is less than $1/|T|$, where*

(a) $L_{\max} = 2 \log \sigma$ when $\sigma \leq |T|$;
(b) $L_{\max} = \frac{2\sigma \log |T|}{|T|}$ when $\sigma \geq |T|$.

Proof. See the full version of the paper [50].

COMPLETING THE PROOF OF THEOREM 1. From Lemmas 3 and 4, we obtain by the union bound that

$$\Pr\left[\Theta_{\mathrm{id}} \in \mathsf{BadT}\right] \leq \frac{2(m-1)\sigma + 1}{|T|} + f(\sigma), \tag{4}$$

with $f(\sigma)$ as in the statement of Theorem 1. Combining (4) with Lemmas 1 and 2 (taking $\varepsilon_1 = 0$), we obtain the same upper bound for δ (defined by (2)) as for $\Pr\left[\Theta_{\mathrm{id}} \in \mathsf{BadT}\right]$. Finally, Eq. (1) yields the result.

VARIANTS. In the full version of the paper [50], we describe two variants of CTRT, a purely nonce-based one and a purely IV-based one.

---

[12] The successor of the tweak T is $\mathsf{Inc}(T)$.
[13] Note that the adversary must commit to the length ℓ_i of the chain *before* knowing the initial bin IV_i since it first makes the query (N_i, M_i) and only then receives the answer (IV_i, C_i).

5 The PWC and EPWC Message Authentication Codes

In this section, we describe two related modes for message authentication, PWC (*Parallel Wegman-Carter*) and EPWC (*Encrypted PWC*). Let $\mathcal{K}$ and $\mathcal{T}$ be two sets, and let $\widetilde{E}$ be a tweakable block cipher with key space $\mathcal{K}$, tweak space[14] $\mathcal{T}' = \{2, \dots, 5\} \times \mathcal{T}$, and domain $\mathcal{X} = \{0, 1\}^n$. Let Inc be a cyclic permutation of $\mathcal{T}$. From $\widetilde{E}$, we construct two nonce-based keyed functions, PWC$[\widetilde{E}]$ and EPWC$[\widetilde{E}]$, both with key space $\mathcal{K}$, nonce-space $\mathcal{N} = \mathcal{X} = \{0, 1\}^n$, domain $\mathcal{D} = \mathcal{A} \times \mathcal{M}$, where $\mathcal{A} = \mathcal{M} = \{0, 1\}^*$,[15] and range $\mathcal{Y} = \mathcal{X} = \{0, 1\}^n$, as defined in Fig. 3 and illustrated on top of Fig. 1 (for PWC, just omit the final call to $\widetilde{E}_K^{4,0}$). We will prove that both PWC$[\widetilde{E}]$ and EPWC$[\widetilde{E}]$ are 2^n-secure as nonce-based MAC and nonce-based PRF, and that EPWC$[\widetilde{E}]$ is moreover a birthday bound-secure PRF in the nonce-misuse scenario.

The PWC construction follows the Wegman-Carter paradigm [14,56,60] by combining a xor-universal hash function H inspired from PMAC [13,52] applied to (A, M), and a pseudorandom function F applied to the nonce N. This pseudorandom function is constructed from $\widetilde{E}$ by summing two independent pseudorandom permutations in order to obtain security beyond the birthday bound [42]. The EPWC construction is simply PWC with an additional layer of encryption to provide nonce-misuse resistance.

Before stating and proving the security results for (E)PWC, we focus on how to obtain the BBB-secure pseudorandom function F from $\widetilde{E}$. A straightforward way would be to "put the nonce in the tweak", e.g.,

$$F_K'(N) = \widetilde{E}_K^{6,N}(0).$$

This would result in a uniformly random value for each new nonce, but this is only possible when the intended nonce space is smaller than the effective tweak space $\mathcal{T}$ of $\widetilde{E}$. In order to allow the nonce length to be as large as the block length of $\widetilde{E}$, we use instead the "sum-of-PRPs" construction by defining (the exact tweak prefixes are unimportant)

$$F_K(N) = \widetilde{E}_K^{2,0}(N) \oplus \widetilde{E}_K^{2,1}(N). \tag{5}$$

The pseudorandomness of this construction has been well studied. Assuming that $\widetilde{E}_K^{2,0}$ and $\widetilde{E}_K^{2,1}$ are perfectly random and independent permutations, Lucks [42, Theorem 5] showed that an information-theoretic adversary trying to distinguish F_K from a random function $\rho : \{0, 1\}^n \to \{0, 1\}^n$ within q queries has an advantage upper bounded by $q^3/2^{2n-1}$ (see also [16]). Better bounds were proposed in three different papers: Bellare and Impagliazzo [8] proved that the advantage is upper bounded by $\mathcal{O}(n)(q/2^n)^{1.5}$, while Patarin proved in two different

---

[14] We use a set of prefixes which is disjoint from the set used for the CTRT mode in order to later combine the two modes smoothly.

[15] When constructing an AE scheme, it is more convenient to directly define a vector-input MAC, rather than a string-input MAC that must later be transformed to handle vectors of strings, as required for an AE scheme.

1 **algorithm** $\text{PWC}[\widetilde{E}]_K(N, A, M)$ $\boxed{\text{EPWC}[\widetilde{E}]_K(N, A, M)}$

2    $\ell_a := |A|/n$

3    $\ell_m := |M|/n$

4   parse A as $A_1\|\cdots\|A_{\ell_a}$, with $|A_1| = \ldots = |A_{\ell_a-1}| = n$, $1 \le |A_{\ell_a}| \le n$

5   parse M as $M_1\|\cdots\|M_{\ell_m}$, with $|M_1| = \ldots = |M_{\ell_m-1}| = n$, $1 \le |M_{\ell_m}| \le n$

6    $\text{auth} := \widetilde{E}_K^{2,0}(N) \oplus \widetilde{E}_K^{2,1}(N)$

7   **if** $\ell_a > 0$ **then**

8   **for** $i = 1$ **to** $\ell_a - 1$ **do**

9            $\text{auth} := \text{auth} \oplus \widetilde{E}_K^{2,i+1}(A_i)$

10   **if** $|A_{\ell_a}| = n$ **then**

11            $\text{auth} := \text{auth} \oplus \widetilde{E}_K^{2,\ell_a+1}(A_{\ell_a})$

12   **else**

13            $\text{auth} := \text{auth} \oplus \widetilde{E}_K^{3,\ell_a+1}(A_{\ell_a}10^*)$

14   **if** $\ell_m > 0$ **then**

15   **for** $i = 1$ **to** $\ell_m - 1$ **do**

16            $\text{auth} := \text{auth} \oplus \widetilde{E}_K^{4,i}(M_i)$

17   **if** $|M_{\ell_m}| = n$ **then**

18            $\text{auth} := \text{auth} \oplus \widetilde{E}_K^{4,\ell_m}(M_{\ell_m})$

19   **else**

20            $\text{auth} := \text{auth} \oplus \widetilde{E}_K^{5,\ell_m}(M_{\ell_m}10^*)$

21    $\boxed{\text{tag} := \widetilde{E}_K^{4,0}(\text{auth})}$

22   **return tag**

Fig. 3. Definition of the PWC and EPWC modes, using a TBC $\widetilde{E} \in \text{TBC}(\mathcal{K}, \mathcal{T}', \mathcal{X})$ with $\mathcal{T}' = \{2,\ldots,5\} \times \mathcal{T}$ and $\mathcal{X} = \{0,1\}^n$. The boxed statement only applies to EPWC. For notational simplicity, we identify $\mathcal{T}$ with $\{0,\ldots,|\mathcal{T}|-1\}$ and $\text{Inc}^i(0)$ with i.

ways [47,49] an upper bound $\mathcal{O}(q/2^n)$. However, in all three cases the exact $\mathcal{O}(\cdot)$ function was left unspecified and the upper bound was not explicitly worked out. For the sake of concreteness, we propose the following optimistic conjecture.

Conjecture 1. There is an absolute constant C such the advantage of any adversary trying to distinguish the sum of two independent random permutations of $\mathcal{X}$ from a random function from $\mathcal{X}$ to $\mathcal{X}$ within q queries is at most $Cq/|\mathcal{X}|$.

The security of PWC and EPWC is captured by Theorems 2 and 3 below. We denote by $t_{\text{PWC}}(\sigma)$, resp. $t_{\text{EPWC}}(\sigma)$, an upper bound on the time needed to compute $\text{PWC}[\widetilde{E}]$, resp. $\text{EPWC}[\widetilde{E}]$ on inputs of total (AD + message) length at most σ blocks of n bits when calls to $\widetilde{E}_K$ cost unit time.

Theorem 2 (PRF-Security of PWC and EPWC). *Let* $\widetilde{E} \in \text{TBC}(\mathcal{K}, \mathcal{T}', \mathcal{X})$ *with* $\mathcal{X} = \{0,1\}^n$ *and* $\mathcal{T}' = \{2,\ldots,5\} \times \mathcal{T}$, *and assume Conjecture 1. Let* A *be a* (q, m, ℓ, σ, t)-*adversary against the PRF-security of* $\text{PWC}[\widetilde{E}]$, *resp.* $\text{EPWC}[\widetilde{E}]$, with $\ell \leq |\mathcal{T}| - 2$. Then there exists an absolute constant C and an adversary A', resp. A'', against the TPRP-security of $\widetilde{E}$, making at most $\sigma + 2q$, resp. $\sigma + 3q$ oracle queries and running it time at most $t + t_{\mathsf{PWC}}(\sigma)$, resp. $t + t_{\mathsf{EPWC}}(\sigma)$, such that

(a) if A is nonce-respecting ($m = 1$), then

$$\mathbf{Adv}_{\mathsf{PWC}[\widetilde{E}]}^{\mathrm{nPRF}}(\mathsf{A}) \leq \mathbf{Adv}_{\widetilde{E}}^{\mathrm{TPRP}}(\mathsf{A}') + \frac{Cq}{|\mathcal{X}|};$$

$$\mathbf{Adv}_{\mathsf{EPWC}[\widetilde{E}]}^{\mathrm{nPRF}}(\mathsf{A}) \leq \mathbf{Adv}_{\widetilde{E}}^{\mathrm{TPRP}}(\mathsf{A}'') + \frac{Cq}{|\mathcal{X}|};$$

(b) if A is allowed to repeat nonces ($m > 1$), then

$$\mathbf{Adv}_{\mathsf{EPWC}[\widetilde{E}]}^{\mathrm{PRF}}(\mathsf{A}) \leq \mathbf{Adv}_{\widetilde{E}}^{\mathrm{TPRP}}(\mathsf{A}'') + \frac{q^2}{|\mathcal{X}|}.$$

Proof. Fix a $(q, m, |\mathcal{T}| - 2, \sigma, t)$-adversary A against the PRF-security of $\mathsf{PWC}[\widetilde{E}]$ or $\mathsf{EPWC}[\widetilde{E}]$, trying to distinguish the construction from a random function $R \leftarrow_{\$} \mathsf{Func}(\mathcal{N} \times \mathcal{D}, \mathcal{Y})$, where $\mathcal{D} = \mathcal{A} \times \mathcal{M}$. We start by proving (a), assuming A is nonce-respecting ($m = 1$). First, slightly abusing the notation, let us see (E)PWC as a construction based on an arbitrary tweakable permutation, identifying (E)PWC$[\widetilde{E}]_K$ with (E)PWC$[\widetilde{E}_K]$. We start by replacing $\widetilde{E}_K$ in the security experiment by a uniformly random tweakable permutation $\widetilde{P}$. One can see $\mathsf{A}^{\mathsf{PWC}[\cdot]}$, resp. $\mathsf{A}^{\mathsf{EPWC}[\cdot]}$, as an adversary A', resp. A'' against the TPRP-security of $\widetilde{E}$, making at most $\sigma + 2q$, resp. $\sigma + 3q$ queries to its oracle (since a query of ℓ_i blocks to PWC, resp. EPWC, costs $\ell_i + 2$, resp. $\ell_i + 3$ calls to $\widetilde{E}$) and running in time at most $t' = t + t_{\mathsf{PWC}}(\sigma)$, resp. $t' = t + t_{\mathsf{EPWC}}(\sigma)$, so that

$$\mathbf{Adv}_{\mathsf{PWC}[\widetilde{E}]}^{\mathrm{nPRF}}(\mathsf{A}) \leq \mathbf{Adv}_{\widetilde{E}}^{\mathrm{TPRP}}(\mathsf{A}') + \delta, \tag{6}$$

$$\mathbf{Adv}_{\mathsf{EPWC}[\widetilde{E}]}^{\mathrm{nPRF}}(\mathsf{A}) \leq \mathbf{Adv}_{\widetilde{E}}^{\mathrm{TPRP}}(\mathsf{A}'') + \delta, \tag{7}$$

where

$$\delta = \left| \Pr\left[\widetilde{P} \leftarrow_{\$} \mathsf{TP}(\mathcal{T}', \mathcal{X}) : \mathsf{A}^{(\mathsf{E})\mathsf{PWC}[\widetilde{P}]} = 1 \right] \right.$$

$$\left. - \Pr\left[R \leftarrow_{\$} \mathsf{Func}(\mathcal{N} \times \mathcal{D}, \mathcal{Y}) : \mathsf{A}^R = 1 \right] \right|. \tag{8}$$

In order to upper bound δ, we abstract the high-level structure of $(\mathsf{E})\mathsf{PWC}[\widetilde{P}]$ as follows. First, we see how A and M are handled as applying a keyed hash function (the key being $\widetilde{P}$) to the pair (A, M), viz.

$$H_{\widetilde{P}}(A, M) \stackrel{\text{def}}{=} \left(\bigoplus_{i=1}^{\ell_a - 1} \widetilde{P}^{2, i+1}(A_i) \right) \oplus \widetilde{P}^{2/3, \ell_a + 1}(A_{\ell_a}(10^*))$$

$$\oplus \left(\bigoplus_{i=1}^{\ell_m - 1} \widetilde{P}^{4, i}(M_i) \right) \oplus \widetilde{P}^{4/5, \ell_m}(M_{\ell_m}(10^*)). \tag{9}$$

We also define a pseudorandom function (again with key $\widetilde{P}$) as

$$F_{\widetilde{P}}(N) \stackrel{\text{def}}{=} \widetilde{P}^{2,0}(N) \oplus \widetilde{P}^{2,1}(N). \tag{10}$$

Then, $(\mathsf{E})\mathsf{PWC}[\widetilde{P}]$ can be written

$$\mathsf{PWC}[\widetilde{P}](N, A, M) = F_{\widetilde{P}}(N) \oplus H_{\widetilde{P}}(A, M), \tag{11}$$

$$\mathsf{EPWC}[\widetilde{P}](N, A, M) = \widetilde{P}^{4,0}\Big(F_{\widetilde{P}}(N) \oplus H_{\widetilde{P}}(A, M)\Big), \tag{12}$$

which should make it clear that the PWC construction follows the Wegman-Carter paradigm [60] with an additional layer of encryption for EPWC (note that the three sets of tweaks used in F, H, and for the final encryption call are disjoint, so that these three building blocks are independent).

We start by showing that the hash function family $(H_{\widetilde{P}})$, with $\widetilde{P} \in \mathsf{TP}(\mathcal{T}', \mathcal{X})$, is xor-universal, i.e., for any two distinct inputs (A, M), (A', M'), and any $X \in \mathcal{X} = \{0,1\}^n$, the probability, over the random draw of $\widetilde{P} \leftarrow_\$ \mathsf{TP}(\mathcal{T}', \mathcal{X})$, that

$$H_{\widetilde{P}}(A, M) \oplus H_{\widetilde{P}}(A', M') = X, \tag{13}$$

is less than $1/|\mathcal{X}|$. Assume that $A \neq A'$ (the reasoning is similar if $A = A'$ and $M \neq M'$), let $\ell_a = |A|/n$ and $\ell_a' = |A'|/n$, and assume $wlog$ that $\ell_a \geq \ell_a'$. Denote $A = A_1\|\cdots\|A_{\ell_a}$ and $A' = A_1'\|\cdots\|A_{\ell_a'}'$. Assume first that $\ell_a > \ell_a'$. Then (13) is equivalent to

$$\widetilde{P}^{2/3,\ell_a+1}(A_{\ell_a}(10^*)) = Z,$$

where Z is independent of permutations $\widetilde{P}^{2,\ell_a+1}$ and $\widetilde{P}^{3,\ell_a+1}$, hence the probability is exactly $1/|\mathcal{X}|$. Assume now that $\ell_a = \ell_a'$. There is necessarily an index $i \leq \ell_a$ such that $A_i \neq A_i'$. If $i < \ell_a$, then (13) is equivalent to

$$\widetilde{P}^{2,i+1}(A_i) \oplus \widetilde{P}^{2,i+1}(A_i') = Z,$$

where Z is a value potentially depending on $\widetilde{P}$ for tweaks different from $(2, i+1)$. Since the probability of this equality (over the random draw of $\widetilde{P}^{2,i+1}$) is either 0 when $Z = 0$ or exactly $1/|\mathcal{X}|$ when $Z \neq 0$, it follows that the condition is met with probability at most $1/|\mathcal{X}|$ in that case. Similarly, if $i = \ell_a$, then (13) is equivalent to

$$\widetilde{P}^{2/3,i+1}(A_i(10^*)) \oplus \widetilde{P}^{2/3,i+1}(A_i'(10^*)) = Z,$$

where Z is a value potentially depending on $\widetilde{P}$ for tweaks different from $(2, i+1)$ and $(3, i+1)$. Again, the condition is met with probability at most $1/|\mathcal{X}|$ in that case. This concludes the proof that H is xor-universal.

As a second step, we replace $F_{\widetilde{P}}$ by a uniformly random function ρ from $\mathcal{N} = \{0,1\}^n$ to $\mathcal{X} = \{0,1\}^n$. Let $\mathsf{PWC}'[\rho, \widetilde{P}]$, resp. $\mathsf{EPWC}'[\rho, \widetilde{P}]$ be defined as in (11), resp. (12), except that $F_{\widetilde{P}}$ is replaced by a call to ρ. Since A is nonce-respecting, then both $\mathsf{PWC}'[\rho, \widetilde{P}]$ and $\mathsf{EPWC}'[\rho, \widetilde{P}]$ are perfectly indistinguishable from Rand (this is obvious for PWC′, while for EPWC′ this follows from the fact that applying any fixed permutation to uniformly random values yields uniformly random values). Hence, it remains to upper bound A's advantage in distinguishing (E)PWC[$\widetilde{P}$] from (E)PWC′[$\rho, \widetilde{P}$]. By a straightforward hybrid argument, this is exactly the advantage of an adversary A‴ simulating H (and $\widetilde{P}^{4,0}$ for EPWC) in distinguishing $F_{\widetilde{P}}$ from ρ within at most q queries (since each query to the construction translates in exactly one query to the function applied to the nonce). Using Conjecture 1, this advantage is upper bounded by $Cq/|\mathcal{X}|$. Combining this with (6), resp. (7), we obtain the result.

We then prove (b), assuming A is allowed to repeat nonces ($m > 1$). Exactly as before, one has

$$\mathbf{Adv}^{\mathrm{PRF}}_{\mathsf{EPWC}[\widetilde{E}]}(\mathsf{A}) \leq \mathbf{Adv}^{\mathrm{TPRP}}_{\widetilde{E}}(\mathsf{A}'') + \delta,$$

with δ defined as in (8). We now see EPWC[$\widetilde{P}$] as a construction based on a universal hash function applied to (N, A, M) followed by a PRF. More precisely, let

$$H'_{\widetilde{P}}(N, A, M) = F_{\widetilde{P}}(N) \oplus H_{\widetilde{P}}(A, M),$$

with H and F as defined in resp. (9) and (10). Then

$$\mathsf{EPWC}[\widetilde{P}](N, A, M) = \widetilde{P}^{4,0}(H'_{\widetilde{P}}(N, A, M)).$$

It is easy to adapt the proof that H is xor-universal to show that H' is also xor-universal (hence, in particular, universal, which is all we need here). The remaining of the proof is now standard [57], and we only sketch it. We first replace $\widetilde{P}^{4,0}$ in EPWC[$\widetilde{P}$] by a uniformly random function $\rho : \mathcal{X} \to \mathcal{X}$, and denote EPWC″[$\rho, \widetilde{P}$] the resulting construction. By the PRP-PRF switching lemma, A can distinguish EPWC[$\widetilde{P}$] from EPWC″[$\rho, \widetilde{P}$] with advantage at most $q^2/(2|\mathcal{X}|)$, and because H' is universal, it can distinguish EPWC″[$\rho, \widetilde{P}$] from Rand with advantage at most $q^2/(2|\mathcal{X}|)$. The result follows.   □

Theorem 3 (MAC-Security of PWC and EPWC). *Let* $\widetilde{E} \in \mathsf{TBC}(\mathcal{K}, \mathcal{T}', \mathcal{X})$ *with* $\mathcal{X} = \{0, 1\}^n$ *and* $\mathcal{T}' = \{2, \ldots, 5\} \times \mathcal{T}$, *and assume Conjecture 1. Let* B *be a nonce-respecting* $(q, 1, \ell, \sigma, t)$-*adversary against the MAC-security of* PWC[$\widetilde{E}$], *resp.* EPWC[$\widetilde{E}$], *with* $\ell \leq |\mathcal{T}| - 2$. *Then there exists an absolute constant* C *and an adversary* B′, *resp.* B″, *against the TPRP-security of* $\widetilde{E}$, *making at most* $\sigma + 2q$, *resp.* $\sigma + 3q$ *oracle queries and running it time at most* $t + t_{\mathsf{PWC}}(\sigma)$, *resp.* $t + t_{\mathsf{EPWC}}(\sigma)$, *such that*

$$\mathbf{Adv}^{\mathrm{nMAC}}_{\mathsf{PWC}[\widetilde{E}]}(\mathsf{B}) \leq \mathbf{Adv}^{\mathrm{TPRP}}_{\widetilde{E}}(\mathsf{B}') + \frac{(C+1)q}{|\mathcal{X}|};$$

$$\mathbf{Adv}^{\mathrm{nMAC}}_{\mathsf{EPWC}[\widetilde{E}]}(\mathsf{B}) \leq \mathbf{Adv}^{\mathrm{TPRP}}_{\widetilde{E}}(\mathsf{B}'') + \frac{(C+1)q}{|\mathcal{X}|}.$$

Proof. The proof is standard and deferred to the full version of the paper [50].   □

Remark 1. While it is in principle possible to save one encryption call in the EPWC construction by keeping the final AD or message block unencrypted as in the standard PMAC construction [13,52], we avoid this to ensure that static AD always gets treated the same, independently of the message. Indeed, applying this optimization would result in a construction where the final block of AD should be treated differently depending on whether the message is empty or not. Handling the AD independently of the message allows to precompute

$$\text{auth}' = \bigoplus_{i=1}^{\ell_a - 1} \widetilde{P}^{2,i+1}(A_i) \oplus \widetilde{P}^{2/3, \ell_a + 1}(A_{\ell_a}(10^*))$$

and to process the nonce and the message later (in particular, when the AD is static, auth' need not be recomputed each time).

Remark 2. In the nonce-misuse scenario, there is a simple birthday attack against $\text{EPWC}[\widetilde{E}]$ as soon as the adversary can repeat a single nonce twice. The attack proceeds as follows: simply query $\text{EPWC}[\widetilde{E}]_K$ for roughly $2^{n/2}$ pairs (N_i, A, M) with distinct nonces and the same AD and message until a collision occurs on the outputs for two nonces N_1 and N_2. Clearly, a collision on the MACs implies that $F_K(N_1) = F_K(N_2)$ (where F_K is given by (5)). Hence, the adversary can now query $Y = \text{EPWC}[\widetilde{E}]_K(N_1, A', M')$ for a new pair $(A', M') \neq (A, M)$. Then Y is a valid forgery for (N_2, A', M'). It remains an open problem to design a nonce-based MAC scheme ensuring graceful degradation of security with the maximal number of nonce repetitions.

6 The SCT Mode

6.1 The NSIV Construction

In this section, we present the nAE mode SCT and analyze its security. We first describe a generic composition method named NSIV, which defines a nAE scheme from a nonce-based keyed function and an nivE scheme. The NSIV construction results from a small (but important from a security viewpoint) modification to the (generic) SIV construction [55]. While in SIV the encryption part is purely IV-based, NSIV relies on a combined nonce- and IV-based encryption (nivE) scheme, the nonce being used as input *both* to the keyed function and the nivE scheme. This is the only difference with SIV, where the nonce is only given as input to the keyed function.

More formally, let F be a nonce-based keyed function with key-space $\mathcal{K}_1$, nonce space $\mathcal{N}$, domain $\mathcal{D} = \mathcal{A} \times \mathcal{M}$, and range $\mathcal{Y}$, and $\Pi = (\mathcal{K}_2, \mathcal{N}, \mathcal{IV}, \mathcal{M}, \text{Enc}, \text{Dec})$ be a nivE scheme. Fix a regular function $\text{Conv} : \mathcal{Y} \to \mathcal{IV}$. We define the nAE scheme[16] $\text{NSIV}[F, \Pi] = (\mathcal{K}, \mathcal{N}, \mathcal{A}, \mathcal{M}, \text{Enc}, \text{Dec})$ with key-space $\mathcal{K} = \mathcal{K}_1 \times \mathcal{K}_2$ as specified on Fig. 4.

---

[16] Our formalization of an nAE scheme in Definition 4 assumes that the ciphertext is a binary string, whereas in our description, $\text{NSIV}[F, \Pi]$.Enc returns a pair (C, tag). We assume some implicit encoding of this pair into a single binary string.

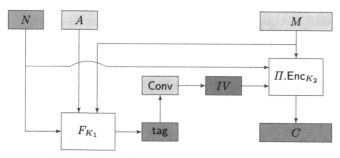

```
1  algorithm NSIV[F, Π].Enc_{K_1,K_2}(N, A, M)
2      tag := F_{K_1}(N, A, M)
3      IV := Conv(tag)
4      C := Π.Enc_{K_2}(N, IV, M)
5      return (C, tag)
6
7  algorithm NSIV[F, Π].Dec_{K_1,K_2}(N, A, C, tag)
8      IV := Conv(tag)
9      M := Π.Dec_{K_2}(N, IV, C)
10     tag' := F_{K_1}(N, A, M)
11     if tag' = tag then return M else return ⊥
```

Fig. 4. The NSIV construction, defining a nAE scheme from a nonce-based keyed function $F : \mathcal{K}_1 \times \mathcal{N} \times \mathcal{D} \to \mathcal{Y}$ where $\mathcal{D} = \mathcal{A} \times \mathcal{M}$ and a nivE scheme $\Pi = (\mathcal{K}_2, \mathcal{N}, \mathcal{IV}, \mathcal{M}, \mathsf{Enc}, \mathsf{Dec})$. Function Conv is a regular function from $\mathcal{Y}$ to $\mathcal{IV}$.

The security of NSIV$[F, \Pi]$ is given by Theorem 4 below. We assume that $\mathcal{A} = \mathcal{M} = \{0,1\}^*$ for convenience, but this restriction can be lifted easily. We denote by $t_\Pi(\sigma)$ an upper bound on the time needed for computing Π.Enc or Π.Dec on inputs of total message length at most σ blocks of n bits, and we assume that computing Conv(tag) or sampling uniformly from Conv$^{-1}(IV)$ takes negligible time for any tag $\in \mathcal{Y}$ and $IV \in \mathcal{IV}$. The proof of this theorem is similar to the security proof of SIV, and deferred to the full version of the paper [50].

Theorem 4 (Security of NSIV). *Let* $F : \mathcal{K}_1 \times \mathcal{N} \times \mathcal{D} \to \mathcal{Y}$, *where* $\mathcal{D} = \mathcal{A} \times \mathcal{M}$, *be a nonce-based keyed function,* $\Pi = (\mathcal{K}_2, \mathcal{N}, \mathcal{IV}, \mathcal{M}, \mathsf{Enc}, \mathsf{Dec})$ *be a nivE scheme, and* Conv $: \mathcal{Y} \to \mathcal{IV}$ *be a regular function. Let* A *be a* (q, m, ℓ, σ, t)-*adversary against* NSIV$[F, \Pi]$. *Then, letting* $t' = t + t_\Pi(\sigma)$, *the following holds:*

(a) if A *is allowed to repeat nonces* $(m > 1)$, *then there exists a* (q, m, ℓ, σ, t')-*adversary* A' *against* Π *and a* (q, q, ℓ, σ, t')-*adversary* A'' *against the PRF-security of* F *such that*

$$\mathbf{Adv}^{\mathrm{AE}}_{\mathsf{NSIV}[F,\Pi]}(\mathsf{A}) \leq \mathbf{Adv}^{\mathrm{ivE}}_{\Pi}(\mathsf{A}') + \mathbf{Adv}^{\mathrm{PRF}}_{F}(\mathsf{A}'') + \frac{q}{|\mathcal{Y}|};$$

(b) if A *is nonce-respecting (m = 1), then there exists a $(q, 1, \ell, \sigma, t')$-adversary* A' *against Π and $(q, 1, \ell, \sigma, t')$-adversaries* A'' *and* A''' *against respectively the PRF- and the MAC-security of F, all nonce-respecting, such that*

$$\mathbf{Adv}^{\text{nAE}}_{\text{NSIV}[F,\Pi]}(\mathsf{A}) \leq \mathbf{Adv}^{\text{nivE}}_{\Pi}(\mathsf{A}') + \mathbf{Adv}^{\text{nPRF}}_{F}(\mathsf{A}'') + \mathbf{Adv}^{\text{nMAC}}_{F}(\mathsf{A}''').$$

6.2   From NSIV to SCT

The $\mathsf{SCT}[\widetilde{E}]$ mode is simply $\mathsf{NSIV}[F, \Pi]$ where F is instantiated with $\mathsf{EPWC}[\widetilde{E}]$ and Π is instantiated with $\mathsf{CTRT}[\widetilde{E}]$. Additionally, in order to be able to use the same key for calls to $\widetilde{E}$ both in EPWC and in CTRT, we use tweak separation to ensure that all calls to $\widetilde{E}$ in EPWC and in CTRT are independent. The resulting construction is illustrated in Fig. 1. Combining Theorem 4 with Theorems 1, 2 and 3, we finally obtain the following result for the security of SCT.[17] We denote by $t_{\mathsf{SCT}}(\sigma)$ an upper bound on the time needed for computing $\mathsf{SCT}[\widetilde{E}].\mathsf{Enc}_K$ or $\mathsf{SCT}[\widetilde{E}].\mathsf{Dec}_K$ on inputs of total (AD + message) length at most σ blocks of n bits when calls to $\widetilde{E}_K$ cost unit time.

Theorem 5 (Security of SCT). *Let $\widetilde{E} \in \mathsf{TBC}(\mathcal{K}, \mathcal{T}', \mathcal{X})$ with $\mathcal{X} = \{0, 1\}^n$, $\mathcal{T} = \{1, \ldots, 5\} \times \mathcal{T}$, and $|\mathcal{T}| \geq 8$. Let Conv be a regular[18] function from $\mathcal{X}$ to $\mathcal{T}$. Assume Conjecture 1 and let $f(\sigma)$ be defined as in Theorem 1. Let* A *be a (q, m, ℓ, σ, t)-adversary against $\mathsf{SCT}[\widetilde{E}]$ with $\ell \leq |\mathcal{T}| - 2$. Then there exists an absolute constant C and an adversary* A' *against the TPRP-security of $\widetilde{E}$, making at most $\sigma + 3q$ oracle queries and running in time at most $t + t_{\mathsf{SCT}}(\sigma)$, such that*

(a) if A *is allowed to repeat nonces in encryption queries (m > 1), then*

$$\mathbf{Adv}^{\text{AE}}_{\mathsf{SCT}[\widetilde{E}]}(\mathsf{A}) \leq \mathbf{Adv}^{\text{TPRP}}_{\widetilde{E}}(\mathsf{A}') + \frac{2(m-1)\sigma + 1}{|\mathcal{T}|} + f(\sigma) + \frac{q^2 + q}{|\mathcal{X}|};$$

(b) if A *is nonce-respecting (m = 1), then*

$$\mathbf{Adv}^{\text{nAE}}_{\mathsf{SCT}[\widetilde{E}]}(\mathsf{A}) \leq \mathbf{Adv}^{\text{TPRP}}_{\widetilde{E}}(\mathsf{A}') + \frac{1}{|\mathcal{T}|} + f(\sigma) + \frac{(2C + 1)q}{|\mathcal{X}|}.$$

Acknowledgements. The authors would like to thank Jérémy Jean and Ivica Nikolic for their remarks on early designs. The first author is supported by the Singapore National Research Foundation Fellowship 2012 (NRF-NRFF2012-06).

---

[17] In more details, it is more convenient to prove Theorem 5 by first replacing $\widetilde{E}_K$ by a uniformly random tweakable permutation, and then applying Theorems 1, 2, and 3 for a perfect TBC.

[18] Note that this regularity condition imposes $|\mathcal{T}| \leq |\mathcal{X}|$. However, when $\mathcal{T} > |\mathcal{X}|$, the security bounds of CTRT and EPWC do not depend on the tweak length (only the maximal message length does). Hence, one can always use a subset of tweaks of size $|\mathcal{X}|$ in case $|\mathcal{T}| > |\mathcal{X}|$.

References

1. CAESAR: Competition for Authenticated Encryption: Security, Applicability, and Robustness. http://competitions.cr.yp.to/caesar.html
2. Abed, F., Fluhrer, S., Forler, C., List, E., Lucks, S., McGrew, D., Wenzel, J.: Pipelineable on-line encryption. In: Cid, C., Rechberger, C. (eds.) FSE 2014. LNCS, vol. 8540, pp. 205–223. Springer, Heidelberg (2015)
3. AlFardan, N.J., Paterson, K.G., Lucky thirteen: breaking the TLS and DTLS record protocols. In: Security and Privacy - SP 2013, pp. 526–540 (2013)
4. Andreeva, E., Bogdanov, A., Luykx, A., Mennink, B., Mouha, N., Yasuda, K.: How to securely release unverified plaintext in authenticated encryption. In: Sarkar, P., Iwata, T. (eds.) ASIACRYPT 2014. LNCS, vol. 8873, pp. 105–125. Springer, Heidelberg (2014)
5. Andreeva, E., Bogdanov, A., Luykx, A., Mennink, B., Tischhauser, E., Yasuda, K.: Parallelizable and authenticated online ciphers. In: Sako, K., Sarkar, P. (eds.) ASIACRYPT 2013, Part I. LNCS, vol. 8269, pp. 424–443. Springer, Heidelberg (2013)
6. Asharov, G., Naor, M., Segev, G., Shahaf, I., Encryption, S.S.: Optimal locality in linear space via two-dimensional balanced allocations. IACR Cryptology ePrint Archive, Report 2016/251 (2016). To appear at STOC 2016
7. Bellare, M., Desai, A., Jokipii, E., Rogaway, P.: A concrete security treatment of symmetric encryption. In: FOCS 1997, pp. 394–403 (1997)
8. Bellare, M., Impagliazzo, R.: A tool for obtaining tighter security analyses of pseudorandom function based constructions, with applications to PRP to PRF conversion. IACR Cryptology ePrint Archive, Report 1999/024 (1999)
9. Bellare, M., Kilian, J., Rogaway, P.: The security of the cipher block chaining message authentication code. J. Comput. Syst. Sci. **61**(3), 362–399 (2000)
10. Bellare, M., Namprempre, C.: Authenticated encryption: relations among notions and analysis of the generic composition paradigm. In: Okamoto, T. (ed.) ASIACRYPT 2000. LNCS, vol. 1976, pp. 531–545. Springer, Heidelberg (2000)
11. Bellare, M., Rogaway, P.: Encode-then-encipher encryption: how to exploit nonces or redundancy in plaintexts for efficient cryptography. In: Okamoto, T. (ed.) ASIACRYPT 2000. LNCS, vol. 1976, pp. 317–330. Springer, Heidelberg (2000)
12. Bellare, M., Rogaway, P., Wagner, D.: The EAX mode of operation. In: Roy, B., Meier, W. (eds.) FSE 2004. LNCS, vol. 3017, pp. 389–407. Springer, Heidelberg (2004)
13. Black, J.A., Rogaway, P.: A block-cipher mode of operation for parallelizable message authentication. In: Knudsen, L.R. (ed.) EUROCRYPT 2002. LNCS, vol. 2332, pp. 384–397. Springer, Heidelberg (2002)
14. Brassard, G.: On computationally secure authentication tags requiring short secret shared keys. In: CRYPTO 1982, pp. 79–86 (1982)
15. Chen, S., Steinberger, J.: Tight security bounds for key-alternating ciphers. In: Nguyen, P.Q., Oswald, E. (eds.) EUROCRYPT 2014. LNCS, vol. 8441, pp. 327–350. Springer, Heidelberg (2014)
16. Cogliati, B., Lampe, R., Patarin, J.: The indistinguishability of the XOR of k permutations. In: Cid, C., Rechberger, C. (eds.) FSE 2014. LNCS, vol. 8540, pp. 285–302. Springer, Heidelberg (2014)
17. Dodis, Y., Steinberger, J.: Domain extension for MACs beyond the birthday barrier. In: Paterson, K.G. (ed.) EUROCRYPT 2011. LNCS, vol. 6632, pp. 323–342. Springer, Heidelberg (2011)

18. Duong, T., Rizzo, J.: Here come the ⊕ Ninjas. Unpublished manuscript (2011). https://bug665814.bugzilla.mozilla.org/attachment.cgi?id=540839
19. Ferguson, N.: Collision attacks on OCB. Unpublished manuscript (2002). http://www.cs.ucdavis.edu/~rogaway/ocb/fe02.pdf
20. Ferguson, N., Lucks, S., Schneier, B., Whiting, D., Bellare, M., Kohno, T., Callas, J., Walker, J.: The Skein hash function family. SHA3 Submission to NIST (Round 3) (2010)
21. Fleischmann, E., Forler, C., Lucks, S.: McOE: a family of almost foolproof on-line authenticated encryption schemes. In: Canteaut, A. (ed.) FSE 2012. LNCS, vol. 7549, pp. 196–215. Springer, Heidelberg (2012)
22. Gligor, V.D., Donescu, P.: Fast encryption and authentication: XCBC encryption and XECB authentication modes. In: Matsui, M. (ed.) FSE 2001. LNCS, vol. 2355, p. 92. Springer, Heidelberg (2002)
23. Goldreich, O., Goldwasser, S., Micali, S.: How to construct random functions. J. ACM **33**(4), 792–807 (1986)
24. Grosso, V., Leurent, G., Standaert, F.-X., Varici, K., Durvaux, F., Gaspar, L., Kerckhof, S.: SCREAM and iSCREAM. Submitted to CAESAR (2014)
25. Gueron, S., Lindell, Y.: GCM-SIV: full nonce misuse-resistant authenticated encryption atunder one cycle per byte. In: ACM CCS 2015, pp. 109–119 (2015)
26. Hoang, V.T., Krovetz, T., Rogaway, P.: Robust authenticated-encryption AEZ and the problem that it solves. In: Oswald, E., Fischlin, M. (eds.) EUROCRYPT 2015. LNCS, vol. 9056, pp. 15–44. Springer, Heidelberg (2015)
27. Hoang, V.T., Reyhanitabar, R., Rogaway, P., Vizár, D.: Online authenticated-encryption and its nonce-reuse misuse-resistance. In: Gennaro, R., Robshaw, M. (eds.) CRYPTO 2015. LNCS, vol. 9215, pp. 493–517. Springer, Heidelberg (2015)
28. Iwata, T.: New blockcipher modes of operation with beyond the birthday bound security. In: Robshaw, M. (ed.) FSE 2006. LNCS, vol. 4047, pp. 310–327. Springer, Heidelberg (2006)
29. Iwata, T.: Authenticated encryption mode for beyond the birthday bound security. In: Vaudenay, S. (ed.) AFRICACRYPT 2008. LNCS, vol. 5023, pp. 125–142. Springer, Heidelberg (2008)
30. Jean, J., Nikolic, I., Peyrin, T.: Deoxys v1. Submitted to the CAESAR competition (2014)
31. Jean, J., Nikolic, I., Peyrin, T.: Joltik v1. Submitted to the CAESAR competition (2014)
32. Jean, J., Nikolic, I., Peyrin, T.: KIASU v1. Submitted to the CAESAR competition (2014)
33. Jean, J., Nikolic, I., Peyrin, T.: Tweaks and keys for blockciphers: the TWEAKEY framework. In: Sarkar, P., Iwata, T. (eds.) ASIACRYPT 2014. LNCS, vol. 8874, pp. 274–288. Springer, Heidelberg (2014)
34. Jutla, C.S.: Encryption modes with almost free message integrity. In: Pfitzmann, B. (ed.) EUROCRYPT 2001. LNCS, vol. 2045, pp. 529–544. Springer, Heidelberg (2001)
35. Jutla, C.S.: Encryption modes with almost free message integrity. J. Cryptol. **21**(4), 547–578 (2008). Earlier version at EUROCRYPT 2001
36. Katz, J., Yung, M.: Characterization of security notions for probabilistic private-key encryption. J. Cryptol. **19**(1), 67–95 (2006)
37. Krawczyk, H.: The order of encryption and authentication for protecting communications (or: how secure is SSL?). In: Kilian, J. (ed.) CRYPTO 2001. LNCS, vol. 2139, pp. 310–331. Springer, Heidelberg (2001)

38. Krovetz, T., Rogaway, P.: The software performance of authenticated-encryption modes. In: Joux, A. (ed.) FSE 2011. LNCS, vol. 6733, pp. 306–327. Springer, Heidelberg (2011)

39. Lampe, R., Seurin, Y.: Tweakable blockciphers with asymptotically optimal security. In: Moriai, S. (ed.) FSE 2013. LNCS, vol. 8424, pp. 133–152. Springer, Heidelberg (2014)

40. Landecker, W., Shrimpton, T., Terashima, R.S.: Tweakable blockciphers with beyond birthday-bound security. In: Safavi-Naini, R., Canetti, R. (eds.) CRYPTO 2012. LNCS, vol. 7417, pp. 14–30. Springer, Heidelberg (2012)

41. Liskov, M., Rivest, R.L., Wagner, D.: Tweakable block ciphers. In: Yung, M. (ed.) CRYPTO 2002. LNCS, vol. 2442, pp. 31–46. Springer, Heidelberg (2002)

42. Lucks, S.: The sum of PRPs is a secure PRF. In: Preneel, B. (ed.) EUROCRYPT 2000. LNCS, vol. 1807, pp. 470–484. Springer, Heidelberg (2000)

43. McGrew, D.A., Viega, J.: The security and performance of the Galois/counter mode (GCM) of operation. In: Canteaut, A., Viswanathan, K. (eds.) INDOCRYPT 2004. LNCS, vol. 3348, pp. 343–355. Springer, Heidelberg (2004)

44. Mennink, B.: Optimally secure tweakable blockciphers. In: Leander, G. (ed.) FSE 2015. LNCS, vol. 9054, pp. 428–448. Springer, Heidelberg (2015)

45. Minematsu, K.: Beyond-birthday-bound security based on tweakable block cipher. In: Dunkelman, O. (ed.) FSE 2009. LNCS, vol. 5665, pp. 308–326. Springer, Heidelberg (2009)

46. Namprempre, C., Rogaway, P., Shrimpton, T.: Reconsidering generic composition. In: Nguyen, P.Q., Oswald, E. (eds.) EUROCRYPT 2014. LNCS, vol. 8441, pp. 257–274. Springer, Heidelberg (2014)

47. Patarin, J.: A proof of security in $O(2^n)$ for the Xor of two random permutations. In: Safavi-Naini, R. (ed.) ICITS 2008. LNCS, vol. 5155, pp. 232–248. Springer, Heidelberg (2008)

48. Patarin, J.: The "Coefficients H" technique. In: Avanzi, R.M., Keliher, L., Sica, F. (eds.) SAC 2008. LNCS, vol. 5381, pp. 328–345. Springer, Heidelberg (2009)

49. Patarin, J.: Security in $O(2^n)$ for the Xor of two random permutations: proof with the standard H technique. IACR Cryptology ePrint Archive, Report 2013/368 (2013)

50. Peyrin, T., Seurin, Y.: Counter-in-Tweak: authenticated encryption modes for tweakable block ciphers. Full version of this paper. http://eprint.iacr.org/2015/1049

51. Rogaway, P.: Authenticated-encryption with associated-data. In: ACM CCS 2002, pp. 98–107 (2002)

52. Rogaway, P.: Efficient instantiations of tweakable blockciphers and refinements to modes OCB and PMAC. In: Lee, P.J. (ed.) ASIACRYPT 2004. LNCS, vol. 3329, pp. 16–31. Springer, Heidelberg (2004)

53. Rogaway, P.: Nonce-based symmetric encryption. In: Roy, B., Meier, W. (eds.) FSE 2004. LNCS, vol. 3017, pp. 348–359. Springer, Heidelberg (2004)

54. Rogaway, P., Bellare, M., Black, J.: OCB: a block-cipher mode of operation for efficient authenticated encryption. ACM Trans. Inf. Syst. Secur. 6(3), 365–403 (2003)

55. Rogaway, P., Shrimpton, T.: A provable-security treatment of the key-wrap problem. In: Vaudenay, S. (ed.) EUROCRYPT 2006. LNCS, vol. 4004, pp. 373–390. Springer, Heidelberg (2006)

56. Shoup, V.: On fast and provably secure message authentication based on universal hashing. In: Koblitz, N. (ed.) CRYPTO 1996. LNCS, vol. 1109, pp. 313–328. Springer, Heidelberg (1996)

57. Shoup, V.: Sequences of games: a tool for taming complexity in security proofs. IACR Cryptology ePrint Archive, Report 2004/332 (2004)
58. Shrimpton, T., Terashima, R.S.: A modular framework for building variable-input-length tweakable ciphers. In: Sako, K., Sarkar, P. (eds.) ASIACRYPT 2013, Part I. LNCS, vol. 8269, pp. 405–423. Springer, Heidelberg (2013)
59. Vaudenay, S.: Security flaws induced by CBC padding - applications to SSL, IPSEC, WTLS ... In: Knudsen, L.R. (ed.) EUROCRYPT 2002. LNCS, vol. 2332, pp. 534–546. Springer, Heidelberg (2002)
60. Wegman, M.N., Carter, L.: New hash functions and their use in authentication and set equality. J. Comput. Syst. Sci. $22(3)$, 265–279 (1981)
61. Whiting, D., Housley, R., Ferguson, N.: Counter with CBC-MAC (CCM). Submission to NIST (2002). http://csrc.nist.gov/groups/ST/toolkit/BCM/documents/proposedmodes/ccm/ccm.pdf
62. Yasuda, K.: A new variant of PMAC: beyond the birthday bound. In: Rogaway, P. (ed.) CRYPTO 2011. LNCS, vol. 6841, pp. 596–609. Springer, Heidelberg (2011)

XPX: Generalized Tweakable Even-Mansour with Improved Security Guarantees

Bart Mennink[✉]

Department of Electrical Engineering, ESAT/COSIC, KU Leuven and iMinds,
Leuven, Belgium
bart.mennink@esat.kuleuven.be

Abstract. We present XPX, a tweakable blockcipher based on a single permutation P. On input of a tweak $(t_{11}, t_{12}, t_{21}, t_{22}) \in \mathcal{T}$ and a message m, it outputs ciphertext $c = P(m \oplus \Delta_1) \oplus \Delta_2$, where $\Delta_1 = t_{11}k \oplus t_{12}P(k)$ and $\Delta_2 = t_{21}k \oplus t_{22}P(k)$. Here, the tweak space $\mathcal{T}$ is required to satisfy a certain set of trivial conditions (such as $(0,0,0,0) \notin \mathcal{T}$). We prove that XPX with any such tweak space is a strong tweakable pseudorandom permutation. Next, we consider the security of XPX under related-key attacks, where the adversary can freely select a key-deriving function upon every evaluation. We prove that XPX achieves various levels of related-key security, depending on the set of key-deriving functions and the properties of $\mathcal{T}$. For instance, if $t_{12}, t_{22} \neq 0$ and $(t_{21}, t_{22}) \neq (0,1)$ for all tweaks, XPX is XOR-related-key secure. XPX generalizes Even-Mansour (EM), but also Rogaway's XEX based on EM, and various other tweakable blockciphers. As such, XPX finds a wide range of applications. We show how our results on XPX directly imply related-key security of the authenticated encryption schemes Prøst-COPA and Minalpher, and how a straightforward adjustment to the MAC function Chaskey and to keyed Sponges makes them provably related-key secure.

Keywords: XPX · XEX · Even-Mansour · Tweakable blockcipher · Related-key security · Prøst · COPA · Minalpher · Chaskey · Keyed sponges

1 Introduction

Even-Mansour Blockcipher. A blockcipher $E : \mathcal{K} \times \{0,1\}^n \rightarrow \{0,1\}^n$ is a function that is a permutation on $\{0,1\}^n$ for every key $k \in \mathcal{K}$. The simplest way of designing a blockcipher is the Even-Mansour construction [23,24]: it is built on top of a single n-bit permutation P:

$$\mathrm{EM}_{k_1,k_2}(m) = P(m \oplus k_1) \oplus k_2. \tag{1}$$

See also Fig. 1. In the classical indistinguishability security model, this construction achieves security up to approximately $2^{n/2}$ queries, both for the case where the keys are independent [23,24] as well as for the case where $k_1 = k_2$ [22]. On the

© International Association for Cryptologic Research 2016
M. Robshaw and J. Katz (Eds.): CRYPTO 2016, Part I, LNCS 9814, pp. 64–94, 2016.
DOI: 10.1007/978-3-662-53018-4_3

downside, this construction clearly does not achieve security against related-key distinguishers that may freely choose an offset δ to transform the key. Indeed, for any $\delta \neq 0$, we have $\mathrm{EM}_{k_1,k_2}(m) = \mathrm{EM}_{k_1 \oplus \delta, k_2}(m \oplus \delta)$. Recently, Farshim and Procter [25] and Cogliati and Seurin [17] reconsidered the security of Even-Mansour in the related-key security model. The former considered the case of $k_1 = k_2$, and derived minimal conditions on the set of key-deriving functions such that EM is related-key secure. The latter showed that if $k_1 = \gamma_1(k)$ and $k_2 = \gamma_2(k)$ for two almost perfect nonlinear permutations γ_1, γ_2 [45], the construction is XOR-related-key secure. Karpman showed how to transform related-key distinguishing attacks on EM to key recovery attacks [28].

Even though our focus is on the single-round Even-Mansour (1), we briefly elaborate on its generalization, the iterated $r \geq 1$ round Even-Mansour construction:

$$\mathrm{EM}[r]_{k_1,\ldots,k_{r+1}}(m) = P_r(\cdots P_1(m \oplus k_1) \cdots \oplus k_r) \oplus k_{r+1},$$

where $P_1, \ldots, P_r$ are n-bit permutations. It has been proved that this construction tightly achieves $\mathcal{O}(2^{rn/(r+1)})$ security in the single-key indistinguishability model [9,13,14,30,50]. It has furthermore been analyzed in the chosen-key indifferentiability model [2,31], the known-key indifferentiability model [4,18], and the related-key indistinguishability model [17,25]. As our work centers around the 1-round Even-Mansour of (1), we will not discuss these results in detail; we refer to Cogliati and Seurin [17] for a recent and complete discussion of the state of the art.

Tweakable Blockciphers. A tweakable blockcipher $\widetilde{E} : \mathcal{K} \times \mathcal{T} \times \{0,1\}^n \to \{0,1\}^n$ generalizes over E by ways of an additional parameter, the tweak $t \in \mathcal{T}$. The tweak is a public parameter which brings additional flexibility to the cipher. In more detail, $\widetilde{E}$ is a family of permutations on $\{0,1\}^n$, indexed by $(k,t) \in \mathcal{K} \times \mathcal{T}$. Liskov et al. [34] formalized the principle of tweakable blockciphers, and introduced two modular constructions based on a classical blockcipher. One of their proposals is the following:

$$\mathrm{LRW}_{k,h}(t,m) = E_k(m \oplus h(t)) \oplus h(t),$$

where h is a universal hash function taken from a family of hash functions H. This construction is proven to achieve security up to $2^{n/2}$ queries. Rogaway [48] introduced XEX: it generalizes over LRW by eliminating the universal hash function (and thus by halving the key size) and by replacing it by an efficient tweaking mechanism based on E_k. In more detail, he suggested the use of masking $\Delta = \mathbf{x}_1^{\alpha_1} \cdots \mathbf{x}_\ell^{\alpha_\ell} E_k(N)$ for some pre-defined generators $\mathbf{x}_1, \ldots, \mathbf{x}_\ell \in \mathrm{GF}(2^n)$ (Fig. 2):

$$\mathrm{XEX}_k((\alpha_1, \ldots, \alpha_\ell, N), m) = E_k(m \oplus \Delta) \oplus \Delta. \tag{2}$$

If the generators and the tweak space are defined such that the $\mathbf{x}_1^{\alpha_1} \cdots \mathbf{x}_\ell^{\alpha_\ell}$ are unique and unequal to 1 for all tweaks, XEX achieves birthday bound security

[40,48]. Along with XEX, Rogaway also considered XE, its cousin which only masks the inputs to E and achieves PRP instead of SPRP security. Here, ℓ is usually a small number, and the generators and the tweak space are defined in such a way that adjusting the tweak is very cheap. For instance, practical applications with $n = 128$ often take $\ell \leq 3$ and $(\mathbf{x}_1, \mathbf{x}_2, \mathbf{x}_3) = (2, 3, 7)$, and an allowed tweak space would be $[1, 2^{n/2}] \times [0, 10] \times [0, 10] \times \{0, 1\}^n$. Chakraborty and Sarkar [11] generalized XEX to word-based powering-up, and more recently Granger et al. [27] presented a generalization to constant-time LFSR-based masking.

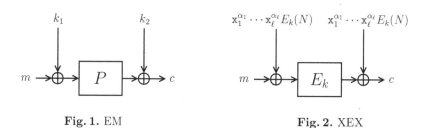

Fig. 1. EM   **Fig. 2.** XEX

Sasaki et al. [49] recently introduced the "Tweakable Even-Mansour" (TEM) for the purpose of the Minalpher authenticated encryption scheme. TEM is a variant of XEX with E_k replaced by a public permutation P:

$$\text{TEM}_k((\alpha_1, \ldots, \alpha_\ell, N), m) = P(m \oplus \Delta) \oplus \Delta, \tag{3}$$

where $\Delta = \mathbf{x}_1^{\alpha_1} \cdots \mathbf{x}_\ell^{\alpha_\ell}(k\|N \oplus P(k\|N))$ for some generators $\mathbf{x}_1, \ldots, \mathbf{x}_\ell \in \text{GF}(2^n)$. (The masking is in fact slightly different, but adjusted for the sake of presentation; cf. Sect. 6.3 for the details.) Independently, Cogliati et al. [15] considered the generalization of LRW to the permutation-based setting. The contribution by Granger et al. [27], Masked EM or MEM, is in fact a generalization of TEM to masking $\Delta = f_1^{\alpha_1} \circ \cdots \circ f_\ell^{\alpha_\ell} \circ P(k\|N)$ for some LFSRs $f_1, \ldots, f_\ell : \{0, 1\}^n \to \{0, 1\}^n$, but their goal is merely to achieve improved efficiency rather than to achieve improved security.

These constructions all achieve approximately birthday bound security, and extensive research has been performed on achieving beyond birthday bound security for tweakable blockciphers [32,33,35,36,41,47]. Because this is out of scope for this article, we will not go into detail; we refer to Mennink [36] and Cogliati and Seurin [16] for a recent and complete discussion of the state of the art.

Application of Tweakable Blockciphers. Tweakable blockciphers find a wide spectrum of applications, most importantly in the area of authenticated encryption and message authentication. For instance, XEX has been originally introduced for the authenticated encryption scheme OCB2 and the message authentication code PMAC [48], and its idea has furthermore been adopted in 18 out of 57 initial submissions to the CAESAR [10] competition for the design of a new authenticated encryption scheme: Deoxys, Joltik, KIASU, and SCREAM use a dedicated tweakable blockcipher; AEZ, CBA, COBRA, COPA, ELmD, iFeed, Marble, OCB, OTR, POET, and SHELL are (in-)directly inspired by XE or XEX; OMD transforms XE to a random function setting; and Minalpher uses TEM. Finally, the Prøstsubmission is simply a permutation P, which is (among others) plugged into COPA and OTR in an Even-Mansour mode. We note that OTR internally uses XE, while COPA uses XEX with $N = 0$ (see also Sect. 6.2).

Related-Key Security of XEX and TEM. XEX resists related-key attacks if the underlying blockcipher is sufficiently related-key secure. However, this premise is not necessarily true if Even-Mansour is plugged into XEX, as is done in Prøst-COPA and Prøst-OTR. In fact, Dobraunig et al. [21] derived a related-key attack on Prøst-OTR. This attack uses that the underlying XE-with-EM construction is not secure under related-key attacks, and it ultimately led to the withdrawal of Prøst-OTR. The attack exploits the nonce N that is used in the masking. Karpman [28] generalized the attack to a key recovery attack. Because COPA uses XEX without nonce (hence with $N = 0$), the attack of Dobraunig et al. does not seem to be directly applicable to Prøst-COPA. Nevertheless, it is unclear whether a variant of it generalizes to Prøst-COPA.

1.1   Our Contribution

We present the tweakable blockcipher XPX. It can be seen as a natural generalization of TEM as well as of XEX with integrated Even-Mansour, and due to its generality it has direct implications for various schemes in literature. In more detail, XPX is a tweakable blockcipher based on an n-bit permutation P. It has a key space $\{0,1\}^n$, a tweak space $\mathcal{T} \subseteq (\{0,1\}^n)^4$ (see below), and a message space $\{0,1\}^n$. It is defined as

$$\mathrm{XPX}_k((t_{11}, t_{12}, t_{21}, t_{22}), m) = P(m \oplus \Delta_1) \oplus \Delta_2, \tag{4}$$

with $\Delta_1 = t_{11}k \oplus t_{12}P(k)$ and $\Delta_2 = t_{21}k \oplus t_{22}P(k)$. Note that XPX boils down to the original Even-Mansour blockcipher by taking $\mathcal{T}_{\mathrm{EM}} = \{(1,0,1,0)\}$. It also generalizes XEX based on Even-Mansour and with $N = 0$, by defining $\mathcal{T}_{\mathrm{XEX}}$ to be a tweak space depending on $(\alpha_1, \ldots, \alpha_\ell)$, and similarly captures TEM and MEM to a certain degree (cf. Sect. 3 for the details).

Valid Tweak Sets. Obviously, XPX is not secure for any possible tweak space $\mathcal{T}$. For instance, if $(0,0,0,0) \in \mathcal{T}$, the scheme is trivially insecure. Also, if $(1,0,0,1) \in \mathcal{T}$, an attacker can easily distinguish by observing that $\mathrm{XPX}_k((1,0,0,1),0) = 0$. Therefore, it makes sense to limit the tweak space in some way, and we define the notion of valid tweak spaces. This condition eliminates the trivial cases (such as above two) and allows us to focus on the "interesting" tweaks. We remark that $\mathcal{T}_{\mathrm{EM}}$ and $\mathcal{T}_{\mathrm{XEX}}$ are valid tweak spaces.

Single-Key Security. As a first step, we consider the security of XPX in the traditional single-key indistinguishability setting, and we prove that if $\mathcal{T}$ is a valid set, then XPX achieves strong PRP (SPRP) security up to about $2^{n/2}$ queries. The proof is performed in the ideal permutation model, and uses Patarin's H-coefficient technique [46] which has found recent adoption in, among others, generic blockcipher analysis [13,14,17,19,35,36] and security of message authentication algorithms [5,20,39,43].

Related-Key Security. Next, we consider the security of XPX in the related-key setting, where for every query, the adversary can additionally choose a function to transform the key. We focus on the following two types of key-deriving function sets:

- $\varPhi_{\oplus}$: the set of functions that transform k to $k \oplus \delta$, for any offset δ;
- $\varPhi_{P\oplus}$: the set of functions that transform k to $k \oplus \delta$, *or* that transform $P(k)$ to $P(k) \oplus \delta$, for any offset δ.

The first set, $\varPhi_{\oplus}$, has been formally introduced alongside the formal specification of related-key security by Bellare and Kohno [6]. It is the most logical choice, given that the maskings in XPX itself are XORed into the state. We remark that Cogliati and Seurin [17] also use $\varPhi_{\oplus}$ in their related-key analysis of Even-Mansour. The second set, $\varPhi_{P\oplus}$, is a natural generalization of $\varPhi_{\oplus}$, noting that the masks in XPX are of the form $t_{i1}k \oplus t_{i2}P(k)$. For the case of $\varPhi_{P\oplus}$, we assume that the underlying permutation is available for the key-deriving functions. Albrecht et al. [1] showed how to generalize the setting of Bellare and Kohno [6] to primitive-dependent key-deriving functions. In this work, we consider the related-key security for XPX in a security model that is a straightforward generalization of the models of Bellare and Kohno and Albrecht et al. to tweakable blockciphers.

For the two key-deriving sets $\varPhi_{\oplus}$ and $\varPhi_{P\oplus}$, we show that XPX achieves the following levels of related-key security:

if $\mathcal{T}$ is valid, and for all tweaks:	security	rk
$t_{12} \neq 0$	PRP	$\varPhi_{\oplus}$
$t_{12}, t_{22} \neq 0$ and $(t_{21}, t_{22}) \neq (0,1)$	SPRP	$\varPhi_{\oplus}$
$t_{11}, t_{12} \neq 0$	PRP	$\varPhi_{P\oplus}$
$t_{11}, t_{12}, t_{21}, t_{22} \neq 0$	SPRP	$\varPhi_{P\oplus}$

In brief, if $P(k)$ does not drop from the masking Δ_1 (resp. maskings Δ_1, Δ_2) the scheme achieves PRP (resp. SPRP) related-key security under $\varPhi_{\oplus}$. To achieve related-key security under $\varPhi_{P\oplus}$, we require that this condition holds for both k and $P(k)$. The requirement "$(t_{21}, t_{22}) \neq (0,1)$" is technically equivalent to the requirement for XEX that $\mathbf{x}_1^{\alpha_1} \cdots \mathbf{x}_\ell^{\alpha_\ell} \neq 1$ for all tweaks: if the conditions were violated, both schemes can be attacked in a similar way.

The proof for related-key security is again performed using the H-coefficient technique, but various difficulties arise, mostly due to the fact that we pursue stronger security requirements and that we aim to minimize the number of conditions we put on the tweaks.

1.2 Applications

XPX as described in (4) appears in many constructions or modes (either directly or indirectly), and can be used to argue related-key security for these modes. We exemplify this for authenticated encryption and for message authentication codes.

Firstly, Prøst-COPA is related-key secure for both key-deriving function sets $\Phi_\oplus$ and $\Phi_{P\oplus}$. The crux behind this observation is that the XEX-with-EM evaluations in Prøst-COPA are in fact XPX evaluations with $t_{11}, t_{12}, t_{21}, t_{22} \neq 0$ for all tweaks. (Recall that EM itself is not related-key secure and this result cannot be shown by straightforward reduction.) A similar observation can be made for Minalpher, with an additional technicality that the key k in TEM is not of full size. Due to the structural differences between the masking approaches of XPX and MEM [27], multiplication versus influence via function evaluation, the proof techniques are technically incompatible. Nonetheless, it is of interest to combine our results with the observations from [27], improving *both the security and the efficiency* of existing modes.

Secondly, we consider the Chaskey permutation-based MAC function by Mouha et al. [42,43]. We first note that the proof of [43] is implicitly using XPX with a tweak space of size $|\mathcal{T}| = 3$. Next, we introduce Chaskey', an adjustment of Chaskey that uses permuted key $P(k)$ instead of k, which achieves XOR-related-key security. Similar findings can be made for keyed Sponges.

It may be of interest to generalize XPX to the case where the maskings are performed using universal hash functions, e.g., $\Delta_i = h_1(t_{i1}) \oplus h_2(t_{i2})$. This generalization may, however, in certain settings be less efficient as one evaluation of the permutation is traded for two hash function evaluations.

1.3 Outline

Section 2 introduces preliminary notation as well as the security models targeted in this work. XPX is introduced in Sect. 3. In Sect. 4, the notion of valid tweak spaces is defined and justified. XPX is analyzed for the various security models in Sect. 5. We apply the results on XPX to authenticated encryption in Sect. 6 and to MACs in Sect. 7.

2 Preliminaries

By $\{0,1\}^n$ we denote the set of bit strings of length n. Let $\mathrm{GF}(2^n)$ be the field of order 2^n. We identify bit strings from $\{0,1\}^n$ and finite field elements in $\mathrm{GF}(2^n)$. This is done by representing a string $a = a_{n-1}a_{n-2}\cdots a_1 a_0 \in \{0,1\}^n$ as polynomial $a(\mathrm{x}) = a_{n-1}\mathrm{x}^{n-1} + a_{n-2}\mathrm{x}^{n-2} + \cdots + a_1\mathrm{x} + a_0 \in \mathrm{GF}(2^n)$ and vice versa. There is additionally a one-to-one correspondence between $[0, 2^n - 1]$

and $\{0,1\}^n$, by considering $a(2) \in [0, 2^n-1]$. For $a, b \in \{0,1\}^n$, we define addition $a \oplus b$ as addition of the polynomials $a(\mathbf{x}) + b(\mathbf{x}) \in \mathrm{GF}(2^n)$. Multiplication $a \otimes b$ is defined with respect to the irreducible polynomial $f(\mathbf{x})$ used to represent $\mathrm{GF}(2^n)$: $a(\mathbf{x}) \cdot b(\mathbf{x}) \bmod f(\mathbf{x})$.

For integers $a \geq b \geq 1$, we denote by $(a)_b = a(a-1) \cdots (a-b+1) = \frac{a!}{(a-b)!}$ the falling factorial power. If $\mathcal{M}$ is some set, $m \xleftarrow{\$} \mathcal{M}$ denotes the uniformly random drawing of m from $\mathcal{M}$. The size of $\mathcal{M}$ is denoted by $|\mathcal{M}|$. By $\mathsf{Perm}(\mathcal{M})$ we denote the set of all permutations on $\mathcal{M}$.

A blockcipher $E : \mathcal{K} \times \mathcal{M} \to \mathcal{M}$ is a function such that for every key $k \in \mathcal{K}$, the mapping $E_k(\cdot) = E(k, \cdot)$ is a permutation on $\mathcal{M}$. For fixed k its inverse is denoted by $E_k^{-1}(\cdot)$. A tweakable blockcipher $\widetilde{E}$ is a function $\widetilde{E} : \mathcal{K} \times \mathcal{T} \times \mathcal{M} \to \mathcal{M}$ such that for every $k \in \mathcal{K}$ and tweak $t \in \mathcal{T}$, the mapping $\widetilde{E}_k(t, \cdot) = \widetilde{E}(k, t, \cdot)$ is a permutation on $\mathcal{M}$. Like before, its inverse is denoted by $\widetilde{E}_k^{-1}(\cdot, \cdot)$. Denote by $\widetilde{\mathsf{Perm}}(\mathcal{T}, \mathcal{M})$ the set of tweakable permutations, i.e., the set of all families of permutations on $\mathcal{M}$ indexed with $t \in \mathcal{T}$.

Note that a blockcipher is a special case of a tweakable blockcipher with $|\mathcal{T}| = 1$, and hence it suffices to restrict our analysis to tweakable blockciphers. In this work, we target the design of a tweakable blockcipher $\widetilde{E}$ from an underlying permutation P, which is modeled as a perfectly random permutation $P \xleftarrow{\$} \mathsf{Perm}(\mathcal{M})$. In Sect. 2.1 we describe the single-key security model and in Sect. 2.2 the related-key security model. We give a description of Patarin's technique for bounding distinguishing advantages in Sect. 2.3.

2.1 Single-Key Security Model

Consider a tweakable blockcipher $\widetilde{E} : \mathcal{K} \times \mathcal{T} \times \mathcal{M} \to \mathcal{M}$ based on a random permutation $P \xleftarrow{\$} \mathsf{Perm}(\mathcal{M})$. Let $\widetilde{\pi} \xleftarrow{\$} \widetilde{\mathsf{Perm}}(\mathcal{T}, \mathcal{M})$ be an ideal tweakable permutation. The single-key security of $\widetilde{E}$ is informally captured by a distinguisher $\mathcal{D}$ that has adaptive oracle access to either $(\widetilde{E}_k, P)$, for some secret key $k \xleftarrow{\$} \mathcal{K}$, or $(\widetilde{\pi}, P)$. The distinguisher always has two-directional access to P. It may or may not have two-directional access to the construction oracle ($\widetilde{E}_k$ or $\widetilde{\pi}$) depending on whether we consider PRP or strong PRP security. The distinguisher is computationally unbounded, deterministic, and it never makes duplicate queries.

Security Definitions. More formally, we define the PRP security of $\widetilde{E}$ based on P as

$$\mathbf{Adv}_{\widetilde{E}}^{\mathrm{prp}}(\mathcal{D}) = \left| \mathbf{Pr}\left[\mathcal{D}^{\widetilde{E}_k, P^{\pm}} = 1\right] - \mathbf{Pr}\left[\mathcal{D}^{\widetilde{\pi}, P^{\pm}} = 1\right]\right|,$$

and the strong PRP (SPRP) security of $\widetilde{E}$ based on P as

$$\mathbf{Adv}_{\widetilde{E}}^{\mathrm{sprp}}(\mathcal{D}) = \left| \mathbf{Pr}\left[\mathcal{D}^{\widetilde{E}_k^{\pm}, P^{\pm}} = 1\right] - \mathbf{Pr}\left[\mathcal{D}^{\widetilde{\pi}^{\pm}, P^{\pm}} = 1\right]\right|,$$

where the probabilities are taken over the random selections of $k \xleftarrow{\$} \mathcal{K}$, $P \xleftarrow{\$}$ Perm$(\mathcal{M})$, and $\widetilde{\pi} \xleftarrow{\$} \widetilde{\mathsf{Perm}}(\mathcal{T}, \mathcal{M})$. For $q, r \geq 0$, we define by

$$\mathbf{Adv}_{\widetilde{E}}^{\mathrm{(s)prp}}(q, r) = \max_{\mathcal{D}} \mathbf{Adv}_{\widetilde{E}}^{\mathrm{(s)prp}}(\mathcal{D})$$

the security of $\widetilde{E}$ against any single-key distinguisher $\mathcal{D}$ that makes q queries to the construction oracle ($\widetilde{E}_k$ or $\widetilde{\pi}_k$) and r queries to the primitive oracle.

2.2 Related-Key Security Model

We generalize the security definitions of Sect. 2.1 to related-key security using the theoretical framework of Bellare and Kohno [6] and Albrecht et al. [1]. The generalization is similar to the one of Cogliati and Seurin [17] with the difference that tweakable blockciphers are considered (and that we consider more general key-deriving functions).

Related-Key Oracle. In related-key attacks, the distinguisher may query its construction oracle not just on $\widetilde{E}_k$, but on $\widetilde{E}_{\varphi(k)}$ for some function φ chosen by the distinguisher. This function may vary for the different construction queries, but should come from a pre-described set. Let Φ be a set of key-deriving functions (a *KDF-set*). For a tweakable blockcipher $\widetilde{E} : \mathcal{K} \times \mathcal{T} \times \mathcal{M} \to \mathcal{M}$, we define a *related-key oracle* $\mathsf{RK}[\widetilde{E}] : \mathcal{K} \times \Phi \times \mathcal{T} \times \mathcal{M} \to \mathcal{M}$ as

$$\mathsf{RK}[\widetilde{E}](k, \varphi, t, m) = \mathsf{RK}[\widetilde{E}]_k(\varphi, t, m) = \widetilde{E}_{\varphi(k)}(t, m).$$

For fixed φ its inverse is denoted $\mathsf{RK}[\widetilde{E}]_k^{-1}(\varphi, t, c) = \widetilde{E}_{\varphi(k)}^{-1}(t, c)$. Denote by $\widetilde{\mathsf{RK\text{-}Perm}}(\Phi, \mathcal{T}, \mathcal{M})$ the set of tweakable related-key permutations, i.e., the set of all families of permutations on $\mathcal{M}$ indexed with $(\varphi, t) \in \Phi \times \mathcal{T}$.

Security Definitions. For a KDF-set Φ, we define the related-key (strong) PRP (RK-(S)PRP) security of $\widetilde{E}$ based on P as

$$\mathbf{Adv}_{\Phi, \widetilde{E}}^{\mathrm{rk\text{-}prp}}(\mathcal{D}) = \left| \mathbf{Pr}\left[\mathcal{D}^{\mathsf{RK}[\widetilde{E}]_k, P^{\pm}} = 1 \right] - \mathbf{Pr}\left[\mathcal{D}^{\widetilde{\mathsf{RK}\pi}, P^{\pm}} = 1 \right] \right|,$$

$$\mathbf{Adv}_{\Phi, \widetilde{E}}^{\mathrm{rk\text{-}sprp}}(\mathcal{D}) = \left| \mathbf{Pr}\left[\mathcal{D}^{\mathsf{RK}[\widetilde{E}]_k^{\pm}, P^{\pm}} = 1 \right] - \mathbf{Pr}\left[\mathcal{D}^{\widetilde{\mathsf{RK}\pi}^{\pm}, P^{\pm}} = 1 \right] \right|,$$

where the probabilities are taken over the random selections of $k \xleftarrow{\$} \mathcal{K}$, $P \xleftarrow{\$}$ Perm$(\mathcal{M})$, and $\widetilde{\mathsf{RK}\pi} \xleftarrow{\$} \widetilde{\mathsf{RK\text{-}Perm}}(\Phi, \mathcal{T}, \mathcal{M})$. For $q, r \geq 0$, we define by

$$\mathbf{Adv}_{\Phi, \widetilde{E}}^{\mathrm{rk\text{-}(s)prp}}(q, r) = \max_{\mathcal{D}} \mathbf{Adv}_{\Phi, \widetilde{E}}^{\mathrm{rk\text{-}(s)prp}}(\mathcal{D})$$

the security of $\widetilde{E}$ against any related-key distinguisher $\mathcal{D}$ that makes q queries to the construction oracle ($\mathsf{RK}[\widetilde{E}]_k$ or $\widetilde{\mathsf{RK}\pi}$) and r queries to the primitive oracle.

Note that we have opted to design the ideal world to behave independently for each φ. This only increases the adversarial success probability in comparison with earlier models: if for some $k \in \mathcal{K}$ there exist two distinct $\varphi, \varphi' \in \Phi$ such that $\varphi(k) = \varphi'(k)$ with non-negligible probability, $\widetilde{\mathsf{RK}\pi}_k$ behaves as two independent tweakable permutations for these two key-deriving functions but $\mathsf{RK}[\widetilde{E}]_k$ does not. In this case, $\mathcal{D}$ can easily distinguish (it corresponds to the collision-resistance property in [6]). We remark that, by using this approach, related-key security can be seen as a specific case of tweakable blockcipher security.

Key-Deriving Functions. Note that for $\Phi_{\mathrm{id}} = \{\varphi : k \mapsto k\}$, we simply have $\mathbf{Adv}^{\mathrm{rk\text{-}(s)prp}}_{\Phi_{\mathrm{id}}, \widetilde{E}}(\mathcal{D}) = \mathbf{Adv}^{\mathrm{(s)prp}}_{\widetilde{E}}(\mathcal{D})$, and we will sometimes view single-key security as related-key security under KDF-set Φ_{id}. Two other KDF-sets we consider in this work are the following:

$$
\begin{aligned}
\Phi_\oplus &= \{\varphi_\delta : k \mapsto k \oplus \delta \mid \delta \in \mathcal{K}\}, \\
\Phi_{P\oplus} &= \{\varphi_{\delta,\epsilon} : k \mapsto P^{-1}(P(k) \oplus \epsilon) \oplus \delta \mid \delta, \epsilon \in \mathcal{K}, \delta = 0 \vee \epsilon = 0\}.
\end{aligned}
\tag{5}
$$

We regularly simply write $\delta \in \Phi_\oplus$ to say that $\varphi_\delta \in \Phi_\oplus$, and similarly write $(\delta, \epsilon) \in \Phi_{P\oplus}$ to say that $\varphi_{\delta,\epsilon} \in \Phi_{P\oplus}$.[1]

Note that every $\varphi_\delta \in \Phi_\oplus$ satisfies $\varphi_\delta = \varphi_{\delta,0} \in \Phi_{P\oplus}$, and hence $\Phi_\oplus \subseteq \Phi_{P\oplus}$ by construction. The side condition "$\delta = 0 \vee \epsilon = 0$" for $\Phi_{P\oplus}$ deserves an additional explanation. In our scheme XPX, the in- and outputs will be masked using the values $(k, P(k))$. A function $\varphi_\delta \in \Phi_\oplus$ (or, equivalently, $\varphi_{\delta,0} \in \Phi_{P\oplus}$) transforms these values to $(k \oplus \delta, P(k \oplus \delta))$. The set $\Phi_{P\oplus}$ generalizes the strength of the attacker by also transforming $P(k)$ under XOR. In more detail, for any ϵ, $\varphi_{0,\epsilon} \in \Phi_{P\oplus}$ transforms $(k, P(k))$ to $(P^{-1}(P(k) \oplus \epsilon), P(k) \oplus \epsilon)$. From a theoretical point, it may be of interest to drop the side condition from $\Phi_{P\oplus}$. This would, however, make the security analysis of XPX much more complicated and technically demanding.

2.3   Patarin's Technique

We use the H-coefficient technique by Patarin [46] and Chen and Steinberger [14], and we introduce it for our definitions of related-key security. Recall that these definitions simplify to single-key security by using KDF-set Φ_{id}.

Let $P \xleftarrow{\$} \mathsf{Perm}(\mathcal{M})$, and $\widetilde{\mathsf{RK}\pi} \xleftarrow{\$} \widetilde{\mathsf{Perm}}(\Phi, \mathcal{T}, \mathcal{M})$. Let $k \xleftarrow{\$} \mathcal{K}$ and $\widetilde{E} : \mathcal{K} \times \mathcal{T} \times \mathcal{M} \to \mathcal{M}$ be a tweakable blockcipher based on P. Consider any fixed deterministic distinguisher $\mathcal{D}$ for the RK-(S)PRP security of $\widetilde{E}$. It has access to either the *real world* $\mathcal{O}_{\mathrm{re}} = (\mathsf{RK}[\widetilde{E}]^{(\pm)}_k, P^\pm)$ or the *ideal world* $\mathcal{O}_{\mathrm{id}} = (\widetilde{\mathsf{RK}\pi}^{(\pm)}, P^\pm)$ and its goal is to distinguish both. Here, the distinguisher has inverse query access to the construction oracle if and only if we are considering *strong* PRP security (hence

---

[1] $\Phi_{P\oplus}$ could alternatively be written as the set of functions $\varphi_{b,\delta} : k \mapsto (k \oplus \delta$ (if $b = 0$) or $P^{-1}(P(k) \oplus \delta)$ (if $b = 1$)). We have opted for the writeup in (5) to make the appearance of the key relation (δ or ϵ) more explicit.

the parentheses around $\pm$). The information that $\mathcal{D}$ learns from the interaction with $\mathcal{O}_{re}/\mathcal{O}_{id}$ is collected in a view v. Denote by X_{re} (resp. X_{id}) the probability distribution of views when interacting with $\mathcal{O}_{re}$ (resp. $\mathcal{O}_{id}$). Let $\mathcal{V}$ be the set of all attainable views, i.e., views that occur in the ideal world with non-zero probability.

Lemma 1 (Patarin's Technique). *Let $\mathcal{D}$ be a deterministic distinguisher. Consider a partition $\mathcal{V} = \mathcal{V}_{good} \cup \mathcal{V}_{bad}$ of the set of attainable views. Let $0 \le \varepsilon \le 1$ be such that for all $v \in \mathcal{V}_{good}$,*

$$\mathbf{Pr}\left[X_{re} = v\right] \ge (1 - \varepsilon)\mathbf{Pr}\left[X_{id} = v\right]. \tag{6}$$

Then, the distinguishing advantage satisfies $\mathbf{Adv}(\mathcal{D}) \le \varepsilon + \mathbf{Pr}\left[X_{id} \in \mathcal{V}_{bad}\right]$.

A proof of this lemma is given in [13,14,38]. The idea of the technique is that only few views are significantly more likely to appear in $\mathcal{O}_{id}$ than in $\mathcal{O}_{re}$. In other words, the ratio (6) is close to 1 for all but the "bad" views. Note that taking a large $\mathcal{V}_{bad}$ implies a higher $\mathbf{Pr}\left[X_{id} \in \mathcal{V}_{bad}\right]$, while a small $\mathcal{V}_{bad}$ implies a higher ε. The definition of what views are "bad" is thus a tradeoff between the two terms.

Let $v_C = \{(\varphi_1, t_1, m_1, c_1), \ldots, (\varphi_q, t_q, m_q, c_q)\}$ be a view on a construction oracle. We say that a tweakable related-key permutation $\widetilde{RK\pi} \in \widetilde{Perm}(\varPhi, \mathcal{T}, \mathcal{M})$ *extends* v_C, denoted $\widetilde{RK\pi} \vdash v_C$, if $\widetilde{RK\pi}(\varphi, t, m) = c$ for each $(\varphi, t, m, c) \in v_C$. Note that if $\widetilde{E} : \mathcal{K} \times \mathcal{T} \times \mathcal{M} \to \mathcal{M}$ is a tweakable blockcipher and $k \in \mathcal{K}$, then $RK[\widetilde{E}]_k \in \widetilde{Perm}(\varPhi, \mathcal{T}, \mathcal{M})$ and the definition reads $RK[\widetilde{E}]_k \vdash v_C$. Similarly, if $v_P = \{(x_1, y_1), \ldots, (x_r, y_r)\}$ is a primitive view, we say that a permutation $P \in Perm(\mathcal{M})$ *extends* v_P, denoted $P \vdash v_P$, if $P(x) = y$ for each $(x, y) \in v_P$.

3 XPX

Let P be any n-bit permutation. We present the tweakable blockcipher XPX that has a key space $\{0, 1\}^n$, a tweak space $\mathcal{T} \subseteq (\{0, 1\}^n)^4$, and a message and ciphertext space $\{0, 1\}^n$. Formally, XPX $: \{0, 1\}^n \times \mathcal{T} \times \{0, 1\}^n \to \{0, 1\}^n$ is defined as

$$\text{XPX}_k((t_{11}, t_{12}, t_{21}, t_{22}), m) = P(m \oplus \varDelta_1) \oplus \varDelta_2, \text{ where } \varDelta_1 = t_{11}k \oplus t_{12}P(k),$$
$$\text{and } \varDelta_2 = t_{21}k \oplus t_{22}P(k). \tag{7}$$

XPX is depicted in Fig. 3. The design is general in that $\mathcal{T}$ can (still) be any set, and we highlight two examples.

- **Even-Mansour.** XPX meets the single-key Even-Mansour construction (1) by fixing $\mathcal{T} = \{(1, 0, 1, 0)\}$. More generally, if $|\mathcal{T}| = 1$, we are simply considering an ordinary (not a tweakable) blockcipher;

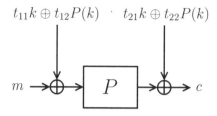

$$t_{11}k \oplus t_{12}P(k) \qquad t_{21}k \oplus t_{22}P(k)$$

Fig. 3. XPX

- **XEX with Even-Mansour.** XPX covers XEX based on Even-Mansour with $N = 0$ by taking

$$\mathcal{T} = \left\{ \begin{array}{l} (\mathbf{x}_1^{\alpha_1} \cdots \mathbf{x}_\ell^{\alpha_\ell} \oplus 1, \mathbf{x}_1^{\alpha_1} \cdots \mathbf{x}_\ell^{\alpha_\ell}, \\ \mathbf{x}_1^{\alpha_1} \cdots \mathbf{x}_\ell^{\alpha_\ell} \oplus 1, \mathbf{x}_1^{\alpha_1} \cdots \mathbf{x}_\ell^{\alpha_\ell}) \end{array} \middle| (\alpha_1, \ldots, \alpha_\ell) \in \mathbb{I}_1 \times \cdots \times \mathbb{I}_\ell \right\},$$

where $\mathbf{x}_1, \ldots, \mathbf{x}_\ell$ and tweak space $\mathbb{I}_1 \times \cdots \times \mathbb{I}_\ell$ are as described in Sect. 1. In this case, $(\alpha_1, \ldots, \alpha_\ell)$ is in fact the "real" tweak, and $(t_{11}, t_{12}, t_{21}, t_{22})$ is a function of $(\alpha_1, \ldots, \alpha_\ell)$.

Further applications follow in Sects. 6 and 7. Obviously, XPX does not achieve security for all choices of $\mathcal{T}$; e.g., if $(1, 0, 1, 1) \in \mathcal{T}$, then we have

$$\mathrm{XPX}_k((1, 0, 1, 1), 0) = k. \qquad (8)$$

In Sect. 4, we derive a minimal set of conditions on $\mathcal{T}$ to make the XPX construction meaningful. Then, in Sect. 5 we prove that XPX is secure in various settings, from single-key (S)PRP security to RK-SPRP security for the key-deriving function sets of Sect. 2.2.

4   Valid Tweak Sets

To eliminate trivial cases such as (8), we define a set of minimal conditions $\mathcal{T}$ needs to satisfy in order for XPX to achieve a reasonable level of security. In more detail, we define the notion of a *valid* tweak space $\mathcal{T}$. After the definition we present its rationale. We give some example of valid tweak spaces in Sect. 4.1, and show that XPX is insecure if $\mathcal{T}$ is invalid in Sect. 4.2.

Definition 1. *We say that $\mathcal{T}$ is valid if:*

(i) *For any $(t_{11}, t_{12}, t_{21}, t_{22}) \in \mathcal{T}$ we have $(t_{11}, t_{12}) \neq (0, 0)$ and $(t_{21}, t_{22}) \neq (0, 0)$;*

(ii) *For any distinct $(t_{11}, t_{12}, t_{21}, t_{22}), (t'_{11}, t'_{12}, t'_{21}, t'_{22}) \in \mathcal{T}$ we have $(t_{11}, t_{12}) \neq (t'_{11}, t'_{12})$ and $(t_{21}, t_{22}) \neq (t'_{21}, t'_{22})$;*

(iii) *If $(1, 0, t_{21}, t_{22}) \in \mathcal{T}$ for some t_{21}, t_{22}:*

    (a) $t_{21} \neq 0$ and $t_{22} \neq 1$;

(b) For any other $(t'_{11}, t'_{12}, t'_{21}, t'_{22}) \in \mathcal{T}$ and $b \in \{0, 1\}$ we have

$$t'_{11} \neq t'_{12} t_{21} (t_{22} \oplus 1)^{-1} \oplus b \text{ and } t'_{22} \neq t'_{21} t_{21}^{-1} (t_{22} \oplus 1) \oplus b;$$

(c) For any distinct $(t'_{11}, t'_{12}, t'_{21}, t'_{22}), (t''_{11}, t''_{12}, t''_{21}, t''_{22}) \in \mathcal{T}$ we have

$$t'_{12} \oplus t''_{12} \neq (t'_{11} \oplus t''_{11}) t_{21}^{-1} (t_{22} \oplus 1) \text{ and } t'_{22} \oplus t''_{22} \neq (t'_{21} \oplus t''_{21}) t_{21}^{-1} (t_{22} \oplus 1);$$

(iv) If $(t_{11}, t_{12}, 0, 1) \in \mathcal{T}$ for some t_{11}, t_{12}:
(a) $t_{12} \neq 0$ and $t_{11} \neq 1$;
(b) For any other $(t'_{11}, t'_{12}, t'_{21}, t'_{22}) \in \mathcal{T}$ and $b \in \{0, 1\}$ we have

$$t'_{11} \neq t'_{12} t_{12}^{-1} (t_{11} \oplus 1) \oplus b \text{ and } t'_{22} \neq t'_{21} t_{12} (t_{11} \oplus 1)^{-1} \oplus b;$$

(c) For any distinct $(t'_{11}, t'_{12}, t'_{21}, t'_{22}), (t''_{11}, t''_{12}, t''_{21}, t''_{22}) \in \mathcal{T}$ we have

$$t'_{11} \oplus t''_{11} \neq (t'_{12} \oplus t''_{12}) t_{12}^{-1} (t_{11} \oplus 1) \text{ and } t'_{21} \oplus t''_{21} \neq (t'_{22} \oplus t''_{22}) t_{12}^{-1} (t_{11} \oplus 1).$$

Conditions (i) and (ii) are basic requirements, in essence guaranteeing that the input to and output of the underlying permutation P is always masked. Conditions (iii) and (iv) are more obscure but are in fact necessary to prevent the key from being leaked. The presence of conditions (iii-a) and (iv-a) is justified by equation (8), but even beyond that, an evaluation $\mathrm{XPX}_k((1, 0, t_{21}, t_{22}), 0)$ for some $t_{21} \neq 0$ and $t_{22} \neq 1$ leaks the value $t_{21} k \oplus (t_{22} \oplus 1) P(k)$ and additional conditions are required.

4.1 Examples of Valid Tweak Spaces

Due to our quest for a minimal definition of valid tweak spaces, Definition 1 is a bit hard to parse. Fortunately, conditions (iii) and (iv) often turn out to be trivially satisfied, as we will show in the next examples.

Example 1. Consider a tweak space $\mathcal{T}$ where all tweaks are of the form $(t_{11}, 0, t_{21}, 0)$ for $t_{11}, t_{21} \neq 0$. The tweak space is valid if and only if

– every t_{11} appears at most once;
– every t_{21} appears at most once.

Concretely, condition (i) of Definition 1 is satisfied as $t_{11}, t_{21} \neq 0$; condition (ii) is enforced by above two simplified conditions; conditions (iii) and (iv) turn out to hold trivially for the specific type of tweaks. This example corresponds to XPX with $\Delta_1 = t_{11} k$ and $\Delta_2 = t_{21} k$, and covers, among others, the Even-Mansour construction. Interestingly, by putting $t_{11} = t_{21} =: t$, XPX corresponds to Cogliati et al. [15]'s tweakable Even-Mansour construction with universal hash function $h_k(t) = k \cdot t$.

Example 2. Consider a tweak space $\mathcal{T}$ where all tweaks are of the form $(0, t_{12}, 0, t_{22})$ for $t_{12}, t_{22} \neq 0$. The tweak space is valid if and only if

– every t_{12} appears at most once;
– every t_{22} appears at most once.

This example corresponds to XPX with $\Delta_1 = t_{12}P(k)$ and $\Delta_2 = t_{22}P(k)$, and it is the symmetrical equivalent of Example 1.

Example 3. Consider a tweak space $\mathcal{T}$ where all tweaks $(t_{11}, t_{12}, t_{21}, t_{22})$ satisfy $t_{11}, t_{12}, t_{21}, t_{22} \neq 0$. The tweak space is valid if and only if

– every (t_{11}, t_{12}) appears at most once;
– every (t_{21}, t_{22}) appears at most once.

As in Example 1, condition (i) of Definition 1 is satisfied as $t_{11}, t_{12}, t_{21}, t_{22} \neq 0$; condition (ii) is enforced by above two simplified conditions; conditions (iii) and (iv) turn out to hold trivially for the specific type of tweaks. This example covers, among others, XEX with Even-Mansour, noticing that XEX requires that $(\alpha_1, \dots, \alpha_\ell) \neq (0, \dots, 0)$ [48].

4.2   Minimality of Definition 1

In below proposition, we show that XPX is insecure whenever $\mathcal{T}$ is invalid. We remark that the second part of condition (ii) and the entire condition (iv) are not strictly needed for PRP security and only apply to SPRP security. We nevertheless included them for completeness.

Proposition 1. *Let $n \geq 1$ and let $\mathcal{T} \subseteq (\{0,1\}^n)^4$ an* invalid *set. We have*

$$\mathbf{Adv}_{\mathrm{XPX}}^{\mathrm{sprp}}(5, 2) \geq 1 - 1/(2^n - 1).$$

Proof. We consider conditions (i), (ii), and (iii) separately. Condition (iv) is symmetrically equivalent to (iii), and omitted.

Condition (i). Assume, w.l.o.g., that $(0, 0, t_{21}, t_{22}) \in \mathcal{T}$ for some t_{21}, t_{22}. For any $m \in \{0,1\}^n$ we have $\mathrm{XPX}_k((0, 0, t_{21}, t_{22}), m) \oplus P(m) = t_{21}k \oplus t_{22}P(k)$. Making these two queries for two different messages $m \neq m'$ gives a collision with probability 1. For a random $\widetilde{\pi}$ this happens with probability at most $1/(2^n - 1)$. Thus, if condition (i) is violated, $\mathbf{Adv}_{\mathrm{XPX}}^{\mathrm{sprp}}(2, 2) \geq 1 - 1/(2^n - 1)$. The analysis for $(t_{11}, t_{12}, 0, 0) \in \mathcal{T}$ is equivalent.

Condition (ii). Assume, w.l.o.g., that $(t_{11}, t_{12}, t_{21}, t_{22}), (t_{11}, t_{12}, t'_{21}, t'_{22}) \in \mathcal{T}$ for some $(t_{21}, t_{22}) \neq (t'_{21}, t'_{22})$. For any m,

$$\mathrm{XPX}_k((t_{11}, t_{12}, t_{21}, t_{22}), m) \oplus \mathrm{XPX}_k((t_{11}, t_{12}, t'_{21}, t'_{22}), m)$$
$$= (t_{21} \oplus t'_{21})k \oplus (t_{22} \oplus t'_{22})P(k).$$

Making these queries for two different messages $m \neq m'$ gives a collision with probability 1. For a random $\widetilde{\pi}$ this happens with probability at most $1/(2^n - 1)$. Thus, if condition (ii) is violated, $\mathbf{Adv}_{\mathrm{XPX}}^{\mathrm{sprp}}(4, 0) \geq 1 - 1/(2^n - 1)$.

Condition (iii-a). Suppose $(1, 0, t_{21}, t_{22}) \in \mathcal{T}$ for some t_{21}, t_{22}. By construction, $\text{XPX}_k((1, 0, t_{21}, t_{22}), 0) = t_{21}k \oplus (t_{22} \oplus 1)P(k)$. If $t_{21} = 0$ or $t_{22} = 1$, this value leaks k or $P(k)$. By making one additional invocation of $P^{\pm}$ the other value is learned as well, giving the distinguisher both $(k, P(k))$. For arbitrary $m \neq 0$, the distinguisher now queries $\text{XPX}_k((1, 0, t_{21}, t_{22}), m) = c$ and $P(m \oplus k) = y$ and verifies whether $c = y \oplus t_{21}k \oplus t_{22}P(k)$. For a random $\tilde{\pi}$ this happens with probability at most $1/(2^n - 1)$. Thus, if condition (iii-a) is violated, $\textbf{Adv}^{\text{sprp}}_{\text{XPX}}(2, 2) \geq 1 - 1/(2^n - 1)$.

Condition (iii-b). Suppose $(1, 0, t_{21}, t_{22}) \in \mathcal{T}$ for some t_{21}, t_{22}, and assume $t_{21} \neq 0$ and $t_{22} \neq 1$ (otherwise, the attack of (iii-a) applies). Suppose there is a $(t'_{11}, t'_{12}, t'_{21}, t'_{22}) \in \mathcal{T}$ such that $t'_{22} = t'_{21}t_{21}^{-1}(t_{22} \oplus 1) \oplus b$ for some $b \in \{0, 1\}$. This is without loss of generality, as the other case is symmetric and the attack applies by reversing all queries for tweak $(t'_{11}, t'_{12}, t'_{21}, t'_{22})$. We first consider case $b = 0$, case $b = 1$ is treated later.

For $b = 0$: firstly, the attacker queries $\text{XPX}_k((1, 0, t_{21}, t_{22}), 0)$ to receive $c = t_{21}k \oplus (t_{22} \oplus 1)P(k)$. Fix any $c' \in \{0, 1\}^n$, and query $\text{XPX}_k^{-1}((t'_{11}, t'_{12}, t'_{21}, t'_{22}), c')$ to receive $m' = t'_{11}k \oplus t'_{12}P(k) \oplus P^{-1}(\text{inp}')$ where $\text{inp}' = c' \oplus t'_{21}k \oplus t'_{22}P(k)$. Eliminating $P(k)$ using c gives

$$\text{inp}' = c' \oplus t'_{22}(t_{22} \oplus 1)^{-1}c \oplus \left(t'_{21} \oplus t'_{22}(t_{22} \oplus 1)^{-1}t_{21}\right)k = c' \oplus t'_{22}(t_{22} \oplus 1)^{-1}c,$$

where we use the violation of property (iii-b). Therefore,

$$m' \oplus P^{-1}(c' \oplus t'_{22}(t_{22} \oplus 1)^{-1}c) = t'_{11}k \oplus t'_{12}P(k).$$

This equation is independent of the choice of c'. Making these queries for two different ciphertexts $c' \neq c''$ gives a collision with probability 1. For a random $\tilde{\pi}$ this happens with probability at most $1/(2^n - 1)$. Thus, if condition (iii-b) is violated with $b = 0$, $\textbf{Adv}^{\text{sprp}}_{\text{XPX}}(3, 2) \geq 1 - 1/(2^n - 1)$.

For $b = 1$: in this case we specifically consider $c' = t'_{21}t_{21}^{-1}c$, and have

$$\begin{aligned} \text{inp}' &= t'_{21}t_{21}^{-1}c \oplus t'_{21}k \oplus t'_{22}P(k) \\ &= \left(t'_{21}t_{21}^{-1}(t_{22} \oplus 1) \oplus t'_{22}\right)P(k) = P(k), \end{aligned}$$

using that $c = t_{21}k \oplus (t_{22} \oplus 1)P(k)$ and the violation of property (iii-b). Therefore,

$$\begin{pmatrix} t_{21} & t_{22} \oplus 1 \\ t'_{11} \oplus 1 & t'_{12} \end{pmatrix} \begin{pmatrix} k \\ P(k) \end{pmatrix} = \begin{pmatrix} c \\ m' \end{pmatrix},$$

If this matrix is singular, it implies that $m' = \text{const} \cdot c$ with $\text{const} = t_{21}^{-1}(t'_{11} \oplus 1) = (t_{22} \oplus 1)^{-1}t'_{12}$. For a random tweakable permutation this happens with probability at most $1/2^n$. On the other hand, if it is non-singular, this reveals k and $P(k)$.

For arbitrary $m \neq 0$, the distinguisher now queries $\text{XPX}_k((1, 0, t_{21}, t_{22}), m'') = c''$ and $P(m'' \oplus k) = y$ and verifies whether $c'' = y \oplus t_{21}k \oplus t_{22}P(k)$. For a random $\tilde{\pi}$ this happens with probability at most $1/(2^n - 1)$. Thus, if condition (iii-b) is violated with $b = 1$, $\textbf{Adv}^{\text{sprp}}_{\text{XPX}}(3, 1) \geq 1 - 1/(2^n - 1)$.

Condition (iii-c). Suppose $(1, 0, t_{21}, t_{22}) \in \mathcal{T}$ for some t_{21}, t_{22}, and assume $t_{21} \neq 0$ and $t_{22} \neq 1$ (otherwise, the attack of (iii-a) applies). Suppose there are $(t'_{11}, t'_{12}, t'_{21}, t'_{22}), (t''_{11}, t''_{12}, t''_{21}, t''_{22}) \in \mathcal{T}$ such that $t'_{22} \oplus t''_{22} = (t'_{21} \oplus t''_{21}) t_{21}^{-1}(t_{22} \oplus 1)$. This is without loss of generality, as the other case is symmetric and the attack applies by reversing all queries for tweaks $(t'_{11}, t'_{12}, t'_{21}, t'_{22}), (t''_{11}, t''_{12}, t''_{21}, t''_{22})$. Firstly, the attacker makes queries $\mathrm{XPX}_k((1, 0, t_{21}, t_{22}), 0)$ to receive $c = t_{21}k \oplus (t_{22} \oplus 1)P(k)$. Now, fix any $c' \in \{0, 1\}^n$, and query

- $\mathrm{XPX}_k^{-1}((t'_{11}, t'_{12}, t'_{21}, t'_{22}), c')$ to receive $m' = t'_{11}k \oplus t'_{12}P(k) \oplus P^{-1}(\mathrm{inp'})$ where $\mathrm{inp'} = c' \oplus t'_{21}k \oplus t'_{22}P(k)$;
- $\mathrm{XPX}_k^{-1}((t''_{11}, t''_{12}, t''_{21}, t''_{22}), c' \oplus (t'_{21} \oplus t''_{21})t_{21}^{-1}c)$ to receive $m'' = t''_{11}k \oplus t''_{12}P(k) \oplus P^{-1}(\mathrm{inp''})$ where $\mathrm{inp''} = c' \oplus (t'_{21} \oplus t''_{21})t_{21}^{-1}c \oplus t''_{21}k \oplus t''_{22}P(k)$.

Plugging c into $\mathrm{inp'}$ and $\mathrm{inp''}$ gives

$$\mathrm{inp''} = c' \oplus t'_{21}k \oplus \left(t''_{22} \oplus (t'_{21} \oplus t''_{21})t_{21}^{-1}(t_{22} \oplus 1)\right)P(k)$$
$$= c' \oplus t'_{21}k \oplus t'_{22}P(k) = \mathrm{inp'},$$

where we use the violation of property (iii-c). Therefore,

$$m' \oplus m'' = (t'_{11} \oplus t''_{11})k \oplus (t'_{12} \oplus t''_{12})P(k).$$

This equation is independent of the choice of c'. Making these queries for two different ciphertexts $c' \neq c''$ gives a collision with probability 1. For a random $\tilde{\pi}$ this happens with probability at most $1/(2^n - 1)$. Thus, if condition (iii-c) is violated, $\mathbf{Adv}_{\mathrm{XPX}}^{\mathrm{sprp}}(5, 0) \geq 1 - 1/(2^n - 1)$.

Conclusion. In any case, a distinguishing attack with success probability at least $1 - 1/(2^n - 1)$ can be performed in at most 5 construction queries and 2 primitive queries. □

5   Security of XPX

In this section, we analyze the security of XPX in various security models. We will focus on valid $\mathcal{T}$ only. Theorem 1 captures all security levels for the three key-deriving function sets of (5).

Theorem 1. *Let $n \geq 1$ and let $\mathcal{T} \subseteq (\{0, 1\}^n)^4$ be a valid set.*

(a) We have

$$\mathbf{Adv}_{\mathrm{XPX}}^{\mathrm{prp}}(q, r) \leq \mathbf{Adv}_{\mathrm{XPX}}^{\mathrm{sprp}}(q, r) \leq \frac{(q+1)^2 + 2q(r+1) + 2r}{2^n}.$$

(b) If for all $(t_{11}, t_{12}, t_{21}, t_{22}) \in \mathcal{T}$ we have $t_{12} \neq 0$, then

$$\mathbf{Adv}_{\Phi_\oplus, \mathrm{XPX}}^{\mathrm{rk\text{-}prp}}(q, r) \leq \frac{\frac{7}{2}q^2 + 4qr}{2^n - q}.$$

(c) *If for all* $(t_{11}, t_{12}, t_{21}, t_{22}) \in T$ *we have* $t_{12}, t_{22} \neq 0$ *and* $(t_{21}, t_{22}) \neq (0, 1)$, *then*

$$\mathbf{Adv}_{\Phi_{\oplus},\mathrm{XPX}}^{\mathrm{rk\text{-}sprp}}(q, r) \leq \frac{\frac{7}{2}q^2 + 4qr}{2^n}.$$

(d) *If for all* $(t_{11}, t_{12}, t_{21}, t_{22}) \in T$ *we have* $t_{11}, t_{12} \neq 0$, *then*

$$\mathbf{Adv}_{\Phi_{P\oplus},\mathrm{XPX}}^{\mathrm{rk\text{-}prp}}(q, r) \leq \frac{4q^2 + 4qr}{2^n - q}.$$

(e) *If for all* $(t_{11}, t_{12}, t_{21}, t_{22}) \in T$ *we have* $t_{11}, t_{12}, t_{21}, t_{22} \neq 0$, *then*

$$\mathbf{Adv}_{\Phi_{P\oplus},\mathrm{XPX}}^{\mathrm{rk\text{-}sprp}}(q, r) \leq \frac{4q^2 + 4qr}{2^n}.$$

In Sect. 5.1, we prove that the conditions T are minimal, meaning that the security proof cannot go through if the conditions are omitted. The proof of Theorem 1(a) is given in Sect. 5.2. The proofs of Theorem 1(b-c) and (d-e) are given in the full version [37].

5.1   Minimality of the Conditions of Theorem 1

We show that the conditions we put on T in Theorem 1 are minimal, in the sense that XPX can be broken if the conditions are omitted. For the validity condition on T, this is already justified by Proposition 1. Below proposition considers the remaining conditions on T put by parts (b)-(e) of Theorem 1.

Proposition 2. *Let* $n \geq 1$ *and let* $T \subseteq (\{0, 1\}^n)^4$ *a valid set.*

(a) *If* $(t_{11}, 0, t_{21}, t_{22}) \in T$ *for some* t_{11}, t_{21}, t_{22}, *then*

$$\mathbf{Adv}_{\Phi_{\oplus},\mathrm{XPX}}^{\mathrm{rk\text{-}prp}}(4, 0) \geq 1 - 1/(2^n - 1).$$

(b) *If* $(t_{11}, t_{12}, t_{21}, 0) \in T$ *or* $(t_{11}, t_{12}, 0, 1) \in T$ *for some* t_{11}, t_{12}, t_{21}, *then*

$$\mathbf{Adv}_{\Phi_{\oplus},\mathrm{XPX}}^{\mathrm{rk\text{-}sprp}}(4, 0) \geq 1 - 1/(2^n - 1).$$

(c) *If* $(0, t_{12}, t_{21}, t_{22}) \in T$ *for some* t_{12}, t_{21}, t_{22}, *then*

$$\mathbf{Adv}_{\Phi_{P\oplus},\mathrm{XPX}}^{\mathrm{rk\text{-}prp}}(4, 0) \geq 1 - 1/(2^n - 1).$$

(d) *If* $(t_{11}, t_{12}, 0, t_{22}) \in T$ *for some* t_{11}, t_{12}, t_{22}, *then*

$$\mathbf{Adv}_{\Phi_{P\oplus},\mathrm{XPX}}^{\mathrm{rk\text{-}sprp}}(4, 0) \geq 1 - 1/(2^n - 1).$$

Proof. We consider the four cases separately.

Case (b). Suppose $(t_{11}, 0, t_{21}, t_{22}) \in T$ for some t_{11}, t_{21}, t_{22}. Fix any $\delta \neq \delta'$ and any $m \in \{0, 1\}^n$. The attacker makes the following queries:

- $XPX_k(\delta, (t_{11}, 0, t_{21}, t_{22}), m)$ to receive $c = t_{21}(k \oplus \delta) \oplus t_{22}P(k \oplus \delta) \oplus P(\text{inp})$ where $\text{inp} = m \oplus t_{11}(k \oplus \delta)$;
- $XPX_k(\delta', (t_{11}, 0, t_{21}, t_{22}), m \oplus t_{11}(\delta \oplus \delta'))$ to receive $c' = t_{21}(k \oplus \delta') \oplus t_{22}P(k \oplus \delta') \oplus P(\text{inp}')$ where $\text{inp}' = m \oplus t_{11}(\delta \oplus \delta') \oplus t_{11}(k \oplus \delta')$.

By construction, $\text{inp}' = \text{inp}$, and thus

$$c \oplus c' = t_{21}(\delta \oplus \delta') \oplus t_{22}\big(P(k \oplus \delta) \oplus P(k \oplus \delta')\big).$$

This equation is independent of the choice of m. Making these queries for two different messages $m \neq m'$ gives a collision with probability 1. For a random $\widetilde{RK}\pi$ this happens with probability at most $1/(2^n - 1)$. Thus, $\mathbf{Adv}^{\text{rk-prp}}_{\Phi_\oplus, \text{XPX}}(4, 0) \geq 1 - 1/(2^n - 1)$.

Case (c). If $(t_{11}, t_{12}, t_{21}, 0) \in \mathcal{T}$ for some t_{11}, t_{12}, t_{21} the attack is the inverse of the one for case (b). Now, suppose $(t_{11}, t_{12}, 0, 1) \in \mathcal{T}$ for some t_{11}, t_{12}. The attacker makes the following queries:

- $XPX_k^{-1}(0, (t_{11}, t_{12}, 0, 1), 0)$ to receive $m = (t_{11} \oplus 1)k \oplus t_{12}P(k)$;
- $XPX_k(0, (t_{11}, t_{12}, 0, 1), m \oplus \delta)$ for $\delta \neq 0$ to receive

$$\begin{aligned} c_\delta &= P(k) \oplus P(m \oplus \delta \oplus t_{11}k \oplus t_{12}P(k)) \\ &= P(k) \oplus P(k \oplus \delta). \end{aligned}$$

Now, fix any m' and query

- $XPX_k(\delta, (t_{11}, t_{12}, 0, 1), m')$ to receive $c' = P(m' \oplus t_{11}(k \oplus \delta) \oplus t_{12}P(k \oplus \delta)) \oplus P(k \oplus \delta)$;
- $XPX_k(0, (t_{11}, t_{12}, 0, 1), m' \oplus t_{11}\delta \oplus t_{12}c_\delta)$ to receive $c'' = P(m' \oplus t_{11}\delta \oplus t_{12}c_\delta \oplus t_{11}k \oplus t_{12}P(k)) \oplus P(k)$.

These queries satisfy $c' \oplus c'' = c_\delta$. For a random $\widetilde{RK}\pi$ this happens with probability at most $1/(2^n - 1)$. Thus, $\mathbf{Adv}^{\text{rk-sprp}}_{\Phi_\oplus, \text{XPX}}(4, 0) \geq 1 - 1/(2^n - 1)$.

Case (d). Suppose $(0, t_{12}, t_{21}, t_{22}) \in \mathcal{T}$ for some t_{12}, t_{21}, t_{22}. Fix any $\delta \neq \delta'$ and any $m \in \{0, 1\}^n$. The attacker makes the following queries:

- $XPX_k((0, \delta), (0, t_{12}, t_{21}, t_{22}), m)$ to receive $c = t_{21}P^{-1}(P(k) \oplus \delta) \oplus t_{22}(P(k) \oplus \delta) \oplus P(\text{inp})$ where $\text{inp} = m \oplus t_{12}(P(k) \oplus \delta)$;
- $XPX_k((0, \delta'), (0, t_{12}, t_{21}, t_{22}), m \oplus t_{12}(\delta \oplus \delta'))$ to receive $c' = t_{21}P^{-1}(P(k) \oplus \delta') \oplus t_{22}(P(k) \oplus \delta') \oplus P(\text{inp}')$ where $\text{inp}' = m \oplus t_{12}(\delta \oplus \delta') \oplus t_{12}(P(k) \oplus \delta')$.

By construction, $\text{inp}' = \text{inp}$, and thus

$$c \oplus c' = t_{21}\big(P^{-1}(P(k) \oplus \delta) \oplus P^{-1}(P(k) \oplus \delta')\big) \oplus t_{22}(\delta \oplus \delta').$$

This equation is independent of the choice of m. Making these queries for two different messages $m \neq m'$ gives a collision with probability 1. For a random $\widetilde{RK}\pi$ this happens with probability at most $1/(2^n - 1)$. Thus, $\mathbf{Adv}^{\text{rk-prp}}_{\Phi_{P\oplus}, \text{XPX}}(4, 0) \geq 1 - 1/(2^n - 1)$.

Case (e). The attack is the inverse of the one for case (d). $\square$

5.2 Proof of Theorem 1(a)

Note that $\mathbf{Adv}^{\mathrm{prp}}_{\mathrm{XPX}}(q,r) \leq \mathbf{Adv}^{\mathrm{sprp}}_{\mathrm{XPX}}(q,r)$ holds by construction, and we will focus on bounding the latter. The proof is a generalization of the proofs of Even-Mansour [5,15,22,23,25,43], but difficulties arise due to the tweaks.

Let $k \xleftarrow{\$} \{0,1\}^n$, $P \xleftarrow{\$} \mathrm{Perm}(\{0,1\}^n)$, and $\widetilde{\pi} \xleftarrow{\$} \widetilde{\mathrm{Perm}}(\mathcal{T},\{0,1\}^n)$. Consider any fixed deterministic distinguisher $\mathcal{D}$ for the SPRP security of XPX. In the real world it has access to (XPX_k, P), and in the ideal world to $(\widetilde{\pi}, P)$. It makes q construction queries which are summarized in view $v_1 = \{((t_{11},t_{12},t_{21},t_{22})_1, m_1, c_1), \ldots, ((t_{11},t_{12},t_{21},t_{22})_q, m_q, c_q)\}$. It additionally makes r queries to P, summarized in a view $v_2 = \{(x_1,y_1), \ldots, (x_r,y_r)\}$. As $\mathcal{D}$ is deterministic this properly summarizes the conversation.

To suit the analysis, we generalize our oracles by providing $\mathcal{D}$ with extra data. How these extra data look like, depends on whether or not $\mathcal{T}$ contains tweak tuple $(1,0,\bar{t}_{21},\bar{t}_{22})$ or $(\bar{t}_{11},\bar{t}_{12},0,1)$.[2] Because of their dedicated treatment, we will always refer to these tweak tuples with overlines. As $\mathcal{T}$ is valid, and more specifically by condition (iii-b), at most one of the two tweaks is in $\mathcal{T}$, but it may as well be that none of these is allowed.

More formally, *before* $\mathcal{D}$'s interaction with the oracles, we reveal forward construction query $((1,0,\bar{t}_{21},\bar{t}_{22}),0,\bar{c})$ or inverse construction query $((\bar{t}_{11},\bar{t}_{12},0,1),\bar{m},0)$, depending on whether one of the two tweaks is in $\mathcal{T}$, and store the resulting tuple in view v_0. If none of the two tweaks is in $\mathcal{T}$, we simply have $|v_0| = 0$.

Then, *after* $\mathcal{D}$'s interaction with its oracles but before $\mathcal{D}$ makes its final decision, we reveal $v_k = \{(k, k^\star)\}$. In the real world, k is the key used for encryption and $k^\star = P(k)$. In the ideal world, $k \xleftarrow{\$} \{0,1\}^n$ will be a randomly drawn dummy key and $k^\star$ will be defined based on k and v_0. If $|v_0| = 0$, then $k^\star \xleftarrow{\$} \{0,1\}^n$. Otherwise, it is the unique[3] value that satisfies

$$
\begin{aligned}
\bar{t}_{21}k \oplus (\bar{t}_{22} \oplus 1)k^\star = \bar{c} & \quad \text{if } v_0 = \{((1,0,\bar{t}_{21},\bar{t}_{22}),0,\bar{c})\}, \text{ or} \\
(\bar{t}_{11} \oplus 1)k \oplus \bar{t}_{12}k^\star = \bar{m} & \quad \text{if } v_0 = \{((\bar{t}_{11},\bar{t}_{12},0,1),\bar{m},0)\}.
\end{aligned} \tag{9}
$$

Clearly, these disclosures are without loss of generality as they may only *help* the distinguisher. The complete view is denoted $v = (v_0, v_1, v_2, v_k)$. Recall that $\mathcal{D}$ is assumed not to make any repeat queries, and hence $v_0 \cup v_1$ and v_2 do not contain any duplicate elements. Note that v_k may collide with v_2, but this will be captured as a bad event.

Throughout, we consider attainable views only. Recall that a view is attainable if it can be obtained in the ideal world. For $v_0 \cup v_1$, this is the case if and only if for any distinct i, i' such that $(t_{11},t_{12},t_{21},t_{22})_i = (t_{11},t_{12},t_{21},t_{22})_{i'}$, we have $m_i \neq m_{i'}$ and $c_i \neq c_{i'}$. For v_2 the condition is equivalent: there should

---

[2] Indeed, if (for instance) $(1,0,\bar{t}_{21},\bar{t}_{22}) \in \mathcal{T}$, a construction query $((1,0,\bar{t}_{21},\bar{t}_{22}),0)$ will reveal $\bar{c} = \bar{t}_{21}k \oplus (\bar{t}_{22} \oplus 1)P(k)$ and a special analysis is needed.

[3] Because $\mathcal{T}$ is valid, $\bar{t}_{21}, \bar{t}_{22} \oplus 1 \neq 0$ in the former case and $\bar{t}_{11} \oplus 1, \bar{t}_{12} \neq 0$ in the latter.

be no two distinct $(x, y), (x', y') \in v_2$ such that $x = x'$ or $y = y'$. Attainability implies for v_k that $k^\star$ satisfies (9) if $|v_0| = 1$.

We say that a view v is *bad* if one of the following conditions holds:

BV_1 : for some $(x, y) \in v_2$ and $(k, k^\star) \in v_k$:

$\quad \mathsf{BV}_{1a}$: $k = x$, or

$\quad \mathsf{BV}_{1b}$: $k^\star = y$, or

BV_2 : for some $((t_{11}, t_{12}, t_{21}, t_{22}), m, c) \in v_1$, $(x, y) \in v_2 \cup v_k$, and $(k, k^\star) \in v_k$:

$\quad \mathsf{BV}_{2a}$: $m \oplus t_{11}k \oplus t_{12}k^\star = x$, or

$\quad \mathsf{BV}_{2b}$: $c \oplus t_{21}k \oplus t_{22}k^\star = y$, or

BV_3 : for some distinct $((t_{11}, t_{12}, t_{21}, t_{22}), m, c), ((t'_{11}, t'_{12}, t'_{21}, t'_{22}), m', c') \in v_0 \cup v_1$

$\quad$ and $(k, k^\star) \in v_k$:

$\quad \mathsf{BV}_{3a}$: $m \oplus t_{11}k \oplus t_{12}k^\star = m' \oplus t'_{11}k \oplus t'_{12}k^\star$, or

$\quad \mathsf{BV}_{3b}$: $c \oplus t_{21}k \oplus t_{22}k^\star = c' \oplus t'_{21}k \oplus t'_{22}k^\star$.

Note that every tuple in $v_0 \cup v_1$ uniquely corresponds to an evaluation of the underlying P, namely via (7) where v_k is used as key material. The above conditions cover all cases where two different tuples in v collide at their P evaluation. In more detail, BV_1 covers the case where $v_k = \{(k, k^\star)\}$ collides with a tuple in v_2, BV_2 the case where a tuple in v_1 collides with a tuple in $v_2 \cup v_k$, and BV_3 the case where two tuples in $v_0 \cup v_1$ collide with each other. Note that two different tuples in v_2 never collide (by construction), and that the case of a tuple of v_0 colliding with v_2 is implicitly covered in BV_1. The only remaining case, v_0 colliding with v_k, is not required to be a bad event, as this is the exact way v_k is defined.

In accordance with Patarin's technique (Lemma 1), we derive an upper bound on $\mathbf{Pr}\,[X_{\mathrm{id}} \in \mathcal{V}_{\mathrm{bad}}]$ in Lemma 2, and in Lemma 3 we will prove that $\varepsilon = 0$ works for good views.

Lemma 2. *For Theorem 1(a), we have* $\mathbf{Pr}\,[X_{\mathrm{id}} \in \mathcal{V}_{\mathrm{bad}}] \leq \frac{(q+1)^2 + 2q(r+1) + 2r}{2^n}$.

Proof. Consider a view v in the ideal world $(\tilde{\pi}, P)$. We will essentially compute

$$\mathbf{Pr}\,[\mathsf{BV}_1 \vee \mathsf{BV}_2 \vee \mathsf{BV}_3] \leq \mathbf{Pr}\,[\mathsf{BV}_1] + \mathbf{Pr}\,[\mathsf{BV}_2 \mid \neg\mathsf{BV}_1] + \mathbf{Pr}\,[\mathsf{BV}_3]. \qquad (10)$$

We have $k \xleftarrow{\$} \{0, 1\}^n$. If $|v_0| = 0$, we would also have $k^\star \xleftarrow{\$} \{0, 1\}^n$. If $|v_0| = 1$, the value $k^\star$ is defined based on v_0. In fact, the probability that a transcript is bad is largest in case $|v_0| = 1$ and we consider this case only (the derivation for $|v_0| = 0$ is in fact a simplification of the below one). Without loss of generality, $v_0 = \{((\bar{t}_{11}, \bar{t}_{12}, 0, 1), \bar{m}, 0)\}$, where $\bar{t}_{11} \neq 1$ and $\bar{t}_{12} \neq 0$ by validity of $\mathcal{T}$. By (9), we have

$$k^\star = \bar{t}_{12}^{-1}\big((\bar{t}_{11} \oplus 1)k \oplus \bar{m}\big).$$

At a high level, we will prove that all bad events become a condition on k once $k^\star$ gets replaced using this equation. We will use validity of $\mathcal{T}$ (and more specifically point (iv)) to show that these are non-trivial conditions (i.e., k never cancels out).

Condition BV_1. Condition BV_{1a} is clearly satisfied with probability $r/2^n$. Regarding BV_{1b}, we have r choices for $(x, y) \in v_2$, and k is a bad key if

$$k = (\bar{t}_{11} \oplus 1)^{-1}(\bar{t}_{12}y \oplus \bar{m}),$$

where we use that $\bar{t}_{11} \neq 1$. This happens with probability at most $r/2^n$. Therefore, $\mathbf{Pr}\,[BV_1] \leq 2r/2^n$.

Condition BV_2. Consider any choice of $((t_{11}, t_{12}, t_{21}, t_{22}), m, c) \in v_1$ and $(x, y) \in v_2 \cup v_k$. Regarding BV_{2a}, it is set if

$$t_{11}k \oplus t_{12}\bar{t}_{12}^{-1}\big((\bar{t}_{11} \oplus 1)k \oplus \bar{m}\big) = x \oplus m.$$

This translates to

$$
\begin{aligned}
\big(t_{11} \oplus t_{12}\bar{t}_{12}^{-1}(\bar{t}_{11} \oplus 1) \oplus 1\big)k &= m \oplus t_{12}\bar{t}_{12}^{-1}\bar{m} && \text{if } (x,y) = (k, k^\star) \in v_k, \\
\big(t_{11} \oplus t_{12}\bar{t}_{12}^{-1}(\bar{t}_{11} \oplus 1)\big)k &= x \oplus m \oplus t_{12}\bar{t}_{12}^{-1}\bar{m} && \text{if } (x,y) \in v_2.
\end{aligned}
$$

Here, we use that $\neg BV_1$ holds. Now, if $(t_{11}, t_{12}, t_{21}, t_{22}) = (\bar{t}_{11}, \bar{t}_{12}, 0, 1)$, we necessarily have $m \neq \bar{m}$ as v does not contain any duplicate elements. Then, the key is bad with probability 0 if $(x, y) = (k, k^\star) \in v_k$ and with probability $1/2^n$ otherwise. If $(t_{11}, t_{12}, t_{21}, t_{22}) \neq (\bar{t}_{11}, \bar{t}_{12}, 0, 1)$, the factor in front of k is nonzero as $\mathcal{T}$ is valid (condition (iv-b)), and k satisfies this equation with probability $1/2^n$. Concluding, BV_{2a} is set with probability at most $q(r+1)/2^n$. Regarding BV_{2b}, it is set if

$$t_{21}k \oplus t_{22}\bar{t}_{12}^{-1}\big((\bar{t}_{11} \oplus 1)k \oplus \bar{m}\big) = y \oplus c.$$

As before, this translates to

$$
\begin{aligned}
\big(t_{21} \oplus (t_{22} \oplus 1)\bar{t}_{12}^{-1}(\bar{t}_{11} \oplus 1)\big)k &= c \oplus (t_{22} \oplus 1)\bar{t}_{12}^{-1}\bar{m} && \text{if } (x,y) = (k, k^\star) \in v_k, \\
\big(t_{21} \oplus t_{22}\bar{t}_{12}^{-1}(\bar{t}_{11} \oplus 1)\big)k &= y \oplus c \oplus t_{22}\bar{t}_{12}^{-1}\bar{m} && \text{if } (x,y) \in v_2.
\end{aligned}
$$

The remainder of the analysis is the same, showing that BV_{2b} is set with probability at most $q(r+1)/2^n$. Therefore, $\mathbf{Pr}\,[BV_2] \leq 2q(r+1)/2^n$.

Condition BV_3. Consider any two distinct $((t_{11}, t_{12}, t_{21}, t_{22}), m, c)$, $((t'_{11}, t'_{12}, t'_{21}, t'_{22}), m', c') \in v_0 \cup v_1$. If $(t_{11}, t_{12}, t_{21}, t_{22}) = (t'_{11}, t'_{12}, t'_{21}, t'_{22})$, then necessarily $m \neq m'$ and $c \neq c'$ and BV_3 cannot be satisfied. Otherwise, we have $(t_{11}, t_{12}) \neq (t'_{11}, t'_{12})$ and $(t_{21}, t_{22}) \neq (t'_{21}, t'_{22})$ because of valid $\mathcal{T}$. Plugging $k^\star$ into the equation of BV_{3a} gives

$$\big(t_{11} \oplus t'_{11} \oplus (t_{12} \oplus t'_{12})\bar{t}_{12}^{-1}(\bar{t}_{11} \oplus 1)\big)k = m \oplus m' \oplus (t_{12} \oplus t'_{12})\bar{t}_{12}^{-1}\bar{m}.$$

As before, $t_{11} \oplus t'_{11} \oplus (t_{12} \oplus t'_{12})\bar{t}_{12}^{-1}(\bar{t}_{11} \oplus 1) \neq 0$: if (t_{11}, t_{12}) or (t'_{11}, t'_{12}) equals $(\bar{t}_{11}, \bar{t}_{12})$ this is due to validity of $\mathcal{T}$ point (iv-b), and otherwise due to point (iv-c).

Therefore, k satisfies this equation with probability $1/2^n$. Thus, BV_{3a} is set with probability at most $\binom{q+1}{2}/2^n$. Regarding BV_{3b}, we similarly find

$$(t_{21} \oplus t'_{21} \oplus (t_{22} \oplus t'_{22})\bar{t}_{12}^{-1}(\bar{t}_{11} \oplus 1))k = c \oplus c' \oplus (t_{22} \oplus t'_{22})\bar{t}_{12}^{-1}\bar{m},$$

and BV_{3b} is set with probability at most $\binom{q+1}{2}/2^n$. Therefore, $\mathbf{Pr}\,[\mathsf{BV}_3] \le 2\binom{q+1}{2}/2^n \le (q+1)^2/2^n$.

Conclusion. Using (10), we have $\mathbf{Pr}\,[X_{\mathrm{id}} \in \mathcal{V}_{\mathrm{bad}}] \le \frac{(q+1)^2 + 2q(r+1) + 2r}{2^n}$. This completes the proof. $\qquad\square$

Lemma 3. *For Theorem 1(a), we have* $\mathbf{Pr}\,[X_{\mathrm{re}} = v] \ge \mathbf{Pr}\,[X_{\mathrm{id}} = v]$ *for any good transcript* $v \in \mathcal{V}_{\mathrm{good}}$.

Proof. For the computation of $\mathbf{Pr}\,[X_{\mathrm{re}} = v]$ and $\mathbf{Pr}\,[X_{\mathrm{id}} = v]$, it suffices to compute the *fraction of oracles* that could result in view v. Recall that we assume that $\mathcal{D}$ never makes redundant queries, and particularly that $v_0 \cup v_1$ consists of $|v_0| + q$ distinct oracle queries.

In the real world, k will always be a randomly drawn key. The tuples $v_0 \cup v_1$ are construction evaluations and the tuples $v_1 \cup v_k$ are direct permutation evaluations. If $|v_0| = 0$, all of these tuples define a unique P-evaluation, $q + r + 1$ in total. This is because of the fact that we consider good transcripts. If $|v_0| = 1$, the P-evaluations by v_0 and v_k are the same, but apart from that all tuples define unique P-evaluations. So also in this case, we have $q + r + 1$ P-evaluations. Therefore,

$$\mathbf{Pr}\,[X_{\mathrm{re}} = v] = \mathbf{Pr}\left[k' \xleftarrow{\$} \{0,1\}^n \; : \; k' = k\right] \cdot$$
$$\mathbf{Pr}\left[P \xleftarrow{\$} \mathrm{Perm}(\mathcal{M}) \; : \; \mathrm{XPX}_k^P \vdash v_0 \cup v_1 \wedge P \vdash v_2 \cup v_k\right]$$
$$= \frac{1}{2^n} \cdot \frac{1}{(2^n)_{q+r+1}}.$$

For the analysis in the ideal world, we group the tuples in $v_0 \cup v_1$ according to the tweak value. Formally, for $t = (t_{11}, t_{12}, t_{21}, t_{22}) \in \mathcal{T}$, we define

$$\#_t = |\{(t, m, c) \in v_0 \cup v_1 \mid m, c \in \{0,1\}^n\}|.$$

The computation of $\mathbf{Pr}\,[X_{\mathrm{id}} = v]$ now differs depending on whether $|v_0| = 0$ or $|v_0| = 1$. If $|v_0| = 0$:

$$\mathbf{Pr}\,[X_{\mathrm{id}} = v \wedge |v_0| = 0] = \mathbf{Pr}\left[k', k^{\star\prime} \xleftarrow{\$} \{0,1\}^n \; : \; k' = k \wedge k^{\star\prime} = k^\star\right] \cdot$$
$$\mathbf{Pr}\left[\widetilde{\pi} \xleftarrow{\$} \widetilde{\mathrm{Perm}}(\mathcal{T}, \mathcal{M}) \; : \; \widetilde{\pi} \vdash v_1\right] \cdot$$
$$\mathbf{Pr}\left[P \xleftarrow{\$} \mathrm{Perm}(\mathcal{M}) \; : \; P \vdash v_2\right]$$
$$= \frac{1}{2^{2n}} \cdot \frac{1}{\prod_t (2^n)_{\#_t}} \cdot \frac{1}{(2^n)_r}, \quad \text{where } \sum_t \#_t = q.$$

If $|v_0| = 1$:

$$\mathbf{Pr}\left[X_{\mathrm{id}} = v \wedge |v_0| = 1\right] = \mathbf{Pr}\left[k' \xleftarrow{\$} \{0,1\}^n \; : \; k' = k\right] \cdot$$
$$\mathbf{Pr}\left[\widetilde{\pi} \xleftarrow{\$} \widetilde{\mathsf{Perm}}(\mathcal{T}, \mathcal{M}) \; : \; \widetilde{\pi} \vdash v_0 \cup v_1\right] \cdot$$
$$\mathbf{Pr}\left[P \xleftarrow{\$} \mathsf{Perm}(\mathcal{M}) \; : \; P \vdash v_2\right]$$
$$= \frac{1}{2^n} \cdot \frac{1}{\prod_t (2^n)_{\#_t}} \cdot \frac{1}{(2^n)_r}, \text{ where } \sum_t \#_t = q+1.$$

In either case,

$$\mathbf{Pr}\left[X_{\mathrm{id}} = v\right] \leq \frac{1}{2^n} \cdot \frac{1}{\prod_t (2^n)_{\#_t}} \cdot \frac{1}{(2^n)_r}, \text{ where } \sum_t \#_t = q+1$$
$$\leq \frac{1}{2^n} \cdot \frac{1}{(2^n)_{q+r+1}}$$
$$= \mathbf{Pr}\left[X_{\mathrm{re}} = v\right],$$

where we use that $(a)_{b_1}(a)_{b_2} \geq (a)_{b_1+b_2}$. This completes the proof. $\square$

6 Application to Authenticated Encryption

We will show how XPX applies to the Prøst-COPA [3,29] and Minalpher [49] authenticated encryption schemes. Before doing so, we briefly discuss the security model.

6.1 Security Model

Authenticated encryption covers the case where both privacy and authenticity of data is required. In more detail, an authenticated encryption scheme consists of an encryption function Enc and a decryption function Dec. Enc gets as input a key, nonce, associated data, and message, and outputs a ciphertext and a tag. Dec gets as input a key, nonce, associated data, ciphertext, and tag, and it either outputs a message (if the authentication is correct) or a dedicated $\bot$ symbol.

Let $\mathsf{AE} = (\mathsf{Enc}, \mathsf{Dec})$ be an authenticated encryption scheme, and let $\mathcal{P}$ be an idealized primitive upon which AE is based, if any (note that if AE is based on a blockcipher, $\mathcal{P}$ is non-existent). Let k be a randomly drawn key. Let $\$$ be a function with the same interface as E_k, but that returns fresh and random answers to every query. Let $\bot$ be a function that outputs $\bot$ on every query. We define the privacy of AE based on $\mathcal{P}$ as

$$\mathbf{Adv}_{\mathsf{AE}}^{\mathrm{priv}}(\mathcal{D}) = \left|\mathbf{Pr}\left[\mathcal{D}^{\mathsf{Enc}_k, \mathcal{P}^{\pm}} = 1\right] - \mathbf{Pr}\left[\mathcal{D}^{\$, \mathcal{P}^{\pm}} = 1\right]\right|,$$

and the authenticity of AE based on $\mathcal{P}$ as

$$\mathbf{Adv}_{\mathsf{AE}}^{\mathrm{auth}}(\mathcal{D}) = \left|\mathbf{Pr}\left[\mathcal{D}^{\mathsf{Enc}_k, \mathsf{Dec}_k, \mathcal{P}^{\pm}} = 1\right] - \mathbf{Pr}\left[\mathcal{D}^{\mathsf{Enc}_k, \bot, \mathcal{P}^{\pm}} = 1\right]\right|.$$

In both definitions, some conditions on $\mathcal{D}$ may apply (such as the nonce-respecting condition). For $q, \ell, \sigma, r \geq 0$, we define by

$$\mathbf{Adv}_{\mathsf{AE}}^{\mathrm{priv/auth}}(q, \ell, \sigma, r) = \max_{\mathcal{D}} \mathbf{Adv}_{\mathsf{AE}}^{\mathrm{priv/auth}}(\mathcal{D})$$

the security of AE against any distinguisher $\mathcal{D}$ that makes q queries to the construction oracle, each of length at most ℓ and of total size σ, and r queries to the primitive oracle.

So far, the model is in the single-key setting, But it generalizes to related-key security straightforwardly (the way Sect. 2.2 generalizes Sect. 2.1). We denote the corresponding related-key security definitions by

$$\mathbf{Adv}_{\Phi,\mathsf{AE}}^{\mathrm{rk\text{-}priv/auth}}(\mathcal{D}) \text{ and } \mathbf{Adv}_{\Phi,\mathsf{AE}}^{\mathrm{rk\text{-}priv/auth}}(q, \ell, \sigma, r),$$

where Φ is some key-deriving function set.

6.2 Prøst-COPA

COPA is an authenticated encryption scheme by Andreeva et al. [3]. COPA for integral message is depicted in Fig. 4 (we refer to [3] for the general case). At its core, it is using a blockcipher E in XEX mode (2) with masks $\Delta = 2^\alpha 3^\beta 7^\gamma E_k(0)$, where (α, β, γ) is the tweak coming from tweak space $\{0, \ldots, \ell\} \times \{0, \ldots, 5\} \times \{0, 1\} \backslash \{(0, 0, 0)\} = \mathcal{T}_{\mathrm{COPA}}$.[4]

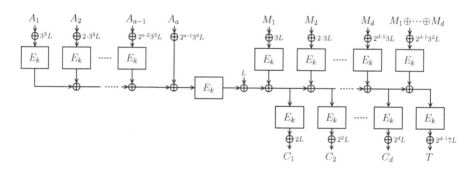

Fig. 4. COPA for integral data. Here, $L = E_k(0)$.

Before discussing the related-key security of COPA, we quickly revisit the original security proof at a high level. Consider an attacker against COPA that has resources (q, ℓ, σ, r). As a first step, all XEX evaluations in COPA are replaced with a random tweakable permutation $\widetilde{\pi} \xleftarrow{\$} \widetilde{\mathsf{Perm}}(\mathcal{T}_{\mathrm{COPA}}, \{0, 1\}^n)$.

---

[4] The fact that $(0, 0, 0) \notin \mathcal{T}_{\mathrm{COPA}}$ is important, cf. Rogaway [48] and Minematsu [40] who describe an attack on XEX if $(0, 0, 0)$ were permitted.

This step costs us $\mathbf{Adv}_{\mathrm{XEX}}^{\mathrm{sprp}}(2(\sigma+q),r)$. Next, COPA with ideal tweakable permutation is proven to achieve privacy up to bound $A_{\mathrm{priv}}(q,\ell,\sigma) = \frac{\sigma^2}{2^n} + \frac{(\ell+2)(q-1)^2}{2^n}$ and authenticity up to bound $A_{\mathrm{auth}}(q,\ell,\sigma) = \frac{(\sigma+q)^2}{2^n} + \frac{(\ell+2)(q-1)^2}{2^n} + \frac{2q}{2^n}$. Thus:

$$\mathbf{Adv}_{\mathrm{COPA}}^{\mathrm{priv/auth}}(q,\ell,\sigma,r) \leq \mathbf{Adv}_{\mathrm{XEX}}^{\mathrm{sprp}}(2(\sigma+q),r) + A_{\mathrm{priv/auth}}(q,\ell,\sigma).$$

The step towards RK-security of COPA is quite straightforward, noting that an attacker against COPA with ideal tweakable *related-key* permutation has no benefit over an attacker against COPA with ideal tweakable (non-related-key) permutation.

Theorem 2 (RK-security of COPA). *Let Φ be any KDF-set. We have*

$$\mathbf{Adv}_{\Phi,\mathrm{COPA}}^{\mathrm{rk\text{-}priv/auth}}(q,\ell,\sigma,r) \leq \mathbf{Adv}_{\Phi,\mathrm{XEX}}^{\mathrm{rk\text{-}sprp}}(2(\sigma+q),r) + A_{\mathrm{priv/auth}}(q,\ell,\sigma).$$

Proof. Consider an attacker against COPA that has resources (q,ℓ,σ,r). As a first step, all XEX evaluations in COPA are replaced with a random tweakable related-key permutation $\widetilde{\mathrm{RK}\pi} \xleftarrow{\$} \widetilde{\mathrm{RK\text{-}Perm}}(\Phi,\mathcal{T}_{\mathrm{COPA}},\{0,1\}^n)$. This step costs $\mathbf{Adv}_{\Phi,\mathrm{XEX}}^{\mathrm{rk\text{-}sprp}}(2(\sigma+q),r)$. It remains to consider COPA with $\widetilde{\mathrm{RK}\pi}$. However, as $\widetilde{\mathrm{RK}\pi}$ instantiates an ideal permutation for every different related-key function, every new related-key function instantiates a completely independent instance of COPA. Formally, assume the adversary queries COPA for s different key-deriving functions, $\varphi_1,\ldots,\varphi_s$, where φ_i is used with total resources (q_i,ℓ,σ_i). These all instantiate independent versions of COPA, contributing $A_{\mathrm{priv/auth}}(q_i,\ell,\sigma_i)$ to the bound, totaling to

$$\sum_{i=1}^{s} A_{\mathrm{priv/auth}}(q_i,\ell,\sigma_i) \leq A_{\mathrm{priv/auth}}(q,\ell,\sigma),$$

using that $q_i \geq 1$, $\sum_{i=1}^{s} q_i = q$, and $\sum_{i=1}^{s} \sigma_i = \sigma$. The bound then applies to all adversaries. $\qquad\square$

Prøst-COPA [29], in turn, uses the Prøst permutation in Even-Mansour mode. In other words, Prøst-COPA does not simply use XEX, but XPX with tweak space

$$\mathcal{T}_{\mathrm{Prøst}} = \left\{ \begin{array}{l} (2^\alpha 3^\beta 7^\gamma \oplus 1, 2^\alpha 3^\beta 7^\gamma, \\ 2^\alpha 3^\beta 7^\gamma \oplus 1, 2^\alpha 3^\beta 7^\gamma) \end{array} \middle| (\alpha,\beta,\gamma) \in \mathcal{T}_{\mathrm{COPA}} \right\}. \tag{11}$$

Taking any of the KDF-sets $\Phi \in \{\Phi_\oplus, \Phi_{P\oplus}\}$ of (5), we find:

Corollary 1 (RK-security of Prøst-COPA). *For Φ being $\Phi_\oplus$ or $\Phi_{P\oplus}$ of (5), we have*

$$\mathbf{Adv}_{\Phi,\mathrm{Prøst\text{-}COPA}}^{\mathrm{rk\text{-}priv/auth}}(q,\ell,\sigma,r) \leq \frac{16(\sigma+q)^2 + 8(\sigma+q)r}{2^n} + A_{\mathrm{priv/auth}}(q,\ell,\sigma).$$

Proof. The proof of Theorem 2 generalizes to Prøst-COPA straightforwardly, where $\mathbf{Adv}_{\Phi,\mathrm{XEX}}^{\mathrm{rk\text{-}sprp}}(2(\sigma+q),r)$ gets replaced with $\mathbf{Adv}_{\Phi,\mathrm{XPX}}^{\mathrm{rk\text{-}sprp}}(2(\sigma+q),r)$. This XPX is instantiated using tweak space $\mathcal{T}_{\mathrm{Prøst}}$ of (11), which is valid and satisfies $t_{11}, t_{12}, t_{21}, t_{22} \neq 0$ for any $(t_{11}, t_{12}, t_{21}, t_{22}) \in \mathcal{T}_{\mathrm{Prøst}}$ (note that $(\alpha, \beta, \gamma) = (0,0,0)$ is excluded). Therefore, Theorem 1(c) applies for $\Phi = \Phi_\oplus$ and Theorem 1(e) for $\Phi = \Phi_{P\oplus}$. In the worst case, we find that

$$\mathbf{Adv}_{\Phi,\mathrm{XPX}}^{\mathrm{rk\text{-}sprp}}(2(\sigma+q),r) \leq \frac{16(\sigma+q)^2 + 8(\sigma+q)r}{2^n},$$

completing the proof.   □

Note that if Prøst-COPA were not to use Prøst permutation in Even-Mansour mode, but if it simply had $E = P$, then the resulting XPX construction would have tweak space

$$\mathcal{T}_{\mathrm{Prøst}'} = \left\{ (0, 2^\alpha 3^\beta 7^\gamma, 0, 2^\alpha 3^\beta 7^\gamma) \mid (\alpha, \beta, \gamma) \in \mathcal{T}_{\mathrm{COPA}} \right\}.$$

This tweak space does not satisfy the conditions of Theorem 1(e) and we can only argue the related-key security of Prøst-COPA under $\Phi_\oplus$.

6.3   Minalpher

Minalpher is an authenticated encryption scheme by Sasaki et al. [49]. Minalpher for integral message is depicted in Fig. 5 (we refer to [49] for the general case). At its core, it is using tweakable Even-Mansour TEM of (3): an evaluation of an n-bit permutation with masks[5] $\Delta = 2^\alpha 3^\beta \big(k\|\mathsf{flag}\|N \oplus P(k\|\mathsf{flag}\|N)\big)$, where $(\alpha, \beta, \mathsf{flag}, N)$ is the tweak coming from tweak space

$$\big(\{0, \dots, \ell\} \times \{0, 1, 2\}\big) \setminus \{(0,0)\} \times \{\mathsf{flag_m}, \mathsf{flag_{ad}}, \mathsf{flag_{mac}}\} \times \{0,1\}^{n/2-s} = \mathcal{T}_{\mathrm{Minalpher}}.$$

Here, the key k is of size $n/2$ bits, the flag of size s bits, and the nonce N of size $n/2 - s$ bits.

The authors prove, among others, that $\mathbf{Adv}_{\mathrm{TEM}}^{\mathrm{sprp}}(q,r) \leq \mathcal{O}((q+r)^2/2^n + (q+r)/2^{n/2})$. The extra term $\mathcal{O}((q+r)/2^{n/2})$ is new compared to Theorem 1(a), and is caused by the shorter key size. A bit of thought reveals that, because the tweaks $\mathsf{flag}\|N$ are concatenated to k instead of XORed with k, the results of Theorem 1(b-e) generalize to TEM. Here, again, the specific key length needs to be taken into account. In [49], the designers prove that if the underlying TEM is sufficiently strong, Minalpher is a secure authenticated encryption scheme. In a similar fashion as Theorem 2 and Corollary 1, a generalization of Theorem 1(b-e) can be used to argue the related-key security of Minalpher.

---

[5] The original specification uses a generator y instead of 2.

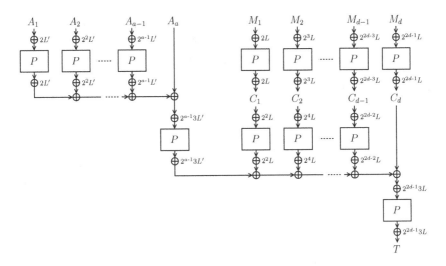

Fig. 5. Minalpher for integral data. Here, $L' = k\|\mathsf{flag}\|0 \oplus P(k\|\mathsf{flag}\|0)$ and $L = k\|\mathsf{flag}\|N \oplus P(k\|\mathsf{flag}\|N)$

7 Application to MAC

Various novel MAC functions, such as the keyed Sponges [5,7,12,26,39,44] and Chaskey [42,43], consist of a sequential application of a permutation, where the key is used to mask the state. We discuss an application of the analysis of XPX to Chaskey in detail, and explain how similar reasoning applies to keyed Sponges. We first briefly discuss the security model.

7.1 Security Model

A MAC function is expected to guarantee authenticity. However, we consider a different security model, namely PRF security. More formally, let MAC be a MAC function that gets as input a key and message, and outputs a tag. Let $\mathcal{P}$ be an idealized primitive upon which MAC is based (optional, for instance a blockcipher or permutation). Let k be a randomly drawn key. Let $\$$ be a function with the same interface as MAC, but that returns fresh and random answers to every query. We define the PRF security of MAC based on $\mathcal{P}$ as

$$\mathbf{Adv}^{\mathrm{prf}}_{\mathsf{MAC}}(\mathcal{D}) = \left| \mathbf{Pr}\left[\mathcal{D}^{\mathsf{MAC}_k,\mathcal{P}^{\pm}} = 1 \right] - \mathbf{Pr}\left[\mathcal{D}^{\$,\mathcal{P}^{\pm}} = 1 \right] \right|.$$

For $q, \ell, \sigma, r \geq 0$, we define by

$$\mathbf{Adv}^{\mathrm{prf}}_{\mathsf{MAC}}(q, \ell, \sigma, r) = \max_{\mathcal{D}} \mathbf{Adv}^{\mathrm{prf}}_{\mathsf{MAC}}(\mathcal{D})$$

the security of MAC against any distinguisher $\mathcal{D}$ that makes q queries to the construction oracle, each of length at most ℓ and of total size σ, and r queries to the primitive oracle.

As before, the definition generalizes to related-key security straightforwardly, and we denote the corresponding related-key security definitions by

$$\mathbf{Adv}_{\Phi,\mathsf{MAC}}^{\text{rk-prf}}(\mathcal{D}) \text{ and } \mathbf{Adv}_{\Phi,\mathsf{MAC}}^{\text{rk-prf}}(q,\ell,\sigma,r),$$

where Φ is some key-deriving function set.

7.2   Chaskey

Chaskey is a permutation-based MAC function by Mouha et al. [42,43]. We consider a small adjustment, called Chaskey′, that processes the initialized state with an evaluation of the permutation. Chaskey and Chaskey′ without final truncation are depicted in Fig. 6.

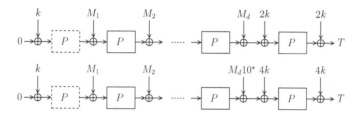

Fig. 6. Chaskey′ for integral messages (top) and fractional messages (bottom). The dashed P's are absent in the original Chaskey.

Mouha et al. [43] proved the security of Chaskey (without the first evaluation of P). It consists of the idea that XORing the key k twice in-between every two consecutive P evaluations gives a blockcipher-based Chaskey using Even-Mansour constructions $m \mapsto P(m \oplus k) \oplus k$, $m \mapsto P(m \oplus 3k) \oplus 2k$, and $m \mapsto P(m \oplus 5k) \oplus 4k$. The security of Chaskey boils down to the advantage of a distinguisher in distinguishing these three constructions from three ideal permutations, an advantage the authors dub the "3PRP" security. This 3PRP security is effectively equivalent to the PRP security of XPX with tweak space $\{(1,0,1,0),(3,0,2,0),(5,0,4,0)\} = \mathcal{T}_{\text{Chaskey}}$, and we find:[6]

$$\mathbf{Adv}_{\text{Chaskey}}^{\text{prf}}(q,\ell,\sigma,r) \leq \mathbf{Adv}_{\text{XPX}}^{\text{prp}}(\sigma,r) + \frac{2\sigma^2}{2^n}.$$

Now, for Chaskey′, the idea is to XOR $P(k) \oplus P(k)$ everywhere in-between two consecutive P evaluations *except for the first two*. In this case, Chaskey′ would simply be using XPX with tweak space

$$\{(0,1,0,1),(2,1,2,0),(4,1,4,0)\} = \mathcal{T}_{\text{Chaskey}'}.$$

Note that $\mathcal{T}_{\text{Chaskey}'}$ satisfies the conditions of Theorem 1(b). Similarly to Theorem 2 and Corollary 1, we directly obtain:

---

[6] The authors of [43] effectively consider MAC security instead of PRF security, but the analysis carries over.

Corollary 2 (RK-security of Chaskey'). *For $\Phi_\oplus$ of (5), we have*

$$\mathbf{Adv}^{\text{rk-prf}}_{\Phi_\oplus,\text{Chaskey}'}(q,\ell,\sigma,r) \leq \frac{\frac{7}{2}\sigma^2 + 4\sigma r}{2^n - \sigma} + \frac{2\sigma^2}{2^n}.$$

7.3 Keyed Sponge

Following [7,12], Andreeva et al. [5] formalized two Sponges: the inner-keyed Sponge and the outer-keyed Sponge. Gaži et al. [26] generalized these results (among others) to full-state absorption. This construction, to some extent, resembles the Donkey Sponge construction [8]. Mennink et al. [39] considered the full-state Sponge and full-state Duplex. In a similar fashion as the analysis of Sect. 7.2, the inner-keyed Sponge [5], the Donkey Sponge [8], and the full-state Sponge and Duplex [39] can be adjusted to achieve related-key security.

Acknowledgments. This work was supported in part by the Research Council KU Leuven: GOA TENSE (GOA/11/007), and in part by COST Action "Cryptography for Secure Digital Interaction." Bart Mennink is a Postdoctoral Fellow of the Research Foundation – Flanders (FWO). The author would like to thank the DTU Compute team and the anonymous reviewers of CRYPTO 2016 for their comments and suggestions.

References

1. Albrecht, M.R., Farshim, P., Paterson, K.G., Watson, G.J.: On cipher-dependent related-key attacks in the ideal-cipher model. In: Joux, A. (ed.) FSE 2011. LNCS, vol. 6733, pp. 128–145. Springer, Heidelberg (2011)
2. Andreeva, E., Bogdanov, A., Dodis, Y., Mennink, B., Steinberger, J.P.: On the indifferentiability of key-alternating ciphers. In: Canetti, R., Garay, J.A. (eds.) CRYPTO 2013, Part I. LNCS, vol. 8042, pp. 531–550. Springer, Heidelberg (2013)
3. Andreeva, E., Bogdanov, A., Luykx, A., Mennink, B., Tischhauser, E., Yasuda, K.: Parallelizable and authenticated online ciphers. In: Sako, K., Sarkar, P. (eds.) ASIACRYPT 2013, Part I. LNCS, vol. 8269, pp. 424–443. Springer, Heidelberg (2013)
4. Andreeva, E., Bogdanov, A., Mennink, B.: Towards understanding the known-key security of block ciphers. In: Moriai, S. (ed.) FSE 2013. LNCS, vol. 8424, pp. 348–366. Springer, Heidelberg (2014)
5. Andreeva, E., Daemen, J., Mennink, B., Van Assche, G.: Security of keyed sponge constructions using a modular proof approach. In: Leander, G. (ed.) FSE 2015. LNCS, vol. 9054, pp. 364–384. Springer, Heidelberg (2015)
6. Bellare, M., Kohno, T.: A theoretical treatment of related-key attacks: RKA-PRPs, RKA-PRFs. In: Biham, E. (ed.) EUROCRYPT 2003. LNCS, vol. 2656, pp. 491–506. Springer, Heidelberg (2003)
7. Bertoni, G., Daemen, J., Peeters, M., Van Assche, G.: On the security of the keyed sponge construction. In: Symmetric Key Encryption Workshop (SKEW 2011) (2011)
8. Bertoni, G., Daemen, J., Peeters, M., Van Assche, G.: Permutation-based encryption, authentication and authenticated encryption. In: Directions in Authenticated Ciphers (DIAC 2012) (2012)

9. Bogdanov, A., Knudsen, L.R., Leander, G., Standaert, F.-X., Steinberger, J., Tischhauser, E.: Key-alternating ciphers in a provable setting: encryption using a small number of public permutations. In: Pointcheval, D., Johansson, T. (eds.) EUROCRYPT 2012. LNCS, vol. 7237, pp. 45–62. Springer, Heidelberg (2012)

10. CAESAR: Competition for Authenticated Encryption: Security, Applicability, and Robustness, May 2015. http://competitions.cr.yp.to/caesar.html

11. Chakraborty, D., Sarkar, P.: A general construction of tweakable block ciphers and different modes of operations. In: Lipmaa, H., Yung, M., Lin, D. (eds.) Inscrypt 2006. LNCS, vol. 4318, pp. 88–102. Springer, Heidelberg (2006)

12. Chang, D., Dworkin, M., Hong, S., Kelsey, J., Nandi, M.: A keyed sponge construction with pseudorandomness in the standard model. In: NIST's 3rd SHA-3 Candidate Conference 2012 (2012)

13. Chen, S., Lampe, R., Lee, J., Seurin, Y., Steinberger, J.: Minimizing the two-round Even-Mansour cipher. In: Garay, J.A., Gennaro, R. (eds.) CRYPTO 2014, Part I. LNCS, vol. 8616, pp. 39–56. Springer, Heidelberg (2014)

14. Chen, S., Steinberger, J.: Tight security bounds for key-alternating ciphers. In: Nguyen, P.Q., Oswald, E. (eds.) EUROCRYPT 2014. LNCS, vol. 8441, pp. 327–350. Springer, Heidelberg (2014)

15. Cogliati, B., Lampe, R., Seurin, Y.: Tweaking Even-Mansour ciphers. In: Gennaro, R., Robshaw, M. (eds.) CRYPTO 2015, Part I. LNCS, vol. 9215, pp. 493–517. Springer, Heidelberg (2015)

16. Cogliati, B., Seurin, Y.: Beyond-birthday-bound security for tweakable Even-Mansour ciphers with linear tweak and key mixing. In: Iwata, T., Cheon, J.H. (eds.) ASIACRYPT 2015. LNCS, vol. 9453, pp. 134–158. Springer, Heidelberg (2015)

17. Cogliati, B., Seurin, Y.: On the provable security of the iterated Even-Mansour cipher against related-key and chosen-key attacks. In: Oswald, E., Fischlin, M. (eds.) EUROCRYPT 2015. LNCS, vol. 9056, pp. 584–613. Springer, Heidelberg (2015)

18. Cogliati, B., Seurin, Y.: Strengthening the known-key security notion for block ciphers. In: FSE 2016. LNCS, Springer, Heidelberg (2016, to appear)

19. Dai, Y., Lee, J., Mennink, B., Steinberger, J.: The security of multiple encryption in the ideal cipher model. In: Garay, J.A., Gennaro, R. (eds.) CRYPTO 2014, Part I. LNCS, vol. 8616, pp. 20–38. Springer, Heidelberg (2014)

20. Datta, N., Nandi, M.: ELmD v1.0, submission to CAESAR competition (2014)

21. Dobraunig, C., Eichlseder, M., Mendel, F.: Related-key forgeries for Prøst-OTR. In: Leander, G. (ed.) FSE 2015. LNCS, vol. 9054, pp. 282–296. Springer, Heidelberg (2015)

22. Dunkelman, O., Keller, N., Shamir, A.: Minimalism in cryptography: the Even-Mansour scheme revisited. In: Pointcheval, D., Johansson, T. (eds.) EUROCRYPT 2012. LNCS, vol. 7237, pp. 336–354. Springer, Heidelberg (2012)

23. Even, S., Mansour, Y.: A construction of a cipher from a single pseudorandom permutation. In: Matsumoto, T., Imai, H., Rivest, R.L. (eds.) ASIACRYPT 1991. LNCS, vol. 739, pp. 201–224. Springer, Heidelberg (1993)

24. Even, S., Mansour, Y.: A construction of a cipher from a single pseudorandom permutation. J. Cryptol. **10**(3), 151–162 (1997)

25. Farshim, P., Procter, G.: The related-key security of iterated Even-Mansour ciphers. In: Leander, G. (ed.) FSE 2015. LNCS, vol. 9054, pp. 342–363. Springer, Heidelberg (2015)

26. Gaži, P., Pietrzak, K., Tessaro, S.: The exact PRF security of truncation: tight bounds for keyed sponges and truncated CBC. In: Gennaro, R., Robshaw, M. (eds.) CRYPTO 2015 Part I. LNCS, vol. 9215, pp. 368–387. Springer, Heidelberg (2015)

27. Granger, R., Jovanovic, P., Mennink, B., Neves, S.: Improved masking for tweakable blockciphers with applications to authenticated encryption. In: Fischlin, M., Coron, J.-S. (eds.) EUROCRYPT 2016. LNCS, vol. 9665, pp. 263–293. Springer, Heidelberg (2016)

28. Karpman, P.: From distinguishers to key recovery: improved related-key attacks on Even-Mansour. In: López, J., Mitchell, C.J. (eds.) ISC 2015. LNCS, vol. 9290, pp. 177–188. Springer, Heidelberg (2015)

29. Kavun, E., Lauridsen, M., Leander, G., Rechberger, C., Schwabe, P., Yalçın, T.: Prøst v1, submission to CAESAR competition (2014)

30. Lampe, R., Patarin, J., Seurin, Y.: An asymptotically tight security analysis of the iterated even-mansour cipher. In: Wang, X., Sako, K. (eds.) ASIACRYPT 2012. LNCS, vol. 7658, pp. 278–295. Springer, Heidelberg (2012)

31. Lampe, R., Seurin, Y.: How to construct an ideal cipher from a small set of public permutations. In: Sako, K., Sarkar, P. (eds.) ASIACRYPT 2013, Part I. LNCS, vol. 8269, pp. 444–463. Springer, Heidelberg (2013)

32. Lampe, R., Seurin, Y.: Tweakable blockciphers with asymptotically optimal security. In: Moriai, S. (ed.) FSE 2013. LNCS, vol. 8424, pp. 133–152. Springer, Heidelberg (2014)

33. Landecker, W., Shrimpton, T., Terashima, R.S.: Tweakable blockciphers with beyond birthday-bound security. In: Safavi-Naini, R., Canetti, R. (eds.) CRYPTO 2012. LNCS, vol. 7417, pp. 14–30. Springer, Heidelberg (2012)

34. Liskov, M., Rivest, R.L., Wagner, D.: Tweakable block ciphers. In: Yung, M. (ed.) CRYPTO 2002. LNCS, vol. 2442, pp. 31–46. Springer, Heidelberg (2002)

35. Mennink, B.: Optimally secure tweakable blockciphers. In: Leander, G. (ed.) FSE 2015. LNCS, vol. 9054, pp. 428–448. Springer, Heidelberg (2015)

36. Mennink, B.: Optimally secure tweakable blockciphers. Cryptology ePrint Archive, report 2015/363, full version of [35] (2015)

37. Mennink, B.: XPX: Generalized tweakable Even-Mansour with improved security guarantees. Cryptology ePrint Archive, report 2015/476, full version of this paper (2015)

38. Mennink, B., Preneel, B.: On the XOR of multiple random permutations. In: Malkin, T., Kolesnikov, V., Lewko, A.B., Polychronakis, M. (eds.) ACNS 2015. LNCS, vol. 9092, pp. 619–634. Springer, Heidelberg (2015)

39. Mennink, B., Reyhanitabar, R., Vizár, D.: Security of full-state keyed sponge and duplex: applications to authenticated encryption. In: Iwata, T., Cheon, J.H. (eds.) ASIACRYPT 2015. LNCS, vol. 9453, pp. 465–489. Springer, Heidelberg (2015)

40. Minematsu, K.: Improved security analysis of XEX and LRW modes. In: Biham, E., Youssef, A.M. (eds.) SAC 2006. LNCS, vol. 4356, pp. 96–113. Springer, Heidelberg (2007)

41. Minematsu, K.: Beyond-birthday-bound security based on tweakable block cipher. In: Dunkelman, O. (ed.) FSE 2009. LNCS, vol. 5665, pp. 308–326. Springer, Heidelberg (2009)

42. Mouha, N.: Chaskey: a MAC algorithm for microcontrollers - status update and proposal of Chaskey-12. Cryptology ePrint Archive, report 2015/1182 (2015)

43. Mouha, N., Mennink, B., Van Herrewege, A., Watanabe, D., Preneel, B., Verbauwhede, I.: Chaskey: an efficient MAC algorithm for 32-bit microcontrollers. In: Joux, A., Youssef, A. (eds.) SAC 2014. LNCS, vol. 8781, pp. 306–323. Springer, Heidelberg (2014)

44. Naito, Y., Yasuda, K.: New bounds for keyed sponges with extendable output: Independence between capacity and message length. In: FSE 2016. LNCS, Springer, Heidelberg (2016, to appear)

45. Nyberg, K., Knudsen, L.R.: Provable security against differential cryptanalysis. In: Brickell, E.F. (ed.) CRYPTO 1992. LNCS, vol. 740, pp. 566–574. Springer, Heidelberg (1993)

46. Patarin, A.: A proof of security in $O(2^n)$ for the Xor of two randompermutations. In: Safavi-Naini, R. (ed.) ICITS 2008. LNCS, vol. 5155, pp. 232–248. Springer, Heidelberg (2008)

47. Procter, G.: A note on the CLRW2 tweakable block cipher construction. Cryptology ePrint Archive, report 2014/111 (2014)

48. Rogaway, P.: Efficient instantiations of tweakable blockciphers and refinements to modes OCB and PMAC. In: Lee, P.J. (ed.) ASIACRYPT 2004. LNCS, vol. 3329, pp. 16–31. Springer, Heidelberg (2004)

49. Sasaki, Y., Todo, Y., Aoki, K., Naito, Y., Sugawara, T., Murakami, Y., Matsui, M., Hirose, S.: Minalpher v1, submission to CAESAR competition (2014)

50. Steinberger, J.: Improved security bounds for key-alternating ciphers via Hellinger distance. Cryptology ePrint Archive, report 2012/481 (2012)

Indifferentiability of 8-Round Feistel Networks

Yuanxi Dai and John Steinberger[⊠]

Institute for Interdisciplinary Information Sciences,
Tsinghua University, Beijing, China
dyx13@mails.tsinghua.edu.cn, jpsteinb@gmail.com

Abstract. We prove that a balanced 8-round Feistel network is indifferentiable from a random permutation, improving on previous 10-round results by Dachman-Soled et al. and Dai et al. Our simulator achieves security $O(q^8/2^n)$, similarly to the security of Dai et al. For further comparison, Dachman-Soled et al. achieve security $O(q^{12}/2^n)$, while the original 14-round simulator of Holenstein et al. achieves security $O(q^{10}/2^n)$.

Keywords: Feistel network · Block ciphers

1 Introduction

For many cryptographic protocols the only known analyses are in a so-called *ideal primitive model*. In such a model, a cryptographic component is replaced by an idealized information-theoretic counterpart (e.g., a random oracle takes the part of a hash function, or an ideal cipher substitutes for a concrete blockcipher such as AES) and security bounds are given as functions of the query complexity of an information-theoretic adversary with oracle access to the idealized primitive. Early uses of such ideal models include Winternitz [33], Fiat and Shamir [19] (see proof in [28]) and Bellare and Rogaway [2], with such analyses rapidly proliferating after the latter paper.

Given the popularity of such analyses a natural question that arises is to determine the relative "power" of different classes of primitives and, more precisely, whether one class of primitives can be used to "implement" another. E.g., is a random function always sufficient to implement an ideal cipher, in security games where oracle access to the ideal cipher/random function is granted to all parties? The challenge of such a question is partly definitional, since the different primitives have syntactically distinct interfaces. (Indeed, it seems that it was not immediately obvious to researchers that such a question made sense at all [7].)

A sensible definitional framework, however, was proposed by Maurer et al. [23], who introduce a simulation-based notion of *indifferentiability*. This framework allows to meaningfully discuss the instantiation of one ideal primitive by a syntactically different primitive, and to compose such results. (Similar simulation-based definitions appear in [4,5,26,27].) Coron et al. [7] are early adopters of the framework, and give additional insights.

Informally, given ideal primitives Z and Q, a construction C^Q (where C is some stateless algorithm making queries to Q) is *indifferentiable* from Z if there

M. Robshaw and J. Katz (Eds.): CRYPTO 2016, Part I, LNCS 9814, pp. 95–120, 2016.
DOI: 10.1007/978-3-662-53018-4_4

exists a simulator S (a stateful, randomized algorithm) with oracle access to Z such that the pair (C^Q, Q) is statistically indistinguishable from the pair (Z, S^Z). The more efficient the simulator, the lower its query complexity, and the better the statistical indistinguishability, the more practically meaningful the result.

The present paper focuses on the natural question of implementing a permutation from one or more random functions (a small number of distinct random functions can be emulated by a single random function with a slightly larger domain) such that the resulting construction is indifferentiable from a random permutation. This means building a permutation $C : \{0,1\}^{m(n)} \to \{0,1\}^{m(n)}$ where

$$C = C[F_1, \ldots, F_r]$$

depends on a small collection of random functions $F_1, \ldots, F_r : \{0,1\}^n \to \{0,1\}^n$ such that the vector of $r + 1$ oracles

$$(C[F_1, \ldots, F_r], F_1, \ldots, F_r)$$

is statistically indistinguishable from a pair

$$(Z, S^Z)$$

where $Z : \{0,1\}^{m(n)} \to \{0,1\}^{m(n)}$ is a random permutation from $m(n)$ bits to $m(n)$ bits, for some efficient simulator S. Thus, in this case, the simulator emulates the random functions $F_1, \ldots, F_r$, and it must use its oracle access to Z to invent answers that make the (fake) random functions $F_1, \ldots, F_r$ look "compatible" with Z, as if Z where really $C[F_1, \ldots, F_r]$. (On the other hand, the simulator does not know what queries the distinguisher might be making to Z.) Here $m(n)$ is polynomially related to n: concretely, the current paper discusses a construction with $m = 2n$.

The construction $C[F_1, \ldots, F_r]$ that we consider in this paper, and as considered in previous papers with the same goal as ours (see discussion below), is an r-round (balanced, unkeyed) *Feistel network*. To wit, given arbitrary functions $F_1, \ldots, F_r : \{0,1\}^n \to \{0,1\}^n$, we define a permutation

$$C[F_1, \ldots, F_r] : \{0,1\}^{2n} \to \{0,1\}^{2n}$$

by the following application: for an input $(x_0, x_1) \in \{0,1\}^{2n}$, values $x_2, \ldots, x_{r+1}$ are defined by setting

$$x_{i+1} = x_{i-1} \oplus F_i(x_i) \tag{1}$$

for $i = 1, \ldots, r$; then $(x_r, x_{r+1}) \in \{0,1\}^{2n}$ is the output of C on input (x_0, x_1). One can observe that C is a permutation since x_{i-1} can be computed from x_i and x_{i+1}, by (1). The value r is the number of *rounds* of the Feistel network. (See, e.g., Fig. 1.)

The question of showing that a Feistel network with a sufficient number of rounds is indifferentiable from a random permutation already has a growing history. Coron et al. [9] show that an r-round Feistel network cannot be

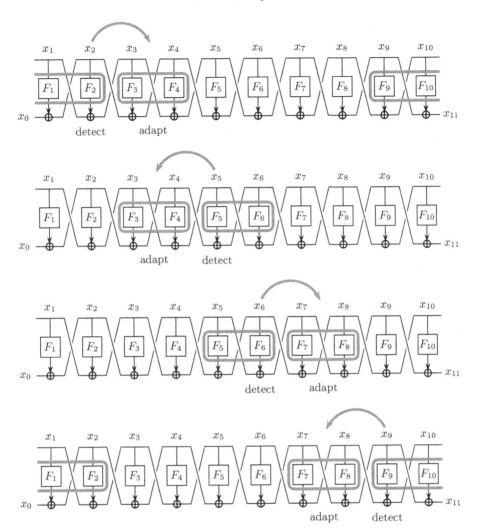

Fig. 1. A sketch of the 10-round simulator from [11] (and also Seurin's 10-round simulator). Rounds 5 and 6 form one detect zone; rounds 1, 2, 9 and 10 form another detect zone; rounds 3 and 4 constitute the left adapt zone, 7 and 8 constitute the right adapt zone; red arrows point from the position where a path is detected (a.k.a., "pending query") to the adapt zone for that path. (Color figure online)

indifferentiable from a random permutation for $r \leq 5$, due to explicit attacks. They also give a proof that indifferentiability is achieved at $r = 6$, but this latter result was found to have a serious flaw by Holenstein et al. [20], who could only prove, as a replacement, that indifferentiability is achieved at $r = 14$ rounds. At the same time, Holenstein et al. found a flaw in the proof of indifferentiability of a 10-round simulator of Seurin's [31] (a simplified alternative to the

6-round simulator of [9]), after which Seurin himself found an explicit attack against his own simulator, showing that the proof could not be patched [32]. More recently, Dachman-Soled et al. [10] and the authors of the present paper [11] have presented independent indifferentiability proofs at 10 rounds.

In [11] we achieve slightly better security than other proofs ($O(q^8/2^n)$, compared to $O(q^{10}/2^n)$ for Holenstein et al. and $O(q^{12}/2^n)$ for Dachman-Soled et al.), and their work also introduces an interesting "last-in-first-out" simulator paradigm. In fact, the simulator of [11] is essentially Seurin's (flawed) 10-round simulator, only with "first-in-first-out" path completion replaced by "last-in-first-out" path completion. This change, as it turns out, is sufficient to repair the flaw discovered by Holenstein et al. [20].

In the current work we prove that an 8-round Feistel network is indifferentiable from a random permutation. The security, query complexity and runtime of our 8-round simulator are $O(q^8/2^n)$, $O(q^4)$ and $O(q^4)$ respectively, just like our previous 10-round simulator [11]. (The query complexity of previous simulators of Dachman-Soled et al. and Holenstein et al. can apparently be reduced to $O(q^4)$ as well with suitable optimizations [11], though higher numbers are quoted in the original papers.) In fact our work closely follows the ideas [11], and is obtained by making a number of small optimizations to that simulator in order to reduce it to 8 rounds. It remains open whether 6 or 7 rounds might suffice for indifferentiability.

Concerning our optimizations, more specifically, in [11,20,31] the "outer detect zone" requires four-out-of-four queries in order to trigger a path completion (the outer detect zone consists of four rounds, these being rounds 1, 2 and $r - 1$, r). In the current paper, we optimize by always making the outer detect zone trigger a path completion as soon as possible, i.e., by completing a path whenever three-out-of-four matching queries occur in the outer detect zone. (This is similar to an idea of Dachman-Soled et al. [10].) By detecting a little earlier in this fashion, we can move the "adapt zones" on either side by one position towards the left and right edges of the network, effectively removing one round at either end, but this creates a fresh difficulty, as two of the four different types of paths detected by the outer detect zone cannot make use of the new translated adapt zones because the translated adapt zones overlap with the query that triggers the path. For these two types of paths (which are triggered by queries at round 2 or at round $r - 1$), we use a brand new adapt zone instead, consisting of the middle two rounds of the network. (Rounds 4 and 5, in our 8-round design.) This itself creates another complication, since an adapted query should not trigger a path completion, lest the proof blow up, and since the "middle detect zone" is traditionally made up of rounds 4 and 5 precisely. We circumvent this problem with a fresh trick: We split the middle detect zone into two separate overlapping zones, each of which has *three* rounds: rounds 3, 4, 5 for one zone, rounds 4, 5, 6 for the other; after this change, adapted queries at rounds 4, 5 (and as argued within the proof) do not trigger either of the middle detect zones. The simulator's "termination argument" is slightly affected by the presence of two separate middle detect zones, but not much: one can observe

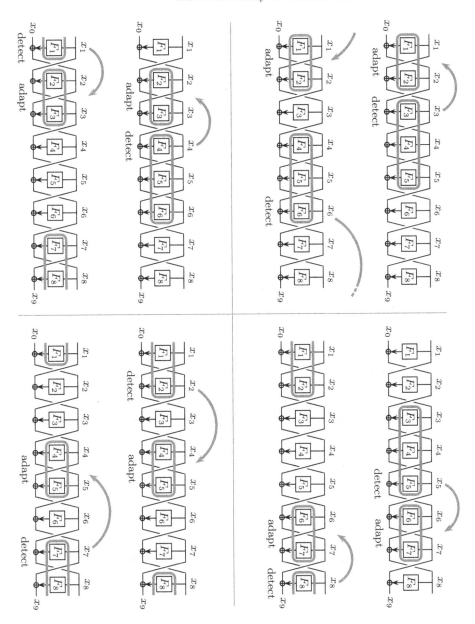

Fig. 2. A sketch of our 8-round simulator drawn in the same style as Fig. 1. Red groups of three queries are detect zones; when a query completing a detect zone (a.k.a., "pending query") occurs at one of the endpoints of the zone, a path completion is triggered; the adapt zone for that path completion is shown in blue; the four quadrants correspond to the four possible adapt zones. (The adapt zone at positions F_1, F_2 in the upper right quadrant could equivalently be moved to F_7, F_8.) (Color figure online)

that neither type of middle path detection adds queries at rounds 4 and 5, even though paths triggered by one middle detect zone can trigger a path in the other middle detect zone. Hence, the original termination argument of Coron et al. [9] (used in [11,14,20] and in many other places since) goes through practically unchanged.

The resulting 8-round simulator ends up having a highly symmetric structure: It can be abstracted as having four detect zones of three consecutive rounds each, with two "inner zones" (rounds 3, 4, 5 and 4, 5, 6) and two "outer zones" (rounds 1, 2, 8 and 1, 7, 8); each detect zone of three consecutive rounds detects "at either end" (e.g., the detect zone with rounds 3, 4, 5 detects at rounds 3 and 5, etc.); the upshot is that each of rounds 1, . . . , 8 ends up being a detection point for exactly one of the four three-round detect zones. We refer to Fig. 2 in Sect. 3. A much more leisurely description of our simulator can be found in Sect. 3.

OTHER RELATED WORK. Before [9], Dodis and Puniya [13] investigated the indifferentiability of Feistel networks in the so-called *honest-but-curious* model, which is incomparable to the standard notion of indifferentiability. They found that in this case, a super-logarithmic number of rounds is sufficient to achieve indifferentiability. Moreover, [9] later showed that super-logarithmically many rounds are also necessary.

Besides Feistel networks, the indifferentiability of many other types of constructions (and particularly hash functions and compression functions) have been investigated. More specifically on the blockcipher side, [1] and [21] investigate the indifferentiability of key-alternating ciphers (with and without an idealized key scheduler, respectively). In a recent eprint note, Dodis et al. [14] investigate the indifferentiability of substitution-permutation networks, treating the S-boxes as independent idealized permutations. Moreover, the "LIFO" design philosophy of [11]—that also carries over to this work—is partly inspired by the latter simulator, as explained in [11].

It should be recalled that indifferentiability does not apply to a cryptographic game for which the adversary is stipulated to come from a special class that does not contain the computational class to which the simulator belongs (the latter class being typically "probabilistic polynomial-time"). See [29].

Finally, Feistel networks have been the subject of a very large body of work in the secret-key (or "indistinguishability") setting, such as in [22,24,25,30] and the references therein.

2   Definitions and Main Result

FEISTEL NETWORKS. Let $r \geq 0$ and let $F_1, \ldots, F_r : \{0,1\}^n \to \{0,1\}^n$. Given values $x_0, x_1 \in \{0,1\}^n$ we define values $x_2, \ldots, x_{r+1}$ by

$$x_{i+1} = F_i(x_i) \oplus x_{i-1}$$

for $1 \leq i \leq r$. As noted in the introduction, the application

$$(x_0, x_1) \to (x_r, x_{r+1})$$

defines a permutation of $\{0,1\}^{2n}$. We let

$$\Psi[F_1, \ldots, F_r]$$

denote this permutation. We say that Ψ is an r-*round Feistel network* and that F_i is the i-*th round function* of Ψ.

In this paper, whenever a permutation is given as an oracle, our meaning is that both forward and inverse queries can be made to the permutation. This applies in particular to Feistel networks.

INDIFFERENTIABILITY. A *construction* is a stateless deterministic algorithm that evaluates by making calls to an external set of *primitives*. The latter are functions that conform to a syntax that is specified by the construction. Thus $\Psi[F_1, \ldots, F_r]$ can be seen as a construction with primitives $F_1, \ldots, F_r$. In the general case we notate a construction C with oracle access to a set of primitives Q as C^Q.

A primitive is *ideal* if it is drawn uniformly at random from the set of all functions meeting the specified syntax. A *random function* $F : \{0,1\}^n \to \{0,1\}^n$ is a particular case of an ideal primitive. Such a function is drawn uniformly at random from the set of all functions of domain $\{0,1\}^n$ and of range $\{0,1\}^n$.

A *simulator* is a stateful randomized algorithm that receives and answer queries, possibly being given oracles of its own. We assume that a simulator is initialized to some default state (which constitutes part of the simulator's description) at the start of each experiment. A simulator S with oracle access to an ideal primitive Z is notated as S^Z.

A *distinguisher* is an algorithm that initiates a query-response session with a set of oracles, that has a limited total number of queries, and that outputs 0 or 1 when the query-response session is over. In our case distinguishers are information-theoretic; this implies, in particular, that the distinguisher can "know by heart" the (adaptive) sequence of questions that will maximize its distinguishing advantage. In particular, one may assume without loss of generality that a distinguisher is deterministic.

Indifferentability seeks to determine when a construction C^Q, where Q is a set of ideal primitives, is "as good as" an ideal primitive Z that has the same syntax (interface) as C^Q. In brief, there must exist a simulator S such that having oracle access to the pair (C^Q, Q) (often referred to as the "real world") is indistinguishable from the pair (Z, S^Z) (often referred to as the "simulated world").

In more detail we refer to the following definition, which is due to Maurer et al. [23].

Definition 1. A construction C with access to a set of ideal primitives Q is (t_S, q_S, ε)-*indifferentiable* from an ideal primitive Z if there exists a simulator $S = S(q)$ such that

$$\Pr\left[D^{C^Q, Q} = 1\right] - \Pr\left[D^{Z, S^Z} = 1\right] \le \varepsilon$$

for every distinguisher D making at most q queries in total, and such that S runs in total time t_S and makes at most q_S queries to Z. Here t_S, q_S and ε are functions of q, and the probabilities are taken over the randomness in Q, Z, S and (if any) in D.

As indicated, we allow S to depend on q.[1] The notation

$$D^{C^Q,Q}$$

indicates that D has oracle access to C^Q as well as to each of the primitives in the set Q. We also note that the oracle

$$S^Z$$

offers one interface for D to query for each of the primitives in Q; however the simulator S is "monolithic" and treats each of these queries with knowledge of the others.

Thus, S's job is to make Z look like C^Q by inventing appropriate answers for D's queries to the primitives in Q. In order to do this, S requires oracle access to Z. On the other hand, S doesn't know which queries D is making to Z.

Informally, C^Q is *indifferentiable* from Z if it is (t_S, q_S, ε)-indifferentiable for "reasonable" values of t_S, q_S and for ε negligibly small in the security parameter n. The value q_S in Definition 1 is called the *query complexity* of the simulator.

In our setting C will be the 8-round Feistel network Ψ and Q will be the set $\{F_1, \ldots, F_8\}$ of round functions, with each round function being an independent random function. Consequently, Z (matching C^Q's syntax) will be a random permutation from $\{0,1\}^{2n}$ to $\{0,1\}^{2n}$, queriable (like C^Q) in both directions; this random permutation is notated P in the body of the proof.

MAIN RESULT. The following theorem is our main result. In this theorem, Ψ plays the role of the construction C, while $\{F_1, \ldots, F_8\}$ (where each F_i is an independent random function) plays the role of Q, the set of ideal primitives called by C.

Theorem 1. *The Feistel network $\Psi[F_1, \ldots, F_8]$ is (t_S, q_S, ε)-indifferentiable from a random 2n-bit to 2n-bit permutation with $t_S = O(q^4)$, $q_S = 32q^4 + 8q^3$ and $\varepsilon = 7400448q^8/2^n$. Moreover, these bounds hold even if the distinguisher is allowed to make q queries to each of its 9 ($= 8 + 1$) oracles.*

The simulator that we use to establish Theorem 1 is described in the next section. The proof of Theorem 1 can be found in the full version of this paper [12].

MISCELLANEOUS NOTATIONS. We write $[k]$ for the set $\{1, \ldots, k\}$, $k \in \mathbb{N}$.

---

[1] This introduces a small amount of non-uniformity into the simulator, but which seems not to matter in practice. While in our case the dependence of S on q is made mainly for the sake of simplicity and could as well be avoided (with a more convoluted proof and a simulator that runs efficiently only with high probability), we note, interestingly, that there is one indifferentiabiliy result that we are aware of—namely that of [16]—for which the simulator crucially needs to know the number of distinguisher queries in advance.

The symbol $\perp$ denotes an uninitialized or null value and can be taken to be synonymous with a programming language's **null** value, though we reserve the latter for uninitialized object fields. If T is a table, moreover, we write $x \in T$ to mean that $T(x) \neq \perp$. Correspondingly, $x \notin T$ means $T(x) = \perp$.

3   High-Level Simulator Overview

In this section we give a somewhat non-technical overview of our 8-round simulator which, like [20] and [11], is a modification of a 10-round simulator by Seurin [31].

ROUND FUNCTION TABLES. We recall that the simulator is responsible for 8 interfaces, i.e., one for each of the rounds functions. These interfaces are available to the adversary through a single function, named

$$F$$

and which takes two inputs: an integer $i \in [8]$ and an input $x \in \{0,1\}^n$.

Correspondingly to these 8 interfaces, the simulator maintains 8 tables, notated $F_1, \ldots, F_8$, whose fields are initialized to $\perp$: initially, $F_i(x) = \perp$ for all $x \in \{0,1\}^n$, all $i \in [8]$. (Hence we note that F_i is no longer the name of a round function, but the name of a table. The i-th round function is now $F(i, \cdot)$.) The table F_i encodes "what the simulator has decided so far" about the i-th round function. For instance, if $F_i(x) = y \neq \perp$, then any subsequent distinguisher query of the form $F(i, x)$ will simply return $y = F_i(x)$. Entries in the tables $F_1, \ldots, F_8$ are not overwritten once they have been set to non-$\perp$ values.

THE $2n$-BIT RANDOM PERMUTATION. Additionally, the distinguisher and the simulator both have oracle access to a random permutation on $2n$ bits, notated

$$P$$

and which plays the role of the ideal primitive Z in Definition 1. Thus P accepts an input of the form $(x_0, x_1) \in \{0,1\}^n \times \{0,1\}^n$ and produces an output $(x_8, x_9) \in \{0,1\}^n \times \{0,1\}^n$. P's inverse P^{-1} is also available as an oracle to both the distinguisher and the simulator.

DISTINGUISHER INTUITION AND COMPLETED PATHS. One can think of the distinguisher as checking the consistency of the oracles $F(1, \cdot)$, ..., $F(8, \cdot)$ with P/P^{-1}. For instance, the distinguisher could choose random values $x_0, x_1 \in \{0,1\}^n$, construct the values $x_2, \ldots, x_9$ by setting

$$x_{i+1} \leftarrow F(i, x_i) \oplus x_{i-1}$$

for $i = 2, \ldots, 9$, and finally check if $(x_8, x_9) = P(x_0, x_1)$. (In the real world, this will always be the case; if the simulator is doing its job, it should also be the case in the simulated world.) In this case we also say that the values

$$x_1, \ldots, x_8$$

queried by the distinguisher form a *completed path*. (The definition of a "completed path" will be made more precise in the next section.)

It should be observed that the distinguisher has multiple options for completing paths; e.g., "left-to-right" (as above), "right-to-left" (starting from values x_8, x_9 and evaluating the Feistel network backwards), "middle-out" (starting with some values x_i, x_{i+1} in the middle of the network, and growing a path outwards to the left and to the right), "outward-in" (starting from the endpoints x_0, x_1, x_8, x_9 and going right from x_0, x_1 and left from x_8, x_9), etc. Moreover, the distinguisher can try to reuse the same query for several different paths, can interleave the completion of several paths in a complex manner, and so on.

To summarize, and for the purpose of intuition, one can picture the distinguisher as trying to complete all sorts of paths in a convoluted fashion in order to confuse and/or "trap" the simulator in a contradiction.

THE SIMULATOR'S DILEMMA. Clearly a simulator must to some extent detect which paths a distinguisher is trying to complete, and "adapt" the values along these paths such as to make the (simulated) Feistel network compatible with P. Concerning the latter, one can observe that a pair of missing consecutive queries is sufficient to adapt the two ends of a path to one another; thus if, say,

$$x_0, x_1, x_4, x_5, x_6, x_7, x_8, x_9$$

are values such that

$$F_i(x_i) \neq \perp$$

for $i \in \{1, 4, 5, 6, 7, 8\}$, and such that

$$x_{i+1} = x_{i-1} \oplus F_i(x_i)$$

for $i \in \{5, 6, 7, 8\}$, and such that

$$P(x_0, x_1) = (x_8, x_9)$$

and such that

$$F_2(x_2) = F_3(x_3) = \perp$$

where $x_2 := x_0 \oplus F_1(x_1)$, $x_3 := F_4(x_4) \oplus x_5$, then by making the assignments

$$F_2(x_2) \leftarrow x_1 \oplus x_3 \tag{2}$$
$$F_3(x_3) \leftarrow x_2 \oplus x_4 \tag{3}$$

the simulator turns $x_1, \ldots, x_8$ into a completed path that is compatible with P. In such a case, we say that the simulator *adapts a path*. The values $F_2(x_2)$ and $F_3(x_3)$ are also said to be *adapted*.

In general, however, if the simulator always waits until the last minute (e.g., until only two adjacent undefined queries are left) before adapting a path,

it can become caught in an over-constrained situation whereby several different paths request different adapted values for the same table entry. Hence, it is
usual for simulators to give themselves a "safety margin" and to pre-emptively
complete paths some time in advance. When pre-emptively completing a path,
typical simulators sample all but two values along the path randomly, while
"adapting" the last two values as above.

It should be emphasized that our simulator, like previous simulators
[9, 20, 31], makes no distinction between a non-null value $F_i(x_i)$ that is non-null
because the distinguisher has made the query $F(i, x_i)$ or that is non-null because
the simulator has set the value $F_i(x_i)$ during a pre-emptive path completion.
(Such a distinction seems tricky to leverage, particularly since the distinguisher
can know a value $F_i(x_i)$ without making the query $F(i, x_i)$, simply by knowing adjacent values and by knowing how the simulator operates.) Moreover, the
simulator routinely calls its own interface

$$F(\cdot, \cdot)$$

during the process of path completion, and it should be noted that our simulator,
again like previous simulators, makes no difference between distinguisher calls
to F and its own calls to F.

One of the basic dilemmas, then, is to decide at which point it is worth it
to complete a path; if the simulator waits too long, it is prone to finding itself
in an over-constrained situation; if it is too trigger-happy, on the other hand, it
runs the danger of creating out-of-control chain reactions of path completions,
whereby the process of completing a path sets off another path, and so on. We
refer to the latter problem (that is, avoiding out-of-control chain reactions) as
the problem of *simulator termination*.

SEURIN'S 10-ROUND SIMULATOR. Our 8-round simulator is based on "tweaking"
a previous 10-round simulator of ours [11] which is itself based on Seurin's (flawed)
10-round simulator [31]. Unfortunately (and after some failed efforts of ours to find
shortcuts) it seems that the best way to understand our 8-round simulator is to
start back with Seurin's 10-round simulator, followed by the modifications of [11]
and by the "tweaks" that bring the network down to 8 rounds.

In a nutshell, Seurin's simulator completes a path for *every* pair of values
(x_5, x_6) such that $F_5(x_5)$ and $F_6(x_6)$ are defined, as well as for every 4-tuple of
values

$$x_1, \ x_2, \ x_9, \ x_{10}$$

such that

$$F_1(x_1), \ F_2(x_2), \ F_9(x_9), \ F_{10}(x_{10})$$

are all defined, and such that

$$P(x_0, x_1) = (x_{10}, x_{11})$$

where $x_0 := F_1(x_1) \oplus x_2$, $x_{11} := x_9 \oplus F_{10}(x_{10})$. By virtue of this, rounds 5 and
6 are called the *middle detect zone* of the simulator, while rounds 1, 2, 9, 10

are called the *outer detect zone*. (Thus whenever a detect zone "fills up" with matching queries, a path is completed.) Paths are adapted either at positions 3, 4 or else at positions 7, 8, as depicted in Fig. 1.

In a little more detail, a function call $F(5, x_5)$ for which $F_5(x_5) = \perp$ triggers a path completion for every value x_6 such that $F_6(x_6) \neq \perp$; such paths are adapted at positions 3 and 4. Symmetrically, a function call $F(6, x_6)$ for which $F_6(x_6) = \perp$ triggers a path completion for every value x_5 such that $F_5(x_5) \neq \perp$; such paths are adapted at positions 7 and 8. For the outer detect zone, a call $F(2, x_2)$ such that $F_2(x_2) = \perp$ triggers a path completion for every tuple of values x_1, x_9, x_{10} such that $F_1(x_1)$, $F_9(x_9)$ and $F_{10}(x_{10})$ are defined, and such that the constraints listed above are satisfied (verifying these constraints thus requires a call to P or P^{-1}); such paths are adapted at positions 3, 4. Paths that are symmetrically triggered by a query $F(9, x_9)$ are adapted at positions 7, 8. Function calls to $F(2, \cdot)$, $F(5, \cdot)$, $F(6, \cdot)$ and $F(9, \cdot)$ are the only ones to trigger path completions. (Indeed, one can easily convince oneself that sampling a new value $F_1(x_1)$ or $F_{10}(x_{10})$ can only trigger the outer detect zone with negligible probability; hence, this possibility is entirely ignored by the simulator.) To summarize, in all cases the completed path is adapted at positions that are immediately *next to* the query that triggers the path completion.

To more precisely visualize the process of path completion, imagine that a query

$$F(2, x_2)$$

has just triggered the second type of path completion, for some corresponding values x_1, x_9 and x_{10}; then Seurin's simulator (which would immediately lazy sample the value $F_2(x_2)$ even before checking if this query triggers any path completions) would (a) make the queries

$$F(8, x_8), \dots, F(6, x_6), F(5, x_5)$$

to itself in that order, where $x_{i-1} := F_i(x_i) \oplus x_{i+1} = F(i, x_i) \oplus x_{i+1}$ for $i = 9, \dots, 6$, and (b) adapt the values $F_3(x_3)$, $F_4(x_4)$ as in (2), (3) where $x_3 := x_1 \oplus F_2(x_2)$, $x_4 := F_5(x_5) \oplus x_6$. In general, some subset of the table entries

$$F_8(x_8), \dots, F_5(x_5)$$

(and more exactly, a prefix of this sequence) may be defined even before the queries $F(8, x_8)$, $\dots$, $F(5, x_5)$ are made. The crucial fact to argue, however, is that $F_3(x_3) = F_4(x_4) = \perp$ right before these table entries are adapted.

Extending this example a little, say moreover that $F_6(x_6) = \perp$ at the moment when the above-mentioned query

$$F(6, x_6)$$

is made. This will trigger another path completion for every value x_5^* such that $F_5(x_5^*) \neq \perp$ at the moment when the query $F(6, x_6)$ occurs. Analogously, such a path completion would proceed by making (possibly redundant) queries

$$F(4, x_4^*), \dots, F(1, x_1^*), F(10, x_{10}^*), F(9, x_9^*)$$

for values $x_4^*, \ldots, x_1^*, x_0^*, x_{11}^*, x_{10}^*, x_9^*$ that are computed in the obvious way (with a query to P to go from (x_0^*, x_1^*) to (x_{10}^*, x_{11}^*), where $x_0^* := F_1(x_1^*) \oplus x_2^*$), before adapting the path at positions 7, 8. The crucial fact to argue would again be that $F_7(x_7^*) = F_8(x_8^*) = \perp$ when the time comes to adapt these table values, where $x_8^* := F_9(x_9^*) \oplus x_{10}^*$, $x_7^* := x_5^* \oplus F_6(x_6)$.

In Seurin's simulator, moreover, paths are completed on a first-come-first-serve (or FIFO[2]) basis: while paths are "detected" immediately when the query that triggers the path completion is made, this information is shelved for later, and the actual path completion only occurs after all previously detected paths have been completed. In our example, for instance, the path triggered by the query $F(2, x_2)$ would be adapted before the path triggered by the query $F(6, x_6)$. The imbroglio of semi-completed paths is rather difficult to keep track of, however, and indeed Seurin's simulator was later found to suffer from a real "bug" related to the simultaneous completion of multiple paths [20,32].

MODIFICATIONS OF[11]. For the following discussion, we will say that x_2, x_5 constitute the *endpoints* of a path that is adapted at positions 3, 4; likewise, x_6, x_9 constitute the *endpoints* of a path that is adapted at positions 7, 8. Hence, the endpoints of a path are the two values that flank the adapt zone. We say that an endpoint x_i is *unsampled* if $F_i(x_i) = \perp$ and *sampled* otherwise. Succinctly, the philosophy espoused in [11] is to not sample the endpoints of a path until right before the path is about to be adapted or, even more succinctly, "to sample randomness at the moment it is needed". This essentially results in two main differences with Seurin's simulator, which are (i) changing the order in which paths are completed and (ii) doing "batch adaptations" of paths, i.e., adapting several paths at once, for paths that happen to share endpoints.

To illustrate the first point, return to the above example of a query

$$F(2, x_2)$$

that triggers a path completion of the second type with respect to some values x_1, x_9, x_{10}. Then by definition

$$F_2(x_2) = \perp$$

at the moment when the call $F(2, x_2)$ is made. Instead of immediately sampling $F_2(x_2)$, as in the original simulator, this value is kept "pending" (the technical term is "pending query") until it comes time to adapt the path. Moreover, and keeping the notations from the previous example, note that the query

$$F(6, x_6)$$

will not result in $F_6(x_6)$ being immediately lazy sampled either (assuming, that is, $F_6(x_6) = \perp$) as long as there is at least one value x_5^* such that $F_5(x_5^*) \neq \perp$, since in such a case x_6 is the endpoint of a path-to-be-completed (namely, the path which we notated as $x_1^*, \ldots, x_5^*, x_6, x_7^*, \ldots, x_{10}^*$ above), and, according to the

---

[2] FIFO: First-In-First-Out. LIFO: Last-In-First-Out.

new policy, this endpoint must be kept unsampled until that path is adapted. In particular, the value $x_5 = F_6(x_6) \oplus x_7$ from the "original" path *cannot be computed* until the "secondary" path containing x_5^* and x_6 has been completed (or even more: until *all* secondary paths triggered by the query F $(6, x_6)$ have been completed). In other words, the query $F(6, x_6)$ "holds up" the completion of the first path. In practical terms, paths that are detected during the completion of another path take precedence over the original path, so that path completion becomes a LIFO process.

Implicitly, the requirement that both endpoints of a path *remain* unsampled until further notice means that both endpoints are *initially* unsampled. For the "starting" endpoint of the path (i.e., where the path is detected) this is obvious, since the path cannot be triggered otherwise, while for the "far" endpoint of the path one can argue that it holds with high probability.

As for "batch adaptations" the intuitive idea is that paths that share unsampled endpoints must be adapted (and in particular have their endpoints lazy sampled) simultaneously. In this event, the group of paths that are collectively sampled[3] and adapted will be an equivalence class in the transitive closure of the relation "shares an endpoint with". Note that paths adapted at 3, 4 can only share their endpoints[4] with other paths adapted at 3, 4, while paths adapted at 7, 8 can only share their endpoints with other paths adapted at 7, 8. Hence the paths in such an equivalence class will, in particular, all have the same adapt zone. Moreover, the batch adaptation of such a group of paths cannot happen at any point in time, but must happen when the group of paths is "stable": none of the endpoints of the paths in the group should currently be a trigger for a path completion that has not yet been detected, or that has started to complete but that has not yet reached its far endpoint. It so turns out, moreover, that the topological structure of such an equivalence class (with endpoints as nodes and paths as edges) will be a tree with all but negligible probability, simplifying many aspects of the simulator and of the proof.

While this describes the (simple) high-level idea of batch adaptations, the implementation details are more tedious. In fact, at this point it is useful to focus on these details.

FURTHER DETAILS: PENDING QUERIES, TREES, ETC. Keeping with the 10-round simulator of [11], if a query $F(i, x_i)$ occurs with $F_i(x_i) = \bot$ and $i \in \{2, 5, 6, 9\}$ the simulator creates a so-called *pending query* at that position, and for that value of x_i. (Strictly speaking, the pending query is the pair (i, x_i).) One can think of a pending query as a kind of "beacon" that periodically[5] checks for new paths to trigger, as per the rules of Fig. 1. E.g., a pending query

$$(2, x_2)$$

---

[3] In this context we use the verb "sampled" as a euphemism for "have their endpoints sampled".

[4] Recall that the endpoints of a path with adapt zone 3, 4 are x_2 and x_5, and that the endpoints of a path with adapt zone 7, 8 are x_6 and x_9.

[5] The simulator is not multi-threaded, but this metaphor is still helpful.

will trigger a new path to complete for any tuple of values x_1, x_9, x_{10} such that (same old!)

$$F_1(x_1) \neq \bot, F_9(x_9) \neq \bot, F_{10}(x_{10}) \neq \bot$$

and such that

$$P(x_0, x_1) = (x_{10}, x_{11})$$

where $x_0 := F_1(x_1) \oplus x_2$, $x_{11} := x_9 \oplus F_{10}(x_{10})$. The tuple of queries x_1, x_9, x_{10} is also called a *trigger* for the pending query $(2, x_2)$. For a pending query $(9, x_9)$, a trigger is a tuple x_1, x_2, x_{10} subject to the symmetric constraints. For a pending query $(5, x_5)$, a trigger is any value x_6 such that $F_6(x_6) \neq \bot$, and likewise any value x_5 such that $F_5(x_5) \neq \bot$ is a trigger for any pending query $(6, x_6)$. We note that a pending query *triggers* a path when there exists a *trigger* for the pending query. Hence there is the word "trigger" has two slightly different uses (as a noun and as a verb).

We differentiate the endpoints of a path according to which one triggered the path: the pending query that triggered the path is called the *origin* of the path, while the other endpoint (if and when present) is the *terminal* of the path.

While pending queries are automatically created each time a function call $F(i, x_i)$ occurs with $F_i(x_i) = \bot$ and with $i \in \{2, 5, 6, 9\}$, the simulator also has a separate mechanism[6] at its disposal for directly creating pending queries without calling $F(\cdot, \cdot)$ by this mechanism. In particular, whenever the simulator reaches the terminal of a path, the simulator turns the terminal into a pending query.

In short: (i) all path endpoints are pending queries, so long as the path has not been sampled and adapted; (ii) pending queries keep triggering paths as long as there are paths to trigger.

For the following, we will use the following extra terminology from [11]:

- A path is *ready* when it has been extended to the terminal, and the terminal has been made pending.
- A ready path with endpoints 2, 5 is called a "$(2, 5)$-path", and a ready path with endpoints 6, 9 is called a "$(6, 9)$-path".
- Two ready paths are *neighbors* if they share an endpoint; let a *neighborhood* be an equivalence class of ready paths under the transitive closure of the neighbor relation. We note that a neighborhood consists either of all $(2, 5)$-paths or consists all of $(6, 9)$-paths.
- A pending query is *stable* if it has no "new" triggers (that is, no triggers for which the simulator hasn't already started to complete a path), and if paths already triggered by the pending query are ready.
- A neighborhood is *stable* if all the endpoints of all the paths that it contains are stable.

A neighborhood can be visualized as a graph with a node for each endpoint and an edge for each ready path. As mentioned above, these neighborhoods actually turn out to be trees with high probability. (The simulator aborts otherwise.)

---

[6] This might sound a bit ad-hoc right now, but it actually corresponds to the most natural way of programming the simulator, as will become clearer in the technical simulator overview.

We will thus speak of a $(2,5)$-*tree* for a neighborhood consisting of $(2,5)$-paths and of a $(6,9)$-*tree* for a neighborhood consisting of $(6,9)$-paths. Moreover, the simulator uses an actual tree *data structure* to keep track of each (i,j)-tree under completion, thus adding further structure to the simulation process.

To summarize, when a query $F(i, x_i)$ triggers a path completion, the simulator starts growing a tree that is "rooted" at the pending query (i, x_i); for other endpoints of paths in this tree (i.e., besides (i, x_i)), the simulator "plants" a pending query at that endpoint without making a call to $F(\cdot, \cdot)$, which pending query tests for further paths to complete, and which may thus cause the tree to grow even larger, etc. If and when the tree becomes stable, the simulator samples all endpoints of all paths in the tree, and adapts all these paths.[7]

The growth of a $(2,5)$-tree may at any moment be interrupted by the apparition of a new $(6,9)$-tree (specifically, when a query to $F(6, \cdot)$ or $F(9, \cdot)$ triggers a new path completion), in which case the $(2,5)$-tree is put "on hold" while the $(6,9)$-tree is grown, sampled and adapted; vice-versa, a $(6,9)$-tree may be interrupted by the apparition of a new $(2,5)$-tree. In this fashion, a "stack of trees" that alternates between $(2,5)$- and $(6,9)$-trees is created. Any tree that is not the last tree on the stack contains a non-ready path (the one, that is, that was interrupted by the next tree on the stack), and so is not stable. For this reason, in fact, the only tree that can become stable at a given moment is the last tree on the stack.

We also note that in certain cases (and more specifically for pending queries at positions 5 and 6), trees higher up in the stack can affect the stability of nodes of trees lower down in the stack: a node that used to be stable loses its stability after a higher-up tree has been created, sampled and adapted. Hence, the simulator always re-checks all nodes of a tree "one last time" before deeming a tree stable, after a tree stops growing—and such a check will typically, indeed,

---

[7] In more detail, when a tree becomes stable the simulator lazy samples

$$F_i(x_i)$$

for every endpoint (a.k.a., pending query) in the tree. Then if the tree is, say, a $(2,5)$-tree, the simulator can compute the values

$$x_3 := x_1 \oplus F_2(x_2)$$
$$x_4 := F_5(x_5) \oplus x_6$$

and set

$$F_3(x_3) := x_2 \oplus x_4$$
$$F_4(x_4) := x_3 \oplus x_5$$

for each path in the tree. If two paths "collide" by having the same value of x_3 or x_4 the simulator aborts. Likewise the simulator aborts if either $F_3(x_3) \neq \perp$ or $F_4(x_4) \neq \perp$ for some path, before adapting those values. We call this two-step process "sampling and adapting" the $(2,5)$-tree. The process of sampling and adapting a $(6,9)$-tree is analogous.

uncover new paths to complete that weren't there before. Moreover, because the factor that determines when these new paths will be adapted is the timestamp of the *pending query* to which they are attached, rather than the timestamp of the *actual last query* that completed a trigger for this pending query, it is a matter of semantic debate whether the simulator of [11] is really "LIFO" or not. (But conceptually at least, it seems safe to think of the simulator as LIFO.)

STRUCTURAL VS. CONCEPTUAL CHANGES. Of the main changes introduced in [11] to Seurin's simulator, one can note that "batch adaptations" are in some sense a conceptual convenience. Indeed, one way or another every non-null value

$$F_j(x_j)$$

for $j \notin \{3, 4, 7, 8\}$ ends up being randomly and independently sampled in their simulator, as well as in Seurin's; so one might as well load a random value into $F_j(x_j)$ as soon as the query $F(j, x_j)$ is made, as in Seurin's original simulator, as long as we take care to keep on completing paths in the correct order. While correct, this approach is conceptually less convenient, because the "freshness" of the random value $F_j(x_j)$ is harder to argue when that randomness is needed (e.g., to argue that adapted queries do not collide, etc.). In fact, our 10-round simulator is an interesting case where the search for a syntactically convenient usage of randomness naturally leads to structural changes that turn out to be critical for correctness.

One should note that the idea of batch adaptations already appears explicitly in the simulator of [14], which, indeed, formed part of the inspiration for [11]. In [14], however, batch adaptations are purely made for conceptual convenience.

Readers seeking more concrete insights can also consult Seurin's attack against his own 10-round simulator [32] and check this attack fails under the LIFO path completion just outlined.

THE 8-ROUND SIMULATOR. In the 10-round simulator, the outer detect zone is in some sense unnecessarily large: for any set of four matching queries that complete the outer detect zone, the simulator can "see" the presence of matching queries already by the third query.

To wit, say the distinguisher chooses random values x_0, x_1, makes the query

$$(x_{10}, x_{11}) \leftarrow P(x_0, x_1)$$

to P, then queries $F(1, x_1)$ and $F(10, x_{10})$. At this point, even if the simulator knows that the values x_1 and x_{10} are related by some query to P, the simulator has no hope of finding *which* query to P, because there are exponentially many possibilities to try for x_0 and/or x_{11}. However, as soon as the distinguisher makes either of the queries

$$F(2, x_2) \qquad \text{or} \qquad F(9, x_9)$$

where $x_2 := x_0 \oplus F(1, x_1)$, $x_9 := F(10, x_{10}) \oplus x_{11}$, then the simulator has enough information to draw a connection between the queries being made at the left- and right-hand sides of the network. (E.g., if the query $F(2, x_2)$ is made, the simulator can compute x_0 from $F_1(x_1)$ and x_2, can call $P(x_0, x_1)$, and recognize, in P's output, the value x_{10} for which it has already answered a query.) More generally, anytime the distinguisher makes three-out-of-four matching queries in the 10-round outer detect zone, the simulator has enough information to reverse-engineer the relevant query to P/P^{-1} and, thus, to see a connection between the queries being made at either side of the network.

This observation (which is also made by Dachman-Soled et al. [10], though our work is independent of theirs) motivates the division of the 4-round outer detect zone into two separate outer detect zones of three (consecutive) rounds each. In the eight-round simulator, then, these two three-round outer detect zones are made up of rounds 1, 2, 8 and rounds 1, 7, 8, respectively. Both of these detect zones detect "at the edges" of the detect zone. I.e., the 1, 7, 8 detect zone might trigger a path completion through queries to $F(7, \cdot)$ and $F(1, \cdot)$, whereas the 1, 2, 8 detect zone might trigger a path completion through queries to $F(2, \cdot)$ or to $F(8, \cdot)$. (Once again the possibility of "completing" a detect zone by a query at the middle of the detect zone is ignored because this event has negligible chance of occuring.)

E.g., a query

$$F(7, x_7)$$

such that $F_7(x_7) = \perp$ and for which there exists values x_0, x_1, x_8 such that $F_8(x_8) \neq \perp$, $F_1(x_1) \neq \perp$, and such that $P^{-1}(x_8, x_9) = (x_0, x_1)$ where $x_9 = x_7 \oplus F_8(x_8)$ would trigger the 1, 7, 8 detect zone, and produce a path completion. Similarly, a query

$$F(1, x_1)$$

such that $F_1(x_1) = \perp$ and for which there exists values x_0, x_7, x_8 such that $F_7(x_7) \neq \perp$, $F_8(x_8) \neq \perp$, and such that $P^{-1}(x_8, x_9) = (x_0, x_1)$ where $x_9 = x_7 \oplus F_8(x_8)$ would trigger the 1, 7, 8 detect zone as well.

When a path is detected at position 1 or at position 8, we can respectively adapt the path at positions 2, 3 or at positions 6, 7—i.e., we adapt the path in an adapt zone that is immediately adjacent to the position that triggered the path completion, as in the[8] 10-round simulator. However, for paths detected at positions 2 and 7, the same adapt zones cannot be used, and we find it more convenient to adapt the path at rounds 4, 5, as depicted in the bottom left quadrant of Fig. 1.

To keep the proof manageable, however, one of the imperatives is that an "adapted" query should not trigger a new path completion. If we kept the middle detect zone as rounds 4, 5 only (by analogy with the 10-round simulator, where the middle detect zone consists of rounds 5 and 6), then the queries that we adapt at rounds 4, 5 would trigger new path completions of themselves—a mess! However, this problem can be avoided by splitting the middle detect zone into two *enlarged* middle detect zones of three rounds each: one middle detect zone consisting of rounds 3, 4, 5 and one consisting of rounds 4, 5, 6. As before, each of these zones detects "at the edges". After this change, bad dreams are

---

[8] Henceforth, "the" 10-round simulator refers to the simulator of [11].

dissipated, and the 8-round simulator recovers essentially the same functioning as the 10-round simulator. The sum total of detect and adapt zones, including which adapt zone is used for paths detected at which point, is shown in Fig. 2.

The 8-round simulator utilizes the same "pending query" mechanism as the 10-round simulator. In particular, now, each query

$$F(j, x_j)$$

with $F_j(x_j) = \perp$ creates a new pending query (j, x_j), because paths are now detected at all positions, and each pending query will detect for paths as depicted[9] in Fig. 2, with there being exactly one type of "trigger" for each position j. A path triggered by a pending query is first extended to a designated terminal (the "other" endpoint of the path), the position of which is a function of the pending query that triggered the path (this position is shortly to be discussed), which becomes a new pending query of its own, etc. As in the 10-round simulator, the simulator turns the terminal into a pending query without making a call to $F(\cdot, \cdot)$.

For the 10-round simulator, we recall that the possible endpoint positions of a path are 2, 5 and 6, 9. The 8-round simulator has more variety, as the endpoints of a path do not always directly flank the adapt zone for that path. Specifically:

- paths detected at positions 1 and 4, as in the top left quadrant of Fig. 2, have endpoints 1, 4; before such paths are adapted, they include only the values x_1, x_4, x_5, x_6, x_7, x_8
- paths detected at positions 3 and 6, as in the top right quadrant of Fig. 2, have endpoints 3, 6; before such paths are adapted, they include only the values x_3, x_4, x_5, x_6
- paths detected at positions 2 and 7, as in the bottom left quadrant of Fig. 2, have endpoints 2, 7; before such paths are adapted, they include only the values x_1, x_2, x_7, x_8
- paths detected at positions 5 and 8, as in the bottom right quadrant of Fig. 2, have endpoints 5, 8; before such paths are adapted, they include only the values x_1, x_2, x_3, x_4, x_5, x_8

Hence, paths with endpoints 1, 4 or 5, 8 are familiar from the 10-round simulator. (Being the analogues, respectively, of paths with endpoints 2, 5 or 6, 9.) On the other hand, paths with endpoints 3, 6 or 2, 7 are shorter, containing only four values before adaptation takes place. As in the 10-round simulator, we speak of an "(i, j)-path" for paths with endpoints i, j. We also say that a path is *ready* once it has reached both its endpoints and these have been turned into pending queries, and that two ready paths are *neighbors* if they share an endpoint.

---

[9] To solidify things with some examples, a "trigger" for a pending query $(5, x_5)$ is a pair values of x_3, x_4 such that $F_3(x_3) \neq \perp$, $F_4(x_4) \neq \perp$ and such that $x_3 \oplus F_4(x_4) = x_5$, corresponding to the rightmost, bottommost diagram of Fig. 2; a "trigger" for a pending query $(1, x_1)$ is pair of values x_7, x_8 such that $F_7(x_7) \neq \perp$, $F_8(x_8) \neq \perp$, and such that $P^{-1}(x_8, x_9) = (*, x_1)$ where $x_9 := x_7 \oplus F_8(x_8)$, corresponding to the leftmost, topmost diagram of Fig. 2. Etc.

Since, by virtue of the endpoint positions, a $(1,4)$-path can only share an endpoint with a $(1,4)$-path, a $(2,7)$-path can only share an endpoint with a $(2,7)$-path, a $(3,6)$-path can only share an endpoint with $(3,6)$-path, and a $(5,8)$-path can only share an endpoint with a $(5,8)$-path, neighborhoods (which are the transitive closure of the neighbor relation) are always comprised of the same kind of (i,j)-path. As in the 10-round simulator, these neighborhoods are actually topological trees, giving rise, thus, to "$(1,4)$-trees", "$(2,7)$-trees", "$(3,6)$-trees" and "$(5,8)$-trees". Given this, the 8-round simulator functions entirely analogously to the 10-round simulator, only with more different types of paths and of trees (which does not make an important difference) and with a slightly modified mechanism for adapting $(2,7)$- and $(3,6)$-trees, which are the trees for which the path endpoints are not directly adjacent to the adapt zone (which does not make an important difference either).

Concerning the latter point, when a $(2,7)$- or $(3,6)$-tree is adapted, some additional queries have to be lazy sampled for each path before reaching the adapt zone. (In the case of a $(3,6)$-tree, each path even requires a query to P^{-1}.) But because the endpoints of each path are lazy sampled as the first step of the batch adaptation process, there is negligible chance that these extra queries will trigger a new path completion. So for those queries the 8-round simulator directly lazy samples the tables F_i without even calling its own $F(\cdot, \cdot)$ interface.

As a small piece of trivia (since it doesn't really matter to the simulator), one can check, for instance, that a $(1,4)$-tree may be followed either by a $(2,7)$-, $(3,6)$-, or a $(5,8)$-tree on the stack—i.e., while making a $(1,4)$-path ready, we may trigger any of the other three types of paths—and symmetrically the growth of a $(5,8)$-tree may be interrupted by any of the three other types of trees. On the other hand, $(2,7)$-trees and $(3,6)$-trees have shorter paths, and in fact when such trees are grown *no* queries to $F(\cdot, \cdot)$ are made, which means that such trees never see their growth interrupted by anything. In other words, a $(3,6)$- or $(2,7)$-tree will only appear as the last tree in the tree stack, if at all.

Overall, it is imperative that pending queries be kept *unsampled* until the relevant tree becomes stable, and is adapted. In particular, the simulator must not overwrite the pending queries of trees lower down in the tree stack while working on the current tree.

In fact, and like [11], our simulator *cannot* overwrite pending queries because it keeps a list of all pending queries, and aborts rather than overwrite a pending query. Nonetheless, one must show that the chance of such an event is negligible. The analysis of this bad event is lengthy but also straightforward. Briefly, this bad event can only occur if ready and non-ready paths arrange to form a certain type of cycle, and the occurence of such cycles can be reduced to the occurence of a few different "local" bad events whose (negligible) probabilities are easily bounded.

THE TERMINATION ARGUMENT. The basic idea of Coron et al.'s [9] termination argument (which only needs to be lightly adapted for our use) is that each path detected in one of the outer detect zones is associated with high probability to a P-query previously made by the distinguisher. Since the distinguisher only has q

queries total, this already implies that the number of path completions triggered by the outer detect zones is at most q with high probability.

Secondly, whenever a path is triggered by one of the middle detect zones, this path completion does not add any new entries to the tables F_4, F_5. Hence, only two mechanisms add entries to the tables F_4 and F_5: queries directly made by the distinguisher and path completions triggered by the outer detect zones. Each of these accounts for at most q table entries in each of F_4, F_5, so that the tables F_4, F_5 do not exceed size $2q$. But *every* completed path corresponds to a *unique* pair of entries in F_4, F_5. (I.e., no two completed paths have the same x_4 *and* the same x_5.) So the total number of paths ever completed is at most $(2q)^2 = 4q^2$.

FURTHER DETAILS. A more technical description of the simulator and the pseudocode are given in the full version of this paper [12].

4 Proof Overview

In this section we give an overview of the proof for Theorem 1, using the simulator described in Sect. 3 as the indifferentiability simulator. Details of the proof are given in the full version [12].

In order to prove that our simulator successfully achieves indifferentiability as defined by Definition 1, we need to upper bound the advantage of any distinguisher, as well as the time and query complexity of the simulator; the latter is related to the *termination argument* for our simulator, already sketched at the end of the last section.

GAME SEQUENCE. Our proof uses a sequence of five games, G_1, ..., G_5, with G_1 being the simulated world and G_5 being the real world. Every game offers the same interface to the distinguisher, consisting of functions F, P and P^{-1}.

In the simulated world G_1, P and P^{-1} are answered by an oracle according to a random permutation and its inverse; the simulator, as described in Sect. 3, is in charge of answering distinguisher queries to F.

The randomness used in the experiment is read from *explicit random tapes*, similar to [20]. In particular, the random permutation is encoded by a tape $p : \{0,1\}^{2n} \to \{0,1\}^{2n}$ (whose inverse is accessible via p^{-1}). The simulator has access to 8 tapes f_1, ..., f_8, which are independent uniform random mappings from $\{0,1\}^n$ to $\{0,1\}^n$; when randomly sampling the value of a query (i, x_i), the simulator reads the value of $f_i(x_i)$ and sets $F_i(x_i) \leftarrow f_i(x_i)$. Since each query is sampled at most once, the tape entry hasn't been read before and its value is uniformly and independently distributed in $\{0,1\}^n$. Note that each f_i encodes a random function, and f_i will be used as the i-th round function in the real world G_5.

A brief synopsis of the changes that occur in the games is as follows:

In G_2: The simulator triggers a path at the outer detect zones only if the distinguisher has issued the permutation query in the path.

Recall that the outer detect zones consist of rounds 1, 7, 8 or of rounds 1, 2, 8; to check whether three queries in an outer detect zone are in the same path, the simulator has to call $P(x_0, x_1)$ or $P^{-1}(x_8, x_9)$ in G_1. In G_2, instead of calling $P^{(-1)}$, the simulator performs a "peek" operation that accesses the query history of the permutation oracle[10]; the path is triggered only if the permutation query is in the history and the three queries are in the same path. Then, the simulator queries P or P^{-1} only when completing a triggered path; therefore, if a permutation query is issued by the simulator, the path containing the permutation query must have been completed (and cannot be triggered again).

Although the change may result in "false negatives" when detecting triggered paths, such false negatives remain unlikely as long as the simulator is efficient.

In G_3: The simulator adds a number of checks that may cause it to abort in places where it did not abort in G_2. Some of these checks cannot be included in G_1 because they also (like the modifications in G_2) involve the simulator "peeking" at the distinguisher's queries to the permutation oracle.

The checks in G_3 are added to catch "bad events" at the earliest possible stage, i.e., at the moment after the relevant randomness has been sampled. If these checks pass, we can show that the simulator will not abort at further key points down the line, such as by attempting to overwrite an existing entry $F_i(x_i)$ or by attempting to overwrite a pending query. (I.e., the checks in G_3 are sufficient conditions for the execution to maintain a "good" structure.)

In G_4: The most important transition occurs in this game, as the oracles P, P^{-1} no longer rely on the random permutation tape $p : \{0,1\}^{2n} \to \{0,1\}^{2n}$, but instead evaluate an 8-round Feistel network using the random tapes $f_1, \ldots, f_8$ (i.e., the same ones used by the simulator) as round functions. Apart from this change to P, P^{-1}, the simulator remains identical.

In G_5: This is the real world, meaning that $F(i, x)$ directly returns the value $f_i(x)$. In particular G_5 never aborts, unlike the previous four games.

Definition 2. The *advantage* of a distinguisher D at distinguishing games G_i and G_j is defined as

$$\Delta_D(G_i, G_j) = \Pr_{G_i}[D^{F,P,P^{-1}} = 1] - \Pr_{G_j}[D^{F,P,P^{-1}} = 1] \tag{4}$$

where the probabilities are taken over the coins of the relevant game as well as over D's coins, if any.

As G_5 never aborts, and as the distinguisher's job is to maximize

$$\Delta_D(G_1, G_5) = \Pr_{G_1}[D^{F,P,P^{-1}} = 1] - \Pr_{G_5}[D^{F,P,P^{-1}} = 1],$$

we can assume without loss of generality that D outputs 1 if the game aborts. In particular, since G_2 is identical to G_3 except for the possibility that G_3 may abort when G_2 does not, we then have

$$\Delta_D(G_2, G_5) \leq \Delta_D(G_3, G_5)$$

---

[10] This operation cannot be included in G_1 as the original simulator is not allowed to see the distinguisher's permutation queries.

so we can upper bound $\Delta_D(G_1, G_5)$ as

$$\begin{aligned} \Delta_D(G_1, G_5) &\leq \Delta_D(G_1, G_2) + \Delta_D(G_2, G_5) \\ &\leq \Delta_D(G_1, G_2) + \Delta_D(G_3, G_5) \\ &\leq \Delta_D(G_1, G_2) + \Delta_D(G_3, G_4) + \Delta_D(G_4, G_5) \end{aligned}$$

by the triangle inequality. Hence, the proof focuses on the individual transitions $G_1 \to G_2$, $G_3 \to G_4$ and $G_4 \to G_5$.

G_1-G_2 TRANSITION. The two games (with the same random tapes) are visibly different only if a path is triggered in G_1 but not in G_2, i.e., when a path triggered by the outer detect zones in G_1 contains a permutation query that hasn't been issued in G_2. In this case (and assuming that G_1 and G_2 are being "run" simultaneously on the same distinguisher D, with the same random tapes $f_1, \ldots, f_8 : \{0,1\}^n \to \{0,1\}^n$ and $p : \{0,1\}^{2n} \to \{0,1\}^{2n}$) we say that G_1 and G_2 *diverge*.

More precisely, one can show that divergence occurs if and only if the simulator makes a certain call of the form[11] "CheckP$^+$ (x_1, x_2, x_8)" such that (i) $p(x_0, x_1)$ is unread in G_2, where[12] $x_0 = F_1(x_1) \oplus x_2$, and (ii) the first n bits of the unread value $p(x_0, x_1)$ are equal to x_8. In fact, this notion can be defined with respect to the execution of G_2 *alone* (i.e., without examining the execution of G_1) which then makes it straightforward to examine. In more detail, the probability of divergence of occuring in G_2 is obtained by a simple union bound over all calls to the CheckP$^+$/CheckP$^-$ procedures, where the number of such calls is upper bounded thanks to the (previously established) simulator efficiency. (Indeed, the proof actually starts off by arguing various efficiency metrics in G_1 and G_2.)

G_3-G_4 TRANSITION. For this transition, a randomness mapping argument is used, as introduced by [20]. We also take advantage of some refinements introduced by [1,14]. Specifically, following [1], we use "footprints" and eschew the use of a "two-way random function"; and following [14], we additionally cancel the probabilities of abort in G_4 in separate transitions from G_3 to G_4 and from G_4 to G_5 in order to avoid double-counting these probabilities.

As usual, the randomness mapping argument consists of two steps: bounding the abort probability in G_3, and mapping the randomness of non-aborting executions of G_3 to the randomness of "matching" executions of G_4. Bounding the abort probability in G_3 is the more technically involved of the two steps.

A small novelty that we introduce also concerns the randomness mapping argument. Specifically, a randomness map needs to be defined with respect to a distinguisher D that "completes" all paths (that contain a permutation query issued by D). Making the assumption that D completes all paths is without loss of generality, but costs a multiplicative factor in the number of queries that is

---

[11] Or a call "CheckP$^-$ (x_1, x_7, x_8)" subject to symmetric conditions.

[12] The fact that CheckP$^+$ (x_1, x_2, x_8) is called in the first place implies that $F_1(x_1) \neq \perp$ in either game.

equal to the number of rounds—potentially annoying! However, we note that if D is allowed q queries to each of its $r + 1$ oracles (the permutation plus the r rounds functions), then the assumption that D completes all paths can be made at the cost of only doubling the number of D's queries. Moreover, there is no real cost in giving D the power to query each of its oracles q times, since most proofs effectively allow this anyway.

G_4-G_5 TRANSITION. A non-aborting execution of G_4 is identical to an execution of G_5 with the same random tape, so the advantage in distinguishing between these two games is upper bounded by the simulator's abort probability in G_4.

Acknowledgments. Yuanxi Dai was supported by the National Basic Research Program of China Grant 2011CBA00300, 2011CBA00301, the National Natural Science Foundation of China Grant 61033001, 61361136003. John Steinberger was funded by National Basic Research Program of China Grant 2011CBA00300, 2011CBA00301, the National Natural Science Foundation of China Grant 61373002, 61361136003, and by the China Ministry of Education grant number 20121088050.

References

1. Andreeva, E., Bogdanov, A., Dodis, Y., Mennink, B., Steinberger, J.P.: On the indifferentiability of key-alternating ciphers. In: Canetti, R., Garay, J.A. (eds.) CRYPTO 2013, Part I. LNCS, vol. 8042, pp. 531–550. Springer, Heidelberg (2013)
2. Bellare, M., Rogaway, P.: Random oracles are practical: a paradigm for designing efficient protocols. In: Proceedings of the 1st ACM Conference on Computer and Communications Security, pp. 62–73 (1993)
3. Bertoni, G., Daemen, J., Peeters, M., Van Assche, G.: On the indifferentiability of the sponge construction. In: Smart, N.P. (ed.) EUROCRYPT 2008. LNCS, vol. 4965, pp. 181–197. Springer, Heidelberg (2008)
4. Canetti, R.: Security and composition of multi-party cryptographic protocols. J. Cryptol. **13**(1), 143–202 (2000)
5. Canetti, R.: Universally composable security: a new paradigm for cryptographic protocols. In: Proceedings of 42nd IEEE Symposium on Foundations of Computer Science (FOCS), pp. 136–145 (2001)
6. Chen, S., Steinberger, J.: Tight security bounds for key-alternating ciphers. In: Nguyen, P.Q., Oswald, E. (eds.) EUROCRYPT 2014. LNCS, vol. 8441, pp. 327–350. Springer, Heidelberg (2014)
7. Coron, J.-S., Dodis, Y., Malinaud, C., Puniya, P.: Merkle-Damgård revisited: how to construct a hash function. In: Shoup, V. (ed.) CRYPTO 2005. LNCS, vol. 3621, pp. 430–448. Springer, Heidelberg (2005)
8. Coron, J.-S., Dodis, Y., Mandal, A., Seurin, Y.: A domain extender for the ideal cipher. In: Micciancio, D. (ed.) TCC 2010. LNCS, vol. 5978, pp. 273–289. Springer, Heidelberg (2010)
9. Coron, J.-S., Patarin, J., Seurin, Y.: The random oracle model and the ideal cipher model are equivalent. In: Wagner, D. (ed.) CRYPTO 2008. LNCS, vol. 5157, pp. 1–20. Springer, Heidelberg (2008)

10. Dachman-Soled, D., Katz, J., Thiruvengadam, A.: 10-round feistel is indifferentiable from an ideal cipher. In: Fischlin, M., Coron, J.-S. (eds.) EUROCRYPT 2016. LNCS, vol. 9666, pp. 649–678. Springer, Heidelberg (2016). doi:10.1007/978-3-662-49896-5_23

11. Dai, Y., Steinberger, J.: Indifferentiability of 10-round Feistel networks. IACR ePrint Archive, Technical Report 2015/874 (2015)

12. Dai, Y., Steinberger, J.: Indifferentiability of 8-round Feistel networks. IACR ePrint Archive, Technical Report 2015/1069 (2015)

13. Dodis, Y., Puniya, P.: On the relation between the ideal cipher and the random oracle models. In: Halevi, S., Rabin, T. (eds.) TCC 2006. LNCS, vol. 3876, pp. 184–206. Springer, Heidelberg (2006)

14. Dodis, Y., Liu, T., Stam, M., Steinberger, J.: On the indifferentiability of confusion-diffusion networks. IACR ePrint Archive, Technical Report 2015/680 (2015)

15. Dodis, Y., Reyzin, L., Rivest, R.L., Shen, E.: Indifferentiability of permutation-based compression functions and tree-based modes of operation, with applications to MD6. In: Dunkelman, O. (ed.) FSE 2009. LNCS, vol. 5665, pp. 104–121. Springer, Heidelberg (2009)

16. Dodis, Y., Ristenpart, T., Steinberger, J., Tessaro, S.: To hash or not to hash again? (In)Differentiability results for H^2 and HMAC. In: Canetti, R., Safavi-Naini, R. (eds.) CRYPTO 2012. LNCS, vol. 7417, pp. 348–366. Springer, Heidelberg (2012)

17. Feistel, H.: Cryptographic coding for data-bank privacy. IBM Technical report RC-2827, 18 March 1970

18. Feistel, H., Notz, W.A., Lynn Smith, J.: Some cryptographic techniques for machine-to-machine data communications. IEEE Proc. **63**(11), 1545–1554 (1975)

19. Fiat, A., Shamir, A.: How to prove yourself: practical solutions to identification and signature problems. In: Odlyzko, A.M. (ed.) CRYPTO 1986. LNCS, vol. 263, pp. 186–194. Springer, Heidelberg (1987)

20. Holenstein, T., Künzler, R., Tessaro, S.: The equivalence of the random oracle model and the ideal cipher model, revisited. In: Fortnow, L., Vadhan, S.P. (eds.) Proceedings of the 43rd ACM Symposium on Theory of Computing, STOC 2011, San Jose, CA, USA, pp. 89–98. ACM, 6–8 June 2011

21. Lampe, R., Seurin, Y.: How to construct an ideal cipher from a small set of public permutations. In: Sako, K., Sarkar, P. (eds.) ASIACRYPT 2013, Part I. LNCS, vol. 8269, pp. 444–463. Springer, Heidelberg (2013)

22. Luby, M., Rackoff, C.: How to construct pseudorandom permutations and pseudorandom functions. SIAM J. Comput. **17**(2), 373–386 (1988)

23. Maurer, U.M., Renner, R.S., Holenstein, C.: Indifferentiability, impossibility results on reductions, and applications to the random oracle methodology. In: Naor, M. (ed.) TCC 2004. LNCS, vol. 2951, pp. 21–39. Springer, Heidelberg (2004)

24. Naor, M., Reingold, O.: On the construction of pseudorandom permutations: Luby-Rackoff revisited. J. Cryptol. **12**(1), 29–66 (1999). Preliminary Version: STOC 1997

25. Patarin, J.: Security of balanced and unbalanced Feistel schemes with linear non equalities. IACR ePrint Arxiv, Technical Report 2010/293 (2010)

26. Pfitzmann, B., Waidner, M.: Composition and integrity preservation of secure reactive systems. In: 7th ACM Conference on Computer and Communications Security, pp. 245–254. ACM Press (2000)

27. Pfitzmann, B., Waidner, M.: A model for asynchronous reactive systems and its application to secure message transmission. Technical report 93350, IBM Research Division, Zürich (2000)

28. Pointcheval, D., Stern, J.: Security proofs for signature schemes. In: Maurer, U.M. (ed.) EUROCRYPT 1996. LNCS, vol. 1070, pp. 387–398. Springer, Heidelberg (1996)
29. Ristenpart, T., Shacham, H., Shrimpton, T.: Careful with composition: limitations of the indifferentiability framework. In: Paterson, K.G. (ed.) EUROCRYPT 2011. LNCS, vol. 6632, pp. 487–506. Springer, Heidelberg (2011)
30. Hoang, V.T., Rogaway, P.: On generalized Feistel networks. In: Rabin, T. (ed.) CRYPTO 2010. LNCS, vol. 6223, pp. 613–630. Springer, Heidelberg (2010)
31. Seurin, Y.: Primitives et protocoles cryptographiques à sécurité prouvée. Ph.D. thesis, Université de Versailles Saint-Quentin-en-Yvelines, France (2009)
32. Seurin, Y.: A note on the indifferentiability of the 10-round Feistel construction. http://yannickseurin.free.fr/pubs/Seurin_note_ten_rounds.pdf
33. Winternitz, R.: A secure one-way hash function built from DES. In: Proceedings of the IEEE Symposium on Information Security and Privacy, pp. 88–90. IEEE Press (1984)

EWCDM: An Efficient, Beyond-Birthday Secure, Nonce-Misuse Resistant MAC

Benoît Cogliati[1] and Yannick Seurin[2(✉)]

[1] University of Versailles, Versailles, France
benoitcogliati@hotmail.fr
[2] ANSSI, Paris, France
yannick.seurin@m4x.org

Abstract. We propose a nonce-based MAC construction called EWCDM (*Encrypted Wegman-Carter with Davies-Meyer*), based on an almost xor-universal hash function and a block cipher, with the following properties: (i) it is simple and efficient, requiring only two calls to the block cipher, one of which can be carried out in parallel to the hash function computation; (ii) it is provably secure beyond the birthday bound when nonces are not reused; (iii) it provably retains security up to the birthday bound in case of nonce misuse. Our construction is a simple modification of the Encrypted Wegman-Carter construction, which is known to achieve only (i) and (iii) when based on a block cipher. Underlying our new construction is a new PRP-to-PRF conversion method coined *Encrypted Davies-Meyer*, which turns a pair of secret random permutations into a function which is provably indistinguishable from a perfectly random function up to at least $2^{2n/3}$ queries, where n is the bit-length of the domain of the permutations.

Keywords: Wegman-Carter MAC · Davies-Meyer construction · Nonce-misuse resistance · Beyond-birthday-bound security

1 Introduction

WEGMAN-CARTER MACs. A *Message Authentication Code* (MAC) is a fundamental symmetric-key primitive that allows a sender to authenticate messages by computing tags that can be verified by the receiver (the sender and the receiver sharing a common secret key). Many MACs are based on some underlying cryptographic primitive such as a block cipher (e.g., CBC-MAC [BKR00]) or a hash function (e.g., HMAC [BCK96]). A different approach, pioneered by Wegman and Carter [WC81] (building on earlier work by Gilbert *et al.* [GMS74]), first treats the message M with an almost xor-universal (AXU) hash function[1] H (i.e., a fast, *combinatorial* primitive rather than a slow, *cryptographic* one) and masks the result with a one-time pad, resulting in *information-theoretically*

---

[1] An AXU hash function is a keyed function with the property that for any two distinct inputs, the probability over the draw of a random key that the outputs have a specific difference is small.

© International Association for Cryptologic Research 2016
M. Robshaw and J. Katz (Eds.): CRYPTO 2016, Part I, LNCS 9814, pp. 121–149, 2016.
DOI: 10.1007/978-3-662-53018-4_5

secure authentication. Since sharing a one-time pad for each message to authenticate is not very practical, one can instead use a pseudorandom function F, as first proposed by Brassard [Bra82], allowing the sender and the receiver to share a short secret K rather than a long list of one-time pads. The mask for each new message is then generated pseudorandomly by applying F_K to a *nonce* N, a value used at most once. This reintroduces a cryptographic primitive (and hence a computational assumption), but only for treating a small nonce rather than a potentially long message. The resulting nonce-based MAC, that we simply call the *Wegman-Carter* (WC) construction, is

$$\mathsf{WC}[F, H]_{K, K_h}(N, M) = F_K(N) \oplus H_{K_h}(M),$$

where K is the key for the pseudorandom function F, K_h is the key for the AXU hash function H, N is the nonce, and M is the message.[2]

The WC construction enjoys a very strong provable security bound when nonces are never reused. Assuming that F is perfect (i.e., F_K is a uniformly random function), any adversary seeing at most q_m honestly generated tags and making at most q_v verification queries (i.e., forgery attempts) succeeds with probability at most εq_v, where ε is the maximal differential probability of H, namely

$$\varepsilon = \max_{X \neq X', Y} \Pr\left[H_{K_h}(X) \oplus H_{K_h}(X') = Y\right],$$

the probabilities being taken over the random draw of the hashing key K_h. When F is not perfect, there is an additional term accounting for its insecurity as a PRF (more precisely, this corresponds to the best advantage an adversary can achieve in distinguishing F_K from a uniformly random function within $q_m + q_v$ queries).

Many AXU hash functions have been proposed for instantiating this construction, most of them based on polynomial hashing [Kra94, Rog95, Sho96, HK97, BHK+99, Ber00, KR00, KVW04, MV04, Ber05c]. See [Ber07] for more references and a comprehensive survey of polynomial hashing. Universal hash functions can also be constructed from a block cipher (e.g. by using the CBC mode with prefix-free encoding [BR05, BPR05]), but in that case the provable maximal differential probability depends on the PRP-security of the block cipher (hence, this yields "computational" rather than "statistical" universal hash functions).

NONCE-MISUSE RESISTANCE. Despite the advantages just mentioned (efficiency and excellent security bound), the WC construction has one major shortcoming: it is very vulnerable to *nonce-misuse*. If a nonce is repeated even a single time, consequences can be catastrophic [Jou06, HP08]. For example, in the case of polynomial universal hashing, this can lead to a complete recovery of the hashing key, which allows universal forgeries. To remedy this nonce-misuse problem, the simplest option, which has been known for long, is to apply the PRF to the output of the hash function. For instance, if the PRF takes $2n$-bit inputs,

---

[2] Here and in all the following, we assume to fix ideas that the outputs of the PRF and the hash function are n-bit strings and the group operation is bitwise xor; this can be easily adapted to any other abelian group.

one can define the tag as $F_K(N\|H_{K_h}(M))$; this construction was analyzed by Black *et al.* [BHK+99, BC09]. If F takes only n-bit inputs, one can instead apply the PRF with an independent key to the output of the WC construction, thereby defining the tag as

$$F_{K'}\big(F_K(N) \oplus H_{K_h}(M)\big). \tag{1}$$

If one gets rid of the nonce, simply defining the tag as $F_K(H_{K_h}(M))$, one obtains a stateless MAC but the security bound includes an extra "birthday-type" term εq_m^2.

BEYOND-BIRTHDAY-BOUND SECURITY. There is another obstacle which can prevent concrete implementations from enjoying the strong security bound promised by the WC construction: pseudorandom functions are not always readily available, and it is common to use a pseudorandom *permutation* instead, or in other words to replace F with a block cipher E. However, as first pointed out by Shoup [Sho96], this causes the proven security bound to drop to the so-called birthday bound. Indeed, a random permutation can be distinguished from a random function within q queries with advantage roughly $q^2/2^n$. For resource-constrained environments, where lightweight cryptographic primitives based on block ciphers with 64-bit blocks are likely to be implemented, this means that security insurance is lost after 2^{32} queries, which is often unacceptable, especially when refreshing keys regularly is excluded.

A first solution to overcome the birthday bound while using only a block cipher is to use a *randomized* construction. However, existing schemes either require very strong properties from the block cipher such as the ideal cipher model [JJV02] or resistance to related-key attacks [JL04], or require a relatively large amount of randomness (at least $3n$ bits for the MACRX construction of [BGK99]). The beyond-birthday-bound secure construction named MAC-R2 of Minematsu [Min10] uses a random n-bit IV per message and bears resemblance to the construction proposed in this paper, but it requires four calls to the underlying block cipher. (Jumping ahead, our new construction requires only two calls.) Moreover, reliable randomness might not always be available in some environments, and it might sometimes be easier to maintain a state.

Another option is to implement F_K in construction (1) from a block cipher E using a so-called *PRP-to-PRF conversion method* [BKR98, HWKS98] with beyond-birthday-bound security. (On the other hand, it is easy to see that the outer PRF $F_{K'}$ can be directly implemented by a block cipher without security loss.) Perhaps the simplest such method is the "xor" construction $E_{K_1}(N) \oplus E_{K_2}(N)$, or its close single-key variant $E_K(N\|0) \oplus E_K(N\|1)$, which have been analyzed in a number of papers [BI99, Luc00, Pat08a, Pat13, CLP14]. However, all known methods require at least two block cipher calls; taking into account the outer encryption layer, this amounts to three block cipher calls for the whole construction. Is it possible to do better?

OUR CONTRIBUTION. We propose a new nonce-based MAC based on a AXU hash function and a block cipher with the following properties:

(i) it is simple and efficient, requiring only two calls to the underlying block cipher, one of which can be carried out in parallel to the hash function computation;
(ii) it provably provides security *beyond the birthday bound* when nonces are never reused;
(iii) it provably retains security up to the birthday bound in case of nonce misuse.

Property (ii) ensures that the scheme is highly secure in the nominal use case where nonces are never repeated, while property (iii) acts as a "safety net" if anything goes wrong with nonces.

Our starting point is what we call the Encrypted Wegman-Carter construction, which is simply construction (1) where the outer PRF layer is replaced by a block cipher, viz.

$$E_{K'}\big(F_K(N) \oplus H_{K_h}(M)\big). \qquad (2)$$

As already briefly explained, this construction enjoys the same security bound as the (unencrypted) WC construction when nonces are never repeated, and is moreover nonce-misuse resistant up to the birthday bound. Replacing F_K by a simple block cipher call causes the security bound to drop to the birthday bound even when nonces are not repeated, while using a PRP-to-PRF conversion method with security beyond the birthday bound results in at least three block cipher calls in total for the resulting construction.

Our main observation is that one can overcome the birthday bound in the nonce-respecting scenario by instantiating F_K using "only" the Davies-Meyer (DM) construction. The DM construction is the easiest way to turn a block cipher into a keyed function.[3] Given a block cipher $E : \mathcal{K} \times \{0,1\}^n \to \{0,1\}^n$, the DM construction based on E is simply

$$\mathsf{DM}[E]_K(N) = E_K(N) \oplus N.$$

Note that this PRF construction is *not* secure beyond the birthday bound: given black-box access to a function $f : \{0,1\}^n \to \{0,1\}^n$, a distinguisher can simply query $f(N_i)$ for roughly $2^{n/2}$ distinct values N_i and look for collisions in values $f(N_i) \oplus N_i$. When f is a uniformly random function this will happen with good probability, whereas when $f = \mathsf{DM}[E]_K$ this cannot happen. However, this attack is not possible anymore if one encrypts the output of the DM construction.

Using the DM construction to instantiate F_K in construction (2) results in a MAC construction based only on E and H, which we call *Encrypted Wegman-Carter with Davies-Meyer* (EWCDM) construction, depicted on Fig. 1 and defined as

$$E_{K'}\big(E_K(N) \oplus N \oplus H_{K_h}(M)\big). \qquad (3)$$

---

[3] Traditionally, the DM construction is rather seen as a way to turn a block cipher into an (unkeyed) compression function.

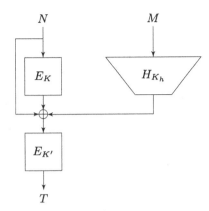

Fig. 1. The "Encrypted Wegman-Carter with Davies-Meyer" construction.

Our main result is that the EWCDM construction is secure up to roughly $2^{2n/3}$ MAC queries and 2^n verification queries against nonce-respecting adversaries (while against nonce-misusing adversaries it still enjoys birthday-bound security) (Table 1). We stress that this does not hold for the (unencrypted) Wegman-Carter construction with Davies-Meyer: if tags are computed as

$$T = E_K(N) \oplus N \oplus H_{K_h}(M),$$

then the resulting MAC scheme is only provably secure up to the birthday bound against nonce-respecting adversaries.[4] Hence, the outer encryption layer $E_{K'}$ turns out to be *twice* useful: for providing nonce-misuse resistance on one hand, and for cheaply enhancing security against nonce-respecting adversaries beyond the birthday bound on the other hand.

We believe that our new construction would be an elementary and easy-to-implement way to enhance the security of widely deployed authentication or authenticated encryption schemes such as Poly1305-AES [Ber05c] or GCM [MV04] (in particular, note that this can be done in a black-box way on top of an existing implementation of those schemes). The main cost would be some additional latency due to the extra block cipher call, but depending on the context this might be tolerable.

PROOF TECHNIQUE. At the heart of construction (3) is a novel PRP-to-PRF conversion method: namely, if we make abstraction for a moment of the hash of the message M, and if we simply denote P and P' in place of E_K and $E_{K'}$, we obtain a function of the nonce defined as

$$F(N) = P'(P(N) \oplus N).$$

---

[4] Indeed, the outputs of this construction can be distinguished from random simply by querying the MAC oracle for tags T_i with the same message and roughly $2^{n/2}$ distinct nonces N_i, and looking for collisions in $T_i \oplus N_i$.

For obvious reasons, we call this the *Encrypted Davies-Meyer* (EDM) construction. The main part of the proof consists in proving that this is a secure PRF up to $2^{2n/3}$ adversarial queries. (We prove this as a standalone result in the full version of the paper [CS16]; this constitutes a good warm-up for the reader before the more complicated security proof of the EWCDM construction in Sect. 4.) However, since the hash of the message is "intermingled" within the EDM construction, it does not seem possible to first prove that the outputs of the MAC oracle are indistinguishable from random, and then handle verification queries (as is usually done for proving the security of the standard Wegman-Carter construction; see Theorem 1 in Sect. 3.1). Note that one cannot hope either to prove security beyond the birthday bound by a sequence of games that would start by replacing the DM construction $E_K(N) \oplus N$ by a uniformly random function.

Hence, it seems that any proof aiming at security beyond the birthday bound must handle MAC queries *and* verification queries both at the same time. For this, we employ the H-coefficients technique, which has been introduced by Patarin [Pat90, Pat91, Pat08b] and which recently regained attention since Chen and Steinberger used it to analyze the iterated Even-Mansour cipher [CS14]. This technique gives a kind of "systematic" way to upper bound the statistical distance between the answers of two interactive systems and is typically used to prove (information-theoretic) pseudorandomness of constructions such as Feistel networks. To the best of our knowledge, this is the first time the H-coefficients technique is used for proving the security of a MAC (i.e., unpredictability rather than pseudorandomness).

MORE RELATED WORK. This paper focuses on nonce-based (hence stateful) MACs, but there is also an important line of work aiming at constructing stateless and deterministic MACs secure beyond the birthday bound. However, existing constructions [Yas10, Yas11, DS11, ZWSW12] are far more complex than the one presented in this paper. We mainly mentioned works related to provable security; there is also a large number of papers (motivated by the analysis of the widely deployed GCM mode [MV04]) investigating attacks against polynomial hash-based MACs [Fer05, HP08, Saa12, PC15, ABBT15].

OPEN PROBLEMS. We prove the security of the EWCDM construction in the nonce-respecting scenario up to $2^{2n/3}$ MAC queries, but we conjecture that security actually holds up to close to 2^n queries (a similar conjecture holds for the Encrypted Davies-Meyer construction).

The EWCDM construction uses two distinct keys for the two calls to the block cipher; a natural question is whether security beyond the birthday bound also holds when the same key is used. We believe this to be true, but likely cumbersome to prove. The corresponding question regarding the Encrypted Davies-Meyer construction is even more intriguing: How many queries are required to distinguish $P(x \oplus P(x))$ from a random function? It might well be that this construction is secure up to close to 2^n queries, which would yield the first optimally secure PRP-to-PRF conversion method which uses a single permutation (unlike $P_1(x) \oplus P_2(x)$) and does not shrink the domain (unlike $P(x\|0) \oplus P(x\|1)$).

Table 1. Proven security bounds (omitting constants and the term accounting for the PRP-security of the underlying block cipher) for the Wegman-Carter construction $\mathsf{WC}[E, H]$, the Encrypted Wegman-Carter construction $\mathsf{EWC}[E, H]$, and the new Encrypted Wegman-Carter with Davies-Meyer construction $\mathsf{EWCDM}[E, H]$.

	Nonce-respecting	Nonce-misusing
$\mathsf{WC}[E, H]$	$(q_m + q_v)^2/2^n + \varepsilon q_v$	—
$\mathsf{EWC}[E, H]$	$(q_m + q_v)^2/2^n + \varepsilon q_v$	$(q_m + q_v)^2/2^n + \varepsilon(q_m + q_v)^2$
$\mathsf{EWCDM}[E, H]$	$q_m^{3/2}/2^n + \varepsilon q_m + q_v/2^n + \varepsilon q_v$	$(q_m + q_v)^2/2^n + \varepsilon(q_m + q_v)^2$

Finally, it would be interesting to investigate how the security of EWCDM is affected by tag truncation. We believe that the only change to be made to the bound of Theorem 3 is to replace the term $6q_v/2^n$ by a term $O(q_v/2^\ell)$, where ℓ is the length of the truncated tag, but this remains to be proven.

ORGANIZATION. We first establish the notation and recall standard security definitions in Sect. 2. In Sect. 3, we recall the previous security results on the Wegman-Carter and the Encrypted Wegman-Carter constructions, and describe our new EWCDM construction. We then prove the security of EWCDM in the nonce-respecting scenario in Sect. 4 and in the nonce-misusing scenario in Sect. 5. We also analyze the Encrypted Davies-Meyer PRP-to-PRF conversion method in the full version of the paper [CS16].

2   Preliminaries

BASIC NOTATION. Given a non-empty set $\mathcal{X}$, we denote $X \leftarrow_\$ \mathcal{X}$ the draw of an element X from $\mathcal{X}$ uniformly at random. The set of all functions from $\mathcal{X}$ to $\mathcal{Y}$ is denoted $\mathsf{Func}(\mathcal{X}, \mathcal{Y})$, and the set of all permutations of $\mathcal{X}$ is denoted $\mathsf{Perm}(\mathcal{X})$. The set of binary strings of length n is denoted $\{0, 1\}^n$. The set of all functions from $\{0, 1\}^n$ to $\{0, 1\}^n$ is simply denoted $\mathsf{Func}(n)$, and the set of all permutations of $\{0, 1\}^n$ is simply denoted $\mathsf{Perm}(n)$. For integers $1 \leq b \leq a$, we will write $(a)_b = a(a - 1) \cdots (a - b + 1)$ and $(a)_0 = 1$ by convention. Note that the probability that a random permutation $P \leftarrow_\$ \mathsf{Perm}(n)$ satisfies q equations $P(X_i) = Y_i$ for distinct X_i's and distinct Y_i's is exactly $1/(2^n)_q$.

PRFs AND BLOCK CIPHERS. A keyed function with key space $\mathcal{K}$, domain $\mathcal{X}$, and range $\mathcal{Y}$ is a function $F : \mathcal{K} \times \mathcal{X} \to \mathcal{Y}$. We denote $F_K(X)$ for $F(K, X)$. A (q, t)-adversary against F is an algorithm A with oracle access to a function from $\mathcal{X}$ to $\mathcal{Y}$, making at most q oracle queries, running in time at most t, and outputting a single bit. The advantage of A in breaking the PRF-security of F is defined as

$$\mathbf{Adv}_F^{\mathrm{PRF}}(\mathsf{A}) = \left| \Pr\left[K \leftarrow_\$ \mathcal{K} : \mathsf{A}^{F_K} = 1\right] - \Pr\left[R \leftarrow_\$ \mathsf{Func}(\mathcal{X}, \mathcal{Y}) : \mathsf{A}^R = 1\right] \right|.$$

A block cipher with key space $\mathcal{K}$ and domain $\mathcal{X}$ is a mapping $E : \mathcal{K} \times \mathcal{X} \to \mathcal{X}$ such that for any key $K \in \mathcal{K}$, $X \mapsto E(K, X)$ is a permutation of $\mathcal{X}$. We denote

$E_K(X)$ for $E(K, X)$. A (q, t)-adversary against E is an algorithm A with oracle access to a permutation of $\mathcal{X}$, making at most q oracle queries, running in time at most t, and outputting a single bit. The advantage of A in breaking the PRP-security of E is defined as

$$\mathbf{Adv}_E^{\mathrm{PRP}}(\mathsf{A}) = \left| \Pr\left[K \xleftarrow{\$} \mathcal{K} : \mathsf{A}^{E_K} = 1 \right] - \Pr\left[P \xleftarrow{\$} \mathsf{Perm}(\mathcal{X}) : \mathsf{A}^P = 1 \right] \right|.$$

Note that we do not need the strongest "two-sided" version of PRP-security (where the adversary also has access to a decryption oracle) since all constructions considered in this paper only use the forward (encryption) direction of the underlying block cipher.

MACs. Given four non-empty sets $\mathcal{K}$, $\mathcal{N}$, $\mathcal{M}$, and $\mathcal{T}$, a nonce-based keyed function with key space $\mathcal{K}$, nonce space $\mathcal{N}$, message space $\mathcal{M}$ and range $\mathcal{T}$ is simply a function $F : \mathcal{K} \times \mathcal{N} \times \mathcal{M} \to \mathcal{T}$. Stated otherwise, it is a keyed function whose domain is a cartesian product $\mathcal{N} \times \mathcal{M}$. We denote $F_K(N, M)$ for $F(K, N, M)$.

Definition 1 (Nonce-Based MAC). *Let $\mathcal{K}$, $\mathcal{N}$, $\mathcal{M}$, and $\mathcal{T}$ be non-empty sets. Let $F : \mathcal{K} \times \mathcal{N} \times \mathcal{M} \to \mathcal{T}$ be a nonce-based keyed function. For $K \in \mathcal{K}$, let Ver_K be the verification oracle which takes as input a triple $(N, M, T) \in \mathcal{N} \times \mathcal{M} \times \mathcal{T}$ and returns 1 ("accept") if $F_K(N, M) = T$, and 0 ("reject") otherwise. A (q_m, q_v, t)-adversary against the MAC-security of F is an adversary A with oracle access to the two oracles F_K and Ver_K for $K \in \mathcal{K}$, making at most q_m "MAC" queries to its first oracle and at most q_v "verification" queries to its second oracle, and running in time at most t. We say that A forges if any of its queries to Ver_K returns 1. The advantage of A against the MAC-security of F is defined as*

$$\mathbf{Adv}_F^{\mathrm{MAC}}(\mathsf{A}) = \Pr\left[K \xleftarrow{\$} \mathcal{K} : \mathsf{A}^{F_K, \mathsf{Ver}_K} \text{ forges} \right],$$

where the probability is also taken over the random coins of A, if any. The adversary is not allowed to ask a verification query (N, M, T) if a previous query (N, M) to F_K returned T. The adversary is said nonce-respecting if it never repeats a nonce $N \in \mathcal{N}$ in its queries to the first oracle F_K.

We say that an adversary is *nonce-misusing* if it does not abide to the rule of non-repeating nonces. The MAC-security of F in face of nonce-misusing adversaries is defined exactly as above, and can be rephrased as the standard (i.e., not nonce-based) MAC-security of a keyed function with domain $\mathcal{N} \times \mathcal{M}$.

AXU HASH FUNCTIONS. We will need the following definition of an almost xor-universal (AXU) hash function.

Definition 2 (ε-AXU Hash Function). *Let $\mathcal{K}_h$, $\mathcal{X}$ and $\mathcal{Y}$ be three non-empty sets and $\varepsilon > 0$. A keyed function $H : \mathcal{K}_h \times \mathcal{X} \to \mathcal{Y}$ is said to be ε-AXU if for any distinct $X, X' \in \mathcal{X}$ and any $Y \in \mathcal{Y}$,*

$$\Pr\left[K_h \xleftarrow{\$} \mathcal{K}_h : H_{K_h}(X) \oplus H_{K_h}(X') = Y \right] \leq \varepsilon.$$

3   Wegman-Carter MAC Constructions

3.1   The Standard Wegman-Carter Construction

We recall the standard Wegman-Carter construction [WC81] of a nonce-based MAC from an ε-AXU hash function and a PRF. Let $\mathcal{K}$, $\mathcal{K}_h$, and $\mathcal{M}$ be non-empty sets. Let $F : \mathcal{K} \times \{0,1\}^n \rightarrow \{0,1\}^n$ be a keyed function and $H : \mathcal{K}_h \times \mathcal{M} \rightarrow \{0,1\}^n$ be an ε-AXU hash function. The Wegman-Carter construction based on F and H is the nonce-based keyed function with key space $\mathcal{K} \times \mathcal{K}_h$, nonce space $\{0,1\}^n$, message space $\mathcal{M}$, and range $\{0,1\}^n$ defined by

$$\mathsf{WC}[F,H]_{K,K_h}(N,M) = F_K(N) \oplus H_{K_h}(M).$$

We recall the classical security result for this construction [WC81] and sketch the proof for completeness. Here and in all the following, t_H is an upper bound on the time needed to compute $H_{K_h}(M)$ for any key $K_h \in \mathcal{K}_h$ and any message $M \in \mathcal{M}$.

Theorem 1. *Let F and H be as above. Then for any (q_m, q_v, t)-nonce-respecting adversary A against the MAC-security of $\mathsf{WC}[F,H]$, there exists a $(q_m + q_v, t')$-adversary A' against the PRF-security of F, where $t' = O(t + (q_m + q_v)t_H)$, such that*

$$\mathbf{Adv}^{\mathrm{MAC}}_{\mathsf{WC}[F,H]}(\mathsf{A}) \le \mathbf{Adv}^{\mathrm{PRF}}_F(\mathsf{A}') + \varepsilon q_v.$$

Proof. Fix a (q_m, q_v, t)-nonce-respecting adversary A. Consider the WC construction where F_K is replaced by a uniformly random function R, and let δ be the advantage of A against this new construction. By a straightforward hybrid argument, there is an adversary A', making at most $q_m + q_v$ oracle queries, and running in time $O(t + (q_m + q_v)t_H)$, such that

$$\mathbf{Adv}^{\mathrm{MAC}}_{\mathsf{WC}[F,H]}(\mathsf{A}) \le \mathbf{Adv}^{\mathrm{PRF}}_F(\mathsf{A}') + \delta.$$

The answers $R(N) \oplus H_{K_h}(M)$ of the MAC oracle are now uniformly random and independent from K_h. Consider the i-th verification query (N', M', T') of the adversary. If N' never appeared in the MAC queries of the adversary, then T' is valid with probability 2^{-n}. If $N' = N$ for some previous MAC query (N, M) that returned T, then the verification query is valid iff

$$R(N') \oplus H_{K_h}(M') = T' \Leftrightarrow H_{K_h}(M) \oplus H_{K_h}(M') = T \oplus T',$$

which happens with probability at most ε by definition of an ε-AXU hash function. (If $M = M'$, then one must have $T \ne T'$ by definition of the security experiment, and the forgery cannot be valid.) Since for an ε-AXU hash function with range $\{0,1\}^n$ one has $\varepsilon \ge 2^{-n}$, in all cases the forgery is valid with probability at most ε. By a union bound over the q_v verification queries, one has $\delta \le \varepsilon q_v$, which concludes the proof. $\qquad \square$

Assume now that F is a family of *permutations* of $\{0,1\}^n$, or in other words, a block cipher, that we denote E. Then E can be distinguished from a random function with q queries and advantage roughly $q^2/2^n$ by simply looking for collisions in its outputs. In other words, by the PRP-PRF switching Lemma [BR06], the best upper bound one can hope to prove for the PRF-advantage of adversary A' appearing in Theorem 1, assuming that E is a secure *PRP*, is

$$\mathbf{Adv}_E^{\mathrm{PRF}}(\mathsf{A}') \le \mathbf{Adv}_E^{\mathrm{PRP}}(\mathsf{A}') + \frac{(q_m + q_v)^2}{2^{n+1}},$$

so that the security bound for the resulting construction $\mathsf{WC}[E, H]$ now has a birthday-type term. Bernstein [Ber05a, Ber05b] proved a better (but still of birthday-type) bound: as long as $q_m \le 2^{n/2}$, the adversary can forge with probability at most $C\varepsilon q_v$, for some small constant C (in all practical cases, $C \le 2$). Note that the distinguishing attack against E does not seem to translate into a forgery attack against the MAC scheme, and it might be possible to improve the security bound under additional assumptions on H and E.

3.2   Nonce-Misuse Resistance and the Encrypted Wegman-Carter Construction

In general, the standard Wegman-Carter construction of the previous section does not offer any security against nonce-misusing adversaries. Consider for example the case where H is a polynomial-based hash function. Then any adversary who gets two tags T and T' for two different messages M and M' generated with the same nonce knows that $H_{K_h}(M) \oplus H_{K_h}(M') \oplus T \oplus T' = 0$. The left hand side is a polynomial in K_h whose coefficients depend on M, M', T and T', and K_h is a root of this polynomial. Even though its degree can be quite high, this is often enough to mount devastating attacks. This weakness was one of the main criticism against the GCM authenticated encryption mode [MV04], whose authentication relies on the standard Wegman-Carter construction [Jou06].

The classical way to remedy this situation and achieve nonce-misuse resistance for Wegman-Carter MACs is to apply an extra PRF layer to the output of the construction. When this additional layer is a block cipher, one obtains what we call the *Encrypted Wegman-Carter* (EWC) construction. Let $F : \mathcal{K} \times \{0,1\}^n \to \{0,1\}^n$ be a keyed function, $E : \mathcal{K}' \times \{0,1\}^n \to \{0,1\}^n$ be a block cipher, and $H : \mathcal{K}_h \times \mathcal{M} \to \{0,1\}^n$ be an ε-AXU hash function. Then the EWC construction based on F, E, and H has key space $\mathcal{K} \times \mathcal{K}' \times \mathcal{K}_h$, nonce space $\{0,1\}^n$, message space $\mathcal{M}$, and range $\{0,1\}^n$, and is defined by

$$\mathsf{EWC}[F, E, H]_{K,K',K_h}(N, M) = E_{K'}\big(\mathsf{WC}[F, H]_{K,K_h}(N, M)\big)$$
$$= E_{K'}\big(F_K(N) \oplus H_{K_h}(M)\big).$$

One can straightforwardly verify that the security of this construction against nonce-respecting adversaries does not depend on E and that the upper bound of Theorem 1 still holds. For nonce-misusing adversaries, one has the following (the proof is omitted since it is exactly the same, *mutatis mutandis*, as the proof of Theorem 4 of Sect. 5).

Theorem 2. *Let F, E and H be as above. Then for any (q_m, q_v, t)-nonce-misusing adversary A against the MAC-security of $\mathsf{EWC}[F, E, H]$, there exists a $(q_m + q_v, t')$-adversary A' against the PRF-security of F and a $(q_m + q_v, t'')$-adversary A'' against the PRP-security of E, where $t', t'' = O(t + (q_m + q_v)t_H)$, such that*

$$\mathbf{Adv}^{\mathrm{MAC}}_{\mathsf{EWC}[F, E, H]}(\mathsf{A}) \leq \mathbf{Adv}^{\mathrm{PRF}}_{F}(\mathsf{A}') + \mathbf{Adv}^{\mathrm{PRP}}_{E}(\mathsf{A}'') + \frac{2(q_m + q_v)^2}{2^n} + \frac{(q_m + q_v)^2 \varepsilon}{2}.$$

It is tempting to implement F from E. The simplest way to do so is simply to let $F = E$, thereby obtaining the construction (overloading notation $\mathsf{EWC}[\cdot]$)

$$\mathsf{EWC}[E, H]_{K, K', K_h}(N, M) = E_{K'}\big(E_K(N) \oplus H_{K_h}(M)\big).$$

However, the resulting MAC suffers from the same birthday-bound type problem against nonce-respecting adversaries as the unencrypted Wegman-Carter MAC $\mathsf{WC}[E, H]$ of Sect. 3.1. As already mentioned in introduction, it is possible to use a PRP-to-PRF conversion method to obtain security beyond the birthday bound, but using the best known constructions yields a MAC that makes at least three calls to the underlying block cipher. Our goal is to reduce the number of block cipher calls to two, which seems to be the minimum to achieve both security beyond the birthday bound and nonce-misuse resistance.

3.3   The New Construction EWCDM

The main contribution of this paper is to propose a much simpler solution that allows to get beyond the birthday bound, namely using the Davies-Meyer (DM) construction which turns a block cipher $E : \mathcal{K} \times \{0, 1\}^n \to \{0, 1\}^n$ into a keyed function as

$$\mathsf{DM}[E]_K(N) = E_K(N) \oplus N.$$

Using the DM construction based on E to instantiate F in $\mathsf{EWC}[F, E, H]$ results in a MAC construction based only on E and H, which we call *Encrypted Wegman-Carter with Davies-Meyer* (EWCDM) construction and denote $\mathsf{EWCDM}[E, H]$, illustrated on Fig. 1 and defined as follows:

$$\mathsf{EWCDM}[E, H]_{K, K', K_h}(N, M) \overset{\mathrm{def}}{=} \mathsf{EWC}[\mathsf{DM}[E], E, H]_{K, K', K_h}(N, M)$$
$$= E_{K'}\big(E_K(N) \oplus N \oplus H_{K_h}(M)\big).$$

As already explained in introduction, the DM construction is *not* PRF-secure beyond the birthday bound. Still, our main result, that we state and prove in the next section, is that the EWCDM construction is secure up to roughly $2^{2n/3}$ MAC queries and 2^n verification queries against nonce-respecting adversaries (while against nonce-misusing adversaries it still enjoys birthday-bound security).

The security proof entails an analysis of what we call the *Encrypted Davies-Meyer* (EDM) PRP-to-PRF conversion method, which turns two independent permutations P and P' of $\{0, 1\}^n$ into a function of $\{0, 1\}^n$ to $\{0, 1\}^n$ defined as

$$\mathsf{EDM}[P, P'](N) = P'(P(N) \oplus N).$$

By "stripping off" from the security proof of EWCDM all details related to the hash function and verification queries, one can extract a proof that the EDM construction is a secure PRF up to $2^{2n/3}$ adversarial queries. We do so in the full version of the paper [CS16], and the reader might want to read this simpler proof before proceeding to Sect. 4. However, as already explained in introduction, it does not seem possible to prove the MAC-security of the EWCDM construction in a modular way from the PRF-security of the EDM construction.

Finally, note that adding the hash of the message to the output of the EDM construction (rather than "in the middle") would result in a construction secure up to $2^{2n/3}$ queries against nonce-respecting adversaries, but insecure against nonce-misusing ones since it is just an instantiation of the standard WC construction of Sect. 3.1 (with the EDM construction as PRF).

4   Nonce-Respecting Security of EWCDM

4.1   Statement of the Result and Overview of the Proof

In all the following, we simply denote $\Pi[E, H]$ the EWCDM construction based on block cipher E and AXU hash function H. Our main security result is as follows.

Theorem 3. *Let $\mathcal{M}$, $\mathcal{K}$ and $\mathcal{K}_h$ be non-empty sets. Let $E : \mathcal{K} \times \{0,1\}^n \to \{0,1\}^n$ be a block cipher and $H : \mathcal{K}_h \times \mathcal{M} \to \{0,1\}^n$ be an ε-AXU hash function. Then for any (q_m, q_v, t)-nonce-respecting adversary A against the MAC-security of $\Pi[E, H]$ with $q_m^{3/2} \le 2^n/4$ and $q_v \le 2^n/4$, there exists a $(q_m + q_v, t')$-adversary A' against the PRP-security of E, where $t' = O(t + (q_m + q_v)t_H)$, such that*

$$\mathbf{Adv}_{\Pi[E,H]}^{\mathrm{MAC}}(\mathsf{A}) \le 2\mathbf{Adv}_E^{\mathrm{PRP}}(\mathsf{A}') + \frac{5q_m^{3/2}}{2^n} + \frac{\varepsilon q_m}{2} + \frac{6q_v}{2^n} + \varepsilon q_v.$$

Hence, assuming $\varepsilon \simeq 2^{-n}$, the EWCDM construction is secure up to $q_m \simeq 2^{2n/3}$ MAC queries and $q_v \simeq 2^n$ verification queries.

In the remaining of the section, we prove Theorem 3. We fix a (q_m, q_v, t)-nonce-respecting adversary A against the MAC-security of $\Pi[E, H]$ and we let

$$\delta = \mathbf{Adv}_{\Pi[E,H]}^{\mathrm{MAC}}(\mathsf{A}).$$

As specified in Definition 1, adversary A has access to a MAC oracle $\Pi[E, H]_{K,K',K_h}$ and a verification oracle Ver_{K,K',K_h} for a randomly drawn key tuple (K, K', K_h).

The first step of the proof is standard and consists in replacing E_K and $E_{K'}$ by two random and independent permutations P and P', both in the MAC and in the verification oracle (in other words, we replace the block cipher E by the *perfect cipher* E^* whose key space is the set of all permutations of $\{0,1\}^n$). Let $\Pi[E^*, H]$ denote the resulting construction. It is easy to show that there exists an adversary against the PRP-security of E, making at most $q_m + q_v$ oracle queries and runnig in time at most $O(t + (q_m + q_v)t_H)$, such that

$$\delta \leq 2\mathbf{Adv}_E^{\mathrm{PRP}}(\mathsf{A}') + \mathbf{Adv}_{\Pi[E^*,H]}^{\mathrm{MAC}}(\mathsf{A}). \tag{4}$$

(We replace successively E_K and $E_{K'}$ by a random permutation, each time constructing an hybrid PRP-adversary, and we consider the best of the two adversaries). Our goal is now to upper bound

$$\delta^* \stackrel{\mathrm{def}}{=} \mathbf{Adv}_{\Pi[E^*,H]}^{\mathrm{MAC}}(\mathsf{A})$$
$$= \Pr\left[(P, P') \leftarrow_{\$} \mathsf{Perm}(n)^2, K_h \leftarrow_{\$} \mathcal{K}_h : \mathsf{A}^{\Pi[P,P',H_{K_h}],\mathsf{Ver}[P,P',H_{K_h}]} \text{ forges}\right],$$

where, overloading the notation, $\Pi[P, P', H_{K_h}]$ denotes the construction $\Pi[E^*, H]$ instantiated with permutations P, P', and hashing key K_h and $\mathsf{Ver}[P, P', H_{K_h}]$ denotes the corresponding verification oracle.

It will be more convenient to express δ^* as a *distinguishing* advantage. Namely, let Rand denote a perfectly random oracle with domain $\{0,1\}^n \times \mathcal{M}$ and range $\{0,1\}^n$, and Rej be an oracle with inputs in $\{0,1\}^n \times \mathcal{M} \times \{0,1\}^n$ which always returns 0 ("reject"). Since the adversary cannot forge (i.e., have the right oracle return 1) when interacting with $(\mathsf{Rand}, \mathsf{Rej})$, we have

$$\delta^* = \Pr\left[\mathsf{A}^{\Pi[P,P',H_{K_h}],\mathsf{Ver}[P,P',H_{K_h}]} \text{ forges}\right] - \Pr\left[\mathsf{A}^{\mathsf{Rand},\mathsf{Rej}} \text{ forges}\right].$$

Consider now an adversary D which queries a pair of oracles $(\mathcal{O}_1, \mathcal{O}_2)$ and outputs a bit β, which we denote $\mathsf{D}^{\mathcal{O}_1,\mathcal{O}_2} = \beta$. (We will refer to such an adversary as a *distinguisher*.) Say that such an adversary is *non-trivial* if it never makes a query (N, M, T) to its right (verification) oracle if a previous query (N, M) to its left (MAC) oracle returned T. Then

$$\delta^* \leq \max_{\mathsf{D}} \Pr\left[\mathsf{D}^{\Pi[P,P',H_{K_h}],\mathsf{Ver}[P,P',H_{K_h}]} = 1\right] - \Pr\left[\mathsf{D}^{\mathsf{Rand},\mathsf{Rej}} = 1\right], \tag{5}$$

where the maximum is taken over non-trivial adversaries. (This follows easily by considering the particular D which runs A and outputs 1 iff A successfully forges.) Hence, we see that δ^* cannot be larger than the advantage of the best non-trivial distinguisher between the two pairs of oracles $(\Pi[P, P', H_{K_h}], \mathsf{Ver}[P, P', H_{K_h}])$ and $(\mathsf{Rand}, \mathsf{Rej})$.[5] This formulation of the problem now allows us to use the H-coefficients technique [Pat08b, CS14], as we explain in more details below.

THE H-COEFFICIENTS TECHNIQUE. From now on, we fix a non-trivial distinguisher D interacting either with the *real world* $(\Pi[P, P', H_{K_h}], \mathsf{Ver}[P, P', H_{K_h}])$ for uniformly random permutations (P, P') and a random hashing key K_h, or with the *ideal world* $(\mathsf{Rand}, \mathsf{Rej})$, making at most q_m queries to its left (MAC)

---

[5] While a verification query answered by 1 constitutes an obvious distinguishing criterion between the two worlds, a more advanced adversary might also use the small difference between the distributions of the answers of the left (MAC) oracle.

oracle and at most q_v queries to its right (verification) oracle, and outputting a single bit. We let

$$\mathbf{Adv}(\mathsf{D}) = \Pr\left[\mathsf{D}^{\Pi[P,P',H_{K_h}],\mathsf{Ver}[P,P',H_{K_h}]} = 1\right] - \Pr\left[\mathsf{D}^{\mathsf{Rand},\mathsf{Rej}} = 1\right].$$

We assume that D is computationally unbounded (and hence *wlog* deterministic) and that it never repeats a query. Let

$$\tau_m = \left((N_1, M_1, T_1), \ldots, (N_{q_m}, M_{q_m}, T_{q_m})\right)$$

be the list of MAC queries of D and corresponding answers. Let also

$$\tau_v = \left((N'_1, M'_1, T'_1, b_1), \ldots, (N'_{q_v}, M'_{q_v}, T'_{q_v}, b_{q_v})\right)$$

be the list of verification queries of D and corresponding answers (with $b_i \in \{0, 1\}$). The pair (τ_m, τ_v) constitutes the *queries transcript* of the attack. For convenience, we slightly modify the security experiment by revealing to the distinguisher (after it made all its queries but before it outputs its decision bit) the hashing key K_h if we are in the real world, or a uniformly random "dummy" key K_h if we are in the ideal world (this is obviously *wlog* since the distinguisher can ignore this additional piece of information). All in all, the *transcript* of the attack is the triplet $\tau = (\tau_m, \tau_v, K_h)$. We will often simply name a tuple $(N, M, T) \in \tau_m$ a *MAC query*, and a tuple $(N', M', T', b) \in \tau_v$ a *verification query*.

A transcript τ is said *attainable* (with respect to distinguisher D) if the probability to obtain this transcript in the ideal world is non-zero. In particular, note that for an attainable transcript $\tau = (\tau_m, \tau_v, K_h)$, any verification query $(N'_i, M'_i, T'_i, b_i) \in \tau_v$ is such that $b_i = 0$.[6] We denote Θ the set of attainable transcripts. We also denote X_{re}, resp. X_{id}, the probability distribution of the transcript τ induced by the real world, resp. the ideal world. The main lemma of the H-coefficients technique is the following one (see e.g. [CS14] or [CLL+14] for the proof).

Lemma 1. *Fix a distinguisher* D*. Let* $\Theta = \Theta_{\mathrm{good}} \sqcup \Theta_{\mathrm{bad}}$ *be a partition of the set of attainable transcripts. Assume that there exists* ε_1 *such that for any* $\tau \in \Theta_{\mathrm{good}}$*, one has[7]*

$$\frac{\Pr[X_{\mathrm{re}} = \tau]}{\Pr[X_{\mathrm{id}} = \tau]} \geq 1 - \varepsilon_1,$$

and that there exists ε_2 *such that* $\Pr[X_{\mathrm{id}} \in \Theta_{\mathrm{bad}}] \leq \varepsilon_2$*. Then* $\mathbf{Adv}(\mathsf{D}) \leq \varepsilon_1 + \varepsilon_2$*.*

The remaining of the proof of Theorem 3 is structured as follows: in Sect. 4.2, we define bad transcripts and upper bound their probability in the ideal world; in Sect. 4.3, we analyze good transcripts and prove that they are almost as likely in the real and the ideal world. Theorem 3 follows easily by combining Eqs. (4) and (5) above, Lemmas 1, 2 and 3 proven below.

---

[6] Hence, some transcripts are attainable in the real world but not in the ideal world. While this is unusual (in most H-coefficients-based proofs, the set of transcripts attainable in the real world is a subset of those attainable in the ideal world), this is not a problem for Lemma 1 to hold.

[7] Recall that for an attainable transcript, one has $\Pr[X_{\mathrm{id}} = \tau] > 0$.

4.2   Definition and Probability of Bad Transcripts

We start by defining bad transcripts. We say that a MAC query $(N_i, M_i, T_i) \in \tau_m$ is *collisioning* if there exists another MAC query $(N_j, M_j, T_j) \in \tau_m$ with $j \neq i$ such $T_i = T_j$, otherwise we say it is *non-collisioning*.

Definition 3. *We say that an attainable transcript* $\tau = (\tau_m, \tau_v, K_h)$ *is* bad *if one of the following conditions is met:*

(i) the number of collisioning MAC queries in τ_m is more than $\sqrt{q_m}$;
(ii) there exists two distinct MAC queries (N_i, M_i, T_i) and (N_j, M_j, T_j) in τ_m such that
$$\begin{cases} T_i = T_j \\ N_i \oplus H_{K_h}(M_i) = N_j \oplus H_{K_h}(M_j); \end{cases}$$

(iii) there exists a MAC query $(N_i, M_i, T_i) \in \tau_m$ and a verification query $(N'_j, M'_j, T'_j, b_j) \in \tau_v$ such that
$$\begin{cases} N_i = N'_j \\ T_i = T'_j \\ H_{K_h}(M_i) = H_{K_h}(M'_j). \end{cases}$$

We denote Θ_{bad}, resp. Θ_{good} the set of bad, respectively good transcripts.

We quickly comment on these three conditions. Condition (i) captures the case where there are too many tag collisions and will be needed when lower bounding the probability of getting a good transcript in the real world. Condition (ii) can only happen in the ideal world and hence allows to trivially distinguish; in the real world, if $N_i \oplus H_{K_h}(M_i) = N_j \oplus H_{K_h}(M_j)$, then, since $N_i \neq N_j$ because the adversary is assumed nonce-respecting, one necessarily has

$$P(N_i) \oplus N_i \oplus H_{K_h}(M_i) \neq P(N_j) \oplus N_j \oplus H_{K_h}(M_j)$$

which implies $T_i \neq T_j$ by applying P' to both sides of the inequality. Similarly, condition (iii) can only happen in the ideal world since in the real world, if $N_i = N'_j$, $T_i = T'_j$, and $H_{K_h}(M_i) = H_{K_h}(M'_j)$, one should have $b_j = 1$ (while $b_j = 0$ in the ideal world).

We now upper bound the probability to get a bad transcript in the ideal world.

Lemma 2. *For any integers q_m and q_v, one has*

$$\Pr\left[X_{\text{id}} \in \Theta_{\text{bad}}\right] \leq \frac{q_m^{3/2}}{2^n} + \frac{\varepsilon q_m}{2} + \varepsilon q_v.$$

Proof. We upper bound the probabilities of the three conditions in turn. We denote Θ_i the set of attainable transcript that satisfy the i-th condition. Recall that, in the ideal world, K_h is drawn independently from the queries transcript.

CONDITIONS (i) AND (ii). We will deal with conditions (i) and (ii) together, using the fact that

$$\Pr\left[X_{\mathrm{id}} \in \Theta_1 \vee X_{\mathrm{id}} \in \Theta_2\right] \leq \Pr\left[X_{\mathrm{id}} \in \Theta_1\right] + \Pr\left[X_{\mathrm{id}} \in \Theta_2 \,|\, X_{\mathrm{id}} \notin \Theta_1\right].$$

Since the adversary does not make useless queries, its MAC queries are distinct. In the ideal world, the values T_i for $i \in \{1, \ldots, q_m\}$ are then simply chosen uniformly and independently at random from $\{0,1\}^n$. We introduce the random variable

$$C = \left|\left\{(i,j) \in \{1, \ldots, q_m\}^2, i \neq j, T_i = T_j\right\}\right|.$$

The number of collisioning MAC queries is always lower than C. Note that

$$\mathbb{E}[C] = \sum_{1 \leq i \leq q_m} \sum_{\substack{1 \leq j \leq q_m \\ i \neq j}} \Pr\left[T_i = T_j\right] \leq \frac{q_m^2}{2^n}.$$

By Markov's inequality,

$$\Pr\left[X_{\mathrm{id}} \in \Theta_1\right] \leq \Pr\left[C \geq \sqrt{q_m}\right] \leq \frac{q_m^{3/2}}{2^n}.$$

Assume now that $X_{\mathrm{id}} \notin \Theta_1$, i.e., τ_m is such that the number of collisioning MAC queries is lower than $\sqrt{q_m}$. Recall that K_h is chosen independently from τ_m in the ideal world. Fix any (i,j) such that $i \neq j$ and $T_i = T_j$. Since the number of collisioning MAC queries is lower than $\sqrt{q_m}$, there are at most $q_m/2$ such pairs of queries. Then, since H is ε-AXU, one has

$$\Pr\left[K_h \leftarrow_\$ \mathcal{K}_h : N_i \oplus H_{K_h}(M_i) = N_j \oplus H_{K_h}(M_j)\right] \leq \varepsilon$$

and, by summing over the at most $q_m/2$ such pairs of queries, one has

$$\Pr\left[X_{\mathrm{id}} \in \Theta_2 \,|\, X_{\mathrm{id}} \notin \Theta_1\right] \leq \frac{\varepsilon q_m}{2}.$$

Hence,

$$\Pr\left[X_{\mathrm{id}} \in \Theta_1 \cup \Theta_2\right] \leq \frac{q_m^{3/2}}{2^n} + \frac{\varepsilon q_m}{2}.$$

CONDITION (iii). We consider any verification query $(N_j', M_j', T_j', b_j) \in \tau_v$ and upper bound the probability that condition (iii) is satisfied for this particular query. Since the adversary is nonce-respecting, there is at most one MAC query (N_i, M_i, T_i) such that $N_i = N_j'$. We distinguish two cases:

- If the verification query comes after the MAC query, then since the distinguisher is non-trivial, either $T_i \neq T_j'$, or $M_i \neq M_j'$. In the former case, condition (iii) cannot be satisfied, while in the latter case, the probability over the random draw of K_h that $H_{K_h}(M_i) \oplus H_{K_h}(M_j') = 0$ is at most ε.
- If the MAC query comes after the verification query, then T_i is random and independent from T_j' and the probability that $T_i = T_j'$ is 2^{-n}.

Since for an ε-AXU hash function with range $\{0,1\}^n$ one has $\varepsilon \geq 2^{-n}$, we see that in all cases condition (iii) is met with probability at most ε. Thus, by summing over every verification query, one has

$$\Pr\left[X_{\mathrm{id}} \in \Theta_3\right] \leq \varepsilon q_v.$$

The Lemma follows by an union bound over all conditions. $\square$

4.3   Analysis of Good Transcripts

We now analyze good transcripts and prove the following lemma.

Lemma 3. *Assume that* $q_m^{3/2} \leq 2^n/4$ *and* $q_v \leq 2^n/4$. *Then, for any good transcript* τ, *one has*

$$\frac{\Pr\left[X_{\mathrm{re}} = \tau\right]}{\Pr\left[X_{\mathrm{id}} = \tau\right]} \geq 1 - \frac{4q_m^{3/2}}{2^n} - \frac{6q_v}{2^n}.$$

Let $\tau = (\tau_m, \tau_v, K_h)$ be a good transcript. Since in the ideal world the MAC oracle is perfectly random and the verification always rejects, one simply has

$$\Pr[X_{\mathrm{id}} = \tau] = \frac{1}{|\mathcal{K}_h| \cdot (2^n)^{q_m}}. \tag{6}$$

We must now lower bound the probability of getting τ in the real world. We say that a pair of permutations (P, P') is compatible with τ_m if

$$\forall i \in \{1, \ldots, q_m\}, \; \Pi[P, P', H_{K_h}](N_i, M_i) = T_i,$$

and we say that it is compatible with τ_v if

$$\forall i \in \{1, \ldots, q_v\}, \; \Pi[P, P', H_{K_h}](N_i', M_i') \neq T_i'.$$

We simply say that (P, P') is compatible with τ if it is compatible with τ_m and τ_v. We denote $\mathsf{Comp}(\tau_m)$, $\mathsf{Comp}(\tau_v)$, and $\mathsf{Comp}(\tau)$ the set of pairs of permutations that are compatible with respectively τ_m, τ_v, and τ. Then one can easily check (see for example [CS14] for a detailed explanation) that

$$\Pr[X_{\mathrm{re}} = \tau] = \frac{1}{|\mathcal{K}_h|} \cdot \Pr\left[(P, P') \leftarrow_\$ \mathsf{Perm}(n)^2 : (P, P') \in \mathsf{Comp}(\tau)\right]. \tag{7}$$

MAC QUERIES TRANSCRIPT. We will first consider the probability that a random pair (P, P') is compatible with the MAC queries transcript τ_m. To ease the notation, we reorder the transcript as follows. Let r be the number of distinct tags T appearing in MAC queries. Then we rewrite the transcript so that all queries with the same tag are consecutive, so that the MAC queries transcript (that we still denote τ_m) is now

$$\begin{aligned}
\tau_m = (&(N_{1,1}, M_{1,1}, T_1), \ldots, (N_{1,q_1}, M_{1,q_1}, T_1), \\
&(N_{2,1}, M_{2,1}, T_2), \ldots, (N_{2,q_2}, M_{2,q_2}, T_2), \\
&\ldots, \\
&(N_{r,1}, M_{r,1}, T_r), \ldots, (N_{r,q_r}, M_{r,q_r}, T_r)),
\end{aligned}$$

where $T_1, \ldots, T_r$ are distinct and $\sum_{i=1}^{r} q_i = q_m$.

Our goal is now to lower bound the probability that a random pair of permutations (P, P') satisfies

$$\forall i \in \{1, \ldots, r\}, \forall j \in \{1, \ldots, q_i\}, \ P'\big(P(N_{i,j}) \oplus N_{i,j} \oplus H_{K_h}(M_{i,j})\big) = T_i.$$

For this, we will consider the possible "internal" values $Z_i = (P')^{-1}(T_i)$. We say that a tuple $\mathbf{Z} = (Z_1, \ldots, Z_r)$ of distinct values in $\{0,1\}^n$ is *good* if

(a) all q_m values $Z_i \oplus N_{i,j} \oplus H_{K_h}(M_{i,j})$ for $i \in \{1, \ldots, r\}$, $j \in \{1, \ldots, q_i\}$ are distinct;

(b) for every verification query $(N', M', T', b) \in \tau_v$ such that $N' = N_{i,j}$ and $T' = T_k$ for some $i \in \{1, \ldots, r\}$, $j \in \{1, \ldots, q_i\}$, and $k \in \{1, \ldots, r\}$ with $k \neq i$, one has

$$Z_i \oplus H_{K_h}(M_{i,j}) \oplus H_{K_h}(M') \neq Z_k.$$

Property (a) is needed since the values $Z_i \oplus N_{i,j} \oplus H_{K_h}(M_{i,j})$ are the images by P of the (distinct) nonces $N_{i,j}$. Property (b) will be needed later when lower bounding the probability that (P, P') is compatible with the verification transcript τ_v.

Given a good tuple $\mathbf{Z}$, the probability, for a randomly drawn pair (P, P'), that

$$\begin{cases} \forall i \in \{1, \ldots, r\}, \forall j \in \{1, \ldots, q_i\}, \ P(N_{i,j}) = Z_i \oplus N_{i,j} \oplus H_{K_h}(M_{i,j}), \\ \forall i \in \{1, \ldots, r\}, \ P'(Z_i) = T_i \end{cases} \quad (8)$$

is exactly

$$\frac{1}{(2^n)_{q_m}(2^n)_r}. \quad (9)$$

(This is simply the probability that P satisfies $q_1 + \ldots + q_r = q_m$ equations and P' satisfies r equations.)

It remains to lower bound the number $N_{\mathbf{Z}}$ of good tuples $\mathbf{Z}$, which can be done as follows. First, note that by definition of a good transcript, for any $i \in \{1, \ldots, r\}$, the values $Z_i \oplus N_{i,j} \oplus H_{K_h}(M_{i,j})$ for $1 \leq j \leq q_i$ are distinct since otherwise condition (ii) defining a bad transcript would be fulfilled (without that, good tuples $\mathbf{Z}$ would not exist). In the following, for $i, k \in \{1, \ldots, r\}$ with $k < i$, we denote $q'_{i,k}$ the number of verification queries $(N', M', T', b) \in \tau_v$ such that either $N' = N_{i,j}$ for some $j \in \{1, \ldots, q_i\}$ and $T' = T_k$, or $N' = N_{k,j}$ for some $j \in \{1, \ldots, q_k\}$ and $T' = T_i$. Note that since a verification query can count for at most one pair (i, k), one has

$$\sum_{i=2}^{r} \sum_{k=1}^{i-1} q'_{i,k} \leq q_v. \quad (10)$$

Then,

– there are at least 2^n possibilities for Z_1;

– once Z_1 is fixed, there are at least $2^n - 1 - q_2 q_1 - q'_{2,1}$ possibilities for Z_2 since Z_2 must be different from the following values:

- Z_1,
- $Z_1 \oplus N_{1,j} \oplus H_{K_h}(M_{1,j}) \oplus N_{2,j'} \oplus H_{K_h}(M_{2,j'})$ for all $j \in \{1, \ldots, q_1\}$ and all $j' \in \{1, \ldots, q_2\}$ (in order for property (a) to be fulfilled),
- $Z_1 \oplus H_{K_h}(M_{1,j}) \oplus H_{K_h}(M')$ for every verification query $(N', M', T', b) \in \tau_v$ such that $N' = N_{1,j}$ for some $j \in \{1, \ldots, q_1\}$ and $T' = T_2$, and $Z_1 \oplus H_{K_h}(M_{2,j}) \oplus H_{K_h}(M')$ for every verification query $(N', M', T', b) \in \tau_v$ such that $N' = N_{2,j}$ for some $j \in \{1, \ldots, q_2\}$ and $T' = T_1$, which amounts to at most $q'_{2,1}$ values (in order for property (b) to be fulfilled);

– once $Z_1, \ldots, Z_i$ are fixed, there are at least $2^n - i - q_{i+1} \sum_{k=1}^{i} q_k - \sum_{k=1}^{i} q'_{i+1,k}$ possibilities for Z_{i+1} since Z_{i+1} must be different from the following values:

- $Z_1, \ldots, Z_i$,
- $Z_k \oplus N_{k,j} \oplus H_{K_h}(M_{k,j}) \oplus N_{i+1,j'} \oplus H_{K_h}(M_{i+1,j'})$ for all $k \in \{1, \ldots, i\}$, all $j \in \{1, \ldots, q_k\}$, and all $j' \in \{1, \ldots, q_{i+1}\}$,
- $Z_k \oplus H_{K_h}(M_{k,j}) \oplus H_{K_h}(M')$ for every verification query $(N', M', T', b) \in \tau_v$ such that $N' = N_{k,j}$ for some $k \in \{1, \ldots, i\}$, $j \in \{1, \ldots, q_k\}$ and $T' = T_{i+1}$, and $Z_k \oplus H_{K_h}(M_{i+1,j}) \oplus H_{K_h}(M')$ for every verification query $(N', M', T', b) \in \tau_v$ such that $N' = N_{i+1,j}$ for some $j \in \{1, \ldots, q_{i+1}\}$ and $T' = T_k$ for some $k \in \{1, \ldots, i\}$, which amounts to at most $\sum_{k=1}^{i} q'_{i+1,k}$ values.

Hence, the number of good tuples $\mathbf{Z} = (Z_1, \ldots, Z_r)$ is at least

$$N_{\mathbf{Z}} \geq \prod_{i=0}^{r-1} \left(2^n - i - q_{i+1} \sum_{k=1}^{i} q_k - \sum_{k=1}^{i} q'_{i+1,k} \right). \tag{11}$$

VERIFICATION QUERIES TRANSCRIPT. From now on, we fix a good tuple $\mathbf{Z}$. We will now lower bound the probability that a random pair (P, P') is compatible with the verification transcript τ_v, conditioned on (P, P') satisfying the set of Eq. (8). (Recall that P is then fixed on q_m values and P' is fixed on r values.) For this, it will be easier to upper bound the probability that (P, P') is *not* compatible with τ_v, i.e., that there exists $(N', M', T', b) \in \tau_v$ such that

$$P'\big(P(N') \oplus N' \oplus H_{K_h}(M')\big) = T'. \tag{12}$$

Fix any verification query $(N', M', T', b) \in \tau_v$. We say that it is *nonce-fresh*, resp. *tag-fresh*, if N', resp. T' does not appear in the MAC queries transcript τ_m.[8] We consider four possible cases.

---

[8] We stress that this freshness definition is with respect to the entire MAC queries transcript τ_m, independently of when the verification query was actually made by the distinguisher.

- *Case 1: the verification query is both nonce-fresh and tag-fresh.* Then $P(N')$ is random and two sub-cases can occur: if $P(N') \oplus N' \oplus H_{K_h}(M') \in \mathbf{Z}$, Eq. (12) cannot be satisfied since the query is tag-fresh; on the other hand, if $P(N') \oplus N' \oplus H_{K_h}(M') \notin \mathbf{Z}$, Eq. (12) is satisfied with probability $1/(2^n - r)$ over the choice of P'. Hence, over the choice of (P, P'), Eq. (12) is satisfied with probability at most

$$\frac{1}{2^n - r} \leq \frac{1}{2^n - q_m}.$$

- *Case 2: the verification query is nonce-fresh, but not tag-fresh.* Then there exists $(N, M, T) \in \tau_m$ such that $T = T'$. Let $Z = (P')^{-1}(T)$ (this value is well defined since we assume Eq. (8) are satisfied). Then Eq. (12) is satisfied iff

$$P(N') = Z \oplus N' \oplus H_{K_h}(M'),$$

hence with probability exactly $1/(2^n - q_m)$ since the query is nonce-fresh and N' does not appear in Eq. (8).

- *Case 3: the verification query is tag-fresh, but not nonce-fresh.* Then there exists a unique $(N, M, T) \in \tau_m$ such that $N' = N$, so that $P(N')$ is fixed by Eq. (8). If $P(N') \oplus N' \oplus H_{K_h}(M') \in \mathbf{Z}$, then Eq. (12) cannot be satisfied since the query is tag-fresh. If $P(N') \oplus N' \oplus H_{K_h}(M') \notin \mathbf{Z}$, then Eq. (12) is satisfied with probability

$$\frac{1}{2^n - r} \leq \frac{1}{2^n - q_m}.$$

- *Case 4: the verification query is neither nonce-fresh nor tag-fresh.* Then there exists a unique $(N_{i,j}, M_{i,j}, T_i) \in \tau_m$ such that $N' = N_{i,j}$ and $(N_k, M_k, T_k) \in \tau_m$ (with possibly $k = i$) such that $T' = T_k$. If $k = i$, then Eq. (12) cannot be satisfied since otherwise one would have

$$P(N') \oplus N' \oplus H_{K_h}(M') = (P')^{-1}(T_i) = P(N_{i,j}) \oplus N_{i,j} \oplus H_{K_h}(M_{i,j}),$$

which implies $H_{K_h}(M') = H_{K_h}(M_{i,j})$ and condition (iii) defining a bad transcript would be fulfilled. On the other hand, if $k \neq i$, then Eq. (12) being satisfied would imply

$$P(N') \oplus N' \oplus H_{K_h}(M') = (P')^{-1}(T_k) = Z_k$$
$$\Rightarrow P(N_{i,j}) \oplus N_{i,j} \oplus H_{K_h}(M') = Z_k$$
$$\Rightarrow Z_i \oplus H_{K_h}(M_{i,j}) \oplus H_{K_h}(M') = Z_k,$$

and this would contradict property (b) of a good tuple $\mathbf{Z}$. Hence, by definition of a good transcript and a good tuple $\mathbf{Z}$, we see that Eq. (12) cannot be satisfied in that case.

Summarizing, we see that for any verification query, Eq. (12) is satisfied with probability at most $1/(2^n - q_m)$. By a union bound over the q_v verification queries, we obtain that

$$\Pr\left[(P, P') \in \mathsf{Comp}(\tau_v) \mid (P, P') \text{ satisfies Eq. (8)}\right] \geq 1 - \frac{q_v}{2^n - q_m}. \qquad (13)$$

SUMMING UP. We can now lower bound the probability that a random pair (P, P') is compatible with τ, that we denote

$$\mathsf{p}(\tau) \overset{\text{def}}{=} \Pr\left[(P, P') \leftarrow_{\$} \mathsf{Perm}(n)^2 : (P, P') \in \mathsf{Comp}(\tau)\right].$$

Namely, summing over all good tuples $\mathbf{Z}$, and using (9), (11), and (13), we have

$$
\begin{aligned}
\mathsf{p}(\tau) &\geq N_{\mathbf{Z}} \times \Pr\left[(P, P') \text{ satisfies Eq. (8)}\right] \\
&\quad \times \Pr\left[(P, P') \in \mathsf{Comp}(\tau_v) \,\middle|\, (P, P') \text{ satisfies Eq. (8)}\right] \\
&\geq \frac{\prod_{i=0}^{r-1}\left(2^n - i - q_{i+1}\sum_{k=1}^{i} q_k - \sum_{k=1}^{i} q'_{i+1,k}\right)}{(2^n)_{q_m}(2^n)_r}\left(1 - \frac{q_v}{2^n - q_m}\right).
\end{aligned}
$$

This, in turn, allows us to lower bound the ratio of the probabilities to obtain τ in the real and the ideal world, namely combining (6) and (7) with the equation above, we have

$$
\frac{\Pr\left[X_{\mathrm{re}} = \tau\right]}{\Pr\left[X_{\mathrm{id}} = \tau\right]} \geq \underbrace{\frac{(2^n)^{q_m}\prod_{i=0}^{r-1}\left(2^n - i - q_{i+1}\sum_{k=1}^{i} q_k - \sum_{k=1}^{i} q'_{i+1,k}\right)}{(2^n)_{q_m}(2^n)_r}}_{A}
$$

$$
\times \left(1 - \frac{q_v}{2^n - q_m}\right). \tag{14}
$$

We focus on term A, that we can rewrite

$$
A = \prod_{i=0}^{q_m-1}\left(1 + \frac{i}{2^n - i}\right)\prod_{i=0}^{r-1}\left(1 - \underbrace{\frac{q_{i+1}\sum_{k=1}^{i} q_k}{2^n - i}}_{a_i} - \underbrace{\frac{\sum_{k=1}^{i} q'_{i+1,k}}{2^n - i}}_{b_i}\right). \tag{15}
$$

The following "Bonferroni-type" inequality will be useful to further lower bound A.

Lemma 4. *Let $r \geq 1$ be an integer and $(a_i)_{0 \leq i \leq r-1}$ and $(b_i)_{0 \leq i \leq r-1}$ be positive reals such that $a_i \leq 1/2$ and $b_i \leq 1/2$ for all $i \in \{0, \ldots, r-1\}$. Then*

$$
\prod_{i=0}^{r-1}(1 - a_i - b_i) \geq \prod_{i=0}^{r-1}(1 - a_i)\prod_{i=0}^{r-1}(1 - 2b_i).
$$

Proof. The proof is by induction. We first prove it for $r = 1$. One has

$$
(1 - a_0)(1 - 2b_0) = 1 - a_0 - 2b_0 + 2a_0 b_0 = 1 - a_0 - b_0 - \underbrace{b_0(1 - 2a_0)}_{\geq 0} \leq 1 - a_0 - b_0.
$$

Assume that the result holds for $r \geq 1$. Then

$$\prod_{i=0}^{r}(1 - a_i) \prod_{i=0}^{r}(1 - 2b_i) = \prod_{i=0}^{r-1}(1 - a_i) \prod_{i=0}^{r-1}(1 - 2b_i) \times \underbrace{(1 - a_r)(1 - 2b_r)}_{\geq 0}$$

$$\leq \prod_{i=0}^{r-1}(1 - a_i - b_i) \times (1 - a_r - b_r - b_r(1 - 2a_r))$$

$$= \prod_{i=0}^{r}(1 - a_i - b_i) - \underbrace{b_r(1 - 2a_r) \prod_{i=0}^{r-1}(1 - a_i - b_i)}_{\geq 0}$$

$$\leq \prod_{i=0}^{r}(1 - a_i - b_i).$$

The result holds for $r + 1$ and the lemma follows.   □

We can apply this lemma to the r.h.s. of (15). Indeed, for any $i \in \{0, \ldots, r - 1\}$, one has $q_{i+1} \leq \sqrt{q_m}$ (as otherwise condition (i) of a bad transcript would be met), and $q_m^{3/2} \leq 2^n/4$ by assumption, so that

$$a_i \overset{\text{def}}{=} \frac{q_{i+1} \sum_{k=1}^{i} q_k}{2^n - i} \leq \frac{q_{i+1} \sum_{k=1}^{i} q_k}{2^n - q_m} \leq \frac{2q_m^{3/2}}{2^n} \leq \frac{1}{2},$$

Moreover, by (10) and the assumption that $q_v \leq 2^n/4$, one has

$$b_i \overset{\text{def}}{=} \frac{\sum_{k=1}^{i} q'_{i+1,k}}{2^n - i} \leq \frac{\sum_{k=1}^{i} q'_{i+1,k}}{2^n - q_m} \leq \frac{2q_v}{2^n} \leq \frac{1}{2}.$$

Hence,

$$A \geq \prod_{i=0}^{q_m-1}\left(1 + \frac{i}{2^n - i}\right) \prod_{i=0}^{r-1}\left(1 - \frac{q_{i+1}\sum_{k=1}^{i} q_k}{2^n - i}\right) \prod_{i=0}^{r-1}\left(1 - \frac{2\sum_{k=1}^{i} q'_{i+1,k}}{2^n - i}\right)$$

$$\geq \prod_{i=0}^{q_m-1}\left(1 + \frac{i}{2^n - i}\right) \prod_{i=0}^{r-1}\left(1 - \frac{q_{i+1}\sum_{k=1}^{i} q_k}{2^n - i}\right)\left(1 - \frac{2\sum_{i=0}^{r-1}\sum_{k=1}^{i} q'_{i+1,k}}{2^n - q_m}\right)$$

$$\geq \underbrace{\prod_{i=0}^{q_m-1}\left(1 + \frac{i}{2^n - i}\right) \prod_{i=0}^{r-1}\left(1 - \frac{q_{i+1}\sum_{k=1}^{i} q_k}{2^n - i}\right)}_{A'}\left(1 - \frac{2q_v}{2^n - q_m}\right), \qquad (16)$$

where for the last inequality we used (10).

In order to further lower bound A', we need to distinguish collisioning MAC queries from non-collisioning ones. Up to reordering the MAC queries transcript, we assume that non-collisioning queries come first, and we let $s \in \{0, \ldots, r\}$ be the integer such that $q_i = 1$ for $i \in \{1, \ldots, s\}$, and $q_i > 1$ for $i \in \{s+1, \ldots, r\}$. Note that since the transcript is good, one has

$$\sum_{i=s+1}^{r} q_i \leq \sqrt{q_m} \tag{17}$$

as otherwise condition (i) of a bad transcript would be fulfilled. Then

$$
\begin{aligned}
A' &\geq \prod_{i=0}^{q_m-1}\left(1+\frac{i}{2^n-i}\right)\prod_{i=0}^{s-1}\left(1-\frac{q_{i+1}\sum_{k=1}^{i}q_k}{2^n-i}\right)\prod_{i=s}^{r-1}\left(1-\frac{q_{i+1}\sum_{k=1}^{i}q_k}{2^n-i}\right) \\
&= \prod_{i=0}^{q_m-1}\left(1+\frac{i}{2^n-i}\right)\prod_{i=0}^{s-1}\left(1-\frac{i}{2^n-i}\right)\prod_{i=s}^{r-1}\left(1-\frac{q_{i+1}\sum_{k=1}^{i}q_k}{2^n-i}\right) \\
&\geq \prod_{i=0}^{q_m-1}\left(1-\frac{i^2}{(2^n-i)^2}\right)\prod_{i=s}^{r-1}\left(1-\frac{q_{i+1}q_m}{2^n-i}\right) \\
&\geq \prod_{i=0}^{q_m-1}\left(1-\frac{i^2}{(2^n-q_m)^2}\right)\prod_{i=s}^{r-1}\left(1-\frac{q_{i+1}q_m}{2^n-q_m}\right) \\
&\geq \left(1-\frac{q_m^3}{3(2^n-q_m)^2}\right)\left(1-\frac{q_m\sum_{i=s+1}^{r}q_i}{2^n-q_m}\right) \\
&\geq \left(1-\frac{4q_m^3}{3\cdot 2^{2n}}\right)\left(1-\frac{2q_m^{3/2}}{2^n}\right), \tag{18}
\end{aligned}
$$

where for the last inequality we used (17) and $q_m \leq 2^n/2$.

Combining (14), (16), and (18), we finally obtain (using $q_m \leq 2^n/2$ once again)

$$\frac{\Pr\left[X_{\mathrm{re}} = \tau\right]}{\Pr\left[X_{\mathrm{id}} = \tau\right]} \geq 1 - \frac{4q_m^3}{3\cdot 2^{2n}} - \frac{2q_m^{3/2}}{2^n} - \frac{6q_v}{2^n}.$$

Lemma 3 follows using $q_m^3/2^{2n} \leq q_m^{3/2}/2^n$ by our assumption that $q_m^{3/2} \leq 2^n/4$.

5   Nonce-Misuse Security of EWCDM

In this section, we consider the security of the EWCDM construction when the adversary is allowed to repeat nonces. In this setting, PRF-security implies MAC-security, hence we can simply consider the EWCDM construction as a function with domain $\mathcal{N} \times \mathcal{M}$ and study its pseudorandomness. Our result on the PRF-security of the EWCDM construction is as follows.

Lemma 5. *Let $\mathcal{M}$, $\mathcal{K}$ and $\mathcal{K}_h$ be non-empty sets. Let $E : \mathcal{K} \times \{0,1\}^n \to \{0,1\}^n$ be a block cipher and $H : \mathcal{K}_h \times \mathcal{M} \to \{0,1\}^n$ be an ε-AXU hash function. Then for any (q,t)-(nonce-misusing) adversary A against the PRF-security of*

$\Pi[E, H]$, there exists a (q, t')-adversary A' against the PRP-security of E, where $t' = O(t + qt_H)$, such that

$$\mathbf{Adv}_{\Pi[E,H]}^{\mathrm{PRF}}(\mathsf{A}) \leq 2\mathbf{Adv}_E^{\mathrm{PRP}}(\mathsf{A}') + \frac{q^2}{2^n} + \frac{q^2 \varepsilon}{2}.$$

The corresponding MAC-security can easily be deduced from Lemma 5 using the following generic result of Bellare *et al.* [BGM04, Proposition 7.3].

Lemma 6. *Let* F *be a keyed function with output length* n. *Then for any* (q_m, q_v, t)-*adversary* A *against the MAC-security of* F, *there exists a* (q_m+q_v, t')-*adversary* A' *against the PRF-security of* F, *where* $t' = O(t)$, *such that*

$$\mathbf{Adv}_F^{\mathrm{MAC}}(\mathsf{A}) \leq \mathbf{Adv}_F^{\mathrm{PRF}}(\mathsf{A}') + \frac{q_v}{2^n}.$$

Combining Lemmas 5 and 6, we obtain the following theorem (absorbing the $q_v/2^n$ term into $(q_m + q_v)^2/2^n$).

Theorem 4. *Let* $\mathcal{M}$, $\mathcal{K}$ *and* $\mathcal{K}_h$ *be non-empty sets. Let* $E : \mathcal{K} \times \{0, 1\}^n \to \{0, 1\}^n$ *be a block cipher and* $H : \mathcal{K}_h \times \mathcal{M} \to \{0, 1\}^n$ *be an* ε-*AXU hash function. Then for any* (q_m, q_v, t)-*nonce-misusing adversary* A *against the MAC-security of* $\Pi[E, H]$, *there exists a* $(q_m + q_v, t')$-*adversary* A' *against the PRP-security of* E, *where* $t' = O(t + (q_m + q_v)t_H)$, *such that*

$$\mathbf{Adv}_{\Pi[E,H]}^{\mathrm{MAC}}(\mathsf{A}) \leq 2\mathbf{Adv}_E^{\mathrm{PRP}}(\mathsf{A}') + \frac{2(q_m + q_v)^2}{2^n} + \frac{(q_m + q_v)^2 \varepsilon}{2}.$$

The proof of Lemma 5 is standard (indeed, the construction, seen as a keyed function with domain $\mathcal{N} \times \mathcal{M}$, follows the classical "hash-then-PRF" paradigm). We include it below for completeness.

PROOF OF LEMMA 5. Fix a (q, t)-adversary A against the PRF-security of $\Pi[E, H]$. The first step of the proof consists in replacing E_K and $E_{K'}$ by two uniformly random and independent permutations P and P'. It is easy to show that there is an adversary A' making at most q queries and running in time at most $t' = O(t + qt_H)$ such that

$$\mathbf{Adv}_{\Pi[E,H]}^{\mathrm{PRF}}(\mathsf{A}) \leq 2\mathbf{Adv}_E^{\mathrm{PRP}}(\mathsf{A}') + \mathbf{Adv}_{\Pi[E^*,H]}^{\mathrm{PRF}}(\mathsf{A}), \tag{19}$$

where E^* denotes the perfect cipher on $\{0, 1\}^n$. Then, we use the PRP/PRF switching lemma [BR06] to replace the random permutations P and P' by two independent and uniformly random functions R and R', obtaining

$$\mathbf{Adv}_{\Pi[E^*,H]}^{\mathrm{PRF}}(\mathsf{A}') \leq \frac{q^2}{2^n} + \mathbf{Adv}_{\Pi[F^*,H]}^{\mathrm{PRF}}(\mathsf{A}), \tag{20}$$

where F^* denotes the perfect keyed function from $\{0, 1\}^n$ to $\{0, 1\}^n$ (i.e., the keyed function with key space $\mathsf{Func}(n)$).

It remains to upper bound the PRF-advantage of A against $\Pi[F^*, H]$. For this, we use the H-coefficients technique. The adversary must distinguish between two worlds:

- the real world in which it interacts with $\Pi[R, R', H]$ where R and R' are two uniformly and independently drawn functions from $\{0,1\}^n$ to $\{0,1\}^n$;
- the ideal world in which it receives independent and uniformly random answers.

Let $\tau_m = ((N_1, M_1, T_1), \ldots, (N_q, M_q, T_q))$ be the list of all queries of A and the corresponding answers. In order to have a simple description of bad transcripts, we reveal to the adversary at the end of the experiment the key K_h and the function R if we are in the real world, while in the ideal world we simply draw a dummy key $K_h \leftarrow_\$ K_h$ and a function R independently from the answers of the oracle. All in all, the transcript of the interaction of A with its oracle is a tuple $\tau = (\tau_m, K_h, R)$ and, in this case, a transcript is said attainable (with respect to an adversary A) if the probability to obtain it in the ideal world is non-zero. We denote Θ the set of attainable transcripts. We also denote X_{re}, resp. X_{id}, the probability distribution of the transcript τ induced by the real world, resp. the ideal world.

We start by defining the set of bad transcripts.

Definition 4. *We say that an attainable transcript $\tau = (\tau_m, K_h, R)$ is bad if there exists distinct queries $(N, M, T), (N', M', T') \in \tau_m$ such that*

$$R(N) \oplus N \oplus H_{K_h}(M) = R(N') \oplus N' \oplus H_{K_h}(M').$$

Otherwise we say that τ is good. We denote Θ_{bad}, resp. Θ_{good}, the set of bad, resp. good transcripts.

We first upper bound the probability to get a bad transcript in the ideal world.

Lemma 7.

$$\Pr\left[X_{\mathrm{id}} \in \Theta_{\mathrm{bad}}\right] \le \frac{q^2 \varepsilon}{2}.$$

Proof. Let τ_m be any attainable query transcript. Recall that, in the ideal world, the key K_h and the function R are drawn uniformly at random and independently from the query transcript τ_m. Fix any pair of distinct queries $(N, M, T), (N', M', T')$. Two cases can occur:

- $M \ne M'$: then the probability, over the random draw of K_h and R, that $R(N) \oplus N \oplus H_{K_h}(M) = R(N') \oplus N' \oplus H_{K_h}(M')$ is lower than ε by the ε-AXU property of H;
- $M = M'$: then, since we assume that the adversary never makes redundant queries, $N \ne N'$ and the probability that $R(N) \oplus N = R(N') \oplus N'$ is $1/2^n \le \varepsilon$.

By summing over every possible pair of queries, one gets the result.   □

We then analyze good transcripts.

Lemma 8. *For any good transcript τ, one has*

$$\frac{\Pr\left[X_{\mathrm{re}} = \tau\right]}{\Pr\left[X_{\mathrm{id}} = \tau\right]} = 1.$$

Proof. Let $\tau = (\tau_m, K_h, R)$ be a good transcript. One has

$$\Pr\left[X_{\mathrm{id}} = \tau\right] = \frac{1}{|\mathcal{K}_h|} \cdot \frac{1}{|\mathsf{Func}(n)|} \cdot \frac{1}{(2^n)^q}$$

since, in the ideal world, the oracle is perfectly random and the key K_h and the function R are chosen uniformly at random and independently from the query transcript.

We say that a function $R' \in \mathsf{Func}(n)$ is compatible with the transcript τ if $R'(R(N_i) \oplus N_i \oplus H_{K_h}(M_i)) = T_i$ for all $i \in \{1, \ldots, q\}$. Let $\mathsf{Comp}(\tau)$ be the set of all compatible functions R'. Then it is easy to see that

$$\Pr\left[X_{\mathrm{re}} = \tau\right] = \frac{1}{|\mathcal{K}_h|} \cdot \frac{1}{|\mathsf{Func}(n)|} \cdot \Pr\left[R' \leftarrow_{\$} \mathsf{Func}(n) : R' \in \mathsf{Comp}(\tau)\right].$$

Since τ is a good transcript, the values $R(N_i) \oplus N_i \oplus H_{K_h}(M_i)$ are distinct. Hence

$$\Pr\left[R' \leftarrow_{\$} \mathsf{Func}(n) : R' \in \mathsf{Comp}(\tau)\right] = \frac{1}{(2^n)^q}$$

and therefore $\Pr\left[X_{\mathrm{re}} = \tau\right] = \Pr\left[X_{\mathrm{id}} = \tau\right]$. $\square$

Combining Lemmas 1, 7, and 8, one obtains

$$\mathbf{Adv}_{\Pi[F^*, H]}^{\mathrm{PRF}}(\mathsf{A}) \leq \frac{q^2 \varepsilon}{2}. \tag{21}$$

Lemma 5 finally follows from Eqs. (19), (20), and (21).

Acknowledgments. Many thanks to Thomas Peyrin. This paper stemmed from discussions with him, and he took part to the early stages of this work.

References

[ABBT15] Abdelraheem, M.A., Beelen, P., Bogdanov, A., Tischhauser, E.: Twisted polynomials and forgery attacks on GCM. In: Oswald, E., Fischlin, M. (eds.) EUROCRYPT 2015. LNCS, vol. 9056, pp. 762–786. Springer, Heidelberg (2015)

[BC09] Black, J., Cochran, M.: MAC reforgeability. In: Dunkelman, O. (ed.) FSE 2009. LNCS, vol. 5665, pp. 345–362. Springer, Heidelberg (2009)

[BCK96] Bellare, M., Canetti, R., Krawczyk, H.: Keying hash functions for message authentication. In: Koblitz, N. (ed.) CRYPTO 1996. LNCS, vol. 1109, pp. 1–15. Springer, Heidelberg (1996)

[Ber00] Bernstein, D.J.: Floating-point arithmetic and message authentication. Unpublished manuscript (2000). http://cr.yp.to/papers.html#hash127

[Ber05a] Bernstein, D.J.: Stronger security bounds for permutations. Unpublished manuscript (2005). http://cr.yp.to/papers.html#permutations

[Ber05b] Bernstein, D.J.: Stronger security bounds for Wegman-Carter-Shoup authenticators. In: Cramer, R. (ed.) EUROCRYPT 2005. LNCS, vol. 3494, pp. 164–180. Springer, Heidelberg (2005)

[Ber05c] Bernstein, D.J.: The Poly1305-AES message-authentication code. In: Gilbert, H., Handschuh, H. (eds.) FSE 2005. LNCS, vol. 3557, pp. 32–49. Springer, Heidelberg (2005)

[Ber07] Bernstein, D.J.: Polynomial evaluation and message authentication. Unpublished manuscript (2007). http://cr.yp.to/papers.html#pema

[BGK99] Bellare, M., Goldreich, O., Krawczyk, H.: Stateless evaluation of pseudorandom functions: security beyond the birthday barrier. In: Wiener, M. (ed.) CRYPTO 1999. LNCS, vol. 1666, p. 270. Springer, Heidelberg (1999)

[BGM04] Bellare, M., Goldreich, O., Mityagin, A.: The power of verification queries in message authentication and authenticated encryption. IACR Cryptology ePrint Archive, Report 2004/309 (2004). http://eprint.iacr.org/2004/309

[BHK+99] Black, J., Halevi, S., Krawczyk, H., Krovetz, T., Rogaway, P.: UMAC: fast and secure message authentication. In: Wiener, M. (ed.) CRYPTO 1999. LNCS, vol. 1666, p. 216. Springer, Heidelberg (1999)

[BI99] Bellare, M., Impagliazzo, R.: A tool for obtaining tighter security analyses of pseudorandom function based constructions, with applications to PRP to PRF conversion. IACR Cryptology ePrint Archive, Report 1999/024 (1999). http://eprint.iacr.org/1999/024

[BKR98] Bellare, M., Krovetz, T., Rogaway, P.: Luby-Rackoff backwards: increasing security by making block ciphers non-invertible. In: Nyberg, K. (ed.) EUROCRYPT 1998. LNCS, vol. 1403, pp. 266–280. Springer, Heidelberg (1998)

[BKR00] Bellare, M., Kilian, J., Rogaway, P.: The security of the cipher block chaining message authentication code. J. Comput. Syst. Sci. **61**(3), 362–399 (2000)

[BPR05] Bellare, M., Pietrzak, K., Rogaway, P.: Improved security analyses for CBC MACs. In: Shoup, V. (ed.) CRYPTO 2005. LNCS, vol. 3621, pp. 527–545. Springer, Heidelberg (2005)

[BR05] Black, J., Rogaway, P.: CBC MACs for arbitrary-length messages: the three-key constructions. J. Cryptol. **18**(2), 111–131 (2005)

[BR06] Bellare, M., Rogaway, P.: The security of triple encryption and a framework for code-based game-playing proofs. In: Vaudenay, S. (ed.) EUROCRYPT 2006. LNCS, vol. 4004, pp. 409–426. Springer, Heidelberg (2006). http://eprint.iacr.org/2004/331

[Bra82] Brassard, G.: On computationally secure authentication tags requiring short secret shared keys. In: Chaum, D., Rivest, R.L., Sherman, A.T. (eds.) CRYPTO 1982, pp. 79–86. Plenum Press, New York (1982)

[CLL+14] Chen, S., Lampe, R., Lee, J., Seurin, Y., Steinberger, J.: Minimizing the two-round even-mansour cipher. In: Garay, J.A., Gennaro, R. (eds.) CRYPTO 2014, Part I. LNCS, vol. 8616, pp. 39–56. Springer, Heidelberg (2014). http://eprint.iacr.org/2014/443

[CLP14] Cogliati, B., Lampe, R., Patarin, J.: The indistinguishability of the XOR of k permutations. In: Cid, C., Rechberger, C. (eds.) FSE 2014. LNCS, vol. 8540, pp. 285–302. Springer, Heidelberg (2015)

[CS14]   Chen, S., Steinberger, J.: Tight security bounds for key-alternating ciphers. In: Nguyen, P.Q., Oswald, E. (eds.) EUROCRYPT 2014. LNCS, vol. 8441, pp. 327–350. Springer, Heidelberg (2014). http://eprint.iacr.org/2013/222

[CS16]   Cogliati, B., Seurin, Y.: EWCDM: an efficient, beyond-birthday secure, nonce-misuse resistant MAC. Full version of this paper. http://eprint.iacr.org/2016/525

[DS11]   Dodis, Y., Steinberger, J.: Domain extension for MACs beyond the birthday barrier. In: Paterson, K.G. (ed.) EUROCRYPT 2011. LNCS, vol. 6632, pp. 323–342. Springer, Heidelberg (2011)

[Fer05]   Ferguson, N.: Authentication weaknesses in GCM. Comments Submitted to NIST Modes of Operation Process (2005). http://csrc.nist.gov/groups/ST/toolkit/BCM/documents/comments/CWC-GCM/Ferguson2.pdf

[GMS74] Gilbert, E.N., MacWilliams, F.J., Sloane, N.J.A.: Codes which detect deception. Bell Syst. Tech. J. **53**(3), 405–424 (1974)

[HK97]   Halevi, S., Krawczyk, H.: MMH: software message authentication in the Gbit/second rates. In: Biham, E. (ed.) FSE 1997. LNCS, vol. 1267, pp. 172–189. Springer, Heidelberg (1997)

[HP08]   Handschuh, H., Preneel, B.: Key-recovery attacks on universal hash function based MAC algorithms. In: Wagner, D. (ed.) CRYPTO 2008. LNCS, vol. 5157, pp. 144–161. Springer, Heidelberg (2008)

[HWKS98] Hall, C., Wagner, D., Kelsey, J., Schneier, B.: Building PRFs from PRPs. In: Krawczyk, H. (ed.) CRYPTO 1998. LNCS, vol. 1462, pp. 370–389. Springer, Heidelberg (1998)

[JJV02]   Jaulmes, É., Joux, A., Valette, F.: On the security of randomized CBC-MAC beyond the birthday paradox limit: a new construction. In: Daemen, J., Rijmen, V. (eds.) FSE 2002. LNCS, vol. 2365, pp. 237–251. Springer, Heidelberg (2002)

[JL04]   Jaulmes, É., Lercier, R.: FRMAC, a fast randomized message authentication code (2004). http://eprint.iacr.org/2004/166

[Jou06]   Joux, A.: Authentication failures in NIST version of GCM. Comments Submitted to NIST Modes of Operation Process (2006). http://csrc.nist.gov/groups/ST/toolkit/BCM/documents/comments/800-38-Series-Drafts/GCM/Joux_comments.pdf

[KR00]   Krovetz, T., Rogaway, P.: Fast universal hashing with small keys and no preprocessing: the PolyR construction. In: Won, D. (ed.) ICISC 2000. LNCS, vol. 2015, pp. 73–89. Springer, Heidelberg (2001)

[Kra94]   Krawczyk, H.: LFSR-based hashing and authentication. In: Desmedt, Y.G. (ed.) CRYPTO 1994. LNCS, vol. 839, pp. 129–139. Springer, Heidelberg (1994)

[KVW04] Kohno, T., Viega, J., Whiting, D.: CWC: a high-performance conventional authenticated encryption mode. In: Roy, B., Meier, W. (eds.) FSE 2004. LNCS, vol. 3017, pp. 408–426. Springer, Heidelberg (2004)

[Luc00]   Lucks, S.: The sum of PRPs is a secure PRF. In: Preneel, B. (ed.) EUROCRYPT 2000. LNCS, vol. 1807, pp. 470–484. Springer, Heidelberg (2000)

[Min10]   Minematsu, K.: How to Thwart birthday attacks against MACs via small randomness. In: Hong, S., Iwata, T. (eds.) FSE 2010. LNCS, vol. 6147, pp. 230–249. Springer, Heidelberg (2010)

[MV04]   McGrew, D.A., Viega, J.: The security and performance of the Galois/counter mode (GCM) of operation. In: Canteaut, A., Viswanathan, K. (eds.) INDOCRYPT 2004. LNCS, vol. 3348, pp. 343–355. Springer, Heidelberg (2004)

[Pat90] Patarin, J.: Pseudorandom permutations based on the DES scheme. In: Cohen, G.D., Charpin, P. (eds.) EUROCODE 1990. LNCS, vol. 514, pp. 193–204. Springer, Heidelberg (1991)

[Pat91] Patarin, J.: New results on pseudorandom permutation generators based on the DES scheme. In: Feigenbaum, J. (ed.) CRYPTO 1991. LNCS, vol. 576, pp. 301–312. Springer, Heidelberg (1992)

[Pat08a] Patarin, J.: A proof of security in $O(2^n)$ for the XOR of two random permutations. In: Safavi-Naini, R. (ed.) ICITS 2008. LNCS, vol. 5155, pp. 232–248. Springer, Heidelberg (2008). http://eprint.iacr.org/2008/010

[Pat08b] Patarin, J.: The "Coefficients H" technique. In: Avanzi, R.M., Keliher, L., Sica, F. (eds.) SAC 2008. LNCS, vol. 5381, pp. 328–345. Springer, Heidelberg (2009)

[Pat13] Patarin, J.: Security in $O(2^n)$ for the XOR of two random permutations: proof with the standard H technique. IACR Cryptology ePrint Archive, Report 2013/368 (2013). http://eprint.iacr.org/2013/368

[PC15] Procter, G., Cid, C.: On weak keys and forgery attacks against polynomial-based MAC schemes. In: Moriai, S. (ed.) FSE 2013. LNCS, vol. 8424, pp. 287–304. Springer, Heidelberg (2014)

[Rog95] Rogaway, P.: Bucket hashing and its application to fast message authentication. In: Coppersmith, D. (ed.) CRYPTO 1995. LNCS, vol. 963, pp. 29–42. Springer, Heidelberg (1995)

[Saa12] Saarinen, M.-J.O.: Cycling attacks on GCM, GHASH and other polynomial MACs and hashes. In: Canteaut, A. (ed.) FSE 2012. LNCS, vol. 7549, pp. 216–225. Springer, Heidelberg (2012)

[Sho96] Shoup, V.: On fast and provably secure message authentication based on universal hashing. In: Koblitz, N. (ed.) CRYPTO 1996. LNCS, vol. 1109, pp. 313–328. Springer, Heidelberg (1996)

[WC81] Wegman, M.N., Carter, L.: New hash functions and their use in authentication and set equality. J. Comput. Syst. Sci. **22**(3), 265–279 (1981)

[Yas10] Yasuda, K.: The sum of CBC MACs is a secure PRF. In: Pieprzyk, J. (ed.) CT-RSA 2010. LNCS, vol. 5985, pp. 366–381. Springer, Heidelberg (2010)

[Yas11] Yasuda, K.: A new variant of PMAC: beyond the birthday bound. In: Rogaway, P. (ed.) CRYPTO 2011. LNCS, vol. 6841, pp. 596–609. Springer, Heidelberg (2011)

[ZWSW12] Zhang, L., Wu, W., Sui, H., Wang, P.: 3kf9: enhancing 3GPP-MAC beyond the birthday bound. In: Wang, X., Sako, K. (eds.) ASIACRYPT 2012. LNCS, vol. 7658, pp. 296–312. Springer, Heidelberg (2012)

Asymmetric Cryptography and Cryptanalysis I

A Subfield Lattice Attack on Overstretched NTRU Assumptions

Cryptanalysis of Some FHE and Graded Encoding Schemes

Martin Albrecht[1]([⊠]), Shi Bai[2], and Léo Ducas[3]

[1] Information Security Group, Royal Holloway, University of London, London, UK
martin.albrecht@royalholloway.ac.uk
[2] ENS de Lyon, Laboratoire LIP (U. Lyon, CNRS, ENSL, INRIA, UCBL),
Lyon, France
shih.bai@gmail.com
[3] Cryptology Group, CWI, Amsterdam, The Netherlands
ducas@cwi.nl

Abstract. The subfield attack exploits the presence of a subfield to solve overstretched versions of the NTRU assumption: norming the public key h down to a subfield may lead to an easier lattice problem and any sufficiently good solution may be lifted to a short vector in the full NTRU-lattice. This approach was originally sketched in a paper of Gentry and Szydlo at Eurocrypt'02 and there also attributed to Jonsson, Nguyen and Stern. However, because it does not apply for small moduli and hence NTRUEncrypt, it seems to have been forgotten. In this work, we resurrect this approach, fill some gaps, analyze and generalize it to any subfields and apply it to more recent schemes. We show that for significantly larger moduli — a case we call overstretched — the subfield attack is applicable and asymptotically outperforms other known attacks.

This directly affects the asymptotic security of the bootstrappable homomorphic encryption schemes LTV and YASHE which rely on a mildly overstretched NTRU assumption: the subfield lattice attack runs in sub-exponential time $2^{O(\lambda/\log^{1/3}\lambda)}$ invalidating the security claim of $2^{\Theta(\lambda)}$. The effect is more dramatic on GGH-like Multilinear Maps: this attack can run in polynomial time without *encodings of zero* nor the *zero-testing parameter*, yet requiring an additional quantum step to recover the secret parameters exactly.

M. Albrecht—Supported by EPSRC grant EP/L018543/1 "Multilinear Maps in Cryptography".

S. Bai—Supported by ERC Starting Grant ERC-2013-StG-335086-LATTAC.

L. Ducas—Supported by a grant from CWI from budget for public-private-partnerships and by a grant from NXP Semiconductors through the European Union's H2020 Programme under grant agreement number ICT-645622 (PQCRYPTO) and ICT-644209 (HEAT).

The full version of the paper is available on http://eprint.iacr.org/2016/127.

© International Association for Cryptologic Research 2016
M. Robshaw and J. Katz (Eds.): CRYPTO 2016, Part I, LNCS 9814, pp. 153–178, 2016.
DOI: 10.1007/978-3-662-53018-4_6

We also report on practical experiments. Running LLL in dimension 512 we obtain vectors that would have otherwise required running BKZ with block-size 130 in dimension 8192. Finally, we discuss concrete aspects of this attack, the condition on the modulus q to guarantee full immunity, discuss countermeasures and propose open questions.

Keywords: Subfield lattice attack · Overstretched NTRU · FHE · Graded encoding schemes

1   Introduction

Lattice-based cryptography relies on the presumed hardness of lattice problems such as the shortest vector problem (SVP) and its variants. For efficiency, many practical lattice-based cryptosystems are based on assumptions on structured lattices such as the NTRU lattice. Introduced by Hoffstein et al. [HPS96, HPS98], the NTRU assumption states that it is hard to find a short vector in the $\mathcal{R}$-module

$$\Lambda_h^q = \{(x, y) \in \mathcal{R}^2 \text{ s.t. } hx - y = 0 \bmod q\}$$

with the promise that a very short solution — the private key — (f, g) exists. The ring $\mathcal{R} = \mathbb{Z}[X]/(P(X))$ is a polynomial ring of rank n over $\mathbb{Z}$, typically a circular convolution ring $(P(X) = X^n - 1)$ or the ring of integers in a cyclotomic number field $(P(X) = \Phi_m(X)$ and $n = \phi(m))$.

Following the pioneer scheme NTRUENCRYPT [HPS98], the NTRU assumption has been re-used in various cryptographic constructions such as signatures schemes [HHGP+03, DDLL13], fully homomorphic encryption [LTV12, BLLN13] and a candidate construction for cryptographic multi-linear maps [GGH13a, LSS14, ACLL15]. After two decades of cryptanalysis, the NTRUENCRYPT scheme remains essentially unbroken, and is one of the fastest candidates for the public-key cryptosystems in the post-quantum era.

Coppersmith and Shamir [CS97] noticed that recovering a short enough vector, may it be different from the actual secret key (f, g), may be sufficient for an attack and claimed that the celebrated LLL algorithm of Lenstra et al. [LLL82] would lead to such an attack. However, it turned out [HPS98] that for sufficiently large dimension n, a much stronger lattice reduction is required and that the NTRUENCRYPT is asymptotically secure. Meanwhile, parameters have been updated to take account for progress in lattice reduction algorithms and potential quantum speed-ups [HPS+15].

Other types of attacks have been considered, such as Odlyzko's meet-in-the-middle attack described in [HSW06]. In practice, the best known algorithm for attacking NTRU lattices is the combined lattice-reduction and meet-in-the-middle attack of Howgrave-Graham [HG07]. Asymptotically, a slightly sub-exponential attack against the ternary-NTRU problem was proposed by Kirchner and Fouque [KF15], with a heuristic complexity $2^{\Theta(n/\log\log q)}$, which is to our knowledge the only sub-exponential attack when q is polynomial in n.

It is typically assumed that NTRU lattices are essentially as intractable as unstructured lattices with similar parameters[1], but without the structure of $\mathcal{R}$-module.

In the present work, we consider the application of lattice reduction in a *subfield* to attack the NTRU assumption for large moduli q. This subfield lattice attack is asymptotically faster than the direct lattice attack as soon as q is super-polynomial, and may also be relevant for polynomially-sized q. We call the problem[2] considered in this work "overstretched NTRU" to distinguish it from the original NTRU parameter choices, which remain secure.

Asymptotics. The subfield attack leads to solving overstreched NTRU instances in time complexity $\mathsf{poly}(n) \cdot 2^{\Theta(\beta)}$ with $\beta/\log \beta = \Theta\left(n \log n / \log^2 q\right)$ when ever the relative degree parameter $r = \Theta(\log q/\log n)$ is greater than 1. In comparison, the direct lattice attack required setting $\beta/\log \beta = \Theta\left(n/\log q\right)$.

We are mostly concerned with overstretched NTRU assumptions when q is super-polynomial in n, in which case the best known attacks are already sub-exponential in n. For cryptographic relevance, we will therefore state all our asymptotics in terms of what was previously thought as the security parameter λ: given $q = q(\lambda)$ we constrain $n = n(\lambda)$ so that the previously best known attack requires exponential time $2^{\Theta(\lambda)}$. In this cryptographic metric, the subfield lattice attack is sub-exponential as soon as q is super-polynomial, and gets polynomial for larger parameters $q = 2^{\tilde{\Theta}(\lambda)} = 2^{\tilde{\Theta}(\sqrt{n})}$.

Our Contribution. In this work, we resurrect[3] the subfield lattice attack sketched in [GS02, Sec.6], attributed to Gentry, Szydlo, Jonsson, Nguyen and Stern. It consists of norming down the secret key to a subfield, running lattice reduction in the subfield to solve a smaller, potentially easier lattice problem and lifting the solution back to the full field.

While the original sketch [GS02] only considered the maximal real subfield, we naturally generalize it to any subfield. We also spell out a different lifting step from arbitrary subfields and prove it applicable even if only an approximation of the normed-down key is found.

We then show that this algorithm solves the overstretched NTRU problem in sub-exponential time when the modulus q is quasi-polynomial in the security parameter λ and in polynomial time when the modulus q is super-exponential in λ (equivalently, $q = 2^{\tilde{\Theta}(\sqrt{n})}$). Applying this algorithm, we show that it gives a subexponential attack on parameter choices for NTRU-based FHE schemes [LTV12,BLLN13] which were believed secure previously. We also show that this algorithm enables new attacks on GGH-like graded encoding

---

[1] Volume, dimension and length of unusually short vectors.

[2] The NTRU problem has also been recently been referred to as DSPR (Decisional Small Polynomial Ratio), but we prefer its historical name for fair attribution of this invention.

[3] A preliminary version of this work qualified the attack considered in this work as new. We are grateful to John Schanck for pointing us to this prior art.

schemes [GGH13a, LSS14, ACLL15]. These attacks lead to subexponential classical and polynomial-time quantum attacks on GGH-like constructions but do
not require encodings of zero nor do they use the zero-testing parameter in contrast to previous work [HJ15].

We also report on experimental results for the subfield lattice attack which
show that the attack is meaningful in practice. Using LLL in dimension 512 we
have obtained vectors that would have required running BKZ with block-size
about 130 in dimension 8192. We refer the reader to the full version of this work
for the experimental results.

Related Work. As mentioned above, a variant of the attack considered in this
work was sketched in [GS02]. Moreover, the Gentry-Szydlo algorithm from the
same work, which allows to reconstruct an element a given the ideal (a) as well as
the Gram element $a\bar{a}$, i.e. the norm $N_{K/K^+}(a)$ of a relatively to the real subfield,
can be seen as a subfield attack. It lead to an attack of the NSS scheme [HPS01]
in which the Gram element $a\bar{a}$ was leaked as the covariance of a certain function
of the signatures. The Gentry-Szydlo algorithm was recently revisited [LS14].

This attack is very similar in spirit to an attack of Gentry [Gen01] against the
NTRU-composite assumption which tackles NTRU problems over rings $\mathcal{R}$ that
can be written as direct products $\mathcal{R} \simeq \mathcal{R}_1 \times \mathcal{R}_2$. More specifically [Gen01] targets
circulant convolution rings $\mathbb{Z}[X]/(X^n - 1) \simeq \mathbb{Z}[X]/(X^{n_1} - 1) \times \mathbb{Z}[X]/(X^{n_2} - 1)$
where $n = n_1 n_2$. Under such condition, there exists a projection $\pi : \mathcal{R} \to \mathcal{R}_1$
that is a ring homomorphism, and he showed that this projection could only
increase the Euclidean length of secret polynomials by a factor $\sqrt{n_2}$. This makes
this attack very powerful (even when the modulus q is quite small). Because this
projection is a ring homomorphism, this approach is not limited to NTRU and
would also apply to Ring-SIS or Ring-LWE.

In some sense, the line of work by Lauter et al. [ELOS15, EHL14, CLS15]
against skewed[4] variants of Ring-LWE falls in this framework, with a direct
factorization of the rings $\mathcal{R}$ modulo q: $(\mathcal{R}/q\mathcal{R}) \simeq (\mathcal{R}_1/q\mathcal{R}_1) \times (\mathcal{R}_2/q\mathcal{R}_2)$. As
already noted in [Gen01], this requires the — seemingly sporadic — property
that the projection map $\pi_q : (\mathcal{R}/q\mathcal{R}) \to (\mathcal{R}_1/q\mathcal{R}_1)$ induces only a manageable
geometric distortion. Similar ideas are being explored to attack schemes based
on certain quasi-cyclic binary codes in work [Loi14, LJ14, HT15].

In comparison, this work tackles NTRU when the ring $\mathcal{R}$ equals $\mathcal{O}_\mathbb{K}$ (the ring
of integer of a number field $\mathbb{K}$) and therefore cannot be a direct product; and
when $\mathbb{K}$ admits proper subfields. Due to the aforementioned attack of [Gen01],
direct product rings are now avoided for lattice-based cryptography, and the typical choice is to use the ring of integers of a cyclotomic number field of the form
$\mathcal{R} = \mathcal{O}_{\mathbb{Q}(\omega_m)} = \mathbb{Z}[\omega_m]$. This setting allows to argue worst-case hardness of certain
problems (Ring-SIS [Mic02], Ideal-LWE [SSTX09], later improved and renamed
to Ring-LWE [LPR10]). Yet all those number fields admit proper subfields

---

[4] It was recently shown that these attacks were in fact made possible by an improper
choice of a very skewed error distributions leading to several noise-free linear equations [CIV16, Pei16].

(at least, the maximal real subfield). Instead of using a projection map π, this attack exploits a relative norm map $N_{\mathbb{K}/\mathbb{L}} : \mathcal{O}_{\mathbb{K}} \to \mathcal{O}_{\mathbb{L}}$, which is only a multiplicative map. This induces a significant yet manageable blow-up on the Euclidean length of secret polynomials and requires a large modulus q. This seems to also limit this attack to the NTRU setting.

Our work is also strongly inspired by the the logarithm-subfield strategy of Bernstein [Ber14], which anticipated other works towards a logarithm attack [CGS14, CDPR16]. While the presence of subfields was in the end not necessary for the recovery of short generators of principal ideals in cyclotomic rings, we show in this work that, indeed, the presence of proper subfields can be exploited in other specific set-ups.

Concurrently and independently to this work, Cheon, Jeong and Lee also investigated subfield attacks on GGH-like graded encoding schemes in work [CJL16]. The general approach is very similar to the one adopted in this work. In [CJL16], however, the trace map is utilised instead of the norm and the result is only presented for the case of powers-of-two cyclotomic rings. Despite using the trace map — which is linear — they obtain a growth of the secret that is similar to ours: multiplicative. For example, when the relative degree of $\mathbb{K}$ over $\mathbb{L}$ is $r = 2$, the trace map $\text{Tr}_{\mathbb{K}/\mathbb{L}}$ sends g/f to $g/f + \bar{g}/\bar{f} = (g\bar{f} + \bar{g}f)/f\bar{f}$ where $\bar{\cdot}$ denotes the adequate automorphism. For comparison, the norm $N_{\mathbb{K}/\mathbb{L}}$ sends g/f to $g\bar{g}/f\bar{f}$. Using the norm map is therefore slightly better when both f, g have the same size (the numerator is smaller by a factor $\approx \sqrt{r}$); but the trace map could be very advantageous when $g \gg f$. Furthermore, Cheon, Jeong and Lee achieve better results for GGH-like graded encoding schemes by making use of the zero-testing parameter which leads to a polynomial-time classical attack for large levels of multilinearity κ.

Outline. Section 2 gives preliminaries on the geometry of NTRU lattices and a brief introduction of the lattice reduction algorithms. Section 3 then presents the subfield lattice attack with its asymptotic performance analyzed in Subsect. 3.4. In Sect. 4, we apply this attack to the FHE and MLM constructions proposed in recent literature. In Sect. 5, we report experimental results for the subfield lattice attack. Finally, Sect. 6 presents the conclusions and suggests directions for future research.

2   Preliminaries

Vectors are presented in row vectors. The notation $[\cdot]_q$ denotes reduction modulo an integer q.

2.1   Number Fields and Subfields

We assume some familiarity with basic algebraic number theory. The reader may refer to [Sam70] for an introduction on the topic.

Let $\mathbb{K}$ be a number field of degree $n = [\mathbb{K} : \mathbb{Q}]$ over $\mathbb{Q}$, and assume $\mathbb{K}$ is a Galois extension of $\mathbb{Q}$ with the Galois group G. The fundamental theorem of Galois Theory states an one-to-one correspondence between the subgroups G' of G and the subfields $\mathbb{L}$ of $\mathbb{K}$ with G' being the subgroup of G fixing $\mathbb{L}$. Let therefore $\mathbb{L}$ be a subfield of $\mathbb{K}$ and G' be the subgroup of G fixing $\mathbb{L}$, and denote $n' = [\mathbb{L} : \mathbb{Q}]$, $r = [\mathbb{K} : \mathbb{L}]$ (so $r = n/n'$). The number fields $\mathbb{K}$, $\mathbb{L}$ and therefore the degrees n, n' and relative degree r are fixed in the rest of this work.

The relative norm $\mathrm{N}_{\mathbb{K}/\mathbb{L}} : \mathbb{K} \to \mathbb{L}$ (resp. relative trace $\mathrm{Tr}_{\mathbb{K}/\mathbb{L}} : \mathbb{K} \to \mathbb{L}$) is a multiplicative (resp. an additive) map defined by

$$\mathrm{N}_{\mathbb{K}/\mathbb{L}} : a \mapsto \prod_{\psi \in G'} \psi(a), \quad \text{resp.} \quad \mathrm{Tr}_{\mathbb{K}/\mathbb{L}} : a \mapsto \sum_{\psi \in G'} \psi(a). \tag{1}$$

The canonical inclusion $\mathbb{L} \subset \mathbb{K}$ will be written explicitly as $L : \mathbb{L} \to \mathbb{K}$. The ring of integers of $\mathbb{K}$ and $\mathbb{L}$ are denoted by $\mathcal{O}_{\mathbb{K}}$ and $\mathcal{O}_{\mathbb{L}}$.

A number field of degree n admits n embeddings –i.e. field morphisms– to the complex numbers. Writing $\mathbb{K} = \mathbb{Q}(X)/(P(X))$ for some monic irreducible polynomial P, and letting $\alpha_1, \ldots, \alpha_n \in \mathbb{C}$ be the distinct complex roots of P, each embedding $e_i : \mathbb{K} \to \mathbb{C}$ consists of evaluating $a \in \mathbb{K}$ at a root α_i, formally $e_i : a \mapsto a(\alpha_i)$. The Galois group acts by permutation on the set of embeddings.

Cyclotomic Number Field. We denote by ω_m an arbitrary primitive m-th root of unity. For cryptanalytic purposes, we are mostly interested in the case when $\mathbb{K} = \mathbb{Q}(\omega_m)$ is the m-th cyclotomic number field; But we may also want to instantiate the attack for subfields $\mathbb{L}$ of $\mathbb{K}$ that are not necessarily cyclotomic number fields.

The number field $\mathbb{L} = \mathbb{Q}(\omega_m)$ has degree $n = \phi(m)$, and has a Galois group isomorphic to $\mathbb{Z}_m^*$: explicitly $i \in \mathbb{Z}_m^*$ corresponds to the automorphism $\psi_i : \omega_m \mapsto \omega_m^i$. Any number field $\mathbb{Q}(\omega_{m'})$ for $m'|m$ is a subfield of $\mathbb{Q}(\omega_m)$, but there are other proper subfields. In particular, the maximal real subfield $\mathbb{Q}(\omega_m + \bar{\omega}_m)$ is a proper subfield of degree $n/2$, and more generally, $\mathbb{K} = \mathbb{Q}(\omega_m)$ admits a subfield of degree n' for any divisor $n'|n$.[5]

We recall (see [Was97], Theorem 2.6) that the ring of integers $\mathcal{O}_{\mathbb{K}}$ of $\mathbb{K} = \mathbb{Q}(\omega_m)$ is exactly $\mathbb{Z}[\omega_m]$.

2.2 Coprimality in $\mathcal{O}_{\mathbb{L}}$

To argue below that we can lift solutions in the subfield to the full field, we rely on two randomly chosen elements in $\mathcal{O}_{\mathbb{L}}$ being coprime. We use density results to estimate such probability. The density of coprime pairs of ideals [Sit10] and elements [FM14] in $\mathcal{O}_{\mathbb{L}}$ is $1/\zeta_{\mathbb{L}}(2)$ where $\zeta_{\mathbb{L}}$ denotes the Dedekind zeta function over $\mathbb{K}$.

---

[5] For example, 7 is prime, so $\mathbb{Q}(\omega_7)$ admits no cyclotomic number fields as proper subfields, yet it admits two proper subfields: $\mathbb{Q}(\omega_7 + \bar{\omega}_7)$ of degree 3 and $\mathbb{Q}(\omega_7 + \omega_7^2 + \omega_7^4)$ of degree 2.

We consider ζ_L for cyclotomic number fields $\mathbb{K} = \mathbb{Q}(\omega_m)$ where $m = p^k$ for some prime p. The next lemma shows that $\lim_{k \to \infty} \zeta_L(s) = 1/(1 - p^{-s})$ for real $s > 3/2$.

Lemma 1. *Let $\mathbb{L}$ be a cyclotomic number field $\mathbb{Q}(\omega_{m'})$ for $m' = p^k$. Then for any real $s > 3/2$ we have*

$$\lim_{k \to \infty} \zeta_L(s) = 1/(1 - p^{-s}).$$

In particular $\lim_{k \to \infty} \zeta_L(2) = 4/3$ for cyclotomic number fields of conductor $m' = 2^k$.

Proof. Please refer to the full version of this work for the proof. □

Further, we numerically approximated $\zeta_L^{-1}(2)$ for $\mathbb{L} = \mathbb{Q}[x]/(x^n + 1)$ for $n = 128$ and $n = 256$ by computing the first 2^{22} terms of the Dirichlet series of the Dedekind zeta function for $\mathbb{L}$ and then evaluated the truncated series at 2. In both cases we get a density ≈ 0.75.

We stress that our pairs f', g' are random elements obtained as relative norms $N_{\mathbb{K}/\mathbb{L}}(f), N_{\mathbb{K}/\mathbb{L}}(g)$ of random *short* f and g, and under the additional condition that f is invertible modulo q. However, our experiments indicate that $3/4$ is a good approximation of the actual probability of coprimality. Additionally, it seems that this requirement is an artifact of our proof, as experiments succeeded even when those elements had a common factor.

2.3 Euclidean Geometry

The number field $\mathbb{K}$ (or $\mathbb{L}$) is viewed as a Euclidean $\mathbb{Q}$-vector space by endowing it with the inner product

$$\langle a, b \rangle = \sum_e e(a)\bar{e}(b) \tag{2}$$

where e ranges over all the n (or n') embeddings $\mathbb{K} \to \mathbb{C}$. This defines a Euclidean norm denoted by $\| \cdot \|$. In addition to the Euclidean norm, we will make use of the operator norm $| \cdot |$ defined by:

$$|a| = \sup_{x \in \mathbb{K}^*} \|ax\|/\|x\|. \tag{3}$$

It is easy to check that the operator norm $|a|$ of a equals to the maximal absolute complex embedding of a:

$$|a| = \max_e |e(a)| \tag{4}$$

where e ranges over all the embeddings $e : \mathbb{K} \to \mathbb{C}$. We note that if $\omega \in \mathbb{K}$ is a root of unity, then $|\omega| = 1$. The operator's norm is sub-multiplicative: $|ab| \le |a| |b|$, and we have the inequality $|a| \le \|a\|$. The Euclidean norm and the operator norm are invariant under automorphisms $\psi : \mathbb{K} \mapsto \mathbb{K}$,

$$\|a\| = \|\psi(a)\|, \quad |a| = |\psi(a)| \tag{5}$$

since the group of automorphisms acts by permutation on the set of embeddings. One also verifies that $\|L(a)\|^2 = r\|a\|^2$ and $|L(a)| = |a|$ for all $a \in \mathbb{L}$. Additionally, the algebraic norm can be bounded in term of geometric norms:

$$N_{\mathbb{K}/\mathbb{Q}}(a) \leq |a|^n \leq \|a\|^n. \tag{6}$$

The inner product (and therefore the Euclidean norm) are extended in a coefficient-wise manner to vectors of $\mathbb{K}^d$: $\langle (a_1, \ldots, a_d), (b_1, \ldots, b_d) \rangle = \sum \langle a_i, b_i \rangle$.

Definition 1. *A distribution $\mathcal{D}$ over $\mathbb{K}^d$ is said to be isotropic of variance $\sigma^2 \geq 0$ if, for any $y \in \mathbb{K}^d$ it hold that*

$$\mathbb{E}_{x \leftarrow \mathcal{D}} \left[\langle x, y \rangle^2 \right] = \sigma^2 \|y\|^2$$

where $\mathbb{E}[\cdot]$ denotes the expectation of a random variable.

Remark. In most theoretical work, the distributions of secrets or errors are spherical discrete Gaussian distribution over $\mathcal{O}_{\mathbb{K}}$ which are isotropic —up to negligible statistical distance. For simplicity, some practically oriented work instead chose random ternary coefficients. In the typical power-of-two case cyclotomic case, such distribution is isotropic of variance $2n/3$. Yet, for more general choices $\mathbb{K} = \mathbb{Q}(\omega_m)$, in the worse case (when m is composed of many small distinct prime factor), this may induce up to quasi-polynomial distortion $n^{\log(n)}$ (see [LPR10]). Such choice of set-up should only marginally affect our asymptotic results.

2.4 $\mathcal{O}_{\mathbb{K}}$ Modules and Lattices

To avoid confusion, we shall speak of the rank of $\mathcal{O}_{\mathbb{K}}$-modules and of $\mathbb{K}$-vectors-spaces when $\mathbb{K} \neq \mathbb{Q}$, and restrict the term of dimension to $\mathbb{Z}$-modules and $\mathbb{Q}$-vector spaces.

The dimension $\dim(\Lambda)$ of a lattice Λ is the dimension over $\mathbb{Q}$ of the $\mathbb{Q}$-vector space it spans[6]. We recall that the minimal distance of a lattice Λ is defined as $\lambda_1(\Lambda) = \min_{v \in \Lambda \setminus \{0\}} \|v\|$. Also, the volume of a lattice $\mathrm{Vol}(\Lambda)$ is defined as the square root of the absolute determinant of the Gram matrix of any basis $\{b_1 \ldots b_{\dim(\Lambda)}\}$ of Λ $\mathrm{Vol}(\Lambda) = \sqrt{\det([\langle b_i, b_j \rangle]_{i,j})}$. For any set of $\mathbb{Q}$-linearly independent vectors $\{v_1, \ldots, v_{\dim(\Lambda)}\} \subset \Lambda$, we have the inequality:

$$\mathrm{Vol}(\Lambda) \leq \prod \|v_i\|. \tag{7}$$

The rank of an $\mathcal{O}_{\mathbb{K}}$ module $M \subset \mathbb{K}^d$ can be defined as the rank over $\mathbb{K}$ of the $\mathbb{K}$ vector-space it spans, but it does not necessarily equal the size of a minimal set of $\mathcal{O}_{\mathbb{K}}$-generators[7]. The Euclidean vector space structure of $\mathbb{K}^d$ allows to view any discrete $\mathcal{O}_{\mathbb{K}}$-module $M \subset \mathbb{K}^d$ as a lattice. The discriminant $\Delta_{\mathbb{K}}$ of a

---

[6] Or equivalently, the size of a minimal sets of $\mathbb{Z}$-generators, since $\mathbb{Z}$ is a principal ideal domain.

[7] Non-principal ideals of $\mathbb{K}$ being a counter-example.

number field relates to the volume of its ring of integers $\sqrt{|\Delta_{\mathbb{K}}|} = \mathrm{Vol}(\mathcal{O}_{\mathbb{K}})$. More generally, we have the identity:

$$\mathrm{Vol}(a\mathcal{O}_{\mathbb{K}}) = \mathrm{N}_{\mathbb{K}/\mathbb{Q}}(a)\sqrt{|\Delta_{\mathbb{K}}|}. \tag{8}$$

This gives rise to a lower bound on the volume $\mathcal{O}_{\mathbb{K}}$-modules of rank 1 in term of its minimal distance:

Lemma 2. *Let $M \subset \mathbb{K}^d$ be a discrete $\mathcal{O}_{\mathbb{K}}$-module of rank 1. It follows that* $\mathrm{Vol}(M) \leq \lambda_1(M)^n \sqrt{|\Delta_{\mathbb{K}}|}$.

Proof. Without loss of generality, we may assume that $d = 1$ (by constructing a $\mathbb{K}$-linear isometry $\iota : \mathrm{Span}_{\mathbb{K}}(M) \to \mathbb{K} \otimes_{\mathbb{Q}} \mathbb{R}$). Let $a \in \mathbb{K} \otimes_{\mathbb{Q}} \mathbb{R}$ be a shortest vector of M, we have $M \supset a\mathcal{O}_{\mathbb{K}}$, therefore $\mathrm{Vol}(M) \leq \mathrm{Vol}(a\mathcal{O}_{\mathbb{K}}) = \mathrm{N}_{\mathbb{K}/\mathbb{Q}}(a)\sqrt{|\Delta_{\mathbb{K}}|}$, and we conclude noting that $\mathrm{N}_{\mathbb{K}/\mathbb{Q}}(a) \leq \|a\|^n$. □

2.5 NTRU Assumption

Let us first describe the NTRU problem as follows.

Definition 2 (NTRU **problem, a.k.a. DSPR**). *The NTRU problem is defined by four parameters: a ring $\mathcal{R}$ (of rank n and endowed with an inner product), a modulus q, a distribution $\mathcal{D}$, and a target norm τ. Precisely, NTRU$(\mathcal{R}, q, \mathcal{D}, \tau)$ is the problem of, given $h = [gf^{-1}]_q$ (conditioned on f being invertible mod q) for $f, g \leftarrow \mathcal{D}$, finding a vector $(x, y) \in \mathcal{R}^2$ such that $(x, y) \neq (0, 0)$ mod q and of Euclidean norm less than $\tau\sqrt{2n}$ in the lattice*

$$\Lambda_h^q = \{(x, y) \in \mathcal{R}^2 \ s.t. \ hx - y = 0 \bmod q\}. \tag{9}$$

We may abuse notation and denote NTRU$(\mathcal{R}, q, \sigma, \tau)$ *for* NTRU$(\mathcal{R}, q, \mathcal{D}, \tau)$ *where $\mathcal{D}$ is any reasonable isotropic distribution of variance σ^2.*

Note that NTRU$(\mathcal{R}, q, \sigma, \sigma)$ is essentially the problem of recovering the secret key (f, g). Yet, in many cases, solving NTRU$(\mathcal{R}, q, \sigma, \tau)$ for some $\tau > \sigma$ is enough to break NTRU-like cryptosystems.

The NTRU *lattice Λ_h^q.* The lattice Λ_h^q defined by the instance $h \leftarrow$ NTRU$(\mathcal{O}_{\mathbb{K}}, q, \sigma, \tau)$ has dimension $2n$ and volume $\mathrm{Vol}(\mathcal{R})^2 q^n$. Consequently, if h were to be uniformly random, the Gaussian heuristic predicts that the shortest vectors of Λ_h^q have norm $\mathrm{Vol}(\mathcal{R})^{1/n}\sqrt{nq/\pi e}$. Therefore, whenever $\sigma < \mathrm{Vol}(\mathcal{R})^{1/n}\sqrt{q/2\pi e}$, the lattice Λ_h^q admits an *unusually short vector*. This vector is not formally a unique shortest vector: for example, if $\mathbb{K} = \mathbb{Q}(\omega_m)$, $\mathcal{R} = \mathcal{O}_{\mathbb{K}}$, all rotations $(\omega_m^i f, \omega_m^i g)$ of that vector have the same norm.

Target Parameter τ for Attacks. Because no solution would be expected if h was uniformly random, note that solving $h \leftarrow$ NTRU$(\mathcal{R}, q, \sigma, \tau)$ for $\tau < \mathrm{Vol}(\mathcal{R})^{1/n}\sqrt{q/2\pi e}$ already constitutes a distinguishing attack on the NTRU problem. As we discuss in Sect. 4, solving NTRU for such τ would break the FHE scheme based on NTRU from [LTV12] and typical parameter choices for the scheme presented in [BLLN13].

2.6   Lattice Reduction Algorithms

Lattice reduction algorithms have been studied for many years in work such as [LLL82, Sch87, GN08, HPS11]. From a theoretical perspective, one of the best lattice reduction algorithm is the slide reduction algorithm from [GN08].

Theorem 1 ([GN08]). *There is an algorithm that, given $\epsilon > 0$, the basis B of a lattice L of dimension d, and performing at most*

$$\mathrm{poly}(d, 1/\epsilon, \mathrm{bitsize}(B))$$

many operations and calls to an SVP oracle in dimension β, outputs a vector $v \in L$ whose length satisfies the following bounds:

– *the approximation-factor bound:*

$$\|v\| \le ((1 + \epsilon)\gamma_\beta)^{\frac{d-\beta}{\beta-1}} \cdot \lambda_1(L) \tag{10}$$

where $\lambda_1(L)$ is the length of a shortest vector in L and $\gamma_\beta \approx \beta$ is the β-dimensional Hermite constant.
– *the Hermite-factor bound:*

$$\|v\| \le ((1 + \epsilon)\gamma_\beta)^{\frac{d-1}{2\beta-2}} \cdot \mathrm{Vol}(L)^{1/d} \tag{11}$$

Alternatively, one may use the BKZ algorithm [Sch87] and its terminated variant [HPS11]. Similar to slide reduction, the terminated BKZ performs at most $\mathrm{poly}(d, 1/\epsilon, \mathrm{bitsize}(B))$ many operations and calls to an SVP oracle in dimension β; and outputs a vector $v \in L$ whose length has order $\beta^{\Theta(n/\beta)} \cdot \mathrm{Vol}(L)^{1/d}$. Using [Lov87, p. 25], the terminated BKZ also provides an algorithm to find an approximated shortest vector of length $\beta^{\Theta(n/\beta)} \cdot \lambda_1(L)$ in similar time.

It is well known [CN11] that in practice lattice reduction algorithms achieve much shorter results and are more efficient, but the approximation and Hermite factors remain of the order of $\beta^{\Theta(n/\beta)}$ asymptotically, for a computational cost in $\mathrm{poly}(\lambda) \cdot 2^{\Theta(\beta)}$. We will use such estimate in the following analysis.

3   The Subfield Lattice Attack

The subfield lattice attack works in three steps. First, we map the NTRU instance to the chosen subfield, then we apply lattice reduction, and finally we lift the solution to the full field. We first describe the three steps of the attacks in Sects. 3.1, 3.2 and 3.3. In Sect. 3.4, we then analyze the asymptotic performances compared to direct reduction in the full field for cryptographically relevant asymptotic parameters.

We are given an instance $h \leftarrow \mathsf{NTRU}(\mathcal{O}_\mathbb{K}, q, \sigma, \tau)$, and $(f, g) \in \mathcal{O}_\mathbb{K}$ is the associated secret. We wish to recover a short vector of Λ_h^q.

3.1 Norming Down

We define $f' = N_{\mathbb{K}/\mathbb{L}}(f)$, $g' = N_{\mathbb{K}/\mathbb{L}}(g)$, and $h' = N_{\mathbb{K}/\mathbb{L}}(h)$. The subfield attack follows from the following observation: (f', g') is a vector of $\Lambda_{h'}^q$, and depending on the parameters it may be an unusually short one.

Lemma 3. *Let $f, g \in \mathcal{O}_{\mathbb{K}} \otimes_{\mathbb{Q}} \mathbb{R}$ be sampled from continuous spherical Gaussians of variance σ^2. For any constant $c > 0$, there exists a constant C, such that,*

$$\|g'\| \le (\sigma n^C)^r, \quad \|f'\| \le (\sigma n^C)^r, \quad |f'| \le (\sigma n^C)^r, \quad |f'^{-1}| \le (n^C/\sigma)^r$$

except with probability $O(n^{-c})$.

Proof. For all embeddings $e : \mathbb{K} \mapsto \mathbb{C}$, it simultaneously holds that

$$\sigma/n^C \le |e(f)| \le \sigma n^C \tag{12}$$

except with polynomially small probability $O(n^{-c})$. Once this is established, the conclusion follows using the invariant $|\psi(a)| = |a|$ since $f' = \prod \psi(f)$, where ψ ranges over r automorphisms of $\mathbb{K}$.

To prove inequality (12), note that for each embedding e, the $\Re(e(f))$ and $\Im(e(f))$ follow a Gaussian distribution of parameter $\Theta(n)\sigma$. Classical tails inequality gives the upper bound $|e(f)| \le \sigma n^C$. For the lower bound, we remark that the probability density function of a Gaussian of parameter $\Theta(n)\sigma$ is bounded by $1/(\Theta(n)\sigma)$. This implies that the probability that a sample falls in the range $\frac{1}{\Theta(n)\sigma}[-\epsilon, \epsilon]$ is less than 2ϵ. It remains to choose $\epsilon = \Theta(n^{-c-1})$ which gives the conclusion by the union-bound. $\square$

In this work, we assume that Lemma 3 holds also for all reasonable distributions considered in cryptographic constructions.

Heuristic 1. *For any m and any $f, g \in \mathcal{O}_{\mathbb{K}}$ with reasonable isotropic distribution of variance σ^2, and any constant $c > 0$, there exists a constant C, such that,*

$$\|g'\| \le (\sigma n^C)^r, \quad \|f'\| \le (\sigma n^C)^r, \quad |f'| \le (\sigma n^C)^r, \quad |f'^{-1}| \le (n^C/\sigma)^r$$

except with probability $O(n^{-c})$.

3.2 Lattice Reduction in the Subfield

We now apply a lattice reduction algorithm with block-size β to the lattice $\Lambda_{h'}^q$, and according to the approximation factor bound (10) we obtain a vector $(x', y') \in \Lambda_{h'}^q$ of norm:

$$\|(x', y')\| \le \beta^{\Theta(2n'/\beta)} \cdot \lambda_1(\Lambda_{h'}^q) \le \beta^{\Theta(n/\beta r)} \cdot \|(f', g')\| \tag{13}$$

$$\le \beta^{\Theta(n/\beta r)} \cdot (n\sigma)^{\Theta(r)}. \tag{14}$$

Next, we argue that if the vector (x', y') is short enough, then it must be an $\mathcal{O}_{\mathbb{K}}$-multiple of (f', g'). In turn, this will allow us to lift (x', y') to a short vector in the full lattice Λ_h^q.

Theorem 2. *Let $f', g' \in \mathcal{O}_\mathbb{L}$ be such that $\langle f' \rangle$ and $\langle g' \rangle$ are coprime ideals and that $h'f' = g' \bmod q\mathcal{O}_\mathbb{L}$ for some $h' \in \mathcal{O}_\mathbb{L}$. If $(x', y') \in \Lambda_{h'}^q$ has length satisfying*

$$\|(x', y')\| < \frac{q}{\|(f', g')\|} \tag{15}$$

then $(x', y') = v(f', g')$ for some $v \in \mathcal{O}_\mathbb{L}$.

Proof. We first prove that that $B = \{(f', g'), (F', G')\}$ is a basis of the $\mathcal{O}_\mathbb{L}$-module $\Lambda_{h'}^q$ for some $(F', G') \in \mathcal{O}_\mathbb{L}^2$. The argument is adapted from [HHGP+03], Sect. 4.1 By coprimality, there exists (F', G') such that $f'G' - g'F' = q \in \mathcal{O}_\mathbb{L}$. We note that:

$$f'(F', G') - F'(f', g') = (0, q);$$
$$g'(F', G') - G'(f', g') = (-q, 0);$$
$$[f'^{-1}]_q(f', g') = (1, h') \bmod q.$$

That is, the module M generated by B contains $q\mathcal{O}_\mathbb{L}^2$ and $(1, h')$: we have proved that $\Lambda_{h'}^q \subset M$. Because $\det_\mathbb{L}(B) = f'G' - g'F' = q = \det_\mathbb{L}(\{(1, h'), (0, q)\})$ we have $\mathrm{Vol}(M) = |\Delta_\mathbb{L}|q^{n'} = \mathrm{Vol}(\Lambda_{h'}^q)$ and therefore $M = \Lambda_{h'}^q$.

We denote $\Lambda = (f', g')\mathcal{O}_\mathbb{L}$ and Λ^* the projection of $(F', G')\mathcal{O}_\mathbb{L}$ orthogonally to Λ. Let s^* of length λ_1^* be a shortest vector of Λ^*. We will conclude using the fact that any vector of $\Lambda_{h'}^q$ of length less than λ_1^* must belong to the sublattice Λ. It remains to give an lower bound for λ_1^*.

We will rely on the identity $\mathrm{Vol}(\Lambda) \cdot \mathrm{Vol}(\Lambda^*) = \mathrm{Vol}(\Lambda_{h'}^q) = |\Delta_\mathbb{L}|q^{n'}$. By Lemma 2, we have

$$\mathrm{Vol}(\Lambda) \leq |\Delta_\mathbb{L}|^{1/2}\|(f', g')\|^{n'} \quad \text{and} \quad \mathrm{Vol}(\Lambda^*) \leq |\Delta_\mathbb{L}|^{1/2}\|s^*\|^{n'}. \tag{16}$$

We deduce that $\lambda_1^* = \|s^*\| \geq q/\|(f', g')\|$. Therefore, the hypothesis (15) ensures that $\|(x', y')\| < \lambda_1^*$, and we conclude that $(x', y') \in \Lambda = (f', g')\mathcal{O}_\mathbb{L}$. $\qquad\square$

We note that according to Heuristic 1, the length condition of Theorem 2 are satisfied asymptotically when

$$\beta^{\Theta(n/\beta r)} \cdot (n\sigma)^{\Theta(r)} \leq q. \tag{17}$$

The probability of satisfying the coprimality condition for random f', g' is discussed in Sect. 2.2, where we argue it to be larger than a constant. On the other hand, experiments (cf. Sect. 5) show that the co-primality condition does not seems necessary in practice for the subfield lattice attack to succeed.

The partial conclusion is that, one may recover non-trivial information about f and g — namely, a small multiple of (f', g') — by solving an NTRU instance in a subfield. Depending on the parameters, this new problem is potentially easier since the dimension $n' = n/r$ of $\mathcal{O}_\mathbb{L}$ is significantly smaller than the dimension $2n$ of the full lattice Λ_h^q.

3.3   Lifting the Short Vector

It remains to lift the solution from the sub-ring $\mathcal{O}_\mathbb{L}$ to $\mathcal{O}_\mathbb{K}$. Simply compute the vector (x, y) where

$$x = L(x') \quad \text{and} \quad y = L(y') \cdot h / L(h') \bmod q \tag{18}$$

where $L : \mathbb{L} \to \mathbb{K}$ is the canonical inclusion map of $\mathbb{L} \subset \mathbb{K}$.

Recall from Theorem 2 that $(x', y') = v(f', g')$. We set $\tilde{f} = L(f')/f$, $\tilde{g} = L(g')/g$ and $\tilde{h} = L(h')/h$. Note that $\tilde{f}, \tilde{g}$ and $\tilde{h}$ are integers of $\mathbb{K}$. We rewrite

$$x = L(v) \cdot \tilde{f} \cdot f \bmod q.$$
$$y = L(v) \cdot L(g')/\tilde{h} = L(v) \cdot g\tilde{g}/\tilde{h} \bmod q$$
$$= L(v) \cdot \tilde{f} \cdot g \bmod q.$$

That is, under condition (17) we have found a short multiple of (f, g):

$$(x, y) = u \cdot (f, g) \in \Lambda_h^q \quad \text{with } u = L(v) \cdot \tilde{f} \in \mathcal{O}_\mathbb{K}$$
$$\|(x, y)\| \leq |v| \cdot |f|^{r-1} \cdot \|(f, g)\|$$
$$\leq |x'| \cdot |f'^{-1}| \cdot |f|^{r-1} \cdot \|(f, g)\|$$
$$\leq \beta^{\Theta(n/\beta r)} \cdot (n\sigma)^{\Theta(r)}.$$

The first inequality is established by writing $\tilde{f}$ as the product of $r - 1$ many $\psi(f)$ where the ψ's are automorphisms of $\mathbb{K}$. The second inequality decomposes $v = x'/f'$, and the last follows from Lemma 3 or Heuristic 1.

Not only we have found a short vector of Λ_h^q, but also have the guarantee that it is an $\mathcal{O}_\mathbb{K}$-multiple of the secret key (f, g). This second property will prove useful to mount attacks on the graded encoding schemes [GGH13a].

3.4   Asymptotic Performance

For the subfield attack to be successful, we require

$$\sqrt{q} = \beta^{\Theta(2n/(\beta\,r))} \cdot \lambda_1(\Lambda_{h'}^q) = \beta^{\Theta(2n/(\beta\,r))} \cdot n^{\Theta(r)}$$

when $\sigma = \mathsf{poly}(n)$. Hence, asymptotically we get

$$\frac{\beta}{\log \beta} = \Theta \left(\frac{4\,n}{r \log q - 2\,r^2 \log n} \right),$$

where we require $r \log q - 2\,r^2 \log n > 0$. Setting $r = 1$ roughly recovers the lattice attack in the full field. Setting $r = \log q/(4 \log n)$ minimizes the expression.

We illustrate the complexity for two extreme cases, where all parameters are expressed in term of a security parameter λ, and are such that the previously best known attack required time greater than 2^λ. Additionally, it is assumed that $\mathbb{K}$ contains adequate subfields so that a subfield $\mathbb{L}$ of the desired relative degree r exists. This condition is satisfied asymptotically for the typical choice $\mathbb{K} = \mathbb{Q}(\omega_{2^k})$.

In the first case, we set $q = 2^{\tilde{\Theta}(\lambda)}$, and the subfield attack is polynomial in the security parameter. For the second case, we show that as soon as q gets super-polynomial, the subfield attack can be made sub-exponential.

Remark. Our analysis does not rule out that the attack may even be relevant even for polynomial gaps q/σ: it could be that it remains exponential but with a better constant than the direct attack.

Exponential and super-exponential q. We set:

$$n = \Theta(\lambda^2 \log^2 \lambda), \quad q = \exp(\Theta(\lambda \log^2 \lambda)), \quad \sigma = \mathsf{poly}(\lambda). \tag{19}$$

Complexity of the Direct Lattice Attack. With such parameters, using 2^λ operations, we argue that one may not find any vector shorter than $\lambda_1(q\mathcal{O}_{\mathbb{K}}) = q\sqrt{n}$. Indeed, one may run lattice reduction up to block-size $\beta = \Theta(\lambda)$. Either from approximation bound or Hermite bound, the vector found should not be shorter than:

$$\beta^{\Theta(n/\beta)} = \exp\left(\Theta(\lambda^2 \log^3(\lambda)/\lambda)\right) > \lambda_1(q\mathcal{O}_{\mathbb{K}}). \tag{20}$$

We verify that having such choice of super-quadratic n makes the Kirchner-Fouque [KF15] attack at least exponential in λ : $\exp(\Theta(n/\log\log q)) = \exp(\Theta(\lambda^2 \log^2(\lambda)/\log \lambda)) > \exp(\Theta(\lambda))$.

Complexity of the Subfield Attack. In contrast, the same parameters allow the subfield attack to recover a vector of norm less than $\sqrt{q}$ in polynomial time: set $r = \Theta(\lambda)$ and $\beta = \Theta(\log \lambda)$. Then, the vector found will have norm

$$\beta^{\Theta(n/\beta r)} \cdot n^{\Theta(r)} = \exp\left(\Theta\left(\frac{\lambda^2 \log \lambda \log \log \lambda}{\lambda \log \lambda} + \lambda \log \lambda\right)\right) \tag{21}$$

$$= \exp\left(\Theta(\lambda \log \lambda \log \log \lambda)\right) < \sqrt{q}. \tag{22}$$

Similarly, setting $n = \Theta(\lambda^2)$, $q = \exp(\Theta(\lambda))$, $\beta = \Theta(\log^{1+\varepsilon} \lambda)$, $r = \Theta(\lambda/(\log \lambda \log \log \lambda))$ leads to a quasi-polynomial version of the subfield attack for exponential q.

Quasi-polynomial q. We set

$$n = \Theta(\lambda \log^\varepsilon \lambda \log \log(\lambda)), \quad q = \exp(\Theta(\log^{1+\varepsilon}\lambda)), \quad \sigma = \mathsf{poly}(\lambda). $$

Complexity of the Direct Lattice Attack. With such parameters, using 2^λ operations, we argue that one may not find any vector shorter than $\lambda_1(q\mathcal{O}_{\mathbb{K}}) = q\sqrt{n}$. Indeed, one may run lattice reduction up to block-size $\beta = \Theta(\lambda)$. Either from approximation bound or Hermite bound, the vector found should not be shorter than:

$$\beta^{\Theta(n/\beta)} = \exp\left(\Theta\left(\log^{1+\varepsilon}\lambda \log \log \lambda\right)\right) > \lambda_1(q\mathcal{O}_{\mathbb{K}}). \tag{23}$$

We verify that having such choice of super-linear n makes the Kirshner and Fouque [KF15] attack at least exponential in λ: $\exp(\Theta(n/\log\log q)) = \exp(\Theta(\lambda\log^\varepsilon \lambda\log\log \lambda/\log\log^{1+\varepsilon}\lambda)) > \exp(\Theta(\lambda))$.

Complexity of the Subfield Attack. In contrast, the same parameters allow the subfield attack to recover a vector of norm less than $\sqrt{q}$ in sub-exponential time $\exp(\lambda/\log^{\varepsilon/3}\lambda)$: set $r = \Theta(\log^{2\varepsilon/3}\lambda)$ and $\beta = \Theta(\lambda/\log^{\varepsilon/3}\lambda)$. Then, the vector found will have norm

$$\beta^{\Theta(n/\beta r)} \cdot n^{\Theta(r)} = \exp\left(\Theta\left(\frac{\log^{1+\frac{4}{3}\epsilon}(\lambda)\log\log(\lambda)}{\log^{\frac{2}{3}\epsilon}(\lambda)} + \log^{1+2/3\epsilon}(\lambda)\right)\right)$$

$$= \exp\left(\Theta\left(\log^{1+2/3\varepsilon}(\lambda)\log\log(\lambda)\right)\right) < \sqrt{q}. \tag{24}$$

4 Applications

We apply this attack to the FHE and MLM constructions from the literature and show that it necessitates to increase parameters for these schemes to remain secure at level λ. In the cryptographic context, we typically have $\mathbb{K} = \mathbb{Q}(\omega_m)$, m a power of 2, and speak of the ring $\mathcal{R} = \mathbb{Z}_q[X]/(X^n+1) \simeq \mathcal{O}_\mathbb{K}$ endowed with the cannonical inner product of its coefficients vector. The ring isomorphism $\mu : \mathcal{R} \to \mathcal{O}_\mathbb{K}$ is a scaled isometry: $\|\mu(x)\| = \sqrt{n}\|x\|$. This normalization is quite convenient, for example $\|1_\mathcal{R}\| = 1$.

4.1 Fully Homomorphic Encryption

NTRU-like schemes are used to realise fully homomorphic encryption starting with the LTV scheme from [LTV12]; the scheme was optimized and implemented in [DHS15].

LTV is motivated by [SS11] which shows that under certain choices of parameters the security of an NTRU-like scheme can be reduced to security of Ring-LWE. That is, [SS11] shows that if f and g have norms $> \sqrt{q} \cdot \mathrm{poly}(\lambda)$, then $h = [g/f]_q \in \mathbb{Z}_q[X]/(X^n+1)$ — with n a power of two — is statistically indistinguishable from a uniformly sampled element. Note that under this choice of parameters the subfield lattice attack does not apply.

However, this choice of parameters rules out even performing one polynomial multiplication and hence the schemes in [LTV12, DHS15] are based on an additional assumption that $[g/f]_q$ is computationally indistinguishable from random even when f and g are small. This assumption — which essentially states that Decisional-NTRU is hard — is called the Decisional Small Polynomial Ratio assumption (DSPR) in [LTV12]. Note that this work shows that DSPR does not hold in the presence of subfields and an overstretched NTRU assumption.

LTV can evaluate circuits of depth $L = \mathcal{O}(n^\varepsilon / \log n)$ for $q = 2^{n^\varepsilon}$ with $\varepsilon \in (0, 1)$ and its decryption circuit can be implemented in depth $(\mathcal{O}\log \log q + \log n)$. This implies

$$\log(n^{\varepsilon+1}) < n^\varepsilon / \log n,$$
$$\log(n^{\varepsilon+1}) < \log q / \log n,$$

i.e. that q must be super-polynomial in n to realise fully homomorphic encryption from LTV.

A scale-invariant variant of the scheme in [LTV12] called YASHE was proposed in [BLLN13]. This variant does not require the DSPR assumption by reducing the noise growth during multiplication. This allows f and g to be sampled from a sufficiently wide Gaussian, such that the reduction in [SS11] goes through. Sampling f and g this way allows to evaluate circuits of depth $L = (\mathcal{O}\log q / (\log \log q + \log n))$ [BLLN13, Theorem 2] for $\mathbb{Z}_2$ being the plaintext space.

On the other hand, setting the bounds on f, g to $\|f\|_\infty = \|g\|_\infty = B_{key} = 1$, the plaintext space to $\mathbb{Z}_2$ via $t = 2$, the multiplicative expansion factor of the ring to $\delta = n$ by assuming n is a power of two and $w = O(1)$, then the multiplicative expansion factor of YASHE is $(\mathcal{O}n^2)$. For correctness, it is required that the noise be less than $q/4$. Hence, to evaluate a circuit of depth L, YASHE requires $q/4 > (\mathcal{O}n^{2L})$ or $L = \mathcal{O}(\log q / \log n)$ under this choice of parameters. As a consequence, YASHE is usually instantiated with f and g very short, cf. [LN14].

Following [BV11, Lemma 4.5], Appendix H of [BLLN13] shows that YASHE is bootstrapable if it can evaluat depth $L = \mathcal{O}(\log \log q + \log n)$ circuits. For $\|f\|_\infty = \|g\|_\infty = B_{key} = 1$ this implies

$$\log \log q + \log(n) < \log q / \log n,$$
$$\log(n \log q) < \log q / \log n,$$

i.e. q must be super-polynomial in n for YASHE to provide fully homomorphic encryption.

To establish a target size, recall that NTRU-like encryption of a binary message $\mu \in \mathbb{Z}_2$ is given by $c = h \cdot e_1 + e_2 + \mu \lfloor q/2 \rfloor$ for random errors of variance ς^2. To decrypt from a solution (F, G) to the instance $h \leftarrow \mathsf{NTRU}(\mathcal{R}, q, \sigma, \tau)$, simply compute $Fc = G \cdot e_1 + F \cdot e_2 + F \cdot \mu \lfloor q/2 \rfloor$. The error term $G \cdot e_1 + F \cdot e_2$ will have entries of magnitudes $\varsigma\tau\sqrt{n}$ which we require to be $< q/2$ to decrypt correctly. Hence, we require $F, G < q/(2\varsigma\sqrt{n})$. In [LTV12, BLLN13] like in other FHE schemes, ς is chosen to be bounded by a very small, constant value.

In [CS15] several Ring-based FHE schemes are compared. For comparability amongst the considered schemes and performance, the authors chose the coefficients of f, g from $\{-1, 0, 1\}$ with the additional guarantee that only 64 coefficients are non-zero in f or g. Then, to establish hardness they assume that an adversary who can find an element $< q$ in a q-ary lattice with dimension m and volume q^n wins for all schemes considered. Now, to achieve security against lattice attacks, the root Hermite factor δ_0 in $q = \delta_0^m q^{n/m}$ should be small enough, where "small enough" depends on which prediction for lattice reduction is used. In [DHS15] the same approach is used to pick parameters, but for a slightly smaller target norm of $q/4$.

The attack presented in this work results in a subexponential attack in the security parameter λ for LTV and YASHE, if L is sufficiently large to enable fully homomorphic encryption and if n is chosen to be minimal such that a lattice attack on the full field does not succeed. Set

$$q = \exp\left(\Theta\left((\epsilon + 1)\log^2 n\right)\right)$$

to satisfy correctness. Now, to rule out lattice attacks on the full field set $n = \Theta\left(\lambda \log \lambda \log \log^2 \lambda\right)$. Hence, for $\beta = \lambda$ we have

$$\beta^{\Theta(n/\beta)} > \sqrt{q},$$
$$\Theta\left(\log^2 \lambda \log \log^2 \lambda\right) > \Theta\left(\log^2 \lambda\right).$$

For the subfield attack, pick $\beta = \Theta\left(\lambda/\log^{1/3}\lambda\right)$ and $r = \Theta\left(\log^{2/3}\lambda\right)$ and we get

$$\beta^{\Theta(n/\beta r)} \cdot n^{\Theta(r)} < \sqrt{q},$$
$$\Theta\left(\log^{\frac{5}{3}}\lambda \log \log^2 \lambda\right) < \Theta\left(\log^2 \lambda\right).$$

4.2 Graded Encoding Schemes

In [GGH13a] a candidate construction for graded encoding schemes approximating multilinear maps was proposed. The GGH construction was improved in [LSS14] and implemented and improved further in [ACLL15]. In these schemes, short elements $m_i \in \mathbb{Z}[X]/(X^n + 1)$ are encoded as $[(r_i \cdot g + m_i)/z]_q \in \mathcal{R}/q\mathcal{R}$ for some r_i, g with norms of size $\mathsf{poly}(\lambda)$ and some random z. For correctness, the latest improvements [ACLL15] require a modulus $q = \mathsf{poly}(\lambda)^\kappa$, where κ is the multi-linearity level. The subfield attack is therefore applicable in subexponential time for any $\kappa = \log^\epsilon \lambda$, according to Sect. 3.4, and would become polynomial for $\kappa > \Theta(\lambda \log \lambda)$. In practice, the fact that the constants in the exponent $q = \lambda^{\Theta(\kappa)}$ is quite large could make this attack quite powerful even for small degrees of multi-linearity.

While initially these constructions permitted the inclusion of encodings of zero ($m_i = 0$) to achieve multilinear maps, it was shown that these encodings break security [HJ15]. Without such encodings, the construction still serves as building-block for realizing Indistinguishability Obfuscation [GGH+13b].

To estimate parameters, [ACLL15] proceeds as follows[8]. Given encodings $x_0 = [(r_0 \cdot g + m_0)/z]_q$ and $x_1 = [(r_1 \cdot g + m_1)/z]_q$ for unknown $m_0, m_1 \neq 0$ we may consider the NTRU lattice Λ_h^q where $h = [x_0/x_1]_q$. This lattice contains a short vector $(r_0 \cdot g + m_0, r_1 \cdot g + m_1)$. In [ACLL15] all elements of norm

---

[8] The attack is attributed to Steven Galbraith in [ACLL15].

$\approx \|r_0 \cdot g + m_0\| = \sigma_1^*$ are considered "interesting" and recovering any such element is considered an attack. This is motivated by the fact that if an attacker recovers $r_0 \cdot g + m_0$ exactly, then it can recover z. This completely breaks the scheme.

The subfield lattice attack does not yield the vector $(r_0 \cdot g + m_0, r_1 \cdot g + m_1)$ exactly but only a relatively small multiple of it $u(r_0 \cdot g + m_0, r_1 \cdot g + m_1)$. We provide two approaches to completely break the scheme from this small multiple. The first approach consists of solving a principal ideal problem and leads to a quantum polynomial-time and classical subexponential attack. The second approach relies on a statistical leak using the Gentry-Szydlo algorithm [GS02, LS14], but is just outside reach with our current tools [GGH13a]. This approach is arguably worrisome, and the authors of [GGH13a] spent significant efforts to rule this approach out completely.

We remark that unlike previous cryptanalysis advances of multi-linear maps [HJ15] this attack does not rely either on the zero testing parameter, neither on encodings of zero. Our cryptanalytic result therefore impacts all applications of multilinear maps, from multi-party key exchange to jigsaw puzzles and Indistinguishability Obfuscation [GGH+13b]. For completeness, we note that the CLT construction [CLT13] of Graded Encoding Schemes is also subject to a quantum polynomial-time attack, because it relies on the hardness of factoring large integers.

The Principal Ideal Problem and Short Generator Recovery. The problem of recovering a short principal ideal generator from any generator received a lot of attention recently, and a series of works has lead to subexponential classical and polynomial-time quantum attacks against principal ideal lattices [EHKS14, CGS14, CDPR16, BS16]. Precisely, given the ideal $\mathfrak{I} = \langle g \rangle$, Biasse and Song [BS16] showed how to recover an arbitrary generator ug of $\mathfrak{I}$ in quantum polynomial time, extending the recent breakthrough of Eisentrager et al. [EHKS14] on quantum algorithms over large degree number fields. Such results were conjectured already in a note of Cambell et al. [CGS14], where a classical polynomial time algorithm is also suggested to recover the original g from ug (namely, LLL in the log-unit lattice). The correctness of a similar algorithm was formally established using analytical number theory by Cramer et al. [CDPR16].

In combination with this subfield lattice attack, this directly implies a polynomial quantum attack. Indeed, the subfield lattice attack allows to recover $u(r_0 \cdot g + m_0)$ for some relatively short u. Repeating this attack several time, and obtaining $u(r_0 \cdot g + m_0)$ for various u eventually leads to the reconstruction of the ideal $\langle r_0 \cdot g + m_0 \rangle$. Because $r_0 \cdot g + m_0$ follows exactly a discrete Gaussian distribution, the approach sketched above can be applied, and reveals $r_0 \cdot g + m_0$ exactly, and therefore z.

In conclusion, for any degree of multi-linearity κ, the subfield attack can be complemented with a quantum polynomial step to a complete break. Alternatively, when $\kappa = O(\lambda^c)$ for any $c < 1/2$, — leading according to the previous best known attacks to a choice of dimension $n = \tilde{\Theta}(\lambda^{1+c})$ — the $2^{\tilde{O}(n^{2/3})}$ algorithms of Biasse and Biasse and Fiecker [Bia14,BF14] combined lead to a classical attack in time sub-exponential in λ.

The Statistical Attack. This attack consists in recovering $u\bar{u}$ and $\langle u \rangle$ and using the Gentry-Szydlo algorithm [GS02,LS14] to recover u.

To recover $\langle u \rangle$, note that we are given $u(a_0, a_1)$. We will assume that $\langle a_0 \rangle, \langle a_1 \rangle$ are coprime with constant probability, cf. Sect. 2.2. Under this assumption, $\langle u \rangle$ can be recovered as $\langle u \rangle = \langle ua_0 \rangle + \langle ua_1 \rangle$.[9]

To recover more information on u, we can compute $ua_0 \cdot [x_i/x_0]_q = ua_i$ for other $i > 1$, and the equation hold over $\mathcal{R}$ because u and a_i are small. For $i > 1$, a_i is a independent of u and follows a spherical Gaussian of parameter σ. It follows that the variance of ua_i leaks $u\bar{u}$: $\mathbb{E}[ua_i \cdot \overline{ua_i}] = \sigma^2 u\bar{u}$.

Given polynomially many samples x_i one can therefore recover $u\bar{u}$ up to a $1 + 1/\mathsf{poly}(\lambda)$ approximation factor. The original attack of Gentry-Szydlo algorithm [GS02,LS14] requires the exact knowledge of $u\bar{u}$ that could be obtained by rounding when u has poly-sized coefficient. However, the u provided by the subfield lattice attack is much larger. In [GGH13a] this algorithm is revisited and extended to when $u\bar{u}$ is only known up to a $1 + (\log n)^{-\Theta(\log n)}$ approximation factor.

In conclusion, with the current algorithmic tools this approach is asymptotically inapplicable if we assume only a polynomial number of available samples, but only barely so. This raises the question of how to improve the tolerance of the Gentry-Szydlo algorithm[10]. Yet, because $(\log n)^{\Theta(\log n)}$ is arguably not so large, it is unclear whether this approach is really infeasible in practice.

We concur with the decision made in [GGH13a], to attempt to rule out such an attack by design even if it is not yet known how to fully exploit it.

5 Experimental Verification

Please refer to the full version of this work for experiments.

6 Conclusions

Practicality of the Attack. The largest instance we broke in practice is for the set of parameter $n = 2^{12}$ and $q \approx 2^{190}$. Choosing a relative degree $r = 16$, the attack required to run LLL in dimension 512, which took about 120 hours, single-threaded, using SAGE [Dev15] and FPLLL [ABC+]. The direct, full field lattice reduction attack, according to root-Hermite-factor based predictions [CN11],

---

[9] Note that the subfield lattice attack may be tweaked to obtain a triplet $u(a_0, a_1, a_2)$ (or more) increasing the probability to recover $\langle u \rangle$.

[10] Asymptotically, the natural idea of replacing LLL by slightly stronger lattice reduction does not seems to help, but should help in practice. The quasi-polynomial factor relates to a number theoretic heuristic. See Sect. 7.6 of [GGH13a].

would have required running BKZ in block-size ≈ 130, and in dimension 8192, which is hardly feasible with the current state-of-the art [CN11] (requiring more than 2^{70} CPU cycles). We conclude that the subfield attack proposed in this work is not only theoretical but also practical.

Obstructions to Concrete Predictions. We are currently unable to predict precisely how a given set of parameters would be affected, for example to predict the power of this attack against concrete parameter choices of NTRU-based FHE [LTV12, BLLN13] and Multilinear Maps [GGH13a].

There are two issues for those predictions. The first issue is that we make use of LLL/BKZ in the approximation-factor regime, not in the Hermite-factor regime. While the behavior of LLL/BKZ is quite well modeled in the latter regime, we are not aware of precise models for the former for NTRU lattices. Unlike the Hermite-factor regime, this case could very well be influenced by the presence of many short vectors rather than just a few.

The second issue is that we do not know the actual size of the shortest vector of $\Lambda_{h'}^q$: all we know is that it is no larger than (f', g'). In several cases in the experiments we found vectors $(x', y') = v(f', g')$ that were actually shorter than (f', g')— the tentative root-approximation factor α is less than 1. One may expect that (f', g') may still be (or close to) the shortest vector for small relative degree r as it is the shortest with high probability in the full field (i.e. when $r = 1$).

Immunity of NTRU Encryption and BLISS Signature Schemes. If q is small enough, then the attacks should become inapplicable, even with the smallest possible relative dimension $r = 2$. Precisely, if (f', g') is not an unusually short vector of $\Lambda_{h'}^q$, then there is little hope that any lattice reduction strategy would lead to information on this vector. Quantitatively, this perfect immunity happens when $\|(f', g')\| \approx \sqrt{2} \cdot \sigma^2 \cdot n' > \sqrt{n'q/\pi e}$. This was the case of the old parameter of NTRU as discussed in [Gen01], which lead this attack being discarded. This is not the case of all the parameters of NTRUENCRYPT [HPS+15] and BLISS [DDLL13], for which (f', g') is sometime unusually short vector, but not by a very large factor. Numerical values are given in Table 1.

Table 1. Vulnerability factor for some parameters of NTRUENCRYPT [HPS+15] and BLISS [DDLL13].

Scheme	n	q	σ	$\sqrt{n'q/\pi e}$	/ $(\sqrt{2}\sigma^2 n')$	$= F$
NTRU-743	743	2048	0.82	298.7	/ 349.8	= 0.85
NTRU-401	401	2048	0.82	219.6	/ 189.5	= 1.16
BLISS-I	512	12289	0.55	607.0	/ 108.6	= 5.59
BLISS-IV	512	12289	0.83	607.0	/ 249.8	= 2.43

When the vulnerability factor F is less then 1, the parameters achieve perfect immunity. When F is greater than 1, the subfield attack consist informally of solving "unusual-SVP" in dimension $2n' = n$, where the unusually short solutions are a factor F shorter than predicted by the Gaussian Heuristic.

According to this table, NTRU-743 should be perfectly immune to the subfield lattice attacks. For other parameters, it seems likely, despite imperfect immunity, that the subfield lattice attack will be more costly than the full attack, but calls for further study, especially for BLISS-I.

Note that the perfect immunity to this attack is achieved asymptotically around $\sigma \approx \Theta(q^{1/4})$, parameter for which h does not have enough entropy to be statistically close to random. For comparison, it was shown that for $\sigma = \omega(q^{1/2})$, h is statistically close to uniform [SS11]. We note that $\sigma > \Theta(q^{1/4})$ could provide enough entropy for the normed-down public key h' to be almost uniform. It would be interesting to see if the proof of [SS11] can be adapted to h'.

Recommendations. Even if credible predictions were to be made, we strongly discourage basing a cryptographic scheme on a set-up to which this attack is applicable. Indeed, it is quite likely that the performance of the attack may be improved in several ways. For example, after having found several subfield solutions $(x', y') = v(f', g')$, it is possible to run a lattice reduction algorithm in the lattice $(f', g') \cdot \mathcal{O}_\mathbb{L}$ of dimension n' rather than $2n'$ to obtain significantly shorter vectors. Additionally, the lifting step may also be improved in the case where $\mathcal{O}_\mathbb{L}$ is a real subfield using the Gentry-Syzdlo algorithm [GS02, LS14] to obtain shorter vector in the full field (i.e. recovering x from $N_{\mathbb{K}/\mathbb{L}}(x)$). More generally, one may recover x from $N_{\mathbb{K}/\mathbb{L}}(x)$ even when $\mathbb{L}$ isn't the real subfield of $\mathbb{K}$: assuming (x) is prime, it can be recovered as a factor of $N_{\mathbb{K}/\mathbb{L}}(x)$, which then leads to x via a short generator recovery; as mentioned before, both steps are now known to be classically sub-exponential or even polynomial for quantum computers [Bia14, EHKS14, CGS14, BS16, CDPR16].

Evaluating concrete security against regular lattice attacks is already a difficult exercise, and leaving open additional algebraic and statistical attack opportunities will only make security assessment intractable. We therefore recommend that this set-up — NTRU assumption, presence of subfields, large modulus — be considered insecure.

Designing Immune Rings. We believe that our work further motivates the design and the study of number fields without subfields to fit for the lattice-based cryptographic purposes, as already recommended in [Ber14]. Even for assumptions that are not directly affected by this attack (Ring-SIS [Mic02], Ideal-LWE [SSTX09], Ring-LWE [LPR10]), it could be considered desirable to have efficient fallback options ready to use, in case subfields induce other unforeseen weaknesses. While this work does not suggest an immediate threat to the Ring-SIS and Ring-LWE, such a precaution is not unreasonable.

An interesting option has been suggested in [Ber14] to use rings of the form $\mathbb{Z}[X]/(X^p - X - 1)$. The design rationale seems to be that $\mathbb{Q}[X]/(X^p - X - 1)$ has a reasonable expansion factor[11] which is often needed for the correctness in cryptographic schemes, but is a non Galois extension with a very large Galois group

---

[11] Multiplication of two small elements remains reasonably small.

for its splitting field, which is intended to hinder algebraic handles. In particular it contains no proper subfields. This leads to the design of the NTRUPrime encryption scheme [BCLvV16]. We note that the security of this scheme is not supported by a worst-case hardness argument. If such an argument is desired then we note that the *search version* of Ideal/Ring-LWE is supported by worst-case hardness for *any choices of number field*, and this is actually sufficient to achieve provable CPA-secure encryption, as already proved by Stehlé et al. [SSTX09].

Open Problems. Another natural option would be to choose p as a safe prime[12] and to work with the ring of integer of the *totally real* number field $\mathbb{K} = \mathbb{Q}(\zeta_p + \bar{\zeta}_p)$. The field remains Galois, and its automorphism group may still allow a quantum worst-case (Ideal-SVP) to average-case (Ring-LWE) reduction a-la [LPR10] thanks to a generalization of the search to decision step presented in [CLS15]. Nevertheless the Galois group has prime order $(p-1)/2$, it has no proper subgroups, and $\mathbb{K}$ has no proper subfields.

But working with $\mathbb{K} = \mathbb{Q}(\zeta_p + \bar{\zeta}_p)$ has a drawback: the class number $h(\mathbb{K}) = h_p^+$ seems quite small (see [Was97, Table 4 pp. 421]), and this makes the worst-case ISVP problem solvable in quantum polynomial time for approximation factors $2^{\tilde{O}(\sqrt{n})}$ as proved in [CDPR16,BS16]: the reduction of [LPR10] is vacuous for such parameters.

This raises the question of whether NTRU and Ring-LWE are actually strictly harder than ISVP in the underlying number field, whether algorithms for ISVP in $\mathbb{K}$ can be lifted to modules over $\mathbb{K}$ as used in NTRU, Ideal-LWE or Ring-LWE. In this regard, overstretched NTRU, and Ideal/Ring-LWE with large approximation factors over the ring $\mathbb{Z}(\zeta_p + \bar{\zeta}_p)$ are very interesting cryptanalytic target: despite those rings not being used in any proposed schemes so far, such an attack will teach us a great deal on the asymptotic security of ideal-lattice based cryptography.

Acknowledgments. We are grateful to Alice Silverberg, and to the participant of the Conference on Mathematics of Cryptography for enlightening talks and discussions. We thank Dan J. Bernstein, Ronald Cramer, Jeffrey Hoffstein, Hendrik W. Lenstra, John Schanck and Damien Stehlé for helpful discussions and comments.

We thank the PSMN (Pôle Scientifique de Modélisation Numérique, Lyon, France) for providing computing facilities.

References

[ABC+] Albrecht, M., Bai, S., Cadé, D., Pujol, X., Stehlé, D.: fpLLL-4.0, a floating-point LLL implementation. https://github.com/dstehle/fplll

---

[12] A safe prime p is an odd prime such that $(p-1)/2$ is also a prime. The terminology relates to weaknesses in RSA and Discrete Logarithm Problem introduced by the smoothness of $p-1$ [Pol74].

[ACLL15] Albrecht, M.R., Cocis, C., Laguillaumie, F., Langlois, A.: Implementing candidate graded encoding schemes from ideal lattices. In: Iwata, T., et al. (eds.) ASIACRYPT 2015. LNCS, vol. 9453, pp. 752–775. Springer, Heidelberg (2015). doi:10.1007/978-3-662-48800-3_31

[BCLvV16] Bernstein, D.J., Chuengsatiansup, C., Lange, T., van Vredendaal, C.: NTRU prime. Cryptology ePrint Archive, Report 2016/461 (2016). http://eprint.iacr.org/

[Ber14] Bernstein, D.: A subfield-logarithm attack against ideal lattices, Febuary 2014. http://blog.cr.yp.to/20140213-ideal.html

[BF14] Biasse, J.-F., Fieker, C.: Subexponential class group, unit group computation in large degree number fields. LMS J. Comput. Math. **17**(Suppl. A), 385–403 (2014)

[Bia14] Biasse, J.-F.: Subexponential time relations in the class group of large degree number fields. Adv. Math. Commun. **8**(4), 407–425 (2014)

[BLLN13] Bos, J.W., Lauter, K., Loftus, J., Naehrig, M.: Improved security for a ring-based fully homomorphic encryption scheme. In: Stam, M. (ed.) IMACC 2013. LNCS, vol. 8308, pp. 45–64. Springer, Heidelberg (2013)

[BS16] Biasse, J.-F., Song, F.: Efficient quantum algorithms for computing class groups and solving the principal ideal problem in arbitrary degree number fields. In: 27th ACM-SIAM Symposium on Discrete Algorithms (SODA 2016) (2016)

[BV11] Brakerski, Z., Vaikuntanathan, V.: Efficient fully homomorphic encryption from (standard) LWE. In: Ostrovsky, R. (ed.) 52nd FOCS, pp. 97–106. IEEE Computer Society Press, October 2011

[CDPR16] Cramer, R., Ducas, L., Peikert, C., Regev, O.: Recovering short generators of principal ideals in cyclotomic rings. In: Fischlin, M., Coron, J.-S. (eds.) EUROCRYPT 2016. LNCS, vol. 9666, pp. 559–585. Springer, Heidelberg (2016). doi:10.1007/978-3-662-49896-5_20

[CG13] Canetti, R., Garay, J.A. (eds.): CRYPTO 2013. LNCS, vol. 8042. Springer, Heidelberg (2013)

[CGS14] Campbell, P., Groves, M., Shepherd, D.: Soliloquy: a cautionary tale. In: ETSI 2nd Quantum-Safe Crypto Workshop (2014). http://docbox.etsi.org/Workshop/2014/201410_CRYPTO/S07_Systems_and_Attacks/S07_Groves_Annex.pdf

[CIV16] Castryck, W., Iliashenko, I., Vercauteren, F.: Provably weak instances of ring-LWE revisited. In: Fischlin, M., Coron, J.-S. (eds.) EUROCRYPT 2016. LNCS, vol. 9665, pp. 147–167. Springer, Heidelberg (2016). doi:10.1007/978-3-662-49890-3_6

[CJL16] Cheon, J.H., Jeong, J., Lee, C.: An algorithm for NTRU problems and cryptanalysis of the GGH multilinear map without an encoding of zero. Cryptology ePrint Archive, Report 2016/139 (2016). http://eprint.iacr.org/

[CLS15] Chen, H., Lauter, K., Stange, K.E.: Attacks on search RLWE. Cryptology ePrint Archive, Report 2015/971 (2015). http://eprint.iacr.org/2015/971

[CLT13] Coron, J.-S., Lepoint, T., Tibouchi, M.: Practical multilinear maps over the integers. In: Canetti, R., Garay, J.A. (eds.) [CG13], pp. 476–493

[CN11] Chen, Y., Nguyen, P.Q.: BKZ 2.0: better lattice security estimates. In: Lee, D.H., Wang, X. (eds.) ASIACRYPT 2011. LNCS, vol. 7073, pp. 1–20. Springer, Heidelberg (2011)

[CS97]   Coppersmith, D., Shamir, A.: Lattice attacks on NTRU. In: Fumy, W. (ed.) EUROCRYPT 1997. LNCS, vol. 1233, pp. 52–61. Springer, Heidelberg (1997)

[CS15]   Costache, A., Smart, N.P.: Which ring based somewhat homomorphic encryption scheme is best? Cryptology ePrint Archive, Report 2015/889 (2015). http://eprint.iacr.org/2015/889

[DDLL13] Ducas, L., Durmus, A., Lepoint, T., Lyubashevsky, V.: Lattice signatures and bimodal gaussians. In: Canetti, R., Garay, J.A. (eds.) [CG13], pp. 40–56

[Dev15]   The Sage Developers: Sage Mathematics Software (2015). http://www.sagemath.org

[DHS15]   Doröz, Y., Yin, H., Sunar, B.: Homomorphic AES evaluation using the modified LTV scheme. Des. Codes Crypt. **80**(2), 333–358 (2016). http://dx.doi.org/10.1007/s10623-015-0095-1

[EHKS14] Eisenträger, K., Hallgren, S., Kitaev, A., Song, F.: A quantum algorithm for computing the unit group of an arbitrary degree number field. In: Proceedings of the 46th Annual ACM Symposium on Theory of Computing, pp. 293–302. ACM (2014)

[EHL14]   Eisenträger, K., Hallgren, S., Lauter, K.: Weak instances of PLWE. In: Joux, A., Youssef, A. (eds.) SAC 2014. LNCS, vol. 8781, pp. 183–194. Springer, Heidelberg (2014)

[ELOS15] Elias, Y., Lauter, K.E., Ozman, E., Stange, K.E.: Provably weak instances of ring-LWE. In: Gennaro, R., Robshaw, M. (eds.) [GR15], pp. 63–92

[FM14]   Ferraguti,   A.,   Micheli,   G.:   On   the   Mertens-Cesàro theorem   for   number   fields.   Bull.   Aust.   Math.   Soc. **93**(2),   199–210   (2016).   doi:10.1017/S0004972715001288. http://journals.cambridge.org/article_S0004972715001288

[Gen01]   Gentry, C.: Key recovery and message attacks on NTRU-composite. In: Pfitzmann, B. (ed.) [Pfi01], pp. 182–194

[GGH13a] Garg, S., Gentry, C., Halevi, S.: Candidate multilinear maps from ideal lattices. In: Johansson, T., Nguyen, P.Q. (eds.) EUROCRYPT 2013. LNCS, vol. 7881, pp. 1–17. Springer, Heidelberg (2013)

[GGH+13b] Garg, S., Gentry, C., Halevi, S., Raykova, M., Sahai, A., Waters, B.: Candidate indistinguishability obfuscation and functional encryption for all circuits. In: 54th FOCS, pp. 40–49. IEEE Computer Society Press, October 2013

[GN08]   Gama, N., Nguyen, P.Q.: Finding short lattice vectors within Mordell's inequality. In: Ladner, R.E., Dwork, C. (eds.) 40th ACM STOC, pp. 207–216. ACM Press, May 2008

[GR15]   Gennaro, R., Robshaw, M. (eds.): CRYPTO 2015. LNCS, vol. 9215. Springer, Heidelberg (2015)

[GS02]   Gentry, C., Szydlo, M.: Cryptanalysis of the revised NTRU signature scheme. In: Knudsen, L.R. (ed.) EUROCRYPT 2002. LNCS, vol. 2332, pp. 299–320. Springer, Heidelberg (2002)

[HG07]   Howgrave-Graham, N.: A hybrid lattice-reduction and meet-in-the-middle attack against NTRU. In: Menezes, A. (ed.) CRYPTO 2007. LNCS, vol. 4622, pp. 150–169. Springer, Heidelberg (2007)

[HHGP+03] Hoffstein, J., Howgrave-Graham, N., Pipher, J., Silverman, J.H., Whyte, W.: NTRUSIGN: digital signatures using the NTRU lattice. In: Joye, M.

(ed.) CT-RSA 2003. LNCS, vol. 2612, pp. 122–140. Springer, Heidelberg (2003)

[HJ15]   Hu, Y., Jia, H.: Cryptanalysis of GGH map. Cryptology ePrint Archive, Report 2015/301 (2015). http://eprint.iacr.org/2015/301

[HPS96]   Hoffstein, J., Pipher, J., Silverman, J.H.: NTRU: a new high speed public key cryptosystem. In: Draft Distributed at Crypto 1996 (1996). http://web.securityinnovation.com/hubfs/files/ntru-orig.pdf

[HPS98]   Hoffstein, J., Pipher, J., Silverman, J.H.: NTRU: a ring-based public key cryptosystem. In: Buhler, J.P. (ed.) ANTS 1998. LNCS, vol. 1423, pp. 267–288. Springer, Heidelberg (1998)

[HPS01]   Hoffstein, J., Pipher, J., Silverman, J.H.: NSS: an NTRU lattice-based signature scheme. In: Pfitzmann, B. (ed.) [Pfi01], pp. 211–228

[HPS11]   Hanrot, G., Pujol, X., Stehlé, D.: Analyzing blockwise lattice algorithms using dynamical systems. In: Rogaway, P. (ed.) CRYPTO 2011. LNCS, vol. 6841, pp. 447–464. Springer, Heidelberg (2011)

[HPS+15]   Hoffstein, J., Pipher, J., Schanck, J.M., Silverman, J.H., Whyte, W., Zhang, Z.: Choosing parameters for NTRUEncrypt. Cryptology ePrint Archive, Report 2015/708 (2015). http://eprint.iacr.org/2015/708

[HSW06]   Hoffstein, J., Silverman, J.H., Whyte, W.: Meet-in-the-middle attack on an ntru private key, 2006. Technical report, NTRU Cryptosystems, Report #04, July 2006. http://www.ntru.com

[HT15]   Hauteville, A., Tillich, J.-P.: New algorithms for decoding in the rank metric and an attack on the LRPC cryptosystem. In: IEEE International Symposium on Information Theory, ISIT 2015, pp. 2747–2751 (2015)

[KF15]   Kirchner, P., Fouque, P.-A.: An improved BKW algorithm for LWE with applications to cryptography and lattices. In: Gennaro, R., Robshaw, M. (eds.) [GR15], pp. 43–62

[LJ14]   Löndahl, C., Johansson, T.: Improved algorithms for finding low-weight polynomial multiples in $f_2[x]$ and some cryptographic applications. Des. Codes Crypt. **73**(2), 625–640 (2014)

[LLL82]   Lenstra, A.K., Lenstra Jr., H.W., Lovász, L.: Factoring polynomials with rational coefficients. Math. Ann. **261**(4), 515–534 (1982)

[LN14]   Lepoint, T., Naehrig, M.: A comparison of the homomorphic encryption schemes FV and YASHE. In: Pointcheval, D., Vergnaud, D. (eds.) AFRICACRYPT. LNCS, vol. 8469, pp. 318–335. Springer, Heidelberg (2014)

[Loi14]   Loidreau, P.: On cellular codes and their cryptographic applications. In: ACCT, Fourteenth International Workshop on Algebraic and Combinatorial Coding Theory, pp. 234–239 (2014)

[Lov87]   Lovasz, L.: An Algorithmic Theory of Numbers, Graphs and Convexity. CBMS-NSF Regional Conference Series in Applied Mathematics. Society for Industrial and Applied Mathematics, Philadelphia (1987)

[LPR10]   Lyubashevsky, V., Peikert, C., Regev, O.: On ideal lattices and learning with errors over rings. In: Gilbert, H. (ed.) EUROCRYPT 2010. LNCS, vol. 6110, pp. 1–23. Springer, Heidelberg (2010)

[LS14]   Lenstra, H.W., Silverberg, A.: Revisiting the Gentry-Szydlo algorithm. In: Garay, J.A., Gennaro, R. (eds.) CRYPTO 2014, Part I. LNCS, vol. 8616, pp. 280–296. Springer, Heidelberg (2014)

[LSS14]   Langlois, A., Stehlé, D., Steinfeld, R.: GGHLite: more efficient multilinear maps from ideal lattices. In: Nguyen, P.Q., Oswald, E. (eds.)

EUROCRYPT 2014. LNCS, vol. 8441, pp. 239–256. Springer, Heidelberg (2014)

[LTV12] López-Alt, A., Tromer, E., Vaikuntanathan, V.: On-the-fly multiparty computation on the cloud via multikey fully homomorphic encryption. In: Karloff, H.J., Pitassi, T. (eds.) 44th ACM STOC, pp. 1219–1234. ACM Press, May 2012

[Mic02] Micciancio, D.: Generalized compact knapsacks, cyclic lattices, and efficient one-way functions from worst-case complexity assumptions. In: 43rd FOCS, pp. 356–365. IEEE Computer Society Press, November 2002

[Pei16] Peikert, C.: How (not) to instantiate ring-LWE. Cryptology ePrint Archive, Report 2016/351 (2016). http://eprint.iacr.org/

[Pfi01] Pfitzmann, B. (ed.): EUROCRYPT 2001. LNCS, vol. 2045. Springer, Heidelberg (2001)

[Pol74] Pollard, J.M.: Theorems on factorization and primality testing. In: Math-ematical Proceedings of the Cambridge Philosophical Society, vol. 76, no. 03, pp. 521–528 (1974)

[Sam70] Samuel, P.: Algebraic Theory of Numbers. Hermann, Paris (1970)

[Sch87] Schnorr, C.-P.: A hierarchy of polynomial time lattice basis reduction algorithms. Theor. Comput. Sci. **53**, 201–224 (1987)

[Sit10] Sittinger, B.D.: The probability that random algebraic integers are relatively r-prime. J. Number Theory **130**(1), 164–171 (2010)

[SS11] Stehlé, D., Steinfeld, R.: Making NTRU as secure as worst-case problems over ideal lattices. In: Paterson, K.G. (ed.) EUROCRYPT 2011. LNCS, vol. 6632, pp. 27–47. Springer, Heidelberg (2011)

[SSTX09] Stehlé, D., Steinfeld, R., Tanaka, K., Xagawa, K.: Efficient public key encryption based on ideal lattices. In: Matsui, M. (ed.) ASIACRYPT 2009. LNCS, vol. 5912, pp. 617–635. Springer, Heidelberg (2009)

[Was97] Washington, L.C.: Introduction to Cyclotomic Fields. Graduate Texts in Mathematics. Springer, New York (1997)

A Practical Cryptanalysis
of the Algebraic Eraser

Adi Ben-Zvi[1], Simon R. Blackburn[2]([⊠]), and Boaz Tsaban[1]

[1] Department of Mathematics, Bar-Ilan University, Ramat Gan 5290002, Israel
[2] Department of Mathematics, Royal Holloway University of London,
Egham TW20 0EX, Surrey, UK
s.blackburn@rhul.ac.uk

Abstract. We present a novel cryptanalysis of the Algebraic Eraser primitive. This key agreement scheme, based on techniques from permutation groups, matrix groups and braid groups, is proposed as an underlying technology for ISO/IEC 29167-20, which is intended for authentication of RFID tags. SecureRF, the company owning the trademark Algebraic Eraser, markets it as suitable in general for lightweight environments such as RFID tags and other IoT applications. Our attack is practical on standard hardware: for parameter sizes corresponding to claimed 128-bit security, our implementation recovers the shared key using less than 8 CPU hours, and less than 64 MB of memory.

1 Introduction

The Algebraic Eraser[TM] is a key agreement scheme using techniques from non-commutative group theory. It was announced by Anshel, Anshel, Goldfeld and Lemieaux in 2004; the corresponding paper [1] appeared in 2006. The Algebraic Eraser is defined in a very general fashion: various algebraic structures (monoids, groups and actions) need to be specified in order to be suitable for implementation. Anshel *et al.* provide most of this extra information, and name this concrete realisation of the Algebraic Eraser the *Colored Burau Key Agreement Protocol (CBKAP)*. This concrete representation involves a novel blend of finite matrix groups and permutation groups with infinite braid groups. A company, SecureRF, owns the trademark to the Algebraic Eraser, and is marketing this primitive as suitable for low resource environments such as RFID tags and Internet of Things (IoT) applications. The primitive is proposed as an underlying technology for ISO/IEC 29167-20, and work on this standard is taking place in ISO/IEC JTC 1/SC 31/WG 7. The company has also presented the primitive to the Internet Research Task Force's Crypto Forum Research Group (IRTF CFRG), with a view towards standardisation. IoT is a growth area, where current widely-accepted public key techniques struggle to operate due to tight efficiency constraints. It is likely that solutions which are efficient enough for these applications will become widely deployed, and the nature of these applications make system changes after deployment difficult. Thus, it is vital to scrutinise the security of primitives such as the Algebraic Eraser

© International Association for Cryptologic Research 2016
M. Robshaw and J. Katz (Eds.): CRYPTO 2016, Part I, LNCS 9814, pp. 179–189, 2016.
DOI: 10.1007/978-3-662-53018-4_7

early in the standardisation process, to ensure only secure primitives underpin standardised protocols.

In a presentation to the NIST Workshop in Lightweight Cryptography in 2015, SecureRF claims a security level of 2^{128} for their preferred parameter sizes, and compares the speed of their system favourably with an implementation of a key agreement protocol based on the NIST recommended [14] elliptic curve K-283. The company reports [3] a speed-up by a factor of 45–150, compared to elliptic curve key agreement at 128-bit security levels. It claims that the computational requirements of the Algebraic Eraser scales linearly with the security parameter, in contrast to the quadratic scaling of elliptic-curve-based key agreement.

Related Work. The criteria for choosing some global parameters of the scheme (namely certain subgroups C and D of matrices over a finite field, and certain subgroups A and B of a certain infinite semidirect product of groups) are not given in [1], and have not been made available by SecureRF. In the absence of this information, it is reasonable to proceed initially with a cryptanalysis on the basis that these parameters are chosen in a generic fashion. All previous cryptanalyses have taken this approach.

Myasnikov and Ushakov [13] provide a heuristic length-based attack on the CBKAP that works for the parameter sizes originally suggested [1]. However, Gunnells [10] reports that this attack quickly becomes unsuccessful as parameter sizes grow; the parameter sizes proposed by SecureRF make this attack impractical[1]. Kalka et al. [11] provide an efficient cryptanalysis of the CBKAP for arbitrary parameter sizes. The attack uses the public key material of Alice and the messages exchanged between Alice and Bob to derive an equivalent to the secret random information generated by Bob, which then compromises the shared key, and so renders the scheme insecure. In particular, the techniques of [11] will succeed when the global parameters are chosen generically.

SecureRF uses proprietary distributions for global parameters, so the cryptanalysis of [11] attack does not imply that the CBKAP as implemented is insecure[2]. Indeed, Goldfeld and Gunnells [9] show that by choosing the subgroup C carefully one step of the attack of [11] does not recover the information required to proceed, and so this attack does not succeed when parameters are generated in this manner.

Our Contribution. There are no previously known attacks on the CBKAP for the proposed parameter sizes, provided the parameters are chosen to resist the attack of [11]. The present paper describes a new attack on the CBKAP that

---

[1] There is an analogy with the development of RSA here: the size of primes (200 digits) proposed in the original article [15] was made obsolete by improvements in integer factorisation algorithms [4].

[2] The analogy with RSA continues: factorisation of a randomly chosen integer n is much easier than when n is a product of two primes of equal size, which is why the latter is used in RSA.

does not assume any structure on the subgroup C. Thus, a careful choice of the subgroup C will have no effect on the applicability of our attack, and so the proposed security measure offered by Goldfeld and Gunnells [9] to the attack of [11] is bypassed.

The earlier cryptanalyses of CBKAP ([11,13]) attempt to recover parts of Alice's or Bob's secret information. The attack presented here recovers the shared key directly from Alice's public key and the messages exchanged between Alice and Bob.

SecureRF have kindly provided us with sets of challenge parameters of the full recommended size, and our implementation is successful in recovering the shared key in all cases. Our (non-optimised) implementation recovers the common key in under 8 h of computation, and thus the security of the system is *much* less than the 2^{128} level claimed for these parameter sizes. The attack scales well with size, so increasing parameter sizes will not provide a solution to the security problem for the CBKAP.

Conclusion and Recommendation. Because our attack efficiently recovers the shared key of the CBKAP for recommended parameter sizes, using parameters provided by SecureRF, we believe the results presented here cast serious doubt on the suitability of the Algebraic Eraser for the applications proposed. We recommend that the primitive in its current form should not be used in practice, and that full details of any revised version of the primitive should be made available for public scrutiny in order to ensure a rigorous security analysis.

Recent Developments. Since the first version of this paper was posted, there have been two recent developments. Firstly, authors from SecureRF have posted [2] a response to our attack, concentrating in the main on the implications for the related ISO standard and providing some preliminary thoughts on how they might redesign the primitive. Until the details are finalised, it is too soon to draw any conclusions on the security of any redesigned scheme, though there have already been some discussions on Cryptography Stack Exchange [8]. Secondly, Blackburn and Robshaw [6] have posted a paper that cryptanalyses the ISO standard itself, rather then the more general underlying Algebraic Eraser primitive.

Structure of the Paper. The remainder of the paper is organised as follows. Sections 2 and 3 establish notation, and describe the CBKAP. We describe a slightly more general protocol than the CBKAP, as our attack naturally generalises to a larger setting. We describe our attack in Sect. 4. In Sect. 5 we describe the results of our implementations and provide a short conclusion.

2   Notation

This section establishes notation for the remainder of the paper. We closely follow the notation from [11], which is in turn mainly derived from the notation in [1], though we do introduce some new terms.

Let $\mathbb{F}$ be a finite field of small order (e.g., $|\mathbb{F}| = 256$) and let n be a positive integer (e.g., $n = 16$). Let S_n be the symmetric group on the set $\{1, 2, \ldots, n\}$, and let $\mathrm{GL}_n(\mathbb{F})$ be the group of invertible $n \times n$ matrices with entries in $\mathbb{F}$.

Let M be a subgroup of $\mathrm{GL}_n(\mathbb{F}(t_1, \ldots, t_n))$, where the elements t_i are algebraically independent commuting indeterminates. Indeed, we assume that the group M is contained in the subgroup of $\mathrm{GL}_n(\mathbb{F}(t_1, \ldots, t_n))$ of matrices whose determinant can be written as $a\mathbf{t}$ for some non-zero element $a \in \mathbb{F}$ and some, possibly empty, word $\mathbf{t}$ in the elements t_i and their inverses. Let $\overline{M}$ be the subgroup of $\mathrm{GL}_n(\mathbb{F}(t_1, \ldots, t_n))$ generated by permuting the indeterminates of elements of M in all possible ways.

Fix non-zero elements $\tau_1, \ldots, \tau_n \in \mathbb{F}$. Define the homomorphism $\varphi \colon \overline{M} \to \mathrm{GL}_n(\mathbb{F})$ to be the evaluation map, computed by replacing each indeterminate t_i by the corresponding element τ_i. Our assumption on the group M means that φ is well defined.

The group S_n acts on $\overline{M}$ by permuting the indeterminates t_i. Let $\overline{M} \rtimes S_n$ be the semidirect product of $\overline{M}$ and S_n induced by this action. More concretely, if we write $^g a$ for the action of an element $g \in S_n$ on an element $a \in \overline{M}$, then the elements of $\overline{M} \rtimes S_n$ are pairs (a, g) with $a \in \overline{M}$ and $g \in S_n$, and group multiplication is given by

$$(a, g)(b, h) = (a\, ^g b, gh)$$

for all $(a, g), (b, h) \in \overline{M} \rtimes S_n$.

Let C and D be subgroups of $\mathrm{GL}_n(\mathbb{F})$ that *commute elementwise*: $cd = dc$ for all $c \in C$ and $d \in D$. The CBKAP specifies that C is a subgroup consisting of all invertible matrices of the form $\ell_0 + \ell_1 \kappa + \cdots + \ell_r \kappa^r$ where κ is a fixed matrix, $\ell_i \in \mathbb{F}$ and $r \geq 0$. So C is the group of units in the $\mathbb{F}$-algebra generated by κ. Moreover, the CBKAP specifies that $D = C$. But we do not assume anything about the forms of C and D in this paper, other than the fact that they commute.

Let $\Omega = \mathrm{GL}_n(\mathbb{F}) \times S_n$ and let $\widehat{S}_n = \overline{M} \rtimes S_n$. We have two actions on Ω. Firstly, there is the right action of the group $\widehat{S}_n$ on Ω via a map

$$* \colon \Omega \times \widehat{S}_n \to \Omega,$$

as defined in [1,11]. So

$$(s, g) * (b, h) = (s\varphi(^g b), gh)$$

for all $(s, g) \in \Omega$ and all $(b, h) \in \widehat{S}_n$. Secondly, there is a left action of the group $\mathrm{GL}_n(\mathbb{F})$ on Ω via the map

$$\bullet \colon \mathrm{GL}_n(\mathbb{F}) \times \Omega \to \Omega$$

given by matrix multiplication:

$$x \bullet (s, g) = (xs, g)$$

for all $x \in \mathrm{GL}_n(\mathbb{F})$ and all $(s, g) \in \Omega$. Note that for all $x \in \mathrm{GL}_n(\mathbb{F})$, all $\omega \in \Omega$ and all $\widehat{g} \in \widehat{S}_n$ we have that

$$(x \bullet \omega) * \widehat{g} = x \bullet (\omega * \widehat{g}).$$

Also note that the left action is $\mathbb{F}$-linear, in the sense that if $x \in \mathrm{GL}_n(\mathbb{F})$ can be written in the form

$$x = \sum_{i=1}^{r} \ell_i c_i$$

for some $c_i \in \mathrm{GL}_n(\mathbb{F})$ and $\ell_i \in \mathbb{F}$, then for all $(s, g) \in \Omega$ we have

$$x \bullet (s, g) = \sum_{i=1}^{r} \ell_i (c_i \bullet (s, g)).$$

To interpret the right hand side of the equality above: the subset of Ω whose second component is a fixed element of S_n is naturally an $\mathbb{F}$-vector space, where addition and scalar multiplication takes place in the first component only.

Finally, let A and B be subgroups of $\widehat{S}_n$ that *-commute: for all $(a, g) \in A$, $(b, h) \in B$ and $\omega \in \Omega$,

$$(\omega * (a, g)) * (b, h) = (\omega * (b, h)) * (a, g).$$

3   The CBKAP Protocol

3.1   Overview

The CBKAP is unusual in that the parties executing it, Alice and Bob, use different parts of the public key in their computations: neither party needs to know all of the public key. The security model assumes that one party's public key material is known to the adversary: say Alice's public key material is known, but Bob's 'public' key (which is better thought of as part of his private key material) is not revealed. The adversary, Eve, receives just Alice's public information, and the messages sent over the insecure channel. Security means that Eve cannot feasibly compute any significant information about K. The attack in [11] works in this model. The same is true for the attack we describe below.

In a typical proposed application, the protocol might be used to enable a low-power device, such as an RFID tag, to communicate with a central server. Data on an RFID tag is inherently insecure, as is system-wide data. So the above security model is realistic (and conservative) for these application settings.

3.2 The Protocol

Public parameters (for Alice) include the parameters n, $\mathbb{F}$, M, $\tau_1, \ldots, \tau_n$, C and A. The groups M, C and A are specified by their generating sets. For efficiency reasons, the generators of the group A are written as words in a certain standard generating set for the group $\widehat{S}_n$. We discuss this further in Sect. 5, but see the TTP algorithm in [1] for full information. It is assumed that Eve knows the parameters n, $\mathbb{F}$, M, $\tau_1, \ldots, \tau_n$, C and A. Bob needs to know the groups B and D, rather than the groups A and C. Eve does not need to know the subgroups B and D for our attack to work.

We write e for the identity element of S_n. We write I_n for the identity matrix in $\mathrm{GL}_n(\mathbb{F})$, and write $1 = (I_n, e) \in \Omega$.

Alice chooses elements $c \in C$ and $\widehat{g} = (a, g) \in A$. She computes the product

$$c \bullet 1 * \widehat{g} = (c\varphi(a), g) \in \Omega$$

and sends it to Bob over an insecure channel.

Bob, who knows the groups B and D, chooses elements $d \in D$ and $\widehat{h} = (b, h) \in B$. He computes the product

$$d \bullet 1 * \widehat{h} = (d\varphi(b), h) \in \Omega$$

and sends it to Alice over the insecure channel.

Note that $cd = dc$ because $c \in C$ and $d \in D$, and the groups C and D commute elementwise. Thus,

$$
\begin{aligned}
d \bullet (c \bullet 1 * \widehat{g}) * \widehat{h} &= (dc) \bullet (1 * \widehat{g}) * \widehat{h} \\
&= (cd) \bullet (1 * \widehat{g}) * \widehat{h} \\
&= (cd) \bullet (1 * \widehat{h}) * \widehat{g} \\
&\quad (\text{as } \widehat{g} \in A, \widehat{h} \in B, \text{ and } A \text{ and } B * \text{-commute}) \\
&= c \bullet (d \bullet 1 * \widehat{h}) * \widehat{g}.
\end{aligned}
$$

The common key K is defined by

$$K = d \bullet (c \bullet 1 * \widehat{g}) * \widehat{h} = c \bullet (d \bullet 1 * \widehat{h}) * \widehat{g}.$$

Alice can compute the key K using the right hand expression in the equation above; Bob can compute K by computing the middle expression.

4 The Proposed Attack

Eve, the adversary, sees all public information, and also sees the elements $(p, g) := c \bullet 1 * \widehat{g} \in \Omega$ and $(q, h) := d \bullet 1 * \widehat{h} \in \Omega$ that are transmitted between Alice and Bob. Eve's goal is to compute the shared key. Rather than attempting to compute Alice's private key material c and $\widehat{g}$, or Bob's private key material d and $\widehat{g}$, our attack will recover the shared key directly.

An overview of our attack is as follows. We first argue that the group C can be replaced by a 'linearised' version of C: this makes it easier to test membership in C. We then show that Eve does not need to compute Alice's or Bob's secret information in order to derive the shared key: more limited information suffices. (This information is specified in Eqs. (1) and (2) below.) Finally, we show how Eve can compute this information.

For a group H of $n \times n$ matrices over a field $\mathbb{F}$, we write $\mathrm{Alg}(H)$ for the $\mathbb{F}$-algebra generated by H [5]. So $\mathrm{Alg}(H)$ is the set of all $\mathbb{F}$-linear combinations of matrices in H. We write $\mathrm{Alg}^*(H)$ for the set of all invertible matrices in $\mathrm{Alg}(H)$.

The groups C proposed in the CBKAP satisfy $C = \mathrm{Alg}^*(C)$. More generally, we may assume that this is always the case. To see this, first note $\mathrm{Alg}^*(C)$ and D commute elementwise since every element of $\mathrm{Alg}^*(C)$ is a linear combination of elements in C. Thus, C may be replaced by $\mathrm{Alg}^*(C)$ to obtain a valid new instance of the protocol. Moreover, since $C \subseteq \mathrm{Alg}^*(C)$ the new instance of the protocol is more general than the original protocol: Alice can choose her matrix c from the larger group $\mathrm{Alg}^*(C)$. So if we successfully recover the common key in *every* new instance of the protocol, we can successfully recover the common key in the original instance.

Thus, from now on, we assume that $C = \mathrm{Alg}^*(C)$. Let $\kappa_1, \kappa_2, \ldots, \kappa_r \in C$ be a basis for $\mathrm{Alg}(C)$. Such a basis is not difficult to compute, using standard techniques. Our assumption means that any invertible linear combination of the matrices κ_i lies in C.

Let $P \trianglelefteq A$ be the *pure subgroup of* A, defined by

$$P = \{ (\alpha, g) \in A : g = e \}.$$

Then $\varphi(P)$ is a subgroup of $\mathrm{GL}_n(\mathbb{F})$. Consider the subgroup $\mathrm{Alg}^*(\varphi(P))$ of $\mathrm{GL}_n(\mathbb{F})$. Concretely, an element $\alpha' \in \mathrm{Alg}^*(\varphi(P))$ is an invertible matrix of the form

$$\alpha' = \sum_{i=1}^{k} \ell_i \varphi(\alpha_i)$$

where $k \geq 0$, $\ell_i \in \mathbb{F}$ and $(\alpha_i, e) \in P$.

Suppose that Eve finds elements $\tilde{c} \in C$, $\alpha' \in \mathrm{Alg}^*(\varphi(P))$ and $(\tilde{a}, g) \in \widehat{S}_n$ such that

$$(p, g) = \tilde{c} \bullet (\alpha', e) * (\tilde{a}, g). \tag{1}$$

Moreover, suppose that Eve can find an element $(\alpha_i, e) \in P$ and $\ell_i \in \mathbb{F}$ such that

$$\sum_{i=1}^{k} \ell_i \varphi(\alpha_i) = \alpha'. \tag{2}$$

Then Eve can compute the common key, as follows. Firstly, she computes the matrix

$$\beta' = \sum_{i=1}^{k} \ell_i \varphi({}^h \alpha_i).$$

This computation is possible for Eve, since h is part of the message $(q, h) = (d\varphi(b), h) \in \Omega$ transmitted from Bob to Alice. Now, $(\alpha_i, e) \in P \le A$, and so (α_i, e) *-commutes with all elements in B. Thus,

$$(q\varphi(^h\alpha_i), h) = d \bullet 1 * (b, h) * (\alpha_i, e) = d \bullet 1 * (\alpha_i, e) * (b, h).$$

Eve then computes $\tilde{c} \bullet (q\beta', h) * (\tilde{a}, g)$. We claim that this is equal to the common key K. To see this, first note that

$$(q\beta', h) = \sum_{i=1}^{k} \ell_i (q\varphi(^h\alpha_i), h)$$

$$= \sum_{i=1}^{k} \ell_i (d \bullet 1 * (\alpha_i, e) * (b, h))$$

$$= \sum_{i=1}^{k} \ell_i (d\varphi(\alpha_i)\varphi(b), h)$$

$$= (d \sum_{i=1}^{k} \ell_i \varphi(\alpha_i)\varphi(b), h)$$

$$= (d\alpha'\varphi(b), h)$$

$$= d \bullet (\alpha', e) * (b, h).$$

Hence

$$\tilde{c} \bullet (q\beta', h) * (\tilde{a}, g) = \tilde{c} \bullet d \bullet (\alpha', e) * (b, h) * (\tilde{a}, g)$$

$$= d \bullet \tilde{c} \bullet (\alpha', e) * (b, h) * (\tilde{a}, g)$$

$$(\text{since } \tilde{c} \in C \text{ and } d \text{ centralises } C)$$

$$= d \bullet \tilde{c} \bullet (\alpha', e) * (\tilde{a}, g) * (b, h)$$

$$(\text{as } (\tilde{a}, g) \in A \text{ and } (b, h) \in B \; * \text{-commute})$$

$$= d \bullet (p, g) * \hat{h}$$

$$= d \bullet (c \bullet 1 * \hat{g}) * \hat{h}$$

$$= K.$$

So it suffices to show that Eve can find elements α_i, ℓ_i, $\tilde{a}$, $\tilde{c}$ and α' so that Eqs. (1) and (2) are satisfied.

Precomputation stage: Find the α_i. Eve computes a collection of elements (α_i, e) such that the matrices $\varphi(\alpha_i)$ form a basis of $\mathrm{Alg}(\varphi(P))$. Once this is done, any $\alpha \in \mathrm{Alg}^*(\varphi(P))$ can easily be written in the form (2). Eve does not need to know the messages (p, g) and (q, h) in this stage, so this stage can be carried out as a precomputation. Eve proceeds as follows.

Eve generates, as in [11], short products (a', g') of generators of A such that the order r of the permutation g' is small (n or less), and computes

$\alpha_1 = (a', g')^r = (a'', e)$. She repeats this procedure to generate $\alpha_2, \alpha_3 \ldots$. (Eve may also take products of some of the previously generated elements (α_1, e), $(\alpha_2, e), \ldots, (\alpha_{i-1}, e)$ to define (α_i, e).) Eve stops when the dimension of the $\mathbb{F}$-linear span of the matrices $\varphi(\alpha_i)$ stops growing, and fixes a linearly independent subset of these matrices.

At the end of this process (relabelling after throwing linearly dependent elements $\varphi(\alpha_i)$ away), Eve has $\alpha_1, \alpha_2, \ldots \alpha_r$ such that $\varphi(\alpha_1), \varphi(\alpha_2), \ldots, \varphi(\alpha_r)$ are a basis for a subspace V of $\mathrm{Alg}(\varphi(P))$. Indeed, we expect (with high probability) that $V = \mathrm{Alg}(\varphi(P))$. We assume that this is true from now on.

Stage 1: Find $\tilde{a}$. Find a product of generators in A whose second component is equal to g, using the method in [11]. Let $(\tilde{a}, g)$ be this product. Define $\gamma \in \mathrm{GL}_n(\mathbb{F})$ by

$$(\gamma, e) = (p, g) * (\tilde{a}, g)^{-1}.$$

Stage 2: Find $\tilde{c}$. Recall that Eve knows $\kappa_1, \kappa_2, \kappa_3, \ldots, \kappa_r \in C$ that form a basis of $\mathrm{Alg}(C)$. She finds (see below) field elements $x_1, x_2, \ldots, x_r \in \mathbb{F}$ such that

$$\gamma^{-1}(x_1\kappa_1 + x_2\kappa_2 + \cdots + x_r\kappa_r) \in V, \text{ and} \tag{3}$$

$$x_1\kappa_1 + x_2\kappa_2 + \cdots + x_r\kappa_r \text{ is invertible.} \tag{4}$$

Set $\tilde{c} = x_1\kappa_1 + x_2\kappa_2 + \cdots + x_r\kappa_r$. Since $\tilde{c}$ is an invertible element of $\mathrm{Alg}(C)$, we see that $\tilde{c} \in C$.

To find a solution to Eqs. (3) and (4), Eve randomly generates solutions x_i that satisfy (3), which is easy, as the conditions are linear. She stops when (4) is also satisfied. We claim that the proportion of solutions to (3) that satisfy (4) is bounded below by $1 - n/|\mathbb{F}|$, which is a non-trivial proportion for the parameters that are proposed. The claim follows by applying the Invertibility Lemma [16, Lemma 9], which states that the proportion of invertible matrices in any $\mathbb{F}$-subspace of matrices over $\mathbb{F}$ is at least $1 - (n/|\mathbb{F}|)$, provided that the subspace contains at least one invertible matrix. We note that the elements of the form $x_1\kappa_1 + x_2\kappa_2 + \cdots + x_r\kappa_r$ that satisfy (3) are a subspace of matrices. So it remains to show that there exists an invertible element of this form. But let $x_1, x_2, \ldots, x_r \in \mathbb{F}$ be such that $x_1\kappa_1 + x_2\kappa_2 + \cdots + x_r\kappa_r = c$. The elements x_i exist since $c \in C \subseteq \mathrm{Alg}(C)$. Clearly $\tilde{c} = c$ is invertible. Moreover (3) holds, because we may show that $\gamma^{-1}c \in \varphi(P) \subseteq V$ as follows. Firstly,

$$(\gamma, e) = (p, g) * (\tilde{a}, g)^{-1} = (c\varphi(a), g) * ({}^{g^{-1}}(\tilde{a}^{-1}), g^{-1}) = (c\varphi(a)\varphi(\tilde{a}^{-1}), e),$$

so $\gamma = c\varphi(a)\varphi(\tilde{a}^{-1})$ and therefore $\gamma^{-1}c = \varphi(\tilde{a})\varphi(a)^{-1}$. And secondly, we see that $\varphi(\tilde{a})\varphi(a)^{-1} = \varphi(\tilde{a}a^{-1}) \in \varphi(P)$, since

$$(\tilde{a}, g)(a, g)^{-1} = (\tilde{a}, g)({}^{g^{-1}}a^{-1}, g^{-1}) = (\tilde{a}a^{-1}, e)$$

and $(\tilde{a}, g), (a, g) \in A$.

Stage 3: The remaining parameters. Eve sets $\alpha' = \tilde{c}^{-1}\gamma$. Since $(\alpha')^{-1} \in V$, we see that α' (being a power of $(\alpha')^{-1}$) also lies in V. So Eve can easily calculate coefficients ℓ_i such that

$$\sum_{i=1}^{k} \ell_i \varphi(\alpha_i) = \alpha'.$$

Hence, Eq. (2) holds. We may also verify that Eq. (1) holds:

$$\tilde{c} \bullet (\alpha', e) * (\tilde{a}, g) = \tilde{c} \bullet (\tilde{c}^{-1}\gamma, e) * (\tilde{a}, g) = (\gamma, e) * (\tilde{a}, g)$$
$$= ((p, g) * (\tilde{a}, g)^{-1}) * (\tilde{a}, g) = (p, g).$$

5   Experiments and Conclusion

We have implemented our attack in Magma [7], running on one 2 GHz core of a multi-core server. We used 5 sets of actual challenge parameters kindly provided by SecureRF. These parameters all used the values $|\mathbb{F}| = 256$ and $n = 16$. The subgroup A is specified by a generating set; each generator for A is given as a word of length approximately 650 (notice the large parameter setting!) in the generating set $\mathcal{X} = \{(x_i(t), s_i) : 1 \le i \le n - 1\}$ for $\overline{M} \ltimes S_n$ defined in [1]. In all 5 cases, our attack terminated successfully, producing the exact shared key. Our attack used less than 64 MB of memory, and terminated in less than 8 h. We would like to emphasise that our code is far from being optimised; we estimate an improvement in CPU time by a significant factor in an optimised version of the attack.

Let B_n be a braid group on n strands, and let $\sigma_1, \sigma_2, \ldots, \sigma_{n-1}$ be the Artin generators for B_n. (See, for example, [12] for an introduction to braid groups.) There is a homomorphism $\psi \colon B_n \to \overline{M} \ltimes S_n$ such that $\psi(\sigma_i) = (x_i, s_i)$ for $1 \le i \le n - 1$, which gives rise to the coloured Burau representation. Thus we could (and did) use standard routines for computing with braids in B_n, rather than dealing with words in $\mathcal{X}$ directly.

The most computationally intensive part of the attack is the computing of $\varphi(a)$ where $(a, g) \in \overline{M} \ltimes S_n$ is given as a word in the generators of A. The long length of the generators in A as words in $\mathcal{X}$ is the cause of difficulty here; we were computing with words of length approximately 20,000 in Stage 1 of our attack.

To decide when the precomputation stage should terminate, we use the criterion that the $\mathbb{F}$-dimension of the algebra generated by the matrices $\varphi(\alpha_i)$ should not grow when 4 generators (α_i, e) in a row are considered.

Not surprisingly, this attack is highly parallelisable. We did not exploit this fact since for the actual parameters a single CPU core sufficed.

It remains open how to immunise the Algebraic Eraser against the presented cryptanalysis. The only hope seems to be to make the problem of expressing a permutation as a short product of given permutations difficult, by working with very carefully chosen distributions. However, for the intended applications, the computational constraints necessitate small values of n. In this case, Schreier–Sims methods solve this problem efficiently, no matter how the permutations are used. See the discussion around [11, Table 4].

Acknowledgement. The authors would like to thank Arkadius Kalka for providing code from the earlier attack [11] on the Algebraic Eraser, and for explaining how to use this code. The authors would also like to thank Martin Albrecht for various excellent editorial suggestions.

References

1. Anshel, I., Anshel, M., Goldfeld, D., Lemieux, S.: Key agreement, the algebraic eraser^{TM} and lightweight cryptography. Contemp. Math. **418**, 1–34 (2006)
2. Anshel, I., Atkins, D., Goldfeld, D., Gunnells, P.: Defeating the Ben-Zvi, Blackburn, and Tsaban attack on the algebraic eraser. IACR ePrint 2016/044
3. Atkins, D., Gunnells, P.: Algebraic Eraser^{TM}: a lightweight, efficient asymmetric key agreement protocol for use in no-power, low-power, and IoT devices. In: NIST Lightweight Cryptography Workshop, 20 July 2015. http://www.nist.gov/itl/csd/ct/lwc_workshop2015.cfm
4. Bahr, F., Boehm, M., Franke, J., Kleinjung, T.: RSA200. http://www.crypto-world.com/announcements/rsa200.txt
5. Ben-Zvi, A., Kalka, A., Tsaban, B.: Cryptanalysis via algebraic spans. IACR ePrint 2014/041
6. Blackburn, S.R., Robshaw, M.J.B.: On the security of the algebraic eraser tag authentication protocol. In: Manulis, M., Sadeghi, A.-R., Schneider, S. (eds.) ACNS 2016. LNCS, vol. 9696, pp. 3–17. Springer, Heidelberg (2016). doi:10.1007/978-3-319-39555-5_1
7. Bosma, W., Cannon, J., Playoust, C.: The Magma algebra system. I. The user language. J. Symb.Comput. **24**, 235–265 (1997)
8. Cryptography Stack Exchange: Status of Algebraic Eraser Key Exchange? http://crypto.stackexchange.com/questions/30644/status-of-algebraic-eraser-key-exchange. Accessed 5 Feb 2016
9. Goldfeld, D., Gunnells, P.: Defeating the Kalka-Teicher-Tsaban linear algebra attack on the algebraic eraser. arXiv:1202.0598, February 2012
10. Gunnells, P.: On the cryptanalysis of the generalized simultaneous conjugacy search problem and the security of the algebraic eraser. arXiv:1105.1141, May 2011
11. Kalka, A., Teicher, M., Tsaban, B.: Short expressions of permutations as products and cryptanalysis of the algebraic eraser. Adv. Appl. Math. **49**, 57–76 (2012)
12. Kassel, C., Turaev, V.: Braid Groups. Graduate Texts in Mathematics, vol. 247. Springer, New York (2008)
13. Myasnikov, A., Ushakov, A.: Cryptanalysis of the Anshel-Anshel-Goldfeld-Lemieux key agreement protocol. Groups Complex. Cryptol. **1**, 63–75 (2009)
14. National Institute of Stadards and Technology: Digital Signature Standard (DSS). Federal Information Processing Standards Publication FIPS PUB **186**–4, July 2013. http://nvlpubs.nist.gov/nistpubs/FIPS/NIST.FIPS.186-4.pdf
15. Rivest, R., Shamir, A., Adleman, L.: A method for obtaining digital signatures and public-key cryptosystems. Commun. ACM **21**, 120–126 (1978)
16. Tsaban, B.: Polynomial-time solutions of computational problems in noncommutative algebraic cryptography. J. Cryptol. **28**, 601–622 (2015)

Lattice-Based Fully Dynamic Multi-key FHE with Short Ciphertexts

Zvika Brakerski[1]([✉]) and Renen Perlman[2]

[1] Weizmann Institute of Science, Rehovot, Israel
zvika.brakerski@weizmann.ac.il
[2] Tel-Aviv University, Tel Aviv, Israel
renenperlman@mail.tau.ac.il

Abstract. We present a multi-key fully homomorphic encryption scheme that supports an *unbounded* number of homomorphic operations for an *unbounded* number of parties. Namely, it allows to perform arbitrarily many computational steps on inputs encrypted by an a-priori unbounded (polynomial) number of parties. Inputs from new parties can be introduced into the computation dynamically, so the final set of parties needs not be known ahead of time. Furthermore, the length of the ciphertexts, as well as the space complexity of an atomic homomorphic operation, grow only *linearly* with the current number of parties.

Prior works either supported only an a-priori bounded number of parties (López-Alt, Tromer and Vaikuntanthan, STOC '12), or only supported single-hop evaluation where all inputs need to be known before the computation starts (Clear and McGoldrick, Crypto '15, Mukherjee and Wichs, Eurocrypt '16). In all aforementioned works, the ciphertext length grew at least quadratically with the number of parties.

Technically, our starting point is the LWE-based approach of previous works. Our result is achieved via a careful use of Gentry's bootstrapping technique, tailored to the specific scheme. Our hardness assumption is that the scheme of Mukherjee and Wichs is circular secure (and thus bootstrappable). A leveled scheme can be achieved under standard LWE.

1 Introduction

In 1978, Rivest et al. [RAD78] envisioned an encryption scheme where it is possible to publicly convert an encryption of a message x into an encryption of $f(x)$ for any f, thus enabling private outsourcing of computation. It took over 30 years for the first realization of this so called *fully homomorphic encryption* (FHE) to materialize in Gentry's breakthrough work [Gen09b, Gen09a], but since then progress has been consistent and rapid. López-Alt et al. [LTV12] considered an extension of this vision into the *multi-key* setting, where it is possible to

Z. Brakerski—Supported by the Israel Science Foundation (Grant No. 468/14), the Alon Young Faculty Fellowship, Binational Science Foundation (Grant No. 712307) and Google Faculty Research Award.

R. Perlman—This work took place at the Weizmann Institute of Science as a part of the Young Weizmann Scholars program for outstanding undergraduate students.

M. Robshaw and J. Katz (Eds.): CRYPTO 2016, Part I, LNCS 9814, pp. 190–213, 2016.
DOI: 10.1007/978-3-662-53018-4_8

compute on encrypted messages even if they were not encrypted using the same key. In multi-key FHE, a public evaluator takes ciphertexts encrypted under different keys, and evaluates arbitrary functions on them. The resulting ciphertext can then be decrypted using the collection of keys of all parties involved in the computation. Note that the security of the encryption scheme compels that all keys need to be used for decryption.

In the dream version of multi-key FHE, each user generates keys for itself and encrypts messages at will. Any third party can then perform an arbitrary computation on any set of encryptions by any set of users. The resulting ciphertext is attributed to the set of users whose ciphertexts were used to create it, and the collection of all of their secret keys is required in order to decrypt it. Most desirable is a *fully dynamic setting* where nothing at all about the parties needs to be known ahead of time: not their identities, not their number and not the order in which they will join the computation. In particular, outputs of previous evaluations can be used as inputs to new evaluations regardless of whether they correspond to the same set of users, intersecting sets or disjoint sets. In short, *any* operation can be performed on *any* ciphertext at *any* point in time, and of course while maintaining ciphertext compactness. However, as we explain below, existing solutions fall short of achieving this functionality.

Multi-key FHE can be useful in various situations involving multiple parties that do not coordinate ahead of time, but only after the fact. In [LTV12], the main motivation is performing *on-the-fly multiparty computation (MPC)* where various parties wish to use a cloud server (or some other untrusted third party) to perform some computation without revealing their private inputs and while having minimal interaction with the server. In this setting, the parties first send their encrypted input to the cloud, which performs the homomorphic operation and sends the output to the parties. The parties then execute a multiparty computation protocol for joint decryption. A similar approach was used by Mukherjee and Wichs [MW16] to construct a 2-round MPC protocol in the common random string setting. We note that one could also consider using this primitive in simpler situations, such as ones where the respective secret keys are being sold after the fact, and the owner of subset of keys can decrypt the output of the respective computation by himself.

As we mentioned, multi-key FHE was introduced by [LTV12] who also introduced the first candidate scheme, building upon the NTRU encryption scheme. Their candidate was almost fully dynamic, except an upper bound on the maximal number of participants in a computation had to be known at the time a key is generated. In particular, to support computation amongst N parties, the bit-length of a ciphertext in their scheme grew with $N^{1+1/\epsilon}$, where $\epsilon < 1$ is a parameter related to the security of the scheme. They were able to support arbitrarily complex computation through use of bootstrapping, but this required a circular security assumption.

The next step forward was by Clear and McGoldrick [CM15] who were motivated by the question of constructing identity based FHE. As a stepping stone, they were able to construct a multi-key FHE scheme based on the hardness of the learning with errors (LWE) problem, which is related to the hardness of certain short vector problems (such as GapSVP, SIVP) in worst case lattices. Their scheme was simplified by Mukherjee and Wichs [MW16] who used it to introduce low-round MPC protocols. In their schemes, they focused on the *single-hop* setting, where the collection of input ciphertexts and the function to be computed are known ahead of time. The dynamic setting where users can join the computation and the function is determined on the fly was not considered. Furthermore, their solution produced ciphertexts whose bit-length grew with N^2, where N is the number of users in the computation. Lastly, their construction requires that all users share common public parameters (a common random string).

1.1   Our Results

As described above, great progress had been made in the study of multi-key FHE, but still much was left to be desired. In particular, coping with the *fully dynamic* setting where no information about the participating parties needs to be known at key generation. This will allow maximum versatility in use of the scheme. The second issue is the ciphertext length and more broadly the space complexity of homomorphic evaluation. Previous works all hit the same barrier of N^2 growth in the ciphertext. Implementations of (single key) FHE already mention the space complexity as a major bottleneck in the usability of the scheme [HS14, HS15], and therefore reducing the overhead in this context is important for making multi-key FHE applicable in the future. Another interesting open problem is removing the requirement for common parameters from the [CM15, MW16] solution.

We address two of the three aforementioned problems by presenting a *fully dynamic* multi-key FHE scheme, with $O(N)$ *ciphertext expansion*, and $O(N)$ space complexity for an atomic homomorphic operation (e.g. evaluating a single gate), where N is the number of parties whose ciphertexts have been introduced into the computation so far (since we are fully dynamic, we allow inputs from more parties to join later). This, in turn, can be used to improve the space complexity of the parties in the MPC protocols of [LTV12, MW16]. Our construction still requires common public parameters as in [CM15, MW16].

In terms of hardness assumption, we are comparable with previous works. We can rely on the hardness of the learning with errors (LWE) problem with a slightly super-polynomial modulus to achieve a *leveled* solution, where only a-priori depth bounded circuits can be evaluated. This restriction can be lifted towards achieving a fully dynamic scheme by making an additional circular security assumption.

We stress that our scheme is not by itself practical. We use the bootstrapping machinery in a way that introduces fair amounts of overhead into the evaluation process. The goal of this work, rather, is to indicate that the theoretical boundaries of multi-key FHE, and open the door for further optimizations bringing solutions closer to the implementable world.

1.2   Our Techniques

Gentry, Sahai and Waters [GSW13] proposed an FHE scheme ("the GSW scheme") with the following properties (we use notation due to Alperin-Sheriff and Peikert [AP14]). A ciphertext is represented as a matrix $\mathbf{C}$ over $\mathbb{Z}_q$, and the secret key is a (row) vector $\mathbf{t}$ such that $\mathbf{t}\mathbf{C} = \mathbf{e} + \mu\mathbf{t}\mathbf{G}$, where $\mu \in \{0,1\}$ is the encrypted message, $\mathbf{e}$ is a low-norm noise vector and $\mathbf{G}$ is a special gadget matrix. So long as the norm of $\mathbf{e}$ is small enough, μ can be retrieved from the ciphertext matrix using the secret key $\mathbf{t}$. In order to homomorphically multiply two ciphertexts $\mathbf{C}_1, \mathbf{C}_2$, compute $\mathbf{C}_1 \cdot \mathbf{G}^{-1}(\mathbf{C}_2)$, where $\mathbf{G}^{-1}(\cdot)$ is an efficiently computable function that operates column-by-column, and whose output is always low-norm. It had been shown by [CM15, MW16] that GSW can be augmented with multi-key features if all parties use common public parameters (which are just a random string) and if the encryption procedure changes as follows. In [CM15, MW16], after encrypting the message with GSW, the randomness that had been used for the encryption is itself encrypted using fresh randomness. The new ciphertext thus contains the matrix $\mathbf{C}$ along with its encrypted randomness $\overrightarrow{\mathcal{R}}$. They show that given a set of N public keys, and a ciphertext of this form under one of these keys, it is possible to obtain a new ciphertext $\widehat{\mathbf{C}}$, which is essentially an $N \times N$ block matrix, with each block having the size of a single-key ciphertext. This new $\widehat{\mathbf{C}}$ encrypts the same message μ as the original $\mathbf{C}$, but under the secret key $\hat{\mathbf{t}}$ which is the concatenation of all N secret keys. In other words, $\hat{\mathbf{t}}\widehat{\mathbf{C}} = \hat{\mathbf{e}} + \mu\hat{\mathbf{t}}\widehat{\mathbf{G}}$, where $\widehat{\mathbf{G}}$ is an expanded gadget matrix (a block matrix with the old $\mathbf{G}$ on its diagonal). This means that given a collection of N ciphertexts, one can expand all of them to correspond to the same $\hat{\mathbf{t}}$ and perform homomorphic operations.

It is clear from the above description that there is an inherent obstacle in adapting this approach to the fully dynamic setting. Indeed, when the expand operation creates the new $\widehat{\mathbf{C}}$, it does not create the respective encrypted randomness $\overrightarrow{\mathcal{R}}$, and therefore one cannot continue to perform homomorphic operations with newly introduced parties. Our first contribution is noticing that this can be resolved by using Gentry's *bootstrapping* technique [Gen09b]. Indeed, previous works [LTV12, MW16] used bootstrapping in order to go from limited amount of homomorphism to full homomorphism. The bootstrapping principle is that so long as a scheme can homomorphically evaluate (a little more than) its own decryption circuit, it can be made to evaluate any circuit. To get the strongest version of the theorem, one needs to further assume that the scheme is *circular secure* (can securely encrypt its own secret key). The idea is to use the encrypted secret key as an input to the function that decrypts a ciphertext and performs an atomic operation on it. Let c_1, c_2 be some ciphertexts and consider the function $h_{c_1,c_2}(x) = (\mathsf{Dec}_x(c_1) \text{ NAND } \mathsf{Dec}_x(c_2))$. This function takes an input, interprets this input as a secret key, uses it to decrypt c_1, c_2 and performs the NAND operation on the messages. Computing this function on a non-encrypted secret key would output $h(\mathsf{sk}) = (\mu_1 \text{ NAND } \mu_2)$. Therefore, performing it homomorphically on the encrypted secret key will result in an encryption of $(\mu_1 \text{ NAND } \mu_2)$. This allows to continue to evaluate the circuit gate by gate. In the context of multi-key FHE, each party only needs to encrypt its own secret key. Then, the multi-key functionality will allow to compute the joint key out of the individual keys and proceed as above.

We therefore consider a new scheme by modifying the MW scheme [MW16] as follows. We append to the public key an encryption of the secret key (using MW encryption which consists of a GSW encryption of the secret key, and an encryption of the respective randomness). Our encryption algorithm is plain GSW encryption - no need to encrypt the randomness. We show that ciphertext expansion can still be achieved here. This is because in bootstrapping, the ciphertexts storing the messages are not the ones upon which homomorphic evaluation is performed. Rather, the input to the homomorphic evaluation is always the encryption of the secret key. We take this approach another step forward and consider ciphertexts c_1, c_2 s.t. c_1 is encrypted under some set of public keys T_1 corresponding to concatenated secret key $\hat{t}_1$, and c_2 is encrypted under a set T_2 corresponding to $\hat{t}_2$. The public keys in T_1, T_2 contain an encryption of the individual secret keys, which in turn can be expanded to an encryption of $\hat{t}_1, \hat{t}_2$ under a key $\hat{t}$ which corresponds to the union of the sets $T_1 \cup T_2$. This will allow us to perform homomorphic evaluation of the bootstrapping function $h(x_1, x_2) = (\mathsf{Dec}_{x_1}(c_1) \ \mathsf{NAND} \ \mathsf{Dec}_{x_2}(c_2))$ and obtain an encrypted output respective to $T_1 \cup T_2$. This process can be repeated as many times as we want (we make sure that we have sufficient homomorphic capacity to evaluate the function $h(\cdot)$ for any polynomial number of parties N). We note that in order for this solution to be secure, encrypting the secret key under the public key in MW needs to be secure (which translates to a circular security assumption on MW). We need to make this hardness assumption explicitly, in addition to the hardness of LWE. However, as in the single-key setting, one can generate a chain of secret keys encrypting one another and obtain a *leveled* scheme that only supports evaluation of circuits up to a predefined depth bound. This can be done while relying on the hardness of LWE alone, which translates to the hardness of approximation of $\mathsf{GapSVP}, \mathsf{SIVP}$ in worst-case lattices.

We thus explained how to achieve a fully dynamic multi-key FHE scheme, but so far the length of the ciphertexts was inherited from the MW scheme, and grew quadratically with N. Examining the bootstrapping solution carefully, it seems that the ciphertext length problem might have a simple solution. We notice that the decryption procedure of the GSW scheme (and thus also of the MW scheme), only computes the inner product of the secret key with a single column derived from the ciphertext matrix. In fact, the rightmost column of the ciphertext matrix will do. In a way, the rest of the ciphertext matrix is only there to allow for the homomorphic evaluation. Therefore, we can amend the previous approach, and after performing each atomic operation, we can just toss out the resulting matrix, except the last column. The length of this last column only grows with N and not N^2, and it is sufficient for the subsequent bootstrapping steps. However, we view this as a very minor victory, since in order to perform homomorphic operations via bootstrapping, we will actually need to expand the encryptions of the secret key to size $N \times N$ and evaluate the bootstrapping procedure with these mammoth matrices. Our goal is to save not only on the communication complexity but also on the memory requirement of the homomorphic evaluation process. As we mentioned above, memory requirement is the main bottleneck in current implementations of FHE.

To reduce the space complexity, we first observe that in some cases, the $N \times N$ block matrices are actually quite sparse. In fact, the expand operation from [CM15, MW16] generates very sparse matrices, where only $(2N - 1)$ of the N^2 blocks are non-zero. Thus the output of expand can be represented using only $O(N)$ space. However, this neat property disappears very quickly as homomorphic operations are performed and ciphertexts are multiplied by one another. It seems hard to shrink a matrix in mid-computation back into $O(N)$ size. The next idea is to incorporate into the scheme the sequentialization method of Brakerski and Vaikuntanathan [BV14]. Their motivation was to reduce the noise accumulation in the (single-key) bootstrapping procedure, and they did this by converting the decryption circuit into a *branching program*. A branching program contains a sequence of steps (polynomially many, in our case), where in each step a local state is being updated through interaction with one of the input bits. We recall that in our case, the input bits are the (expanded) encryptions of the bits of the secret keys. The newly expanded ciphertexts have length $O(N)$, so we only need to worry about the encryption of the running state. However, since homomorphic evaluation is of the form $\mathbf{C}_1 \cdot \mathbf{G}^{-1}(\mathbf{C}_2)$, it is sufficient to hold the last column of $\mathbf{C}_2$ (and thus of $\mathbf{G}^{-1}(\mathbf{C}_2)$) in order to obtain the last column of the product. Therefore, so long as we make sure that the encrypted state of the computation is always the right-hand operand in the multiplication (which can be done in branching program evaluation), we can perform the entire computation with $O(N)$ space in total. A subtle point in this use of branching programs is that the representation of the program itself could have size $\text{poly}(N)$, so just writing it would require more space than we could save anyway. To address this issue, we notice that the construction of the branching program in Barrington's theorem can be performed "on the fly". We can hold a state of size proportional to the depth of the circuit we wish to evaluate and produce the layers of the branching program one by one. Thus we can produce a layer, evaluate it, proceed to the next one etc., all without exceeding our space limit.

1.3 Other Related Work

In an independent and concurrent work, Peikert and Shiehian [PS16] also addressed the problem of constructing multi-hop multi-key FHE. They show that multi-hop homomorphism can be achieved "natively" without using bootstrapping. This is done by utilizing new properties of the [CM15, MW16] multi-key scheme and incorporating them with those of GSW-style encryption. Without bootstrapping, they are unable to achieve a fully dynamic scheme, and require the total number of users and the maximal computation depth to be known ahead of time. Their techniques are very different from ours and we believe that it is a significant contribution to the study of multi-key FHE.

2 Preliminaries

Matrices are denoted by bold-face capital letters, and vectors are denoted by bold-face small letters. For vectors $\mathbf{v} = (v_i)_{i \in [n]}$, we let $\mathbf{v}[i]$ be the i'th element of the vector, for every $i \in [n]$. Similarly, for a matrix $\mathbf{M} = (m_{i,j})_{i \in [n], j \in [m]}$, we let $\mathbf{M}[i, j]$ be the i'th element of the j'th column, for every $i \in [n]$ and $j \in [m]$. Sequences of matrices $\mathbf{M}_1, \ldots, \mathbf{M}_\ell$ are denoted by $\overrightarrow{\mathcal{M}}$. We let $\overrightarrow{\mathcal{M}}[k]$ be the k'th matrix, for every $k \in [\ell]$. To avoid cluttering of notation, we regard to vectors in the same way as we do to matrices and do not denote row vectors with the transposed symbol. The standard rules of matrix arithmetics should be applied to vectors the same as they do for matrices. The vectors of the standard basis are denoted by $\{\mathbf{u}_i\}_i$, the dimension will be clear from the context.

All logarithms are taken to base 2, unless otherwise specified. We let $\mathbb{Z}_q$ be the ring of integers modulo q. Normally we associate $x \in \mathbb{Z}_q$ with the value $y \in (-q/2, q/2] \cap \mathbb{Z}$ s.t. $y = x \pmod{q}$. We denote $\ell_q = \lceil \log q \rceil$ and recall that an element in $\mathbb{Z}_q$ can be represented by a string in $\{0, 1\}^{\ell_q}$, by default x will be represented using two's complement representation of the aforementioned representative y. For a distribution ensemble $\chi = \chi(\lambda)$ over $\mathbb{Z}$, and integer bounds $B_\chi = B_\chi(\lambda)$ we say that χ is B-bounded if $Pr_{x \leftarrow \chi}[|x| > B] = 0$.

For $x \in \mathbb{Z}_q$ we define $|x| = \arg\min_{y = x \pmod{q}} |y|$ (this function does not have all properties of standard absolute value, but the triangle inequality still holds). Further, we will denote $\|\mathbf{v}\|_\infty = \max_i |\mathbf{v}[i]|$.

We let λ denote a security parameter. When we speak of a negligible function $\mathrm{negl}(\lambda)$, we mean a function that decays faster than $1/\lambda^c$ for any constant $c > 0$ and sufficiently large values of λ. When we say that an event happens with overwhelming probability, we mean that it happens with probability at least $1 - \mathrm{negl}(\lambda)$ for some negligible function $\mathrm{negl}(\lambda)$.

2.1 Homomorphic Encryption and Bootstrapping

We now define fully homomorphic encryption and introduce Gentry's bootstrapping theorem. Our definitions are mostly taken from [BV11, BGV12], and adapted to the setting where multiple users can share the same public parameters.

A homomorphic (public-key) encryption scheme HE = (HE.Setup, HE.Keygen, HE.Enc, HE.Dec, HE.Eval) is a 5-tuple of PPT algorithms as follows (λ is the security parameter):

- **Setup** params←HE.Setup(1^λ): Outputs the public parametrization params of the system.
- **Key generation** (pk, sk)←HE.Keygen(params): Outputs a public encryption key pk and a secret decryption key sk.
- **Encryption** c←HE.Enc(pk, μ): Using the public key pk, encrypts a single bit message $\mu \in \{0, 1\}$ into a ciphertext c.
- **Decryption** μ←HE.Dec(sk, c): Using the secret key sk, decrypts a ciphertext c to recover the message $\mu \in \{0, 1\}$.

– **Homomorphic evaluation** $\widehat{c} \leftarrow \mathsf{HE.Eval}(\mathcal{C}, (c_1, \ldots, c_\ell), \mathsf{pk})$: Using the public key pk, applies a circuit $\mathcal{C} : \{0,1\}^\ell \rightarrow \{0,1\}$ to $c_1, \ldots, c_\ell$, and outputs a ciphertext $\widehat{c}$.

A homomorphic encryption scheme is said to be secure if it is semantically secure.

Full homomorphism is defined next. We distinguish between single-hop and multi-hop homomorphism as per [GHV10]. Loosely speaking, single-hop homomorphism supports only a single evaluation of a circuit on ciphertexts. Conversely, in the multi-hop case, one can continue evaluating circuits (on not necessarily fresh) ciphertexts, as long as they decrypt correctly.

Definition 1 (Compactness and Full Homomorphism). *A scheme* HE *is single-hop fully homomorphic, if for any efficiently computable circuit* $\mathcal{C}$ *and any set of inputs* $\mu_1, \ldots, \mu_\ell$, *letting* $\mathsf{params} \leftarrow \mathsf{HE.Setup}(1^\lambda)$, $(\mathsf{pk}, \mathsf{sk}) \leftarrow \mathsf{HE.Keygen}(\mathsf{params})$ *and* $c_i \leftarrow \mathsf{HE.Enc}(\mathsf{pk}, \mu_i)$, *it holds that*

$$\Pr\left[\mathsf{HE.Dec}(\mathsf{sk}, \mathsf{HE.Eval}(\mathcal{C}, (c_1, \ldots, c_\ell), \mathsf{pk})) \neq \mathcal{C}(\mu_1, \ldots, \mu_\ell)\right] = \mathrm{negl}(\lambda),$$

The scheme is multi-hop fully homomorphic if for any circuit $\mathcal{C}$ *and any set of* ciphertexts $c_1, \ldots, c_\ell$, *letting* $\mathsf{params} \leftarrow \mathsf{HE.Setup}(1^\lambda)$, $(\mathsf{pk}, \mathsf{sk}) \leftarrow \mathsf{HE.Keygen}(\mathsf{params})$ *and* $\mu_i \leftarrow \mathsf{HE.Dec}(\mathsf{sk}, c_i)$, *it holds that*

$$\Pr\left[\mathsf{HE.Dec}(\mathsf{sk}, \mathsf{HE.Eval}(\mathcal{C}, (c_1, \ldots, c_\ell), \mathsf{pk})) \neq \mathcal{C}(\mu_1, \ldots, \mu_\ell)\right] = \mathrm{negl}(\lambda).$$

A fully homomorphic encryption scheme is compact if its decryption circuit is independent of the evaluated function. The scheme is leveled fully homomorphic if it takes 1^L *as additional input in key generation, and can only evaluate depth* L *Boolean circuits (this notion usually only refers to single-hop schemes).*

Gentry's bootstrapping theorem shows how to go from limited amount of homomorphism to full homomorphism. This method has to do with the *augmented decryption circuit* and, in the case of pure fully homomorphism, relies on the *weak circular security* property of the scheme.

Definition 2. *Consider a homomorphic encryption scheme* HE. *Let* $(\mathsf{sk}, \mathsf{pk})$ *be properly generated keys and let* $\mathcal{C}$ *be the set of properly decryptable ciphertexts. Then the set of augmented decryption functions,* $\{f_{c_1, c_2}\}_{c_1, c_2 \in \mathcal{C}}$ *is defined by* $f_{c_1, c_2}(x) = \overline{\mathsf{HE.Dec}_x(c_1) \wedge \mathsf{HE.Dec}_x(c_2)}$. *Namely, the function that uses its input as secret key, decrypts* c_1, c_2 *and returns the NAND of the results.*

Definition 3. *A public key encryption scheme* PKE *is said to be* weakly circular secure *if it is secure even against an adversary who gets encryptions of the bits of the secret key.*

The bootstrapping theorem is thus as follows.

Theorem 1 (Bootstrapping [Gen09b, Gen09a]). *A scheme that can homomorphically evaluate its family of augmented decryption circuits can be transformed into a leveled fully homomorphic encryption scheme with the same decryption circuit, ciphertext space and public key.*

Furthermore, if the aforementioned scheme is also weak circular secure, then it can be made into a pure fully homomorphic encryption scheme.

2.2 Multi-key Homomorphic Encryption

A homomorphic encryption scheme is *multi-key* if it can evaluate circuits on ciphertexts encrypted under different public keys. To decrypt an evaluated ciphertext, the algorithm uses the secret keys of all parties whose ciphertexts took part in the computation.

A multi-key homomorphic encryption scheme MKHE = (.Setup, MKHE.Keygen, MKHE.Enc, MKHE.Dec, MKHE.Eval) is a 5-tuple of PPT algorithms as follows:

- **Setup** params←.Setup(1^λ): Outputs the public parametrization params of the system.
- **Key generation** (pk, sk)←MKHE.Keygen(params): Outputs a public encryption key pk and a secret decryption key sk.
- **Encryption** c←MKHE.Enc(pk, μ): Using the public key pk, encrypts a single bit message $\mu \in \{0,1\}$ into a ciphertext c.
- **Decryption** μ←MKHE.Dec(($sk_1, \ldots, sk_N$), c): Using the sequence of secret keys ($sk_1, \ldots, sk_N$), decrypts a ciphertext c to recover the message $\mu \in \{0,1\}$.
- **Homomorphic evaluation** $\hat{c}$←MKHE.Eval($\mathcal{C}$, ($c_1, \ldots, c_\ell$), ($pk_1, \ldots, pk_N$)): Using the sequence of public keys ($pk_1, \ldots, pk_N$), applies a circuit $\mathcal{C} : \{0,1\}^\ell \to \{0,1\}$ to $c_1, \ldots, c_\ell$, where each ciphertext c_j is evaluated under a sequence of public keys $T_j \subset \{pk_1, \ldots, pk_N\}$ (we assume that T_j is implicit in c_i). Upon termination, outputs a ciphertext $\hat{c}$.

In the multi-key setting we define the notion of *fully dynamic* scheme as a generalization of multi-hop homomorphism in the single-key setting. A formal definition follows.

Definition 4 (Fully Dynamic Multi-Key FHE). *A scheme* MKHE *is fully dynamic multi-key FHE, if the following holds. Let* $N = N_\lambda$ *be any polynomial in the security parameter, let* $\mathcal{C} = \mathcal{C}_\lambda$ *be a sequence of circuits, set* params←.Setup(1^λ) *and* (pk_i, sk_i)←MKHE.Keygen(params) *for every* $i \in [N]$, *and let* $\hat{c}_j$ *be such that* MKHE.Dec(($sk_{j,1}, \ldots, sk_{j,s}$), $\hat{c}_j$) $= \mu_j$, *where* $\{sk_{j,i}\}_{j,i} \subseteq \{sk_1, \ldots, sk_N\}$. *Then*

$$\Pr\left[\mathsf{MKHE.Dec}((sk_1, \ldots, sk_N), \mathsf{MKHE.Eval}(\mathcal{C}, (\hat{c}_1, \ldots, \hat{c}_\ell), (pk_1, \ldots, pk_N)))\right.$$
$$\left. \neq \mathcal{C}(\mu_1, \ldots, \mu_\ell)\right] = \mathrm{negl}(\lambda)$$

The scheme is compact if its decryption circuit is independent of the evaluated function and its size is $\mathrm{poly}(\lambda, N)$ *for some fixed polynomial.*

2.3 Barrington's Theorem and an On-the-Fly Variant

We define the computational model of permutation branching programs, and cite the fundamental theorem of Barrington connecting them to depth bounded computation. Finally, we note a corollary from Barrington's construction, which allows to compute the branching program "on-the-fly", layer by layer, keeping only small state.

Definition 5. *A permutation branching program Π with ℓ variables, width k and length L is a collection of L tuples $(p_{0,t}, p_{1,t})_{t \in [L]}$ called the instructions and a function* var : $[L] \rightarrow [\ell]$. *Each tuple is composed of a pair of permutations $p_{0,t}, p_{1,t} : [k] \rightarrow [k]$. The program takes as input a binary vector $\mathbf{x} = (x_1, \ldots, x_\ell) \in \{0,1\}^\ell$, and outputs a bit $b \in \{0,1\}$. The execution of the Π is as follows: the program keeps a state integer $s \in \{1, \ldots, k\}$, initially $s_0 = 1$. On every step $t = 1, \ldots, L$, the next state is determent recursively using the t'th instruction:*

$$s_t := p_{x_{var(t)}, t}(s_{t-1})$$

In other words, $s_t := p_{0,t}(s_{t-1})$ if $x_{var(t)} = 0$, and otherwise $s_t := p_{1,t}(s_{t-1})$. Finally, after the L'th iteration, the branching programs outputs 1 if and only if $s_L = 1$.

Theorem 2 (Barrington's Theorem [Bar89]). *Every Boolean NAND circuit Ψ that acts on ℓ inputs and has depth d can be computed by a width-5 permutation branching program Π of length 4^d. Given the description of the circuit Ψ, the description of the branching program Π can be computed in $poly(\ell, 4^d)$ time.*

In order to state our corollary for on-the-fly branching programs, we will require the following definition.

Definition 6 (Predecessor Function for Circuit). *Let Ψ be a circuit as in Theorem 2. The predecessor function of Ψ, denoted $\mathsf{Pred}_\Psi(i)$, is defined with respect to some arbitrary labeling of the gates of Ψ, where the label of the output gate is always 0, and input gates are labeled by the index of the variable. Given a label i for a gate, $\mathsf{Pred}_\Psi(i)$ returns (j_1, j_2) which are the labels of the wires feeding this gate.*

We can now define the on-the-fly variant of Barrington's Theorem.

Corollary 1 (Barrington On-The-Fly). *There exists a uniform machine* BPOTF *that, given access to a predecessor function Pred_Ψ of a depth d circuit, outputs the layers $(p_{0,t}, p_{1,t})$ of the branching program from Theorem 2 in order for $t = 1, \ldots, L$. Each layer takes time $O(d)$ to produce, and the total space used by* BPOTF *is $O(d)$.*

Proof. This corollary is implicit in the proof of Barrington's Theorem. It will be convenient to refer to the proof as it appears in [Vio09]. Essentially, the branching program in Barrington's theorem is produced recursively, where every node in the circuit, starting with the output gate, applies the branching program generation procedure recursively on its left hand predecessor, right hand, left hand again, right hand again. Each recursive call is parameterized by an element in the group S_5 which is passed on as a recursive parameter. Once an input node is reached, the respective layer in the branching program can be produced based on the S_5 element and the identity of the variable.

We conclude that the computation of the branching program is a traversal of the DAG representing the circuit, and at each point in time there is one path in the graph that is active, and each nodes on that path need to maintain a state of size $O(1)$. Thus the total space required is $O(d)$ and the time to produce the next layer is at most $O(d)$.

2.4   Learning with Errors and the Gadget Matrix

The Learning with Errors (LWE) problem was introduced by Regev [Reg05] as a generalization of "learning parity with noise" [BFKL93, Ale03]. We now define the decisional version of LWE.

Definition 7 (Decisional LWE *(DLWE)* [Reg05]**).** *Let λ be the security para-meter, $n = n(\lambda)$, $m = m(\lambda)$, and $q = q(\lambda)$ be integers and $\chi = \chi(\lambda)$ be a prob-ability distribution over $\mathbb{Z}$ bounded by $B_\chi = B_\chi(\lambda)$. The $\text{DLWE}_{n,q,\chi}$ problem states that for all $m = \text{poly}(n)$, letting $\mathbf{A} \leftarrow \mathbb{Z}_q^{n \times m}$, $\mathbf{s} \leftarrow \mathbb{Z}_q^n$, $\mathbf{e} \leftarrow \chi^m$, and $\mathbf{u} \leftarrow \mathbb{Z}_q^m$, the following distributions are computationally indistinguishable:*

$$(\mathbf{A}, \mathbf{s}\mathbf{A} + \mathbf{e}) \stackrel{c}{\approx} (\mathbf{A}, \mathbf{u})$$

There are known quantum (Regev [Reg05]) and classical (Peikert [Pei09]) reductions between $\text{DLWE}_{n,q,\chi}$ and approximating short vector problems in lat-tices. Specifically, these reductions take χ to be a discrete Gaussian distribution $D_{\mathbb{Z},\alpha q}$ for some $\alpha < 1$. We write $\text{DLWE}_{n,q,\alpha}$ to indicate this instantiation. We now state a corollary of the results of [Reg05, Pei09, MM11, MP12]. These results also extend to additional forms of q (see [MM11, MP12]).

Corollary 2 ([Reg05, Pei09, MM11, MP12]**).** *Let $q = q(n) \in \mathbb{N}$ be either a prime power $q = p^r$, or a product of co-prime numbers $q = \prod q_i$ such that for all i, $q_i = \text{poly}(n)$, and let $\alpha \geq \sqrt{n}/q$. If there is an efficient algorithm that solves the (average-case) $\text{DLWE}_{n,q,\alpha}$ problem, then:*

- *There is an efficient quantum algorithm that solves $\text{GapSVP}_{\tilde{O}(n/\alpha)}$ (and $\text{SIVP}_{\tilde{O}(n/\alpha)}$) on any n-dimensional lattice.*
- *If in addition $q \geq \tilde{O}(2^{n/2})$, there is an efficient classical algorithm for $\text{GapSVP}_{\tilde{O}(n/\alpha)}$ on any n-dimensional lattice.*

Recall that GapSVP_γ is the (promise) problem of distinguishing, given a basis for a lattice and a parameter d, between the case where the lattice has a vector shorter than d, and the case where the lattice doesn't have any vector shorter than $\gamma \cdot d$. SIVP is the search problem of finding a set of "short" vectors. The best known algorithms for GapSVP_γ ([Sch87]) require at least $2^{\tilde{\Omega}(n/\log \gamma)}$ time. We refer the reader to [Reg05, Pei09] for more information.

Lastly, we derive the following corollary which will allow us to choose the LWE parameters for our scheme. It follows immediately by taking χ to be a discrete Gaussian with parameter $B/\omega(\sqrt{\log n})$ with rejection sampling rejecting all samples bigger than B (which only happens with negligible probability).

Corollary 3. *For any function $B = B(n)$ there exists a B-bounded distribution $\chi = \chi(n)$ such that for all q it holds that $\mathrm{DLWE}_{n,q,\chi}$ is at least as hard as the quantum hardness of GapSVP_γ, SIVP_γ for $\gamma = \tilde{O}(nq/B)$, and also the classical hardness of GapSVP_γ if $q \geq \tilde{O}(2^{n/2})$.*

We now define the gadget matrix [MP12, AP14] that plays an important role in our construction. Our definition is a slight variant on definitions from previous works.

Definition 8. *Let $m = n \cdot (\ell_q + c)$ for some $c \in \mathbb{N}$, and define the "gadget matrix"*

$$\mathbf{G}_{n,m} = (\mathbf{0}^c \mid \mathbf{g}) \otimes \mathbf{I}_n \in \mathbb{Z}_q^{n \times m} ,$$

where $\mathbf{g} = (1, 2, 4, \ldots, 2^{\ell_q - 1}) \in \mathbb{Z}_q^{\ell_q}$. We will also refer to this gadget matrix as the "powers-of-two" matrix. We define the inverse function $\mathbf{G}_{n,m}^{-1} : \mathbb{Z}_q^{n \times m'} \to \{0,1\}^{m \times m'}$ which expands each entry $a \in \mathbb{Z}_q$ of the input matrix into a column of size $c + \ell_q$ consisting of the bits of the binary representation of a with leading zeros. We have the property that for any matrix $\mathbf{A} \in \mathbb{Z}_q^{n \times m'}$, it holds that $\mathbf{G}_{n,m} \cdot \mathbf{G}_{n,m}^{-1}(\mathbf{A}) = \mathbf{A}$. We note that we sometimes omit the subscripts when they are clear from the context.

3   Building Blocks from Previous Works

3.1   Noise Level of Matrices and Vectors

Let n, q be natural numbers and let $m \geq n \cdot \ell_q$ be s.t. $\mathbf{G} = \mathbf{G}_{n,m} \in \mathbb{Z}_q^{n \times m}$ (where $\mathbf{G}$ is the gadget matrix from Definition 8). Throughout this section we consider matrices $\mathbf{C} \in \mathbb{Z}_q^{n \times m}$ (to be interpreted as ciphertexts), vectors $\mathbf{t} \in \mathbb{Z}_q^n$ (secret keys), and bits $\mu \in \{0, 1\}$ (plaintexts). Starting with [GSW13], a number of recent homomorphic encryption schemes, and in particular the ones we will consider as a basis for our construction, have the property that $\mathbf{C}$ encrypts μ if and only if

$$\mathbf{t}\mathbf{C} = \mu\mathbf{t}\mathbf{G} + \mathbf{e} , \tag{1}$$

for a sufficiently low-norm *noise vector* $\mathbf{e}$. We would like to keep track of the noise level in the ciphertext throughout homomorphic evaluation. We therefore define the noise level as follows. (Recall that we defined absolute value and norm of elements in $\mathbb{Z}_q$ in the beginning of Sect. 2.)

Definition 9. *The noise level of $\mathbf{C}$ with respect to $(\mathbf{t}, \mu)$ is the infinity norm of the noise vector:*

$$\mathsf{noise}_{(\mathbf{t},\mu)}(\mathbf{C}) = \|\mathbf{t}\mathbf{C} - \mu\mathbf{t}\mathbf{G}\|_\infty .$$

For a vector $\mathbf{c} \in \mathbb{Z}_q^n$, we define:

$$\mathsf{noise}_{(\mathbf{t},\mu)}(\mathbf{c}) = \left| \langle \mathbf{t}, \mathbf{c} \rangle - 2^{\ell_q - 1}\mu \right| .$$

One or both subscripts are sometimes omitted when they are clear from the context.

We note that since $\mathbf{G}^{-1}(2^{\ell_q-1}\mathbf{u}_n) = \mathbf{u}_m$ (where $\mathbf{u}_i$ is the ith unit vector), then $\mathbf{c} = \mathbf{C} \cdot \mathbf{G}^{-1}(2^{\ell_q-1}\mathbf{u}_n)$ is simply the last (mth) column of $\mathbf{C}$ and furthermore for such $\mathbf{c}$ it holds that

$$\mathsf{noise}_{(\mathbf{t},\mu)}(\mathbf{c}) = \mathsf{noise}_{(\mathbf{t},\mu)}(\mathbf{C} \cdot \mathbf{G}^{-1}(2^{\ell_q-1}\mathbf{u}_n)) \leq \mathsf{noise}_{(\mathbf{t},\mu)}(\mathbf{C}) \ .$$

The following are basic properties of the noise vector that had been established in previous works. These are used to establish the homomorphic properties of the respective encryption scheme.

Lemma 1 ([GSW13, AP14]). *The noise in negation, addition and multiplication is bounded as follows.* ***Negation:*** *For all $\mu \in \{0,1\}$ it holds that*

$$\mathsf{noise}_{(\mathbf{t},1-\mu)}(\mathbf{G} - \mathbf{C}) = \mathsf{noise}_{(\mathbf{t},\mu)}(\mathbf{C}) \ .$$

Addition: *If $\mu_1, \mu_2 \in \{0,1\}$ are such that $\mu_1 \cdot \mu_2 = 0$ (i.e. not both are 1) then*

$$\mathsf{noise}_{(\mathbf{t},\mu_1+\mu_2)}(\mathbf{C}_1 + \mathbf{C}_2) \leq \mathsf{noise}_{(\mathbf{t},\mu_1)}(\mathbf{C}_1) + \mathsf{noise}_{(\mathbf{t},\mu_2)}(\mathbf{C}_2) \ .$$

Multiplication: *For all $\mu_1, \mu_2 \in \{0,1\}$ it holds that*

$$\mathsf{noise}_{(\mathbf{t},\mu_1\mu_2)}(\mathbf{C}_1\mathbf{G}^{-1}(\mathbf{C}_2)) \leq m \cdot \mathsf{noise}_{(\mathbf{t},\mu_1)}(\mathbf{C}_1) + \mu_1 \cdot \mathsf{noise}_{(\mathbf{t},\mu_2)}(\mathbf{C}_2) \ .$$

We will also require an almost identical variant about the noise content in vectors. We add a proof for the sake of completeness.

Lemma 2. *Let n, q be integers, let $\mathbf{t} \in \mathbb{Z}_q^n$ be such that $\mathbf{t}[n] = 1$ and $\mathbf{c}_1, \mathbf{c}_2 \in \mathbb{Z}_q^n$, further let $\mathbf{C}_1 \in \mathbb{Z}^{n \times m}$. Recall the definition of noise content (Definition 9). Then:* ***Negation:*** *For all $\mu \in \{0,1\}$ it holds that*

$$\mathsf{noise}_{(\mathbf{t},1-\mu)}(2^{\ell_q-1}\mathbf{u}_n - \mathbf{c}) = \mathsf{noise}_{(\mathbf{t},\mu)}(\mathbf{c}) \ .$$

Addition: *If $\mu_1, \mu_2 \in \{0,1\}$ are such that $\mu_1 \cdot \mu_2 = 0$ (i.e. not both are 1) then*

$$\mathsf{noise}_{(\mathbf{t},\mu_1+\mu_2)}(\mathbf{c}_1 + \mathbf{c}_2) \leq \mathsf{noise}_{(\mathbf{t},\mu_1)}(\mathbf{c}_1) + \mathsf{noise}_{(\mathbf{t},\mu_2)}(\mathbf{c}_2) \ .$$

Multiplication: *For all $\mu_1, \mu_2 \in \{0,1\}$ it holds that*

$$\mathsf{noise}_{(\mathbf{t},\mu_1\mu_2)}(\mathbf{C}_1\mathbf{G}_{n,m}^{-1}(\mathbf{c}_2)) \leq m \cdot \mathsf{noise}_{(\mathbf{t},\mu_1)}(\mathbf{C}_1) + \mu_1 \cdot \mathsf{noise}_{(\mathbf{t},\mu_2)}(\mathbf{c}_2) \ .$$

Proof. We simply compute:
Negation:

$$\begin{aligned}
\mathsf{noise}_{(\mathbf{t},1-\mu)}&(2^{\ell_q-1}\mathbf{u}_n - \mathbf{c}) \\
&= \left| \langle \mathbf{t}, 2^{\ell_q-1}\mathbf{u}_n - \mathbf{c} \rangle - 2^{\ell_q-1}(1-\mu) \right| \\
&= \left| 2^{\ell_q-1} - \langle \mathbf{t}, \mathbf{c} \rangle - 2^{\ell_q-1}(1-\mu) \right| \\
&= \left| \langle \mathbf{t}, \mathbf{c} \rangle - 2^{\ell_q-1}\mu \right| \\
&= \mathsf{noise}_{(\mathbf{t},\mu)}(\mathbf{c})
\end{aligned}$$

Addition:

$$\text{noise}_{(t,\mu_1+\mu_2)}(c_1 + c_2)$$
$$= \left| \langle t, c_1 + c_2 \rangle - 2^{\ell_q-1}(\mu_1 + \mu_2) \right|$$
$$\leq \left| \langle t, c_1 \rangle - 2^{\ell_q-1}\mu_1 \right| + \left| \langle t, c_2 \rangle - 2^{\ell_q-1}\mu_2 \right|$$
$$= \text{noise}_{(t,\mu_1)}(c_1) + \text{noise}_{(t,\mu_2)}(c_2)$$

Multiplication: Define $e_1 = tC_1 - \mu_1 tG$.

$$\text{noise}_{(t,\mu_1\mu_2)}(C_1 G_{n,m}^{-1}(c_2))$$
$$= \left| \langle t, C_1 G_{n,m}^{-1}(c_2) \rangle - 2^{\ell_q-1}(\mu_1\mu_2) \right|$$
$$= \left| \langle \mu_1 tG_{n,m} + e_1, G_{n,m}^{-1}(c_2) \rangle - 2^{\ell_q-1}(\mu_1\mu_2) \right|$$
$$= \left| \mu_1 \langle t, c_2 \rangle + \langle e_1, G_{n,m}^{-1}(c_2) \rangle - 2^{\ell_q-1}(\mu_1\mu_2) \right|$$
$$\leq \mu_1 \left| \langle t, c_2 \rangle - 2^{\ell_q-1}\mu_2 \right| + m \left\| e_1 \right\|_\infty$$
$$= m \cdot \text{noise}_{(t,\mu_1)}(C_1) + \mu_1 \cdot \text{noise}_{(t,\mu_2)}(c_2)$$

The following corollary, which follows from an analysis performed in [BV11], states that if a vector has sufficiently small noise, then μ can be recovered using a shallow boolean circuit. We also observe that the predecessor function of this circuit (Definition 6) is succinctly computable.

Corollary 4 ([BV11]). *Define* $\text{Threshold}(t, c) = \arg\min_{\mu \in \{0,1\}} \text{noise}_{t,\mu}(c)$. *Then if* $\text{noise}_{(t,\mu)}(c) < q/8$ *for some* μ, *then* $\text{Threshold}(t, c) = \mu$. *Furthermore, there exists a depth* $O(\log(n \log(q)))$ *boolean circuit that computes* Threshold, *and there exists a size* $\text{polylog}(n \log q)$ *circuit that computes the predecessor function* $\text{Pred}_{\text{Threshold}}$ *as per Definition 6.*

Proof. Since $2^{\ell_q-1} > q/4$, it follows that if $\text{noise}_{(t,\mu)}(c) < q/8$ then it must me that $\text{noise}_{(t,1-\mu)}(c) > q/8$ and the first part follows.

The boolean circuit that computes Threshold is essentially the same as the decryption circuit described in [BV11, Lemma 4.5]. This is a circuit that first performs an addition, then computes modulo q and finally a threshold function. Specifically we do the following. For an integer $a \in \mathbb{Z}_q$ we let $\text{Bit}_i(a)$ denote the ith bit of a. Note that:

$$\langle t, c \rangle - 2^{\ell_q-1}\mu = \sum_{i=1}^{n} \sum_{j=1}^{\ell_q} (t[i] \cdot 2^j) \cdot \text{Bit}_j(c[i]) - 2^{\ell_q-1}\mu \ .$$

Therefore, $\langle t, c \rangle - 2^{\ell_q-1}\mu$ is a summation of $n\ell_q + 1$ integers. This can be done in depth $O(\log(n\ell_q)) = O(\log(n \log(q)))$, using a 3-to-2 addition tree. In order to take modulo q, we subtract, in parallel, all possible multiples of q and check if the result is in $\mathbb{Z}_q$. Since there are at most $O(n\ell_q)$ possible multiplies, this can be done, using a selection tree, in depth $O(\log(n \log(q)))$ again. Finally, to compute the threshold function, we compare the values of $\text{noise}_{(t,0)}(c)$ and $\text{noise}_{(t,1)}(c)$. This can be done in depth $O(\log(n \log(q)))$ as well, leading to a total depth of $O(\log(n \log(q)))$, as desired.

Note that the circuit of addition of two ℓ_q-bits integers can be computed using a uniform machine with logarithmic space, taking ℓ_q as input. Therefore, the circuit of summation of $n\ell_q + 1$ integers can also be computed by such a machine, taking ℓ_q, n as input. Similarly, each of the components of the circuit of Threshold can be computed in such manner. Since there is a constant number of components, wired sequentially, we can compute each component using logarithmic space, and reuse that space to compute the following component. Therefore, we can compute the circuit of Threshold using a uniform machine with logarithmic space, taking n, ℓ_q as input. In particular, such a machine can also compute the $\mathsf{Pred}_{\mathsf{Threshold}}$ function.

3.2 A Single-Hop Multi-key Homomorphic Encryption Scheme

The scheme below is essentially a restatement of the scheme of [MW16], which in turns relies on [GSW13, CM15].

- SHMK.Setup(1^λ): Generate the parameters (n, q, χ, B_χ) such that $\mathrm{DLWE}_{n-1,q,\chi}$ holds (note that there is freedom in the choice of parameters here, so the scheme can be instantiated in various parameter ranges), recall that we denote $\ell_q = \lceil \log q \rceil$, and choose m s.t. $n|m$ and $m \geq n\ell_q + \omega(\log \lambda)$. Finally, choose a matrix uniformly at random $\mathbf{B} \xleftarrow{\$} \mathbb{Z}_q^{(n-1) \times m}$ and output:

$$\mathsf{params} := (q, n, m, \chi, B_\chi, \mathbf{B})$$

- SHMK.Keygen(params): Sample uniformly at random $\mathbf{s} \xleftarrow{\$} \mathbb{Z}_q^{n-1}$, set the secret key as follows:

$$\mathsf{sk} := \mathbf{t} = (-\mathbf{s}, 1) \in \mathbb{Z}_q^n$$

Sample a noise vector $\mathbf{e} \xleftarrow{\$} \chi^m$ and compute:

$$\mathbf{b} := \mathbf{s}\mathbf{B} + \mathbf{e} \in \mathbb{Z}_q^m \qquad \mathbf{A} := \begin{pmatrix} \mathbf{B} \\ \mathbf{b} \end{pmatrix} \in \mathbb{Z}_q^{n \times m}$$

Finally, set and output:

$$\mathsf{pk} := \mathbf{A}$$

- SHMK.PreEnc(pk, μ): This is a "pre-encryption" algorithm that will be used as an auxiliary procedure in the actual encryption algorithm. Sample uniformly at random $\mathbf{R} \xleftarrow{\$} \{0,1\}^{m \times m}$. Set and output the encryption:

$$\mathbf{C} := \mathbf{A}\mathbf{R} + \mu\mathbf{G}_{n,m} \in \mathbb{Z}_q^{n \times m}$$

- SHMK.Enc(pk, μ): Sample uniformly at random $\mathbf{R} \xleftarrow{\$} \{0,1\}^{m \times m}$. Compute:

$$\mathbf{C} := \mathbf{A}\mathbf{R} + \mu\mathbf{G} \in \mathbb{Z}_q^{n \times m} \qquad \overrightarrow{\mathcal{R}}[j,k] \leftarrow \mathsf{SHMK.PreEnc}(\mathsf{pk}, \mathbf{R}[j,k]) \in \mathbb{Z}_q^{n \times m}$$

Set the ciphertext to be the encrypted massage along with the vector of encryptions and output it $(\mathbf{C}, \overrightarrow{\mathcal{R}}) \in \mathbb{Z}_q^{n \times m} \times (\mathbb{Z}_q^{n \times m})^{m \times m}$. We note that for the most part, the internal structure of $\overrightarrow{\mathcal{R}}$ will not be too important for our purposes.

- SHMK.Dec$((\mathsf{sk}_1, \ldots, \mathsf{sk}_N), \widehat{\mathbf{C}})$: Let $\hat{\mathbf{c}} = \widehat{\mathbf{C}} \cdot \mathbf{G}_{nN,mN}^{-1}(2^{\ell_q - 1}\mathbf{u}_{nN})$ be the last column of $\widehat{\mathbf{C}}$. Denote $\mathbf{t}_i = \mathsf{sk}_i$ for every $i \in [N]$ and $\hat{\mathbf{t}} = (\mathbf{t}_1, \ldots, \mathbf{t}_N)$. Compute and output:

$$\mu' := \mathsf{Threshold}(\hat{\mathbf{t}}, \hat{\mathbf{c}}) \ ,$$

  where $\mathsf{Threshold}(\cdot, \cdot)$ is as defined in Corollary 4.
- SHMK.Extend$(c, (\mathsf{pk}_1, \ldots, \mathsf{pk}_N))$: Takes as input a ciphertext $c = (\mathbf{C}, \vec{\mathcal{R}})$ and a tuple of public keys $\mathsf{pk}_1, \ldots, \mathsf{pk}_N$. It outputs a new ciphertext $\widehat{\mathbf{C}} \in \mathbb{Z}_q^{nN \times mN}$, however $\widehat{\mathbf{C}}$ is represented as an $N \times N$ block matrix containing blocks of size $\mathbb{Z}_q^{n \times m}$, and only $2N - 1$ of these blocks are non-zero. See Lemma 3 below for the properties of the extension procedure.

Lemma 3 ([MW16]). *Let $N \in \mathbb{N}$, $\mu \in \{0, 1\}$, $(\mathsf{sk}_j, \mathsf{pk}_j) = \mathsf{SHMK.Keygen}(1^\lambda)$ for all $j \in [N]$. Let $c = \mathsf{SHMK.Enc}(\mathsf{pk}_i, \mu)$ for some $i \in [N]$ (we assume that i is given implicitly).*
Then $\mathsf{SHMK.Extend}(c, (\mathsf{pk}_1, \ldots, \mathsf{pk}_N))$ runs in $N \cdot \mathrm{poly}(\lambda)$ time and outputs a succinct description of the block matrix $\widehat{\mathbf{C}}$ containing all but $(2N - 1)$ nonzero blocks. Furthermore, denoting $\mathbf{t}_j = \mathsf{sk}_j$ and $\hat{\mathbf{t}} = (\mathbf{t}_1, \ldots, \mathbf{t}_N)$, it holds that $\mathsf{noise}_{\hat{\mathbf{t}}, \mu}(\widehat{\mathbf{C}}) \leq (m^4 + m)B_\chi$, with probability 1.

Since the extended ciphertexts satisfy Eq. (1) with respect to the concatenated secret keys, and since the noise of a fresh extended ciphertext is small, we can preform homomorphic operations, as described in Sect. 3.1.

The following lemma asserts the security of the scheme (we note that security-wise, this scheme is an immediate extension of Regev's encryption scheme [Reg05]).

Lemma 4 ([MW16]). *The scheme SHMK is semantically secure under the $\mathrm{DLWE}_{n-1, q, \chi}$ assumption.*

4   Our Fully Dynamic Multi-key FHE Scheme

We now describe our fully dynamic multi-key FHE scheme FDMK. We start by presenting the setup, key generation, encryption and decryption (without evaluation), which will be a slight variation on the bootstrappable version of the single-hop scheme SHMK from Sect. 3.2. The only difference is that a ciphertext here is only a small fragment of a ciphertext in SHMK, and the public keys are augmented with encryptions of the secret key as required for bootstrapping. Then, in Sect. 4.1 we describe our evaluation procedure using branching programs.

We note that the security of our scheme requires making a circular security assumption on SHMK. We describe a leveled version of our scheme that only requires the hardness of learning with errors in Appendix A.

– FDMK.Setup(1^λ): Generate the parameters such that $\text{DLWE}_{n-1,q,\chi}$ holds, where $n := n(\lambda)$ is the lattice dimension parameter; $\chi := \chi(\lambda)$ is a $B_\chi :=$ $B_\chi(\lambda)$ bounded error distribution; and a modulus $q := B_\chi 2^{\omega(\log \lambda)}$. Denote $\ell_q = \lceil \log q \rceil$, set $m := n\ell_q + \omega(\log \lambda)$ such that $n|m$, and sample uniformly at random $\mathbf{B} \xleftarrow{\$} \mathbb{Z}_q^{(n-1)\times m}$. Set and output:

$$\text{params} := (q, n, m, \chi, B_\chi, \mathbf{B})$$

– FDMK.Keygen(params): Generate a secret key as in the scheme SHMK - sample uniformly at random $\mathbf{s} \xleftarrow{\$} \mathbb{Z}_q^{n-1}$, set and output:

$$\text{sk} := \mathbf{t} = (-\mathbf{s}, 1) \in \mathbb{Z}_q^n$$

Then, to generate the public key - sample a noise vector $\mathbf{e} \xleftarrow{\$} \chi^m$ and compute:

$$\mathbf{b} := \mathbf{s}\mathbf{B} + \mathbf{e} \in \mathbb{Z}_q^m \qquad \mathbf{A} := \begin{pmatrix} \mathbf{B} \\ \mathbf{b} \end{pmatrix} \in \mathbb{Z}_q^{n\times m}$$

Next, encrypt the secret key bit-by-bit according to the SHMK scheme. Let $\text{Bit}_i(\text{sk})$ denote the ith bit of sk. For every $i \in [n \cdot \ell_q]$ compute:

$$\overrightarrow{\mathcal{S}}[i] \leftarrow \text{SHMK.Enc}(\mathbf{A}, \text{Bit}_i(\text{sk}))$$

Set the public key to:

$$\text{pk} := (\mathbf{A}, \overrightarrow{\mathcal{S}}) \ .$$

We note that the $\overrightarrow{\mathcal{S}}$ part is only used for homomorphic evaluation and not for encryption.

– FDMK.Enc(pk, μ): Sample uniformly at random $\mathbf{r} \xleftarrow{\$} \{0, 1\}^m$. Set and output the encryption:

$$\mathbf{c} := \mathbf{A}\mathbf{r} + \mu 2^{\ell_q - 1}\mathbf{u}_n \in \mathbb{Z}_q^n$$

where $\mathbf{u}_i$ is the ith standard basis vector.

– FDMK.Dec($(\text{sk}_1, \ldots, \text{sk}_N), c$) : Parse each secret key as $\mathbf{t}_i := \text{sk}_i$ for every $i \in [N]$. Concatenate the secret keys and set $\widehat{\mathbf{t}} := (\mathbf{t}_1, \ldots, \mathbf{t}_N) \in \mathbb{Z}_q^{nN}$. Compute and output

$$\mu' := \text{Threshold}(\widehat{\mathbf{t}}, \mathbf{c}) \ ,$$

where $\text{Threshold}(\cdot, \cdot)$ is as defined in Corollary 4.

The following lemma states the security of our scheme, which follows immediately from that of SHMK.

Lemma 5. *The scheme* FDMK *is semantically secure if* SHMK *with the same* DLWE *parameters is weakly circular secure.*

Proof. Note that the ciphertexts in the FDMK scheme are the last column of the ciphertexts in the SHMK scheme. In particular it could be computed deterministically out of the ciphertexts in the SHMK scheme. Since the public key in the FDMK is generated as in the SHMK scheme, along with encryption of the secret keys, the semantic security follows from the weak circular security of the SHMK scheme.

The choice of DLWE parameters and the lattice approximation factors they induce is discussed when we present our leveled scheme that does not require circular security. See Appendix A and in particular Lemma 9 and the discussion thereafter.

4.1 Homomorphic Evaluation

Following [BV14], we evaluate the augmented NAND circuits needed for bootstrapping by converting them into branching programs. We recall the definition of a branching program, Barrington's Theorem and our on-the-fly variant of the theorem from Sect. 2.3.

We would like to evaluate a branching program homomorphically. Since the message space of our scheme is binary, it would be more convenient to keep a binary state vector $\mathbf{v} \in \{0,1\}^5$, rather than an integer state $s \in \{1,\dots,5\}$. We keep the following invariant: $\mathbf{v}_t[i] = 1 \Leftrightarrow s_t = i$. To do so we initialize $\mathbf{v}_0[1] = 1$ and $\mathbf{v}_0[i] = 0$ for $i \in \{2,\dots,5\}$. In the iterative stage we evaluate using the recursive formula:

$$\mathbf{v}_t[i] = 1 \Leftrightarrow p_{t,x_{\mathrm{var}(t)}}(s_{t-1}) = i$$

We can rewrite it to get an iterative formula: for every $1 \le i \le 5$, $\mathbf{v}_t[i] = 1$ if and only if

$$\mathbf{v}_{t-1}\left[p_{t,0}^{-1}(i)\right] = 1 \text{ and } x_{\mathrm{var}(t)} = 0 \qquad \text{or} \qquad \mathbf{v}_{t-1}\left[p_{t,1}^{-1}(i)\right] = 1 \text{ and } x_{\mathrm{var}(t)} = 1$$

Equivalently:

$$\begin{aligned}
\mathbf{v}_t[i] :&= \mathbf{v}_{t-1}\left[p_{t,0}^{-1}(i)\right] \cdot \left(1 - x_{\mathrm{var}(t)}\right) + \mathbf{v}_{t-1}\left[p_{t,1}^{-1}(i)\right] \cdot x_{\mathrm{var}(t)} \\
&= \mathbf{v}_{t-1}\left[\alpha_{t,i}\right] \cdot \left(1 - x_{\mathrm{var}(t)}\right) + \mathbf{v}_{t-1}\left[\beta_{t,i}\right] \cdot x_{\mathrm{var}(t)}
\end{aligned} \qquad (2)$$

where $\alpha_{t,i} = p_{t,0}^{-1}(i), \beta_{t,i} = p_{t,1}^{-1}(i)$ are constant parameters derived from the description of the circuit. After the L'th iteration, we accept if and only if $s_L = 1$, that is we output $\mathbf{v}_L[1]$, following from the kept invariant.

The main idea of [BV14] is to convert the circuit into a branching program, and compute it homomorphically by evaluating the above formula (Eq. (2)). We adapt the use of branching programs to reduce the space complexity of evaluation. Specifically, we show that our scheme is bootstrapable using space linear in the number of the participating parties as follows.

- FDMK.NAND$((c_1,c_2),(\mathsf{pk}_1,\dots,\mathsf{pk}_N))$: Let $T_1, T_2 \subset \{\mathsf{pk}_1,\dots,\mathsf{pk}_N\}$ be the sequences of public keys under which c_1 and c_2 are encrypted, respectively. Let $S_1, S_2 \subset \{\mathsf{sk}_1,\dots,\mathsf{sk}_N\}$ be the respective secret keys, and let

$\mu_j = \mathsf{FDMK.Dec}(S_j, c_j)$ for $j = 1, 2$. Set $\mathcal{C}$ to be the description of the circuit

$$\mathcal{C}(x, y) = \mathsf{NAND}(\mathsf{FDMK.Dec}_x(c_1), \mathsf{FDMK.Dec}_y(c_2)) \ .$$

We would like to construct an on-the-fly branching program for $\mathcal{C}$ using the algorithm BPOTF from Corollary 1. To this end we observe that due to the properties of $\mathsf{Threshold}$ (see Corollary 4), the predecessor function $\mathsf{Pred}_\mathcal{C}$ can be computed by a circuit of size $\mathrm{polylog}(nN \log q)$. Given access to $\mathsf{Pred}_\mathcal{C}$, we have that $\mathsf{BPOTF}^{\mathsf{Pred}_\mathcal{C}}$ uses space $O(\log(nN \log q))$ and generates the branching program $\Pi = (\mathrm{var}, (p_{t,0}, p_{t,1})_{i \in [L]})$ that corresponds to $\mathcal{C}$, layer by layer. We will execute $\mathsf{BPOTF}^{\mathsf{Pred}_\mathcal{C}}$ lazily, every time we will produce a layer of the branching program and evaluate it homomorphically, so that $\mathsf{BPOTF}^{\mathsf{Pred}_\mathcal{C}}$ should not require more than $O(\log(Nn \log q))$ space at any point in time. We execute Π on $(\overrightarrow{\mathcal{S}}_i, \overrightarrow{\mathcal{S}}_j)_{i \in S_1, j \in S_2}$ as follows:

- **Initialization:**
  * Set the state vector:

  $$\overrightarrow{\mathbf{w}}_0 := (2^{\ell_q - 1} \mathbf{u}_{nN}, \mathbf{0}, \mathbf{0}, \mathbf{0}, \mathbf{0})$$

  * Initialize $\mathsf{BPOTF}^{\mathsf{Pred}_\mathcal{C}}$.
- **Iterative Step:** For every step $t = 1, \ldots, L$ do:
  * Compute the constants $\alpha_{t,i}, \beta_{t,i}$ by running $\mathsf{BPOTF}^{\mathsf{Pred}_\mathcal{C}}$ to obtain the next layer of the branching program

  $$((\alpha_{t,1}, \ldots, \alpha_{t,5}), (\beta_{t,1}, \ldots, \beta_{t,5}), \mathrm{var}(t))$$

  * For every $i = 1, \ldots, 5$ homomorphically compute the encryption of the next state:

  $$\mathbf{w}_{t,i} := \mathbf{X}^c_{\mathrm{var}(t)} \cdot \mathbf{G}^{-1}_{nN, mN} \left(\mathbf{w}_{t-1, \alpha_{t,i}} \right) + \mathbf{X}_{\mathrm{var}(t)} \cdot \mathbf{G}^{-1}_{nN, mN} \left(\mathbf{w}_{t-1, \beta_{t,i}} \right)$$

  where $\mathbf{X}_{\mathrm{var}(t)} := \mathsf{SHMK.Extend}(\overrightarrow{\mathcal{S}}_{\mathrm{var}(t)}, (\mathsf{pk}_1, \ldots, \mathsf{pk}_N))$ is the extended encryption of the secret key $\overrightarrow{\mathcal{S}}_{\mathrm{var}(t)}$ and $\mathbf{X}^c_{\mathrm{var}(t)} := \mathbf{G}_{nN, mN} - \mathbf{X}_{\mathrm{var}(t)}$ is its complement.
- Finally output the ciphertext $\mathbf{w}_{L,1}$.

Lemma 6. *For every* $N \in \mathrm{poly}(\lambda)$, *and every messages* $\mu_1, \mu_2 \in \{0, 1\}$, *let* $\mathsf{params} \leftarrow \mathsf{FDMK.Setup}(1^\lambda)$ *and* $(\mathsf{pk}_i, \mathsf{sk}_i) \leftarrow \mathsf{FDMK.Keygen}(\mathsf{params})$ *for every* $i \in [N]$. *Let* c_1, c_2 *be ciphertexts such that for some sequence of secret keys* $S_1, S_2 \subset \{\mathsf{sk}_1, \ldots, \mathsf{sk}_N\}$ *it holds that* $\mathsf{FDMK.Dec}(c_j, S_j) = \mu_j$ *for* $j = 1, 2$. *Then the following holds:*

$$\mathsf{FDMK.Dec}(\mathsf{FDMK.NAND}((c_1, c_2), (\mathsf{pk}_1, \ldots, \mathsf{pk}_N)), (\mathsf{sk}_1, \ldots, \mathsf{sk}_N)) = \mathsf{NAND}(\mu_1, \mu_2)$$

for every large enough value of λ.

To prove the correctness we first prove the following lemma:

Lemma 7. *For every $t = 0, \ldots, L$ and every $i = 1, \ldots, 5$ the following holds:*

$$\text{noise}_{\widehat{\text{sk}}, \mathbf{v}_t[i]}(\mathbf{w}_{t,i}) \leq 2t(m^5 + m^2)NB_\chi$$

where $\widehat{\text{sk}} := (\text{sk}_1, \ldots, \text{sk}_N)$ is the concatenation of the secret keys.

Proof. Denote $x = (S_1, S_2)$, recall that the input for the branching program Π is the encryption of the secret keys $(\vec{S}_i, \vec{S}_j)_{i \in S_1, j \in S_2}$.
We proof by induction on t:
The claim clearly holds for the case $t = 0$, by the way we defined $\mathbf{w}_0$. Assume that the hypothesis holds for $t' < t$. Note that by definition of $\mathbf{w}_t$ it follows that

$$\text{noise}_{\widehat{\text{sk}}, \mathbf{v}_t[i]}(\mathbf{w}_{t,i}) =$$

$$\text{noise}_{\widehat{\text{sk}}, \mathbf{v}_t[i]}\left(\mathbf{X}^c_{\text{var}(t)} \cdot \mathbf{G}^{-1}_{nN,mN}\left(\mathbf{w}_{t-1,\alpha_{t,i}}\right) + \mathbf{X}_{\text{var}(t)} \cdot \mathbf{G}^{-1}_{nN,mN}\left(\mathbf{w}_{t-1,\beta_{t,i}}\right)\right)$$

Following from Lemma 2, and the definition of $\mathbf{v}_t[i]$ in Eq. (2) we get:

$$\text{noise}_{\widehat{\text{sk}}, \mathbf{v}_t[i]}(\mathbf{w}_{t,i}) =$$

$$(1 - x_{\text{var}(t)}) \cdot \text{noise}_{\widehat{\text{sk}}, \mathbf{v}_{t-1}[\alpha_{t,i}]}\left(\mathbf{w}_{t-1,\alpha_{t,i}}\right) + mN \cdot \text{noise}_{\widehat{\text{sk}}, \widehat{\text{sk}}_{\text{var}(t)}}\left(\mathbf{X}_{\text{var}(t)}\right)$$

$$+ x_{\text{var}(t)} \cdot \text{noise}_{\widehat{\text{sk}}, \mathbf{v}_{t-1}[\beta_{t,i}]}\left(\mathbf{w}_{t-1,\beta_{t,i}}\right) + mN \cdot \text{noise}_{\widehat{\text{sk}}, \widehat{\text{sk}}_{\text{var}(t)}}\left(\mathbf{X}_{\text{var}(t)}\right)$$

Note that either $x_{\text{var}(t)} = 0$ or $1 - x_{\text{var}(t)} = 0$, thus, using the induction's hypothesis:

$$(1 - x_{\text{var}(t)}) \cdot \text{noise}_{\widehat{\text{sk}}, \mathbf{v}_{t-1}[\alpha_{i,t}]}\left(\mathbf{w}_{t-1,\alpha_{t,i}}\right)$$

$$+ x_{\text{var}(t)} \cdot \text{noise}_{\widehat{\text{sk}}, \mathbf{v}_{t-1}[\beta_{i,t}]}\left(\mathbf{w}_{t-1,\beta_{t,i}}\right)$$

$$\leq \max\left(\text{noise}_{\widehat{\text{sk}}, \mathbf{v}_{t-1}[\alpha_{i,t}]}\left(\mathbf{w}_{t-1,\alpha_{t,i}}\right), \text{noise}_{\widehat{\text{sk}}, \mathbf{v}_{t-1}[\beta_{i,t}]}\left(\mathbf{w}_{t-1,\beta_{t,i}}\right)\right)$$

$$\leq 2(t-1)(m^5 + m^2)NB_\chi$$

Following Lemma 3, $\text{noise}_{\widehat{\text{sk}}, \widehat{\text{sk}}_{\text{var}(t)}}\left(\mathbf{X}_{\text{var}(t)}\right) = (m^4 + m)$. Putting it all together:

$$\text{noise}_{\widehat{\text{sk}}, \mathbf{v}_t[i]}(\mathbf{w}_{t,i}) \leq 2(t-1)(m^5 + m^2)NB_\chi + 2(m^4 + m)mN$$

$$= 2t(m^5 + m^2)NB_\chi$$

Proof. (Proof of Lemma 6). Using the correctness of Barrington's Theorem (Theorem 2), we only need to prove that $\text{noise}_{\widehat{\text{sk}}, \text{NAND}(\mu_1, \mu_2)}(\mathbf{w}_{L,1}) = \text{noise}_{\widehat{\text{sk}}, \mathbf{v}_L[1]}(\mathbf{w}_{L,1}) < q/8$. Indeed, using the previous lemma, $\text{noise}_{\widehat{\text{sk}}, \mathbf{v}_L[1]}(\mathbf{w}_{L,1}) \leq 2L(m^5 + m^2)NB_\chi$. Using Lemma 4, bounding the depth of decryption $d = d_{\text{Threshold}} = O(\log(nN\log(q))) = O(\log \lambda + \log N) = O(\log \lambda)$, since $n, N, \log q = \text{poly}(\lambda)$. Therefore, $L = 4^{d+1} = \text{poly}(\lambda)$. And so, $\text{noise}_{\widehat{\text{sk}}, \mathbf{v}_L[1]}(\mathbf{w}_{L,1}) = \text{poly}(\lambda) \cdot B_\chi$ which is less than $q/8$ by our choice of parameters.

We note that if a bound on N was known a priori, then we could choose a polynomial modulus $q = \text{poly}(N, \lambda) = \text{poly}(\lambda)$.

Lemma 8. *The algorithm* FDMK.NAND *can be computed using* $N\mathrm{poly}(n \log q)$ *space, where* N *is the number of parties involved in the computation.*

Proof. As we saw above, the computation of Pred_C takes space $\mathrm{polylog}(Nn \log q)$ and BPOTF requires space $\log(Nn \log q)$. At every point in the evaluation of the branching program we only hold 5 ciphertexts, each of which is a vector of dimension nN over $\mathbb{Z}_q$. At each step in the computation we apply SHMK.Extend to blow up an encryption of a secret-key bit into a sparse $N \times N$ block matrix that can be represented by $(2N-1)$ matrices in $\mathbb{Z}_q^{n \times m}$. We perform matrix-vector multiplication and vector-vector addition between this matrix and vectors, which can all be performed in $N \cdot \mathrm{poly}(n \log q)$ space. The lemma follows.

A   Leveled Multi-key Fully Homomorphic Encryption

In this section we give a leveled multi-key homomorphic scheme LMK. The scheme is leveled with respect to the total depth of computation. Namely, the summation over all hops of the depth of the evaluated circuit. The scheme is a modified version of the FDMK construction described in Sect. 4. Moreover, it relies only on the hardness of LWE and not on any circular security assumption. We achieve this, as in [Gen09b] and followup works, by generating a sequence of secret keys and encrypting each of them under the next. To evaluate each gate, we bootstrap using the next secret key.

- LMK.Setup($1^\lambda, 1^D$): Generate the parameters such that $\mathrm{DLWE}_{n-1,q,\chi}$ holds, where $n := n(\lambda)$ is the lattice dimension parameter; $\chi := \chi(\lambda)$ is a $B_\chi :=$ $B_\chi(\lambda)$ bounded error distribution; and a modulus $q := B_\chi 2^{\omega(\log \lambda)}$. Denote $\ell_q = \lceil \log q \rceil$, set $m := n\ell_q + \omega(\log \lambda)$ such that $n|m$ and sample uniformly at random $\mathbf{B} \xleftarrow{\$} \mathbb{Z}_q^{(n-1) \times m}$. Set and output:

$$\mathsf{params} := (D, q, n, m, \chi, B_\chi, \mathbf{B})$$

- LMK.Keygen(params): Generate a sequence of $(D + 1)$ pairs of keys as in the SHMK scheme: for every $d \in [D+1]$ sample uniformly at random $\mathbf{s}^{(d)} \xleftarrow{\$} \mathbb{Z}_q^{n-1}$ and set:

$$\mathsf{sk}^{(d)} := \mathbf{t}^{(d)} = (-\mathbf{s}^{(d)}, 1)$$

Also sample a noise vector $\mathbf{e}^{(d)} \xleftarrow{\$} \chi^m$ and compute:

$$\mathsf{pk}^{(d)} := \mathbf{A}^{(d)} = \begin{pmatrix} \mathbf{B} \\ \mathbf{s}^{(d)}\mathbf{B} + \mathbf{e}^{(d)} \end{pmatrix} \in \mathbb{Z}_q^{n \times m}$$

Next, encrypt bit-by-bit each secret key $\mathsf{sk}^{(d)}$ using the sequentially following public key $\mathsf{pk}^{(d+1)}$. Let $\mathsf{Bit}_j(\mathsf{sk}^{(d)})$ denote the jth bit of $\mathsf{sk}^{(d)}$. For every for every $d \in [D]$ and $j \in [n \cdot \ell_q]$ compute:

$$\overrightarrow{\mathcal{S}}^{(d)}[j] \leftarrow \mathsf{SHMK.Enc}(\mathsf{pk}^{(d+1)}, \mathsf{Bit}_j(\mathsf{sk}^{(d)}))$$

Finally, let the secret key be the last secret key of the sequence:

$$\mathsf{sk} := \mathsf{sk}^{(D+1)}$$

And let the public key be the first public key of the sequence, along with the encryptions:

$$\mathsf{pk} := \left(\mathsf{pk}^{(1)}, \vec{\mathcal{S}}^{(1)}, \ldots, \vec{\mathcal{S}}^{(D)} \right)$$

- LMK.Enc(pk, μ): Sample uniformly at random $\mathbf{r} \xleftarrow{\$} \{0,1\}^m$. Set and output the encryption:

$$\mathbf{c} := \mathbf{Ar} + \mu 2^{\ell-1} \mathbf{u}_n \in \mathbb{Z}_q^n$$

- LMK.Dec(($\mathsf{sk}_1, \ldots, \mathsf{sk}_N$), c): Assume w.l.o.g that c is the product of evaluating a depth D circuit on newly encrypted ciphertexts (otherwise apply additional "dummy" homomorphic operations). Parse each secret key as $\mathbf{t}_i = \mathbf{t}_i^{(D+1)} :=$ sk_i for every $i \in [N]$. Concatenate the secret keys and set $\widehat{\mathbf{t}} := (\mathbf{t}_1, \ldots, \mathbf{t}_N) \in \mathbb{Z}_q^{nN}$. Compute and output

$$\mu' := \mathsf{Threshold}(\widehat{\mathbf{t}}, \mathbf{c})$$

Evaluation is done similar to as described in Sect. 4.1. Consider evaluating a depth D circuit, and assume that the circuit is leveled, i.e. the gates of the circuit can be divided into D sets (levels) such that the first set is only fed by input gates, and set $d + 1$ is only fed by the outputs of gates in level d. Clearly every circuit can be made leveled by adding dummy gates. Homomorphic evaluation maintains the invariant that an encryption w.r.t some set of users T at the input of level d is encrypted under their set of $\mathsf{sk}^{(d)}$ keys. The computation of the gate is therefore performed w.r.t the set of $\mathsf{sk}_i^{(d)}$ keys. Then for bootstrapping, we use the $\mathcal{S}_i^{(d)}$ key switching parameters, which allow us to produce an encryption of the bits of the $\mathsf{sk}^{(d)}$ keys under the $\mathsf{sk}^{(d+1)}$ keys, so the invariant is preserved. The noise growth is the same as analyzed in Lemma 7.

Security follows from LWE by a standard hybrid argument.

Lemma 9. *The scheme described above is secure under the* DLWE$_{n-1,q,\chi}$ *assumption.*

Following Corollary 3, we can assume DLWE$_{n-1,q,\chi}$ by assuming the quantum hardness of either GapSVP$_\gamma$ or SIVP$_\gamma$ for $\gamma = \tilde{O}(nq/B_\chi)$. For bootstrapping to go through, we require that $q/B_\chi = 2^{\omega(\log n)}$ (in fact, it is sufficient to have $q/B_\chi = \mathrm{poly}(n, N)$, but we do not want to assume an upper bound on N, except that it's some polynomial). As usual, we can scale q, B_χ so long as their ratio remains large enough. This allows us to achieve better efficiency by scaling B_χ, q to be smallest possible, or achieving classical security by scaling q to be exponential in n, and B_χ appropriately.

References

[Ale03] Alekhnovich, M.: More on average case vs approximation complexity. In: Proceedings of 44th Symposium on Foundations of Computer Science (FOCS 2003), Cambridge, MA, USA, pp. 298–307. IEEE Computer Society, 11–14 October 2003

[AP14] Alperin-Sheriff, J., Peikert, C.: Faster bootstrapping with polynomial error. In: Garay and Gennaro [GG14], pp. 297–314

[Bar89] Mix Barrington, D.A.: Bounded-width polynomial-size branching programs recognize exactly those languages in nc^1. J. Comput. Syst. Sci. **38**(1), 150–164 (1989)

[BFKL93] Blum, A., Furst, M.L., Kearns, M., Lipton, R.J.: Cryptographic primitives based on hard learning problems. In: Stinson, D.R. (ed.) CRYPTO 1993. LNCS, vol. 773, pp. 278–291. Springer, Heidelberg (1994)

[BGV12] Brakerski, Z., Gentry, C., Vaikuntanathan, V.: (Leveled) fully homomorphic encryption without bootstrapping. In: Goldwasser, S. (ed.) ITCS, pp. 309–325. ACM (2012). Invited to ACM Transactions onComputation Theory

[BV11] Brakerski, Z., Vaikuntanathan, V.: Efficient fully homomorphic encryption from (standard) LWE. In: Ostrovsky, R. (ed.) FOCS, pp. 97–106. IEEE (2011). https://eprint.iacr.org/2011/344.pdf

[BV14] Brakerski, Z., Vaikuntanathan, V.: Lattice-based FHE as secure as PKE. In: Naor, M. (ed.) Innovations in Theoretical Computer Science, ITCS 2014, Princeton, NJ, USA, pp. 1–12. ACM, 12–14 January 2014

[CM15] Clear, M., McGoldrick, C.: Multi-identity and multi-key leveled FHE from learning with errors. In: Gennaro, R., Robshaw, M. (eds.) CRYPTO 2015. LNCS, vol. 9216, pp. 630–656. Springer, Heidelberg (2015)

[Gen09a] Gentry, C.: A fully homomorphic encryption scheme. Ph.D. thesis, Stanford University (2009)

[Gen09b] Gentry, C.: Fully homomorphic encryption using ideal lattices. In: STOC, pp. 169–178 (2009)

[GG14] Garay, J.A., Gennaro, R. (eds.): CRYPTO 2014, Part I. LNCS, vol. 8616. Springer, Heidelberg (2014)

[GHV10] Gentry, C., Halevi, S., Vaikuntanathan, V.: i-hop homomorphic encryption and Rerandomizable Yao circuits. In: Rabin, T. (ed.) CRYPTO 2010. LNCS, vol. 6223, pp. 155–172. Springer, Heidelberg (2010)

[GSW13] Gentry, C., Sahai, A., Waters, B.: Homomorphic encryption from learning with errors: conceptually-simpler, asymptotically-faster, attribute-based. In: Canetti, R., Garay, J.A. (eds.) CRYPTO 2013, Part I. LNCS, vol. 8042, pp. 75–92. Springer, Heidelberg (2013)

[HS14] Halevi, S., Shoup, V.: Algorithms in HElib. In: Garay and Gennaro [GG14], pp. 554–571

[HS15] Halevi, S., Shoup, V.: Bootstrapping for HElib. In: Oswald, E., Fischlin, M. (eds.) EUROCRYPT 2015. LNCS, vol. 9056, pp. 641–670. Springer, Heidelberg (2015)

[LTV12] López-Alt, A., Tromer, E., Vaikuntanathan, V.: On-the-fly multiparty computation on the cloud via multikey fully homomorphic encryption. In: Karloff, H.J., Pitassi, T. (eds.) Proceedings of 44th Symposium on Theory of Computing Conference, STOC 2012, New York, NY, USA, pp. 1219–1234. ACM, 19–22 May 2012

[MM11] Micciancio, D., Mol, P.: Pseudorandom knapsacks and the sample complexity of LWE search-to-decision reductions. In: Rogaway, P. (ed.) CRYPTO 2011. LNCS, vol. 6841, pp. 465–484. Springer, Heidelberg (2011)

[MP12] Micciancio, D., Peikert, C.: Trapdoors for lattices: simpler, tighter, faster, smaller. In: Pointcheval, D., Johansson, T. (eds.) EUROCRYPT 2012. LNCS, vol. 7237, pp. 700–718. Springer, Heidelberg (2012)

[MW16] Mukherjee, P., Wichs, D.: Two round multiparty computation via multi-key FHE. In: Fischlin, M., Coron, J.-S. (eds.) EUROCRYPT 2016. LNCS, vol. 9666, pp. 735–763. Springer, Heidelberg (2016). doi:10.1007/978-3-662-49896-5_26

[Pei09] Peikert, C.: Public-key cryptosystems from the worst-case shortest vector problem: extended abstract. In: Proceedings of 41st Annual ACM Symposium on Theory of Computing, STOC 2009, Bethesda, MD, USA, pp. 333–342, 31 May - 2 June 2009

[PS16] Peikert, C., Shiehian, S.: Multi-key FHE from LWE (revisited). IACR Cryptology ePrint Archive 2016:196 (2016)

[RAD78] Rivest, R., Adleman, L., Dertouzos, M.: On data banks and privacy homomorphisms. In: Foundations of Secure Computation, pp. 169–177. Academic Press, Cambridge (1978)

[Reg05] Regev, O.: On lattices, learning with errors, random linear codes, and cryptography. In: Proceedings of 37th Annual ACM Symposium on Theory of Computing, Baltimore, MD, USA, pp. 84–93, 22–24 May 2005

[Sch87] Schnorr, C.-P.: A hierarchy of polynomial time lattice basis reduction algorithms. Theor. Comput. Sci. **53**, 201–224 (1987)

[Vio09] Viola, E.: Barrington's Theorem - Lecture Notes (2009). Scribe by Zhou, F. http://www.ccs.neu.edu/home/viola/classes/gems-08/lectures/le11.pdf

Cryptography with Auxiliary Input and Trapdoor from Constant-Noise LPN

Yu Yu[1,2,3,4]([⊠]) and Jiang Zhang[2]

[1] Department of Computer Science and Engineering,
Shanghai Jiao Tong University, Shanghai, China
yuyuathk@gmail.com
[2] State Key Laboratory of Cryptology, P.O. Box 5159, Beijing 100878, China
jiangzhang09@gmail.com
[3] State Key Laboratory of Information Security,
Institute of Information Engineering,
Chinese Academy of Sciences, Beijing 100093, China
[4] Westone Cryptologic Research Center, Beijing 100070, China

Abstract. Dodis, Kalai and Lovett (STOC 2009) initiated the study of the Learning Parity with Noise (LPN) problem with (static) exponentially hard-to-invert auxiliary input. In particular, they showed that under a new assumption (called Learning Subspace with Noise) the above is quasi-polynomially hard in the high (polynomially close to uniform) noise regime.

Inspired by the "sampling from subspace" technique by Yu (eprint 2009/467) and Goldwasser et al. (ITCS 2010), we show that standard LPN can work in a mode (reducible to itself) where the constant-noise LPN (by sampling its matrix from a random subspace) is robust against sub-exponentially hard-to-invert auxiliary input with comparable security to the underlying LPN. Plugging this into the framework of [DKL09], we obtain the same applications as considered in [DKL09] (i.e., CPA/CCA secure symmetric encryption schemes, average-case obfuscators, reusable and robust extractors) with resilience to a more general class of leakages, improved efficiency and better security under standard assumptions.

As a main contribution, under constant-noise LPN with certain sub-exponential hardness (i.e., $2^{\omega(n^{1/2})}$ for secret size n) we obtain a variant of the LPN with security on poly-logarithmic entropy sources, which in turn implies CPA/CCA secure public-key encryption (PKE) schemes and oblivious transfer (OT) protocols. Prior to this, basing PKE and OT on constant-noise LPN had been an open problem since Alekhnovich's work (FOCS 2003).

1 Introduction

LEARNING PARITY WITH NOISE. The computational version of learning parity with noise (LPN) assumption with parameters $n \in \mathbb{N}$ (length of secret) and

© International Association for Cryptologic Research 2016
M. Robshaw and J. Katz (Eds.): CRYPTO 2016, Part I, LNCS 9814, pp. 214–243, 2016.
DOI: 10.1007/978-3-662-53018-4_9

$0 < \mu < 1/2$ (noise rate) postulates that for any $q = \mathsf{poly}(n)$ (number of queries) it is computationally infeasible for any probabilistic polynomial-time (PPT) algorithm to recover the random secret $\mathbf{x} \xleftarrow{\$} \{0,1\}^n$ given $(\mathbf{A}, \mathbf{A} \cdot \mathbf{x} + \mathbf{e})$, where $\mathbf{a}$ is a random $q \times n$ Boolean matrix, $\mathbf{e}$ follows $\mathcal{B}_\mu^q = (\mathcal{B}_\mu)^q$, $\mathcal{B}_\mu$ denotes the Bernoulli distribution with parameter μ (i.e., $\Pr[\mathcal{B}_\mu = 1] = \mu$ and $\Pr[\mathcal{B}_\mu = 0] = 1 - \mu$), '$\cdot$' denotes matrix vector multiplication over $\mathrm{GF}(2)$ and '$+$' denotes bitwise addition over $\mathrm{GF}(2)$. The decisional version of LPN simply assumes that $(\mathbf{A}, \mathbf{A} \cdot \mathbf{x} + \mathbf{e})$ is pseudorandom. The two versions are polynomially equivalent [4,8,34].

HARDNESS OF LPN. The computational LPN problem represents a well-known NP-complete problem "decoding random linear codes" [6] whose worst-case hardness is well-investigated. LPN was also extensively studied in learning theory, and it was shown in [21] that an efficient algorithm for LPN would allow to learn several important function classes such as 2-DNF formulas, juntas, and any function with a sparse Fourier spectrum. Under a constant noise rate, the best known LPN solvers [9,39] require time and query complexity both $2^{O(n/\log n)}$. The time complexity goes up to $2^{O(n/\log\log n)}$ when restricted to $q = \mathsf{poly}(n)$ queries [40], or even $2^{O(n)}$ given only $q = O(n)$ queries [42]. Under low noise rate $\mu = n^{-c}$ (for constant $0 < c < 1$), the best attacks [5,7,12,38,48] solve LPN with time complexity $2^{O(n^{1-c})}$ and query complexity $q = O(n)$ or more[1]. The low-noise LPN is mostly believed a stronger assumption than constant-noise LPN. In noise regime $\mu = O(1/\sqrt{n})$, LPN can be used to build public-key encryption (PKE) schemes and oblivious transfer (OT) protocols (more discussions below). Quantum algorithms are not known to have any advantages over classic ones in solving LPN, which makes LPN a promising candidate for "post-quantum cryptography". Furthermore, LPN enjoys simplicity and is more suited for weak-power devices (e.g., RFID tags) than other quantum-secure candidates such as LWE [46].

CRYPTOGRAPHY IN minicrypt. LPN was used as a basis for building lightweight authentication schemes against passive [29] and even active adversaries (e.g. [32,34], see [1] for a more complete literature). Kiltz et al. [37] and Dodis et al. [18] constructed randomized MACs from LPN, which implies a two-round authentication scheme with man-in-the-middle security. Lyubashevsky and Masny [41] gave an more efficient three-round authentication scheme from LPN (without going through the MAC transformation) and recently Cash, Kiltz, and Tessaro [13] reduced the round complexity to 2 rounds. Applebaum et al. [3] used LPN to construct efficient symmetric encryption schemes with certain key-dependent message (KDM) security. Jain et al. [30] constructed an efficient perfectly binding string commitment scheme from LPN. We refer to a recent survey [45] on the current state-of-the-art about LPN.

---

[1] We are not aware of any non-trivial time-query tradeoff results to break low-noise LPN in time $2^{o(n^{1-c})}$ even with super-polynomial number of queries.

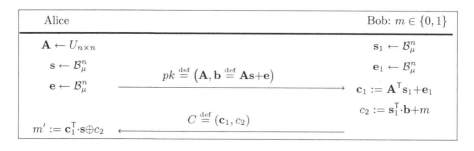

Fig. 1. A two-pass protocol by which Bob transmits a message bit m to Alice with passive security and noticeable correctness (for proper choice of μ), where Bob receives $m' = m + (\mathbf{s}_1^\top \cdot \mathbf{e}) + (\mathbf{e}_1^\top \cdot \mathbf{s})$.

CRYPTOGRAPHY BEYOND minicrypt. Alekhnovich [2] constructed the first (CPA secure) public-key encryption scheme from LPN with noise rate[2] $\mu = 1/\sqrt{n}$. By plugging the correlated products approach of [47] into Alekhnovich's CPA secure PKE scheme, Döttling et al. [20] constructed the first CCA secure PKE scheme from low-noise LPN. After observing that the complexity of the scheme in [20] was hundreds of times worse than Alekhnovich's original scheme, Kiltz et al. [36] proposed a neat and more efficient CCA secure construction by adapting the techniques from LWE-based encryption in [44] to the case of LPN. More recently, Döttling [19] constructed a PKE with KDM security. All the above schemes are based on LPN of noise rate $O(1/\sqrt{n})$. To see that noise rate $1/\sqrt{n}$ is inherently essential for PKE, we illustrate the (weakly correct) single-bit PKE protocol by Döttling et al. [20] in Figure 1, which is inspired by the counterparts based on LWE [23,46]. First, the decisional $\mathsf{LPN}_{\mu,n}$ assumption implies that $(\mathbf{A}, \mathbf{A}\mathbf{x} + \mathbf{e})$ is pseudorandom even when $\mathbf{x}$ is drawn from $X \sim \mathcal{B}_\mu^n$ (instead of $X \sim U_n$), which can be shown by a simple reduction [20]. Second, the passive security of the protocol is straightforward as $(pk, \mathbf{c}_1)$ is pseudorandom even when concatenated with the Goldreich-Levin[3] hardcore bit $\mathbf{s}_1^\top \cdot \mathbf{b}$ (replacing $\mathbf{b}$ with U_n by a hybrid argument). The final and most challenging part is correctness, i.e., m' needs to correlate with m at least noticeably. It is not hard to see for $n\mu^2 = O(1)$ and $\mathbf{e}, \mathbf{s} \leftarrow \mathcal{B}_\mu^n$ we have $\Pr[\langle \mathbf{e}, \mathbf{s} \rangle = 0] \geq 1/2 + \Omega(1)$, and thus noise rate $\mu = O(1/\sqrt{n})$ seems an inherent barrier[4] for the PKE to be correct.

---

[2] More precisely, Alekhnovich's PKE is based on a variant called the Exact LPN whose noise vector is sampled from $\chi_{\mu q}^q$ for $\mu = \frac{1}{\sqrt{n}}$ (i.e., uniform random distribution over q-bit strings of Hamming weight μq), which is implied by LPN with noise rate μ.

[3] Typically (in the context of one-way functions), the Goldreich-Levin Theorem [25] assumes a uniformly random secret $\mathbf{s}$, which is however not necessary. A Markov argument suggests that $\mathbf{s}$ can follow any polynomial-time sampleable distribution, as long as f on $\mathbf{s}$ is hard to invert.

[4] In fact, $\mu = O(\sqrt{\log n / n})$ is sufficient to have a noticeable correctness, i.e., $1/2 + 1/\mathsf{poly}(n)$, but known PKE constructions avoid the strong noise by assuming noise rate $n^{-1/2}$ [36] or even lower rate $n^{-1/2-\epsilon}$ [2,20].

The scheme is "weak" in the sense that correctness is only $1/2 + \Omega(1)$ and it can be transformed into a standard CPA scheme (that encrypts multiple-bit messages with overwhelming correctness) using standard techniques (e.g., [15,20]). Notice a correct PKE scheme (with certain properties) yields also a (weak form of) 2-round oblivious transfer protocol against honest-but-curious receiver. Suppose that Alice has a choice $i \in \{0,1\}$, and she samples pk_i with trapdoor $\mathbf{s}$ (as described in the protocol) and a uniformly random pk_{1-i} without trapdoor. Upon receiving pk_0 and pk_1, Bob uses the scheme to encrypt two bits σ_0 and σ_1 under pk_0 and pk_1 respectively, and sends them to Alice. Alice can then recover σ_i and but knows nothing about σ_{1-i}. David et al. [16] constructed a universally composable OT under LPN with noise rate $1/\sqrt{n}$. Therefore, basing PKE (and OT) on LPN with noise rate $\mu = n^{-1/2+\epsilon}$ (and ideally a constant $0 < \mu < 1/2$) remains an open problem for the past decade.

LPN WITH AUXILIARY INPUT. Despite being only sub-exponentially secure, LPN is known to be robust against any constant-fraction of static linear leakages, i.e., for any constant $0 < \alpha < 1$ and any $f(\mathbf{x}; \mathbf{Z}) = (\mathbf{Z}, \mathbf{Zx})$ it holds that

$$\left(f(\mathbf{x}), \mathbf{A}, \mathbf{Ax} + \mathbf{e} \right) \overset{c}{\sim} \left(f(\mathbf{x}), \mathbf{A}, U_q \right), \tag{1}$$

where $\mathbf{Z}$ is any $(1-\alpha)n \times n$ matrix (that can be sampled in polynomial time and independent of $\mathbf{A}$). The above can be seen by a change of basis so that the security is reducible from the LPN assumption with the same noise rate on a uniform secret of size αn. Motivated by this, Dodis, Kalai and Lovett [17] further conjectured that LPN is secure against any polynomial-time computable f such that 1) $\mathbf{x}$ given $f(\mathbf{x})$ has average min-entropy αn; or even 2) any f that is $2^{-\alpha n}$-hard-to-invert for PPT algorithms (see Definition 2 for a formal definition). Note the distinction between the two types of leakages: the former f is a lossy function and the latter can be even injective (the leakage $f(\mathbf{x})$ may already determine $\mathbf{x}$ in an information theoretical sense). However, they didn't manage to prove the above claim (i.e., LPN with auxiliary input) under standard LPN. Instead, they introduced a new assumption called Learning Subspace with Noise (LSN) as below, where the secret to be learned is the random subspace $\mathbf{V}$.

Assumption 1 (The LSN Assumption [17]). *For any constant $\beta > 0$, there exists a polynomial $p = p_\beta(n)$ such that for any polynomial $q = \mathsf{poly}(n)$ the following two distributions are computationally indistinguishable:*

$$\left((\mathbf{a}_1, \mathbf{Va}_1 + U_n^{(1)} E_1), \cdots, (\mathbf{a}_q, \mathbf{Va}_q + U_n^{(q)} E_q) \right) \overset{c}{\sim} \left((\mathbf{a}_1, U_n^{(1)}), \cdots, (\mathbf{a}_q, U_n^{(q)}) \right),$$

where $\mathbf{V} \sim U_{n \times \beta n}$ is a random $n \times \beta n$ matrix, $\mathbf{a}_1, \cdots, \mathbf{a}_q$ are vectors i.i.d. to $U_{\beta n}$, and $E_1, \cdots, E_q$ are Boolean variables (determining whether the respective noise is uniform randomness or nothing) i.i.d. to $\mathcal{B}_{1-\frac{1}{p}}$.

Then, the authors of [17] showed that LSN with parameters β and $p_\beta = p_\beta(n)$ implies the decisional LPN (as in (1)) under noise rate $\mu = (\frac{1}{2} - \frac{1}{4p_\beta})$ holds with

$2^{-\alpha n}$-hard-to-invert auxiliary input (for any constant $\alpha > \beta$). Further, this yields many interesting applications such as CPA/CCA secure symmetric encryption schemes, average-case obfuscators for the class of point functions, reusable and robust extractors, all remain secure with exponentially hard-to-invert auxiliary input (see [17] for more details). We note that [17] mainly established the feasibility about cryptography with auxiliary input, and there remain issues to be addressed or improved. First, to counteract $2^{-\alpha n}$-hard-to-invert auxiliary input one needs to decide in advance the noise rate noise rate $1/2 - 1/4p_\beta$ (recall the constraint $\beta < \alpha$). Second, Raz showed that for any constant β, $p_\beta = n^{\Omega(1)}$ is necessary (otherwise LSN can be broken in polynomial-time) and even with $p_\beta = n^{\Theta(1)}$ there exist quasi-polynomial attacks (see the full version of [17] for more discussions about Raz's attacks). Therefore, the security reduction in [17] is quite loose. As the main end result of [17], one needs a high-noise LPN for $\mu = 1/2 - 1/\mathsf{poly}(n)$ (and thus low efficiency due to the redundancy needed to make a correct scheme) only to achieve quasi-polynomial security (due to Raz's attacks) against $2^{-\alpha n}$-hard-to-invert leakage for some constant α (i.e., not any exponentially hard-to-invert leakage). Third, LSN is a new (and less well-studied) assumption and it was left as an open problem in [17] whether the aforementioned cryptographic applications can be based on the hardness of standard LPN, ideally admitting more general class of leakages, such as sub-exponentially or even quasi-polynomially hard-to-invert auxiliary input.

THE MAIN OBSERVATION. Yu [49] introduced the "sampling from subspace" technique to prove the above "LPN with auxiliary input" conjecture under standard LPN but the end result of [49] was invalid due to a flawed intermediate step. A similar idea was also used by Goldwasser et al. [26] in the setting of LWE, where the public matrix was drawn from a (noisy) random subspace. Informally, the observation (in our setting) is that, the decisional LPN with constant noise rate $0 < \mu < 1/2$ implies that for any constant $0 < \alpha < 1$, any 2^{-2n^α}-hard-to-invert f and any $q' = \mathsf{poly}(n)$ it holds that

$$(f(\mathbf{x}), \mathbf{A}', \mathbf{A}' \cdot \mathbf{x} + \mathbf{e}) \overset{c}{\sim} (f(\mathbf{x}), \mathbf{A}', U_{q'}), \qquad (2)$$

where $\mathbf{x} \sim U_n{}^5$, $\mathbf{e} \sim \mathcal{B}_\mu^{q'}$, and $\mathbf{A}'$ is a $q' \times n$ matrix with rows sampled from a random subspace of dimension $\lambda = n^\alpha$. Further, if the underlying LPN is $2^{\omega(n^{\frac{1}{1+\beta}})}$-hard[6] for any constant $\beta > 0$, then by setting $\lambda = \log^{1+\beta} n$, (2) holds for any $q' = \mathsf{poly}(n)$ and any $2^{-2\log^{1+\beta} n}$-hard-to-invert f. The rationale is that distribution $\mathbf{A}'$ can be considered as the multiplication of two random matrices $\mathbf{A} \overset{\$}{\leftarrow} \{0,1\}^{q' \times \lambda}$ and $\mathbf{V} \overset{\$}{\leftarrow} \{0,1\}^{\lambda \times n}$, i.e., $\mathbf{A}' \sim (\mathbf{A} \cdot \mathbf{V})$, where $\mathbf{V}$ constitutes the basis of the λ-dimensional random subspace and $\mathbf{A}$ is the random coin for sampling from $\mathbf{V}$. Unlike the LSN assumption whose subspace $\mathbf{V}$ is secret, the $\mathbf{V}$ and

---

[5] We assume $\mathbf{x} \sim U_n$ to be in line with [17], but actually our results hold for any efficiently sampleable $\mathbf{x}$ as long as $\mathbf{x}$ given $f(\mathbf{x})$ is $2^{-2\lambda}$-hard-to-invert.

[6] Informally, we say that a cryptographic scheme/problem Π is T-secure/hard if every probabilistic adversary of time (and query, if applicable) complexity T achieve advantage no more than $1/T$ in breaking/solving Π.

$\mathbf{A}$ in (2) are public coins (implied by $\mathbf{A}'$, see Remark 1). We have by the associative law $\mathbf{A}' \cdot \mathbf{x} = \mathbf{A}(\mathbf{V} \cdot \mathbf{x})$ and by the Goldreich-Levin theorem $\mathbf{V} \cdot \mathbf{x}$ is a pseudo-random secret (even conditioned on $\mathbf{V}$ and $f(\mathbf{x})$), and thus (2) is reducible from the standard decisional LPN on noise rate μ, secret size λ and query complexity q'. Concretely, assume that the LPN problem is $2^{\omega(n^{3/4})}$-hard then by setting $\lambda = n^{2/3}$ (resp., $\lambda = \log^{4/3} n$) we have that (2) is $2^{\Omega(n^{1/2})}$-secure (resp., $n^{\omega(1)}$-secure) with any auxiliary input that is $2^{-2n^{2/3}}$-hard (resp., $2^{-2\log^{4/3} n}$-hard) to invert. Plugging (2) into the framework of [17] we obtain the same applications (CPA/CCA secure symmetric encryption schemes, average-case obfuscators for point functions, reusable and robust extractors) under standard (constant-noise) LPN with improved efficiency (as the noise is constant rather than polynomially close to uniform) and tighter security against sub-exponentially (or even quasi-polynomially) hard-to-invert auxiliary input.

PKE FROM CONSTANT-NOISE LPN. More surprisingly, we show a connection from "LPN with auxiliary input" to "basing PKE on (constant-noise) LPN". The feasibility can be understood by the single-bit weak PKE in Fig. 1 with some modifications: assume that LPN is $2^{\omega(n^{\frac{1}{2}})}$-hard (i.e., $\beta = 1$), then for $\lambda = \log^2 n/4$ we have that (2) holds on any $\mathbf{x} \sim X$ with min-entropy $\mathbf{H}_\infty(X) \geq \log^2 n/2$. Therefore, by replacing the uniform matrix $\mathbf{A}$ with $\mathbf{A}' \sim (U_{n \times \lambda} \cdot U_{\lambda \times n})$, and sampling $\mathbf{s}, \mathbf{s}_1 \leftarrow X$ and $\mathbf{e}, \mathbf{e}_1 \leftarrow \mathcal{B}_\mu^n$ for constant μ and $X \sim \chi_{\log n}^n$ [7], we get that $\mathbf{s}_1^\mathsf{T} \mathbf{e}$ and $\mathbf{e}_1^\mathsf{T} \mathbf{s}$ are both $(1/2 + 1/\mathsf{poly}(n))$-biased to 0 independently, and thus the PKE scheme has noticeable correctness. We then transform the weak PKE into a full-fledged CPA secure scheme, where the extension is not trivial (more than a straightforward parallel repetition plus error-correction codes). In particular, neither $X \sim \chi_{\log n}^n$ or $X \sim \mathcal{B}_{\log n/n}^n$ can guarantee security and correctness simultaneously and thus additional ideas are needed (more details deferred to Sect. 4.3).

PKE WITH CCA SECURITY. Once we have a CPA scheme based on constant-noise LPN, we can easily extend it to a CCA one by using the techniques in [20], and thus suffer from the same performance slowdown as that in [20]. A natural question is whether we can construct a simpler and more efficient CCA scheme as that in [36]. Unfortunately, the techniques in [36] do not immediately apply to the case of constant-noise LPN. The reason is that in order to employ the ideas from the LWE-based encryption scheme [44], the scheme in [36] has to use a variant of LPN (called knapsack LPN), and the corresponding description key is exactly the secret of some knapsack LPN instances. Even though there is a polynomial time reduction [43] from the LPN problem to the knapsack LPN problem, such a reduction will map the noise distribution of the LPN problem into the secret distribution of the knapsack LPN problem. If we directly apply the techniques in [36], the resulting scheme will not have any guarantee of correctness

---

[7] Recall that for $m \ll n$ we have by Stirling's approximation that $\binom{n}{m} \approx n^m/m!$ and thus $\chi_{\log n}^n$ (uniform distribution over n-bit strings of Hamming weight $\log n$) is of min-entropy roughly $\log^2 n - \log n \log \log n \geq \log^2 n/2$.

because the corresponding decryption key follows the Bernoulli distribution with constant parameter μ. Recall that for the correctness of our CPA secure PKE scheme, the decryption key cannot simply be chosen from either $\chi_{\log n}^n$ or $\mathcal{B}_{\log n/n}^n$. Fortunately, based on several new observations and some new technical lemmas, we mange to adapt the idea of [36,44] to construct a simpler and efficient CCA secure PKE scheme from constant-noise LPN.

OT FROM CONSTANT-NOISE LPN. PKE and OT are incomparable in general [24]. But if the considered PKE scheme has some additional properties, then we can build OT protocol from it in a black-box way [24]. Gertner et al. [24] showed that if the public key of some CPA secure PKE scheme can be indistinguishably sampled (without knowing the corresponding secret key) from the public key distribution produced by honestly running the key generation algorithm, then we can use it to construct an OT protocol with honest parties (and thus can be transformed into a standard OT protocol by using zero-knowledge proof). It is easy to check that our CPA secure PKE scheme satisfies this property under the LPN assumption. Besides, none of the techniques used in transforming Alekhnovich's CPA secure PKE scheme into a universally composable OT protocol [16] prevent us from obtaining a universally composable OT protocol from our CPA secure PKE scheme. In summary, our results imply that there exists (universally composable) OT protocol under constant-noise LPN assumption. We omit the details, and refer to [16,24] for more information.

2   Preliminaries

NOTATIONS AND DEFINITIONS. We use capital letters (e.g., X, Y) for random variables and distributions, standard letters (e.g., x, y) for values, and calligraphic letters (e.g. $\mathcal{X}$, $\mathcal{E}$) for sets and events. Vectors are used in the column form and denoted by bold lower-case letters (e.g., $\mathbf{a}$). We treat matrices as the sets of its column vectors and denoted by bold capital letters (e.g., $\mathbf{A}$). The support of a random variable X, denoted by $\mathsf{Supp}(X)$, refers to the set of values on which X takes with non-zero probability, i.e., $\{x : \Pr[X = x] > 0\}$. For set $\S$ and binary string s, $|\S|$ denotes the cardinality of $\S$ and $|s|$ refers to the Hamming weight of s. We use $\mathcal{B}_\mu$ to denote the Bernoulli distribution with parameter μ, i.e., $\Pr[\mathcal{B}_\mu = 1] = \mu$, $\Pr[\mathcal{B}_\mu = 0] = 1 - \mu$, while $\mathcal{B}_\mu^q$ denotes the concatenation of q independent copies of $\mathcal{B}_\mu$. We use χ_i^n to denote a uniform distribution over $\{e \in \{0,1\}^n : |e| = i\}$. We denote by $\mathcal{D}_\lambda^{n_1 \times n} \overset{\text{def}}{=} (U_{n_1 \times \lambda} \cdot U_{\lambda \times n})$ to be a matrix distribution induced by multiplying two random matrices. For $n, q \in \mathbb{N}$, U_n (resp., $U_{q \times n}$) denotes the uniform distribution over $\{0,1\}^n$ (resp., $\{0,1\}^{q \times n}$) and independent of any other random variables in consideration, and $f(U_n)$ (resp., $f(U_{q \times n})$) denotes the distribution induced by applying function f to U_n (resp., $U_{q \times n}$). $X \sim D$ denotes that random variable X follows distribution D. We use $s \leftarrow S$ to denote sampling an element s according to distribution S, and let $s \overset{\$}{\leftarrow} \S$ denote sampling s uniformly from set $\S$.

ENTROPY NOTIONS. For $0 < \mu < 1/2$, the binary entropy function is defined as $\mathbf{H}(\mu) \overset{\text{def}}{=} \mu \log(1/\mu) + (1 - \mu) \log(1/(1 - \mu))$. We define the Shannon entropy and min-entropy of a random variable X respectively, i.e.,

$$\mathbf{H}_1(X) \overset{\text{def}}{=} \sum_{x \in \mathsf{Supp}(X)} \Pr[X = x] \log \frac{1}{\Pr[X = x]}, \quad \mathbf{H}_\infty(X) \overset{\text{def}}{=} \min_{x \in \mathsf{Supp}(X)} \log(1/\Pr[X = x]).$$

Note that $\mathbf{H}_1(\mathcal{B}_\mu) = \mathbf{H}(\mu)$. The average min-entropy of a random variable X conditioned on another random variable Z is defined as

$$\mathbf{H}_\infty(X|Z) \overset{\text{def}}{=} -\log \left(\mathbb{E}_{z \leftarrow Z} \left[2^{-\mathbf{H}_\infty(X|Z=z)} \right] \right).$$

INDISTINGUISHABILITY AND STATISTICAL DISTANCE. We define the (t,ε)- *computational distance* between random variables X and Y, denoted by $X \underset{(t,\varepsilon)}{\sim} Y$, if for every probabilistic distinguisher $\mathcal{D}$ of running time t it holds that

$$| \Pr[\mathcal{D}(X) = 1] - \Pr[\mathcal{D}(Y) = 1] | \leq \varepsilon.$$

The *statistical distance* between X and Y, denoted by $\mathsf{SD}(X, Y)$, is defined by

$$\mathsf{SD}(X, Y) \overset{\text{def}}{=} \frac{1}{2} \sum_x |\Pr[X = x] - \Pr[Y = x]|.$$

Computational/statistical indistinguishability is defined with respect to distribution ensembles (indexed by a security parameter). For example, $X \overset{\text{def}}{=} \{X_n\}_{n \in \mathbb{N}}$ and $Y \overset{\text{def}}{=} \{Y_n\}_{n \in \mathbb{N}}$ are computationally indistinguishable, denoted by $X \overset{c}{\sim} Y$, if for every $t = \mathsf{poly}(n)$ there exists $\varepsilon = \mathsf{negl}(n)$ such that $X \underset{(t,\varepsilon)}{\sim} Y$. X and Y are statistically indistinguishable, denoted by $X \overset{s}{\sim} Y$, if $\mathsf{SD}(X, Y) = \mathsf{negl}(n)$.

SIMPLIFYING NOTATIONS. To simplify the presentation, we use the following simplified notations. Throughout, n is the security parameter and most other parameters are functions of n, and we often omit n when clear from the context. For example, $q = q(n) \in \mathbb{N}$, $t = t(n) > 0$, $\epsilon = \epsilon(n) \in (0,1)$, and $m = m(n) = \mathsf{poly}(n)$, where poly refers to some polynomial.

Definition 1 (Learning Parity with Noise). *The **decisional** $\mathsf{LPN}_{\mu,n}$ problem (with secret length n and noise rate $0 < \mu < 1/2$) is hard if for every $q = \mathsf{poly}(n)$ we have*

$$(\mathbf{A}, \mathbf{A} \cdot \mathbf{x} + \mathbf{e}) \overset{c}{\sim} (\mathbf{A}, U_q), \tag{3}$$

*where $q \times n$ matrix $\mathbf{A} \sim U_{q \times n}$, $\mathbf{x} \sim U_n$ and $\mathbf{e} \sim \mathcal{B}_\mu^q$. The **computational** $\mathsf{LPN}_{\mu,n}$ problem is hard if for every $q = \mathsf{poly}(n)$ and every PPT algorithm $\mathcal{D}$ we have*

$$\Pr[\mathcal{D}(\mathbf{A}, \mathbf{A} \cdot \mathbf{x} + \mathbf{e}) = \mathbf{x}] = \mathsf{negl}(n), \tag{4}$$

where $\mathbf{A} \sim U_{q \times n}$, $\mathbf{x} \sim U_n$ *and* $\mathbf{e} \sim \mathcal{B}_\mu^q$.

LPN WITH SPECIFIC HARDNESS. *We say that the decisional (resp., computational)* LPN$_{\mu,n}$ *is T-hard if for every* $q \leq T$ *and every probabilistic adversary of running time* T *the distinguishing (resp., inverting) advantage in* (3) *(resp.,* (4)*) is upper bounded by* $1/T$.

Definition 2 (Hard-to-Invert Function). *Let* n *be the security parameter and let* $\kappa = \omega(\log n)$. *A polynomial-time computable function* $f : \{0,1\}^n \to \{0,1\}^l$ *is* $2^{-\kappa}$*-hard-to-invert if for every PPT adversary* $\mathcal{A}$

$$\Pr_{\mathbf{x} \sim U_n} [\ \mathcal{A}(f(\mathbf{x})) = \mathbf{x}\ ] \ \leq\ 2^{-\kappa}.$$

Lemma 1 (Union Bound). *Let* $\mathcal{E}_1, \cdots, \mathcal{E}_l$ *be any (not necessarily independent) events such that* $\Pr[\mathcal{E}_i] \geq (1 - \epsilon_i)$ *for every* $1 \leq i \leq l$, *then we have*

$$\Pr[\ \mathcal{E}_1 \wedge \cdots \wedge \mathcal{E}_l\ ] \ \geq\ 1 - (\epsilon_1 + \cdots + \epsilon_l).$$

We will use the following (essentially the Hoeffding's) bound on the Hamming weight of a high-noise Bernoulli vector.

Lemma 2. *For any* $0 < p < 1/2$ *and* $\delta \leq (\frac{1}{2} - p)$, *we have*

$$\Pr[\ |\mathcal{B}_\delta^q| > (\frac{1}{2} - \frac{p}{2})q\ ] < \exp^{-\frac{p^2 q}{8}}.$$

3   Learning Parity with Noise with Auxiliary Input

3.1   Leaky Sources and (Pseudo)randomness Extraction

We define below two types of leaky sources and recall two technical lemmas for (pseudo)randomness extraction from the respective sources, where $\mathbf{x}$ for TYPE-II source is assumed to be uniform only for alignment with [17] (see Footnote 3).

Definition 3 (Leaky Sources). *Let* $\mathbf{x}$ *be any random variable over* $\{0,1\}^n$ *and let* $f : \{0,1\}^n \to \{0,1\}^l$ *be any polynomial-time computable function.* $(\mathbf{x}, f(\mathbf{x}))$ *is called an* (n,κ) *TYPE-I (resp., TYPE-II) leaky source if it satisfies condition 1 (resp., condition 2) below:*

1. **Min-entropy leaky sources.** $\mathbf{H}_\infty(\mathbf{x}|f(\mathbf{x})) \geq \kappa$ *and* $f(\mathbf{x})$ *is polynomial-time sampleable.*
2. **Hard-to-invert leaky sources.** $\mathbf{x} \sim U_n$ *and* f *is* $2^{-\kappa}$*-hard-to-invert.*

Lemma 3 (Goldreich-Levin Theorem [25]). *Let* n *be a security parameter, let* $\kappa = \omega(\log n)$ *be polynomial-time computable from* n, *and let* $f : \{0,1\}^n \to \{0,1\}^l$ *be any polynomial-time computable function that is* $2^{-\kappa}$*-hard-to-invert. Then, for any constant* $0 < \beta < 1$ *and* $\lambda = \lceil \beta \kappa \rceil$, *it holds that*

$$(f(\mathbf{x}), \mathbf{V}, \mathbf{V} \cdot \mathbf{x}) \stackrel{c}{\sim} (f(\mathbf{x}), \mathbf{V}, U_\lambda),$$

where $\mathbf{x} \sim U_n$ *and* $\mathbf{V} \sim U_{\lambda \times n}$ *is a random* $\lambda \times n$ *Boolean matrix.*

Lemma 4 (Leftover Hash Lemma [28]**).** *Let* $(X, Z) \in \mathcal{X} \times \mathcal{Z}$ *be any joint random variable with* $\mathbf{H}_\infty(X|Z) \geq k$, *and let* $\mathcal{H} = \{h_\mathbf{V} : \mathcal{X} \to \{0,1\}^l, \mathbf{V} \in \{0,1\}^s\}$ *be a family of universal hash functions, i.e., for any* $x_1 \neq x_2 \in \mathcal{X}$, $\Pr_{\mathbf{V} \xleftarrow{\$} \{0,1\}^s}[h_\mathbf{V}(x_1) = h_\mathbf{V}(x_2)] \leq 2^{-l}$. *Then, it holds that*

$$\mathsf{SD}\Big((Z, \mathbf{V}, h_\mathbf{V}(X)), (Z, \mathbf{V}, U_l) \Big) \leq 2^{l-k},$$

where $\mathbf{V} \sim U_s$.

3.2 The Main Technical Lemma and Immediate Applications

Inspired by [26,49], we state a technical lemma below where the main difference is that we sample from a random subspace of sublinear-sized dimension (rather than linear-sized one [49] or from a noisy subspace in the LWE setting [26]).

Theorem 1 (LPN with Hard-to-Invert Auxiliary Input). *Let* n *be a security parameter and let* $0 < \mu < 1/2$ *be any constant. Assume that the decisional* $\mathsf{LPN}_{\mu,n}$ *problem is hard, then for every constant* $0 < \alpha < 1$, $\lambda = n^\alpha$, $q' = \mathsf{poly}(n)$, *and every* $(n, 2\lambda)$ *TYPE-I or TYPE-II leaky source* $(\mathbf{x}, f(\mathbf{x}))$, *we have*

$$(f(\mathbf{x}), \mathbf{A}', \mathbf{A}' \cdot \mathbf{x} + \mathbf{e}) \stackrel{c}{\sim} (f(\mathbf{x}), \mathbf{A}', U_{q'}), \tag{5}$$

where $\mathbf{e} \sim \mathcal{B}_\mu^{q'}$, *and* $\mathbf{A}' \sim \mathcal{D}_\lambda^{q' \times n}$ *is a* $q' \times n$ *matrix, i.e.,* $\mathbf{A}' \sim (\mathbf{A} \cdot \mathbf{V})$ *for random matrices* $\mathbf{A} \xleftarrow{\$} \{0,1\}^{q' \times \lambda}$ *and* $\mathbf{V} \xleftarrow{\$} \{0,1\}^{\lambda \times n}$.

Furthermore, if the $\mathsf{LPN}_{\mu,n}$ *problem is* $2^{\omega(n^{\frac{1}{1+\beta}})}$-*hard for any constant* $\beta > 0$ *and any superconstant hidden by* $\omega(\cdot)$ *then the above holds for any* $\lambda = \Theta(\log^{1+\beta} n)$, *any* $q' = \mathsf{poly}(n)$ *and any* $(n, 2\lambda)$ *TYPE-I/TYPE-II leaky source.*

Remark 1 (Closure Under Composition). The random subspace $\mathbf{V}$ and the random coin $\mathbf{A}$ can be public as well, which is seen from the proof below but omitted from (5) to avoid redundancy (since they are implied by $\mathbf{A}'$). That is, there exists a PPT Simu such that $(\mathbf{A}', \mathsf{Simu}(\mathbf{A}'))$ is $2^{-\Omega(n)}$-close to $(\mathbf{A}', (\mathbf{A}, \mathbf{V}))$. Therefore, (5) can be written in an equivalent form that is closed under composition, i.e., for any $q' = \mathsf{poly}(n)$ and $l = \mathsf{poly}(n)$

$$\Big(f(\mathbf{x}), \mathbf{V}, \big(\mathbf{A}_i, (\mathbf{A}_i \cdot \mathbf{V}) \cdot \mathbf{x} + \mathbf{e}_i \big)_{i=1}^l \Big) \stackrel{c}{\sim} \Big(f(\mathbf{x}), \mathbf{V}, \big(\mathbf{A}_i, U_{q'}^{(i)} \big)_{i=1}^l \Big),$$

where $\mathbf{A}_1, \cdots, \mathbf{A}_l \xleftarrow{\$} \{0,1\}^{q' \times \lambda}$, $\mathbf{e}_1, \cdots, \mathbf{e}_l \sim \mathcal{B}_\mu^{q'}$ and $\mathbf{V} \xleftarrow{\$} \{0,1\}^{\lambda \times n}$. This will be a useful property for constructing symmetric encryption schemes w.r.t. hard-to-invert auxiliary input (see more details in [17]).

Proof of Theorem 1. We have by the assumption of $(\mathbf{x}, f(\mathbf{x}))$ and Lemma 3 or Lemma 4 that

$$(f(\mathbf{x}), \mathbf{V}, \mathbf{V}\cdot\mathbf{x}) \overset{c}{\sim} (f(\mathbf{x}), \mathbf{V}, \mathbf{y})$$
$$\Rightarrow (f(\mathbf{x}), (\mathbf{A}, \mathbf{V}), (\mathbf{A}\cdot\mathbf{V})\cdot\mathbf{x}+\mathbf{e}) \overset{c}{\sim} (f(\mathbf{x}), (\mathbf{A}, \mathbf{V}), \mathbf{A}\cdot\mathbf{y}+\mathbf{e}).$$

where $\mathbf{y}\sim U_\lambda$. Next, consider T-hard decisional $\mathsf{LPN}_{\mu,\lambda}$ problem on uniform secret $\mathbf{y}$ of length λ (instead of n), which postulates that for any $q' \leq T$

$$(\mathbf{A}, \mathbf{A}\cdot\mathbf{y}+\mathbf{e}) \underset{T,1/T}{\sim} (\mathbf{A}, U_{q'})$$
$$\Rightarrow (f(\mathbf{x}), (\mathbf{A}, \mathbf{V}), \mathbf{A}\cdot\mathbf{y}+\mathbf{e}) \underset{T-\mathsf{poly}(n),\ 1/T}{\sim} (f(\mathbf{x}), (\mathbf{A}, \mathbf{V}), U_{q'}).$$

Under the LPN assumption with standard asymptotic hardness (i.e., $T = \lambda^{\omega(1)}$) and by setting parameter $\lambda = n^\alpha$ we have $T = n^{\omega(1)}$, which suffices for our purpose since for any $q' = \mathsf{poly}(n)$, any PPT adversary wins the above distinguishing game with advantage no greater than $n^{-\omega(1)}$. In case that $\mathsf{LPN}_{\mu,\lambda}$ is $2^{\omega(n^{\frac{1}{1+\beta}})}$-hard, substitution of $\lambda = \Theta(\log^{1+\beta} n)$ into $T = 2^{\omega(\lambda^{\frac{1}{1+\beta}})}$ also yields $T = n^{\omega(1)}$. Therefore, in both cases the above two ensembles are computationally indistinguishable in security parameter n. The conclusion then follows by a triangle inequality. $\qquad\square$

A COMPARISON WITH [17]. The work of [17] proved results similar to Theorem 1. In particular, [17] showed that the LSN assumption with parameters β and $p = \mathsf{poly}_\beta(n)$ implies LPN with $2^{-\alpha n}$-hard auxiliary input (for constant $\alpha > \beta$), noise rate $\mu = 1/2 - 1/4p$ and quasi-polynomial security (in essentially the same form as (5) except for a uniform matrix $\mathbf{A}'$). In comparison, by sampling $\mathbf{A}'$ from a random subspace of sublinear dimension $\lambda = n^\alpha$ (for $0 < \alpha < 1$), constant-noise LPN implies that (5) holds with $2^{-\Omega(n^\alpha)}$-hard auxiliary input, constant noise and comparable security to the underlying LPN. Furthermore, assume constant-noise LPN with $2^{\omega(n^{\frac{1}{1+\beta}})}$-hardness (for constant $\beta > 0$), then (2) holds for $2^{-\Omega(\log^{1+\beta})}$-hard auxiliary input, constant noise and quasi-polynomial security.

IMMEDIATE APPLICATIONS. This yields the same applications as considered in [17], such as CPA/CCA secure symmetric encryption schemes, average-case obfuscators for point functions, reusable and robust extractors, all under standard (constant-noise) LPN with improved efficiency (by bringing down the noise rate) and tighter security against sub-exponentially (or even quasi-polynomially) hard-to-invert auxiliary input. The proofs simply follow the route of [17] and can be informally explained as: the technique (by sampling from random subspace) implicitly applies pseudorandomness extraction (i.e., $\mathbf{y} = \mathbf{V} \cdot \mathbf{x}$) so that the rest of the scheme is built upon the security of $(\mathbf{A}, \mathbf{A}\mathbf{y} + \mathbf{e})$ on secret $\mathbf{y}$ (which is pseudorandom even conditioned on the leakage), and thus the task is essentially to obtain the aforementioned applications from standard LPN (without auxiliary input). In other words, our technique allows to transform any applications based on constant-noise LPN into the counterparts with auxiliary input under the same assumption. Therefore, we only sketch some applications in the full version of this work and refer to [17] for the redundancy.

4 CPA Secure PKE from Constant-Noise LPN

We show a more interesting application, namely, to build public-key encryption schemes from constant-noise LPN, which has been an open problem since the work of [2]. We refer to Appendix A.2 for standard definitions of public-key encryption schemes, correctness and CPA/CCA security.

4.1 Technical Lemmas

We use the following technical tool to build PKE scheme from constant-noise LPN. It would have been an immediate corollary of Theorem 1 for sub-exponential hard LPN on squared-logarithmic min-entropy sources (i.e., $\beta = 1$), except for the fact that the leakage is also correlated with noise. Notice that we lose the "closure under composition" property by allowing leakage to be correlated with noise, and thus our PKE scheme will avoid this property.

Theorem 2 (LPN on Squared-Log Entropy). *Let n be a security parameter and let $0 < \mu < 1/2$ be any constant. Assume that the computational $\mathsf{LPN}_{\mu,n}$ problem is $2^{\omega(n^{\frac{1}{2}})}$-hard (for any superconstant hidden by $\omega(\cdot)$), then for every $\lambda = \Theta(\log^2 n)$, $q' = \mathsf{poly}(n)$, and every polynomial-time sampleable $\mathbf{x} \in \{0,1\}^n$ with $\mathbf{H}_\infty(\mathbf{x}) \geq 2\lambda$ and every probabilistic polynomial-time computable function $f : \{0,1\}^{n+q'} \times \mathcal{Z} \to \{0,1\}^{O(\log n)}$ with public coin Z, we have*

$$\big(f(\mathbf{x},\mathbf{e};Z), Z, \mathbf{A}', \mathbf{A}'{\cdot}\mathbf{x} + \mathbf{e}\big) \stackrel{c}{\sim} \big(f(\mathbf{x},\mathbf{e};Z), Z, \mathbf{A}', U_{q'}\big),$$

where noise vector $\mathbf{e} \sim \mathcal{B}_\mu^{q'}$ and $q' \times n$ matrix $\mathbf{A}' \sim \mathcal{D}_\lambda^{q' \times n}$.

Proof sketch. It suffices to adapt the proof of Theorem 1 as follows. First, observe that (by the chain rule of min-entropy)

$$\mathbf{H}_\infty(\mathbf{x}|f(\mathbf{x},\mathbf{e};Z), Z, \mathbf{e}) \geq \mathbf{H}_\infty(\mathbf{x}|Z, \mathbf{e}) - O(\log n) = \mathbf{H}_\infty(\mathbf{x}) - O(\log n) \geq 2\lambda - O(\log n).$$

For our convenience, write $\mathbf{A}' \sim (\mathbf{A} \cdot \mathbf{V})$ for $\mathbf{A} \sim U_{q' \times \lambda}$, $\mathbf{V} \sim U_{\lambda \times n}$, and let $\mathbf{y}, \mathbf{r} \sim U_\lambda$. Then, we have by Lemma 4

$$(f(\mathbf{x},\mathbf{e};Z), Z, \mathbf{e}, \mathbf{V}, \mathbf{V}{\cdot}\mathbf{x}) \stackrel{s}{\sim} (f(\mathbf{x},\mathbf{e};Z), Z, \mathbf{e}, \mathbf{V}, \mathbf{y})$$
$$\Rightarrow (f(\mathbf{x},\mathbf{e};Z), Z, (\mathbf{A}{\cdot}\mathbf{V}), (\mathbf{A}{\cdot}\mathbf{V}){\cdot}\mathbf{x}+\mathbf{e}) \stackrel{s}{\sim} (f(\mathbf{x},\mathbf{e};Z), Z, (\mathbf{A}{\cdot}\mathbf{V}), \mathbf{A}{\cdot}\mathbf{y}+\mathbf{e}).$$

Next, $2^{\omega(\lambda^{\frac{1}{2}})}$-hard computational $\mathsf{LPN}_{\mu,\lambda}$ problem with secret size λ postulates that for any $q' \leq 2^{\omega(\lambda^{\frac{1}{2}})} = n^{\omega(1)}$ (recall $\lambda = \Theta(\log^2 n)$) and any probabilistic $\mathcal{D}$, $\mathcal{D}'$ of running time $n^{\omega(1)}$

$$\Pr[\ \mathcal{D}'(\mathbf{A},\ \mathbf{A}{\cdot}\mathbf{y}{+}\mathbf{e}) = \mathbf{y}\ ] \ =\ n^{-\omega(1)}$$
$$\Rightarrow \Pr[\ \mathcal{D}'(f(\mathbf{x},\mathbf{e};Z), Z, \mathbf{A}, \mathbf{A}{\cdot}\mathbf{y}{+}\mathbf{e}) = \mathbf{y}\ ] \ =\ n^{-\omega(1)}$$
$$\Rightarrow (f(\mathbf{x},\mathbf{e};Z), Z, \mathbf{A}, \mathbf{A}{\cdot}\mathbf{y}{+}\mathbf{e}, \mathbf{r}, \mathbf{r}^{\mathsf{T}}{\cdot}\mathbf{y}) \overset{c}{\sim} (f(\mathbf{x},\mathbf{e};Z), Z, \mathbf{A}, \mathbf{A}{\cdot}\mathbf{y}{+}\mathbf{e}, \mathbf{r}, U_1)$$
$$\Rightarrow (f(\mathbf{x},\mathbf{e};Z), Z, \mathbf{A}, \mathbf{A}{\cdot}\mathbf{y}{+}\mathbf{e}) \overset{c}{\sim} (f(\mathbf{x},\mathbf{e};Z), Z, \mathbf{A}, U_{q'})$$
$$\Rightarrow (f(\mathbf{x},\mathbf{e};Z), Z, (\mathbf{A}{\cdot}\mathbf{V}), \mathbf{A}{\cdot}\mathbf{y}{+}\mathbf{e}) \overset{c}{\sim} (f(\mathbf{x},\mathbf{e};Z), Z, (\mathbf{A}\cdot\mathbf{V}), U_{q'}),$$

where the first implication is trivial since Z is independent of $(\mathbf{A},\mathbf{y},\mathbf{e})$ and any $O(\log n)$ bits of leakage affects unpredictability by a fact of $\mathsf{poly}(n)$, the second step is the Goldreich-Levin theorem [25] with $\mathbf{r} \sim U_\lambda$, and the third implication uses the sample-preserving reduction from [4] and is reproduced as Lemma 18. The conclusion follows by a triangle inequality.

We will use Lemma 5 to estimate the noise rate of an inner product between Bernoulli-like vectors .

Lemma 5. *For any $0 < \mu \leq 1/8$ and $\ell \in \mathbb{N}$, let $E_1, \cdots, E_\ell$ be Boolean random variables i.i.d. to $\mathcal{B}_\mu$, then $\Pr[\ \bigoplus_{i=1}^{\ell} E_i = 0\ ] > \frac{1}{2} + 2^{-(4\mu\ell+1)}$.*

Proof. We complete the proof by Fact 1 and Fact 2

$$\Pr[\ \bigoplus_{i=1}^{\ell} E_i = 1\ ] = \frac{1}{2}(1 - (1-2\mu)^\ell) < \frac{1}{2}(1 - 2^{-4\mu\ell}) = \frac{1}{2} - 2^{-(4\mu\ell+1)}.$$

Fact 1 (Piling-Up Lemma). *For $0 < \mu < 1/2$ and random variables $E_1, E_2, \cdots, E_\ell$ that are i.i.d. to $\mathcal{B}_\mu$ we have $\bigoplus_{i=1}^{\ell} E_i \sim \mathcal{B}_\sigma$ with $\sigma = \frac{1}{2}(1 - (1-2\mu)^\ell)$.*

Fact 2 (Mean Value Theorem). *For any $0 < x \leq 1/4$ we have $1 - x > 2^{-2x}$.*

We recall the following facts about the entropy of Bernoulli-like distributions. In general, there's no closed formula for binomial coefficient, but an asymptotic estimation like Fact 3 already suffices for our purpose, where the binary entropy function can be further bounded by Fact 4 (see also Footnote 11).

Fact 3 (Asymptotics for Binomial Coefficients (e.g.[27], p.492). *For any $0 < \mu < 1/2$, and any $n \in \mathbb{N}$ we have $\binom{n}{\mu n} = 2^{n\mathbf{H}(\mu) - \frac{\log n}{2} + O(1)}$.*

Fact 4. *For any $0 < \mu < 1/2$, we have $\mu \log(1/\mu) < \mathbf{H}(\mu) < \mu(\log(1/\mu) + \frac{3}{2})$.*

4.2 Weakly Correct 1-bit PKE from Constant-Noise LPN

As stated in Theorem 2, for any constant $0 < \mu < 1/2$, $2^{\omega(n^{\frac{1}{2}})}$-hard $\mathsf{LPN}_{\mu,n}$ implies that $(\mathbf{A}' \cdot \mathbf{x} + \mathbf{e})$ is pseudorandom conditioned on $\mathbf{A}'$ for $\mathbf{x} \sim X$ with squared-log entropy, where the leakage due to f can be omitted for now as it is only needed for CCA security. If there exists X satisfying the following three conditions at the same time then the 1-bit PKE as in Fig. 1 instantiated with the square matrix $\mathbf{A}' \leftarrow \mathcal{D}_\lambda^{n \times n}$, $\mathbf{s}, \mathbf{s}_1 \leftarrow X$ and $\mathbf{e}, \mathbf{e}_1 \leftarrow \mathcal{B}_\mu^n$ will be secure and noticeably correct (since $\mathbf{s}_1^\mathsf{T} \mathbf{e}$ and $\mathbf{e}_1^\mathsf{T} \mathbf{s}$ are both $(1/2 + 1/\mathsf{poly}(n))$-biased to 0 independently).

1. **(Efficiency)** $X \in \{0,1\}^n$ can be sampled in polynomial time.
2. **(Security)** $\mathbf{H}_\infty(X) = \Omega(\log^2 n)$ as required by Theorem 2.
3. **(Correctness)** $|X| = O(\log n)$ such that $\Pr[\langle X, \mathcal{B}_\mu^n \rangle = 0] \geq 1/2 + 1/\mathsf{poly}(n)$.

Note that any distribution $X \in \{0,1\}^n$ satisfying $|X| = O(\log n)$ implies that $\mathbf{H}_\infty(X) = O(\log^2 n)$ (as the set $\{\mathbf{x} \in \{0,1\}^n : |\mathbf{x}| = O(\log n)\}$ is of size $2^{O(\log^2 n)}$), so the job is to maximize the entropy of X under constraint $|X| = O(\log n)$. The first candidate seems $X \sim \mathcal{B}_{\mu'}^n$ for $\mu' = \Theta(\frac{\log n}{n})$, but it does not meet the security condition because the noise rate μ' is so small that a Chernoff bound only ensures (see Lemma 19) that $\mathcal{B}_{\mu'}^n$ is $(2^{-O(\mu' n)} = 1/\mathsf{poly}(n))$-close to having min-entropy $\Theta(n\mathbf{H}(\mu')) = \Theta(\log^2 n)$. In fact, we can avoid the lower-tail issue by letting $X \sim \chi_{\log n}^n$, namely, a uniform distribution of Hamming weight exact $\log n$, which is of min-entropy $\Theta(\log^2 n)$ by Fact 3. Thus, $X \sim \chi_{\log n}^n$ is a valid option to obtain a single-bit PKE with noticeable correctness.

4.3 CPA Secure PKE from Constant-Noise LPN

Unlike [20] where the extension from the weak single-bit PKE to a fully correct scheme is almost immediate (by a parallel repetition and using error correcting codes), it is not trivial to amplify the noticeable correctness of the single-bit scheme to an overwhelming probability, in particular, the scheme instantiated with distribution $X \sim \chi_{\log n}^n$ would no longer work. To see the difficulty, we define below our CPA secure scheme $\Pi_X = (\mathsf{KeyGen}, \mathsf{Enc}, \mathsf{Dec})$ that resembles the counterpart for low-noise LPN (e.g., [15,20]), where distribution X is left undefined (apart from the entropy constraint).

Distribution X: X is a polynomial-time sampleable distribution satisfying $\mathbf{H}_\infty(X) = \Omega(\log^2 n)$ and we set $\lambda = \Theta(\log^2 n)$ such that $2\lambda \leq \mathbf{H}_\infty(X)$.
$\mathsf{KeyGen}(1^n)$: Given a security parameter 1^n, it samples matrix $\mathbf{A} \sim \mathcal{D}_\lambda^{n \times n}$, column vectors $\mathbf{s} \sim X$, $\mathbf{e} \sim \mathcal{B}_\mu^n$, computes $\mathbf{b} = \mathbf{As} + \mathbf{e}$ and sets $(pk, sk) := ((\mathbf{A}, \mathbf{b}), \mathbf{s})$.
$\mathsf{Enc}_{pk}(\mathbf{m})$: Given the public key $pk = (\mathbf{A}, \mathbf{b})$ and a plaintext $\mathbf{m} \in \{0,1\}^n$, Enc_{pk} chooses
$$\mathbf{S}_1 \sim (X^{(1)}, \cdots, X^{(q)}) \in \{0,1\}^{n \times q}, \mathbf{E}_1 \sim \mathcal{B}_\mu^{n \times q}$$

where $X^{(1)}, \cdots, X^{(q)}$ are i.i.d. to X. Then, it outputs $C = (\mathbf{C}_1, \mathbf{c}_2)$ as cipher-text, where

$$\mathbf{C}_1 := \mathbf{A}^\top \mathbf{S}_1 + \mathbf{E}_1 \in \{0,1\}^{n \times q},$$
$$\mathbf{c}_2 := \mathbf{S}_1^\top \mathbf{b} + \mathbf{G} \cdot \mathbf{m} \in \{0,1\}^q,$$

and $\mathbf{G} \in \{0,1\}^{q \times n}$ is a generator matrix for an efficiently decodable code (with error correction capacity to be defined and analyzed in Sect. 4.4).
$\mathsf{Dec}_{sk}(\mathbf{C}_1, \mathbf{c}_2)$: On secret key $sk = \mathbf{s}$, ciphertext $(\mathbf{C}_1, \mathbf{c}_2)$, it computes

$$\tilde{\mathbf{c}}_0 := \mathbf{c}_2 - \mathbf{C}_1^\top \mathbf{s} = \mathbf{G} \cdot \mathbf{m} + \mathbf{S}_1^\top \mathbf{e} - \mathbf{E}_1^\top \mathbf{s}$$

and reconstructs $\mathbf{m}$ from the error $\mathbf{S}_1^\top \mathbf{e} - \mathbf{E}_1^\top \mathbf{s}$ using the error correction property of $\mathbf{G}$.

We can see that the CPA security of Π_X, for any X with $\mathbf{H}_\infty(X) = \Omega(\log^2 n)$, follows from applying Theorem 2 twice (once for replacing the pubic key $\mathbf{b}$ with uniform randomness, and again together with the Goldreich Levin Theorem for encrypting a single bit) and a hybrid argument (to encrypt many bits).

Theorem 3 (CPA Security). *Assume that the decisional* $\mathsf{LPN}_{\mu,n}$ *problem is* $2^{\omega(n^{\frac{1}{2}})}$-*hard for any constant* $0 < \mu < 1/2$, *then* Π_X *is IND-CPA secure.*

4.4   Which X Makes a Correct Scheme?

$X \sim \chi^n_{\log n}$ MAY NOT WORK. To make a correct scheme, we need to upper bound $|\mathbf{S}_1^\top \mathbf{e} - \mathbf{E}_1^\top \mathbf{s}|$ by $q(1/2 - 1/\mathsf{poly}(n))$, but in fact we do not have any useful bound even for $|\mathbf{S}_1^\top \mathbf{e}|$. Recall that $\mathbf{S}_1^\top$ is now a $q \times n$ matrix and parse $\mathbf{S}_1^\top \mathbf{e}$ as Boolean random variables $W_1, \cdots, W_q$. First, although every W_i satisfies $\Pr[W_i = 0] \geq 1/2 + 1/\mathsf{poly}(n)$, they are not independent (correlated through $\mathbf{e}$). Second, if we fix any $|\mathbf{e}| = \Theta(n)$, all $W_1, \cdots, W_q$ are now independent conditioned on $\mathbf{e}$, but then we could no longer guarantee that $\Pr[W_i = 0 | \mathbf{e}] \geq 1/2 + \mathsf{poly}(n)$ as $\mathbf{S}_1$ follows $(\chi^n_{\log n})^q$ rather than $(\mathcal{B}^n_{\log n/n})^q$. Otherwise said, condition #3 (as in Sect. 4.2) is not sufficient for overwhelming correctness. We introduced a tailored version of Bernoulli distribution (with upper/lower tails chopped off).

Definition 4 (Distribution $\widetilde{\mathcal{B}}^n_{\mu_1}$). *Define* $\widetilde{\mathcal{B}}^n_{\mu_1}$ *to be distributed to* $\mathcal{B}^n_{\mu_1}$ *conditioned on* $(1 - \frac{\sqrt{6}}{3})\mu_1 n \leq |\mathcal{B}^n_{\mu_1}| \leq 2\mu_1 n$. *Further, we define an* $n \times q$ *matrix distribution, denoted by* $(\widetilde{\mathcal{B}}^n_{\mu_1})^q$, *where every column is i.i.d. to* $\widetilde{\mathcal{B}}^n_{\mu_1}$.

$\widetilde{\mathcal{B}}^n_{\mu_1}$ IS EFFICIENTLY SAMPLEABLE. $\widetilde{\mathcal{B}}^n_{\mu_1}$ can be sampled in polynomial-time with exponentially small error, e.g., simply sample $\mathbf{e} \leftarrow \mathcal{B}^n_{\mu_1}$ and outputs $\mathbf{e}$ if $(1 - \frac{\sqrt{6}}{3})\mu_1 n \leq |\mathbf{e}| \leq 2\mu_1 n$. Otherwise, repeat the above until such $\mathbf{e}$ within the Hamming weight range is obtained or the experiment failed (then output $\bot$ in this case) for a predefined number of times (e.g., n).

$\widetilde{\mathcal{B}}^n_{\mu_1}$ IS OF MIN-ENTROPY $\Omega(\log^2 n)$. For $\mu_1 = \Omega(\log n/n)$, it is not hard to see that $\widetilde{\mathcal{B}}^n_{\mu_1}$ is a convex combination of $\chi^n_{(1 - \frac{\sqrt{6}}{3})\mu_1 n}, \cdots, \chi^n_{2\mu_1 n}$, and thus of min-entropy $\Omega(\log^2 n)$ by Fact 3.

Therefore, Π_X when instantiated with $X \sim \widetilde{\mathcal{B}}^n_{\mu_1}$ is CPA secure by Theorem 4, and we proceed to the correctness of the scheme.

Lemma 6. *For constants* $\alpha > 0$, $0 < \mu \leq 1/10$ *and* $\mu_1 = \alpha \log n/n$, *let* $\mathbf{S}_1 \sim (\widetilde{\mathcal{B}}^n_{\mu_1})^q$, $\mathbf{e} \sim \mathcal{B}^n_\mu$, $\mathbf{E}_1 \sim \mathcal{B}^{n \times q}_\mu$ *and* $\mathbf{s} \sim \widetilde{\mathcal{B}}^n_{\mu_1}$, *we have*

$$\Pr\left[\ |\mathbf{S}_1^\mathsf{T}\mathbf{e} - \mathbf{E}_1^\mathsf{T}\mathbf{s}| \leq (\frac{1}{2} - \frac{1}{2n^{3\alpha/2}})q \ \right] \ \geq \ 1 - 2^{-\Omega(n^{-3\alpha}q)}.$$

Proof. It is more convenient to consider $|\mathbf{S}_1^\mathsf{T}\mathbf{e} - \mathbf{E}_1^\mathsf{T}\mathbf{s}|$ conditioned on $|\mathbf{e}| \leq 1.01\mu n$ (except for a $2^{-\Omega(n)}$-fraction) and $|\mathbf{s}| \leq 2\mu n$. We have by Lemmas 7 and 8 that $\mathbf{S}_1^\mathsf{T}\mathbf{e}$ and $\mathbf{E}_1^\mathsf{T}\mathbf{s}$ are identical distributed to $\mathcal{B}^q_{\delta_1}$ and $\mathcal{B}^q_{\delta_2}$ respectively, where $\delta_1 \leq 1/2 - n^{-\alpha/2}$ and $\delta_2 \leq 1/2 - n^{-\alpha}/2$. Thus, $(\mathbf{S}_1^\mathsf{T}\mathbf{e} - \mathbf{E}_1^\mathsf{T}\mathbf{s})$ follows $\mathcal{B}^q_\delta$ for $\delta \leq 1/2 - n^{-3\alpha/2}$ by the Piling-up lemma, and then we complete the proof with Lemma 2.

CONCRETE PARAMETERS. Enc_{pk} simply uses a generator matrix $\mathbf{G} : \{0,1\}^{q \times n}$ that efficiently corrects up to a $(1/2 - n^{-3\alpha/2}/2)$-fraction of bit flipping errors, which exists for $q = O(n^{3\alpha+1})$ (e.g., [22]). We can now conclude the correctness of the scheme since every encryption will be correctly decrypted with overwhelming probability and thus so is the event that polynomially many of them occur simultaneously (even when they are not independent, see Lemma 1).

Theorem 4 (Correctness). *Let* $0 < \mu \leq 1/10$ *and* $\alpha > 0$ *be any constants, let* $q = \Theta(n^{3\alpha+1})$ *and* $\mu_1 = \alpha \log n/n$, *and let* $X \sim \widetilde{\mathcal{B}}^n_{\mu_1}$. *Assume that the decisional* $\mathsf{LPN}_{\mu,n}$ *problem is* $2^{\omega(n^{\frac{1}{2}})}$*-hard, then* Π_X *is a correct scheme.*

Lemma 7. *For any* $0 < \mu \leq 1/10$, $\mu_1 = O(\log n/n) \leq 1/8$ *and any* $\mathbf{e} \in \{0,1\}^n$ *with* $|\mathbf{e}| \leq 1.01\mu n$,
$$\Pr[\langle\widetilde{\mathcal{B}}^n_{\mu_1}, \mathbf{e}\rangle = 0] \geq 1/2 + 2^{-\frac{\mu_1 n}{2}}.$$

Proof. Denote by $\mathcal{E}$ the event $(1 - \frac{\sqrt{6}}{3})\mu_1 n \leq |\mathcal{B}^n_{\mu_1}| \leq 2\mu_1 n$ and thus $\Pr[\mathcal{E}] \geq (1 - 2\exp^{-\mu_1 n/3})$ by the Chernoff bound. We have by Lemma 5

$$\frac{1}{2} + 2^{-(4.04\mu\mu_1 n+1)} \leq \Pr[\langle\mathcal{B}^n_{\mu_1}, \mathbf{e}\rangle = 0]$$
$$\leq \Pr[\mathcal{E}] \cdot \Pr[\langle\widetilde{\mathcal{B}}^n_{\mu_1}, \mathbf{e}\rangle = 0] \ + \ \Pr[\neg\mathcal{E}] \cdot \Pr[\langle\mathcal{B}^n_{\mu_1}, \mathbf{e}\rangle = 0 | \neg\mathcal{E}]$$
$$\leq \Pr[\langle\widetilde{\mathcal{B}}^n_{\mu_1}, \mathbf{e}\rangle = 0] \ + \ \Pr[\neg\mathcal{E}].$$

For $0 < \mu \leq 1/10$, $\Pr[\langle\widetilde{\mathcal{B}}^n_{\mu_1}, \mathbf{e}\rangle = 0] \geq 1/2 + 2^{-(4.04\mu\mu_1 n+1)} - 2\exp^{-\mu_1 n/3} > 1/2 + 2^{-\mu_1 n/2}$.

Lemma 8. *For any* $0 < \mu \leq 1/8$, $\mu_1 = O(\log n/n)$, *and any* $\mathbf{s} \in \{0,1\}^n$ *with* $|\mathbf{s}| \leq 2\mu_1 n$, *we have by Lemma 5*

$$\Pr[\langle\mathcal{B}^n_\mu, \mathbf{s}\rangle = 0] \ \geq \ 1/2 + 2^{-(8\mu\mu_1 n+1)} \ \geq \ 1/2 + 2^{-(\mu_1 n+1)}.$$

5   CCA-Secure PKE from Constant-Noise LPN

In this section, we show how to construct CCA-secure PKE from constant-noise LPN. Our starting point is the construction of a tag-based PKE against selective tag and chosen ciphertext attacks from LPN, which can be transformed into a standard CCA-secure PKE by using known techniques [11,35]. We begin by first recalling the definitions of tag-based PKE.

5.1   Tag-Based Encryption

A tag-based encryption (TBE) scheme with tag-space $\mathcal{T}$ and message-space $\mathcal{M}$ consists of three PPT algorithms $\mathcal{TBE} = (\mathsf{KeyGen}, \mathsf{Enc}, \mathsf{Dec})$. The randomized key generation algorithm KeyGen takes the security parameter n as input, outputs a public key pk and a secret key sk, denoted as $(pk, sk) \leftarrow \mathsf{KeyGen}(1^n)$. The randomized encryption algorithm Enc takes pk, a tag $\mathbf{t} \in \mathcal{T}$, and a plaintext $\mathbf{m} \in \mathcal{M}$ as inputs, outputs a ciphertext C, denoted as $C \leftarrow \mathsf{Enc}(pk, \mathbf{t}, \mathbf{m})$. The deterministic algorithm Dec takes sk and C as inputs, outputs a plaintext $\mathbf{m}$, or a special symbol $\perp$, which is denoted as $\mathbf{m} \leftarrow \mathsf{Dec}(sk, \mathbf{t}, C)$. For correctness, we require that for all $(pk, sk) \leftarrow \mathsf{KeyGen}(1^n)$, any tag $\mathbf{t}$, any plaintext $\mathbf{m}$ and any $C \leftarrow \mathsf{Enc}(pk, \mathbf{t}, \mathbf{m})$, the equation $\mathsf{Dec}(sk, \mathbf{t}, C) = \mathbf{m}$ holds with overwhelming probability.

We consider the following game between a challenger $\mathcal{C}$ and an adversary $\mathcal{A}$ given in [35].

Init. The adversary $\mathcal{A}$ takes the security parameter n as inputs, and outputs a target tag $\mathbf{t}^*$ to the challenger $\mathcal{C}$.

KeyGen. The challenger $\mathcal{C}$ computes $(pk, sk) \leftarrow \mathsf{KeyGen}(1^n)$, gives the public key pk to the adversary $\mathcal{A}$, and keeps the secret key sk to itself.

Phase 1. The adversary $\mathcal{A}$ can make decryption queries for any pair $(\mathbf{t}, C)$ for any polynomial time, with a restriction that $\mathbf{t} \neq \mathbf{t}^*$, and the challenger $\mathcal{C}$ returns $\mathbf{m} \leftarrow \mathsf{Dec}(sk, \mathbf{t}, C)$ to $\mathcal{A}$ accordingly.

Challenge. The adversary $\mathcal{A}$ outputs two equal length plaintexts $\mathbf{m}_0, \mathbf{m}_1 \in \mathcal{M}$. The challenger $\mathcal{C}$ randomly chooses a bit $b^* \xleftarrow{\$} \{0, 1\}$, and returns the challenge ciphertext $C^* \leftarrow \mathsf{Enc}(pk, \mathbf{t}^*, \mathbf{m}_{b^*})$ to the adversary $\mathcal{A}$.

Phase 2. The adversary can make more decryption queries as in Phase 1.

Guess. Finally, $\mathcal{A}$ outputs a guess $b \in \{0, 1\}$. If $b = b^*$, the challenger $\mathcal{C}$ outputs 1, else outputs 0.

Advantage. $\mathcal{A}$'s advantage is defined as $\mathsf{Adv}_{\mathcal{TBE}, \mathcal{A}}^{\text{ind-stag-cca}}(1^n) \overset{\text{def}}{=} |\Pr[b = b^*] - \frac{1}{2}|$.

Definition 5 (IND-sTag-CCA). *We say that a TBE scheme $\mathcal{TBE}$ is IND-sTag-CCA secure if for any PPT adversary $\mathcal{A}$, its advantage $\mathsf{Adv}_{\mathcal{TBE}, \mathcal{A}}^{\text{ind-stag-cca}}(1^n)$ is negligible in n.*

For our convenience, we will use the following corollary, which is essentially a q-fold[8] (transposed) version of Theorem 2 with $q' = n$ and 2 bits of linear leakage (rather than $O(\log n)$ bits of arbitrary leakage) per copy. Following [36], the leakage is crucial for the CCA security proof.

Corollary 1. *Let n be a security parameter and let $0 < \mu < 1/2$ be any constant. Assume that the computational* $\mathsf{LPN}_{\mu,n}$ *problem is $2^{\omega(n^{\frac{1}{2}})}$-hard (for any super-constant hidden by $\omega(\cdot)$). Then, for every $\mu_1 = \Omega(\log n/n)$ and $\lambda = \Theta(\log^2 n)$ such that $2\lambda \leq \mathbf{H}_\infty(\widetilde{\mathcal{B}}_{\mu_1}^n)$, and every $q = \mathsf{poly}(n)$, we have*

$$ \left(\, (\mathbf{S}_0^\mathsf{T}\mathbf{e}, \mathbf{E}_0^\mathsf{T}\mathbf{s}), \mathbf{e}, \mathbf{s}, \mathbf{A}, \mathbf{S}_0^\mathsf{T}\mathbf{A} + \mathbf{E}_0^\mathsf{T} \,\right) \overset{c}{\sim} \left(\, (\mathbf{S}_0^\mathsf{T}\mathbf{e}, \mathbf{E}_0^\mathsf{T}\mathbf{s}), \mathbf{e}, \mathbf{s}, \mathbf{A}, U_{q\times n} \,\right), $$

where the probability is taken over $\mathbf{S}_0 \sim (\widetilde{\mathcal{B}}_{\mu_1}^n)^q$, $\mathbf{E}_0 \sim \mathcal{B}_\mu^{n\times q}$, $\mathbf{A} \sim \mathcal{D}_\lambda^{n\times n}$, $U_{q\times n}$, $\mathbf{s} \leftarrow \widetilde{\mathcal{B}}_{\mu_1}^n$, $\mathbf{e} \leftarrow \mathcal{B}_\mu^n$ and internal random coins of the distinguisher.

5.2 Our Construction

Our construction is built upon previous works in [36,44]. A couple of modifications are made to adapt the ideas of [36,44], which seems necessary due to the absence of a meaningful knapsack version for our LPN (with poly-log entropy and non-uniform matrix). Let n be the security parameter, let $\alpha > 0$, $0 < \mu \leq 1/10$ be any constants, let $\mu_1 = \alpha \log n/n$, $\beta = (\frac{1}{2} - \frac{1}{n^{3\alpha}})$, $\gamma = (\frac{1}{2} - \frac{1}{2n^{3\alpha/2}})$ and choose $\lambda = \Theta(\log^2 n)$ such that $2\lambda \leq \mathbf{H}_\infty(\widetilde{\mathcal{B}}_{\mu_1}^n)$. Let the plaintext-space $\mathcal{M} = \{0,1\}^n$, and let $\mathbf{G} \in \{0,1\}^{q\times n}$ and $\mathbf{G}_2 \in \{0,1\}^{\ell\times n}$ be the generator matrices that can correct at least βq and $2\mu\ell$ bit flipping errors in the codeword respectively, where $q = O(n^{6\alpha+1})$, $\ell = O(n)$ and we refer to [22] and [33] for explicit constructions of the two codes respectively. Let the tag-space $\mathcal{T} = \mathbb{F}_{2^n}$. We use a matrix representation $\mathbf{H}_\mathbf{t} \in \{0,1\}^{n\times n}$ for finite field elements $\mathbf{t} \in \mathbb{F}_{2^n}$ [10,14,36] such that $\mathbf{H}_0 = \mathbf{0}$, $\mathbf{H}_\mathbf{t}$ is invertible for any $\mathbf{t} \neq \mathbf{0}$, and $\mathbf{H}_{\mathbf{t}_1} + \mathbf{H}_{\mathbf{t}_2} = \mathbf{H}_{\mathbf{t}_1+\mathbf{t}_2}$. Our TBE scheme $\mathcal{TBE}$ is defined as follows:

KeyGen(1^n): Given a security parameter n, first uniformly choose matrices $\mathbf{A} \overset{\$}{\leftarrow} \mathcal{D}_\lambda^{n\times n}$, $\mathbf{C} \overset{\$}{\leftarrow} \mathcal{D}_\lambda^{\ell\times n}$, $\mathbf{S}_0, \mathbf{S}_1 \overset{\$}{\leftarrow} (\widetilde{\mathcal{B}}_{\mu_1}^n)^q$ and $\mathbf{E}_0, \mathbf{E}_1 \overset{\$}{\leftarrow} \mathcal{B}_\mu^{n\times q}$. Then, compute $\mathbf{B}_0 = \mathbf{S}_0^\mathsf{T}\mathbf{A} + \mathbf{E}_0^\mathsf{T}$, $\mathbf{B}_1 = \mathbf{S}_1^\mathsf{T}\mathbf{A} + \mathbf{E}_1^\mathsf{T} \in \{0,1\}^{q\times n}$, and set $(pk, sk) = (\,(\mathbf{A}, \mathbf{B}_0, \mathbf{B}_1, \mathbf{C}), (\mathbf{S}_0, \mathbf{S}_1))$.
Enc($pk, \mathbf{t}, \mathbf{m}$): Given the public key $pk = (\mathbf{A}, \mathbf{B}_0, \mathbf{B}_1, \mathbf{C})$, a tag $\mathbf{t} \in \mathbb{F}_{2^n}$, and a plaintext $\mathbf{m} \in \{0,1\}^n$, randomly choose

$$ \mathbf{s} \overset{\$}{\leftarrow} \widetilde{\mathcal{B}}_{\mu_1}^n, \mathbf{e}_1 \overset{\$}{\leftarrow} \mathcal{B}_\mu^n, \mathbf{e}_2 \overset{\$}{\leftarrow} \mathcal{B}_\mu^\ell, \mathbf{S}_0', \mathbf{S}_1' \overset{\$}{\leftarrow} (\widetilde{\mathcal{B}}_{\mu_1}^n)^q, \mathbf{E}_0', \mathbf{E}_1' \overset{\$}{\leftarrow} \mathcal{B}_\mu^{n\times q} $$

---

[8] Please do not confuse q' with q, where q' is the number of samples in LPN (see Theorem 2) and is set to n (for a square matrix), and q is the number of parallel repetitions of LPN on independent secrets and noise vectors.

and define

$$\mathbf{c} := \mathbf{As} + \mathbf{e}_1 \qquad\qquad\qquad \in \{0,1\}^n$$
$$\mathbf{c}_0 := (\mathbf{GH_t} + \mathbf{B}_0)\mathbf{s} + (\mathbf{S}_0')^{\mathsf{T}}\mathbf{e}_1 - (\mathbf{E}_0')^{\mathsf{T}}\mathbf{s} \in \{0,1\}^q$$
$$\mathbf{c}_1 := (\mathbf{GH_t} + \mathbf{B}_1)\mathbf{s} + (\mathbf{S}_1')^{\mathsf{T}}\mathbf{e}_1 - (\mathbf{E}_1')^{\mathsf{T}}\mathbf{s} \in \{0,1\}^q$$
$$\mathbf{c}_2 := \mathbf{Cs} + \mathbf{e}_2 + \mathbf{G}_2\mathbf{m} \qquad\qquad \in \{0,1\}^\ell.$$

Finally, return the ciphertext $C = (\mathbf{c}, \mathbf{c}_0, \mathbf{c}_1, \mathbf{c}_2)$.

$\mathsf{Dec}(sk, \mathbf{t}, C)$: Given the secret key $sk = (\mathbf{S}_0, \mathbf{S}_1)$, tag $\mathbf{t} \in \mathbb{F}_{2^n}$ and ciphertext $C = (\mathbf{c}, \mathbf{c}_0, \mathbf{c}_1, \mathbf{c}_2)$, first compute

$$\tilde{\mathbf{c}}_0 := \mathbf{c}_0 - \mathbf{S}_0^{\mathsf{T}}\mathbf{c} = \mathbf{GH_t s} + (\mathbf{S}_0' - \mathbf{S}_0)^{\mathsf{T}}\mathbf{e}_1 + (\mathbf{E}_0 - \mathbf{E}_0')^{\mathsf{T}}\mathbf{s}.$$

Then, reconstruct $\mathbf{b} = \mathbf{H_t s}$ from the error $(\mathbf{S}_0' - \mathbf{S}_0)^{\mathsf{T}}\mathbf{e}_1 + (\mathbf{E}_0 - \mathbf{E}_0')^{\mathsf{T}}\mathbf{s}$ by using the error correction property of $\mathbf{G}$, and compute $\mathbf{s} = \mathbf{H_t}^{-1}\mathbf{b}$. If it holds that

$$\underbrace{|\mathbf{c} - \mathbf{As}|}_{=\mathbf{e}_1} \leq 2\mu n \wedge \underbrace{|\mathbf{c}_0 - (\mathbf{GH_t} + \mathbf{B}_0)\mathbf{s}|}_{=(\mathbf{S}_0')^{\mathsf{T}}\mathbf{e}_1 - (\mathbf{E}_0')^{\mathsf{T}}\mathbf{s}} \leq \gamma q \wedge \underbrace{|\mathbf{c}_1 - (\mathbf{GH_t} + \mathbf{B}_1)\mathbf{s}|}_{=(\mathbf{S}_1')^{\mathsf{T}}\mathbf{e}_1 - (\mathbf{E}_1')^{\mathsf{T}}\mathbf{s}} \leq \gamma q$$

then reconstruct $\mathbf{m}$ from $\mathbf{c}_2 - \mathbf{Cs} = \mathbf{G}_2\mathbf{m} + \mathbf{e}_2$ by using the error correction property of $\mathbf{G}_2$, else let $\mathbf{m} = \bot$. Finally, return the decrypted result $\mathbf{m}$.

Remark 2. As one can see, the matrix $\mathbf{S}_1$ in the secret key $sk = (\mathbf{S}_0, \mathbf{S}_1)$ can also be used to decrypt the ciphertext, i.e., compute $\tilde{\mathbf{c}}_1 := \mathbf{c}_1 - \mathbf{S}_1^{\mathsf{T}}\mathbf{c} = \mathbf{GH_t s} + (\mathbf{S}_1' - \mathbf{S}_1)^{\mathsf{T}}\mathbf{e}_1 + (\mathbf{E}_1 - \mathbf{E}_1')^{\mathsf{T}}\mathbf{s}$ and recover $\mathbf{s}$ from $\tilde{\mathbf{c}}_1$ by using the error correction property of $\mathbf{G}$. Moreover, the check condition

$$|\mathbf{c} - \mathbf{As}| \leq 2\mu n \wedge |\mathbf{c}_0 - (\mathbf{GH_t} + \mathbf{B}_0)\mathbf{s}| \leq \gamma q \wedge |\mathbf{c}_1 - (\mathbf{GH_t} + \mathbf{B}_1)\mathbf{s}| \leq \gamma q$$

guarantees that the decryption results are the same when we use either $\mathbf{S}_0$ or $\mathbf{S}_1$ in the decryption. This fact seems not necessary for the correctness, but it is very important for the security proof. Looking ahead, it allows us to switch the "exact decryption key" between $\mathbf{S}_0$ and $\mathbf{S}_1$.

Correctness and Equivalence of the Secret Keys $\mathbf{S}_0, \mathbf{S}_1$. In the following, we show that for appropriate choice of parameters, the above scheme $\mathcal{TBE}$ is correct, and has the property that both $\mathbf{S}_0$ and $\mathbf{S}_1$ are equivalent in terms of decryption.

– The correctness of the scheme requires the following:
  1. $|(\mathbf{S}_0' - \mathbf{S}_0)^{\mathsf{T}}\mathbf{e}_1 + (\mathbf{E}_0 - \mathbf{E}_0')^{\mathsf{T}}\mathbf{s}| \leq \beta q$ (to let $\mathbf{G}$ reconstruct $\mathbf{s}$ from $\tilde{\mathbf{c}}_0$).
  2. $|\mathbf{c} - \mathbf{As}| \leq 2\mu n \wedge |\mathbf{c}_0 - (\mathbf{GH_t} + \mathbf{B}_0)\mathbf{s}| \leq \gamma q \wedge |\mathbf{c}_1 - (\mathbf{GH_t} + \mathbf{B}_1)\mathbf{s}| \leq \gamma q$.
  3. $|\mathbf{e}_2| \leq 2\mu\ell$ (such that $\mathbf{G}_2$ can reconstruct $\mathbf{m}$ from $\mathbf{c}_2 - \mathbf{Cs} = \mathbf{G}_2\mathbf{m} + \mathbf{e}_2$).
– For obtaining CCA security, we also need to show that $\mathbf{S}_0$ and $\mathbf{S}_1$ have the same decryption ability except with negligible probability, namely,
  1. If $|\mathbf{c} - \mathbf{As}| \leq 2\mu n \wedge |\mathbf{c}_0 - (\mathbf{GH_t} + \mathbf{B}_0)\mathbf{s}| \leq \gamma q$, then $\mathbf{G}$ can reconstruct $\mathbf{s}$ from a code within bounded error $|(\mathbf{S}_0' - \mathbf{S}_0)\mathbf{e}_1 + (\mathbf{E}_0 - \mathbf{E}_0')\mathbf{s}| \leq \beta q$.

2. If $|\mathbf{c} - \mathbf{As}| \leq 2\mu n \wedge |\mathbf{c}_1 - (\mathbf{GH}_t + \mathbf{B}_1)\mathbf{s}| \leq \gamma q$, then $\mathbf{G}$ can reconstruct s from a code within bounded error $|(\mathbf{S}_1' - \mathbf{S}_1)\mathbf{e}_1 + (\mathbf{E}_1 - \mathbf{E}_1')\mathbf{s}| \leq \beta q$.

It suffices to show that each Hamming weight constraint above holds (with overwhelming probability) individually and thus polynomially many of them hold simultaneously (with overwhelming probability as well) by Lemma 1. First, Chernoff bound guarantees that $\Pr[|\mathbf{e}_1| \leq 2\mu n] = 1 - 2^{-\Omega(n)}$ and $\Pr[|\mathbf{e}_2| \leq 2\mu\ell] = 1 - 2^{-\Omega(\ell)}$. Second, for $i \in \{0,1\}$ the bound $|(\mathbf{S}_i')^\mathsf{T}\mathbf{e}_1 - (\mathbf{E}_i')^\mathsf{T}\mathbf{s}| \leq \gamma q$ is ensured by Lemma 6 and we further bound $|(\mathbf{S}_i' - \mathbf{S}_i)\mathbf{e}_1 + (\mathbf{E}_i - \mathbf{E}_i')\mathbf{s}| \leq \beta q$ with Lemma 9 below (proof similar to Lemma 6 and thus deferred to Appendix B).

Lemma 9. *For constants $\alpha > 0$, $0 < \mu \leq 1/10$ and $\mu_1 = \alpha \log n/n$, let $\mathbf{S}$ and $\mathbf{S}'$ be i.i.d. to $(\widetilde{\mathcal{B}}_{\mu_1}^n)^q$, $\mathbf{E}$ and $\mathbf{E}'$ be i.i.d. to $\mathcal{B}_\mu^{n \times q}$, $\mathbf{s} \sim \widetilde{\mathcal{B}}_{\mu_1}^n$ and $\mathbf{e} \sim \mathcal{B}_\mu^n$. Then,*

$$\Pr\left[\, |(\mathbf{S}' - \mathbf{S})^\mathsf{T}\mathbf{e} + (\mathbf{E} - \mathbf{E}')^\mathsf{T}\mathbf{s}| \leq (\frac{1}{2} - \frac{1}{n^{3\alpha}})q \,\right] \geq 1 - 2^{-\Omega(n^{-6\alpha}q)}.$$

Security of the TBE Scheme. We now show that under the LPN assumption, the above scheme $\mathcal{TBE}$ is IND-sTag-CCA secure in the standard model.

Theorem 5. *Assume that the decisional $\mathsf{LPN}_{\mu,n}$ problem is $2^{\omega(n^{\frac{1}{2}})}$-hard for any constant $0 < \mu \leq 1/10$, then our TBE scheme $\mathcal{TBE}$ is IND-sTag-CCA secure.*

Proof. Let $\mathcal{A}$ be any PPT adversary that can attack our TBE scheme $\mathcal{TBE}$ with advantage ϵ. We show that ϵ must be negligible in n. We continue the proof by using a sequence of games, where the first game is the real IND-sTag-CCA security game, while the last is a random game in which the challenge ciphertext is independent from the choices of the challenge plaintexts. Since any PPT adversary $\mathcal{A}$'s advantage in a random game is exactly 0, the security of $\mathcal{TBE}$ can be established by showing that $\mathcal{A}$'s advantage in any two consecutive games are negligibly close.

Game 0. The challenger $\mathcal{C}$ honestly runs the adversary $\mathcal{A}$ with the security parameter n, and obtains a target tag $\mathbf{t}^*$ from $\mathcal{A}$. Then, it simulates the IND-sTag-CCA security game for $\mathcal{A}$ as follows:

KeyGen. First uniformly choose matrices $\mathbf{A} \xleftarrow{\$} \mathcal{D}_\lambda^{n \times n}, \mathbf{C} \xleftarrow{\$} \mathcal{D}_\lambda^{\ell \times n}, \mathbf{S}_0, \mathbf{S}_1 \xleftarrow{\$} (\widetilde{\mathcal{B}}_{\mu_1}^n)^q$ and $\mathbf{E}_0, \mathbf{E}_1 \xleftarrow{\$} \mathcal{B}_\mu^{n \times q}$. Then, compute $\mathbf{B}_0 = \mathbf{S}_0^\mathsf{T}\mathbf{A} + \mathbf{E}_0^\mathsf{T}, \mathbf{B}_1 = \mathbf{S}_1^\mathsf{T}\mathbf{A} + \mathbf{E}_1^\mathsf{T} \in \{0,1\}^{q \times n}$. Finally, $\mathcal{C}$ sends $pk = (\mathbf{A}, \mathbf{B}_0, \mathbf{B}_1, \mathbf{C})$ to the adversary $\mathcal{A}$, and keeps $sk = (\mathbf{S}_0, \mathbf{S}_1)$ to itself.
Phase 1. After receiving a decryption query $(\mathbf{t}, (\mathbf{c}, \mathbf{c}_0, \mathbf{c}_1, \mathbf{c}_2))$ from the adversary $\mathcal{A}$, the challenger $\mathcal{C}$ directly returns $\perp$ to $\mathcal{A}$ if $\mathbf{t} = \mathbf{t}^*$. Otherwise, it first computes

$$\tilde{\mathbf{c}}_0 := \mathbf{c}_0 - \mathbf{S}_0^\mathsf{T}\mathbf{c} = \mathbf{GH}_t\mathbf{s} + (\mathbf{S}_0' - \mathbf{S}_0)^\mathsf{T}\mathbf{e}_1 + (\mathbf{E}_0 - \mathbf{E}_0')^\mathsf{T}\mathbf{s}.$$

Then, it reconstruct $\mathbf{b} = \mathbf{H_t s}$ from the error $(\mathbf{S}_0' - \mathbf{S}_0)^\top \mathbf{e}_1 + (\mathbf{E}_0 - \mathbf{E}_0')^\top \mathbf{s}$ by using the error correction property of $\mathbf{G}$, and compute $\mathbf{s} = \mathbf{H_t}^{-1}\mathbf{b}$. If

$$|\mathbf{c} - \mathbf{As}| \leq 2\mu n \wedge |\mathbf{c}_0 - (\mathbf{GH_t} + \mathbf{B}_0)\mathbf{s}| \leq \gamma q \wedge |\mathbf{c}_1 - (\mathbf{GH_t} + \mathbf{B}_1)\mathbf{s}| \leq \gamma q$$

is true, reconstruct M from $\mathbf{c}_2 - \mathbf{Cs} = \mathbf{G}_2\mathbf{m} + \mathbf{e}_2$ by using the error correction property of $\mathbf{G}_2$, else let $\mathbf{m} = \bot$. Finally, return the decrypted result $\mathbf{m}$ to the adversary $\mathcal{A}$.

Challenge. After receiving two equal length plaintexts $\mathbf{m}_0, \mathbf{m}_1 \in \mathcal{M}$ from the adversary $\mathcal{A}$, the challenger $\mathcal{C}$ first randomly chooses a bit $b^* \xleftarrow{\$} \{0,1\}$, and

$$\mathbf{s} \xleftarrow{\$} \widetilde{\mathcal{B}}_{\mu_1}^n, \mathbf{e}_1 \xleftarrow{\$} \mathcal{B}_\mu^n, \mathbf{e}_2 \xleftarrow{\$} \mathcal{B}_\mu^\ell, \mathbf{S}_0', \mathbf{S}_1' \xleftarrow{\$} (\widetilde{\mathcal{B}}_{\mu_1}^n)^q, \mathbf{E}_0', \mathbf{E}_1' \xleftarrow{\$} \mathcal{B}_\mu^{n \times q}$$

Then, it defines

$$\begin{aligned}
\mathbf{c}^* &:= \mathbf{As} + \mathbf{e}_1 & \in \{0,1\}^n \\
\mathbf{c}_0^* &:= (\mathbf{GH}_{t^*} + \mathbf{B}_0)\mathbf{s} + (\mathbf{S}_0')^\top\mathbf{e}_1 - (\mathbf{E}_0')^\top\mathbf{s} \in \{0,1\}^q \\
\mathbf{c}_1^* &:= (\mathbf{GH}_{t^*} + \mathbf{B}_1)\mathbf{s} + (\mathbf{S}_1')^\top\mathbf{e}_1 - (\mathbf{E}_1')^\top\mathbf{s} \in \{0,1\}^q \\
\mathbf{c}_2^* &:= \mathbf{Cs} + \mathbf{e}_2 + \mathbf{G}_2\mathbf{m}_{b^*} & \in \{0,1\}^\ell,
\end{aligned}$$

and returns the challenge ciphertext $(\mathbf{c}^*, \mathbf{c}_0^*, \mathbf{c}_1^*, \mathbf{c}_2^*)$ to the adversary $\mathcal{A}$.

Phase 2. The adversary can adaptively make more decryption queries, and the challenger $\mathcal{C}$ responds as in Phase 1.

Guess. Finally, $\mathcal{A}$ outputs a guess $b \in \{0,1\}$. If $b = b^*$, the challenger $\mathcal{C}$ outputs 1, else outputs 0.

EVENT. Let F_i be the event that $\mathcal{C}$ outputs 1 in Game i for $i \in \{0, 1, \ldots, 6\}$.

Lemma 10. $|\Pr[F_0] - \frac{1}{2}| = \epsilon$.

Proof. This lemma immediately follows the fact that $\mathcal{C}$ honestly simulates the attack environment for $\mathcal{A}$, and only outputs 1 if and only if $b = b^*$.

Game 1. This game is identical to Game 0 except that the challenger $\mathcal{C}$ changes the key generation phase as follows:

KeyGen. First uniformly choose matrices $\mathbf{A} \xleftarrow{\$} \mathcal{D}_\lambda^{n \times n}, \mathbf{C} \xleftarrow{\$} \mathcal{D}_\lambda^{\ell \times n}, \mathbf{S}_0, \mathbf{S}_1 \xleftarrow{\$} (\widetilde{\mathcal{B}}_{\mu_1}^n)^q, \mathbf{E}_0, \mathbf{E}_1 \xleftarrow{\$} \mathcal{B}_\mu^{n \times q}$, and $\mathbf{B}_1' \xleftarrow{\$} \{0,1\}^{q \times n}$. Then, compute $\mathbf{B}_0 = \mathbf{S}_0^\top\mathbf{A} + \mathbf{E}_0^\top, \mathbf{B}_1 = \mathbf{S}_1^\top\mathbf{A} + \mathbf{E}_1^\top \in \{0,1\}^{q \times n}$. Finally, $\mathcal{C}$ sends $pk = (\mathbf{A}, \mathbf{B}_0, \mathbf{B}_1', \mathbf{C})$ to the adversary $\mathcal{A}$, and keeps $sk = (\mathbf{S}_0, \mathbf{S}_1)$ to itself.

Lemma 11. *If the decisional $\mathsf{LPN}_{\mu,n}$ problem is $2^{\omega(n^{\frac{1}{2}})}$-hard, then we have $|\Pr[F_1] - \Pr[F_0]| \leq \mathsf{negl}(n)$.*

Proof. Since the only difference between Game 0 and Game 1 is that $\mathcal{C}$ replaces $\mathbf{B}_1 = \mathbf{S}_1^\mathsf{T}\mathbf{A} + \mathbf{E}_1^\mathsf{T} \in \{0,1\}^{q\times n}$ in Game 0 with a randomly chosen $\mathbf{B}_1' \xleftarrow{\$} \{0,1\}^{q\times n}$ in Game 1. we have that Game 0 and Game 1 are computationally indistinguishable for any PPT adversary $\mathcal{A}$ by our assumption and Corollary 1. This means that $|\Pr[F_1] - \Pr[F_0]| \leq \mathsf{negl}(n)$ holds.

Game 2. This game is identical to Game 1 except that the challenger $\mathcal{C}$ changes the key generation phase as follows:

KeyGen. First uniformly choose matrices $\mathbf{A} \xleftarrow{\$} \mathcal{D}_\lambda^{n\times n}, \mathbf{C} \xleftarrow{\$} \mathcal{D}_\lambda^{\ell\times n}, \mathbf{S}_0, \mathbf{S}_1 \xleftarrow{\$}$ $(\widetilde{\mathcal{B}}_{\mu_1}^n)^q, \mathbf{E}_0, \mathbf{E}_1 \xleftarrow{\$} \mathcal{B}_\mu^{n\times q}$, and $\mathbf{B}_1'' \xleftarrow{\$} \{0,1\}^{q\times n}$. Then, compute $\mathbf{B}_0 = \mathbf{S}_0^\mathsf{T}\mathbf{A} + \mathbf{E}_0^\mathsf{T}, \mathbf{B}_1 = \mathbf{S}_1^\mathsf{T}\mathbf{A} + \mathbf{E}_1^\mathsf{T} \in \{0,1\}^{q\times n}$ and $\mathbf{B}_1' = \mathbf{B}_1'' - \mathbf{G}\mathbf{H}_{t^*}$. Finally, $\mathcal{C}$ sends $pk = (\mathbf{A}, \mathbf{B}_0, \mathbf{B}_1', \mathbf{C})$ to the adversary $\mathcal{A}$, and keeps $sk = (\mathbf{S}_0, \mathbf{S}_1)$ to itself.
Challenge. After receiving two equal length plaintexts $\mathbf{m}_0, \mathbf{m}_1 \in \mathcal{M}$ from the adversary $\mathcal{A}$, the challenger $\mathcal{C}$ first randomly chooses a bit $b^* \xleftarrow{\$} \{0,1\}$, and

$$\mathbf{s} \xleftarrow{\$} \widetilde{\mathcal{B}}_{\mu_1}^n, \mathbf{e}_1 \xleftarrow{\$} \mathcal{B}_\mu^n, \mathbf{e}_2 \xleftarrow{\$} \mathcal{B}_\mu^\ell, \mathbf{S}_0', \mathbf{S}_1' \xleftarrow{\$} (\widetilde{\mathcal{B}}_{\mu_1}^n)^q, \mathbf{E}_0', \mathbf{E}_1' \xleftarrow{\$} \mathcal{B}_\mu^{n\times q}$$

Then, it defines

$$\begin{aligned}
\mathbf{c}^* &:= \mathbf{A}\mathbf{s} + \mathbf{e}_1 & \in \{0,1\}^n \\
\mathbf{c}_0^* &:= (\mathbf{G}\mathbf{H}_{t^*} + \mathbf{B}_0)\mathbf{s} + (\mathbf{S}_0')^\mathsf{T}\mathbf{e}_1 - (\mathbf{E}_0')^\mathsf{T}\mathbf{s} \in \{0,1\}^q \\
\mathbf{c}_1^* &:= (\mathbf{G}\mathbf{H}_{t^*} + \mathbf{B}_1)\mathbf{s} + (\mathbf{S}_1')^\mathsf{T}\mathbf{e}_1 - (\mathbf{E}_1')^\mathsf{T}\mathbf{s} \in \{0,1\}^q \\
\mathbf{c}_2^* &:= \mathbf{C}\mathbf{s} + \mathbf{e}_2 + \mathbf{G}_2\mathbf{m}_{b^*} & \in \{0,1\}^\ell,
\end{aligned}$$

and returns the challenge ciphertext $(\mathbf{c}^*, \mathbf{c}_0^*, \mathbf{c}_1^*, \mathbf{c}_2^*)$ to the adversary $\mathcal{A}$.

Lemma 12. $\Pr[F_2] = \Pr[F_1]$.

Proof. Because of $\mathbf{B}_1'' \xleftarrow{\$} \{0,1\}^{q\times n}$, we have that $\mathbf{B}_1' = \mathbf{B}_1'' - \mathbf{G}\mathbf{H}_{t^*}$ is also uniformly distributed over $\{0,1\}^{q\times n}$. This means that the public key in Game 2 has the same distribution as that in Game 1. In addition, since $\mathbf{S}_1 \xleftarrow{\$} (\widetilde{\mathcal{B}}_{\mu_1}^n)^q$ and $\mathbf{E}_1 \xleftarrow{\$} \mathcal{B}_\mu^{n\times q}$ are chosen from the same distribution as $\mathbf{S}_1'$ and $\mathbf{E}_1'$ respectively. By the fact that $\mathbf{B}_1 = \mathbf{S}_1^\mathsf{T}\mathbf{A} + \mathbf{E}_1^\mathsf{T} \in \{0,1\}^{q\times n}$ is not included in the public key $pk = (\mathbf{A}, \mathbf{B}_0, \mathbf{B}_1', \mathbf{C})$ (and thus $\mathcal{A}$ has no information about $\mathbf{S}_1$ and $\mathbf{E}_1$ before the challenge phase), we have that the challenge ciphertext in Game 2 also has the same distribution as that in Game 1. In all, Game 2 is identical to Game 1 in the adversary's view. Thus, we have $\Pr[F_2] = \Pr[F_1]$.

Game 3. This game is identical to Game 2 except that the challenger $\mathcal{C}$ changes the key generation phase as follows:

KeyGen. First uniformly choose matrices $\mathbf{A} \xleftarrow{\$} \mathcal{D}_\lambda^{n\times n}, \mathbf{C} \xleftarrow{\$} \mathcal{D}_\lambda^{\ell\times n}, \mathbf{S}_0, \mathbf{S}_1 \xleftarrow{\$}$ $(\widetilde{\mathcal{B}}_{\mu_1}^n)^q$, and $\mathbf{E}_0, \mathbf{E}_1 \xleftarrow{\$} \mathcal{B}_\mu^{n\times q}$. Then, compute $\mathbf{B}_0 = \mathbf{S}_0^\mathsf{T}\mathbf{A} + \mathbf{E}_0^\mathsf{T}, \mathbf{B}_1 = \mathbf{S}_1^\mathsf{T}\mathbf{A} + \mathbf{E}_1^\mathsf{T} \in \{0,1\}^{q\times n}$ and $\mathbf{B}_1' = \mathbf{B}_1 - \mathbf{G}\mathbf{H}_{t^*}$. Finally, $\mathcal{C}$ sends $pk = (\mathbf{A}, \mathbf{B}_0, \mathbf{B}_1', \mathbf{C})$ to the adversary $\mathcal{A}$, and keeps $sk = (\mathbf{S}_0, \mathbf{S}_1)$ to itself.

Lemma 13. *If the decisional* $\mathsf{LPN}_{\mu,n}$ *problem is* $2^{\omega(n^{\frac{1}{2}})}$*-hard, then* $|\Pr[F_3] - \Pr[F_2]| \le \mathsf{negl}(n)$.

Proof. Since the only difference between Game 2 and Game 3 is that $\mathcal{C}$ replaces the randomly chosen $\mathbf{B}_1'' \xleftarrow{\$} \{0,1\}^{q \times n}$ in Game 2 with $\mathbf{B}_1 = \mathbf{S}_1^\top \mathbf{A} + \mathbf{E}_1^\top \in \{0,1\}^{q \times n}$ in Game 3, by our assumption and Corollary 1 we have that Game 2 and Game 3 are computationally indistinguishable for any PPT adversary $\mathcal{A}$ seeing $(\mathbf{S}_1^\top \mathbf{e}_1, \mathbf{E}_1^\top \mathbf{s})$ in the challenge ciphertext. This means that $|\Pr[F_3] - \Pr[F_2]| \le \mathsf{negl}(n)$ holds.

Remark 3. Note that for the challenge ciphertext $(\mathbf{c}, \mathbf{c}_0^*, \mathbf{c}_1^*, \mathbf{c}_2^*)$ in Game 3, we have that $\mathbf{c}_1^* := (\mathbf{GH}_{t_1^*} + \mathbf{B}_1')\mathbf{s} + \mathbf{S}_1^\top \mathbf{e}_1 - \mathbf{E}_1^\top \mathbf{s} = \mathbf{S}_1^\top \mathbf{c}$.

Game 4. This game is identical to Game 3 except that the challenger $\mathcal{C}$ answers the decryption queries by using $\mathbf{S}_1$ instead of $\mathbf{S}_0$.

Lemma 14. $|\Pr[F_4] - \Pr[F_3]| \le \mathsf{negl}(n)$.

Proof. This lemma directly follows from the fact that both $\mathbf{S}_0$ and $\mathbf{S}_1$ have equivalent decryption ability except with negligible probability.

Game 5. This game is identical to Game 4 except that the challenger $\mathcal{C}$ changes the key generation phase and the challenge phase as follows:

KeyGen. First uniformly choose matrices $\mathbf{A} \xleftarrow{\$} \mathcal{D}_\lambda^{n \times n}, \mathbf{C} \xleftarrow{\$} \mathcal{D}_\lambda^{\ell \times n}, \mathbf{S}_0, \mathbf{S}_1 \xleftarrow{\$} (\tilde{\mathcal{B}}_{\mu_1}^n)^q$, and $\mathbf{E}_0, \mathbf{E}_1 \xleftarrow{\$} \mathcal{B}_\mu^{n \times q}$. Then, compute $\mathbf{B}_0 = \mathbf{S}_0^\top \mathbf{A} + \mathbf{E}_0^\top, \mathbf{B}_1 = \mathbf{S}_1^\top \mathbf{A} + \mathbf{E}_1^\top \in \{0,1\}^{q \times n}$, $\mathbf{B}_0' = \mathbf{B}_0 - \mathbf{GH}_{t^*}$ and $\mathbf{B}_1' = \mathbf{B}_1 - \mathbf{GH}_{t^*}$. Finally, $\mathcal{C}$ sends $pk = (\mathbf{A}, \mathbf{B}_0', \mathbf{B}_1', \mathbf{C})$ to the adversary $\mathcal{A}$, and keeps $sk = (\mathbf{S}_0, \mathbf{S}_1)$ to itself.
Challenge. After receiving two equal length plaintexts $\mathbf{m}_0, \mathbf{m}_1 \in \mathcal{M}$ from the adversary $\mathcal{A}$, the challenger $\mathcal{C}$ first randomly chooses a bit $b^* \xleftarrow{\$} \{0,1\}$, and $\mathbf{s} \xleftarrow{\$} \tilde{\mathcal{B}}_{\mu_1}^n, \mathbf{e}_1 \xleftarrow{\$} \mathcal{B}_\mu^n$ and $\mathbf{e}_2 \xleftarrow{\$} \mathcal{B}_\mu^\ell$. Then, it defines

$$
\begin{aligned}
\mathbf{c}^* &:= \mathbf{As} + \mathbf{e}_1 && \in \{0,1\}^n \\
\mathbf{c}_0^* &:= (\mathbf{GH}_{t^*} + \mathbf{B}_0')\mathbf{s} + \mathbf{S}_0^\top \mathbf{e}_1 - \mathbf{E}_0^\top \mathbf{s} = \mathbf{S}_0^\top \mathbf{c}^* && \in \{0,1\}^q \\
\mathbf{c}_1^* &:= (\mathbf{GH}_{t^*} + \mathbf{B}_1')\mathbf{s} + \mathbf{S}_1^\top \mathbf{e}_1 - \mathbf{E}_1^\top \mathbf{s} = \mathbf{S}_1^\top \mathbf{c}^* && \in \{0,1\}^q \\
\mathbf{c}_2^* &:= \mathbf{Cs} + \mathbf{e}_2 + \mathbf{G}_2 \mathbf{m}_{b^*} && \in \{0,1\}^\ell,
\end{aligned}
$$

and returns the challenge ciphertext $(\mathbf{c}, \mathbf{c}_0^*, \mathbf{c}_1^*, \mathbf{c}_2^*)$ to the adversary $\mathcal{A}$.

Lemma 15. *If the decisional* $\mathsf{LPN}_{\mu,n}$ *problem is* $2^{\omega(n^{\frac{1}{2}})}$*-hard, then we have that* $|\Pr[F_5] - \Pr[F_4]| \le \mathsf{negl}(n)$.

Proof. One can easily show this lemma holds by using similar proofs from Lemma 10 to Lemma 14. We omit the details.

Game 6. This game is identical to Game 5 except that the challenger $\mathcal{C}$ changes the challenge phase as follows:

Challenge. After receiving two equal length plaintexts $\mathbf{m}_0, \mathbf{m}_1 \in \mathcal{M}$ from the adversary $\mathcal{A}$, the challenger $\mathcal{C}$ first randomly chooses $b^* \xleftarrow{\$} \{0,1\}, \mathbf{u} \xleftarrow{\$} \{0,1\}^n$ and $\mathbf{v} \xleftarrow{\$} \{0,1\}^\ell$. Then, it defines

$$
\begin{aligned}
\mathbf{c}^* &:= \mathbf{u} & \in \{0,1\}^n \\
\mathbf{c}_0^* &:= \mathbf{S}_0 \mathbf{c}^* & \in \{0,1\}^q \\
\mathbf{c}_1^* &:= \mathbf{S}_1 \mathbf{c}^* & \in \{0,1\}^q \\
\mathbf{c}_2^* &:= \mathbf{v} + \mathbf{G}_2 \mathbf{m}_{b^*} & \in \{0,1\}^\ell,
\end{aligned}
$$

and returns the challenge ciphertext $(\mathbf{c}, \mathbf{c}_0^*, \mathbf{c}_1^*, \mathbf{c}_2^*)$ to the adversary $\mathcal{A}$.

Lemma 16. *If the decisional* $\mathsf{LPN}_{\mu,n}$ *problem is* $2^{\omega(n^{\frac{1}{2}})}$*-hard, then we have that* $|\Pr[F_6] - \Pr[F_5]| \leq \mathsf{negl}(n)$.

Proof. Since the only difference between Game 5 and Game 6 is that $\mathcal{C}$ replaces $\mathbf{c}^* = \mathbf{As} + \mathbf{e}_1$ and $\mathbf{c}_2^* = \mathbf{Cs} + \mathbf{e}_2 + \mathbf{G}_2 \mathbf{m}_{b^*}$ in Game 5 with $\mathbf{c}^* := \mathbf{u}$ and $\mathbf{c}_2^* := \mathbf{v} + \mathbf{G}_2 \mathbf{m}_{b^*}$ in Game 6, where $\mathbf{u} \xleftarrow{\$} \{0,1\}^n$ and $\mathbf{v} \xleftarrow{\$} \{0,1\}^\ell$, by our assumption and Corollary 1 we have that Game 5 and Game 6 are computationally indistinguishable for any PPT adversary $\mathcal{A}$. Obviously, we have that $|\Pr[F_6] - \Pr[F_5]| \leq \mathsf{negl}(n)$ holds.

Lemma 17. $\Pr[F_6] = \frac{1}{2}$.

Proof. This claim follows from the fact that the challenge ciphertext $(\mathbf{c}, \mathbf{c}_0^*, \mathbf{c}_1^*, \mathbf{c}_2^*)$ in Game 6 perfectly hides the information of $\mathbf{m}_{b^*}$.

In all, by Lemma 10 $\sim$ Lemma 17, we have that $\epsilon = |\Pr[F_0] = \frac{1}{2}| \leq \mathsf{negl}(n)$. This completes the proof of Theorem 5.

Acknowledgments. Yu Yu was supported by the National Basic Research Program of China Grant No. 2013CB338004, the National Natural Science Foundation of China Grant (Nos. 61472249, 61572192, 61572149 and U1536103), Shanghai excellent academic leader funds (No. 16XD1400200) and International Science & Technology Cooperation & Exchange Projects of Shaanxi Province (2016KW-038). Jiang Zhang is supported by the National Basic Research Program of China under Grant No. 2013CB338003 and the National Natural Science Foundation of China under Grant Nos. U1536205, 61472250 and 61402286.

A   Definitions and Security Notions

A.1   Symmetric-Key Encryption Schemes with Auxiliary Input

Definition 6 (Symmetric-Key Encryption Schemes). *A symmetric-key encryption scheme* Π *is a tuple* (KeyGen,Enc,Dec) *with message space* $\mathcal{M}$, *such that*

- KeyGen(1^n) *is a PPT algorithm that takes a security-parameter* 1^n *and outputs a symmetric key* k.

- $\mathsf{Enc}_k(m)$ is a PPT algorithm that encrypts a message $m \in \mathcal{M}$ under key k and outputs a ciphertext c.
- $\mathsf{Dec}_k(c)$ is a deterministic polynomial-time algorithm that decrypts a ciphertext c using key k and outputs a plaintext m.

Definition 7 (Correctness). *We say that a symmetric-key encryption scheme $\Pi = ($KeyGen, Enc, Dec$)$ is correct, if it holds for every plaintext $m \in \mathcal{M}$ that*

$$\Pr_{k \leftarrow \mathsf{KeyGen}(1^n)} [\ \mathsf{Dec}_k(\mathsf{Enc}_k(m)) \neq m\ ] = \mathsf{negl}(n).$$

Definition 8 (IND-CPA/IND-CCA SKE w.r.t. Auxiliary Input). *For $X \in \{CPA, CCA\}$, a symmetric-key encryption scheme $\Pi = ($KeyGen,Enc,Dec$)$ is IND-X secure w.r.t. sub-exponentially hard-to-invert auxiliary input if there exists a constant $0 < \alpha < 1$ such that for any PPT adversary $\mathcal{A}$, any $2^{-\Omega(n^\alpha)}$-hard-to-invert function f*

$$\Pr[\ \mathsf{SKE}^X_{\Pi,f,\mathcal{A}}(1^n, \alpha) = 1\ ] \leq \frac{1}{2} + \mathsf{negl}(n),$$

where $\mathsf{SKE}^{cpa}_{\Pi,f,\mathcal{A}}(1^n,\alpha)$ is the IND-CPA indistinguishability experiment defined as below:

1. *On $k \leftarrow \mathsf{KeyGen}(1^n)$, the adversary takes as input 1^n, $f(k)$, and is given oracle access to Enc_k. Then, he outputs a pair of messages m_0 and m_1 of the same length.*
2. *A random bit $b \overset{\$}{\leftarrow} \{0,1\}$ is sampled, and then a challenge ciphertext $c \leftarrow \mathsf{Enc}_k(m_b)$ is computed and given to $\mathcal{A}$.*
3. *$\mathcal{A}$ continues to have oracle access to Enc_k and finally outputs $b' \in \{0,1\}$.*
4. *The experiment outputs 1 if $b' = b$, and 0 otherwise.*

and $\mathsf{SKE}^{cca}_{\Pi,f,\mathcal{A}}(1^n,\alpha)$ is the IND-CCA indistinguishability experiment defined as below:

1. *On $k \leftarrow \mathsf{KeyGen}(1^n)$, the adversary takes as input 1^n, $f(k)$, and is given oracle access to Enc_k and Dec_k. Then, he outputs a pair of messages m_0 and m_1 of the same length.*
2. *A random bit $b \overset{\$}{\leftarrow} \{0,1\}$ is sampled, and then a challenge ciphertext $c \leftarrow \mathsf{Enc}_k(m_b)$ is computed and given to $\mathcal{A}$.*
3. *$\mathcal{A}$ continues to have oracle access to Enc_k and Dec_k (with the exception that decryption for challenge ciphertext is not allowed) and finally outputs $b' \in \{0,1\}$.*
4. *The experiment outputs 1 if $b' = b$, and 0 otherwise.*

A.2   Public-Key Encryption Schemes

Definition 9 (Public-key Encryption Schemes). *A public key encryption scheme Π is a tuple $($KeyGen,Enc,Dec$)$ with message space $\mathcal{M}$, such that*

- KeyGen(1^n) *is a PPT algorithm that takes a security-parameter 1^n and outputs a pair of public and private keys (pk,sk).*
- Enc$_{pk}$(m) *is a PPT algorithm that encrypts message $m \in \mathcal{M}$ under public key pk and outputs a ciphertext c.*
- Dec$_{sk}$(c) *is a deterministic polynomial-time algorithm that decrypts a ciphertext c using secret key sk and outputs a plaintext m (or $\perp$).*

Definition 10 (Correctness). *We say that a public-key encryption scheme $\Pi = $ (KeyGen,Enc,Dec) is correct, if it holds for every plaintext $m \in \mathcal{M}$ that*

$$\Pr_{(pk,sk) \leftarrow \text{KeyGen}(1^n)} [\text{ Dec}_{sk}(\text{Enc}_{pk}(m)) \neq m] = \text{negl}(n).$$

Definition 11 (IND-CPA/IND-CCA PKE). *For $X \in \{CPA, CCA\}$, a public-key encryption scheme $\Pi = $ (KeyGen,Enc,Dec) is IND-X secure if for any PPT adversary $\mathcal{A}$*

$$\Pr[\text{ PKE}_{\Pi,\mathcal{A}}^{X}(1^n) = 1] \leq \frac{1}{2} + \text{negl}(n),$$

where PKE$_{\Pi,\mathcal{A}}^{cpa}(1^n)$ is the IND-CPA indistinguishability experiment defined as below:

1. *On $(pk, sk) \leftarrow$ KeyGen(1^n), the adversary takes as input 1^n and pk. Then, he outputs a pair of messages m_0 and m_1 of the same length.*
2. *A random bit $b \xleftarrow{\$} \{0,1\}$ is sampled, and then a challenge ciphertext $c \leftarrow$ Enc$_{pk}(m_b)$ is computed and given to $\mathcal{A}$.*
3. *$\mathcal{A}$ continues his computation and finally outputs $b' \in \{0,1\}$.*
4. *The experiment outputs 1 if $b' = b$, and 0 otherwise.*

and PKE$_{\Pi,\mathcal{A}}^{cca}(1^n)$ is the IND-CCA indistinguishability experiment defined as below:

1. *On $(pk, sk) \leftarrow$ KeyGen(1^n), the adversary takes as input 1^n and pk, and is given oracle access to Dec$_{sk}$. Then, he outputs a pair of messages m_0 and m_1 of the same length.*
2. *A random bit $b \xleftarrow{\$} \{0,1\}$ is sampled, and then a challenge ciphertext $c \leftarrow$ Enc$_{pk}(m_b)$ is computed and given to $\mathcal{A}$.*
3. *$\mathcal{A}$ continues to have oracle access to Dec$_{sk}$ (with the exception that decryption for challenge ciphertext is not allowed) and finally outputs $b' \in \{0,1\}$.*
4. *The experiment outputs 1 if $b' = b$, and 0 otherwise.*

B   Facts, Lemmas, Inequalities and Proofs Omitted

Lemma 18 (Sample-Preserving Reduction). *For the same assumptions and notations as in the proof of Theorem 2, we have*

$$(f(\mathbf{x}, \mathbf{e}; Z), Z, \mathbf{A}, \ \mathbf{A} \cdot \mathbf{y} + \mathbf{e}, \mathbf{r}^{\mathsf{T}}, \mathbf{r}^{\mathsf{T}} \cdot \mathbf{y}) \overset{c}{\sim} (f(\mathbf{x}, \mathbf{e}; Z), Z, \mathbf{A}, \ \mathbf{A} \cdot \mathbf{y} + \mathbf{e}, \mathbf{r}^{\mathsf{T}}, U_1)$$
$$\Rightarrow (f(\mathbf{x}, \mathbf{e}; Z), Z, \mathbf{A}, \ \mathbf{A} \cdot \mathbf{y} + \mathbf{e}) \overset{c}{\sim} (f(\mathbf{x}, \mathbf{e}; Z), Z, \mathbf{A}, U_{q'}).$$

Proof. Assume for contradiction that there exists a polynomial $p(\cdot)$ and a PPT distinguisher $\mathcal{D}$ such that

$$\Pr[\mathcal{D}(f(\mathbf{x}, \mathbf{e}; Z), Z, \mathbf{A}, \mathbf{A}\cdot\mathbf{y}+\mathbf{e}) = 0] - \Pr[\mathcal{D}(f(\mathbf{x}, \mathbf{e}; Z), Z, \mathbf{A}, U_{q'}) = 0] \geq 1/p(n)$$

for infinitely many n's and we recall that $\mathbf{y}, \mathbf{r} \sim U_\lambda$. Given input $(z_1, z, \mathbf{A}, \mathbf{A}\cdot\mathbf{y} + \mathbf{e}, \mathbf{r}^\mathsf{T})$, we use an efficient $\mathcal{D}'$ (which invokes $\mathcal{D}$) to predict the Goldreich-Levin hardcore bit $\mathbf{r}^\mathsf{T}\cdot\mathbf{y}$ with non-negligible probability (and thus a contradiction to the assumption). $\mathcal{D}'$ chooses a random $\mathbf{u} \overset{\$}{\leftarrow} \{0,1\}^{q'}$, computes a new $q' \times n$ Boolean matrix $\tilde{\mathbf{A}} = \mathbf{A} - \mathbf{u} \cdot \mathbf{r}^\mathsf{T}$, applies $\mathcal{D}$ on $(z_1, z, \tilde{\mathbf{A}}, \mathbf{A}\mathbf{y} + \mathbf{e})$ and outputs his answer. Note that $\tilde{\mathbf{A}} \sim U_{q' \times n}$ and $\mathbf{A}\mathbf{y} + \mathbf{e} = \tilde{\mathbf{A}}\mathbf{y} + \mathbf{e} + \mathbf{u} \cdot \mathbf{r}^\mathsf{T}\mathbf{y}$. Therefore, when $\mathbf{r}^\mathsf{T}\mathbf{y} = 0$ we have $(z_1, z, \tilde{\mathbf{A}}, \mathbf{A}\mathbf{y} + \mathbf{e})$ follows $(f(\mathbf{x}, \mathbf{e}; Z), Z, \tilde{\mathbf{A}}, \tilde{\mathbf{A}}\cdot\mathbf{y}+\mathbf{e})$ and for $\mathbf{r}^\mathsf{T}\mathbf{y} = 1$ it is distributed according to $(f(\mathbf{x}, \mathbf{e}; Z), Z, \tilde{\mathbf{A}}, U_{q'})$.

$$\begin{aligned}
&\Pr[\mathcal{D}'(f(\mathbf{x}, \mathbf{e}; Z), Z, \tilde{\mathbf{A}}, \tilde{\mathbf{A}}\mathbf{y} + \mathbf{e}, \mathbf{r}^\mathsf{T}) = \mathbf{r}^\mathsf{T}\cdot\mathbf{y}] \\
&= \Pr[\mathbf{r}^\mathsf{T}\cdot\mathbf{y} = 0] \cdot \Pr[\mathcal{D}'(f(\mathbf{x}, \mathbf{e}; Z), Z, \tilde{\mathbf{A}}, \tilde{\mathbf{A}}\mathbf{y} + \mathbf{e}, \mathbf{r}^\mathsf{T}) = 0 \mid \mathbf{r}^\mathsf{T}\cdot\mathbf{y} = 0] \\
&\quad + \Pr[\mathbf{r}^\mathsf{T}\cdot\mathbf{y} = 1] \cdot \Pr[\mathcal{D}'(f(\mathbf{x}, \mathbf{e}; Z), Z, \tilde{\mathbf{A}}, \tilde{\mathbf{A}}\mathbf{y} + \mathbf{e}, \mathbf{r}^\mathsf{T}) = 1 \mid \mathbf{r}^\mathsf{T}\cdot\mathbf{y} = 1] \\
&= \frac{1}{2}\Big(\Pr[\mathcal{D}(f(\mathbf{x}, \mathbf{e}; Z), Z, \mathbf{A}, \mathbf{A}\cdot\mathbf{y}+\mathbf{e}) = 0] \\
&\quad + 1 - \Pr[\mathcal{D}(f(\mathbf{x}, \mathbf{e}; Z), Z, \mathbf{A}, U_{q'}) = 0]\Big) \\
&\geq \frac{1}{2} + \frac{1}{2p(n)},
\end{aligned}$$

which completes the proof.

Proof of Lemma 9 Consider $|(\mathbf{S}_0' - \mathbf{S}_0)^\mathsf{T}\mathbf{e} + (\mathbf{E}_0 - \mathbf{E}_0')^\mathsf{T}\mathbf{s}|$ conditioned on any $|\mathbf{e}| \leq 1.01\mu n$ (except for a $2^{-\Omega(n)}$-fraction) and $|\mathbf{s}| \leq 2\mu n$. We have by Lemma 7 and Lemma 8 that $\mathbf{S}_0^\mathsf{T}\mathbf{e}$, $\mathbf{S}_0'^\mathsf{T}\mathbf{e}$ are i.i.d. to $\mathcal{B}_{\delta_1}^q$, and $\mathbf{E}_0^\mathsf{T}\mathbf{s}$, $\mathbf{E}_0'^\mathsf{T}\mathbf{s}$ are i.i.d. to $\mathcal{B}_{\delta_2}^q$, where $\delta_1 \leq 1/2 - n^{-\alpha/2}$ and $\delta_2 \leq 1/2 - n^{-\alpha}/2$. Thus, $((\mathbf{S}_0' - \mathbf{S}_0)^\mathsf{T}\mathbf{e} + (\mathbf{E}_0 - \mathbf{E}_0')^\mathsf{T}\mathbf{s})$ follows $\mathcal{B}_\delta^q$ for $\delta \leq 1/2 - 2n^{-3\alpha}$ by the Piling-up lemma, and then we complete the proof with Lemma 2.

Lemma 19 (Flattening Shannon Entropy). *For any $n \in \mathbb{N}$, $0 < \mu < 1/2$ and any constant $0 < \Delta < 1$, there exists some random variable $W \in \{0,1\}^n$ such that $\mathbf{H}_\infty(W) \geq (1 - \Delta)n\mathbf{H}(\mu)$ and $\mathsf{SD}(\mathcal{B}_\mu^n, W) \leq 2^{-\Omega(\mu n)}$.*

References

1. Related work on LPN-based authentication schemes. http://www.ecrypt.eu.org/lightweight/index.php/HB
2. Alekhnovich, M.: More on average case vs approximation complexity. In: 44th Annual Symposium on Foundations of Computer Science, pp. 298–307. IEEE, Cambridge, Massachusetts, October 2003
3. Applebaum, B., Cash, D., Peikert, C., Sahai, A.: Fast cryptographic primitives and circular-secure encryption based on hard learning problems. In: Halevi, S. (ed.) CRYPTO 2009. LNCS, vol. 5677, pp. 595–618. Springer, Heidelberg (2009)

4. Applebaum, B., Ishai, Y., Kushilevitz, E.: Cryptography with constant input locality. In: Menezes, A. (ed.) CRYPTO 2007. LNCS, vol. 4622, pp. 92–110. Springer, Heidelberg (2007)

5. Becker, A., Joux, A., May, A., Meurer, A.: Decoding random binary linear codes in $2^n/20$: how $1 + 1 = 0$ improves information set decoding. In: Pointcheval, D., Johansson, T. (eds.) EUROCRYPT 2012. LNCS, vol. 7237, pp. 520–536. Springer, Heidelberg (2012)

6. Berlekamp, E., McEliece, R.J., van Tilborg, H.: On the inherent intractability of certain coding problems. IEEE Trans. Inf. Theor. **24**(3), 384–386 (1978)

7. Bernstein, D.J., Lange, T., Peters, C.: Smaller decoding exponents: ball-collision decoding. In: Rogaway, P. (ed.) CRYPTO 2011. LNCS, vol. 6841, pp. 743–760. Springer, Heidelberg (2011)

8. Blum, A., Furst, M.L., Kearns, M., Lipton, R.J.: Cryptographic primitives based on hard learning problems. In: Stinson, D.R. (ed.) CRYPTO 1993. LNCS, vol. 773, pp. 278–291. Springer, Heidelberg (1994)

9. Blum, A., Kalai, A., Wasserman, H.: Noise-tolerant learning, the parity problem, and the statistical query model. J. ACM **50**(4), 506–519 (2003)

10. Boneh, D., Canetti, R., Halevi, S., Katz, J.: Chosen-ciphertext security from identity-based encryption. SIAM J. Comput. **36**(5), 1301–1328 (2006). http://dx.doi.org/10.1137/S009753970544713X

11. Canetti, R., Halevi, S., Katz, J.: Chosen-ciphertext security from identity-based encryption. In: Cachin, C., Camenisch, J.L. (eds.) EUROCRYPT 2004. LNCS, vol. 3027, pp. 207–222. Springer, Heidelberg (2004)

12. Canteaut, A., Chabaud, F.: A new algorithm for finding minimum-weight words in a linear code: application to mceliece's cryptosystem and to narrow-sense BCH codes of length 511. IEEE Trans. Inf. Theor. **44**(1), 367–378 (1998)

13. Cash, D., Kiltz, E., Tessaro, S.: Two-round man-in-the-middle security from LPN. In: Kushilevitz, E., et al. (eds.) TCC 2016-A. LNCS, vol. 9562, pp. 225–248. Springer, Heidelberg (2016). doi:10.1007/978-3-662-49096-9_10

14. Cramer, R., Damgård, I.: On the amortized complexity of zero-knowledge protocols. In: Halevi, S. (ed.) CRYPTO 2009. LNCS, vol. 5677, pp. 177–191. Springer, Heidelberg (2009)

15. Damgård, I., Park, S.: How practical is public-key encryption based on lpn and ring-lpn? Cryptology ePrint Archive, Report 2012/699. http://eprint.iacr.org/2012/699 (2012)

16. David, B., Dowsley, R., Nascimento, A.C.A.: Universally composable oblivious transfer based on a variant of LPN. In: Gritzalis, D., Kiayias, A., Askoxylakis, I. (eds.) CANS 2014. LNCS, vol. 8813, pp. 143–158. Springer, Heidelberg (2014)

17. Dodis, Y., Kalai, Y.T., Lovett, S.: On cryptography with auxiliary input. In: Mitzenmacher, M. (ed.) STOC. pp. 621–630. ACM (2009)

18. Dodis, Y., Kiltz, E., Pietrzak, K., Wichs, D.: Message authentication, revisited. In: Pointcheval, D., Johansson, T. (eds.) EUROCRYPT 2012. LNCS, vol. 7237, pp. 355–374. Springer, Heidelberg (2012)

19. Döttling, N.: Low noise lpn: Kdm secure public key encryption and sample amplification. In: Katz, J. (ed.) PKC 2015. LNCS, vol. 9020, pp. 604–626. Springer, Heidelberg (2015)

20. Döttling, N., Müller-Quade, J., Nascimento, A.C.A.: IND-CCA secure cryptography based on a variant of the lpn problem. In: Wang, X., Sako, K. (eds.) ASIACRYPT 2012. LNCS, vol. 7658, pp. 485–503. Springer, Heidelberg (2012)

21. Feldman, V., Gopalan, P., Khot, S., Ponnuswami, A.K.: New results for learning noisy parities and halfspaces. In: 47th Symposium on Foundations of Computer Science, pp. 563–574. IEEE, Berkeley, CA, USA, 21–24 October 2006

22. Forney, D.: Concatenated Codes. MIT Press, Cambridge (1966)

23. Gentry, C., Peikert, C., Vaikuntanathan, V.: Trapdoors for hard lattices and new cryptographic constructions. In: Ladner, R.E., Dwork, C. (eds.) Proceedings of the 40th Annual ACM Symposium on Theory of Computing, pp. 197–206. ACM, Victoria, BC, Canada, 17–20 May 2008

24. Gertner, Y., Kannan, S., Malkin, T., Reingold, O., Viswanathan, M.: The relationship between public key encryption and oblivious transfer. In: Proceedings of 41st Annual Symposium on Foundations of Computer Science, 2000, pp. 325–335 (2000)

25. Goldreich, O., Levin, L.A.: A hard-core predicate for all one-way functions. In: Johnson[31] , pp. 25–32

26. Goldwasser, S., Kalai, Y., Peikert, C., Vaikuntanathan, V.: Robustness of the learning with errors assumption. In: Innovations in Theoretical Computer Science, ITCS 2010, pp. 230–240. Tsinghua University Press (2010)

27. Graham, R.L., Knuth, D.E., Patashnik, O.: Concrete Mathematics: A Foundation for Computer Science, 2nd edn. Addison-Wesley Longman Publishing Co., Inc, Boston (1994)

28. Håstad, J., Impagliazzo, R., Levin, L., Luby, M.: Construction of pseudorandom generator from any one-way function. SIAM J. Comput. **28**(4), 1364–1396 (1999)

29. Hopper, N.J., Blum, M.: Secure human identification protocols. In: Boyd, C. (ed.) ASIACRYPT 2001. LNCS, vol. 2248, pp. 52–66. Springer, Heidelberg (2001)

30. Jain, A., Krenn, S., Pietrzak, K., Tentes, A.: Commitments and efficient zero-knowledge proofs from learning parity with noise. In: Wang, X., Sako, K. (eds.) ASIACRYPT 2012. LNCS, vol. 7658, pp. 663–680. Springer, Heidelberg (2012)

31. Johnson, D.S. (ed.): Proceedings of the Twenty First Annual ACM Symposium on Theory of Computing. Seattle, Washington, 15–17 May 1989

32. Juels, A., Weis, S.A.: Authenticating pervasive devices with human protocols. In: Shoup, V. (ed.) CRYPTO 2005. LNCS, vol. 3621, pp. 293–308. Springer, Heidelberg (2005)

33. Justesen, J.: A class of constructive asymptotically good algebraic codes. IEEE Trans. Info. Theor. **18**(5), 652–656 (1972)

34. Katz, J., Shin, J.S.: Parallel and concurrent security of the HB and HB$^+$ protocols. In: Vaudenay, S. (ed.) EUROCRYPT 2006. LNCS, vol. 4004, pp. 73–87. Springer, Heidelberg (2006)

35. Kiltz, E.: Chosen-ciphertext security from tag-based encryption. In: Halevi, S., Rabin, T. (eds.) TCC 2006. LNCS, vol. 3876, pp. 581–600. Springer, Heidelberg (2006)

36. Kiltz, E., Masny, D., Pietrzak, K.: Simple chosen-ciphertext security from low-noise LPN. In: Krawczyk, H. (ed.) PKC 2014. LNCS, vol. 8383, pp. 1–18. Springer, Heidelberg (2014)

37. Kiltz, E., Pietrzak, K., Cash, D., Jain, A., Venturi, D.: Efficient authentication from hard learning problems. In: Paterson, K.G. (ed.) EUROCRYPT 2011. LNCS, vol. 6632, pp. 7–26. Springer, Heidelberg (2011)

38. Kirchner, P.: Improved generalized birthday attack. Cryptology ePrint Archive, Report 2011/377 (2011). http://eprint.iacr.org/2011/377

39. Levieil, É., Fouque, P.-A.: An improved LPN algorithm. In: De Prisco, R., Yung, M. (eds.) SCN 2006. LNCS, vol. 4116, pp. 348–359. Springer, Heidelberg (2006)

40. Lyubashevsky, V.: The parity problem in the presence of noise, decoding random linear codes, and the subset sum problem. In: Chekuri, C., Jansen, K., Rolim, J.D.P., Trevisan, L. (eds.) APPROX 2005 and RANDOM 2005. LNCS, vol. 3624, pp. 378–389. Springer, Heidelberg (2005)

41. Lyubashevsky, V., Masny, D.: Man-in-the-Middle secure authentication schemes from LPN and weak PRFs. In: Canetti, R., Garay, J.A. (eds.) CRYPTO 2013, Part II. LNCS, vol. 8043, pp. 308–325. Springer, Heidelberg (2013)

42. May, A., Meurer, A., Thomae, E.: Decoding Random Linear Codes in $\tilde{\mathcal{O}}(2^{0.054n})$. In: Lee, D.H., Wang, X. (eds.) ASIACRYPT 2011. LNCS, vol. 7073, pp. 107–124. Springer, Heidelberg (2011)

43. Micciancio, D., Mol, P.: Pseudorandom knapsacks and the sample complexity of LWE search-to-decision reductions. In: Rogaway, P. (ed.) CRYPTO 2011. LNCS, vol. 6841, pp. 465–484. Springer, Heidelberg (2011)

44. Micciancio, D., Peikert, C.: Trapdoors for lattices: simpler, tighter, faster, smaller. In: Pointcheval, D., Johansson, T. (eds.) EUROCRYPT 2012. LNCS, vol. 7237, pp. 700–718. Springer, Heidelberg (2012)

45. Pietrzak, K.: Cryptography from learning parity with noise. In: Bieliková, M., Friedrich, G., Gottlob, G., Katzenbeisser, S., Turán, G. (eds.) SOFSEM 2012. LNCS, vol. 7147, pp. 99–114. Springer, Heidelberg (2012)

46. Regev, O.: On lattices, learning with errors, random linear codes, and cryptography. In: Gabow, H.N., Fagin, R. (eds.) STOC, pp. 84–93. ACM (2005)

47. Rosen, A., Segev, G.: Chosen-ciphertext security via correlated products. In: Reingold, O. (ed.) TCC 2009. LNCS, vol. 5444, pp. 419–436. Springer, Heidelberg (2009)

48. Stern, J.: A method for finding codewords of small weight. In: 3rd International Colloquium on Coding Theory and Applications, pp. 106–113 (1988)

49. Yu, Y.: The LPN problem with auxiliary input. (withdrawn) see historical versions at. http://eprint.iacr.org/2009/467

Cryptography in Theory and Practice

The Multi-user Security of Authenticated Encryption: AES-GCM in TLS 1.3

Mihir Bellare[⊠] and Björn Tackmann

Department of Computer Science and Engineering,
University of California San Diego, La Jolla, USA
{mihir,btackmann}@eng.ucsd.edu

Abstract. We initiate the study of multi-user (mu) security of authenticated encryption (AE) schemes as a way to rigorously formulate, and answer, questions about the "randomized nonce" mechanism proposed for the use of the AE scheme GCM in TLS 1.3. We (1) Give definitions of mu ind (indistinguishability) and mu kr (key recovery) security for AE (2) Characterize the intent of nonce randomization as being improved mu security as a defense against mass surveillance (3) Cast the method as a (new) AE scheme RGCM (4) Analyze and compare the mu security of both GCM and RGCM in the model where the underlying block cipher is ideal, showing that the mu security of the latter is indeed superior in many practical contexts to that of the former, and (5) Propose an alternative AE scheme XGCM having the same efficiency as RGCM but better mu security and a more simple and modular design.

1 Introduction

Traditionally, security definitions were single-user, meaning there was a single target key. Consideration of the multi-user setting began with public-key encryption [3]. In this setting, there are many users, each with their own key, and the target is to violate security under *some* key. This is, first, simply more realistic, reflecting real usage, but is now even more relevant from the mass-surveillance perspective. This paper initiates a study of the multi-user security of authenticated encryption. Our motivation comes from TLS 1.3.

AE. The form of authenticated encryption (AE) we consider is nonce-based [28]. The encryption algorithm AE.Enc takes key K, nonce N, message M and header H to deterministically return a ciphertext $C \leftarrow \mathsf{AE.Enc}(K, N, M, H)$. The requirement formalized in [28] is to provide privacy of M, and authenticity of both M and H, as long as a nonce is not re-used. The formalization refers to only one target key, meaning is in the single user (su) setting.

There are many AE schemes (provably) meeting this security requirement. One simple way to obtain them is via generic composition of privacy-only encryption schemes with MACs [5,26]. There are also dedicated schemes such as OCB [22,29,31], CCM [11] and GCM [12,24]. The last, with AES, is used in TLS 1.3.

© International Association for Cryptologic Research 2016
M. Robshaw and J. Katz (Eds.): CRYPTO 2016, Part I, LNCS 9814, pp. 247–276, 2016.
DOI: 10.1007/978-3-662-53018-4_10

MULTI-USER SECURITY OF AE. We formalize multi-user (mu) security of an authenticated-encryption scheme AE. The game picks an adversary-determined number u of independent target keys $K_1, \ldots, K_u$. The adversary gets an encryption oracle that takes an index $i \in [1..u]$, a message, nonce and header, and returns either an encryption of these under K_i or a random string of the same length. It also gets a verification oracle that takes i, a ciphertext, nonce and header, and indicates whether or not decryption is valid. Security is required as long as the adversary does not re-use a nonce *for a particular user*. That is, it is fine to obtain encryptions under the same nonce for different keys, just not under the same key. When $u = 1$, we get a definition equivalent to (but formulated slightly differently from) the (single-user) definition of [28].

Besides this usual goal (which we call indistinguishability), we also formalize a mu key-recovery goal. Again the game picks target keys $K_1, \ldots, K_u$ and gives the adversary an encryption oracle. This time time the latter is always true, meaning it takes an index $i \in [1..u]$, a message, nonce and header, and returns an encryption of these under K_i. The adversary also gets a verification oracle, and, to win, must find one of the target keys. A key-recovery attack is much more damaging than a distinguishing attack, and is the threat of greatest concern to practioners. Key recovery security is usually dismissed by theoreticians as being implied by indistinguishability, but this view misses the fact that the quantitative security of a scheme, in terms of bounds on adversary advantage, can be very different for the two metrics, making it worthwhile to consider key recovery security separately and additionally.

We give our definitions in the ideal-cipher model. (Standard-model definitions follow because this is just the special case where scheme algorithms and adversaries make no queries to the ideal cipher.) For all the schemes we consider, the assumption that the underlying blockcipher is a PRP suffices to prove security. The reason we use the ideal-cipher model is that adversary queries to the ideal cipher give a clear and rigorous way to measure the offline computation being performed in an attack. Also in some cases we get better bounds.

Multi-user security is not qualitatively different from single-user security. A hybrid argument shows that the latter implies the former. But the two could be quantitatively quite different, and this has important practical implications. In the hybrid reduction, there is a loss of a factor u in adversary advantage. Thus, the mu advantage of an adversary could be as much as u times its su advantage. This is the worst case. But it could be a lot less, degrading much more slowly with u. This would be better.

AE IN TLS 1.3. As the protocol underlying `https`, TLS is the basis for secure communication on the Internet, used millions of times a day. The existing versions up to TLS 1.2 have however been subject to many attacks. The effort to create a new and (hopefully) better version, TLS 1.3, is currently underway. TLS (of whatever version) begins with a *handshake*. This is an authenticated key exchange that establishes a shared session key, called the traffic secret, between client and server. This step will not be our concern. After the handshake, data is authenticated and encrypted within the so-called *record layer*, using an authenticated encryption scheme AE that is keyed by a key K derived from the traffic secret. The currently proposed choice of AE is AES-GCM.

The most natural way to use AE in the record layer is directly, meaning the data message M is simply encrypted via $C \leftarrow \mathsf{AE.Enc}(K, N, M, H)$, where N is a nonce (in TLS 1.3 this is a sequence number that is known to the receiver) and H is the header. This is not what TLS 1.3 proposes. Instead, they randomize the nonce, computing $C \leftarrow \mathsf{AE.Enc}(K, N \oplus L, M, H)$, where the randomizer L is also derived from the traffic secret. (It is thus known to the receiver, enabling decryption.) Why do this? Brian Smith gave the following motivation on the TLS 1.3 mailing list [33]:

> ... massively parallel attacks on many keys at once seem like the most promising way to break AES-128. It seems bad to have popular endpoints encrypting the same plaintext block with the same nonce with different keys. That seems like exactly the recipe for making such attacks succeed. It seems like it would be better, instead, to require that the initial nonces to be calculated from the key block established during key agreement ... This ... should prevent any such massively-parallel attack from working.

In this paper, we aim to understand and formalize the threat alluded to here, and then assess to what extent one can prove that nonce-randomization guarantees security. In particular, we suggest that the formal cryptographic goal underlying nonce randomization and Smith's comment is improved multi-user security. In our model, the "massively parallel attack" is a key-search attack that finds the GCM key of some user out of u target users —here we are referring to the basic GCM scheme, in the absence of nonce randomization— in time $2^\kappa / u$ where κ is the key length of the underlying block cipher, $\kappa = 128$ for AES. The attack picks some N, M, H and for each $i \in [1..u]$ obtains from its encryption oracle the encryption C_i of these quantities under K_i. Now, it goes through all possible κ-bit keys L, for each computing $C_L \leftarrow \mathsf{AE.Enc}(L, N, M, H)$, and returning L if $C_L = C_i$ for some i. Note that the attack needs a single computation of AE.Enc for each L, not one per user, which is why the running time is $2^\kappa / u$. Given NSA computing capabilities, the fear of the TLS 1.3 designers is that this attack may be feasible for them for large u, and thus a mass-surveillance threat. Nonce randomization is a candidate way to circumvent the attack. The question this raises is whether nonce randomization works. To answer this in a rigorous way, we abstract out a (new) AE scheme and then use our definitions of mu security.

RGCM. In TLS 1.3, nonce randomization is viewed as a way to use GCM in the record layer. We take a different perspective. We view the method as defining a new AE scheme that we call RGCM. In this scheme, the randomizer is part of the key. This view is appropriate because the randomizer was derived from the traffic secret just like the base key, and has the security necessary to be used as a key, and the randomizer is also static across the session, just like the base key. While GCM has a key whose length is the key length κ of the underlying block cipher ($\kappa = 128$ for AES), RGCM has a key of length $\kappa + \nu$, where ν is the length of the randomizer ($\nu = 96$ for GCM in TLS 1.3). Nonces are assumed to also have length ν so that xoring the nonce with the randomizer makes sense.

RESULTS. With this perspective, we are looking at two AE schemes, GCM and RGCM. We can now divorce ourselves of TLS details and analyze them as AE schemes to determine the quantitative mu security of both. The number p of adversary queries to the ideal cipher is the central parameter, capturing the offline computational effort of the adversary. As before u is the number of users, and we let m denote the total number of bits encrypted, meaning the sum of the lengths of all messages in queries.

Let us first discuss mu security under key recovery, where the picture is clearer. Roughly, we show that key recovery for GCM needs $p = 2^{\kappa}/u$ while for RGCM it needs $p = 2^{\kappa+\nu}/um$. We expect m to be quite a bit less than 2^{ν} —in the current schemes, $\nu = 96$— so the effort to recover a key is significantly higher for RGCM than for GCM. This says that nonce randomization works, meaning it does increase mu security as targeted by the TLS 1.3 designers, at least for key recovery.

For mu-ind security, the picture is more complex. We distinguish the case of passive attacks, where the adversary does not query its verification oracle, and active attacks, where it does. In the passive case, RGCM still emerges as superior, but in the active case, the two schemes become comparable. Also, our bounds in the ind case are complex, and interesting terms get swamped by collision terms. We stress that the bounds here may not be tight, so the picture we are seeing could reflect limitations of our analysis techniques rather than the inherent security of the schemes. Obtaining better (and ideally tight) bounds is an interesting open question.

XGCM. Even if under some metrics superior to GCM, RGCM performs considerably worse than expected from an AE with key length $\kappa + \nu$, and the natural question is, why not use some standard scheme or construction paradigm rather than "roll your own" with RGCM? The most obvious choice is AES256-GCM. Our analysis of GCM shows that AES256-GCM has good enough mu security, simply due to the larger key size. However, AES256-GCM is slower than AES-RGCM, and a scheme using AES itself would be preferable. We suggest and analyze XGCM, derived simply as GCM with the blockcipher $E: \{0,1\}^{\kappa} \times \{0,1\}^{\lambda} \to \{0,1\}^{\lambda}$ replaced by $EX: \{0,1\}^{\kappa+\lambda} \times \{0,1\}^{\lambda} \to \{0,1\}^{\lambda}$, defined by $EX(K\|L, X) = L \oplus E(K, L \oplus X)$. This transform of a blockcipher uses the Even-Mansour technique [13]. It was suggested by Rivest as a key-extension method for DES and first analyzed by Kilian and Rogaway [19]. Our analysis implies that, with AES parameters, the mu security of XGCM is better than that of RGCM. Its performance is however essentially the same as that of GCM or RGCM. While it would be a viable alternative for AES-RGCM in TLS 1.3, it does require non-black-box changes to the implementation of AES-GCM, whereas for AES-RGCM the change is only the randomization of the nonce input.

RELATED WORK. GCM was proposed by McGrew and Viega (MV) [24] and standardized by NIST as [12]. MV [24] prove single-user security assuming PRP-security of the underlying blockcipher. While the original scheme allows variable-length nonces [24], IOM [18] showed that the security proof of MV was flawed in this case and the claimed security bounds did not hold. They provide a corrected proof, which was later improved by NOMI [27]. In this paper we only consider fixed-length nonces. We prove security in the mu setting in the ideal cipher model.

Key-recovery security of symmetric encryption schemes was defined in [30] for the single-user, privacy-only setting. We extend their definition to the mu, authenticated encryption setting.

BMMRT [1] and FGMP [14] analyze the record layer of TLS 1.3 relative to the goal of providing a secure channel, under an appropriate formalization of the latter. These works assume that AES-GCM is a secure AE scheme. Our work is not attempting to analyze the record layer. It is analyzing the security of GCM and RGCM as stand-alone AE schemes, with emphasis on their mu security.

We are seeing increased interest in multi-user security, further reflected in this paper. BCK [4] considered mu security for PRFs as an intermediate step in the analysis of the cascade construction. Multi-user security of PRFs and PRPs (blockciphers) has been further considered in [2,25,34]. The first work that highlighted mu security as a goal and targeted quantitative security improvements seems to have been BBM [3], the primitive here being public-key encryption. Multi-user security for signatures was considered by GMS [16] and has been the subject of renewed interest in [8,20]. Further works involving multi-user security include [9,10,17], and, in the cryptanalytic context, [15].

2   Preliminaries

We let ε denote the empty string. If Z is a string then $|Z|$ denotes its length and $Z[1..i]$ denotes bits 1 through i of Z. If X is a finite set, we let $x \leftarrow_\$ X$ denote picking an element of X uniformly at random and assigning it to x. Algorithms may be randomized unless otherwise indicated. Running time is worst case. If A is an algorithm, we let $y \leftarrow A(x_1, \ldots; r)$ denote running A with random coins r on inputs $x_1, \ldots$ and assigning the output to y. We let $y \leftarrow_\$ A(x_1, \ldots)$ be the result of picking r at random and letting $y \leftarrow A(x_1, \ldots; r)$. We let $[A(x_1, \ldots)]$ denote the set of all possible outputs of A when invoked with inputs $x_1, \ldots$.

We use the code-based game-playing framework of BR [6]. (See Fig. 1 for an example.) By $\Pr[G]$ we denote the probability that the execution of game G results in the game returning true. In games, integer variables, set variables and boolean variables are assumed initialized, respectively, to 0, the empty set, and false.

A family of functions F: F.Keys $\times$ F.Dom $\to$ F.Rng is a two-argument function that takes a key K in the key space F.Keys, an input x in the domain F.Dom and returns an output $F(K, x)$ in the range F.Rng. In the ROM, F takes an oracle RO. We say F has key length F.kl if F.Keys $= \{0,1\}^{F.kl}$; output length F.ol if F.Rng $= \{0,1\}^{F.ol}$; and input length F.il if F.Dom $= \{0,1\}^{F.il}$.

We say that F: $\{0,1\}^{F.kl} \times \{0,1\}^{F.il} \to \{0,1\}^{F.ol}$ is a *block cipher* if F.il = F.ol and $F(K, \cdot): \{0,1\}^{F.il} \to \{0,1\}^{F.ol}$ is a permutation for each K in $\{0,1\}^{F.kl}$. We denote by $F^{-1}(K, \cdot)$ the inverse of $F(K, \cdot)$.

Let $\mathsf{H}\colon \mathsf{H.Keys} \times (\{0,1\}^* \times \{0,1\}^*) \to \{0,1\}^{\mathsf{H.ol}}$ be a family of functions with domain $\mathsf{H.Dom} = \{0,1\}^* \times \{0,1\}^*$. Let $\epsilon\colon \mathbb{N} \times \mathbb{N} \to [0,1]$ be a function. Somewhat extending [21], we say that H is ϵ-*almost XOR-universal* if for all distinct $(M_1, H_1), (M_2, H_2) \in \mathsf{H.Dom}$ and all $s \in \{0,1\}^{\mathsf{H.ol}}$, we have

$$\Pr[\mathsf{H}(hk, (M_1, H_1)) \oplus \mathsf{H}(hk, (M_2, H_2)) = s \; : \; hk \leftarrow\!\!\!{}_\$ \; \mathsf{H.Keys}]$$
$$\leq \epsilon(\max(|M_1|, |M_2|), \max(|H_1|, |H_2|)) \,.$$

3 Multi-user Security of Symmetric Encryption

We consider symmetric encryption in a multi-user setting. We give two definitions of security. The first, an indistinguishability-style definition, extends Rogaway's single-user definition [28] to the multi-user setting, and represents a very strong requirement. We also define security against key recovery, representing the goal the attacker would most like to achieve and the most common target of cryptanalysis. We will see that the security bounds for these notions can differ. Since our analyses will be in the ideal-cipher model, the definitions are given directly in that model.

SYNTAX. A symmetric encryption scheme AE specifies a deterministic encryption algorithm $\mathsf{AE.Enc}\colon \{0,1\}^{\mathsf{AE.kl}} \times \mathsf{AE.NS} \times \{0,1\}^* \times \{0,1\}^* \to \{0,1\}^*$ that takes a key $K \in \{0,1\}^{\mathsf{AE.kl}}$, a nonce $N \in \mathsf{AE.NS}$, a message $M \in \{0,1\}^*$ and a header $H \in \{0,1\}^*$ to return a ciphertext $C \leftarrow \mathsf{AE.Enc}^{\mathsf{E},\mathsf{E}^{-1}}(K,N,M,H)$ $\in \{0,1\}^{\mathsf{AE.cl}(|M|)}$. Here $\mathsf{AE.kl} \in \mathbb{N}$ is the key length of the scheme, $\mathsf{AE.NS}$ is the nonce space and $\mathsf{AE.cl}\colon \mathbb{N} \to \mathbb{N}$ is the ciphertext length function. The oracles represent a cipher $\mathsf{E}\colon \{0,1\}^{\mathsf{AE.ckl}} \times \{0,1\}^{\mathsf{AE.bl}} \to \{0,1\}^{\mathsf{AE.bl}}$ and its inverse E^{-1}. In the security games this cipher will be chosen at random, meaning be ideal. We view the key length $\mathsf{AE.ckl}$ and block length $\mathsf{AE.bl}$ of the cipher as further parameters of AE itself. Also specified is a deterministic decryption algorithm $\mathsf{AE.Dec}\colon \{0,1\}^{\mathsf{AE.kl}} \times \mathsf{AE.NS} \times \{0,1\}^* \times \{0,1\}^* \to \{0,1\}^* \cup \{\perp\}$ that takes K,N,C,H and returns $M \leftarrow \mathsf{AE.Dec}^{\mathsf{E},\mathsf{E}^{-1}}(K,N,C,H) \in \{0,1\}^* \cup \{\perp\}$. Correctness requires that $\mathsf{AE.Dec}(K,N,\mathsf{AE.Enc}(K,N,M,H),H) = M$ for all $M,H \in \{0,1\}^*$, all $N \in \mathsf{AE.NS}$ and all $K \in \{0,1\}^{\mathsf{AE.kl}}$.

INDISTINGUISHABILITY SECURITY. We extend Rogaway's definition of indistinguishability security for authenticated encryption [28], which is in the single-user setting, to the multi-user setting. The formalization is based on game $\mathbf{G}^{\mathrm{mu\text{-}ind}}_{\mathsf{AE}}(A)$ of Fig. 1, associated to encryption scheme AE and adversary A. The game initially samples a random bit challenge b, with $b = 1$ indicating it is in "real" mode and $b = 0$ that it is in "ideal" mode. As per our conventions noted in Sect. 2, the sets U, V are assumed initialized to the empty set, and the integer v is assumed initialized to 0. Now the adversary A has access to an oracle NEW that creates new user instances. A also has access to an encryption oracle ENC that takes a user instance identifier i, a nonce $N \in \mathsf{AE.NS}$, a message M, and a header H. The oracle either returns a uniformly random bit string of length $\mathsf{AE.cl}$ that depends only on the length of M (for $b = 0$), or an encryption under

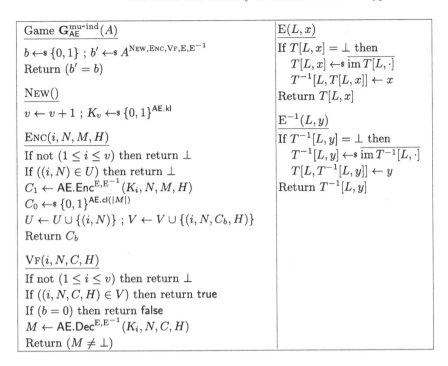

Fig. 1. Game defining multi-user indistinguishability security of symmetric encryption scheme AE in the ideal-cipher model.

AE.Enc using the key of user i (for $b = 1$). The oracle checks that A does not re-use nonces for a user instance, and that it is invoked only for user instances that exist. Analogously, there is a verification oracle VF that takes user instance i, nonce $N \in$ AE.NS, ciphertext C, and header H. Oracle VF always accepts ciphertexts generated by ENC for the same i, N, and H, rejects all other ciphertexts for $b = 0$, and uses the decryption algorithm AE.Dec to check the validity of the ciphertext for $b = 1$. As a last step, the adversary outputs a bit b' that can be viewed as a guess for b. The advantage of adversary A in breaking the mu-ind security of AE is defined as $\text{Adv}_{\text{AE}}^{\text{mu-ind}}(A) = 2\Pr[\mathbf{G}_{\text{AE}}^{\text{mu-ind}}(A)] - 1$.

The ideal-cipher oracles E and E^{-1} are given to the adversary, the encryption algorithm and the decryption algorithm, where the inputs are $L \in \{0,1\}^{\text{AE.ckl}}$ and $x, y \in \{0,1\}^{\text{AE.bl}}$. The oracles are defined using lazy sampling. The description of game $\mathbf{G}_{\text{AE}}^{\text{mu-ind}}$ in Fig. 1 uses some notation that we introduce here and use also elsewhere. First of all, $T[\cdot, \cdot]$ describes a map $\{0,1\}^* \times \{0,1\}^* \to \{0,1\}^*$ that is initially $\perp$ everywhere, with new values defined during the game. By $\text{im}T[\cdot, \cdot]$ we denote the set $\{z \in \{0,1\}^* : \exists x, y \in \{0,1\}^* \text{ with } T[x,y] = z\}$ and by $\text{supp}T[\cdot, \cdot]$ the set $\{(x,y) \in \{0,1\}^* \times \{0,1\}^* : T[x,y] \neq \perp\}$. Both terms are also used in the obvious sense in settings where one of the inputs is fixed. (In Fig. 1, this input is L.) Finally, for a subset $A \subset B$, the notation $\overline{A}$ refers to the complement $B \setminus A$ in B. We use this notation in places when the superset B is clear from the context. (In Fig. 1, the set B is $\{0,1\}^{\text{AE.bl}}$.)

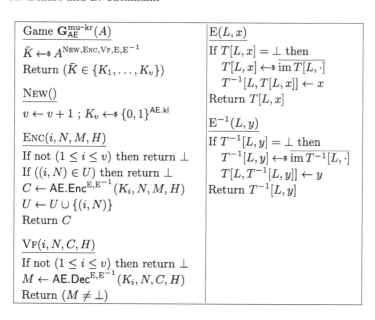

Fig. 2. Game defining multi-user key-recovery security of symmetric encryption scheme AE in the ideal-cipher model.

Definitions of mu security for authenticated encryption in the standard model are obtained as a special case, namely by restricting attention to schemes and adversaries that do not make use of the E and E^{-1} oracles.

One can further strengthen the security of the above ind definition by considering *nonce-misuse resistance* as defined by Rogaway and Shrimpton [32]. This requires changing the condition $(i, N) \in U$ in oracle ENC to only prevent queries where nonce *and* message (or even nonce, message, and header) are repeated. We do not use such a stronger definition in this work because GCM does not achieve it.

We say that an adversary is passive if it makes no queries to its VF oracle. In some cases we will get better bounds for passive adversaries.

Rogaway's definition of indistinguishability security for authenticated encryption (in the su setting) [28] gives the adversary a decryption oracle, while we give it a verification oracle. The latter is simpler and our definition can be shown equivalent to one with a decryption oracle by the technique of BN [5].

KEY-RECOVERY SECURITY. The qualitatively weaker requirement of key-recovery security can sometimes be established with better bounds than ind, which is of practical importance since violating key recovery is much more damaging that violating ind. The formalization is based on game $\mathbf{G}_{\mathsf{AE}}^{\mathrm{mu\text{-}kr}}(A)$ of Fig. 2, associated to encryption scheme AE and adversary A. The goal of the adversary A is simply to output the key of any honest user. It again has access to oracles NEW, ENC, VF, E, and E^{-1}. Oracles ENC and VF are defined to always return the values as determined by the scheme AE. Adversary A wins if it outputs any

CAU.Enc$^{\mathrm{E,E}^{-1}}(K, N, M, H)$	CAU.Dec$^{\mathrm{E,E}^{-1}}(K, N, T\|C, H)$
$\ell \leftarrow \lceil |M|/\lambda \rceil$	$\ell \leftarrow \lceil |M|/\lambda \rceil - 1$
$M_1\| \ldots \|M_\ell \leftarrow M$ // block length λ	$C_1\| \ldots \|C_\ell \leftarrow C$ // block length λ
$r \leftarrow |M_\ell|$ // last block length	$r \leftarrow |C_\ell|$ // last block length
$G \leftarrow \mathrm{E}(K, 0^\lambda)$; $Y \leftarrow N\|0^{\lambda-\nu-1}1$	$G \leftarrow \mathrm{E}(K, 0^\lambda)$; $Y \leftarrow N\|0^{\lambda-\nu-1}1$
For $i = 1$ to $\ell - 1$	$T' \leftarrow \mathrm{H}(G, H, C) \oplus \mathrm{E}(K, Y)$
$\quad C_i \leftarrow M_i \oplus \mathrm{E}(K, Y + i)$	If $T \neq T'$ then return $\perp$
$C_\ell \leftarrow M_\ell \oplus \mathrm{msb}_r(\mathrm{E}(K, Y + \ell))$	For $i = 1$ to $\ell - 1$
$C \leftarrow C_1\| \ldots \|C_\ell$	$\quad M_i \leftarrow C_i \oplus \mathrm{E}(K, Y + i)$
$T \leftarrow \mathrm{H}(G, H, C) \oplus \mathrm{E}(K, Y)$	$M_\ell \leftarrow C_\ell \oplus \mathrm{msb}_r(\mathrm{E}(K, Y + \ell))$
Return $T\|C$	Return $M_1\| \ldots \|M_\ell$

Fig. 3. Encryption scheme CAU $=$ **CAU**$[\mathrm{H}, \kappa, \lambda, \nu]$. **Left:** Encryption algorithm CAU.Enc. **Right:** Decryption algorithm CAU.Dec.

one of the keys that was generated using the NEW oracle. The advantage of A in breaking the mu-kr security of AE is defined as $\mathsf{Adv}_{\mathsf{AE}}^{\mathrm{mu\text{-}kr}}(A) = \Pr[\mathbf{G}_{\mathsf{AE}}^{\mathrm{mu\text{-}kr}}(A)]$.

4   The Schemes

We present a symmetric encryption scheme we call CAU, for Counter-Mode with a AXU hash function. GCM is a special case. This allows us to divorce our results and analyses from some details of GCM (namely, the particular, polynomial-evaluation based hash function) making them both simpler and more general.

The TLS Working Group introduced a specific usage mode of GCM in recent draft versions of TLS 1.3 in which material, obtained in the handshake key derivation phase, is used to mask the nonce. We take a different perspective and view this as a new symmetric encryption scheme whose generalized version we specify here as RCAU. Finally we specify XCAU, our own variant that better achieves the same goals.

CAU. Let $\kappa, \lambda, \nu \geq 1$ be integers such that $\nu \leq \lambda - 2$, where κ is referred to as the *cipher key length*, λ as the *block length* and ν as the *nonce length*. Let H: $\{0,1\}^\lambda \times (\{0,1\}^* \times \{0,1\}^*) \to \{0,1\}^\lambda$ be an ϵ-XOR universal hash function. We associate to these the symmetric encryption scheme CAU $=$ **CAU**$[\mathrm{H}, \kappa, \lambda, \nu]$ —here **CAU** is a transform taking $\mathrm{H}, \kappa, \lambda, \nu$ and returning a symmetric encryption scheme that we are denoting CAU— whose encryption and decryption algorithms are specified in Fig. 3. The scheme has key length CAU.kl $= \kappa$, cipher key length CAU.ckl $= \kappa$ and block length CAU.bl $= \lambda$. It has nonce space CAU.NS $= \{0,1\}^\nu$ and ciphertext length function CAU.cl$(\cdot)$ defined by CAU.cl$(m) = m + \lambda$. Explanations follow.

The algorithms CAU.Enc and CAU.Dec are given access to oracles that represent a cipher E: $\{0,1\}^\kappa \times \{0,1\}^\lambda \to \{0,1\}^\lambda$ and its inverse E^{-1}. In the security games the cipher will be chosen at random, meaning be ideal. In practice, it will be instantiated by a block cipher, usually AES.

CAU is an encrypt-then-mac scheme [5]. Encryption is counter-mode of the block cipher. The MAC is a Carter-Wegman MAC based on the AXU function family H. Some optimizations are performed over and above generic encrypt-then-mac to use the same key for both parts. The name stands for "Counter Almost Universal."

In the description of Fig. 3, the plaintext M is first partitioned into $\ell = \lceil |M|/\lambda \rceil$ plaintext blocks $M_1, \ldots, M_\ell$. The first $\ell - 1$ blocks have length λ. The final block M_ℓ has length $1 \leq r \leq \lambda$. The value G defined as $E(K, 0^\lambda)$ is later used as a key for the hash function H. The loop then computes the counter mode encryption. Here and in the rest of the paper we use the following notation. If Z is a λ bit string and $j \geq 0$ is an integer then we let

$$Z + j = Z[1..\nu] \| \langle 1 + j \rangle \tag{1}$$

where $\langle 1 + j \rangle$ is the representation of the integer $(1 + j) \bmod 2^{\lambda - \nu}$ as a $(\lambda - \nu)$-bit string. Thus, in the scheme, $Y + i = N \| \langle 1 + i \rangle$. Function msb_n, which is needed to compute the final and possibly incomplete ciphertext block C_ℓ, maps a string of length $\geq n$ to its n-bit prefix. The final step in the scheme is then to compute the function H on H and $C = C_1 \| \ldots \| C_\ell$ and xor it to the output of the block cipher on input Y. To simplify the technical descriptions in our proofs, we define the ciphertext as consisting of the tag prepended to the output of the counter-mode encryption.

GCM, as proposed by McGrew and Viega [24] and standardized by NIST [12], is obtained by instantiating the block cipher with AES, so that $\lambda = \kappa = 128$. The nonce length (in the standardized version) is $\nu = 96$. The hash function H is based on polynomial evaluation. The specifics do not matter for us. For our security analysis, all we need is that H is an ϵ-almost XOR-universal hash function (according to our definition of Sect. 2) for some $\epsilon \colon \mathbb{N} \times \mathbb{N} \to [0, 1]$. McGrew and Viega [24, Lemma 2] show that H has this property for $\epsilon(m, n) = (\lceil m/\lambda \rceil + \lceil n/\lambda \rceil + 1)/2^\lambda$.

CAU has fixed-length nonces, reflecting the standardized version of GCM in which $\nu = 96$. While the original scheme allows variable-length nonces [24], IOM [18] showed that the original security proof was flawed for variable-length nonces and the claimed security bounds did not hold.

RCAU. The TLS Working Group introduced a specific usage mode of GCM in recent draft versions of TLS 1.3 to prevent the scheme from evaluating the block cipher on the same inputs in each session. This countermeasure is described as computing an additional ν bits of key material in the key derivation phase, and using these to mask the ν-bit nonce given to GCM.

In order to analyze the effectiveness of this countermeasure, we take a different perspective, casting the method as specifying a new symmetric encryption scheme in which the mask becomes part of the key. Formally, as before, let $\kappa, \lambda, \nu \geq 1$ be integers representing the cipher key length, block length and

$\text{RCAU.Enc}^{\mathrm{E},\mathrm{E}^{-1}}(K\|L,N,M,H)$	$\text{RCAU.Dec}^{\mathrm{E},\mathrm{E}^{-1}}(K\|L,N,T\|C,H)$
$\ell \leftarrow \lceil |M|/\lambda \rceil$	$\ell \leftarrow \lceil |M|/\lambda \rceil - 1$
$M_1\|\dots\|M_\ell \leftarrow M$ // block length λ	$C_1\|\dots\|C_\ell \leftarrow C$ // block length λ
$r \leftarrow |M_\ell|$ // last block length	$r \leftarrow |C_\ell|$ // last block length
$G \leftarrow \mathrm{E}(K,0^\lambda)$; $Y \leftarrow (N\oplus L)\|0^{\lambda-\nu-1}1$	$G \leftarrow \mathrm{E}(K,0^\lambda)$; $Y \leftarrow (N\oplus L)\|0^{\lambda-\nu-1}1$
For $i = 1$ to $\ell - 1$	$T' \leftarrow \mathrm{H}(G,H,C)\oplus \mathrm{E}(K,Y)$
$\quad C_i \leftarrow M_i \oplus \mathrm{E}(K,Y+i)$	If $T \neq T'$ then return $\bot$
$C_\ell \leftarrow M_\ell \oplus \mathrm{msb}_r(\mathrm{E}(K,Y+\ell))$	For $i = 1$ to $\ell - 1$
$C \leftarrow C_1\|\dots\|C_\ell$	$\quad M_i \leftarrow C_i \oplus \mathrm{E}(K,Y+i)$
$T \leftarrow \mathrm{H}(G,H,C)\oplus \mathrm{E}(K,Y)$	$M_\ell \leftarrow C_\ell \oplus \mathrm{msb}_r(\mathrm{E}(K,Y+\ell))$
Return $T\|C$	Return $M_1\|\dots\|M_\ell$

Fig. 4. Encryption scheme RCAU = **RCAU**$[\mathrm{H},\kappa,\lambda,\nu]$. **Left:** Encryption algorithm RCAU.Enc. **Right:** Decryption algorithm RCAU.Dec.

$\text{XCAU.Enc}^{\mathrm{E},\mathrm{E}^{-1}}(K\|L,N,M,H)$	$\text{XCAU.Dec}^{\mathrm{E},\mathrm{E}^{-1}}(K\|L,N,T\|C,H)$
$\ell \leftarrow \lceil |M|/\lambda \rceil$	$\ell \leftarrow \lceil |M|/\lambda \rceil - 1$
$M_1\|\dots\|M_\ell \leftarrow M$ // block length λ	$C_1\|\dots\|C_\ell \leftarrow C$ // block length λ
$r \leftarrow |M_\ell|$ // last block length	$r \leftarrow |C_\ell|$ // last block length
$G \leftarrow L\oplus \mathrm{E}(K,L)$; $Y \leftarrow N\|0^{\lambda-\nu-1}1$	$G \leftarrow L\oplus \mathrm{E}(K,L)$; $Y \leftarrow N\|0^{\lambda-\nu-1}1$
For $i = 1$ to $\ell - 1$	$T' \leftarrow \mathrm{H}(G,H,C)\oplus L\oplus \mathrm{E}(K,L\oplus Y)$
$\quad C_i = M_i \oplus L\oplus \mathrm{E}(K,L\oplus(Y+i))$	If $T \neq T'$ then return $\bot$
$C_\ell \leftarrow M_\ell \oplus \mathrm{msb}_r(L\oplus \mathrm{E}(K,L\oplus(Y+\ell)))$	For $i = 1$ to $\ell - 1$
$C \leftarrow C_1\|\dots\|C_\ell$	$\quad M_i \leftarrow C_i \oplus L\oplus \mathrm{E}(K,L\oplus(Y+i))$
$T \leftarrow \mathrm{H}(G,H,C)\oplus L\oplus \mathrm{E}(K,L\oplus Y)$	$M_\ell \leftarrow C_\ell\oplus\mathrm{msb}_r(L\oplus\mathrm{E}(K,L\oplus(Y+\ell)))$
Return $T\|C$	Return $M_1\|\dots\|M_\ell$

Fig. 5. Encryption scheme XCAU = **XCAU**$[\mathrm{H},\kappa,\lambda,\nu]$. **Left:** Encryption algorithm XCAU.Enc. **Right:** Decryption algorithm XCAU.Dec.

nonce length, where $\nu \leq \lambda - 2$. Let $\mathrm{H}: \{0,1\}^\lambda \times (\{0,1\}^* \times \{0,1\}^*) \to \{0,1\}^\lambda$ be an ϵ-XOR universal hash function. We associate to these the symmetric encryption scheme RCAU = **RCAU**$[\mathrm{H},\kappa,\lambda,\nu]$ whose encryption and decryption algorithms are specified in Fig. 4. The scheme has key length RCAU.kl = $\kappa + \nu$, cipher key length RCAU.ckl = κ and block length RCAU.bl = λ. It has nonce space RCAU.NS = $\{0,1\}^\nu$ and ciphertext length function RCAU.cl$(\cdot)$ defined by RCAU.cl$(m) = m + \lambda$. Note that the key length is $\kappa + \nu$, while that of CAU was κ. The definition of $Y + i$ is as per (1), so $Y + i = (N\oplus L)\|\langle 1 + i\rangle$.

XCAU. We suggest a different scheme to achieve the multi-user security goal targeted by RCAU. Recall that if $\mathrm{E}: \{0,1\}^\kappa \times \{0,1\}^\lambda \to \{0,1\}^\lambda$ is a block cipher than $\mathrm{EX}: \{0,1\}^{\kappa+\lambda} \times \{0,1\}^\lambda \to \{0,1\}^\lambda$ is the block cipher defined by $\mathrm{EX}(K\|L,X) = L\oplus\mathrm{E}(K,L\oplus X)$. This can be viewed as strengthening E using an Even-Mansour technique [13]. This was suggested by Rivest as a key-extension method for DES and first analyzed by Kilian and Rogaway [19]. We then simply use EX in place of E in the basic CAU. Formally, as before, let $\kappa, \lambda, \nu \geq 1$ be integers representing the cipher key length, block length and nonce length, where $\nu \leq \lambda - 2$. Let H: $\{0,1\}^\lambda \times (\{0,1\}^* \times \{0,1\}^*) \rightarrow \{0,1\}^\lambda$ be an ϵ-XOR universal hash function. We associate to these the symmetric encryption scheme XCAU = **XCAU**$[H, \kappa, \lambda, \nu]$ whose encryption and decryption algorithms are specified in Fig. 5. The scheme has key length XCAU.kl = $\kappa + \lambda$, cipher key length XCAU.ckl = κ and block length XCAU.bl = λ. It has nonce space XCAU.NS = $\{0,1\}^\nu$ and ciphertext length function XCAU.cl$(\cdot)$ defined by XCAU.cl$(m) = m + \lambda$. Note that the key length is $\kappa + \lambda$, while that of RCAU was $\kappa + \nu$. The definition of $Y + i$ is as per (1), so $Y + i = N \| \langle 1 + i \rangle$.

Our analysis of this scheme builds on the work of Kilian and Rogaway, but analyzes the construction directly in the multi-user setting. We believe that the bounds can be further improved along the lines of Mouha and Luykx's work [25], but this does not affect the terms we are most interested in.

5   Key-Recovery Security

The multi-user security differences between the schemes are most easily seen in the case of security against key recovery, so we start there.

5.1   Security of CAU

We show that the multi-user kr advantage scales linearly in the number of adversarial evaluations of the ideal cipher (corresponding to offline evaluations of the blockcipher in practice) and the number of user instances. We give both an upper bound (security proof) and lower bound (attack) on the kr-advantage to show this, beginning with the former.

Theorem 1. *Let* $\kappa, \lambda, \nu \geq 1$ *be such that* $\nu \leq \lambda - 2$. *Let* H: $\{0,1\}^\lambda \times (\{0,1\}^* \times \{0,1\}^*) \rightarrow \{0,1\}^\lambda$ *be a family of functions. Let* CAU = **CAU**$[H, \kappa, \lambda, \nu]$. *Let* A *be an adversary that makes at most* u *queries to its* NEW *oracle and* p *queries to its* E *and* E^{-1} *oracles. Then*

$$\mathsf{Adv}^{\mathrm{mu\text{-}kr}}_{\mathrm{CAU}}(A) \leq \frac{u(p+1)}{2^\kappa}.$$

Proof. We use the code-based game-playing technique of BR [6]. Without loss of generality, we assume that the adversary A does not input invalid user identifiers $i \notin \{1, \ldots, v\}$ to ENC or VF, and does not re-use nonces in encryption queries. We also assume that A does not verify correct ciphertexts they obtained from ENC at its VF oracle. These restrictions allow us to simplify the descriptions of the games, and any arbitrary adversary A can be translated into an adversary A' that adheres to these restrictions and makes at most the same number of queries as A. Our proof proceeds in a sequence of games.

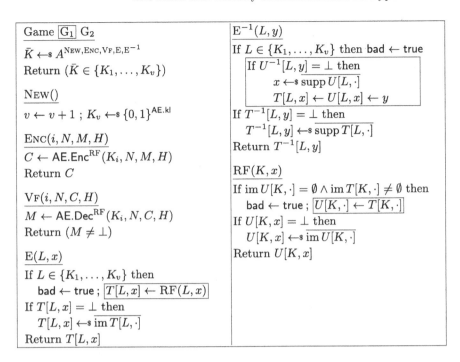

Fig. 6. Intermediate games for decoupling the oracles E/E^{-1} and RF in the proof of Theorem 1.

The first step in the proof is to rewrite game $G_{CAU}^{mu\text{-}kr}(A)$ syntactically by introducing an additional oracle RF that implements the forward evaluation of the ideal cipher for the algorithms CAU.Enc and CAU.Dec. This is sufficient as encryption and decryption in CAU never query E^{-1}. We call this game G_0, but do not explicitly describe as it is obtained easily from $G_{CAU}^{mu\text{-}kr}(A)$.

We then rewrite the game in the form of G_1, which is described in Fig. 6 and basically obtained by a syntactic modification of the oracles E, E^{-1}, and RF. In more detail, oracle RF samples the ideal cipher for the keys used in the encryption using the map $U[\cdot, \cdot]$. The oracles E and E^{-1} are adapted such that, for keys used in the game, they sample the map $T[\cdot, \cdot]$ consistently with $U[\cdot, \cdot]$. We introduce a flag bad that is set when the adversary A queries one of the oracles E or E^{-1} with a key that is also used in the oracle RF.

The next game G_2 modifies the way in which the responses for the E, E^{-1}, and RF oracles are determined. In particular, we break the consistency between E and E^{-1} on the one hand, and RF on the other hand, by sampling the oracle responses independently. Since all changes appear only after bad has been set, we can relate the games using the Fundamental Lemma from Bellare and Rogaway [6] and proceed by bounding the probability of setting bad. This probability is in fact bounded by $up/2^\kappa$. As all computations while bad is not set are

Adversary A_{u,p_e}
For $i = 1$ to u do
NEW ; $C_i \leftarrow \text{ENC}(i, 0^\nu, 0^{2\lambda}, \varepsilon)$
For $j = 1$ to $p_e/2$ do
$y \leftarrow \text{E}([j]_\lambda, 0^\nu \| 0^{\lambda-\nu-2} \| 10)$ // $[j]_\lambda$ is the encoding of integer j as a λ-bit string
$y' \leftarrow \text{E}([j]_\lambda, 0^\nu \| 0^{\lambda-\nu-2} \| 11)$
If $(\exists i : C_i[(\lambda+1)..2\lambda] = y$ and $C_i[(2\lambda+1)..3\lambda] = y')$ then return $[j]_\lambda$

Fig. 7. Adversary A_{u,p_e} used in Theorem 2.

independent of the values $K_1, \ldots, K_u$, the maximal probability of the adversary to guess one of these uniformly random values is $u/2^\kappa$ in each of its p queries to E and E^{-1}.

The keys in G_2 only serve as labels, the game is independent of their actual values. The only remaining step is to compute the probability of guessing any one of the u keys that are chosen at random without collision, which is also incorporated into the advantage. In more detail:

$$\text{Adv}_{\text{CAU}}^{\text{mu-kr}}(A) = \Pr\left[\mathbf{G}_{\text{CAU}}^{\text{mu-kr}}(A)\right] = \Pr[\text{G}_0] = \Pr[\text{G}_1]$$

$$\leq \Pr[\text{G}_2] + \frac{up}{2^\kappa} \leq \frac{u}{2^\kappa} + \frac{up}{2^\kappa} = \frac{u(p+1)}{2^\kappa},$$

which concludes the proof.   □

Next we show that the security bound proven in Theorem 1 is (almost) tight. We describe an attack (adversary) that achieves the described bound up to a (for realistic parameters small) factor. The adversary is shown in Fig. 7. It is parameterized by a number u of users and an (even) number p_e of queries to E. It first encrypts a short message $0^{2\lambda}$ for each of the u users. Next, it queries the E oracle on the value $0^{\lambda-2}10$, the first block that is used for masking actual plaintext, for up to p_e different keys. As soon as it finds a matching key L for the first block, it simply evaluates $\text{E}(L, 0^{\lambda-2}11)$ and checks for consistency with the second block. If the check succeeds, the adversary outputs the key L, otherwise it tries further keys.

The described attack strategy extends to any application of CAU in which the nonces used in the scheme are the same in each session. As TLS 1.3 uses the sequence number to compute the nonces, a version without the nonce randomization technique would be susceptible to this attack.

Theorem 2. *Let* $\kappa, \lambda, \nu \geq 1$ *be such that* $\nu \leq \lambda - 2$. *Let* $\text{H}: \{0,1\}^\lambda \times (\{0,1\}^* \times \{0,1\}^*) \to \{0,1\}^\lambda$ *be a family of functions. Let* $\text{CAU} = \mathbf{CAU}[\text{H}, \kappa, \lambda, \nu]$. *Let* $u \geq 1$ *be an integer and* $p_e \geq 2$ *an even integer. Associate to them the adversary* A_{u,p_e} *described in Fig. 7, which makes* u *queries to* NEW, $q_e = u$ *queries to* ENC *of length* 2λ *bits, no queries to* VF, p_e *queries to* E, *and no queries to* E^{-1}. *Then*

$$\text{Adv}_{\text{CAU}}^{\text{mu-kr}}(A_{u,p_e}) \geq \mu \cdot \left(1 - e^{-\frac{p_e u}{2^{\kappa+1}}}\right)$$

where

$$\mu = \left(1 - \frac{u(u-1)}{2^{\kappa+1}}\right) \cdot \left(1 - \frac{u(2^{\kappa} - u)}{2^{\lambda}(2^{\lambda} - 1)}\right).$$

This means that the advantage of A_{u,p_e} scales (almost) linearly with the number of users, and in fact, for values u, p_e such that $up_e/2^{\kappa+1} \leq 1$, the advantage is lower bounded by $\mu \cdot (1 - 1/e) \cdot \frac{p_e u}{2^{\kappa+1}}$. The proof we give can be improved in terms of tightness, for instance, we allow the attack to completely fail if only a single collision occurs between honest users' keys. In particular the factor $(1 - u(u-1)/2^{\kappa+1})$ could be improved especially for large u.

Proof (Theorem 2). The probability for any of the $u = q_e$ keys to collide is at most $u(u-1)/2^{\kappa+1}$. In the subsequent steps we compute the probabilities based on the assumption that no user keys generated within NEW collide, which is correct with probability at least $1 - u(u-1)/2^{\kappa+1}$. In more detail, given that we have no collisions of user keys, the adversary uses at least $p_e/2$ attempts to guess any one of $u = q_e$ (uniformly random, without collision) keys from a set of size 2^{κ}. The probability for each honest user's key to be among the adversary's guesses is $p_e/2^{\kappa+1}$, and so the overall probability for any one of the adversary's attempts to succeed is

$$1 - \left(1 - \frac{p_e}{2^{\kappa+1}}\right)^u \geq 1 - e^{-\frac{p_e u}{2^{\kappa+1}}}.$$

We still need to bound the probability of false positives, that is, keys that were not sampled in a NEW oracle but coincide with the block cipher outputs, and therefore lead to a wrong guess: The probability that the ideal cipher for a specific "wrong" key (out of $2^{\kappa} - u$) coincides with the ideal cipher for each of the u "correct" keys on both inputs $0^{\nu}\|0^{\lambda-\nu-2}\|10$ and $0^{\nu}\|0^{\lambda-\nu-2}\|11$ is $2^{-\lambda}(2^{\lambda}-1)^{-1}$. The existence of such a colliding key can be bounded using the Union Bound to be at most $u(2^{\kappa} - u)/(2^{\lambda}(2^{\lambda} - 1))$, so the probability that no such collision exists is at least $1 - u(2^{\kappa} - u)/(2^{\lambda}(2^{\lambda} - 1))$. Overall, we obtain the stated bound. □

Evaluating the formula for realistic values for GCM in TLS 1.3, we set $\kappa = 128$. We allow the adversary to make $p_e = 2^{64}$ evaluations of the block cipher. We estimate the number of TLS sessions per day as 2^{40}, which leaves us a security margin of roughly 2^{24}. While this means that on expectation the attack still needs 2^{24} days to recover a single key, it is important to recall that this estimate is obtained under the strong assumption that AES behaves like an ideal cipher.

5.2 Security of RCAU

RCAU aims to avoid the attack strategy described in Sect. 5.1 by randomizing the nonce before it is used in the block cipher. Here we assess whether the measure succeeds, again first upper bounding adversary advantage via a proof, then lower bounding it via an attack.

In contrast to the bound for CAU, the bound for RCAU depends on more parameters. This is caused by the more intricate "decoupling" of the E/E^{-1} and RF oracles.

Theorem 3. *Let $\kappa, \lambda, \nu \geq 1$ be such that $\nu \leq \lambda - 2$. Let $\mathsf{H}: \{0,1\}^\lambda \times (\{0,1\}^* \times \{0,1\}^*) \to \{0,1\}^\lambda$ be a family of functions. Let $\mathsf{RCAU} = \mathbf{RCAU}[\mathsf{H}, \kappa, \lambda, \nu]$. Let A be an adversary that makes at most u queries to its NEW oracle, q_e queries to its ENC oracle with messages of length at most ℓ_{bit} bits, q_v queries to its VF oracle of length at most $\ell_{\mathrm{bit}} + \lambda$ bits, p_e queries to its E oracle, and p_i queries to its E^{-1} oracle. Then*

$$\mathsf{Adv}^{\mathrm{mu\text{-}kr}}_{\mathsf{RCAU}}(A) \leq \frac{2up(\ell_{\mathrm{blk}}(q_e + q_v) + 1)}{2^{\kappa+\nu}} + \frac{up(\ell_{\mathrm{blk}}(q_e + q_v) + 1)}{2^\kappa(2^\lambda - p)}$$

$$+ \frac{up(\ell_{\mathrm{blk}}(q_e + q_v) + 1)}{2^\kappa(2^\lambda - q_e - q_v)} + \frac{p_i + u}{2^\kappa}, \quad (2)$$

where $\ell_{\mathrm{blk}} = \lceil \ell_{\mathrm{bit}}/\lambda \rceil + 1$.

Proof. As in Theorem 1, we restrict our attention to adversaries A that do not use invalid user identifiers, that do not re-use nonces, and that do not verify ciphertexts obtained from the ENC oracle. As in the proof of Theorem 1, we now aim at "decoupling" the oracles $\mathrm{E}/\mathrm{E}^{-1}$ and RF, but this time we have to be cautious: we cannot just "give up" when the adversary "guesses" one of the users keys in calls to $\mathrm{E}/\mathrm{E}^{-1}$; this would ruin our bound. The first step is as above to introduce an auxiliary map $U[\cdot, \cdot]$ in addition to $T[\cdot, \cdot]$, but keep the maps synchronized. The change from $\mathbf{G}^{\mathrm{mu\text{-}kr}}_{\mathsf{RCAU}}(A)$ to G_0 is therefore only syntactic. Intuitively, the lazy sampling of the block cipher is now performed using both maps, where $T[\cdot, \cdot]$ is filled in calls to E and E^{-1}, and $U[\cdot, \cdot]$ is filled in RF. The oracles make sure that the maps stay consistent.

In game G_1, described in detail in Fig. 8, we first change the way in which the responses are sampled, but still in an equivalent way, namely we first attempt to sample consistently only for $T[\cdot, \cdot]$ and then check for consistency with $U[\cdot, \cdot]$. If this fails, we set bad $\leftarrow$ true and re-sample with the correct distribution. Additionally, we set bad $\leftarrow$ true whenever we need to answer for either $T[\cdot, \cdot]$ or $U[\cdot, \cdot]$ and the answer is already defined by the respective other map. Game G_1 is equivalent to G_0. The proof is further complicated by the fact that RCAU derives the key for H as $\mathrm{E}(K, 0^\lambda)$ and this query is therefore not randomized. As a consequence, we have to treat the queries with value 0^λ independently of the other queries, and keep the maps $T[\cdot, 0^\lambda]$ and $U[\cdot, 0^\lambda]$ consistent for the next proof steps.

In game G_2 we modify the behavior of the oracles E, E^{-1}, and RF to not re-sample to avoid inconsistencies with the other oracles. Also, we do not enforce consistency between $T[K, \cdot]$ and $U[K, \cdot]$ for values that are defined already in one of the maps; we sample a fresh value in a map independently of whether the point is already defined in the other map. As both modifications occur only after the flag bad has been set, we can use the Fundamental Lemma to relate the advantages of an adversary in games G_1 and G_2.

To bound the probability for the flag bad to be set in games G_1 or G_2, respectively, we begin with the following observation: As long as bad is not set, each row $T[K, \cdot]$ or $U[K, \cdot]$ for a specific key K is sampled without collisions within this row, but independently of any other row, and also mutually independent

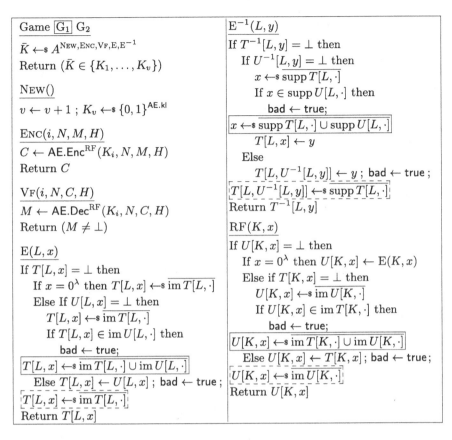

Fig. 8. Intermediate games for decoupling the oracles E/E^{-1} and RF in the proof of Theorem 3.

between $T[K, \cdot]$ and $U[K, \cdot]$. This is the case because the only other way of defining a value for those maps is either re-sampling or copying from the other map; in both cases we set the flag bad. Furthermore, we observe that all operations that occur before the flag bad is set are independent of the actual values of the keys $K_1, \ldots, K_u$. Given these insights, we now analyze the probabilities for setting the bad flag at the different code points, first for E and E^{-1}:

- The probability of enforcing re-sampling in E or E^{-1} is analyzed as follows: For a particular key $\bar{K} \in \{K_1, \ldots, K_u\}$ for which m blocks have been defined through queries to RF, the probability of sampling a value that collides is at most $m/(2^\lambda - p)$, as we choose uniformly from $2^\lambda - p$ values. The expected number of blocks for the key L in the query is $u(\ell_{\mathrm{blk}}(q_e + q_v) + 1)/2^\kappa$, which leads to an overall probability of $u(\ell_{\mathrm{blk}}(q_e + q_v) + 1)/(2^\kappa(2^\lambda - p))$ for each query.

- The probability of enforcing that a value be copied (that is, the final "Else" statement becomes active) in E is bounded by $u(\ell_{\text{blk}}(q_e + q_v) + 1)/2^{\kappa+\nu}$ for each of the p queries. This is computed analogously to above: executing the "Else" statement means that the adversary guessed a combination of a κ-bit key and a ν-bit mask value.
- Finally, the probability for copying a value in E^{-1} is bounded by the term $1/2^{-\kappa}$. The reason is that it corresponds to guessing a the key for a specific user.

We obtain the bounds $up(\ell_{\text{blk}}(q_e + q_v) + 1)/(2^{\kappa}(2^\lambda - p))$, $up_e(\ell_{\text{blk}}(q_e + q_v) + 1)/2^{\kappa+\nu}$, and $p_i/2^{-\kappa}$ as the adversary makes at most p_e queries to E, p_i queries to E^{-1}, and $p = p_e + p_i$ queries accumulated.

We proceed by analyzing the probabilities for RF analogously:

- With respect to enforcing re-sampling in RF, for a key L for which m blocks have been defined, the probability of sampling a colliding value is $m/(2^\lambda - q_e - q_v)$. This leads to an overall probability of at most $p/(2^{\kappa}(2^\lambda - q_e - q_v))$.
- The probability of enforcing that a value be copied (that is, the final "Else" statement becomes active) in RF is bounded by $u(\ell_{\text{blk}}(q_e + q_v) + 1)p/2^{\kappa+\nu}$. The reason is that for a particular key L for which m blocks have been defined through queries to E and E^{-1}, the probability that an query to RF as done by $\text{CAU.Enc}^{\text{RF}}$ uses the same input is bounded by $m/2^\nu$. This leads to a probability of $p/2^{\kappa+\nu}$.

Since the encryption and decryption algorithms overall make $u(\ell_{\text{blk}}(q_e + q_v) + 1)$ queries to RF, we obtain the bounds $up(\ell_{\text{blk}}(q_e + q_v) + 1)/(2^{\kappa}(2^\lambda - q_e - q_v))$ and $up(\ell_{\text{blk}}(q_e + q_v) + 1)/2^{\kappa+\nu}$.

Finally, as in G_2 the oracles E and E^{-1} are independent of the oracle RF that is used in RCAU, the probability of guessing a key is $u/2^\kappa$. All these terms together comprise the bound in the theorem statement.   □

For realistic parameters, the bound in Theorem 3 means that the "best" attack for passive adversaries is now the inversion of a block observed while eavesdropping. In contrast to the attack analyzed in Sect. 5.1, this attack does not scale in the mass surveillance scenario, because the adversary has to target one specific ciphertext block.

In more detail, the adversary strategy A analyzed in the below lemma and specified in detail in Fig. 9 proceeds as follows. First obtain an encryption of $0^{2\lambda}$ from an honest user. Then brute-force the key by decrypting the first ciphertext block using E^{-1}, checking whether the output satisfies the structure $N\|0^{\lambda-\nu-2}10$. In case this structure is observed, verify the key by checking if the next block is consistent with an evaluation of E with the same key and plaintext $N\|0^{\lambda-\nu-2}11$.

Since the described attack strategy applies independently of how the nonces are chosen (prior to the randomization) as long as the value is predictable, the lower bound also applies to the scheme as used in the latest draft of TLS 1.3.

Adversary A_{p_i}

NEW ; $C \leftarrow \text{ENC}(1, 0^\nu, 0^{2\lambda}, \varepsilon)$

For $j = 1$ to $p_i/2$ do

$\quad y \leftarrow \text{E}^{-1}([j]_\lambda, C[(\lambda+1)..2\lambda])$ $/\!/$ $[j]_\lambda$ means encoding as λ-bit string

$\quad$ If $\exists N \in \{0,1\}^\nu : y = N\|0^{\lambda-\nu-2}10$ then

$\quad\quad$ If $\text{E}^{-1}([j]_\lambda, C[(2\lambda+1)..3\lambda]) = N\|0^{\lambda-\nu-2}10$ then

$\quad\quad\quad$ Return $[j]_\lambda$

Fig. 9. Adversary A_{p_i} used in Theorem 4.

Theorem 4. *Let $\kappa, \lambda, \nu \geq 1$ be such that $\nu \leq \lambda - 2$. Let $\mathsf{H}: \{0,1\}^\lambda \times (\{0,1\}^* \times \{0,1\}^*) \to \{0,1\}^\lambda$ be a family of functions. Let $\mathsf{RCAU} = \mathbf{RCAU}[\mathsf{H}, \kappa, \lambda, \nu]$. Let $p_i \geq 2$ an even integer and the adversary A_{p_i} as described in Fig. 9, which makes 1 query to each NEW and ENC (the latter of length 2λ bits), no queries to VF, p_i queries to E^{-1}, and no queries to E. Then*

$$\mathsf{Adv}^{\text{mu-kr}}_{\mathsf{RCAU}}(A_{p_i}) \geq \mu \cdot p_i \cdot 2^{-\kappa-1},$$

with

$$\mu = 1 - \frac{(2^\kappa - 1)2^\nu}{2^\lambda(2^\lambda - 1)}.$$

Proof. Let K_1 be the key sampled during the invocation of NEW in the game. The probability for the block cipher on a key $K \neq K_1$ to satisfy the first condition is $2^{\nu-\lambda}$, since in the first invocation of E^{-1} the value is sampled uniformly at random and $\lambda - \nu$ bits have to match. The second invocation of E^{-1} has to lead to the correct outcome $N\|0^{\lambda-\nu-2}11$, the value is drawn uniformly at random from the remaining $2^\lambda - 1$ values not equal to the outcome of the first query. There are 2^κ keys, so by the Union Bound the probability of any key $K \neq K_1$ to lead to an admissible pattern on the first two blocks is bounded by $(2^\kappa - 1)2^\nu/(2^\lambda(2^\lambda - 1))$.

In the event that no key $K \neq K_1$ satisfies the above condition, this advantage of adversary A_{p_i} is simply the probability of guessing a uniformly random key of κ bits in $p_i/2$ attempts, as for each key A_{p_i} spends at most 2 queries. This completes the proof. $\qquad\square$

The attack analyzed in Theorem 4 is considerably harder to mount than the one analyzed in Theorem 2, because the queries in the Theorem 2 attack can be preprocessed and apply to all observed communication sessions equally, whereas in the Theorem 4 attack the queries have to be made for a particular session under attack. Still, in the following Sect. 5.3, we show that at low computational cost for the honest parties, the Theorem 4 attack can be made considerably harder.

5.3   Security of XCAU

The term $p_i/2^\kappa$ in the bound for RCAU originates in the fact that only the input of the block cipher is masked, and inversion queries by the adversaries are

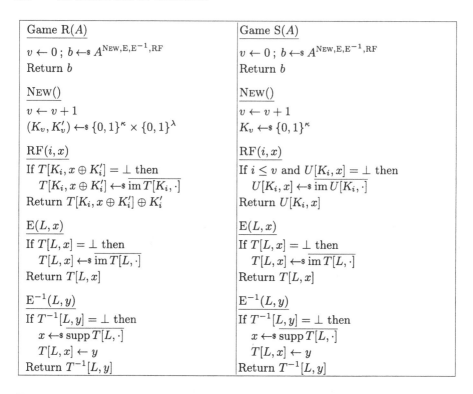

Fig. 10. Multi-user security for block-cipher key extension. **Left:** Game giving the adversary access to the actual construction. **Right:** Game giving the adversary access to an independent ideal cipher.

not hindered. In the scheme XCAU, an advantage beyond the randomization of the input to derive the hash function key is that the output of the block cipher is masked, which restricts the power of inversion queries to the block cipher considerably.

Our analysis of XCAU is based on combining the analysis of DESX-like input and output whitening in a multi-user setting, and then prove the security of XCAU along the lines of Theorem 8. We first prove a multi-user bound for the DESX-like construction. The security goal is described by the games in Fig. 10.

Theorem 5. *Let A be an adversary that makes at most u queries to its* NEW, *q_2 queries to its* RF *oracle per user, p queries to its* E *oracle and* E^{-1} *oracles. Then*

$$|\Pr[R(A)] - \Pr[S(A)]| \leq \frac{u \cdot q_2 \cdot p}{2^{\lambda+\kappa+1}}.$$

Proof. We introduce two intermediate games G_0 and G_1 in Fig. 11. Game G_0 is equivalent to game $R(A)$; the introduction of the additional map $U[\cdot, \cdot]$ is only syntactic as we make sure that it stays consistent with $T[\cdot, \cdot]$ throughout. We also modify the procedures for sampling new values for the maps $U[\cdot, \cdot]$ and $T[\cdot, \cdot]$

such that first we sample a new value such that it is consistent only with the respective map, then check whether it is consistent with the other map, and re-sample consistently if we determine that it is not. In G_1, the map $U[\cdot, \cdot]$ is completely independent of the map $T[\cdot, \cdot]$. Both G_0 and G_1 set the flag bad on occasions where the sampling creates inconsistencies between $U[\cdot, \cdot]$ and $T[\cdot, \cdot]$.

The probability of setting the bad flag in G_2 and G_3 can be bounded as follows. We first observe that besides the bad flag, G_3 is equivalent to S. For both G_2 and G_3, as long as bad is not set, all outputs are uniformly distributed among the values that are valid for the respective oracle and key. Moreover, xoring K_i' to all inputs or outputs modifies each concrete permutation; however, the distribution of a uniformly random permutation remains unchanged by this operation. Following the definition of Maurer [23], this means that both games G_2 and G_3 with the respective flags bad are *conditionally equivalent* to the game S. (In other words, *conditioned on* bad = false, the outputs of the games are distributed exactly as in S.)

Subsequently, we can employ Maurer's result [23, Theorem 1] to bound the distinguishing advantage between G_2 and G_3 by the advantage of the best *non-adaptive* distinguisher. As the adversary makes at most p queries to its E and E^{-1} oracles, and uq_2 queries to its RF oracle, there are $u \cdot q_2 \cdot p$ possible combinations of queries that may provoke the flag bad to be set, and each case appears with probability $2^{-\lambda - \kappa - 1}$. We conclude the proof via the Union Bound.   □

Analogously to the previous results on CAU and RCAU, we now analyze the key-recovery security of XCAU.

Theorem 6. *Let* $\kappa, \lambda, \nu \geq 1$ *be such that* $\nu \leq \lambda - 2$. *Let* H: $\{0,1\}^\lambda \times (\{0,1\}^* \times \{0,1\}^*) \rightarrow \{0,1\}^\lambda$ *be a family of functions. Let* XCAU = **XCAU**[H, κ, λ, ν]. *Let* A *be an adversary that makes at most* u *queries to its* NEW *oracle,* q_e *queries to its* ENC *oracle with messages of length at most* ℓ_{bit} *bits,* q_v *queries to its* VF *oracle with messages of length at most* $\ell_{\text{bit}} + \lambda$ *bits, and* p *queries to its* E *and* E^{-1} *oracles. Assume furthermore that* $q_e \leq 2^\nu$, *and* $\ell_{\text{bit}} \leq \lambda(2^{\lambda - \nu} - 2)$. *Then, with* $\ell_{\text{blk}} = \lceil \ell_{\text{bit}}/\lambda \rceil + 1$,

$$\text{Adv}_{\text{XCAU}}^{\text{mu-kr}}(A) \leq \frac{up(\ell_{\text{blk}}(q_e + q_v) + 1)}{2^{\lambda + \kappa + 1}} + \frac{u}{2^\kappa}.$$

Proof. As in Theorems 1 and 3, we restrict our attention to adversaries A that do not use invalid user identifiers, that do not re-use nonces, and that do not verify ciphertexts obtained from the ENC oracle. The first step in this proof is to rewrite the game as G_0 in the same way as in the previous proofs; the scheme is changed to use the oracle RF that is, however, kept consistent with E and E^{-1}. The game is described in Fig. 12.

The next game G_1 is again a syntactic modification from G_0. The change is that we replace XCAU, which uses the original block cipher and applies the input and output whitening for the block cipher as a part of the encryption and decryption procedures, by CAU instantiated with a block cipher with key length $\lambda + \kappa$. Consequently, we rewrite the oracle RF to perform the input and output whitening.

Game $\boxed{G_0}$ $\overline{\boxed{G_1}}$

$v \leftarrow 0$; $b \leftarrow\!\!{\scriptstyle\$}\; A^{\mathrm{New},\mathrm{E},\mathrm{E}^{-1},\mathrm{RF}}$
Return b

$\underline{\mathrm{New}}$

$v \leftarrow v + 1$
$(K_v, K'_v) \leftarrow\!\!{\scriptstyle\$}\; \{0,1\}^\kappa \times \{0,1\}^\lambda$

$\underline{\mathrm{RF}(i,x)}$

If $U[K_i, x \oplus K'_i] = \bot$ then
   If $T[K_i, x \oplus K'_i] = \bot$ then
      $U[K_i, x \oplus K'_i] \leftarrow\!\!{\scriptstyle\$}\; \overline{\mathrm{im}\, U[K_i, \cdot]}$
      If $U[K_i, x \oplus K'_i] \in \mathrm{im}\, T[K_i, \cdot]$ then
         $\mathsf{bad} \leftarrow \mathsf{true}$; $\boxed{U[K_i, x \oplus K'_i] \leftarrow\!\!{\scriptstyle\$}\; \overline{\mathrm{im}\, U[K_i, \cdot] \cup \mathrm{im}\, T[K_i, \cdot]}}$
   Else
      $U[K_i, x \oplus K'_i] \leftarrow T[K_i, x \oplus K'_i]$; $\mathsf{bad} \leftarrow \mathsf{true}$
      $\overline{U[K_i, x \oplus K'_i] \leftarrow\!\!{\scriptstyle\$}\; \overline{\mathrm{im}\, U[K_i, \cdot]}}$
Return $U[K_i, x \oplus K'_i] \oplus K'_i$

$\underline{\mathrm{E}(L,x)}$

If $T[L,x] = \bot$ then
   If $U[L,x] = \bot$ then
      $T[L,x] \leftarrow\!\!{\scriptstyle\$}\; \overline{\mathrm{im}\, T[L, \cdot]}$
      If $T[L,x] \in U[L, \cdot]$ then
         $\mathsf{bad} \leftarrow \mathsf{true}$; $\boxed{T[L,x] \leftarrow\!\!{\scriptstyle\$}\; \overline{\mathrm{im}\, T[L, \cdot] \cup \mathrm{im}\, U[L, \cdot]}}$
   Else $T[L,x] \leftarrow U[L,x]$; $\mathsf{bad} \leftarrow \mathsf{true}$; $\overline{T[L,x] \leftarrow\!\!{\scriptstyle\$}\; \overline{\mathrm{im}\, T[L, \cdot]}}$
Return $T[L,x]$

$\underline{\mathrm{E}^{-1}(L,y)}$

If $T^{-1}[L,y] = \bot$ then
   If $U^{-1}[L,y] = \bot$ then
   $x \leftarrow\!\!{\scriptstyle\$}\; \overline{\mathrm{supp}\, T[L, \cdot]}$
      If $x \in \mathrm{supp}\, U[L, \cdot]$ then
         $\mathsf{bad} \leftarrow \mathsf{true}$; $\boxed{x \leftarrow\!\!{\scriptstyle\$}\; \overline{\mathrm{supp}\, T[L, \cdot] \cup \mathrm{supp}\, U[L, \cdot]}}$
   Else $x \leftarrow U^{-1}[L,y]$; $\mathsf{bad} \leftarrow \mathsf{true}$; $\overline{x \leftarrow\!\!{\scriptstyle\$}\; \overline{\mathrm{supp}\, T[L, \cdot]}}$
   $T[L,x] \leftarrow y$
Return $T^{-1}[L,y]$

Fig. 11. Modification of the sampling algorithm. In G_0, the values are sampled to keep consistency between $U[\cdot, \cdot]$ and $T[\cdot, \cdot]$, with the flag bad set if attempted independent sampling leads to inconsistencies. In G_1, the maps $U[\cdot, \cdot]$ and $T[\cdot, \cdot]$ are sampled independently, making RF an independent ideal cipher.

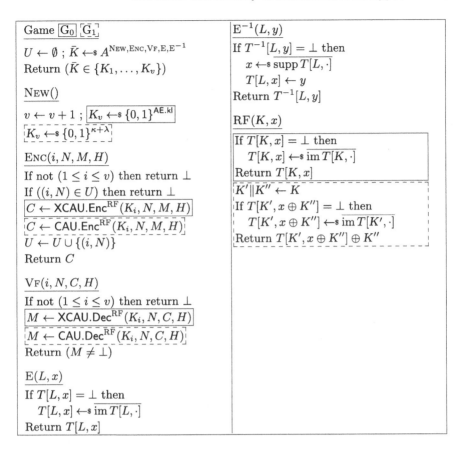

Fig. 12. Games that intuitively correspond to the security of AES-XCAU (G$_0$) as well as AESX-CAU (G$_1$).

In the next game G$_2$, the oracles E and E^{-1}, and the oracle RF are based on different maps $T[\cdot, \cdot]$ (for E and E^{-1}) and $U[\cdot, \cdot]$ (for RF), but the oracles are defined to keep them consistent. This is achieved by first sampling them independently, but then re-sampling in case an inconsistency occurs. Should that be the case, the flag bad is set. Apart from this flag, games G$_1$ and G$_2$ are equivalent. We do not describe the game G$_2$ explicitly, but remark that it is obtained by verbatim replacement of the oracles E, E^{-1}, and RF in game G$_1$ by the ones described in game G$_0$ in Fig. 11. In the next game G$_3$, the re-sampling procedure keeping the oracles consistent is abandoned, which means that the oracles RF and E together with E^{-1} are independent. Like G$_2$, game G$_3$ is obtained by replacing the oracles E, E^{-1}, and RF by the ones in game G$_1$ in Fig. 11.

The probability of setting the bad flag in G$_2$ and G$_3$ can be bounded using Theorem 5. More technically, we describe an adversary $B = B(A)$ that emulates oracles to A as follows: Queries NEW, E, and E^{-1} by B are responded by B

performing the same query in its game. Queries ENC and DEC are responded by B emulating the respective oracles using the oracle RF in its game to evaluate CAU.Enc and CAU.Dec. The view of A is the same in G_2 and in the game $R(B(A))$, and in G_3 and the game $S(B(A))$, respectively. The numbers of queries u to the NEW oracle and p to the E and E^{-1} oracles are preserved by B. At most q_e queries of length at most ℓ_{bit} to ENC and at most q_v queries of length at most $\ell_{bit} + \lambda$ to VF translate into at most $\ell_{blk}(q_e + q_v) + 1$ queries to RF in the game played by B. Using Theorem 5, this means that the probability of setting bad can be bounded by $up(\ell_{blk}(q_e + q_v) + 1)/2^{\lambda+\kappa+1}$.

All that remains to be done is bounding the probability of A guessing any key in G_3. As in this game, similarly to the previous proofs, the keys used to reference values in $U[\cdot, \cdot]$ is only used as an index to the table and is unrelated to all values that A observes in the game, the guessing probability is at most $u/2^\kappa$. This concludes the proof. □

6 Indistinguishability Security

In this section we prove the multi-user indistinguishability security bounds for CAU, RCAU, and XCAU, all in the ideal cipher model. All proofs in this section are deferred to the full version of this paper [7].

6.1 Preparation: A Lemma on CAU

We begin with a multi-user analysis of CAU which models the block cipher as a uniform random permutation and is useful in the subsequent proofs. The analysis is related to the ones of MV [24], IOM [18], and NOMI [27], with the main difference that they proved single-user security, while we directly prove multi-user security. We formalize the random-permutation model using our game $G_{CAU}^{mu\text{-}ind}$ while considering only adversaries that do not make use of the oracles E and E^{-1}.

Lemma 7. *Let $\kappa, \lambda, \nu \geq 1$ be such that $\nu \leq \lambda - 2$. Let $H: \{0,1\}^\lambda \times (\{0,1\}^* \times \{0,1\}^*) \to \{0,1\}^\lambda$ be an ϵ-almost XOR-universal hash function, for some $\epsilon: \mathbb{N} \times \mathbb{N} \to [0,1]$. Let $CAU = \mathbf{CAU}[H, \kappa, \lambda, \nu]$. Let A be an adversary that makes at most u queries to its NEW oracle, q_e queries to its ENC oracle with messages of length at most ℓ_{bit} bits, and q_v queries to its VF oracle with messages of length at most $\ell_{bit} + \lambda$ bits,[1]. In particular, A does not use the E and E^{-1} oracles. Assume furthermore that $q_e \leq 2^\nu$ and $\ell_{bit} \leq \lambda(2^{\lambda-\nu} - 2)$. Then*

$$\mathsf{Adv}_{CAU}^{mu\text{-}ind}(A) \leq \frac{u(u-1)}{2^{\kappa+1}} + \frac{u(\ell_{blk}(q_e + q_v) + 1)^2\cdot}{2^{\lambda+1}} + uq_v \cdot \epsilon(\ell_{bit}, \ell_{head}),$$

for $\ell_{blk} = \lceil \ell_{bit}/\lambda \rceil + 1$ and where the AEAD headers are restricted to ℓ_{head} bits.

---

[1] The ciphertext contains an λ-bit MAC tag, so the length of the contained plaintext is ℓ_{bit} bits.

6.2 Security of CAU

We now prove the multi-user indistinguishability security of plain CAU in the ideal-cipher model.

Theorem 8. *Let* $\kappa, \lambda, \nu \geq 1$ *be such that* $\nu \leq \lambda - 2$. *Let* $\mathsf{H} \colon \{0,1\}^\lambda \times (\{0,1\}^* \times \{0,1\}^*) \rightarrow \{0,1\}^\lambda$ *be an* ϵ*-almost XOR-universal hash function, for some* $\epsilon \colon \mathbb{N} \times \mathbb{N} \rightarrow [0,1]$. *Let* $\mathsf{CAU} = \mathbf{CAU}[\mathsf{H}, \kappa, \lambda, \nu]$. *Let* A *be an adversary that makes at most* u *queries to its* NEW *oracle,* q_e *queries to its* ENC *oracle with messages of length at most* ℓ_{bit} *bits,* q_v *queries to its* VF *oracle with messages of length at most* $\ell_{\mathrm{bit}} + \lambda$ *bits, and* p *queries to its* E *and* E^{-1} *oracles. Assume furthermore that* $q_e \leq 2^\nu$ *and* $\ell_{\mathrm{bit}} \leq \lambda(2^{\lambda-\nu} - 2)$. *Then*

$$\mathsf{Adv}_{\mathsf{CAU}}^{\mathrm{mu\text{-}ind}}(A) \leq \frac{up}{2^\kappa} + \frac{u(\ell_{\mathrm{blk}}(q_e + q_v) + 1)^2 \cdot}{2^{\lambda+1}} + \frac{u(u-1)}{2^{\kappa+1}} + uq_v \cdot \epsilon(\ell_{\mathrm{bit}}, \ell_{\mathrm{head}}),$$

for $\ell_{\mathrm{blk}} = \lceil \ell_{\mathrm{bit}}/\lambda \rceil + 1$ *and where the AEAD headers are restricted to* ℓ_{head} *bits.*

The first term originates from the advantage of the adversary in guessing a user's key in a query to the ideal cipher. This term grows linearly in the number of honest sessions, and it also grows linearly in the number of adversary calls to the ideal cipher. We show below in Theorem 2 that a term of this size is inevitable by proving the effectiveness of an attack. The second term stems from a PRF/PRP-switching in the proof of counter mode. The third term stems from a potential collision of honest-user keys, and the final term from the authentication using the AUH-based MAC.

6.3 Security of RCAU

In terms of bounds for RCAU, we first show a simple corollary proving that the same bounds as for CAU also apply for RCAU. This follows immediately by a reduction that randomizes the nonces.

Corollary 9. *Let* $\kappa, \lambda, \nu \geq 1$ *be such that* $\nu \leq \lambda - 2$. *Let* $\mathsf{H} \colon \{0,1\}^\lambda \times (\{0,1\}^* \times \{0,1\}^*) \rightarrow \{0,1\}^\lambda$ *be an* ϵ*-almost XOR-universal hash function, for some* $\epsilon \colon \mathbb{N} \times \mathbb{N} \rightarrow [0,1]$. *Let* $\mathsf{RCAU} = \mathbf{RCAU}[\mathsf{H}, \kappa, \lambda, \nu]$. *Let* A *be an adversary that makes at most* u *queries to its* NEW *oracle,* q_e *queries to its* ENC *oracle with messages of length at most* ℓ_{bit} *bits,* q_v *queries to its* VF *oracle with messages of length* $\ell_{\mathrm{bit}} + \lambda$ *bits,* p_e *queries to its* E *oracle,* p_i *queries to its* E^{-1} *oracle, and* $p = p_e + p_i$. *Assume furthermore that* $q_e \leq 2^\nu$, *and* $\ell_{\mathrm{bit}} \leq \lambda(2^{\lambda-\nu} - 2)$. *(For brevity we write* $q = q_e + q_v$.) *Then*

$$\mathsf{Adv}_{\mathsf{RCAU}}^{\mathrm{mu\text{-}ind}}(A) \leq \frac{up}{2^\kappa} + \frac{u(\ell_{\mathrm{blk}}(q_e + q_v) + 1)^2 \cdot}{2^{\lambda+1}} + \frac{u(u-1)}{2^{\kappa+1}} + uq_v \cdot \epsilon(\ell_{\mathrm{bit}}, \ell_{\mathrm{head}}), \quad (3)$$

for $\ell_{\mathrm{blk}} = \lceil \ell_{\mathrm{bit}}/\lambda \rceil + 1$ *and where the AEAD headers are restricted to* ℓ_{head} *bits.*

We prove a stronger bound for the advantage of a *passive* adversary that does not use its VF oracle in a non-trivial way. The bound differs from the one proven above significantly: we show that for *passive* adversaries we can replace the term $up/2^\kappa$ in the bound for CAU by terms that are smaller for realistic parameters. The proof does, however, not extend to *active* adversaries that make use of the VF oracle: In fact, RCAU evaluates the block cipher, in each session, on the fixed value 0^λ to obtain the key for H, and our analysis of the authenticity guarantee requires that this key be uniformly random. This requirement is of course not fulfilled if the adversary evaluated the block cipher on the value 0^λ for the respective key.

In the result for RCAU, we explicitly distinguish between the numbers for evaluation p_e and inversion p_i queries for the block cipher, with $p = p_e + p_i$.

Theorem 10. *Let $\kappa, \lambda, \nu \geq 1$ be such that $\nu \leq \lambda - 2$. Let $H: \{0,1\}^\lambda \times (\{0,1\}^* \times \{0,1\}^*) \to \{0,1\}^\lambda$ be a family of functions. Let RCAU = $\mathbf{RCAU}[H, \kappa, \lambda, \nu]$. Let A be an adversary that makes at most u queries to its NEW oracle, q_e queries to its ENC oracle with messages of length at most ℓ_{bit} bits, q_v queries to its VF oracle with messages of length $\ell_{\mathrm{bit}} + \lambda$ bits, p_e queries to its E oracle, p_i queries to its E^{-1} oracle, and $p = p_e + p_i$. Assume furthermore that $q_e \leq 2^\nu$, and $\ell_{\mathrm{bit}} \leq \lambda(2^{\lambda-\nu} - 2)$. (For brevity we write $q = q_e + q_v$.) Then*

$$\mathsf{Adv}^{\mathrm{mu\text{-}ind}}_{\mathsf{RCAU}}(A) \leq \frac{u(\ell_{\mathrm{blk}}q_e + 1)^2}{2^{\lambda+1}} + \frac{up(\ell_{\mathrm{blk}}q_e + 1)}{2^{\kappa+\nu-1}}$$

$$+ \frac{up(\ell_{\mathrm{blk}}q_e + 1)}{2^\kappa(2^\lambda - p)} + \frac{up(\ell_{\mathrm{blk}}q_e + 1)}{2^\kappa(2^\lambda - q_e)} + \frac{2p_i + u(u-1)}{2^{\kappa+1}}, \quad (4)$$

for $\ell_{\mathrm{blk}} = \lceil \ell_{\mathrm{bit}}/\lambda \rceil + 1$, and for an adversary A making $q_v = 0$ verification queries.

In comparison with the bound proven in Theorem 8, the major difference in Eq. (4) is that the term $up/2^\kappa$ is replaced by the four terms $up(\ell_{\mathrm{blk}}(q_e + q_v) + 1)/2^{\kappa+\nu}$, $up(\ell_{\mathrm{blk}}(q_e + q_v) + 1)/2^\kappa(2^\lambda - u)$, $up(\ell_{\mathrm{blk}}(q_e + q_v) + 1)/2^\kappa(2^\lambda - q_e - q_v)$ and $p_i/2^\kappa$. This is an improvement because for the values used in TLS 1.3 it is reasonable to assume $\ell_{\mathrm{blk}}(q_e + q_v) + 1 \ll 2^{96}$ as well as $q_e + q_v, p \ll 2^{96}$, and the term $p_i/2^\kappa$ does not scale with u. Unfortunately, our proof does not support a similar statement for active attacks.

We stress that the term $up/2^\kappa$ in Eq. (3) does, unlike the one in Theorem 8, not immediately corresponds to a matching attack on the use of the scheme within the TLS protocol. The reason is that such an attack would require sending a great amount of crafted ciphertexts within the TLS session, but TLS tears down a session and discards the keys after the first failure in MAC verification. Therefore, it is conceivable that the scheme as used within TLS achieves considerably better security against active attacks than our above bound suggests. Moreover, such an attack would be inherently *active* and not suitable for mass surveillance.

6.4   Security of XCAU

To analyze the indistinguishability security of XCAU, we combine the results of Theorem 5 and Lemma 7. The proof is almost the same as the one for Theorem 8, but the step of "decoupling" the E/E^{-1} and RF oracles makes use of the results in Theorem 5. Most notably and in contrast to RCAU, the bound does not contain a term of the type $p_i/2^\kappa$, and applies to active adversaries as well.

Theorem 11. *Let $\kappa, \lambda, \nu \geq 1$ be such that $\nu \leq \lambda - 2$. Let $H: \{0,1\}^\lambda \times (\{0,1\}^* \times \{0,1\}^*) \to \{0,1\}^\lambda$ be an ϵ-almost XOR-universal hash function, for some $\epsilon: \mathbb{N} \times \mathbb{N} \to [0,1]$. Let $\mathsf{XCAU} = \mathbf{XCAU}[H, \kappa, \lambda, \nu]$. Let A be an adversary that makes at most u queries to its NEW oracle, q_e queries to its ENC oracle with messages of length at most ℓ_{bit} bits, q_v queries to its VF oracle with messages of length at most $\ell_{\mathrm{bit}} + \lambda$ bits, and p queries to its E and E^{-1} oracles. Assume furthermore that $q_e \leq 2^\nu$ and $\ell_{\mathrm{bit}} \leq \lambda(2^{\lambda-\nu} - 2)$. Then*

$$\mathsf{Adv}_{\mathsf{XCAU}}^{\mathrm{mu\text{-}ind}}(A) \leq \frac{up(\ell_{\mathrm{blk}}(q_e + q_v) + 1)}{2^{\lambda+\kappa+1}} + \frac{up(\ell_{\mathrm{blk}}(q_e + q_v) + 1)^2}{2^{\lambda+1}}$$

$$+ uq_v \cdot \epsilon(\ell_{\mathrm{bit}}, \ell_{\mathrm{head}}) + \frac{u(u-1)}{2^{\kappa+1}},$$

for $\ell_{\mathrm{blk}} = \lceil \ell_{\mathrm{bit}}/\lambda \rceil + 1$, and with headers of length at most $\lambda\ell_{\mathrm{head}}$ bits.

7   Conclusion

TLS 1.2 is the most widely used cryptographic protocol in the Internet, but due to issues with both performance and security, it will soon be replaced by its successor, TLS 1.3. Given that the bulk of Internet traffic will likely be protected by TLS 1.3 in the next years, it is extremely important that the security of the protocol is well-understood. Facing the threat of mass surveillance and the expected great number of TLS 1.3 sessions, the TLS Working Group has introduced a nonce-randomization technique to improve the resilience of TLS 1.3 against such attacks.

We show that the proposed technique can be understood as a key-length extension for AE; it essentially extends the 128-bit key of AES-GCM to a 224-bit key. We first describe the authenticated encryption CAU (Counter mode Almost Universal) as an abstraction of GCM. We then describe the scheme with randomized nonces as its variant RCAU and analyze it in the multi-user setting, where we show that it improves the resilience against (passive) mass surveillance as intended by the designers. We also show, however, that the AE does not perform as well as one might expect from an AE with a 224-bit key, especially in presence of active attacks. One alternative would be to simply increase the key size by, e.g., switching to an AES-256-based mode; this achieves better security but also impacts performance.

We suggest a new encryption mode that we call XCAU. The mode uses an additional 128-bit key (256 bits in total) to randomize the inputs and outputs of the block cipher (here AES) as in DESX. The mode is almost as efficient as the mode RCAU used in TLS 1.3, only adding two 128-bit xor operations for each call to the block cipher over plain CAU, our abstraction for GCM. We show that, still, its security is improved over RCAU in two ways. The security bounds we prove for security of XCAU *against active attacks* scale significantly better in the number u of users than those for RCAU, this stems mostly from the fact that *all* inputs to the block cipher are randomized. Furthermore, the whitening of the block-cipher output allows to remove the (for realistic parameters largest) term $p_i/2^\kappa$ from the security bound. (It should be noted, however, that this term is not worrisome for realistic parameters.) The fact that the implementation of XCAU, in contrast to that of RCAU, requires non-black-box changes to the libraries implementing CAU, however, makes adoption in the currently developed standard TLS 1.3 difficult.

Acknowledgments. Bellare was supported in part by NSF grants CNS-1526801 and CNS-1228890, ERC Project ERCC FP7/615074 and a gift from Microsoft. Tackmann was supported in part by the Swiss National Science Foundation (SNF) via Fellowship No. P2EZP2_155566 and by NSF grant CNS-1228890.

References

1. Badertscher, C., Matt, C., Maurer, U., Rogaway, P., Tackmann, B.: Augmented secure channels and the goal of the TLS 1.3 record layer. In: AU, M.-H., et al. (eds.) ProvSec 2015. LNCS, vol. 9451, pp. 85–104. Springer, Heidelberg (2015). doi:10.1007/978-3-319-26059-4_5
2. Bellare, M., Bernstein, D.J., Tessaro, S.: Hash-function based PRFs: AMAC and its multi-user security. In: Fischlin, M., Coron, J.-S. (eds.) EUROCRYPT 2016. LNCS, vol. 9665, pp. 566–595. Springer, Heidelberg (2016). doi:10.1007/978-3-662-49890-3_22
3. Bellare, M., Boldyreva, A., Micali, S.: Public-key encryption in a multi-user setting: security proofs and improvements. In: Preneel, B. (ed.) EUROCRYPT 2000. LNCS, vol. 1807, pp. 259–274. Springer, Heidelberg (2000)
4. Bellare, M., Canetti, R., Krawczyk, H.: Pseudorandom functions revisited: the cascade construction and its concrete security. In: 37th FOCS, pp. 514–523. IEEE Computer Society Press, October 1996
5. Bellare, M., Namprempre, C.: Authenticated encryption: relations among notions and analysis of the generic composition paradigm. In: Okamoto, T. (ed.) ASIACRYPT 2000. LNCS, vol. 1976, pp. 531–545. Springer, Heidelberg (2000)
6. Bellare, M., Rogaway, P.: The security of triple encryption and a framework for code-based game-playing proofs. In: Vaudenay, S. (ed.) EUROCRYPT 2006. LNCS, vol. 4004, pp. 409–426. Springer, Heidelberg (2006)
7. Bellare, M., Tackmann, B.: The multi-user security of authenticated encryption: AES-GCM in TLS 1.3. Cryptology ePrint Archive, Report 2016/564 (2016). http://eprint.iacr.org/
8. Bernstein, D.J.: Multi-user Schnorr security, revisited. Cryptology ePrint Archive, Report 2015/996 (2015). http://eprint.iacr.org/2015/996
9. Boyarsky, M.K.: Public-key cryptography and password protocols: the multi-user case. In: ACM CCS 1999, pp. 63–72. ACM Press, November 1999

10. Dodis, Y., Lee, P.J., Yum, D.H.: Optimistic fair exchange in a multi-user setting. In: Okamoto, T., Wang, X. (eds.) PKC 2007. LNCS, vol. 4450, pp. 118–133. Springer, Heidelberg (2007)
11. Dworkin, M.: Recommendation for block cipher modes of operation: the CCM mode for authentication and confidentiality. NIST Special, Publication 800-38C, May 2004
12. Dworkin, M.: Recommendation for block cipher modes of operation: Galois/Counter Mode (GCM) and GMAC. NIST Special, Publication 800-38D, November 2007
13. Even, S., Mansour, Y.: A construction of a cipher from a single pseudorandom permutation. J. Cryptol. **10**(3), 151–162 (1997)
14. Fischlin, M., Günther, F., Marson, G.A., Paterson, K.G.: Data is a stream: security of stream-based channels. In: Gennaro, R., Robshaw, M. (eds.) CRYPTO 2015. LNCS, vol. 9216, pp. 545–564. Springer, Heidelberg (2015)
15. Fouque, P.-A., Joux, A., Mavromati, C.: Multi-user collisions: applications to discrete logarithm, even-mansour and PRINCE. In: Sarkar, P., Iwata, T. (eds.) ASIACRYPT 2014. LNCS, vol. 8873, pp. 420–438. Springer, Heidelberg (2014)
16. Galbraith, S., Malone-Lee, J., Smart, N.P.: Public key signatures in the multi-user setting. Inf. Process. Lett. **83**(5), 263–266 (2002)
17. Huang, Q., Yang, G., Wong, D.S., Susilo, W.: Efficient optimistic fair exchange secure in the multi-user setting and chosen-key model without random oracles. In: Malkin, T. (ed.) CT-RSA 2008. LNCS, vol. 4964, pp. 106–120. Springer, Heidelberg (2008)
18. Iwata, T., Ohashi, K., Minematsu, K.: Breaking and repairing GCM security proofs. In: Safavi-Naini, R., Canetti, R. (eds.) CRYPTO 2012. LNCS, vol. 7417, pp. 31–49. Springer, Heidelberg (2012)
19. Kilian, J., Rogaway, P.: How to protect DES against exhaustive key search (an analysis of DESX). J. Cryptol. **14**(1), 17–35 (2001)
20. Kiltz, E., Masny, D., Pan, J.: Optimal security proofs for signatures from identification schemes. Cryptology ePrint Archive, Report 2016/191 (2016). http://eprint.iacr.org/
21. Krawczyk, H.: LFSR-based hashing and authentication. In: Desmedt, Y.G. (ed.) CRYPTO 1994. LNCS, vol. 839, pp. 129–139. Springer, Heidelberg (1994)
22. Krovetz, T., Rogaway, P.: The software performance of authenticated-encryption modes. In: Joux, A. (ed.) FSE 2011. LNCS, vol. 6733, pp. 306–327. Springer, Heidelberg (2011)
23. Maurer, U.M.: Indistinguishability of random systems. In: Knudsen, L.R. (ed.) EUROCRYPT 2002. LNCS, vol. 2332, pp. 110–132. Springer, Heidelberg (2002)
24. McGrew, D.A., Viega, J.: The security and performance of the Galois/Counter Mode (GCM) of operation. In: Canteaut, A., Viswanathan, K. (eds.) INDOCRYPT 2004. LNCS, vol. 3348, pp. 343–355. Springer, Heidelberg (2004)
25. Mouha, N., Luykx, A.: Multi-key security: the even-mansour construction revisited. In: Gennaro, R., Robshaw, M.J.B. (eds.) CRYPTO 2015. LNCS, vol. 9215, pp. 209–223. Springer, Heidelberg (2015)
26. Namprempre, C., Rogaway, P., Shrimpton, T.: Reconsidering generic composition. In: Nguyen, P.Q., Oswald, E. (eds.) EUROCRYPT 2014. LNCS, vol. 8441, pp. 257–274. Springer, Heidelberg (2014)
27. Niwa, Y., Ohashi, K., Minematsu, K., Iwata, T.: GCM security bounds reconsidered. In: Leander, G. (ed.) FSE 2015. LNCS, vol. 9054, pp. 385–407. Springer, Heidelberg (2015)

28. Rogaway, P.: Authenticated-encryption with associated-data. In: Atluri, V. (ed.) ACM CCS 2002, pp. 98–107. ACM Press, November 2002

29. Rogaway, P.: Efficient instantiations of tweakable blockciphers and refinements to modes OCB and PMAC. In: Lee, P.J. (ed.) ASIACRYPT 2004. LNCS, vol. 3329, pp. 16–31. Springer, Heidelberg (2004)

30. Rogaway, P., Bellare, M.: Robust computational secret sharing and a unified account of classical secret-sharing goals. In: Ning, P., di Vimercati, S.D.C., Syverson, P.F. (eds.) ACM CCS 2007, pp. 172–184. ACM Press, October 2007

31. Rogaway, P., Bellare, M., Black, J., Krovetz, T.: OCB: a block-cipher mode of operation for efficient authenticated encryption. In: ACM CCS 2001, pp. 196–205. ACM Press, November 2001

32. Rogaway, P., Shrimpton, T.: A provable-security treatment of the key-wrap problem. In: Vaudenay, S. (ed.) EUROCRYPT 2006. LNCS, vol. 4004, pp. 373–390. Springer, Heidelberg (2006)

33. Smith, B.: Pull request: removing the AEAD explicit IV. Mail to IETF TLS Working Group, March 2015

34. Tessaro, S.: Optimally secure block ciphers from ideal primitives. In: Iwata, T., et al. (eds.) ASIACRYPT 2015. LNCS, vol. 9453, pp. 437–462. Springer, Heidelberg (2015). doi:10.1007/978-3-662-48800-3_18

A Modular Treatment of Cryptographic APIs: The Symmetric-Key Case

Thomas Shrimpton[1](✉), Martijn Stam[2], and Bogdan Warinschi[2]

[1] University of Florida, Gainesville, USA
teshrim@ufl.edu
[2] University of Bristol, Bristol, UK
{csxms,csxbw}@bris.ac.uk

Abstract. Application Programming Interfaces (APIs) to cryptographic tokens like smartcards and Hardware Security Modules (HSMs) provide users with commands to manage and use cryptographic keys stored on trusted hardware. Their design is mainly guided by industrial standards with only informal security promises.

In this paper we propose cryptographic models for the security of such APIs. The key feature of our approach is that it enables modular analysis. Specifically, we show that a secure cryptographic API can be obtained by combining a secure API for key-management together with secure implementations of, for instance, encryption or message authentication. Our models are the first to provide such compositional guarantees while considering realistic adversaries that can adaptively corrupt keys stored on tokens. We also provide a proof of concept instantiation (from a deterministic authenticated-encryption scheme) of the key-management portion of cryptographic API.

1 Introduction

Key management, i.e. the secure creation, storage, backup, and destruction of keys, has long been identified as a major challenge in all practical uses of cryptography. To achieve high levels of security, in practice one commonly relies on physical protection: store cryptographic keys inside a tamper-resistant device, called a *cryptographic token*, and only allow access to the keys (e.g. for performing cryptographic operations) indirectly through an Application Programming Interface (API). Tokens are widely deployed in practice and range from smart cards and USB sticks to powerful Hardware Security Modules (HSMs). They are used to generate and store keys for certification authorities, to accelerate SSL/TLS connections and they form the backbone of interbank communication networks.

A user with access to the token may use the API to perform—securely on the token—cryptographic operations, such as encryption or authentication of user-provided data, using the stored keys. A key feature of such APIs is their support for key management across tokens. We focus on *wrapping*, the mechanism to transport keys between devices by encrypting them under an already shared key.

© International Association for Cryptologic Research 2016
M. Robshaw and J. Katz (Eds.): CRYPTO 2016, Part I, LNCS 9814, pp. 277–307, 2016.
DOI: 10.1007/978-3-662-53018-4_11

Finally, the API prevents insecure or unauthorized use of keys, typically based on attributes and policies. Through their APIs, the overall distributed architecture provides an increased level of security for keys, simplifies access control through flexible key-management, and enables modular application development.

The design and analysis of key-management APIs mainly follows industrial standards, notably PCKS#11 [23], that are geared towards specifying functionality and interoperability. The standards typically lack a clearly defined security goal, let alone a rigorous analysis that any security claim is reasonably met. As a result, proper deployment relies strongly on best practices (undocumented in the public domain); moreover, tokens are subject to regular successful attacks [2–4,7]. This raises the question whether the security of cryptographic APIs can be captured and compartmentalized, taking into account the reality that some keys will leak.

The main symmetric operation employed in key-management, namely the key-wrapping primitive, is fairly well understood through appropriate models and efficient implementations [15,21,22]. However, the security of the overall design of cryptographic APIs is a far more complicated problem, that only recently received attention [5,17,18]. None of the existing models is entirely satisfactory: they are either too specific [5,17]; underspecified while imposing unnecessary restrictions on how PKCS#11 can be used [18]; or avoid the highly relevant issue of adaptive key corruption [5,17]. We provide a more in-depth comparison later in the paper (Sect. 6). Our model naturally and unsurprisingly shares various modelling choices with past work: We keep track of the information concerning which key encrypts which key using a graph; we maintain information about keys, handles and attributes in a similar way. Our focus is on the modular analysis where the key-management component can be analyzed separately from the cryptographic schemes that use the keys, and all of this in a reasonable corruption model.

Our Contributions. We give a formal syntax and security model for cryptographic APIs, reflecting concepts distilled from PKCS#11. We have aimed for a level of abstraction that allows for common deployment "best practices" (e.g. hierachical layering of managed keys based upon their intended use), without being overly tied to any particular implementation. Our formalism captures the core *symmetric* functionalities exposed by cryptographic APIs. Specifically, management and exporting/importing of cryptographic keys via the API; and cryptographic operations (e.g. encryption) performed under the managed keys, on behalf of applications requesting these operations via the API.

To foster modular analysis, we establish security goals for the key-management system (the abstract "back end" whose state is affected by key-management API calls). These goals are agnostic toward the particular cryptographic operations the keys will support. The primitives underlying the cryptographic operations exposed by the API are also are treated abstractly, as are their corresponding security notions.

Our key technical result shows that—as one would hope and expect—composing a secure key-management system with a secure primitive yields a secure overall system, provided certain conditions are met. Remarkably, our composition result holds while allowing adaptive corruptions of managed keys; we discuss later how we overcome the well-documented difficulties associated to merging composition and adaptive security in a single framework.

We also show how to instantiate a secure key-management system based upon deterministic authenticated-encryption (DAE). The DAE primitive was previously proposed as a method for secure "key-wrapping", loosely the symmetric encryption of key K_1 (and its associated data) under another key K_2, ostensibly for the purpose of transporting K_1 between devices that share K_2. We build upon this functionality to deliver a (minimal) secure key-management component of a cryptographic API, specifically one with hierarchical layering of keys. Below, we discuss these contributions in greater detail.

Our Syntax and Security Model. Our syntax for a cryptographic API abstractly captures the following abilities: (1) to create keys with specified attributes on a named token; (2) to wrap, and subsequently unwrap, a managed key for external transport between tokens; (3) to transport keys directly from one token to another (without (un)wrapping); and (4) to run (non-key-management) cryptographic primitives on user-provided inputs, under user-indicated keys. These operations are all subject to the policy enforced by the token. We include this dependency on the policy in our model, but leave it unspecified.

The security model exposes these capabilities to adversaries who, speaking informally, attempt to "break" the token by sequences of API calls. In particular, an adversary can create, wrap, and unwrap keys as it wishes, and use these keys in the supported cryptographic primitives. The realistic multi-token setting is captured by allowing the adversary to cause direct transfers of keys between tokens (modeling secure injection of a single key into several tokens, say during the manufacturing process, or by security officers), and by allowing it to corrupt individual keys adaptively. The latter capability models the real possibility that some keys leak, due to for instance partial security breaches or successful cryptanalysis.

Security with respect to our model will demand that all managed keys that are not compromised (directly, or indirectly by clever API calls) can be used securely by the cryptographic primitives. Our focus on the exported primitives, instead of individual keys, highlights the raison d'être of cryptographic tokens: they should guarantee the security of the operations performed with the keys that they store.

One salient feature of our model is its generality. Instead of providing a model only for say an encryption API, we work with an abstract (symmetric) cryptographic primitive. In brief, we start with an abstract security definition for arbitrary (symmetric) primitives and lift it to the setting of APIs. Our general treatment has the benefit that the resulting security definition can be instantiated for APIs that export a large class of symmetric primitives (including all of the usual ones).

Composition Theorem. The main technical contribution of this paper is a modular treatment for cryptographic APIs. As a first step, we isolate the core common component shared by cryptographic tokens, namely key-management, and provide a separate security model for it. Essentially, we define a key-management API (or KM-API, in short) to be a cryptographic API that allows only key-management operations. We define security of a KM-API to mean that any key that is not trivially compromised (directly or indirectly) is indistinguishable from random.

Next, we show how to compose a KM-API and an arbitrary (abstract) primitive. We require common sense syntactic restrictions to ensure the composition is meaningful (e.g. that the space of keys managed by the KM-API fits the one of the symmetric primitives). More importantly, the design that we propose requires that each key is used for either key-management or for keying the primitive, but not for both. Many of the existent attacks on APIs are the result of careless enforcement of this separation of key roles. Technically, we enforce this requirement via a mechanism from the PKCS#11 standard—the security of our construction essentially confirms the validity of this mechanism.

In a nutshell, to each key an attribute is associated making the key either external or not.

We ensure the attribute has the desired effect by requiring that it is *sticky*. This notion formalizes an integrity property for attributes informally defined by PKCS#11. It guarantees that once set, the value of an attribute cannot be changed. The following theorem establishes the security of our design, allowing for the two components to be designed and analyzed separately.

Theorem 1 (Informal). *If CA is a secure KM-API and P is a secure primitive then the composition of CA and P (as above) is a secure cryptographic API that exports P.*

Importantly, the composition theorem is for a setting where adversaries can *adaptively* corrupt keys. Our models rely on game-based definitions, which is the main tool that we use to reconcile composition and adaptive corruption, two features that raise well-known problems in settings based on simulation [6,20].

Construction Based on Deterministic Authenticated Encryption. We show that a secure KM-API can be built upon DAE schemes. In particular, we show that when the wrapping and unwrapping functionalities are implemented by a secure DAE scheme, one can securely instantiate a KM-API for an abstract "back end" that enforces hierachical layering of keys. Keys at the lowest layer of the hierarchy are used only to key the cryptographic primitives (we call these *external* keys), and keys above this are used only to wrap keys at lower layers (we call these *internal* keys). Whether a key is external or internal is specified in that key's attributes. To wrap external key K_1 under internal key K_2, the encryption algorithm of a DAE scheme is used, and the attributes of key K_2 serve as the associated data. (Of course, the KM-API only allows calling applications to indicate which keys are to be involved, not the actual key values.) The design of our proposed API ensures that the API policy will enforce layering.

Extensions. Our ultimate goal is to provide usable security models that should facilitate the analysis of security tokens in realistic scenarios. In this paper, for simplicity we restricted attention to the symmetric aspect of APIs only; moreover our security definition for cryptographic APIs only concerns the primitives they export. We do not address other properties that can be enforced by token policies, e.g. that internal policies may restrict operations to authenticated users that log-in to the token. Such policies play important roles in the logic of applications that rely on tokens. Nonetheless, we believe our model provides a suitable starting point for further extension. Indeed, we already incorporate attributes and use a very simple policy to enforce the security of our composition. We leave identifying and formalizing the intended semantics for other PKCS#11 attributes and extending to public-key functionality as an open problem.

2   Cryptographic Primitives

In this section we provide an abstract framework for *cryptographic primitives* that captures common goals such as encryption and message authentication. Our abstraction is tailored specifically for its subsequent use in defining (Sect. 3) and constructing (Sect. 4) cryptographic APIs. Thus, while our abstraction is rather general the choices regarding to what to abstract and what to make explicit in our framework are strongly motivated by the later context.

Standard notions of encryption and authentication (e.g., IND-CPA and EUF-CMA) are usually defined based on a single key and corruption of this single key is seldom considered: it typically renders the game trivial (either the adversary wins easily, or winning is information theoretically impossible). Adding explicit corruption to the single-key security model facilitates moving to the multi-key scenario (that is needed in the more general API setting). There are also true multi-key definitions in the literature (e.g. for key-dependent message security), but for technical reasons we require a modular multi-key definition that is induced by a single-key one.

Syntax. A primitive P is defined by a pair of stateless, randomized algorithms (Kg_P, Alg_P). Algorithm Kg_P takes as input some parameter pm and generates a key from some set $Keys_{pm}$; here the distribution may depend on the parameter (e.g. which key length to use). Algorithm Alg_P implements the functionality of the primitive, taking as input both a key and a primary input in, and producing an output out. Without loss of generality, the definition of the primitive requires only a single formal algorithm. If some functionality is naturally implemented using several algorithms (e.g. one for encryption and decryption each) these can all be "packed" inside Alg_P by tagging the input to Alg_P with a label that indicates which of the natural algorithms is to be executed. This means that our framework also captures a situation where multiple "types" of primitives (e.g. both encryption and a MAC) need to be supported, as all relevant algorithms can be neatly packed in the single Alg_P (for which several distinct security notions can be defined, e.g. one for confidentiality and one for authenticity).

Correctness. Correctness is usually defined as a requirement on a sequence of calls that involve the algorithms that define a primitive. For instance encrypting an arbitrary message and subsequently decrypting the ciphertext (under the same key) should return the original message. Definition 1 captures this idea in the context of arbitrary primitives. For generality, the definition is formulated in a setting with multiple keys.

We consider an adversary that can create keys of its choice for the primitive (using oracle NEW), and can invoke the algorithms of the primitive, via oracle ALG using the index i of a key K_i. The experiment maintains a list tr that records the execution trace: the occurrence of triple (i, x, y) in the trace indicates that $\mathsf{Alg_P}$ was invoked on key K_i, with input x and returning y. The correctness of P is captured by a predicate $\mathsf{corr_P}$ applied to the execution trace. Usually $\mathsf{corr_P}$ will be monotone: initially, for the empty trace, it will be true and, once set to false, it will remain false.

game $\mathbf{Exp_P^{corr_P}}(\mathcal{A})$:	**oracle** NEW(K):	**oracle** ALG(j, in):
$i \leftarrow 0, \mathsf{tr} \leftarrow [\,]$	if $K \notin$ Keys then	if $j > i$ then return $\frac{\iota}{\iota}$
$\mathcal{A}^{\mathrm{NEW},\mathrm{ALG}}$	return $\frac{\iota}{\iota}$	$y \leftarrow_{\$} \mathsf{Alg_P}(K_j, \mathsf{in})$
return $\mathsf{corr_P}(\mathsf{tr})$	$i{+}{+}$	$\mathsf{tr} \leftarrow \mathsf{tr} :: (j, \mathsf{in}, \mathsf{out})$
	$K_i \leftarrow K$	return out
	return K_i	

Fig. 1. The experiment $\mathbf{Exp_P^{corr_P}}(\mathcal{A})$ (with oracles) to define correctness for primitive $\mathsf{P} = (\mathsf{Kg_P}, \mathsf{Alg_P})$.

Definition 1. *Let* $(\mathsf{Kg_P}, \mathsf{Alg_P})$ *implement a primitive* P *with* Keys $\supseteq$ $\bigcup_{\mathsf{pm}}[\mathsf{Keys_{pm}}]$. *Let* $\mathsf{corr_P}$ *be a correctness predicate and* $\mathcal{A}$ *an adversary, then the incorrectness advantage of* $\mathcal{A}$ *against* $(\mathsf{Kg_P}, \mathsf{Alg_P})$ *with respect to* $\mathsf{corr_P}$ *is defined as*

$$\mathbf{Adv_P^{corr_P}}(\mathcal{A}) = \mathbf{Pr}\left[\mathbf{Exp_P^{corr_P}}(\mathcal{A}) = \mathsf{false}\right]$$

for the experiment $\mathbf{Exp_P^{corr_P}}(\mathcal{A})$ *as given in Fig. 1.*

We call P *correct with respect to* $\mathsf{corr_P}$ *iff for all (terminating) adversaries the advantage is* 0.

Security. Next, we introduce a formalism for specifying security notions for symmetric primitives. We first consider the case of a single key (which we associate with index 1) and then extend the formalism to the case of multiple keys.

Single-Key Scenario. A security notion for primitive P is given by four algorithms $\mathsf{sec} = (\mathsf{setup}, \mathsf{chal_0}, \mathsf{chal_1}, \mathsf{chal_{aux}})$. Informally, these algorithms define two experiments $\mathbf{Exp_P^{1sec(pm)\text{-}0}}$ and $\mathbf{Exp_P^{1sec(pm)\text{-}1}}$ which characterize security in terms of an adversary that tries to distinguish between the two. Both experiments maintain a state st initialized via the algorithm setup.

In experiment $\mathbf{Exp}_P^{\mathsf{1sec(pm)}-b}(\mathcal{A})$ (b is either 0 or 1), the adversary has access to the algorithm Alg_P only indirectly through its challenge oracle CHAL$_b$ and the auxiliary oracle CHAL$_{\mathsf{aux}}$. The behavior of these oracles is defined by the algorithm chal$_x$ (for the relevant $x \in \{0, 1, \mathsf{aux}\}$) which has both access to the game's state and oracle access to the actual primitive algorithm Alg_P. Our formalization generalizes many of the standard definitions for security of cryptographic primitives, where an adversary needs to distinguish between two "worlds" (modeled here by oracles CHAL$_b$ with $b = 0, 1$). For example, to define indistinguishability under chosen-plaintext attack for probabilistic symmetric encryption schemes, we would instantiate oracle CHAL$_b$ with a left-right oracle that receives a pair of messages m_0, m_1, checks that they have the same length, and returns an encryption of m_b. Oracle CHAL$_{aux}$ would allow the adversary to see encryptions of whatever messages it wants. Security under chosen-ciphertext attacks can be captured by letting oracle CHAL$_{aux}$ also answer decryption queries.

Without loss of generality we assume chal$_x$ makes at most one call to Alg_P. The state of the game allows the algorithm chal$_x$ to suppress or modify the output of Alg_P, for instance to avoid the decryption of a challenge ciphertext being made available directly to the adversary. Of course, how a sequence of calls to a chal$_b$ and chal$_{\mathsf{aux}}$ interact with each other is specific to the security game.

Our model allows the adversary to corrupt the secret key. The distinction between the *algorithm* chal$_b$ and the *oracle* CHAL$_b$ as an interface to chal$_b$ allows us to deal with corruptions explicitly: if the key is corrupted, the interface CHAL$_b$ will suppress the output of the algorithm chal$_b$. We record if the key is used in some challenge oracle CHAL$_b$ in set H and record its corruption using set C and then prevent trivial wins by appropriate checks.

Multi-Key Scenario. When utilized within tokens, primitives are effectively in a multi-key setting. Looking ahead, our definition for cryptographic API security essentially bootstraps from the security of primitives in a standalone scenario as described above to when used in this more complex scenarios.

3   Cryptographic APIs

A cryptographic API is an interface between an untrusted usage environment, and a trusted environment that stores cryptographic objects (e.g. keys) and carries out cryptographic operations (e.g. encryption). In practice, the trusted environment is instantiated as a hardware token, or a hardware security module; we will simply refer to the trusted environment as the token. A user may request, via the cryptographic API, that the token carry out cryptographic operations on the user's behalf. In typical scenarios, the user will also control which key or keys are to be used, by specifying one or more handles to these keys. However, the cryptographic value of the key (as stored on the token) should remain hidden from the user and the outside world in general.

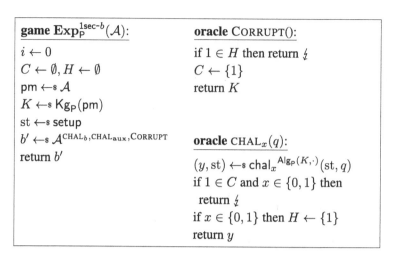

Fig. 2. The experiments $\mathbf{Exp}_{P}^{\mathrm{1sec(pm)}-b}(\mathcal{A})$ for the single key security notion 1sec defined by the tuple (setup, chal$_0$, chal$_1$, chal$_{\mathrm{aux}}$) for primitive $\mathsf{P} = (\mathsf{Kg}_\mathsf{P}, \mathsf{Alg}_\mathsf{P})$. The single key in the system has implicit index 1 and is generated using parameter pm selected by the adversary from a set of possible parameters.

To protect the confidentiality and proper usage of exported and imported keys, tokens employ key wrapping and unwrapping mechanisms. Oftentimes there are multiple tokens in the same cryptologic ecosystem. In this case, keys may be exported from one trusted token to another (via the API). Thus our abstraction includes (a minimal set of) explicit key management functions, and an interface to use some specific cryptographic primitive.

The ultimate goal of a cryptographic API is the correct and secure implementation of some cryptographic primitive and our main target in this section are appropriate definitions of correctness (Definition 3) and security (Definition 5). These definitions build on the abstract notion of a cryptographic primitive from Sect. 2.

As explained in the introduction, the focal point of this paper is on those aspects associated to key-management, shared by cryptographic APIs. For instance, when wrapping a key, the expectation is that after unwrapping the original, wrapped key emerges (correctness) and that this key has not leaked (security), e.g. as a result of the wrapping. We provide a separate set of notions relevant for the key management part of a cryptographic API (Definitions 4, 6, and 7).

3.1   Modeling and Syntax

Tokens, Handles, Keys, and Attributes. Formally, we model a token t as having some abstract state $s \in \mathsf{States}$ plus a number of associated handles. For simplicity, we assume the token identity t is a (unique) natural number and let the token's initial state consist of this identity only. When API calls to the token are being made, its state might evolve arbitrarily.

Handles are part of some set Handles (that itself can be thought of as some fixed, finite subset of $\{0,1\}^*$). Each handle belongs to a unique token, identified by $\mathsf{tkn}(h)$, and points to an actual cryptographic key value, denoted $h.\mathsf{key}$. Since the key will be stored on the token $\mathsf{tkn}(h)$, the value represented by $h.\mathsf{key}$ depends on the token's state. Since this state is not static, $h.\mathsf{key}$ could change over time. The different notation, $\mathsf{tkn}(h)$ versus $h.\mathsf{key}$, captures the distinction between immutable properties associated to a handle (possibly for bookkeeping within a cryptographic game) and changeable quantities that are associated to it directly by the API.

The association between a handle and a cryptographic key is annotated by an attribute, denoted $h.\mathsf{attr}$. For instance, an attribute could indicate that the handle points to a 128-bit AES key to be used in some specific mode of operation only, say CBC-MAC.

Like the key, the attribute will be stored on the cryptographic token (and could change over time). We will assume that $h.\mathsf{attr} \in$ Attributes, where Attributes is some fixed set of possible attributes. Note that the abstraction to a single attribute only is without loss of generality, as one can capture say the more traditional setting of many Boolean attributes by a single attribute (in this case a true/false vector).

Our model is purposefully abstract, but it is worth bearing in mind typical implementations as used in practice. For instance, PKCS #11 reliance on 'objects' implies that a token's state will contain a mapping between handles and key–attribute pairs, plus additional information that helps the token to maintain the security policy. Thus, for most APIs it will be possible to write the state explicitly in the form $s = (\tilde{s}, (h \mapsto (\mathsf{key}, a))_h)$, where for each handle h, the mapping $h \mapsto (\mathsf{key}, a)$ indicates the associated key and attribute pair (so $h.\mathsf{key} = \mathsf{key}$ and $h.\mathsf{attr} = a$), and the state $\tilde{s}$ contains a snapshot of the token's past I/O only (which in principle could be made public without compromising core cryptographic security).

The Application Programming Interface (API). Each token runs an API that allows the outside world to interface with the keys present on the token. Definition 2 lists the procedures supported by our abstract API. Intuitively, each of the API procedures has a clearly specified objective. For instance, there is an API call $\mathsf{CA.new}(t, a)$ that is *supposed* to create a new key on the token t and returns a fresh handle h such that $h.\mathsf{key}$ is this newly generated key and $h.\mathsf{attr} = a$. Here freshness is global and means that the handle does not yet occur elsewhere, so that a handle can uniquely be associated to a token (explicitly embedding the token identity in the handle could facilitate global freshness). While the syntax thus guarantees uniqueness of the handles returned by the API calls, there is no guarantee that API calls behave as intended (other than possibly implied by the correctness properties introduced later).

Definition 2. *A cryptographic API* CA *exporting a primitive* P *(cf. Sect. 2) is defined by the following tuple of algorithms.*

- $h \leftarrow_{\$} \mathsf{CA.new}(t, a)$ *creates and returns a fresh handle on token* t, *so* $\mathsf{tkn}(h) = t$; *the intention is that* $h.\mathsf{attr} = a$ *and* $h.\mathsf{key}$ *is a newly generated key, drawn from some set* Keys *according to a distribution that could for instance depend on* a.
- $h \leftarrow_{\$} \mathsf{CA.create}(t, \mathsf{key}, a)$ *creates and returns a fresh handle on token* t, *so* $\mathsf{tkn}(h) = t$; *the intention is that* $h.\mathsf{attr} = a$ *and* $h.\mathsf{key} = \mathsf{key}$.
- $w \leftarrow_{\$} \mathsf{CA.wrap}(h_1, h_2)$ *takes as input two handles and runs on the first handle's token* $\mathsf{tkn}(h_1)$. *It returns some* $w \in \mathsf{CWraps}$, *where* CWraps *is the space of all wraps. Supposedly* w *is a wrap of* $h_2.\mathsf{key}$ *tied to* $h_2.\mathsf{attr}$ *under key* $h_1.\mathsf{key}$.
- $\bar{h} \leftarrow_{\$} \mathsf{CA.unwrap}(h, w, a)$ *takes as input a handle to use for unwrapping, a wrap and an attribute string. If unwrapping succeeds, a fresh handle* $\bar{h}$ *is created on* $\mathsf{tkn}(h)$ *and returned. The intention is that* $\bar{h}.\mathsf{attr} = a$ *and* $\bar{h}.\mathsf{key}$ *equals the key that was wrapped under* $h.\mathsf{key}$.
- $\mathsf{out} \leftarrow_{\$} \mathsf{CA.alg}(h, \mathsf{in})$ *intends to evaluate the primitive* $\mathsf{Alg_P}$ *on key* $h.\mathsf{key}$ *and input* in, *returning* out.

Any call may result in an API error $\perp_{\mathrm{api}}$. *An API for key-management only may omit the procedure* CA.alg.

All of the above commands, but the CA.create, reflect the typical interface available to the user of a token. We use CA.create as an abstraction of (often non-cryptographic) mechanisms for transferring keys from one token to another. For example, in the production phase the same cryptographic key may be injected in several devices (which are to be used by the same company).

The procedures of the API directly manipulate the state of *one* token only, where the relevant token is either made explicit by the API call (CA.new and CA.create), or it follows from the handles involved (e.g. $\mathsf{CA.wrap}(h_1, h_2)$ can affect the state of $\mathsf{tkn}(h_1)$). We could make this manipulation explicit by keeping track of the token's state as input and output of each of the API's procedures etc. For readability, we keep the state of the token implicit and only stress that the commands may not depend on, or modify, the state of *another* token.

Policies and Attributes. To protect the security of the keys the API will enforce a policy. For instance, an API may forbid usage of a key intended for authentication to be used for encryption. To indicate that an operation is not allowed, an API call can return a policy error message (distinct from possible error messages resulting for instance from decrypting an invalid ciphertext). For simplicity, we will model all possible policy errors with a single[1] symbol $\perp_{\mathrm{api}}$.

We will not give a formal definition of what constitutes a policy. Actually, the level of abstraction of our model makes it somewhat cumbersome to pin down an exact, yet general concept of a policy. In a practical, multi-token setting, the use of attributes is useful to enforce consistent yet efficient implementation of a policy across tokens. We will see a concrete example of this in Sect. 4 (see Definition 8).

---

[1] An extension of our model could consider a more fine-grained level of errors, identifying why an operation results in an error.

An API can also use the token's state for this decision (e.g. to prevent wrapping a sensitive key under a key that is somehow deemed insecure or to avoid circularity). For instance, a token could keep track of all the calls (with responses) ever made to it (note that, with the exception of the key value of CA.create queries, this information can all be made public). If only a single token exists, this leads to a complete history of the API's use, which suffices to implement (albeit inefficiently) a meaningful security policy (cf. [5]).

Enforcing Meaning. So far our syntax does not formally give any guarantees that h.key and h.attr are used by the API in an explicit, meaningful way. The generality of our notion of state would allow an API to for instance declare some key as h.key but in fact use a completely different cryptographic value throughout. The KSW definitions, which use a similar abstract state as our work, share this problem, but leave it unaddressed.

Since working completely abstractly (e.g. making no assumptions on states) seems to easily lead to difficulties without obvious gains we make explicit assumptions regarding the implementations. Our upcoming correctness notion deals with the wrapping mechanism as a means to transfer keys from token to token. Notice that wrapping involves h.key where h is the 'source' handle, and only implicitly involves the associated key. Since, we would like to reflect that the actual key is transferred we need to make explicit the assumption that wraps are linked to actual cryptographic keys. Along similar lines, we make explicit the assumption that the cryptographic operations exported by the API make use of actual keys. The assumption is useful to define and analyze the composition between an API for key-management with actual primitives. Moreover, we will use the attributes to create a policy separating keys that can be used by the primitive and those that cannot. This slight loss of generality enables simpler definitions and analysis and still reflects virtually all designs commonly used in practice.

3.2   Correctness of a Cryptographic API

In this section we present a definition of correctness for a cryptographic API. Much of the discussion and formalization is relevant to the latter sections where we define security since both for correctness and security we explain how to lift the definitions of Sect. 2 from primitives to primitives exported by the APIs.

The main difficulty is an important difference between the interfaces that an adversary has against a primitive and against a primitive exported by an API. In Sect. 2, primitive correctness is modeled as a predicate on the execution trace of an adversary, where the trace keeps track of both the keys that are generated and of the cryptographic operations that the adversary executes with these keys. Crucially, the trace only included the indexes of the keys and not their cryptographic values. In contrast, an adversary against the API refers to the underlying keys using the handles provided by the API. Notice that the difference goes further, in that several handles may point to the same cryptographic key.

To bridge this gap we introduce a mapping that associates to each handle some index. The map idx that we introduce reflects the idea that handles with same index have associated the same cryptographic key.[2] Formally, when defining the oracles used by an adversary to interact with a cryptographic API, we explicitly keep track of the indexes associated to handles—we explain below our modeling. We then lift the definitions from primitives to primitives exported by APIs by (essentially) replacing the handles with their associated index in the execution trace. We detail below our approach.

Indexing Handles by Equivalence Classes. To each handle h we will assign an index $\mathsf{idx}(h) \in \mathbb{N}$ as soon as the handle is created following some key-management operation. This indexing induces an equivalence relation: two handles h_1 and h_2 are equivalent iff $\mathsf{idx}(h_1) = \mathsf{idx}(h_2)$. We aim to ensure that if two handles are expected to have the same associated cryptographic key then they should belong to the same class. Notice that we aim to maintain this property globally, i.e. the mapping handles to indexes is "system wide" and is not restricted to one particular token.

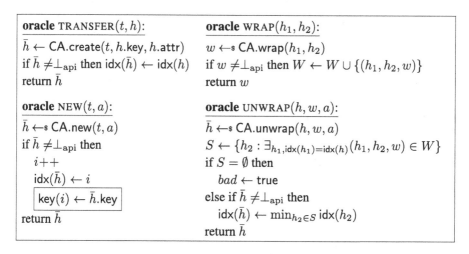

Fig. 3. Oracles used in experiments $\mathbf{Exp}^{\mathsf{corr}_\mathsf{P}}_{\mathsf{CA}[\mathsf{P}]}(\mathcal{A})$ and $\mathbf{Exp}^{\mathsf{corr}_\mathsf{km}}_{\mathsf{CA}}(\mathcal{A})$ that define the correctness of a crypto API CA. The boxed line is only relevant for the experiment involving $\mathsf{corr}_\mathsf{km}$.

Formal Definitions. Our formal definitions of correctness for key-management (Fig. 5) and primitive-exporting APIs (Fig. 4) use the oracles in Fig. 3 to model the interaction of an adversary with the API via key-management commands. Each oracle reflects the behavior of the API and contains the bookkeeping that

---

[2] Notice that the converse implication is not necessarily true.

we do to maintain and assign equivalence classes to handles. The games where these oracles are used maintain a global variable i (initially 0) that counts the number of equivalence classes.

The only way to create a new equivalence class is through the NEW oracle: whenever oracle NEW is called successfully (i.e. does not return $\perp_{\mathrm{api}}$) we increment i and assign it as the index of the handle that is returned. Handles can be added to the equivalence classes through calling either the TRANSFER or the UNWRAP oracle.

The TRANSFER oracle is used, as explained earlier, for bootstrapping purposes: to create a wrap on one token and then unwrap it on another one, the two tokens already need to contain the same key. Oracle TRANSFER models this ability: handle $\bar{h}$ pointing to the transferred key has the same index as h which points to the original key.

Dealing with handles created via unwrapping requires some more bookkeeping. We use set W (initially empty) to maintain all wraps created by the WRAP oracle, together with the handles involved: we add (h_1, h_2, w) to W if w was the result of wrapping (the key associated to) h_2 under (the key associated to) h_1. When calling UNWRAP(h, w, a) we use W to test whether w was created by wrapping some h_2 under a handle h_1 equivalent to h (the set S contains all such h_2). If this is the case (S is not empty), then the newly returned handle is equivalent to h_2 and is therefore assigned the same index. In case a wrap w was created multiple times, the lowest applicable index is used (if the key-management component is secure, it should not be possible to create identical wraps under non-equivalent handles). If S is empty, the wrap w is adversarially generated and, since we do not wish to consider dishonest adversaries for defining the correctness of a cryptographic API, we set the flag bad to force an adversarial loss.

Valid Traces. The calls that the adversary makes to the algorithm CA.alg (through its oracle ALG) are recorded in a similar way as done in the experiment for primitive correctness (Fig. 1). To account for the possibility that the same key is used via equivalent handles, we identify the key used in the cryptographic operation by the *index* of the handle. For an ALG call, this derived index neatly matches the index of the algorithm used in our multi-key primitive definition.

Definition 3. *Let API* CA[P] *implement a primitive* P *and let* corr$_P$ *be a correctness predicate. Then the incorrectness advantage of* $\mathcal{A}$ *against* CA[P] *with respect to* corr$_P$ *is defined as*

$$\mathbf{Adv}^{\mathrm{corr}_P}_{\mathsf{CA[P]}}(\mathcal{A}) = \mathbf{Pr}\left[\mathbf{Exp}^{\mathrm{corr}_P}_{\mathsf{CA[P]}}(\mathcal{A}) = \mathsf{false}\right]$$

for the experiment $\mathbf{Exp}^{\mathrm{corr}_P}_{\mathsf{CA[P]}}(\mathcal{A})$ *as given in Fig. 4. We call* CA[P] *correct with respect to* corr$_P$ *iff for all (terminating) adversaries the advantage is* 0.

Note that correctness only really implies consistency, it does not incorporate robustness. There is no guarantee that a successfully wrapped key can in fact be

game $\mathbf{Exp}^{\mathsf{corrp}}_{\mathsf{CA[P]}}(\mathcal{A})$:	**oracle** $\mathrm{ALG}(h, \mathsf{in})$:
$W \leftarrow \emptyset, i \leftarrow 0, bad \leftarrow \mathsf{false}$	$\mathsf{out} \leftarrow\!\!{\scriptstyle\$}\ \mathsf{CA.alg}(h, \mathsf{in})$
$\mathrm{tr} \leftarrow []$	if $\mathsf{out} \neq \perp_{\mathrm{api}}$ then
$\mathcal{A}^{\mathrm{ALG}, \mathcal{O}}$	$\mathrm{tr} \leftarrow \mathrm{tr} :: (\mathsf{idx}(h), \mathsf{in}, \mathsf{out})$
return $\mathsf{corr_P}(\mathrm{tr}) \vee bad$	return out

Fig. 4. The experiment $\mathbf{Exp}^{\mathsf{corrp}}_{\mathsf{CA[P]}}(\mathcal{A})$ for defining the correctness of a crypto API CA that exports primitive P based on correctness predicate $\mathsf{corr_P}$. An adversary additionally has access to the oracles $\mathcal{O}$ given in Fig. 3.

unwrapped at all, or that a primitive API call will result in an evaluation of the primitive. In both cases, the policy might well result in $\perp_{\mathrm{api}}$, in which case the correctness game effectively ignores the output of the corresponding call. As an extreme example, the cryptographic API that *always* returns $\perp_{\mathrm{api}}$ is considered correct.

3.3   Correctness of an API's Key Management

For the correctness definition above, we only looked directly at the final primitive calls, ignoring the cryptographic key values. However, intuitively if two handles are equivalent, one might expect that the associated cryptographic keys are identical. This intuition is captured by the experiment described in Fig. 5, where an adversary tries to find a handle pointing to a key distinct from the key associated to the handle's index.

game $\mathbf{Exp}^{\mathsf{corr_{km}}}_{\mathsf{CA}}(\mathcal{A})$:
$W \leftarrow \emptyset, i \leftarrow 0, bad \leftarrow \mathsf{false}$
$h \leftarrow\!\!{\scriptstyle\$}\ \mathcal{A}^{\mathcal{O}}$
return $h.\mathsf{key} = \mathsf{key}(\mathsf{idx}(h)) \vee bad$

Fig. 5. The experiment $\mathbf{Exp}^{\mathsf{corr_{km}}}_{\mathsf{CA}}(\mathcal{A})$ for defining the correctness of the key management component of a cryptographic API CA. An adversary has access to the oracles $\mathcal{O}$ given in Fig. 3.

Definition 4 (Correctness of the Key Management). *Let* CA *be a key management API and* $\mathcal{A}$ *an adversary. Then the advantage of* $\mathcal{A}$ *against* CA*'s key correctness is defined as*

$$\mathbf{Adv}^{\mathsf{corr_{km}}}_{\mathsf{CA}}(\mathcal{A}) = \mathbf{Pr}\left[\mathbf{Exp}^{\mathsf{corr_{km}}}_{\mathsf{CA}}(\mathcal{A}) = \mathsf{false}\right]$$

for the experiment $\mathbf{Exp}^{\mathsf{corr_{km}}}_{\mathsf{CA}}(\mathcal{A})$ *as given in Fig. 5. We call* CA *key-correct iff for all (terminating) adversaries the advantage is* 0.

Note that correctness of the key management component of a cryptographic API does not relate to the attribute. For deployed systems, it is common that equivalent handles are associated using different attributes; moreover, these attributes might change over time. Nonetheless, some attributes should not easily be changed by an adversary. For example, it should not be possible to change an attribute that declares a key as "sensitive" (a PKCS#11 term).

This relates to the well-known notion of stickyness, for which we provide a formal definition later on (Definition 7).

Cryptographic Key Wrap Assumption. Definition 2 mentions that CA.wrap is supposed to wrap h_2.key tied to h_2.attr under key h_1.key. Implicitly, this assumes that knowledge of both w and h_1.key suffices to determine h_2.key as well. For most schemes used in practice this is indeed the case, however it does not follow logically from our abstract syntax (even when taking into account correctness of the key management component).[3] Assumption 2 formalizes the idea that an honestly, successfully generated wrap $w \leftarrow$ CA.wrap(h_1, h_2) contains sufficient information to recover the wrapped key h_2.key, provided one knows the actual key h_1.key used for wrapping, and the attributes h_1.attr, h_2.attr associated to the handles in the wrapping command.

Henceforth, we will restrict our attention to schemes satisfying the key wrap assumption (which has direct consequences for the security notion we consider in the upcoming sections).

Assumption 2 (Key Wrap Assumption). *A cryptographic API* CA *satisfies the key wrap assumption iff there exists an extractor U that extracts keys from wraps. Specifically, for all $w \leftarrow$ CA.wrap(h_1, h_2), $w \neq \perp_{\mathrm{api}}$ with, at the time of calling, $\mathrm{key}_1 = h_1$.key and $\mathrm{key}_2 = h_2$.key it holds that $U(w, \mathrm{key}_1, h_1.\mathrm{attr}, h_2.\mathrm{attr})$ outputs key_2 with probability 1.*

3.4   Security of a Cryptographic API

We will consider three types of security. Our primary concern is the security of the exported primitive (Definition 5), of secondary concern are security of keys managed internally by the API (Definition 6) and the integrity of the attributes (Definition 7). The various security experiments to define these notions rely on a set of common oracles, given in Fig. 6. With the exception of CORRUPT and ATTRIB, the oracles match those for the correctness game (as given in Fig. 3), but with more elaborate internal bookkeeping, whose reasoning is explained below. The oracles NEW and UNWRAP contain a macro initclass that will be defined depending on the game.

---

[3] As an example, a scheme could effectively share the key over multiple wraps, where unwrapping fails (outputs $\perp_{\mathrm{api}}$) unless sufficient shares (wraps) have been received: no single wrap will allow extraction of the key.

oracle NEW(t, a):

$\bar{h} \leftarrow_\$ \mathsf{CA.new}(t, a)$
if $\bar{h} \neq \perp_{\mathsf{api}}$ then
  $i{+}{+}$
  $\mathsf{idx}(\bar{h}) \leftarrow i$
  $V \leftarrow V \cup \{i\}$
  initclass
return $\bar{h}$

oracle CORRUPT(h):

$C \leftarrow C \cup \{\mathsf{idx}(h)\}$
return $h.\mathsf{key}$

oracle TRANSFER(t, h):

$\bar{h} \leftarrow \mathsf{CA.create}(t, h.\mathsf{key}, h.\mathsf{attr})$
if $\bar{h} \neq \perp_{\mathsf{api}}$ then $\mathsf{idx}(\bar{h}) \leftarrow \mathsf{idx}(h)$
return $\bar{h}$

oracle ATTRIB(h):

return $h.\mathsf{attr}$

oracle WRAP(h_1, h_2):

$w \leftarrow_\$ \mathsf{CA.wrap}(h_1, h_2)$
If $w \neq \perp_{\mathsf{api}}$ then
  $W \leftarrow W \cup \{(h_1, h_2, w)\}$
  $E \leftarrow E \cup \{(\mathsf{idx}(h_1), \mathsf{idx}(h_2))\}$
return w

oracle UNWRAP(h, w, a):

$\bar{h} \leftarrow_\$ \mathsf{CA.unwrap}(h, w, a)$
if $\bar{h} \neq \perp_{\mathsf{api}}$ then
  $S \leftarrow \{h_2 : (h_1, h_2, w) \in W$ where
      $\mathsf{idx}(h_1) = \mathsf{idx}(h)\}$
  if $S \neq \emptyset$ then $\mathsf{idx}(\bar{h}) \leftarrow \min_{h_2 \in S} \mathsf{idx}(h_2)$
  else if $\mathsf{idx}(h) \in L(C)$ then $\mathsf{idx}(\bar{h}) \leftarrow 0$
  else
    $i{+}{+}, \mathsf{idx}(\bar{h}) \leftarrow i, V \leftarrow V \cup \{i\}$
    initclass
return $\bar{h}$

Fig. 6. Oracles common to the security experiments $\mathbf{Exp}_{\mathsf{CA[P]}}^{\mathsf{sec}\text{-}b}(\mathcal{A})$, $\mathbf{Exp}_{\mathsf{CA}}^{\mathsf{km}\text{-}b}(\mathcal{A})$, and $\mathbf{Exp}_{\mathsf{CA}}^{\mathsf{sticky}}(\mathcal{A})$, for a crypto API CA that exports $\mathsf{P} = (\mathsf{Kg_P}, \mathsf{Alg_P})$. The macro initclass is defined separately for each of the experiments.

Corrupt and Compromised Handles. We have explained earlier in the context of the *correctness* game for APIs that "honest" wrap/unwrap queries induce an equivalence relation on handles, and how the equivalence class of a handle can be represented (and maintained) by an index. For defining the *security* of APIs we also have to take into account adversaries that may be actively trying to subvert the system. In addition to dishonest API calls (e.g. asking for unwrappings of adversarially created wraps), we will also model *corruptions* of handles. When an adversary corrupts a handle, the associated cryptographic key is returned to the adversary. Note that the API itself is *not* aware of corruptions. Moreover, corruptions and (dishonest API) calls tend to reinforce each other, which we model by compromised handles, namely those handles for which an adversary can reasonably be assumed to know the corresponding key. The notion of corrupt and compromised handles is based on ideas similar to those used by Cachin and Chandran [5], and Kremer et al. [18].

Corruptions. The premise of cryptographic APIs is that keys should be kept secret and are stored securely—an adversary does not have access to cryptographic keys. Yet, in practice keys that are initially stored securely on HSMs might be exported to weaker tokens that can be breached physically (e.g. by means of side-channel analysis or fault injection). As a result, the adversary can learn these keys. Such leakage of keys is modeled by corruptions: an adversary can issue a corruption request of a handle to learn the associated key. In general, one cannot guarantee security for handles that have been corrupted (cf. the primitive's security game). Moreover, corruption of a handle automatically leads to corruption of the equivalence class of that handle (as equivalent handles are presumed to point to identical cryptographic keys). We let C be the set of indices corresponding to handles that have been corrupted directly by the adversary through making a corruption query.

Compromised Handles. An adversary could issue a query WRAP(h_1, h_2), receiving a wrap $w \neq \perp_{\text{api}}$ as a result. Subsequent corruption of h_1 might then also compromise h_2. Indeed, Assumption 2 states that knowledge of a wrapping key suffices to unwrap (and learn) a wrapped key, making the compromise of h_2 inevitable. Thus, the corruption of a small set of keys could lead to the compromise of a much larger set.

We let $L(C)$ be the set of indices corresponding to compromised handles (where $C \subseteq L(C)$). To identify precisely the set $L(C)$ of compromised equivalence classes, we keep track of which key (handle) wraps which key by means of a directed graph (V, E). The vertices of the graph are defined by the equivalence classes associated to the handles (so a subset of the natural numbers). There is an edge from i to j iff for some handles h_1, h_2 with $\text{idx}(h_1) = i$ and $\text{idx}(h_2) = j$ the adversary has issued a query WRAP(h_1, h_2), receiving $w \neq \perp_{\text{api}}$ as a result. For a given graph (V, E) and corrupted set $C \subseteq V$, we define $L(C)$ as the set of all vertices that can be reached from C (including C itself).

Dishonest Wraps. Since a wrap is just a bitstring, an adversary can try to unwrap some w that has not been produced by the API itself (i.e., $S = \emptyset$ in UNWRAP(h, w, a)). If unwrapping succeeds and returns a fresh handle, the security game needs to associate this handle to some equivalence class. We will consider two options.

Firstly, the unwrapping could have been performed under a handle that has not been compromised (intuitively, this corresponds to a wrapping forgery). In that case, the handle returned by the unwrapping will be assumed to create a new equivalence class. Technically, w is now a wrap of a handle in this new class i under $\text{idx}(h)$, yet we do not add a corresponding edge $(\text{idx}(h), i)$ to E. Adding this edge would have resulted in the new class being compromised as a result of the corruption of $\text{idx}(h)$, so that an adversary could no longer win the primitive game based on the newly introduced equivalence class. Since the new class is effectively the result of a successfully forged wrap (as $S = \emptyset$), we prefer the stronger definition (i.e. without adding an edge to E) where an adversary might benefit from a forged wrap.

Secondly, the unwrapping could have been called using a compromised handle. Since the adversary knows the key corresponding to the compromised handle, creation of such wraps is likely feasible; moreover, the adversary can be assumed to know the key corresponding to the handle being returned. To simplify matters, we will use the equivalence class 0 for all handles that result from unwrapping under compromised handles. The set C of corrupt handles initially contains the class 0. The index class 0 is special as there are no correctness guarantees for it: if $\mathsf{idx}(h_1) = \mathsf{idx}(h_2) = 0$, it is quite possible that $h_1.\mathsf{key} \neq h_2.\mathsf{key}$.

Incorporating the Primitive's Security Game. Intuitively, an adversary breaks a cryptographic API, exporting a primitive P, if and only if he manages to win the primitive's security game. Formally, in order to express an adversary's advantage against the cryptographic API in terms of the abstract security game for the primitive itself, we would need to interpret an adversary's actions against a cryptographic API as that of an adversary directly playing the abstract primitive game.

As in the correctness game, we use the equivalence to associate handles in the API game with keys in the primitive game. Whenever a new equivalence class is created, the API game creates a new instance of the primitive game by calling $\mathsf{st}[i] \leftarrow\!\!\!{}^{\$}\ \mathsf{setup}()$ (the macro initclass takes care of this).

For the API's challenge oracle we want to draw on the challenge algorithms from the primitive game. These algorithms themselves expect an oracle that implements the primitive. In the API's game the challenge oracle can use the API primitive interface. If the API outputs $\perp_{\mathsf{api}}$ we suppress the output of the challenge oracle and regard the challenge call as not having taken place in the primitive's game (note that the call might still have had an effect on the API's state).

As in the multi-key primitive game, at the end of the game we check whether the adversary caused a breach by challenging on corrupt (or in this case compromised) key or not. As mentioned before, an alternative (and stronger) formulation would maintain $L(C) \cap H = \emptyset$ as invariant by suppressing any query that would cause a breach of the invariant (possibly allowing for those queries that the API already caught). However, our formalism is easier to specify and simplifies an already complex model without materially changing its meaning.

Note that if a cryptographic API exports several different primitives, each with their own security notion, one can consider several security notions for the cryptographic API. One could modify the $\mathsf{chal}_{\mathsf{aux}}$ algorithm to model joint security.

Definition 5. *Let API* CA[P] *export primitive* P *and let* sec = (setup, chal$_0$, chal$_1$, chal$_{\mathsf{aux}}$) *be a security notion for* P. *Then the advantage of an adversary* $\mathcal{A}$ *against* CA[P] *is defined by*

$$\mathbf{Adv}_{\mathsf{CA[P]}}^{\mathsf{sec}}(\mathcal{A}) = \left| \mathbf{Pr}\left[\mathbf{Exp}_{\mathsf{CA[P]}}^{\mathsf{sec}\text{-}0}(\mathcal{A}) = 1 \right] - \mathbf{Pr}\left[\mathbf{Exp}_{\mathsf{CA[P]}}^{\mathsf{sec}\text{-}1}(\mathcal{A}) = 1 \right] \right| ,$$

for the experiments $\mathbf{Exp}_{\mathsf{CA[P]}}^{\mathsf{sec}\text{-}b}(\mathcal{A})$ *as given in Fig. 7.*

game $\mathbf{Exp}^{\text{sec-}b}_{\text{CA[P]}}(\mathcal{A})$:	**oracle** $\text{CHAL}_x(h,q)$:
$i \leftarrow 0$	$j \leftarrow \text{idx}(h)$
$H \leftarrow \emptyset, C \leftarrow \{0\}$	Run $(y, \text{st}[j]) \leftarrow \text{chal}_x^{\tilde{\mathcal{O}}(\cdot)}(\text{st}[j], q)$
$W \leftarrow \emptyset, V \leftarrow \emptyset, E \leftarrow \emptyset$	where $\tilde{\mathcal{O}}(\text{in})$ is simulated as follows
$b' \leftarrow \mathcal{A}^{\mathcal{O}, \text{CHAL}_b, \text{CHAL}_{\text{aux}}}$	out $\leftarrow \text{CA.alg}(h, \text{in})$
if $H \cap L(C) \neq \emptyset$ then return 0	if out $= \perp_{\text{api}}$ then abort chal_b
else return b'	by setting $y \leftarrow \text{\textbrokenbar}$
	leaving $\text{st}[j]$ unchanged
macro initclass:	if $x \in \{0,1\} \wedge y \neq \text{\textbrokenbar}$ then $H \leftarrow H \cup$
$\text{st}[i] \leftarrow_{\$} \text{setup}()$	$\{j\}$
	return y

Fig. 7. The security experiment $\mathbf{Exp}^{\text{sec-}b}_{\text{CA[P]}}(\mathcal{A})$ for a crypto API CA that exports $\mathsf{P} = (\mathsf{Kg_P}, \mathsf{Alg_P})$ with security notion $\overset{.}{\text{sec}} = (\text{setup}, \text{chal}_0, \text{chal}_1, \text{chal}_{\text{aux}})$. The adversary additionally has access to the oracles defined in Fig. 6 (which is where the macro initclass is used).

3.5 Security of an API's Key Management

When concentrating on the security of the exported primitive, we ignored confidentiality of cryptographic keys and authenticity of associated attributes. However, for the key-management component of a cryptographic API these are important properties and we capture these with Definitions 6 and 7, respectively.

We define the security of a key-management API via the experiment $\mathbf{Exp}^{\text{km-}b}_{\text{CA}}(\mathcal{A})$ as given in Fig. 8. Here, the goal of an adversary is to distinguish real keys managed by the API from fake ones generated at random. As usual, we capture this idea via a challenge oracle parametrized by a bit b which the adversary needs to determine. When called with handle h as input, the oracle returns the real key associated with h or a fake key (depending on b). In the process the adversary controls the key-management API which we model via the oracles in Fig. 6. We impose only minimal restrictions to prevent trivial wins. As before, we assume that for all compromised handles, the adversary knows the corresponding (real) key, making a win trivial (we can exclude these wins at the end by imposing that $H \cap L(C) = \emptyset$ as before).

Moreover, notice that under the key wrap assumption (Assumption 2), if a handle has been used to wrap another key, an adverary may easily distinguish between the key and a random one by unwrapping: the operation would always succeed with the real key and would fail with the fake one. We call an index *tainted* if one of the keys with that index is compromised, or has been used in a wrapping operation.

We write $T(C)$ for the class of tainted indexes: the adversary loses (the experiment returns 0) if it challenges a key that belongs to a tainted class.

Definition 6. *Let API* CA *be a key management API. Then the advantage of an adversary* $\mathcal{A}$ *against* CA *is defined by*

$$\mathbf{Adv}_{CA}^{km}(\mathcal{A}) = \left| \Pr\left[\mathbf{Exp}_{CA}^{km\text{-}0}(\mathcal{A}) = 1\right] - \Pr\left[\mathbf{Exp}_{CA}^{km\text{-}1}(\mathcal{A}) = 1\right] \right| ,$$

for the experiments $\mathbf{Exp}_{CA}^{km\text{-}b}(\mathcal{A})$ *as given in Fig. 8.*

game $\mathbf{Exp}_{CA}^{km\text{-}b}(\mathcal{A})$:	**oracle** $\mathrm{CHAL}_b(h)$:
$i \leftarrow 0$	$j \leftarrow \mathsf{idx}(h)$
$H \leftarrow \emptyset, C \leftarrow \{0\}$	$H \leftarrow H \cup \{j\}$
$W \leftarrow \emptyset, V \leftarrow \emptyset, E \leftarrow \emptyset$	$y \leftarrow \mathsf{fake}(j)$
$b' \leftarrow \mathcal{A}^{\mathcal{O}}$	if $b = 1$ then $y \leftarrow h.\mathsf{key}$
if $H \cap T(C) \neq \emptyset$ then return 0	return y
else return b'	
	macro initclass:
	$\mathsf{fake}(i) \leftarrow_{\$} \mathsf{Kg}(a)$

Fig. 8. The security experiment $\mathbf{Exp}_{CA}^{km\text{-}b}(\mathcal{A})$ for a key-management API CA, relative to generator Kg. The adversary additionally has access to the oracles defined in Fig. 6.

Our notion of secure key management differs from existing ones, e.g. KSW describe a fake game where the challenge key is not directly revealed, but instead wraps based on fake keys are given to an adversary. We believe that our notion is the natural one: in the key agreement literature (including KEMs) distinguishing between real and random keys is standard. Our notion of secure key management has some beneficial implications: indistinguishability of keys (privacy) implies correctness, to some extent.[4] See the full version of this paper.

Remark 1. A useful observation is that key-management security with respect to adversaries that make polynomially many challenge queries can be reduced via a hybrid argument to security against an adversary that makes a single challenge query. Specifically, for any adversary $\mathcal{A}$ that makes q_c challenge queries, there is an adversary $\mathcal{B}$ that makes a single challenge query so that $\mathbf{Adv}_{CA}^{km}(\mathcal{A}) \leq q_c \cdot \mathbf{Adv}_{CA}^{km}(\mathcal{B})$.

3.6 Stickyness: Attribute Security

In general, attributes associated to a handle may evolve over time. For instance, an attribute might indicate whether its handle has been used to perform a wrap operation or not. Initially this will not be true, but once it has occurred, it will be and should remain true. Existing API attacks show the importance of the

---

[4] For information theoretic adversaries the lemma is worthless.

integrity of critical parts of the attribute (e.g. to prevent a handle from being used for two conflicting purposes). In PKCS# 11 parlance, a binary attribute is sticky iff it cannot be unset. We model this by a stickyness game defined for an arbitrary predicate over the attribute space. Our notion of stickyness allows no change whatsoever (i.e. a predicate that is initially not set will have to remain unset). Note that, as expected, there are no guarantees for index 0.

---

game $\mathbf{Exp}_{\mathsf{CA}}^{\pi\text{-sticky}}(\mathcal{A})$:   **macro** initclass:

$i \leftarrow 0, C \leftarrow 0$ $\mathsf{pred}(i) \leftarrow \pi(a)$

$W \leftarrow \emptyset, V \leftarrow \emptyset, E \leftarrow \emptyset$

$h^* \leftarrow \mathcal{A}^\mathcal{O}$

if $0 < \mathsf{idx}(h^*) \leq i$ then

    return $\pi(h^*.\mathsf{attr}) \neq \mathsf{pred}(\mathsf{idx}(h^*))$

else return false

Fig. 9. Oracles for defining experiment $\mathbf{Exp}_{\mathsf{CA}}^{\pi\text{-sticky}}(\mathcal{A})$ for the partial authenticity of attributes in a cryptographic API. The adversary additionally has access to the oracles defined in Fig. 6.

Definition 7. *Let* CA *be a cryptographic API with attribute space* Attributes. *Let* $\pi :$ Attributes $\to \{0,1\}$ *be a predicate on the attribute space. Then the advantage of an adversary $\mathcal{A}$ against the stickyness of π equals* $\mathbf{Pr}\left[\mathbf{Exp}_{\mathsf{CA}}^{\pi\text{-sticky}}(\mathcal{A}) = \mathsf{true}\right]$ *with the experiment as given in Fig. 9.*

In the next section we exhibit one particular predicate which specifies whether the key is intended for key management or for other cryptographic operations. These two possibilities are modelled through a predicate external applied to attributes: the predicate is set if the key is intended for cryptographic operations other than key management.

Remark 2. In this section we have defined secrecy of keys via indistinguishability from random ones. This may seem like a questionable choice, since API keys are usually used to accomplish some cryptographic task, and any such use immediately gives rise to a distinguishing attack. This result can be understood by drawing a useful analogy with the area of key-exchange protocols. There, security is also defined via indistinguishability, even though keys are used later to achieve some other task (i.e. implement a secure channel). The composition of a good key exchange with a secure implementation of secure channels should yield a secure channel establishment protocol.

Similarly, one should understand the model of this section as a steppingstone towards the modular analysis of cryptographic APIs of the next section. There, we show how to combine a key-management API secure in the sense defined in this section with arbitrary (symmetric) primitives to yield a secure cryptographic API. The security of the latter is defined by asking that all of the cryptographic tasks implemented by the cryptographic API meet their (standard) game-based security notion.

4   The Power of Key Management

In this section we show how to compose, generically, a key-management API with an arbitrary primitive. First we identify some compatibility conditions that permit the composition of the two components. Informally, these require that the keys of the API are of one of two types. *Internal* keys are used exclusively for key management (i.e. wrapping other keys). *External* keys are used exclusively for keying the primitives exported by the API. Whether a key is internal or external follows from the attribute associated to the handle through a predicate external. Below, we write h.external for the value of the external predicate associated to handle h.

Definition 8. *Let* CA $=$ (CA.init, CA.new, CA.create, CA.key, CA.wrap, CA.unwrap) *be a key-management API and let* P $= (\mathsf{Kg_P}, \mathsf{Alg_P})$ *be the implementation of an arbitrary primitive with key space* Keys. *We say that* CA *and* P *are compatible if:*

1. *there exists an easy to compute predicate* external *on the attribute space* Attributes, *denoted* h.external *for a particular handle* h;
2. *if* h_1.external $=$ true *(at call time) then both* CA.wrap(h_1, h_2) *and* CA.unwrap(h_1, w) *return* $\perp_{\mathrm{api}}$;
3. *if* h.external $=$ true *then* h.key $\in$ Keys.

An abstract primitive P and a compatible key-management API CA can be composed in a generic fashion by exploiting the predicate external, leading to a cryptographic API [CA; P] as formalized in Definition 9 below. Correctness of [CA; P] follows from correctness of its two constituent components (Theorem 3). Our main composition result (Theorem 4) states that if both components are secure and, additionally, the external predicate is sticky, then the composition yields a secure cryptographic API exporting P. We formalize our construction in the following definition.

Definition 9 (Construction of [CA; P]). *Let* CA *be a key management API defined by algorithms* (CA.init, CA.new, CA.create, CA.key, CA.wrap, CA.unwrap), *and let* P $= (\mathsf{Kg_P}, \mathsf{Alg_P})$ *be a compatible primitive. We define the composition of key management API* CA *and the primitive* P *as*

$$[\mathsf{CA}; \mathsf{P}] = (\mathsf{CA.init}, \mathsf{CA.new}, \mathsf{CA.create}, \mathsf{CA.key}, \mathsf{CA.wrap}, \mathsf{CA.unwrap}, \mathsf{CA.alg})$$

where CA.alg(h, x) *simply returns* $\mathsf{Alg_P}(h.\mathsf{key}, x)$ *if* h.external $=$ true *and returns* $\perp_{\mathrm{api}}$ *otherwise (note that a call* CA.alg(h, x) *does not change the API's state).*

The following theorem states that if the components are correct, the result of the composition is also correct.

Theorem 3 (Correctness of $[CA; P]$). *Let* CA *be a key-management API and let* $P = (Kg_P, Alg_P)$ *be a compatible primitive with correctness notion defined by the predicate* $corr_P$. *Then*

$$\mathbf{Adv}^{corr_P}_{[CA;P]} \leq \mathbf{Adv}^{corr_P}_{CA} + \mathbf{Adv}^{corr_P}_P$$

Then correctness of both CA *and* P *implies correctness of* $[CA; P]$.

Proof. Consider the game $\mathbf{Exp}^{corr_P}_{[CA;P]}$ with CA.alg specified for the construction at hand. The resulting oracle ALG is specified in Fig. 10 (without the boxed statement). Adding the boxed statement provides an identical game, unless at some point $h.\mathsf{key} \neq \mathsf{key}(idx(h))$. This event is exactly the event that triggers a win in the key-management's correctness game (the *bad* flags in the cryptographic API game and the key management game coincide). Furthermore, when considering the overall correctness game using ALG with boxed statement included, a win can be syntactically mapped to a win in the primitive's correctness game, concluding the proof.   □

---

oracle $\text{ALG}(h, \mathsf{in})$:

if $h.\mathsf{external} = \mathsf{true}$ then

    $\mathsf{key} \leftarrow h.\mathsf{key}$

    $\boxed{\text{if } idx(h) \neq 0 \text{ then } \mathsf{key} \leftarrow \mathsf{key}(idx(h))}$

    $\mathsf{out} \leftarrow \mathsf{Alg}_P(\mathsf{key}, \mathsf{in})$

    $\mathsf{tr} \leftarrow \mathsf{tr} :: (idx(h), \mathsf{in}, \mathsf{out})$

else

    $\mathsf{out} \leftarrow \perp_{\mathsf{api}}$

return out

Fig. 10. Crucial oracle hop for the $[CA; P]$ correctness proof.

Compatibility of a key-management API CA and a primitive (Kg_P, Alg_P) only involved the set from which the primitive's keys are drawn. While for correctness this suffices, for security the way keys are distributed matters as well. Recall that Kg_P takes as input a parameter pm, whereas a NEW call to the key-management API takes as input an attribute a. Let a2pm be a function that maps attributes to parameters (or ξ). Let Kg take in an attribute such that for all attributes a for which the predicate external is set, it holds that $Kg(a) = Kg_P(a2pm(a))$ (i.e. the output distributions of the two algorithms match).

The following theorem establishes that composing a secure key management API with a compatible secure primitive yields a secure cryptographic API. The proof of the theorem is in the full version of the paper.

Theorem 4 (Security of [CA; P]). *Let* CA *be a key-management API and let* P = (Kg$_P$, Alg$_P$) *be a compatible primitive with security notion defined by the tuple of algorithms* (setup, chal$_0$, chal$_1$, chal$_{aux}$). *Then for any adversary* $\mathcal{A}$ *against the security of the cryptographic API* [CA; P], *there exist efficient reductions* $\mathcal{B}_1$, $\mathcal{B}_2$, *and* $\mathcal{B}_3$ *such that*

$$\mathbf{Adv}^{sec}_{[CA;P]}(\mathcal{A}) \leq 2\mathbf{Adv}^{sticky}_{CA}(\mathcal{B}_1) + q_e \left(4\mathbf{Adv}^{sticky}_{CA}(\mathcal{B}_1) + 2\mathbf{Adv}^{km}_{CA}(\mathcal{B}_2) + \mathbf{Adv}^{sec}_{P}(\mathcal{B}_3) \right)$$

where $\mathbf{Adv}^{sticky}_{CA}$ *refers to* external, $\mathbf{Adv}^{km}_{CA}$ *is relative to* Kg *defined above, and* q_e *is an upper bound on the number of non-zero index classes that ever contain a handle with attribute* external *set (in the game played by* $\mathcal{A}$).

Remark 3. To avoid being tied down to a particular cryptographic interface, we have developed an abstract framework for arbitrary security games. One nice side-effect of our choice is that we can treat (modularly) settings where APIs leak "fingerprints" of their external keys via their attributes. Specifically, we can treat these fingerprints as an additional functionality of the abstract primitive (instead of an attribute). Obviously the actual primitive needs to be proven secure in the presence of fingerprints.

5   Instantiating a KM-API

We now show how to instantiate a KM-API from a DAE scheme. This KM-API enforces a "leveled" key hierarchy. The bottom level will contain keys for one or more (unspecified) cryptographic primitives. The top level will contain keys for a DAE scheme. Our KM-API will enforce the following policy: top-level keys may only be used to (un)wrap keys at the bottom level, and bottom-level keys may not (un)wrap any key. Intuitively, keys on the bottom should only be used with their associated cryptographic primitive.

DAE Schemes. A deterministic authenticated encryption scheme (DAE) is a tuple $\Pi = (\mathcal{K}, \mathcal{E}, \mathcal{D})$. The first component $\mathcal{K} \subseteq \{0,1\}^*$ is the set of encryption keys. The encryption algorithm $\mathcal{E}$ and decryption algorithm $\mathcal{D}$ both take an input in $\mathcal{K} \times \{0,1\}^* \times \{0,1\}^*$ and return either a string or a distinguished value $\bot$. We write $\mathcal{E}^V_K(X)$ for $\mathcal{E}(K, V, X)$ and $\mathcal{D}^V_K(Y)$ for $\mathcal{D}(K, V, Y)$. We assume there are an *associated data space* $\mathcal{V} \subseteq \{0,1\}^*$ and a *message space* $\mathcal{X} \subseteq \{0,1\}^*$, such that $X \in \mathcal{X} \Rightarrow \{0,1\}^{|X|} \subset \mathcal{X}$ and $\mathcal{E}^V_K(X) \in \{0,1\}^*$ iff $V \in \mathcal{V}$ and $X \in \mathcal{X}$.

Our convention is that $\mathcal{E}^V_K(\bot) = \mathcal{D}^V_K(\bot) = \bot$ for all $K \in \mathcal{K}, V \in \mathcal{V}$. We require that $\mathcal{D}$ and $\mathcal{E}$ are each others inverses on their range excluding $\bot$: for all $K \in \mathcal{K}, V \in \mathcal{V}, Y \in \{0,1\}^*$, if there is an X such $\mathcal{E}^V_K(X) = Y$ then we require that $\mathcal{D}^V_K(Y) = X$ (correctness), moreover if no such X exists, then we require that $\mathcal{D}^V_K(Y) = \bot$ (tidyness).   •

We require $\mathcal{E}$ to be length-regular with *stretch* $\tau \colon \mathbb{N} \times \mathbb{N} \to \mathbb{N}$, meaning that for all $K \in \mathcal{K}, V \in \mathcal{V}, X \in \mathcal{X}$ it holds that $|\mathcal{E}^V_K(X)| = |X| + \tau(|V|, |X|)$. Consequently, ciphertext lengths can only on the lengths of V and X.

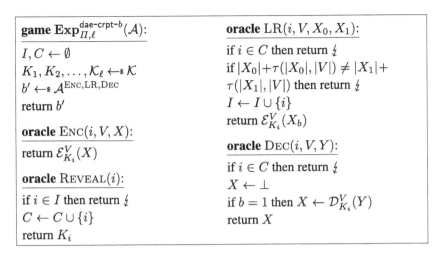

Fig. 11. The experiments $\mathbf{Exp}_{\Pi}^{\ell\text{-dae-crpt-}b}(\mathcal{A})$ for defining left-or-right DAE security with adaptive key-corruption. To prevent trivial wins, we make the following assumptions on the adversary: (1) it does not ask (i, V, Y) to its DEC-oracle if some previous ENC-oracle query (i, V, X) returned Y, or if some previous LR-oracle query (i, V, X_0, X_1) returned Y; (2) it does not ask (i, V, X) to its ENC-oracle if some previous DEC-oracle query (i, V, Y) returned X; (3) if (i, V, X) is ever asked to the ENC-oracle, then no query of the form $(i, V, X, \cdot)$ or $(i, V, \cdot, X)$ is ever made to the LR-oracle, and vice versa.

DAE Scheme Security. For integer $\ell \geq 1$ we define the advantage of adversary $\mathcal{A}$ in the *ℓ-key left-or-right DAE with corruptions* experiment as

$$\mathbf{Adv}_{\Pi}^{\ell\text{-dae-crpt}}(\mathcal{A}) = \left| \mathbf{Pr}\left[\mathbf{Exp}_{\Pi}^{\ell\text{-dae-crpt-0}}(\mathcal{A}) = 1 \right] - \mathbf{Pr}\left[\mathbf{Exp}_{\Pi}^{\ell\text{-dae-crpt-1}}(\mathcal{A}) = 1 \right] \right|,$$

where the probability is over the experiment in Fig. 11 and the coins of $\mathcal{A}$. Without loss of generality, we assume that the adversary does not repeat any query, and that it does not ask queries that are outside of the implied domains of its oracles.

As a special case of this, we also define the advantage of adverary $\mathcal{A}$ in the *ℓ-key left-or-right DAE* experiment as

$$\mathbf{Adv}_{\Pi}^{\ell\text{-dae}}(\mathcal{A}) = \left| \mathbf{Pr}\left[\mathbf{Exp}_{\Pi}^{\ell\text{-dae-0}}(\mathcal{A}) = 1 \right] - \mathbf{Pr}\left[\mathbf{Exp}_{\Pi}^{\ell\text{-dae-1}}(\mathcal{A}) = 1 \right] \right|,$$

where $\mathbf{Exp}_{\Pi}^{\ell\text{-dae-}b}$ is defined by modifying Fig. 11 to no longer include the ENC or REVEAL oracles, the sets I, C, and any references to these. The applicable restrictions on adversarial behavior carry over.

We note that this notion differs from the DAE security notion first given by Rogaway and Shrimpton [22]. We use a left-or-right version, more along the lines of Gennaro and Halevi [15] because it suits our needs better.

A standard "hybrid argument" provides a proof of the following theorem, along with the description of the claimed adversary $\mathcal{B}$. We omit this proof.

Theorem 5. [1-key left-or-right DAE implies ℓ-key left-or-right DAE with corruptions.] *Fix an integer $\ell \geq 1$. Let $\Pi = (\mathcal{K}, \mathcal{E}, \mathcal{D})$ be a DAE scheme with associated-data space $\mathcal{V}$, message space $\mathcal{X}$, and ciphertext-expansion function e. Let $\mathcal{A}$ be an adversary compatible with the ℓ-key DAE advantage notion. Let $\mathcal{A}$ ask q_i LR-queries of the form $(i, \cdot, \cdot, \cdot)$ and p_i DEC-queries of the form $(i, \cdot, \cdot)$, and have time-complexity t. Then there is an adversary $\mathcal{B}$ that makes black-box use of $\mathcal{A}$ such that*

$$\mathbf{Adv}_{\Pi}^{\ell\text{-dae-crpt}}(\mathcal{A}) \leq \ell\,\mathbf{Adv}_{\Pi}^{1\text{-dae}}(\mathcal{B})$$

where $\mathcal{B}$ asks at most $\max_i\{q_i\}$ LR-queries and $\max_i\{p_i\}$ DEC-queries.

Building a KM-API from a DAE Scheme. Assume that there exists an easy to compute predicate external on the attribute space Attributes $\subseteq \{0,1\}^*$, and assume that sampling attributes for which the predicate holds, respectively does not hold, both are easy. Recall that, as before, for a particular handle h, we use the shorthand h.external for the predicate evaluated on h.attr.

Let $\mathsf{Kg_P}$ be the key generation for some primitive with key space Keys and let pm be a function that maps attributes to parameters (or $\notin$). Let $\Pi = (\mathcal{K}, \mathcal{E}, \mathcal{D})$ be a DAE-scheme with associated-data space $\mathcal{V} = $ Attributes and message-space $\mathcal{X}$ that contains Keys. Define Kg: Attributes $\rightarrow$ Keys $\cup\, \mathcal{K}$ to be the algorithm that, on input an attribute a that satisfies external, samples from Keys according to $\mathsf{Kg_P}(\mathsf{pm}(a))$ and otherwise samples uniformly from $\mathcal{K}$.

Before specifying the algorithms that comprise our KM-API, let us detail our assumptions on the state of tokens with its scope. We assume that all tokens have state of the form $s = (\tilde{s}, (h \mapsto (\mathsf{key}, a))_h)$, where for each handle h, the mapping $h \mapsto (\mathsf{key}, a)$ indicates the associated key and attribute pair (so h.key = key and h.attr $= a$), and the state $\tilde{s}$ contains a snapshot of the token's past I/O only. Let fresh be a mechanism that creates fresh (unique) handles on a per token basis.

With all of this established, the algorithms of our KM-API are defined as follows:

- CA.new(t, a): Create a fresh handle h on token t by calling fresh(t). Sample $K \leftarrow\!\!{}^{\$}\,\mathsf{Kg}(a)$ and update the state of token t to reflect the new mapping $h \mapsto (K, a)$. Return h.
- CA.create(t, K, a): Create a fresh handle h on token t by calling fresh(t) and update the state on token t to reflect the new mapping $h \mapsto (K, a)$. Return h.
- CA.wrap(h_1, h_2): If h_1.external $\lor\, \neg h_2$.external then return $\perp_{\text{api}}$. Otherwise, $w \leftarrow \mathcal{E}_{h_1.\mathsf{key}}^{h_2.\mathsf{attr}}(h_2.\mathsf{key})$. Return w.
- CA.unwrap(h, w, a): If h.external return $\perp_{\text{api}}$. Compute $K \leftarrow \mathcal{D}_{h.\mathsf{key}}^{a}(w)$. If $K = \perp$ then return $\perp_{\text{api}}$. Otherwise, create a fresh handle $\bar{h}$ and update the state on token $\mathsf{tkn}(h)$ to reflect the new mapping $\bar{h} \mapsto (K, a)$. Return $\bar{h}$.

Theorem 6. *Fix a nonempty set Keys. Let $\Pi = (\mathcal{K}, \mathcal{E}, \mathcal{D})$ be a DAE-scheme with associated-data space $\mathcal{V} = $ Attributes and message-space $\mathcal{X}$ that contains Keys. Let CA be the KM-API just described, and let $\mathcal{A}$ be an efficient KM-API adversary asking a single challenge query. Let q_n be the number of NEW-oracle* *queries made by* $\mathcal{A}$ *in its execution, and let* $\ell \leq q_n$ *be the number of these that produce an internal key. Then there exist efficient adversaries* $\mathcal{B}, \mathcal{F}$ *for the* ℓ-*key DAE with corruptions experiment such that*

$$\mathbf{Adv}_{\mathsf{CA}}^{\mathsf{km}}(\mathcal{A}) \leq 2\mathbf{Adv}_{\Pi}^{\ell\text{-dae-crpt}}(\mathcal{F}) + (q_n - \ell)\mathbf{Adv}^{\ell\text{-dae-crpt}}\Pi(\mathcal{B})$$

6   Related Work

Symbolic Models for API Security. Given that many attacks against APIs rely on logical flaws rather than weak cryptography a large body of work addresses their security using symbolic models. The first set of attacks were discovered by Longley and Rigby [19], Bond [3], and Clulow [7]. More recently, Cortier et al. [8], Delaune et al. [13], and Bortolozzo et al. [4] uncovered further vulnerabilities by using automated tools. Security models and proofs of security include the work of Courant and Monin who use Coq to analyze a variant of IBM CCA API [11] and Cortier et al. [8]. Fröschle and Steel [14] and Cortier and Steel [9] analyze a fragment of the of PKCS#11 standard. Newer models consider key-management that employs asymmetric cryptographic [12] and revocation of keys [10]. While symbolic models are suitable for finding attacks, security proofs are less meaningful—in particular they do not a priori imply security with respect to the types of stronger computational models that we develop in this paper.

The Cachin–Chandran Model [5]. This is the first computational security model for a cryptographic API. The model is based on a particular design that relies on a centrally located server which keeps track of all key-management operations (how realistic the presence of such a server is in the distributed environment in which tokens typically operate is unclear).

The security model is intrinsically stated in terms of this suggested implementation of an API by hardwiring into the syntax of what constitutes an API their specific implementation choices (e.g. how and when certain attributes change, how and what information the overall internal state of the token should maintain). Clearly this severely restricts the model's applicability. For example, the security of the wrap scheme is hardwired into the model and essentially demands that the wrap operation be implemented with a probabilistic scheme—schemes employing a deterministic wrapping mechanism would be ruled insecure under the model (in particular our key-management scheme is not captured as our tokens need not keep track, internally, of the attributes associated to keys).

From a security perspective, just like in our model, the adversary has access to the full interface of the token and aims to break the cryptographic functionality that the token provides. Yet, there are three aspects—we believe shortcomings—of the Cachin–Chandran model on which our model significantly improves. Firstly, as stated already, the Cachin–Chandran model rules out (either explicitly or by implication) some very reasonable and secure implementations of a cryptographic API. Secondly, aliasing issues caused by the possibility that a key can have multiple distinct handles pointing to it are sidestepped in the Cachin–Chandran model (essentially, unwrapping of wraps that were not previously created is not permitted). Finally, the corruption model considered in the Cachin–Chandran model is restricted to users, which implies that an adversary can then act on that user's behalf. However, there are no further implications to the experiment, as acting on behalf of a user does *not* give access to any keys. Consequently, the notion of corrupted or compromised *keys* is absent in the Cachin–Chandran model.

Our model makes only minimal assumptions about the inner-workings of a cryptographic API (it allows but certainly does not impose a central server for the implementation). Our security model carefully keeps track of the equivalence classes on handles that the wrap/unwrap operations give rise to. More importantly, we explicitly allow for adaptive corruption of API keys and demand that any other key that is not directly or indirectly affected by corruptions stays secure.

The Kremer–Steel–Warinschi Model [18]. The KSW computational model[5] fixes some of the shortcomings of the Cachin–Chandran model. In particular, it presents definitions for the syntax and security of an encryption-exporting API not driven by any particular implementation and allowing adaptive corruption of keys. The syntax is single token and the security requirements imposed are incompatible with PKCS#11 implementations: all attributes need to be sticky, whereas PKCS#11 mandates that some attributes change during operations. Interestingly, while the Cachin-Chandran model imposes that wrapping be implemented with a probabilistic encryption scheme, the modelling choice adopted by the KSW model enforces wrapping to be deterministic. Perhaps worse, the high level of abstraction led to underspecified, malformed definitions.[6]

In contrast, we consider a multi-token environment and only surface minimal assumptions that avoid the underspecification in the KSW model. Our security notion is more relaxed. For example, for key-management APIs we only demand that the application keys are secret, which allows for both probabilistic and deterministic solutions to the key-wrap problem. Crucially, we show that our notion of security for key-management APIs is composable, whereas no such result is known to hold for the KSW model.

Universally Composable Key-Management [17]. This paper is, in spirit, closest to ours. It aims to provide a compositional framework where key-management can be analyzed separately from the other cryptographic operations that tokens

---

[5] The paper also introduces two other related models: an idealized and a symbolic one.

[6] For example, the formalization crucially relies on the notation $s[\mathsf{key}(h) \overset{\$}{\mapsto} k_0]$ which indicates some state s in which $\mathsf{key}(h)$ has been replaced with the randomly chosen k_0. However, given the abstract notion of state, it is unclear what this state change even means. For instance, if another handle points to the same key, does that handle's key also get affected? Is the state change persistent? Is k_0 drawn anew each time?.

may export. The formalization relies on the universal composability framework (as refined by Hofheinz and Shoup [16]) and consists of an ideal key-management functionality which, as usual, should be emulated by a secure implementation. The framework naturally encompasses multi-token scenarios which are simply distributed implementations of the functionality and should guarantee the desired guarantee: the implementation can replace the functionality in any other scenario.

Since the underlying definitional framework relies on simulation, the model does not tolerate well adversaries that adaptively corrupt keys (we discuss this issue below), so the adversary is only allowed static corruptions. An additional issue is that in simulation based settings keys cannot be freely passed around between functionalities. The solution adopted here employs a cumbersome capability-based mechanism to model the interaction between key-management and other cryptographic operations. The key-management functionality is not fully agnostic of the primitive in which the managed keys are to be used. Furthermore, the key-management functionality has hardwired a wrapping algorithm (which needs to be deterministic, authenticated and secure against related-key attacks).

We avoid all of these shortcomings. Our construction is mostly oblivious to the primitive in which keys are used and allows various instantiations where wrapping can be either probabilistic or deterministic. Our use of game-based definitions enables the proof of the composition theorem even with adaptive corruptions.

Computationally Sound API Analysis. Recently, Scerri and Stanley-Oakes have proposed an approach for the analysis of key-management APIs [24] using the framework of Bana and Comon-Lundh [1]. This framework allows to model and reason about cryptographic systems using a high-level of abstraction and then use a general theorem that links the results with security in a standard computational model. The approach used by Scerri and Stanley-Oakes is similar to ours in that they treat the key-management component of APIs separately and retrieve the security of the overall API through a composition theorem that considers the use of API keys in symmetric encryption. That work provides a more detailed treatment of API policies and benefits from the simple, axiomatic way of reasoning about security of protocols. The main drawback is that the adversary is only allowed a constant number of queries to the API.

7   Conclusion

We propose models that capture the core security guarantees that cryptographic and key-management APIs should provide. Our treatment is general, in that we do not consider a particular primitive (or primitives) but rely on an abstraction that allows multiple instantiations. Our work opens several interesting research avenues. We currently treat policies abstractly, and only indicate their influence on tokens as part of our execution model. It would be interesting to investigate further additional guarantees for tokens that relate to secure policy enforcement. For example, useful policies may attempt to ensure that certain keys are used only by certain users and only for a restricted set of operations. Such guarantees can be defined and analyzed in an extension of our model that incorporates the notion of token users and formalizes the type of restrictions envisioned by the policy. In this paper we consider only the management of symmetric keys. It would be useful to extend our treatment to include private keys for the asymmetric cryptographic primitives that are part of a standard PKCS#11 interface.

Acknowledgements. This work was supported in part by European Union Seventh Framework Programme (FP7/2007-2013) grant agreement 609611 (PRACTICE), and ERC Advanced Grant ERC-2010AdG-267188-CRIPTO. It was also supported by National Science Foundation grant CNS-1319061.

References

1. Bana, G., Comon-Lundh, H.: Towards unconditional soundness: computationally complete symbolic attacker. In: Degano, P., Guttman, J.D. (eds.) POST 2012 (ETAPS 2012). LNCS, vol. 7215, pp. 189–208. Springer, Heidelberg (2012)
2. Bardou, R., Focardi, R., Kawamoto, Y., Simionato, L., Steel, G., Tsay, J.-K.: Efficient padding oracle attacks on cryptographic hardware. In: Safavi-Naini, R., Canetti, R. (eds.) CRYPTO 2012. LNCS, vol. 7417, pp. 608–625. Springer, Heidelberg (2012)
3. Bond, M.: Attacks on cryptoprocessor transaction sets. In: Koç, Ç.K., Naccache, D., Paar, C. (eds.) CHES 2001. LNCS, vol. 2162, pp. 220–234. Springer, Heidelberg (2001)
4. Bortolozzo, M., Centenaro, M., Focardi, R., Steel, G.: Attacking and fixing PKCS#11 security tokens. In: Al-Shaer, E., Keromytis, A.D., Shmatikov, V. (eds.) ACM CCS 2010, pp. 260–269. ACM Press, October 2010
5. Cachin, C., Chandran, N.: A secure cryptographic token interface. In: Proceedings of 22th IEEE Computer Security Foundations Symposium (CSF 2009), pp. 141–153. IEEE Computer Society Press (2009)
6. Canetti, R., Feige, U., Goldreich, O., Naor, M.: Adaptively secure multi-party computation. In: 28th ACM STOC, pp. 639–648. ACM Press, May 1996
7. Clulow, J.: On the security of PKCS#11. In: Walter, C.D., Koç, Ç.K., Paar, C. (eds.) CHES 2003. LNCS, vol. 2779, pp. 411–425. Springer, Heidelberg (2003)
8. Cortier, V., Keighren, G., Steel, G.: Automatic analysis of the security of XOR-based key management schemes. In: Grumberg, O., Huth, M. (eds.) TACAS 2007. LNCS, vol. 4424, pp. 538–552. Springer, Heidelberg (2007)
9. Cortier, V., Steel, G.: A generic security API for symmetric key management on cryptographic devices. In: Backes, M., Ning, P. (eds.) ESORICS 2009. LNCS, vol. 5789, pp. 605–620. Springer, Heidelberg (2009)
10. Cortier, V., Steel, G., Wiedling, C.: Revoke and let live: a secure key revocation api for cryptographic devices. In: Yu, T., Danezis, G., Gligor, V.D. (eds.) ACM CCS 2012, pp. 918–928. ACM Press, October 2012
11. Courant, J., Monin, J.F.: Defending a bank with a proof assistant. In: WITS, pp. 87–98 (2006)

12. Daubignard, M., Lubicz, D., Steel, G.: A secure key management interface with asymmetric cryptography. In: Abadi, M., Kremer, S. (eds.) POST 2014 (ETAPS 2014). LNCS, vol. 8414, pp. 63–82. Springer, Heidelberg (2014)

13. Delaune, S., Kremer, S., Steel, G.: Formal analysis of PKCS#11. In: Proceedings of 21th IEEE Computer Security Foundations Symposium (CSF 2008), pp. 331–344. IEEE Computer Society Press (2008)

14. Fröschle, S., Steel, G.: Analysing PKCS#11 key management APIs with unbounded fresh data. In: Degano, P., Viganò, L. (eds.) ARSPA-WITS 2009. LNCS, vol. 5511, pp. 92–106. Springer, Heidelberg (2009)

15. Gennaro, R., Halevi, S.: More on key wrapping. In: Jacobson Jr., M.J., Rijmen, V., Safavi-Naini, R. (eds.) SAC 2009. LNCS, vol. 5867, pp. 53–70. Springer, Heidelberg (2009)

16. Hofheinz, D., Shoup, V.: GNUC: a new universal composability framework. J. Cryptol. **28**(3), 423–508 (2015)

17. Kremer, S., Künnemann, R., Steel, G.: Universally composable key-management. In: Crampton, J., Jajodia, S., Mayes, K. (eds.) ESORICS 2013. LNCS, vol. 8134, pp. 327–344. Springer, Heidelberg (2013)

18. Kremer, S., Steel, G., Warinschi, B.: Security for key management interfaces. In: Proceedings of 24th IEEE Computer Security Foundations Symposium (CSF 2011), pp. 266–280. IEEE Computer Society Press (2011)

19. Longley, D., Rigby, S.: An automatic search for security flaws in key management schemes. Comput. Secur. **11**(1), 75–89 (1992)

20. Nielsen, J.B.: Separating random oracle proofs from complexity theoretic proofs: the non-committing encryption case. In: Yung, M. (ed.) CRYPTO 2002. LNCS, vol. 2442, pp. 111–126. Springer, Heidelberg (2002)

21. Osaki, Y., Iwata, T.: Further more on key wrapping. IEICE Trans. **95–A**(1), 8–20 (2012)

22. Rogaway, P., Shrimpton, T.: A provable-security treatment of the key-wrap problem. In: Vaudenay, S. (ed.) EUROCRYPT 2006. LNCS, vol. 4004, pp. 373–390. Springer, Heidelberg (2006)

23. RSA Security Inc: PKCS#11: cryptographic token interface standard, June 2004

24. Scerri, G., Stanley-Oakes, R.: Analysis of key wrapping APIs: generic policies, computational security. In: Proceedings of 29th IEEE Computer Security Foundations Symposium (CSF 2016). IEEE Computer Society Press (2016)

Encryption Switching Protocols

Geoffroy Couteau[1]($\boxtimes$), Thomas Peters[2], and David Pointcheval[1]

[1] ENS, CNRS, INRIA, PSL Research University, Paris, France
geoffroy.couteau@ens.fr
[2] UCLouvain, ICTEAM, Louvain-la-Neuve, Belgium
thomas.peters@uclouvain.be

Abstract. We formally define the primitive of *encryption switching protocol* (ESP), allowing to switch between two encryption schemes. Intuitively, this two-party protocol converts given ciphertexts from one scheme into ciphertexts of the same messages under the other scheme, for any polynomial number of *switches*, in any direction. Although ESP is a special kind of two-party computation protocol, it turns out that ESP implies general two-party computation (2-PC) under natural conditions. In particular, our new paradigm is tailored to the evaluation of functions over rings. Indeed, assuming the compatibility of two additively and multiplicatively homomorphic encryption schemes, switching ciphertexts makes it possible to efficiently reconcile the two internal laws. Since no such pair of public-key encryption schemes appeared in the literature, except for the non-interactive case of fully homomorphic encryption which still remains prohibitive in practice, we build the first multiplicatively homomorphic ElGamal-like encryption scheme over $(\mathbb{Z}_n, \times)$ as a complement to the Paillier encryption scheme over $(\mathbb{Z}_n, +)$, where n is a strong RSA modulus. Eventually, we also instantiate secure ESPs between the two schemes, in front of malicious adversaries. This enhancement relies on a new technique called *refreshable twin ciphertext pool*, which we show being of independent interest. We additionally prove this is enough to argue the security of our general 2-PC protocol against malicious adversaries.

1 Introduction

The development of the Internet witnessed the explosive growth of the amount of available data. We now live in an era of big data in which there is an always increasing need for efficient tools to store and manipulate huge quantities of information. While most companies now outsource their data to get an arbitrarily large storage capacity with efficient access, manipulating data in the Cloud raises many security issues. Secure multi-party computation (MPC) has thus gained tremendous importance by providing privacy-preserving tools allowing manipulations of sensitive inputs.

T. Peters—Work done while being at ENS under the ERC CryptoCloud Project.

M. Robshaw and J. Katz (Eds.): CRYPTO 2016, Part I, LNCS 9814, pp. 308–338, 2016.
DOI: 10.1007/978-3-662-53018-4_12

Secure Two-Party and Multiparty Computation. Secure two-party computation (2-PC) targets the following problem: Alice and Bob, modeled as probabilistic polynomial-time algorithms, wish to jointly compute a public function f of their respective inputs x and y, while keeping them private. We will focus on the case where Alice only gets the final result $f(x, y)$, while Bob should learn nothing, but this is not really a loss of generality. To this end, they perform an *interactive protocol*, that is expected to be *correct* (*i.e.*, the final output of the protocol is indeed $f(x, y)$) and *private* (*i.e.*, no one can learn from his own view any information that he could not have deduced from his input, and the outcome $f(x, y)$ for Alice). Secure multiparty computation is the natural extension of this problem to more than two players. Two kinds of adversarial behaviors are usually considered: *semi-honest* adversaries (*a.k.a.* honest-but-curious) follow the specifications of the protocol and try to get as much information as possible from the transcript, while *malicious* adversaries might deviate from these specifications in any way to gain more information.

Starting with the seminal work of Yao [41], there have been a vast amount of publications targeting secure two-party and multiparty computation. Today's most efficient schemes are based on various paradigms, such as secret sharing with preprocessing (e.g. TinyOT [32], SPDZ [11], MiniMac [12]), oblivious transfers [1], garbled circuits [28], or homomorphic encryption [10]. In addition, there are several hybrid constructions which combine various approaches (e.g. garbled circuit and homomorphic encryption in [21], secret sharing and garbled circuits in [13]). Most of those schemes are very efficient when the circuit to be computed is of low depth. However, when high-depth circuits are involved, the efficiency drops down: protocols based on secret sharing, oblivious transfers, partially homomorphic encryption, or garbled circuits have a communication proportional to the depth of the circuit. At the exception of the latter one, they also have a round complexity proportional to the depth of the circuit. This can be avoided with somewhat homomorphic encryption, but as soon as the circuit has a high depth, the players will have to rely on prohibitively expensive bootstrapping procedures. In the honest-but-curious setting, hybrid protocols might provide efficient solutions in some particular cases (although they will still suffer from comparable downsides in general, as they combine approaches which do all have such downsides). However, enhancing hybrid protocols *efficiently* to security against malicious adversaries is highly non-trivial, due to the lack of a common structure between the various elements manipulated in those protocols; in fact, [13,21] do only consider the honest-but curious setting.

Switching Between Homomorphic Schemes. The existence of very efficient MPC protocols for circuits containing a large number of additions, and few multiplications, suggests that multiplications might be way more expensive than additions. However, there exist encryption schemes which are *multiplicatively homomorphic*, the most famous one being the ElGamal encryption scheme [14]. In such cryptosystems, multiplications come essentially for free, but additions cannot be performed (unless a fully homomorphic scheme is used). Therefore, a natural way to design a MPC protocol in which multiplications would not incur a significant overhead compared to additions would be to *combine* a multiplicative cryptosystem with an additive cryptosystem: multiplications would be performed homomorphically on multiplicative ciphertexts, and additions on additive ciphertexts. The missing ingredient in such a protocol is a procedure to *convert* a multiplicative (resp. additive) ciphertext into an additive (resp. multiplicative) ciphertext encrypting the same plaintext: an *encryption switching protocol*.

To our knowledge, three papers have considered switching between ciphertexts under different homomorphic schemes in the past. The concept was initially introduced in [17], where the authors propose a variant of the ElGamal encryption scheme to work over $\mathbb{Z}_n^*$, together with a protocol to switch between this scheme and the Paillier scheme. In [40], a trusted software is used to switch between various homomorphic schemes. In a recent unpublished paper [27], the authors propose methods to switch from the ElGamal scheme to the Paillier scheme, to evaluate DNF formulae.

As [40] relies on a trusted software, it cannot be compared to our work, which does not make this assumption. Moreover, we found both [17,27] to be flawed: in [17], a variant of the ElGamal encryption scheme is proposed; however, the public key of the scheme contains a square root β of unity with Jacobi symbol -1. But then, computing $\gcd(\beta - 1, n)$ gives a non-trivial factor of n. Hence, the scheme leaks the factorization of the modulus. In [27], the following variant of the ElGamal scheme is proposed: to encrypt $m \in \mathbb{Z}_n^*$, pick a random scalar r in $\mathbb{Z}_n$ and output $(g^r \bmod n, mh^r \bmod n)$, where g is a square ($g = 16$ in the article) and h is g^x for some secret key x. Given a ciphertext (c_0, c_1), any player can compute the Jacobi symbol of c_0 and c_1, and check whether they are equal or different. The former case corresponds to the Jacobi symbol of m being 1, while the latter case corresponds to the Jacobi symbol of m being -1: the scheme leaks the Jacobi symbol of the plaintext, which contradicts the semantic security, at least in $\mathbb{Z}_n^*$.

Indeed, constructing a multiplicatively homomorphic variant of the ElGamal encryption scheme that is still semantically secure over $\mathbb{Z}_n^*$ (and *a fortiori* over $\mathbb{Z}_n$) turns out to be a non-trivial task.

Our Contribution. In this work, we formally define *encryption switching protocol* (ESP), which allows two players to interactively and obliviously convert an encryption of a message m with a cryptosystem Π_1 to an encryption of the same message with a cryptosystem Π_2, provided that m lies in the intersection of the plaintext spaces of the cryptosystems. To instantiate this primitive, we introduce (and formally prove the security of) a new multiplicatively homomorphic variant of the ElGamal encryption scheme whose plaintext space is $\mathbb{Z}_n^*$. To our knowledge, our scheme is the first secure construction of a multiplicatively homomorphic IND-CPA encryption scheme over $\mathbb{Z}_n^*$ and might be of independent interest. We extend our variant of the ElGamal cryptosystem to a space which is "nearly" equal to $\mathbb{Z}_n$, in a sense that we formally define. We then construct *encryption switching protocols* between our new scheme and the Paillier encryption scheme. Our ESPs (between the two encryption schemes, in both directions) have a *constant* communication (counted as a number of group elements), and their security relies on standard assumptions (the decisional composite residuosity, the decisional Diffie-Hellman, and the quadratic residuosity assumptions). In addition to its application to two-party computation, which will be outlined afterward, we believe that the primitive of ESP is of theoretical interest on its own.

To demonstrate the generality of our approach, we construct a generic two-party computation protocol over a ring $(\mathscr{R}, \oplus, \otimes)$ assuming the existence of homomorphic cryptosystems for each law, $\oplus$ and $\otimes$, and encryption switching protocols. We formally prove that our generic protocol achieves the standard security notions for two-party computation. Our new paradigm is particularly suited for high depth circuits.

We then turn our attention to the malicious setting. The natural way to provide security against malicious adversaries is to ask each player to prove, using a zero-knowledge proof, that he behaved honestly. However, ESPs can be seen as hybrid protocols, as they combine primitives with very different structures (in our case, the ElGamal scheme and the Paillier scheme). As is often the case in hybrid schemes, the lack of a common algebraic structure between the schemes prevents us from using standard zero-knowledge proofs. We tackle this issue by introducing a new technique for zero-knowledge, which we call a *refreshable twin-ciphertext pool*. In addition to providing an efficient way to enhance the security of ESPs to the malicious setting, we show that our new technique allows us to improve over several classical zero-knowledge proofs, such as proofs of knowledge of a double logarithm, or proof of primality of a committed value, which is of independent interest.

A nice feature of our two-party computation paradigm is that it is in fact *sufficient* to instantiate it with an ESP secure against malicious adversaries for the full generic two-party computation protocol to be secure against malicious adversaries.

Related Work. We already mentioned (and argued the insecurity of) [17,27] which design methods for switching between homomorphic schemes, and [40], which relies on a trusted software to achieve a comparable goal. Fully homomorphic encryption (FHE), gathering both additive and multiplicative homomorphic properties in a single encryption scheme, has been a long standing open problem until the seminal work of Gentry [18]. It relies on a somewhat homomorphic encryption scheme, that allows to perform a bounded number of operations, and a technique called bootstrapping to remove this bound. Our work can be seen as a similar line of work, using homomorphic encryption schemes (HEs) to perform an unlimited number of *specific* operations, and then relying on a *switching* technique to replace one HE by another one to get access to other specific operations. However, a fundamental difference is that the bootstrapping is a *non-interactive* technique, while our encryption switching protocols are interactive.

We stress that our ESP primitive makes use of shared decryption keys to obliviously decrypt and re-encrypt under the other encryption scheme, with a similar public key. This is totally different from proxy re-encryption, where the proxy knows a key to convert a ciphertext under one key into a ciphertext under another independent key. For instance, disclosure of secret key of one encryption scheme in our realization breaks the semantic security of the other one too.

Preliminaries. Because of lack of space, basics on classical tools are postponed to the full version [6] (as well as the optimizations and detailed proofs), and the reader is recommended to refer to it for more details. But in short, a public-key encryption scheme Π is defined by the four algorithms (Setup, KeyGen, Enc, Dec), where the two first generate the global parameters and the keys, and the two others encrypt and decrypt. If nothing else is specified we assume that a correctly encrypted message is always returned back by the decryption algorithm. We denote $\mathcal{M}$ the message space.

Throughout this paper, κ denotes the security parameter. The notation $x \xleftarrow{\$} S$ indicates that x is sampled uniformly at random from the finite set S. We write $a = b \bmod n$ to specify that $a = b$ in $\mathbb{Z}_n$ and we write $a \leftarrow [b \bmod n]$ to affect the smallest non-negative integer to a so that $a = b \bmod n$.

2 Two-Party Computation from ESPs

We introduce a theoretical framework for alternating between different encryption schemes: the new primitive of *encryption switching protocol* (ESP) allows to switch a ciphertext under an encryption scheme into a ciphertext of the same message under the other encryption scheme without damaging their semantic security. We define this primitive as a 2-party protocol and we show that secure ESP implies secure general 2-party computation under natural conditions. This is the first main contribution of the paper.

2.1 Definitions

Definition 1 (Twin-Ciphertext Pair). *For $i = 1, 2$, let Π_i be an encryption scheme (Setup$_i$, KeyGen$_i$, Enc$_i$, Dec$_i$) with plaintext space $\mathcal{M}_i$. A twin-ciphertext pair (c_1, c_2) is a pair of ciphertexts so that:*

1. *c_1 is an encryption of $m_1 \in \mathcal{M}_1$ under Π_1;*
2. *c_2 is an encryption of $m_2 \in \mathcal{M}_2$ under Π_2;*
3. *$m_1 = m_2$ (which in turn belongs to $\mathcal{M}_1 \cap \mathcal{M}_2$).*

Given an encryption c of a message $m \in \mathcal{M}_1 \cap \mathcal{M}_2$, under one of the two above encryption schemes, we will say that any ciphertext c' which does encrypt m under the other encryption scheme is a twin ciphertext *of c.*

On the other hand, if c and c' encrypt the same m under the same encryption scheme, they are said *equivalent*. Informally, given a ciphertext c of a plaintext m under one of the two above encryption schemes, an *encryption switching protocol* (ESP) describes how users A and B, sharing the decryption key, can interact to construct a twin ciphertext of c. This is of course under the restriction that the plaintext m lies in the intersection of the two message spaces. We focus on two encryption schemes that use common Setup and KeyGen algorithms for generating the global parameters and the keys[1].

Definition 2 (Encryption Switching Protocol). *For $i = 1, 2$, let Π_i be an encryption scheme* (Setup, KeyGen, Enc_i, Dec_i). *An encryption switching protocol (ESP) between Π_1 and Π_2, noted $\Pi_1 \rightleftharpoons \Pi_2$, is a tuple* (Share, Switch):

Share(pk, sk) *given the common keys* sk *and* pk *of both schemes, it outputs a secret sharing* $(\mathsf{sk}_A, \mathsf{sk}_B)$ *of* sk *and updates* pk *if necessary. The party A (resp. B) is intended to be given* sk_A *(resp. sk_B);*
Switch$_{\mathsf{par}}((\mathsf{pk}, \mathsf{sk}_A, c), (\mathsf{pk}, \mathsf{sk}_B, c))$ *is an interactive protocol in the direction* par $\in$ $\{1 \rightarrow 2, 2 \rightarrow 1\}$ *which, from a ciphertext c under the source encryption scheme, jointly computes a twin ciphertext c' of c under the target encryption scheme or outputs $\bot$ (in case of problems during the protocol execution).*

Correctness. An ESP $\Pi_1 \rightleftharpoons \Pi_2 =$ (Share, Switch) is *correct* if both Π_1 and Π_2 are correct encryption schemes, and for any pp $\leftarrow$ Setup(1^κ), any keys (pk, sk) $\leftarrow$ KeyGen(pp), any key shares (pk, sk_A, sk_B) $\leftarrow$ Share(pk, sk), any message $m \in \mathcal{M}_1 \cap \mathcal{M}_2$, and any $c_i \leftarrow \mathsf{Enc}_i(\mathsf{pk}_i, m)$ for $i = 1, 2$,

$$\mathsf{Dec}_2(\mathsf{sk}, \mathsf{Switch}_{1 \rightarrow 2}((\mathsf{pk}, \mathsf{sk}_A, c_1), (\mathsf{pk}, \mathsf{sk}_B, c_1))) = m,$$
$$\mathsf{Dec}_1(\mathsf{sk}, \mathsf{Switch}_{2 \rightarrow 1}((\mathsf{pk}, \mathsf{sk}_A, c_2), (\mathsf{pk}, \mathsf{sk}_B, c_2))) = m,$$

always hold. Ciphertexts on messages in the intersection of the two plaintext spaces are called *switchable*.

2.2 Security Notions

We expect ESP not to break the IND-CPA security of the encryption schemes, even in front of malicious adversaries: the adversary $\mathscr{A}$ is given pk, but since it plays against Alice or Bob it can choose either sk_B or sk_A, respectively. Then, even interacting with an oracle that emulates the other party as an honest player, $\mathscr{A}$ should not be able to break IND-CPA security of neither Π_1 nor Π_2. Let us more formally define this security notion.

Definition 3 ($\mathscr{O}_A$ and $\mathscr{O}_B$ Oracles). *For appropriate keys* (pk, sk_A, sk_B), *we denote the stateful oracle $\mathscr{O}_A(i \rightarrow j, c, \mathsf{Flow})$ that emulates the honest player A: it provides the answers A would send back upon receiving the flow* Flow *when running the protocol* Switch$_{i \rightarrow j}((\mathsf{pk}, \mathsf{sk}_A, c), (\mathsf{pk}, \mathsf{sk}_B, c))$. *We similarly define the oracle $\mathscr{O}_B$ that emulates the honest player B. A special flow 'Start' is used to initialize the protocol.*

---

[1] In any case, we could just take the concatenation of the outputs of the algorithms of the two schemes.

In our target application of 2-PC, these oracles will not be available on any input, but on controlled ciphertexts only. Hence our following security notion.

Definition 4 (ESP Security). *An encryption switching protocol $\Pi_1 \rightleftharpoons \Pi_2$ is* ***secure*** *if it is strongly sound and zero-knowledge (see below).*

The soundness property guarantees that no malicious player can successfully force the outcome of Switch not to be a twin ciphertext of the input, when the input is indeed a switchable ciphertext. The *strong* requirement means that the soundness holds even if the adversary is also given the whole secret key sk (or both sk_A and sk_B), instead of just one of the two shares.

Definition 5 (Strong Soundness). *An encryption switching protocol $\Pi_1 \rightleftharpoons \Pi_2$ is* ***strongly sound****, if it is strongly sound for A and strongly sound for B. The scheme is strongly sound for B, if for any $\mathsf{pp} \leftarrow \mathsf{Setup}(1^\kappa)$, any keys $(\mathsf{pk}, \mathsf{sk}) \leftarrow \mathsf{KeyGen}(\mathsf{pp})$, any secret key shares $(\mathsf{pk}, \mathsf{sk}_A, \mathsf{sk}_B) \leftarrow \mathsf{Share}(\mathsf{pk}, \mathsf{sk})$, for all PPT adversary $\mathscr{A}$ playing the role of A, the success*

$$\mathsf{Succ}_B^{esp\text{-}sound}(\mathscr{A}) = \Pr[BadSwitch| \mathscr{A}^{\mathscr{O}_B(\cdot,\cdot,\cdot)}(\mathsf{pk}, \mathsf{sk}_A, \mathsf{sk})]$$

is negligible, where the event BadSwitch is raised when a full protocol execution of Switch *with $\mathscr{O}_B$ on a switchable input ciphertext c successfully outputs $c^\star$ which is not a twin ciphertext of c. (In a non-strong version of soundness the adversary is only given $(\mathsf{pk}, \mathsf{sk}_A)$.) We denote $\mathsf{Succ}^{esp\text{-}sound}(\kappa, t)$ the maximal success an adversary can get against A or B within time t.*

The zero-knowledge property guarantees that no information leaks about the secret key shares to a malicious player when switches are performed on switchable ciphertexts: its view can be simulated without any additional information than its own secret share.

Definition 6 (Zero-Knowledge). *An encryption switching protocol $\Pi_1 \rightleftharpoons \Pi_2$ is* ***zero-knowledge****, if it is zero-knowledge for A and zero-knowledge for B. The scheme is zero-knowledge for B if there exist two efficient simulators, $\mathscr{Sim}_B^{\mathsf{share}}$ and $\mathscr{Sim}_B^{\mathsf{ESP}}$ of* Share *and the oracle $\mathscr{O}_B$ respectively, with the following property: for any $\mathsf{pp} \leftarrow \mathsf{Setup}(1^\kappa)$, any keys $(\mathsf{pk}, \mathsf{sk}) \leftarrow \mathsf{KeyGen}(\mathsf{pp})$, any secret key shares $(\mathsf{pk}, \mathsf{sk}_A, \mathsf{sk}_B) \leftarrow \mathsf{Share}(\mathsf{pk}, \mathsf{sk})$ or simulated shares $(\mathsf{pk}', \mathsf{sk}'_A) \leftarrow \mathscr{Sim}_B^{\mathsf{share}}(\mathsf{pk})$, and for any PPT adversary $\mathscr{A}$ playing the role of A, the advantage*

$$\mathsf{Adv}_B^{esp\text{-}zk}(\mathscr{A}) = \big| \Pr[1 \leftarrow \mathscr{A}^{\mathscr{O}_B(\cdot,\cdot,\cdot,\cdot)}(\mathsf{pk}, \mathsf{sk}_A)] - \Pr[1 \leftarrow \mathscr{A}^{\mathscr{Sim}_B(\cdot,\cdot,\cdot,\cdot)}(\mathsf{pk}', \mathsf{sk}'_A)] \big|$$

is negligible, where the adversary $\mathscr{A}$ is given unbounded access to either the simulator $\mathscr{Sim}_B$ or the stateful oracle $\mathscr{O}'_B$ described below, with the restriction that input ciphertexts $(c, \bar{c})$ to $\mathscr{Sim}_B$ or $\mathscr{O}'_B$ are twin ciphertexts:

Oracle $\mathscr{O}'_B(i{\rightarrow}j, c, \bar{c}, \mathsf{Flow})$: *on input a direction $i{\rightarrow}j$, a ciphertext c under the encryption scheme Π_i, a ciphertext $\bar{c}$ under the encryption scheme Π_j, and a message flow* Flow*, ignores $\bar{c}$ and runs $\mathscr{O}_B(i{\rightarrow}j, c, \mathsf{Flow})$;*

Simulator $\mathscr{S}im_B(i{\rightarrow}j, c, \bar{c}, \mathsf{Flow})$: *on the same inputs as above, emulates the output an honest player B would answer upon receiving the flow* Flow *when running the protocol* $\mathsf{Switch}_{i{\rightarrow}j}((\mathsf{pk}, \mathsf{sk}_A, c), (\mathsf{pk}, \mathsf{sk}_B, c))$, *without* sk_B *but possibly with* sk_A, *and forcing the output to be a ciphertext* $\bar{c}'$ *equivalent to* $\bar{c}$ *(i.e., a ciphertext* $\bar{c}'$ *such that* $\mathsf{Dec}(\mathsf{sk}, \bar{c}) = \mathsf{Dec}(\mathsf{sk}, \bar{c}'))$.

If the adversary $\mathscr{A}$ *can be unbounded,* $\Pi_1 \rightleftharpoons \Pi_2$ *is statistically zero-knowledge. We denote* $\mathsf{Adv}^{\mathsf{esp\text{-}zk}}(\kappa, t)$ *the maximal advantage an adversary can get against* A *or* B *within time* t.

At a high level, Definition 4 says that (misbehaving) players A and B separately gain no information on the plaintexts even if they can switch the ciphertexts between Π_1 and Π_2. In that sense, switching ciphertexts is a special kind of two-party computation. It is pretty clear that a secure ESP on appropriate encryption schemes allows to build two-party protocols in $\mathscr{M}_1 \cap \mathscr{M}_2$.

2.3   Computational Equality

Let us consider an adversary $\mathscr{A}$ which can efficiently sample messages in both the intersection of the message spaces $\mathscr{M}_1 \cap \mathscr{M}_2$ and their symmetric difference $\mathscr{M}_1 \oplus \mathscr{M}_2 = (\mathscr{M}_1 \cup \mathscr{M}_2) \backslash (\mathscr{M}_1 \cap \mathscr{M}_2)$. A simple observation shows that a secure ESP could not be safe to use inside a larger protocol, even in front of a passive adversary, since the switching protocol does not provide any guarantee on non-switchable ciphertexts, that encrypt messages outside $\mathscr{M}_1 \cap \mathscr{M}_2$. They could help to distinguish ciphertexts. More generally, we would like Switch not to help for distinguishing *switchable* ciphertexts from *non-switchable* ciphertexts, which would break the IND-CPA security with the Switch oracle.

A solution could be a restriction on the choice of the ciphertexts asked to the Switch oracles, so that the plaintexts lie in $\mathscr{M}_1 \cap \mathscr{M}_2$. But this would not be strong enough for practical purpose, since there is no reason that it cannot happen during a complex evaluation. We thus define the following additional property, to be satisfied by the message spaces, with the common public key pk as auxiliary input:

Definition 7 (Computational Equality). *Let* $(\mathscr{M}_1, \mathscr{M}_2, \mathsf{aux})$ *be two sets and some additional information.* $\mathscr{M}_1$ *and* $\mathscr{M}_2$ *are computationally equal given auxiliary input* aux *if, for any adversary* $\mathscr{A}$, *its success probability for outputting a message in the symmetric difference* $\mathscr{M}_1 \oplus \mathscr{M}_2$, *denoted* $\mathsf{Succ}^{\mathsf{comp\text{-}eq}}(\mathscr{A}) = \Pr[m \leftarrow \mathscr{A}(\mathscr{M}_1, \mathscr{M}_2, \mathsf{aux}) : m \in \mathscr{M}_1 \oplus \mathscr{M}_2]$, *is negligible.*

We have defined the security of ESP for switchable inputs and, informally, the computational equality will guarantee that non-switchable inputs are quite unlikely during the execution of a protocol involving ESPs.

2.4   Ring-Homomorphic Encryption Schemes

Toward our aim of getting two-party computation protocols from ESP, our goal is to design two encryption schemes on a ring structure $(\mathscr{R}, \oplus, \otimes)$, where the encryption algorithms are homomorphic on the plaintexts (under either $\oplus$ or $\otimes$) and on the random coins (with an appropriate group law $\odot$ over the randomness space R which may differ in every case), using the combinations $\boxplus$ and $\boxtimes$ of the ciphertexts:

$$
\begin{aligned}
\mathscr{E}_\oplus(m_1; r_1) \boxplus \mathscr{E}_\oplus(m_2; r_2) &= \mathscr{E}_\oplus(m_1 \oplus m_2; r_1 \odot r_2) \\
\mathscr{E}_\otimes(m_1; r_1) \boxtimes \mathscr{E}_\otimes(m_2; r_2) &= \mathscr{E}_\otimes(m_1 \otimes m_2; r_1 \odot r_2)
\end{aligned}
\tag{1}
$$

In particular, this implies that we can maul any ciphertext of m into a ciphertext of $R \otimes m$, for a *known* R, with an appropriate operation $\bullet$ in each case (and the appropriate operation $\cdot$ on the random coins) on the ciphertexts:

$$
R \bullet \mathscr{E}_\oplus(m; r) = \mathscr{E}_\oplus(R \otimes m; R \cdot r) \quad R \bullet \mathscr{E}_\otimes(m; r) = \mathscr{E}_\otimes(R \otimes m; R \cdot r). \tag{2}
$$

Note that we explicitly choose $\boxplus$, $\boxtimes$ and $\bullet$ to be deterministic functions, so that any local homomorphic evaluation on ciphertexts leads to the same ciphertext result. Note also that the existence of $\boxplus$ and $\boxtimes$ implies the stability of the plaintexts spaces of $\mathscr{E}_\oplus()$ and $\mathscr{E}_\otimes()$, under $\oplus$ and $\otimes$ respectively.

2.5   General Secure Two-Party Computation

The reason of designing ESP is to take advantage of the nice (homomorphic) properties of the two schemes which may not be available in a single *efficient* encryption scheme. When additions $\oplus$ are required, we use ciphertexts under the additively homomorphic encryption scheme *w.r.t.* $\boxplus$, and when multiplications $\otimes$ and exponentiations are needed, we convert the operands into the other multiplicatively homomorphic encryption scheme *w.r.t.* $\boxtimes$. In other words, ESP aims at reconciling additively and multiplicatively homomorphic schemes, to jointly compute the encryption of $f(x, y)$, for any public function f over $(\mathscr{R}, \oplus, \otimes)$, on encryptions of x and y. Below, we consider two-party computation which reveals the result to a single party only (Alice).

Secure 2-PC. More formally, assuming only Alice gets the outcome, the security game of such a privacy-preserving evaluation is the following one: The adversary against Bob chooses its input x and the possible inputs y_0, y_1 for Bob, with the additional restriction that $f(x, y_0) = f(x, y_1)$ (otherwise the outcome would reveal Bob's actual input value); It gets the encryption of x and the encryption of y_b for a random bit $b \xleftarrow{\$} \{0, 1\}$; At the end of the joint evaluation with Bob, it should try to guess b, and thus Bob's actual input value. If the adversary plays the role of Bob against Alice, then it chooses its input y and the possible inputs x_0, x_1 for Alice but without any additional restriction. When no adversary can guess b in any of the two games (against Alice or Bob), with non-negligible advantage, we say that the 2-PC protocol is *input-indistinguishable*. This is formally defined in the full version [6].

   Since we assume that Alice receives the outcome of the 2-PC in our design we also assume that Alice and Bob are able to decrypt ciphertexts *from their shares*. Without loss of generality, we assume that Π_2 admits a 2-party decryption (as detailed in the full version [6]) so that only Alice gets the plaintexts. A rigorous construction Π_{2PC} is proposed in the full version [6], using a secure ESP between homomorphic encryption schemes over computationally-equal message spaces, following the above intuition, leads to the next result.

Theorem 8. *Let Π_1 and Π_2 be IND-CPA (complementary) homomorphic encryption schemes over a ring $(\mathscr{R}, \oplus, \otimes)$, whose message spaces are computationally equal, equipped with a secure ESP, $\Pi_1 \rightleftharpoons \Pi_2 = (\text{Share}, \text{Switch})$, so that Π_2 admits a 2-party decryption for A from the same key shares output by Share and which is statistically sound and zero-knowledge, then the Π_{2PC} protocol is an input-indistinguishable 2-PC for any function f over $(\mathscr{R}, \oplus, \otimes)$.*

We stress that this theorem is for the malicious setting: if the ESP protocols (and the 2-party decryption) are secure against malicious adversaries, the Π_{2PC} protocol is secure against malicious adversaries, without any additional zero-knowledge proofs.

Intuition. Our approach for Π_{2PC} consists in starting from ciphertexts of x and y, and to switch to the appropriate encryption scheme in order to be able to make operations through the homomorphic property, until the encryption of the result is reached. The rationale of the *computational-equality property* for the message spaces, with the public key as auxiliary input, is the following one: on encryptions of valid inputs x and y_b, the evaluation of the encryption of $f(x, y_b)$ follows a deterministic path of switches and public homomorphic operations on the ciphertexts. In the honest-but-curious setting, the sequences of involved plaintexts is indeed determined by x and y_b, and in the malicious setting, the soundness property ensures that the same happens. Then, if all the ciphertexts are switchable, using the simulators from the zero-knowledge property of the ESP leads to the privacy of the computation: no information leaks on b. If a ciphertext happens to be non-switchable with non-negligible probability during the computation, simply generating the sequences of plaintexts from (x, y_0) and from (x, y_1) would efficiently generate an element in the symmetric difference: we need this to be intractable. Eventually, the outcome of the protocol is recovered by performing 2-party decryption.

Sketch of the Proof. The structure of the proof follows a sequence of indistinguishable games from the real game with (x, y_0), between the adversary and a simulator emulating the challenger using $b = 0$ with all the secret information to the real game with (x, y_1), and so using $b = 1$. We consider the output guess b', which should remain the same. The first games consist of a preparation for replacing y_0 by y_1. We indeed cannot apply the semantic security of the encryption schemes yet since the decryption keys are known to the simulator. But first, with the computational-equality property, we can guarantee that all the input ciphertexts of the ESPs are switchable. Then, with the soundness of the ESPs, we know that the outputs of the ESPs are twin ciphertexts. Actually, we need here the *strong* flavor of soundness since the secret key is still known. Again we apply the soundness of the final 2-party decryption to guarantee the correct decryption (since the decryption key is still known, we require the *statistical* soundness, but a *strong* flavor would be enough too). Now that we know all the input-output pairs of the internal primitives (ESPs and decryption) are correct, we can safely replace the honest emulation using the secret key by the simulators without the secret key, thanks to the zero-knowledge property. So, the secret key is not required anymore, and we can replace y_0 by y_1, applying the IND-CPA security game to the first encryption scheme. We also have to propagate to the outputs of the ESPs, using again the IND-CPA security game of the other encryption scheme. This is done sequentially, with hybrid games, to end with a game where the input is (x, y_1) and all the intermediate ciphertexts are consistent. We can then move back to the honest emulation (and not the simulators for the ESPs and the decryption) using the secret key. The full construction is described and formally proven secure in the full version [6].

Our Next Goal. Three properties must be satisfied to securely evaluate functions over a ring: the *homomorphism* of the encryption schemes, the *security* of the ESPs and the *computational equality* of the messages spaces. Instantiating these building blocks would allow us to achieve our second objective: building an *efficient* two-party computation over a ring as a realistic alternative to standard methods, particularly for arithmetic functions with a high multiplicative depth. After discussing some applications of ESPs, we provide a first step toward our goal by designing a secure ESP to switch between two homomorphic encryption schemes over $\mathbb{Z}_n^*$.

3 Applications

In this section, we motivate our paradigm for two-party computation with some concrete examples involving high-depth circuits.

Private Disjointness Testing (PDT). Two players, Alice and Bob, holding respective databases $A = (a_i)_{i \leq a}$ and $B = (b_i)_{i \leq b}$, wish to know whether their databases have at least one common element or not, and nothing more. The state-of-the-art solution to PDT is [42], which solves the problem with complexity $O\left((a+b)^2\right)$ (counting group elements).

   A natural way to solve the PDT is to view the items of A as the roots of a polynomial $P(X) = \sum_{i=0}^{a} \alpha_i X^i$. Alice and Bob perform an interactive protocol which outputs $u = r \prod_{i=1}^{b} P(b_i)$ to Alice, where r is a uniformly random value picked by Bob. If this value is 0, then one of the $P(b_i)$'s is zero, which means that one of the b_i's is in A. However, the circuit computing u is of depth $O(\log b)$, hence most 2-PC protocols computing this circuit are not constant round. Using carefully constructed circuits such as the sort-compare-shuffle circuit of [22] (adapted to the case of PDT), the (constant-round) garbled circuit approach transmits $O(\kappa \ell(a+b) \log(a+b) + \kappa b M(\kappa))$ bits, where ℓ is the size of the items in A and B and $M(\kappa)$ the circuit size of modular multiplication (multiplications are performed modulo a κ-bit value to avoid integer multiplication while maintaining statistical correctness).

Our framework allows us to design a linear-communication constant-round protocol for the private disjointness test:

1. Alice builds the polynomial $P = \sum \alpha_i X^i$ so that $P(a_i) = 0$ for $i \le a$, and sends $(C_i = \mathscr{E}_\oplus(\alpha_i))_i$;
2. Bob computes and sends $D_i \leftarrow \boxplus_j b_i^j \bullet C_i = \mathscr{E}_\oplus(P(b_i))$ for $i \le b$;
3. They perform b ESPs in parallel to get $(D_i' = \mathscr{E}_\otimes(P(b_i)))_{i \le b}$;
4. Bob picks $r \xleftarrow{\$} \mathbb{Z}_n$ and computes $E \leftarrow r \bullet \boxtimes_i D_i' = \mathscr{E}_\otimes(r \times \prod P(b_i))$.
5. Alice and Bob jointly decrypt the ciphertext, Bob gets the result and checks whether the plaintext is zero or not.

The total communication complexity of this protocol is $a+b+2$ ciphertexts and b parallel ESPs. With constant size ESPs (as we will construct in the following), this gives a total communication of $O(a+b)$ in constant round. We want to stress that this does not mean that, for concrete parameters, this approach will necessarily beat the best super-linear garbled circuits for PDT; however, garbled circuits have enjoyed decades of optimizations, and given its asymptomatic complexity, our new approach seems worth considering for further investigations and could benefit from numerous optimizations. Note also that hybrid frameworks (such as [21]) can also provide linear-communication constant-round solutions, but unlike these protocols, our approach is easily enhanced to the malicious setting: in a high level, items 1 and 2 are secure from [7] and the next items are secure against malicious adversaries if so are the ESPs performing the switches (and Sect. 6 provides an efficient technique to achieve this security).

Oblivious Multivariate Polynomial Evaluation (OMPE). This is the natural extension of oblivious polynomial evaluation [31] over multivariate polynomials [39]. Once an ESP is available, constructing an OMPE protocol is straightforward (we use the notations of [39]). Unlike previous solutions, it keeps the degree d of P hidden.

- Alice holds an N-variate polynomial P of degree d with M monomials;
- Bob holds $(x_1, \cdots, x_N)$ and sends $(\mathscr{E}_\otimes(x_i))_{i \le N}$;
- Alice computes all the M monomials of $P(x_1, \cdots, x_N)$ encrypted under $\mathbb{Z}_n^*$-EG, due to the multiplicativity;
- Alice and Bob perform M parallel ESPs on the encrypted monomials to get the M additively encrypted monomials, and then get $\mathscr{E}_\oplus(P(x_1, \cdots, x_N))$;
- Alice and Bob jointly decrypt it, so that Bob (or both) gets $P(x_1, \cdots, x_N)$.

Our OMPE protocol transmits $O((N + M) \log n)$ bits, to be compared with $O(Nd\kappa^2)$ for [39]. In addition, our protocol can be adapted to the case of multivariate polynomials whose most compact representation is not their canonical form; for example, if the polynomial is of the form $\prod_i \sum_j X_j^{\delta_{ij}}$, extending it to its canonical form would result in an expression with exponentially many terms.

Instead, the polynomial can be directly evaluated from this compact form: first using the multiplicative homomorphism to evaluate the $X_i^{\delta_{ij}}$'s, they switch to perform the sums, and then switch again to perform the final product. Several applications of OMPE are discussed in [39], such as testing whether the union of two sets of vectors are of full rank which has applications in linear secret sharing schemes, where the secret can be recovered when a full rank set of vectors is known; the players can determine whether they could recover the secret together without revealing their set. We get a more efficient *Full-Rank Test* protocol.

4 An Encryption Switching Protocol over $\mathbb{Z}_n^*$

For the internal laws on the plaintexts in $\mathbb{Z}_n$ we keep the usual notations $+$ and $\times$ (or $\cdot$ and even nothing), but we still use the notations of the Sect. 2.4 for the external operations on the ciphertexts and the relations on the random coins.

In order to complete the Paillier encryption scheme, that is additively homomorphic in $\mathbb{Z}_n$, we build an ElGamal variant that is multiplicatively homomorphic in $\mathbb{Z}_n^*$, both for the same RSA modulus n. The security of our new variant relies on the DDH assumption in $\mathbb{J}_n$, the (maximal) cyclic subgroup of $\mathbb{Z}_n^*$ whose elements have a Jacobi symbol equal to $+1$, and the QR assumption in $\mathbb{Z}_n^*$ (see the full version [6] for more details about the structure of the ring $\mathbb{Z}_n$). In order to build a secure encryption switching protocol, we need an additional property from the two encryption schemes: they can be *randomized*. An encryption scheme $\mathscr{E}$ is randomizable if there exists an efficient algorithm Rand such that for every message m and every random coins $r \in \mathsf{R}$:

$$\{\mathscr{E}(m;r') \mid r' \xleftarrow{\$} \mathsf{R}\} \equiv \{\mathsf{Rand}(\mathscr{E}(m;r),r') \mid r' \xleftarrow{\$} \mathsf{R}\} \qquad (3)$$

where $\equiv$ denotes the computational/statistical/perfect indistinguishability of the two distributions. For the sake of simplicity, we will denote $\mathsf{Rand}(C)$ the probabilistic algorithm which picks r uniformly at random and returns $\mathsf{Rand}(C;r)$.

We now recall basic computational assumptions and an implication to $\mathbb{J}_n$, then we review the Paillier encryption which also admits a verifiable 2-party decryption algorithm (where either the two players, or one player only, get the result) and we introduce our new ElGamal encryption schemes. Finally, we show how to switch between these schemes from encryptions over $\mathbb{Z}_n^*$.

4.1 Computational Assumptions

The security of our protocols will rely on the following standard assumptions:

- The DDH (Decisional Diffie-Hellman) assumption in a cyclic group $\mathbb{G} = \langle g \rangle$ of order q states that, given (g^a, g^b) for $a, b \xleftarrow{\$} \mathbb{Z}_q$, g^{ab} is indistinguishable from a random element in $\mathbb{G}$.
- The QR (Quadratic Residuosity) assumption in $\mathbb{Z}_n^*$, for an RSA modulus n, states that a random element in QR_n (square in $\mathbb{Z}_n^*$) is indistinguishable from a random element in $\mathbb{J}_n$ (element of $\mathbb{Z}_n^*$ with Jacobi symbol $+1$).

- The DCR (Decisional Composite Residuosity) assumption in $\mathbb{Z}_{n^2}^*$, for an RSA modulus n, states that a random n-th power in $\mathbb{Z}_{n^2}^*$ is indistinguishable from a random element in $\mathbb{Z}_{n^2}^*$.

The DDH assumption is usually assumed to hold in large prime-order subgroups of $\mathbb{Z}_p^*$. In the following, $n = pq$ is a **strong RSA modulus** if $p = 2p' + 1$ and $q = 2q' + 1$ are safe primes (with both p' and q' also prime). With such a modulus n, DDH is also a reasonable assumption in QR_n, since the order is $p'q'$ (see the full version [6] for more details). Adding the QR assumption in $\mathbb{Z}_n^*$, this makes the DDH assumption in $\mathbb{J}_n$ (of order $2p'q'$) reasonable too:

Theorem 9. *When $n = pq$ is a strong RSA modulus, the DDH assumption in $\mathbb{J}_n$ is implied by the DDH assumption in both the large prime-order subgroups of $\mathbb{Z}_p^*$ and $\mathbb{Z}_q^*$ and the QR assumption in $\mathbb{Z}_n^*$. (The proof is in the full version [6].)*

However, given $m \in \mathbb{Z}_n^*$, computing Jacobi symbol $J_n(m)$ is easy and then the DDH assumption does not hold in $\mathbb{Z}_n^*$ which, in addition, is non cyclic.

4.2 $\mathbb{Z}_n$-P: The Paillier Encryption Scheme on $\mathbb{Z}_n$

For the Paillier encryption scheme (denoted $\mathbb{Z}_n$-P), we will use the notation $\mathcal{E}_\oplus(\cdot)$ since this will be our additively homomorphic encryption scheme. It implicitly uses the strong RSA modulus $n = pq$, and we denote $\lambda = \lambda(n) = (p-1)(q-1)/2$, the maximal order of an element of $\mathbb{Z}_n^*$. One can note that $\lambda = (n-1)/2 + (2 - (p+q))/2$ is statistically close to $(n-1)/2$ or $n/2$ if we consider the Euclidean division (we will abuse this notation $n/2$ in the following).

The Paillier Cryptosystem. In [33], Paillier proposed an encryption scheme $\mathbb{Z}_n$-P for a modulus $\mathsf{pk} = n$ as public key, and $\mathsf{sk} = d \leftarrow [\lambda^{-1} \bmod n] \times \lambda \bmod n\lambda$ as secret key: $\mathbb{Z}_n$-P.Enc$(\mathsf{pk}, m; r)$, for a message $m \in \mathbb{Z}_n$ and random coins r in $\mathbb{Z}_n^*$, outputs $c = (1 + n)^m \cdot r^n \bmod n^2$; $\mathbb{Z}_n$-P.Dec(sk, c) returns $m = ([c^d \bmod n^2] - 1)/n$. (See details in the full version [6]). This scheme is IND-CPA under the DCR assumption over $\mathbb{Z}_{n^2}^*$, and it is additively homomorphic in $\mathbb{Z}_n$. It satisfies Eq. (1), $\boxplus$ being the multiplication in $\mathbb{Z}_{n^2}^*$. The randomization algorithm Rand is given by $\mathbb{Z}_n$-P.Rand$(c; r) = c \cdot r^n \bmod n^2$, for any random coins r in $\mathbb{Z}_n^*$.

2-Party Paillier Decryption. In this section, we briefly recall the semi-honest case where players are honest-but-curious. The reader can refer to the full version [6] for more details and a description in the malicious case which makes use of classical zero-knowledge proofs.

We assume that a trusted dealer generates the key shares for the two parties, Alice and Bob (distributed key generation can be found in [20]). The dealer generates random $d_A, d_B \in \mathbb{Z}_{n\lambda}$ subject to $d_A + d_B = d \bmod n\lambda$ defined above. Then, Alice gets d_A and Bob gets d_B.

In order to allow Bob to decrypt the ciphertext C, Alice computes and sends $C_A \leftarrow C^{d_A} \bmod n^2$, which allows Bob to get the plaintext $m \leftarrow ([C_A \times C^{d_B} \bmod$

$n^2] - 1)/n$. Note that we do intentionally not disclose m to Alice in general. But this is perfectly symmetric if one wants Alice to get the result instead of Bob.

The **correctness** of this protocol is straightforward. Let us show that it is statistically **zero-knowledge**: To emulate Alice in front of a curious Bob, we first pick d_B in $\mathbb{Z}_{n^2/2}$ instead of $\mathbb{Z}_{n\lambda}$ (since $n/2$ is statistically close to λ) and we give it to Bob. The simulator with input (m, d_B) sends $C_A \leftarrow (1 + n \cdot m) \times C^{-d_B}$, which enforces the decryption to m for Bob. This simulation is statistically indistinguishable from a real execution when C does indeed encrypt m. No emulation of Bob is needed as he does not send any message.

4.3 $\mathbb{Z}_n^*$-EG: An ElGamal Variant in $\mathbb{Z}_n^*$

The ElGamal Cryptosystem. In [14], ElGamal proposed the famous encryption scheme that applies in any cyclic group $\mathbb{G} = \langle g \rangle$ of order q, in which the DDH assumption holds: for a secret scalar $\mathsf{sk} = x \xleftarrow{\$} \mathbb{Z}_q$, the public key is $\mathsf{pk} = h \leftarrow g^x$: $\mathsf{Enc}(\mathsf{pk}, m; r)$, for a message $m \in \mathbb{G}$ and random coins r in $\mathbb{Z}_q$, outputs $c = (c_0 = g^r, c_1 = h^r \cdot m)$; $\mathsf{Dec}(\mathsf{sk}, c)$ returns $m = c_1/c_0^x$.

This scheme is IND-CPA under the DDH assumption over $\mathbb{G}$, and it is multiplicatively homomorphic in $\mathbb{G}$. ElGamal encryption scheme satisfies Eq. (1), $\boxtimes$ being the component-wise multiplication in $\mathbb{G}^2$. The randomization algorithm Rand is given by $\mathsf{Rand}(c; r) = (c_0 \cdot g^r, c_1 \cdot h^r)$, for any random coins r in $\mathbb{Z}_q$. The 2-party decryption protocol is quite similar to the above Paillier one.

In the following, we will essentially use QR_n-EG and $\mathbb{J}_n$-EG, the ElGamal encryption schemes in QR_n and $\mathbb{J}_n$ respectively.

Extension to $\mathbb{Z}_n^*$. However, our main goal is to extend the ElGamal encryption scheme to $\mathbb{Z}_n^*$. The global parameters contain the strong RSA modulus n, with a generator g of $\mathbb{J}_n$. The global setup and the algorithms are described on Fig. 1.

Description. Since the larger space that ElGamal can securely encrypt is $\mathbb{J}_n$, in order to encrypt a message $m \in \mathbb{Z}_n^*$, we have to split m into two parts, $m_1, m_2 \in \mathbb{J}_n$: given $\chi \in \mathbb{Z}_n^* \setminus \mathbb{J}_n$, a natural encoding is $m_1 = J_n(m) = (-1)^a$ and $m_2 = \chi^a m$, with an appropriate integer a. But, even if $\{\pm 1\}$ could be seen as a subgroup of $\mathbb{J}_n$, $\psi : \mathbb{Z}_2 \times \mathbb{J}_n \mapsto \mathbb{Z}_n^*$, $\psi(a, m) = \chi^{-a} m$ is not an homomorphism when the order of χ is not 2. But we cannot leave in the clear[2] a square root of 1 lying in $\mathbb{Z}_n^* \setminus \mathbb{J}_n$ (as done in [17]). However, for a generator g of $\mathbb{J}_n$, we can instead encode m with $m_1 = g^a$ and $m_2 = \chi^{-a} m$ for any integer a such that $J_n(m) = (-1)^a$, and encrypt m_2 into (C_0, C_1) using $\mathbb{J}_n$-EG, and appending m_1 in clear. The intricate point in the decryption phase will be to reconstruct χ^a from $m_1 = g^a$: if one defines $v = [p^{-1} \bmod q] \cdot p \bmod n$ and $\chi \leftarrow (1 - v) \cdot g^{t_p} + v \cdot g^{t_q} \bmod n$, for even t_p and odd t_q randomly drawn in $\mathbb{Z}_\lambda$, then $\chi \in \mathbb{Z}_n^* \setminus \mathbb{J}_n$. In addition, from $m_1 = g^a$, one gets χ^a as $(1 - v) m_1^{t_p} + v m_1^{t_q} \bmod n$. The complete description of the scheme is described on Fig. 1.

---

[2] Given two square roots of the same element with distinct Jacobi symbols allows efficiently factoring n.

Setup and Key Generation

- The main strong RSA modulus n:
  - p, q two safe primes, $n \leftarrow pq$;
  - $g_0 \xleftarrow{\$} \mathbb{Z}_n^*$, $g \leftarrow -g_0^2$ (a generator of $\mathbb{J}_n$, of order λ);
  - $d \leftarrow [\lambda^{-1} \bmod n] \cdot \lambda \bmod n\lambda$: $d = 0 \bmod \lambda$ and $d = 1 \bmod n$;
  - $v \leftarrow [p^{-1} \bmod q] \cdot p \bmod n$: $v = 0 \bmod p$ and $v = 1 \bmod q$;
  - an even $t_p \xleftarrow{\$} \mathbb{Z}_\lambda$ and an odd $t_q \xleftarrow{\$} \mathbb{Z}_\lambda$: $\chi \leftarrow (1 - v) \cdot g^{t_p} + v \cdot g^{t_q} \bmod n$;
  - $s \xleftarrow{\$} \mathbb{Z}_\lambda$, and set $g_1 \leftarrow g^s \bmod n$ (for $\mathbb{J}_n$-EG).
- The additional modulus N:
  - P, Q two strong primes, $N \leftarrow PQ$ (such that $N > (2 + 2^{\kappa+1})n^2$);
  - $D \leftarrow [\Lambda^{-1} \bmod N] \cdot \Lambda \bmod N\Lambda$, where Λ is the order of $\mathbb{J}_N$.
- Keys: $\mathsf{pk} \leftarrow (n, g, \chi, g_1, N)$ and $\mathsf{sk} \leftarrow (d, v, t_p, t_q, s, D)$.
- Partial keys: $(d_A, v_A, t_{pA}, t_{qA}, s_A, D_A) \xleftarrow{\$} \mathbb{Z}_{n\lambda} \times \mathbb{Z}_n \times \mathbb{Z}_\lambda^3 \times \mathbb{Z}_{N\Lambda}$, and $d_B \leftarrow d - d_A \bmod n\lambda$, $v_B \leftarrow v - v_A \bmod n$, $t_{pB} \leftarrow t_p - t_{pA} \bmod \lambda$, $t_{qB} \leftarrow t_q - t_{qA} \bmod \lambda$, $s_B \leftarrow s - s_A \bmod \lambda$, and $D_B \leftarrow D - D_A \bmod N\Lambda$.

$\mathscr{E}_\otimes(\cdot) = \mathbb{Z}_n^*$-EG: ElGamal Encryption Scheme in $\mathbb{Z}_n^*$

$\mathsf{Enc}(\mathsf{pk}, m)$: On input $m \in \mathbb{Z}_n^*$, compute $(m_1, m_2) \leftarrow (g^a, \chi^{-a}m) \in \mathbb{J}_n^2$ for $a \xleftarrow{\$} \mathbb{Z}_{n/2}$, so that $J_n(m) = (-1)^a$. Then, choose $r \xleftarrow{\$} \mathbb{Z}_{n/2}$ and compute $C \leftarrow \mathbb{J}_n\text{-EG.Enc}(m_2; r) = (c_0 = g^r, c_1 = m_2 g_1^r)$.
Return the ciphertext $c \leftarrow \mathscr{E}_\otimes(m; r) = (C = (c_0, c_1), m_1)$.

$\mathsf{Rand}(\mathsf{pk}, c)$: Parse $c = (C = (c_0, c_1), m_1)$, choose $r_1 \xleftarrow{\$} \mathbb{Z}_{n/2}$ and $r_2 \xleftarrow{\$} \mathbb{Z}_{n/4}$, output $c' \leftarrow (C' = (g^{r_1} \cdot c_0, \chi^{-2r_2} g_1^{r_1} \cdot c_1), g^{2r_2} \cdot m_1)$.

$\mathsf{Dec}(\mathsf{sk}, c)$: Parse $c = (C = (c_0, c_1), m_1)$ and check whether $J_n(c_1) = 1$. If not, return $\perp$, otherwise compute $m_2 \leftarrow \mathbb{J}_n\text{-EG.Dec}(C) = c_1/c_0^s$ in $\mathbb{Z}_n^*$ and then $m_0 \leftarrow (1 - v) \cdot m_1^{t_p} + v \cdot m_1^{t_q} \bmod n$.
Return $m \leftarrow m_0 m_2 \bmod n$.

$\mathscr{E}_\oplus(\cdot) = \mathbb{Z}_n$-P: Paillier Encryption Scheme on $\mathbb{Z}_n$

$\mathsf{Enc}(\mathsf{pk}, m)$: given $m \in \mathbb{Z}_n$, for a random $r \xleftarrow{\$} \mathbb{Z}_n^*$, output $c \leftarrow (1 + n)^m \cdot r^n \bmod n^2$.

$\mathsf{Rand}(\mathsf{pk}, c)$: choose $r \xleftarrow{\$} \mathbb{Z}_n^*$, output $c' \leftarrow r^n \cdot c \bmod n^2$.

$\mathsf{Dec}(\mathsf{sk}, c)$: return $m \leftarrow ([c^d \bmod n^2] - 1)/n$.

Fig. 1. Setup and encryption schemes in $\mathbb{Z}_n^*$

Properties. The **correctness** follows from the Chinese Remainder Theorem: by construction, $\chi \leftarrow (1 - v) \cdot g^{t_p} + v \cdot g^{t_q} \bmod n$, with v such that $v = 0 \bmod p$ and $v = 1 \bmod q$, then, $\chi = g^{t_p} \bmod p$ (so that $\chi \in \mathsf{QR}_p$) and $\chi = g^{t_q} \bmod q$ (so that $\chi \notin \mathsf{QR}_q$). Then, from $m_0 \leftarrow (1 - v)m_1^{t_p} + vm_1^{t_q} \bmod n$, we also have $m_0 = g^{at_p} = \chi^a \bmod p$ and $m_0 = g^{at_q} = \chi^a \bmod q$, and so $m_0 = \chi^a \bmod n$. Hence, $m_0 \cdot m_2 \bmod n$ is indeed the plaintext m in $\mathbb{Z}_n^*$.

The **multiplicative homomorphism** comes from the fact that a does not need to be in $\mathbb{Z}_2$, but just has to satisfy $(-1)^a = J_n(m)$ to make both m_1 and m_2 in $\mathbb{J}_n$. If one multiplies two ciphertexts c and c', of m and m' respectively, one gets $(g^{r+r'}, \chi^{-a-a'} mm' \cdot g_1^{r+r'}, g^{a+a'}) = (g^{r''}, \chi^{-a''} mm' \cdot g_1^{r''}, g^{a''})$, which is statistically indistinguishable from a direct encryption of mm' since $\mathbb{Z}_{n/2}$ is statistically close to $\mathbb{Z}_\lambda$.

As usual, the **randomization** just consists in multiplying by a ciphertext of $m = 1$, and so with any random encoding of 1: $(m_1 = g^{2a}, m_2 = \chi^{-2a})$. Hence, on input a ciphertext $C = (C_0, C_1, \alpha)$ and two random integers (r_1, r_2), $\mathsf{Rand}(C; r_1, r_2)$ outputs $C' \leftarrow (g^{r_1} \cdot C_0, \chi^{-2r_2} \cdot g_1^{r_1} \cdot C_1, g^{2r_2} \cdot \alpha)$. Note that this algorithm returns a ciphertext in which both the random coins and the encoding of the plaintext are uniform, hence this is a perfect randomization algorithm.

Security. A ciphertext $c = (C = (c_0, c_1), m_1)$ contains m_1 in clear but m_2 is encrypted using $\mathbb{J}_n$-EG. While m_1 encodes the Jacobi symbol of the plaintext m (if m_1 is a square, $m \in \mathbb{J}_n$ and if m_1 is not a square, $m \in \mathbb{Z}_n^* \setminus \mathbb{J}_n$), under the QR assumption in $\mathbb{Z}_n^*$, it is infeasible to distinguish squares from non-squares in $\mathbb{J}_n$: m_1 does not leak anything. The choice of χ is completely independent from the $\mathbb{J}_n$-EG decryption key. This means that the IND-CPA security of the scheme just relies on the DDH assumption in $\mathbb{J}_n$ (Theorem 9) and the QR assumption in $\mathbb{Z}_n^*$.

4.4 $\mathbb{Z}_n^*$-ESP: Encryption Switching Protocols on $\mathbb{Z}_n^*$

For an ESP, the general approach consists of four steps: Alice first randomizes the ciphertext, Bob gets the decryption and then re-encrypts it under the second encryption scheme, and Alice eventually de-randomizes it. Figure 2 contains the full description of the two protocols, from $\mathbb{Z}_n$-P to $\mathbb{Z}_n^*$-EG and from $\mathbb{Z}_n^*$-EG to $\mathbb{Z}_n$-P. The former is easy because of the simple 2-party $\mathbb{Z}_n$-P decryption. The latter requires a more intricate 2-party $\mathbb{Z}_n^*$-EG decryption, that needs to interactively compute χ^a from g^a. It requires a second Paillier encryption scheme in $\mathbb{Z}_{N^2}^*$ for a larger modulus $N > (2 + 2^{\kappa+1})n^2$ to make the computations in $\mathbb{Z}$ but masking the number of loops in the reduction modulo n.

Proof of Security of $\mathbb{Z}_n^*$-ESP. About the **correctness**, C encrypts m, C_A' encrypts R_A^{-1}, and C_A encrypts $x = R_A \cdot m$, in both directions. Then $x \bullet C_A'$ is a ciphertext of m under the second encryption scheme. In the multiplicative to additive direction, this is a bit more intricate, but $A_3 = -B^{t_{pA}}B_1 + B^{t_{qA}}B_2 = -B^{t_p} + B^{t_q}$ and $A_4 = A_1B_1 + A_2B_2 = (1 - v_A)B^{t_p} + v_A B^{t_q}$, hence E_5 and E_6 contain encryption of $B' \times (v_B(B^{t_q} - B^{t_p}) + ((1 - v_A)B^{t_p} + v_A B^{t_q} + kn)) = B' \times ((1-v)B^{t_p} + vB^{t_p})$. But as already remarked, $(1-v)B^{t_p} + vB^{t_p} = \chi^{a+r_1} \bmod n$ if $\alpha = g^a$. Hence, the plaintext $m_6 = \chi^a$, and x is the expected value. (The blinding factor kn added in E_6, which masks the number of reductions modulo n, disappears in the end.)

About the **zero-knowledge**, the full and detailed proof in the honest-but-curious setting of Theorem 10 can be found in the full version [6]. But in short, the proof is done in two steps, for Alice and for Bob. For each player, we exhibit a simulator which, essentially, generates the key share of its opponent from the public key without having any information on the key of the player it emulates, and is given for each switch a target output of the protocol. The simulator forces the output of the switch to be a re-randomization of its target output. He does so by sending random ciphertexts instead of correct ciphertexts and computing some

---

2-Party $\mathsf{ESP}^{\times}_{+}$ from $C = \mathscr{E}_{\oplus}(m)$ into $C' = \mathscr{E}_{\otimes}(m)$

$R_A \xleftarrow{\$} \mathbb{Z}_n^*, C'_A \leftarrow \mathscr{E}_{\otimes}(R_A^{-1})$
$C_A \leftarrow \mathbb{Z}_n\text{-P.Rand}(R_A \bullet C)$
$C_1 \leftarrow C_A^{d_A} \bmod n^2$

$\xrightarrow{\quad C'_A, C_A, C_1 \quad}$ $x \leftarrow ([C_1 \times C_A^{d_B} \bmod n^2] - 1)/n$

$\xleftarrow{\quad C' \quad}$ $C' \leftarrow \mathbb{Z}_n^*\text{-EG.Rand}(x \bullet C'_A)$

2-Party $\mathsf{ESP}^{+}_{\times}$ from $C = \mathscr{E}_{\otimes}(m)$ into $C' = \mathscr{E}_{\oplus}(m)$

$R_A \xleftarrow{\$} \mathbb{Z}_n^*, C'_A \leftarrow \mathscr{E}_{\oplus}(R_A^{-1})$
$C_A \leftarrow \mathbb{Z}_n^*\text{-EG.Rand}(R_A \bullet C)$
$\quad = (C_0, C_1, \alpha)$
$C_2 \leftarrow C_0^{s_A}$

$\xrightarrow{\quad C'_A, C_A, C_2 \quad}$ $C_3 \leftarrow C_0^{s_B}, \beta \leftarrow C_1/C_2 C_3$
$\qquad r_1 \xleftarrow{\$} \mathbb{Z}_{n/2}, B \leftarrow \alpha g^{r_1}, B' \leftarrow \chi^{-r_1}$
$\qquad (u_1, u_2) \leftarrow ([v_B B' \bmod n], [B' \bmod n])$

$A_1 \leftarrow (1 - v_A)B^{t_{p_A}}$
$A_2 \leftarrow v_A B^{t_{q_A}}$ $\xleftarrow{\quad B, B_1, B_2 \quad}$ $(B_1, B_2) \leftarrow (B^{t_{p_B}}, B^{t_{q_B}})$
$A_3 \leftarrow -B^{t_{p_A}} B_1 + B^{t_{q_A}} B_2$
$A_4 \leftarrow A_1 B_1 + A_2 B_2$
$(r_3, r_4, k) \xleftarrow{\$} \mathbb{Z}_N^{*2} \times \mathbb{Z}_{2^{\kappa+1}n}$
$E_3 \leftarrow \mathbb{Z}_N\text{-P.Enc}(A_3; r_3)$

$E_4 \leftarrow \mathbb{Z}_N\text{-P.Enc}(A_4; r_4)$ $\xrightarrow{\quad E_3, E_4 \quad}$ $r_5 \xleftarrow{\$} \mathbb{Z}_N^*$

$\xleftarrow{\quad E_5 \quad}$ $E_5 \leftarrow E_3^{u_1} E_4^{u_2} \times r_5^N \bmod N^2$
$E_6 \leftarrow \mathbb{Z}_N\text{-P.Rand}(kn \boxplus E_5)$

$F_6 \leftarrow E_6^{D_A} \bmod N^2$ $\xrightarrow{\quad E_6, F_6 \quad}$ $m_6 \leftarrow ([F_6 E_6^{D_B} \bmod N^2] - 1)/N$
$\qquad x \leftarrow \beta[m_6 \bmod n] \bmod n$

$\xleftarrow{\quad C' \quad}$ $C' \leftarrow \mathbb{Z}_n\text{-P.Rand}(x \bullet C'_A)$

Fig. 2. Interactive protocols for encryption switching in $\mathbb{Z}_n^*$

intermediate values using either its input or its target output (both being a twin-ciphertext pair). The Paillier scheme with a second larger modulus N is necessary to hide some redundancy in the flows sent by the player that a simulator could not have sampled without the knowledge of the keys. The full proof involves several subtleties (typically, ensuring that indistinguishability between two situations involving values over $\mathbb{J}_n$ is implied by the DDH assumption over QR_n).

Theorem 10. *When instantiated with the Paillier encryption scheme and the $\mathbb{Z}_n^*$-EG encryption scheme, both over $\mathbb{Z}_n^*$, the $\mathbb{Z}_n^* - ESP$ are zero-knowledge under the DDH assumption in QR_n, the QR assumption in $\mathbb{Z}_n^*$, the DCR assumption over $\mathbb{Z}_n^*$, and the DCR assumption over $\mathbb{Z}_N^*$.*

Using our two complementary homomorphic schemes and $\mathbb{Z}_n^* - ESP$ allows to evaluate functions over $\mathbb{Z}_n^*$, but no information leaks only if no intermediate computation will evaluate to 0 during the protocol. This is the goal of the next section to extend the message space of our ElGamal variant to $\mathbb{Z}_n^* \cup \{0\}$, which can be shown to be computationally equal to $\mathbb{Z}_n$.

5   An Encryption Switching Protocol over the Ring $\mathbb{Z}_n$

In order to allow computations over encrypted data in the full ring $(\mathbb{Z}_n, +, \times)$, we need to extend $\mathbb{Z}_n^*$-EG to a message space that is computationally equal to $\mathbb{Z}_n$. To this aim, we just have to handle zero. This will indeed make the two sets $\mathcal{M}_1 = \mathbb{Z}_n$ and $\mathcal{M}_2 = \mathbb{Z}_n^* \cup \{0\}$ computationally equal: finding an element in the symmetric difference provides a non-trivial non-invertible element, which breaks the factorization of n.

In the following, we use the notation $\mathscr{E}_\otimes(\cdot)$ for our above $\mathbb{Z}_n^*$-EG, and still $\mathscr{E}_\oplus(\cdot)$ for the Paillier encryption scheme $\mathbb{Z}_n$-P, both homomorphic on $(\mathbb{Z}_n^*, \times)$ and $(\mathbb{Z}_n, +)$ respectively, with the same strong RSA modulus n. We will also denote QR_n-EG and QR_n-EG$'$, two ElGamal encryption schemes over QR_n, and so with additional secret keys s_2, s_3, and $g_2 = g^{2s_2}$, $g_3 = g^{2s_3}$. QR_n-EG and QR_n-EG$'$ are clearly homomorphic in $(\mathsf{QR}_n, \times)$, and the IND-CPA security just relies on the DDH assumption in QR_n, which is independent of the factorization of n. Note however that QR_n-EG$'$ will be used as an extractable commitment and not an encryption scheme: the secret key s_3 is not kept by anybody (excepted the simulator in the security proof).

5.1 $\mathbb{Z}_n$-EG: Zero-Handling ElGamal Encryption Scheme in $\mathbb{Z}_n$

The global setup and the algorithms are represented in Fig. 3, but our $\mathbb{Z}_n$-EG encryption scheme essentially uses $\mathbb{Z}_n^*$-EG to encrypt $m + b$, where $b = 0$ if $m \neq 0$ and $b = 1$ otherwise, in $C_1 \leftarrow \mathscr{E}_\otimes(m + b)$, and is completed with two ciphertexts of b: $C_2 \leftarrow \mathsf{QR}_n$-EG.Enc$(T^b)$ and $C_3 \leftarrow \mathsf{QR}_n$-EG$'$.Enc$(T'^b)$, with two random squares T and T'.

The decryption algorithm is in two steps: one first decrypts C_2 to check whether the plaintext is 1, in which case $b = 0$ and so C_1 can be decrypted to get m, otherwise $b = 1$ and so one does not need to decrypt C_1 since $m = 0$. The purpose of C_3 will be for the simulation of the ESP (and namely of the *encrypted zero-test*, see below, in which the simulator is given a twin-ciphertext pair). This is reason why the decryption key s_3 will just be known to the simulator.

Properties. This scheme is correct, although the decryption is only statistically correct since the random square T can be equal to 1 with negligible probability. Since this is a combination of ElGamal encryption schemes, the resulting scheme is also IND-CPA. The 2-party decryption algorithms of $\mathbb{Z}_n^*$-EG and QR_n-EG immediately give rise to a 2-party decryption algorithm for $\mathbb{Z}_n$-EG: this is in two steps, as above, since the decryption of C_2 leads to either 1 or a random value.

Homomorphism. The multiplicativity of $\mathbb{Z}_n^*$-EG makes this scheme homomorphic until a zero is involved. And thanks to the absorbing property of random values T, it also captures the absorbing property of the zero value in the ring $\mathbb{Z}_n$: the multiplication is thus performed component-wise. In Fig. 3, we propose a randomization algorithm. One could note that C_1 will keep track of the operations performed on the ciphertexts when the global ciphertext encrypts zero, even

---

Setup and Key Generation
- The main strong RSA modulus n:
  - p, q two safe primes, $n \leftarrow pq$;
  - $g_0 \xleftarrow{\$} \mathbb{Z}_n^*$, $g \leftarrow -g_0^2$ (a generator of $\mathbb{J}_n$, of order λ);
  - $d \leftarrow [\lambda^{-1} \bmod n] \cdot \lambda \bmod n\lambda$: $d = 0 \bmod \lambda$ and $d = 1 \bmod n$;
  - $v \leftarrow [p^{-1} \bmod q] \cdot p \bmod n$: $v = 0 \bmod p$ and $v = 1 \bmod q$;
  - an even $t_p \xleftarrow{\$} \mathbb{Z}_\lambda$ and an odd $t_q \xleftarrow{\$} \mathbb{Z}_\lambda$: $\chi \leftarrow (1 - v) \cdot g^{t_p} + v \cdot g^{t_q} \bmod n$;
  - $s \xleftarrow{\$} \mathbb{Z}_\lambda$, and set $g_1 \leftarrow g^s \bmod n$ (for $\mathbb{J}_n$-EG).
  - $s_2, s_3 \xleftarrow{\$} \mathbb{Z}_{\lambda/2}^2$, and set $g_2 \leftarrow g^{2s_2} \bmod n$ (for QR_n-EG) and $g_3 \leftarrow g^{2s_3} \bmod n$ (for QR_n-EG').
- The additional modulus N:
  - P, Q two strong primes, $N \leftarrow PQ$ (such that $N > (2 + 2^{\kappa+1})n^2$);
  - $D \leftarrow [\Lambda^{-1} \bmod N] \cdot \Lambda \bmod N\Lambda$, where Λ is the order of $\mathbb{J}_N$.
- Keys: $\mathsf{pk} \leftarrow (n, g, \chi, g_1, g_2, g_3, N)$ and $\mathsf{sk} \leftarrow (d, v, t_p, t_q, s, s_2, D)$.
- Partial keys: $(d_A, v_A, t_{pA}, t_{qA}, s_A, s_{2A}, D_A) \xleftarrow{\$} \mathbb{Z}_{n\lambda} \times \mathbb{Z}_n \times \mathbb{Z}_\lambda^3 \times \mathbb{Z}_{\lambda/2} \times \mathbb{Z}_{N\Lambda}$, and $d_B \leftarrow d - d_A \bmod n\lambda$, $v_B \leftarrow v - v_A \bmod n$, $t_{pB} \leftarrow t_p - t_{pA} \bmod \lambda$, $t_{qB} \leftarrow t_q - t_{qA} \bmod \lambda$, $s_B \leftarrow s - s_A \bmod \lambda$, $s_{2B} \leftarrow s_2 - s_{2A} \bmod \lambda/2$, and $D_B \leftarrow D - D_A \bmod N\Lambda$.

$\mathscr{E}_\otimes^0(\cdot) = \mathbb{Z}_n$**-EG: ElGamal Encryption Scheme in $\mathbb{Z}_n$**

$\mathsf{Enc}(\mathsf{pk}, m)$: On input $m \in \mathbb{Z}_n$, if $m = 0$, then set $b = 1$ else set $b = 0$. Then, choose $T, T' \xleftarrow{\$} \mathsf{QR}_n$ and compute $C_1 \leftarrow \mathscr{E}_\otimes(m + b)$, $C_2 \leftarrow \mathsf{QR}_n\text{-EG.Enc}(T^b)$, $C_3 \leftarrow \mathsf{QR}_n\text{-EG'.Enc}(T'^b)$.
Return the ciphertext $C = \mathscr{E}_\otimes^0(m) = (C_1, C_2, C_3)$.

$\mathsf{Rand}(\mathsf{pk}, C = (C_1, C_2, C_3))$: Choose random $r_2, r_3 \xleftarrow{\$} \mathbb{Z}_{n/4}$, and compute $C_1' \leftarrow \mathbb{Z}_n^*\text{-EG.Rand}(C_1)$, $C_2' \leftarrow \mathsf{QR}_n\text{-EG.Rand}(C_2^{r_2})$, and $C_3' \leftarrow \mathsf{QR}_n\text{-EG'.Rand}(C_3^{r_3})$. Output $C' \leftarrow (C_1', C_2', C_3')$.

$\mathsf{Dec}(\mathsf{sk}, C)$: Parse $C = (C_1, C_2, C_3)$ and first decrypt $T'' \leftarrow \mathsf{QR}_n\text{-EG.Dec}(C_2)$. If $T'' = \bot$, return $\bot$; if $T'' = 1$, return 0; otherwise compute $m \leftarrow \mathscr{D}_\otimes(C_1)$ and return m.

$\mathscr{E}_\oplus(\cdot) = \mathbb{Z}_n$**-P: Paillier Encryption Scheme on $\mathbb{Z}_n$**

$\mathsf{Enc}(\mathsf{pk}, m)$: given $m \in \mathbb{Z}_n$, for a random $r \xleftarrow{\$} \mathbb{Z}_n^*$, compute $c \leftarrow (1+n)^m \cdot r^n \bmod n^2$. Output $c \in \mathbb{Z}_{n^2}^*$;

$\mathsf{Rand}(\mathsf{pk}, c)$: choose $r \xleftarrow{\$} \mathbb{Z}_n^*$, output $c' \leftarrow r^n \cdot c \bmod n^2$.

$\mathsf{Dec}(\mathsf{sk}, c)$: return $m \leftarrow ([c^d \bmod n^2] - 1)/n$.

Fig. 3. Setup and encryption schemes in $\mathbb{Z}_n$

after randomization. We will limit the decryption of C_1 only if C_2 contains 1, and then C_1 contains the plaintext, independent of the previous steps.

Computational Equality of Message Spaces. The message space of $\mathbb{Z}_n$-EG is now $\mathbb{Z}_n^* \cup \{0\}$, which is computationally equal to $\mathbb{Z}_n$, the message space of the Paillier encryption scheme: elements in the symmetric difference are non-trivial multiples of p or q, which lead to the factorization of the modulus n.

5.2   Encrypted Zero Test

To switch between encryption schemes over $\mathbb{Z}_n$, we have to obliviously detect the zeroes during the switch; this will be done by a sub-protocol, the *encrypted zero-test* (EZT). An EZT is a protocol in which two players share a decryption key, with an encryption C of a message m as input, and wish to get an encryption C' of a bit b as output, where $b = 1$ if $m = 0$, and $b = 0$ otherwise. An EZT is *zero-knowledge* if there is an efficient simulator for each player which is indistinguishable from an honest player, and runs on input (C, C'), where C' is a twin ciphertext of C, without the knowledge of the share of the secret key of the player it emulates, but just the share of the other player.

We stress that the EZT takes as input a Paillier ciphertext C of a message m and outputs a Paillier ciphertext of b, that is 1 if $m = 0$ and 0 otherwise. However, for our ESP protocols, the simulators of the ESP are given twin-ciphertext pairs (the input C of the ESP and an expected output C'), the simulator of the EZT can also take advantage of such a pair: thanks to C_3 in C' and the trapdoor s_3, the simulator can learn the value of b.

Various protocols have been proposed for this functionality (or closely related functionalities), such as [19,30,43]. Garbled circuits for testing the equality of strings, as proposed in [25], can also be used to construct an EZT with a better communication: given a ciphertext C encrypting a plaintext m,

- Alice picks $x \xleftarrow{\$} \mathbb{Z}_n$ and sends $C_A \xleftarrow{\$} \mathsf{Rand}(C \boxplus x) = \mathsf{Enc}(mx)$ to Bob. Both players jointly decrypt C_A; Bob gets the result y. Let $x' \leftarrow x \bmod 2^\kappa$ (that Alice computes) and $y' \leftarrow y \bmod 2^\kappa$ (that Bob computes).
- Let $f(u, v)$ be the function which returns 1 if $u = v$, and 0 else. Alice picks $b_A \xleftarrow{\$} \{0, 1\}$ and builds a garbled circuit computing $b_A \mathsf{\ xor\ } f(x', y')$. Using [25], the resulting circuit has 2κ gates.
- Bob gets a bit b_B from evaluating the garbled circuit with Yao's protocol. He sends an encryption C_B of b_B to Alice.
- Alice outputs $C' \leftarrow \mathsf{Rand}(b_A \boxplus C_B \boxminus (2b_A) \bullet C_B) = \mathsf{Enc}(b_A + b_B - 2b_A b_B) = \mathsf{Enc}(b_A \oplus b_B) = \mathsf{Enc}(f(x', y'))$.

The correctness follows from the fact that $x' = y'$ implies, with overwhelming probability, that $x - y = m = 0$, which is the plaintext of C.

Figure 4 sums up the efficiency the protocol of [25], and of the protocol of [30], which is the most efficient solution based on homomorphic encryption. Both protocols involve three rounds of on-line communication.

Authors	Preprocessing	Communication	Assumptions
[30]	2κ ciphertexts	3 ciphertexts	DCR
[25]	$8\kappa^2$ bits	κ^2 bits + κ oblivious transfers	Oblivious transfers

Fig. 4. Comparison of two EZT protocols

5.3 Encryption Switching Protocols on $\mathbb{Z}_n$

Our ESP on $\mathbb{Z}_n$ is described on Fig. 5, where the double arrows indicate an execution of an interactive sub-protocol, either a $\mathbb{Z}_n^*$-ESP or an EZT, and com is any extractable commitment (for the simulation). In the full version [6], we prove:

Theorem 11. *When instantiated with the $\mathbb{Z}_n$-P and $\mathbb{Z}_n$-EG encryption schemes, both over $\mathbb{Z}_n$, if both $\mathbb{Z}_n^* - ESP$ and EZT are zero-knowledge and if com is hiding, then the $\mathbb{Z}_n - ESP$ given in Fig. 5 is zero-knowledge under the DDH assumption in QR_n, the QR assumption in $\mathbb{Z}_n^*$ and the DCR assumption over $\mathbb{Z}_n^*$.*

Instantiating the $\mathbb{Z}_n^* - ESP$ with our construction of Sect. 4, we additionally require the DCR assumption over $\mathbb{Z}_N^*$.

6 Security Against Malicious Adversaries

In the previous sections, we built two homomorphic encryption schemes with zero-knowledge ESPs that achieve our goal of secure two-party computation from ESP: namely, at the end of the ESP executions, the *semi-honest* users do not know more than before about the input plaintexts. To prevent malicious behaviors, and move from the semi-honest setting to the malicious setting, additional validity checks are required. They are performed with zero-knowledge proofs.

Indeed, the last step toward secure ESPs is to ensure its soundness, so that malicious adversaries will not gain more information than an honest adversary would do. Moreover, the use of the simulators of the additional proofs will preserve the zero-knowledge of our ESPs. In this section, we provide this final building block.

Unfortunately, ESPs are essentially *non-arithmetic* protocols, and namely the internal decryptions and re-encryptions. Hence, ensuring honest behavior might require garbled circuits-based zero-knowledge proofs such as [15,23,35], or cut-and-choose techniques, both at a very high computational cost, but also from the communication point of view, which we cannot afford.

In this section, we present a more efficient technique for such zero-knowledge proofs, based on a particular pre-processing phase. We first explain how a *pool of random twin-ciphertext pairs* allows designing efficient (amortized) proofs of honest behavior in our ESPs.

6.1 Refreshable Twin-Ciphertext Pool

First, we set up a perfectly hiding commitment scheme $\mathsf{com}_\oplus$ over a group of order n: let k be a small integer such that $t \leftarrow 2kn + 1$ is prime. Let (g_t, h_t) be two generators of the subgroup of $\mathbb{Z}_t^*$ of order n. On input $m \in \mathbb{Z}_n$ and a randomness $r \in \mathbb{Z}_n$, the scheme outputs $\mathsf{com}_\oplus(m; r) = g_t^m h_t^r$.

2-Party $\mathsf{ESP}^{\times}_{+}$ from $C = \mathscr{E}_{\oplus}(m)$ into $(C_{1\times}, C_{2\times}, C_{3\times}) = \mathscr{E}^0_{\otimes}(m)$

Alice gets $C_{\mathsf{EZT}} = \mathscr{E}_{\oplus}(b)$ $\qquad \xleftarrow{\quad C_{\mathsf{EZT}} \leftarrow \mathsf{EZT}(C) \quad}$ $\qquad$ Bob gets nothing

$C_{1+} \leftarrow C \boxplus C_{\mathsf{EZT}}$ $\qquad \xrightarrow{\quad C_{1+} \quad}$

Alice gets $C_{1\times}$ $\qquad \xleftarrow{\quad C_{1\times} \leftarrow \mathbb{Z}^*_n\text{-}\mathsf{ESP}(C_{1+}) \quad}$ $\qquad$ Bob gets $C_{1\times}$

$T, T', R_2, R_3, k \xleftarrow{\$} \mathsf{QR}_n$
$c \leftarrow \mathsf{com}(k)$
$C_{2+} \leftarrow \mathscr{E}_{\oplus}(1) \boxplus (T-1) \bullet C_{\mathsf{EZT}}$
$C_{3+} \leftarrow \mathscr{E}_{\oplus}(1) \boxplus (T'-1) \bullet C_{\mathsf{EZT}}$
$C_2 \leftarrow \mathbb{Z}_n\text{-}\mathsf{P.Rand}(R_2 \bullet C_{2+})$
$C'_2 \leftarrow \mathsf{QR}_n\text{-}\mathsf{EG.Enc}(R_2^{-1})$
$C_3 \leftarrow \mathbb{Z}_n\text{-}\mathsf{P.Rand}(k \cdot R_3 \bullet C_{3+})$
$C'_3 \leftarrow \mathsf{QR}_n\text{-}\mathsf{EG'.Enc}(R_3^{-1})$
$D_2 \leftarrow C_2^{d_A} \bmod n^2$
$D_3 \leftarrow C_3^{d_A} \bmod n^2$ $\qquad \xrightarrow{\quad c, C_2, C'_2, D_2, C_3, C'_3, D_3 \quad}$ $\quad x_2 \leftarrow ([C_2^{d_B} D_2 \bmod n^2] - 1)/n$
$\qquad\qquad\qquad\qquad x_3 \leftarrow ([C_3^{d_B} D_3 \bmod n^2] - 1)/n$
$\qquad\qquad\qquad\qquad C_{2\times} \leftarrow \mathsf{QR}_n\text{-}\mathsf{EG.Rand}(x_2 \bullet C'_2)$

$C''_{3\times} \leftarrow k^{-1} \bullet C'_{3\times}$ $\qquad \xleftarrow{\quad C_{2\times}, C'_{3\times} \quad}$ $\quad C'_{3\times} \leftarrow \mathsf{QR}_n\text{-}\mathsf{EG'.Rand}(x_3 \bullet C'_3)$
$k' \xleftarrow{\$} \mathbb{Z}_{n/4}$
$C_{3\times} \leftarrow \mathsf{QR}_n\text{-}\mathsf{EG'.Rand}(C''^{k'}_{3\times})$ $\qquad \xrightarrow{\quad C_{3\times} \quad}$

2-Party $\mathsf{ESP}^{+}_{\times}$ from $(C_{1\times}, C_{2\times}, C_{3\times}) = \mathscr{E}^0_{\otimes}(m)$ into $C' = \mathscr{E}_{\oplus}(m)$

Alice gets C_{1+} $\qquad \xleftarrow{\quad C_{1+} \leftarrow \mathbb{Z}^*_n\text{-}\mathsf{ESP}(C_{1\times}) \quad}$ $\qquad$ Bob gets C_{1+}

$R_2 \xleftarrow{\$} \mathsf{QR}_n,$
$k' \xleftarrow{\$} \mathbb{Z}_{n/4}$
$C_2 \leftarrow \mathsf{QR}_n\text{-}\mathsf{EG.Rand}(R_2 \bullet C^{k'}_{2\times})$
$= (c_0, c_1)$
$C'_2 \leftarrow \mathscr{E}_{\oplus}(R_2^{-1})$
$d_1 \leftarrow c_0^{s_{2A}}$ $\qquad \xrightarrow{\quad C_2, C'_2, d_1 \quad}$ $\quad x_2 \leftarrow c_1/d_1 c_0^{s_{2B}}$
$\qquad\qquad\qquad\qquad k \xleftarrow{\$} \mathbb{Z}^*_n$
$\qquad \xleftarrow{\quad C'_{2+} \quad}$ $\quad C'_{2+} \leftarrow k \bullet (x_2 \bullet C'_2 \boxminus \mathscr{E}_{\oplus}(1))$

Alice gets nothing $\qquad \xleftarrow{\quad C_{\mathsf{EZT}} \leftarrow \mathsf{EZT}(C'_{2+}) \quad}$ $\quad$ Bob gets $C_{\mathsf{EZT}} = \mathscr{E}_{\oplus}(b')$
$\qquad\qquad\qquad\qquad (\rho_0, \rho_1) \xleftarrow{\$} \mathbb{Z}^2_n$
$\qquad\qquad\qquad\qquad B_1 \leftarrow C_{1+} \boxplus \mathscr{E}_{\oplus}(\rho_0)$
$\qquad\qquad\qquad\qquad B_2 \leftarrow \mathscr{E}_{\oplus}(\rho_1) \boxplus C_{\mathsf{EZT}}$
$\qquad\qquad\qquad\qquad B_3 \leftarrow \rho_0 \bullet C_{\mathsf{EZT}}$
$\qquad\qquad\qquad\qquad B_4 \leftarrow \rho_1 \bullet C_{1+}$
$\qquad\qquad\qquad\qquad B_5 \leftarrow \mathscr{E}_{\oplus}(\rho_0 \rho_1)$
$\qquad\qquad\qquad\qquad B_6 \leftarrow B_3 \boxplus B_4 \boxplus B_5$
$\qquad\qquad\qquad\qquad A'_1 \leftarrow B_1^{d_B} \bmod n^2$
$A_1 \leftarrow ([B_1^{d_A} A'_1 \bmod n^2] - 1)/n$ $\qquad \xleftarrow{\quad B_1, B_2, A'_1, A'_2, B_6 \quad}$ $\quad A'_2 \leftarrow B_2^{d_B} \bmod n^2$
$A_2 \leftarrow ([B_2^{d_A} A'_2 \bmod n^2] - 1)/n$
$C' \leftarrow \mathscr{E}_{\oplus}(A_1 \cdot A_2) \boxminus B_6$ $\qquad \xrightarrow{\quad C' \quad}$

Fig. 5. Interactive Encryption Switching in $\mathbb{Z}_n$

Pre-processing Random Twin-Ciphertext Pairs. Our starting point is a protocol that allows a prover to convince a verifier that two ciphertexts, from two different cryptosystems, do indeed encrypt the same value. This means they form a *twin-ciphertext pair*. Such a proof will be denoted TCP, for *Twin-Ciphertext Proof*. It comes at a cost of the cut-and-choose technique and thus requires $O(\kappa)$ communication. However, we show in the full version [6] how to amortize ℓ TCPs using only a *single* cut-and-choose protocol, for any arbitrarily large ℓ. It relies on the techniques developed by Groth and Bayer on generalized Pedersen commitments [3,34]. But we use a new zero-knowledge proof on multi-exponentiation with committed base, of independent interest: we can create a pool of ℓ proven twin-ciphertext pairs in $O(\ell + \kappa)$. We then show several applications to speed-up various zero-knowledge arguments.

In order to generate an arbitrary number of twin-ciphertext pairs $(C_i, C_i')_i = (\mathcal{E}_\oplus(m_i), \mathcal{E}_\otimes(m_i))_i$ of random plaintexts m_i, under two homomorphic encryption schemes, we first show how to generate a first pair: Alice has a pair of ciphertexts $(C, C') = (\mathcal{E}_\oplus(m, r), \mathcal{E}_\otimes(m', s))$ for which she knows both the plaintexts and the random coins. She wants to prove that $m = m'$ to Bob:

- Alice generates κ twin-ciphertext pairs $(C_i, C_i')_i = (\mathcal{E}_\oplus(\mu_i; r_i), \mathcal{E}_\otimes(\mu_i; s_i))_i$, for values μ_i picked at random over $\mathbb{Z}_n$, and commits to those pairs (using any commitment scheme);
- Bob sends a challenge $c = c_1 \cdots c_\kappa \overset{\$}{\leftarrow} \{0,1\}^\kappa$;
- Alice opens the κ commitments on the twin-ciphertext pairs, and for each $i \le \kappa$, she sends
  - the plaintext μ_i and the random coins (r_i, s_i), if $c_i = 0$;
  - the ratio $R_i = m/\mu_i$ and the random coins $\rho_i \leftarrow (R_i \cdot r_i) \odot ((-1) \cdot r)$ according to the additive case, and $\sigma_i \leftarrow (R_i \cdot s_i) \odot ((-1) \cdot s)$ according to the multiplicative case — using the notations from Sect. 2.4;
- Bob checks the openings of commitments and
  - either checks the validity of (C_i, C_i') with μ_i and the random coins;
  - or computes $D_i = R_i \bullet C_i \boxplus (-1) \bullet C$ and $D_i' = R_i \bullet C_i' \boxtimes (-1) \bullet C'$, according to the relations (1) and (2). Bod then checks whether both $D_i = \mathcal{E}_\oplus(0; \rho_i)$ and $D_i' = \mathcal{E}_\otimes(1; \sigma_i)$ hold.

We prove the security of TCP in the full version [6].

Using $com_\oplus$ Instead of Paillier. The Paillier encryption scheme $\mathcal{E}_\oplus()$ in the twin-ciphertext proof can be replaced by the above perfectly hiding commitment scheme $com_\oplus : (m; r) \mapsto g_t^m h_t^r$, that is also additively homomorphic. But then the proofs become arguments. Alice generates κ pairs, each pair consisting of an additive commitment and a $\mathbb{Z}_n^*$-EG ciphertext, and the rest of the proof is exactly the same. We keep using the notation $\mathcal{E}_\oplus()$ below.

Efficient Online TCP. Let us assume that we have already proven that a *random* twin-ciphertext pair $P_i = (\mathcal{E}_\oplus(m_i; r_i), \mathcal{E}_\otimes(m_i; s_i))$ is correct. When one wants to perform a TCP during a protocol on a new twin-ciphertext pair $P = (\mathcal{E}_\oplus(m; r), \mathcal{E}_\otimes(m; s))$, it is enough to reveal some relations between the random coins of the pairs P and P_i, in order to show that the plaintexts are co-linear: if one of them is correct, so is the other. And this can be done without disclosing m (as m_i is random, disclosing m/m_i will not reveal m). Thereby, all our protocols are described in the following model: first, in a pre-processing phase, a large pool of random twin-ciphertext pairs are generated and proven correct with a batch argument. Then, in the on-line phase, each time a TCP is required, a twin-ciphertext pair from the pool is used and the player performs a cheap co-linearity proof. This proof *consumes* P_i and ensures the correctness of the switch.

Refreshing the Twin-Ciphertext Pool. The expected number of TCPs might not be known to the players; however, once a pool of twin-ciphertext pairs has been set up, the same batch technique that we describe in the full version [6] can be used to generate ℓ new random twin-ciphertext pairs, while consuming a single pair of the pool. The batch argument transmits $O(\ell + \kappa)$ group elements but does not rely on cut-and-choose, hence cut-and-choose in only needed *once*, when generating the very first element of the pool.

6.2   Zero-Knowledge Proofs

The pool of twin-ciphertext pairs allows the players to perform TCPs efficiently. Apart from TCPs, the zero-knowledge proofs needed to enhance ESPs to the malicious setting are classical protocols. For zero-knowledge proofs involving the decryption keys, we have to add the corresponding *verification keys* in the public key: first, we pick $h_0, r_0 \xleftarrow{\$} \mathbb{Z}_n^*$, $R_0 \xleftarrow{\$} \mathbb{Z}_N^*$ and set $h \leftarrow -h_0^2$, then we add $(h_p, h_q) = (h^{t_p}, h^{t_q})$ and $(u, U) \leftarrow ((1+n) \cdot r_0^{2n} \bmod n^2, (1+N) \cdot R_0^{2N} \bmod N^2) \in$ $\mathsf{QR}_{n^2} \times \mathsf{QR}_{N^2}$ to the public key. The latter pair satisfies $u^d = 1 + n \bmod n^2$ and $U^D = 1 + N \bmod N^2$. Second, we set up the commitment scheme $\mathsf{com}_\oplus$ previously described. Each time a player performs computations, he commits to the operands if they are not already encrypted, and proves his honest behavior, with a zero-knowledge proof, using the above elements in the public key. Note that in our generic 2-PC from ESP, the switches run either sequentially or in parallel, but they are never intertwined; hence, we do not need to use zero-knowledge proofs secure in the concurrent setting (which would be less efficient).

Range Proofs. In the multiplicative to additive direction, a second Paillier encryption scheme is used, with a different modulus N. The plaintext space of this scheme is large enough to ensure that no modular reduction occurs during computations over input ciphertexts encrypting values in $\{0, \cdots, n-1\}$. Thereby, it is necessary to prove that these values are indeed in that range, which is handled by range proofs. The method of [4] provides an efficient (constant-communication) proof. Hence, we first have to commit to the encrypted value, using a generator of a space with a different modulus $n' > n$ whose factorization is unknown as in [8,16]: the plaintext is thus committed over $\mathbb{Z}_{\lambda(n')}$, a space of unknown order. Then, equality between the encrypted value (over $\mathbb{Z}_N$) and the committed value (over $\mathbb{Z}_{\lambda(n')}$, whose order is unknown) can be proven using [9], and the range proof is performed on the committed value. The soundness of this proof relies on the *strong RSA assumption* [2,16] modulo $n' > n$. We stress that it is necessary that the factorization of n' is not known by anyone nor shared between the parties, as the strong soundness requirement of our proof of 2-PC states that the adversary is given the full secret key; hence, unlike [9], we *cannot* use $n' = N$. Note also that n' can be taken way smaller than N (which has to be large enough to allow for some multiplications without overflow).

We stress that the need of the strong RSA assumption in the security of our constant-size on-line ESP comes from the range proof, only. To date, there is no constant-size range proof over $\mathbb{Z}_n$ in the literature whose soundness does not rely on this assumption.

Classical Zero-Knowledge Proofs. Appart from TCP and range proofs, all the proofs are classical zero-knowledge proofs *á la* Schnorr [36]: Proof of re-randomization of ciphertexts and proof of correct computation of $R \bullet C$ given $\mathsf{com}_\oplus(R)$ are just proofs of exponentiations to the same power either in the same groups or in two groups which one order ($\mathbb{J}_n$) is unknown. It is also easy to generate $\mathscr{E}_\otimes(m^{-1})$ from $\mathscr{E}_\otimes(m)$, just inverting all the components in $\mathbb{J}_n$.

6.3   Ensuring Honest Behavior in ESP Protocols

Let us illustrate how TCP are used on the $\mathbb{Z}_n^*$-ESP from Paillier to $\mathbb{Z}_n^*$-EG (see Fig. 2): Alice sends (C_A, C_A') and commits to R_A: $c \leftarrow \mathsf{com}_\oplus(R_A)$. She makes a TCP on the pair $(c, C_A'^{-1})$, and a classical proof of product to show that C_A encrypts the product of the value committed in c and the value encrypted in C; combined together, those proofs are enough to ensure Alice's honest behavior. Additional proofs (including range proofs) are needed in the other direction as it is a more complex case.

Sketch of the Proof of Security. In our full proof of security of the semi-honest protocols, we have already included the generation of the verification keys by the simulator. Hence, the enhanced protocol only adds zero-knowledge proofs and perfectly hiding commitments to the semi-honest proof. The zero-knowledge property of these proofs states that a simulator can fake them, *i.e.* convince a verifier of the truth of the associated statement, even if the statement is not true. Thereby, we add the two following games to our game-based proof of security in the semi-honest model, right after the very first game (in which the simulator plays honestly, using the secret keys):

1. from this game, each time the simulator is asked to perform a zero-knowledge proof, it fakes it instead. This game is indistinguishable from the honest game due to the zero-knowledge property of the proofs;
2. from this game, each time the simulator has to commit, the simulator sends a uniformly random commitment. As $\mathsf{com}_\oplus$ is perfectly hiding, this game is perfectly indistinguishable from the previous one.

The rest of the game-based proof is exactly the same as the semi-honest proof.

6.4   From Secure ESP to Secure 2-PC

We stress that our 2-PC protocol is made fully secure as soon as ESPs is secure against malicious adversaries (as well as the 2-party decryption procedure at the end of the protocol). This comes from the fact, that apart from ESPs and the final decryption, all the operations are *local* homomorphic evaluations of *public* functions on ciphertexts known by both players. The homomorphic operations themselves are deterministic and performed by both players.

A sequence of public operations is followed by either an ESP or, at the very end of the protocol, a 2-party decryption. From the strong soundness of the ESP, any honest player is guaranteed that his output of the ESP is necessarily a twin ciphertext of his input, or an abort is triggered. Similarly, in the 2-party decryption protocol, an honest player is guaranteed that the output is the plaintext of his input ciphertext, unless an error is raised.

Therefore, an honest player that correctly performs his (local) homomorphic evaluations is also guaranteed of the correct evaluation of the switches and of the decryption: the final answer is necessarily correct, or an error is raised in case of a misbehaving partner.

6.5   Exponential Relations Among Committed Values

We describe several applications of our preprocessing technique. In the following applications, we use a commitment scheme $\text{com}(\cdot)$ we assume to be additively homomorphic. This can either be $\mathscr{E}_\oplus(\cdot)$ (perfectly binding) or the previous $\text{com}_\oplus(\cdot)$ (perfectly hiding).

Proof of Knowledge of an Exponential Relation over Committed Values. A prover has sent a tuple $(C_a = \text{com}(a), C_b = \text{com}(b), d)$ and wishes to prove that $b = a^d$. Let C_a' and C_b' be twin ciphertexts under $\mathscr{E}_\otimes()$ of C_a and C_b; the prover sends them and proves them with two TCPs. Both players can compute $D \leftarrow \boxtimes^d C_a' = \mathscr{E}_\otimes(a^d)$. She then proves that D encrypts the same value as C_b', which can be done since she knows all the random coins.

Extension to the Case of a Committed Exponent. Let us now suppose d has also been committed in $C_d = \text{com}(d)$. The prover sends $C_a' = (C_{a0}', C_{a1}', \alpha)$, C_b', and C_d', twin ciphertexts of C_a and C_d respectively, and proves them with two TCPs. The prover computes $D = (D_0, D_1, D_2) \leftarrow \boxtimes^d C_a' = \mathscr{E}_\otimes(a^d)$, and proves its knowledge of (b, r) such that $C_b = \text{com}(b; r)$, $D_0 = (C_{a0}')^d$, $D_1 = (C_{a1}')^d$, and $D_2 = \alpha^d$. She then proves that C_b' and D encrypt the same plaintext.

Proof of Knowledge of a Double Logarithm (or Double Decker Exponentiation). In this case, the prover wants to prove her knowledge of x that satisfies $X = g^{(h^x)}$, for public values (g, h, X). Such proofs are required for example in some publicly verifiable secret sharing schemes [38], in group signature or group encryption [24]. Let $n = pq$ be an RSA modulus such that $\pi = 2n + 1$

is a prime. Let g be a generator of a subgroup of $\mathbb{Z}_\pi^*$ of order n. Let h be a generator of $\mathbb{J}_n$ and x be an element of $\mathbb{Z}_\lambda$. The prover computes $H \leftarrow h^x$, $C \leftarrow \mathbb{Z}_n\text{-P.Enc}(H)$, and $C' \leftarrow \mathbb{J}_n\text{-EG.Enc}(H)$. She sends (C, C'), proves that she knows the discrete log of this encrypted value in C' (a classical Schnorr-like proof), and makes a TCP on (C, C'). She then proves her knowledge of H and r such that $X = g^H \bmod \pi$ and $C = \mathbb{Z}_n\text{-P.Enc}(H; r)$ (again a Schnorr-like proof).

Proof that a Committed Value Is Prime. In [5], the authors design a zero-knowledge proof that a committed value is a product of two safe primes, which has applications in numerous RSA-based protocols. The idea is the following: to prove that a committed number π is a prime, one proves that it passed each step of the Lehmann's primality test [26,37], *i.e.* commit to κ random numbers (in an interactive way to ensure they are random) and for each of the κ random committed a, prove that $a^{(\pi-1)/2} = \pm 1 \bmod \pi$ by committing to each bit of $(\pi - 1)/2$, and by using a zero-knowledge proof for each step of the square-and-multiply algorithm. This can be done way more efficiently with our above proof of knowledge of an exponential relation over a committed exponent, enhanced using the technique from [29] to work modulo a committed value. Overall, we improve [5] by a factor of $O(\log(\pi))$. In typical applications, π will be 1024 to 2048 bit-long.

Acknowledgments. We thank Fabrice Ben Hamouda for the fruitful discussions on the ElGamal variant. This work was supported in part by the European Research Council under the European Community's Seventh Framework Programme (FP7/2007-2013 Grant Agreement no. 339563 – CryptoCloud). The second author is supported by the F.R.S-FNRS as a postdoctoral researcher.

References

1. Asharov, G., Lindell, Y., Schneider, T., Zohner, M.: More efficient oblivious transfer extensions with security for malicious adversaries. In: Oswald, E., Fischlin, M. (eds.) EUROCRYPT 2015. LNCS, vol. 9056, pp. 673–701. Springer, Heidelberg (2015)

2. Barić, N., Pfitzmann, B.: Collision-free accumulators and fail-stop signature schemes without trees. In: Fumy, W. (ed.) EUROCRYPT 1997. LNCS, vol. 1233, pp. 480–494. Springer, Heidelberg (1997)

3. Bayer, S., Groth, J.: Efficient zero-knowledge argument for correctness of a shuffle. In: Pointcheval, D., Johansson, T. (eds.) EUROCRYPT 2012. LNCS, vol. 7237, pp. 263–280. Springer, Heidelberg (2012)

4. Boudot, F.: Efficient proofs that a committed number lies in an interval. In: Preneel, B. (ed.) EUROCRYPT 2000. LNCS, vol. 1807, pp. 431–444. Springer, Heidelberg (2000)

5. Camenisch, J.L., Michels, M.: Proving in zero-knowledge that a number is the product of two safe primes. In: Stern, J. (ed.) EUROCRYPT 1999. LNCS, vol. 1592, pp. 107–122. Springer, Heidelberg (1999)

6. Couteau, G., Peters, T., Pointcheval, D.: Encryption switching protocols. Cryptology ePrint Archive, Report 2015/990 (2015). http://eprint.iacr.org/2015/990

7. Dachman-Soled, D., Malkin, T., Raykova, M., Yung, M.: Efficient robust private set intersection. In: Abdalla, M., Pointcheval, D., Fouque, P.-A., Vergnaud, D. (eds.) ACNS 2009. LNCS, vol. 5536, pp. 125–142. Springer, Heidelberg (2009)

8. Damgård, I.B., Fujisaki, E.: A statistically-hiding integer commitment scheme based on groups with hidden order. In: Zheng, Y. (ed.) ASIACRYPT 2002. LNCS, vol. 2501, pp. 125–142. Springer, Heidelberg (2002)

9. Damgård, I.B., Jurik, M.: Client/server tradeoffs for online elections. In: Naccache, D., Paillier, P. (eds.) PKC 2002. LNCS, vol. 2274, pp. 125–140. Springer, Heidelberg (2002)

10. Damgård, I.B., Nielsen, J.B.: Universally composable efficient multiparty computation from threshold homomorphic encryption. In: Boneh, D. (ed.) CRYPTO 2003. LNCS, vol. 2729, pp. 247–264. Springer, Heidelberg (2003)

11. Damgård, I., Pastro, V., Smart, N., Zakarias, S.: Multiparty computation from somewhat homomorphic encryption. In: Safavi-Naini, R., Canetti, R. (eds.) CRYPTO 2012. LNCS, vol. 7417, pp. 643–662. Springer, Heidelberg (2012)

12. Damgård, I., Zakarias, S.: Constant-overhead secure computation of Boolean circuits using preprocessing. In: Sahai, A. (ed.) TCC 2013. LNCS, vol. 7785, pp. 621–641. Springer, Heidelberg (2013)

13. Demmler, D., Schneider, T., Zohner, M.: ABY-a framework for efficient mixed-protocol secure two-party computation. In: Network and Distributed System Security, NDSS (2015)

14. ElGamal, T.: A public key cryptosystem and a signature scheme based on discrete logarithms. IEEE Tran. Inf. Theory 31, 469–472 (1985)

15. Frederiksen, T.K., Nielsen, J.B., Orlandi, C.: Privacy-free garbled circuits with applications to efficient zero-knowledge. Cryptology ePrint Archive, Report 2014/598 (2014). http://eprint.iacr.org/2014/598

16. Fujisaki, E., Okamoto, T.: Statistical zero knowledge protocols to prove modular polynomial relations. In: Kaliski Jr., B.S. (ed.) CRYPTO 1997. LNCS, vol. 1294, pp. 16–30. Springer, Heidelberg (1997)

17. Gavin, G., Minier, M.: Oblivious multi-variate polynomial evaluation. In: Roy, B., Sendrier, N. (eds.) INDOCRYPT 2009. LNCS, vol. 5922, pp. 430–442. Springer, Heidelberg (2009)

18. Gentry, C.: Fully homomorphic encryption using ideal lattices. In: Mitzenmacher, M. (ed.) 41st ACM STOC, pp. 169–178. ACM Press, May/June 2009

19. Gentry, C., Halevi, S., Jutla, C., Raykova, M.: Private database access with HE-over-ORAM architecture. Cryptology ePrint Archive, Report 2014/345 (2014). http://eprint.iacr.org/2014/345

20. Hazay, C., Mikkelsen, G.L., Rabin, T., Toft, T.: Efficient RSA key generation and threshold paillier in the two-party setting. In: Dunkelman, O. (ed.) CT-RSA 2012. LNCS, vol. 7178, pp. 313–331. Springer, Heidelberg (2012)

21. Henecka, W., Kögl, S., Sadeghi, A.R., Schneider, T., Wehrenberg, I.: TASTY: tool for automating secure two-party computations. In: Al-Shaer, E., Keromytis, A.D., Shmatikov, V. (eds.) ACM CCS 2010, pp. 451–462. ACM Press, October 2010

22. Huang, Y., Evans, D., Katz, J.: Private set intersection: are garbled circuits better than custom protocols? In: NDSS 2012. The Internet Society, February 2012

23. Jawurek, M., Kerschbaum, F., Orlandi, C.: Zero-knowledge using garbled circuits: how to prove non-algebraic statements efficiently. Cryptology ePrint Archive, Report 2013/073 (2013). http://eprint.iacr.org/2013/073

24. Kiayias, A., Tsiounis, Y., Yung, M.: Group encryption. Cryptology ePrint Archive, Report 2007/015 (2007). http://eprint.iacr.org/2007/015
25. Kolesnikov, V., Schneider, T.: Improved garbled circuit: free XOR gates and applications. In: Aceto, L., Damgård, I., Goldberg, L.A., Halldórsson, M.M., Ingólfsdóttir, A., Walukiewicz, I. (eds.) ICALP 2008, Part II. LNCS, vol. 5126, pp. 486–498. Springer, Heidelberg (2008)
26. Kranakis, E.: Primality and Cryptography. Wiley, Hoboken (1986)
27. Lim, H.W., Tople, S., Saxena, P., Chang, E.C.: Faster secure arithmetic computation using switchable homomorphic encryption. Cryptology ePrint Archive, Report 2014/539 (2014). http://eprint.iacr.org/2014/539
28. Lindell, Y., Pinkas, B.: Secure two-party computation via cut-and-choose oblivious transfer. In: Ishai, Y. (ed.) TCC 2011. LNCS, vol. 6597, pp. 329–346. Springer, Heidelberg (2011)
29. Lipmaa, H.: On diophantine complexity and statistical zero-knowledge arguments. Cryptology ePrint Archive, Report 2003/105 (2003). http://eprint.iacr.org/2003/105
30. Lipmaa, H., Toft, T.: Secure equality and greater-than tests with sublinear online complexity. In: Fomin, F.V., Freivalds, R., Kwiatkowska, M., Peleg, D. (eds.) ICALP 2013, Part II. LNCS, vol. 7966, pp. 645–656. Springer, Heidelberg (2013)
31. Naor, M., Pinkas, B.: Oblivious polynomial evaluation. SIAM J. Comput. 35, 1254–1281 (2006)
32. Nielsen, J.B., Nordholt, P.S., Orlandi, C., Burra, S.S.: A new approach to practical active-secure two-party computation. In: Safavi-Naini, R., Canetti, R. (eds.) CRYPTO 2012. LNCS, vol. 7417, pp. 681–700. Springer, Heidelberg (2012)
33. Paillier, P.: Public-key cryptosystems based on composite degree residuosity classes. In: Stern, J. (ed.) EUROCRYPT 1999. LNCS, vol. 1592, pp. 223–238. Springer, Heidelberg (1999)
34. Pedersen, T.P.: Non-interactive and information-theoretic secure verifiable secret sharing. In: Feigenbaum, J. (ed.) CRYPTO 1991. LNCS, vol. 576, pp. 129–140. Springer, Heidelberg (1992)
35. Ranellucci, S., Tapp, A., Zakarias, R.W.: Efficient generic zero-knowledge proofs from commitments. Cryptology ePrint Archive, Report 2014/934 (2014). http://eprint.iacr.org/2014/934
36. Schnorr, C.-P.: Efficient identification and signatures for smart cards. In: Quisquater, J.-J., Vandewalle, J. (eds.) EUROCRYPT 1989. LNCS, vol. 434, pp. 688–689. Springer, Heidelberg (1990)
37. Solovay, R., Strassen, V.: A fast monte-carlo test for primality. SIAM J. Comput. 6(1), 84–85 (1977)
38. Stadler, M.A.: Publicly verifiable secret sharing. In: Maurer, U.M. (ed.) EUROCRYPT 1996. LNCS, vol. 1070, pp. 190–199. Springer, Heidelberg (1996)
39. Tassa, T., Jarrous, A., Ben-Ya'akov, Y.: Oblivious evaluation of multivariate polynomials. J. Math. Cryptol. 7, 1–29 (2013)
40. Tople, S., Shinde, S., Chen, Z., Saxena, P.: AUTOCRYPT: enabling homomorphic computation on servers to protect sensitive web content. In: Sadeghi, A.R., Gligor, V.D., Yung, M. (eds.) ACM CCS 2013, pp. 1297–1310. ACM Press, November 2013
41. Yao, A.C.C.: How to generate and exchange secrets (extended abstract). In: 27th FOCS, pp. 162–167. IEEE Computer Society Press, October 1986

42. Ye, Q., Wang, H., Pieprzyk, J., Zhang, X.-M.: Efficient disjointness tests for private datasets. In: Mu, Y., Susilo, W., Seberry, J. (eds.) ACISP 2008. LNCS, vol. 5107, pp. 155–169. Springer, Heidelberg (2008)
43. Yu, C.-H., Yang, B.-Y.: Probabilistically correct secure arithmetic computation for modular conversion, zero test, comparison, MOD and exponentiation. In: Visconti, I., De Prisco, R. (eds.) SCN 2012. LNCS, vol. 7485, pp. 426–444. Springer, Heidelberg (2012)

Compromised Systems

Message Transmission with Reverse Firewalls—Secure Communication on Corrupted Machines

Yevgeniy Dodis[1], Ilya Mironov[2], and Noah Stephens-Davidowitz[1(✉)]

[1] Department of Computer Science, New York University, New York, USA
noahsd@gmail.com
[2] Google, Menlo Park, USA

Abstract. Suppose Alice wishes to send a message to Bob privately over an untrusted channel. Cryptographers have developed a whole suite of tools to accomplish this task, with a wide variety of notions of security, setup assumptions, and running times. However, almost all prior work on this topic made a seemingly innocent assumption: that Alice has access to a trusted computer with a proper implementation of the protocol. The Snowden revelations show us that, in fact, powerful adversaries can and will corrupt users' machines in order to compromise their security. And, (presumably) accidental vulnerabilities are regularly found in popular cryptographic software, showing that users cannot even trust implementations that were created honestly. This leads to the following (seemingly absurd) question: "Can Alice securely send a message to Bob even if she cannot trust her own computer?!"

Bellare, Paterson, and Rogaway recently studied this question. They show a strong impossibility result that in particular rules out even semantically secure public-key encryption in their model. However, Mironov and Stephens-Davidowitz recently introduced a new framework for solving such problems: reverse firewalls. A secure reverse firewall is a third party that "sits between Alice and the outside world" and modifies her sent and received messages so that *even if the her machine has been corrupted*, Alice's security is still guaranteed. We show how to use reverse firewalls to sidestep the impossibility result of Bellare et al., and we achieve strong security guarantees in this extreme setting.

Indeed, we find a rich structure of solutions that vary in efficiency, security, and setup assumptions, in close analogy with message transmission in the classical setting. Our strongest and most important result shows a protocol that achieves interactive, concurrent CCA-secure message transmission with a reverse firewall—i.e., CCA-secure message transmission on a possibly compromised machine! Surprisingly, this protocol is quite efficient and simple, requiring only four rounds and a small

Y. Dodis—Partially supported by gifts from VMware Labs and Google, and NSF grants 1319051, 1314568, 1065288, 1017471.

N. Stephens-Davidowitz—Partially supported by National Science Foundation under Grant No. CCF-1320188. Any opinions, findings, and conclusions or recommendations expressed in this material are those of the authors and do not necessarily reflect the views of the National Science Foundation.

M. Robshaw and J. Katz (Eds.): CRYPTO 2016, Part I, LNCS 9814, pp. 341–372, 2016.
DOI: 10.1007/978-3-662-53018-4_13

342 Y. Dodis et al.

constant number of public-key operations for each party. It could easily be used in practice. Behind this result is a technical composition theorem that shows how key agreement with a sufficiently secure reverse firewall can be used to construct a message-transmission protocol with its own secure reverse firewall.

1 Introduction

We consider perhaps the simplest, most fundamental problem in cryptography: secure message transmission, in which Alice wishes to send a plaintext message to Bob without leaking the plaintext to an eavesdropper. Of course, this problem has a rich history, and it is extremely well-studied with a variety of different setup assumptions and notions of security (e.g., [4]). There are many beautiful solutions, based on symmetric-key encryption, public-key encryption, key agreement, etc.

However, in the past few years, it has become increasingly clear that the real world presents many vulnerabilities that are not captured by the security models of classical cryptography. The revelations of Edward Snowden show that the United States National Security Agency successfully gained access to secret information by extraordinary means, including subverting cryptographic standards [3,34] and intercepting and tampering with hardware on its way to users [23]. Meanwhile, many (apparently accidental) security flaws have been found in widely deployed pieces of cryptographic software, leaving users completely exposed [12–14,25,28]. Due to the complexity of modern cryptographic software, such vulnerabilities are extremely hard to detect in practice, and, ironically, cryptographic modules are often the easiest to attack, as attackers can often use cryptographic mechanisms to mask their activities or opportunistically hide their communications within encrypted traffic. This has led to a new direction for cryptographers (sometimes called "post-Snowden cryptography"), which in our context is summarized by the following (seemingly absurd) question: "How can Alice and Bob possibly communicate securely when an eavesdropper might have corrupted their computers?!"

Motivated by such concerns, Bellare, Paterson, and Rogaway consider the problem of securely encrypting a message when the encrypting party might be compromised [7]. They consider the case in which the corrupted party's behavior is indistinguishable from that of an honest implementation. Even in this setting, their main result shows that even a relatively weak adversary can break any scheme that "non-trivially uses randomness." (They also provide a nice deterministic symmetric-key solution, which we use as a subprotocol in the sequel.) In particular, it is easy to see that a semantically secure public-key message transmission is impossible in their framework. (See [5] for an analysis of weaker notions of security for public-key encryption in this setting.)

1.1   Reverse Firewalls

Due to the strong restriction proved in [7], we consider a relaxation of their model in which we allow for an additional party, a *(cryptographic) reverse firewall* (RF) as recently introduced by Mironov and Stephens-Davidowitz [31]. We provide formal definitions in Sect. 2.1, but since RFs are quite a new concept (and they can be rather confusing at first), we now provide a high-level discussion of some of the salient aspects of the reverse-firewall framework.

A reverse firewall for Alice is an autonomous intermediary that modifies the messages that Alice's machine sends and receives. The hope is that the protocol equipped with a reverse firewall can provide meaningful security guarantees for Alice *even if her own machine is compromised.* As we explain in detail below, the firewall is *untrusted* in the sense that it shares no secrets with Alice, and in general we expect Alice to place no more trust in the firewall than she does in the communication channel itself.

More concretely, Mironov and Stephens-Davidowitz start by considering an arbitrary cryptographic protocol that satisfies some notions of functionality (i.e., correctness) and security.[1] For example, perhaps the simplest non-trivial case is semantically secure message transmission from Alice to Bob, which has the functionality requirement that Bob should receive the correct plaintext message from Alice and the security requirement that a computationally bounded adversary "should not learn anything about Alice's plaintext message" from the transcript of a run of the protocol. Formally, we can model this functionality by providing Alice with an input plaintext and requiring Bob's output to match this, and we can model semantic security by a standard indistinguishability security game.

A reverse firewall for Alice in some protocol *maintains functionality* if the protocol "with the firewall in the middle" achieves the same functionality as the original protocol. E.g., in the case of message transmission, Bob should still receive Alice's message—his output should still match Alice's input. More interestingly, the firewall *preserves security* if the protocol with the firewall is secure even when we replace Alice's computer with some arbitrarily corrupted party. For example, a reverse firewall for Alice preserves semantic security of message transmission if a computationally bounded adversary "learns nothing about Alice's plaintext message" from the transcript of messages sent between the firewall and Bob, *regardless of how Alice behaves.* E.g., the firewall may rerandomize the messages that Alice sends in a way that makes them indistinguishable from random from the adversary's perspective, regardless of Alice's original message. (We analyze such protocols in a stronger setting in Sect. 3, and in a different setting in Appendix B.)

Note that it also makes sense to consider reverse firewalls for the receiver, Bob. For example, consider a protocol in which Bob first sends his public key

---

[1] The notion of functionality in [31] is quite simple, and it should not be confused with the much more complicated concept of functionality used in the universal composability framework. Formally, Mironov and Stephens-Davidowitz define a functionality requirement as any condition on the output of the parties that may depend on the input, and in practice, these requirements are straightforward.

to Alice, and Alice responds with an encryption of her message under this key. Clearly, if Bob's computer is corrupted in such a protocol, this can compromise security, even if Alice behaves properly. In such a protocol, a firewall for Bob might rerandomize his public key. Of course, to maintain correctness, this firewall must also intercept Bob's incoming messages and convert ciphertexts under this rerandomized key to encryptions under Bob's original key. (Again, see Sect. 3 for a formal treatment of such protocols.)

A key feature of protocols with reverse firewalls, as defined in [31], is that they should be functional and secure *both* with the reverse firewall *and* without it. I.e., there should be a well-defined *underlying protocol* between Alice and Bob that satisfies classical functionality and security requirements. This is one important difference between reverse firewalls and some similar models, such as the mediated model [1] and divertible protocols [9,10,32], and it comes with a number of benefits. ([31] contains a thorough comparison of many different related models.) First, it means that these protocols can be implemented and used without worrying about whether reverse firewalls are present—one protocol works regardless; we simply obtain additional security guarantees with an RF.

Second, and more importantly, this definitional choice provides an elegant solution to a natural concern about reverse firewalls: What happens when the firewall itself is corrupted? Of course, if *both* Alice's own machine and her firewall are compromised, then we cannot possibly hope for security. But, if Alice's own implementation is correct and the firewall has been corrupted, then we can view the firewall as "part of" the adversary in the firewall-free protocol between Alice and Bob. Since this underlying protocol must itself be secure, it trivially remains secure in the presence of a corrupted firewall.[2] This is why we can say that the firewall is trusted no more than the communication channel.

Of course, the advantage of using a firewall comes when Alice's machine is corrupted but the firewall is implemented correctly, in which case the firewall provides Alice with a security guarantee that she could not have had otherwise. In short, *the firewall can only help.* (See Table 1.) In fact, we even require firewalls to be "stackable," so that arbitrarily many firewalls may be deployed, and security is guaranteed as long as *either* (1) Alice's own machine is uncorrupted; or (2) at least one of these firewalls is implemented correctly and honestly.

Finally, it is convenient to identify a class of *functionality-maintaining corruptions*: compromised implementations that are "technically legal" in the sense that they may deviate arbitrarily from the protocol, as long as they do not break its functionality. Some of our reverse firewalls are only secure against this type of corruption. (This model is introduced by [31], and the authors call security against unrestricted compromise *strong* security.) We emphasize that, while this restricted class of compromised implementations is not ideal, it is still quite large. In particular, all of the real-world compromises mentioned above fall into this category [3,12–14,23,25,28,34,40], as do essentially all other forms of compromise considered in prior work, such as backdoored PRNGs [21],

---

[2] Technically, this statement only holds if the underlying protocol is secure against active adversaries.

Table 1. Security of a protocol with a secure reverse firewall for Alice in various different scenarios.

Alice	Alice's RF	Secure?	
Honest	Honest	✔	Trivial
Honest	Compromised	✔	Underlying protocol's security
Compromised	Honest	✔	RF's security
Compromised	Compromised	✗	Everyone is compromised!

Algorithm Substitution Attacks [7], subliminal channels [39], etc. (We discuss functionality-maintaining corruption in our setting in more detail in the full version [22]. See [31] for a detailed discussion of the general reverse framework.)

1.2 Our Results

In this section, we walk through the results that we obtain in different settings, starting with simpler cases and working our way up to our stronger results. In what follows, Alice is always the sender and Bob is always the receiver of the message. All of our security notions apply to the concurrent setting, in which the adversary may instantiate many runs of the protocol simultaneously. (The proofs are in the full version [22].)

The Symmetric-Key Setting. In the first and simplest scenario, Alice and Bob have a shared secret key. (See Appendix A.) Quite naturally, Alice might want to use a symmetric-key encryption scheme to communicate with Bob. Using a standard scheme (e.g., AES-CBC) would, however, expose her to a number of "algorithm-substitution attacks" (what we call corruption or compromise) described by Bellare, Paterson, and Rogaway [7], such as IV-replacement or a biased-ciphertext attack. To defend against such attacks, Bellare et al. propose using a clever solution: a *deterministic* encryption scheme based on either a counter or a nonce. We briefly consider this case, observing that their solution corresponds to a one-round protocol in our model (in which the firewall simply lets messages pass unaltered).

Unfortunately, we show that strong security (i.e., security against corrupted implementations of Alice that are not necessarily functionality-maintaining) is not achievable without using (less efficient) public-key primitives, even in the reverse-firewalls framework. This provides further motivation to study reverse firewalls in the public-key setting.

Rerandomizable Encryption. As we mentioned earlier, the simplest nontrivial reverse firewall in the public-key setting uses CPA-secure rerandomizable public-key encryption. In particular, Alice can send her plaintext encrypted under Bob's public key, and Alice's reverse firewall can simply rerandomize this ciphertext. We observe that this folklore technique works in our setting. In Sect. 3, we present a generalization of this idea that does not require any public-key infrastructure, by having Bob send his public key as a first message. Following [31], we observe that Bob can have a reverse firewall for such a protocol that rerandomizes his key (and converts Alice's ciphertext from an encryption under the rerandomized key to an encryption under Bob's original key). We therefore show a simple two-round protocol with a reverse firewall for each party.

While such protocols are simple and elegant, they have two major drawbacks. First, they are only secure against passive adversaries (we will return to this issue soon). Second, and arguably more importantly, such protocols require the computation of public-key operations on the entire plaintext. Since plaintexts are often quite long and public-key operations tend to be much slower than symmetric-key operations, it is much faster in practice to use public-key operations to transmit a (relatively short) key for a symmetric-key encryption scheme and then to send the plaintext encrypted under this symmetric key. There are two general methods for transmitting this key in the classical setting: hybrid encryption and key agreement.

Failure of Hybrid Encryption. Unfortunately, hybrid encryption does not buy us anything in the reverse-firewalls framework. Recall that in a hybrid encryption scheme, Alice selects a uniformly random key rk for a symmetric-key scheme and sends rk encrypted under Bob's public key together with the encryption of her message under the symmetric-key scheme with key rk. We might naively hope that we can build a reverse firewall for such a scheme by simply applying the "rerandomizing firewall" to the "public-key part" of Alice's ciphertext. But, this does not work because of the attack in which a corrupted implementation of Alice chooses a "bad key" rk^* with which to encrypt the message. The "bad key" rk^* might be known to an adversary; might be chosen so that the ciphertext takes a specific form that leaks some information; or might otherwise compromise Alice's security. So, intuitively, a reverse firewall in such a scheme must be able to rerandomize the key rk, and it therefore must be able to convert an encryption under some key rk into an encryption of the same plaintext under some new key rk'. Unfortunately, we show that any such "key-malleable" symmetric-key encryption scheme implies public-key encryption. Therefore, it cannot be faster than public-key encryption and is useless for our purposes.

Key Agreement. Recall that a key-agreement protocol allows Alice and Bob to jointly select a secret key over an insecure channel. Security requires that the resulting key is indistinguishable from random to an eavesdropper. Such a protocol is often used in conjunction with symmetric-key encryption in the classical setting, where it is justified by composition theorems relating the security of the message-transmission protocol to the underlying key-agreement protocol. Indeed, we give an analogous result (Theorem 2) that works in our setting, showing that a carefully designed key-agreement protocol with sufficiently secure reverse firewalls can be combined with symmetric-key encryption to produce an efficient CPA-secure message-transmission protocol with secure reverse firewalls.

This motivates the study of key-agreement protocols with secure reverse firewalls. As a first attempt at constructing such an object, we might try to somehow rerandomize the messages in the celebrated Diffie-Hellman key-agreement protocol, in which Alice first sends the message g^a, Bob then sends g^b, and the shared key is g^{ab}. (See Fig. 8.) Here, we run into an immediate problem. Since the firewall must maintain correctness, no matter what message A^* the firewall sends to Bob, it must be the case that the final key is A^{*b}, where b is chosen by Bob. But, this allows a corrupt implementation of Bob to influence the key. For example, Bob can repeatedly resample b until, say, the first bit of the key A^{*b} is zero, thus compromising the security. It is easy to see that this problem is not unique to Diffie-Hellman and in fact applies to *any* protocol in which "a party can learn what the key will be before sending a message that influences the key."

So, to truly prevent any party from having any control over the final key, we use a three-round protocol in which Bob sends a *commitment of g^b* as his first message. Alice then sends g^a, and Bob then opens his commitment.[3] Of course, the commitment scheme that Bob uses must itself be rerandomizable and malleable, so that the firewall can both rerandomize the commitment itself *and* the committed group element. Fortunately, we show that a very simple scheme, a natural variant of the Pedersen commitment, actually suffices.

Since this simple protocol is unauthenticated, it cannot be secure against active adversaries. While passive security might be sufficient in some settings (powerful adversaries are known to passively gather large amounts of web traffic [23]), it would be much better to achieve security against active adversaries. We address this next.

CCA-Security From Key Agreement. We attempt to construct a reverse firewall that preserves CCA-security (i.e., security against active adversaries that may "feed" Alice and Bob arbitrary adversarial messages and read Bob's output). In this setting, we again prove a generic composition theorem, which shows that it suffices to find a key-agreement protocol with a reverse firewall that satisfies certain security properties. In analogy with the passive setting, after agreeing to a key with Bob, Alice can use symmetric-key encryption to send the actual plaintext message. The resulting protocol is CCA-secure, and Alice's reverse firewall preserves this security. (See Theorem 4.)

To instantiate this scheme, we must construct a key-agreement protocol that is secure against active adversaries and has a reverse firewall that preserves this security. Unfortunately, many of the common techniques used in classical key-agreement protocols (or even protocols that are secure against key control) are useless here. In particular, most key-agreement protocols that achieve security

---

[3] We note that the problem that we face here is very similar to the problem of key control, and our solution is similar to solutions used in the key-control literature. See, e.g., [18].

against active adversaries do so by essentially having both parties sign the transcript at the end of the protocol. Intuitively, this allows the parties to know if the adversary has tampered with any messages, so that they will never agree to a key if a man in the middle has modified their messages. But in our setting, we actually *want* the firewall to be able to modify the parties' messages. We therefore need to somehow find some unique information that the parties can use to confirm that they have agreed to the same key without preventing the firewall from modifying the key. Furthermore, we need the firewall to be able to check these signatures, so that it can block invalid messages. Therefore, our primary technical challenge in this context is to find a protocol with some string that (1) uniquely identifies the key; (2) does not leak the key; (3) respects the firewall's changes to the parties' messages; and (4) is efficiently computable from the transcript. And, of course, the protocol must be secure against active adversaries, even though it is in some sense "designed to help a man in the middle."

In spite of these challenges, we construct a protocol with a reverse firewall for each party that preserves security against active adversaries. Remarkably, our protocol achieves this extremely strong notion of security with only four rounds and relatively short messages, and the parties themselves (including the firewall) only need to perform a small constant number of operations. This compares quite favorably with protocols that are currently implemented in practice (which of course are completely insecure in our setting), and we therefore believe that this protocol can and should be implemented and used in the real world.

This surprising solution, which we describe in detail in Sect. 5.1, uses hashed Diffie-Hellman (similar in spirit to [27]) and bilinear maps. We also use unique signatures to prevent the signatures from becoming a channel themselves.

Rerandomizable Encryption and Active Adversaries. Finally, we return to the question of (necessarily less efficient) protocols based on rerandomizable encryption, but now in the setting of active adversaries. We show how to achieve CCA-security in a single round using rerandomizable RCCA-secure encryption [4]. (See Appendix B.) Indeed, we show that such a primitive is actually *equivalent to* a one-round protocol with a firewall that preserves CCA-security. Such schemes are fairly well-studied, and solutions exist [24,36]. But, our work leads to an interesting open question. Currently known schemes are rerandomizable in the sense that the rerandomization of any valid ciphertext is indistinguishable from a fresh ciphertext, even with access to a decryption oracle. We ask whether these schemes can be made "strongly rerandomizable," in the sense that the same is true even for *invalid* ciphertexts. (See Appendix B for the formal definition.) We show the weaker notion of rerandomizability is equivalent to protocols with firewalls that are secure against functionality-maintaining corruption, while strong rerandomizability is equivalent to security against arbitrary corruption.

1.3   Related Work

Message transmission in the classical setting (i.e., without reverse firewalls) is of course extremely well-studied, and a summary of such work is beyond the scope of this paper. We note, however, that our security definitions for message transmission protocols follow closely Dodis and Fiore [20].

There have been many different approaches to cryptography in the presence of compromise. [31] contains a thorough discussion of many of these (though they naturally do not mention the many relevant papers that appeared simultaneously with or after their publication, such as [2,5,6,15,21,37,38]). In particular, [31] contains a detailed comparison of the reverse-firewall framework with many prior models, showing that RFs generalize much of the prior work on insider attacks and various related notions. Here, we focus on works whose setting or techniques are most similar to our own.

Our work can be viewed as a generalization of that of Bellare, Paterson, and Rogaway [7] in a number of directions. We consider multi-round protocols in which the parties might not share secret keys, and we consider arbitrarily compromised adversaries. In order to get around the very strong restrictions proved in [7], we use the RF framework of [31]. Our techniques are therefore quite different. However, we do use the deterministic encryption scheme of Bellare et al. as part of two of our protocols. (See Appendix A.)

Our work is closely related to Mironov and Stephens-Davidowitz [31], which introduces the reverse-firewalls framework. [31] demonstrate feasibility of this framework by constructing reverse firewalls for parties participating in Oblivious Transfer and Secure Function Evaluation protocols—very strong cryptographic primitives. The fact that such strong primitives can be made secure in this model is quite surprising and bodes well for the reverse-firewalls framework. However, these protocols are very inefficient and therefore mostly of theoretical interest. And, while the primitives considered in [31] have very strong functionality, the security notions that they achieve are rather weak (e.g., security in the semi-honest model). To fulfill the promise of reverse firewalls, we need to consider protocols of more practical importance. We construct much more efficient protocols for widely used primitives with very strong security guarantees. Naturally, we inherit some of the techniques of [31], but we also develop many new ideas.

Bellare and Hoang [5] build on [7] in a different direction, showing how to build deterministic and hedged public-key encryption schemes that are secure against randomness subversion and Algorithm Substitution Attacks. Essentially, they show public-key encryption schemes that are secure even when the sender is compromised, provided that (1) the type of compromise is restricted; and (2) the plaintext itself comes from a high-entropy distribution. These notions of security are much weaker than ours, but they achieve them without the use of an RF. (Recall that [7] implies that semantically secure public-key encryption secure against Algorithm Substitution Attacks is not possible in the classical model.)

Recently, Ateniese, Magri, and Venturi studied reverse firewalls for signature schemes and showed a number of clever solutions [2]. Their work can be considered as complementary to ours, as we are concerned with privacy, while they consider authentication. We also note that our more advanced key-agreement scheme uses unique signatures, and we implicitly rely on the fact that unique signatures have a (trivial) reverse firewall. Indeed, the more general primitive of rerandomizable signatures that Ateniese et al. consider would also suffice for our purposes and might be more efficient in practice.

Our frequent use of rerandomization to "sanitize" messages is very similar to much of the prior work on subliminal channels [9,10,16,17,39], divertible protocols [9,10,32], collusion-free protocols [1,29], etc—particularly the elegant work of Blaze, Bleumer, and Strauss [9] and Alwen, shelat, and Visconti [1]. Again, we refer the reader to [31] for a thorough discussion of these models and their relationship to the reverse-firewall framework.

Finally, we note that some of our study of key agreement is similar to work on key-agreement protocols secure against active insiders, and the study of key control (e.g., [18,26,35]). These works consider key-agreement protocols involving at least three parties, in which one or more of the participants wishes to maliciously fix the key or otherwise subvert the security of the protocol. Some of the technical challenges that we encounter are similar to those encountered in the key control literature, and indeed, the simple commitment-based protocol that we present in Sect. 4.1 can be viewed as a simple instantiation of some of the known (more sophisticated) solutions to the key-control problem (see, e.g., [18]). However, since prior work approached this problem from a different perspective—with three or more parties and without reverse firewalls—our more technical solutions presented in Sect. 5.1 are quite different. In particular, almost all prior work on key agreement focuses on creating protocols that produce a "non-malleable" key, whereas our protocols need some type of malleability specifically to allow the firewall to rerandomize the key. Perhaps surprisingly, we accomplish this without sacrificing security, and our techniques might therefore be of independent interest.

2   Definitions

2.1   Reverse Firewalls

We use the definition of reverse firewalls from [31] (and we refer the reader to [31] for further discussion of the reverse-firewall framework). A reverse firewall $\mathcal{W}$ is a stateful algorithm that maps messages to messages. For a party A and reverse firewall $\mathcal{W}$, we define $\mathcal{W} \circ A$ as the "composed" party in which $\mathcal{W}$ is applied to the messages that A receives before A "sees them" and the messages that A sends before they "leave the local network of A." $\mathcal{W}$ has access to all public parameters, but not to the private input of A or the output of A. (We can think of $\mathcal{W}$ as an "active router" that sits at the boundary between Alice's private network and the outside world and modifies Alice's incoming and outgoing messages.) We repeat all relevant definitions from [31] below, and we add two new ones.

As in [31], we assume that a cryptographic protocol comes with some functionality or correctness requirements $\mathcal{F}$ and security requirements $\mathcal{S}$. (For example, a functionality requirement $\mathcal{F}$ might require that Alice and Bob output the same thing at the end of the protocol. A security requirement $\mathcal{S}$ might ask that no efficient adversary can distinguish between the transcript of the protocol and a uniformly random string.) Throughout, we use $\bar{A}$ to represent arbitrary adversarial implementations of party A and $\tilde{A}$ to represent functionality-maintaining implementations of A (i.e., implementations of A that still satisfy the functionality requirements of the protocol). For a protocol $\mathcal{P}$ with party A, we write $\mathcal{P}_{A\to\tilde{A}}$ to represent the protocol in which the role of party A is replaced by party $\tilde{A}$.

We are only interested in firewalls that themselves maintain functionality. In other words, the *composed party* $\mathcal{W} \circ A$ should not break the correctness of the protocol. (Equivalently, $\mathcal{P}_{A\to\mathcal{W}\circ A}$ should satisfy the same functionality requirements as the underlying protocol $\mathcal{P}$.) We follow [31] in requiring something slightly stronger—reverse firewalls should be "stackable", so that many reverse firewalls composed in series $\mathcal{W}\circ\cdots\circ\mathcal{W}\circ A$ still do not break correctness.

Definition 1 (Reverse firewall). *A reverse firewall $\mathcal{W}$ maintains functionality $\mathcal{F}$ for party A in protocol $\mathcal{P}$ if protocol $\mathcal{P}$ satisfies $\mathcal{F}$, the protocol $\mathcal{P}_{A\to\mathcal{W}\circ A}$ satisfies $\mathcal{F}$, and the protocol $\mathcal{P}_{A\to\mathcal{W}\circ\cdots\circ\mathcal{W}\circ A}$ also satisfies $\mathcal{F}$. (I.e., we can compose arbitrarily many reverse firewalls without breaking functionality.)*

Of course, a firewall is not interesting unless it provides some benefit. The most natural reason to deploy a reverse firewall is to *preserve* the security of a protocol, even in the presence of compromise. The below definition (which again follows [31]) captures this notion by asking that the protocol obtained by replacing party A with $\mathcal{W} \circ \tilde{A}$ for an arbitrary corrupted party $\tilde{A}$ still achieves some notion of security. For example, when we consider message transmission, we will want the firewall to guarantee Alice's privacy against some adversary, even when Alice's own computer has been corrupted.

Definition 2 (Security preservation). *A reverse firewall strongly preserves security $\mathcal{S}$ for party A in protocol $\mathcal{P}$ if protocol $\mathcal{P}$ satisfies $\mathcal{S}$, and for any polynomial-time algorithm $\bar{A}$, the protocol $\mathcal{P}_{A\to\mathcal{W}\circ\bar{A}}$ satisfies $\mathcal{S}$. (I.e., the firewall can guarantee security even when an adversary has tampered with A.)*

A reverse firewall preserves security $\mathcal{S}$ for party A in protocol $\mathcal{P}$ satisfying functionality requirements $\mathcal{F}$ if protocol $\mathcal{P}$ satisfies $\mathcal{S}$, and for any polynomial-time algorithm $\tilde{A}$ such that $\mathcal{P}_{A\to\tilde{A}}$ satisfies $\mathcal{F}$, the protocol $\mathcal{P}_{A\to\mathcal{W}\circ\tilde{A}}$ satisfies $\mathcal{S}$. (I.e., the firewall can guarantee security even when an adversary has tampered with A, provided that the tampered implementation does not break the functionality of the protocol.)

For technical reasons, we will also need a new definition not present in [31]. We wish to show generic composition theorems, allowing us to construct a message-transmission protocol with secure reverse firewall from any key-agreement protocol with its own firewalls. In order to accomplish this, we will need the notion of *detectable failure*. Essentially, a protocol fails detectably if we can distinguish between transcripts of valid runs of the protocol and invalid transcripts. For simplicity, we assume that honest parties always output $\perp$ when they receive a malformed message (e.g., when a message that should be a pair of group elements is not a pair of group elements). While the general notion of validity is a bit technical, we will use it in very straightforward ways. (E.g., transcripts will be valid if and only if a commitment is properly opened and a certain signature is valid.)

Definition 3 (Valid transcripts). *A sequence of bits r and private input I generate transcript T in protocol P if a run of the protocol P with input I in which the parties' coin flips are taken from r results in the transcript T. A transcript T is a valid transcript for protocol P if there is a sequence r and private input I generating T such that no party outputs $\perp$ at the end of the run. (Here, we assume that the public input is part of the transcript.) A protocol has unambiguous transcripts if for any valid transcript T, there is no possible input I and coins r generating T that results in a party outputting $\perp$. (In other words, a valid transcript never results from a failed run of the protocol.)*

Definition 4 (Detectable failure). *A reverse firewall W detects failure for party A in protocol P if*

- *$P_{A \to W \circ A}$ has unambiguous transcripts;*
- *the firewall W outputs the special symbol $\perp$ when run on any transcript that is not valid for $P_{A \to W \circ A}$; and*
- *there is a polynomial-time deterministic algorithm that decides whether a transcript T is valid for $P_{A \to W \circ A}$.*

We will also need the notion of *exfiltration resistance*, introduced in [31]. Intuitively, a reverse firewall is exfiltration resistant if "no corrupt implementation of Alice can leak information through the firewall." We say that it is exfiltration resistant for Alice against Bob if Alice cannot leak information to Bob through the firewall, and we say that it is exfiltration resistant against eavesdroppers (or just exfiltration resistant) if Alice cannot leak information through the firewall to an adversary that is only given access to the protocol transcript.

The second definition below (which uses the notion of valid transcripts) is new to this paper and is necessary for our composition theorems.

$$
\begin{array}{l}
\textbf{proc. } \mathsf{LEAK}(P, \mathsf{A}, \mathsf{B}, W, \lambda) \\
(\bar{\mathsf{A}}, \bar{\mathsf{B}}, I) \leftarrow \mathcal{E}(1^\lambda) \\
b \xleftarrow{\$} \{0, 1\} \\
\text{IF } b = 1, \mathsf{A}^* \leftarrow W \circ \bar{\mathsf{A}} \\
\text{ELSE, } \mathsf{A}^* \leftarrow W \circ \mathsf{A} \\
T^* \leftarrow P_{\mathsf{A} \to \mathsf{A}^*, \mathsf{B} \to \bar{\mathsf{B}}}(I) \\
b^* \leftarrow \mathcal{E}(T^*, S_{\bar{\mathsf{B}}}) \\
\texttt{OUTPUT } (b = b^*)
\end{array}
$$

Fig. 1. $\mathsf{LEAK}(P, \mathsf{A}, \mathsf{B}, W, \lambda)$, the exfiltration resistance security game for a reverse firewall W for party A in protocol P against party B with input I. $\mathcal{E}$ is the adversary, λ the security parameter, $S_{\bar{B}}$ the state of $\bar{B}$ after the run of the protocol, I valid input for P, and T^* is the transcript resulting from a run of the protocol $mathcal P_{\mathsf{A} \to \mathsf{A}^*, \mathsf{B} \to \bar{B}}$ with input I.

Definition 5 (Exfiltration resistance). *A reverse firewall is* exfiltration resistant *for party* A *against party* B *in protocol* $\mathcal{P}$ *satisfying functionality* $\mathcal{F}$ *if no PPT algorithm* $\mathcal{E}$ *with output circuits* $\widetilde{\mathsf{A}}$ *and* $\widetilde{\mathsf{B}}$ *such that* $\mathcal{P}_{\mathsf{A}\to\widetilde{\mathsf{A}}}$ *and* $\mathcal{P}_{\mathsf{B}\to\widetilde{\mathsf{B}}}$ *satisfy* $\mathcal{F}$ *has non-negligible advantage in* $\mathsf{LEAK}(\mathcal{P}, \mathsf{A}, \mathsf{B}, \mathcal{W}, \lambda)$. *If* B *is empty, then we simply say that the firewall is* exfiltration resistant.

A reverse firewall is exfiltration resistant *for party* A *against party* B *in protocol* $\mathcal{P}$ *with valid transcripts if no PPT algorithm* $\mathcal{E}$ *with output circuits* $\widetilde{\mathsf{A}}$ *and* $\widetilde{\mathsf{B}}$ *such that* $\mathcal{P}_{\mathsf{A}\to\widetilde{\mathsf{A}}}$ *and* $\mathcal{P}_{\mathsf{B}\to\widetilde{\mathsf{B}}}$ *produce valid transcripts for* $\mathcal{P}$ *has non-negligible advantage in* $\mathsf{LEAK}(\mathcal{P}, \widetilde{\mathsf{A}}, \widetilde{\mathsf{B}}, \mathcal{W}, \lambda)$. *If* B *is empty, then we simply say that the firewall is* exfiltration resistant with valid transcripts.

A reverse firewall is strongly exfiltration resistant *for party* A *against party* B *in protocol* $\mathcal{P}$ *if no PPT adversary* $\mathcal{E}$ *has non-negligible advantage in* $\mathsf{LEAK}(\mathcal{P}, \mathsf{A}, \mathsf{B}, \mathcal{W}, \lambda)$. *If* B *is empty, then we say that the firewall is* strongly exfiltration resistant.

2.2 Message-Transmission Protocols

A *message-transmission protocol* is a two-party protocol in which one party, Alice, is able to communicate a plaintext message to the other party, Bob. (For simplicity, we only formally model the case in which Alice wishes to send a single plaintext to Bob per run of the protocol, but this of course naturally extends to a more general case in which Alice and Bob wish to exchange many plaintext messages.) We consider two notions of security for such messages. First, we consider *CPA security*, in which the adversary must distinguish between the transcript of a run of the protocol in which Alice communicates the plaintext m_0 to Bob and the transcript with which Alice communicates m_1 to Bob, where m_0 and m_1 are adversarially chosen plaintexts. (Even in this setting, we allow the adversary to start many concurrent runs of the protocol with adaptively chosen plaintexts.) Our strongest notion of security is *CCA security* in which the adversary may "feed" the parties any messages and has access to a decryption oracle. Our security definitions are similar in spirit to [20], but adapted for our setting.

Session Ids. Throughout this paper, we consider protocols that may be run concurrently many times between the same two parties. In order to distinguish one run of a protocol from another, we therefore "label" each run with a unique session id, denoted sid. We view sid as an implicit part of every message, and we often ignore sid when it is not important. Our parties and firewalls are stateful, and we assume that the parties and the firewall maintain a list of the relevant session ids, together with any information that is relevant to continue the run of the protocol corresponding to sid (such as the number of messages sent so far, any values that need to be used later in the protocol, etc.). We typically suppress explicit reference to these states. In our security games, the adversary may choose the value sid for each run of the protocol, provided that each party has a unique run for each session sid. (In fact, it does not even make sense for the adversary to use the same sid for two different runs of the protocol with the same party, as this party will necessarily view any calls with the same sid as corresponding to a single run of the protocol. However, as is clear from our security games, an active adversary may maintain two separate runs of a protocol with two different parties but the same sid.) In practice, sid can be a simple counter or any other nonce (perhaps together with any practical information necessary for communication, such as IP addresses). We note in passing that, in the setting of reverse firewalls, a counter is preferable to, e.g., a random nonce to avoid providing a channel through sid, but such concerns are outside of our model and the scope of this paper.

The definition below makes the above formal and provides us with some useful terminology.

Definition 6 (Message-transmission protocol). *A* message-transmission protocol *is a two-party protocol in which one party, Alice, receives as input a plaintext m from some plaintext space* $\mathcal{M}$. *The protocol is* correct *if for any input* $m \in \mathcal{M}$, *Bob's output is always m.*

We represent the protocol by four algorithms $\mathcal{P} = (\mathsf{setup}, \mathsf{next_A},$ $\mathsf{next_B}, \mathsf{return_B})$. setup *takes as input* 1^λ, *where* λ *is the security parameter, and returns the starting states for each party,* S_A, S_B, *which consist of both private input,* σ_A *and* σ_B *respectively, and public input* π. *Each party's* next *procedure is a stateful algorithm that takes as input* sid *and an incoming message, updates the party's state, and returns an outgoing message. The* return_B *procedure takes as input Bob's state* S_B *and* sid *and returns Bob's final output.*

We say that a message-transmission protocol is

- unkeyed *if* setup *does not return any private input* σ_A *or* σ_B;
- singly keyed *if* setup *returns private input* σ_B *for Bob but none for Alice;*
- publicly keyed *if* setup *returns private input for both parties* σ_A *and* σ_B, *but these private inputs are independently distributed; and*
- privately keyed *if* setup *returns private input for both parties whose distributions are dependent.*

When we present protocols, we will drop the formality of defining explicit functions $\mathcal{P} = (\mathsf{setup}, \mathsf{next_A}, \mathsf{next_B}, \mathsf{return_B})$ and states for the parties, preferring instead to use diagrams as in Fig. 4. But, this formulation is convenient for our security definitions. In particular, we present the relevant subprocedures for our security games in Fig. 2. An adversary plays the game depicted in Fig. 2 by first calling initialize (receiving as output π) and then making various calls to the other subprocedures. Each time it calls a subprocedure, it receives any output from the procedure. The game ends when the adversary calls finalize, and the adversary wins if and only if the output of finalize is one.

The below definitions capture formally the intuitive notions of security that we presented above. In particular, the CPA security definition allows the adversary to start arbitrarily many concurrent runs of the protocol with adversarial input, but it does not allow the adversary to change the messages sent by the

---

proc. initialize(1^λ)
$(\sigma_A, \sigma_B, \pi) \xleftarrow{\$} \text{setup}(1^\lambda)$
$S_A \leftarrow (\sigma_A, \pi);\ S_B \leftarrow (\sigma_B, \pi)$
sid* $\leftarrow \perp$; compromised $\leftarrow$ false
$b \xleftarrow{\$} \{0,1\}$
OUTPUT π

proc. finalize(b^*)
IF $b = b^*$, RETURN 1
ELSE, RETURN 0

proc. start-run(sid, m)
IF sid $\notin S_A$, S_A.add(sid, m)

proc. start-challenge(sid, m_0, m_1)
IF sid $\notin S_A$ AND sid* $= \perp$,
    sid* $\leftarrow$ sid
    S_A.add(sid, m_b)

proc. get-next$_A$(sid, M)
IF compromised,
    OUTPUT $\perp$
OUTPUT next$_A$(S_A, sid, M)

proc. get-next$_B$(sid, M)
IF compromised,
    OUTPUT $\perp$
OUTPUT next$_B$(S_B, sid, M)

proc. get-output$_B$(sid)
IF sid $=$ sid* OR compromised,
    OUTPUT $\perp$
OUTPUT return$_B$(S_B, sid)

proc. get-secrets
compromised $\leftarrow$ true
OUTPUT (σ_A, σ_B)

Fig. 2. Procedures used to define security for message-transmission protocol $\mathcal{P} =$ (setup, next$_A$, next$_B$). An adversary plays this game by first calling initialize and then making various oracle calls. The game ends when the adversary calls finalize, and the output of finalize is one if the adversary wins and zero otherwise.

two parties or to send its own messages. We also define forward secrecy, which requires that security hold even if the parties' secret keys may be leaked to the adversary.

Definition 7 (Message-transmission security). *A message-transmission protocol is called*

- chosen-plaintext secure *(CPA-secure) if no PPT adversary has non-negligible advantage in the game presented in Fig. 2 when* get-next$_A$(sid, M) *and* get-next$_B$(sid, M) *output* $\perp$ *unless this is the first* get-next *call with this* sid *or* M *is the output from the previous* get-next$_A$ *call with the same* sid *or the previous* get-next$_B$ *with the same* sid *respectively (i.e., the adversary is passive); and*
- chosen-ciphertext secure *(CCA-secure) if no PPT adversary has non-negligible advantage in the game presented in Fig. 2 with access to all oracles.*

We say that the protocol is chosen-plaintext (resp. chosen-ciphertext) secure without forward secrecy *if the above holds without access to the* get-secrets *oracle.*

We note that it does not make sense to consider chosen-ciphertext security when Bob may be corrupted. In this case, the output of get-output$_B$ could be arbitrary. (Note that the firewall can potentially "sanitize" Bob's *messages*, but it of course does not have access to his *output*.) We therefore only consider firewalls that preserve CPA security for Bob.

2.3 Key Agreement

Key-agreement protocols will play a central role in our constructions, so we now provide a definition of key agreement that suffices for our purposes. Our notion of key agreement closely mirrors the definitions from the previous section.

Definition 8 (Key agreement). *A key-agreement protocol is represented by five algorithms,* $\mathcal{P} = (\mathsf{setup}, \mathsf{next_A}, \mathsf{next_B}, \mathsf{return_A}, \mathsf{return_B})$. setup *takes as input* 1^λ, *where* λ *is the security parameter and returns the starting states for each party,* S_A, S_B, *which consists of public input* π *and the private input for each party* σ_A *and* σ_B. *Each party's* next *procedure is a stateful algorithm that takes as input* sid *and an incoming message, updates the party's state, and returns an outgoing message. Each party's* return *procedure takes as input the relevant party's state and* sid *and returns the party's final output from some key space* $\mathcal{K}$ *or* $\perp$. *We also allow* auxiliary *input* aux *to be added to Alice's state before the first message of a protocol is sent.*

The protocol is correct *if Alice and Bob always output the same thing at the end of the run of a protocol for any random coins and auxiliary input* aux.

We say that a key-agreement protocol is

– unkeyed *if* setup *does not return any private input* σ_A *or* σ_B;
– singly keyed *if* setup *returns private input* σ_B *for Bob but no private input* σ_A *for Alice; and*
– publicly keyed *if* setup *returns private input for both parties* σ_A *and* σ_B.

Definition 9 (Key-agreement security). *A key-agreement protocol is*

– secure against passive adversaries *if no probabilistic polynomial-time adversary has non-negligible advantage in the game presented in Fig. 3 when* $\mathsf{get\text{-}next_A}(\mathsf{sid}, M)$ *and* $\mathsf{get\text{-}next_B}(\mathsf{sid}, M)$ *output* $\perp$ *unless this is the first* $\mathsf{get\text{-}next}$ *call with this* sid *or* M *is the output from the previous* $\mathsf{get\text{-}next_B}$ *call with the same* sid *or the previous* $\mathsf{get\text{-}next_A}$ *call with the same* sid *respectively (i.e., the adversary is passive);*
– secure against active adversaries for Alice *if no probabilistic polynomial-time algorithm has non-negligible advantage in the game presented in Fig. 3 without access to the* $\mathsf{get\text{-}output_A}$ *oracle;*
– secure against active adversaries for Bob *if no probabilistic polynomial-time algorithm has non-negligible advantage in the game presented in Fig. 3 without access to the* $\mathsf{get\text{-}output_B}$ *oracle; and*
– secure against active adversaries *if it is secure against active adversaries for both Bob and Alice; and*

proc. initialize(1^λ)
$(\sigma_A, \sigma_B, \pi) \xleftarrow{\$} \text{setup}(1^\lambda)$
$S_A \leftarrow (\sigma_A, \pi)$
$S_B \leftarrow (\sigma_B, \pi)$
$\text{sid}^* \leftarrow \perp$
compromised $\leftarrow$ false
$b \xleftarrow{\$} \{0, 1\}$
OUTPUT π

proc. finalize(b^*)
IF $b = b^*$,
    RETURN 1
ELSE, RETURN 0

proc. start-run(sid, aux)
IF sid $\notin S_A$,
    S_A.add(sid, aux)

proc. start-challenge(sid, aux)
IF $\text{sid}^* = \perp$ AND sid $\notin S_A$,
    $\text{sid}^* \leftarrow$ sid
    $\mathcal{R}_{\text{sid}^*} \xleftarrow{\$} \mathcal{K}$
    S_A.add(sid, aux)

proc. get-next$_A$(sid, M)
IF NOT compromised,
    OUTPUT next$_A$(S_A, sid, M)

proc. get-next$_B$(sid, M)
IF NOT compromised,
    OUTPUT next$_B$(S_B, sid, M)

proc. get-output$_A$(sid)
IF compromised, OUTPUT $\perp$
IF sid $=$ sid* AND $b = 0$,
    IF return$_A$(S_A, sid) $= \perp$, OUTPUT $\perp$
    ELSE, OUTPUT R_{sid}
ELSE, OUTPUT return$_A$(S_A, sid)

proc. get-output$_B$(sid)
IF compromised, OUTPUT $\perp$
IF sid $=$ sid* AND $b = 0$,
    IF return$_B$(S_B, sid) $= \perp$, OUTPUT $\perp$
    ELSE, OUTPUT R_{sid}
ELSE, OUTPUT return$_B$(S_B, sid)

proc. get-secrets
compromised $\leftarrow$ true
OUTPUT (σ_A, σ_B)

Fig. 3. Procedures used to define security for key-agreement protocol $\mathcal{P} =$ (setup, next$_A$, next$_B$, return$_A$, return$_B$). An adversary plays this game by first calling initialize and then making various oracle calls. The game ends when the adversary calls finalize, and the output of finalize is one if the adversary wins and zero otherwise. We suppress the auxiliary input aux when it is irrelevant.

– authenticated for Bob *if no probabilistic polynomial-time algorithm playing the game presented in Fig. 3 can output a valid transcript with corresponding session id* sid *unless* return$_B$(S_B, sid) $\neq \perp$ *or* compromised $=$ true. *(I.e., it is hard to find a valid transcript unless Bob returns a key.) Furthermore, if the transcript is valid and* get-output$_A$(sid) $\neq \perp$ *then* get-output$_A$(sid) $=$ return$_B$(sid). *(I.e., if the transcript is valid and Alice outputs a key, then Bob outputs the same key.)*

Note that these definitions are far from standard. In particular, in the case of active adversaries, we define security for Alice in terms of the keys that *Bob* outputs and security for Bob in terms of the keys that *Alice* outputs. This may seem quite counterintuitive. But, in our setting, we are worried that Alice may be corrupted. In this case, we cannot hope to restrict Alice's output after she receives invalid messages. (The firewall can modify Alice's *messages*, but not her *output*.) So, the best we can hope for is that the firewall prevents a tampered implementation of Alice (together with an active adversary) from "tricking" Bob into returning an insecure key.

3   A Two-Round Protocol from Rerandomizable Encryption

We first consider the simple case of CPA-secure two-round schemes in which the first message is a public key chosen randomly by Bob and the second message is an encryption of Alice's plaintext under this public key. Figure 4 shows the protocol.

In order to provide a reverse firewall for Alice in this protocol, the encryption scheme must be *rerandomizable*. In order to provide a reverse firewall for Bob, the scheme must be *key malleable*. Intuitively, a scheme is key malleable if a third party can "rerandomize" a public key and map ciphertexts under the "rerandomized" public key to ciphertexts under the original public key. We include formal definitions in the full version, and we observe there that ElGamal encryption suffices [22].

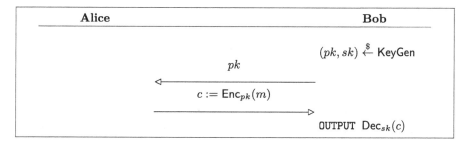

Fig. 4. Two round message-transmission protocol using the public-key encryption scheme (KeyGen, Enc, Dec).

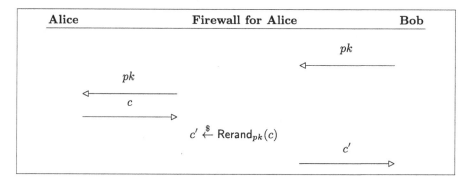

Fig. 5. Reverse firewall for Alice for the protocol shown in Fig. 5 that works if the encryption scheme is rerandomizable.

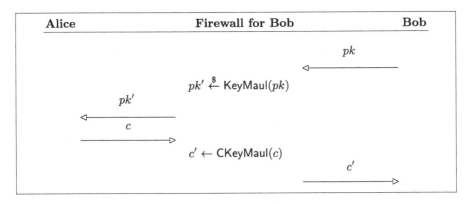

Fig. 6. Reverse firewall for Bob for the protocol shown in Fig. 4 that works if the encryption scheme is key malleable. We suppress the randomness r used as input to KeyMaul and CKeyMaul.

If the underlying encryption scheme in Fig. 4 is rerandomizable, then we can build a reverse firewall for Alice as in Fig. 5. If it is key malleable, then we can build a reverse firewall for Bob as in Fig. 6. The following theorem shows that this protocol and its reverse firewalls are secure. The simple proof is included in the full version [22].

Theorem 1. *The unkeyed message-transmission protocol shown in Fig. 4 is CPA-secure if the underlying encryption scheme is semantically secure. If the scheme is also (strongly) rerandomizable, then the reverse firewall shown in Fig. 5 (strongly) preserves security for Alice and (strongly) resists exfiltration for Alice. If the scheme is key malleable, then the reverse firewall shown in Fig. 6 maintains functionality for Bob, strongly preserves Bob's security, and strongly resists exfiltration for Bob against Alice.*

3.1 Hybrid encryption fails.

A major drawback of the above scheme is that it requires public-key operations of potentially very long plaintexts, which can be very inefficient in practice. A common solution in the classical setting is to use *hybrid encryption*, in which $\mathsf{Enc}_{pk}(m)$ is replaced by $(\mathsf{Enc}_{pk}(rk), \mathsf{SEnc}_{rk}(m))$, where SEnc is some suitable symmetric-key encryption scheme and rk is a freshly chosen uniformly random key for SEnc. However, if we simply replace the public-key encryption in Fig. 4 with the corresponding hybrid-key encryption scheme, then this fails spectacularly. For example, a tampered version of Alice $\widetilde{\mathsf{A}}$ can choose some *fixed* secret key rk^* and send the message $(\mathsf{Enc}_{pk}(rk^*), \mathsf{SEnc}_{rk^*}(m))$. If rk^* is a valid key, then $\widetilde{\mathsf{A}}$ maintains functionality, but an adversary that knows rk^* can of course read any messages that Alice sends.

So, in order for such a protocol to have a secure reverse firewall, the RF must⁻ be able to maul the encrypted key $\mathsf{Enc}_{pk}(rk)$ into $\mathsf{Enc}_{pk}(rk')$ for some

rk' and then convert the encrypted plaintext $\mathsf{SEnc}_{rk}(m))$ into an encryption under this new key $\mathsf{SEnc}_{rk'}(m))$. In particular, the symmetric-key scheme must be "key malleable." Unfortunately, such a scheme implies public-key encryption. Therefore, our supposed "symmetric-key" scheme actually must use public-key primitives. So, hybrid encryption buys us nothing. (In the full version [22], we give a proof sketch.)

Remark (Informal). *Any key-malleable symmetric-key encryption scheme implies public-key encryption.*

4   A Solution Using Key Agreement

In this section, we remain in the setting in which neither Alice nor Bob has a public key, so we are still interested in CPA security. (We address CCA security in the next section.) The protocol from Sect. 3 works, but it requires a public-key operation on the plaintext, which may be very long. In practice, this can be very inefficient. And, we showed in Sect. 3.1 that one common solution to this problem in the classical setting, hybrid encryption, seems hopeless with reverse firewalls because it allows Alice to *choose* a key rk that will be used to encrypt the plaintext—thus allowing a tampered version of Alice to "choose a bad key."

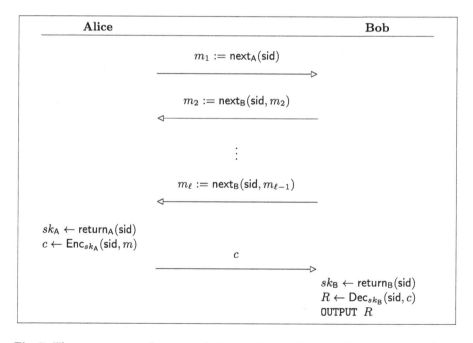

Fig. 7. The message-transfer protocol obtained by combining a key-agreement scheme (setup, next_A, next_B, $\mathsf{return}_\mathsf{A}$, $\mathsf{return}_\mathsf{B}$) and a nonce-based encryption scheme, (Enc, Dec).

So, we instead consider an alternative common solution to this efficiency problem: key agreement followed by symmetric-key encryption. (See Fig. 7.) As in Appendix A, we use a nonce-based encryption scheme with unique ciphertexts. We can view this as a modification of hybrid encryption in which "Alice and Bob together choose the key rk" that will be used to encrypt the plaintext. More importantly from our perspective, the messages that determine the key will go through the firewall. As an added benefit, once a key is established, Alice can use it to efficiently send multiple messages, not just one, without any additional public-key operations (though we do not model this here).

The composition theorem below shows that this protocol can in fact have a reverse firewall for both parties, provided that the key-agreement protocol itself has a reverse firewall that satisfies some suitable security requirements. See the full version for the proof. In the next section, we construct such a protocol.

Theorem 2 (Composition theorem for CPA security). *Let $\mathcal{W}_A$ and $\mathcal{W}_B$ be reverse firewalls in the underlying key-agreement protocol in Fig. 7 for Alice and Bob respectively. Let $\mathcal{W}_A^*$ be the firewall for Alice in the full protocol in Fig. 7 obtained by applying $\mathcal{W}_A$ to the key-agreement messages and then letting the last message through if $\mathcal{W}_A$ does not output $\bot$ and replacing the last message by $\bot$ otherwise. Let $\mathcal{W}_B^*$ be the firewall for Bob in the full protocol in Fig. 7 obtained by applying $\mathcal{W}_B$ to the key-agreement messages and simply letting the last message through. Then,*

1. *the protocol in Fig. 7 is CPA-secure if the underlying key-agreement protocol is secure against passive adversaries and the underlying nonce-based encryption scheme is CPA-secure;*
2. *$\mathcal{W}_B^*$ preserves CPA security if $\mathcal{W}_B$ preserves security of the key-agreement protocol; and*
3. *$\mathcal{W}_A^*$ preserves CPA security if the encryption scheme has unique ciphertexts and $\mathcal{W}_A$ preserves semantic security and is exfiltration resistant against Bob.*

Finally, we note that strong security preservation is not possible for this protocol (at least for Alice). (We include a proof sketch in the full version [22].)

Remark 1 (Informal). There is no reverse firewall for Alice in the protocol illustrated in Fig. 7 that maintains functionality and strongly preserves Alice's security.

4.1   Key Agreement Secure Against Passive Adversaries

Theorem 2 motivates the study of unkeyed key-agreement protocols with reverse firewalls that preserve security against passive adversaries. In the classical setting (i.e., without reverse firewalls), the canonical example is the celebrated key-agreement protocol of Diffie and Hellman [19], shown in Fig. 8, whose security follows immediately from the hardness of DDH over the base group G. We use this as an example to illustrate the basic idea of a reverse firewall in the key-agreement setting.

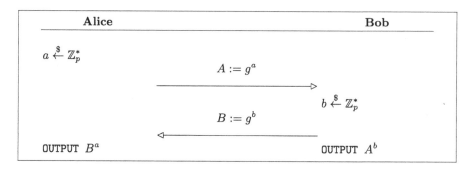

Fig. 8. Diffie-Hellman key agreement over a group G of prime order p with generator g.

Diffie-Hellman key agreement has a simple reverse firewall for Alice, which raises both messages to a single random power, $\alpha \in \mathbb{Z}_p^*$. We present this reverse firewall in Fig. 9. Note that this firewall effectively replaces Alice's message with a uniformly random message. Security then follows from the security of the underlying protocol, since the transcript and resulting key in the two cases are distributed identically.

But, this protocol cannot have a reverse firewall that maintains correctness and preserves security for Bob, as we described in the introduction. In particular, we run the risk that one party has the ability to selectively reject keys.

To solve this problem, we add an additional message to the beginning of the protocol in which Bob commits to the message that he will send later. Of course, in order to permit a secure firewall, the commitment scheme itself must be both malleable (so that the firewall can rerandomize the underlying message that Bob has committed to, mapping a commitment of B to a commitment of B^α) and rerandomizable (so that the randomness used by Bob to commit and open will not leak any information about his message). To achieve our strongest level of security,

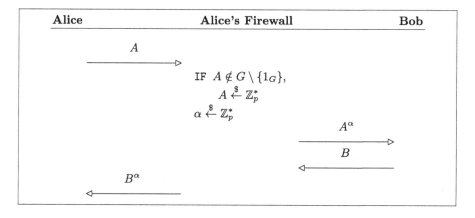

Fig. 9. Reverse firewall for Alice in the protocol from Fig. 8.

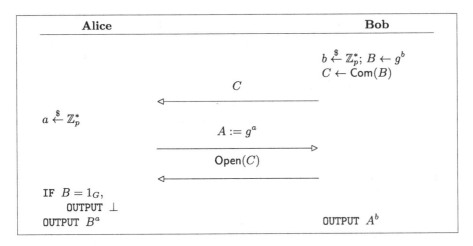

Fig. 10. A variant of Diffie-Hellman key agreement over a group G of prime order p with public generator g. $(\mathsf{Com}, \mathsf{Open})$ is a commitment scheme.

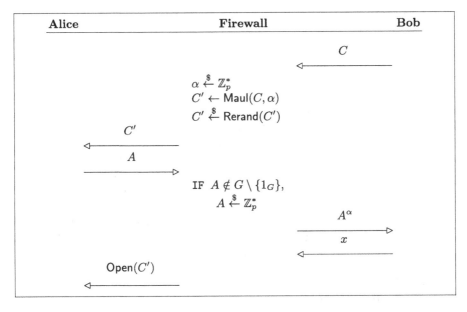

Fig. 11. Reverse firewall for either Alice or Bob in the protocol from Fig. 8. $\mathsf{Maul}(C, \alpha)$ takes a commitment $C = \mathsf{Com}(B)$ and converts it into a commitment of B^α. $\mathsf{Rerand}(C)$ takes a commitment $C = \mathsf{Com}(B)$ and converts it into a uniformly random commitment of B. We assume that a rerandomized and mauled commitment can be opened with access to an opening of the original commitment (and the randomness used in the rerandomization and mauling).

we also need the scheme to be statistically hiding and for each commitment to have a unique opening for a given message. (These requirements are easily met in practice. For example, a simple variant of the Pedersen commitment suffices [33]. For completeness, we present such a scheme in the full version [22].) The protocol is shown in Fig. 10. In Fig. 11, we present a single reverse firewall for this protocol that happens to work for either party. (Each party would need to deploy its own version of this firewall to guarantee its own security. It just happens that each party's firewall would have the same "code.")

In the full version, we prove the following theorem [22].

Theorem 3. *The protocol in Fig. 10 is secure against passive adversaries if DDH is hard in G. The reverse firewall W in Fig. 11 is functionality maintaining. If the commitment scheme is statistically hiding, then W preserves security for Alice and is strongly exfiltration resistant against Bob. If the commitment scheme is computationally binding, then W is exfiltration resistant for Bob against Alice and preserves security for Bob. W also detects failure for both parties.*

5   CCA-security Using Key Agreement

In the setting of the previous section, with no public-key infrastructure, it is trivially impossible to achieve CCA-security. (An adversary can simply "pretend to be Bob" and read Alice's plaintext.) In this section, we show that a CCA-secure message-transmission protocol with reverse firewalls does in fact exist in the publicly keyed setting. In particular, below, we give the CCA analogue of Theorem 2, showing that a key-agreement protocol that is secure against active adversaries and has sufficiently secure reverse firewalls together with a symmetric-key encryption scheme suffices. As in the previous section, this key-agreement-based protocol has the additional benefit that it is efficient, in the sense that it does not apply public-key operations to the plaintext. In Sect. 5.1, we construct a key-agreement protocol that suffices.

We now present our stronger composition theorem. (Recall that Bob's reverse firewall can only preserve CPA security. Such a firewall is already given by Theorem 2, so we do not repeat this here.) See the full version for the proof [22].

Theorem 4 (Composition theorem for CCA security). *Define $\mathcal{W}_A$ and $\mathcal{W}_A^*$ as in Theorem 2. Then,*

1. *the protocol in Fig. 7 is CCA-secure if the underlying key-agreement protocol is secure against active adversaries for Alice and the underlying nonce-based encryption scheme is CCA-secure; and*
2. *$\mathcal{W}_A^*$ preserves CCA-security if the encryption scheme has unique ciphertexts, the key-agreement protocol is authenticated for Bob, and $\mathcal{W}_A$ preserves security for Alice, is exfiltration resistant against Bob with valid transcripts, and detects failure.*

5.1   Key Agreement Secure Against Active Adversaries

Theorem 4 motivates the study of key-agreement protocols with reverse firewalls that preserve security against active adversaries. In the classical setting, the common solution is essentially for each of the parties to sign the transcript of this run of the protocol. However, this solution does not work in our setting because it is important for us that messages *can* be altered without breaking functionality, so that the firewall can rerandomize messages when necessary.

Of course, while we want to allow for the possibility that Alice and Bob see a different transcript but still output a key, we still want them to agree on the key itself. This leads to the idea of signing some deterministic function of the key, so that the signatures can be used to verify that the parties share the same key without necessarily requiring them to share the same transcript. This is the heart of our solution.

We also have to worry that the signatures themselves can provide channels, allowing tampered versions of the parties to leak some information. We solve this by using a unique signature scheme, as defined by [30]. (See [2] for a thorough analysis of signatures in the context of reverse firewalls and corrupted implementations, including alternative ways to implement signatures that would suffice for our purposes.)

Furthermore, in order for our firewall to fail detectably, it has to be able to check the signature itself—so that it can distinguish a valid transcript from an invalid one. So, we would like the parties to sign a deterministic function of g^{ab} that is efficiently computable given only access to g^a and g^b. This leads naturally to the use of a symmetric bilinear map $e : G \times G \to G_T$. The parties then sign $e(g^a, g^b)$. Of course, g^{ab} is no longer indistinguishable from random in the presence of a bilinear map. But, it can be hard to compute. So, we apply a cryptographic hash function H to g^{ab} in order to extract the final key $H(g^{ab})$.

We now provide two definitions to make this precise.

Definition 10 (Unique Signatures). *A* unique signature scheme *is a triple of algorithms* (KeyGen, USig, Ver). KeyGen *takes as input* 1^λ *where* λ *is the security parameter and outputs a public key* pk *and a private key* sk. Sig *takes as input the secret key* sk *and a plaintext* m *and outputs a signature* τ. Ver *takes as input the public key* pk, *a signature* τ *and a message* m *and outputs either* true *or* false. *A signature scheme is* correct *if* $\text{Ver}_{pk}(\text{USig}_{sk}(m), m) = \text{true}$. *It is* unique *if for each plaintext* m *and public key* pk, *there is a unique signature* τ *such that* $\text{Ver}_{pk}(\tau, m) = \text{true}$.

A signature scheme is secure against adaptive chosen-message existential-forgery attacks *if no adversary with access to the public key and a signature oracle can produce a valid signature not returned by the oracle.*

We will need to use a group with a symmetric bilinear map in which the following variant of the computational Diffie-Hellman assumption holds.

Definition 11 (Any-base CDH). *For a group* G *of order* p, *we say that* any-base CDH is hard in G *if no probabilistic polynomial-time adversary taking input*

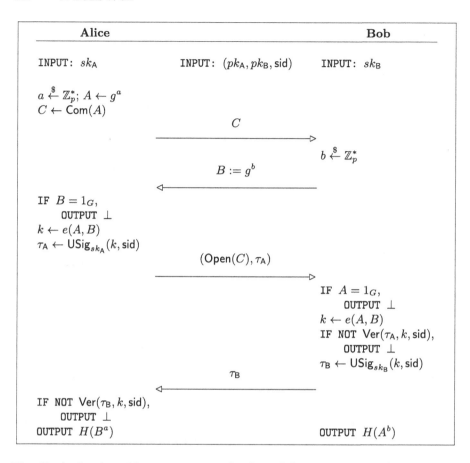

Fig. 12. Authenticated key agreement with a firewall for both parties. USig is a unique signature.

(g, g^a, g^b) where $g \xleftarrow{\$} G$ and $(a, b) \xleftarrow{\$} \mathbb{Z}_p^2$ has non-negligible probability of returning (h^a, h^b, h^{ab}) for some element $h \in G \setminus \{1_G\}$.

We now present our protocol in Fig. 12 with a reverse firewall for both parties in Fig. 13. It requires a unique signature scheme (USig, Ver) with public keys pk_A for Alice and pk_B for Bob and corresponding secret keys sk_A and sk_B respectively, a base group G with generator g in which any-base CDH is hard, a target group G_T, and a non-trivial bilinear map between the two groups $e : G \times G \to G_T$. We also need a function $H : G \to \{0, 1\}^\ell$ for some polynomially large ℓ that extracts hardcore bits from CDH. Presumably a standard cryptographic hash function will work. For simplicity, we model H as a random oracle, but we note that the proof can be modified to apply to any function H such that $(g^a, g^b, H(g^{ab}))$ is indistinguishable from random. (See [27] for a discussion of such functions.) We stress again that this protocol is remarkably efficient, and we think that it can

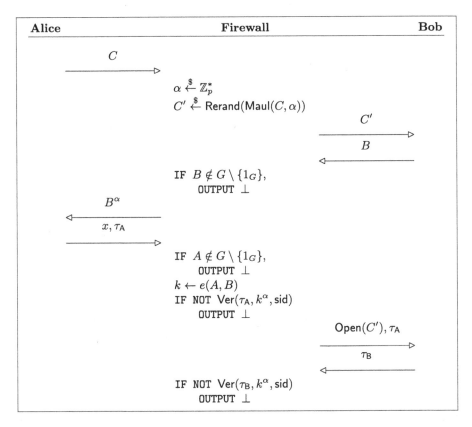

Fig. 13. Reverse firewall for either Alice or Bob in the protocol from Fig. 12. C is a commitment of the group element A. $\mathsf{Maul}(C, \alpha)$ takes a commitment $C = \mathsf{Com}(A)$ and converts it into a commitment of A^α. $\mathsf{Rerand}(C)$ takes a commitment $C = \mathsf{Com}(A)$ and converts it into a uniformly random commitment of A. We assume that a rerandomized and mauled commitment can be opened with access to an opening of the original commitment.

and should be used in practice. The proof of the following theorem is in the full version [22].

Theorem 5. *The protocol shown in Fig. 12 is authenticated for Bob and secure against active adversaries if the signature scheme is secure and any-base CDH is hard in G. The reverse firewall $\mathcal{W}$ shown in Fig. 13 preserves security against active adversaries for Alice, preserves authenticity, is exfiltration resistant for Alice against Bob with valid transcripts, and detects failure for Alice. $\mathcal{W}$ also preserves security against active adversaries for Bob, is exfiltration resistant for Bob against Alice with valid transcripts, and detects failure for Bob.*

References

1. Alwen, J., Shelat, A., Visconti, I.: Collusion-free protocols in the mediated model. In: Wagner, D. (ed.) CRYPTO 2008. LNCS, vol. 5157, pp. 497–514. Springer, Heidelberg (2008)
2. Ateniese, G., Magri, B., Venturi, D.: Subversion-resilient signature schemes. In: CCS (2015)
3. Ball, J., Borger, J., Greenwald, G.: Revealed: how US and UK spy agencies defeat internet privacy and security. Guardian Weekly, September 2013
4. Bellare, M., Desai, A., Pointcheval, D., Rogaway, P.: Relations among notions of security for public-key encryption schemes. In: Krawczyk, H. (ed.) CRYPTO 1998. LNCS, vol. 1462, p. 26. Springer, Heidelberg (1998)
5. Bellare, M., Hoang, V.T.: Resisting randomness subversion: fast deterministic and hedged public-key encryption in the standard model. In: Oswald, E., Fischlin, M. (eds.) EUROCRYPT 2015. LNCS, vol. 9057, pp. 627–656. Springer, Heidelberg (2015)
6. Bellare, M., Jaeger, J., Kane, D.: Mass-surveillance without the state: strongly undetectable algorithm-substitution attacks. In: CCS (2015)
7. Bellare, M., Paterson, K.G., Rogaway, P.: Security of symmetric encryption against mass surveillance. In: Garay, J.A., Gennaro, R. (eds.) CRYPTO 2014, Part I. LNCS, vol. 8616, pp. 1–19. Springer, Heidelberg (2014). Full version: [8]
8. Bellare, M., Paterson, K.G., Rogaway, P.: Security of symmetric encryption against mass surveillance. Cryptology ePrint Archive, report 2014/438 (2014). http://eprint.iacr.org/
9. Blaze, M., Bleumer, G., Strauss, M.J.: Divertible protocols and atomic proxy cryptography. In: Nyberg, K. (ed.) EUROCRYPT 1998. LNCS, vol. 1403, pp. 127–144. Springer, Heidelberg (1998)
10. Burmester, M., Desmedt, Y.G.: All Languages in NP have divertible zero-knowledge proofs and arguments under cryptographic assumptions. In: Damgård, I.B. (ed.) EUROCRYPT 1990. LNCS, vol. 473, pp. 1–10. Springer, Heidelberg (1991)
11. Canetti, R., Krawczyk, H., Nielsen, J.B.: Relaxing chosen-ciphertext security. In: Boneh, D. (ed.) CRYPTO 2003. LNCS, vol. 2729, pp. 565–582. Springer, Heidelberg (2003)
12. Vulnerability summary for CVE-2014-1260 ('Heartbleed'), April 2014. http://cve.mitre.org/cgi-bin/cvename.cgi?name=CVE-2014-1260
13. Vulnerability summary for CVE-2014-1266 ('goto fail'), February 2014. http://cve.mitre.org/cgi-bin/cvename.cgi?name=CVE-2014-1266
14. Vulnerability summary for CVE-2014-6271 ('Shellshock'), September 2014. http://cve.mitre.org/cgi-bin/cvename.cgi?name=CVE-2014-6271
15. Degabriele, J.P., Farshim, P., Poettering, B.: A more cautious approach to security against mass surveillance. In: Leander, G. (ed.) FSE 2015. LNCS, vol. 9054, pp. 579–598. Springer, Heidelberg (2015)
16. Desmedt, Y.G.: Abuses in cryptography and how to fight them. In: Goldwasser, S. (ed.) CRYPTO 1988. LNCS, vol. 403, pp. 375–389. Springer, Heidelberg (1990)
17. Desmedt, Y.: Subliminal-free sharing schemes. In: Information Theory (1994)
18. Desmedt, Y.G., Pieprzyk, J., Steinfeld, R., Wang, H.: A non-malleable group key exchange protocol robust against active insiders. In: Katsikas, S.K., López, J., Backes, M., Gritzalis, S., Preneel, B. (eds.) ISC 2006. LNCS, vol. 4176, pp. 459–475. Springer, Heidelberg (2006)

19. Diffie, W., Hellman, M.: New directions in cryptography. IEEE Trans. Inf. Theor. **22**(6), 644–654 (1976)
20. Dodis, Y., Fiore, D.: Interactive encryption and message authentication. In: Abdalla, M., De Prisco, R. (eds.) SCN 2014. LNCS, vol. 8642, pp. 494–513. Springer, Heidelberg (2014)
21. Dodis, Y., Ganesh, C., Golovnev, A., Juels, A., Ristenpart, T.: A formal treatment of backdoored pseudorandom generators. In: Oswald, E., Fischlin, M. (eds.) EUROCRYPT 2015. LNCS, vol. 9056, pp. 101–126. Springer, Heidelberg (2015)
22. Dodis, Y., Mironov, I., Stephens-Davidowitz, N.: Message transmission with reverse firewalls–secure communication on corrupted machines. Cryptology ePrint Archive, report 2015/548 (2015). http://eprint.iacr.org/2015/548
23. Greenwald, G.: No Place to Hide: Edward Snowden the N.S.A. and the U.S. Surveillance State. Metropolitan Books, New York (2014)
24. Groth, J.: Rerandomizable and replayable adaptive chosen ciphertext attack secure cryptosystems. In: Naor, M. (ed.) TCC 2004. LNCS, vol. 2951, pp. 152–170. Springer, Heidelberg (2004)
25. Juniper vulnerability (2015). https://kb.juniper.net/InfoCenter/index?page=content&id=JSA10713
26. Katz, J., Shin, J.S.: Modeling insider attacks on group key-exchange protocols. In: CCS (2005)
27. Kiltz, E.: Chosen-ciphertext secure key-encapsulation based on Gap Hashed Diffie-Hellman. In: PKC (2007)
28. Lenstra, A.K., Hughes, J.P., Augier, M., Bos, J.W., Kleinjung, T., Wachter, C.: Public keys. In: Safavi-Naini, R., Canetti, R. (eds.) CRYPTO 2012. LNCS, vol. 7417, pp. 626–642. Springer, Heidelberg (2012)
29. Lepinksi, M., Micali, S., Shelat, A.: Collusion-free protocols. In: STOC (2005)
30. Micali, S., Rabin, M., Vadhan, S.: Verifiable random functions. In: FOCS (1999)
31. Mironov, I., Stephens-Davidowitz, N.: Cryptographic reverse firewalls. In: Oswald, E., Fischlin, M. (eds.) EUROCRYPT 2015. LNCS, vol. 9057, pp. 657–686. Springer, Heidelberg (2015)
32. Okamoto, T., Ohta, K.: Divertible zero knowledge interactive proofs and commutative random self-reducibility. In: Quisquater, J.-J., Vandewalle, J. (eds.) EURO-CRYPT 1989. LNCS, vol. 434, pp. 134–149. Springer, Heidelberg (1990)
33. Pedersen, T.P.: Non-interactive and information-theoretic secure verifiable secret sharing. In: Feigenbaum, J. (ed.) CRYPTO 1991. LNCS, vol. 576, pp. 129–140. Springer, Heidelberg (1992)
34. Perlroth, N., Larson, J., Shane, S.: National Security Agency able to foil basic safeguards of privacy on Web. The New York Times, September 2013
35. Pieprzyk, J., Wang, H.: Key control in multi-party key agreement protocols. In: Workshop on Coding, Cryptography and Combinatorics (2003)
36. Prabhakaran, M., Rosulek, M.: Rerandomizable RCCA encryption. In: Menezes, A. (ed.) CRYPTO 2007. LNCS, vol. 4622, pp. 517–534. Springer, Heidelberg (2007)
37. Russell, A., Tang, Q., Yung, M., Zhou, H.-S.: Cliptography: clipping the power of kleptographic attacks. Cryptology ePrint Archive, report 2015/695 (2015). https://eprint.iacr.org/2015/695
38. Schneier, B., Fredrikson, M., Kohno, T., Ristenpart, T.: Surreptitiously weakening cryptographic systems. Technical report, IACR Cryptology ePrint Archive, 2015: 97 (2015). http://eprint.iacr.org/2015/97
39. Simmons, G.: The prisoners' problem and the subliminal channel. In: Chaum, D. (ed.) CRYPTO 1983, pp. 51–67. Springer, New York (1984)
40. https://www.us-cert.gov/ncas/alerts/TA15-051A. February 2015

A   The symmetric-key setting

Here, we consider the setting in which Alice and Bob share a private key. We observe that a one-round protocol due to Bellare, Paterson, and Rogaway provides a solution that does not even need a reverse firewall [7]. We also use this scheme elsewhere to build protocols that do not rely on shared private keys. We first define nonce-based encryption.

Definition 12. (Nonce-based encryption). *A* nonce-based symmetric-key encryption scheme *is a pair of deterministic algorithms* (Enc, Dec). Enc *takes as input a key from a key space* $\mathcal{K}$, *a nonce from a nonce space* $\mathcal{N}$, *and a plaintext from a plaintext space* $\mathcal{M}$ *and outputs a ciphertext from a ciphertext space* $\mathcal{C}$. Dec *takes as input a key, a nonce, and a ciphertext and returns a plaintext or the special symbol* $\perp$. *The scheme is* correct *if for any key* sk, *nonce* r, *and plaintext* m, $\mathsf{Dec}(r, \mathsf{Enc}_{sk}(r, m)) = m$.

Such a scheme is CPA *secure if no probabilistic polynomial-time adversary can distinguish between* $\mathsf{Enc}_{sk}(r^*, m_0)$ *and* $\mathsf{Enc}_{sk}(r^*, m_1)$ *with non-negligible advantage where* r^*, m_0, *and* m_1 *are adversarially chosen when given access to an encryption oracle that outputs* $\perp$ *unless given a unique nonce* r. *It is* CCA-secure *if no probabilistic polynomial-time has non-negligible advantage when also given access to a decryption oracle that outputs* $\perp$ *if* $r = r^*$.

Such a scheme (Enc, Dec) *has* unique ciphertexts *if for any key* sk, *message* m, *and nonce* r, *there is exactly one ciphertext* c *such that* $\mathsf{Dec}(r, c) = m$.

Theorem 6. *Let* (Enc, Dec) *be a nonce-based symmetric-key encryption scheme. Then, if the encryption scheme is CPA-secure (resp. CCA-secure) the one-round protocol in which Alice sends* $\mathsf{Enc}_{sk}(\mathsf{sid}, m)$ *and Bob returns* $\mathsf{Dec}_{sk}(\mathsf{sid}, m)$ *is a CPA-secure (resp. CCA-secure) one-round privately keyed message-transmission protocol without forward secrecy. If the encryption scheme has unique ciphertexts, then the "trivial" reverse firewall that simply passes Alice's messages to Bob unchanged preserves security and is exfiltration resistant against Bob.*

See, e.g., [7] for formal analysis and the construction of such a scheme. The key thing to note from our perspective is that, as Bellare et al. observe, the fact that the encryption scheme has unique ciphertexts implies that any tampered version of Alice that maintains functionality necessarily behaves identically to honest Alice. The next theorem shows that we essentially cannot do any better for a one-round protocol without using public-key primitives. The proof is in the full version.

Theorem 7. *There is a black-box reduction from semantically secure public-key encryption to CPA-secure symmetric-key encryption with at least four possible plaintexts and a reverse firewall that strongly preserves CPA security.*

Of course, the primary drawbacks of this approach are that it requires Alice and Bob to share a secret key and that it does not offer forward secrecy.

B   Less efficient one-round protocols

Finally, we show one-round protocols in the singly keyed setting, in which Bob has a public-key/private-key pair, but Alice does not. These are essentially the single-round analogue of the two-round protocol presented in Fig. 4 in Sect. 3. They can also be thought of as the natural extension of public-key encryption schemes to the reverse firewall setting. (In particular, these protocols are not efficient, in the sense that they use public-key operations on the plaintext, which may be very long.) Indeed, we show that the existence of such a protocol is equivalent to the existence of rerandomizable encryption, and we show how to achieve CCA-security (though not forward secrecy).

B.1   One-round CPA-secure protocols

The next theorem shows that one-round CPA-secure protocols with reverse firewalls are equivalent to rerandomizable public-key encryption. The proof is in the full version [22].

Theorem 8. *Any (strongly) rerandomizable semantically secure public-key encryption scheme implies a one-round CPA-secure singly keyed message-transmission protocol without forward security with a reverse firewall that (strongly) preserves security and (strongly) resists exfiltration. Conversely, any one-round CPA-secure message-transmission protocol with a reverse firewall that (strongly) preserves security implies a (strongly) rerandomizable semantically secure public-key encryption scheme.*

B.2   A one-round CCA-secure protocol

To extend this idea to stronger notions of security, we need the underlying encryption scheme to satisfy stronger notions of security. A natural candidate is CCA security. However, CCA-secure encryption schemes cannot be rerandomizable, so we need a slightly weaker notion. RCCA security, as defined by [11], suffices, and rerandomizable RCCA-secure schemes do exist (see, e.g., [24,36]), though they are relatively inefficient. (They are not *strongly* rerandomizable; their rerandomization procedures do not work on invalid ciphertexts.) We present the RCCA security game in Fig. 14. In addition, we need a rerandomized ciphertext to be indistinguishable from a valid encryption *even with access to a decryption oracle.* Figure 15 and the definition below makes this precise.

Definition 13. *An encryption scheme is* RCCA *secure if no probabilistic polynomial-time adversary $\mathcal{E}$ has non-negligible advantage in the game presented in Fig. 14. It is* RCCA *rerandomizable if there exists an algorithm* Rerand *with access to the public key such that for any ciphertext c with $\mathsf{Dec}(c) \neq \bot$, $\mathsf{Dec}(\mathsf{Rerand}(c)) = \mathsf{Dec}(c)$ and no probabilistic polynomial-time adversary $\mathcal{E}$ has non-negligible advantage in the game presented in Fig. 15 when we require that $\mathsf{Dec}(c_i) \neq \bot$. It is* strongly RCCA-rerandomizable *if the previous statement holds even if $\mathsf{Dec}(c_i) = \bot$.*

proc. IND-RCCA(λ)

$(pk, sk) \xleftarrow{\$} \mathsf{KeyGen}(1^\lambda)$
$(m_0, m_1) \leftarrow \mathcal{E}^{\mathcal{O}_1}(pk)$
$b \xleftarrow{\$} \{0, 1\}; C^* \xleftarrow{\$} \mathsf{Enc}_{pk}(m_b)$
$b^* \leftarrow \mathcal{E}^{\mathcal{O}_2}(\sigma, c^*)$
OUTPUT $(b = b^*)$

proc. $\mathcal{O}_1(c)$
OUTPUT $\mathsf{Dec}_{sk}(c)$

proc. $\mathcal{O}_2(c)$
$m \leftarrow \mathsf{Dec}_{sk}(c)$
IF $m = m_0$ OR $m = m_1$,
    OUTPUT Challenge
ELSE,
    OUTPUT m

Fig. 14. The RCCA security game.

proc. IND-RCCA(λ)

$(pk, sk) \xleftarrow{\$} \mathsf{KeyGen}(1^\lambda)$
$(c_0, c_1) \leftarrow \mathcal{E}^{\mathcal{O}_1}(pk)$
$b \xleftarrow{\$} \{0, 1\}; c^* \xleftarrow{\$} \mathsf{Rerand}_{pk}(c_b)$
$b^* \leftarrow \mathcal{E}^{\mathcal{O}_2}(c^*)$
OUTPUT $(b = b^*)$

proc. $\mathcal{O}_1(c)$
OUTPUT $\mathsf{Dec}_{sk}(c)$

proc. $\mathcal{O}_2(c)$
$m \leftarrow \mathsf{Dec}_{sk}(c)$
IF $m = \mathsf{Dec}_{sk}(c_0)$ OR $m = \mathsf{Dec}_{sk}(c_1)$,
    OUTPUT Challenge
ELSE,
    OUTPUT m

Fig. 15. The RCCA rerandomization game.

The below theorem is the CCA analogue of Theorem 8. The proof is in the full version [22].

Theorem 9. *Any (strongly) RCCA-rerandomizable, RCCA-secure encryption scheme implies a one-round CCA-secure singly keyed message-transmission protocol without forward security with a reverse firewall that (strongly) preserves security and (strongly) resists exfiltration.*

Big-Key Symmetric Encryption: Resisting Key Exfiltration

Mihir Bellare[1]([⊠]), Daniel Kane[1], and Phillip Rogaway[2]

[1] Department of Computer Science and Engineering,
University of California, San Diego, USA
`mihir@eng.ucsd.edu`
[2] Department of Computer Science, University of California, Davis, USA
`http://cseweb.ucsd.edu/~mihir/`
`http://cseweb.ucsd.edu/~dakane/`
`http://web.cs.ucdavis.edu/~rogaway/`

Abstract. This paper aims to move research in the bounded retrieval model (BRM) from theory to practice by considering symmetric (rather than public-key) encryption, giving efficient schemes, and providing security analyses with sharp, concrete bounds. The threat addressed is malware that aims to exfiltrate a user's key. Our schemes aim to thwart this by using an enormously long key, yet paying for this almost exclusively in storage cost, not speed. Our main result is a general-purpose lemma, the *subkey prediction lemma*, that gives a very good bound on an adversary's ability to guess a (modest length) subkey of a big-key, the subkey consisting of the bits of the big-key found at random, specified locations, after the adversary has exfiltrated partial information about the big-key (e.g., half as many bits as the big-key is long). We then use this to design a new kind of key encapsulation mechanism, and, finally, a symmetric encryption scheme. Both are in the random-oracle model. We also give a less efficient standard-model scheme that is based on universal computational extractors (UCE). Finally, we define and achieve hedged BRM symmetric encryption, which provides authenticity in the absence of leakage.

1 Introduction

This paper is concerned with the possibility of mass surveillance by APTs. An APT (Advanced Persistent Threat) is malware that resides on your system and attempts to exfiltrate your key. (This means that it aims to communicate your key to its home base, probably using your system's network connection.) How might one protect against this? One answer is: by strengthening system security to the point that we eliminate APTs. Unfortunately, this approach seems out of reach. Indeed, the Snowden revelations show that the NSA (through TAO, their Tailored Access Operations unit) and others have sophisticated system penetration capabilities that they use to plant APTs. Another answer is provided by the bounded retrieval model (BRM) [2,3,17,20,23], namely to make secret keys so big that their undetected exfiltration is difficult.

© International Association for Cryptologic Research 2016
M. Robshaw and J. Katz (Eds.): CRYPTO 2016, Part I, LNCS 9814, pp. 373–402, 2016.
DOI: 10.1007/978-3-662-53018-4_14

So far, BRM research has been largely theoretical and foundational. Our intent is to move it towards being a plausible countermeasure to mass surveillance. This involves the following. First, we treat symmetric rather than asymmetric encryption. Second, we focus on simple, efficient schemes. Third, we provide security analyses that are strong and fully concrete (no hidden constants), giving good numerical bounds on security.

Our main technical contribution is a very good upper bound on the probability of predicting a subset of random positions in a large key in the presence of leakage. We then give a big-key encapsulation mechanism, and thence a big-key symmetric encryption scheme, both efficient and in the random-oracle model (ROM) [12]. Let us now look at all this in more detail.

THE BRM. The BRM evolved through a series of works [2,3,17,20,23]. It is part of the broader area of Leakage-Resilient Cryptography [1,25,34] and is also related to the Bounded Storage Model [16,29]. A survey by Alwen et al. [4] explains that

> If an attacker hacks into a remote system (or infects it with some malware) it may ... be infeasible/impractical for the attacker to download "too much" data (say, more than 10 GBytes).

In such a setting, making the key very big, say 1 TByte, ensures that the adversary obtains limited information about it. The idea was echoed by Adi Shamir at the RSA 2013 conference in a somewhat broader context of secrets that are not necessarily keys, and with specific reference to APTs. He said:

> We have to think in a totally different way about how we are going to protect computer systems assuming there are APTs inside already which cannot be detected. Is everything lost? I claim that not: there are many things that you can do, because the APT is basically going to have a very, very narrow pipeline to the outside world. ... I would like, for example, all the small data to become big data, just in terms of size. I want that the secret of the Coco-Cola company to be kept not in a tiny file of one kilobyte, which can be exfiltrated easily by an APT ····. I want that file to be a terabyte, which cannot be [easily] exfiltrated.

The problem we aim to solve is how to effectively utilize a key K whose length k is big in the presence of an adversary that has some information about K. In the BRM, the APT is modeled as a function that takes K as input and returns a string L of $\ell < k$ bits, the *leakage*, where ℓ is some parameter; for example, $\ell = k/10$ corresponding to the assumption that the adversary can't exfiltrate more than 100 GBytes of information about a 1 TByte key. In the BRM, effective utilization imposes two requirements, one on security and the other on efficiency. The former is that security must be maintained in the presence of the leakage. The latter, called *locality*, is that the scheme's algorithms may access only a very small part of the key. Without this, working with a big-key would be too inefficient.

ADW [3] give BRM schemes for authenticated key-exchange and public-key identification. ADNSWW [2] give BRM schemes for public-key encryption. (Here K is the secret decryption key.) The latter construction is based on identity-based hash proof systems, which are instantiated via bilinear maps, lattices and quadratic residuosity.

OVERVIEW. We treat symmetric encryption in the BRM. We refer to it, synonymously, as big-key symmetric encryption, to emphasize the use of a large key. The k-bit key K of a big-key symmetric encryption scheme SE is shared between sender and receiver. Algorithm SE.Enc maps K and a plaintext to a ciphertext, while SE.Dec maps K and a ciphertext to a plaintext, both making few accesses to K.

Our scheme is simple, efficient, and easily described. It is parameterized by the number of probes p into the k-bit key K, for example, $p = 500$. To encrypt message M, pick a random R, apply the random oracle RO to it to get a sequence of probes $\mathbf{p}[1], \ldots, \mathbf{p}[p] \in [1..k]$ into K, and let $J = K[\mathbf{p}[1]] \ldots K[\mathbf{p}[p]]$ be the corresponding bits of K. Next, obtain a short, conventional key K by applying RO to J. Finally, encrypt M under K with a conventional symmetric encryption scheme to get a ciphertext C. Return (R, C) as the ciphertext of the big-key scheme.

This scheme is derived via a modular framework with three steps, each involving its own definition, problem and analysis, namely *subkey prediction*, *encapsulation* and *encryption*. We will discuss them in turn below. Then we describe two extensions of the basic scheme sketched above, namely, a standard-model variant based on UCE, and the idea of *hedged* big-key encryption.

SUBKEY PREDICTION. Our core technical contribution concerns the *subkey-prediction problem*. We consider a game parameterized by the length k of the big-key, a number p of random probes into it, and a bound ℓ on the leakage. The game selects a k-bit big-key K at random, and the adversary is given the result $L \in \{0,1\}^\ell$ of applying an arbitrary leakage function to K. This L represents the exfiltrated information, and, as a special case, could consist of ℓ bits of K. Now the game picks random indices $\mathbf{p}[1], \ldots, \mathbf{p}[p] \in [1..k]$, called probes. These probes are also given to the adversary. We ask the adversary, given the leakage and probes, to predict, in its entirety, the induced p-bit *subkey* $J = K[\mathbf{p}[1]]K[\mathbf{p}[2]] \cdots K[\mathbf{p}[p]]$ found at those spots. We ask: how well can the adversary do at this? That is, we want to know the adversary's maximum probability, denoted $\mathsf{Adv}^{\mathrm{skp}}_{k,p,1}(\ell)$ in our formalization of Sect. 3, of guessing J as a function of k, ℓ, and p.

The analysis turns out to be surprisingly technical. One might think that there is no better strategy than to leak some ℓ bits of K, for example the first ℓ, and then predict J in the obvious way. (If $\mathbf{p}[i] \in [1..\ell]$ then $K[\mathbf{p}[i]]$ is known from the leakage, else guess it.) We give a counter-example showing that this is not the best strategy, and one can do better using an error-correcting code. We then show that, roughly, the best strategy for the adversary is to select the leakage function so that the pre-images of any point under this function are sandwiched between adjacent Hamming balls. In Sect. 3 we formalize and prove this and then use it to show that

$$\mathsf{Adv}^{\mathrm{skp}}_{k,p,1}(\ell) \approx 2^{-p \cdot w(\ell/k)} \quad \text{for} \ \ w(\lambda) = -\lg(1 - H_2^{-1}(1 - \lambda)),$$

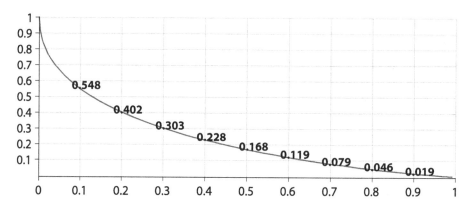

Fig. 1. Subkey predictability. The x-axis indicates the fraction $\lambda = \ell/k$ of the bits adversarially exfiltrated from the big-key. The corresponding point $w(\lambda)$ on the y-axis then indicates how many bits of unpredictability are achieved from each random probe (e.g., $w(0.5) \approx 0.168$). In particular, the adversary's ability to guess the contents of a p-bit probe are about $2^{-p \cdot w(\lambda)}$. Results apply to large k and modest p.

where $H_2(x) = -x \lg(x) - (1-x) \lg(1-x)$ is the binary entropy function and $H_2^{-1}(1-\lambda)$ is the smaller of the two possible inverses of $1-\lambda$ under H_2. The salient point is that the probability decreases exponentially in the number of probes p, with the factor in the exponent depending on the fraction $\lambda = \ell/k$ of the bits of the big-key that are leaked. See Fig. 1 for a plot of $w(\lambda)$ as a function of λ.

Related settings are analyzed by several lemmas in the literature, notably NZ [30, Lemma 11], Vadhan [35, Lemma 9] and ADW [3, Lemma A.3]. We are not aware of any direct way of applying the first two to get bounds on subkey prediction probability. (They do give bounds on what in Sect. 3 we call the restricted subkey prediction probability.) They also involve hidden constants that make it hard to obtain the concrete bounds needed to estimate security in usage. In contrast, the elegant lemma of ADW [3, Lemma A.3] can be directly applied to bound $\mathsf{Adv}_{k,p,1}^{\mathrm{skp}}(\ell)$, and it gives a concrete bound with no hidden constants. However, the bound obtained in this way is much inferior to ours, as we now illustrate.

In Sect. 3 we show that [3, Lemma A.3] implies $\mathsf{Adv}_{k,p,1}^{\mathrm{skp}}(\ell) \leq 2^{-c}$ for c given by Eq. (8), namely $c = p(k-\ell-5)/(2k \lg(2k)+3p)$. Figure 2 compares the bounds obtained by our result (column "New") with the ones obtained via [3, Lemma A.3] (column "Old"). We see for example that for k being 1 TB and $\ell/k = 0.1$, for 500 probes, we show that $\mathsf{Adv}_{k,p,1}^{\mathrm{skp}}(\ell) \approx 2^{-274}$ while the prior bound would be only $2^{-5.1}$. Other entries show similarly large gaps for other parameter values. Another way to compare is, for a certain fixed k, ℓ, to ask how many probes are needed to get 256 bits of security, meaning have $\mathsf{Adv}_{k,p,1}^{\mathrm{skp}}(\ell) \leq 2^{-256}$. According to Fig. 2, for k being 1 TB and $\ell/k = 0.1$, our result says that 468 probes suffice, while the prior result would require us to use 24954 probes. The difference in

p	New	Old
250	137	3.3
500	274	6.6
1000	548	13
234	**128**	3.1
468	**256**	6.2
9642	5284	**128**
19284	10567	**256**

1 GB, 10% leak
$(k = 8 \cdot 10^9,\ \ell = 0.1k)$

p	New	Old
250	42	1.8
500	84	3.7
1000	168	7.4
762	**128**	5.6
1523	**256**	11
17536	2919	**128**
34711	5837	**256**

1 GB, 50% leak
$(k = 8 \cdot 10^9,\ \ell = 0.5k)$

p	New	Old
250	137	2.6
500	274	5.1
1000	548	10
234	**128**	2.4
468	**256**	4.8
12477	6837	**128**
24954	13674	**256**

1 TB, 10% leak
$(k = 8 \cdot 10^{12},\ \ell = 0.5k)$

p	New	Old
250	42	1.4
500	84	2.8
1000	168	5.7
762	**128**	4.3
1523	**256**	8.7
22458	3777	**128**
44916	7553	**256**

1 TB, 50% leak
$(k = 8 \cdot 10^{12},\ \ell = 0.5k)$

Fig. 2. Numerical examples, comparisons with ADW [3, Lemma A.3]. The "New" and "Old" columns show approximate values of x for which $\mathsf{Adv}^{\mathrm{skp}}_{k,p,1}(\ell) \leq 2^{-x}$ using our results and those of ADW [3, Lemma A.3], respectively. The first two tables use a key size k of 1 GB; the rest, 1 TB. In each case, we consider leakage restricted to 10 % or 50 % of the key length. The first column gives the number of probes p. In each table, the first three rows represent natural probe counts $p \in \{250, 500, 1000\}$ while the remaining four rows are determined by asking how many probes would be needed to get either 128-bit or 256-bit security according to each of the bounds.

efficiency is dramatic, meaning that reliance on the prior bounds would translate to a significant loss of practical efficiency for big-key symmetric encryption.

ENCAPSULATION. Building on the above, we provide a general tool, XKEY, for using big-keys in symmetric settings. XKEY takes in a key K and a random *selector* R, which is a short string (like 128–256 bits). It returns a *derived key* $K = \mathsf{XKEY}(K, R)$, which has conventional length (like 128 bits) and can be used in a conventional scheme. We formalize the goal of XKEY, which we call *big-key encapsulation*. It asks that the derived key is indistinguishable from a random string of the same length, even given the selector and leakage on the big-key. This is reminiscent of a classical key-encapsulation mechanism (KEM) as defined by CS [18], yet it is also very different, since we are in the presence of leakage and in the symmetric (rather than asymmetric) setting. Additionally, in the ROM, not only does the adversary have access to the random oracle RO, but, also, *the leakage function can also itself invoke the random oracle*. This is crucial, as otherwise it is easy to give an example of a scheme that is secure in the ROM yet insecure when the random oracle is instantiated. This element increases the technical difficulty of our security proof.

Given K and R, our XKEY algorithm applies the random oracle RO to R to specify probes $\mathbf{p}[1], \ldots, \mathbf{p}[p] \in [1..k]$ into the big-key K. It lets $J = K[\mathbf{p}[1]] \, K[\mathbf{p}[2]] \cdots K[\mathbf{p}[p]]$ be the corresponding subkey. By subkey unpredictability, J is unpredictable, but it is not guaranteed to be indistinguishable from random. XKEY further applies RO to $R\|J$ to obtain the derived key K. Theorem 12 says that this derived key is indistinguishable from random even to an adversary that sees multiple encapsulations and gets leakage about K. The theorem gives a concrete bound on the adversary advantage. The proof has two steps, first addressing the ability of the leakage function to use the random oracle by a coin-fixing argument, and then reducing to subkey prediction via a game sequence.

BIG-KEY ENCRYPTION. In Sect. 5, we define and achieve big-key symmetric encryption. Our definition ensures indistinguishability in the presence of leakage on the big-key, the leakage again allowed to depend on the random oracle. To encrypt a message M under $\boldsymbol{K}$, we pick R at random, obtain a session key $K = \mathsf{XKEY}(\boldsymbol{K}, R)$, and output as ciphertext a pair (R, C) where C is an encryption of M under K with a base, conventional symmetric encryption scheme, for example an AES mode of operation. Theorem 13 shows that this achieves our definition of big-key encryption privacy assuming that XKEY achieves our definition of encapsulation security and the base scheme meets a standard privacy definition for symmetric encryption. The scheme is very efficient. Relative to the base scheme, the added communications cost is small (transmission of R) and the added computation cost is also small (one XKEY operation).

STANDARD-MODEL SCHEME. A variant of our scheme, still quite efficient, can be proven secure in the standard model. We can focus on encapsulation, since our reduction of encryption to the latter does not use a random oracle. We consider a variant $\mathsf{XKEY2}$ of XKEY where the selector $R = (I, \mathbf{p})$ is a key I for a UCE (Universal Computation Extractor) H [8] together with a sequence of probes $\mathbf{p}[1], \ldots, \mathbf{p}[p] \in [1..k]$ into $\boldsymbol{K}$. Given $\boldsymbol{K}$ and this selector, $\mathsf{XKEY2}$ lets $J = \boldsymbol{K}[\mathbf{p}[1]] \ldots \boldsymbol{K}[\mathbf{p}[p]]$ be the corresponding bits of $\boldsymbol{K}$ and obtains subkey K by applying $\mathsf{H}(I, \cdot)$ to J. Efficiency is the same as for our ROM scheme, but the ciphertext of the encryption scheme is longer because the selector, which is included in the ciphertext, is longer. Theorem 14 proves security assuming H is $\mathsf{UCE}[\mathcal{S}^{\mathrm{sup}}]$-secure, namely UCE-secure for statistically unpredictable sources. This version of UCE, from [8,15], has been viable and has been used in many applications. We use our subkey unpredictability bound in a crucial way, to prove statistical unpredictability of the source constructed in the reduction.

AUTHENTICITY AND HEDGED BIG-KEY ENCRYPTION. Our big-key encryption schemes provide privacy. What about authenticity—meaning big-key authenticated encryption (AE)? ADW [3] remark that secure signatures are not possible in the BRM. The same attack applies to rule out big-key AE. Namely, the leakage function can simply compute and leak a valid ciphertext. We take the view that authenticity is important in normal usage but the target of mass surveillance is violating privacy, not authenticity. Accordingly, we suggest *hedged* big-key encryption. Encryption would continue to provide, in the presence of leakage, the guarantees of our above-discussed schemes. Additionally, in the absence of leakage, the same scheme should provide AE, meeting a standard and strong formalizations of the latter [10]. Throughout all this, the scheme must remain true to the local efficiency requirement of the BRM. In Sect. 7 we give a simple way to turn a privacy-only big-key scheme into a hedged one while preserving locality.

DISCUSSION. We clarify some assumptions and limitations of the BRM and our work. In the BRM, leakage (exfiltration) on the big-key is assumed to occur once, at the beginning. The leakage function cannot depend on ciphertexts. This is true in most models of leakage-resilient cryptography, but leakage after encryption has been considered [26] and it would be interesting to extend this to big-key symmetric encryption. We assume encryption code is trusted. Algorithm substitution attacks [11] consider the case of untrusted encryption code. Whether any defense against ASAs is possible in the big-key setting remains open. Finally we assume the availability of trusted randomness in the encryption process.

Our schemes view the big-key as a string over $\{0,1\}$, so each probe draws one bit of the big-key. More generally, we could view the big-key as a vector over $\{0,1\}^b$ where b is some block or word length, for example $b = 8$ or $b = 32$. Each probe would then result in a b-bit string. This could increase efficiency and ease of implementation of the scheme. Our current subkey prediction lemma addresses only the $b = 1$ case. One can apply [3, Lemma A.3] to get a bound for larger b, but we would expect that an extension of our subkey prediction Lemma would yield better bounds. Obtaining such an extension is an interesting open question.

RELATED WORK. In Maurer's bounded storage model [29], parties have access to a public source of randomness that transmits a sequence $X_1, X_2, \ldots$ of high min-entropy strings. Parties are limited in storage and the goal is information theoretic security. Symmetric encryption in this setting is studied in [5,6,24,28, 29,35]. However, in this information-theoretic setting one can derive only one session key from each output of the source and the ability to encrypt multiple messages relies on the expectation of receiving a continuous stream of strings from the source. In contrast, in the BRM setting, the big-key is static, and, in the presence of leakage on it, we want to encrypt an arbitrary number of messages without changing the key.

Di Crescenzo et al. [20] and Dziembowski [23] independently introduced the bounded retrieval model (BRM), where the adversary has a bounded amount of information on the data stored by the users. See the excellent survey of Alwen et al. [4] for history and results in this setting. Here we touch on only a few examples. DLW [20] design password protocols for the setting where the "password file" stored by the server is huge but the amount of information the adversary can get about it is limited, yet the server is efficient. Dziembowski [23] considers malicious code that can exfiltrate only a limited portion of a long key before it is sanitized. Dziembowski's aim is to achieve entity authentication and session-key distribution. Like us, the author works in the ROM. His symmetric key-derivation scheme, and its analysis, are similar to ours. Following up on this work, CDDLLW [17] provide a general paradigm for achieving intrusion-resilient authenticated key exchange (AKE), as well as a solution in the standard (as opposed to RO) model. Alwen et al. [3] design authenticated key agreement protocols in the public-key setting where the secret key is huge but the public key is small and security must be maintained in the presence of bounded leakage on the secret key. ADNSWW [2] construct public-key encryption schemes in this model.

Predating the BRM, Kelsey and Schneier consider an authentication scheme in which a user with a large-memory token authenticates itself by providing XORs of randomly specified subsets of its bits [27]. An adversary who manages to exfiltrate only some of the bits of the device will be unable to subsequently impersonate the token. Dagon, Lee, and Lipton consider the problem of securely storing a ciphertext encrypted in a weak password on a device that's subject to adversarial attack [19]. They create a long ciphertext for a short plaintext where partial knowledge of the ciphertext will frustrate dictionary attacks.

Lu [28] and Vadhan [35] construct locally computable extractors. These yield big-key encapsulation schemes, but with the limitation that one can only obtain a small, bounded number of encapsulated keys from one big-key. (Encapsulated keys are statistically close to random, so after a few derivations, the entropy of the big-key is exhausted.) This is not sufficient for big-key symmetric encryption, where, with the big-key in place, we want to encrypt an arbitrary number of messages. XKEY in this light yields a locally computable *computational* extractor in the ROM. (The computational element is that the number of queries to the random oracle is limited. In asymptotic terms, it is a polynomial.) It uses the random oracle as a hardcore function following [12] to be able to encapsulate an unbounded number of keys under a single big-key. Our UCE-based encapsulation scheme XKEY2 similarly yields a standard-model locally-computable computational extractor. One might also view XKEY and XKEY2 as reusable locally-computable extractors following [2,17,23]. Reusability of extractors also aims to address deriving multiple subkeys and arose in [14,21,31].

A condenser [22,32,33] is a min-entropy extractor. Our subkey prediction Lemma can be viewed as building a BRM (or locally computable) condenser for a random source. The algorithm is to simply return the subkey J given by random probes into the big-key.

One could obtain a big-key symmetric encryption scheme by adapting the asymmetric BRM scheme of ADNSWW [2]. Our schemes are much more efficient.

2   Notation

NOTATION. For integers $a \leq b$ we let $[a..b] = \{a, \ldots, b\}$. If $\mathbf{x}$ is a vector then $|\mathbf{x}|$ denotes its length and $\mathbf{x}[i]$ denotes its i-th coordinate. (For example if $\mathbf{x} = (10, 00, 1)$ then $|\mathbf{x}| = 3$ and $\mathbf{x}[2] = 00$.) We let ε denote the empty vector, which has length 0. If $0 \leq i \leq |\mathbf{x}|$ then we let $\mathbf{x}[1..i] = (\mathbf{x}[1], \ldots, \mathbf{x}[i])$, this being ε when $i = 0$. We let S^n denote the set of all length n vectors over the set S and we let S^* denote the set of all finite-length vectors over the set S. Strings are treated as the special case of vectors over $\{0, 1\}$. Thus, if x is a string then $|x|$ is its length, $x[i]$ is its i-th bit, $x[1..i] = x[1]...x[i]$, ε is the empty string, $\{0, 1\}^n$ is the set of n-bit strings and $\{0, 1\}^*$ the set of all strings. If K is a k-bit string and $\mathbf{p}$ is a p-vector over $[1..k]$ then we let $K[\mathbf{p}] = K[\mathbf{p}[1]]K[\mathbf{p}[2]] \cdots K[\mathbf{p}[p]]$ denote the length-p string consisting of the bits of K in the positions indicated by $\mathbf{p}$. For example, if $K = 01010101$ and $\mathbf{p} = (1, 8, 2, 2)$ then $K[\mathbf{p}] = 0111$.

If X is a finite set, we let $x \twoheadleftarrow X$ denote picking an element of X uniformly at random and assigning it to x. Algorithms may be randomized unless otherwise indicated. Running time is worst case. If A is an algorithm, we let $y \leftarrow A(x_1, \cdots; r)$ denote running A with random coins r on inputs $x_1, \cdots$ and assigning the output to y. We let $y \twoheadleftarrow A(x_1, \cdots)$ be the result of picking r at random and letting $y \leftarrow A(x_1, \cdots; r)$. We let $[A(x_1, \cdots)]$ denote the set of all possible outputs of A when invoked with inputs $x_1, \cdots$. We denote by $\mathsf{Func}[a, b]$ the set of all functions f from $\{0, 1\}^a$ to $\{0, 1\}^b$.

We use the code-based game-playing framework [13] (see Fig. 3 for an example). By $\Pr[\mathbf{G}]$ we denote the probability that game $\mathbf{G}$ returns true. Uninitialized boolean variables, sets and integers are assume initialized to false, the empty set and 0, respectively.

3   The Subkey Prediction Lemma

Suppose an adversary computes 1 TByte of information L about a random 2 TByte key $\boldsymbol{K}$. Afterward, we challenge the adversary to identify the 128-bit substring K whose bits are those found at some 128 random locations of $\boldsymbol{K}$. We tell the adversary those locations. How well can the adversary do at this game? This section introduces what we call the *subkey prediction* game to formalize this question and answer it.

THE SUBKEY-PREDICTION GAME. Let $k, \ell, p, q \geq 1$ be integers with $k \geq \ell$. We call these values the *key length*, the *leakage length*, the *probe length*, and the *iteration count*. Let $\mathsf{Lk}: \{0, 1\}^k \rightarrow \{0, 1\}^\ell$ be a function, the *leakage function*. We associate to these values and an adversary $\mathcal{A}$ the *subkey prediction game* $\mathbf{G}_{k,p,q}^{\mathrm{skp}}(\mathcal{A}, \mathsf{Lk})$ depicted in the left panel of Fig. 3. (Ignore the other game for now). The game picks a random key $\boldsymbol{K}$ (the big-key) of length k and computes *leakage* $L \leftarrow \mathsf{Lk}(\boldsymbol{K})$. It then picks q random probes $\mathbf{p}_1, \ldots, \mathbf{p}_q$, each consisting of p random indexes into the key $\boldsymbol{K}$. The adversary $\mathcal{A}$ must guess one of the strings $\boldsymbol{K}[\mathbf{p}_i]$ given leakage L and probe locations $\mathbf{p}_1, \ldots, \mathbf{p}_q$. Any one will do. It outputs a guess J and the game returns true if and only if J is one of the values $\boldsymbol{K}[\mathbf{p}_i]$. The adversary needn't identify which subkey it has guessed. Now define

$$\mathsf{Adv}_{k,p,q}^{\mathrm{skp}}(\mathcal{A}, \mathsf{Lk}) = \Pr[\mathbf{G}_{k,p,q}^{\mathrm{skp}}(\mathcal{A}, \mathsf{Lk})]$$

$$\mathsf{Adv}_{k,p,q}^{\mathrm{skp}}(\mathsf{Lk}) = \max_{\mathcal{A}} \mathsf{Adv}_{k,p,q}^{\mathrm{skp}}(\mathcal{A}, \mathsf{Lk})$$

$$\mathsf{Adv}_{k,p,q}^{\mathrm{skp}}(\ell) = \max_{\mathsf{Lk} \in \mathsf{Func}[k,\ell]} \mathsf{Adv}_{k,p,q}^{\mathrm{skp}}(\mathsf{Lk})$$

The first is the probability that the game returns true, that is, the probability that $\mathcal{A}$ wins the game. For the second definition the maximum is over all adversaries $\mathcal{A}$, regardless of running time. For the third definition the maximum is over all leakage functions $\mathsf{Lk}: \{0, 1\}^k \rightarrow \{0, 1\}^\ell$ that return ℓ bits. This last advantage function measures the best possible prediction probability when the

Game $\mathbf{G}^{\mathrm{skp}}_{k,p,q}(\mathcal{A},\mathsf{Lk})$	Game $\mathbf{G}^{\mathrm{skp1}}_{k,p}(\mathcal{A},\mathcal{K})$
$\boldsymbol{K} \twoheadleftarrow \{0,1\}^k$; $L \leftarrow \mathsf{Lk}(\boldsymbol{K})$	$\boldsymbol{K} \twoheadleftarrow \mathcal{K}$
for $i \leftarrow 1,\ldots,q$ do $\mathbf{p}_i \twoheadleftarrow [1..k]^p$	$\mathbf{p} \twoheadleftarrow [1..k]^p$
$J \leftarrow \mathcal{A}(L,\mathbf{p}_1,\ldots,\mathbf{p}_q)$	$J \leftarrow \mathcal{A}(\mathbf{p})$
return $\left(J \in \{\boldsymbol{K}[\mathbf{p}_1],\ldots,\boldsymbol{K}[\mathbf{p}_q]\}\right)$	return $\left(J = \boldsymbol{K}[\mathbf{p}]\right)$

Fig. 3. Subkey-prediction game (left) and restricted subkey-prediction game (right).

leakage is restricted to ℓ bits and the adversary is arbitrary. We are interested in upper bounding this last advantage as a function of k, ℓ, p, q.

The idea here is that an adversary has some information about $\boldsymbol{K}$, but it is limited to $\mathsf{Lk}(\boldsymbol{K})$, which is ℓ bits. This leaves $k - \ell$ bits of average min-entropy in $\boldsymbol{K}$. We are trying to extract it in a particular and efficient way, by taking a random subset of positions of $\boldsymbol{K}$. We are asking "only" for unpredictability, and will then apply a random oracle to get random-looking bits.

FROM MANY QUERIES TO ONE. One simple observation is that we can reduce the case of q probes to the case of a single probe via the union bound:

Lemma 1. *Let k, ℓ, p, q be integers with $k \geq \ell$. Then*

$$\mathsf{Adv}^{\mathrm{skp}}_{k,p,q}(\ell) \leq q \cdot \mathsf{Adv}^{\mathrm{skp}}_{k,p,1}(\ell) \quad \blacksquare$$

Proof (Lemma 1). Given adversary $\mathcal{A}$ we let $\mathcal{A}_1$ be the adversary defined as follows. On input $L, \mathbf{p}$ it picks $g \twoheadleftarrow [1..q]$ and $\mathbf{p}_j \twoheadleftarrow [1..k]^p$ for $j \in [1..q] \setminus \{g\}$. It lets $\mathbf{p}_g \leftarrow \mathbf{p}$ and returns $J \twoheadleftarrow \mathcal{A}(L, \mathbf{p}_1, \ldots \mathbf{p}_q)$. By a union bound we have $\mathsf{Adv}^{\mathrm{skp}}_{k,p,q}(\mathcal{A}, \mathsf{Lk}) \leq q \cdot \mathsf{Adv}^{\mathrm{skp}}_{k,p,1}(\mathcal{A}_1, \mathsf{Lk})$ and the lemma follows. $\blacksquare$

We retain the q-probe definition because this is what we will use in applications, but Lemma 1 allows us to focus on the $q = 1$ case for the remainder of this section.

CONNECTION TO HAMMING BALLS. Given k, ℓ, let us ask which leakage function $\mathsf{Lk}: \{0,1\}^k \to \{0,1\}^\ell$ maximizes the advantage. At first glance it's hard to imagine there's a strategy better than the greedy one of leaking the first ℓ bits of $\boldsymbol{K}$. Specifically let $\mathsf{Lk}_\ell: \{0,1\}^k \to \{0,1\}^\ell$ be defined by $\mathsf{Lk}_\ell(\boldsymbol{K}) = \boldsymbol{K}[1] \cdots \boldsymbol{K}[\ell]$, the function that returns the first ℓ bits of its input. Then a natural conjecture is that $\mathsf{Adv}^{\mathrm{skp}}_{k,p,1}(\ell) = \mathsf{Adv}^{\mathrm{skp}}_{k,p,1}(\mathsf{Lk}_\ell)$, meaning that the subkey prediction advantage is maximized by Lk_ℓ. Indeed this was our first guess.

This conjecture, however, is false. We now give a counter-example that shows this. Besides indicating the subtleties in the problem, it makes a connection with error-correcting codes and Hamming balls that will underly our eventual results and bound. Here is the counter-example. Let $k = 7$ and $\ell = 4$. Let $p = 1$. Then

$$\mathsf{Adv}^{\mathrm{skp}}_{k,1,1}(\mathsf{Lk}_\ell) = (4/7)(1) + (3/7)(1/2) = 11/14 .$$

Now consider the following alternative $\mathsf{Lk}\colon \{0,1\}^7 \to \{0,1\}^4$ for which we will show that $\mathsf{Adv}^{\mathrm{skp}}_{k,1,1}(\mathsf{Lk}) = 7/8 > 11/14$. Let $W_1,\dots,W_{16} \in \{0,1\}^7$ be the codewords of the Hamming $(7,4)$ code. (See Wikipedia article "Hamming(7,4)" for the definition.) This code has message length $\ell = 4$ and codeword length $k = 7$, so that it describes an (injective) encoding function $E\colon \{0,1\}^4 \to \{W_1,\dots,W_{16}\} \subseteq \{0,1\}^7$. It has minimum distance 3 and corrects one error. Let B_i be the set of all 7-bit strings whose Hamming distance from W_i is ≤ 1. Then $|B_i| = 8$ and the sets $B_1,\dots,B_{16}$ are a partition of $\{0,1\}^7$. The decoding function $D\colon \{0,1\}^7 \to \{0,1\}^4$, given $\boldsymbol{K}$, finds the unique i such that $\boldsymbol{K} \in B_i$. It then returns message $L = E^{-1}(W_i)$. The leakage function Lk is simply D; namely, given the big-key $\boldsymbol{K}$, it returns its decoding $D(\boldsymbol{K})$. Now the adversary $\mathcal{A}$ receives (L,p_1) where $p_1 \in [1..7]$ is the random probe into $\boldsymbol{K}$ chosen by the game and $L = \mathsf{Lk}(\boldsymbol{K}) = D(\boldsymbol{K})$, and it wants to predict $\boldsymbol{K}[p_1]$. Adversary $\mathcal{A}$ lets $W = E(L)$ and returns $W[p_1]$ as its guess. Then $\mathsf{Adv}^{\mathrm{skp}}_{k,1,1}(\mathcal{A},\mathsf{Lk}) = 7/8$ because if $W = W_i$ then 7 of the 8 strings in B_i have $W_i[p_1]$ as their p_1-th bit. So

$$\mathsf{Adv}^{\mathrm{skp}}_{k,1,1}(\mathsf{Lk}) \ge 7/8 > 11/14 = \mathsf{Adv}^{\mathrm{skp}}_{k,1,1}(\mathsf{Lk}_\ell) .$$

SUBKEY-PREDICTION BOUND. We now give our upper bound on $\mathsf{Adv}^{\mathrm{skp}}_{k,p,1}(\ell)$ as a function of k,ℓ,p. Let $w_{\mathrm{H}}(x)$ denote the Hamming weight of string x. Let $\mathcal{B}_k(r) = \{\, x \in \{0,1\}^k : w_{\mathrm{H}}(x) \le r \,\}$ be the Hamming ball of radius r with center 0^k and let $B_k(r) = |\mathcal{B}_k(r)|$ be its size. Then

$$B_k(r) = \sum_{i=0}^{r} \binom{k}{i} . \tag{1}$$

Now define the *radius* $\mathsf{rd}_k(N)$ as the largest integer r such that $B_k(r) \le N$, and let

$$
\begin{aligned}
G_{k,p}(N) & \\
&= \frac{1}{N} \sum_{i=0}^{\mathsf{rd}_k(N)} \binom{k}{i} \left(1 - \frac{i}{k}\right)^p + \frac{N - B_k(\mathsf{rd}_k(N))}{N} \left(1 - \frac{1 + \mathsf{rd}_k(N)}{k}\right)^p \quad (2) \\
&\le \frac{1}{N} \sum_{i=0}^{\min(1+\mathsf{rd}_k(N),k)} \binom{k}{i} \left(1 - \frac{i}{k}\right)^p . \quad (3)
\end{aligned}
$$

The following says this function provides an upper bound on the subkey prediction probability:

Theorem 2 (Subkey-prediction bound). *Let k,ℓ,p be integers, $k \ge \ell$. Then*

$$\mathsf{Adv}^{\mathrm{skp}}_{k,p,1}(\ell) \le G_{k,p}(2^{k-\ell}) \ \blacksquare \tag{4}$$

Before we prove Theorem 2, let's try to understand the growth rate of the bound for parameters of interest.

ESTIMATES. The $G_{k,p}(N)$ formula itself is somewhat intractable. To use Theorem 2 it is easier to work with the following approximation that we used in Sect. 1. The approximation is very good. For $\lambda \in [0,1]$ we let

$$w(\lambda) = -\lg(1 - H_2^{-1}(1 - \lambda))$$

where lg is the logarithm to base two, H_2 is the binary entropy function defined for $x \in [0,1]$ by $H_2(x) = -x\ln(x) - (1-x)\ln(1-x)$ and $H_2^{-1}(1-\lambda)$ returns the smaller of the two values x satisfying $H_2(x) = 1 - \lambda$. Then for $\ell \leq k$ we have

$$G_{k,p}(2^{k-\ell}) \approx 2^{-p \cdot w(\ell/k)} . \tag{5}$$

We now justify this. Let us define

$$G_{k,p,r}^*(N) = \frac{1}{N} \sum_{i=0}^{r} \binom{k}{i} \left(1 - \frac{i}{k}\right)^p . \tag{6}$$

The following lemma gives both a lower bound and an upper bound on $G_{k,p,r}^*(N)$ that are within a factor two of each other, showing that the estimate of the lemma is tight to within a small constant. The proof is in [9].

Lemma 3. *Let $N, k, r, p \geq 1$ be integers with $r \leq k$. Let $\theta = r/k$ and $\gamma = 2\binom{k}{r}/N$. Suppose: (1) $p \leq 0.27 \cdot k$ and (2) $\theta \leq 0.22$. Let $\alpha = -\lg(1-\theta) > 0$. Then*

$$\frac{\gamma}{2} \cdot 2^{-\alpha \cdot p} \leq G_{k,p,r}^*(N) \leq \gamma \cdot 2^{-\alpha \cdot p} \; \blacksquare$$

Letting $r^* = \mathsf{rd}_k(2^{k-\ell})$, from Eq. (3) we have $G_{k,p}(2^{k-\ell}) \approx G_{k,p,r^*}^*(2^{k-\ell})$, which by Lemma 3 is $\approx 2^{-\alpha \cdot p}$ where $\alpha = -\lg(1 - r^*/k)$. So to justify Eq. (5) we need to show that $\alpha \approx w(\ell/k)$. For this we use the well-known estimate $2^{k-\ell} \approx B_k(r^*) \approx 2^{k \cdot H_2(r^*/k)}$ to get $1 - \ell/k \approx H_2(r^*/k)$ or $r^*/k \approx H_2^{-1}(1 - \ell/k)$. Thus $-\lg(1 - r^*/k) \approx -\lg(1 - H_2^{-1}(1 - \ell/k)) = w(\ell/k)$ as desired.

RESTRICTED SUBKEY-PREDICTION. Our proof of Theorem 2 will rely on an analysis of the *restricted* subkey-prediction game $\mathbf{G}_{k,p}^{\mathrm{skp1}}(\mathcal{A}, \mathcal{K})$ shown in the right panel of Fig. 3. The game is associated to integers k, p, adversary $\mathcal{A}$ and a set $\mathcal{K} \subseteq \{0,1\}^k$. Our results concerning this game are also of independent interest because we obtain bounds that are *tight*. In this game, the big-key $\boldsymbol{K}$ is drawn from $\mathcal{K}$ rather than from $\{0,1\}^k$, and there is no leakage. Also, there is only one probe. The rest is the same as in the subkey prediction game. Intuitively, one can think of $\mathcal{K}$ as being $\mathsf{Lk}^{-1}(L)$ for some particular L received by $\mathcal{A}$ in game $\mathbf{G}^{\mathrm{skp}}$, so that the new game, $\mathbf{G}^{\mathrm{skp1}}$, effectively represents the view of $\mathcal{A}$ in the prior game at the point it receives leakage L. Now define

$$\mathsf{Adv}_{k,p}^{\mathrm{skp1}}(\mathcal{A}, \mathcal{K}) = \Pr[\mathbf{G}_{k,p}^{\mathrm{skp1}}(\mathcal{A}, \mathcal{K})]$$

$$\mathsf{Adv}_{k,p}^{\mathrm{skp1}}(\mathcal{K}) = \max_{\mathcal{A}} \mathsf{Adv}_{k,p}^{\mathrm{skp1}}(\mathcal{A}, \mathcal{K})$$

$$\mathsf{Adv}_{k,p}^{\mathrm{skp1}}(N) = \max_{\mathcal{K} \subseteq \{0,1\}^k, \, |\mathcal{K}| = N} \mathsf{Adv}_{k,p}^{\mathrm{skp1}}(\mathcal{K})$$

The first advantage is the probability that $\mathcal{A}$ wins the game. In the second, the maximum is over all adversaries $\mathcal{A}$, regardless of running time. In the third, the maximum is over all sets $\mathcal{K} \subseteq \{0,1\}^k$ that have size $|\mathcal{K}| = N$.

MAIN LEMMAS AND PROOF OF THEOREM. Theorem 2 is obtained via two main lemmas. Here we state them and show how they yield the theorem. We will then prove the lemmas. The first main lemma reduces the task of upper bounding the subkey prediction probability $\mathsf{Adv}_{k,p,1}^{\mathrm{skp}}(\ell)$ to the task of bounding the special subkey prediction probability $\mathsf{Adv}_{k,p}^{\mathrm{skp1}}(N)$ for $N = 2^{k-\ell}$:

Lemma 4. *Let* $k, \ell, p \geq 1$ *be integers with* $k \geq \ell$. *Then*

$$\mathsf{Adv}_{k,p,1}^{\mathrm{skp}}(\ell) \ \leq \ \mathsf{Adv}_{k,p}^{\mathrm{skp1}}(2^{k-\ell}) \ \blacksquare \tag{7}$$

The proof of this lemma, given below, involves a definition of concavity for discrete functions and a lemma saying where such functions attain their maximum. Then a particular function $F_{k,p}$ we define is shown to meet this definition of concavity, and Lemma 4 results. The second main lemma characterizes the special subkey prediction probability:

Lemma 5. *Suppose* $1 \leq N \leq 2^k$ *and* $p \geq 1$. *Then*

$$\mathsf{Adv}_{k,p}^{\mathrm{skp1}}(N) = G_{k,p}(N) \ \blacksquare$$

We note that Lemma 5 is an equality, not a bound. We are able to say *exactly* what is the special subkey prediction probability for a given value of N. Lemma 5 is obtained by showing that for a given N, the maximum of $\mathsf{Adv}_{k,p}^{\mathrm{skp1}}(\mathcal{K})$ over sets $\mathcal{K}$ of size N, occurs for a set that is *monotone* and sandwiched in between two adjacent Hamming balls. Monotone means that if a string is in the set, so is any string obtained by flipping one bits to zero bits on the original string. For monotone sets, it is quite easy to estimate the optimal advantage. All this put together will lead to Lemma 5.

The proof of Theorem 2 is immediate from these two lemmas, which we still need to prove. But first, with the above, we are in a position to compare with prior work.

COMPARISON WITH PRIOR WORK. Lemmas related to subkey and restricted subkey prediction have been given by NZ [30, Lemma 11], Vadhan [35, Lemma 9] and ADW [3, Lemma A.3]. Briefly, the first two don't give bounds on subkey prediction. They do give bounds on restricted subkey prediction but they are hard to use due to hidden constants, and these bound are inferior to Lemma 4 since the latter is tight. The elegant lemma of ADW [3, Lemma A.3], however, not only applies directly to subkey prediction but also gives a concrete bound with no hidden constants. The difference here, as quantified in Sect. 1, is that the bound is much inferior to that of Theorem 2, translating to a significant loss of practical efficiency for big-key symmetric encryption.

NZ [30, Lemma 11] considers drawing a string $\boldsymbol{K}$ from $\{0,1\}^k$ according to a distribution D with min-entropy δ. Like us, for a random probe $\mathbf{p} \in [1..k]^p$, they then consider $\boldsymbol{K}[\mathbf{p}]$. The lemma specifies ϵ, δ' such that for all but an ϵ fraction of the $\mathbf{p}$'s, the distribution $\boldsymbol{K}[\mathbf{p}]$ is within statistical distance ϵ of a δ' source. Unlike our work there is no leakage. Their setting does not capture subkey prediction but it does capture restricted subkey prediction game, which corresponds to the distribution D that puts a uniform probability on $\mathcal{K}$ and 0 probability on points outside it. However, the formulas in [30, Lemma 11] make ϵ very large for the parameters of interest to us, and also have un-specified constants. In contrast Lemma 5 gives a tight bound with no unspecified constants. The same remains true with Vadhan [35, Lemma 9]. (Here the hidden constant is an exponent in the statistical distance.) The values of the constants can in principle, of course, be obtained from the proofs, but since our bound of Lemma 4 for restricted subkey prediction is tight, we would not see an improvement. (And the indication from what follows, where concrete bounds are given, is that the bounds would be substantially worse than ours anyway.)

ADW [3, Lemma A.3] can, however, be directly applied to get a bound on $\mathsf{Adv}^{\mathrm{skp}}_{k,p,1}(\ell)$. Referring to their lemma, let random variable X represent the big-key $\boldsymbol{K}$, and let experiment $\mathcal{E}_1$ represent the leakage. Their t corresponds to our p. (We think the N in their lemma is a typo. It should be t.) Then $\tilde{\mathrm{H}}_\infty(X) = k - \ell$. The lemma then implies that $\mathsf{Adv}^{\mathrm{skp}}_{k,p,1}(\ell) \leq 2^{-c}$ as long as $k - \ell \geq 2ck \lg(2k)/p + 3c + 5$. This translates to

$$c \leq \frac{p(k - \ell - 5)}{2k \lg(2k) + 3p} \, . \tag{8}$$

This is the formula used for Fig. 2.

We note that ADW [3, Lemma A.3] considers a more general setting than subkey prediction. We are saying that in this special setting, we can get much better bounds. In the big-key context, an important case where their bound applies but ours does not is when we work over blocks rather than bits. Here $\boldsymbol{K}$ is a k-vector over $\{0,1\}^b$ for some block length parameter b, so that each probe draws a b-bit block. This setting is more convenient for implementations of big-key symmetric encryption. Giving better bounds for this setting is an interesting open question.

PROOF OF LEMMA 5. Recall that $w_{\mathrm{H}}(x)$ denotes the Hamming weight of string x. For equal-length strings $x, y \in \{0,1\}^*$, write $x \preceq y$ if $x[i] \leq y[i]$ —that is, if $x[i] = 1$ then $y[i] = 1$— for every $i \in [1..|x|]$. Write $x \prec y$ if $x \preceq y$ and $x \neq y$. Then $\preceq$ is a partial order on $\{0,1\}^k$, satisfying the required conditions, namely it is reflexive, anti-symmetric and transitive. If $U \subseteq \{0,1\}^*$ is a finite set then we let $w_{\mathrm{H}}(U) = \sum_{u \in U} w_{\mathrm{H}}(u)$ be the Hamming weight of U. Call a subset $\mathcal{K} \subseteq \{0,1\}^k$ monotone if for all $x, y \in \{0,1\}^k$ we have that if $x \prec y$ and $y \in \mathcal{K}$ then $x \in \mathcal{K}$. The following lemma says that the maximum restricted subkey prediction advantage occurs for a set $\mathcal{K}$ that is monotone. The proof is in [9].

Lemma 6. *Suppose $1 \leq N \leq 2^k$ and $p \geq 1$. Then there is a monotone $\mathcal{K} \subseteq \{0,1\}^k$ such that $|\mathcal{K}| = N$ and*

$$\mathsf{Adv}_{k,p}^{\mathrm{skp1}}(N) = \mathsf{Adv}_{k,p}^{\mathrm{skp1}}(\mathcal{K}) \quad \blacksquare$$

Now define the function $g_{k,p}$, taking input a set $\mathcal{K} \subseteq \{0,1\}^k$, via

$$g_{k,p}(\mathcal{K}) = \frac{1}{|\mathcal{K}|} \sum_{x \in \mathcal{K}} \left(1 - \frac{w_{\mathrm{H}}(x)}{k} \right)^p . \tag{9}$$

A nice property of monotone sets is that we can easily compute the maximal advantage on them, via the function just defined:

Lemma 7. *Suppose $1 \leq N \leq 2^k$ and $p \geq 1$. Suppose $\mathcal{K} \subseteq \{0,1\}^k$ is a monotone set of size N. Then*

$$\mathsf{Adv}_{k,p}^{\mathrm{skp1}}(\mathcal{K}) = g_{k,p}(\mathcal{K}) \quad \blacksquare$$

Proof (Lemma 7). Since $\mathcal{K}$ is monotone, the best strategy for the adversary given $\mathbf{p}$ is to simply return $K = 0^k$ as the guess. Letting $\mathcal{A}$ denote the adversary that does this, we now want to evaluate $\mathsf{Adv}_{k,p}^{\mathrm{skp1}}(\mathcal{A}, \mathcal{K})$. This is just the probability that if x is chosen at random from $\mathcal{K}$ and $\mathbf{p}[1], \ldots, \mathbf{p}[p]$ are chosen at random from $[1..k]$ then $x[\mathbf{p}[j]] = 0$ for all $j \in [1..p]$. This probability is $g_{k,p}(\mathcal{K})$. $\blacksquare$

We will now further characterize the sets which attain the maximum. We say that $\mathcal{K} \subseteq \{0,1\}^k$ is sandwiched between Hamming balls if there is an r such that $\mathcal{B}_k(r) \subseteq \mathcal{K} \subset \mathcal{B}_k(r+1)$. Note in this case it must be that $r = \mathrm{rd}_k(N)$ where N is the size of $\mathcal{K}$. The proof of the following, which exploits Lemma 6, is in [9].

Lemma 8. *Suppose $1 \leq N \leq 2^k$ and $p \geq 1$. Let $r = \mathrm{rd}_k(N)$. Then there is a monotone, size N set $\mathcal{K}$ such that $\mathcal{B}_k(r) \subseteq \mathcal{K} \subset \mathcal{B}_k(r+1)$ and*

$$\mathsf{Adv}_{k,p}^{\mathrm{skp1}}(N) = \mathsf{Adv}_{k,p}^{\mathrm{skp1}}(\mathcal{K}) \quad \blacksquare$$

We are now in a position to prove our second main lemma.

Proof (Lemma 5). Let $r = \mathrm{rd}_k(N)$. By Lemma 8 there is a monotone, size N set $\mathcal{K}$ such that $\mathcal{B}_k(r) \subseteq \mathcal{K} \subset \mathcal{B}_k(r+1)$ and $\mathsf{Adv}_{k,p}^{\mathrm{skp1}}(N) = \mathsf{Adv}_{k,p}^{\mathrm{skp1}}(\mathcal{K})$. But

$$\mathsf{Adv}_{k,p}^{\mathrm{skp1}}(\mathcal{K}) = g_{k,p}(\mathcal{K}) = G_{k,p}(N)$$

where the first equality is by Lemma 7 and the second is because $\mathcal{B}_k(r) \subseteq \mathcal{K} \subset \mathcal{B}_k(r+1)$. $\blacksquare$

PROOF OF LEMMA 4. We now prove the first main lemma. We begin with a general result about the maximization of discrete concave functions.

The standard definition of concavity of functions applies to continuous functions. Here we provide a definition for functions defined on a discrete domain that allows us to prove a lemma we will use later. Proceeding, suppose $F: \mathbb{Z}_M \to \mathbb{R}$. We say that F is concave if $F(a+1) - F(a) \leq F(b+1) - F(b)$ for all $a, b \in \mathbb{Z}_{M-1}$ satisfying $a \geq b$. Now suppose $t, m \geq 1$ are integers with $1 \leq m \leq t$. Then we let

$$S(M, m, t) = \{ (x_1, \ldots, x_m) \in \mathbb{Z}_M^m : x_1 + \cdots + x_m = t \} .$$

Define $F^m: \mathbb{Z}_M^m \to \mathbb{R}$ by $F^m(x_1, \ldots, x_m) = F(x_1) + \cdots + F(x_m)$. The proof of the following is in [9].

Lemma 9. *Suppose $F: \mathbb{Z}_M \to \mathbb{R}$ is concave. Suppose $1 \leq m \leq t$ are integers such that m divides t and $t/m \in \mathbb{Z}_M$. Then*

$$\max_{(x_1, \ldots, x_m) \in S(M, m, t)} F^m(x_1, \ldots, x_m) = m \cdot F(t/m) \ \blacksquare$$

That is, the maximum of F^m over $S(M, m, t)$ is attained when all inputs have the same value $t/m \in \mathbb{Z}_M$, and is thus equal to $m \cdot F(t/m)$. To apply this in our setting, we introduce the function $F_{k,p}: \mathbb{Z}_{2^k+1} \to \mathbb{R}$ defined by

$$F_{k,p}(N) = \frac{N}{2^k} \cdot \mathsf{Adv}_{k,p}^{\mathrm{skp1}}(N) . \tag{10}$$

Rewriting the claim of Lemma 4 in terms of the function $F_{k,p}$, our aim is to show that $\mathsf{Adv}_{k,p,1}^{\mathrm{skp}}(\ell) \leq 2^\ell \cdot F_{k,p}(2^{k-\ell})$. The following says that the function $F_{k,p}$ meets our definition of concavity. The proof is in [9].

Lemma 10. *Let $k, p \geq 1$ be integers. Then the function $F_{k,p}$ is concave.* $\blacksquare$

We now show how to obtain our first main lemma.

Proof (Lemma 4). Let $M = 2^k + 1$, $m = 2^\ell$ and $t = 2^k$. Now we have

$$\mathsf{Adv}_{k,p,1}^{\mathrm{skp}}(\ell) = \max_{\mathsf{Lk}} \left(\sum_L \frac{|\mathsf{Lk}^{-1}(L)|}{2^k} \cdot \max_{\mathcal{A}} \Pr[\, \mathbf{G}_{k,p,1}^{\mathrm{skp}}(\mathcal{A}, \mathsf{Lk}) \mid \mathsf{Lk}(\boldsymbol{K}) = L \,] \right)$$

$$= \max_{\mathsf{Lk}} \left(\sum_L \frac{|\mathsf{Lk}^{-1}(L)|}{2^k} \cdot \mathsf{Adv}_{k,p}^{\mathrm{skp1}}(\mathsf{Lk}^{-1}(L)) \right)$$

$$\leq \max_{(N_1, \ldots, N_m) \in S(M, m, t)} \sum_{i=1}^m F_{k,p}(N_i)$$

$$= \max_{(N_1, \ldots, N_m) \in S(M, m, t)} F_{k,p}^m(N_1, \ldots, N_m)$$

$$= m \cdot F_{k,p}(2^{k-\ell}) \tag{11}$$

$$= m \cdot \frac{2^{k-\ell}}{2^k} \cdot \mathsf{Adv}_{k,p}^{\mathrm{skp1}}(2^{k-\ell}) = \mathsf{Adv}_{k,p}^{\mathrm{skp1}}(2^{k-\ell}) . \tag{12}$$

Equation (11) is justified as follows. Lemma 10 says that $F_{k,p}$ is concave. So we can apply Lemma 9, and here $t/m = 2^{k-\ell}$. Equation (12) is by Eq. (10) and because $m = 2^{\ell}$. ∎

4   Encapsulating a Key

In this section we introduce *big-key encapsulation*. A scheme for this aim lets a user encapsulate a random, conventional-length key K using a big-key $\boldsymbol{K}$. We speak of "encapsulation" instead of "encryption" because the user never selects a value K to encrypt: rather, a random R is chosen and this value, together with the big-key $\boldsymbol{K}$, determines a derived key $K = \mathsf{KEY}(\boldsymbol{K}, R)$. A user can transmit R to a party that knows $\boldsymbol{K}$ and in this way name an induced key K. While the aim is similar to a KEM (key encapsulation mechanism) [18], there are also many differences, which we will later discuss.

DEFINITIONS. A big-key encapsulation algorithm is a deterministic algorithm KEY that, given strings $\boldsymbol{K} \in \{0,1\}^k$ and $R \in \{0,1\}^r$, returns a string $K = \mathsf{KEY}(\boldsymbol{K}, R) \in \{0,1\}^\kappa$. We call $\boldsymbol{K}$, R, and K the *big-key*, *selector*, and *derived key*, respectively. Their lengths, being part of the signature of KEY, are numbers associated to it. Since we will be working in the RO model, we allow the encapsulation algorithm to depend on an oracle RO. We write such a function as a superscript, $K = \mathsf{KEY}^{\mathrm{RO}}(\boldsymbol{K}, R)$, if we want to emphasize its presence.

The security requirement for an encapsulation algorithm captures the idea that a derived key K should be indistinguishable from a uniform random string even when accompanied by R and some bounded amount of leakage from the big-key $\boldsymbol{K}$. This is formalized via the game $\mathbf{G}_{\mathsf{KEY}}^{\mathrm{key}}(\mathcal{A})$ on the left of Fig. 4. The game is associated to encapsulation algorithm KEY and adversary $\mathcal{A}$. The game is in the ROM, oracle RO taking (x, l) and returning a random l-bit string. Not only do algorithms and the adversary have access to RO, but, importantly, so does the leakage function $\mathsf{LK}^{\mathrm{RO}}\colon \{0,1\}^k \to \{0,1\}^\kappa$, now called an *oracle* leakage function to emphasize this.

In its first stage, the adversary specifies the leakage function it wants. It then makes a sequence of DERIVE calls, each providing an (R, K) pair that is either *real*—the derived key was determined by running KEY with a random selector— or *random*—the derived key is uniformly random. Which of these possibilities occurs depends on a bit b chosen at the beginning of the game. The adversary must guess that bit. We let $\mathsf{Adv}_{\mathsf{KEY}}^{\mathrm{key}}(\mathcal{A}) = 2 \Pr[\mathbf{G}_{\mathsf{KEY}}^{\mathrm{key}}(\mathcal{A})] - 1$ be its advantage in doing so.

DISCUSSION. A big-key encapsulation algorithm is in some ways similar to a conventional key encapsulation mechanism [18]. But there are many differences. First, we are in the symmetric setting instead of the asymmetric setting. Second, we are considering security under leakage, so the leakage function, chosen by the adversary, comes into the picture. Finally, we have chosen a syntax under which we do not have a separate decapsulation algorithm, preferring to surface the coins R across the encapsulation algorithm's interface.

Game $\mathbf{G}_{\mathsf{KEY}}^{\mathrm{key}}(\mathcal{A})$	DERIVE( )	RO(x,l)
$b \twoheadleftarrow \{0,1\}$	$R \twoheadleftarrow \{0,1\}^r$	if not $T[x,l]$ then
$\mathbf{K} \twoheadleftarrow \{0,1\}^k$	if $(b=0)$ then $K \twoheadleftarrow \{0,1\}^\kappa$	$\quad T[x,l] \twoheadleftarrow \{0,1\}^l$
$(\mathsf{LK}, \sigma) \twoheadleftarrow \mathcal{A}^{\mathrm{RO}}( )$	else $K \leftarrow \mathsf{KEY}^{\mathrm{RO}}(\mathbf{K}, R)$	return $T[x,l]$
$L \leftarrow \mathsf{LK}^{\mathrm{RO}}(\mathbf{K})$	return (R, K)	
$b' \twoheadleftarrow \mathcal{A}^{\mathrm{DERIVE, RO}}(L, \sigma)$		
return $(b' = b)$		

Fig. 4. Game for defining the security of a big-key key encapsulation algorithm $\mathsf{KEY} \colon \{0,1\}^k \times \{0,1\}^r \to \{0,1\}^\kappa$

One may ask why we let the adversary encapsulate an arbitrary number of conventional keys, using its DERIVE oracle q times; wouldn't a hybrid argument show that providing a single derived key is equivalent, up to a factor of q? In fact, such a hybrid argument fails because the single-query adversary has no means to simulate the real derived keys. For this reason, it is important to consider multiple keys.

We explain the importance of giving the leakage function access to RO. Otherwise, the scheme $\mathsf{KEY}^{\mathrm{RO}}(\mathbf{K}, R) = \mathrm{RO}(\mathbf{K}, \kappa)$ is secure. Yet, in practice, when RO is instantiated with a concrete hash function $H \colon \{0,1\}^* \to \{0,1\}^\kappa$, this scheme is certainly not secure because the leakage function can return $H(\mathbf{K})$. (We note that in this example the scheme does not make efficient use of $\mathbf{K}$ as we want, but this is an orthogonal issue to security.) Once the leakage function has access to RO, it too can return $\mathrm{RO}(\mathbf{K}, \kappa)$, preventing this and other similar schemes from being deemed secure.

ENCAPSULATION SCHEME XKEY. Let $k, \kappa, r \geq 1$ be integers, and, for convenience, assume k is a power of two. Given the results of Sect. 3, a natural big-key encapsulation algorithm is as follows. On input the big-key $\mathbf{K} \in \{0,1\}^k$, algorithm KEY picks random probes $\mathbf{p} \in [1..k]^p$ and computes the induced subkey $J = \mathbf{K}[\mathbf{p}]$. While this subkey is unpredictable, up to the bounds we have seen, it is not indistinguishable from random bits, so cannot, by itself, function as the derived key K. So, instead, the algorithm would let the derived key be $K \leftarrow \mathrm{RO}(J, \kappa)$ where κ is the desired length for the derived key.

While the above might sound reasonable, there are two problems with the scheme—one with regard to security and the other with regard to efficiency. We explain these and then present a scheme that resolves them.

The security problem is that, we can only show security for the scheme just described if the leakage function does not have access to the oracle RO. But we have argued that it *must* have such access, and in that case the proof breaks down and it is not clear if the scheme is secure. A simple remedy is to have the scheme pick a random string and include it in the scope of the RO used to determine the subkey. The proof can then exploit the fact that leakage function can't depend on this (not yet chosen) value.

This still leaves the efficiency issue, which is that the probe $\mathbf{p}$ is quite long, a total of $p \lg(k)$ bits. The number of components p of $\mathbf{p}$ is fairly large and grows with the fraction of bits potentially leaked; it will typically be 100–1000. If $\mathbf{K}$ is 1 TByte then $k \approx 2^{43}$ and we're using up a few kilobytes to communicate the selector $\mathbf{p}$, which is unpleasant.

Since we're working in the ROM, an easy solution is to obtain $\mathbf{p}$ by applying RO to a short seed. Conveniently, the same random choices needed for this remedy can be used for the security problem as well. The resulting encapsulation algorithm $\mathsf{XKEY}_{k,\kappa,p,r}: \{0,1\}^k \times \{0,1\}^r \to \{0,1\}^\kappa$ is shown in Fig. 5 and described below.

Algorithm $\mathsf{XKEY}^{\mathrm{RO}}_{k,\kappa,p,r}(\mathbf{K}, R)$

for $i \leftarrow 1, \ldots, p$ do $\mathbf{p}[i] \leftarrow \mathrm{RO}(\langle R, i, 0 \rangle, \lg(k))$
$J \leftarrow \mathbf{K}[\mathbf{p}]$; $K \leftarrow \mathrm{RO}(\langle R, J, 1 \rangle, \kappa)$; Return K

Fig. 5. Encapsulation algorithm XKEY. Given a length-k big-key $\mathbf{K}$ and a length-r selector R, the algorithm returns a length-κ subkey K. The value p, a security parameter, specifies the number of probes into the big-key.

The XKEY encapsulation algorithm picks or is provided a random selector R of length r. It picks the ith probe into $\mathbf{K}$ not directly at random, but by applying the random oracle RO to a string encoding R, i, and 0. In defining $\mathbf{p}[i]$ we interpret a $\lg(k)$-bit string as an integer in $[1..k]$ in the natural way. The encapsulation algorithm then lets the subkey J be the positions of $\mathbf{K}$ indicated by the probes. The derived key K is obtained by applying the RO to a string encoding J, R, and 0. The third component in each encoding $\langle \cdot, \cdot, \cdot \rangle$ is for domain separation.

Theorem 12 will establish security of XKEY, providing concrete bounds. Those bounds indicate that $r = |R|$ can be chosen to be quite small, like 128–256. Since only this value is transmitted when XKEY is used, bandwidth overhead in small and independent of p.

ENHANCED SUBKEY-PREDICTION GAME. Towards the proof of Theorem 12 it is convenient to consider an enhanced version of the subkey-prediction problem. The security goal reflects two changes over our earlier subkey-prediction game $\mathbf{G}^{\mathrm{skp}}$. First, the leakage function is not fixed but dynamically chosen by an adversary. Second, that choice may depend on the RO, which the leakage function itself may depend on. These issues, particularly the second, lead to design choices embedded in XKEY and make proofs more challenging. We must now revisit sample predictability, formulate it in the extended setting just discussed, and then show that security in this new setting is implied by security in the basic one. Afterwards, we will be in a position to prove security for XKEY.

As in the basic sample predictability problem, let $k, \ell, p, q \geq 1$ be integers with $\ell \leq k$. Consider game $\mathbf{G}_{k,p,q}^{\text{skp2}}(\mathcal{B})$ defined in Fig. 6. In its first stage, the adversary returns an oracle leakage function Lk. We say that $\mathcal{B}$ leaks (or exfiltrates) ℓ bits if $\mathsf{LK}: \{0,1\}^k \rightarrow \{0,1\}^\ell$. Adversary $\mathcal{B}$ also returns state σ representing information that could be known to LK and will be passed to the adversary's second stage. The game picks the big-key $\boldsymbol{K}$, computes the leakage as $\mathsf{LK}^{\text{RO}}(\boldsymbol{K})$ and picks probe vectors $\mathbf{p}_1, \ldots, \mathbf{p}_q$. These implicitly determine subkeys $K_1, \ldots, K_q$ where $K_i = \boldsymbol{K}[\mathbf{p}_i]$. In its second stage, the adversary $\mathcal{B}$ gets the leakage L and probe vectors as before, but now additionally gets σ. Its task is to guess one of the subkeys, and the game returns true if it does this correctly. We let $\mathsf{Adv}_{k,p,q}^{\text{skp2}}(\mathcal{B}) = \Pr[\mathbf{G}_{k,p,q}^{\text{skp2}}(\mathcal{B})]$ be its advantage.

Game $\mathbf{G}_{k,p,q}^{\text{skp2}}(\mathcal{B})$	$\text{RO}(x, l)$
$(\mathsf{LK}, \sigma) \leftarrow \mathcal{B}^{\text{RO}}$	if not $T[x, l]$ then $T[x, l] \twoheadleftarrow \{0,1\}^l$
$\boldsymbol{K} \twoheadleftarrow \{0,1\}^k$; $L \leftarrow \mathsf{LK}^{\text{RO}}(\boldsymbol{K})$	return $T[x, l]$
for $i \leftarrow 1, \ldots, q$ do $\mathbf{p}_i \twoheadleftarrow [1..k]^p$	
$K \twoheadleftarrow \mathcal{B}^{\text{RO}}(L, \mathbf{p}_1, \ldots, \mathbf{p}_q, \sigma)$	
return $\big(K \in \{\boldsymbol{K}[\mathbf{p}_1], \ldots, \boldsymbol{K}[\mathbf{p}_p]\}\big)$	

Fig. 6. Game defining the "enhanced" subkey-prediction game, $\mathbf{G}^{\text{skp2}}$. The game differs from $\mathbf{G}^{\text{skp}}$ by allowing the adversary to select LK, both the latter and the adversary having access to a random oracle.

Lemma 11. *Let $\ell, k, p, q \geq 1$ be integers with $\ell \leq k$. Let $\mathcal{B}$ be an adversary leaking ℓ bits. Then*

$$\mathsf{Adv}_{k,p,q}^{\text{skp2}}(\mathcal{B}) \leq \mathsf{Adv}_{k,p,q}^{\text{skp}}(\ell) \ \blacksquare \tag{13}$$

The proof uses a fairly standard "coin-fixing" argument in which a predictor adversary uses the "best" choice of random oracle and coins for $\mathcal{B}$. The details follow.

Proof (Lemma 11). Let $\mathbf{H}$ denote the set of all functions H such that $H(\cdot, l): \{0,1\}^* \rightarrow \{0,1\}^l$ for all $l \in \mathbb{N}$. A random oracle is a function drawn at random from $\mathbf{H}$. Regard $\mathcal{B}, \ell, k, p, q$ as fixed and define the function $g: \mathbf{H} \times \{0,1\}^* \rightarrow [0,1]$ as $g(H, \omega) = \Pr[\mathbf{G}(H, \omega)]$ where game $\mathbf{G}(H, \omega)$ is on the left below:

Game $\mathbf{G}(H, \omega)$	Adversary $\mathcal{P}(L, \mathbf{p}_1, \ldots, \mathbf{p}_q)$
$(\mathsf{LK}, \sigma) \leftarrow \mathcal{B}^H(\omega)$; $\boldsymbol{K} \twoheadleftarrow \{0,1\}^k$	$J' \leftarrow \mathcal{B}^{H^*}(L, \mathbf{p}_1, \ldots, \mathbf{p}_q, \sigma^*; \omega^*)$
$L \leftarrow \mathsf{LK}^H(\boldsymbol{K})$	return J'
for $j \leftarrow 1, \ldots, q$ do	
$\quad \mathbf{p}_j[1], \ldots, \mathbf{p}_j[p] \twoheadleftarrow [1..k]$	
$\quad J_j \leftarrow \boldsymbol{K}[\mathbf{p}_j]$	
$J' \leftarrow \mathcal{B}^H(L, \mathbf{p}_1, \ldots, \mathbf{p}_q, \sigma; \omega)$	
return $(J' \in \{J_1, \ldots, J_q\})$	

In this game, the coins of $\mathcal{B}$ are fixed to ω and its random oracle is fixed to H. The probability is only over the choices made in the game. Let $(H^*, \omega^*) \in \mathbf{H} \times \{0,1\}^*$ be such that $g(H^*, \omega^*) \geq g(H, \omega)$ for all $(H, \omega) \in \mathbf{H} \times \{0,1\}^*$, and let $(\mathsf{LK}, \sigma^*) \leftarrow \mathcal{B}^{H^*}(\omega^*)$. Let $\mathsf{Lk} = \mathsf{LK}^{H^*}$. This is a basic (not oracle) leakage function, $\mathsf{Lk}: \{0,1\}^k \rightarrow \{0,1\}^\ell$. Define predictor adversary $\mathcal{P}$ as on the right above. Then

$$\begin{aligned}
\mathsf{Adv}_{k,p,q}^{\mathrm{skp2}}(\mathcal{B}) &= \mathbf{E}_{(H,\omega) \leftarrow \mathbf{H} \times \{0,1\}^*}\left[g(H, \omega)\right] \\
&\leq g(H^*, \omega^*) \\
&= \mathsf{Adv}_{k,p,q}^{\mathrm{skp}}(\mathcal{P}, \mathsf{Lk}) \leq \mathsf{Adv}_{k,p,q}^{\mathrm{skp}}(\ell)
\end{aligned}$$

which yields Eq. (13). ∎

Theorem 12. *Let* $k, \kappa, p, r \geq 1$ *be integers with* k *a power of two. Let* $\mathsf{KEY} = \mathsf{XKEY}_{k,\kappa,p,r}$ *be the big-key encapsulation scheme associated to them as per Fig. 5. Let* $\mathcal{A}$ *be an adversary making at most* q *queries to its* DERIVE *oracle and leaking* ℓ *bits. Assume the number of* RO *queries made by* $\mathcal{A}$ *in its first stage, plus the number made by the oracle leakage function* LK *that it outputs in this stage, is at most* q_1, *and the number of* RO *queries made by* $\mathcal{A}$ *in its second stage is at most* q_2. *Then*

$$\mathsf{Adv}_{\mathsf{KEY}}^{\mathrm{key}}(\mathcal{A}) \leq q_2 \cdot \mathsf{Adv}_{k,p,q}^{\mathrm{skp}}(\ell) + \frac{q \cdot (2q_1 + q - 1)}{2^{r+1}} \quad \blacksquare \qquad (14)$$

Game $\boxed{\mathbf{G}_0}$, $\mathbf{G}_1$

$K \twoheadleftarrow \{0,1\}^k$
for $j \leftarrow 1, \ldots, q$ do
    $\mathbf{R}[j] \twoheadleftarrow \{0,1\}^r$; $\mathbf{p}_j[1], \ldots, \mathbf{p}_j[p] \twoheadleftarrow [1..k]$; $\mathbf{KK}[j] \twoheadleftarrow \{0,1\}^\kappa$; $J_j \leftarrow K[\mathbf{p}_j]$
    For $i = 1, \ldots, j - 1$ do
        If $(\mathbf{R}[j] = \mathbf{R}[i])$ then $(\mathbf{p}_j, J_j) \leftarrow (\mathbf{p}_i, J_i)$; bad $\leftarrow$ true ; $\boxed{\mathbf{KK}[j] \leftarrow \mathbf{KK}[i]}$
$(\mathsf{LK}, \sigma) \twoheadleftarrow \mathcal{A}^{\mathrm{RO}}$; $L \leftarrow \mathsf{LK}^{\mathrm{RO}}(K)$; $c' \twoheadleftarrow \mathcal{A}^{\mathrm{DERIVE,RO}}(L, \sigma)$; Return $(c' = 1)$

Game $\mathbf{G}_2$, $\mathbf{G}_3$, $\mathbf{G}_4$, $\mathbf{G}_5$

$K \twoheadleftarrow \{0,1\}^k$
for $j \leftarrow 1, \ldots, q$ do
    $\mathbf{R}[j] \twoheadleftarrow \{0,1\}^r$; $\mathbf{p}_j[1], \ldots, \mathbf{p}_j[p] \twoheadleftarrow [1..k]$; $\mathbf{KK}[j] \twoheadleftarrow \{0,1\}^\kappa$; $J_j \leftarrow K[\mathbf{p}_j]$
    For $i = 1, \ldots, j - 1$ do
        If $(\mathbf{R}[j] = \mathbf{R}[i])$ then $(\mathbf{p}_j, J_j) \leftarrow (\mathbf{p}_i, J_i)$
stage $\leftarrow 1$; $(\mathsf{LK}, \sigma) \twoheadleftarrow \mathcal{A}^{\mathrm{RO}}$; $L \leftarrow \mathsf{LK}^{\mathrm{RO}}(K)$
stage $\leftarrow 2$; $j \leftarrow 0$; $c' \twoheadleftarrow \mathcal{A}^{\mathrm{DERIVE,RO}}(L, \sigma)$; Return $(c' = 1)$

DERIVE()

$j \leftarrow j + 1$; return $(\mathbf{R}[j], \mathbf{KK}[j])$

Fig. 7. Games for proof of Theorem 12. See Fig. 8 for the RO procedures.

$\mathrm{RO}(x, l)$ // Games $\mathbf{G}_0$, $\mathbf{G}_1$

if not $T[x, l]$ then
  $T[x, l] \twoheadleftarrow \{0, 1\}^l$
  for $j \leftarrow 1, \ldots, q$ do
    for $i \leftarrow 1, \ldots, p$ do
      if $(x = (\mathbf{R}[j], i, 0)$ and $l = \lg(k))$ then $T[x, l] \leftarrow \mathbf{p}_j[i]$
      if $(x = (\mathbf{R}[j], J_j, 1)$ and $l = r)$ then $T[x, l] \leftarrow \mathbf{KK}[j]$
return $T[x, l]$

$\mathrm{RO}(x, l)$ // Games $\boxed{\mathbf{G}_2}$, $\mathbf{G}_3$

if not $T[x, l]$ then
  $T[x, l] \twoheadleftarrow \{0, 1\}^l$
  if (stage = 1) then
    for $j \leftarrow 1, \ldots, q$ do
      for $i \leftarrow 1, \ldots, p$ do
        if $(x = (\mathbf{R}[j], i, 0)$ and $l = \lg(k))$ then bad $\leftarrow$ true ; $\boxed{T[x, l] \leftarrow \mathbf{p}_j[i]}$
        if $(x = (\mathbf{R}[j], J_j, 1)$ and $l = r)$ then bad $\leftarrow$ true ; $\boxed{T[x, l] \leftarrow \mathbf{KK}[j]}$
  if (stage = 2) then
    for $j \leftarrow 1, \ldots, q$ do
      for $i \leftarrow 1, \ldots, p$ do
        if $(x = (\mathbf{R}[j], i, 0)$ and $l = \lg(k)))$ then $T[x, l] \leftarrow \mathbf{p}_j[i]$
        if $(x = (\mathbf{R}[j], J_j, 1)$ and $l = r)$ then $T[x, l] \leftarrow \mathbf{KK}[j]$
return $T[x, l]$

$\mathrm{RO}(x, l)$ // Games $\boxed{\mathbf{G}_4}$, $\mathbf{G}_5$

if not $T[x, l]$ then
  $T[x, l] \twoheadleftarrow \{0, 1\}^l$
  if (stage = 2) then
    for $j \leftarrow 1, \ldots, q$ do
      for $i \leftarrow 1, \ldots, p$ do
        if $(x = (\mathbf{R}[j], i, 0)$ and $l = \lg(k))$ then $T[x, l] \leftarrow \mathbf{p}_j[i]$
        if $(x = (\mathbf{R}[j], J_j, 1)$ and $l = r)$ then bad $\leftarrow$ true ; $\boxed{T[x, l] \leftarrow \mathbf{KK}[j]}$
return $T[x, l]$

Fig. 8. RO procedures for games for proof of Theorem 12.

We note that the bound of Eq. (14) does not depend on the length κ of the output keys.

Proof (Theorem 12). Consider the games of Fig. 7. Their RO procedures are in Fig. 8. Game $\mathbf{G}_0$ includes the boxed code and is equivalent to case of game $\mathbf{G}_{\mathsf{KEY}}^{\mathsf{key}}(\mathcal{A})$ in which $b = 1$. Game $\mathbf{G}_5$ excludes the boxed code and mimics the $b = 0$ case of game $\mathbf{G}_{\mathsf{KEY}}^{\mathsf{key}}(\mathcal{A})$, except that the probability the former returns true is the probability the latter returns false. Let $p_i = \Pr[\mathbf{G}_i]$ for $0 \le i \le 5$. Then

$$
\begin{aligned}
\mathsf{Adv}_{\mathsf{KEY}}^{\mathsf{key}}(\mathcal{A}) & \\
&= \Pr[\mathbf{G}_{\mathsf{KEY}}^{\mathsf{key}}(\mathcal{A}) \mid b = 1] - \left(1 - \Pr[\mathbf{G}_{\mathsf{KEY}}^{\mathsf{key}}(\mathcal{A}) \mid b = 0]\right) \\
&= \Pr[\mathbf{G}_0] - \Pr[\mathbf{G}_5] \\
&= (p_0 - p_1) + (p_1 - p_2) + (p_2 - p_3) + (p_3 - p_4) + (p_4 - p_5) \\
&= (p_0 - p_1) + (p_2 - p_3) + (p_4 - p_5) & (15) \\
&\le \Pr[\mathbf{G}_1 \text{ sets bad}] + \Pr[\mathbf{G}_3 \text{ sets bad}] + \Pr[\mathbf{G}_5 \text{ sets bad}] . & (16)
\end{aligned}
$$

Equation (15) used the fact that $p_1 = p_2$ and $p_3 = p_4$. Equation (16) used the Fundamental Lemma of Game Playing [13]. Now

$$
\Pr[\mathbf{G}_1 \text{ sets bad}] \le \frac{q(q-1)}{2^{r+1}} . \tag{17}
$$

Also

$$
\Pr[\mathbf{G}_3 \text{ sets bad}] \le \frac{q \cdot q_1}{2^r} . \tag{18}
$$

The first estimate of the above bound may be that each RO query can set the first instance of bad with probability $pq/2^r$ and the second with probability $q/2^r$

Adversary $\mathcal{B}^{\mathrm{RO}}$	Adversary $\mathcal{B}^{\mathrm{RO}}(L, \mathbf{p}_1, \ldots, \mathbf{p}_q, (\sigma, T))$

$(\mathsf{LK}, \sigma) \twoheadleftarrow \mathcal{A}^{\mathrm{RO}}$
 return $(\mathsf{LK}, (\sigma, T))$
	for $j \leftarrow 1, \ldots, q$ do
	    $\mathbf{R}[j] \twoheadleftarrow \{0,1\}^r$; $\mathbf{KK}[j] \twoheadleftarrow \{0,1\}^\kappa$
	    for $i = 1, \ldots, j-1$ do
	      If $(\mathbf{R}[j] = \mathbf{R}[i])$ then $\mathbf{p}_j \leftarrow \mathbf{p}_i$

$\underline{\mathrm{ROSIM}(x,l)}$ If not $T[\mathsf{x},\mathsf{l}]$ then $T[x,l] \leftarrow \mathrm{RO}(x,l)$ Return $T[x,l]$	$S \leftarrow \emptyset$; $j \leftarrow 0$; $c' \twoheadleftarrow \mathcal{A}^{\mathrm{DERIVE,ROSIM}}(L, \sigma)$ $J' \twoheadleftarrow S$; return J'
	$\underline{\mathrm{DERIVE}()}$ $j \leftarrow j+1$; return $(\mathbf{R}[j], \mathbf{KK}[j])$

	$\underline{\mathrm{ROSIM}(x,l)}$
	 If not $T[x,l]$ then
	    $T[x,l] \leftarrow \mathrm{RO}(x,l)$
	    for $j \leftarrow 1, \ldots, q$ do
	      for $i \leftarrow 1, \ldots, p$ do
	        if $(x = (\mathbf{R}[j], i, 0)$ and $l = \lg(k))$ then
	          $T[x,l] \leftarrow \mathbf{p}_j[i]$
	 $(R, J, b) \leftarrow x$
	 If $b = 1$ then $S \leftarrow S \cup \{J\}$
	 return $T[x,l]$

Fig. 9. Adversary $\mathcal{B}$ for proof of Theorem 12.

for a bound of $q_1(p+1)q/2^r$. But these different events are mutually exclusive due to the queries including the index i and the domain separation bit, whence Eq. (18). Now we will present an adversary $\mathcal{A}$ such that

$$\Pr[\mathbf{G}_5 \text{ sets bad}] \leq q_2 \cdot \mathsf{Adv}^{\text{skp2}}_{k,p,q}(\mathcal{B}) \ . \tag{19}$$

The theorem follows from Lemma 11. Adversary $\mathcal{B}$ is shown in Fig. 9. In the first stage, $\mathcal{B}$ simulates $\mathcal{A}$'s RO directly via its own RO, keeping track of values in table T, which is passed to the next stage. In the second it does the same, makes sure to map selectors to probes as per game $\mathbf{G}_5$, and also it saves subkey guesses in the set S. Equation (19) follows because $|S| \leq q_2$. $\blacksquare$

5 Big-Key Symmetric Encryption

Here we define and achieve big-key symmetric encryption.

DEFINITIONS. A symmetric encryption scheme SE specifies a key length $\mathsf{SE.kl} \in \mathbb{N}$, an encryption algorithm SE.Enc that given a key K and message M returns a ciphertext, and a deterministic decryption algorithm SE.Dec such that for all K, M we have $\mathsf{SE.Dec}(K, \mathsf{SE.Enc}(K, M)) = M$ with probability one, where the probability is over the coins of SE.Enc. Privacy is formalized by left or right indistinguishability [7] via game $\mathbf{IND}_{\mathsf{SE}}(\mathcal{A})$ on the right of Fig. 10 associated to SE and adversary $\mathcal{A}$. We let $\mathsf{Adv}^{\text{ind}}_{\mathsf{SE}}(\mathcal{A}) = 2\Pr[\mathbf{IND}_{\mathsf{SE}}(\mathcal{A})] - 1$ be the advantage of $\mathcal{A}$ in violating privacy of SE.

Syntactically, a big-key symmetric encryption scheme **SE** continues to be a symmetric encryption scheme as above, specifying **SE.kl**, **SE.Enc** and **SE.Dec**. Two things make it special. First, privacy is measured under leakage on the key. Second, the encryption and decryption algorithms have the "locality" efficiency attribute, which means that in any one execution they access only a small part of the key. Privacy is formalized via game $\mathbf{LIND}_{\mathsf{SE}}(\mathcal{A})$ on the left of Fig. 10 associated to **SE** and adversary $\mathcal{A}$. In its first stage, the adversary, given access to RO, specifies an oracle leakage function $\mathsf{LK} \colon \{0,1\}^{\mathsf{SE.kl}} \to \{0,1\}^{\ell}$ together with state information σ. We refer to ℓ as the number of bits leaked by $\mathcal{A}$. In its second stage, $\mathcal{A}$ gets the leakage $L \leftarrow \mathsf{LK}^{\mathrm{RO}}(\boldsymbol{K})$, the state σ, and access to the challenge encryption oracle ENC, while continuing to have access to RO. To win it must guess the challenge bit b. We let $\mathsf{Adv}^{\text{lind}}_{\mathsf{SE}}(\mathcal{A}) = 2\Pr[\mathbf{LIND}_{\mathsf{SE}}(\mathcal{A})] - 1$ be the advantage of $\mathcal{A}$ in violating privacy of **SE** under leakage. Locality will not be formalized but rather visible in specific constructs. Giving the leakage function access to RO is important for same reason as we discussed in Sect. 4 for key encapsulation, namely that, otherwise, there are trivial ROM schemes that are secure but when the RO is instantiated the resulting scheme is clearly not secure.

BIG-KEY ENCRYPTION SCHEME. Let SE be a symmetric encryption scheme. Let KEY be a key-encapsulation algorithm with big-key length k, randomness length r and derived key length $\kappa = \mathsf{SE.kl}$ (keys output by KEY are suitable for use with SE). We associate to SE, KEY the big-key symmetric encryption scheme

Game $\mathbf{LIND}_{\mathsf{SE}}(\mathcal{A})$	Game $\mathbf{IND}_{\mathsf{SE}}(\mathcal{A})$
$(\mathrm{LK}, \sigma) \twoheadleftarrow \mathcal{A}^{\mathrm{RO}};\ \boldsymbol{K} \twoheadleftarrow \{0,1\}^{\mathsf{SE.kl}}$	$S \twoheadleftarrow \{0,1\}^{\mathsf{SE.kl}}$
$L \leftarrow \mathsf{Lk}^{\mathrm{RO}}(\boldsymbol{K});\ b \twoheadleftarrow \{0,1\}$	$b \twoheadleftarrow \{0,1\}$
$c' \leftarrow \mathcal{A}^{\mathrm{ENC},\mathrm{RO}}(L, \sigma)$	$c' \leftarrow \mathcal{A}^{\mathrm{ENC}}$
Return $(c' = b)$	Return $(c' = b)$
$\underline{\mathrm{ENC}(M_0, M_1)}$	$\underline{\mathrm{ENC}(M_0, M_1)}$
$\overline{C} \twoheadleftarrow \mathsf{SE.Enc}^{\mathrm{RO}}(\boldsymbol{K}, M_b)$	$C \twoheadleftarrow \mathsf{SE.Enc}(S, M_b)$
Return $\overline{C}$	Return C
$\underline{\mathrm{RO}(x, l)}$	
If not $T[x, l]$ then $T[x, l] \twoheadleftarrow \{0,1\}^l$	
Return $T[x, l]$	

Fig. 10. Game defining privacy of symmetric encryption scheme **SE** under leakage, and game defining standard privacy of symmetric encryption scheme SE.

$\mathsf{SE} = \mathsf{BKSE}[\mathsf{SE}, \mathsf{KEY}]$ defined as follows. The key length is $\mathsf{SE.kl} = k$ (the key for **SE** is the same as that for KEY) and the encryption and decryption algorithms are as follows:

Algorithm $\mathsf{SE.Enc}^{\mathrm{RO}}(K, M)$	Algorithm $\mathsf{SE.Dec}^{\mathrm{RO}}(K, \overline{C})$
$R \twoheadleftarrow \{0,1\}^r;\ K \twoheadleftarrow \mathsf{KEY}^{\mathrm{RO}}(K, R)$	$(R, C) \leftarrow \overline{C}$
$C \twoheadleftarrow \mathsf{SE.Enc}(K, M)\ ;\ \overline{C} \leftarrow (R, C)$	$K \leftarrow \mathsf{KEY}^{\mathrm{RO}}(K, R)$
Return $\overline{C}$	$M \leftarrow \mathsf{SE.Dec}(K, C)$
	Return M

Encryption applies the key encapsulation algorithm to the big-key to get a derived key K. The message is encrypted under SE using key K. The locality of this scheme is exactly that of KEY since accesses to the key are done only by KEY. The big-key aspect is similarly inherited from KEY. The following says that our big-key scheme achieves privacy under leakage on the key assuming standard privacy of the base scheme SE and the lror-security of KEY. The proof is in [9].

Theorem 13. *Let SE be a symmetric encryption scheme. Let KEY be a key encapsulation algorithm with big-key length k, randomness length r and derived key length $\kappa = \mathsf{SE.kl}$. Let $\mathsf{SE} = \mathsf{BKSE}[\mathsf{SE}, \mathsf{KEY}]$ be the big-key symmetric encryption scheme associated to them as above. Let $\mathcal{A}$ be an adversary making at most q queries to its ENC oracle and leaking ℓ bits. Then the proof below specifies an adversary $\mathcal{A}_1$ and an adversary $\mathcal{A}_2$ such that*

$$\mathsf{Adv}^{\mathrm{lind}}_{\mathsf{SE}}(\mathcal{A}) \ \leq \ \mathsf{Adv}^{\mathrm{key}}_{\mathsf{KEY}}(\mathcal{A}_2) + q \cdot \mathsf{Adv}^{\mathrm{ind}}_{\mathsf{SE}}(\mathcal{A}_1)\,. \tag{20}$$

Adversary $\mathcal{A}_1$ makes only one query to its ENC oracle and its running time is about that of $\mathcal{A}$. Adversary $\mathcal{A}_2$ makes q queries to its DERIVE oracle and has *running time about that of $\mathcal{A}$. Its first stage is the same as that of $\mathcal{A}$, so it also leaks ℓ bits. In its second stage it makes the same number of RO queries as $\mathcal{A}$ does in its second stage.* ∎

Since $\mathcal{A}_1$ makes only one query to its ENC oracle, a one-time encryption scheme suffices to instantiate SE.

6   Standard-Model Big-Key Encryption

In this section we give a standard-model variant of our scheme whose security relies on UCE. The scheme is as efficient as our ROM one, but ciphertexts are longer.

DEFINITIONS. We recall the UCE framework following [8]. Let H: $\{0,1\}^{H.kl} \times \{0,1\}^{H.il} \to \{0,1\}^{H.ol}$ be a family of functions taking an H.kl-bit key I and H.il-bit input x to a H.ol-bit output $H(I,x)$. Game $\mathbf{G}^{uce}$ of Fig. 11 is associated to H, an adversary $\mathcal{S}$ called the *source*, an adversary $\mathcal{D}$ called the distinguisher, and a number q of keys. (We are using the multi-key version of UCE from [8].) Here $\mathcal{S}$ does not get the keys. It produces leakage M that is passed to $\mathcal{D}$, who does get the keys. We let $\mathsf{Adv}^{uce}_{H,q}(\mathcal{S},\mathcal{D}) = 2\Pr[\mathbf{G}^{uce}_{H,q}(\mathcal{S},\mathcal{D})] - 1$. We can only expect this to be small for sources restricted in some way. We require statistical unpredictability [8,15] of the source's oracle queries. If $\mathcal{P}$ is an adversary called the *predictor*, let $\mathsf{Adv}^{pred}_{\mathcal{S}}(\mathcal{P}) = \Pr[\mathbf{G}^{sp}_{\mathcal{S}}(\mathcal{P})]$ where the game is again in Fig. 11, and let $\mathsf{Adv}^{pred}_{\mathcal{S}} = \max_{\mathcal{P}} \mathsf{Adv}^{pred}_{\mathcal{S}}(\mathcal{P})$. The predictor here is unbounded (corresponding to statistical unpredictability) so the maximum is over all predictors. The assumption, informally, is that if $\mathsf{Adv}^{pred}_{\mathcal{S}}$ is small then so is $\mathsf{Adv}^{uce}_{H,q}(\mathcal{S},\mathcal{D})$ for all efficient $\mathcal{S},\mathcal{D}$. An important element of results is thus to be able to bound $\mathsf{Adv}^{pred}_{\mathcal{S}}$ for the $\mathcal{S}$ constructed by the reduction.

XKEY2. The encapsulation algorithm is specified in Fig. 12. If κ is the desired length of the derived key, k the length of the big-key $\boldsymbol{K}$ (assumed a power of two for simplicity) and p the number of probes then it uses a family of functions

Game $\mathbf{G}^{uce}_{H,q}(\mathcal{S},\mathcal{D})$	Game $\mathbf{G}^{sp}_{\mathcal{S}}(\mathcal{P})$
$b \twoheadleftarrow \{0,1\}$	$Q \leftarrow \emptyset$; $M \twoheadleftarrow \mathcal{S}^{HASH}$; $x \twoheadleftarrow \mathcal{P}(M)$
For $i = 1,\ldots,q$ do $I_i \twoheadleftarrow \{0,1\}^{H.kl}$	Return $(x \in Q)$
$M \twoheadleftarrow \mathcal{S}^{HASH}$; $b' \twoheadleftarrow \mathcal{D}(I_1,\ldots,I_q,M)$	
Return $(b' = b)$	$\underline{HASH(x,j)}$
	If not $T[x,j]$ then $T[x,j] \twoheadleftarrow \{0,1\}^{H.ol}$
$\underline{HASH(x,j)}$	$Q \leftarrow Q \cup \{x\}$; Return $T[x,j]$
If not $T[x,j]$ then	
If $b = 0$ then $T[x,j] \twoheadleftarrow \{0,1\}^{H.ol}$	
Else $T[x,j] \leftarrow H(I_j,x)$	
Return $T[x,j]$	

Fig. 11. Games $\mathbf{G}^{uce}$ and $\mathbf{G}^{sp}$ to define UCE security.

H with H.ol $= \kappa$ and H.il $= p$. The selector is of length $r = $ H.kl $+ p \cdot \lg(k)$ and specifies a key I for H as well as the probe sequence $\mathbf{p}$. The derived key K is then computed as shown. The following theorem says that the scheme works, meaning achieves our notion of encapsulation security. This involves two claims. First is that the key encapsulation advantage can be bounded by the uce advantage of a source-distinguisher pair. But this by itself is not enough. To ensure this uce advantage is small, we also show that the predictability of the source can be bounded. Here we appeal to our bound on sub-key predictability, so that once again the latter emerges as crucial.

Algorithm $\mathsf{XKEY2}_{k,\kappa,p,r}(\mathbf{K}, R)$

$(I, \mathbf{p}) \leftarrow R$; $J \leftarrow \mathbf{K}[\mathbf{p}]$; $K \leftarrow \mathsf{H}(I, J)$; Return K

Fig. 12. Encapsulation algorithm $\mathsf{XKEY2}$. Given a length-k big-key $\mathbf{K}$ and a length-r selector $R = (I, \mathbf{p})$, the algorithm returns a length-κ subkey K.

Theorem 14. Let $k, \kappa, p \geq 1$ be integers and H a family of functions with H.ol $= \kappa$ and H.il $= p$. Let $r = $ H.kl $+ p \cdot \lg(k)$. Let $\mathsf{KEY} = \mathsf{XKEY2}_{k,\kappa,p,r}$ be the big-key key-derivation scheme associated to them as per Fig. 12. Let $\mathcal{A}$ be an adversary making at most q queries to its DERIVE oracle and leaking ℓ bits. The proof specifies a source adversary $\mathcal{S}$ and a distinguisher adversary $\mathcal{D}$ such that

$$\mathsf{Adv}^{\mathrm{key}}_{\mathsf{KEY}}(\mathcal{A}) \leq \mathsf{Adv}^{\mathrm{uce}}_{\mathsf{H},q}(\mathcal{S}, \mathcal{D}) \quad and \quad \mathsf{Adv}^{\mathrm{pred}}_{\mathcal{S}} \leq \mathsf{Adv}^{\mathrm{skp}}_{k,p,q}(\ell) . \qquad (21)$$

Adversary $\mathcal{S}$ makes q queries to its HASH oracle (one per key) and the running times of $\mathcal{S}$ and $\mathcal{D}$ add up to essentially that of $\mathcal{A}$. ∎

Proof (Theorem 14). Adversaries $\mathcal{S}, \mathcal{D}$ are as follows:

Adversary $\mathcal{S}^{\mathrm{HASH}}$	Adversary $\mathcal{D}(I_1, \ldots, I_q, M)$
$\mathbf{K} \twoheadleftarrow \{0,1\}^k$; $(\mathsf{LK}, \sigma) \leftarrow \mathcal{A}()$	$(L, \sigma, \mathbf{p}_1, \ldots, \mathbf{p}_q, K_1, \ldots, K_q) \leftarrow M$
$L \twoheadleftarrow \mathsf{LK}(\mathbf{K})$	$i \leftarrow 0$; $c' \twoheadleftarrow \mathcal{A}^{\mathrm{DERIVE}}(L, \sigma)$
For $i = 1, \ldots, q$ do	Return c'
$\quad \mathbf{p}_i \twoheadleftarrow [1..k]^p$; $J_i \leftarrow \mathbf{K}[\mathbf{p}_i]$	$\underline{\mathrm{DERIVE}()}$
$\quad K_i \leftarrow \mathrm{HASH}(J_i, i)$	$i \leftarrow i + 1$; return $((I_i, \mathbf{p}_i), K_i)$
$M \leftarrow (L, \sigma, \mathbf{p}_1, \ldots, \mathbf{p}_q, K_1, \ldots, K_q)$	
Return M	

Adversary $\mathcal{S}$ itself picks the big-key $\mathbf{K}$ and runs $\mathcal{A}$ to get the leakage function, producing its own leakage M as shown. Adversary $\mathcal{D}$ continues the execution of $\mathcal{A}$, being in a position to answer DERIVE queries because it has $I_1, \ldots, I_q$. If $\mathcal{P}$ is any predictor adversary, the leakage M it gets in its game specifies the output L of the leakage function, the probe sequences, and independent random strings $K_1, \ldots, K_q$, and an oracle query of the source is a subkey, so guessing it is exactly guessing a subkey. It follows that $\mathsf{Adv}^{\mathrm{pred}}_{\mathcal{S}}(\mathcal{P}) \leq \mathsf{Adv}^{\mathrm{skp}}_{k,p,q}(\ell)$. We omit the details. ∎

BIG-KEY ENCRYPTION. We can turn XKEY2 into a big-key encryption scheme via the general transform of Sect. 5. This transform does not introduce a random oracle. (If the big-key key encapsulation mechanism used one, it will inherit it, but will not introduce an additional use.) Thus the result of applying the transform to XKEY2 is a standard-model big-key encryption scheme. It satisfies locality because XKEY2 does. Theorem 13 reduces its security to that of XKEY2, and thus we can conclude by applying Theorem 14. We omit the details.

7   Authenticity and Hedged Big-Key Encryption

In real-world settings we are likely to want authenticated encryption (AE) rather than privacy-only encryption. We should thus ask whether, we can have big-key AE rather than the privacy-only formulation we have now. As discussed in Sect. 1, this is not possible due to the following attack: the adversary simply leaks a valid ciphertext. This is a small amount of leakage, yet violates authenticity.

To overcome this difficulty we suggest to use what we call *hedged* big-key encryption. This provides privacy in the big-key setting we have already defined and achieved; additionally, in the absence of leakage, it provides authenticity. We suggest that this is a good goal because, in the mass-surveillance / APT context, it is privacy that is the main concern, not authenticity; but in the absence of an APT, our concerns would be the usual ones, which include authenticity. Hedged big-key encryption provides both, so that security does not degrade by moving to big keys.

There is a simple and generic way to turn a privacy-only big-key encryption scheme into a hedged big-key encryption scheme. Reserve a small (128-bit, say) portion K of the big-key $\boldsymbol{K}$ as a key for a conventional PRF or MAC. Then use encrypt-then-mac [10]. Namely, big-key encrypt the message under the remaining (big) portion of $\boldsymbol{K}$ to get a ciphertext C, and return (C, T) as the ciphertext for the hedged big-key scheme, where T is the result of applying a PRF, keyed by K, to C. In the absence of leakage, we have authenticated encryption by applying results of [10]. In the presence of leakage, we must assume the small key K is leaked in its entirety, but the big-key privacy-only component will still provide the same privacy as before. Here we use the fact that in the privacy proof of [10], the adversary can be given the PRF (MAC) key.

Acknowledgments. Bellare was supported in part by NSF grants CNS-1526801 and CNS-1228890, a gift from Microsoft corporation and ERC Project ERCC (FP7/615074). Rogaway was supported in part by NSF grants CNS-1228828 and CNS-1314885. We thank Joseph Jaeger for comments and corrections, Wei Dai for helpful discussions, and the CRYPTO 2016 reviewers for their knowledgeable reviews, corrections and pointers to the literature.

References

1. Agrawal, D., Archambeault, B., Rao, J.R., Rohatgi, P.: The EM side-channels. In: Kaliski Jr., B.S., Koç, Ç.K., Paar, C. (eds.) CHES 2002. LNCS, vol. 2523. Springer, Heidelberg (2003)

2. Alwen, J., Dodis, Y., Naor, M., Segev, G., Walfish, S., Wichs, D.: Public-key encryption in the bounded-retrieval model. In: Gilbert, H. (ed.) EUROCRYPT 2010. LNCS, vol. 6110, pp. 113–134. Springer, Heidelberg (2010)

3. Alwen, J., Dodis, Y., Wichs, D.: Leakage-resilient public-key cryptography in the bounded-retrieval model. In: Halevi, S. (ed.) CRYPTO 2009. LNCS, vol. 5677, pp. 36–54. Springer, Heidelberg (2009)

4. Alwen, J., Dodis, Y., Wichs, D.: Survey: leakage resilience and the bounded retrieval model. In: Kurosawa, K. (ed.) Information Theoretic Security. LNCS, vol. 5973, pp. 1–18. Springer, Heidelberg (2010)

5. Aumann, Y., Ding, Y.Z., Rabin, M.O.: Everlasting security in the bounded storage model. IEEE Trans. Inf. Theory 48(6), 1668–1680 (2002)

6. Aumann, Y., Rabin, M.O.: Information theoretically secure communication in the limited storage space model. In: Wiener, M. (ed.) CRYPTO 1999. LNCS, vol. 1666, pp. 65–79. Springer, Heidelberg (1999)

7. Bellare, M., Desai, A., Jokipii, E., Rogaway, P.: A concrete security treatment of symmetric encryption. In: 38th FOCS, pp. 394–403. IEEE Computer Society Press, October 1997

8. Bellare, M., Hoang, V.T., Keelveedhi, S.: Instantiating random oracles via UCEs. In: Canetti, R., Garay, J.A. (eds.) CRYPTO 2013, Part II. LNCS, vol. 8043, pp. 398–415. Springer, Heidelberg (2013)

9. Bellare, M., Kane, D., Rogaway, P.: Big-key symmetric encryption: resisting key exfiltration. Cryptology ePrint Archive, report 2016/541 (2016)

10. Bellare, M., Namprempre, C.: Authenticated encryption: relations among notions and analysis of the generic composition paradigm. In: Okamoto, T. (ed.) ASIACRYPT 2000. LNCS, vol. 1976, pp. 531–545. Springer, Heidelberg (2000)

11. Bellare, M., Paterson, K.G., Rogaway, P.: Security of symmetric encryption against mass Surveillance. In: Garay, J.A., Gennaro, R. (eds.) CRYPTO 2014, Part I. LNCS, vol. 8616, pp. 1–19. Springer, Heidelberg (2014)

12. Bellare, M., Rogaway, P.: Random oracles are practical: a paradigm for designing efficient protocols. In: Ashby, V. (ed.) ACM CCS 1993, pp. 62–73. ACM Press, November 1993

13. Bellare, M., Rogaway, P.: The Security of triple encryption and a framework for code-based game-playing proofs. In: Vaudenay, S. (ed.) EUROCRYPT 2006. LNCS, vol. 4004, pp. 409–426. Springer, Heidelberg (2006)

14. Boyen, X.: Reusable cryptographic fuzzy extractors. In: Atluri, V., Pfitzmann, B., McDaniel, P. (eds.) ACM CCS 2004, pp. 82–91. ACM Press, October 2004

15. Brzuska, C., Farshim, P., Mittelbach, A.: Indistinguishability obfuscation and UCEs: the case of computationally unpredictable sources. In: Garay, J.A., Gennaro, R. (eds.) CRYPTO 2014, Part I. LNCS, vol. 8616, pp. 188–205. Springer, Heidelberg (2014)

16. Cachin, C., Maurer, U.M.: Unconditional security against memory-bounded adversaries. In: Kaliski Jr., B.S. (ed.) CRYPTO 1997. LNCS, vol. 1294, pp. 292–306. Springer, Heidelberg (1997)

17. Cash, D.M., Ding, Y.Z., Dodis, Y., Lee, W., Lipton, R.J., Walfish, S.: Intrusion-resilient key exchange in the bounded retrieval model. In: Vadhan, S.P. (ed.) TCC 2007. LNCS, vol. 4392, pp. 479–498. Springer, Heidelberg (2007)

18. Cramer, R., Shoup, V.: Design and analysis of practical public-key encryption schemes secure against adaptive chosen ciphertext attack. SIAM J. Comput. **33**(1), 167–226 (2003)

19. Dagon, D., Lee, W., Lipton, R.J.: Protecting secret data from insider attacks. In: Patrick, A.S., Yung, M. (eds.) FC 2005. LNCS, vol. 3570, pp. 16–30. Springer, Heidelberg (2005)

20. Di Crescenzo, G., Lipton, R.J., Walfish, S.: Perfectly secure password protocols in the bounded retrieval model. In: Halevi, S., Rabin, T. (eds.) TCC 2006. LNCS, vol. 3876, pp. 225–244. Springer, Heidelberg (2006)

21. Dodis, Y., Kalai, Y.T., Lovett, S.: On cryptography with auxiliary input. In: Mitzenmacher, M. (ed.) 41st ACM STOC, pp. 621–630. ACM Press, May/June 2009

22. Dodis, Y., Ristenpart, T., Vadhan, S.: Randomness condensers for efficiently samplable, seed-dependent sources. In: Cramer, R. (ed.) TCC 2012. LNCS, vol. 7194, pp. 618–635. Springer, Heidelberg (2012)

23. Dziembowski, S.: Intrusion-resilience via the bounded-storage model. In: Halevi, S., Rabin, T. (eds.) TCC 2006. LNCS, vol. 3876, pp. 207–224. Springer, Heidelberg (2006)

24. Dziembowski, S., Maurer, U.M.: Optimal randomizer efficiency in the bounded-storage model. J. Cryptol. **17**(1), 5–26 (2004)

25. Dziembowski, S., Pietrzak, K.: Leakage-resilient cryptography. In: 49th FOCS, pp. 293–302. IEEE Computer Society Press, October 2008

26. Halevi, S., Lin, H.: After-the-fact leakage in public-key encryption. In: Ishai, Y. (ed.) TCC 2011. LNCS, vol. 6597, pp. 107–124. Springer, Heidelberg (2011)

27. Kelsey, J., Schneier, B.: Authenticating secure tokens using slow memory access. In: Proceedings of the USENIX Workshop on Smartcard Technology (Smartcard 1999), 10–11 May 1999, Chicago, Illinois, USA, p. 101. USENIX Association (1999)

28. Lu, C.-J.: Hyper-encryption against space-bounded adversaries from on-line strong extractors. In: Yung, M. (ed.) CRYPTO 2002. LNCS, vol. 2442, pp. 257–271. Springer, Heidelberg (2002)

29. Maurer, U.M.: Conditionally-perfect secrecy and a provably-secure randomized cipher. J. Cryptol. **5**(1), 53–66 (1992)

30. Nisan, N., Zuckerman, D.: Randomness is linear in space. J. Comput. Syst. Sci. **52**(1), 43–52 (1996)

31. Pietrzak, K.: A leakage-resilient mode of operation. In: Joux, A. (ed.) EUROCRYPT 2009. LNCS, vol. 5479, pp. 462–482. Springer, Heidelberg (2009)

32. Raz, R., Reingold, O.: On recycling the randomness of states in space bounded computation. In: 31st ACM STOC, pp. 159–168. ACM Press, May 1999

33. Reingold, O., Shaltiel, R., Wigderson, A.: Extracting randomness via repeated condensing. SIAM J. Comput. **35**(5), 1185–1209 (2006)

34. Shin, S.H., Kobara, K., Imai, H.: Leakage-resilient authenticated key establishment protocols. In: Laih, C.-S. (ed.) ASIACRYPT 2003. LNCS, vol. 2894, pp. 155–172. Springer, Heidelberg (2003)

35. Vadhan, S.P.: Constructing locally computable extractors and cryptosystems in the bounded-storage model. J. Cryptol. **17**(1), 43–77 (2004)

Backdoors in Pseudorandom Number Generators: Possibility and Impossibility Results

Jean Paul Degabriele[1]([✉]), Kenneth G. Paterson[1], Jacob C.N. Schuldt[2], and Joanne Woodage[1]

[1] Royal Holloway, University of London, London, UK
{jean.degabriele,kenny.paterson}@rhul.ac.uk
joanne.woodage.2014@live.rhul.ac.uk
[2] AIST, Tokyo, Japan
jacob.schuldt@aist.go.jp

Abstract. Inspired by the Dual EC DBRG incident, Dodis et al. (Eurocrypt 2015) initiated the formal study of backdoored PRGs, showing that backdoored PRGs are equivalent to public key encryption schemes, giving constructions for backdoored PRGs (BPRGs), and showing how BPRGs can be "immunised" by careful post-processing of their outputs. In this paper, we continue the foundational line of work initiated by Dodis et al., providing both positive and negative results.

We first revisit the backdoored PRG setting of Dodis et al., showing that PRGs can be *more strongly* backdoored than was previously envisaged. Specifically, we give efficient constructions of BPRGs for which, given a single generator output, Big Brother can recover the initial state and, therefore, *all* outputs of the BPRG. Moreover, our constructions are *forward-secure* in the traditional sense for a PRG, resolving an open question of Dodis et al. in the negative.

We then turn to the question of the effectiveness of backdoors in robust PRNGs with input (c.f. Dodis et al., ACM-CCS 2013): generators in which the state can be regularly refreshed using an entropy source, and in which, provided sufficient entropy has been made available since the last refresh, the outputs will appear pseudorandom. The presence of a refresh procedure might suggest that Big Brother could be defeated, since he would not be able to predict the values of the PRNG state backwards or forwards through the high-entropy refreshes. Unfortunately, we show that this intuition is not correct: we are also able to construct robust PRNGs with input that are backdoored in a backwards sense. Namely, given a single output, Big Brother is able to rewind through a number of refresh operations to earlier "phases", and recover all the generator's outputs in those earlier phases.

Finally, and ending on a positive note, we give an impossibility result: we provide a bound on the number of previous phases that Big Brother can compromise as a function of the state-size of the generator: smaller states provide more limited backdooring opportunities for Big Brother.

© International Association for Cryptologic Research 2016
M. Robshaw and J. Katz (Eds.): CRYPTO 2016, Part I, LNCS 9814, pp. 403–432, 2016.
DOI: 10.1007/978-3-662-53018-4_15

1   Introduction

Background: In the wake of the Snowden revelations, the cryptographic research community has begun to realise that it faces a more powerful and insidious adversary than it had previously envisaged: Big Brother, an adversary willing to subvert cryptographic standards and implementations in order to gain an advantage against users of cryptography. The Dual EC DRBG debacle, and subsequent research showing the widespread use of this NIST-standardised pseudorandom generator (PRG) and its security consequences [11], has highlighted that inserting backdoors into randomness-generating components of systems is a profitable, if high-risk, strategy for Big Brother.

The threat posed by the Big Brother adversary brings new research challenges, both foundational and applied. The study of subversion of cryptographic systems — how to undetectably and securely subvert them, and how to defend against subversion — is a central one. Current research efforts to understand various forms of subversion include the study of Algorithm Substitution Attacks (ASAs) [2,6,13,23,28] and that of backdooring of cryptosystems [3,8,11,15]. These lines of research have a long and rich history through topics such as kleptography [34] and subliminal channels [31]. In an ASA, the subversion is specific to a specific *implementation* of a particular algorithm or scheme, whereas in backdooring, the backdoor resides in the specification of the scheme or primitive itself and any implementation faithful to the specification will be equally vulnerable. There is a balancing act at play with these two types of attack: while ASAs are arguably easier to carry out, their impact is limited to a specific implementation, whereas the successful introduction of a backdoor into a cryptographic scheme, albeit ostensibly harder to mount and subsequently conceal, can have much wider impact.

The Importance of Randomness: Many cryptographic processes rely heavily on good sources of randomness, for example, key generation, selection of IVs for encryption schemes and random challenges in authentication protocols, and the selection of Diffie-Hellman exponents. Indeed randomness failures of various kinds have led to serious vulnerabilities in widely deployed cryptographic systems, with a growing literature on such failures [1,7,10,19,21,22,25,27,33]. Furthermore it is well established in the theory of cryptography that the security of most cryptographic tasks relies crucially on the quality of that randomness [16].

Since true random bits are hard to generate without specialised hardware, and such hardware has only recently started to become available on commodity computing platforms,[1] Pseudorandom Generators (PRGs) and Pseudorandom Number Generators with input ("PRNGs with input" for short) are almost universally used in implementations. These generate pseudorandom bits instead of truly random bits; PRNGs with input can also have their state regularly refreshed with fresh entropy, though from a possibly biased source of randomness. Typically, a host

---

[1] See for example https://en.wikipedia.org/wiki/RdRand for a description of Intel's "Bull Mountain" random number generator.

operating system will make PRNGs with input available to applications, with the entropy being gathered from a variety of events, e.g. keyboard or disk timings, or timing of interrupts and other system events; programming libraries typically also provide access to PRG functionality, though of widely varying quality.

Backdooring Randomness: Given the ubiquity of PRGs and PRNGs with input in cryptographic implementations, they constitute the ideal target for maximising the spread and impact of backdoors. This was probably the rationale behind the Dual EC DRBG [11] which is widely believed to have been backdoored by the NSA. Despite this generator's low-speed, known output biases, and known capability to be backdoored (which was pointed out as early as 2007 by Shumow and Ferguson [30]), it managed to be covertly deployed in a range of widely used systems. Such systems continue to be discovered today, more than three years after the original Snowden revelations relating to Dual EC DRBG and project Bullrun.[2] The Dual EC DRBG provides a particularly useful backdoor to Big Brother: given a single output from the generator, its state can be recovered, and all future outputs can be recovered (with moderate computational effort). Protocols like SSL/TLS directly expose PRG outputs in protocol messages, making the Dual EC DRBG exploitable in practice [11].

Formal Analysis of Backdoored PRGs: The formal study of backdoored PRGs (BPRGs) was initiated by Dodis et. al. [15], building on earlier work of Vazirani and Vazirani [32]. Dodis et al. showed that BPRGs are equivalent to public-key encryption (PKE) with pseudorandom ciphertexts (IND\$-CPA-security), provided constructions using PKE schemes and KEMs, and analysed folklore immunisation techniques. Understanding the nature of backdoored primitives together with their capabilities and limitations is an important first step towards finding solutions that will safeguard against backdooring attacks. For instance the equivalence of BPRGs with public key encryption shown in [15] suggests that a PRG based on purely symmetric techniques is less likely to contain a backdoor, since we currently do not know how to build public key encryption from one-way functions.

A basic question that was posed – and partly answered – in [15] is: *to what extent can a PRG be backdoored while at the same time being provably secure?* This question makes perfect sense in the context of subversion via backdooring, where the backdoor resides in the specification of the PRG itself, and where the PRG can be publicly assessed and its security evaluated. The Dual EC DRBG has notable biases which directly rule out any possibility of it being provably secure as a PRG. Nevertheless, in [15] it is noted that by using special encodings of curve points as in [9,24,35], these biases can be eliminated and the Dual EC DRBG can be turned into a provably forward-secure PRG under the DDH assumption.

---

[2] See for example http://www.realworldcrypto.com/rwc2016/program/rwc16-shacham.pdf?attredirects=0\&d=1 for the Dual EC DRBG being used as a backdoor in Juniper networking equipment; see also http://www.theguardian.com/world/2013/sep/05/nsa-gchq-encryption-codes-security for the original reporting on project Bullrun.

Yet the backdoor in the Dual EC DRBG, while relatively powerful and certainly completely undermining security in certain applications like SSL/TLS, has its limitations. In particular, it does not allow Big Brother (who holds the backdoor key) to predict *previous* outputs from a given output but only future ones. The random-seek BPRG construction of [15] provides a stronger type of backdoor: given any single output, it allows Big Brother to recover any past or future output with probability roughly $\frac{1}{4}$. But the random-seek BPRG construction of [15] attains this stronger backdooring at the expense of no longer being a forward-secure PRG (in the usual sense). Indeed, forward-security and the random-seek backdoor property would intuitively seem to be opposing goals, and it is then natural to ask whether this tradeoff is inherent, or whether strong forms of backdooring of forward-secure PRGs *are* possible. If the limitation was inherent, then a proof of forward-security for a PRG would serve to preclude backdoors with the backward-seek feature, so a forward-secure PRG would be automatically immunised, to some extent, against backdoors.

1.1   Our Contributions

In this work we advance understanding of backdoored generators in two distinct directions.

Stronger Backdooring of PRGs: We settle the above open question from [15] in the negative by providing two different constructions of random-seek BPRGs that are provably forward-secure. In fact we demonstrate something substantially stronger:

- Firstly, both of our constructions allow Big Brother to succeed with probability 1 (rather than the 1/4 attained for the random-seek BPRG construction of [15]).
- Secondly, the backdooring is much stronger, in that for both of our BPRG constructions, Big Brother is able to recover the initial state of the BPRG, given only a single output value. This then enables all states and output values to be reconstructed.

Our constructions require a number of cryptographic tools. Unsurprisingly, given the connection between BPRGs and PKE with pseudorandom ciphertexts that was shown in [15], they both make use of the latter primitive. To give a flavour of what lies ahead, we remark that our simplest construction, shown in Fig. 7, uses such a PKE scheme to encrypt its state s, with the resulting ciphertext C forming the generator's output; s is also evolved using a one-way function, to provide forward security. Clearly, Big Brother, with access to a single output and the decryption key, can recover the state s. But we use a *trapdoor* one-way function so that Big Brother can then "unwind" s back to its starting value. For the security proof, we need to use a random oracle applied to s to generate the encryption randomness, making our construction reminiscent of the "Encrypt-with-hash" construction of [5], while for technical reasons, we require the trapdoor one-way function to be lossy [26]. Our second construction is in the standard model and combines, in novel ways, other primitives such as re-randomizable PKE schemes.

Backdooring PRNGs with Input: We then turn our attention to the study of backdoored PRNGs with input (BPRNGs). This is a very natural extension to the study of BPRGs conducted in [15] and continued here, particularly in view of the widespread deployment of PRNGs with input in real systems.

The formal study of PRNGs with input (but without backdooring) commenced with Barak and Halevi's work in [4], later extended in [17,18]. Various security notions have been proposed in the literature for PRNGs with input, namely *resilience, forward security, backward security* and *robustness*. Of these, robustness is the strongest notion. It captures the ability of a generator to both preserve security when its entropy inputs are influenced by an attacker and to recover security after its state is compromised, via refreshing (provided sufficient entropy becomes available to it). Robustness is generally accepted as the *de facto* security target for any new PRNG design, though several widely-deployed PRNGs fail to meet it (see, for example, [12,17]).

Given that we are in the backdooring setting for subversion, in which the full specification of the cryptographic primitive targeted for backdooring is public, any construction can be vetted for security. It is therefore logical to require any BPRNG to be robust. (This is analogous to requiring a BPRG to be forward-secure, or at least, a PRG in the traditional sense.) As such, a BPRNG *cannot* just ignore its entropy inputs and revert to being a PRG. One might then hope that, with additional high entropy inputs being used to refresh the generator state, and with this entropy not being under the direct control of Big Brother (since, otherwise, no security at all is possible), backdooring a PRNG with input might be impossible. This would be a positive result in the quest to defeat backdooring. Unfortunately, we show that this is not the case.

As a warm-up, we show how to adapt the robust PRNG of [17] to make it backdoored. This requires only a simple trick (and some minor changes to the processing of entropy): replace the PRG component of the generator with a BPRG. Given a single output from the generator, this then allows Big Brother to compute *all* outputs from the last refresh operation to the next refresh operation. Yet the generator is still robust.

Much more challenging is to develop a robust PRNG with input in which Big Brother can use his backdoor to "pass through" refresh operations when computing generator outputs. We provide a construction which does just that, see Fig. 11. Our construction is based on the idea of interleaving outputs of a (non-backdoored) PRNG with encryptions of snapshots of that PRNG's state, using an IND\$-CPA secure encryption scheme to ensure pseudorandomness of the outputs. By taking a snapshot of the state whenever it is refreshed and storing a list of the previous k snapshots in the state (for a parameter k), the construction enables Big Brother to recover, with some probability, old output values that were computed as many as k refreshes previously. The actual construction is considerably more complex than this sketch hints, since achieving robustness, in the sense of [17], is challenging when the state has this additional structure. We also sketch variants of this construction that trade state and output size for strength of backdooring.

An Impossibility Result for BPRNGs: We close the paper on a more positive note, providing an impossibility result showing that backdooring in a strong sense cannot be achieved (whilst preserving robustness) without significantly enlarging the state of the generator. More precisely, we show that it is not possible for Big Brother to perform a *state* recovery attack in which he recovers more than some number k of properly refreshed previous states from an output of the generator, when k is large relative to the state-size of the BPRNG. A precise formalisation of our result is contained in Theorem 5.

Note that the backdooring attack here requires more of Big Brother than might be needed in practice, since he may be considered successful if he can recover just one previous state, or a fraction of the previous BPRNG outputs. Our construction shows that backdooring of this kind is certainly possible. Nor does our result say anything about Big Brother's capabilities (or lack thereof) when it comes to recovering *future* states/outputs (after a generator has undergone further high-entropy refresh operations). It is an important open problem to strengthen our impossibility results – and to improve our constructions – to explore the limits of backdooring for PRNGs with input.

2   Preliminaries

2.1   Notation

The set of binary strings of length n is denoted $\{0,1\}^n$ and ε denotes the empty string. For any two binary strings x and y we write $|x|$ to denote the size of x and $x\|y$ to denote their concatenation. For any set U we denote by $u \leftarrow U$ the process of sampling an element uniformly at random from U and assigning it to u. All logs are to base 2.

2.2   Entropy

We recall a number of standard definitions on entropy, statistical distance, and (k,ϵ)-extractors in the full version [14].

Definition 1. *An (k,ϵ)-extractor* Ext $: \{0,1\}^* \times \{0,1\}^v \rightarrow \{0,1\}^w$ *is said to be online-computable on inputs of length p if there exists a pair of efficient algorithms* iterate $: \{0,1\}^p \times \{0,1\}^p \times \{0,1\}^v \rightarrow \{0,1\}^p$, *and* finalize $: \{0,1\}^p \times \{0,1\}^v \rightarrow \{0,1\}^w$ *such that for all inputs $\bar{I} = (I_1,\ldots,I_d)$ where each $I_j \in \{0,1\}^p$, and $d \geq 2$, then after setting $y_1 = I_1$, and $y_j = $ iterate$(y_{j-1}, I_j; A)$ $j = 2,\ldots,d$, it holds that*

$$\mathsf{Ext}(\bar{I}; A) = \mathsf{finalize}(y_d; A).$$

2.3   Cryptographic Primitives

In the full version [14], we recall a number of standard definitions for PKE schemes. Throughout this work we require that PKE schemes be length-regular. For the constructions that follow, we shall require an IND\$-CPA-secure PKE scheme; that is to say a PKE scheme having pseudorandom ciphertexts. We define such schemes formally below. Concrete and efficient examples of such schemes can be obtained by applying carefully constructed encoding schemes to the group elements of ciphertexts in the ElGamal encryption scheme (in which ciphertexts are of the form $(g^R, M \cdot g^{Rx})$ where g generates a group of prime order p in which DDH is hard; $(g^x, x) \leftarrow$ KGen with $x \leftarrow \mathbb{Z}_p$; $R \leftarrow \mathbb{Z}_p$; and M is a message, encoded here as a group element); see for example [9,24,35].

Definition 2. *A PKE scheme* $\mathcal{E} =$ (KGen, Enc, Dec) *is said to be* (t, q, δ)-*IND\$-CPA-secure if for all adversaries* $\mathscr{A}$ *running in time* t *and making at most* q *oracle queries, it holds that* $\mathsf{Adv}_{\mathcal{E}}^{\mathrm{ind\$-cpa}}(\mathscr{A}) \leq \delta$, *where:*

$$\mathsf{Adv}_{\mathcal{E}}^{\mathrm{ind\$-cpa}}(\mathscr{A}) = \left| \Pr\left[(pk, sk) \leftarrow \mathsf{KGen} : \mathscr{A}^{\mathsf{Enc}(pk, \cdot)}(pk) \Rightarrow 1 \right] \right.$$
$$\left. - \Pr\left[(pk, sk) \leftarrow \mathsf{KGen} : \mathscr{A}^{\$(\cdot)}(pk) \Rightarrow 1 \right] \right|$$

and $\$(\cdot)$ *is such that on input a message* M *it returns a random string of size* $|\mathsf{Enc}(pk, M)|$.

It is straightforward to show that if $\mathcal{E}$ is (t, q, δ)-IND\$-CPA-secure, then it is also $(t, q, 2\delta)$-IND-CPA-secure in the usual sense.

We shall also utilise PKEs which are *statistically re-randomizable*; again the ElGamal scheme and its group-element-encoded variants have the required property.

Definition 3. [20] *A* (t, q, δ, ν)-*statistically re-randomizable encryption scheme is a tuple of algorithms* $\mathcal{E} =$ (KGen, Enc, Rand, Dec) *where* (KGen, Enc, Dec) *is a standard PKE scheme and* Rand *is an efficient randomised algorithm such that for all* $(pk, sk) \leftarrow$ KGen *and for all* M, R_0',

$$\Delta(\{\mathsf{Enc}(pk, M; R_0) : R_0 \leftarrow \mathsf{Coins}(\mathsf{Enc})\},$$
$$\{\mathsf{Rand}(\mathsf{Enc}(pk, M; R_0'); R_1) : R_1 \leftarrow \mathsf{Coins}(\mathsf{Rand}) :\}) \leq \nu.$$

That is, the distributions of an honestly generated ciphertext and a ciphertext obtained by applying Rand *to one generated with arbitrary randomness are statistically close. We write* $\mathsf{Rand}(C_0; R_1, \ldots, R_q)$ *to denote the value of* C_q *where* $C_j = \mathsf{Rand}(C_{j-1}; R_j)$ *for* $j = 1, \ldots, q$.

We now define encryption schemes which have the additional property of being *reverse re-randomizable*. It is easy to see that ElGamal encryption and its encoded variants has the required property.

Definition 4. *A* (t, q, δ, ν)-*statistically reverse re-randomizable encryption scheme $\mathcal{E}$ is a tuple of algorithms* $\mathcal{E} = (\mathsf{KGen}, \mathsf{Enc}, \mathsf{Rand}, \mathsf{Rand}^{-1}, \mathsf{Dec})$ *such that:*

- $(\mathsf{KGen}, \mathsf{Enc}, \mathsf{Rand}, \mathsf{Dec})$ *is a* (t, q, δ, ν) *statistically re-randomizable encryption scheme.*
- Rand^{-1} *is an efficient algorithm such that for all* $(pk, sk) \leftarrow \mathsf{KGen}$ *and for all* M, R_0, R_1, *it holds that, if* $C = \mathsf{Enc}(pk, M; R_0)$, *then:*

$$\Pr\left[\mathsf{Rand}^{-1}(\mathsf{Rand}(C; R_1); R_1) = C\right] = 1.$$

Suppose $C_q = \mathsf{Rand}(C_0; R_1, \ldots, R_q)$, *so that* $C_j = \mathsf{Rand}(C_{j-1}; R_j)$ *for* $j = 1, \ldots, q$. *Then, from the above, we know that* $C_{j-1} = \mathsf{Rand}^{-1}(C_j; R_j)$ *for* $1 \leq j \leq q$; *to denote* C_0, *we write* $\mathsf{Rand}^{-1}(C_q; R_1, \ldots, R_q)$.

We recall the definitions of trapdoor one-way permutations, and lossy trapdoor permutations, in the full version [14].

2.4 Pseudorandom Generators

A pseudorandom generator (PRG) takes a small amount of true statistical randomness as an input seed, and outputs arbitrary (polynomial) length bit-strings which are *pseudorandom*. Following [15], we will equip PRGs with a parameter generation algorithm, setup. This allows backdooring to be introduced into the formalism.

Definition 5. *A* PRG *is a triple of algorithms* PRG $= (\mathsf{setup}, \mathsf{init}, \mathsf{next})$, *with associated parameters* $(n, l) \in \mathbb{N}^2$, *defined as follows:*

- $\mathsf{setup} : \{0, 1\}^* \to \{0, 1\}^* \times \{0, 1\}^*$ *takes random coins as input and outputs a pair of parameters* (pp, bk), *where* pp *denotes the public parameter for the generator, and* bk *is the secret backdoor parameter. In a non-backdoored* PRG, *we set* $bk = \perp$.
- $\mathsf{init} : \{0, 1\}^* \times \{0, 1\}^* \to \{0, 1\}^n$ *takes* pp *and random coins as input, and returns an initial state for the* PRG, $s_0 \in \{0, 1\}^n$.
- $\mathsf{next} : \{0, 1\}^* \times \{0, 1\}^n \to \{0, 1\}^l \times \{0, 1\}^n$ *takes* pp *and a state* $s \in \{0, 1\}^n$ *as input, and outputs an output/state pair* $(r, s') \leftarrow \mathsf{next}(pp, s)$ *where* $r \in \{0, 1\}^l$ *is the* PRG*'s output, and* $s' \in \{0, 1\}^n$ *is the updated state.*

Definition 6. *Let* PRG $= (\mathsf{setup}, \mathsf{init}, \mathsf{next})$ *be a* PRG. *Given an initial state* s_0, *we set* $(r_i, s_i) \leftarrow \mathsf{next}(pp, s_{i-1})$ *for* $i = 1, \ldots, q$. *We write* $\mathsf{out}^q(\mathsf{next}(pp, s_0))$ *for the sequence of outputs* $r_1, \ldots, r_q$ *and* $\mathsf{state}^q(\mathsf{next}(pp, s_0))$ *for the sequence of states* $s_1, \ldots, s_q$ *produced by this process.*

Definition 7 (PRG Security). *Let* PRG $= (\mathsf{setup}, \mathsf{init}, \mathsf{next})$ *be a* PRG. *Consider the game* PRG-DIST$_{\mathsf{PRG}}^{\mathscr{A}, q}$ *of Fig. 1 in which the adversary receives either* q *outputs from the* PRG *or* q *random strings of the appropriate size. We define the* PRG *distinguishing advantage of* $\mathscr{A}$ *against* PRG *to be*

$$\mathsf{Adv}_{\mathsf{PRG}}^{\mathsf{dist}}(\mathscr{A}, q) = 2|\Pr\left[\text{PRG-DIST}_{\mathsf{PRG}}^{\mathscr{A}, q} \Rightarrow \mathsf{true}\right] - \frac{1}{2}|.$$

Game PRG-DIST$_{\mathsf{PRG}}^{\mathscr{A},q}$	Game PRG-FWD$_{\mathsf{PRG}}^{\mathscr{A},q}$
$(pp, bk) \leftarrow\!\!\!\leftarrow \mathsf{setup}$	$(pp, bk) \leftarrow\!\!\!\leftarrow \mathsf{setup}$
$s_0 \leftarrow\!\!\!\leftarrow \mathsf{init}(pp)$	$s_0 \leftarrow\!\!\!\leftarrow \mathsf{init}(pp)$
$r_1^0, \ldots, r_q^0 \leftarrow \mathsf{out}^q(\mathsf{next}(pp, s_0))$	$r_1^0, \ldots, r_q^0 \leftarrow \mathsf{out}^q(\mathsf{next}(pp, s_0))$
$r_1^1, \ldots, r_q^1 \leftarrow\!\!\!\leftarrow (\{0,1\}^l)^q$	$r_1^1, \ldots, r_q^1 \leftarrow\!\!\!\leftarrow (\{0,1\}^l)^q$
$b \leftarrow\!\!\!\leftarrow \{0,1\}$	$s_1, \ldots, s_q \leftarrow \mathsf{state}^q(\mathsf{next}(pp, s_0))$
$b' \leftarrow\!\!\!\leftarrow \mathscr{A}(pp, r_1^b, \ldots, r_q^b)$	$b \leftarrow\!\!\!\leftarrow \{0,1\}$
$\mathbf{return}\ (b = b')$	$b' \leftarrow\!\!\!\leftarrow \mathscr{A}(pp, r_1^b, \ldots, r_q^b, s_q)$
	$\mathbf{return}\ (b = b')$

Fig. 1. The games for PRG-DIST$_{\mathsf{PRG}}^{\mathscr{A},q}$ and PRG-FWD$_{\mathsf{PRG}}^{\mathscr{A},q}$.

Definition 8. *A PRG* PRG $=$ (setup, init, next) *is said to be* (t, q, δ)-*secure if for all adversaries* $\mathscr{A}$ *running in time at most* t *it holds that* $\mathsf{Adv}_{\mathsf{PRG}}^{\mathsf{dist}}(\mathscr{A}, q) \leq \delta$.

Definition 9 (PRG Forward Security). *Let* PRG $=$ (setup, init, next) *be a PRG. Consider the game* PRG-FWD$_{\mathsf{PRG}}^{\mathscr{A},q}$ *of Fig. 1 in which the adversary receives either* q *outputs from the* PRG *and the final state, or* q *random strings of the appropriate size and the final state. We define the* PRG *forward-security advantage of* $\mathscr{A}$ *against* PRG *to be*

$$\mathsf{Adv}_{\mathsf{PRG}}^{\mathsf{fwd}}(\mathscr{A}, q) := 2\left|\Pr\left[\mathsf{PRG\text{-}FWD}_{\mathsf{PRG}}^{\mathscr{A},q} \Rightarrow \mathsf{true}\right] - \frac{1}{2}\right|.$$

Definition 10. *A PRG* PRG *is said to be* (t, q, δ)-*FWD-secure if for all adversaries* $\mathscr{A}$ *running in time at most* t *it holds that* $\mathsf{Adv}_{\mathsf{PRG}}^{\mathsf{fwd}}(\mathscr{A}, q) \leq \delta$.

2.5 Backdoored Pseudorandom Generators

The first formal treatment of backdoored PRGs was that of Dodis et al. [15]. Intuitively, a backdoored cryptosystem is a scheme coupled with some secret backdoor information. In the view of an adversary who does not know the backdoor information, the scheme fulfils its usual security definition. However an adversary in possession of the backdoor information will gain some advantage in breaking the security of the cryptosystem. The backdoor attacker is modelled as an algorithm which we call $\mathscr{B}$ (for 'Big Brother'), to distinguish it from an attacker $\mathscr{A}$ whose goal is to break the usual security of the scheme *without* access to the backdoor. Whilst the backdoor attacker $\mathscr{B}$ will be external in the sense that it will only be able to observe public outputs and parameters, the attack is also internalised as the backdoor algorithm is designed alongside, and incorporated into, the scheme.

We define backdoored PRGs (BPRGs) in conjunction with different games BPRNG-TYPE$_{\mathsf{PRG}}^{\mathscr{B},q}$ which capture specific backdooring goals, each game having a corresponding advantage term. The three games considered in [15] are defined in Fig. 2.

Definition 11. *A tuple of algorithms* $\overline{\mathsf{PRG}} = (\mathsf{setup}, \mathsf{init}, \mathsf{next}, \mathscr{B})$ *is defined to be a* $(t, q, \delta, (\mathsf{type}, \epsilon))$-*secure BPRG if:*

- $\mathsf{PRG} = (\mathsf{setup}, \mathsf{init}, \mathsf{next})$ *is a* (t, q, δ)-*secure PRG;*
- $\mathsf{Adv}_{\overline{\mathsf{PRG}}}^{\mathsf{type}}(\mathscr{B}, q) \geq \epsilon.$

Definition 12. *Let* $\overline{\mathsf{PRG}} = (\mathsf{setup}, \mathsf{init}, \mathsf{next}, \mathscr{B})$ *be a BPRG. We define*

- $\mathsf{Adv}_{\overline{\mathsf{PRG}}}^{\mathsf{dist}}(\mathscr{B}, q) := 2|\Pr\left[\mathsf{BPRG\text{-}DIST}_{\overline{\mathsf{PRG}}}^{\mathscr{B},q} \Rightarrow \mathsf{true}\right] - \frac{1}{2}|,$

- $\mathsf{Adv}_{\overline{\mathsf{PRG}}}^{\mathsf{next}}(\mathscr{B}, q) := \Pr\left[\mathsf{BPRG\text{-}NEXT}_{\overline{\mathsf{PRG}}}^{\mathscr{B},q} \Rightarrow \mathsf{true}\right],$

- $\mathsf{Adv}_{\overline{\mathsf{PRG}}}^{\mathsf{rseek}}(\mathscr{B}, q) := \min_{1 \leq i,j, \leq q} \Pr\left[\mathsf{BPRG\text{-}RSEEK}_{\overline{\mathsf{PRG}}}^{\mathscr{B},q}(i,j) \Rightarrow \mathsf{true}\right].$

Game BPRG-DIST$_{\overline{\mathsf{PRG}}}^{\mathscr{B},q}$

$(pp, bk) \twoheadleftarrow \mathsf{setup}$
$s_0 \twoheadleftarrow \mathsf{init}(pp)$
$r_1^0, \ldots, r_q^0 \leftarrow \mathsf{out}^q(\mathsf{next}(pp, s_0))$
$r_1^1, \ldots, r_q^1 \twoheadleftarrow (\{0,1\}^l)^q$
$b \twoheadleftarrow \{0,1\}$
$b^* \leftarrow \mathscr{B}(pp, bk, r_1^0, \ldots, r_q^0)$
return $(b = b^*)$

Game BPRG-NEXT$_{\overline{\mathsf{PRG}}}^{\mathscr{B},q}$

$(pp, bk) \twoheadleftarrow \mathsf{setup}$
$s_0 \twoheadleftarrow \mathsf{init}(pp)$
$r_1^0, \ldots, r_q^0 \leftarrow \mathsf{out}^q(\mathsf{next}(pp, s_0))$
$s_1, \ldots, s_q \leftarrow \mathsf{state}^q(\mathsf{next}(pp, s_0))$
$s_q^* \leftarrow \mathscr{B}(pp, bk, r_1^0, \ldots, r_q^0)$
return $(s_q = s_q^*)$

Game BPRG-RSEEK$_{\overline{\mathsf{PRG}}}^{\mathscr{B},q}(i,j)$

$(pp, bk) \twoheadleftarrow \mathsf{setup}$
$s_0 \twoheadleftarrow \mathsf{init}(pp)$
$r_1^0, \ldots, r_q^0 \leftarrow \mathsf{out}^q(\mathsf{next}(pp, s_0))$
$s_1, \ldots, s_q \leftarrow \mathsf{state}^q(\mathsf{next}(pp, s_0))$
$r_j^* \leftarrow \mathscr{B}(pp, bk, i, j, r_i)$
return $(r_j = r_j^*)$

Fig. 2. Security games for backdooring of PRGs.

In Fig. 2, game BPRG-DIST$_{\overline{\mathsf{PRG}}}^{\mathscr{B},q}$ challenges Big Brother to use the backdoor to break the security of the PRG in the most basic sense of distinguishing real from random outputs. In game BPRG-NEXT$_{\mathsf{PRG}}^{\mathscr{B},q}$, $\mathscr{B}$ aims to recover the current state of the PRG given q consecutive outputs from the generator. This is a far more powerful compromise since it then allows $\mathscr{B}$ to predict all of the generator's future outputs. In the third game, BPRG-RSEEK$_{\mathsf{PRG}}^{\mathscr{B},q}(i,j)$, $\mathscr{B}$ is given only the ith output (rather than q outputs) and index j, and tries to recover the jth output (but not any state).

It is noted in [15] that an adversary $\mathscr{B}$ winning in game BPRG-NEXT$_{\mathsf{PRG}}^{\mathscr{B},q}$ represents a stronger form of backdooring than an adversary $\mathscr{B}$ winning in game BPRG-DIST$_{\mathsf{PRG}}^{\mathscr{B},q}$ for the same parameters, whilst an adversary $\mathscr{B}$ winning in game BPRG-RSEEK$_{\mathsf{PRG}}^{\mathscr{B},q}(i,j)$ may be more or less powerful than one for game BPRG-NEXT$_{\mathsf{PRG}}^{\mathscr{B},q}$ depending on the circumstances. The paper [15] presents constructions of BPRGs that are backdoored in the BPRG-NEXT$_{\mathsf{PRG}}^{\mathscr{B},q}$ and BPRG-RSEEK$_{\mathsf{PRG}}^{\mathscr{B},q}(i,j)$ senses, but does also note that their construction for a scheme of the latter type is *not* forward-secure.

Both for their intrinsic interest, and because they will be needed in our later constructions of backdoored PRNGs with input, we are interested in BPRGs that *are* forward secure against normal adversaries. For a generic type of game BPRNG-TYPE$_{\mathsf{PRG}}^{\mathscr{B},q}$, these are formally defined as follows.

Definition 13. *A tuple of algorithms* $\overline{\mathsf{PRG}} = (\mathsf{setup}, \mathsf{init}, \mathsf{next}, \mathscr{B})$ *is said to be a* $(t,q,\delta,(\mathsf{type},\epsilon))$-*FWD-secure BPRG if:*

- $\mathsf{PRG} = (\mathsf{setup}, \mathsf{init}, \mathsf{next})$ *is a* (t,q,δ)-*FWD-secure PRG;*
- $\mathsf{Adv}_{\mathsf{PRG}}^{\mathsf{type}}(\mathscr{B},q) \geq \epsilon$.

2.6 Pseudorandom Number Generators with Input

Definition 14 (PRNG with Input). *A PRNG with input is a tuple of algorithms* $\mathsf{PRNG} = (\mathsf{setup}, \mathsf{init}, \mathsf{refresh}, \mathsf{next})$ *with associated parameters* $(n,l,p) \in \mathbb{N}^3$, *where:*

- $\mathsf{setup} : \{0,1\}^* \to \{0,1\}^*$ *takes as input some random coins and returns a public parameter pp.*
- $\mathsf{init} : \{0,1\}^* \times \{0,1\}^* \to \{0,1\}^n$ *takes the public parameter pp and some random coins to return an initial state* s_0.
- $\mathsf{refresh} : \{0,1\}^* \times \{0,1\}^n \times \{0,1\}^p \to \{0,1\}^n$ *takes as input the public parameter pp, the current state S, and a sample I from the entropy source, and returns a new state* s'.
- $\mathsf{next} : \{0,1\}^* \times \{0,1\}^n \to \{0,1\}^n \times \{0,1\}^l$ *takes as input the public parameter pp and the current state s, and returns a new state s' together with an output string r.*

Definition 15 (Distribution Sampler). *A distribution sampler* $\mathscr{D}$: $\{0,1\}^* \to \{0,1\}^* \times \{0,1\}^p \times \mathbb{R}^{\geq 0} \times \{0,1\}^*$ *is a probabilistic and possibly stateful algorithm which takes its current state σ as input and returns an updated state*

σ', a sample I, an entropy estimate γ, and some leakage information z about I. The state σ is initialised to the empty string.

A distribution sampler $\mathcal{D}$ is said to be valid up to q_r samples, if for all $j \in \{1, \ldots, q_r\}$ it holds (with probability 1) that:

$$H_\infty(I_j \mid I_1, \ldots, I_{j-1}, I_{j+1}, \ldots, I_{q_r}, z_1, \ldots, z_{q_r}, \gamma_1, \ldots, \gamma_{q_r}) \geq \gamma_j$$

where $(\sigma_i, I_i, \gamma_i, z_i) = \mathcal{D}(\sigma_{i-1})$ for $i \in \{1, \ldots, q_r\}$ and $\sigma_0 = \varepsilon$.

2.7 Security for Pseudorandom Number Generators with Input

We now turn to discussing security definitions for PRNGs with input. We follow [17], with some minor differences noted below.

Definition 16 (Security of PRNG with Input). *With references to the security game shown in Fig. 3, a PRNG with input* $\mathsf{PRNG} = (\mathsf{setup}, \mathsf{init}, \mathsf{refresh}, \mathsf{next})$ *is said to be* $(t, q_r, q_n, q_c, \gamma^*, \epsilon)$*-ROB-secure, for any distribution sampler $\mathcal{D}$ valid up to q_r samples, and any adversary $\mathcal{A}$ running in time at most t, making at most q_r queries to* REF, *q_n queries to* ROR *and a total of q_c queries to* GET *and* SET, *the corresponding advantage in game* $\mathrm{ROB}_{\mathsf{PRNG}, \gamma^*}^{\mathcal{D}, \mathcal{A}}$ *is bounded by ϵ, where*

$$\mathsf{Adv}_{\mathsf{PRNG}}^{\mathrm{rob}}(\mathcal{A}, \mathcal{D}) := 2|\Pr\left[\mathrm{ROB}_{\mathsf{PRNG}, \gamma^*}^{\mathcal{D}, \mathcal{A}} \Rightarrow \mathsf{true}\right] - \frac{1}{2}|.$$

Game $\mathrm{ROB}_{\mathsf{PRNG}, \gamma^*}^{\mathcal{D}, \mathcal{A}}$	REF	ROR	GET
$pp \leftarrow \mathsf{setup}$	$(\sigma, I, \gamma, z) \leftarrow \mathcal{D}(\sigma)$	$(s, r_0) \leftarrow \mathsf{next}(pp, s)$	$c \leftarrow 0$
$\sigma \leftarrow \varepsilon;\ c \leftarrow \infty$	$s \leftarrow \mathsf{refresh}(pp, s, I)$	$r_1 \leftarrow \{0,1\}^l$	**return** s
$s \leftarrow \mathsf{init}(pp)$	$c \leftarrow c + \gamma$	if $c < \gamma^*$	
$b \leftarrow \{0,1\}$	**return** (γ, z)	$c \leftarrow 0$	SET (s^*)
$b' \leftarrow \mathcal{A}^{\mathrm{REF, ROR, GET, SET}}(pp)$		**return** r_0	$c \leftarrow 0$
return $b' = b$		**else return** r_b	$s \leftarrow s^*$

Fig. 3. PRNG with input security game $\mathrm{ROB}_{\mathsf{PRNG}, \gamma^*}^{\mathcal{D}, \mathcal{A}}$.

Our definition here deviates from that in [17] in the following ways.

- We generalise the syntax so as to allow the state to be initialised according to some arbitrary distribution rather than requiring it to be uniformly random. In particular we allow this distribution to depend on pp. This facilitates our backdooring definitions to follow.
- We have removed the NEXT oracle from the model, without any loss of generality (as was shown in [12]).

One of the key insights of [17] is to decompose the somewhat complex notion of robustness into the two simpler notions of PRE and REC security. We recall these definitions below, generalised here to include the init algorithm.

Definition 17 (Preserving and Recovering Security). *Consider the security games described in* Fig. 4. *The PRE security advantage of an adversary $\mathscr{A}$ against a PRNG with input PRNG is defined to be*

$$\mathsf{Adv}^{pre}_{\mathsf{PRNG}}(\mathscr{A}) := 2|\Pr\left[\mathsf{PRE}^{\mathscr{A}}_{\mathsf{PRNG}} \Rightarrow \mathsf{true}\right] - \frac{1}{2}|.$$

The REC security advantage with respect to parameters q_r, γ^* of an adversary/sampler pair $(\mathscr{A}, \mathscr{D})$ against a PRNG with input PRNG is defined to be

$$\mathsf{Adv}^{rec}_{\mathsf{PRNG}}(\mathscr{A}, \mathscr{D}) := 2|\Pr\left[\mathsf{REC}^{\mathscr{D},\mathscr{A},q_r}_{\mathsf{PRNG},\gamma^*} \Rightarrow \mathsf{true}\right] - \frac{1}{2}|.$$

In the REC security game, it is required that $\sum_{j=k+1}^{k+d} \boldsymbol{\gamma}[j] \geq \gamma^*$ for the value d output by $\mathscr{A}$.

Game $\mathsf{PRE}^{\mathscr{A}}_{\mathsf{PRNG}}$	Game $\mathsf{REC}^{\mathscr{D},\mathscr{A},q_r}_{\mathsf{PRNG},\gamma^*}$	SAM
$b \twoheadleftarrow \{0,1\}$	$k \leftarrow 0;\ \sigma[0] \leftarrow \varepsilon$	$k \leftarrow k+1$
$pp \twoheadleftarrow \mathsf{setup}$	$b \twoheadleftarrow \{0,1\}$	**return** $\mathbf{I}[k]$
$s^0 \twoheadleftarrow \mathsf{init}(pp)$	$pp \twoheadleftarrow \mathsf{setup}$	
$\mathbf{I}[1:d] \leftarrow \mathscr{A}(pp)$	**for** $i = 1$ **to** q_r	
for $i = 1$ **to** d	$\quad (\boldsymbol{\sigma}[i], \mathbf{I}[i], \boldsymbol{\gamma}[i], \mathbf{z}[i]) \leftarrow \mathscr{D}(\boldsymbol{\sigma}[i-1])$	
$\quad s^i \leftarrow \mathsf{refresh}(pp, s^{i-1}, \mathbf{I}[i])$	$(s^0, d) \leftarrow \mathscr{A}^{\mathrm{SAM}}(pp, \boldsymbol{\gamma}, \mathbf{z})$	
$(s_0, r_0) \leftarrow \mathsf{next}(pp, s^d)$	**for** $i = k+1$ **to** $k+d$	
$s_1 \twoheadleftarrow \mathsf{init}(pp);\ r_1 \twoheadleftarrow \{0,1\}^l$	$\quad s^i \leftarrow \mathsf{refresh}(pp, s^{i-1}, \mathbf{I}[k+i])$	
$b' \leftarrow \mathscr{A}(pp, s_b, r_b)$	$(s_0, r_0) \leftarrow \mathsf{next}(pp, s^d)$	
return $b' = b$	$s_1 \twoheadleftarrow \mathsf{init}(pp);\ r_1 \twoheadleftarrow \{0,1\}^l$	
	$b' \leftarrow \mathscr{A}(pp, \mathbf{I}[k+d+1:q_r], s_b, r_b)$	
	return $b' = b$	

Fig. 4. PRNG with input security games $\mathsf{PRE}^{\mathscr{A}}_{\mathsf{PRNG}}$ and $\mathsf{REC}^{\mathscr{D},\mathscr{A},q_r}_{\mathsf{PRNG},\gamma^*}$.

Definition 18 (Preserving Security). *A PRNG with input PRNG is said to have (t, ϵ_{pre})-PRE security if for all attackers $\mathscr{A}$ running in time t, it holds that $\mathsf{Adv}^{pre}_{\mathsf{PRNG}}(\mathscr{A}) \leq \epsilon_{pre}$.*

Definition 19 (Recovering Security). *A PRNG with input PRNG is said to have* $(t, q_r, \gamma^*, \epsilon_{rec})$-*REC security if for any attacker* $\mathscr{A}$ *and sampler* $\mathscr{D}$ *valid up to* q_r *samples and running in time* t*, it holds that* $\mathsf{Adv}^{rec}_{\mathsf{PRNG}}(\mathscr{A}, \mathscr{D}) \leq \epsilon_{rec}$.

Informally, preserving security concerns a generator's ability to maintain security (in the sense of having pseudorandom state and output) when the adversary completely controls the entropy source used to refresh the generator but does not compromise its state. Meanwhile, recovering security captures the idea that a generator whose state is set by the adversary should eventually get to a secure state, and start producing pseudorandom outputs, once sufficient entropy has been made available to it. The proof of Theorem 1 can be found in the full version [14].

Theorem 1. *Let* PRNG *be a* PRNG *with input. If* PRNG *has both* (t, ϵ_{pre})-*PRE security, and* $(t, q_r, \gamma^*, \epsilon_{rec})$-*REC security, then* PRNG *is* $((t', q_r, q_n, q_c), \gamma^*, \epsilon)$-*ROB secure where* $t \approx t'$ *and* $\epsilon = q_n(\epsilon_{pre} + \epsilon_{rec})$.

To simplify notation, we will make use of an algorithm, evolve, to generate output values and update the internal state of a PRNG. It takes as input a PRNG with input PRNG = (setup, init, next, refresh), public parameter pp, an initial state s, a refresh pattern $\boldsymbol{rp} = (a_1, b_1, \ldots, a_\rho, b_\rho)$, and a distribution sampler $\mathscr{D}$. The refresh pattern $\boldsymbol{rp}$ denotes a sequence of calls to next and refresh; for each i, a_i denotes the number of consecutive calls to next and b_i denotes the subsequent number of consecutive calls to refresh. More specifically, evolve proceeds as shown in Fig. 5.

$\mathsf{evolve}(\mathsf{PRNG}, pp, s, \boldsymbol{rp}, \mathscr{D})$

---

parse $\boldsymbol{rp}$ **as** $(a_1, b_1, \ldots, a_\rho, b_\rho)$

$\mathcal{S} \leftarrow (); \sigma \leftarrow \varepsilon$

for $i = 1$ **to** ρ

    for $j = 1$ **to** a_i

        $(r, s) \leftarrow \mathsf{next}(pp, s)$

        $\mathcal{S} \leftarrow \mathcal{S} \| (r, s)$

    for $k = 1$ **to** b_i

        $(\sigma, I, \gamma, z) \leftarrow \mathscr{D}(\sigma)$

        $s \leftarrow \mathsf{refresh}(pp, s, I)$

return $\mathcal{S}$

Fig. 5. The evolve algorithm.

The output of evolve is a sequence, $(r_1, s_1, \ldots, r_{q_n}, s_{q_n})$, of PRNG output and state pairs, where $q_n = \sum_{i=1}^{\rho} a_i$. Based on evolve, we define an additional algorithm, out, which takes the same input, runs evolve, and returns only the output values $(r_1, \ldots, r_{q_n})$.

3   Stronger Models and New Constructions for Backdoored Pseudorandom Generators

In this section, we first present two new, strong backdooring security models for PRGs. The stronger of the two implies all the backdooring notions in [15]. We then give two new constructions of BPRGs which achieve our two backdooring notions. In contrast to the strongest constructions in [15], all of our constructions are *forward-secure*.

3.1   Backdoored PRG Security Models

In the first of our two new models, the BPRG is run with initial state s_0 to produce q outputs $r_1, \ldots, r_q$. The Big Brother adversary $\mathscr{B}$ is then given a particular output r_i, and challenged to recover the initial state s_0 of the BPRG. In the second model, the BPRG is again run with initial state s_0 to produce q outputs, one of which is given to $\mathscr{B}$. However $\mathscr{B}$ is now asked to reproduce the remaining $q - 1$ unseen output values. We formalise these two models as games BPRG-FIRST and BPRG-OUT in Fig. 6.

Definition 20. *Let* $\overline{\mathsf{PRG}} = (\mathsf{setup}, \mathsf{init}, \mathsf{next}, \mathscr{B})$ *be a BPRG. We define*

- $\mathsf{Adv}_{\overline{\mathsf{PRG}}}^{\mathrm{first}}(\mathscr{B}, q, i) := \Pr\left[\mathsf{BPRG\text{-}FIRST}_{\overline{\mathsf{PRG}}}^{\mathscr{B},q}(i) \Rightarrow \mathsf{true}\right]$, *and*
- $\mathsf{Adv}_{\overline{\mathsf{PRG}}}^{\mathrm{out}}(\mathscr{B}, q, i) := \Pr\left[\mathsf{BPRG\text{-}OUT}_{\overline{\mathsf{PRG}}}^{\mathscr{B},q}(i) \Rightarrow \mathsf{true}\right]$

Game BPRG-FIRST$_{\overline{\mathsf{PRG}}}^{\mathscr{B},q}(i)$	Game BPRG-OUT$_{\overline{\mathsf{PRG}}}^{\mathscr{B},q}(i)$
$(pp, bk) \twoheadleftarrow \mathsf{setup}$	$(pp, bk) \twoheadleftarrow \mathsf{setup}$
$s_0 \twoheadleftarrow \mathsf{init}(pp)$	$s_0 \twoheadleftarrow \mathsf{init}(pp)$
$r_1, \ldots, r_q \leftarrow \mathsf{out}^q(\mathsf{next}(pp, s_0))$	$r_1, \ldots, r_q \leftarrow \mathsf{out}^q(\mathsf{next}(pp, s_0))$
$s_0^* \leftarrow \mathscr{B}(pp, bk, r_i)$	$r_1^*, \ldots, r_q^* \leftarrow \mathscr{B}(pp, bk, r_i)$
return $(s_0 = s_0^*)$	**return** $((r_1, \ldots r_q) = (r_1^*, \ldots, r_q^*))$

Fig. 6. Backdoored PRG security games BPRG-FIRST and BPRG-OUT.

Discussion. We observe that our first backdooring notion, as formalised in BPRG-FIRST$_{\overline{\mathsf{PRG}}}^{\mathscr{B},q}$ and $\mathsf{Adv}_{\overline{\mathsf{PRG}}}^{\mathrm{first}}(\mathscr{B}, q, i)$, is strictly stronger than the three notions for BPRGs defined in [15] and discussed in Sect. 2.5: it is straightforward to see that any $(t, q, \delta, (\mathsf{first}, \epsilon))$-secure BPRG is also a $(t, q, \delta, (\mathsf{type}, \epsilon))$-secure BPRG for $\mathsf{type} \in \{\mathsf{dist}, \mathsf{state}, \mathsf{rseek}\}$.

Moreover, simple comparison of definitions shows that any $(t, q, \delta, (\mathsf{out}, \epsilon))$-secure BPRG is also a $(t, q, \delta, (\mathsf{type}, \epsilon))$-secure BPRG for $\mathsf{type} \in \{\mathsf{dist}, \mathsf{rseek}\}$.

However, a BPRG backdoored in the out sense need not be backdoored in the state sense, since the latter concerns state prediction rather than output prediction. (And indeed it is easy to construct separating examples for the out and state backdooring notions.)

Since the initial state of a PRG determines all of its output, it is also clear that any $(t, q, \delta, (\text{first}, \epsilon))$-secure BRPG is also a $(t, q, \delta, (\text{out}, \epsilon))$-secure BPRG. However, the converse need not hold, and first backdooring is strictly stronger than out backdooring. To see this, consider $\overline{\text{PRG}}$, a $(t, q, \delta, (\text{out}, \epsilon))$-secure BPRG, and define a modified BRPG $\overline{\text{PRG}}'$ in which the initial state s_0 is augmented to $s_0 \| d$ for $d \leftarrow \{0, 1\}^n$, but where d is not used in any computations and all other algorithms of $\overline{\text{PRG}}$ are left unchanged. In particular, the output produced by $\overline{\text{PRG}}'$ is identical to that of $\overline{\text{PRG}}$. Then it is easy to see that $\overline{\text{PRG}}'$ is a $(t, q, \delta, (\text{out}, \epsilon))$-secure BPRG, but that $\text{Adv}_{\overline{\text{PRG}}'}^{\text{first}}(\mathcal{B}, q, i) \leq 2^{-n}$, since $\mathcal{B}$ can do no better than guessing the n extra bits of state d.

In most attack scenarios, and taking Big Brother's perspective, the ability of $\mathcal{B}$ to compute all unseen output (as in out) is as useful in practice as being able to compute the initial state (as in first), since it is the output values of the BPRG that will be consumed in applications. This makes the out notion a natural and powerful target for constructions of BPRGs. That said, in the sequel we will obtain constructions for the even stronger first setting.

A $(t, q, \delta, (\text{rseek}, \epsilon))$-secure BPRG is also a $(t, q, \delta, (\text{out}, \epsilon^{q-1}))$-secure BPRG, implying an exponential loss in going from rseek backdooring to out backdooring. This means that achieving either first or out backdooring with a high value of ϵ is significantly more powerful than achieving rseek backdooring with the same ϵ.

3.2   Forward-Secure BPRGs in the Random Oracle Model

We present our first construction for a forward-secure BPRG that is backdoored in the first sense in Fig. 7. This construction uses as ingredients an LTDP family and an IND\$-CPA-secure PKE scheme. Its security analysis is in the Random Oracle Model (ROM). It achieves our strongest first notion with $\epsilon = 1$.

The scheme is reminiscent of the "Encrypt-with-Hash" paradigm for constructing deterministic encryption schemes from [5]. At each stage, the generator encrypts its own state s, with randomness derived from hashing s, to produce the next output. The IND\$-CPA-security of the PKE scheme ensures these outputs are pseudorandom. The state s is also transformed by applying a one-way function F at each stage. This is necessary to provide forward security against non-$\mathcal{B}$ adversaries. The function is trapdoored, enabling $\mathcal{B}$ to decrypt an output to recover a state, then reverse the state update repeatedly to recover the initial state, thereby realising first backdooring. For technical reasons that will become apparent in the proof, we require the one-way function F to be a lossy permutation. The proof of the following theorem can be found in the full version [14].

Theorem 2. Let $\mathcal{E} = (\text{KGen}, \text{Enc}, \text{Dec})$ be a (t, q, δ)-IND\$-CPA secure PKE scheme. Let $\text{LTDP} = (\text{G}_0, \text{G}_1, \text{S}, \text{F}, \text{F}^{-1})$ be a family of (n, k, t, ϵ)-lossy trapdoor permutations. Then $\overline{\text{PRG}} = (\text{setup}, \text{init}, \text{next}, \mathcal{B})$ with algorithms as shown in

setup	init (pp)	next (pp, s)	$\mathscr{B}(bk, i, r_i)$
$(pk, sk) \leftarrow \mathsf{KGen}$	$(pk, \mathsf{PK}) \leftarrow pp$	$(pk, \mathsf{PK}) \leftarrow pp$	$(sk, \mathsf{SK}) \leftarrow bk$
$(\mathsf{PK}, \mathsf{SK}) \twoheadleftarrow \mathsf{G}_1$	$s_0 \twoheadleftarrow \mathsf{S}(\mathsf{PK})$	$r \leftarrow \mathsf{Enc}(pk, s; \mathsf{RO}(s))$	$s^*_{i-1} \leftarrow \mathsf{Dec}(sk, r_i)$
$pp \leftarrow (pk, \mathsf{PK})$	**return** (s_0)	$s' \leftarrow \mathsf{F}_{\mathsf{PK}}(s)$	$s^*_0 \leftarrow \mathsf{F}^{-(i-1)}_{\mathsf{SK}}(s^*_{i-1})$
$bk \leftarrow (sk, \mathsf{SK})$		**return** (r, s')	**return** (s^*_0)
return (pp, bk)			

Fig. 7. Construction of a forward-secure BPRG (setup, init, next, $\mathscr{B}$) from an LTDP family $\mathsf{LTDP} = (\mathsf{G}_0, \mathsf{G}_1, \mathsf{S}, \mathsf{F}, \mathsf{F}^{-1})$ and an IND\$-CPA-secure PKE scheme $\mathcal{E} = (\mathsf{KGen}, \mathsf{Enc}, \mathsf{Dec})$.

Fig. 7 is a $(t', q, (2\delta + 3\epsilon + (q+1)2^{-(k-1)}), (\text{first}, 1))$*-FWDsecure BPRG in the ROM, where* $t' \approx t$.

3.3 Standard Model, Forward-Secure BPRGs from Reverse Re-randomizable Encryption

Our second construction dispenses with the ROM and the use of lossy trapdoor permutations, at the expense of requiring as a component an IND\$-CPA-secure reverse re-randomizable PKE scheme (see Definition 4). It is instantiable in the standard model using a variant of the ElGamal encryption scheme. The scheme is again backdoored in the first sense with $\epsilon = 1$.

The scheme, shown in Fig. 8, uses algorithm next′ from a normal (forward-secure) PRG PRG′ to generate the next state s' and a pseudorandom value $>$ using the current state s as a seed. The value $>$ is then used to re-randomise a ciphertext C that encrypts an initial state value s_0, and the 'old' value C is used as the generator's output r. The re-randomisation at each step ensures that the outputs collectively appear pseudorandom to a regular PRG adversary; the fact that PRG′ is forward-secure ensures that the constructed BPRG is too.

Meanwhile, the use of PKE allows $\mathscr{B}$ (who knows the decryption key) to recover s_0 from any of the generator's outputs, run the component generator PRG′ from its starting state s_0, and recover all the values $>$ used for re-randomisation at each step; finally $\mathscr{B}$ can run the re-randomisation process backwards to recover the initial state. The proof of the following theorem can be found in the full version [14].

Theorem 3. *Let* $\mathcal{E} = (\mathsf{Key}, \mathsf{Enc}, \mathsf{Rand}, \mathsf{Rand}^{-1}, \mathsf{Dec})$ *be a* (t, q, δ, ν)*-IND\$-CPA secure reverse re-randomizable encryption scheme, and suppose that* PRG′ $=$ (setup′, init′, next′) *is a* (t, q, ϵ_{fwd})*-secure PRG. Then* PRG $=$ (setup, init, next, $\mathscr{B}$) *as defined in Fig. 8 is a* $(t', q, 6\delta + 2\epsilon_{fwd} + q(q+3)\nu/2, (\text{first}, 1))$*-FWDsecure BPRG, where* $t' \approx t$.

setup	next (pp, S)	$\mathscr{B}(sk, r_i, i)$
$(pk, sk) \leftarrow \mathsf{KGen}$	$(pk, pp') \leftarrow pp$	$C_{i-1}^* \leftarrow r_i$
$(pp', \perp) \leftarrow \mathsf{setup}'$	$(s, C) \leftarrow S$	$s_0^* \leftarrow \mathsf{Dec}_{sk}(C_{i-1})$
$pp \leftarrow (pk, pp')$	$(t, s') \leftarrow \mathsf{next}'(pp', s)$	$(t_1^*, \dots, t_q^*) \leftarrow \mathsf{out}^q(\mathsf{next}'(pp', s_0^*))$
$bk \leftarrow sk$	$C' \leftarrow \mathsf{Rand}(C, t)$	**for** $j = 1, \dots, i-1$
return (pp, bk)	$r \leftarrow C$	$\quad C_{j-1}^* \leftarrow \mathsf{Rand}^{-1}(C_j^*, t_j^*)$
	$S \leftarrow (s', C')$	$S^* \leftarrow (s_0^*, C_0^*)$
init (pp)	**return** (r, S)	**return** (S^*)
$(pk, pp') \leftarrow pp$		
$s_0 \leftarrow \mathsf{init}'(pp')$		
$C_0 \leftarrow \mathsf{Enc}_{pk}(s_0)$		
$S \leftarrow (s_0, C_0)$		
return S		

Fig. 8. Construction of a forward-secure BPRG (setup, init, next, $\mathscr{B}$) from a (t, q, δ, ν)-reverse-re-randomizable IND\$-CPA-secure PKE scheme $\mathcal{E} = (\mathsf{KGen}, \mathsf{Enc}, \mathsf{Dec})$ and a forward-secure PRG PRG' $= (\mathsf{setup}', \mathsf{init}', \mathsf{next}')$.

4   Backdooring PRNGs with Input

In this section, we address the second main theme in our paper: backdooring of PRNGs with input. To begin with, we show a simple construction for a PRNG with input that is both robust and subject to a limited form of backdooring: given a single output, $\mathscr{B}$ can recover the state and all outputs back to the previous refresh and up to the next refresh operations (see Sect. 4.1). We then move on to provide our formal definition for backdoored PRNGs with input (BPRNGs) in Sect. 4.2; this definition demands much more of $\mathscr{B}$, asking him to compute outputs beyond refresh operations, at the same time as asking that the BPRNG remain robust. Finally, in Sect. 4.3, we give a construction for a BPRNG meeting our backdooring notion for PRNGs with input, with various extensions to this construction being described in Sect. 4.4.

4.1   A Simple Backdoored PRNG

Let PRNG $= (\mathsf{setup}, \mathsf{init}, \mathsf{refresh}, \mathsf{next})$ be a ROB-secure PRNG with input. By considering the special case of Game $\mathsf{ROB}_{\mathsf{PRNG}, \gamma^*}^{\mathscr{D}, \mathscr{A}}$ in which the adversary $\mathscr{A}$ makes no SET or REF calls, and one GET call at the conclusion of the game, it is straightforward to see that PRG $= (\mathsf{setup}, \mathsf{init}, \mathsf{next})$ must be a FWD-secure PRG. This suggests that in order to backdoor PRNG, we might try to replace PRG with a BPRG. As long as this implicit BPRG is running without any refreshes, this should enable $\mathscr{B}$ to carry out backdooring.

To make this idea concrete, we present in Fig. 9 a construction of a ROB-secure PRNG with input from a PRG PRG. This scheme is closely based on the PRNG with input from [17]. It utilises an online-computable extractor and a FWD-secure PRG; our main modification is to ensure that repeated next calls are processed via a repeated iteration of a FWD-secure PRG. A proof of robustness for this PRNG with input is easily derived from that of the original construction:

Lemma 1. *Let* Ext : $\{0,1\}^* \times \{0,1\}^v \to \{0,1\}^n$ *be an online-computable* $(\gamma^*, \epsilon_{ext})$-*extractor. Let* PRG $=$ (setup, init, next) *be a* (t, q, ϵ_{prg})-*PRG such that* $s_0 \leftarrow$ init(pp) *is equivalent to* $s_0 \leftarrow \{0,1\}^n$. *Then* $\overline{\text{PRG}}$ $=$ (setup, init, refresh, next) *as shown in Fig. 9 is a* $((t', q_r, q_n, q_c), \gamma^*, q_n(2\epsilon_{prg} + q_r^2 \epsilon_{ext} + 2^{-n+1}))$-*robust PRNG with input, where* $t' \approx t$.

We now simply substitute a FWD-secure BPRG (such as that presented in Theorem 2) for PRG in this construction. Now, during the period between any pair of refresh calls in which the PRNG is producing output, we inherit the backdooring advantage of the BPRG in the new construction. However, the effectiveness of this backdoor is highly limited: as soon as refresh is called, the state of the PRNG is refreshed with inputs, which, if of sufficiently high entropy, will make the state information-theoretically unpredictable. Then $\mathcal{B}$ would need to compromise more output in order to regain his backdooring advantage.

One implication of this construction is that it makes it clear that, when considering stronger forms of backdooring, we must turn our attention to subverting refresh calls in some way.

setup	refresh(pp, s, I)	next(pp, s)
$(pp', \bot) \leftarrow$ setup	**parse** pp **as** (pp', A)	**parse** pp **as** (pp', A)
$A \leftarrow \{0,1\}^v$	**parse** $\bar{s}$ **as** $(s_1, s_2, \text{flgRfrsh})$	**parse** $\bar{s}$ **as** $(s_1, s_2, \text{flgRfrsh})$
$pp \leftarrow (pp', A)$	$s_2 \leftarrow$ iterate($s_2, I; A$)	**if** flgRfrsh $= 1$
$bk \leftarrow \bot$	flgRfrsh $\leftarrow 1$	$U \leftarrow$ finalize($s_2; A$)
return (pp, bk)	$\bar{s} \leftarrow (s_1, s_2, \text{flgRfrsh})$	$s_1 \leftarrow U \oplus s_1$
	return $(\bar{s})$	$s_2 \leftarrow 0^p$
init(pp)		$(s_1, r) \leftarrow$ next($pp'; s_1$)
$s_1 \leftarrow \{0,1\}^n$		flgRfrsh $\leftarrow 0$
$s_2 \leftarrow 0^p$		$\bar{s} \leftarrow (s_1, s_2, \text{flgRfrsh})$
flgRfrsh $\leftarrow 0$		**return** $(\bar{s}, r)$
$\bar{s} \leftarrow (s_1, s_2, \text{flgRfrsh})$		
return $(\bar{s})$		

Fig. 9. Construction of a robust PRNG PRNG from a FWD-secure PRG PRG, based on [17].

4.2   Formal Definition for Backdoored PRNGs with Input

To make our backdooring models for PRNGs with input as strong as possible, we wish to make minimal assumptions about Big Brother's influence, whilst allowing the non-backdoored adversary $\mathscr{A}$, to whom the backdoored schemes must still appear secure, maximum power to compromise the scheme. To this end, we will model $\mathscr{B}$ as a passive observer who is able to capture just one PRNG output, which he is then challenged to exploit. Simultaneously, we demand that the scheme is still secure in the face of a ROB-adversary $\mathscr{A}$, with all the capabilities this allows. Notably, the latter condition also offers the benefit of allowing us to explore the extent to which a guarantee of robustness may act as an immuniser against backdooring.

In our models to follow, we do not allow $\mathscr{B}$ any degree of compromise over the distribution sampler $\mathscr{D}$. This is again to fit with our ethos of making minimal assumptions on $\mathscr{B}$'s capabilities. It strengthens the backdooring model by demanding that the backdoor be effective against *all* samplers $\mathscr{D}$ valid up to q_r samples, including in particular those not under the control of $\mathscr{B}$. We also note that, in the extreme case where $\mathscr{B}$ has complete knowledge of all the inputs used in refresh calls, then $\mathscr{B}$'s view of the evolution of the state is deterministic and the PRNG is reduced to a FWD-secure PRG which is periodically reseeded with correlated values. Thus this restriction on Big Brother's power ensures a clear separation between the results of Sect. 3 and those which follow.

Next consider a PRNG with input which produces its output via a sequence of refresh and next calls. The evolution of the state, and subsequent production of output, is determined not only by the number of such calls, but also by their position in the sequence. To reflect this, each backdooring game below will take as input the specific refresh pattern $\textit{rp}$ which was used to produce the challenge. In line with this, and to reflect the fact that the refresh pattern may impact $\mathscr{B}$'s ability to subvert the scheme, the advantage of $\mathscr{B}$ in our formal definition will be allowed to depend on the refresh pattern $\textit{rp}$.

We present two new backdooring models for PRNGs with input in Fig. 10. In the first game, the PRNG is evolved according to the specified refresh pattern. Big Brother is given an output r_i, and challenged to recover state s_j. In the second game, Big Brother is again given output r_i, but now we ask him to recover a different output value r_j. In both games, Big Brother is additionally given the refresh pattern. Stronger notions can be achieved by considering games in which Big Brother is not given the refresh pattern, but for simplicity, we will consider the games shown in Fig. 10. In Sect. 4.4 we will discuss how our concrete construction of a BPRNG presented in Sect. 4.3 can be extended to the stronger setting in which Big Brother is not given the used refresh pattern. As with the corresponding PRG definitions in Sect. 3.1, a BPRNG backdoored in the state sense is strictly stronger than one backdoored in the out sense.

Definition 21. *A tuple of algorithms* $\overline{\mathsf{PRNG}}$ = *(*setup, init, next, refresh, $\mathscr{B}$*) is said to be a* $(t, q_r, q_n, q_c, \gamma^*, \epsilon, (\textit{type}, \delta))$*-robust BPRNG, where* type $\in$ {state, out}, *if*

Game BPRNG-STATE$_{\overline{\text{PRNG}},\mathscr{D}}^{\mathscr{B}}(\boldsymbol{rp}, i, j)$	Game BPRNG-OUT$_{\overline{\text{PRNG}},\mathscr{D}}^{\mathscr{B}}(\boldsymbol{rp}, i, j)$
$(pp, bk) \leftarrow \text{setup}$	$(pp, bk) \leftarrow \text{setup}$
$s_0 \leftarrow \text{init}(pp)$	$s_0 \leftarrow \text{init}(pp)$
$(r_1, s_1, \ldots, r_{q_n}, s_{q_n})$	$(r_0, \ldots, r_{q_n}) \leftarrow \text{out}(\overline{\text{PRNG}}, pp, s_0, \boldsymbol{rp}, \mathscr{D})$
$\quad \leftarrow \text{evolve}(\overline{\text{PRNG}}, pp, s_0, \boldsymbol{rp}, \mathscr{D})$	$r'_j \leftarrow \mathscr{B}(pp, bk, r_i, i, j, \boldsymbol{rp})$
$s'_j \leftarrow \mathscr{B}(pp, bk, r_i, i, j, \boldsymbol{rp})$	return $(r'_j = r_j)$
return $(s'_j = s_j)$	

Fig. 10. Backdooring security games BPRNG-STATE$_{\overline{\text{PRNG}},\mathscr{D}}^{\mathscr{B}}$ and BPRNG-OUT$_{\overline{\text{PRNG}},\mathscr{D}}^{\mathscr{B}}$ for BPRNGs.

- PRNG $=$ (setup, init, refresh, next) *is a* $(t, q_r, q_n, q_c, \gamma^*, \epsilon)$-*robust* PRNG *with input;*
- *For all refresh patterns* $\boldsymbol{rp} = (a_1, b_1, \ldots, a_\rho, b_\rho)$, *where* a_i, b_i, n *are polynomial in the security parameter, for all distribution samplers* $\mathscr{D}$, *for all* $1 \leq i, j \leq \sum_{\nu=1}^{\rho} a_\nu$, *where* $i \neq j$, *it holds that* $\text{Adv}_{\overline{\text{PRNG}},\mathscr{D}}^{\text{type}}(\boldsymbol{rp}, i, j) \geq \delta(\boldsymbol{rp}, i, j)$ *where*

$$\text{Adv}_{\overline{\text{PRNG}},\mathscr{D}}^{\text{type}}(\boldsymbol{rp}, i, j) := \Pr\left[\text{BPRNG-TYPE}_{\overline{\text{PRNG}},\mathscr{D}}^{\mathscr{B}}(\boldsymbol{rp}, i, j) \Rightarrow \text{true}\right].$$

We note that by replacing the index j with a vector of indices $(j_1, \ldots, j_k)$, we can immediately extend both of the above games to challenge Big Brother to recover multiple outputs and states.

4.3 Backdoored PRNG Construction

In Fig. 11, we present our construction of a BPRNG. The construction makes use of an ordinary non-backdoored PRNG with input, PRNG, and is based on the simple idea of interleaving outputs of PRNG with encryptions of snapshots of the state of PRNG, using an IND\$-CPA secure encryption scheme. By taking a snapshot of the state whenever this is refreshed and storing a list of the previous k snapshots, the construction will enable $\mathscr{B}$ to recover, with reasonable probability, the previous output values that were computed up to k refreshes ago. Of course, this means that the state of the final construction is large compared to that of the PRNG with input used as a component in its construction.

More specifically, the construction maintains a list of ciphertexts, $(C_1, \ldots, C_k)$, encrypting k snapshots of the state of PRNG. A snapshot of the state is taken in the $\overline{\text{next}}$ algorithm of our construction, whenever the previous operation was a refresh. This ensures that if the state is successively refreshed multiple times, only a single snapshot will be stored. To produce an output value, the construction will use the next function of PRNG to compute a seed r which will either be used to directly compute an output value $\bar{r}$ via a pair of PRGs, or used to re-randomize $(C_1, \ldots, C_k)$, which will then be used as $\bar{r}$. The combination of the IND\$-CPA-security of the encryption scheme and the re-randomization will ensure that the output value in the latter case will remain pseudorandom to a regular PRNG adversary. Which of the two different output values the construction will produce is decided based on the seed r.

We prove robustness of the generator by going via preserving and recovering security. To be able to achieve these notions, the ciphertexts $(C_1, \ldots, C_k)$ are re-randomized a second time in $\overline{\mathsf{next}}$ to ensure that the overall state returned by $\overline{\mathsf{next}}$ appears independent of the output value $\bar{r}$. Furthermore, to ensure recovering security, in which the adversary is allowed to maliciously set the state, the construction requires that the validity of ciphertexts can be verified. In particular, we assume the used encryption scheme is equipped with an additional algorithm, invalid, which given a public key pk and a ciphertext C, returns 1 if C is invalid for pk, and 0 if it is valid. This is used to ensure that the state of the construction always contains valid ciphertexts. Additionally, we require the used encryption scheme to satisfy a stronger re-randomization property than was introduced in Sect. 2: the re-randomisation of an adversarially chosen ciphertext should be indistinguishable from the encryption of any message. We will formalize this property below.

For the Big Brother algorithm $\mathscr{B}$ in the construction to be successful, it is required that the output value $\bar{r}_i$ given to $\mathscr{B}$ corresponds to $(C_1, \ldots, C_k)$, and that the output value $\bar{r}_j$ that $\mathscr{B}$ is required to recover corresponds to a value computed directly from the then current state of PRNG. Since the type of the produced output is decided from the output of PRNG and a PRG which are both assumed to be good generators, this will happen with probability close to $1/4$. Furthermore, it is required that the number of refresh periods between $\bar{r}_j$ and $\bar{r}_i$ is less than k. More precisely, for a refresh pattern $\boldsymbol{rp} = (a_1, b_1, \ldots, a_\rho, b_\rho)$, the number of refresh periods PRNG has undergone when $\bar{r}_i$ and $\bar{r}_j$ are produced, are $i_{ref} = \max_\sigma[\sum_{\nu=1}^\sigma a_\nu < i]$ and $j_{ref} = \max_\sigma[\sum_{\nu=1}^\sigma a_\nu < j]$, respectively. If $i_{ref} - j_{ref} < k$, the initial refreshed state used to compute $\bar{r}_j$ will be encrypted in $C_{i_{ref}-j_{ref}+1}$. Hence, all $\mathscr{B}$ has to do is to decrypt and iterate this state $j_{it} = j - \sum_{\nu=1}^{j_{ref}} a_\nu$ times to obtain the seed used to compute $\bar{r}_j$.

The full construction, shown in Fig. 11, is based on a (non-backdoored) (n, l, p)-PRNG with input, PRNG = (setup, init, refresh, next), a pair of PRGs PRG : $\{0,1\}^l \to \{0,1\}^{2ku+1}$ and PRG' : $\{0,1\}^u \to \{0,1\}^{k \times m}$, and a re-randomizable encryption scheme $\mathcal{E}$ = (KGen, Enc, Rand, Dec, invalid) with message space $\{0,1\}^n$, randomness space $\{0,1\}^u$, and ciphertext space $\{0,1\}^m$, and produces a $(k \times m + n + 1, k \times m, p)$-PRNG with input.

Before proving the construction to be robust and backdoored, we formalize the stronger re-randomization property mentioned above. Note that this property is not comparable to the re-randomization definition for PKE given in Sect. 2: that was a statistical notion concerning encryptions of the same message, while, in contrast, the following is a computational notion regarding possibly different messages.

Definition 22. *An encryption scheme $\mathcal{E}$ = (KGen, Enc, Dec) with message space $\{0,1\}^n$ is said to be (t, δ)-strongly re-randomizable, if there exists a polynomial time algorithm* Rand *such that*

- For all $(pk, sk) \leftarrow \mathsf{KGen}$, $M \in \{0,1\}^n$, and $c \leftarrow \mathsf{Enc}(pk, M)$, it holds that

$$\Pr[\mathsf{Dec}_{sk}(\mathsf{Rand}(C)) = M] = 1.$$

- For all adversaries $\mathscr{A}$ with running time t and for all messages $M \in \{0,1\}^n$, it holds that $\mathsf{Adv}_{\mathcal{E}}^{\mathsf{rand}}[(\mathscr{A})] < \delta$, where

$$\mathsf{Adv}_{\mathcal{E}}^{\mathsf{rand}}(\mathscr{A}) = \Big| \Pr\big[(pk, sk) \leftarrow \mathsf{KGen}; b \leftarrow \{0,1\}; C^* \leftarrow \mathscr{A}(pk);$$

$$C_0 \leftarrow \mathsf{Rand}(pk, C^*); C_1 \leftarrow \mathsf{Enc}(pk, M); b' \leftarrow \mathscr{A}(C_b) : b = b'\big] - 1/2 \Big|.$$

In the above, it is required that the output C^* of $\mathscr{A}$ is a valid ciphertext under pk.

It is relatively straightforward to see that ElGamal encryption satisfies the above re-randomization property. Specifically, for a public key $y = g^x$ and a ciphertext $C = (C^1, C^2) = (g^r, M \cdot y^r)$, a re-randomization C_0 of C is obtained by picking random r' and computing $C_0 = (C^1 \cdot g^{r'}, C^2 \cdot y^{r'})$. However, under the DDH assumption, the tuples $(g, g^{r'}, y, y^{r'})$ and $(g, g^{r'}, y, z)$ are indistinguishable, where z is a random group element. Hence, re-randomization of C is indistinguishable from multiplying the components of C with random group elements, which again makes C_0 indistinguishable from two random group elements. Likewise, the encryption of any message M, $C_1 = (g^r, M \cdot y^r)$, is indistinguishable from two random group elements under the DDH assumption, which makes C_0 and C_1 indistinguishable.

The proof of the following theorem appears in the full version [14].

Theorem 4. Let PRG and $\mathsf{PRG'}$ be ϵ_{prg}-secure and ϵ'_{prg}-secure PRGs respectively, and let PRNG be a (t, ϵ_{pre})-PRE and $(t, q_r, \gamma^*, \epsilon_{rec})$-REC secure PRNG with input. Suppose further that $\mathcal{E}$ is a $(t, q_{ind}, \epsilon_{ind})$-IND\$-CPA secure and (t, ϵ_{rand})-strongly re-randomizable encryption scheme. Then $\overline{\mathsf{PRNG}}$ shown in Fig. 11 is a $(t', q_r, q_n, q_c, \gamma^*, \epsilon, (\mathsf{out}, \delta))$-robust BPRNG, where $t' \approx t$,

$$\epsilon = 2q_n(8\epsilon_{ind} + 2\epsilon_{prg} + 2\epsilon'_{prg} + 4k\epsilon_{rand} + 3\epsilon_{pre} + \epsilon_{rec})$$

and

$$\delta(\boldsymbol{rp}, i, j) = \begin{cases} (1/4 - 2\epsilon_{prg} - a(\epsilon_{pre} + \epsilon_{rec})) & \text{if } j \le i \wedge i_{ref} - j_{ref} + 1 \le k \\ 0 & \text{otherwise} \end{cases}$$

where $\boldsymbol{rp} = (a_1, b_1, \ldots, a_\rho, b_\rho)$, $a = \sum_{\nu=1}^{\rho} a_\nu$, $i_{ref} \leftarrow \max_\sigma [\sum_{\nu=1}^{\sigma} a_\nu < i]$, and $j_{ref} \leftarrow \max_\sigma [\sum_{\nu=1}^{\sigma} a_\nu < j]$.

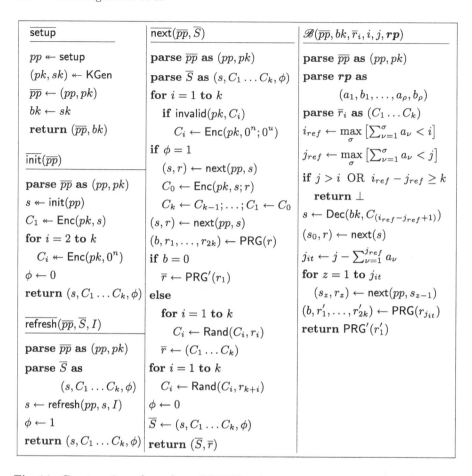

Fig. 11. Construction of a robust BPRNG using as components a re-randomisable PKE scheme $\mathcal{E} = $ (KGen, Enc, Dec, Rand, invalid), a PRNG with input PRNG = (setup, init, refresh, next), and PRGs PRG and PRG'.

4.4   Extensions and Modifications of Our Main Construction

The above construction can be modified and extended to provide slightly different properties. For example, an alternative to storing a snapshot of a refreshed state by rotating the ciphertexts $(C_1, \ldots, C_k)$ as done in line 9 of $\overline{\mathsf{next}}$, would be to choose a random ciphertext to replace. More specifically, the output value r of PRNG computed in line 7 could be stretched to produce a $\log k$ bit value t, and ciphertext C_t would then be replaced with C_0. Note, however, that $\mathcal{B}$ would no longer be able to tell which ciphertext corresponds to which snapshot of the state. This can be addressed if the used encryption scheme is additionally assumed to be additively homomorphic, e.g. like ElGamal encryption, which, using an appropriate group, also satisfies all of the other requirements of the construction. In this case, the construction would be able to maintain an encrypted counter of the number of refresh periods, and, for each snapshot, store an encrypted value corresponding to the number of refresh periods PRNG has undergone before the snapshot was taken. If the ciphertexts containing these values are concatenated with $(C_1, \ldots, C_k)$ to produce the output value $\bar{r}$, then $\mathscr{B}$ obtains sufficient information to derive what state to use to recover a given output value. This yields a construction with slightly different advantage function $\delta(\boldsymbol{rp}, i, j)$ compared to the above construction; instead of a sharp drop to 0 when i and j are separated by k refresh periods, the advantage gradually declines as the distance (in terms of the number of refresh periods) between i and j increases.

The above construction can furthermore be modified to produce shorter output values. Specifically, instead of setting $\bar{r} \leftarrow (C_1, \ldots, C_k)$ in line 16 of $\overline{\text{next}}$, a random ciphertext C_t could be chosen as $\bar{r}$, by stretching the output of PRG in line 11 with an additional $\log k$ bits to produce t. This will reduce the output length from km bits to m bits. However, a similar problem to the above occurs: $\mathscr{B}$ will not be able to tell which snapshot C_t represents. Using a similar solution to the above will increase the output length to $2m$ bits. This modification will essentially reduce the backdooring advantage by a factor of $1/k$ compared to the above construction.

Lastly, we note that the above construction assumes $\mathscr{B}$ receives as input the refresh pattern $\boldsymbol{rp}$. Again, by maintaining encrypted counters for both the number of refresh periods and the number of produced output values for each snapshot, we can obtain an algorithm $\mathscr{B}$ which does not require $\boldsymbol{rp}$ as input, but at the cost of increasing the output size.

All of the above modifications can be shown to be secure using almost identical arguments to the existing security analysis for the above construction.

5 On the Inherent Resistance of PRNGs with Input to Backdoors

In the previous section we have shown a construction, and variations thereof, for a PRNG with input that is backdoored in a powerful sense: from a given output Big Brother can recover prior state and output values past an arbitrary number of refreshes. One can see however that in our constructions, Big Brother's ability to go past refreshes is limited by the size of the state and output of the constructed generator. We now show that this limitation is inherent in any PRNG with input that is robust.

In particular consider the sequence representing the evolution of a PRNG's state, and select a subsequence of states where any two states are separated by consecutive refreshes that in combination have high entropy. Then we will show that the number of such states that Big Brother can predict *simultaneously* with non-negligible probability is limited by the size of the state. Thus if we limit the state size of a robust PRNG, then Big Brother's ability in exploiting any potential backdoors that it may contain must *decrease* as more entropy becomes available to the PRNG.

5.1   An Impossibility Result

We now turn to formalising the preceding claim. In order to simplify the analysis to follow, we focus on a restricted class of distribution samplers. We say that a distribution sampler is well-behaved if it satisfies the following properties:

- It is efficiently sampleable.
- For any i the entropy estimate γ_i of the random variable I_i is fixed, but may vary across different values of i.
- For all $i > 0$ such that $\Pr(\sigma_{i-1}) > 0$ it holds that:

$$\mathrm{H}_\infty\left(I_i \mid I_1, \ldots, I_{i-1}, I_{i+1}, \ldots, I_{q_r}, z_1, \ldots, z_{q_r}, \gamma_1, \ldots, \gamma_{q_r}\right) \geq \gamma_i$$

where $(\sigma_i, I_i, \gamma_i, z_i) = \mathscr{D}(\sigma_{i-1})$ for $i \in \{1, \ldots, q_r\}$ and $\sigma_0 = \varepsilon$.

For any well-behaved distribution sampler $\mathscr{D}$ and any PRNG with input PRNG, let us now consider the experiment of running setup and init to obtain a public paramer pp and an initial state S_0, and then applying a sequence of queries $q_1, \ldots, q_i, \ldots$ where each q_i represents a query to refresh or next. To any query q_i we associate a tuple $(R_i, S_i, \sigma_i, I_i, \gamma_i)$ that represents the outcome of that query. If q_i is a refresh query these variables are set by the outputs of $\mathscr{D}$ and refresh, while R_i is set to ε. If q_i is a next query these variables are set to the outputs of next while γ_i is set to zero, I_i is set to the empty string, and $\sigma_i \leftarrow \sigma_{i-1}$. (Note that we deviate slightly here in the notation we use for the output and state of a PRNG with input: we use R_i and S_i to denote *random variables* and we use r_i and s_i respectively to denote *values* assumed by these random variables.)

Now let the function $f : \mathbb{N} \to \mathbb{N}$ where $f(0) = 0$ identify a subsequence $(R_{f(j)}, S_{f(j)}, \sigma_{f(j)}, I_{f(j)}, \gamma_{f(j)})$. We say that a subsequence is *legitimate* if for all $S_{f(j)}$ there exists $f(j-1) \leq c \leq d \leq f(j)$ such that $\sum_c^d \gamma_i \geq \gamma^*$, and all queries between c and d are refresh queries. For ease of notation we let ϵ denote an upper bound on $\mathrm{Adv}_{\mathsf{PRNG}}^{\mathrm{rob}}(\mathscr{A}, \mathscr{D}') + \frac{1}{2^r}$ over all $\mathscr{D}'$ and all $\mathscr{A}$ in some class of adversaries with restricted sources.

With this notation established, we can state the main theorem of this section as follows:

Theorem 5. *For any PRNG with input* PRNG *having associated parameters* (n, l, p), *any well-behaved distribution sampler* $\mathscr{D}$, *any sequence of queries, any legitimate subsequence identified by the function* f, *any index* j, *and any* $k \in \mathbb{N}$, *it holds that:*

$$\tilde{\mathrm{H}}_\infty\left(\bar{S}'_{f(j)} \mid R_{f(j)+k}, pp\right) \geq \frac{j+1}{2} \log\left(\frac{1}{\epsilon}\right) - \min(n, l).$$

The proof of the theorem can be found in the full version [14].

This theorem deserves some interpretation. On the left-hand-side, $R_{f(j)+k}$ refers to a particular output received by $\mathscr{B}$ and pp to the public parameters. The theorem says that, conditioned on these, the vector of states $\bar{S}'_{f(j)}$ still has large average min-entropy, provided j is sufficiently large. This is because, on the right-hand-side, $\min(n, l)$ is fixed for a given generator, ϵ is small (so $\log\left(\frac{1}{\epsilon}\right)$ is large), and the first term scales linearly with j, thus attaining arbitrarily large values as j increases. This means that it is impossible for $\mathscr{B}$ to compute or guess the state vector with a good success probability. In short, no adversary, irrespective of its computational resources or backdoor information, can recover all the state information represented by the vector $\bar{S}'_{f(j)}$. In addition the result extends easily to the stronger setting where the adversary is given any sequence of outputs following $R_{f(j)}$, since these will depend only on $S_{f(j)}$ and independently sampled future I values. In that case, we simply replace the $R_{f(j)+k}$ term by any sequence of ouputs following $R_{f(j)}$ and $\min(n, l)$ by n.

5.2   Discussion and Open Problems

Theorem 5 concerns *state* recovery attacks against robust PRNGs with input. It seems plausible to us that the result can be strengthened to say something about the impossibility of recovering old outputs, instead of old states. Likewise, the theorem only concerns the impossibility of recovering *old* states from current outputs, but nothing about the hardness of recovering *future* states or outputs (after refreshing) from current outputs. Informally, the strength of the robustness security notion seems to make such a result plausible, since it essentially requires that a PRNG with input cannot ignore its entropy inputs when refreshing. However, we have not yet proved a formal result in this direction. These are problems that we intend to study in our immediate future work. They relate closely to the kind of impossibility result that would be useful in demonstrating the absence of the kind of effective backdooring that $\mathscr{B}$ might prefer to perform.

This result can also be seen as saying that a PRNG with input is, to some extent, intrinsically immunised against backdooring attacks, since $\mathscr{B}$ cannot recover *all* old states once sufficient entropy has been accumulated in the generator. Here the immunisation is a direct consequence of the nature of the primitive. By contrast, for PRGs, the results of [15] concerning immunisation of PRGs require intrusive changes to the PRG, essentially post-processing the generator's output with either a keyed primitive (a PRF) or a hash with relatively strong security (a random oracle or a Universal Computational Extractor). Moreover, our strengthening of the result of [15], via constructions of forward-secure PRGs that are backdoored in the strong first sense, shows that PRGs cannot resist backdooring in general. So some form of external immunisation is inevitable if PRGs are to resist backdooring.

On the other hand, exploring immunisation for PRNGs with input would still be useful, since, as our constructions in Sect. 4 show, it is possible to achieve meaningful levels of backdooring for PRNGs with input. Naively, the immunisation techniques of [15] should work equally well for PRNGs with input as they do for PRGs, since a PRNG with input certainly contains within it an implicit PRG, and if that simpler component is immunised, then so should be the more complex PRNG primitive. Furthermore, it may be that PRNGs with input, being informally *harder* to backdoor, could be immunised by applying less intrusive or less idealised cryptographic techniques.

Acknowledgments. Degabriele and Paterson were supported by EPSRC grant EP/M013472/1 (UK Quantum Technology Hub for Quantum Communications Technologies). Schuldt was supported by JSPS KAKENHI Grant Number 15K16006. Woodage was supported by the EPSRC and the UK government as part of the Centre for Doctoral Training in Cyber Security at Royal Holloway, University of London (EP/K035584/1)

References

1. Abeni, P., Bello, L., Bertacchini, M.: Exploiting DSA-1571: How to break PFS in SSL with EDH, July 2008
2. Ateniese, G., Magri, B., Venturi, D.: Subversion-resilient signature schemes. Cryptology ePrint Archive, Report 2015/517 (2015). http://eprint.iacr.org/2015/517
3. Baignères, T., Delerablée, C., Finiasz, M., Goubin, L., Lepoint, T., Rivain, M.: Trap me if you can - million dollar curve. IACR Cryptology ePrint Archive 2015:1249 (2015)
4. Barak, B., Halevi, S.: A model and architecture for pseudo-random generation with applications to/dev/random. In: Atluri, V., Meadows, C., Juels, A. (eds.) ACM CCS 05, Alexandria, Virginia, USA, 7–11 November 2005, pp. 203–212. ACM Press (2005)
5. Bellare, M., Boldyreva, A., O'Neill, A.: Deterministic and efficiently searchable encryption. In: Menezes, A. (ed.) CRYPTO 2007. LNCS, vol. 4622, pp. 535–552. Springer, Heidelberg (2007)
6. Bellare, M., Paterson, K.G., Rogaway, P.: Security of symmetric encryption against mass surveillance. In: Garay, J.A., Gennaro, R. (eds.) CRYPTO 2014, Part I. LNCS, vol. 8616, pp. 1–19. Springer, Heidelberg (2014)
7. Bernstein, D.J., Chang, Y.-A., Cheng, C.-M., Chou, L.-P., Heninger, N., Lange, T., van Someren, N.: Factoring RSA keys from certified smart cards: coppersmith in the wild. In: Sako, K., Sarkar, P. (eds.) ASIACRYPT 2013, Part II. LNCS, vol. 8270, pp. 341–360. Springer, Heidelberg (2013)
8. Bernstein, D.J., Chou, T., Chuengsatiansup, C., Hülsing, A., Lange, T., Niederhagen, R., van Vredendaal, C.: How to manipulate curve standards: a white paper for the black hat. Cryptology ePrint Archive, Report 2014/571 (2014). http://eprint.iacr.org/2014/571
9. Bernstein, D.J., Hamburg, M., Krasnova, A., Lange, T.: Elligator: elliptic-curve points indistinguishable from uniformrandom strings. In: Sadeghi, A.-R. et al. [29], pp. 967–980
10. Brown, D.R.L.: A weak-randomizer attack on RSA-OAEP with e = 3. Cryptology ePrint Archive, Report 2005/189 (2005). http://eprint.iacr.org/2005/189
11. Checkoway, S., Niederhagen, R., Everspaugh, A., Green, M., Lange, T., Ristenpart, T., Bernstein, D.J., Maskiewicz, J., Shacham, H., Fredrikson, M.: On the practical exploitability of dual EC in TLS implementations. In: Fu, K., Jung, J. (eds.) Proceedings of the 23rd USENIX Security Symposium, San Diego, CA, USA, 20–22 August 2014, pp. 319–335. USENIX Association (2014)

12. Cornejo, M., Ruhault, S.: Characterization of real-life PRNGs under partial state corruption. In: Ahn, G.-J., Yung, M., Li, N. (eds.) ACM CCS 14, Scottsdale, AZ, USA, 3–7 November 2014, pp. 1004–1015. ACM Press (2014)

13. Degabriele, J.P., Farshim, P., Poettering, B.: A more cautious approach to security against mass surveillance. In: Leander, G. (ed.) FSE 2015. LNCS, vol. 9054, pp. 579–598. Springer, Heidelberg (2015)

14. Degabriele, J.P., Paterson, K.G., Schuldt, J.C.N., Woodage, J.: Backdoors in pseudorandom number generators: possibility andimpossibility results. Cryptology ePrint Archive, Report 2016/577 (2016). http://eprint.iacr.org/2016/577

15. Dodis, Y., Ganesh, C., Golovnev, A., Juels, A., Ristenpart, T.: A formal treatment of backdoored pseudorandom generators. In: Oswald, E., Fischlin, M. (eds.) EUROCRYPT 2015. LNCS, vol. 9056, pp. 101–126. Springer, Heidelberg (2015)

16. Dodis, Y., Ong, S.J., Prabhakaran, M., Sahai, A.: On the (im)possibility of cryptography with imperfect randomness. In: 45th FOCS, Rome, Italy, 17–19 October 2004, pp. 196–205. IEEE Computer Society Press (2004)

17. Dodis, Y., Pointcheval, D., Ruhault, S., Vergnaud, D., Wichs, D.: Security analysis of pseudo-random number generators with input: /dev/random is not robust. In: Sadeghi, A.-R., et al. [29], pp. 647–658

18. Dodis, Y., Shamir, A., Stephens-Davidowitz, N., Wichs, D.: How to eat your entropy and have it too – optimal recovery strategies for compromised RNGs. In: Garay, J.A., Gennaro, R. (eds.) CRYPTO 2014, Part II. LNCS, vol. 8617, pp. 37–54. Springer, Heidelberg (2014)

19. Goldberg, I., Wagner, D.: Randomness and the Netscape browser. Dr Dobb's J.-Softw. Tools Prof. Programmer 21(1), 66–71 (1996)

20. Hemenway, B., Libert, B., Ostrovsky, R., Vergnaud, D.: Lossy encryption: constructions from general assumptions and efficient selective opening chosen ciphertext security. In: Lee, D.H., Wang, X. (eds.) ASIACRYPT 2011. LNCS, vol. 7073, pp. 70–88. Springer, Heidelberg (2011)

21. Heninger, N., Durumeric, Z., Wustrow, E., Halderman, J.A.: Mining your Ps, Qs: detection of widespread weak keys in network devices. In: Kohno, T. (ed.) Proceedings of the 21th USENIX Security Symposium, Bellevue, WA, USA, 8–10 August 2012, pp. 205–220. USENIX Association (2012)

22. Lenstra, A.K., Hughes, J.P., Augier, M., Bos, J.W., Kleinjung, T., Wachter, C.: Public keys. In: Safavi-Naini, R., Canetti, R. (eds.) CRYPTO 2012. LNCS, vol. 7417, pp. 626–642. Springer, Heidelberg (2012)

23. Mironov, I., Stephens-Davidowitz, N.: Cryptographic reverse firewalls. In: Oswald, E., Fischlin, M. (eds.) EUROCRYPT 2015. LNCS, vol. 9057, pp. 657–686. Springer, Heidelberg (2015)

24. Möller, B.: A public-key encryption scheme with pseudo-random ciphertexts. In: Samarati, P., Ryan, P.Y.A., Gollmann, D., Molva, R. (eds.) ESORICS 2004. LNCS, vol. 3193, pp. 335–351. Springer, Heidelberg (2004)

25. Mueller, M.: Debian OpenSSL predictable PRNG bruteforce SSH exploit, May 2008

26. Peikert, C., Waters, B.: Lossy trapdoor functions and their applications. In: Ladner, R.E., Dwork, C. (eds.) 40th ACM STOC, Victoria, British Columbia, Canada, 17–20 May 2008, pp. 187–196. ACM Press (2008)

27. Ristenpart, T., Yilek, S.: When good randomness goes bad: virtual machine reset vulnerabilities and hedging deployed cryptography. In: Proceedings of the Network and Distributed System Security Symposium, NDSS 2010, San Diego, California, USA, 28 February–3 March 2010. The Internet Society (2010)

28. Russell, A., Tang, Q., Yung, M., Zhou, H.-S.: Cliptography: clipping the power of kleptographic attacks. Cryptology ePrint Archive, Report 2015/695 (2015). http://eprint.iacr.org/2015/695

29. Sadeghi, A.-R., Gligor, V.D., Yung, M. (eds.) ACM CCS 13, Berlin, Germany, 4–8 November 2013. ACM Press (2013)

30. Shumow, D., Ferguson, N.: On the possibility of a back door in the NIST SP800-90 Dual EC PRNG. Presentation at rump session of CRYPTO 2007 (2007)

31. Simmons, G.J.: The prisoners' problem and the subliminal channel. In: Chaum, D. (ed.) CRYPTO 1983, Santa Barbara, CA, USA, pp. 51–67. Plenum Press, New York (1983)

32. Vazirani, U.V., Vazirani, V.V.: Trapdoor pseudo-random number generators, with applications to protocol design. In: 24th Annual Symposium on Foundations of Computer Science, Tucson, Arizona, USA, 7–9 November 1983, pp. 23–30. IEEE Computer Society (1983)

33. Yilek, S., Rescorla, E., Shacham, H., Enright, B., Savage, S., When private keys are public: results from the 2008 Debian OpenSSL vulnerability. In: Feldmann, A., Mathy, L. (eds.) Proceedings of the 9th ACM SIGCOMM Internet Measurement Conference, IMC 2009, Chicago, Illinois, USA, 4–6 November 2009, pp. 15–27. ACM (2009)

34. Young, A., Yung, M.: Kleptography: using cryptography against cryptography. In: Fumy, W. (ed.) EUROCRYPT 1997. LNCS, vol. 1233, pp. 62–74. Springer, Heidelberg (1997)

35. Young, A., Yung, M.: Relationships between Diffie-Hellman and "index oracles". In: Blundo, C., Cimato, S. (eds.) SCN 2004. LNCS, vol. 3352, pp. 16–32. Springer, Heidelberg (2005)

Symmetric Cryptanalysis

A 2^{70} Attack on the Full MISTY1

Achiya Bar-On$^{(\boxtimes)}$ and Nathan Keller

Department of Mathematics, Bar Ilan University, 52900 Ramat Gan, Israel
abo1000@gmail.com

Abstract. MISTY1 is a block cipher designed by Matsui in 1997. It is widely deployed in Japan, and is recognized internationally as a European NESSIE-recommended cipher and an ISO standard. After almost 20 years of unsuccessful cryptanalytic attempts, a first attack on the full MISTY1 was presented at CRYPTO 2015 by Yosuke Todo. The attack, using a new technique called *division property*, requires almost the full codebook and has time complexity of $2^{107.3}$ encryptions.

In this paper we present a new attack on the full MISTY1. It is based on Todo's division property, along with a variety of refined key-recovery techniques. Our attack requires almost the full codebook (like Todo's attack), but allows to retrieve 49 bits of the secret key in time complexity of only 2^{64} encryptions, and the full key in time complexity of $2^{69.5}$ encryptions.

While our attack is clearly impractical due to its large data complexity, it shows that MISTY1 provides security of only 2^{70} — significantly less than what was considered before.

1 Introduction

MISTY1 [10] is a 64-bit block cipher with 128-bit keys designed in 1997 by Matsui. In 2002, MISTY1 was selected by the Japanese government to be one of the e-government candidate recommended cipher, and since then, it became widely deployed in Japan. MISTY1 also gained recognition outside Japan, when it was selected to the portfolio of European NESSIE-recommended ciphers, and approved as an ISO standard in 2005. Furthermore, the block cipher KASUMI [1] designed as a slight modification of MISTY1 is used in the 3G cellular networks, which makes it one of the most widely used block ciphers today.

MISTY1 has an 8-round recursive Feistel structure, where the round function FO is in itself a 3-round Feistel construction, whose F-function FI is in turn a 3-round Feistel construction using 7-bit and 9-bit invertible S-boxes. The specific choice of S-boxes and the recursive structure ensure provable security against differential and linear cryptanalysis. In order to thwart other types of attacks, after every two rounds an FL function is applied to each of the two halves independently. The FL functions are key-dependent linear functions which play the role of whitening layers.

A. Bar-On – This research was partially supported by the Israeli Ministry of Science, Technology and Space, and by the Check Point Institute for Information Security.

M. Robshaw and J. Katz (Eds.): CRYPTO 2016, Part I, LNCS 9814, pp. 435–456, 2016.
DOI: 10.1007/978-3-662-53018-4_16

In the 18 years since its design, MISTY1 withstood numerous cryptanalytic attempts. More than a dozen of papers analyzed its reduced-round variants (see, e.g., [2,7,13,14]), yet the full 8-round variant seemed completely out of reach. The situation changed when Yosuke Todo presented at CRYPTO'2015 [11] the first attack on the full MISTY1. The attack is based on a new variant of integral cryptanalysis, called *division property*, presented by Todo at EURO-CRYPT'2015. Using the new technique, Todo showed that there exist seven independent integral structures of 2^{63} plaintexts, whose propagation can be traced through 6 rounds of MISTY1. (This should be compared with 4 rounds achieved by the best previously known integral characteristics.) These characteristics allow to break the full MISTY1 with data complexity of $2^{63.994}$ chosen plaintexts and time complexity of $2^{107.3}$ encryptions.

In this paper, we present an improved attack on the full MISTY1, which allows to significantly reduce the time complexity. Our attack uses Todo's division property both in the encryption direction (like Todo used it) and in the decryption direction (which turns out to be more useful than the encryption direction due to key schedule considerations). In addition, we use refined key recovery techniques, including the partial sums technique [6], and a two-dimensional meet-in-the-middle attack (like the one is used in [4]). Our attack has two phases. The first phase requires $2^{64} - 2^{50}$ chosen ciphertexts and allows to recover the equivalent of 49 key bits in time of 2^{64} encryptions (i.e., dominated by the time required for encrypting the plaintexts!). The second phase requires almost all the rest of the codebook and allows to recover all remaining key bits in time of $2^{69.5}$ encryptions. Alternatively, the rest of the key can be found in time of 2^{79} encryptions without additional data. A comparison of our attack with the best previously known attacks on full-round and reduced-round MISTY1 is presented in Table 1.

The paper is organized as follows. In Sect. 2 we describe the structure of MISTY1 and introduce some notations that will be used throughout the paper. The division property is described in Sect. 3, as well as the 6-round integral characteristic based on its propagation. Our improved attack on full MISTY1 is presented in Sect. 4. Finally, in Sect. 5 we summarize the paper.

2   Brief Description of MISTY1

MISTY1 is an 8-round Feistel construction, where the round function, FO, is in itself a variant of a 3-round Feistel construction, defined as follows. The input to FO is divided into two halves. The left one is XORed with a subkey, enters a keyed permutation FI, and the output is XORed with the right half. After the XOR, the two halves are swapped, and the same process (including the swap) is repeated two more times. After that, an additional swap and an XOR of the left half with a subkey is performed (see Fig. 1).

The FI function in itself also has a Feistel-like structure. Its 16-bit input is divided into two unequal parts – one of 9 bits, and the second of 7 bits. The left part (which contains 9 bits) enters an S-box, $S9$, and the output is XORed

Table 1. Summary of the best known single-key attacks on MISTY1

FO rounds	FL layers	Data complexity	Time complexity	Type
7	3	2^{58} KP	$2^{124.4}$	ID attack [7]
7	4	$2^{62.9}$ KP	2^{118}	MZC attack [14]
7	4	$2^{49.7}$ CP	$2^{116.4}$	HOD attack [13]
7	4	$2^{50.1}$ CP	$2^{100.4}$	HOD attack [2]
7	5	$2^{51.45}$ CP	2^{121}	HOD attack [2]
8	5	$2^{63.58}$ CP	2^{121}	IDP attack [11]
8	5	$2^{63.994}$ CP	$2^{107.3}$	IDP attack [11]
8	5	$2^{63.9999}$ CC	2^{64}	IDP attack[†] (Sect. 4)
8	5	$2^{63.9999}$ CC	2^{79}	IDP attack (Sect. 4)
8	5	$2^{64} - 2^{36}$ CPC	$2^{69.5}$	IDP attack (Sect. 4)

ID attack: Impossible Differential attack
HOD attack: Higher Order Differential attack
MZC attack: Multi-Dimensional Zero Correlation attack
IDP attack: Integral attack using division property
[†] Attack recovers 49 key bits
KP: Known Plaintexts; CP: Chosen Plaintexts; CC: Chosen Ciphertexts; CPC: Chosen Plaintexts and Ciphertexts

with the right 7-bit part (after padding the 7-bit value with two zeroes as the most significant bits). The two parts are swapped, the 7-bit part enters a different S-box, $S7$, and the output is XORed with 7 bits out of the 9 of the right part. The two parts are then XORed with a subkey, and swapped again. The 9-bit value again enters $S9$, and the output is XORed with the 7-bit part (after padding). The two parts are then swapped for the last time.

Every two rounds, starting before the first one, each of the two 32-bit halves enters an FL layer. The FL layer is a simple linear transformation. Its input is divided into two halves of 16 bits each, the AND of the left half with a subkey is XORed to the right half, and the OR of the updated right half with another subkey is XORed to the left half. We outline the structure of MISTY1 and its parts in Fig. 1.

The key schedule of MISTY1 takes the 128-bit key, and treats it as eight 16-bit words $K_1, K_2, \ldots, K_8$. From this sequence of words, another sequence of eight 16-bit words is generated, according to the rule $K_i' = FI_{K_{i+1}}(K_i)$.

In each round, seven words are used as the round subkey, and each of the FL functions accepts two subkey words. We give the exact key schedule of MISTY1 in Table 2.

2.1 Notations Used in the Paper

Throughout the paper, we use the following notations for intermediate values during the MISTY1 encryption process.

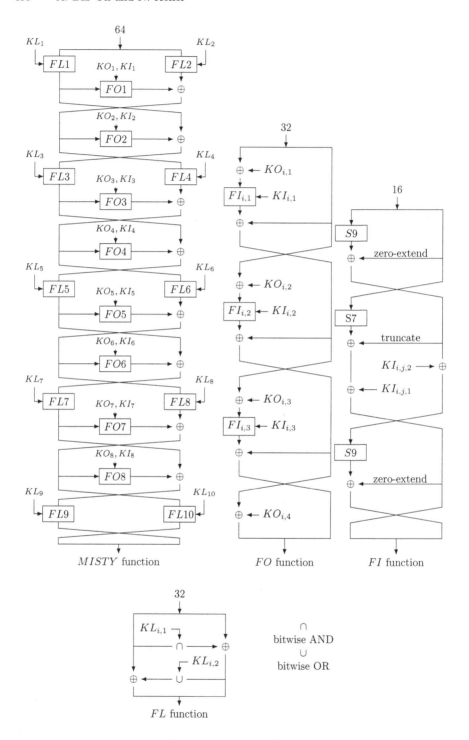

Fig. 1. Outline of MISTY1

Table 2. The Key Schedule of MISTY1

$KO_{i,1}$	$KO_{i,2}$	$KO_{i,3}$	$KO_{i,4}$	$KI_{i,1}$	$KI_{i,2}$	$KI_{i,3}$	$KL_{i,1}$	$KL_{i,2}$
K_i	K_{i+2}	K_{i+7}	K_{i+4}	K'_{i+5}	K'_{i+1}	K'_{i+3}	$K_{\frac{i+1}{2}}$ (odd i) $K'_{\frac{i}{2}+2}$ (even i)	$K'_{\frac{i+1}{2}+6}$ (odd i) $K_{\frac{i}{2}+4}$ (even i)

- The plaintext and the ciphertext are denoted, as usual, by P and $C = E(P)$.
- The input of the i'th round ($1 \le i \le 8$) is denoted by X_i. If we want to emphasize that the intermediate value corresponds to the plaintext P, we denote it by $X_i(P)$.
- For odd rounds, we denote by X'_i the intermediate value after application of the FL functions.
- The output of the FO function of round i is denoted Out_i.
- For any intermediate value Z, $Z[k-l]$ denotes bits from k to l (inclusive) of Z. The special case $Z[i]$ denotes the i'th bit of Z.
- For any intermediate value Z, the right and left halves of Z are denoted by Z_R and Z_L, respectively.
- For any 16-bit key K, the 9 rightmost bits and 7 left most bits of K are denoted by K^R and K^L, respectively.
- The first $S9$ function of $FI_{i,j}$ is denoted by $S9_{i,j,1}$ and its input is denoted by $ES9_{i,j,1}$. Similarly, $S7_{i,j}$, $ES7_{i,j}$ and $S9_{i,j,2}$, $ES9_{i,j,2}$ denote the other S-boxes, $S7$ and the second $S9$, and their inputs in $FI_{i,j}$.

3 Integral Cryptanalysis Using Division Property and Its Application to MISTY1

3.1 Integral Cryptanalysis

Integral cryptanalysis is a powerful cryptanalytic technique presented in [3,8]. The basic idea behind integral cryptanalysis is to trace the encryption process of a structured set of plaintexts, called *integral structure*, during part of the cipher's rounds. Usually, the information on the intermediate values gradually reduces as the encryption proceeds, so that eventually, one can predict only the *sum* of the set of intermediate values in part of the state.

Definition 1. A set of values V is called balanced if $\bigoplus_{x \in V} x = 0$.

An *integral characteristic* predicts that for a structured set V of plaintexts (usually, an affine subspace of the plaintext space), the set of corresponding intermediate values after i rounds is balanced in some bit j. That is,

$$\bigoplus_{x \in V} X_{i+1}[j](x) = 0. \tag{1}$$

Note that unlike differential and linear cryptanalysis, an integral characteristic is combinatorial and not statistical, meaning that (1) holds with probability 1. A characteristic that satisfies (1) is called an i-round characteristic of order $\log_2 |V|$.

An example of an integral characteristic is an 8-order 3-round characteristic of AES presented in [3], which predicts that if V consists of 256 plaintexts that are constant in all bytes but one, and assume all possible values in the remaining byte, then the corresponding set of intermediate values after 3 rounds is balanced in each of the 16 bytes. This property is used in [3] to attack up to 6 rounds of AES with a practical complexity.

Once an integral characteristic is found, it can be used to mount a key-recovery attack. Suppose that for some block cipher $E \colon \{0,1\}^n \to \{0,1\}^n$, there exists an i-round integral characteristic that satisfies (1). Denote by E_1^{-1} the Boolean function that represents the mapping from the ciphertext of E to the intermediate state bit $X_i[j]$ (see Fig. 2). Then Eq. (1) can be rewritten as

$$\bigoplus_{x \in V} E_1^{-1}(E(x)) = 0. \tag{2}$$

(Eq. (2) is called *the attack equation*). The adversary asks for the encryption of several structured sets of plaintexts of the form V, partially decrypts the corresponding ciphertexts (by guessing the key material used in E_1^{-1}), and checks whether Eq. (2) holds.

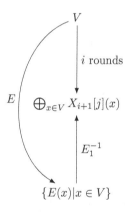

Fig. 2. Outline of the integral attack

3.2 Division Property

The idea behind the division property is to increase the precision of the information on intermediate values traced by the integral attack. As the idea is rather complex in its general form, we present in this section all definitions and notations required for it (mostly taken from [11]), and in the next section its application to MISTY1. For the general *division property* technique, we refer the reader to [12].

For a linear subspace $U \subset \mathbb{F}_2^n$ and a vector $v \in \mathbb{F}_2^n$, we call $V = \{u+v | u \in U\}$ an *affine subspace* of $\mathbb{F}_2^n$. The dimension of V is, as usual, $\log_2 |V|$.

For $u \in \mathbb{F}_2^n$ we denote by w_u the Hamming weight of u (i.e., $w_u = \sum_{i=1}^n u[i]$) and by $\mathbb{S}_k^n = \{u \in \mathbb{F}_2^n : w_u \geq k\}$ the set of all values with Hamming weight larger than (or equal to) k. For $x, u \in \mathbb{F}_2^n$ we define $x^u = \prod_{i=1}^n x[i]^{u[i]}$. In general, for $x = (x_1, x_2, \ldots, x_m), u = (u_1, u_2, \ldots, u_m) \in \mathbb{F}_2^{n_1} \times \mathbb{F}_2^{n_2} \times \cdots \times \mathbb{F}_2^{n_m}$, we write x^u for $\prod_{i=1}^m x_i^{u_i}$ and $\mathbb{S}_{[k_1,k_2,\ldots,k_m]}^{n_1,n_2,\ldots,n_m}$ for the set $\{u \in \mathbb{F}_2^{n_1} \times \cdots \times \mathbb{F}_2^{n_m} : w_{u_i} \geq k_i \text{ for all } i\}$.

Let $\mathbb{X} \subset \mathbb{F}_2^{n_1} \times \cdots \times \mathbb{F}_2^{n_m}$ be a multiset. We say that $\mathbb{X}$ has the division property $\mathcal{D}_{[k_1,k_2,\ldots,k_m]}^{n_1,n_2,\ldots,n_m}$ if

$$\bigoplus_{x \in \mathbb{X}} x^u = 0, \quad \text{for all } u \in (\mathbb{F}_2^{n_1} \times \cdots \times \mathbb{F}_2^{n_m} \setminus \mathbb{S}_{[k_1,\ldots,k_m]}^{n_1,\ldots,n_m})$$

(with no restriction on $\bigoplus_{x \in \mathbb{X}} x^u$ for $u \in \mathbb{S}_{[k_1,\ldots,k_m]}^{n_1,\ldots,n_m}$). We also define

$$\mathbb{S}_{\mathbf{k}_1,\ldots,\mathbf{k}_t}^{n_1,n_2,\ldots,n_m} = \bigcup_{i=1}^{t} \mathbb{S}_{\mathbf{k}_i}^{n_1,n_2,\ldots,n_m}$$

($\mathbf{k}_i \in \mathbb{F}_2^{n_1} \times \cdots \times \mathbb{F}_2^{n_m}$) and define the division property $\mathcal{D}_{\mathbf{k}_1,\ldots,\mathbf{k}_t}^{n_1,n_2,\ldots,n_m}$ similarly to the above definition of $\mathcal{D}_{[k_1,k_2\ldots,k_m]}^{n_1,n_2,\ldots,n_m}$.

Example 2. Any 4-dimensional affine subspace $V \subset \mathbb{F}_2^7$ has the property

$$\bigoplus_{x \in V} x_{i_1} x_{i_2} x_{i_3} = 0$$

for all $1 \leq i_1 < i_2 < i_3 \leq 7$ (where $x_j = x[j]$ denote the jth bit of x). Hence, V has the division property $\mathcal{D}_4^7$.

Example 3. For $1 \leq i \leq m$, let $V_i \subset \mathbb{F}_2^{n_i}$ be an affine subspace of dimension k_i. Then $\mathbb{X} = \{(x_1, \ldots, x_m) \in \mathbb{F}_2^{n_1} \times \cdots \times \mathbb{F}_2^{n_m} : x_i \in V_i\}$ has the division property $\mathcal{D}_{[k_1,k_2\ldots,k_m]}^{n_1,n_2,\ldots,n_m}$.

Notation 4. Let $\mathbf{k}_i, \mathbf{k}_j \in \mathbb{Z}^m$. We write $\mathbf{k}_i \leq \mathbf{k}_j$ if $\mathbf{k}_i[\ell] \leq \mathbf{k}_j[\ell]$ for all $1 \leq \ell \leq m$.

Example 5. Let $\mathbb{X} \subseteq \mathbb{F}_2^7 \times \mathbb{F}_2^7$ be a multiset that has the division property $\mathcal{D}_{[5,0],[1,4],[2,3]}^{7,7}$. Then

$$\bigoplus_{(x,y) \in \mathbb{X}} x_{i_1} \cdots x_{i_t} y_{j_1} \cdots y_{j_s}$$

is unknown if $(5,0) \leq (t,s)$ or $(1,4) \leq (t,s)$ or $(2,3) \leq (t,s)$ and equals 0 otherwise.

Observation 6. If $\mathbf{k}_i \leq \mathbf{k}_j$ ($\iff \mathbb{S}_{\mathbf{k}_i} \subseteq \mathbb{S}_{\mathbf{k}_j}$) then we omit $\mathbf{k}_j$ from $\mathbb{S}_{\mathbf{k}_1,\ldots,\mathbf{k}_t}^{n_1,n_2,\ldots,n_m}$ because $\mathbb{S}_{\mathbf{k}_1,\ldots,\mathbf{k}_t}^{n_1,n_2,\ldots,n_m} = \mathbb{S}_{\mathbf{k}_1,\ldots,\mathbf{k}_{j-1},\mathbf{k}_{j+1},\ldots,\mathbf{k}_t}^{n_1,n_2,\ldots,n_m}$ in this case. For example, we replace $\mathcal{D}_{[5,0],[1,4],[2,4],[2,3]}^{7,7}$ with $\mathcal{D}_{[5,0],[1,4],[2,3]}^{7,7}$.

Remark 7. Let $\mathbb{X}$ be a multiset that has the division property $\mathcal{D}_{\mathbf{k}_1,\ldots,\mathbf{k}_t}^{n_1,n_2,\ldots,n_m}$. If $(2,0,0,\ldots,0) \preceq \mathbf{k}_j$ for some j then

$$\bigoplus_{x \in \mathbb{X}} x_i = 0$$

for $1 \leq i \leq n_1$. In other words, $\mathbb{X}$ is balanced in the n_1 first bits.

For the full details of the propagation of the division property, we refer the interested reader to the original paper in [12].

3.3 An Integral Characteristic of 6-round MISTY1

In [11], Yosuke Todo presented a new integral characteristic for 6-round MISTY1, constructed by tracing the propagation of a division property. Todo showed that if a set V of values after the first FL layer (i.e., a set of X_1' values) has the division property $\mathcal{D}_{[6,2,7,7,2,7,7,2,7,7,2,7]}^{7,2,7,7,2,7,7,2,7,7,2,7}$ then the corresponding set of X_7 values has the division property $\mathcal{D}_{\mathbf{k}_1,\ldots,\mathbf{k}_{132}}^{7,2,7,7,2,7,7,2,7,7,2,7}$, where $\mathbf{k}_1,\ldots,\mathbf{k}_{132}$ is a list of vectors presented in [11]. For our purposes, it is sufficient to know that one of the $\mathbf{k}_i$'s is $[2,0,0,0,0,0,0,0,0,0,0,0]$. In particular, if we take V_{63} to be a 63-dimensional affine subspace of the values X_1' of a specific form, then the following equation holds:

$$\bigoplus_{x \in V_{63}} X_7[57-63](x) = 0. \tag{3}$$

The "specific form" of V_{63} is defined as follows: For every 6-dimensional affine subspace $V_6 \subset \mathbb{F}_2^7$, the set $V_{63} = \{(x_6, x_{57}) : x_6 \in V_6, x_{57} \in \mathbb{F}_2^{57}\}$ has the "correct" form. Note that while there are many options for "correct" V_{63} (since there are many options for V_6), we can construct only 7 independent Eq. 3 equations. For example, define 7 particular V_{63}'s by fixing one of the 7 bits $X_1'[57-63]$ and take all the values in the other bits. Knowing that Eq. 3 holds for these seven V_{63}'s implies that Eq. 3 for all possible V_{63}'s. Therefore, attacks using Eq. 3 exploit up to 7 V_{63}'s simultaneously. Figure 3 illustrates the 6-round integral characteristic.

3.4 Todo's Integral Attack on Full MISTY1

Todo's integral attack on full MISTY1 [11] uses the 6-round integral characteristic described above and has the following steps.

1. Choose V to be one of the "possible" V_{63}'s and ask for the encryption of $2^{63.58}$ chosen plaintexts such that V is included in the set of intermediate X_1' values.
2. As the values of plaintexts that correspond to the values in V depend on the value of 2 key bits used in FL_1, guess those key bits, and for each guess, perform the following steps assuming that Eq. (3) holds.

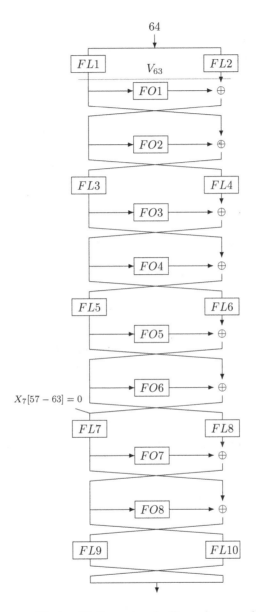

V_{63} is a 63-dimensional affine subspace of the form
$$V_{63} = \{(x_6, x_{57}) : x_6 \in V_6, x_{57} \in \mathbb{F}_2^{57}\}.$$

Fig. 3. A 6-round integral characteristic

(a) Guess the key bits needed to partially decrypt the ciphertexts of V and get the corresponding values $X_7[57 - 63]$. (This step is performed efficiently using the *partial sums* technique.)

(b) Check whether Eq. (3) holds. A wrong key-guess will pass this seven-bit condition with probability 2^{-7}, and thus, $2^{128-7} = 2^{121}$ wrong keys are expected to remain.

3. Check the 2^{121} remaining options by exhaustive search.

The time complexity of the attack is dominated by the last step. Hence, the attack requires $2^{63.58}$ chosen plaintexts and has time complexity of 2^{121} encryptions.

To reduce the time complexity, more V_{63}'s can be used. As we can use only independent subspaces, we can use up to 7 of them. Increasing the number of the V_{63}'s has two effects. On the one hand, there is an increase in the time complexity of the filtering step and in the data complexity. On the other hand, as the filtering of wrong key-guesses is stronger, the time complexity of the exhaustive search step is reduced. The tradeoff between those two effects was considered in [11]. The optimal time complexity is $2^{107.3}$, achieved by using 4 subspaces of the form V_{63} that require $2^{64} - 2^{56} = 2^{63.994}$ chosen plaintexts.

4   Improved Attack on Full MISTY1

In this section we present our improved attack on the full MISTY1. Our attack is based on using Todo's characteristic both in the encryption and the decryption directions and on a mixture of improved key-recovery techniques. First, we discuss application of Todo's characteristic in the decryption direction and present several observations used in the attack. Then we present the first phase of the attack that requires $2^{64} - 2^{50}$ chosen ciphertexts and recovers 49 key bits in time complexity of 2^{64} encryptions. Finally, we present the second phase of the attack that recovers the rest of the key in $2^{69.5}$ time, given almost the entire codebook.

4.1   Using Todo's Characteristic in the Decryption Direction

We observe that while Todo's characteristic (described in Sect. 3.3 and illustrated in Fig. 3) holds in the encryption direction of MISTY1, it can be used in the decryption direction as well. Indeed, since the characteristic exploits only the general structure of MISTY1 and not the exact subkeys used in the encryption process, and as MISTY1 is a Feistel construction, the characteristic holds also for the inverse cipher MISTY1^{-1}, which possesses the same general structure. (It should be noted that MISTY1^{-1} is not exactly equivalent to MISTY1, since the FL functions are not involutions. However, as we verified experimentally, this difference does not affect the applicability of Todo's characteristic.)

Hence, we have the following "dual" claim:

Claim 8. *The equation:*

$$\bigoplus_{x \in V_{63}} X'_3[25 - 31](x) = 0 \tag{4}$$

holds for every 63-dimensional affine subspace V_{63} of the values X_9 of the form $V_{63} = \{(x_{57}, x_6) : x_6 \in V_6, x_{57} \in \mathbb{F}_2^{57}\}$ (where $V_6 \subset \mathbb{F}_2^7$ is a 6-dimensional affine subspace).

Our attack (see Sects. 4.3 and 4.4) takes advantage of Todo's characteristic in both the encryption and the decryption directions, and by that increases the filtering of wrong key-guesses. The reverse 6-round integral characteristic is illustrated in Fig. 4.

4.2 Preliminaries

In this section we give several useful observations exploited in our attack.

Computing the Attack Equation by Independent Calculations. Consider the attack equation $\bigoplus_{x \in V_{63}} X'_3[25 - 31](x) = 0$ (see also Fig. 5a). The last functions applied during the evaluation of the attack equation are the bit-wise AND and OR of FL_4. Since both functions are linear/affine functions, an equivalent function EFL_4 can be used. In EFL_4 the OR operation is replaced by AND operation and those for every value x it holds that $FL_4(x, K'_4, K_6) = EFL_4(x, K'_4, K_6 \oplus 1^{16}) \oplus (K_6, 0^{16})$. The constant $(K_6, 0^{16})$ can be omitted because we sum over even number of values[1]. So, it is sufficient to compute the contributions of Out_1 and $FL_2(P_R)$ independently and then XOR them together. Moreover, the bits $X'_3[25 - 31]$ depend only on bits $9-15, 25-31$ of the input of FL_4. Thus, it is sufficient to compute the contributions of $Out_1[9 - 15, 25 - 31]$ and $FL_2(P_R)[9 - 15, 25 - 31] = FL_2(P_R[9 - 15, 25 - 31])$ independently.

Another independence appears in computing Out_1 given the inputs to $FI_{1,2}$ and to $FI_{1,3}$. (As the computation is up to XORs with key bits, we assume that these bits have already been guessed.) Denoting the inputs to $FI_{1,2}$ and to $FI_{1,3}$ by I_2 and I_3, respectively, the explicit formula for Out_1 is $Out_1 = (FI_{1,2}(I_2) \oplus I_3, FI_{1,2}(I_2) \oplus I_3 \oplus FI_{1,3}(I_3))$. Furthermore, we need only the values in $Out_1[9 - 15, 25 - 31]$ and these values can be found by independent computations of the four S-boxes $S9_{1,2,1}, S7_{1,2}, S9_{1,3,1}, S7_{1,3}$.

As a result, we get the following Observation:

Observation 9. Suppose we want to compute $\bigoplus_{x \in S} X'_3[25 - 31](x)$ for some set S of plaintexts and we already partially encrypted S using known $KL_1, KO_{1,1}, KI_{1,1}$. The rest of the computation can be done as follows. For every $x \in S$ and for every guess of the 14 bits of K'^L_4, K^L_6:

1. Guess the 14 relevant bits of KL_2 (the bits K'^L_3, K^L_5). For each guess, calculate the contribution of $FL_2(P_R[9 - 15, 25 - 31])$ to $\bigoplus_{x \in S} X'_3[25 - 31](x)$.

---

[1] We will write FL even when the meaning is of the equivalent function EFL.

2. Guess the 9 leftmost bits of $KO_{1,2}$. For each guess, calculate the value $ES9_{1,2,1}$ (the bits that enter $S9_{1,2,1}$), encrypt it through $S9_{1,2,1}$, and store in a table its contribution to $\bigoplus_{x \in S} X'_3[25 - 31](x)$.

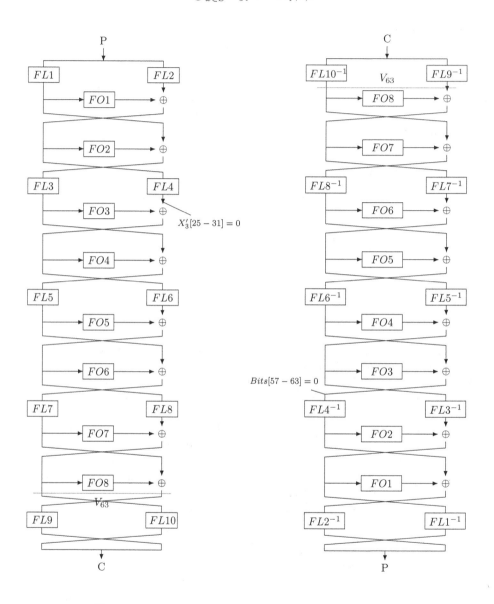

V_{63} is a 63-dimensional affine subspace of the form
$$V_{63} = \{(x_6, x_{57}) : x_6 \in V_6, x_{57} \in \mathbb{F}_2^{57}\}.$$

Fig. 4. A 6-round integral characteristic of MISTY1 in the decryption direction

3. Guess the 7 rightmost bits of $KO_{1,2}$. For each guess, calculate the value $ES7_{1,2}$, encrypt it through $S7_{1,2}$, and store in a table its contribution to $\bigoplus_{x \in S} X_3'[25-31](x)$.
4. Guess the 9 leftmost bits of $KO_{1,3}$. For each guess, calculate the value $ES9_{1,3,1}$, encrypt it through $S9_{1,3,1}$, and store in a table its contribution to $\bigoplus_{x \in S} X_3'[25-31](x)$.
5. Guess the 7 rightmost bits of $KO_{1,3}$. For each guess, calculate the value $ES7_{1,3}$, encrypt it through $S7_{1,3}$, and store in a table its contribution to $\bigoplus_{x \in S} X_3'[25-31](x)$.
6. Finally, $\bigoplus_{x \in S} X_3'[25-31](x)$ is equal to the XOR of the five contributions listed above.

The addition of $KI_{1,2}$ and $KI_{1,3}$ was omitted from the calculation in Observation 9. We give the explanation for this in the next section.

Removing Unnecessary Key-Bits and Equivalent Keys. As observed by Kühn ([9], see also [5]), the structure of FO and FI allows to use equivalent keys. In MISTY1, each $FI_{i,j}$ computation involves 32 key bits – 16 bits of $KO_{i,j}$ and 16 bits of $KI_{i,j}$. These 32 bits can be replaced by an equivalent 25-bit subkey, by "pushing" all key additions forward until they meet a non-linear operation, using the fact that linear operations can be interchanged easily. (The 7 leftmost bits of $KI_{i,j}$ do not meet any S-box of $FI_{i,j}$ and therefore are "pushed" outside $FI_{i,j}$ and added at a later place where they are merged into another subkey.)

Another way to decrease the key material involved in the computations is removing unnecessary key bits. When we compute $\bigoplus_{x \in S} X_3'[25-31](x)$ for a set S of even size, there is no effect of the key bits $KI_{1,2}, KI_{1,3}$ and $KO_{1,4}$ since they all affect $X_3'[25-31]$ in the same (constant) way for every plaintext and their contributions cancel each other. In our attack we consider equivalent keys and remove unnecessary key bits, as described in Fig. 5a.

Partial Sums and Piles Construction. Key guessing becomes more efficient if we combine it with the partial sums technique, presented by Ferguson et al. [6]. Here is an example that illustrates the technique.

Suppose there are 2^n values $\{D_i\}$ after the partial encryption of one FI function in FO_1 (the first round) and we want to calculate the XOR of the outputs from $S9_{1,2,1}$. If two values are equal in the 9 input bits of $S9_{1,2,1}$, then their contributions to the XOR of the output from $S9_{1,2,1}$ cancel each other. Hence, we sort the 2^n values according to those 9 bits, check which of the 2^9 values appears an odd number of times and encrypt through $S9_{1,2,1}$ only them (and each of them only once). Hence, it is sufficient to encrypt 2^9 values instead of 2^n values. We use the term *pile* for a set of the values that appears an odd number of times in the 9 examined bits (so that there are at most 2^9 values in a pile in total), and *constructing pile* for the process of reducing the 2^n inputs to 2^9 values. In general, we define:

Definition 10. Let $S = \{D_i\}$ be a set of n-bit values that are required for some calculation. Assume that the calculation can be done only with the knowledge of $\{D_i \cap \Omega\}$, where Ω is a mask with Hamming weight d and $\cap$ means bitwise AND (i.e., knowing only d of the n bits is sufficient for the calculation). Denote by B the position of those d bits. Constructing pile of size 2^d in B means reducing the number of values from $|S|$ to at most 2^d by considering only the values $\{D_i \cap \Omega\}$ that appear an odd number of times.

4.3   A New Integral Attack on Full MISTY1 – The First Phase

In this section we present the first phase of our attack on the full MISTY1. This phase uses only the "reverse" 6-round characteristic presented in Sect. 4.1. It requires $2^{64} - 2^{50}$ chosen ciphertexts and recovers the equivalent of 49 key bits in time complexity dominated by the decryption of the ciphertexts.

Remark 11. We chose to begin with the "reverse" characteristic due to key scheduling arguments (namely, this allows to exploit the fact that the subkeys KL_1 and KO_1 share 16 key bits).

First, we choose seven independent structures V_{63} of X_9 states, to be used in the integral characteristics exploited by the attack. Then, we choose $2^{64} - 2^{50}$ ciphertexts, structured such that for any value of the subkey KL_{10}, the corresponding intermediate X_9 values contain all seven V_{63} structures. (The way to choose ciphertexts such that this property holds is presented in [11].)

After choosing the ciphertexts, we guess the subkeys K_1 and K_7'. This allows us to identify the right 2^{63} ciphertexts that yield each of the seven chosen V_{63} structures. An efficient procedure for this identification is presented below.

The goal of this phase is to discard wrong key guesses using the attack equation $\bigoplus_{x \in V_{63}} X_3'[25-31](x) = 0$ (derived from the "reverse" 6-round characteristic). We use a meet-in-the-middle (MITM) approach: We split the 7-bit attack equation into two equations, a 4-bit equation and a 3-bit equation. We treat each equation separately and then combine the results, getting an extra filtering by comparing the key bits involved in both equations.

To check whether the attack equation holds, we construct the following piles (see Definition 10 and Fig. 5):

(i) The pile $\alpha_R = \bigoplus_{x \in V_{63}} P_R$ (which is of size 1).
(ii) A pile of size 2^9 in the 9 leftmost bits of B.
(iii) A pile of size 2^7 in the 7 rightmost bits of B.
(iv) A pile of size 2^9 in the 9 leftmost bits of D.
(v) A pile of size 2^7 in the 7 rightmost bits of D.

The total sum $\bigoplus_{x \in V_{63}} X_3'[25-31](x)$ is the sum of the contributions of the five piles. We use the MITM approach once again (thus, getting a two-dimensional MITM attack) by dividing the piles into two sets – piles (i), (ii), and (iii) on the one hand and piles (iv), (v) on the other hand. We then compute the contributions of each set of piles separately and check whether they are equal.

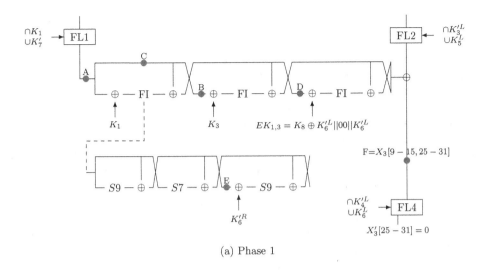

(a) Phase 1

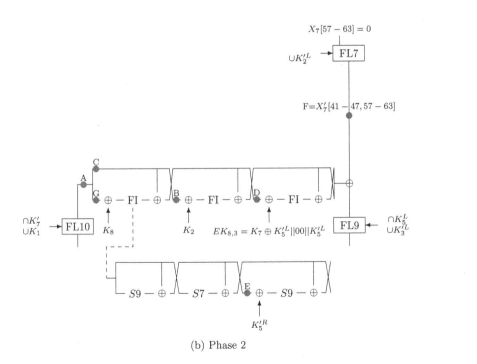

(b) Phase 2

Fig. 5. Reference figures for the attack

For the right key, the contributions must be equal (since the contributions of the five piles sum to zero), while for a wrong key the contributions are equal with probability 2^{-7} (per V_{63} structure). Since we use seven V_{63} structures, we obtain $7 \cdot 7 = 49$ bits of filtering overall.

As mentioned above, the sets of ciphertexts that correspond to the seven V_{63}'s can be computed once K_1 and K'_7 are guessed. Those key bits are indeed guessed at the beginning of this phase but if we identify the V_{63}'s *after* guessing K_1, K'_7 then the time complexity of this identification would be at least $2^{32} \cdot 2^{63} \cdot 7$. We can reduce this time complexity by a precomputation, as follows.

Recall that for each structure V_{63}, the exact 2^{63} ciphertexts that are decrypted to V_{63} are determined by the value of only two key bits (one bit of K_1 and another of K'_7). For each of the 4 options, we identify the 2^{63} ciphertexts $\{C^i\}$, get the corresponding plaintexts $\{P^i\}$ and save $\alpha_R = \bigoplus_i P^i_R$. Additionally, construct a pile of size 2^{32} in P_L and save it. All information we need for the attack is now contained in the computed piles (so that once K_1, K'_7 are guessed, we can continue the attack with the right piles). In this way, the time complexity of the identification step becomes $7 \cdot 4 \cdot 2^{63} = 2^{67.8}$ operations.

The procedure of Phase 1 consists of several steps for each one of the seven V_{63}'s. First, guess K_1 and K'_7, get α_R and the pile of size 2^{32}. Partially encrypt the 2^{32} values through FL_1 to the point A (the time complexity of this step is $2^{32} \cdot 2^{32} = 2^{64}$) and continue as follows:

1. Construct piles of size 2^9 and 2^7 in B (piles (ii), (iii)).
   1.1. Guess the 9 leftmost bits of K_3 and partially encrypt the 2^9 values of pile number (ii) to the point F.
   1.2. Guess the 7 rightmost bits of K_3, encrypt the 2^7 values of pile number (iii) to the point F and calculate their XOR. Note that the XOR of the values is sufficient (instead of the values themselves) because FL_4 is linear. Explicitly, for a set of values S, the equation

$$\bigoplus_{x \in S} FL_4(x) = EFL_4(\bigoplus_{x \in S} x) \oplus \bigoplus_{x \in S} const$$

   holds for the right key (where $const = (K_6, 0^{16})$).
   1.3. Guess $K'_3[12-15], K_5[12-15]$ (8 bits of KL_2) and encrypt $\alpha_R[12-15, 28-31]$ to the point F.
   1.4. XOR the results from the previous steps in F and call the joint XOR J_1.
   1.5. Guess $K'_4[12-15], K_6[12-15]$ (8 bits of KL_4), encrypt J_1 and save in a table T_1 the contribution to $X'_3[28-31]$ (4 bits of the attack equation) with the relevant key (a table of size $2^{9+7+8+8} = 2^{32}$).
2. Partially encrypt the 2^{32} values through $S9_{1,1,1}$ and $S7_{1,1}$ (note that we already know K_1), get partial outputs from FI_1 and add them to the values in C. To construct pile number (iv), we first construct a pile of size 2^{18} corresponding to the 9 leftmost bits in C and the 9 bits in E. Similarly, to construct pile number (v), we first construct a pile of size 2^{16} corresponding to the 7 rightmost bits in C and the 9 bits in E.

2.1. Guess the 9 bits $K_6'^R$, encrypt the 2^{18} and 2^{16} values through FI_1 to construct piles of size 2^9 and 2^7 in D (piles number (iv) and (v)).

2.2. Guess the 9 leftmost bits of $EK_{1,3}$, encrypt the 2^9 values of pile number (iv) to the point F and calculate their XOR.

2.3. Guess the 7 rightmost bits of $EK_{1,3}$, encrypt the 2^7 values of pile number (v) to the point F and calculate their XOR.

2.4. XOR the results from the previous steps in the point F and call the joint XOR J_2.

2.5. Guess $K_4'[12 - 15], K_6[12 - 15]$ (8 bits of KL_4), encrypt J_2, calculate the contribution to $X_3'[28 - 31]$ (4 bits of the attack equation) and search for a collision in T_1 (i.e., a collision both in the contributions and in the common key bits). Save the collision in a table T_2.

2.6. The size of T_2 is $2^{9+7+8+8} \cdot 2^{9+9+7+8} \cdot 2^{-8} \cdot 2^{-4 \cdot 7} = 2^{29}$, since a match must occur in $K_4'[12 - 15], K_6[12 - 15]$ (8 bits of KL_4) and in the contribution to $X_3'[28 - 31]$ (a 4-bit attack equation, for the seven V_{63}'s).

3. Produce T_2' using the 3-bit attack equation $\bigoplus_{x \in V_{63}} X_3'[25 - 27] = 0$ by a similar process. The difference is that we guess $K_4'[9 - 11], K_6[9 - 11]$ instead of $K_4'[12 - 15], K_6[12 - 15]$ and $K_3'[9 - 11], K_5[9 - 11]$ instead of $K_3'[12 - 15], K_5[12 - 15]$. The size of T_2' is $2^{9+7+6+6} \cdot 2^{9+9+7+6} \cdot 2^{-6} \cdot 2^{-3 \cdot 7} = 2^{32}$.

4. There are $9 + 16 + 16 = 41$ shared bit-guesses in T_2 and T_2' (the bits $K_6'^L, K_3, EK_{1,3}$). Search for a collision in those bits and store them in a table T_3. The size of T_3 is $2^{29} \cdot 2^{32} \cdot 2^{-41} = 2^{20}$.

We save in T_3 the corresponding guess of K_1 and K_7', and thus, the size of T_3 is $2^{20} \cdot 2^{32} = 2^{52}$. For each suggestion in T_3, we guess the subkey $K_6'^L$ (seven bits), retrieve the entire subkey K_6', and then:

- Retrieve K_8 from $K_8 \oplus K_6'^L \| 00 \| K_6'^L$ and K_6',
- Retrieve K_7 from K_7' and K_8,
- Retrieve K_6 from K_6' and K_7.
- Compare with the known K_6^L and discard wrong guesses (we guess seven bits of $K_6'^L$ and have a 7-bit condition, so we remain with a table of the same size).
- Retrieve K_4^L from $K_3'^L$ and K_3.

The output of Phase 1 is two tables. The first is the table T_{phase1}, that contains 2^{52} suggestions for five full subkeys K_1, K_3, K_6, K_7', K_8 and four 7-bit subkeys $K_3'^L, K_5^L, K_4'^L, K_4^L$, sorted according to K_1, K_7', K_8. The second is the table T_{phase1}' with the same 2^{52} suggestions, but sorted according to $K_1, K_7', K_3'^L, K_5^L$.

The time complexity of Phase 1 for a single V_{63} and a fixed guess of K_1, K_7' is less than $2^{33.7}$ operations. The calculation is given in Table 3 and composed of the sum of time complexities of the steps involving key guessing and (partial) encryption. Since we use seven V_{63}'s, the total time complexity of Phase 1 (with the precomputation) is bounded by

$$\mathbf{T1} = 2^{67.8} + 7 \cdot 2^{32} \cdot 2^{33.7} = 2^{69.2}$$

operations, which is less than 2^{64} encryptions.

Table 3. Time complexity of Phase 1 for single V_{63} and K_1, K_7' guess

Step	Time complexity	Description
1.1	$2^{18} = 2^9 \cdot 2^9$	Guess 9 bits of K_3 and partially encrypt 2^9 values
1.2	$2^{14} = 2^7 \cdot 2^7$	Guess 7 bits of K_3 and partially encrypt 2^7 values
1.3	$2^8 = 2^8 \cdot 1$	Guess 8 bits of KL_2 and partially encrypt $\alpha_R[12 - 15, 28 - 31]$
1.4	$2^{24} = 2^9 \cdot 2^7 \cdot 2^8$	Calculate J_1
1.5	$2^{32} = 2^{24} \cdot 2^8$	Guess 8 bits of KL_4 and partially encrypt J_1
2.1	$2^{27.3} = 2^9 \cdot (2^{18} + 2^{16})$	Guess 9 bits of $K_6'^R$ and partially encrypt 2^{18} and 2^{16} values
2.2	$2^{27} = 2^{9+9} \cdot 2^9$	Guess 9 bits of $EK_{1,3}$ and partially encrypt 2^9 values
2.3	$2^{23} = 2^{9+7} \cdot 2^7$	Guess 7 bits of $EK_{1,3}$ and partially encrypt 2^7 values
2.4	$2^{25} = 2^9 \cdot 2^9 \cdot 2^7$	Calculate J_2
2.5	$2^{33} = 2^{25} \cdot 2^8$	Guess 8 bits of KL_4 and partially encrypt J_2.
3	$<$ step 1+ step 2	Similar to 1 and 2, using $\oplus X_3'[25 - 27] = 0$
4	$2^{32.2} = 2^{29} + 2^{32}$	Compare two tables of sizes 2^{29} and 2^{32}
Total	$< 2^{33.7}$	

After the first phase is completed, the rest of the key can be found immediately in time complexity of 2^{79} encryptions, by guessing the rest of the key (for each of 2^{52} entries of T_{stage1}, guess 2^{27} key bits $K_4^R, K_5^R, K_3'^R$ and derive a master-key candidate) and checking it by a trial encryption. Of course, this approach does not require additional data. In the following section we show that if more data is available, the time complexity of the second phase can be reduced to $2^{69.5}$ encryptions.

4.4 A New Integral Attack on Full MISTY1 – The Second Phase

The second phase of our attack uses the 6-round characteristic presented in Sect. 3.3 to apply a second filtering to the key suggestions remaining from Phase 1.

The goal of this phase is to discard wrong key guesses that pass Phase 1, using the attack equation $\bigoplus_{x \in V_{63}} X_7[57 - 63](x) = 0$ (derived from the 6-round characteristic).

To check whether the attack equations holds, we construct the following piles (see Fig. 5b):

(i) The pile $\beta_R = \bigoplus_{x \in V_{63}} C_R$ (which is of size 1).
(ii) A pile of size 2^9 in the 9 leftmost bits of B.
(iii) A pile of size 2^7 in the 7 rightmost bits of B.
(iv) A pile of size 2^{16+9} in the 9 leftmost bits of C + 16 bits of D.
(v) A pile of size 2^7 in the 7 rightmost bits of C + 16 bits of D.

The sum $\bigoplus_{x \in V_{63}} X_7[57 - 63](x)$ is the sum of the contributions of each pile separately. As in Phase 1 above, we calculate the contribution of piles number (i), (ii), (iii) and of piles number (iv), (v) separately and check whether the two contributions are equal. For the right key they must be equal, and for a wrong key guess they are equal with probability 2^{-7} (per V_{63} structure). There are 7 optional V_{63}'s and we use only two of them to get $2 \cdot 7 = 14$ bits filtering.

The procedure of Phase 2 consists of several steps (similarly to Phase 1) for each one of the seven V_{63}'s.

First, we ask for the encryption of $2^{64} - 2^{50}$ chosen plaintexts, such that each of the seven chosen V_{63} structures (of X'_1 values) is covered by the texts. (This data requirement, combined with the requirement of Phase 1, makes the data complexity equal to $2^{64} - 2^{36}$ chosen plaintexts and ciphertexts, which is rather close to the entire codebook). Second, we construct pile number (i) and a pile of size 2^{32} in C_L for constructing the other piles.

Then, we guess K_1 and K'_7 and partially decrypt the 2^{32} values through FL_{10} to the point A (the time complexity of this step is negligible compared to the total time complexity). The procedure continues as follows:

1. Construct piles of size 2^9 and 2^7 in B (piles (ii), (iii)).
   1.1. Guess the 9 leftmost bits of K_2 and decrypt the 2^9 values of pile number (ii) to the point F.
   1.2. Guess the 7 rightmost bits of K_2, decrypt the 2^7 values of pile number (iii) to the point F and calculate their XOR.
   1.3. Guess K'^L_3, K^L_5 (14 bits of KL_9 and decrypt the $\beta_R[9 - 15, 25 - 31]$ to the point F.
   1.4. XOR the results from the previous steps in the point F and call the joint XOR J_1.
   1.5. Get the K_1, K'_7, K'^L_3, K^L_5 entry of T'_{stage1}. The entry consists of $2^{52-46} = 2^6$ values for $K_3, K_6, K_8, K'^L_4, K^L_4$. Compute K'_2 from K_2, K_3, decrypt J_1 and save in a table T_1 the contribution to $X'_7[57 - 63]$, along with the relevant key (a table of size $2^{9+7+14+6} = 2^{36}$).

2. To construct pile number (iv), we first construct a pile of size 2^{16+9} that corresponds to the 9 leftmost bits of C + 16 bits of G. Similarly, to construct pile number (v), we first construct a pile of size 2^{16+7} that corresponds to the 7 rightmost bits in C + 16 bits of G.
   2.1. Guess K_8 and construct a pile of size 2^{9+9} that corresponds to the 9 leftmost bits of C + 9 bits of E. Construct a pile of size 2^{9+7} that corresponds to the 7 rightmost bits of C + 9 bits of E.
   2.2. Guess the 9 bits K'^R_5, decrypt the 2^{18} and 2^{16} values of the piles from the previous step through FI_1 to construct piles of size 2^9 and 2^7 in D (piles number (iv) and (v)). For the piles of size 2^9, this step can be performed more efficiently by guessing key bits one by one, as described in [2].
   2.3. Get the K_1, K'_7, K_8 entry of T_{stage1}. The entry consists of $2^{52-48} = 2^4$ values for $K_3, K_6, K'^L_3, K^L_5, K'^L_4, K^L_4$. For each value, guess K^R_5, derive K_5, compute K'_5 (from K_5, K_6) and compare with K'^R_5 that was guessed. We remain with 2^4 values for K'_5 and hence 2^4 values for $EK_{8,3}$.
   2.4. With the known $EK_{8,3}$, decrypt the 2^9 values of pile number (iv) to the point F and calculate their XOR. In addition, decrypt the 2^7 values of pile number (v) to the point F and calculate their XOR.
   2.5. XOR the results from the previous steps at the point F and call the joint XOR J_2.

2.6. Guess $K_2'[9-15]$, decrypt J_2, calculate the contribution to $X_7[57-63]$ and search for a collision in the table T_1 (i.e., collision in the contributions + in the common key bits).

2.7. The expected number of collisions is $2^{9+7+14+6} \cdot 2^{16+9+4+7} \cdot 2^{-7} \cdot 2^{-2 \cdot 7} \cdot 2^{-20} = 2^{31}$, since a match must occur in $K_2'[9-15]$, in the contributions in $X_7[57-63]$ (for two V_{63}'s) and in the entry of T_{stage1}.

3. For each of the 2^{31} suggestions, guess K_4^R, and use the knowledge of $K_4, K_5, K_4'^L$ to get a 7-bit filtering. This yields 2^{65} suggestions for the entire key. Test them with a single plaintext/ciphertext pair. Only $2^{65} \cdot 2^{-64} = 2$ suggestions are expected to remain. Test them with another plaintext/ciphertext pair and find the key.

The time complexity of stage 2 for each structure V_{63} and each guess of K_1, K_7' is less than $2^{42.5}$. The calculation is given in Table 4 and is composed of the sum of the time complexities of the steps involving key guessing and (partial) decryption. Since we use two V_{63}'s, the total time complexity of Phase 2 is bounded by

$$\mathbf{T2} = 2 \cdot 2^{32} \cdot 2^{42.5} = 2^{75.5}$$

operations.

All the operations of Phase 2 are simple operations besides Step 3 that consists of 2^{33} full MISTY1 encryptions (for each K_1, K_7' guess). Thus, the time complexity $\mathbf{T2}$ is $2^{75.5}$ simple operations $+ 2^{65}$ full MISTY1 encryptions. Assuming that each simple operation is comparable to an S-box evaluation, the time complexity of stage 2 (in terms of full encryptions) is $\frac{2^{75.5}}{8 \cdot 9} + 2^{65} = 2^{69.5}$, since MISTY1 has 9 S-boxes in each of its 8 rounds.

Table 4. Time complexity of Phase 2 for single V_{63} and K_1, K_7' guess

Step	Time complexity	Description
1.1.	$2^{18} = 2^9 \cdot 2^9$	Guess 9 bits of K_2 and partially decrypt 2^9 values
1.2.	$2^{14} = 2^7 \cdot 2^7$	Guess 7 bits of K_2 and partially decrypt 2^7 values
1.3.	$2^{14} = 2^{14} \cdot 1$	Guess 14 bits of KL_9 and partially decrypt $\beta_R[9-15, 25-31]$
1.4.	$2^{30} = 2^9 \cdot 2^7 \cdot 2^{14}$	Calculate J_1
1.5.	$2^{36} = 2^{30+6} \cdot 1$	Partially decrypt J_1
2.1.	$2^{41.3} = 2^{16} \cdot (2^{25} + 2^{23})$	Guess K_8 and partially decrypt 2^{25} and 2^{23} values
2.2.	$2^{41.3} = 2^{16+9} \cdot (2^{14} + 2^{16})$	Guess 9 bits of $K_5'^R$ and partially encrypt 2^{18} and 2^{16} values.
2.3.	$2^{38} = 2^{16+9+4+9}$	Get 2^4 values from T_{stage2} and for each value guess 9 bits of K_5^R
2.4.+2.5.	$2^{38.3} = 2^{16+9+4} \cdot (2^9 + 2^7)$	Partially decrypt 2^9 and 2^7 values. Calculate J_2
2.6.	$2^{36} = 2^{16+9+4+7} \cdot 1$	Partially decrypt J_2
3	$2^{33} = 2^{31} \cdot 2^2$	Obtain another 7-bit filtering (upon the attack equation filtering of the **two** V_{63}'s), and then test remaining key suggestions by trial encryptions
Total	$< 2^{42.5}$ simple operations $+ 2^{33}$ full MISTY1 encryptions	

The output of Phase 1 required tables of size 2^{52}. In a naive approach, this is the memory complexity of the attack but maybe it can be reduced.

5 Summary and Conclusions

In this paper we presented a new attack on the full MISTY1. The attack uses Todo's 6-round integral characteristic [11] both in the encryption direction (as it was used by Todo) and in the decryption direction. The attack equations derived from the characteristics provide a filtering for wrong key guesses. Exploiting the filtering efficiently by using partial sums, two-dimensional meet-in-the-middle and other techniques, our attack has time complexity of $2^{69.5}$ encryptions. This is a reduction by a factor of 2^{38} over Todo's attack that has time complexity of $2^{107.3}$ encryptions.

While our attack is clearly impractical due to its high data complexity, it shows that MISTY1 has a rather low security margin, providing only 70 bits of security.

As a problem for further research, it will be interesting to find out whether the data complexity can be reduced. A possible direction for achieving this is finding additional 6-round integral characteristics of a lower order.

References

1. 3rd Generation Partnership Project: Specification of the 3GPP. Confidentiality, Integrity Algorithms - Document 2: KASUMI Specification (Release 6). Technical report 3GPP. TS 35.202 V6.1.0 (2005–2009), September 2005
2. Bar-On, A.: Improved higher-order differential attacks on MISTY1. In: Leander, G. (ed.) FSE 2015. LNCS, vol. 9054, pp. 28–47. Springer, Heidelberg (2015)
3. Daemen, J., Knudsen, L.R., Rijmen, V.: The block cipher SQUARE. In: Biham, E. (ed.) FSE 1997. LNCS, vol. 1267, pp. 149–165. Springer, Heidelberg (1997)
4. Dinur, I., Dunkelman, O., Shamir, A.: Improved attacks on full GOST. In: Canteaut, A. (ed.) FSE 2012. LNCS, vol. 7549, pp. 9–28. Springer, Heidelberg (2012)
5. Dunkelman, O., Keller, N.: Practical-time attacks against reduced variants of MISTY1. Des. Codes Crypt. **76**, 601–627 (2013)
6. Ferguson, N., Kelsey, J., Lucks, S., Schneier, B., Stay, M., Wagner, D., Whiting, D.L.: Improved cryptanalysis of Rijndael. In: Schneier, B. (ed.) FSE 2000. LNCS, vol. 1978, pp. 213–230. Springer, Heidelberg (2001)
7. Jia, K., Li, L.: Improved impossible differential attacks on reduced-round MISTY1. In: Lee, D.H., Yung, M. (eds.) WISA 2012. LNCS, vol. 7690, pp. 15–27. Springer, Heidelberg (2012)
8. Knudsen, L.R., Wagner, D.: Integral cryptanalysis. In: Daemen, J., Rijmen, V. (eds.) FSE 2002. LNCS, vol. 2365, pp. 112–127. Springer, Heidelberg (2002)
9. Kühn, U.: Improved cryptanalysis of MISTY1. In: Daemen, J., Rijmen, V. (eds.) FSE 2002. LNCS, vol. 2365, pp. 61–75. Springer, Heidelberg (2002)
10. Matsui, M.: New block encryption algorithm MISTY. In: Biham, E. (ed.) FSE 1997. LNCS, vol. 1267, pp. 54–68. Springer, Heidelberg (1997)

11. Todo, Y.: Integral cryptanalysis on full MISTY1. In: Gennaro, R., Robshaw, M. (eds.) CRYPTO 2015. LNCS, vol. 9215, pp. 413–432. Springer, Heidelberg (2015)
12. Todo, Yosuke: Structural Evaluation by Generalized Integral Property. In: Oswald, Elisabeth, Fischlin, Marc (eds.) EUROCRYPT 2015. LNCS, vol. 9056, pp. 287–314. Springer, Heidelberg (2015)
13. Tsunoo, Y., Saito, T., Kawabata, T., Nakagawa, H.: Differentials, finding higher order of MISTY1. IEICE Trans. **95**(A(6)), 1049–1055 (2012)
14. Yi, W., Chen, S.: Multidimensional zero-correlation linear attacks on reduced-round MISTY1. In: CoRR (2014). arXiv:1410.4312

Cryptanalysis of the **FLIP** Family of Stream Ciphers

Sébastien Duval[(⊠)], Virginie Lallemand, and Yann Rotella

Inria, Project-team SECRET, Paris, France
{Sebastien.Duval,Virginie.Lallemand,Yann.Rotella}@inria.fr

Abstract. At Eurocrypt 2016, Méaux et al. proposed **FLIP**, a new family of stream ciphers intended for use in Fully Homomorphic Encryption systems. Unlike its competitors which either have a low initial noise that grows at each successive encryption, or a high constant noise, the **FLIP** family of ciphers achieves a low constant noise thanks to a new construction called *filter permutator*.

In this paper, we present an attack on the early version of **FLIP** that exploits the structure of the filter function and the constant internal state of the cipher. Applying this attack to the two instantiations proposed by Méaux et al. allows for a key recovery in 2^{54} basic operations (resp. 2^{68}), compared to the claimed security of 2^{80} (resp. 2^{128}).

Keywords: Stream cipher · Guess-and-determine attack · FLIP · FHE

1 Introduction

One of the challenges of recent years is to create an acceptable system of Fully Homomorphic Encryption (FHE) that would allow users to delegate computations to so-called Cloud Services. While Gentry showed in [6] the theoretic feasibility of such a framework, two main difficulties remain: first, the important computational and memory costs, and second the limited homomorphic capacities.

In order to overcome these limitations, one of the important aspects that have to be lessened is the cost of the evaluation of the symmetric encryption algorithm used in the framework, which mainly depends on the multiplicative depth of the circuit implementing the primitive. Since adapting the AES seems hard [3,4,7], several new symmetric schemes purposed for FHE have been proposed, among which the block cipher LowMC [1] and the stream ciphers Trivium and Kreyvium [2].

At Eurocrypt 2016, Méaux et al. [12] proposed the new stream cipher construction **FLIP** which aims at overcoming some of the drawbacks of previous

Partially supported by the French Agence Nationale de la Recherche through the BRUTUS project under Contract ANR-14-CE28-0015 and by the Commission of the European Communities through the Horizon 2020 program under project number 645622 PQCRYPTO.

© International Association for Cryptologic Research 2016
M. Robshaw and J. Katz (Eds.): CRYPTO 2016, Part I, LNCS 9814, pp. 457–475, 2016.
DOI: 10.1007/978-3-662-53018-4_17

schemes by, among other things, allowing for constant and smaller noise. This achievement was made possible by the use of a new construction that resembles a filter generator but with a constant register that is permuted before entering the filtering function in order to limit the multiplicative depth of the circuit.

This design has been presented in October 2015 by the authors of FLIP at a national workshop [11], and then submitted to Eurocrypt 2016. Our study shows that the concrete instantiations proposed by the designers suffer from several flaws that can be extended to a cryptanalysis. We reported our findings to the authors which led them to change their design after their paper was accepted, in order to resist our attack. A fixed version of the construction is then described in the final version of the Eurocrypt 2016 article entitled *Toward Stream Ciphers for Efficient FHE with Low-Noise Ciphertexts* [12].

In the following, we only deal with the *preliminary version* of the FLIP family of stream ciphers: so everytime that we mention "FLIP" we mean the version presented in [11] and submitted to Eurocrypt 2016 (which differs from the final version of [12]).

This paper is organised as follows. We start by giving a description of the submitted version of the FLIP family of stream ciphers in Sect. 2. Then, we discuss its vulnerabilities against guess-and-determine attacks in Sect. 3 and show how to break the cipher by exploiting these vulnerabilities through an algebraic attack (Sect. 4). The pseudocode of the attack is given in Sect. 5. Our analyses are supported by experiments reported in Sect. 6. The last section concludes this paper.

2   Description of the **FLIP** Family of Stream Ciphers

2.1   General Idea: The Filter Permutator Structure

When it comes to designing a symmetric construction tailored for FHE applications, both block and stream ciphers can be considered, each with advantages and disadvantages, as discussed in [2].

Since the targeted applications use noise-based cryptography (such as lattice-based cryptography), one of the pursued goals is to limit the growth of noise as much as possible, which is equivalent to considering circuits with a limited multiplicative depth. This desirable property, also refered to as a high *homomorphic capacity*, is hard to obtain with block ciphers since the round iterations lead to an output with a large algebraic degree. However, the good point is that the noise is constant per block, which implies that noise does not add any limitation on the number of generated ciphertext blocks. On the other hand, the homomorphic capacity of stream ciphers is usually very high for the first ciphertext bits, but decreases as more bits are generated, imposing to re-initialise the cipher or to use techniques like bootstrapping.

The innovative design of Méaux et al. [12] succeeds in taking the best from both sides and enjoys a very good homomorphic capacity that remains constant with time. Their proposal is a family of stream ciphers named FLIP that is based on the filter generator construction, but drops the register update part

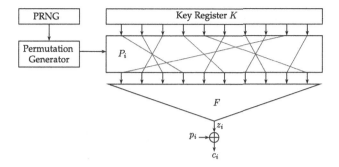

Fig. 1. General structure of the filter permutator construction used for the FLIP family of stream ciphers.

to avoid the algebraic degree increase. Instead, the register bits are permuted before entering the filter function, thus the name *filter permutator*.

Its operational principle is represented in Fig. 1.

It is made up of three main components:

- a register storing the N-bit key K,
- a bit permutation generator, parametrised by a public PseudoRandom Number Generator (PRNG), producing at each clock i an N-bit permutation P_i,
- a filtering (Boolean) function F, generating the keystream bit z_i.

Once the PRNG is initialised using an IV, the master key K is loaded into the register and encryption starts: at each step i, the permutation generator produces a permutation P_i that shuffles the key bits right before they enter the filtering function F. F produces the keystream bit z_i which is XORed to the corresponding plaintext bit p_i and gives the ciphertext c_i.

To recover the plaintext, the same process is used to generate the bits z_i of the keystream that are simply XORed back with the c_i.

Every part of the scheme is public except the key.

2.2   The FLIP Family of Stream Ciphers

After an extensive analysis of the filter permutator construction with respect both to FHE constraints and resistance against most common stream ciphers attacks, the authors chose the concrete instantiation we now describe.

The PRNG used for the permutation generator is defined as a *forward secure* PRNG based on AES-128, and the permutation generator itself is a Knuth shuffle [9], which ensures that all the N-bit permutations have the same probability to be generated (provided that it is used with a random generator).

F is an N-variable Boolean function defined by the direct sum of three specific Boolean functions f_1, f_2 and f_3 that are defined in [13] and that we now recall.

In the following, n and k are positive integers and operations are considered over $\mathbb{F}_2$.

Definition 1 (L-type Function). *The n-th L-type function L_n is the n-variable linear function defined by:*

$$L_n(x_0, \cdots, x_{n-1}) = \sum_{i=0}^{n-1} x_i.$$

Definition 2 (Q-type Function). *The n-th Q-type function Q_n is defined by the $2n$-variable quadratic function:*

$$Q_n(x_0, \cdots, x_{2n-1}) = \sum_{i=0}^{n-1} x_{2i} x_{2i+1}$$

Definition 3 (T-type Function). *The k-th T-type function is the $\frac{k(k+1)}{2}$-variable Boolean function defined by:*

$$T_k(x_0, \cdots, x_{\frac{k(k+1)}{2}-1}) = \sum_{i=1}^{k} \prod_{j=0}^{i-1} x_{j+\sum_{\ell=0}^{i-1} \ell}.$$

For instance the 3rd T-type function is equal to:

$$T_3 = x_0 + x_1 x_2 + x_3 x_4 x_5$$

Each of these types of functions has nice properties according to one or several security criteria (non-linearity, resiliency, algebraic immunity, ...).

The filtering function used in the FLIP family of stream ciphers uses a combination of 3 Boolean functions parameterised by the integers n_1, n_2 and n_3, chosen so that the resulting properties of F are good:

- $f_1(x_0, \cdots, x_{n_1-1}) = L_{n_1}$,
- $f_2(x_{n_1}, \cdots, x_{n_1+n_2-1}) = Q_{n_2/2}$,
- $f_3(x_{n_1+n_2}, \cdots, x_{n_1+n_2+n_3-1}) = T_k$ where k is such that $n_3 = \frac{k(k+1)}{2}$.

F is defined as the direct sum of f_1, f_2 and f_3:

$$F(x_0, \cdots, x_{n_1+n_2+n_3-1}) = L_{n_1} + Q_{n_2/2} + T_k \quad \text{where } n_1 + n_2 + n_3 = N$$

and thus inherits in some measure the good properties of f_1, f_2 and f_3 (see [13]).

The initial analysis performed by Méaux et al. and presented in the submitted version of the construction [13] takes into account the most common attacks on filter generators (which are very similar to filter permutators) and resulted in the selection of the parameters reported in Table 1.

The security analysis detailed in [13] and more precisely the study of weak keys results in an additional limitation on the design which is that the key must be balanced (in the submitted version of the Eurocrypt paper [13] it is stated that: *"Since our N parameter will typically be significantly larger than the bit-security of our filter permutator instances, we suggest to restrict the key space to keys of Hamming weight N/2"*).

Finally, note that the specification document does not give any limit on the number of keystream bits that can be generated under the same key.

Table 1. Parameters of the two concrete instantiations of FLIP with the corresponding complexities of Algebraic Attacks (AA), Fast Algebraic Attacks (FAA) Higher-Order Correlation attacks (HC).

FLIP (n_1, n_2, n_3)	Key size (N)	Security	AI (F)	FAI (F)	res (F)	AA	FAA	HC
FLIP (47,40,105)	192	80	14	15	47	194	88	119
FLIP (87,82,231)	400	128	21	22	87	323	136	180

3 Preliminary Remarks on the Vulnerabilities of the **FLIP** Family of Stream Ciphers

3.1 Attack Scenario and Computation Model

In the following we examine one of the most common attack scenarios considered for stream cipher analysis, which is the known-plaintext scenario: we suppose that we know a part of the plaintext together with the corresponding ciphertext, which implies that we know the value of some bits of the keystream z. Needless to say, our goal is to recover the secret key, which in the case of the FLIP family of stream ciphers is equivalent to recovering the internal state.

To express the performance of our attack, we use the three usual metrics which are time, data and memory complexities. Time complexity (hereafter denoted by C_T) expresses the quantity of operations that the attacker has to perform to execute the attack. In our case, we compute it in the same way as in the specification paper [13] so we count the number of basic operations. Data complexity (C_D) corresponds to the required number of keystream bits and finally memory complexity (C_M) measures the memory (in bits) needed during the attack.

3.2 The **FLIP** Family of Stream Ciphers and Guess-and-Determine Attacks

The attack we propose uses a variant of the guess-and-determine technique. This approach, which seems to have been named first in [5,8], has been extensively used to analyse stream ciphers, starting with the ones submitted to the NESSIE project. The idea is to start by making a hypothesis on the value of some bits of the internal state or of the key (the 'Guess') and to use the information coming from keystream bits to deduce the unknown ones (to 'Determine' them). Most of the time the attack is completed thanks to algebraic techniques.

Two features of the FLIP family of stream ciphers seem to indicate that an attack using guess-and-determine techniques would be efficient: first its fixed internal state and second the definition of its filtering function. More precisely, the fact that the register is not updated implies that a guess of one key/internal state bit at any time would give an information of one bit at any other time. This is different from common stream ciphers for which the update function mixes the internal state bits together, implying that a one-bit information quickly vanishes after some (forward or backward) rounds.

The second feature that seems exploitable for a guess-and-determine attack is the definition of the filtering function F which contains very few monomials of high-degree. This is what we detail now.

3.3   Observations on the Boolean Function F

As reported before, the Boolean function F is made of the direct sum of 3 Boolean functions f_1, f_2 and f_3 which are respectively of L-, Q- and T-type. This definition implies that all the monomials of degree greater than or equal to 3 are present in f_3, which is given by the following formula:

$$f_3(x_{n_1+n_2}, \cdots, x_{n_1+n_2+n_3-1}) = T_k(x_{n_1+n_2}, \cdots, x_{n_1+n_2+n_3-1})$$
$$= \sum_{i=1}^{k} \prod_{j=n_1+n_2}^{n_1+n_2+i-1} x_{j+\sum_{\ell=0}^{i-1} \ell}$$

where k is the algebraic degree of f_3 and is such that $n_3 = \frac{k(k+1)}{2}$.

From this expression, we see that there are $k-2$ monomials of degree greater than or equal to 3 in F, in a total of $n_3 - 3$ variables. Given the multiplicative depth constraint, k has to be low[1], which implies that the T-type function has few monomials and therefore easy to cancel, as we show in the next section.

The core idea of our attack is to notice that since there are few high-degree monomials but a lot of null key bits, there is a high probability that the high-degree monomials of F are cancelled. In these cases, it also means that the keystream bits can be seen as expressions of degree less than or equal to 2 in the non-null key bits.

The attack we perform uses this specificity by doing a slight variant of the guess-and-determine technique: instead of making a hypothesis on the value of key/internal state bits, we guess the indices of some null key bits[2]. We deduce from that the clocks when the keystream bits are an expression of low-degree in the other key bits and build a system from it. Finally we solve the system with linearisation techniques, which in the case of low-degree equations is of reasonable cost.

3.4   Probability of Cancelling all the High-Degree Monomials of F Given that ℓ Input Variables are Null

To figure out the feasibility of such a procedure, we have to evaluate the probability that, given exactly ℓ positions of null bits in K, the expression of the keystream bit z_i is of degree less than or equal to 2 in the remaining key bits[3].

---

[1] To give the order of magnitude, we recall here that the 2 concrete instantiations described in [13] use $k = 14$ and $k = 21$ for respective security of 80 and 128 bits.

[2] As we saw in Sect. 2, we are sure that there are $\frac{N}{2}$ null key bits.

[3] This is what we denote by an *exploitable equation* or *exploitable clock*.

This probability is directly linked to the amount of data that is required to lead to the attack since it determines the amount of keystream bits that an attacker needs such that enough of them are exploitable to construct the system.

From the previous discussion, we know that there are exactly $k - 2$ disjoint monomials of degree greater than or equal to 3 in the expression of $z_i = F(P_i(k_0, k_1, \cdots k_{N-1}))$. Then, if the attacker is only aware of $\ell < k - 2$ zero positions, she won't be able to determine exploitable clocks, which forces $\ell \geq k - 2$, i.e. at minimum one zero bit that could be positioned in each of the high-degree monomials.

First case: if $\ell = k - 2$. The first possibility is to choose the number of null positions equal to the number of high-degree monomials that we want to cancel, i.e. $\ell = k - 2$. In this case, exactly one null bit has to go into each monomial: for instance, if we are looking at a specific monomial of degree d: $x_0 x_1 \cdots x_{d-1}$, it is equivalent to choosing which of the variables is null, so there are d possibilities. From that, we can enumerate the set of valid configurations, which corresponds to choosing one index in each monomial, so since there are 3 possible choices for the monomial of degree 3, 4 possibilities for the one of degree 4 and so on up to the monomial of degree k, there is a total of $3 \times 4 \times 5 \cdots \times k = k!/2$ valid configurations. To obtain the probability, this amount has to be compared with the total number of possibilities for choosing the null positions, which is $\binom{N}{\ell}$ so we have:

$$\mathbb{P}_{\ell=k-2} = \frac{k!/2}{\binom{N}{\ell}}.$$

General case: if $\ell \geq k - 2$. To increase the probability that a clock is exploitable, the attacker can guess more null key bit positions and choose $\ell \geq k - 2$. A first way of computing this probability is:

$$\mathbb{P}_\ell = \frac{\sum_{i_1+i_2+\cdots+i_{k-2}\leq\ell} \binom{3}{i_1}\binom{4}{i_2}\cdots\binom{k}{i_{k-2}}\binom{N-m}{\ell-I}}{\binom{N}{\ell}}$$

where m is the number of variables that occur in the monomials of degree greater than or equal to 3 and $I = i_1 + i_2 + \cdots + i_{k-2}$.

Proof. Suppose that we are given ℓ null bit positions in K. We are interested in the probability that a random permutation P_i shuffles the key bits in a way that the evaluation of F does not contain any monomial of degree greater than or equal to 3. As previously, we count the number of valid configurations among the total number of permutations.

The idea is to list all the possible ways of positioning at least one null bit in each monomial: we set i_1 null bits in the monomial of degree 3, i_2 null bits in the monomial of degree 4, and so on up to i_{k-2} null bits in the monomial of highest degree (k). If we denote by $I = i_1 + i_2 + \cdots + i_{k-2}$ the number of null bits positioned in such a way, we are left with $\ell - I$ null bits to position in the other $N - m$ monomials. To obtain the probability, we have to divide this quantity by the number of ways to position ℓ guesses among N bits. $\qquad\square$

Another way of obtaining the probability is to compute the number of configurations that do not cancel the monomials of degree greater than or equal to 3, which is the complementary probability of the one we are looking for. The advantage is that this complementary can be easily expressed with the inclusion-exclusion principle. Let us denote A_J the event that our guess doesn't cancel the monomials of degrees included in the set J, i.e.

$$A_J \text{ is the event: } \{\forall j \in J, M_j \neq 0\}$$

where M_j is the unique monomial of degree j in T_k.

$\mathbb{P}(A_J)$ is the probability of setting the ℓ bits among the monomials whose degrees are not in J so is equal to:

$$\mathbb{P}(A_J) = \frac{\binom{N - \sum_{j \notin J} j}{\ell}}{\binom{N}{\ell}}$$

Then we can express the probability that our guess yields a polynomial of degree higher than or equal to 3 by:

$$\mathbb{P}\left(\bigcup_{d \in \{3, \cdots, k\}} A_{\{d\}} \right) = \sum_{s=1}^{k-2} \left((-1)^s \sum_{\substack{J \subseteq \{3, \cdots, k\} \\ |J| = s}} \mathbb{P}(A_J) \right)$$

which can be expressed as

$$\mathbb{P}\left(\bigcup_{d \in \{3, \cdots, k\}} A_{\{d\}} \right) = \frac{\sum_{s=1}^{k-2} \left((-1)^s \sum_{\substack{J \subseteq \{3, \cdots, k\} \\ |J| = s}} \binom{N - \sum_{j \notin J} j}{\ell} \right)}{\binom{N}{\ell}}$$

From which we get the expression of the probability that we are looking for:

$$P_\ell = \mathbb{P}\left(\bigcap_{d \in \{3, \cdots, k\}} \overline{A_{\{d\}}} \right) = 1 - \mathbb{P}\left(\bigcup_{d \in \{3, \cdots, k\}} A_{\{d\}} \right)$$

The evaluation of these formulas gives the results reported in Tables 3, 4 and 5 in Appendix, and we will see in the next section that they are good enough to mount an attack. For instance, if we attack the small version[4] of FLIP and do the minimal number of guesses (i.e. $\ell = 12$) we will have a probability of having an exploitable equation of $P_{\ell=12} = 2^{-26.335}$. For the other version[5] and a minimal number of guesses we have $P_{\ell=19} = 2^{-42.382}$.

---

[4] FLIP (47,40,105).
[5] FLIP (87,82,231).

4   Our Attack

4.1   Description

Setting. Since we consider a known-plaintext scenario, we suppose that we are given C_D keystream bits that we denote by z_i, $i = 0, \cdots, C_D - 1$. Additionally, the associated permutations P_i are public so we have expressions of the keystream bits as function of the unknown key bits $k_0, \cdots, k_{N-1}$:

$$z_i = F(P_i(k_0, k_1, \cdots, k_{N-1})) \quad \forall i \geq 0$$

Our attack takes advantage of the two vulnerabilities detailed in the previous section to boil down the key recovery problem to the solving of a linearised system.

First step: initial guess. The first step consists in making a hypothesis on the positions of ℓ null key bits, where $\ell \geq k - 2$. Assuming that these bits are null gives us a simplified expression of z_i in only $N - \ell$ unknowns. Since the key K is balanced, the probability of our guess being right is[6]:

$$\mathbb{P}_{rg} = \frac{\binom{\frac{N}{2}}{\ell}}{\binom{N}{\ell}}$$

Second step: extraction of low-degree equations. The objective of step 2 is to collect equations of low-degree in the unknown key bits. To do so, we look at the expressions of the available z_i and pick up all the equations for which the null key bits cancel the monomials of degree greater than or equal to 3. As seen in previous section, this event is of probability $\mathbb{P}_\ell$.

Third step: solving the system. One of the easiest ways of solving the quadratic system is to use linearisation techniques, which consist in converting the system into a linear one by introducing a new variable for each non-linear monomial that appears. In our specific case, the only non-linear expressions we have to deal with are the monomials of degree 2. Since F takes as input N variables but we guessed ℓ of them, we are left with $N - \ell$ unknown variables, which in the worst case scenario form $\binom{N-\ell}{2}$ monomials of degree two. This implies that once linearised, our converted system will contain

$$v_\ell = N - \ell + \binom{N - \ell}{2}$$

variables.

Assuming that the equations are random, the number of equations that are necessary to give a unique solution (or show a contradiction) is roughly equal to the number of unknowns[7]. This implies that the necessary amount of keystream

---

[6] This probability is slightly smaller than in the case of a random key ($2^{-\ell}$), but the advantage is that as long as we guess $\ell \leq \frac{N}{2}$ we are sure that at least one guess will be correct while it could fail for a random key that does not have enough null bits.

[7] This will be confirmed by our experiments detailed in Sect. 6.

bits that the attacker needs is the product of the number of variables and the inverse of the probability that a z_i is exploitable:

$$C_D = v_\ell \times \frac{1}{\mathbb{P}_\ell}$$

The time complexity is determined by the time to solve the system[8] multiplied by the number of times we have to repeat the guess of ℓ null bit positions before finding a correct one:

$$C_T = v_\ell^3 \times \frac{1}{\mathbb{P}_{rg}}$$

The final memory complexity is dominated by the memory necessary to store the system, so is roughly equal to:

$$C_M = v_\ell^2$$

Tables 3, 4 and 5 in Appendix give the possible trade-offs between time and data complexity for the two versions of FLIP. As we can see, increasing the number of initial guesses ℓ allows to reduce the amount of data necessary to conduct the attack at the cost of an increased time complexity.

4.2   Discussion and Possible Improvements

Data Complexity Reduction. The data complexity can be further improved if, rather than choosing the guesses at random, the attacker chooses them according to the observed permutations. With the PRNG seed being public, at any point in time, she knows all the upcoming permutations so she can deduce a guess that cancels the triangular part for many of the upcoming permutations.

Possibility of Precomputations. Most of the computational cost of the attack lies in the linear system solving. Notice that this linear system depends only on the permutation and the guess, which are all known to the attacker, who can therefore compute the system inversion for several guesses without any knowledge of the keystream. Once she receives the keystream bits, she plugs them into her precomputations to obtain the results. The drawback of this technique is its increase in memory complexity.

Seed Independence. Our attack has the property of being unaffected by a re-initialisation of the system. What we mean here is that a change of the PRNG seed in the middle of the attack will not force the attacker to restart her attack: she can combine the previously obtained equations with the one obtained under the new seed.

Security. The security level of the FLIP family of stream ciphers is at most proportional to $\sqrt{N}$ bits, where N is the key size.

---

[8] Which is v_ℓ^3 for a basic Gaussian elimination or $v_\ell^{2.8}$ with Strassen's algorithm. We will use the first one for simplicity.

Proof. The time complexity of our attack is

$$C_T = v_\ell^3 \times \frac{1}{\mathbb{P}_{rg}}$$

As $\ell \ll N$, one can say that $\mathbb{P}_{rg}$ is roughly equivalent to $2^{-\ell}$. Also, as $v_\ell = N - \ell + \binom{N-\ell}{2}$, we can give an approximation of C_T which is

$$C_T \sim N^6 \times 2^\ell$$

Additionally, the number of guesses we need to perform our attack is the number of monomials of degree greater than or equal to 3 in T_{n_3}. Thus $n_3 = (\ell + 2)(\ell + 3)/2$, so $\ell \sim \sqrt{n_3}$, from which we get:

$$\log C_T \sim \alpha \sqrt{N} \qquad \square$$

Figure 2 represents the evolution of the time complexity of our attack as function of the key size when we consider instances of FLIP of the form FLIP (n_1, n_2, n_3) where $N = n_1 + n_2 + n_3 = 2n_3$ (which is consistent with the parameters proposed in [13]). ℓ is chosen as the minimal number of guesses needed to perform the attack, i.e. $\ell = k - 2$.

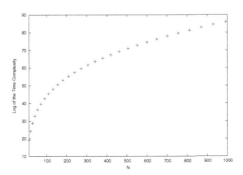

Fig. 2. Evolution of the time complexity as function of the key size N.

Attempt to Cancel the Quadratic Part. Our attack consists in guessing key bits to cancel the triangular part of the filtering function: another possibility would be to cancel the monomials of degree 2 in order to reduce the resistance of the scheme against correlation attacks. We considered this option but our studies showed that the complexity of such an attack would be too high. We also thought of cancelling both quadratic and triangular parts, thus leaving only linear relations, but the data complexity of such an attack makes it less practical.

5   Description of the Algorithm

The main computation part of our attack is a linear system solving over $\mathbb{F}_2$. If the solving detects a contradiction, we deduce that our guess is wrong and we start again with another guess. Otherwise, the guess was right and the solving yields the key. The intuition is that we don't always need a full-rank system to detect a contradiction. We can therefore improve the attack by treating every equation as they come, rather than waiting for a full-rank system.

A pseudocode description of the attack using this improvement is given in Algorithm 1. In this algorithm, an equation will be represented as a $(v_\ell + 1)$-bit word containing 1 where a variable is present in the equation and 0 otherwise. The least significant bit of this representation contains the value of the keystream bit of the equation. We also memorise if equation i is present in the system through the vector $Exists$. If $Exists[i] = 1$, then equation i is in the system.

6   Verification of the Attack on a Toy Version

To support our findings, we implemented our attack on a toy version of the cipher. We reduced the key size to $N = 64$ bits and adapted accordingly the values of the parameters to $n_1 = 14$, $n_2 = 14$ and $n_3 = 36$ (the proportions between the size of the parameters are kept). The filtering function F has algebraic degree 8 and is defined as follows:

$$F(x_0, \cdots x_{63}) = f_1(x_0, \cdots x_{13}) + f_2(x_{14}, \cdots x_{27}) + f_3(x_{28} \cdots x_{63})$$

where:

$$f_1(x_0, \cdots, x_{13}) = L_{14}(x_0, \cdots, x_{13}) = x_0 + x_1 + \cdots + x_{13}$$
$$f_2(x_{14}, \cdots, x_{27}) = Q_7(x_{14}, \cdots, x_{27}) = x_{14}x_{15} + x_{16}x_{17} + \cdots + x_{26}x_{27}$$
$$f_3(x_{28}, \cdots, x_{63}) = T_8(x_{28}, \cdots, x_{63}) = x_{28} + x_{29}x_{30} + x_{31}x_{32}x_{33} + \cdots + x_{56}x_{57} \cdots x_{63}$$

According to our analysis, the parameters of the attacks are the ones described in Table 6: for instance if we decide to make a hypothesis on $\ell = 8$ null indices, the probability that our guess is correct is

$$\mathbb{P}_{rg} = 2^{-8.717}.$$

The probability that a permutation is exploitable is equal to:

$$\mathbb{P}_\ell = 2^{-7.814}$$

and the linearised system depends on $v_\ell = 1596$ variables. We expect that $C_D = 2^{18.454}$ bits are necessary to conduct the analysis and that the attack requires $C_T = 2^{40.638}$ basic operations.

We implemented our own version of this toy instance of FLIP on which we performed our attack with $\ell = 8$ guesses. The statistics we obtain are given in Table 2.

Algorithm 1. FLIP Key recovery

Input: Keystream, PRNG seed
Output: Key
1: SYSTEM ← Vector of v_ℓ null words
2: $Exists$ ← Vector of v_ℓ null bits
3: $KeyNotFoundYet$ ← **true**
4: **while** $KeyNotFoundYet$ **do**
5:    G ← $NewRandomGuess$
6:    $NoContradiction$ ← **true**
7:    N_{eq} ← 0
8:    **while** $NoContradiction$ **and** $N_{eq} \leq v_\ell$ **do**
9:        E ← $NewEquation$
10:       $NewIndex$ ← −1
11:       i ← MSB(E)
12:       **while** $i \leq v_\ell$ **do**
13:           **if** $Exists[i]$ **then**
14:               E ← $E \oplus$ SYSTEM$[i]$
15:               i ← MSB(E)
16:           **else**
17:               **if** $NewIndex = -1$ **then**
18:                   $NewIndex$ ← i
19:               **end if**
20:               i ← Index of the next bit with value 1 starting from index $i + 1$
                  and going towards the LSB
21:           **end if**
22:       **end while**
23:       **for** $j = 1$ to $NewIndex$ **do**
24:           **if** $Exists[i]$ **and** SYSTEM$[i][NewIndex] = 1$ **then**
25:               SYSTEM$[i]$ ← SYSTEM$[i] \oplus E$
26:           **end if**
27:       **end for**
28:       **if** $E = 1$ **then**
29:           $NoContradiction$ ← **false**
30:       **else**
31:           **if** $E \neq 0$ **then**
32:               SYSTEM$[NewIndex]$ ← E
33:               N_{eq} ← $N_{eq} + 1$ {If $E = 0$, the equation is linearly dependent
                  from the first ones but brings no contradiction, we then don't
                  increment N_{eq}}
34:           **end if**
35:       **end if**
36:    **end while**
37:    **if** $NoContradiction$ **then**
38:        Get x_i and $x_i x_j$ and Test if there is a contradiction
39:        **if** There is no contradiction **then**
40:            $KeyNotFoundYet$ ← **false**
41:            $Key = (x_i)_{1 \leq i \leq n}$
42:        **end if**
43:    **end if**
44: **end while**
45: **return** Key

Table 2. Comparison of the experimental results with theory: attack on the toy version FLIP (14,14,36) with a hypothesis on $\ell = 8$ null bit positions (average on 1000 tests, launched on an Intel(R) Xeon(R) CPU W3670 at 3.20 GHz (12 MB cache), and with 8 GB of RAM)

	Guesses	Data generated	Ratio exploited	Elementary Op	Time (sec)
Practice	437.1	$2^{18.455}$	$2^{-7.813}$	$2^{38.588}$	280.93
Theory	$\mathbb{P}_{rg}^{-1} = 420.8$	$C_D = 2^{18.454}$	$\mathbb{P}_\ell = 2^{-7.814}$	$C_T = 2^{40.638}$	∅

Although the equations have a very specific structure, we noticed that they behave like random equations in the following sense: the first linearly dependent equation is only found after generating 1590 equations, which fits with the theory in the case of random equations [10]. However, treating the equations as they come allows us to discard right away any equation that is linearly dependent from the others. This way, we can stop collecting equations as soon as we have as many equations in our system as are variables[9].

As we can see in Table 2, experimental results fit pretty well with the theory.

7   Conclusion

In this paper we presented a cryptanalysis of the FLIP family of stream ciphers. Our attack makes use of the weaknesses of the FLIP structure against guess-and-determine attacks to reduce the degree of the filtering function, after what an algebraic attack suffices to recover the key. We obtained theoretical estimations of the complexity of the attack and an implementation of the attack on a toy version shows that this complexity holds in practice. This attack can be performed in 2^{54} basic operations (resp. 2^{68}), compared to the claimed security of 2^{80} (resp. 2^{128}), and we discussed trade-offs and improvements that can lower this complexity even more. We also underlined that a simple increase of the key size is not an efficient countermeasure as the complexity of the attack doesn't increase much with the key size.

Finally, in view of fixing this attack, one should keep in mind the inherent weakness of the filter permutator construction against guess-and-determine attacks due to its constant register. The biggest issue of the FLIP family of stream ciphers is that its filtering function increases the fragility against guess-and-determine attacks. To strengthen the security of the filter permutator, a possible direction would be to refine its filtering function, for instance by using more high-degree monomials.

---

[9] The experiments show that we discard about 500 equations before we get 1596 independent equations.

References

1. Albrecht, M.R., Rechberger, C., Schneider, T., Tiessen, T., Zohner, M.: Ciphers for MPC and FHE. In: Oswald, E., Fischlin, M. (eds.) EUROCRYPT 2015. LNCS, vol. 9056, pp. 430–454. Springer, Heidelberg (2015)
2. Canteaut, A., Carpov, S., Fontaine, C., Lepoint, T., Naya-Plasencia, M., Paillier, P., Sirdey, R.: How to compress homomorphic ciphertexts. In: FastSoftware Encryption FSE 2016 (to appear). http://eprint.iacr.org/2015/113
3. Coron, J.-S., Lepoint, T., Tibouchi, M.: Scale-invariant fully homomorphic encryption over the integers. In: Krawczyk, H. (ed.) PKC 2014. LNCS, vol. 8383, pp. 311–328. Springer, Heidelberg (2014)
4. Doröz, Y., Hu, Y., Sunar, B.: Homomorphic AES evaluation using NTRU. IACR Cryptology ePrint Archive 2014, 39 (2014). http://eprint.iacr.org/2014/039
5. Ekdahl, P., Johansson, T.: SNOW - a new stream cipher. In: Proceedings of First Open NESSIE Workshop, KU-Leuven, pp. 167–168 (2000)
6. Gentry, C.: Fully homomorphic encryption using ideal lattices. In: Mitzenmacher, M. (ed.) Proceedings of the 41st Annual ACM Symposium on Theory of Computing, STOC 2009, pp. 169–178. ACM (2009)
7. Gentry, C., Halevi, S., Smart, N.P.: Homomorphic evaluation of the AES circuit. In: Safavi-Naini, R., Canetti, R. (eds.) CRYPTO 2012. LNCS, vol. 7417, pp. 850–867. Springer, Heidelberg (2012)
8. Hawkes, P., Rose, G.G.: Exploiting multiples of the connection polynomial in word-oriented stream ciphers. In: Okamoto, T. (ed.) ASIACRYPT 2000. LNCS, vol. 1976, pp. 303–316. Springer, Heidelberg (2000)
9. Knuth, D.E.: The Art of Computer Programming, Volume II: Seminumerical Algorithms. Addison-Wesley, Reading (1969)
10. Lidl, R., Niederreiter, H.: Finite Fields. Cambridge University Press, Cambridge (1983)
11. Méaux, P.: Symmetric Encryption Scheme adapted to FullyHomomorphic Encryption Scheme. In: Journées Codage etCryptographie - JC2 2015 -12ème édition des Journées Codage et Cryptographie du GT C2, 5 au 9octobre 2015, La Londe-les-Maures, France (2015). http://imath.univ-tln.fr/C2/
12. Méaux, P., Journault, A., Standaert, F., Carlet, C.: Towards stream ciphers for efficient fhe with low-noise ciphertexts. In: Fischlin, M., Coron, J. (eds.) EUROCRYPT 2016. LNCS, vol. 9665, pp. 311–343. Springer, Heidelberg (2016). http://eprint.iacr.org/2016/254
13. Méaux, P., Journault, A., Standaert, F.X., Carlet, C.: Towards stream ciphers for efficient FHE with low-noise ciphertexts. Personal communication, October 2015

A Possible Trade-Offs
A.1 FLIP (47,40,105)
See Table 3.

Table 3. Log of the complexities of the attacks as function of the number of initial guesses (ℓ) for the instantiation FLIP (47,40,105)

ℓ	$\mathbb{P}_\ell$	v_ℓ	$\mathbb{P}_{rg}$	C_D	C_T	C_M
12	-26.335	13.992	-12.528	40.326	54.503	27.983
13	-23.049	13.976	-13.627	37.025	55.554	27.951
14	-20.653	13.960	-14.736	34.613	56.615	27.919
15	-18.738	13.943	-15.854	32.682	57.684	27.887
16	-17.141	13.927	-16.982	31.069	58.763	27.854
17	-15.775	13.911	-18.120	29.686	59.852	27.821
18	-14.585	13.894	-19.267	28.480	60.950	27.788
19	-13.536	13.878	-20.425	27.414	62.057	27.755
20	-12.601	13.861	-21.592	26.462	63.175	27.722
21	-11.762	13.844	-22.771	25.606	64.303	27.688
22	-11.004	13.827	-23.960	24.831	65.442	27.654
23	-10.315	13.810	-25.160	24.125	66.591	27.621
24	-9.686	13.793	-26.371	23.479	67.750	27.586
25	-9.110	13.776	-27.593	22.886	68.921	27.552
26	-8.580	13.759	-28.827	22.339	70.103	27.517
27	-8.092	13.741	-30.073	21.833	71.297	27.483
28	-7.640	13.724	-31.331	21.364	72.502	27.448
29	-7.221	13.706	-32.601	20.927	73.720	27.413
30	-6.832	13.689	-33.883	20.520	74.949	27.377
31	-6.469	13.671	-35.179	20.140	76.191	27.342
32	-6.131	13.653	-36.487	19.784	77.446	27.306
33	-5.816	13.635	-37.809	19.450	78.714	27.270
34	$\mathbf{-5.520}$	**13.617**	$\mathbf{-39.145}$	**19.137**	**79.995**	**27.233**
>34	...	...	...	...	>80	...
35	-5.243	13.598	-40.495	18.842	81.290	27.197

A.2 FLIP (87,82,231)
See Tables 4 and 5.

Table 4. Log of the complexities of the attacks as function of the number of initial guesses (ℓ) for the instantiation FLIP (87,82,231)

ℓ	$\mathbb{P}_\ell$	v_ℓ	$\mathbb{P}_{rg}$	C_D	C_T	C_M
19	−42.382	16.151	−19.647	58.533	68.100	32.302
20	−38.522	16.144	−20.721	54.666	69.151	32.287
21	−35.589	16.136	−21.799	51.725	70.206	32.272
22	−33.169	16.128	−22.881	49.298	71.266	32.257
23	−31.097	16.121	−23.967	47.218	72.329	32.241
24	−29.282	16.113	−25.058	45.395	73.397	32.226
25	−27.667	16.105	−26.153	43.772	74.469	32.211
26	−26.214	16.098	−27.253	42.311	75.546	32.195
27	−24.895	16.090	−28.357	40.985	76.627	32.180
28	−23.691	16.082	−29.465	39.773	77.712	32.164
29	−22.584	16.074	−30.578	38.658	78.802	32.149
30	−21.562	16.067	−31.696	37.629	79.896	32.133
31	−20.615	16.059	−32.818	36.674	80.994	32.118
32	−19.734	16.051	−33.944	35.785	82.097	32.102
33	−18.912	16.043	−35.075	34.955	83.205	32.086
34	−18.142	16.035	−36.211	34.178	84.317	32.071
35	−17.421	16.027	−37.352	33.448	85.434	32.055
36	−16.743	16.020	−38.497	32.762	86.556	32.039
37	−16.104	16.012	−39.648	32.116	87.683	32.023
38	−15.502	16.004	−40.803	31.505	88.814	32.007
39	−14.932	15.996	−41.963	30.928	89.950	31.991
40	−14.393	15.988	−43.128	30.381	91.091	31.975
41	−13.883	15.980	−44.298	29.862	92.237	31.959
42	−13.398	15.972	−45.473	29.370	93.388	31.943
43	−12.937	15.964	−46.653	28.901	94.543	31.927
44	−12.499	15.956	−47.838	28.455	95.704	31.911
45	−12.082	15.947	−49.028	28.029	96.870	31.895
46	−11.684	15.939	−50.224	27.624	98.042	31.879
47	−11.305	15.931	−51.425	27.236	99.218	31.862
48	−10.942	15.923	−52.631	26.865	100.400	31.846
49	−10.596	15.915	−53.842	26.511	101.586	31.830

Table 5. Log of the complexities of the attacks as function of the number of initial guesses (ℓ) for the instantiation FLIP (87,82,231)

ℓ	$\mathbb{P}_\ell$	v_ℓ	$\mathbb{P}_{rg}$	C_D	C_T	C_M
50	-10.265	15.907	-55.059	26.171	102.779	31.813
51	-9.948	15.898	-56.282	25.846	103.976	31.797
52	-9.644	15.890	-57.509	25.534	105.180	31.780
53	-9.353	15.882	-58.743	25.235	106.388	31.763
54	-9.074	15.873	-59.982	24.947	107.602	31.747
55	-8.806	15.865	-61.227	24.671	108.822	31.730
56	-8.548	15.857	-62.477	24.405	110.048	31.713
57	-8.301	15.848	-63.734	24.149	111.279	31.697
58	-8.063	15.840	-64.996	23.903	112.516	31.680
59	-7.835	15.831	-66.264	23.666	113.758	31.663
60	-7.614	15.823	-67.538	23.437	115.007	31.646
61	-7.402	15.815	-68.818	23.217	116.262	31.629
62	-7.198	15.806	-70.104	23.004	117.522	31.612
63	-7.001	15.797	-71.397	22.799	118.789	31.595
64	-6.812	15.789	-72.695	22.601	120.062	31.578
65	-6.629	15.780	-74.000	22.409	121.341	31.561
66	-6.452	15.772	-75.311	22.224	122.627	31.543
67	-6.281	15.763	-76.629	22.044	123.918	31.526
68	-6.116	15.754	-77.953	21.871	125.216	31.509
69	-5.957	15.746	-79.284	21.703	126.521	31.491
70	**-5.803**	**15.737**	**-80.621**	**21.540**	**127.832**	**31.474**
>70	...	...	...	...	>128	...
71	-5.655	15.728	-81.965	21.383	129.150	31.457

B Complexities of the Attack on the Toy Version of FLIP
See Table 6.

Table 6. Log of the complexities of the attacks as function of the number of initial guesses (ℓ) for the toy version FLIP (14,14,36)

ℓ	$\mathbb{P}_\ell$	v_ℓ	$\mathbb{P}_{rg}$	C_D	C_T	C_M
6	−11.861	10.741	−6.370	22.601	38.592	21.481
7	−9.436	10.691	−7.528	20.126	39.601	21.382
8	−7.814	10.640	−8.717	18.454	40.638	21.280
9	−6.602	10.589	−9.939	17.191	41.706	21.177
10	−5.649	10.536	−11.197	16.185	42.806	21.072
11	−4.876	10.483	−12.493	15.359	43.941	20.966
12	−4.235	10.428	−13.828	14.663	45.113	20.857
13	−3.696	10.373	−15.207	14.069	46.325	20.746
14	−3.237	10.316	−16.631	13.553	47.580	20.633
15	−2.843	10.259	−18.105	13.102	48.881	20.517
16	−2.503	10.200	−19.632	12.703	50.231	20.399
17	−2.208	10.140	−21.217	12.347	51.636	20.279
18	−1.950	10.078	−22.865	12.028	53.100	20.156
19	−1.723	10.015	−24.581	11.739	54.628	20.031
20	−1.524	9.951	−26.373	11.476	56.227	19.903
21	−1.349	9.886	−28.247	11.235	57.904	19.771
22	−1.194	9.819	−30.214	11.013	59.670	19.637
23	−1.057	9.750	−32.284	10.807	61.534	19.500
24	**−0.935**	**9.679**	**−34.472**	**10.615**	**63.510**	**19.359**
>24	...	...	...	...	>64	...
25	−0.827	9.607	−36.794	10.435	65.616	19.215

Crypto 2016 Award Papers

The Magic of ELFs

Mark Zhandry[1,2(✉)]

[1] MIT, Cambridge, USA
[2] Princeton University, Princeton, USA
mzhandry@princeton.edu

Abstract. We introduce the notion of an *Extremely Lossy Function* (ELF). An ELF is a family of functions with an image size that is tunable anywhere from injective to having a polynomial-sized image. Moreover, for any efficient adversary, for a sufficiently large polynomial r (necessarily chosen to be larger than the running time of the adversary), the adversary cannot distinguish the injective case from the case of image size r.

We develop a handful of techniques for using ELFs, and show that such extreme lossiness is useful for instantiating random oracles in several settings. In particular, we show how to use ELFs to build secure point function obfuscation with auxiliary input, as well as polynomially-many hardcore bits for any one-way function. Such applications were previously known from strong knowledge assumptions — for example polynomially-many hardcore bits were only know from differing inputs obfuscation, a notion whose plausibility has been seriously challenged. We also use ELFs to build a simple hash function with *output intractability*, a new notion we define that may be useful for generating common reference strings.

Next, we give a construction of ELFs relying on the *exponential* hardness of the decisional Diffie-Hellman problem, which is plausible in pairing-based groups. Combining with the applications above, our work gives several practical constructions relying on qualitatively different — and arguably better — assumptions than prior works.

1 Introduction

Hash functions are a ubiquitous tool in cryptography: they are used for password verification, proofs of work, and are central to a variety of cryptographic algorithms including efficient digital signatures and encryption schemes.

Unfortunately, formal justifications of many of the uses of hash functions have been elusive. The trouble stems from the difficulty of even defining what security properties a hash function should satisfy. On one extreme, a hash function can be assumed to have standard security properties such as one-wayness or collision resistance, which are useless for most of the applications above. On the other

M. Zhandry—This work was sponsored by the Defense Advanced Research Projects Agency (DARPA) and the U.S. Army Research Office under contract number W911NF-15-C-0226.

M. Robshaw and J. Katz (Eds.): CRYPTO 2016, Part I, LNCS 9814, pp. 479–508, 2016.
DOI: 10.1007/978-3-662-53018-4_18

extreme, a hash function can be modeled as a truly random function, where it is assumed that an adversary only has black-box access. In the so-called random oracle model (ROM) [5], all of the above applications are secure. However, random oracles clearly do not exist and moreover provably cannot be replaced by any concrete hash function [16]. In this light, it is natural to ask:

What are useful properties of random oracles that can be realized by real-world hash functions.

Some attempts have been made to answer this question; however, many such attempts have serious limitations. For example Canetti et al. [16] propose the notion of *correlation intractability* as a specific feature of random oracles that could potentially have a standard model instantiation. However, they show that for some parameter settings such standard model hash functions cannot exist. The only known positive example [15] relies on extremely strong cryptographic assumptions such as general-purpose program obfuscation. For another example, Bellare et al. [4] define a security property for hash functions called Universal Computational Extractors (UCE), and show that hash functions with UCE security suffice for several uses of the random oracle model. While UCE's present an important step toward understanding which hash function properties might be achievable and which are not, UCE's have several limitations. For example, the formal definition of a UCE is somewhat complicated to even define. Moreover, UCE is not a single property, but a family or "framework" of assumptions. The most general form of UCE is trivially unattainable, and some of the natural restricted classes of UCE have been challenged [7,13]. Therefore, it is unclear which versions of UCE should be trusted and which untrusted.

Similar weaknesses have been shown for other strong assumptions that can be cast as families of assumptions or as knowledge/extracting assumptions, such as extractable one-way functions (eOWFs) [8] and differing inputs obfuscation (diO) [2,12,21]. These weakness are part of a general pattern for strong assumptions such as UCE, eOWFs, and diO that are not specified by a cryptographic game. In particular, these assumptions do not meet standard notions of falsifiability ([22,28]), and are not *complexity assumptions* in the sense of Goldwasser and Kalai [24]. We stress that such knowledge/extracting/framework assumptions are desirable as security *properties*. However, in order to trust that the property actually holds, it should be derived from a "nice" and trusted assumption. Therefore, an important question in this space is the following:

Are there primitives with "nice" (e.g. simple, well-established, game-based, falsifiable, complexity assumption, etc.) security properties that can be used to build hash functions suitable for instantiating random oracles for many protocols.

1.1 Our Work

Our Random Oracle Targets. We aim to base several applications of random oracles on concrete, "nice" assumptions with relatively simple instantiations.

- **Boosting selective to adaptive security.** A trivial application of random oracles is to boost selective to adaptive security in the case of signatures and identity-based encryption. This is done by first hashing the message/identity with the random oracle before signing/generating secret keys. There has been no standard-model security notion for hash functions that allows for this conversion to be secure, though in the case of signatures, chameleon hash functions [27] achieve this conversion with a small tweak.

- **Password hashing.** Another common use of hash functions is to securely store a password in "encrypted" form, allowing for the password to be verified, but hiding the actual password in case the password database is compromised. This use case is a special instance of *point obfuscation* (PO). In the case that there may be side information about the password, we have the notion of *auxiliary input* point obfuscation (AIPO). The only prior constructions of AIPO [9,14] rely on very strong knowledge assumptions. The first is Canetti's [14] strong knowledge variant of the decisional Diffie Hellman (DDH) assumption, whose plausibility has been called into question by a recent work showing it is incompatible with the existence of certain strong forms of obfuscation [7]. The second is a strong knowledge assumption about one-way permutations due to Bitansky and Paneth [9], which is a strengthening of Wee's strong one-way permutation assumption [37]. To the best of our knowledge, the only currently known ways to instantiate the [9] assumption is to make the tautological assumption that a particular one-way permutation is secure. For reasons mentioned above, such tautological knowledge assumptions are generally considered undesirable in cryptography.

- **Generating system parameters.** A natural use case of hash functions is for generating common random strings (crs) in a trusted manner. More specifically, suppose a (potentially untrusted) authority is generating a crs for some protocol. Unfortunately, such a crs may admit a "trapdoor" that allows for breaking whatever protocol is using it (Dual_EC_DRBG is a prominent example of this). In order to ensure to untrusting parties that no trapdoor is known, the authority will generate the crs as an output of the hash function on some input. The authority may have some flexibility in choosing the input; we wish to guarantee that it cannot find an input such that it also knows a trapdoor for the corresponding output. In the random oracle model, this methodology is sound: the authority cannot choose an input so that it knows the trapdoor for the output. However, standard notions of security for hash functions give no guarantees for this setting. We propose (Sect. 5) the notion of *output intractability* as a standard-model security notion that captures this use case. Output intractability is related to, but incomparable with, the notion of correlation intractability mentioned above. As an assumption, our notion of output intractability takes the form of a knowledge assumption on hash functions; no construction based on "nice" assumptions is currently known.

- **Hardcore bits for any one-way function.** A random oracle serves as a good way to extract many hardcore bits for any one-way function. This fact gives rise to a simple public-key encryption scheme from trapdoor permutations. While it is known how to extract many hardcore bits for specific functions [1, 29, 34], extracting many bits for general one-way functions may be useful in settings where we cannot freely choose the function, such as if the function is required to be a trapdoor permutation. Unfortunately, for general one-way functions, the only known way to extract more than a logarithmic number of hardcore bits is to use very strong (and questionable [21]) knowledge assumptions: differing inputs obfuscation [6] (plus one-way functions) or extractable witness PRFs [39]. In the case of *injective* one-way functions, Bellare et al. [6] show that the weaker assumption of *indistiguishability* obfuscation (iO) (plus one-way functions) suffices. While weaker than diO, iO is still one of the strongest assumptions made in cryptography. Either way, the forms of obfuscation required are also completely impractical [3]. Another limitation of prior constructions is that randomness used to sample the hardcore function needs to be kept secret.

- **Instantiating Full Domain Hash (FDH) signatures.** Finally, we consider using random oracles to instantiate the Full Domain Hash (FDH) protocol transforming trapdoor permutations into signatures. Hohenberger et al. [26] show that (indistinguishability) obfuscating a (puncturable) pseudorandom function *composed with the permutation* is sufficient for FDH signatures. However, their proof has two important limitations. First, the resulting signature scheme is only selectively secure. Second, the instantiation depends on the particular trapdoor permutation used, as well as the public key of the signer. Thus, each signer needs a separate hash function, which needs to be appended to the signer's public keys. To use their protocol, everyone will therefore need to publish new keys, even if they already have published keys for the trapdoor permutation.

Our approach. We take a novel approach to addressing the questions above. We isolate a (generally ignored) property of random oracles, namely that random oracles are indistinguishable from functions that are extremely lossy. More precisely, the following is possible in the random oracle model. Given any polynomial time oracle adversary $\mathcal{A}$ and an inverse polynomial δ, we can choose the oracle such that (1) the image size of the oracle is a polynomial r (even for domain/range sizes where a truly random oracle will have exponential image size w.h.p.), and (2) $\mathcal{A}$ cannot tell the difference between such a lossy oracle and a truly random oracle, except with advantage smaller than δ. Note that the tuning of the image size must be done with knowledge of the adversary's running time — an adversary running in time $O(\sqrt{r})$ can with high probability find a collision, thereby distinguishing the lossy function from a truly random oracle. However, by setting $\sqrt{r}$ to be much larger than the adversary's running time, the probability of finding a collision diminishes. We stress that any protocol would still use a truly random oracle and hence not depend on the adversary; the image size tuning would only appear in the security proof. Our observation of this property is inspired by prior works of Boneh and Zhandry [11, 38], who use it for the entirely different goal of giving security proofs in the so-called *quantum* random oracle model (random oracle instantiation was not a goal nor accomplishment of these prior works).

We next propose the notion of an *Extremely Lossy Function (ELF)* as a standard-model primitive that captures this tunable image size property. The definition is related to the notion of a lossy *trapdoor* function due to Peikert and Waters [30], with two important differences: we do not need any trapdoor, giving hope that ELFs could be constructed from symmetric primitives. On the other hand, we need the functions to be much, much more lossy, as standard lossy functions still have exponential image size.

On the surface, extreme lossiness without a trapdoor does not appear incredibly useful, since many interesting applications of standard lossy functions (e.g. CCA-secure public key encryption) require a trapdoor. Perhaps surprisingly, we show that this extremely lossy property, in conjunction with other tools — usually pairwise independence — can in fact quite powerful, and we use this power to give new solutions to each of the tasks above. Our results are as follows:

- (Section 3) We give a practical construction of ELFs assuming the *exponential* hardness of the decisional Diffie-Hellman (DDH) problem: roughly, that the best attack on DDH for groups of order p takes time $O(p^c)$ for some constant c. More generally, our construction can be based on the exponential hardness of the k-Lin problem. Our construction is based on the lossy trapdoor functions due to Peikert and Waters [30] and Freeman et al. [20], though we do not need the trapdoor from those works. Our construction starts from a trapdoor-less version of the DDH-based construction of [20], and iterates it many times at different security levels, together with pairwise independent hashing to keep the output length from growing too large. Having many different security levels allows us to do the following: when switching the function to be lossy, we can do so at a security level that is just high enough to prevent the particular adversary from detecting the switch. Using the exponential DDH assumption, we show that the security level can be set low enough so that (1) the adversary cannot detect the switch, and (2) so that the resulting function has polynomial image size. We note that a couple prior works [10,18] have used a similar technique of combining several "bounded adversary" instances at multiple security levels, and invoking the security of the instance with "just high enough" security. The main difference is that in prior works, "bounded adversary" refers to bounded queries, and the security parameter itself is kept constant across instances; in our work, "bounded adversary" refers to bounding the running time of the adversary, and the security parameter is what is varied across instances.

  Our iteration at multiple security levels is somewhat generic and would potentially apply to other constructions of lossy functions, such as those based on LWE. However, LWE-based constructions of lossy functions are not quite lossy enough for our needs since even "exponentially secure" LWE can be solved in time sub-exponential in the length of the secret.

  The exponential hardness of DDH is plausible on elliptic curve groups — despite over a decade of wide-spread use and cryptanalysis attempts, there are virtually no non-trivial attacks on most elliptic curve groups and the current best attacks on DDH take time $\Omega(p^{1/2})$. In fact, the parameter settings for most real-world uses of the Diffie-Hellman problem are set assuming the Diffie-Hellman problem takes exponential time to solve. If our assumption turns out to be false, it would have significant ramifications as it would suggest that parameters for many cryptosystems in practice are set too aggressively. It would therefore be quite surprising if DDH turned out to *not* be exponentially hard on elliptic curves. While not a true falsifiable assumption in the sense of Naor [28] or Gentry and Wichs [22] due to the adversary being allowed to run in exponential time, the exponential DDH assumption is falsifiable in spirit and naturally fits within the complexity assumption framework of Goldwasser and Kalai [24].

While our ELFs are built from public key tools, we believe such tools are unnecessary and we leave as an interesting open question the problem of constructing ELFs from symmetric or generic tools.

We observe that our construction achieves a public coin notion, which is useful for obtaining public coin hash functions in applications[1].

– We give several different hash function instantiations based on ELFs ranging in complexity and the additional tools used. In doing so, we give new solutions to each of the problems above. Each construction uses the ELFs in different ways, and we develop new techniques for the analysis of these constructions. Thus we give an initial set of tools for using ELFs that we hope to be useful outside the immediate scope of this work.

- The simplest instantiation is just to use an ELF itself as a hash function. Such a function can be used to generically boost selective security to adaptive security in signatures and identity-based encryption by first hashing the message/user identity (more details below).
- (Section 4) The next simplest instantiation is to pre-compose the ELF with a pairwise independent hash function. This function gives rise to a simple (public coin) point function obfuscation (PO). Proving this uses a slight generalization of the "crooked leftover hash lemma" [17].
- (Section 5) A slightly more complicated instantiation is given by *post*-composing and ELF with a k-wise independent function. We show that this construction satisfies our notion of *output intractability*. It is moreover public coin. This construction and analysis can be seen as a generalization of the result of [30] that post-composing a standard lossy function with a pairwise independent hash function gives a collision resistant function, though the details of the analysis are very different.
- (Section 6) We then give an even more complicated construction, though still using ELF's as the only underlying source of cryptographic hardness. The construction roughly follows a common paradigm used in leakage resilience [19]: apply a computational step (in our case, involving ELFs), compress with pairwise independence, and repeat. We note however that the details of the construction and analysis are new to this work.

---

[1] The construction of [20] can also be made public coin by tweaking the generation procedure. However, this necessarily loses the trapdoor, as having a trapdoor and being public coin are incompatible. To the best of our knowledge, however, we are the first to observe this public coin feature.

We demonstrate that our construction is a pseudorandom generator attaining a very strong notion of leakage resilience for the seed. This property strengthens the one-way notion of Bitansky and Paneth [9]. Our construction therefore shows how to instantiate the knowledge properties conjectured in their work using a more traditional-style assumption.

An immediate consequences of our generator requirement is a (public coin) point function obfuscation that is secure even in the presence of auxiliary information (AIPO), which was previously known from either *permutations* satisfying [9]'s one-wayness requirement (our function is *not* a permutation), or from Canetti's strong knowledge variant of DDH [9,14][2]. Our AIPO construction is qualitatively very different from these existing constructions, and when plugging in our ELF construction, again relies on just exponential DDH.

Our generator also immediately gives a family of (public coin) hardcore functions of arbitrary stretch for any one-way function. Unlike the previous obfuscation-based solutions, our is practical, and public coin, and ultimately based on a well-studied game-based assumption.

Our analysis also demonstrates that our ELF-based function can be used in a standard random oracle public key encryption protocol [5].

- In the full version [40], we give an instantiation useful for Full Domain Hash (FDH) signatures which involves obfuscating the composition of an ELF and a (puncturable) pseudorandom function using an indistinguishability obfuscator. Since we use obfuscation as in Hohenberger et al. [26] scheme, this construction is still completely impractical and therefore currently only of theoretical interest. We show that our construction can be used in the FDH protocol, solving some of the limitations in [26]. In particular, by composing with an ELF, we immediately get adaptive security as observed above. Our construction is moreover independent of the permutation (except for the size of the circuit computing it), and is also independent of the signer's public key. Thus, our instantiation is universal and one instantiation can be used by any signer, even using existing keys. Similar to [26], this construction is still required to be secret coin, even if the underlying ELF is public coin.

Warm up: generically boosting selective to adaptive security. To give a sense for our techniques, we show how ELFs can be used to generically boost selective to adaptive security in signatures and identity-based encryption. We demonstrate the case for signatures; the case for identity based encryption is almost identical.

Recall that in selective security for signatures, the adversary commits to a message m^* at the beginning of the experiment before seeing the public key. Then the adversary makes a polynomial q adaptive signing queries on messages $m_1, \ldots, m_q \neq m^*$, receiving signatures $\sigma_1, \ldots, \sigma_q$. Then, the adversary produces

---

[2] One drawback — which is shared with some of the prior constructions — is that we achieve a relaxed notion of correctness where for some sparse "bad" choices of the obfuscation randomness, the outputted program may compute the wrong function.

a forged signature σ^* on m^*, and security states that σ^* is with overwhelming probability invalid for any efficient adversary. Adaptive security, in contrast, allows the adversary to choose m^* potentially *after* the q adaptive queries.

We now convert selective to adaptive security using ELFs: first hash the message using the ELF, and then sign. Adaptive security is proved through a sequence of hybrids. The first is the standard adaptive security game above. Toward contradiction, suppose that the adversary runs in polynomial time t and succeeds in forging a signature on m^* with non-negligible probability ϵ. Let δ be an inverse polynomial that lower bounds ϵ infinitely often. In the second hybrid, the ELF is selected to have polynomial image size r, where $r \geq 2q$ is chosen, say, so that no t-time adversary can distinguish between this ELF and an injective ELF, except with probability at most $\delta/2$. Thus, in this hybrid, the adversary still successfully forges with probability $\epsilon - \delta/2$. This is lower bounded by $\delta/2$ infinitely often, and is therefore non-negligible.

In the next hybrid, at the beginning of the experiment, one of the r image points of the ELF, y^*, is chosen at random[3]. Then we abort the experiment if the adversary's chosen m^* does not hash to y^*; with probability $1/r$, we do not abort[4]. This abort condition is independent of the adversary's view, meaning that we do not abort, and the adversary successfully forges, with probability at least $(\epsilon - \delta/2)/r$, which again is non-negligible. Notice now that y^* can be chosen at the beginning of the experiment. This is sufficient for obtaining an adversary for the selective security of the original signature scheme.

1.2 Complexity Absorption

It may be more reasonable to assume the (sub-)exponential hardness of an existing well-studied problem than to assume such hardness for new and untested problems. Moreover, there might be implementation issues (such as having to re-publish longer keys, see the full version [40] for a setting where this could happen) that make the sub-exponential hardness of certain primitives undesirable.

The application of ELFs to signatures and identity-based encryption above can be seen as an instance of a more general task of *complexity absorption*, where an extra complexity-absorbing primitive (in our case, and ELF) is introduced into the protocol. The original building blocks of the protocol (the underlying signature/identity-based encryption in this case) can be reduced from (sub)exponential security to polynomial security. Meanwhile, the complexity-absorbing primitive may still require exponential hardness as in our case, but hopefully such hardness is a reasonable assumption. Our hardcore function with arbitrary span can also be seen in this light: it is straightforward to extend Goldreich-Levin [23] to a hardcore function of polynomial span for exponentially-secure one-way functions. By introducing an ELF into the hardcore function,

---

[3] The ability to sample a random image point does not follow immediately from our basic ELF definition, though this can be done in our construction.

[4] We also need to abort if any of the m_i do hash to y_i. It is straightforward to show that we still do not abort with probability at least $\frac{1}{2r}$.

the ELF can absorb the complexity required of the one-way function, yielding a hardcore function for *any* one-way function, even one-way functions that are only polynomially secure. Similarly, our random oracle instantiation for Full Domain Hash can also be seen as an instance of complexity absorption.

Thus, our work can be seen as providing an initial set of tools and techniques for the task of complexity absorption that may be useful in other settings where some form of sub-exponential hardness is difficult or impossible to avoid. For example, Rao [32] argues that any proof of adaptive security for multiparty non-interactive key exchange (NIKE) will likely incur an exponential loss. As all current multiparty NIKE protocols are built from multilinear maps or obfuscation, which in turn rely on new, untested (and in many cases broken) hardness assumptions, assuming the sub-exponential security of the underlying primitives to attain adaptive security is undesirable. Hofheinz et al. [25] give a construction in the random oracle model that only has a polynomial loss; our work gives hope that a standard model construction based on ELFs may be possible where the ELF is the only primitive that needs stronger than polynomial hardness.

1.3 Non-black Box Simulation

Our proofs require knowledge of the adversary's running time (and success probability). Thus, they do not make black box use of the adversary. Yet, this is the only non-black box part of our proofs — the reduction does not need to know the description or internal workings of the adversary. This is similar to Goldreich-Levin [23], where only the adversary's success probability is needed. Thus our reductions are nearly black box, while potentially giving means to circumvent black-box impossibilities. For example, proving the security of AIPO is known to require non-black box access to the adversary [9,37], and yet our reduction proves the security of AIPO knowing only the adversary's running time and success probability. We leave it as an interesting open question to see if your techniques can be used to similarly circumvent other black box impossibilities.

1.4 On the Minimal Assumptions Needed to Build ELFs

We show how to construct extremely lossy functions from a specific assumption on elliptic curve groups. One could also hope for generic constructions of ELFs based on existing well-studied primitives. Unfortunately, this appears to be a difficult task, and there are several barriers to constructing ELFs. For example, lossy functions (even standard ones) readily imply collision resistance [30], which cannot be built from one-way functions in a black-box fashion [35]. Rosen and Segev [33] show a similar separation from functions that are secure under correlated products. Pietrzak et al. [31] show that efficiently amplifying lossiness in a black box way is impossible — this suggests that building ELFs from standard lossy functions will be difficult, if not impossible.

Perhaps an even stronger barrier to realizing ELFs from standard assumptions is the following. Our assumption, unfortunately, is about *exponential*-time adversaries, as opposed to typical assumptions about polynomial-time adversaries. One could hope for basing ELFs on standard polynomial assumptions, such as polynomial DDH. However, this would require major breakthroughs in complexity theory. Indeed, lossy and injective modes of an ELF can be distinguished very efficiently using a super-logarithmic amount of non-determinism as follows. Let $D = [2^{\omega(\log m)}]$ where m is the number of input bits to the ELF. In the injective mode, there will be no collisions when the domain is restricted to D. However, in the lossy mode for *any* polynomial image size $r = r(m)$, there is guaranteed to be a collision in D. Points in D can be specified by $\omega(\log m)$ bits. Therefore, we can distinguish the two modes by non-deterministically guessing two inputs in D (using $\omega(\log m)$ bits of non-determinism) and checking that they form a collision. Therefore, if NP restricted to some super-logarithmic amount of non-determinism was solvable in polynomial time, then this algorithm could be made efficient while removing all non-determinism. Such an algorithm would violate ELF security.

Theorem 1. *If ELFs exist, then for any super-logarithmic function t, NP with t bits of non-determinism is not solvable in polynomial time.*

Therefore, it seems implausible to base ELFs on any polynomially-secure primitive, since it is consistent with our current knowledge that NP with, say, $\log^2$ bits of non-determinism is solvable in polynomial time, but polynomially-secure cryptographic primitives exist. This may seem to suggest that ELFs are too strong of a starting point for our applications; to the contrary, we argue that for most of our applications — point functions[5] (Sect. 4), output intractability (Sect. 5), and polynomially-many hardcore bits for any one-way function (Sect. 6) — similar barriers are inherent to the applications. Therefore, this limitation of ELFs is shared with any primitive strong enough to realize the applications.

Therefore, instead of starting from standard polynomially-secure primitives, we may hope to build ELFs generically from, say, an exponentially secure primitive which has a similar limitation. Can we build ELFs from exponentially secure (injective) one-way functions? Exponentially-secure collision resistant hash functions? To what extent do the black-box barriers above extend into the regime of exponential hardness? We leave these as interesting open questions.

2 Preliminaries

Given a distribution $\mathcal{D}$ over a set $\mathcal{X}$, define the support of $\mathcal{D}$, $\text{Supp}(\mathcal{D})$, to be the set of points in $\mathcal{X}$ that occur with non-zero probability. For any $x \in \mathcal{X}$, let $\Pr[\mathcal{D} = x]$ be the probability that $\mathcal{D}$ selects x. For any set $\mathcal{X}$, define $U_{\mathcal{X}}$ to be the uniform distribution on $\mathcal{X}$. Define the collision probability of $\mathcal{D}$ to be $CP(\mathcal{D}) = \Pr[x_1 = x_2 : x_1, x_2 \leftarrow \mathcal{D}] = \sum_{x \in \mathcal{X}} \Pr[\mathcal{D} = x]^2$. Given two distributions $\mathcal{D}_1, \mathcal{D}_2$, define the statistical distance between $\mathcal{D}_1$ and $\mathcal{D}_2$ to be $\Delta(\mathcal{D}_1, \mathcal{D}_2) = \frac{1}{2} \sum_{x \in \mathcal{X}} \left| \Pr[\mathcal{D}_1 = x] - \Pr[\mathcal{D}_2 = x] \right|$. Suppose $\text{Supp}(\mathcal{D}_1) \subseteq \text{Supp}(\mathcal{D}_2)$. Define the

---

[5] The case of point functions is more or less equivalent to a similar result of Wee [37].

Rényi Divergence between $\mathcal{D}_1$ and $\mathcal{D}_2$ to be $RD(\mathcal{D}_1, \mathcal{D}_2) = \sum_{x \in \sup(\mathcal{D}_1)} \frac{\Pr[\mathcal{D}_1 = x]^2}{\Pr[\mathcal{D}_2 = x]}$ [6]. The Rényi divergence is related to the statistical distance via the following lemma:

Lemma 1. *For any distributions $\mathcal{D}_1, \mathcal{D}_2$ over a set $\mathcal{Z}$ such that* $\mathrm{Supp}(\mathcal{D}_1) \subseteq \mathrm{Supp}(\mathcal{D}_2)$, $\Delta(\mathcal{D}_1, \mathcal{D}_2) \leq \frac{1}{2}\sqrt{RD(\mathcal{D}_1, \mathcal{D}_2) - 1}$.

Consider a distribution $\mathcal{H}$ over the set of functions $h : \mathcal{X} \to \mathcal{Y}$. We say that $\mathcal{H}$ is *pairwise independent* if, for any $x_1 \neq x_2 \in \mathcal{X}$, the random variables $\mathcal{H}(x_1)$ and $\mathcal{H}(x_2)$ are independent and identically distributed, though not necessarily uniform. Similarly define k-wise independence. We say that $\mathcal{H}$ has *output distribution* $\mathcal{D}$ if for all x, the random variable $\mathcal{H}(x)$ is identical to $\mathcal{D}$. Finally, we say that $\mathcal{H}$ is *uniform* if it has output distribution $U_{\mathcal{Y}}$ [7]. We will sometimes abuse notation and say that a function h is a pairwise independent function (resp. uniform) if h is drawn from a pairwise independent (resp. uniform) distribution of functions.

We will say that a (potentially probabilistic) algorithm $\mathcal{A}$ outputting a bit b distinguishes two distributions $\mathcal{D}_0, \mathcal{D}_1$ with advantage ϵ if $\Pr[\mathcal{A}(\mathcal{D}_b) : b \leftarrow \{0,1\}] \in \left[\frac{1}{2} - \epsilon, \frac{1}{2} + \epsilon\right]$. This is equivalent to the random variables $\mathcal{A}(\mathcal{D}_0)$ and $\mathcal{A}(\mathcal{D}_1)$ have 2ϵ statistical distance.

Unless otherwise stated, all cryptographic protocols will implicitly take a security parameter λ as input. Moreover, any sets (such as message spaces, ciphertext spaces, etc.) will be implicitly indexed by λ, unless otherwise stated. In this context, when we say that an adversary is efficient, we mean its running time is polynomial in λ. A non-negative function $\epsilon = \epsilon(n)$ is negligible if, for any polynomial $p = p(\lambda)$, $\epsilon < 1/p$ for all sufficiently large λ. When discussing cryptographic protocols, we say that a probability of an event or advantage of an adversary is negligible if it is negligible in λ. Two distributions $\mathcal{D}_0, \mathcal{D}_1$ (implicitly parameterized by λ) are computationally indistinguishable if any efficient algorithm has only negligible distinguishing advantage, and are statistically indistinguishable if the distributions have negligible statistical distance. In the statistical setting, we also sometimes say that $\mathcal{D}_0, \mathcal{D}_1$ are statistically close.

The Crooked Leftover Hash Lemma. Here we state a slight generalization of the "crooked Leftover Hash Lemma" of Dodis and Smith [17]; the proof is in the full version [40] and follows [17].

Lemma 2. *Let H be a distribution on functions $h : \mathcal{X} \to \mathcal{Y}$ that is pairwise independent with output distribution $\mathcal{E}$, for some distribution $\mathcal{E}$ that is possibly non-uniform. Let $\mathcal{D}$ be an arbitrary distribution over $\mathcal{X}$. Then we have that*

$$\Delta\big((H, H(\mathcal{D})), (H, \mathcal{E})\big) \leq \frac{1}{2}\sqrt{CP(\mathcal{D})(|\mathrm{Supp}(\mathcal{E})| - 1)}.$$

---

[6] Often, the Rényi Divergence is defined to be proportional to the logarithm of this quantity. The definition here will be more convenient for our purposes.

[7] Note that the typical use of *pairwise independence* is equivalent to our notion of pairwise independence *plus* uniformity. For our purposes, it will be convenient to separate out the two properties.

3   Extremely Lossy Functions

Here, we define our notion of *extremely lossy functions*, or ELFs. A standard lossy function [30] is intuitively a function family with two modes: an injective mode where the function is injective, and a lossy mode where the image size of the function is much smaller than the domain. The standard security requirement is that no polynomial-time adversary can distinguish the two modes[8].

An ELF is a lossy function with a much stronger security requirement. In the lossy mode, the image size can be taken to be a polynomial r. Clearly, such a lossy mode can be distinguished from injective by an adversary running in time $O(\sqrt{r})$ that simply evaluates the function on $\sqrt{r}$ inputs, looking for a collision. Therefore, we cannot have security against arbitrary polynomial-time attackers. Instead, we require security against r^c-time attackers, for some $c \leq 1/2$. Moreover, we require that r is actually tunable, and can be chosen based on the adversary in question. This means that for *any* polynomial time attacker, we can set the lossy function to have domain r for some polynomial r, and the lossy function will be indistinguishable from injective to that particular attacker (note that the honest protocol will always use the injective mode, and therefore will not depend on the adversary in any way).

Definition 1. *An* extremely lossy function *(ELF) consists of an algorithm* ELF.Gen, *which takes as input integers M and $r \in [M]$. There is no security parameter here; instead, $\log M$ acts as the security parameter.* ELF.Gen *outputs the description of a function $f : [M] \to [N]$ such that:*

- *f is computable in time polynomial in the bit-length of its input, namely $\log M$. The running time is independent of r.*
- *If $r = M$, then f is injective with overwhelming probability (in $\log M$).*
- *For all $r \in [M]$, $|f([M])| \leq r$ with overwhelming probability. That is, the function f has image size at most r.*
- *For any polynomial p and inverse polynomial function δ (in $\log M$), there is a polynomial q such that: for any adversary $\mathcal{A}$ running in time at most p, and any $r \in [q(\log M), M]$, we have that $\mathcal{A}$ distinguishes* ELF.Gen(M, M) *from* ELF.Gen(M, r) *with advantage less than δ. Intuitively, no polynomial-time adversary $\mathcal{A}$ can distinguish an injective from polynomial image size (where the polynomial size depends on the adversary's running time.).*

For some applications, we will need an additional requirement for ELFs:

Definition 2. *An ELF has an* efficiently enumerable *image space if there is a (potentially randomized) procedure running in time polynomial in r and $\log M$ that, given $f \leftarrow$ ELF.Gen(M, r) and r, outputs a polynomial-sized set S of points in $[N]$ such that, with overwhelming probability over the choice of f and the randomness of the procedure, $f([M]) \subseteq S$.*

---

[8] [30] additionally require that, in the injective mode, there is a trapdoor that allows inverting the function. We will not need any such trapdoor.

Definition 3. *An ELF has a efficiently sampleable* image space *if there is a polynomial s and a randomized polynomial time procedure (where "polynomial" means polynomial in r and* $\log M$ *) such that the following holds. Given* $f \leftarrow$ ELF.Gen(M, r) *and r, the procedure outputs a point* $y \in [N]$ *such that with overwhelming probability over the choice of f, the point y has a distribution that places weight at least* $1/s$ *on each image point in* $f([M])$.

Lemma 3. *An ELF is efficiently sampleable iff it is efficiently enumerable.*

Proof. In one direction, we just sample a random element from the polynomial-sized list S, obtaining each image point with probability $1/|S|$. In the other direction, by sampling λs points independently at random, except with negligible probability in λ, the set of sampled points will contain each of the r images. □

The following property will be useful for attaining ELFs with efficiently enumerable/sampleable image spaces:

Definition 4. *An ELF is* regular *if, for all polynomial r, with overwhelming probability over the choice of* $f \leftarrow$ ELF.Gen(M, r), *the distribution* $f(x)$ *for a uniform* $x \leftarrow [M]$ *is statistically close to uniform over* $f([M])$.

Lemma 4. *If an ELF is regular, then it is efficiently sampleable/enumerable.*

Proof. To sample, just apply the ELF to a random point. Notice that the sampled point is guaranteed to be an image point. Thus regularity actually implies a strong notion of enumerability where $S = f([M])$ with overwhelming probability. □

The final ELF property we define is *public coin*.

Definition 5. *An ELF is* public coin *if the description of an injective mode f outputted by* ELF.Gen(M, M) *is simply the random coins used by* ELF.Gen(M, M). *The descriptions of lossy mode f's outputted by* ELF.Gen$(M, r), r < M$ *may (and in fact, must) be a more complicated function of the random coins.*

3.1   Constructing ELFs

We now show how to construct ELFs. Our construction will have two steps: first, we will show that ELFs can be constructed from a weaker primitive called a *bounded adversary* ELF, which is basically and ELF that is only secure against a priori bounded adversaries. Then we show essentially that the DDH-based lossy function of [20], when the group size is taken to be polynomial, satisfies our notion of a bounded-adversary ELF.

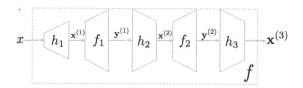

Fig. 1. An example instantiation for $k = 3$.

Bounded Adversary ELFs

Definition 6. *An* bounded adversary *extremely lossy function (bounded ELF) consists of an algorithm* ELF.Gen$'$, *which takes as input integers* M, $r \in [M]$, *and* $b \in \{0, 1\}$. *Here,* b *will indicate whether the function should be lossy, and* r *will specify the lossiness. Similar to regular ELFs, there is no security parameter here; instead,* $\log M$ *acts as the security parameter.* ELF.Gen$'$ *outputs the description of a function* $f : [M] \to [N]$ *such that:*

- f *is computable in time polynomial in the bit-length of its input, namely* $\log M$. *The running time is independent of* r.
- *If* $b = 0$, *then* f *is injective with overwhelming probability (in* $\log M$).
- *For all* $r \in [M]$, *if* $b = 1$, *then* $|f([M])| \leq r$ *with overwhelming probability. That is, the function* f *has image size at most* r.
- *For any polynomial* p *and inverse polynomial function* δ *(in* $\log M$), *there is a polynomial* q *such that: for any adversary* $\mathcal{A}$ *running in time at most* p, *and any* $r \in [q(\log M), M]$, *we have that* $\mathcal{A}$ *distinguishes* ELF.Gen$'(M, r, 0)$ *from* ELF.Gen$'(M, r, 1)$ *with advantage less than* δ.

Intuitively, the difference between a regular ELF and a bounded adversary ELF is that in a regular ELF, r can be chosen dynamically based on the adversary, whereas in a bounded adversary ELF, r must be chosen first, and then security only applies to adversaries whose running time is sufficiently small. In a bounded adversary ELF, the adversary may be able to learn r. We now show that bounded ELFs are sufficient for constructing full ELFs.

Construction 1. *On input* M, r, ELF.Gen *does:*

- *For simplicity, assume* M *is a power of* 2: $M = 2^k$. *Let* $M' = M^3 = 2^{2k}$, *and* $[N]$ *be the co-domain of* ELF.Gen$'$ *on domain* $[M']$.
- *Let* i^* *be the integer such that* $2^{i^*} \in (r/2, r]$. *Set* $b_{i^*} = 1$ *and* $b_i = 0$ *for* $i \neq i^*$
- *For* $i = 1, \ldots, k - 1$, *let* $f_i \leftarrow$ ELF.Gen$'(M', 2^i, b_i)$.
- *For* $i = 2, \ldots, k$, *choose a pairwise independent random* $h_i : [N'] \to [M']$.
- *Choose a pairwise independent random* $h_1 : [M] \to [M']$.
- *Output the function* $f = h_k \circ f_{k-1} \circ h_{k-1} \circ f_{k-2} \circ \cdots \circ f_1 \circ h_1$.

Theorem 2. *If* ELF.Gen$'$ *is a bounded-adversary ELF, then* ELF.Gen *is a (standard) ELF. If* ELF.Gen$'$ *is public coin, then so is* ELF.Gen. *If* ELF.Gen$'$ *is enumerable, then so is* ELF.Gen. *If* ELF.Gen$'$ *is regular, then so is* ELF.Gen.

Proof. First, if $r = M$, then $i^* = k$, and so each of the b_i will be 0. Thus each of the f_i will be injective with overwhelming probability. Fix $h_1, f_i, \ldots, h_{i-1}, f_{i-1}$, and let S_i be the image of $f_{i-1} \circ h_{i-1} \circ f_{k-2} \circ \cdots \circ f_1 \circ h_1$. Since each of the functions h_i have co-domain of size $M' = M^3$, by pairwise independence, h_i will be injective on S_i with overwhelming probability. Thus, with overwhelming probability, the entire evaluation of f will be injective.

Second, if $r < M$, the function f_{i^*} is set to be lossy with image size $2^{i^*} \leq r$. Thus, f will have image size at most r. Third, we need to argue security. Let p be a polynomial and σ be an inverse polynomial (in $\log M$). Let $p' = p + c$ for some c to be determined later. We can think of p', σ as being functions of $\log M' = 3 \log M$. Let q be the polynomial guaranteed by ELF.Gen' for p' and σ. Then we can consider q to be a polynomial in $\log M$. Consider any adversary A for ELF.Gen running in time at most p. Let $r \in (q(\log M), M]$, and let i^* be such that $2^{i^*} \in (r/2, r]$. We construct an adversary A' for ELF.Gen': let f_{i^*} be the f that A' receives, where f_{i^*} is either ELF.Gen$(M, 2^{i^*}, 0)$ or ELF.Gen$(M, 2^{i^*}, 1)$. A' simulates the rest of f for itself, setting $b_i = 0$, $f_i \leftarrow$ ELF.Gen'$(M, 2^i, b_i)$ for $i \neq i^*$ as well as generating the h_i. A' then runs A on the simulated f. Let c be the overhead of this reduction, so that A' runs in time $p + c = p'$. Thus by the bounded-adversary security of ELF.Gen', A' cannot distinguish injective or lossy mode, except with advantage σ. Moreover, if f_{i^*} is generated as ELF.Gen$(M, 2^{i^*}, 0)$, then this corresponds to the injective mode of ELF.Gen, and if f_{i^*} is generated as ELF.Gen$(M, 2^{i^*}, 1)$, then this corresponds to ELF.Gen(M, r). Thus, A' and A have the same distinguishing advantage, and therefore A cannot distinguish the two cases except with probability less than σ.

It remains to show that ELF.Gen inherits some of the properties of ELF.Gen'. Being public coin is trivially inherited. To get a sampler for ELF.Gen, apply the .sampler for ELF.Gen' to the instance f_{i^*} that is lossy, obtaining point $y^{(i^*)}$. Then compute $y = h_k \circ f_{k-1} \circ h_{k-1} \circ f_{k-2} \circ \cdots \circ f_{i^*+1} \circ h_{i^*+1}(y^{(i^*)})$. Since any image of f is necessarily computed as $h_k \circ f_{k-1} \circ h_{k-1} \circ f_{k-2} \circ \cdots \circ f_{i^*+1} \circ h_{i^*+1}(y^{(i^*)})$ for *some* y_{i^*} in the image of f_{i^*}, and all other steps are injective with overwhelming probability, the result $x^{(k+1)}$ will hit each image point of f frequently as well. In the full version [40], we also show that regularity is inherited. □

Instantiation for Bounded Adversary ELFs. Our construction of bounded adversary ELFs is based on the DDH-based lossy *trapdoor* functions of Peikert and Waters [30] and Freeman et al. [20]. We stress that we do not need the trapdoor property of their construction, only the lossy property. Security will be based on the exponential hardness of the decisional Diffie-Hellman problem, or its k-linear generalizations.

Definition 7. *A cryptographic group consists of an algorithm* GroupGen *that takes as input a security parameter λ, and outputs the description of a cyclic group $\mathbb{G}$ of prime order $p \in [2^\lambda, 2 \times 2^\lambda)$, and a generator g for $\mathbb{G}$ such that:*

- *The group operation $\times : \mathbb{G}^2 \to \mathbb{G}$ can be computed in time polynomial in λ.*

- *Exponentiation by elements in $\mathbb{Z}_p$ can be carried out in time polynomial in λ. This follows from the efficient group operation procedure by repeated doubling and the fact that $\log p \leq \lambda + 1$.*
- *The representation of a group element h has size polynomial in λ. This also follows implicitly from the assumption that the group operation is efficient.*

We now introduce some notation. For vectors $\mathbf{v}, \mathbf{w} \in \mathbb{Z}_p^n$, let $\mathbf{v} * \mathbf{w}$ denote the point-wise product of the two vectors. For a matrix $\mathbf{A} \in \mathbb{Z}_p^{m \times n}$, we write $g^\mathbf{A} \in \mathbb{G}^{m \times n}$ to be the $m \times n$ matrix of group elements $g^{A_{i,j}}$. Similarly define $g^\mathbf{w}$ for a vector $\mathbf{w} \in \mathbb{Z}_p^n$. Given a matrix $\hat{\mathbf{A}} \in \mathbb{G}^{m \times n}$ of group elements and a vector $\mathbf{v} \in \mathbb{Z}_p^n$, define $\hat{\mathbf{A}} \cdot \mathbf{v}$ to be $\hat{\mathbf{w}} \in \mathbb{G}^m$ where $\hat{w}_i = \prod_{j=1}^n \hat{A}_{i,j}^{v_j}$. Using this notation, $(g^\mathbf{A}) \cdot \mathbf{v} = g^{\mathbf{A} \cdot \mathbf{v}}$. Therefore, the map $g^\mathbf{A}, \mathbf{v} \mapsto g^{\mathbf{A} \cdot \mathbf{v}}$ is efficiently computable.

Definition 8. *The exponential decisional k-linear assumption (k-eLin) on a cryptographic group specified by* GroupGen *holds if there is a polynomial $q(\cdot, \cdot)$ such that the following is true. For any time bound t and probability ϵ, let $\lambda = \log q(t, 1/\epsilon)$. Then for any adversary $\mathcal{A}$ running in time at most t, the following two distributions are indistinguishable, except with advantage at most ϵ:*

$$(\mathbb{G}, g, g^\mathbf{v}, g^{\mathbf{v}*\mathbf{w}}, g^c) : (\mathbb{G}, g, p) \leftarrow \mathtt{GroupGen}(\lambda), \mathbf{v}, \mathbf{w} \leftarrow \mathbb{Z}_p^k, c \leftarrow \mathbb{Z}_p \ and$$

$$(\mathbb{G}, g, g^\mathbf{v}, g^{\mathbf{v}*\mathbf{w}}, g^{\sum_{i=1}^k w_i} : (\mathbb{G}, g, p) \leftarrow \mathtt{GroupGen}(\lambda), \mathbf{v}, \mathbf{w} \leftarrow \mathbb{Z}_p^k$$

Definition 9. *A cryptographic group is* public coin *if the following holds:*

- *The "description" of $\mathbb{G}, g, p$ is just the random coins sampled by* GroupGen.
- *There is a (potentially redundant) efficiently computable representation of group elements in $\mathbb{G}$ as strings in $\{0,1\}^n$ such that (1) a random string in $\{0,1\}^n$ corresponds to a random element in $\mathbb{G}$, and (2) a random representation of a random element in $\mathbb{G}$ is a random string in $\{0,1\}^n$.*

A plausible candidate for a cryptographic group supporting the k-eLin assumption are groups based on elliptic curves. Despite over a decade or research, essentially no non-trivial attack is known on general elliptic curve groups. Therefore, the k-eLin assumption on these groups appears to be a reasonable assumption. We note that groups based on elliptic curves can be made public coin.

Construction 2. *Our construction is as follows, and will be parameterized by k.* ELF.Gen$_k'(M, r, b)$ *does the following.*

- *Let λ be the largest integer such that $(2 \times 2^\lambda)^k < r$. Run $(\mathbb{G}, g, p) \leftarrow$* GroupGen$(\lambda)$.
- *Assume for simplicity that $M = p^m$ for some integer m. Then associate the domain $[M]$ with $\mathbb{Z}_p^m$. The more general case can be handled by hashing $[M]$ into $\mathbb{Z}_p^m$ for some m using a pairwise independent hash function which is injective with overwhelming probability; we defer the analysis to the full version [40] and here focus on the simple case.*
- *Let $n \geq m$ be such that a random matrix sampled from $\mathbb{Z}_p^{n \times m}$ has rank m with overwhelming probability. For this, it suffices to set $n = 2m$.*

- If $b = 0$, *choose a random matrix of group elements* $g^{\mathbf{A}}$. *If* $b = 1$, *choose a random rank-k matrix* $\mathbf{A} \in \mathbb{Z}_p^{n \times m}$ *and compute* $g^{\mathbf{A}}$.
- *Output the function* $f(x) = \mathbf{A} \cdot x$. *The description of* f *will consist of* $(\mathbb{G}, p, \mathbf{A}, m, n)$.

Theorem 3. *If* GroupGen *is a group where the* k*-eLin assumption holds for some constant* k, *then* ELF.Gen$'_k$ *is a regular bounded adversary ELF. If* GroupGen *is public coin, then so is* ELF.Gen$'$.

Proof. If $\mathbf{A}$ is full rank, then the map $\mathbf{y} \mapsto g^{\mathbf{A} \cdot \mathbf{y}}$ is injective. If $\mathbf{A}$ has rank k, then the map has image size $p^k < r$. For security, we just need to show that the two distributions on $g^{\mathbf{A}}$ are indistinguishable. Note that it is well known that the k-linear assumption implies that it is hard to distinguish $g^{\mathbf{B}}$ for a random $\mathbf{B} \in \mathbb{Z}_p^{k+1,k+1}$ from $g^{\mathbf{B}}$ for a random rank k matrix $\mathbf{B}$ with no loss in the security of the reduction. From here, it is straightforward to show that it is hard to distinguish the full rank and rank k cases of $g^{\mathbf{A}}$, with a loss of a factor of $m - k$. In fact, using the ideas of [36], the loss can even be made logarithmic in m, but we will use m as an upper bound on the loss for simplicity. Let q be the polynomial guaranteed by the k-eLin assumption. Let t be a polynomial and δ an inverse polynomial. Let $q' = 4q(t + u, m/\delta)^k$, where u is the overhead in the reduction from k-eLin to the problem of distinguishing ranks of matrices. Suppose an adversary runs in time t and distinguishes the two distributions on $g^{\mathbf{A}}$ with advantage δ. For any $r \geq q'$, we have that $\lambda \geq r^{1/k}/4 \geq q(t + u, m/\delta)$. This means no $(t + u)$-time adversary can break the k-eLin assumption with advantage greater than δ/m. By our reduction from distinguishing ranks, this means no t-time adversary can distinguish the two cases of $g^{\mathbf{A}}$, except with advantage at most δ, as desired.

Notice that if GroupGen is public coin, we can sample $g^{\mathbf{A}}$ directly in the injective mode since it is just a matrix of random group elements. Finally, note that the function $\mathbf{y} \mapsto g^{\mathbf{A} \cdot \mathbf{y}}$ is *perfectly* regular due to its linear structure. $\square$

Corollary 1. *If there exists a constant* k *and a cryptographic group where the* k*-eLin assumption holds, then there exists an ELF with efficiently sampleable/enumerable image. If the group is public coin, then so is the ELF.*

4   Point Function Obfuscation

A (expanding) random oracle H serves as a good point function obfuscator: to obfuscate the point function $I_x(x') = \begin{cases} 1 & \text{if } x' = x \\ 0 & \text{if } x' \neq x \end{cases}$, simply output $y = H(x)$.

Then to run the "program" on input x', simply check that $H(x') = y$. For any x that is drawn from an source with super-logarithmic min-entropy, an adversary making a polynomial number of queries to H will not be able to determine x from y. Thus, x is hidden to all efficient adversaries.

In this section, we show how to use ELFs to implement a concrete function H for which the strategy above still yields a secure point obfuscation (PO).

Definition 10. *A point obfuscator (PO) is an efficient probabilistic algorithm $\mathcal{O}$ with the following properties:*

- *(Almost Perfect Correctness) On input a point function I_x, with overwhelming probability over the random coins of $\mathcal{O}$, $\mathcal{O}$ outputs the description of a program P that is functionally equivalent to I_x. P must run in time polynomial in the length of x and the security parameter.*
- *(Secrecy) For any distribution $\mathcal{D}$ over a set $\mathcal{X}$ with super-logarithmic min-entropy, the distribution $\mathcal{O}(I_x)$ for $x \leftarrow \mathcal{D}$ is computationally indistinguishable from $\mathcal{O}(I_{x'})$ where $x' \leftarrow U_{\mathcal{X}}$.*

Before giving our construction, we point out that a point obfuscator implies a separation from NP with super-logarithmic non-determinism and P. Thus, any primitive used to build point obfuscation, such as ELFs, must necessarily imply such a separation. This is essentially the same statement as a theorem of Wee [37], and is proved in the full version [40].

Theorem 4. *If Point Obfuscators exist, then for any super-logarithmic function t, NP with t bits of non-determinism is not solvable in polynomial time.*

4.1   The Construction

Construction 3. *Let $\mathcal{X}$ be the desired domain of H. To generate H,*

- *Let $\mathcal{Z}$ be some set such that $|\mathcal{X}|^2/|\mathcal{Z}|$ is negligible, and sample a hash function h from a uniform and pairwise independent function distribution from $\mathcal{X}$ to $\mathcal{Z}$. h will thus be injective with overwhelming probability.*
- *Let $f \leftarrow \mathtt{ELF.Gen}(|\mathcal{Z}|, |\mathcal{Z}|)$ to get an injective-mode f.*
- *Output $H = f \circ h$.*

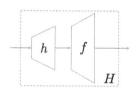

Fig. 2. The function $H = f \circ h$.

Theorem 5. *Assuming* ELF *is a secure ELF, H in Construction 3 gives a secure point obfuscator. If* ELF *is public coin, then so is H.*

Proof. We will actually show something stronger: that the point function obfuscation of x is indistinguishable from an obfuscation of the all-zeros function. In particular, we will show that no efficient adversary can distinguish $y = f(h(x))$

from $y = f(z)$ for a uniformly random z. Notice that by injectivity of f, y has a pre-image under $H = f \circ h$ if and only if $z = f^{-1}(y)$ has a pre-image under h. Since we chose h to be expanding, when we sample z uniformly random, z will have no pre-image with overwhelming probability. Therefore, $y = f(z)$ has no pre-image with overwhelming probability.

The proof involves a sequence of hybrids. Suppose the adversary runs in time t and distinguishes $y = f(h(x))$ from $y = f(z)$ with non-negligible advantage ϵ. This means there is an inverse polynomial δ such that $\epsilon \geq \delta$ infinitely often.

Hybrid 0. This is the honestly generated $y = f(h(x))$ for f drawn in injective mode and x drawn from D.

Hybrid 1. Now, we change f to be lossy. That is, we generate $f \leftarrow$ ELF.Gen$(|\mathcal{Z}|, r)$ where r is chosen so that no adversary running in time t can distinguish this lossy f from an injective f, except with advantage at most $\delta/3$. Thus by ELF security, the adversary cannot distinguish **Hybrid 0** from **Hybrid 1**, except with probability $\delta/3$.

Hybrid 2. Now we change y to be $y = f(z)$ for a random uniform $z \in \mathcal{Z}$. Fix f, and let E be the distribution of y. Then notice that by the pairwise independence and uniformity of h, the composition $H = f \circ h$ is pairwise independent and has output distribution E. Moreover, Supp$(E) \leq r$ is a polynomial. Therefore, by Lemma 2, we see that **Hybrid 1** and **Hybrid 2** are indistinguishable, except with probability $\frac{1}{2}\sqrt{CP(D)(|\text{Supp}(E)| - 1)}$. As long as the collision probability of $\mathcal{X}$ is negligible (which in particular happens when $\mathcal{X}$ has super-logarithmic min-entropy), this quantity will be negligible. In particular, the distinguishing advantage will be less than $\delta/3$.

Hybrid 3. Now we change f to be injective again. The distinguishing advantage between **Hybrid 2** and **Hybrid 3** will be at most $\delta/3$. Notice that **Hybrid 3** is exactly our all-zeros obfuscation. Therefore, **Hybrid 0** and **Hybrid 3** are indistinguishable, except with probability less than δ, meaning $\epsilon < \delta$. This contradicts our assumption about the adversary. $\qquad\square$

In Sect. 6, we will show how to strengthen our construction to get a point obfuscator that is secure even against auxiliary information about the point.

5   Output Intractability

Consider any $k+1$-ary relation R over $\mathcal{Y}^k \times \mathcal{W}$ that is *computationally intractable*: on a random input $\mathbf{y} \in \mathcal{Y}^k$, it is computationally infeasible to find a $w \in \mathcal{W}$ such that $R(\mathbf{y}, w)$ outputs 1. If H is a random oracle, assuming k is a constant, it is computationally infeasible for find a set of distinct inputs $\mathbf{x}$, $x_i \neq x_j \forall i \neq j$, and a $w \in \mathcal{W}$, such that $R(H(\mathbf{x}), w) = 1$. We will now show how to build standard-model hash functions H that achieve the same property.

Definition 11. *A family of hash functions $H : [M] \rightarrow \mathcal{Y}$ is k-ary output intractable if, for any computationally intractable $k + 1$-ary relation $R : \mathcal{Y}^k \times \mathcal{W} \rightarrow \{0, 1\}$, no efficient adversary, given H, can find a set of distinct inputs $\mathbf{x} \in [M]^k$ and an element $w \in \mathcal{W}$, such that $R(H(\mathbf{x}), w) = 1$.*

Note that binary output intractability implies as a special case collision resistance. In the unary case, and if $\mathcal{W}$ is just a singleton set, then output intractability is a special case of *correlation intractability*, which allows the relation to additionally depend on the *input*.

The unary case captures the following use case of hash functions: a given protocol may require a common reference string (crs), but some or all instances of the crs may admit a trapdoor that allows breaking the protocol. Of course, such a trapdoor should be difficult to find for a random crs. To "prove" that the crs is generated so that the generator of the crs does not know a trapdoor, the generator sets the crs to be the output of a public hash function on an arbitrary point. Since the potentially malicious generator does not control the hash function, he should be unable to find an output along with a corresponding trapdoor. Modeling the hash function as a random oracle, this methodology is sound. However, standard notions of security do not prevent the crs generator from choosing the input in such a way so that it knows a trapdoor. Unary output intractability precludes this case. Of course, the hash function itself needs to be set up in a trusted manner; however, once the hash function is set up · and trusted, it can be used to generate arbitrarily many different crs by even untrusted authorities.

We note, however, that the unary case on its own is not very interesting: the family of hash functions H parameterized by a string $y \in \mathcal{Y}$ where $H(x) = y$ for all x is clearly unary intractable. Depending on the application, one may want additional features such as collision resistance, which as noted above is implied by binary output intractability ($k = 2$). Therefore, $k = 2$ and above are likely to be the most interesting settings. In the full version [40], we argue that $k \geq 2$ inherently requires some sort of super-polynomial hardness:

Theorem 6. *If binary output intractable hash functions exist, then for any super-logarithmic function t, NP with t bits of non-determinism is not solvable in polynomial time.*

Trivial impossibility for arbitrary k. We note that no one family of hash functions H can satisfy k-ary output intractability for all k. That is, for different k, a different family will be required. Suppose to the contrary that a family H satisfied k-output intractability for all k. Let t be the size of the circuit computing H. Choose k so that $k \log |\mathcal{Y}| \geq t$. Then with overwhelming probability over the choice of random $\mathbf{y} \in \mathcal{Y}^k$, there is no circuit of size at most t that outputs y_i on input $i \in [k]$. Therefore, let $\mathcal{W}$ be the set of circuits of size at most t, and let $R(\mathbf{y}, C)$ output 1 if and only if $C(i) = y_i$ for each $i \in [k]$. Then R is computationally (in fact statistically) intractable. However, it is trivial to find an $\mathbf{x}, w$ that satisfy $R(H(\mathbf{x}), w) = 1$: set $\mathbf{x} = [k]$ and $w = H$. Therefore, output intractability is violated. We obtain the following:

Theorem 7. *For any family $H : [M] \to \mathcal{Y}$ of hash functions, let t be the description size of H. Then H cannot be output intractable for any $k \geq t/\log |\mathcal{Y}|$.*

In the following, we show that it is nonetheless possible to obtain output intractability for any given constant k. Our functions will be described by strings of length $k(\log|\mathcal{Y}| + \texttt{poly}(\log M))$, which in the case $|\mathcal{Y}| \gg M$ gives a near-optimal relationship between k and t.

5.1 The Construction

Construction 4. *Let $[M]$ be the desired domain of H, and $\mathcal{Y}$ the desired range. To generate H, to the following:*

- *Let $f \leftarrow \texttt{ELF.Gen}(M, M)$ to get an injective-mode f, with codomain $\mathcal{Z}$.*
- *Let g be a k-wise independent and uniform function from $\mathcal{Z}$ to $\mathcal{Y}$.*
- *Output $H = g \circ f$.*

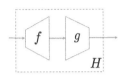

Fig. 3. The function $H = g \circ f$.

Theorem 8. *If* ELF *is a secure ELF with an efficiently enumerable image, then for any constant k the hash function H in Construction 4 is k-ary output intractable. If* ELF *is public coin, then so is H.*

Proof. Suppose toward contradiction that there is an intractable $k + 1$-ary relation R and an adversary $\mathcal{A}$ that on input H finds a set of distinct inputs $\mathbf{x}$ and a value $w \in \mathcal{W}$ such that $R(H(\mathbf{x}), w) = 1$ with non-negligible probability ϵ. Let δ be an inverse polynomial such that $\epsilon \geq \delta$ infinitely often. We will switch to a lossy mode for f so that (1) f has polynomial image size, and (2) no adversary running in time t (for a t to be chosen later) can distinguish the lossy mode from injective, except with probability $\delta/3$. By choosing t to be larger than the running time of $\mathcal{A}$, we have that $\mathcal{A}$ still outputs $\mathbf{x}$ of distinct elements, and a string w, such that $R(H(\mathbf{x}), w) = 1$ with probability $\epsilon - \delta/3$.

We first argue that each of the elements of $f(\mathbf{x})$ are distinct except with probability $\delta/3$. Since this was true in the injective case (since $\mathbf{x}$ is distinct), if this is not true in the lossy case, then the injective and lossy modes could be easily distinguished by an adversary taking slightly more time than $\mathcal{A}$. Let t be this time, so that this distinguisher is impossible. Thus, the adversary succeeds *and* the elements of $f(\mathbf{x})$ are distinct with probability at least $\epsilon - 2\delta/3$. This probability is larger than $\delta/3$ infinitely often, and is therefore non-negligible. Let S be the polynomial-sized set of image points of f. Then in other words, the adversary comes up with an ordered set $\mathbf{z}$ of distinct elements in S, and a string w, such that $R(g(\mathbf{z}), w) = 1$.

Now, note that, for any ordered set $\mathbf{z}$ of k distinct inputs, $g(\mathbf{z})$ is distributed uniformly at random, by the k-wise independence of g. Moreover, it is straightforward, given $\mathbf{z}$ and a vector $\mathbf{y} \in \mathcal{Y}^k$, to sample a random g conditioned on $g(\mathbf{z}) = \mathbf{y}$. Sampling random $\mathbf{y}$, and then g in this way, gives a correctly distributed g.

We now describe an algorithm $\mathcal{B}$ that breaks the intractability of R. $\mathcal{B}$, on input $\mathbf{y} \in \mathcal{Y}^k$, chooses lossy f as above, and then selects k distinct (potential) image points from the image sampling procedure. Let $\mathbf{z}$ be the ordered list of points. Next, it chooses a random g such that $g(\mathbf{z}) = \mathbf{y}$. Finally, it runs $\mathcal{A}$ on the hash function $H = g \circ f$. When $\mathcal{A}$ outputs $\mathbf{x}, w$, if $f(\mathbf{x}) = \mathbf{z}$ (equivalently, $H(\mathbf{x}) = \mathbf{y}$), $\mathcal{B}$ outputs w; otherwise it aborts.

Since $\mathbf{y}$ is hidden from $\mathcal{A}$'s view, g is distributed randomly according to the k-wise independent distribution. Therefore, $\mathcal{A}$ will output a valid w with probability at least $\epsilon - 2\delta/3$. If $\mathcal{B}$'s guess for $\mathbf{z}$ was correct, then w will break the intractability of R on $\mathbf{y}$. Since $\mathbf{z}$ is independent of $\mathcal{A}$'s view, the probability of a good guess is at least $1/s^k$, where $1/s$ is the inverse polynomial lower bound on the probability any image point is selected. Therefore, $\mathcal{B}$ breaks the intractability of R with probability $(\epsilon - 2\delta/3)/s^k$, which is larger than $\delta/3s^k$ infinitely often, and is therefore non-negligible. □

6   Leakage-Resilient PRGs, AIPO and Poly-Many Hardcore Bits

In this section, we use ELFs to give arbitrarily-many hardcore bits for any one-way function, and for constructing point function obfuscation secure in the presence of auxiliary information. Both of these can be seen as special cases of a very strong security requirement for pseudorandom generators.

Definition 12. *A distribution $\mathcal{D}$ on pairs $(x, z) \in \mathcal{X} \times \mathcal{Z}$ is computationally unpredictable if no efficient adversary can guess x given z.*

Definition 13. *A family of pseudorandom generators $H : \mathcal{X} \to \mathcal{Y}$ secure for computationally unpredictable seeds if, for any computationally unpredictable distribution on $(\mathcal{X}, \mathcal{Z})$, no efficient adversary can distinguish $(H, z, H(x))$ from (H, z, S) where $(x, z) \leftarrow \mathcal{D}$ and $S \leftarrow U_{\mathcal{Y}}$.*

Basically, this requirement states that H is a secure pseudorandom generator for arbitrary distributions on the seed, and even remains secure in the presence of arbitrary leakage about the seed, so long as the seed remains *computationally* unpredictable. The only restriction is that the distribution on the seed and the leakage must be chosen independently of H. However, in the absence of other restrictions, this independence between the source $\mathcal{D}$ and function H can easily be seen to be necessary: if z contained a few bits of $H(x)$, then it is trivial to distinguish $H(x)$ from random.

6.1   The Construction

The intuition behind our construction is the following. The usual way of extracting pseudorandomness from computationally unpredictable source is to output a hardcore bit of the source, say using Goldreich-Levin [23]. While this can be used to generate a logarithmic number of pseudorandom bits, security is lost once a super-logarithmic number of hardcore bits have been generated in this way.

In order to get around this logarithmic barrier, we actually compute a *polynomial* number of Goldreich-Levin bits. Of course, we cannot output these in the clear or else the seed can be easily computed by linear algebra. Instead, we scramble the hardcore bits using a sequence of ELFs. We can argue that each of the (scrambled) hardcore bits really is "as good as" random, in the sense that we can replace each bit with a truly random bit before scrambling without detection. To do so, we use the lossiness of the ELFs to argue that, when the ith hardcore bit is incorporated into the scramble, enough information is lost about the previous bits that the ith bit actually still is hardcore. By iterating this for each bit, we replace each one with random. We now give the details.

Construction 5. *Let q be the input length and m be the output length. Let λ be a security parameter. We will consider inputs x as q-dimensional vectors $\mathbf{x} \in \mathbb{F}_2^q$. Let* ELF *be an ELF. Let $M = 2^{m+\lambda+1}$, and let n be the bit-length of the ELF on input $m + 1$. Set $N = 2^n$. Let ℓ be some polynomial in m, λ to be determined later. First, we will construct a function H' as follows.*

Choose random $f_1, \ldots, f_\ell \leftarrow$ ELF.Gen(M, M) where $f_i : [M] \to [N]$, and let $h_1, \ldots, h_{\ell-1} : [N] \to [M/2] = [2^{m+\lambda}]$ and $h_\ell : [N] \to [2^m]$ be pairwise independent and uniform functions. Define $\mathbf{f} = \{f_1, \ldots, f_\ell\}$ and $\mathbf{h} = \{h_1, \ldots, h_\ell\}$. Define $H_i' : \{0, 1\}^i \to [M/2]$ (and $H_\ell' : \{0, 1\}^\ell \to [2^m]$) as follows:

- *$H_0'() = y_1 = 1 \in [2^{m+\lambda}]$*
- *$H_i'(\mathbf{b}_{[1,i-1]}, b_i) = y_{i+1} = h_i(z_i)$ where $y_i \leftarrow H_{i-1}'(\mathbf{b}_{[1,i-1]}), z_i \leftarrow f_i(y_i \| b_i)$.*

Then we set $H' = H_\ell'$. Then to define H, choose a random matrix $\mathbf{R} \in \mathbb{F}_2^{\ell \times q}$. The description of H consists of $\mathbf{f}, \mathbf{h}, \mathbf{R}$. Then set $H(x) = H'(\mathbf{R} \cdot \mathbf{x})$. A diagram of H is given in Fig. 4.

We now prove several important facts about H and H':

Claim. If $\ell \geq m + \lambda$, and if $\mathbf{b}$ is drawn uniformly at random, then $(H', H'(\mathbf{b}))$ is statistically close to (H', R) where R is uniformly random in $[2^m]$.

Proof. We will prove the case $\ell = m + \lambda$, the case of larger ℓ being similar. We will consider $f_1, \ldots, f_\ell$ as being fixed injective functions; since the f_i are injective with overwhelming probability, the claim follows from this case. This means that the composition $h_i \circ f_i$ is pairwise independent, for all i.

Let $d_i(h_1, \ldots, h_i)$ be the collision probability of y_{i+1} when $b_i \ldots, b_i$ are random bits, for fixed $h_1, \ldots, h_i$. Let d_i be the expectation (over $h_1, \ldots, h_i$) of this value. There are two possibilities for a collision at y_{i+1}:

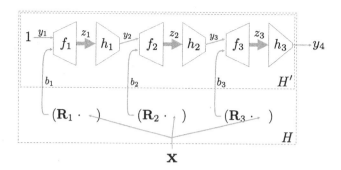

Fig. 4. An example instantiation for $\ell = 3$. Notice that each iteration is identical, except for the final iteration, where h_ℓ has a smaller output.

- There is a collision at y_i and b_i. This happens with half the probability that there is a collision at y_i.
- There is not a collision at y_i and b_i, but $h_i \circ f_i$ maps the two points to the same y_{i+1}. Conditioned on there being no collision at (y_i, b_i), this occurs with probability $\frac{1}{2^{m+\lambda}}$ for $i < \ell$, and $\frac{1}{2^m}$ for $i = \ell$.

Then we have that $d_0 = 1$ and it is straightforward to show the following recurrence for $i < \ell$: $d_i = d_{i-1}\left(\frac{1}{2} - \frac{1}{2\times 2^{m+\lambda}}\right) + \frac{1}{2^{m+\lambda}}$. This recurrence has the form $d_i = a + b d_{i-1}$, which is solved by $d_i = b^i + a\frac{b^i - 1}{b - 1}$. Therefore, with some manipulation we have that $d_i = \frac{2}{2^{m+\lambda}+1} + \frac{(2^{m+\lambda}-1)^{i+1}}{2^{(m+\lambda)i}(2^{m+\lambda}+1)}$. Now, for $i = m+\lambda-1$, this becomes $d_{m+\lambda-1} = \frac{2\left(1+\left(1-2^{-m-\lambda}\right)^{m+\lambda}\right)}{2^{m+\lambda}+1} \leq \frac{4}{2^{m+\lambda}}$. Next, it is straightforward to adapt the above argument to show that $d_{m+\lambda} = d_{m+\lambda-1}\left(\frac{1}{2} - \frac{1}{2\times 2^m}\right) + \frac{1}{2^m} \leq \frac{1}{2^m} + \frac{2}{2^{m+\lambda}}$. Now, the Rényi entropy of $y_{m+\lambda+1}$ is exactly the collision probability, scaled up by a factor of the 2^m. Therefore, the expected Rényi entropy of $y_{m+\lambda+1}$ is at most $1 + \frac{2}{2^\lambda}$. Finally, we relate the Rényi entropy to the statistical distance from uniform using Lemma 1:

$$\mathop{\mathbb{E}}_{h_1,\dots,h_\ell} \Delta(y_{\ell+1}, R) \leq \mathbb{E}\frac{1}{2}\sqrt{RE(y_{\ell+1}) - 1} \leq \frac{1}{2}\sqrt{\mathbb{E}[RE(y_{\ell+1})] - 1} \leq 2^{-(\lambda+1)/2}$$

The statistical distance between $(H', H'(\mathbf{b}))$ and (H', R) is exactly this quantity, which is negligible in λ. □

We will thus set $\ell = m + \lambda$ in our construction of H'. Claim 6.1 will be crucial for our security proof for H. We also show our H is injective (with overwhelming probability) exactly when a truly random function with the same domain and co-domain is injective (with overwhelming probability).

Claim. If $2^{-(m-2q)}$ is negligible (in q), and $\ell \geq m$, then with overwhelming probability H is injective.

Proof. First, note that with overwhelming probability by our choice of $\ell \geq m \geq 2q$, $\mathbf{R}$ is full rank. Next, let $\mathcal{Y}_i$ be the set of possible y_i values as we vary $\mathbf{x}$, and $\mathcal{Z}_i$ be the set of possible z_i values. By the injectivity of f_i, we have that $|\mathcal{Z}_i| \geq |\mathcal{Y}_i|$. Moreover, since h_i is pairwise independent and uniform, with overwhelming probability h_i is injective on $\mathcal{Z}_i$ since $|\mathcal{Z}_i| \leq 2^q$ but the co-domain of h_i has size at least $2^m \gg (2^q)^2$. Therefore $|\mathcal{Y}_{i+1}| = |\mathcal{Z}_i| \geq |\mathcal{Y}_i|$. This means that as we increase i, the image size never decreases (with overwhelming probability).

Now pick q linearly independent rows of $\mathbf{R}$. We will assume that the q rows constitute the first q rows of $\mathbf{R}$; the more general case is handled analogously. By performing an appropriate invertible transformation on the domain, we can assume that these q rows form the identity matrix. Therefore, we can take $b_i = x_i$ for $i \in [q]$. Next, observe that y_i for $i \in [q]$ only depends on the first $i-1$ bits of $\mathbf{x}$. Thus the set of possible pairs $(y_i, b_i) = (y_i, x_i)$ is exactly $\mathcal{Y}_i \times \{0,1\}$, which has size $2|\mathcal{Y}_i|$. By the injectivity of f_i, $|\mathcal{Z}_i| = 2|\mathcal{Y}_i|$. Since $|\mathcal{Y}_{i+1}| = |\mathcal{Z}_i| = 2|\mathcal{Y}_i|$, we have that the image size exactly doubles in each iteration for $i \in [q]$. Once we get to $i = q$, the image size is 2^q, and the remaining iterations do not introduce any collisions. Thus the image size of H is 2^q, meaning H is injective. $\square$

Theorem 9. *If* ELF *is a secure ELF, then H in Construction 5 is a pseudorandom generator secure for computationally unpredictable seeds. If* ELF *is public coin, then so is H.*

Proof. Recall that $H(\mathbf{x}) = H'(\mathbf{R} \cdot \mathbf{x})$, and that $H'(\mathbf{b})$ is (with overwhelming probability over the choice of H') statistically close to random when $\mathbf{b}$ is random. Therefore, it suffices to show that the following distributions are indistinguishable: $(\mathbf{f}, \mathbf{h}, \mathbf{R}, z, H'(\mathbf{R} \cdot \mathbf{x}))$ and $(\mathbf{f}, \mathbf{h}, \mathbf{R}, z, H'(\mathbf{b}))$ for a uniformly random $\mathbf{b}$.

Suppose an adversary $\mathcal{A}$ has non-negligible advantage ϵ in distinguishing the two distributions. Define $\mathbf{b}^{(i)}$ so that the first i bits of $\mathbf{b}^{(i)}$ are equal to the first i bits of $\mathbf{R} \cdot \mathbf{x}$, and the remaining $\ell - i$ bits are chosen uniformly at random independently of $\mathbf{x}$. Define **Hybrid** i to be the case where $\mathcal{A}$ is given the distribution $(\mathbf{f}, \mathbf{h}, \mathbf{R}, z, H'(\mathbf{b}^{(i)}))$.

Then $\mathcal{A}$ distinguishes **Hybrid 0** from **Hybrid** ℓ with probability ϵ. Thus there is an index $i \in [\ell]$ such that the adversary distinguishes **Hybrid** $i-1$ from **Hybird** i with probability at least ϵ/ℓ. Next, observe that since bits $i+1$ through t are random in either case, they can be simulated independently of the challenge. Moreover, $H'(\mathbf{b})$ can be computed given $H'_{i-1}(\mathbf{b}_{[i-1]})$, b_i (be it random or equal to $\mathbf{R}_i, \mathbf{x}$), and the random $b_{i+1}, \ldots, b_\ell$. Thus, we can construct an adversary $\mathcal{A}'$ that distinguishes $\mathbf{R}_i \cdot \mathbf{x}$ from a random b_i — given $(\mathbf{f}, \mathbf{h}, \mathbf{R}_{[i-1]}, z, H'_{i-1}(\mathbf{R}_{[i-1]} \cdot \mathbf{x}), \mathbf{R}_i)$ — with advantage ϵ/ℓ, where $\mathbf{R}_{[i-1]}$ consists of the first $i-1$ rows of $\mathbf{R}$, $\mathbf{R}_i$ is the ith row of $\mathbf{R}$, and b_i is a random bit.

Next, since $\epsilon/3\ell$ is non-negligible, there is an inverse polynomial δ such that $\epsilon/3\ell \geq \delta$ infinitely often. Then, there is a polynomial r such $\mathcal{A}'$ cannot distinguish f_i generated as ELF.Gen(M, r) from the honest f_i generated from ELF.Gen(M, M), except with probability at most δ. This means, if we generate $f_i \leftarrow$ ELF.Gen(M, r), we have that $\mathcal{A}'$ still distinguishes $\mathbf{R}_i \cdot \mathbf{x}$ from a random b_i — given $(\mathbf{f}, \mathbf{h}, \mathbf{R}_{[i-1]}, z, H'_{i-1}(\mathbf{R}_{[i-1]} \cdot \mathbf{x}), \mathbf{R}_i)$ — with advantage $\epsilon' = \epsilon/\ell - 2\delta$.

Put another way, given $(\mathbf{f}, \mathbf{h}, \mathbf{R}_{[i-1]}, z, H'_{i-1}(\mathbf{R}_{[i-1]} \cdot \mathbf{x}), \mathbf{R}_i)$, $\mathcal{A}'$ is able to compute $\mathbf{R}_i \cdot \mathbf{x}$ with probability $\frac{1}{2} + \epsilon'$. Note that $\epsilon' \geq \delta$ infinitely often, and is therefore non-negligible.

Now fix $\mathbf{f}, \mathbf{h}, \mathbf{R}_{[i-1]}$, which fixes H'_{i-1}. Let $y_i = H'_{i-1}(\mathbf{R}_{[i-1]} \cdot \mathbf{x})$. Notice that since $\mathbf{f}, \mathbf{h}$ are fixed, there are at most r possible values for y_i, and recall that r is a polynomial. We now make the following claim:

Claim. Let $\mathcal{D}$ be a computationally unpredictable distribution on $\mathcal{X} \times \mathcal{Z}$. Suppose $T : \mathcal{X} \to \mathcal{R}$ is drawn from a family $\mathcal{T}$ of efficient functions where the size of the image of T is polynomial. Then the following distribution is also computationally unpredictable: $(x, (T, z, T(x)))$ where $T \leftarrow \mathcal{T}$, $(x, z) \leftarrow \mathcal{D}$.

Proof. Suppose we have an efficient adversary $\mathcal{B}$ that predicts x with non-negligible probability γ given $T, z, T(x)$, and suppose T has polynomial image size r. We then construct a new adversary $\mathcal{C}$ that, given x, samples a random T, samples $(x', z') \leftarrow \mathcal{D}$, and sets $a = T(x')$. It then runs $\mathcal{B}(T, z, a)$ to get a string x'', which it outputs. Notice that a is sampled from the same distribution as $T(x)$, so with probability at least $1/r$, $a = T(x)$. In this case, $x'' = x$ with probability γ. Therefore, $\mathcal{C}$ outputs x with probability γ/r, which is non-negligible. ☐

Using Claim 6.1 with $T = H'_{i-1}(\mathbf{R}_{[i-1]} \cdot \mathbf{x})$, we see that $(x, (\mathbf{f}, \mathbf{h}, \mathbf{R}_{[i-1]}, z, H'_{i-1}(\mathbf{R}_{[i-1]} \cdot \mathbf{x})))$ is computationally unpredictable. Moreover, $\mathbf{R}_i \cdot \mathbf{x}$ is a Goldreich and Levin [23] hardcore bit for any computationally unpredictable source. Hence, no efficient adversary can predict $\mathbf{R}_x \cdot \mathbf{x}$ given $(\mathbf{f}, \mathbf{h}, \mathbf{R}_{[i-1]}, z, H'_{i-1}(\mathbf{R}_{[i-1]} \cdot \mathbf{x}), \mathbf{R}_i)$. This contradicts the existence of $\mathcal{A}'$, proving the theorem. ☐

6.2   Applications

Polynomially-many hardcore bits for any one-way function. We see that H immediately gives us a hardcore function of arbitrary stretch for any computationally unpredictable distribution. This includes any one-way function. To the best of our knowledge, this is the first hardcore function of arbitrary stretch based on simple assumptions that applies to general computationally unpredictable sources. In the special case of one-way functions, the only prior constructions are due to Bellare et al. [6] using differing inputs obfuscation (diO), and of Zhandry [39] using extractable witness PRFs. Our construction offers an entirely different approach to constructing hardcore functions with arbitrary stretch, and is based on a very simple primitive.

Strong injective one-way functions. Bitansky and Paneth [9] conjecture the existence of a very strong one-way permutation family. We demonstrate that our function H meets this notion of security. Unfortunately, however, it is only injective, not a permutation.

Definition 14. *A [9] permutation is a family of functions H such that for any computationally unpredictable distribution $\mathcal{D}$, the following two distributions are also unpredictable: $(x,\ (z, H, H(x))\ )$ and $(H(x),\ (z, H)\ )$ where $(x, z) \leftarrow \mathcal{D}$.*

The first property is a generalization of a strong uninvertability assumption of Wee [37]. The second guarantees that if x is unpredictable, then so is $H(x)$. In the full version [40], we show that H satisfies this definition:

Theorem 10. *H constructed above using a secure ELF, when set to be injective as in Claim 6.1, is a [9]-injective one-way function.*

The main application of Bitansky and Paneths [9] assumption is to build auxiliary input point function obfuscation (AIPO). Since H is not a permutation, it cannot be immediately plugged into their construction. Yet, next, we show that going through their construction is unnecessary in our case: we show that our function H gives an AIPO "out of the box" with no additional overhead.

Point function obfuscation with auxiliary input (AIPO). We now show how to achieve full AIPO using just the assumption of ELFs.

Definition 15. *A auxiliary input point obfuscator (AIPO) is an efficient probabilistic algorithm $\mathcal{O}$ that saitsfies the almost perfect correctness requirement of Definition 10, as well as the following secrecy requirement: for any unpredictable distribution $\mathcal{D}$ over pairs $(x, z) \in \mathcal{X} \times \mathcal{Z}$, $(\mathcal{O}(I_x), z)$ and $(\mathcal{O}(I_{x'}), z)$ are computationally indistinguishable, where $(x, z) \leftarrow \mathcal{D}$ and $x' \leftarrow \mathcal{X}$.*

As in Sect. 4, an expanding ideal hash function (random oracle) H gives a very natural AIPO: the obfuscation of a point function I_x is simply $S = H(x)$. Injectivity of H gives (almost perfect) correctness. Moreover, security is easily proved in the random oracle model.

We now show that by choosing H to be as in the construction above, the same is true. In particular, by Claim 6.1, H is injective in the same regime of input/output sizes as a random oracle. For security, we have the following:

Theorem 11. *The obfuscation construction described above is a secure AIPO assuming H is constructed as in Construction 5 using a secure ELF.*

Proof. Note that since H is expanding, if we choose S at random from $[2^m]$, then with overwhelming probability there are no inputs $\mathbf{x}$ that map to S. Therefore, the obfuscated program corresponding to S is just the all-zeros function.

Let $\mathcal{D}$ be any computationally unpredictable source. We thus need to show that the following two distributions are indistinguishable: $(H, z, H(\mathbf{x}))$ and (H, z, S) (where $(\mathbf{x}, z) \leftarrow \mathcal{D}$). This follows immediately from Theorem 9. $\square$

Public key encryption from trapdoor permutations. In the full version [40], we show that our hardcore function can be used in a simple hybrid encryption scheme of Bellare and Rogaway [5].

6.3 Difficulty of Realizing Applications

Since AIPO implies PO, AIPO implies that NP with a super-logarithmic amount of non-determinism cannot be solved in polynomial time. Hence, this separation is inherent to the AIPO application. As an immediately corollary, we also have that our pseudorandom generator definition also implies such a separation. Since our pseudorandom generator definition is essentially equivalent to obtaining hardcore functions of arbitrary span for any unpredictable source, we also see that such a separation is inherent to such hardcore functions. In contrast, this separation does not extend to the special case of hardcore functions for any one-way function. It is consistent with our current knowledge that NP with, say, $\log^2$ bits of non-determinism *is* solvable in polynomial time, and yet there are still hardcore functions of arbitrary stretch for any one-way function. However, in the full version [40], we still demonstrate some barriers to realizing this special case from polynomially-hard primitives.

References

1. Akavia, A., Goldwasser, S., Vaikuntanathan, V.: Simultaneous hardcore bits and cryptography against memory attacks. In: Reingold, O. (ed.) TCC 2009. LNCS, vol. 5444, pp. 474–495. Springer, Heidelberg (2009)
2. Ananth, P., Boneh, D., Garg, S., Sahai, A., Zhandry, M.: Differing-inputs obfuscation and applications. Cryptology ePrint Archive, Report 2013/689 (2013). http://eprint.iacr.org/2013/689
3. Apon, D., Huang, Y., Katz, J., Malozemoff, A.J.: Implementing cryptographic program obfuscation. Cryptology ePrint Archive, Report 2014/779 (2014). http://eprint.iacr.org/2014/779
4. Bellare, M., Hoang, V.T., Keelveedhi, S.: Instantiating random oracles via UCEs. In: Canetti, R., Garay, J.A. (eds.) CRYPTO 2013, Part II. LNCS, vol. 8043, pp. 398–415. Springer, Heidelberg (2013)
5. Bellare, M., Rogaway, P.: Random oracles are practical: a paradigm for designing efficient protocols. In: Ashby, V. (ed.) ACM CCS 1993, pp. 62–73. ACM Press, November 1993
6. Bellare, M., Stepanovs, I., Tessaro, S.: Poly-many hardcore bits for any one-way function and a framework for differing-inputs obfuscation. In: Sarkar, P., Iwata, T. (eds.) ASIACRYPT 2014, Part II. LNCS, vol. 8874, pp. 102–121. Springer, Heidelberg (2014)
7. Bellare, M., Stepanovs, I., Tessaro, S.: Contention in cryptoland: obfuscation, leakage and UCE. In: Kushilevitz, E., et al. (eds.) TCC 2016-A. LNCS, vol. 9563, pp. 542–564. Springer, Heidelberg (2016). doi:10.1007/978-3-662-49099-0_20
8. Bitansky, N., Canetti, R., Paneth, O., Rosen, A.: On the existence of extractable one-way functions. In: Shmoys, D.B. (ed.) 46th ACM STOC, pp. 505–514. ACM Press, May/June 2014
9. Bitansky, N., Paneth, O.: Point obfuscation and 3-round zero-knowledge. In: Cramer, R. (ed.) TCC 2012. LNCS, vol. 7194, pp. 190–208. Springer, Heidelberg (2012)
10. Böhl, F., Hofheinz, D., Jager, T., Koch, J., Seo, J.H., Striecks, C.: Practical signatures from standard assumptions. In: Johansson, T., Nguyen, P.Q. (eds.) EUROCRYPT 2013. LNCS, vol. 7881, pp. 461–485. Springer, Heidelberg (2013)

11. Boneh, D., Zhandry, M.: Secure signatures and chosen ciphertext security in a quantum computing world. In: Canetti, R., Garay, J.A. (eds.) CRYPTO 2013, Part II. LNCS, vol. 8043, pp. 361–379. Springer, Heidelberg (2013)

12. Boyle, E., Chung, K.-M., Pass, R.: On extractability obfuscation. In: Lindell, Y. (ed.) TCC 2014. LNCS, vol. 8349, pp. 52–73. Springer, Heidelberg (2014)

13. Brzuska, C., Farshim, P., Mittelbach, A.: Indistinguishability obfuscation and UCEs: the case of computationally unpredictable sources. In: Garay, J.A., Gennaro, R. (eds.) CRYPTO 2014, Part I. LNCS, vol. 8616, pp. 188–205. Springer, Heidelberg (2014)

14. Canetti, R.: Towards realizing random oracles: hash functions that hide all partial information. In: Kaliski Jr., B.S. (ed.) CRYPTO 1997. LNCS, vol. 1294, pp. 455–469. Springer, Heidelberg (1997)

15. Canetti, R., Chen, Y., Reyzin, L.: On the correlation intractability of obfuscated pseudorandom functions. In: Kushilevitz, E., et al. (eds.) TCC 2016-A. LNCS, vol. 9562, pp. 389–415. Springer, Heidelberg (2016). doi:10.1007/978-3-662-49096-9_17

16. Canetti, R., Goldreich, O., Halevi, S.: The random oracle methodology, revisited (preliminary version). In: 30th ACM STOC, pp. 209–218. ACM Press, May 1998

17. Dodis, Y., Smith, A.: Correcting errors without leaking partial information. In: Gabow, H.N., Fagin, R. (eds.) 37th ACM STOC, pp. 654–663. ACM Press, May 2005

18. Döttling, N., Schröder, D.: Efficient pseudorandom functions via on-the-fly adaptation. In: Gennaro, R., Robshaw, M.J.B. (eds.) CRYPTO 2015, Part I. LNCS, vol. 9215, pp. 329–350. Springer, Heidelberg (2015)

19. Dziembowski, S., Pietrzak, K.: Leakage-resilient cryptography. In: 49th FOCS, pp. 293–302. IEEE Computer Society Press, October 2008

20. Freeman, D.M., Goldreich, O., Kiltz, E., Rosen, A., Segev, G.: More constructions of lossy and correlation-secure trapdoor functions. In: Nguyen, P.Q., Pointcheval, D. (eds.) PKC 2010. LNCS, vol. 6056, pp. 279–295. Springer, Heidelberg (2010)

21. Garg, S., Gentry, C., Halevi, S., Wichs, D.: On the implausibility of differing-inputs obfuscation and extractable witness encryption with auxiliary input. In: Garay, J.A., Gennaro, R. (eds.) CRYPTO 2014, Part I. LNCS, vol. 8616, pp. 518–535. Springer, Heidelberg (2014)

22. Gentry, C., Wichs, D.: Separating succinct non-interactive arguments from all falsifiable assumptions. In: Fortnow, L., Vadhan, S.P. (eds.) 43rd ACM STOC, pp. 99–108. ACM Press, June 2011

23. Goldreich, O., Levin, L.A.: A hard-core predicate for all one-way functions. In: 21st ACM STOC, pp. 25–32. ACM Press, May 1989

24. Goldwasser, S., Tauman Kalai, Y.: Cryptographic assumptions: a position paper. In: Kushilevitz, E., et al. (eds.) TCC 2016-A. LNCS, vol. 9562, pp. 505–522. Springer, Heidelberg (2016). doi:10.1007/978-3-662-49096-9_21

25. Hofheinz, D., Jager, T., Khurana, D., Sahai, A., Waters, B., Zhandry, M.: How to generate and use universal samplers. Cryptology ePrint Archive, Report 2014/507 (2014). http://eprint.iacr.org/2014/507

26. Hohenberger, S., Sahai, A., Waters, B.: Replacing a random oracle: full domain hash from indistinguishability obfuscation. In: Nguyen, P.Q., Oswald, E. (eds.) EUROCRYPT 2014. LNCS, vol. 8441, pp. 201–220. Springer, Heidelberg (2014)

27. Krawczyk, H., Rabin, T.: Chameleon signatures. In: NDSS 2000. The Internet Society, February 2000

28. Naor, M.: On cryptographic assumptions and challenges. In: Boneh, D. (ed.) CRYPTO 2003. LNCS, vol. 2729, pp. 96–109. Springer, Heidelberg (2003)

29. Patel, S., Sundaram, G.S.: An efficient discrete log pseudo random generator. In: Krawczyk, H. (ed.) CRYPTO 1998. LNCS, vol. 1462, p. 304. Springer, Heidelberg (1998)

30. Peikert, C., Waters, B.: Lossy trapdoor functions and their applications. In: Ladner, R.E., Dwork, C. (eds.) 40th ACM STOC, pp. 187–196. ACM Press, May 2008

31. Pietrzak, K., Rosen, A., Segev, G.: Lossy functions do not amplify well. In: Cramer, R. (ed.) TCC 2012. LNCS, vol. 7194, pp. 458–475. Springer, Heidelberg (2012)

32. Rao, V.: Adaptive multiparty non-interactive key exchange without setup in the standard model. Cryptology ePrint Archive, Report 2014/910 (2014). http://eprint.iacr.org/2014/910

33. Rosen, A., Segev, G.: Chosen-ciphertext security via correlated products. In: Reingold, O. (ed.) TCC 2009. LNCS, vol. 5444, pp. 419–436. Springer, Heidelberg (2009)

34. Schrift, A.W., Shamir, A.: The discrete log is very discreet. In: 22nd ACM STOC, pp. 405–415. ACM Press, May 1990

35. Simon, D.R.: Findings collisions on a one-way street: can secure hash functions be based on general assumptions? In: Nyberg, K. (ed.) EUROCRYPT 1998. LNCS, vol. 1403, pp. 334–345. Springer, Heidelberg (1998)

36. Villar, J.L.: Optimal reductions of some decisional problems to the rank problem. In: Wang, X., Sako, K. (eds.) ASIACRYPT 2012. LNCS, vol. 7658, pp. 80–97. Springer, Heidelberg (2012)

37. Wee, H.: On obfuscating point functions. In: Gabow, H.N., Fagin, R. (eds.) 37th ACM STOC, pp. 523–532. ACM Press, May 2005

38. Zhandry, M.: How to construct quantum random functions. In: 53rd FOCS, pp. 679–687. IEEE Computer Society Press, October 2012

39. Zhandry, M.: How to avoid obfuscation using witness PRFs. In: Kushilevitz, E., et al. (eds.) TCC 2016-A. LNCS, vol. 9563, pp. 421–448. Springer, Heidelberg (2016). doi:10.1007/978-3-662-49099-0_16

40. Zhandry, M.: The magic of ELFs. In: Proceedings of CRYPTO (2016). Full version available at the Cryptology ePrint Archives http://eprint.iacr.org/2016/114

Breaking the Circuit Size Barrier for Secure Computation Under DDH

Elette Boyle[1], Niv Gilboa[2], and Yuval Ishai[3(✉)]

[1] IDC Herzliya, Herzliya, Israel
elette.boyle@idc.ac.il
[2] Ben Gurion University, Beersheba, Israel
gilboan@bgu.ac.il
[3] Technion and UCLA, Haifa, Israel
yuvali@cs.technion.ac.il

Abstract. Under the Decisional Diffie-Hellman (DDH) assumption, we present a 2-out-of-2 secret sharing scheme that supports a compact evaluation of branching programs on the shares. More concretely, there is an evaluation algorithm Eval with a single bit of output, such that if an input $w \in \{0,1\}^n$ is shared into (w^0, w^1), then for any deterministic branching program P of size S we have that $\mathsf{Eval}(P, w^0) \oplus \mathsf{Eval}(P, w^1) = P(w)$ except with at most δ failure probability. The running time of the sharing algorithm is polynomial in n and the security parameter λ, and that of Eval is polynomial in S, λ, and $1/\delta$. This applies as a special case to boolean formulas of size S or boolean circuits of depth $\log S$. We also present a public-key variant that enables homomorphic computation on inputs contributed by multiple clients.

The above result implies the following DDH-based applications:

- A secure 2-party computation protocol for evaluating any branching program or formula of size S, where the communication complexity is linear in the input size and only the running time grows with S.
- A secure 2-party computation protocol for evaluating layered boolean circuits of size S with communication complexity $O(S/\log S)$.
- A 2-party *function secret sharing* scheme, as defined by Boyle et al. (Eurocrypt 2015), for general branching programs (with inverse polynomial error probability).
- A 1-round 2-server *private information retrieval* scheme supporting general searches expressed by branching programs.

Prior to our work, similar results could only be achieved using fully homomorphic encryption. We hope that our approach will lead to more practical alternatives to known fully homomorphic encryption schemes in the context of low-communication secure computation.

1 Introduction

In this paper we introduce a simple new technique for low-communication secure computation that can be based on the Decisional Diffie-Hellman (DDH) assumption and avoids the use of fully homomorphic encryption. We start with some relevant background.

© International Association for Cryptologic Research 2016
M. Robshaw and J. Katz (Eds.): CRYPTO 2016, Part I, LNCS 9814, pp. 509–539, 2016.
DOI: 10.1007/978-3-662-53018-4_19

Since the seminal feasibility results of the 1980s [3,8,23,38], a major challenge in the area of secure computation has been to break the "circuit size barrier." This barrier refers to the fact that all classical techniques for secure computation required a larger amount of communication than the size of a boolean circuit representing the function to be computed, even when the circuit is much bigger than the inputs. The circuit size barrier applied not only to general circuits, but also to useful restricted classes of circuits such as boolean formulas (namely, circuits with fan-out 1) or branching programs (a stronger computational model capturing non-uniform logarithmic-space computations). Moreover, the same barrier applied also to secure computation protocols that can rely on a trusted source of correlated randomness, provided that this correlated randomness needs to be *reusable*.

The first significant progress has been made in the context of *private information retrieval* (PIR), where it was shown that for the bit-selection function $f(x, i) = x_i$ it is possible to break the circuit size barrier either in the multi-server model [9,11], where a client holds i and two or more servers hold x, or in the two-party model [27] under standard cryptographic assumptions. However, progress on extending this to other useful computations has been slow, with several partial results [4,10,17,25,30] that do not even cover very simple types of circuits such as general DNF or CNF formulas, let alone more expressive ones such as general formulas or branching programs.[1]

All this has changed with Gentry's breakthrough on fully homomorphic encryption (FHE) [19,33]. FHE enables local computations on encrypted inputs, thus providing a general-purpose solution to the problem of low-communication secure computation. On the down side, even the best known implementations of FHE [24] are still quite slow. Moreover, while there has been significant progress on basing the feasibility of FHE on more standard or different assumptions [7,21,36], the set of cryptographic assumptions on which FHE can be based is still very narrow, and in particular it does not include any of the "traditional" assumptions that were known in the 20th century.

1.1   Our Contribution

Our new approach was inspired by the recent work on *function secret sharing* (FSS) [6]. A (2-party) FSS scheme for a function class $\mathcal{F}$ allows a client to split (a representation of) $f \in \mathcal{F}$ into succinctly described functions f_0 and f_1 such that for any input x we have that $f(x) = f_0(x) + f_1(x)$ (over some Abelian group), but each f_b hides f.

The notion of FSS was originally motivated by applications to m-server PIR and related problems. FSS schemes for simple classes of functions such as point functions were constructed from one-way functions in [6,22]. However, a result

---

[1] In the homomorphic encryption for branching programs from [25] (see also [26]), the size of the encrypted output must grow with the *length* of the branching program. When simulating a boolean formula by a branching program, the length of the branching program is typically comparable to the formula size.

from [6] shows that 2-party FSS for richer circuit classes, from AC^0 and beyond, would imply (together with a mild additional assumption) breaking the circuit size barrier for similar classes.

The idea is that by encrypting the inputs and applying FSS to the function f' that first decrypts the inputs and then computes f, the parties can shift the bulk of the work required for securely evaluating f to local evaluations of f'_0 and f'_1. Thus, breaking the circuit size barrier reduces to securely distributing the generation of f'_0 and f'_1 from f and the secret decryption keys, which can be done using standard secure computation protocols and reused for an arbitrary number of future computations. This was viewed in [6] as a negative result, providing evidence against the likelihood of basing powerful forms of FSS on assumptions that are not known to imply FHE.

We turn the tables by constructing FSS schemes for branching programs under DDH, which implies low-communication secure 2-party computation protocols under DDH.

HOMOMORPHIC SECRET SHARING. For the purpose of presenting our results, it is more convenient to consider a dual version of FSS that can also be viewed as a form of "homomorphic secret sharing," or alternatively a variant of threshold FHE [1,19]. Concretely, a client wants to split a secret input $w \in \{0,1\}^n$ into a pair of shares (w^0, w^1), each of which is sent to a different server. Each individual share should computationally hide w. Each server, holding (a representation of) a function $f \in \mathcal{F}$, can apply an evaluation algorithm to compute $y_b = \mathsf{Eval}(f, w^b)$, so that $y_0 + y_1 = f(w)$. Note that this is precisely the original notion of FSS with the roles of the function and input reversed.[2]

Cast in the this language, our main technical contribution is such a homomorphic secret sharing scheme, based on DDH, with output group $\mathbb{Z}_2$ (or any other $\mathbb{Z}_p$), and the class $\mathcal{F}$ of functions represented by deterministic[3] branching programs. The scheme only satisfies a relaxed form of the above correctness requirement: for every input w and branching program P, the probability of producing local outputs that do not add up to the correct output $P(w)$ is upper bounded by an error parameter $\delta > 0$ which affects the running time of Eval. This probability is over the randomness of the sharing. The running time of the sharing algorithm is $n \cdot \mathsf{poly}(\lambda)$, where λ is a security parameter. The running time of Eval is polynomial in S, λ, and $1/\delta$.

We would like to stress that branching programs are quite powerful and capture many useful real-life computations. In particular, a branching program of size S can simulate any boolean formula of size S or boolean circuit of depth $\log_2 S$, and

---

[2] While one can always switch between the notions by changing the definition of $\mathcal{F}$, for classes $\mathcal{F}$ that contain *universal functions* [13,35] the switch can be done with polynomial overhead without changing $\mathcal{F}$. This will be the case for all function classes considered in this work.

[3] In fact, our construction can handle a larger class of *arithmetic* branching programs over the integers, but correctness only holds as long as all integers involved in intermediate computations are bounded by some fixed polynomial.

polynomial-size branching programs can simulate any computation in the complexity classes NC^1 or (non-uniform) deterministic log-space.

We also present a public-key variant of the homomorphic secret sharing scheme. This variant can be viewed as a threshold homomorphic encryption scheme with secret evaluation keys and additive reconstruction. That is, there is a key generation algorithm that outputs a single public key and a pair of secret evaluation keys. Given the public key, an arbitrary number of clients can encrypt their inputs. Each server, given the public ciphertexts and its secret evaluation key, can locally compute an additive share of the output.

The above results imply the following applications, all based on the DDH assumption alone.

SUCCINCT SECURE COMPUTATION OF BRANCHING PROGRAMS. The general transformation from FSS to secure two-party computation described above can be used to obtain succinct two-party protocols for securely evaluating branching programs with reusable preprocessing. However, the public-key variant of our construction implies simpler and more efficient protocols. The high level approach is similar to that of other low-communication secure protocols from different flavors of FHE [1,19,29], except for requiring secret homomorphic evaluation keys and an additional error-correction sub-protocol. For a two-party functionality with a total of n input bits and $m = m(n)$ output bits, where each output can be computed by a polynomial-size branching program (alternatively, logarithmic space Turing Machine or NC^1 circuit), the protocol can be implemented with a constant number of rounds and $n + m \cdot \text{poly}(\lambda)$ bits of communication, where λ is a security parameter. To reduce the $n \cdot \text{poly}(\lambda)$ cost of a bit-by-bit encryption of the inputs, the protocol employs a hybrid homomorphic encryption technique from [20].

BREAKING THE CIRCUIT SIZE BARRIER FOR "WELL STRUCTURED" CIRCUITS. In the case of evaluating general boolean circuits, we can make the total communication slightly sublinear in the circuit size by breaking the computation into segments of logarithmic depth and homomorphically computing additive shares of the outputs of each segment given additive shares of the inputs. For instance, we can evaluate a *layered* circuit of size S using $O(S/\log S)$ bits of communication (ignoring lower order additive terms; see Sect. 4 for a more precise statement). We employ error-correcting codes with encoding and decoding in NC^1 to ensure that errors introduced by the computation of a segment are corrected before propagating to the next segment.

FUNCTION SECRET SHARING. Using a universal branching program we can reverse the roles of P and w in the above homomorphic secret sharing scheme, obtaining a polynomial-time 2-party FSS scheme for branching programs. Unlike the main definition of FSS from [6] here we can only satisfy a relaxed notion that allows an inverse polynomial error probability. However, the error probability can be made negligible in the context of natural applications. An m-party FSS scheme for *circuits* was recently obtained by Dodis et al. [15] under the Learning with Errors (LWE) assumption, by making use of multi-key FHE [12,28,29].

Our construction gives the first FSS scheme that applies to a rich class of functions and does not rely on FHE.

PRIVATE INFORMATION RETRIEVAL. Following the application of FSS to PIR from [6] with a simple repetition-based error-correction procedure, a consequence of the above result is a 1-round 2-server (computational) PIR scheme in which a client can privately search a database consisting of N documents for existence of a document satisfying a predicate P, where P is expressed as a branching program applied to the document. For instance, any deterministic finite automaton can be succinctly expressed by such a branching program. The length of the query sent to each server is polynomial in the size of the branching program and a computational security parameter, whereas the length of the answer is a statistical security parameter times $\log N$.

1.2   Overview of Techniques

We now describe the main ideas behind our construction. It will be convenient to use the homomorphic secret sharing view: a client would like to share an input w between 2 servers so that the servers, on input P, can locally compute additive shares of $P(w)$.

Let $\mathbb{G}$ be a DDH group of prime order q with generator g. Our construction employs three simple ideas.

The first is that a combination of a threshold version of ElGamal and linear secret sharing allows the servers to locally multiply an encrypted input x with a linearly secret-shared value y, such that the result $z = xy$ is shared multiplicatively between the servers; namely the servers end up with elements $z_i \in \mathbb{G}$ such that the product of the z_i is g^z. This idea alone is already useful, as it gives an $(m-1)$-private m-server protocol for computing any degree-2 polynomial P with small integer coefficients held by the servers on a vector w of small integers held by the client, where the communication complexity in each direction is essentially optimal.

To see how this step is possible, consider a simplified version of the world, where (instead of requiring ElGamal) it held that g^w is a secure encryption of w. In this world, we can secret share input w by giving both servers a copy of the encryption g^w. Then, given an additive secret sharing x_0, x_1 of another value x, the servers can generate a multiplicative sharing of wx, by each computing $(g^w)^{x_b}$. Indeed, $(g^{wx_0}) \cdot (g^{wx_1}) = g^{wx}$. Extending this idea to ElGamal (as, alas, g^w is not a secure encryption) can be done via comparable "linear algebra in the exponent" given additive shares of x as well as for xc, where c is the ElGamal secret key.

What seems to stop us at degree-2 polynomials is the fact that z is now shared *multiplicatively* rather than linearly, so the servers cannot multiply z by a new input encrypted by the client. Moreover, converting multiplicative shares to additive shares seems impossible without the help of the client, due to the intractability of computing discrete logarithms in $\mathbb{G}$. The second, and perhaps most surprising, idea is that if we allow for an inverse polynomial error probability, and *assuming there are only $m = 2$ servers*, the servers can convert multiplicative shares of g^z into linear shares of z without any interaction. For simplicity, suppose $z \in \{0, 1\}$. Taking the inverse of the second server's share, the servers now hold group elements g_0, g_1 such that $g_0 = g_1$ if $z = 0$ and $g_1 = g \cdot g_0$ if $z = 1$. Viewing the action of multiplication by g as a cycle over $\mathbb{Z}_q$, the elements g_0, g_1 are either in identical positions, or g_1 is one step ahead. Conversion is done by picking a pseudo-random δ-sparse[4] subset $\mathbb{G}' \subset \mathbb{G}$ and having each server $b \in \{0, 1\}$ locally find the minimal integer $z_b \geq 0$ such that $g_b \cdot g^{z_b} \in \mathbb{G}'$. The first such z_b is expected to be found in roughly $1/\delta$ steps and if it is not found in $(1/\delta)\log(1/\delta)$ steps, we set $z_b = 0$. The key observation is that except with $O(\delta)$ probability, both searches will find the same point in $\mathbb{G}'$ and the servers end up with integers z_1, z_2 such that $z_1 - z_2 = z$, yielding the desired linear sharing of z.

Once we have a linear sharing of z, we can freely add it with other values that have a similar linear representation. We cannot hope to multiply two linearly shared values, but only to multiply them with another encrypted input. However, in order to perform another such a multiplication, we need additive shares not only of z, but also of zc for the ElGamal key c.

The third idea is that the client can assist the conversion by also providing an encryption of each input w multiplied by the secret key. This introduces two problems: the first is that semantic security may break down given a circular encryption of the secret key, which we handle either by assuming circular security of ElGamal or (with some loss of efficiency) by using the circular-secure variant of Boneh et al. [5] instead of standard ElGamal. A more basic problem is that for the conversion to produce correct results with high probability, the secrets must be small integers, whereas c (and so zc) is a large number. This is handled by providing an encryption of each input x multiplied by each *bit* of the secret key, and applying a linear combination whose coefficients are powers of 2 to the linear shares of the products of x and the bits of the key.

These ideas allow the servers to compute a restricted type of "straight-line programs" on the client's input, consisting of a sequence of instructions that either: load an input into memory, add the values of two memory locations, or multiply a memory location by an input. (Note that we cannot multiply two memory locations, which would allow evaluation of arbitrary circuits.) Such programs can emulate any branching program of size S by a sequence of $O(S)$ instructions.

It is instructive to note that the only limit on the number of instructions performed by the servers is the accumulation of error probabilities. This is analogous to the accumulation of noise in FHE schemes. However, the mechanisms for coping with errors are very different: in the context of known FHE schemes the simplest way of coping with noise is by using larger ciphertexts, whereas here we can reduce the error probability by simply increasing the *running time* of the

---

[4] Ideally, such a sparse subset would include each $g \in \mathbb{G}$ independently with probability δ. To emulate this efficiently we include each $g \in \mathbb{G}$ in $\mathbb{G}'$ if $\phi(g) = 0^{\lceil \log 1/\delta \rceil}$, where ϕ is a pseudorandom function.

servers, without affecting the ciphertext size or the complexity of encryption and decryption at all. We can also further trade running time for succinctness: the share size in our basic construction can be reduced by replacing the binary representation of the secret key with a representation over a larger basis, which leads to a higher homomorphic evaluation time.

The surprising power of local share conversions, initially studied in [14], has already been observed in the related contexts of information-theoretic PIR and locally decodable codes [2,16,39]. However, the type of share conversion employed here is very different in nature, as it is inherently tied to efficient computation rather than information.

Interestingly, our share conversion technique has resemblance to a cryptanalytic technique introduced by van Oorschot and Weiner for the purpose of parallel collision finding [37], where a set of "distinguished points" is used to synchronize two different processors.

1.3 Future Directions

This work gives rise to many natural open questions and future research directions. Can one bootstrap from branching programs to general circuits without relying on FHE? Can similar results be obtained for more than 2 parties? Can similar results be based on other assumptions that are not known to imply FHE? Can the dependence on the error parameter δ be eliminated or improved? To what extent can our protocols be optimized for practical use?

We hope that our approach will lead to faster solutions for some practical use-cases of FHE.

2 Preliminaries

We describe the necessary primitives and assumptions we rely on.

FUNCTION REPRESENTATIONS. We capture a function representation (such as a circuit, formula, or branching program) by an infinite collection $\mathcal{P}$ of bit strings P (called "programs"), each specifying an input length n and an output length m, together with an efficient algorithm Evaluate, such that $y \leftarrow$ Evaluate(P, w) (denoted by shorthand notation "$P(w)$"), for any input $w \in \{0,1\}^n$, defines the output of P on w.

HOMOMORPHIC SECRET SHARING. A (2-party) *Homomorphic Secret Sharing (HSS)* for a class of programs $\mathcal{P}$ consists of algorithms (Share, Eval), where Share$(1^\lambda, (w_1, \dots, w_n))$ splits the input w into a pair of shares (share$_0$, share$_1$), and Eval$(b, \text{share}, P, \delta, \beta)$ homomorphically evaluates P on share, where the correct output is additively shared over $\mathbb{Z}_\beta$ except with error probability δ. When β is omitted it is understood to be $\beta = 2$. We allow Eval to run in time polynomial in its input length and in $1/\delta$ and require that each share$_b$ output by Share keeps w semantically secure.

PUBLIC-KEY VARIANT. We further consider a stronger variant of the homomorphic secret sharing primitive that supports homomorphic computations on inputs contributed by different clients. In fact, what we achieve is stronger: there is a single public key that can be used to encrypt inputs as in a standard public-key encryption scheme. However, similar to the original notion of homomorphic secret sharing (and in contrast to standard homomorphic encryption schemes), homomorphic computations on encrypted inputs are done in a distributed way and require two separate (secret) evaluation keys. As before, we require the reconstruction of the output to be additive.

The corresponding security notion guarantees "semantic"-style secrecy of an encrypted value, given only the evaluation key of a single server. In a setting consisting of two servers and an arbitrary number of clients, the above security notion implies that inputs contributed by a set of uncorrupted clients remain secure even if one of the two servers colludes with all the remaining clients.

Definition 1 (Distributed-Evaluation   Homomorphic   Encryption).
A (2-party) Distributed-Evaluation Homomorphic Encryption (DEHE) for a class of programs $\mathcal{P}$ consists of algorithms (Gen, Enc, Eval) with the following syntax:

- *Gen(1^λ): On input a security parameter 1^λ, the key generation algorithm outputs a public key pk and a pair of evaluation keys $(\mathsf{ek}_0, \mathsf{ek}_1)$.*
- *Enc(pk, w): On a public key pk and a secret input value $w \in \{0,1\}$, the encryption algorithm outputs a ciphertext ct.*
- *Eval(b, ek, $(\mathsf{ct}_1, \ldots, \mathsf{ct}_n), P, \delta, \beta$): On input party index $b \in \{0,1\}$, an evaluation key ek, vector of n ciphertexts, a program $P \in \mathcal{P}$ with n input bits and m output bits, error bound $\delta > 0$, and an integer $\beta \geq 2$, the homomorphic evaluation algorithm outputs $y_b \in \mathbb{Z}_\beta^m$, constituting party b's share of an output $y \in \{0,1\}^m$. When β is omitted it is understood to be $\beta = 2$.*

The algorithms Gen and Enc are PPT algorithms, whereas Eval can run in time polynomial in its input length and in $1/\delta$.
The algorithms (Gen, Enc, Eval) should satisfy the following correctness and security requirements:

- **Correctness:** *There exists a negligible function ν such that for every positive integer λ, input $(w_1, \ldots, w_n) \in \{0,1\}^n$, program $P \in \mathcal{P}$ with input length n, error bound $\delta > 0$, and integer $\beta \geq 2$,*

$$\Pr[(\mathsf{pk}, (\mathsf{ek}_0, \mathsf{ek}_1)) \leftarrow \mathsf{Gen}(1^\lambda);$$
$$(\mathsf{ct}_1, \ldots, \mathsf{ct}_n) \leftarrow (\mathsf{Enc}(\mathsf{pk}, w_1), \ldots, \mathsf{Enc}(\mathsf{pk}, w_n));$$
$$y_b \leftarrow \mathsf{Eval}(b, \mathsf{pk}(\mathsf{ct}_1, \ldots, \mathsf{ct}_n), P, \delta, \beta) \ \forall b \in \{0,1\}:$$
$$y_0 + y_1 = P(w_1, \ldots, w_n)] \geq 1 - \delta - \nu(\lambda),$$

  where addition of y_0 and y_1 is carried out modulo β.
- **Security:** *The two distribution ensembles $C_0(\lambda)$ and $C_1(\lambda)$ are computationally indistinguishable, where $C_w(\lambda)$ is obtained by letting $(\mathsf{pk}, (\mathsf{ek}_0, \mathsf{ek}_1)) \leftarrow \mathsf{Gen}(1^\lambda)$ and outputting $(\mathsf{pk}, \mathsf{ek}_b, \mathsf{Enc}(\mathsf{pk}, w))$.*

2.1 DDH and Circular Security

Definition 2 (DDH). *Let* $\mathcal{G} = \{\mathbb{G}_\rho\}$ *be a set of finite cyclic groups, where* $|\mathbb{G}_\rho| = q$ *and* ρ *ranges over an infinite index set. We use multiplicative notation for the group operation and use* $g \in \mathbb{G}_\rho$ *to denote a generator of* $\mathbb{G}_\rho$. *Assume that there exists an algorithm running in polynomial time in* $\log q$ *that computes the group operation of* $\mathbb{G}_\rho$. *Assume further that there exists a PPT instance generator algorithm* $\mathcal{IG}$ *that on input* 1^λ *outputs an index* ρ *which determines the group* $\mathbb{G}_\rho$ *and a generator* $g \in \mathbb{G}_\rho$. *We say that the Decisional Diffie-Hellman assumption (DDH) is satisfied on* $\mathcal{G}$ *if* $\mathcal{IG}(1^\lambda) = (\rho, g)$ *and for every non-uniform PPT algorithm* $\mathcal{A}$ *and every three random* $a, b, c \in \{0, \ldots, q-1\}$ *we have*

$$|Pr[\mathcal{A}(\rho, g^a, g^b, g^{ab}) = 1] - Pr[\mathcal{A}(\rho, g^a, g^b, g^c) = 1]| < \varepsilon(\lambda),$$

for a negligible function ε. *We will sometimes write* $(\mathbb{G}, g, q) \leftarrow \mathcal{IG}(1^\lambda)$.

A more efficient variant of our construction requires a circular security assumption on the underlying bit encryption scheme, in which an efficient adversary cannot distinguish encryptions of the bits of the secret key from encryptions of 0.

Definition 3 (Circular Security). *We say that a public-key encryption scheme* (Gen, Enc, Dec) *with key length* $\ell(\lambda)$ *and message space containing* $\{0, 1\}$ *is* circular secure *if there exists a negligible function* $\nu(\lambda)$ *for which the following holds for every non-uniform PPT* $\mathcal{A}$:

$$\Pr[(\mathsf{pk}, \mathsf{sk}) \leftarrow \mathsf{Gen}(1^\lambda), b \leftarrow \{0, 1\}, b' \leftarrow \mathcal{A}^{\mathcal{O}_b}(\mathsf{pk}) : \quad b' = b] \leq \frac{1}{2} + \nu(\lambda),$$

where the oracle $\mathcal{O}_b$ *takes no input and outputs the following (where* $\mathsf{sk}^{(i)}$ *denotes the ith bit of* sk*):*

$$(C_1, \ldots, C_\ell), \; where \; \begin{cases} \forall i \in [\ell], C_i \leftarrow \mathsf{Enc}(\mathsf{pk}, 0) & if \; b = 0 \\ \forall i \in [\ell], C_i \leftarrow \mathsf{Enc}(\mathsf{pk}, \mathsf{sk}^{(i)}) & if \; b = 1 \end{cases}.$$

3 Homomorphic Secret Sharing for Branching Programs

In this section, we present constructions of homomorphic secret sharing schemes that enable non-interactive evaluation of a certain class of programs, known as restricted-multiplication straight-line programs. In particular, this class will include deterministic branching programs.

Definition 4 (RMS programs). *The class of* Restricted Multiplication Straight-line (RMS) *programs consists of an arbitrary sequence of the four following instructions, each with a unique identifier* id:

– *Load an input into memory:* $(\mathsf{id}, \hat{y}_j \leftarrow \hat{w}_i)$.
– *Add values in memory:* $(\mathsf{id}, \hat{y}_k \leftarrow \hat{y}_i + \hat{y}_j)$.

- *Multiply value in memory by an input value:* $(\mathsf{id}, \hat{y}_k \leftarrow \hat{w}_i \cdot \hat{y}_j)$.
- *Output value from memory, as element of* $\mathbb{Z}_\beta$: $(\mathsf{id}, \beta, \hat{O}_j \leftarrow \hat{y}_i)$.

Our construction will support homomorphic evaluation of straight-line programs of this form over inputs $w_i \in \mathbb{Z}$, provided that all intermediate computation values in $\mathbb{Z}$ remain "small" (where the required runtime grows with this size bound). Our final result is a public-key variant—i.e., a homomorphic encryption scheme with distributed evaluation (as per Definition 1)—based on DDH, with ciphertext size $O(\ell)$ group elements per input (for ℓ the logarithm of the DDH group size), and where runtime for homomorphic evaluation of an RMS program of size S with intermediate computation values bounded by M is $\mathsf{poly}(\lambda, S, M, 1/\delta)$.

An important sub-procedure of our homomorphic share evaluation algorithms is a local share conversion algorithm DistributedDLog, which intuitively converts a multiplicative secret sharing of g^x to an *additive* secret sharing of the value x, with inverse polynomial probability of error.

In the following subsections, we present: (1) The share conversion procedure DistributedDLog, (2) a simplified version of the homomorphic secret sharing scheme (in the secret-key setting), assuming circular security of ElGamal encryption, (3) the analogous public-key construction, and (4) the final public-key construction based on standard DDH.

3.1 Share Conversion Procedure

We now describe the local share conversion algorithm DistributedDLog, which receives as input a group element $h \in \mathbb{G}$ and outputs an integer w. Loosely speaking, DistributedDLog outputs the distance on the cycle generated by $g \in \mathbb{G}$ between h and the first $z \in \mathbb{G}$ such that a pseudo-random function outputs 0 on z. DistributedDLog is a deterministic algorithm and consequently two invocations of the algorithm with the same element h result in the same output w. Two invocations of the algorithm on inputs h and $h \cdot g^\mu$ for a small μ result, with good probability, in outputs w and $w - \mu$. Therefore, the DistributedDLog algorithm converts a difference of small μ in the cycle generated by g in $\mathbb{G}$ to the same difference over $\mathbb{Z}$.

The detailed description of $\mathsf{DistributedDLog}_{\mathbb{G},g}$ follows. The algorithm is hard-wired with ρ defining a group $\mathbb{G} = \mathbb{G}_\rho$ and a generator $g \in \mathbb{G}$. $\mathsf{DistributedDLog}_{\mathbb{G},g}$ receives as input $h \in \mathbb{G}$, an allowable error probability δ, maximum difference $\mu \in \mathbb{N}$, and a pseudo-random function $\phi : \mathbb{G} \to \{0,1\}^{\log(4\mu/\delta)}$. The difference μ specifies the maximum distance from h, along the cycle that g generates, of an element that is the input to a parallel invocation of $\mathsf{DistributedDLog}_{\mathbb{G},g}$ within a given application.

Proposition 1. *Let λ be a security parameter, let $\mathcal{G} = \{\mathbb{G}_\rho\}$ be a set of finite cyclic groups and let the instance generator algorithm of the set running on input 1^λ return a group $\mathbb{G} = \mathbb{G}_\rho$ with generator g. Let $\delta > 0$, let $\mu \in \mathbb{N}$, let $\mathcal{F}^r$ be a family of PRF defined over $\mathcal{G}$ and let $\phi : \mathbb{G} \to \{0,1\}^{\log(4/\delta)}$ be randomly chosen*

Algorithm 1. DistributedDLog$_{\mathbb{G},g}(h, \delta, \mu, \phi)$

1: Set $z \leftarrow h$, $w \leftarrow 0$.
2: **while** $(\phi(z) \neq 0^{\log(4\mu/\delta)}$ and $w < \frac{4\mu \ln(4/\delta)}{\delta})$ **do**
3:     $z \leftarrow z \cdot g$, $w \leftarrow w + 1$.
4: **end while**
5: Output w.

---

from all the members of $\mathcal{F}^r$ with domain $\mathbb{G}$. Then, for any $h \in \mathbb{G}$ and $\mu' \leq \mu$ we have that

$$\mathsf{DistributedDLog}_{\mathbb{G},g}(h, \delta, \mu, \phi) - \mathsf{DistributedDLog}_{\mathbb{G},g}(h \cdot g^{\mu'}, \delta, \mu, \phi) = \mu'$$

with probability greater than $1 - \delta$.

Proof. The values of ϕ on the sequence of elements traversed by the variable z in an execution of $\mathsf{DistributedDLog}_{\mathbb{G},g}(h, \delta, \mu, \phi)$ can be divided into three cases. The first case is that $\phi(h \cdot g^c) = 0^{\log(4\mu/\delta)}$ for some c in the range $0 \leq c \leq \mu - 1$, the second case is that $\phi(h \cdot g^c) \neq 0^{\log(4\mu/\delta)}$ for $c = 0, \ldots, \mu - 1$, but $\phi(h \cdot g^c) = 0^{\log(4\mu/\delta)}$ for some c in the range $\mu \leq c \leq \frac{4\mu \ln(4/\delta)}{\delta}$ and the last case is that neither of the former occurs, i.e. $\phi(h \cdot g^c) \neq 0^{\log(4\mu/\delta)}$ for every $c = 0, 1, \ldots, \frac{4\mu \ln(4/\delta)}{\delta}$.

In the second case, since $\phi(h \cdot g^c) \neq 0^{\log(4\mu/\delta)}$ for any $c = 0, \ldots, \mu - 1$, the execution of $\mathsf{DistributedDLog}_{\mathbb{G},g}(h, \delta, \mu, \phi)$ returns the smallest $c, \mu \leq c \leq \frac{4\mu \ln(4/\delta)}{\delta}$ such that $\phi(h) = 0^{\log(4\mu/\delta)}$. In $\mathsf{DistributedDLog}_{\mathbb{G},g}(h \cdot g^\mu, \delta, \mu, \phi)$ the variable z ranges over the elements $h \cdot g^\mu, \ldots, h \cdot g^{c+\mu}$ and the return value is $c - \mu$. Therefore, $\mathsf{DistributedDLog}_{\mathbb{G},g}(h, \delta, \mu, \phi) - \mathsf{DistributedDLog}_{\mathbb{G},g}(h \cdot g^\mu, \delta, \mu, \phi) = \mu$. By showing that the second case occurs with probability at least $1 - \delta$ we complete the proof.

If R is a random function then the probability that $R(h \cdot g^c) = 0^{\log(4\mu/\delta)}$ for some $c, 0 \leq c \leq \mu - 1$ is exactly $1 - (1 - \delta/4\mu)^\mu$, which by induction is at most $1 - (1 - \mu\delta/4\mu) = \delta/4$. In addition,

$$\Pr[\forall c \in \{0, \ldots, \frac{4\mu \ln \frac{4}{\delta}}{\delta}\} \ R(hg^c) \neq 0^{\log(4\mu/\delta)}] \leq (1 - \frac{\delta}{4\mu})^{\frac{4\mu \ln(4/\delta)}{\delta}}$$
$$< e^{-\ln 4/\delta} = \frac{\delta}{4}$$

Since ϕ is selected randomly from an appropriate family of pseudo-random functions any non-uniform PPT algorithm can distinguish between R and ϕ with negligible probability. Specifically, the probability of the first case is bound by $|\Pr[\exists c, 0 \leq c \leq \mu - 1 \ \phi(h \cdot g^c) = 0^{\log(4/\delta)}] - \delta/4| < \varepsilon(\delta)$ (for a negligible function $\varepsilon(\delta)$), since otherwise an efficient non-uniform algorithm can distinguish between R and ϕ by testing their value on hard-coded element (h, δ, μ, ϕ) and returning

1 if the result is $0^{\log(4\mu/\delta)}$. Similarly, the probability of the third case is bound by $|\Pr[\forall c \in \{0, \ldots, \frac{4\ln(4/\delta)}{\delta}\} \, \phi(hg^c) \neq 0^{\log(4/\delta)}] - \delta/4| < \varepsilon(\delta)$.

The probability of the second case is therefore at least $1 - \frac{\delta}{2} - 2\varepsilon(\delta)$. $\qquad\square$

3.2 Homomorphic Secret Sharing

We now construct a simple version of the homomorphic secret sharing scheme, using the procedure DistributedDLog as a sub-routine. The resulting scheme will be a "secret-key" version. Further, the security of the scheme will rely on the assumption that ElGamal encryption is circular secure. These restrictions will be removed in the following subsections.

Consider a DDH group $\mathbb{G}$ of prime order q (with λ bits of security) with generator g, and $\ell = \lceil \log_2 q \rceil$. We will use $c = c^{(1)}, \ldots, c^{(\ell)}$ to denote bits of an element $c \in \mathbb{Z}_q$ (i.e., $c = \sum_{i=0}^{\ell-1} 2^i c^{(i+1)}$).

OVERVIEW OF CONSTRUCTION. All values generated within the secret sharing and homomorphic evaluation sit within three "levels." We will maintain notation as in the top portion of Fig. 1. Namely,

Level 1: ElGamal Ciphertexts $[\![w]\!]_c$.
> Initial input values w will be "uploaded" into the homomorphic evaluation system by generating an ElGamal encryption $[\![w]\!]_c$ of the value w with respect to a common secret key c, as well as encryptions $[\![c^{(i)}w]\!]_c$ of each of the products $c^{(i)}w$ for the bits $c^{(i)}$ of the corresponding key c.

Level 2: Additive secret shares $\langle y \rangle$.
> Each value y in memory of the RMS program will be maintained via two sets of additive secret shares: $\langle y \rangle$ itself, and $\langle cy \rangle$ secret sharing the product of y with the ElGamal secret key c of the system. We start with secret shares of this form for each input value (e.g., in the secret-key setting, these will be generated as part of the Share procedure). Then, after each emulated RMS instruction, we will maintain the invariant that each newly computed memory value is stored as secret shares in this fashion.

Level 3: Multiplicative secret shares $\langle\!\langle xy \rangle\!\rangle$.
> Multiplicative secret shares appear only as intermediate values during the execution of homomorphic evaluation (of multiplication), and are then converted back to additive shares via DistributedDLog.

Remark 1 (Valid vs Random). We emphasize that a "valid" encoding (e.g., $[\![x]\!]_c$, $\langle x \rangle$, or $\langle\!\langle x \rangle\!\rangle$) speaks *only* to the correctness of decoding, and does not imply that the encoding is a *random* such encoding (e.g., a randomly sampled ciphertext, or fresh secret shares).

The bottom portion of Fig. 1 describes two pairing operations that constitute *cross-level* computations. The first, MultShares, "multiplies" a level-1 encoding by a level-2 encoding. Namely, it takes as input a level-1 (ElGamal ciphertext) encoding of x under key c, and level-2 (additive secret sharing) encodings of

---

Notation. For $x \in \mathbb{Z}_q$ (or $x \in \mathbb{Z}$ with $x < q$).
Items in which *both* parties receive same value.
- $[\![x]\!]_c = (g, h) \in \mathbb{G}^2$ for which $h/g^c = g^x$. That is, ElGamal ciphertext of x w.r.t. key c.

Items in which each party receives a separate share.
- $\langle x \rangle =$ Additive secret shares $x_1, x_2 \in \mathbb{Z}_q$ for which $x_1 + x_2 = x \in \mathbb{Z}_q$.
- $\langle\!\langle x \rangle\!\rangle =$ "Multiplicative" secret shares $h_1, h_2 \in \mathbb{G}$ for which $h_1 \cdot h_2 = g^x \in \mathbb{G}$.

Pairing Operations.
Let $\phi : \{0,1\}^\lambda \times \mathbb{G} \rightarrow \{0,1\}^\ell$ be a given PRF
- MultShares$\left([\![x]\!]_c, \langle y \rangle, \langle cy \rangle\right) \rightarrow \langle\!\langle xy \rangle\!\rangle$.
  1. Denote $[\![x]\!]_c = (h_1, h_2) \in \mathbb{G}^2$.
  2. Compute $\langle\!\langle xy \rangle\!\rangle = h_2^{\langle y \rangle} h_1^{-\langle cy \rangle}$.
- ConvertShares$(b, \langle\!\langle x \rangle\!\rangle, \text{id}, \delta', M) \rightarrow \langle x \rangle$, with party identifier $b \in \{0,1\}$, nonce id, error parameter δ' and max size bound M.
  1. Denote by $\phi' : \mathbb{G} \rightarrow \{0,1\}^{\lfloor \log(4M/\delta') \rfloor}$ the appropriate prefix output of $\phi(\text{id}, \cdot)$.
  2. Let x_b denote the present party b's share of $\langle\!\langle x \rangle\!\rangle$.
     If $b = 1$, then replace $x_b \leftarrow x_b^{-1}$ // i.e., convert so that $x_0/x_1 = g^x$.
  3. Let $x_b' \leftarrow \mathsf{DistributedDLog}_{\mathbb{G},g}(\langle\!\langle x \rangle\!\rangle, \delta', M, \phi')$.
  4. If $b = 0$, output x_0'. If $b = 1$, output $-x_1'$.   // Output additive shares

---

Fig. 1. Notation for components of the homomorphic secret sharing scheme, and pairing operations for transforming between different components.

y, and of cy (the product of y with the ElGamal secret key), and outputs a level-3 (multiplicative secret sharing) encoding of the product xy. The second, ConvertShares, converts from a level-3 (multiplicative) encoding back down to a level-2 (additive) encoding, with some probability of error, as dictated by given parameters.

Roughly, the intermediate values of homomorphic evaluation will be maintained in level-2 (additive) secret shared form. Any linear combination of such shares can be performed directly. Multiplication between a value in memory and an input value will be performed by performing the MultShares between the input value (encoded in level 1) and the relevant memory value (encoded in level 2). This will yield an encoding of the product, but in level 3 (i.e., as multiplicative shares). To return the computed product back to level 2, the parties will execute the pairing procedure ConvertShares, which essentially runs the DistributedDLog procedure from the previous subsection.

Remark 2 (Variable Types). Note that the relevant values are nearly all elements of $\mathbb{G}$ (e.g., elements of ElGamal ciphertexts) or of $\mathbb{Z}_q$ (e.g., the values cy_i, as well as shares of SubtShare). An important exception to this are the values w_i, y_i, which are interpreted as (small) *integers*. When necessary for computation, we will sometimes perform a type cast back and forth between $\mathbb{Z}$ and $\mathbb{Z}_q$, using the notation $(int)(x) \in \mathbb{Z}$ for $x \in \mathbb{Z}_q$, and $(x \mod q) \in \mathbb{Z}_q$ for $x \in \mathbb{Z}$.

A few calculations provides us with the following two claims on these pairing procedures (see full version of this work for details). A slight subtlety arises in the case of ConvertShares, regarding the PRF. Namely, the share values on which we run ConvertShares are the results of partial computations and previous ConvertShares executions, meaning in particular that they depend on the choice of the sampled PRF ϕ. However, this is not an issue due to two reasons: (1) this dependence is efficiently computable given oracle access to the PRF *outputs*, as specified in the algorithm DistributedDLog, and (2) we will explicitly ensure the PRF is never used on the same input twice, by use of nonces id. In such case, the PRF will still act as a random function in each ConvertShares invocation, and yield the required share conversion correctness guarantee.

Claim (Pairing Operations).

1. MultShares: For all values $x, y \in \mathbb{Z}_q$ and any key $c \in \mathbb{Z}_q$, then on input a valid level-1 encoding $[\![x]\!]_c$ with respect to key c, and valid level-2 encodings $\langle y \rangle, \langle cy \rangle$, the output of MultShares($[\![x]\!]_c, \langle y \rangle, \langle cy \rangle$) is a valid *level-3* encoding $\langle\!\langle xy \rangle\!\rangle$ of the product $xy \in \mathbb{Z}_q$.
2. ConvertShares: For every id $\in \{0,1\}^\lambda, \delta > 0, M \in \mathbb{N}$, PPT $\mathcal{A}$, $\Pr[\phi \leftarrow$ PRFGen(1^λ); $(h, \mu) \leftarrow \mathcal{A}^{\phi(\mathrm{id}', \cdot)}(1^\lambda) : \mu < M \wedge$ ConvertShares($0, h, \mathrm{id}, \delta, M$) + ConvertShares($1, h^{-1}g^\mu, \mathrm{id}, \delta, M$) $\neq \mu] \leq M\delta$, where $\phi(\mathrm{id}', \cdot)$ is an oracle to the PRF ϕ on any input with prefix $\mathrm{id}' \neq \mathrm{id}$.

Homomorphic Secret Sharing Scheme - Share$_{\mathbb{G},g}(1^\lambda, w_1, \ldots, w_n)$
Inputs: 1^λ and input values $w_1, \ldots, w_n \in \mathbb{Z}$.

- Sample a PRF with input $\{0,1\}^n \times \mathbb{G}$ and output $\{0,1\}^\ell$: $\phi \leftarrow$ PRFGen(1^λ).
- Sample an ElGamal secret key: $c \leftarrow \mathbb{Z}_q$.
- For each input w_i, sample the following values:
  1. ElGamal encryptions:
     (a) of $w_i \in \mathbb{Z}$: let $[\![w_i]\!]_c \leftarrow$ Enc(g^c, w_i) $\in \mathbb{G}^2$.
     (b) of $(c^{(t)} \cdot w_i) \in \mathbb{Z}$: i.e., for each $t \in [\ell]$, let $[\![c^{(t)}w_i]\!]_c \leftarrow$ Enc($g^c, c^{(t)}w_i$).
  2. Additive secret sharings:
     (a) of $w_i \in \mathbb{Z}$: let $\langle w_i \rangle \leftarrow$ AdditiveShare($1^\lambda, w_i$).
     (b) of $cw_i \in \mathbb{Z}_q$: let $\langle cw_i \rangle \leftarrow$ AdditiveShare($1^\lambda, cw_i$).
- For each $b \in \{0,1\}$, output share$_b = \left\{ \phi, \left([\![w_i]\!]_c, \left\{ [\![c^{(t)}w_i]\!]_c \right\}_{t \in [\ell]}, \langle w_i \rangle_b, \langle cw_i \rangle_b \right)_{i \in [n]} \right\}$.

Fig. 2. Share generation procedure Share$_{\mathbb{G},g}$ for secret sharing an input w via the homomorphic secret sharing scheme.

We present our secret sharing scheme Share in Fig. 2, and the corresponding homomorphic operations on shares Eval in Fig. 3.

We remark that our combined construction obtains a generalization of the notion of HSS from Sect. 2 both extending beyond the Boolean setting to support arithmetic computations over small integers, and allowing multiple outputs of the

---

Homomorphic Share Evaluation of RMS Programs - $\mathsf{Eval}_{\mathsf{G},g}(b, \mathsf{share}, P, M, \delta)$
Inputs: Homomorphic secret share value share, RMS program description P of size $\leq S$,
plaintext size bound $M \in \mathbb{Z}$, error bound δ.
Take $\delta' = \delta/((\ell+1)MS)$.

Parse share as in Figure 2. Parse P as a sequence of instructions (as in Definition 4);
for each sequential instruction, perform the corresponding sequence of operations:

Instruction $(\mathsf{id}, \hat{y}_j \leftarrow \hat{w}_i)$:
    1: Let $\langle y_j \rangle \leftarrow \langle w_i \rangle$ and $\langle cy_j \rangle \leftarrow \langle cw_i \rangle$, where $\langle w_i \rangle, \langle cw_i \rangle$ are as in share.

Instruction $(\mathsf{id}, \hat{y}_k \leftarrow \hat{y}_i + \hat{y}_j)$:
    1: Compute $\langle y_k \rangle \leftarrow \langle y_i \rangle + \langle y_j \rangle$, directly on the additive shares (over $\mathbb{Z}_q$).
    2: Compute $\langle cy_k \rangle \leftarrow \langle cy_i \rangle + \langle cy_j \rangle$, on the additive shares (over $\mathbb{Z}_q$).

Instruction $(\mathsf{id}, \hat{y}_k \leftarrow \hat{w}_i \cdot \hat{y}_j)$:
    1: Let $[\![w_i]\!]_c$ and $\{[\![c^{(t)}w_i]\!]_c\}_{t \in [\ell]}$ be the ElGamal ciphertexts associated with w_i,
       and let $\langle y_j \rangle$ and $\langle cy_j \rangle$ the additive secret shares associated with y_j.
    2: Compute the pairing $\langle\!\langle w_i y_j \rangle\!\rangle = \mathsf{MultShares}([\![w_i]\!]_c, \langle y_j \rangle, \langle cy_j \rangle)$, as in Figure 1.
    3: Execute Share Conversion: $\langle w_i y_j \rangle = \mathsf{ConvertShares}(b, \langle\!\langle w_i y_j \rangle\!\rangle, (\mathsf{id}, 0), \delta', M, \phi)$,
       as in Figure 1.
    4: **for** $t = 1$ to ℓ **do**   // Repeat above process for each $c^{(t)} \cdot y_k$ in the place of y_k
    5:     Compute $\langle\!\langle c^{(t)} w_i y_j \rangle\!\rangle = \mathsf{MultShares}([\![c^{(t)}w_i]\!]_c, \langle y_j \rangle, \langle cy_j \rangle)$.
    6:     Execute $\langle c^{(t)} w_i y_j \rangle = \mathsf{ConvertShares}(b, \langle\!\langle c^{(t)} w_i y_j \rangle\!\rangle, (\mathsf{id}, t), \delta', M, \phi)$.
    7: **end for**
    8: Compute $\langle cw_i y_j \rangle = \sum_{t=1}^{\ell} 2^{t-1} \langle c^{(t)} w_i y_j \rangle$.
    9: Set $\langle y_k \rangle \leftarrow \langle w_i y_j \rangle$ (from Step 3) and $\langle cy_k \rangle \leftarrow \langle cw_i y_j \rangle$.

Instruction $(\mathsf{id}, \beta, \hat{O}_j \leftarrow \hat{y}_i)$:
    1: Shift $\langle y_i \rangle$ share by rerandomization offset: $\langle y_i \rangle \leftarrow \langle y_i \rangle + \phi(\mathsf{id}, g)$, over $\mathbb{Z}_q$.
       // Note that shifting *both* shares does not change the shared value
    2: Convert share from $\mathbb{Z}_q$ to $\mathbb{Z}_\beta$: i.e., $\langle O_i \rangle \leftarrow \langle y_i \rangle \mod \beta$.
    3: Output $\langle O_i \rangle$.

Fig. 3. Procedures for performing homomorphic operations on secret shares. Note that
we distinguish variables of the straight-line program from the actual values by using $\hat{y}_i$
as opposed to y_i, etc. Here, notation $\langle y \rangle$ is used to represent *this party's* share of the
corresponding subtractive secret shared pair. Evaluation maintains the invariant that
each of the *additive secret shares* $\langle y_i \rangle$ encode the correct current computation value
of $\hat{y}_i$.

program from possibly different groups $\mathbb{Z}_\beta$ (as specified by the program description P). The restricted case of the definition coincides with our construction with
size bound M set to 1 and all program outputs in a fixed group $\mathbb{Z}_\beta$.

Theorem 1 (Homomorphic Secret Sharing). *Assume that ElGamal is circular secure (as per Definition 3). Then the scheme* (Share, Eval) *as specified in
Figs. 2, 3 is a secure homomorphic secret sharing scheme for the class of deterministic branching programs.*

The theorem follows from the next two correctness and security lemmas. In particular, correctness is proved for the broader class of RMS programs with bounded-size intermediate computation values, which captures deterministic branching programs. We provide proof sketches and refer the reader to the full version for a complete analysis.

Lemma 1 (Correctness of Eval). *For every input* $w_1, \ldots, w_n \in \mathbb{Z}_q$ *and every RMS program* P *(as in Definition 4) of size* S *for which* all *intermediate values* $y_i \in \mathbb{Z}$ *in the execution of* P *are bounded by* $|y_i| \leq M$,

$$
\Pr[(\mathsf{share}_0, \mathsf{share}_1) \leftarrow \mathsf{Share}_{\mathbb{G},g}(1^\lambda, w_1, \ldots, w_n) :
$$
$$
\mathsf{Eval}_{\mathbb{G},g}(\mathsf{share}_0, P, M, \delta) + \mathsf{Eval}_{\mathbb{G},g}(\mathsf{share}_1, P, M, \delta)
$$
$$
= P(w_1, \ldots, w_n)] \geq 1 - \delta.
$$

Proof Sketch. We first address the probability of error due to execution of the Share Conversion Procedure ConvertShares. Observe that the homomorphic evaluation of program P performs at most $S(\ell+1)$ executions of the Share Conversion Procedure (with error parameter $\delta' = \delta/(\ell+1)MS$). By Claim 3.2, together with a union bound, this implies that no conversion errors will occur with probability $\delta'(\ell+1)MS = \delta$.

Assume, then, that every ConvertShares execution returns without error. We prove that following invariant is maintained at each step of homomorphic evaluation:

For every memory item $\hat{y}_i$, let $y_i \in \mathbb{Z}$ denote the correct value that should be presently stored in memory. Then the shares $\langle y_i \rangle = (y_i^A, y_i^B) \in \mathbb{Z}_q^2$ and $\langle cy_i \rangle = (v^A, v^B) \in \mathbb{Z}_q^2$ held by the parties satisfy:

(a) $(int)(y_i^A + y_i^B) = y_i \in \mathbb{Z}$ (where addition is in $\mathbb{Z}_q$).
(b) $v^A + v^B = cy_i \in \mathbb{Z}_q$.

Note that this invariant holds vacuously at the start of Eval, as all memory locations are empty, and is preserved directly by our actions in each Load Input Into Memory and Add Values In Memory instructions, due to the structure of the additive secret sharing and since all intermediate values $y_i \in \mathbb{Z}$ in the execution are bounded by $M < q$ (so that relations over $\mathbb{Z}$ align with relations over the shares $\mathbb{Z}_q$).

For each instruction Multiply Input By Memory Value, consider the resulting shares $\langle y_k \rangle$ and $\langle cy_k \rangle$. By Claim 3.2, the shares $\langle\langle w_i y_j \rangle\rangle$ computed via MultShares constitute a valid level-3 sharing of the product $w_i y_j$ (as per Fig. 1). Since we are in the case where ConvertShares does not err, the resulting converted shares $\langle w_i y_j \rangle$ encode exactly the value $w_i y_j \in \mathbb{Z}_q$. Since $w_i y_j$ is an intermediate computation value y_k in the evaluation of P, we have $0 \leq w_i y_j \leq M < q$. Thus, invariant (a) holds. Consider now $\langle cy_k \rangle$. From precisely the same argument (since we are in the case where ConvertShares does not err), we have for each $t \in [\ell]$ that the computed intermediate value $\langle c^{(t)} y_k \rangle$ is a level-2 encoding of the corresponding value $v_t := c^{(t)} y_k \in \mathbb{Z}_q$ (where $y_k = w_i y_j$). Since y_k is an intermediate computation value in P we have $0 \leq y_k \leq M < q$, and so for $c^{(t)} \in \{0, 1\}$, then $0 \leq c^{(t)} y_k \leq M < q$. Thus, it holds $(int)(v_t^A - v_t^B) = c^{(t)} w_i y_j \in \mathbb{Z}$. Combining the respective values over $t \in [\ell]$, it holds $\left(\sum_{t \in [\ell]} 2^t v_t^A \right) + \left(\sum_{t \in [\ell]} 2^t v_t^B \right) = \sum_{t \in [\ell]} 2^t (v_t^A + v_t^B) = \sum_{t \in [\ell]} 2^t c^{(t)} w_i y_j = c w_i y_j \in \mathbb{Z}_q$. Therefore, invariant (b) holds.

Finally, for each Output Memory Value instruction, the shares of both parties are deliberately shifted by a (pseudo-)random value to avoid the bad "edge case" in which the shares y^A, y^B of the desired output value y satisfy $(int) y^B \geq q - M$ and $(int)(y^A) - (int)(y^B) = y - q$ (where locally modding each share by β may disrupt correctness), as opposed to the almost-everywhere case that $(int)(y^A) - (int)(y^B) = y$. The error probability is bounded by $M/q + \mathsf{negl}(\lambda)$, which is negligible.

Lemma 2 (Security of Share). *Based on the assumption that ElGamal is a weakly circular secure encryption scheme (as per Definition 3), then* Share *is a computationally secure secret sharing scheme.*

Proof Sketch. We prove that the distribution of a single party's share resulting from Share is computationally indistinguishable from a distribution that is *independent* of the shared values $w_1, \ldots, w_n$, via three hybrids. In Hybrid 1, we replace all additive secret shares by random values. This yields an identical distribution, by the perfect security of the 2-out-of-2 additive secret sharing scheme. In Hybrid 2, we replace all ciphertexts of the form $\{[\![c^{(t)} w_i]\!]_c\}$ with encryptions of 0, which we address below. In Hybrid 3, we rely on standard semantic security to further replace all ciphertexts of the form $[\![w_i]\!]_c$ with encryptions of 0.

We now demonstrate that an efficient adversary $\mathcal{A}$ who distinguishes between Hybrids 2 and 3 for some choice of inputs $w_1, \ldots, w_n$ can be used to break the circular security of ElGamal. In the circular security challenge, the adversary $\mathcal{B}$ receives an ElGamal public key pk and access to an oracle which provides either vectors of encryptions of the bits of the secret key, or vectors of encryptions of 0. To simulate the ciphertext vectors for $\mathcal{A}$ as either $\{[\![c^{(t)} w_i]\!]_c\}$ (Hybrid 2) or $\{[\![0]\!]_c\}$ (Hybrid 3), $\mathcal{B}$ (with $w_1, \ldots, w_n$ hardcoded) considers two cases. If $w_i = 0$ then anyway $c^{(t)} w_i = 0$, so he simply generates a vector of 0 ciphertexts. For each $w_i \neq 0$, he: queries his circular security oracle, receives a vector of ciphertexts $[\![x_{i,1}]\!]_c, \ldots, [\![x_{i,\ell}]\!]_c$, and then exponentiates each ciphertext (that is, each of the 2 group elements) by the corresponding input w_i. This operation maps ciphertexts of 0 to ciphertexts of 0, and ciphertexts of 1 to ciphertexts of w_i; further, since $w_i \neq 0$ and $\mathbb{Z}_q$ is a field, the resulting distribution of encryption randomness is also equally distributed (i.e., uniform). Therefore, either Hybrid 2 or 3 is exactly simulated, and so the adversary $\mathcal{A}$ enables $\mathcal{B}$ to distinguish, breaking circular security.

3.3   Public-Key HSS for Branching Programs

In the construction of the previous section, secret shares of an input w consisted of ElGamal encryptions $[\![w]\!]_c, \{[\![c^{(t)} w]\!]_c\}_{t \in [\ell]}$ and additive secret shares $\langle w \rangle, \langle cw \rangle$, where c was a (freshly sampled) key for ElGamal. At face value, it would seem that one must know the value of the key c in order to generate these values—meaning, in turn, that homomorphic computation can only be performed on the data of a single user who generates the key c. In this section, we demonstrate that by leveraging the homomorphic properties of ElGamal encryption, we can in fact generate all required values for a secret sharing of w while maintaining security, given only "public key" information independent of the input w. That is, we obtain homomorphic encryption with distributed evaluation, as discussed in Sect. 2.

More formally, we now consider a separate procedure Gen for generating common setup information pk and secret evaluation keys ek_0, ek_1 (which we consider to be given to two servers). Given access to pk, a user can "upload" his input w to the system via Enc. Then, given their respective evaluation keys, two servers can perform non-interactive homomorphic computations on *all* users' inputs via Eval.

In our construction, the algorithm Gen samples an ElGamal key pair, and outputs pk consisting of encryptions $[\![1]\!]_c, \{[\![c^{(t)}]\!]_c\}_{t \in [\ell]}$ and evaluation keys ek_b corresponding to additive secret shares of $\langle c \rangle$. In Enc, a user computes the necessary ciphertexts $[\![w]\!]_c$ and $\{[\![c^{(t)}w]\!]_c\}_{t \in [\ell]}$ for his input w by exponentiating the ciphertexts in pk component-wise by w, (i.e., using multiplicative homomorphism of ElGamal). The final required values $\langle w \rangle, \langle cw \rangle$ can be obtained directly by the servers within Eval by performing the procedure for a homomorphic multiplication between the "input value" w (i.e., given $[\![w]\!]_c, \{[\![c^{(t)}]\!]_c\}_{t \in [\ell]}$) together with "memory value" 1 (i.e., given a trivial sharing $\langle 1 \rangle$ together with $\langle c \rangle$ from ek).

A formal description of the algorithms Gen, Enc, Eval is given in Fig. 4.

Theorem 2 (DEHE). *Assume that ElGamal is circular secure (as per Definition 3). Then the scheme (Gen, Enc, Eval) as given in Fig. 4 is a secure Distributed-Evaluation Homomorphic Encryption scheme for the class of deterministic branching programs.*

Proof. Correctness. To reduce to Theorem 1, it suffices to demonstrate: (1) the values $[\![w]\!]_c, \{[\![c^{(t)}w]\!]_c\}_{t \in [\ell]}$ as generated in Enc are valid level-1 encodings of w and $\{c^{(t)}w\}_{t \in [\ell]}$; and (2) the values $\langle w \rangle, \langle cw \rangle$ as generated in Eval are valid level-2 encodings of w and cw. Property (1): Recall $[\![w]\!]_c$ was obtained as (h_1^w, h_2^w) for (h_1, h_2) a valid level-1 encoding of 1. This means $h_2 h_1^{-c} = g^1$, which implies $(h_2^w)(h_1^w)^{-c} = g^w$, as desired. Same for each $[\![c^{(t)}w]\!]_c$. Property (2): Holds by the correctness of homomorphic RMS multiplication evaluation of Eval as per Theorem 1.

Security. Semantic security of the scheme follows as in Theorem 1, assuming circular security of ElGamal. Namely, the view of a server holding ek_b consists of information theoretically hiding secret shares, ciphertexts of values independent of the secret key, and vectors of ciphertexts encrypting the vector $(c^{(1)}w, \ldots, c^{(\ell)}w)$ for various inputs w. An adversary distinguishing between this view and one consisting of random share elements and ciphertexts of 0 can be used to break the circular security of ElGamal, precisely as in Theorem 1.

Distributed-Evaluation Homomorphic Encryption: Gen, Enc, Eval

$\mathsf{Gen}(1^\lambda)$:

1. Sample a PRF with input $\{0,1\}^n \times \mathbb{G}$ and output $\{0,1\}^\ell$: $\phi \leftarrow \mathsf{PRFGen}(1^\lambda)$.
2. ElGamal Key Setup:
   (a) Sample a DDH-hard group and generator $(\mathbb{G}, g, q) \leftarrow \mathcal{IG}(1^\lambda)$.
   (b) Sample an ElGamal key pair: $c \leftarrow \mathbb{Z}_q$.
3. Sample ElGamal encryptions:
   (a) The constant $1 \in \mathbb{Z}_q$: let $[\![1]\!]_c \leftarrow \mathsf{Enc}(g^c, 1)$.
   (b) The bits of the secret key c: $\forall i \in [\ell]$, let $[\![c_i]\!]_c \leftarrow \mathsf{Enc}(g^c, c_i)$.
4. Sample 2-out-of-2 additive secret sharings:
   (a) The constant $1 \in \mathbb{Z}_q$: $\langle 1 \rangle \leftarrow \mathsf{AdditiveShare}(1)$. // Included for notational convenience
   (b) The bits of the secret key c: $\forall t \in [\ell]$, let $\langle ct \rangle \leftarrow \mathsf{AdditiveShare}(c^{(t)})$.
5. Output $\mathsf{pk} = \left(\mathbb{G}, g, [\![1]\!]_c, \{[\![c^{(t)}]\!]_c\}_{t \in [\ell]} \right)$, $\mathsf{ek}_b = \left(\mathsf{pk}, \langle 1 \rangle, \{\langle c^{(t)} \rangle\}_{t \in [\ell]} \right)$.

$\mathsf{Enc}_{\mathbb{G},g}(\mathsf{pk}, w)$:

1. Parse pk as in Gen above.
2. Compute the following ElGamal ciphertexts:
   (a) Of $w \in \mathbb{Z}$: parse $[\![1]\!]_c = (h_1, h_2)$, and let $[\![w]\!]_c = (h_1^w, h_2^w) \in \mathbb{G}^2$.
   (b) Of $c^{(t)} w \in \mathbb{Z}$: i.e., for each $t \in [\ell]$, parse $[\![c^{(t)}]\!]_c = (h_1^{(t)}, h_2^{(t)})$ let $[\![c^{(t)} w]\!]_c = ((h_1^{(c)})^w, (h_2^{(c)})^w)$.
3. Output $([\![w]\!]_c, \{[\![c^{(t)} w]\!]_c\}_{t \in [\ell]})$.

$\mathsf{Eval}_{\mathbb{G},g}(b, \mathsf{ek}, \mathsf{ct}, P, \delta, M)$:

1. Parse ek as in Gen above; interpret $\hat{1}$ as loaded into memory, via $\langle 1 \rangle, \{\langle c^{(t)} \rangle\}_{t \in [\ell]}$ as given.
2. Parse P as a sequence of instructions (as in Definition 4).
3. For each instruction $(\mathsf{id}, \hat{y}_k \leftarrow \hat{y}_i + \hat{y}_j)$, $(\mathsf{id}, \hat{y}_k \leftarrow \hat{w}_i \cdot \hat{y}_j)$, or $(\mathsf{id}, \hat{O}_j \leftarrow \hat{y}_i)$, perform the corresponding sequence of operations as given in Figure 3.
4. For each instruction $(\mathsf{id}, \hat{y}_j \leftarrow \hat{w}_i)$, execute $(\mathsf{id}, \hat{y}_j \leftarrow \hat{w}_i \cdot \hat{1})$.

Fig. 4. Construction of "public-key" variant of homomorphic secret sharing: i.e., homomorphic encryption with distributed evaluation.

Comparing the complexity of the public-key scheme (Gen, Enc, Eval) to that of the secret-key scheme (Share, Eval) from the previous section, we see that the computation cost to the user for uploading inputs $w_1, \ldots, w_n$ via Enc is essentially equivalent to the cost of sharing the inputs via Share (exponentiating given ciphertexts by the respective inputs in one case, versus encrypting the values directly in the other), but the cost of each "load input" instruction $(\mathsf{id}, \hat{y}_i \leftarrow \hat{w}_i)$ within the homomorphic evaluation now incurs the cost of a multiplication step to generate additive secret shares $\langle w_i \rangle, \langle cw_i \rangle$ given only $\langle c \rangle$ and the uploaded ElGamal ciphertexts associated with w_i, as opposed to being essentially for free for the client to generate $\langle w_i \rangle, \langle cw_i \rangle$ when he knew the values of w_i, cw_i.

3.4 Removing the Circular Security Assumption

We now show how to remove the ElGamal circular security assumption in the construction in the previous section, yielding a scheme that relies solely on DDH. As in the setting of FHE, this can be achieved directly in exchange for a multiplicative blowup of the computation depth in the share size, by considering a *leveled* version of the scheme (i.e., replacing the circular encryptions of bits of c under key c by bits of c_i under key c_{i+1} for a depth-length sequence of keys). However, we now demonstrate an alternative approach, which does *not* require increasing the share size with respect to the size of computation.

Our new construction replaces ElGamal encryption with the ElGamal-like cryptosystem of Boneh, Halevi, Hamburg, and Ostrovsky (BHHO) [5], which is provably circular secure based on DDH. At a high level, BHHO ciphertexts possess an analogous structure of "linear algebra in the exponent," which allows us to mirror the procedure we used with ElGamal for multiplicatively pairing a ciphertext with an additively shared value.

It will be convenient to consider a slightly modified version of the BHHO scheme, given below, in which the message space is a subset of the *exponent* space $\mathbb{Z}_q$ instead of the group $\mathbb{G}$ itself (i.e., the multiplication by message m in standard encryption is replaced by g^m). Since decryption of such scheme requires taking discrete log, efficient decryption will hold for a polynomial-size message space.

Definition 5 (BHHO Encryption [5]). *Let $\mathbb{G}$ be a group of prime order q and g a fixed generator of $\mathbb{G}$. The size of $\mathbb{G}$ is determined by a security parameter λ, in particular, $1/q$ is negligible in λ. The BHHO public-key encryption scheme for polynomial-size message space $\mathsf{Msg} \subset \mathbb{Z}_q$ is as follows:*

- *Key Generation. Let $\ell := \lceil 3\log_2 q \rceil$. Choose random $g_1, \ldots, g_\ell \leftarrow \mathbb{G}$ and a random secret key vector $s = (s_1, \ldots, s_\ell) \leftarrow \{0,1\}^\ell$. Let $h = (g^{s_1} \cdots g_\ell^{s_\ell})^{-1}$ and define the public and secret keys to be*

$$\mathsf{pk_{BHHO}} := (g_1, \ldots, g_\ell, h), \quad \mathsf{sk_{BHHO}} = (g^{s_1}, \ldots, g^{s_\ell}).$$

- *Encryption. To encrypt $m \in \mathsf{Msg}$, choose a random $r \leftarrow \mathbb{Z}_q$ and output the ciphertext $(g_1^r, \ldots, g_\ell^r, h^r \cdot g^m)$.*
- *Decryption. Let $(c_1, \ldots, c_\ell, d)$ be a ciphertext and $\mathsf{sk_{BHHO}} = (v_1, \ldots, v_\ell)$ a secret key. Do:*
  - *Decode the secret key: For $i = 1, \ldots, \ell$, set $s_i \leftarrow 0$ if $v_i = 1$ and $s_i \leftarrow 1$ otherwise.*
  - *Output $m \in \mathsf{Msg}$ for which $g^m = d \cdot (c_1^{s_1} \cdots c_\ell^{s_\ell})$.*

Theorem 3 (Circular Security of BHHO [5]). *Assuming DDH, the BBHO encryption scheme satisfies circular security, as per Definition 3.*

In order to emulate the homomorphic evaluation procedure of the previous sections, there are two steps we must modify:

First, we must provide a means for pairing a BHHO ciphertext of an input w with additive secret sharings of a value x to obtain a multiplicative secret sharing of g^{wx}. For ElGamal this was done given $\langle x \rangle$ and $\langle cx \rangle$, and computing $h_2^{\langle x \rangle} h_1^{-\langle cx \rangle}$. Now, for BHHO, we can perform an analogous "partial decryption" procedure given shares $\langle x \rangle$ and $\{\langle s_i x \rangle\}_{i \in [\ell]}$, for the bits s_i of the BHHO secret key. The corresponding pairing computation is given as MultShares in Fig. 5.

Once we obtain a multiplicative secret sharing of g^{wx}, we can perform the same share-conversion procedure DistributedDLog from the previous sections to

---

DDH-Based Notation and Pairing Operations

Let $s = (s_1, \ldots, s_\ell) \in \{0,1\}^\ell$.

- **Notation:** $[\![x]\!]_s = (g_1', \ldots, g_\ell', h') \in \mathbb{G}^{\ell+1}$ for which $g^x = h' \cdot \prod_{t \in \ell} (g_t')^{s_t}$.

  That is, BHHO ciphertext of x w.r.t. secret key s is the new level-1 encoding.

- **Pairing:** MultShares$\big([\![x]\!]_s, \langle y \rangle, \{\langle s_t y \rangle\}_{t \in [\ell]}\big) \to \langle\!\langle xy \rangle\!\rangle$.

  1. Denote $[\![x]\!]_s = (g_1', \ldots, g_\ell', h') \in \mathbb{G}^{\ell+1}$.
  2. Compute $\langle\!\langle xy \rangle\!\rangle = (h')^{\langle y \rangle} \cdot \prod_{t \in [\ell]} (g_i')^{\langle s_t y \rangle}$.

- **Pairing:** $([\![x]\!]_s)^y$, for ciphertext $[\![x]\!]_s \in \mathbb{G}^{\ell+1}$ and plaintext $y \in \mathbb{Z}_q$.

  1. Denote $[\![x]\!]_s = (g_1', \ldots, g_\ell', h') \in \mathbb{G}^{\ell+1}$.
  2. Output $[\![xy]\!]_s := \big((g_1')^y, \ldots, (g_\ell')^y, (h')^y\big) \in \mathbb{G}^{\ell+1}$.

Fig. 5. Modified DDH-based notation and pairing operations, making use of BHHO encryption [5].

DDH-Based Distributed-Evaluation HE: Gen and Enc Algorithms

Gen(1^λ):

1. Sample a PRF with input $\{0,1\}^n \times \mathbb{G}$ and output $\{0,1\}^\ell$: $\phi \leftarrow$ PRFGen(1^λ).
2. BHHO Key Setup:
   (a) Sample a DDH-hard group and generator $(\mathbb{G}, g, q) \leftarrow \mathcal{IG}(1^\lambda)$.
   (b) Sample a BHHO secret key: $s \leftarrow \{0,1\}^\ell$.
3. Sample BHHO encryptions:
   (a) The constant $1 \in \mathbb{Z}_q$: let $[\![1]\!]_c \leftarrow$ Enc($e, 1$).
   (b) The bits of the secret key s: $\forall i \in [\ell]$, let $[\![s_i]\!]_s \leftarrow$ Enc(e, s_i).
4. Sample 2-out-of-2 additive secret sharings:
   (a) The constant $1 \in \mathbb{Z}_q$: $\langle 1 \rangle \leftarrow$ AdditiveShare(1). // Included for notational convenience
   (b) The bits of the secret key s: $\forall i \in [\ell]$, let $\langle s_i \rangle \leftarrow$ AdditiveShare(s_i).
5. Output pk $= \big(\mathbb{G}, g, [\![1]\!]_s, \{[\![s_i]\!]_s\}_{i \in [\ell]}\big)$, ek$_b = \big($pk, $\langle 1 \rangle, \{\langle s_i \rangle\}_{i \in [\ell]}\big)$.

Enc$_{\mathbb{G}, g}$(pk, w):

1. Compute the following values:
   (a) BHHO encryption of $w \in \mathbb{Z}$: let $[\![w]\!]_s = ([\![1]\!]_s)^w \in \mathbb{G}^{\ell+1}$.
   (b) BHHO encryptions of $(w \cdot s_i) \in \mathbb{Z}$: i.e., for each $i \in [\ell]$, let $[\![s_i w]\!]_s = ([\![s_i]\!]_s)^w$.
2. Output $([\![w]\!]_s, \{[\![s_i w]\!]_s\}_{i \in [\ell]})$.

Fig. 6. DDH-based homomorphic encryption with distributed evaluation, making use of the BHHO cryptosystem.

return to an additive secret sharing of wx (with some error probability δ). But, to be able to perform a future pairing as above, we additionally must generate additive secret sharings $\langle wxs_i \rangle$ for each of the bits s_i of the secret key (analogous to generating $\langle cwx \rangle$ in the ElGamal case). Conveniently, this BHHO task is actually slightly simpler than that for ElGamal: whereas before we had to deal with the large size of the secret key $c \in \mathbb{Z}_q$ by operating on a bit decomposition of c and then reconstructing, here the secret key $(s_1, \ldots, s_\ell)$ is already interpreted as a binary vector. This means we can perform the multiplication steps directly without requiring the decomposition/reconstruction steps.

We remark that BHHO ciphertexts are multiplicatively homomorphic in the same fashion as ElGamal, which allows us to obtain a public-key variant of the secret sharing scheme precisely as in the previous section. The required procedure of modifying a ciphertext of some message x to one encrypting xy given y is explicitly described as $([\![x]\!]_s)^y$ in Fig. 5.

In Fig. 5, we provide the modified notation and pairing procedures for this setting. The remaining notations $\langle x \rangle$, $\langle\!\langle x \rangle\!\rangle$ and pairing operation ConvertShares will remain as in the previous sections (Fig. 1). Given these sub-procedures, we present in Figs. 6 and 7 the corresponding algorithms Gen, Enc, Eval. The resulting share size is roughly λ times larger than in the previous section, as BHHO ciphertexts are $\lambda + 1$ group elements in comparison to ElGamal which is 2. We refer the reader to the full version for a full proof of the following theorem.

Theorem 4 (DEHE from DDH). *Assuming DDH, then the scheme* (Gen, Enc, Eval) *as given in Figs. 6, 7 is a secure Distributed-Evaluation Homomorphic Encryption scheme for the class of deterministic branching programs.*

4   Applications

In this section we describe applications of our homomorphic secret sharing scheme and its public-key variant in the context of secure computation. We restrict attention to security against semi-honest parties; to obtain similar asymptotic efficiency in the presence of malicious parties, one can apply general-purpose compilation techniques [23,30]. For lack of space, formal protocol descriptions and security proofs are postponed to the full version.

4.1   Succinct Protocols for Branching Programs

Our protocols for branching programs can be based either on the weaker HSS primitive via the transformation from [6], or can be built more directly from the public-key variant. We present here the latter approach, which is more direct. For simplicity, we restrict attention to the case of evaluating a *single* branching program P on inputs x_0, x_1 held by Party 0 and Party 1 respectively. This can be extended in a straightforward way to functions with m bits of output that are computed either by m separate branching programs or by a single RMS program.

DDH-Based Homomorphic Evaluation - $\mathsf{Eval}_{\mathbb{G},g}(\mathsf{ek}, b, \mathsf{share}, P, M, \delta)$

Inputs: Homomorphic secret share parameters **params**, shared value **share**, RMS program description P of size $\leq S$, plaintext size bound $M \in \mathbb{Z}$, error bound δ. Take $\delta' = \delta/((k+1)MS)$.

Parse **ek** as in Figure 6, and interpret $\hat{1}$ as loaded into memory, via $\langle 1 \rangle, \{\langle s_i \rangle\}_{i \in [\ell]}$ as given. Parse P as a sequence of instructions (as in Definition 4); for each sequential instruction, perform the corresponding sequence of operations described below.

Instruction $(\mathsf{id}, \hat{y}_j \leftarrow \hat{w}_i)$:
    1: Execute the multiplication operation $(\mathsf{id}, \hat{y}_j \leftarrow \hat{w}_i \cdot \hat{1})$, as described below.

Instruction $(\mathsf{id}, \hat{y}_k \leftarrow \hat{y}_i + \hat{y}_j)$:
    1: Compute $\langle y_k \rangle \leftarrow \langle y_i \rangle + \langle y_j \rangle$, directly on the additive shares (over $\mathbb{Z}_q$).
    2: For each $t \in [\ell]$, compute $\langle s_t y_k \rangle \leftarrow \langle s_t y_i \rangle + \langle s_t y_j \rangle$ (over $\mathbb{Z}_q$).

Instruction $(\mathsf{id}, \hat{y}_k \leftarrow \hat{w}_i \cdot \hat{y}_j)$:
    1: Let $[\![w_i]\!]_s$ and $\{[\![s_t w_i]\!]_s\}_{t \in [\ell]}$ be the BHHO ciphertexts associated with w_i, and $\langle y_j \rangle$ and $\{\langle s_t y_j \rangle\}_{t \in [\ell]}$ the additive secret shares associated with y_j.
    2: Compute $\langle\!\langle w_i y_j \rangle\!\rangle = \mathsf{MultShares}([\![w_i]\!]_s, \langle y_j \rangle, \{\langle s_t y_j \rangle\}_{t \in [\ell]})$, as in Figure 5.
    3: Execute Share Conversion: $\langle w_i y_j \rangle = \mathsf{ConvertShares}(b, \langle\!\langle w_i y_j \rangle\!\rangle, \delta', M, \phi)$.
      // DistributedDLog will yield subtractive shares of $w_i \cdot y_j$
    4: **for** $t = 1$ to ℓ **do**   // Repeat above process for each $s_t \cdot y_k$
    5:     Compute $\langle\!\langle s_t w_i y_j \rangle\!\rangle = \mathsf{MultShares}([\![s_t w_i]\!]_s, \langle y_j \rangle, \{\langle s_t y_j \rangle\}_{t \in [\ell]})$.
    6:     Execute Share Conversion: $\langle s_t w_i y_j \rangle = \mathsf{ConvertShares}(b, \langle\!\langle s_t w_i y_j \rangle\!\rangle, \delta', M, \phi)$.
    7: **end for**
    8: Let $\langle y_k \rangle \leftarrow \langle w_i y_j \rangle$ and $\langle s_t y_k \rangle \leftarrow \langle s_t w_i y_j \rangle$, for each $t \in [\ell]$.

Instruction $(\mathsf{id}, \beta, \hat{O}_j \leftarrow \hat{y}_i)$:
    1: Shift $\langle y_i \rangle$ share by rerandomization offset: $\langle y_i \rangle \leftarrow \langle y_i \rangle + \phi(\mathsf{id}, g)$, over $\mathbb{Z}_q$.
      // Note that shifting *both* shares does not change the shared value
    2: Convert share from $\mathbb{Z}_q$ to $\mathbb{Z}_\beta$: i.e., $\langle O_i \rangle \leftarrow \langle y_i \rangle \mod \beta$.
    3: Output $\langle O_i \rangle$.

Fig. 7. Procedures for performing homomorphic operations on secret shares. Here, notation $\langle y \rangle$ is used to represent *this party's* share of the corresponding subtractive secret shared pair. Evaluation maintains the invariant that each of the additive secret shares $\langle y_i \rangle$ encode the correct current computation value of $\hat{y}_i$.

The simplest protocol proceeds as follows. The two parties run a general-purpose protocol (such as Yao's protocol) to jointly evaluate the key generation Gen. In the end of this sub-protocol, both parties hold a public key pk and each holds a secret evaluation key ek_b. While this step may be expensive, its complexity depends (polynomially) only on the security parameter λ, and moreover the same key setup can be used for evaluating an arbitrary number of branching programs on an arbitrary number of inputs. In this basic version of the protocol, the key generation protocol is the only step that does not make a black-box use of the underlying DDH group.

Next, each party uses $\mathsf{Enc}(\mathsf{pk}, \cdot)$ to encrypt every bit of its input, and sends the encryptions to the other party. Finally, the two parties locally run Eval to generate additive (mod-2) shares of the output $P(x_0, x_1)$. If Eval had negligible error, the parties could simply exchange their shares of the output, since the share sent to Party b is determined by the output and the share computed by Party b.

The fact that Eval has a non-negligible error δ is problematic for two reasons. First, it poses a correctness problem. This can be fixed by setting δ to be a constant (say, $\delta = 1/4$), running σ independent instances of Eval, for a statistical security parameter σ,[5] and outputting the majority value. However, this modification alone will not suffice, because the existence of errors within the homomorphic evaluation is *dependent* on the computation values, and as such the σ output bits may leak information about the inputs. Instead, the parties apply the σ instances of Eval locally (as before), and distribute the reconstruction function (computing majority of XORs) using general-purpose secure computation. This ensures that *only* the correct output is revealed (and no further information) with negligible correctness and secrecy error.

The communication complexity of the above protocol is $n \cdot \mathsf{poly}(\lambda)$, where $n = |x_0| + |x_1|$ is the combined length of the two parties' inputs. This can be improved to $n + \mathsf{poly}(\lambda)$ by using the following hybrid encryption techniques [20]. Let F_r be a pseudorandom function computable in NC^1, which can be based on DDH [31]. Following the key generation phase, each party encrypts a random key r_b for F. Then, instead of separately encrypting each bit of x_b using Enc, Party b simply masks each bit i of its input using $F_{r_b}(i)$ and sends to the other party the encryption of r_b and all of the masked bits. The value of program P on the inputs can now be expressed as the value of a (polynomially larger) publicly known branching program P' on the inputs r_0, r_1, where P' is determined by P and the masked inputs. The evaluation of P' is repeated σ times as before. This yields the following:

Theorem 5. *Under the DDH assumption, there exists a constant-round secure 2-party protocol for evaluating branching programs of size S on inputs (x_0, x_1) of total length n, using $n + \mathsf{poly}(\lambda)$ bits of communication.*

4.2 Breaking the Circuit Size Barrier for "Well Structured" Circuits

We turn to the question of reducing the communication complexity of evaluating a deep boolean circuit C of size S and depth D. We assume for simplicity that the circuit is *layered* in the sense that its S gates can be partitioned into $D + 1$ layers such that the gates from layer i (except input gates) receive their inputs from gates of layer $i - 1$. This can be generalized to a broader class of "well-structured" circuits that captures most instances of circuits that arise naturally.

---

[5] Here we assume that the events of error in different instances of Eval are independent. This can be enforced by using a fresh set of pseudorandom values for each share conversion.

Given a layered circuit as above, we divide the layers into intervals of $\lceil \log S \rceil$ consecutive layers, and pick for every interval the layer that has the smallest number of gates (except for the input layer). Overall, we have at most $D/\log S$ "special" layers, whose total size is at most $S/\log S$. In addition, the output layer is considered the last special layer.

The crucial observation is that each output of a new special layer can be expressed as a circuit of depth $O(\log S)$ applied to values of the previous special layer. The protocol will compute the values of the special layers one at a time, by using the previous protocol for branching programs, except that the reconstruction protocol is only applied in the end. That is, given additive shares of special layer i, each party encrypts his shares and the parties apply Eval on a function (computable by polynomial-size branching programs) that first reconstructs the value and then computes the outputs of special layer $i+1$.

To avoid a multiplicative factor of σ in communication, we need to apply a more efficient error correction procedure for intermediate layers. To this end, we apply an asymptotically good error-correcting code, with encoding and decoding in NC^1, for encoding the values of each special layer. (Many such codes are known to exist; see, e.g., [34]; moreover, by using a Las-Vegas type algorithm for the share conversion it suffices to correct *erasures*.) The computation performed by Eval will start by reconstructing the noisy encoding of layer i (using XOR), then apply a decoder to recover the actual values of layer i, then compute the outputs of layer $i+1$, and then encode these outputs. If the error probability δ of Eval is smaller than the relative error correction radius of the code, the error rate in the encoded output will be within the error-correction radius with overwhelming probability. Thus, we can use a general-purpose protocol for decoding the correct outputs from the shared noisy encoding. This approach yields the following theorem.

Theorem 6. *Under the DDH assumption, there exists a secure 2-party protocol for evaluating any layered boolean circuit of size S, depth D, input length n, and output length m using $O(S/logS) + O(D\sigma/\log D) + n + m \cdot poly(\lambda)$ bits of communication (for σ, λ statistical and computational security parameters).*

4.3 Function Secret Sharing and Generalized PIR

FUNCTION SECRET SHARING. As discussed in the Introduction, homomorphic secret sharing can be viewed as a "dual" notion of *function secret sharing*, as defined in [6]. In a homomorphic secret sharing scheme for a class of programs $\mathcal{P}$, given a share of a secret input w and a public *program* $P \in \mathcal{P}$, one can locally compute a share of $\Pi(w)$. In a function secret sharing (FSS) scheme for function class $\mathcal{F}$, given a share of a secret function (represented by a "program"') and a public *input* x, one can locally compute a share of $f(x)$. In particular, given a homomorphic secret sharing scheme supporting a class of programs $\mathcal{P}$ containing a *universal program* U, one can directly obtain a FSS scheme for $\mathcal{P}$, by secret sharing a *description* of the secret program $P \in \mathcal{P}$, and then shares of the evaluation of P on an input x can be obtained by homomorphically evaluating the universal program $U_x(\cdot)$ on the given shares. If for each program $P \in \mathcal{P}$ the homomorphic secret sharing scheme produces output error on the evaluation of P with probability δ (over the randomness of the secret sharing), then for each input x in the domain of f, the resulting FSS scheme will also yield an output error with probability δ.

Thus, as a corollary of our homomorphic secret sharing scheme, we obtain a DDH-based FSS scheme for branching program with an arbitrary inverse polynomial error. The resulting FSS key size corresponds to the size of a homomorphic secret share of a description of the secret function: namely, a fixed polynomial in the size of the branching program S and security parameter λ.

PRIVATE INFORMATION RETRIEVAL. A motivating application regime of function secret sharing (and thus our homomorphic secret sharing scheme) is that of 2-server private information retrieval (PIR) for expressive query classes [6]. As we demonstrate, such applications can be achieved with *negligible* error even when starting with FSS with inverse-polynomial error δ. Together with our construction of such δ-FSS, this gives us DDH-based 2-server PIR for queries expressed by branching programs. Useful examples include counting or retrieving matches that are specified by conjunction queries or fuzzy match predicates (e.g., requiring that a document contains at least a given threshold of keywords from a given list).

A (standard) FSS scheme for a program class $\mathcal{P}$ can be used to obtain secure 2-server PIR schemes for classes of queries related to $\mathcal{P}$, via three basic steps. For simplicity, we focus our treatment to querying the count of database entries satisfying a (secret) predicate $f \in \mathcal{F}$.[6] (1) The client generates FSS shares P_0, P_1 of the desired query P and sends one share to each server. (2) The servers locally compute, and reply with, the linear combination $\sum_{x \in DB} P_b(x)$ for database DB (where the output group of P, P_0, P_1 is $\mathbb{Z}_N$ for $N = |DB|$). (3) Then, leveraging the linearity of FSS reconstruction, the client can recover the desired output $\sum_{x \in DB} P(x) = \sum_{x \in DB} P_0(x) + \sum_{x \in DB} P_1(x)$. To extend this approach to δ-FSS, we execute several independent parallel instances of the δ-FSS scheme, and compute the *majority* of the resulting execution outputs. Note that here, unlike the secure computation application, there is no danger of releasing the (potentially noisy) outcomes of the parallel executions directly, as no hiding guarantees are required against the PIR client.

5   Examples and Optimizations

We now introduce several optimizations and trade-offs between computation and communication within our HSS construction, and describe their applications within two examples of using homomorphic secret sharing. We first consider a toy example of a client who computes the AND of n bits $x_1, \ldots, x_n$ by homomorphic

---

[6] Using sketching or coding techniques (e.g., [18,32]), this approach can be extended to *recovery* of data entries satisfying a hidden predicate.

secret sharing. We then describe a partial match search in a PIR setting. All the examples and optimizations within this section assume the circular security of ElGamal.

The communication of our homomorphic secret sharing scheme is dominated by the $\ell+1$ ElGamal ciphertexts, or $2(\ell+1)$ group elements encoding each input bit (where $\ell = \log_2 |\mathbb{G}|$ for DDH group $\mathbb{G}$). The computation is dominated by running MultShares and ConvertShares $\ell+1$ times for each product of a memory variable and input variable.

MultShares consists of raising a group element to the power of a secret share, which given a sliding window implementation requires less than $3\ell/2$ group operations. The computation time of ConvertShares with target error δ' and maximum difference between two shares M is dominated by $\frac{4M \ln(4/\delta')}{\delta'}$ group operations. Consider the following optimizations and trade-offs.

1. *Ciphertext description reduction.* The first optimization is heuristically secure (or alternatively, secure in the random oracle model) and uses a PRG $G : \{0,1\}^\ell \to \mathbb{G}^{\ell+1}$ to reduce the communication by almost half. Let $G(\sigma) = (g^r, g^{r_1}, \ldots, g^{r_\ell})$ for a seed $\sigma \in \{0,1\}^\ell$. To encode an input bit, instead of sending $\ell+1$ complete ElGamal ciphertexts, a party will now send (a random) σ and the $\ell+1$ group elements $(g^{rc+w_i}, g^{r_1 c + w_i c_1}, \ldots, g^{r_\ell c + w_i c_\ell})$, corresponding to the *second terms* of the prescribed ciphertexts, using the outputs of $G(\sigma)$ implicitly as the first terms. Given this information, each party can locally generate the full $\ell+1$ ciphertexts, and compute as before.

2. *Modified key representation.* A trade-off reducing communication and increasing computation is possible by changing the representation of the key c from $\sum_{j=0}^{\ell-1} c_j 2^j$ in base 2 to $\sum_{j=0}^{\ell'-1} c'_j B^j$ in base $B = 2^b$ for some $b > 1$. Communication complexity and the *number* of MultShares and ConvertShares are reduced by a factor of b, as the ℓ encryptions of $\{c_j w_i\}_{j \in [\ell]}$ encoding input w_i can be replaced by ℓ' encryptions of $\{c_{j'} w_i\}_{j' \in [\ell']}$. However, in ConvertShares the possible difference M between the shares held by the two parties (equivalently, the size of encoded values) increases from 1 to $B-1$, increasing the computation time by a factor of $B-1$.

3. *Las-Vegas algorithm.* A Las-Vegas type algorithm for share conversion can be used to relax the target error probability and reduce the computation time. ConvertShares potentially induces an error in one of two situations (these are the two error cases in the proof of Proposition 1) which can both be identified by the second party. In the proposed optimization, the second party outputs a flag indicating failure in each of these cases. The client is sure that the result is correct if the second player does not return a failure. Given target error probabilities δ for the whole protocol and, e.g., $1/4$ for a single execution, we require that the number of independent executions of the algorithm γ satisfies $(1/4)^\gamma < \delta$, or $\gamma > \frac{\ln 1/\delta}{\ln 4}$. Note that this optimization may reveal to the client information on intermediate computation values, since errors are input dependent. However, this type of leakage is harmless for applications like PIR.

4. *Breaking computation into chunks.* The final trade-off increases communication and decreases computation by breaking the computation into "chunks" and encrypting (and communicating) the input to each chunk separately. Loosely speaking, if the computation is split into ζ chunks, then the required communication increases by a factor of ζ, and computation is reduced by a factor of ζ because the quadratic overhead in computing n gates is reduced to ζ times computing a quadratic overhead in n/ζ gates. In general this method requires up to ζ communication rounds, but in certain applications (like PIR) it does not require additional interaction.

HOMOMORPHIC n-BIT AND. In the first (toy) example application, the communication complexity is dominated by $2(\ell + 1)n$ group elements, to encode n bits. The operations are $n - 1$ homomorphic evaluations of AND of bits, which amount to less than $n\ell$ applications of MultShares and ConvertShares or a total of less than $\frac{3n\ell^2}{2} + \frac{4n^2\ell \ln 4n/\delta}{\delta}$ group operations. In this example, communication is minimized by using the ciphertext reduction optimization and by representing c in base $B = 2^b$. Communication complexity is about $\frac{(\ell+1)n}{b}$ group elements and computation is dominated by $\frac{3n\ell^2}{2b^2} + \frac{4Bn^2\ell \ln 4n/\delta}{b\delta}$ group operations. Computation is minimized using ciphertext reduction, the Las-Vegas algorithm, and breaking into chunks. The communication complexity is increased by a factor of ζ for each of the $\lceil \frac{\ln 1/\delta}{\ln 4} \rceil$ invocations of the Las Vegas algorithm or $\lceil \frac{\ln 1/\delta}{\ln 4} \rceil \zeta n\ell$ group elements altogether. The computation requires at most

$$\left\lceil \frac{\ln 1/\delta}{\ln 4} \right\rceil \left(\frac{3n\ell^2}{2} + \frac{16n^2\ell \ln(16n/\zeta)}{\zeta^2} \right)$$

group operations altogether.

2PC FORMULA EVALUATION. The second example is a two-party computation of a formula ψ. This application requires the public-key variant of our protocol. The unoptimized version of this protocol is roughly similar in performance to the unoptimized version of homomorphic secret sharing. However, two of the optimizations, ciphertext reduction and the Las-Vegas algorithm do not apply in this case. Communication can be minimized by representing c in base B, reducing communication by $\log B$ and increasing computation by B compared to the unoptimized version. Computation can be minimized by breaking ψ into ζ chunks increasing communication by ζ and reducing computation by ζ compared to the unoptimized version.

Acknowledgements. We thank an anonymous reviewer for pointing out the relevance of [37].

Research done in part while visiting the Simons Institute for the Theory of Computing, supported by the Simons Foundation and by the DIMACS/Simons Collaboration in Cryptography through NSF grant #CNS-1523467. Supported by ERC starting grant 259426.

First author was additionally supported by ISF grant 1709/14 and ERC starting grant 307952. Second author was additionally supported by ISF grant 1638/15, a grant by the BGU Cyber Center, the Israeli Ministry Of Science and Technology Cyber Program and by the European Union's Horizon 2020 ICT program (Mikelangelo project). Third author was additionally supported by ISF grant 1709/14, BSF grant 2012378, a DARPA/ARL SAFEWARE award, NSF Frontier Award 1413955, NSF grants 1228984, 1136174, 1118096, and 1065276. This material is based upon work supported by the Defense Advanced Research Projects Agency through the ARL under Contract W911NF-15-C-0205. The views expressed are those of the author and do not reflect the official policy or position of the Department of Defense, the National Science Foundation, or the U.S. Government.

References

1. Asharov, G., Jain, A., López-Alt, A., Tromer, E., Vaikuntanathan, V., Wichs, D.: Multiparty computation with low communication, computation and interaction via threshold FHE. In: Pointcheval, D., Johansson, T. (eds.) EUROCRYPT 2012. LNCS, vol. 7237, pp. 483–501. Springer, Heidelberg (2012)

2. Beimel, A., Ishai, Y., Kushilevitz, E., Orlov, I.: Share conversion and private information retrieval. In: Proceedings of CCC, pp. 258–268 (2012)

3. Ben-Or, M., Goldwasser, S., Wigderson, A.: Completeness theorems for non-cryptographic fault-tolerant distributed computation (extended abstract). In: Proceedings of STOC, pp. 1–10 (1988)

4. Boneh, D., Goh, E.-J., Nissim, K.: Evaluating 2-DNF formulas on ciphertexts. In: Kilian, J. (ed.) TCC 2005. LNCS, vol. 3378, pp. 325–341. Springer, Heidelberg (2005)

5. Boneh, D., Halevi, S., Hamburg, M., Ostrovsky, R.: Circular-secure encryption from decision Diffie-Hellman. In: Wagner, D. (ed.) CRYPTO 2008. LNCS, vol. 5157, pp. 108–125. Springer, Heidelberg (2008)

6. Boyle, E., Gilboa, N., Ishai, Y.: Function secret sharing. In: Oswald, E., Fischlin, M. (eds.) EUROCRYPT 2015. LNCS, vol. 9057, pp. 337–367. Springer, Heidelberg (2015)

7. Brakerski, Z., Vaikuntanathan, V.: Efficient fully homomorphic encryption from (standard) LWE

8. Chaum, D., Crépeau, C., Damgård, I.: Multiparty unconditionally secure protocols (extended abstract). In: Proceedigs of STOC, pp. 11–19 (1988)

9. Chor, B., Gilboa, N.: Computationally private information retrieval (extended abstract). In: Proceedings of 29th Annual ACM Symposium on the Theory of Computing, pp. 304–313 (1997)

10. Chor, B., Gilboa, N., Naor, M.: Private information retrieval by keywords. IACR Cryptology ePrint Archive 1998:3 (1998)

11. Chor, B., Goldreich, O., Kushilevitz, E., Sudan, M.: Private information retrieval. J. ACM **45**(6), 965–981 (1998)

12. Clear, M., McGoldrick, C.: Multi-identity and multi-key leveled fhe from learning with errors. In: Gennaro, R., Robshaw, M. (eds.) CRYPTO 2015. LNCS, vol. 9216, pp. 630–656. Springer, Heidelberg (2015)

13. Cook, S.A., Hoover, H.J.: A depth-universal circuit. SIAM J. Comput. **14**(4), 833–839 (1985)

14. Cramer, R., Damgård, I.B., Ishai, Y.: Share conversion, pseudorandom secret-sharing and applications to secure computation. In: Kilian, J. (ed.) TCC 2005. LNCS, vol. 3378, pp. 342–362. Springer, Heidelberg (2005)

15. Dodis, Y., Halevi, S., Rothblum, R.D., Wichs, D.: Spooky encryption and its applications. IACR Cryptology ePrint Archive, 2016:272 (2016). To appear in Crypto 2016
16. Efremenko, K.: 3-query locally decodable codes of subexponential length. In: Proceedings of STOC, pp. 39–44 (2009)
17. Feigenbaum, J., Ishai, Y., Malkin, T., Nissim, K., Strauss, M.J., Wright, R.N.: Secure multiparty computation of approximations. In: Orejas, F., Spirakis, P.G., van Leeuwen, J. (eds.) ICALP 2001. LNCS, vol. 2076, pp. 927–938. Springer, Heidelberg (2001)
18. Finiasz, M., Ramchandran, K.: Private stream search at the same communication cost as a regularsearch: role of LDPC codes. In: Proceedings of ISIT, pp. 2556–2560 (2012)
19. Gentry, C.: Fully homomorphic encryption using ideal lattices. In: Proceedings of STOC, pp. 169–178 (2009)
20. Gentry, C., Groth, J., Ishai, Y., Peikert, C., Sahai, A., Smith, A.D.: Using fully homomorphic hybrid encryption to minimize non-interactive zero-knowledge proofs. J. Cryptol. 28(4), 820–843 (2015)
21. Gentry, C., Sahai, A., Waters, B.: Homomorphic encryption from learning with errors: conceptually-simpler, asymptotically-faster, attribute-based. In: Canetti, R., Garay, J.A. (eds.) CRYPTO 2013, Part I. LNCS, vol. 8042, pp. 75–92. Springer, Heidelberg (2013)
22. Gilboa, N., Ishai, Y.: Distributed point functions and their applications. In: Nguyen, P.Q., Oswald, E. (eds.) EUROCRYPT 2014. LNCS, vol. 8441, pp. 640–658. Springer, Heidelberg (2014)
23. Goldreich, O., Micali, S., Wigderson, A.: How to play any mental game or a completeness theorem for protocols with honest majority. In: Proceedings of STOC, pp. 218–229 (1987)
24. Halevi, S., Shoup, V.: Bootstrapping for HElib. In: Oswald, E., Fischlin, M. (eds.) EUROCRYPT 2015. LNCS, vol. 9056, pp. 641–670. Springer, Heidelberg (2015)
25. Ishai, Y., Paskin, A.: Evaluating branching programs on encrypted data. In: Vadhan, S.P. (ed.) TCC 2007. LNCS, vol. 4392, pp. 575–594. Springer, Heidelberg (2007)
26. Kiayias, A., Leonardos, N., Lipmaa, H., Pavlyk, K., Tang, Q.: Optimal rate private information retrieval from homomorphic encryption. PoPETs 2015(2), 222–243 (2015)
27. Kushilevitz, E., Ostrovsky, R.: Replication is NOT needed: SINGLE database, computationally-private information retrieval. In: Proceedings of FOCS 1997, pp. 364–373 (1997)
28. López-Alt, A., Tromer, E., Vaikuntanathan, V.: On-the-fly multiparty computation on the cloud via multikey fully homomorphic encryption. In: Proceedings of STOC 2012, pp. 1219–1234 (2012)
29. Mukherjee, P., Wichs, D.: Two round multiparty computation via multi-key FHE. In: Fischlin, M., Coron, J.-S. (eds.) EUROCRYPT 2016. LNCS, vol. 9666, pp. 735–763. Springer, Heidelberg (2016). doi:10.1007/978-3-662-49896-5_26
30. Naor, M., Nissim, K.: Communication preserving protocols for secure function evaluation. In: Proceedings of STOC, pp. 590–599 (2001)
31. Naor, M., Reingold, O.: Number-theoretic constructions of efficient pseudo-random functions. In: Proceedings of FOCS, pp. 458–467 (997)
32. Ostrovsky, R., Skeith III, W.E.: Private searching on streaming data. In: Shoup, V. (ed.) CRYPTO 2005. LNCS, vol. 3621, pp. 223–240. Springer, Heidelberg (2005)

33. Rivest, R.L., Adleman, L., Dertouzos, M.L.: On data banks and privacy homomorphisms. In: Foundations of Secure Computation, pp. 169–179. Academic, New York (1978)
34. Spielman, D.A.: Linear-time encodable and decodable error-correcting codes. IEEE Trans. Inf. Theory **42**(6), 1723–1731 (1996)
35. Valiant, L.G.: Universal circuits (preliminary report). In: Proceedings of STOC 1976, pp. 196–203 (1976)
36. van Dijk, M., Gentry, C., Halevi, S., Vaikuntanathan, V.: Fully homomorphic encryption over the integers. In: Gilbert, H. (ed.) EUROCRYPT 2010. LNCS, vol. 6110, pp. 24–43. Springer, Heidelberg (2010)
37. van Oorschot, P.C., Wiener, M.J.: Parallel collision search with cryptanalytic applications. J. Cryptol. **12**(1), 1–28 (1999)
38. Yao, A.C.-C.: How to generate and exchange secrets (extended abstract). In: Proceedings of FOCS, pp. 162–167 (1986)
39. Yekhanin, S.: Towards 3-query locally decodable codes of subexponential length. In: Proceedings of STOC, pp. 266–274 (2007)

Algorithmic Number Theory

Extended Tower Number Field Sieve: A New Complexity for the Medium Prime Case

Taechan Kim[1](✉) and Razvan Barbulescu[2]

[1] NTT Secure Platform Laboratories, Tokyo, Japan
taechan.kim@lab.ntt.co.jp
[2] CNRS, Univ Paris 6, Univ Paris 7, Paris, France
razvan.barbulescu@imj-prg.fr

Abstract. We introduce a new variant of the number field sieve algorithm for discrete logarithms in $\mathbb{F}_{p^n}$ called exTNFS. The most important modification is done in the polynomial selection step, which determines the cost of the whole algorithm: if one knows how to select good polynomials to tackle discrete logarithms in $\mathbb{F}_{p^\kappa}$, exTNFS allows to use this method when tackling $\mathbb{F}_{p^{\eta\kappa}}$ whenever $\gcd(\eta, \kappa) = 1$. This simple fact has consequences on the asymptotic complexity of NFS in the medium prime case, where the complexity is reduced from $L_Q(1/3, \sqrt[3]{96/9})$ to $L_Q(1/3, \sqrt[3]{48/9})$, $Q = p^n$, respectively from $L_Q(1/3, 2.15)$ to $L_Q(1/3, 1.71)$ if multiple number fields are used. On the practical side, exTNFS can be used when $n = 6$ and $n = 12$ and this requires to updating the keysizes used for the associated pairing-based cryptosystems.

Keywords: Discrete logarithm problem · Number field sieve · Finite fields · Cryptanalysis

1 Introduction

The discrete logarithm problem (DLP) is at the foundation of a series of public key cryptosystems. Over a generic group of cardinality N, the best known algorithm to solve the DLP has an exponential running time of $O(\sqrt{N})$. However, if the group has a special structure one can design better algorithms, as is the case for the multiplicative group of finite fields $\mathbb{F}_Q = \mathbb{F}_{p^n}$ where the DLP can be solved much more efficiently than in the exponential time. For example, when the characteristic p is small compared to the extension degree n, the best known algorithms have quasi-polynomial time complexity [6,21].

DLP Over Fields of Medium and Large Characteristic. Recall the usual L_Q-notation,

$$L_Q(\ell, c) = \exp\left((c + o(1))(\log Q)^\ell (\log \log Q)^{1-\ell}\right),$$

This work is a merged version of two consecutive works [4,24].

© International Association for Cryptologic Research 2016
M. Robshaw and J. Katz (Eds.): CRYPTO 2016, Part I, LNCS 9814, pp. 543–571, 2016.
DOI: 10.1007/978-3-662-53018-4_20

for some constants $0 \leq \ell \leq 1$ and $c > 0$. We call the characteristic $p = L_Q(\ell_p, c_p)$ medium when $1/3 < \ell_p < 2/3$ and large when $2/3 < \ell_p \leq 1$. We say that a field $\mathbb{F}_{p^n}$ is in the boundary case if $\ell_p = 2/3$.

For medium and large characteristic, in particular when Q is prime, all the state-of-the-art attacks are variants of the number field sieve (NFS) algorithm. Initially used for factoring, NFS was rapidly introduced in the context of DLP [20,32] to target prime fields. One had to wait almost one decade before the first constructions for $\mathbb{F}_{p^n}$ with $n > 1$ were proposed [33], known today [7] as the tower number field sieve (TNFS). This case is important because it is used to choose the key sizes for pairing based cryptosystems. Since 2006 one can cover the complete range of large and medium characteristic finite fields [22]. This latter approach that we denote by JLSV has the advantage to be very similar to the variant used to target prime fields, except for the first step called polynomial selection where two new methods were proposed: $JLSV_1$ and $JLSV_2$.

In the recent years NFS in fields $\mathbb{F}_{p^n}$ with $n > 1$ has become a laboratory where one can push NFS to its limits and test new ideas which are ineffective or impossible in the factorization variant of NFS. Firstly, the polynomial selection methods were supplemented with the generalized Joux-Lercier (GJL) method [5,27], with the Conjugation (Conj) method [5] and the Sarkar-Singh (SS) method [31]. One can see Table 1 for a summary of the consequences of these methods on the asymptotic complexity. In particular, in all these algorithms the complexity for the medium prime case is slightly larger than that of the large prime case.

Table 1. The complexity of each algorithms in the medium and large prime cases. Each cell indicates c if the complexity is $L_Q(1/3, (c/9)^{\frac{1}{3}})$.

$p = L_Q(\ell_p)$	$1/3 < \ell_p < 2/3$	best $\ell_p = 2/3$	$2/3 < \ell_p < 1$
TNFS [7,33]	none	none	64
NFS-JLSV [22]	128	64	64
NFS-(Conj and GJL) [5]	96	48	64
NFS-SS [31]	96	48	64
exTNFS (this article)	48^a	48^a	64

[a]The best complexity is obtained when n has a factor of the appropriate size as specified in Theorem 1.

Secondly, a classical idea which was introduced in the context of factorization is to replace the two polynomials f and g used in NFS by a polynomial f and several polynomials g_i, $i = 1, 2, \ldots$ which play the role of g. All the currently known variants of NFS admit such variants with multiple number fields (MNFS) which have a slightly better asymptotic complexity, as shown in Table 2. The discrete logarithm problem allows to have a case with no equivalent in the factorization context: instead of having a distinguished polynomial f and many sides g_i all the polynomials are interchangeable [8].

Table 2. The complexity of each algorithms using multiple number fields. Each cell indicates an approximation of c if the complexity is $L_Q(1/3, (c/9)^{\frac{1}{3}})$

$p = L_Q(\ell_p)$	$1/3 < \ell_p < 2/3$	best $\ell_p = 2/3$	$2/3 < \ell_p < 1$
MTNFS [7]	none	none	61.93
MNFS-JLSV [8]	122.87	61.93	61.93
MNFS-(Conj and GJL) [30]	89.45	45.00	61.93
MNFS-SS [31]	89.45	45.00	61.93
MexTNFS (this article)	45.00^a	45.00^a	61.93

[a] The best complexity is obtained under the assumption that n has a factor of the appropriate size. See Theorem 1.

Thirdly, when the characteristic p has a special form, as it is the case for fields in several pairing-based cryptosystems, one might speed-up the computations by variants called special number field sieve (SNFS). In Table 3 we list the asymptotic complexity of each algorithm. Once again, the medium characteristic case has been harder than the large characteristic one.

Table 3. The complexity of each algorithms used when the characteristic has a special form (SNFS) Each cell indicates an approximation of c if the complexity is $L_Q(1/3, (c/9)^{\frac{1}{3}})$

$p = L_Q(\ell_p)$	$1/3 < \ell_p < 2/3$	$2/3 < \ell_p < 1$
SNFS-JP [23]	64	32
STNFS [7]	none	32
SexTNFS (this article)	32^a	32

[a] The best complexity is obtained under the assumption that n has a factor of the appropriate size. See Theorem 1.

Our Contributions. Let us place ourselves in the case when the extension degree is composite with relatively prime factors, $n = \eta\kappa$ with $\gcd(\eta, \kappa) = 1$. If the particular cases $\eta = 1$ and $\kappa = 1$ we obtain known algorithms but we don't exclude thses cases from our presentation. The basic idea is to use the trivial equality

$$\mathbb{F}_{p^n} = \mathbb{F}_{(p^\eta)^\kappa}.$$

In the JLSV algorithm, $\mathbb{F}_{p^n}$ is constructed as $\mathbb{F}_p[x]/k(x)$ for an irreducible polynomial $k(x)$ of degree n. In the TNFS algorithm $\mathbb{F}_{p^n}$ is obtained as R/pR where R is a ring of integers of a number field where p is inert. In our construction $\mathbb{F}_{p^n} = R/pR$ as in TNFS and $\mathbb{F}_{p^n} = (R/pR)[x]/(k(x))$ where k is a degree κ irreducible polynomial over $\mathbb{F}_{p^\eta}$.

Interestingly, this construction can be integrated in an algorithm, that we call the extended number field sieve (exTNFS), in which we can target $\mathbb{F}_{p^{\eta\kappa}}$

· with the same complexity as $\mathbb{F}_{P^\kappa}$ for a prime P of the same bitsize as p^η. Hence we obtain complexities for composite extension degrees which are similar in the medium characteristic case to the large characteristic case. This is because our construction lets us to consider the norm of an element from a number field K_f that is 'doubly' extended by $h(t)$ and $f(x)$, i.e. $K_f := \mathbb{Q}(\iota, \alpha_f)$, where ι and α_f denote roots of h and f, respectively. It provides a smaller norm size, which plays an important role during the complexity analysis than when we work with an absolute extension of the same degree.

Since the previous algorithms have an "anomaly" in the case $\ell_p = 2/3$, where the complexity is better than in the large prime case, when n is composite we obtain a better complexity for the medium prime case than in the large prime case.

Overview. We introduce the new algorithm in Sect. 2 and analyse its complexity in Sect. 3. The multiple number field variant and the one dedicated to fields of SNFS characteristic are discussed in Sect. 4. In Sect. 5 we make a precise comparison to the state-of-the-art algorithms at cryptographic sizes before deriving new key sizes for pairings in Sect. 6. We conclude with cryptographic implications of our result in Sect. 7.

2   Extended TNFS

2.1   Setting

Throughout this paper, we target fields $\mathbb{F}_Q$ with $Q = p^n$ where $n = \eta\kappa$ such that $\eta, \kappa \neq 1$, $\gcd(\eta, \kappa) = 1$ and the characteristic p is medium or large, i.e. $\ell_p > 1/3$.

First we select a polynomial $h(t) \in \mathbb{Z}[t]$ of degree η which is irreducible modulo p. We put $R := \mathbb{Z}[t]/h(t)$ and note that $R/pR \simeq \mathbb{F}_{p^\eta}$. Then we select two polynomials f and g with integer coefficients whose reductions modulo p have a common factor $k(x)$ of degree κ which is irreducible over $\mathbb{F}_{p^\eta}$. Our algorithm is unchanged if f and g have coefficients in R because in all the cases we use the number fields K_f (resp. K_g) defined by f (resp. g) above the fraction field of R but this generalization is not needed for the purpose of this paper, except in a MNFS variant.

The conditions on f, g and h yield two ring homomorphisms from $R[x]/f(x)$ (resp. $R[x]/g(x)$) to $(R/pR)/k(x) = \mathbb{F}_{p^{\eta\kappa}}$: in order to compute the reduction of a polynomial in $R[x]$ modulo p then modulo $k(x)$ one can start by reducing modulo f (resp. g) and continue by reducing modulo p and then modulo $k(x)$. The result is the same if we use f as when we use g. Thus one has the commutative diagram in Fig. 1 which is a generalization of the classical diagram of NFS.

After the polynomial selection, the exTNFS algorithm proceeds as all the variants of NFS, following the same steps: relations collection, linear algebra and individual logarithm. Most of these steps are very similar to the TNFS algorithms as we shall explain below.

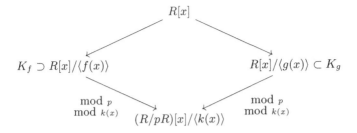

Fig. 1. Commutative diagram of exTNFS. When $R = \mathbb{Z}$ this is the diagram of NFS for non-prime fields. When $k(x) = x - m$ for some $m \in R$ this is the diagram of TNFS. When both $R = \mathbb{Z}$ and $k(x) = x - m$ this is the diagram of NFS.

2.2 Detailed Descriptions

Polynomial Selection.

Choice of h. We have to select a polynomial $h(t) \in \mathbb{Z}[t]$ of degree η which is irreducible modulo p and whose coefficients are as small as possible. As in TNFS we try random polynomials h with small coefficients and factor them in $\mathbb{F}_p[t]$ to test irreducibility. Heuristically, one succeeds after η trials and since $\eta \leq 3^\eta$ we expect to find h such that $\|h\|_\infty = 1$. For a more rigorous description on the existence of such polynomials one can refer to [7].

Next we select f and g in $\mathbb{Z}[x]$ which have a common factor $k(x)$ modulo p of degree κ which remains irreducible over $\mathbb{F}_{p^\eta}$. It is here that we use the condition $\gcd(\eta, \kappa) = 1$ because an irreducible polynomial $k(x) \in \mathbb{F}_p[x]$ remains irreducible over $\mathbb{F}_{p^\eta}$ if and only if $\gcd(\eta, \kappa) = 1$. If one has an algorithm to select f and g in $R[x]$ one might drop this condition, but in this paper f and g have integer coefficients. Thus it is enough to test the irreducibility of $k(x)$ over $\mathbb{F}_p$ and we have the same situation as in the classical variant of NFS for non-prime fields (JLSV): JLSV$_1$, JLSV$_2$, Conjugation method, GJL and Sarkar-Singh. Let us present two of these methods which are important for results of asymptotic complexity.

JLSV$_2$ Method. We briefly describe the polynomial selection introduced in Sect. 3.2 of [22]. One first chooses a monic polynomial $f_0(x)$ of degree κ with small coefficients, which is irreducible over $\mathbb{F}_p$ (and automatically over $\mathbb{F}_{p^\eta}$ because $\gcd(\eta, \kappa) = 1$). Set an integer $W \approx p^{1/(D+1)}$, where D is a parameter determined later subject to the condition $D \geq \kappa$. Then we define $f(x) := f_0(x + W)$. Take the coefficients of $g(x)$ as the shortest vector of an LLL-reduced basis of the lattice L defined by the columns:

$$L := (p \cdot \mathbf{x^0}, \ldots, p \cdot \mathbf{x^\kappa}, \mathbf{f(x)}, \mathbf{xf(x)}, \ldots, \mathbf{x^{D+1-\kappa}f(x)}).$$

Here, $\mathbf{f(x)}$ denotes the vector formed by the coefficients of a polynomial f. Finally, we set $k = f$ then we have

- $\deg(f) = \kappa$ and $\|f\|_\infty = O(p^{\frac{\kappa}{D+1}})$;
- $\deg(g) = D \geq \kappa$ and $\|g\|_\infty = O(p^{\frac{\kappa}{D+1}})$.

Conjugation Method. We recall the polynomial selection method in Algorithm 4 of [5]. First, one chooses two polynomials $g_1(x)$ and $g_0(x)$ with small coefficients such that $\deg g_1 < \deg g_0 = \kappa$. Next one chooses a quadratic, monic, irreducible polynomial $\mu(x) \in \mathbb{Z}[x]$ with small coefficients. If $\mu(x)$ has a root δ in $\mathbb{F}_p$ and $g_0 + \delta g_1$ is irreducible over $\mathbb{F}_p$ (and automatically over $\mathbb{F}_{p^\eta}$ because $\gcd(\eta, \kappa) = 1$), then set $k = g_0 + \delta g_1$. Otherwise, one repeats the above steps until such g_1, g_0, and δ are found. Once it has been done, find u and v such that $\delta \equiv u/v \pmod{p}$ and $u, v \leq O(\sqrt{p})$ using rational reconstruction. Finally, we set $f = \mathrm{Res}_Y(\mu(Y), g_0(x) + Yg_1(x))$ and $g = vg_0 + ug_1$. By construction we have

- $\deg(f) = 2\kappa$ and $\|f\|_\infty = O(1)$;
- $\deg(g) = \kappa$ and $\|g\|_\infty = O(\sqrt{p}) = O(Q^{\frac{1}{2\eta\kappa}})$.

The bound on $\|f\|_\infty$ depends on the number of polynomials $g_0 + \delta g_1$ tested before we find one which is irreducible over $\mathbb{F}_p$. Heuristically this happens on average after 2κ trials. Since there are $3^{2\kappa} > 2\kappa$ choices of g_0 and g_1 of norm 1 we have $\|f\|_\infty = O(1)$.

Relation Collection. The elements of $R = \mathbb{Z}[t]/h(t)$ can be represented uniquely as polynomials of $\mathbb{Z}[t]$ of degree less than $\deg h$.

We proceed as in TNFS and enumerate all the pairs $(a, b) \in \mathbb{Z}[t]^2$ of degree $\leq \eta - 1$ such that $\|a\|_\infty, \|b\|_\infty \leq A$ for a parameter A to be determined. We say that we obtain a relation for the pair (a, b) if

$$N_f(a, b) := \mathrm{Res}_t(\mathrm{Res}_x(a(t) - b(t)x, f(x)), h(t)) \text{ and}$$
$$N_g(a, b) := \mathrm{Res}_t(\mathrm{Res}_x(a(t) - b(t)x, g(x)), h(t))$$

are B-smooth for a parameter B to be determined (an integer is B-smooth if all its prime factors are less than B). If ι denotes a root of h in R our enumeration is equivalent to putting linear polynomials $a(\iota) - b(\iota)x$ in the top of the diagram of Fig. 1.

One can put non-linear polynomials $r(x) \in R[x]$ of degree $\tau - 1$ in the diagram for any $\tau \geq 2$ but this is not necessary in this paper. Indeed, in this paper we enumerate polynomials r to attack $\mathbb{F}_{p^{\kappa\eta}}$ of the same degree as those that one would use to attack $\mathbb{F}_{P^\kappa}$ for a prime $P \approx p^\eta$. It happens that in the large prime case and for the best parameters of the boundary case the optimal value of τ is 2. This determines us to state Lemma 1 only in the case $\tau = 2$ and to write everywhere $r = a(\iota) - b(\iota)x$, but we bear in mind that r could have a larger degree, prove Lemma 2 in Appendix A, use it in the last paragraph of Sect. 4 and write Table 5 for arbitrary values of τ before observing that the optimal value is again $\tau = 2$.

Remark 1. The choice of the polynomials r in the top of the diagram is such that the norm sizes are as small as possible. If one had an algorithm to pinpoint the principal ideals of a number field which have small norms then one would use this algorithm to generate the polynomials r.

As one of the referees notices, the advantage of exTNFS when compared to the classical version of NFS is that our enumeration is less naive. Indeed, since the norms are computed as an iteration of resultants, i.e. $N_f(r(t,x)) = \mathrm{Res}_t(\mathrm{Res}_x(r(t,x), f(x)), h(t))$, we can enumerate polynomials r which make the relative norm $\mathrm{Res}_x(r(t,x), f(x))$ small in some sense, for example we restrict to linear polynomials r.

For each pair (a, b), i.e. $r = a - bx$, one obtains a linear equation where the unknowns are logarithms of elements of the factor base as in the classical variant of NFS for discrete logarithms. But let us define the factor base in our particular case.

Factor Base. Let α_f (resp. α_g) be a root of f in K_f (resp. of g in K_g), the number field it defines over the fraction field of R. Then the norm of $a(\iota) - b(\iota)\alpha_f$ (resp. $a(\iota) - b(\iota)\alpha_g$) over $\mathbb{Q}$ is $\mathrm{Res}_t(\mathrm{Res}_x(a(t) - b(t)x, f(x)), h(t))$ (resp. $\mathrm{Res}_t(\mathrm{Res}_x(a(t) - b(t)x, g(x)), h(t)))$ up to a power of $l(f)$ (resp. $l(g)$), the leading coefficient of f (resp. g). We call factor base the set of prime ideals of K_f and K_g which can occur in the factorization of $a(\iota) - b(\iota)\alpha_f$ and $a(\iota) - b(\iota)\alpha_g$ when both norms are B-smooth. By Proposition 1 in [7] we can give an explicit description of the factor base as $\mathcal{F}(B) := \mathcal{F}_f(B) \bigcup \mathcal{F}_g(B)$ where

$$\mathcal{F}_f(B) = \left\{ \langle \mathfrak{q}, \alpha - \gamma \rangle : \begin{array}{l} \mathfrak{q} \text{ is a prime in } \mathbb{Q}(\iota) \text{ lying over a prime} \\ p \le B \text{ and } f(\gamma) \equiv 0 \pmod{\mathfrak{q}} \end{array} \right\}$$
$$\bigcup \{\text{prime ideals of } K_f \text{ dividing } l(f)\mathrm{Disc}(f)\}.$$

and similarly for $\mathcal{F}_g(B)$.

Schirokauer Maps. If $\langle a(\iota) - b(\iota)\alpha_f \rangle = \prod_{\mathfrak{q} \in \mathcal{F}_f(B)} \mathfrak{q}^{\mathrm{val}_\mathfrak{q}(a(\iota) - b(\iota)\alpha_f)}$ and $\langle a(\iota) - b(\iota)\alpha_g \rangle = \prod_{\mathfrak{q} \in \mathcal{F}_g(B)} \mathfrak{q}^{\mathrm{val}_\mathfrak{q}(a(\iota) - b(\iota)\alpha_g)}$ we write

$$\sum_{\mathfrak{q} \in \mathcal{F}_f(B)} \mathrm{val}_\mathfrak{q}(a(\iota) - b(\iota)\alpha_f) \log \mathfrak{q} + \epsilon_f(a,b) = \sum_{\mathfrak{q} \in \mathcal{F}_g(B)} \mathrm{val}_\mathfrak{q}(a(\iota) - b(\iota)\alpha_g) \log \mathfrak{q} + \epsilon_g(a,b)$$

where the log sign denotes virtual logarithms in the sense of [22,32] and ϵ_f and ϵ_g are correction terms called Schirokauer maps which were first introduced in [32].

The novelty for TNFS and exTNFS with respect to JLSV is that K_f and K_g are constructed as tower extensions instead of absolute extensions. On the other hand, it is more convenient to work on absolute extensions when we compute Schirokauer maps. We solve this problem by computing primitive elements θ_f (resp. θ_g) of $K_f/\mathbb{Q}$ (resp. $K_g/\mathbb{Q}$). For a proof we refer to Sect. 4.3 in [22].

Linear Algebra and Individual Logarithm. These two steps are unchanged with respect to the classical variant of NFS. The linear algebra step, comes after relation collection and consists in solving the linear system over $\mathbb{F}_l$ for some prime factor l of the order of $\mathbb{F}_Q^*$. Using Wiedemann's algorithm this has a quasi-quadratic complexity in the size of the linear system, which is equal to the cardinality of the factor base. In [7] it is shown that the factor base has $(2 + o(1))B/\log B$ elements, so the cost of the linear algebra is $B^{2+o(1)}$.

In the individual logarithm step one writes any desired discrete logarithm as a sum of virtual logarithms of elements in the factor base. Since the step is very similar to the corresponding step in NFS we keep the description for the Appendix.

3   Complexity

The complexity analysis of exTNFS follows the steps of the analysis of NFS in the case of prime fields. It is expected that the stages of the algorithm other than the relation collection and the linear algebra are negligible, hence we select parameters to minimize their cost and afterwards we check that the other stages are indeed negligible.

Let us call T the time spent in average for each polynomial $r \in R[x]$ enumerated in the relation collection stage (in this paper $r = a(\iota) - b(\iota)x$), and let P_f (resp. P_g) be the probability that the norm N_f (resp. N_g) of r with respect to f (resp. g) is B-smooth. The number of polynomials that we test before finding each new relation is on average $1/(P_f P_g)$, so the cost of the relations collection is $\#\mathcal{F}(B)T/(P_f P_g)$.

We make the usual heuristic that the proportion of smooth norms is the same as the proportion of arbitrary positive integers of the same size which are also smooth, so $P_f = \text{Prob}(N_f, B)$ (resp $P_g = \text{Prob}(N_g, B)$) where $\text{Prob}(x, y)$ is the probability that an arbitrary integer less than x is y-smooth. The value of T depends on whether we use a sieving technique or we consider each value and test smoothness with ECM [26]; if we use the latter variant we obtain $T = L_B(1/2, \sqrt{2})(\log Q)^{O(1)}$, so $T = B^{o(1)}$. Using the algorithm of Wiedemann [34] the cost of the linear algebra is $(\#\mathcal{F}(B))^{2+o(1)} = B^{2+o(1)}$. Hence, up to an exponent $1 + o(1)$, we have

$$\text{complexity(exTNFS)} = \frac{B}{\text{Prob}(N_f, B)\text{Prob}(N_g, B)} + B^2. \tag{1}$$

This equation is the same for NFS, TNFS, exTNFS and the corresponding SNFS variants. The differences begin when we look at the size of N_f and N_g which depend on the polynomial selection method. In what follows we instantiate Eq. (1) with various cases and obtain equations which have already been analyzed in the literature.

Lemma 1. *Let h and f be irreducible polynomials over $\mathbb{Z}$ and call $\eta := \deg h$ and $\kappa := \deg(f)$. Let $a(t), b(t) \in \mathbb{Z}[t]$ be polynomials of degree at most $\eta - 1$ with $\|a\|_\infty, \|b\|_\infty \le A$. We put $N_f(a, b) := \text{Res}_t(\text{Res}_x(a(t) - b(t)x, f(x)), h(t))$. Then we have*

1.
$$|N_f(a,b)| < A^{\eta \cdot \kappa} \|f\|_\infty^\eta \|h\|_\infty^{\kappa \cdot (\eta-1)} C(\eta, \kappa), \tag{2}$$

where $C(\eta, \kappa) = (\eta + 1)^{(3\kappa+1)\eta/2}(\kappa + 1)^{3\eta/2}$.

2. Assume in addition that $\|h\|_\infty$ is bounded by an absolute constant H and that $p = L_Q(\ell_p, c)$ for some $\ell_p > 1/3$ and $c > 0$. Then

$$N_f(a,b) \leq E^\kappa \|f\|_\infty^\eta L_Q(2/3, o(1)), \tag{3}$$

where $E = A^\eta$

Proof. 1. This is proven in Theorem 3 in [7].

2. The overhead is bounded as follows

$$\begin{aligned}
\log(\|h\|_\infty^{\kappa(\eta-1)} C(\eta, \kappa)) &\leq \kappa\eta \log H + 3\kappa\eta \log \eta + 3\eta \log \kappa \\
&= O(\log(Q)^{1-\ell_p} (\log\log Q)^{\ell_p}) \\
&= o(1) \log(Q)^{2/3} (\log\log Q)^{1/3}.
\end{aligned}$$

$\square$

If $N_f = L_Q(2/3)$ then we can forget the overhead $L_Q(2/3, o(1))$ as the Canfield-Erdös-Pomerance theorem states that the smoothness probability satisfies, uniformly on x and y in the validity domain,

$$\mathrm{Prob}(x^{1+o(1)}, y) = \mathrm{Prob}(x, y)^{1+o(1)}.$$

. The next statement summarizes our results.

Theorem 1. *(under the classical NFS heuristics) If $Q = p^n$ is a prime power such that*

- $p = L_Q(\ell_p, c_p)$ with $1/3 < \ell_p$;
- $n = \eta\kappa$ such that $\eta, \kappa \neq 1$ and $\gcd(\eta, \kappa) = 1$

then the discrete logarithm over $\mathbb{F}_Q$ can be solved in $L_Q(1/3, C)$ where C and the additional conditions are listed in Table 4.

In the rest of this section we prove this statement. In any case in the table, one shares the conditions $\kappa = o\left(\left(\frac{\log Q}{\log\log Q}\right)^{\frac{1}{3}}\right)$ or $\kappa \leq c\left(\frac{\log Q}{\log\log Q}\right)^{\frac{1}{3}}$ for some constant $c > 0$. These are equivalent to say that $P = p^\eta = L_Q(\ell_P)$ for some $\ell_P \geq 2/3$.

3.1   exTNFS-JLSV₂

In this section we assume that n has a factor κ such that

$$\kappa = o\left(\left(\frac{\log(Q)}{\log\log(Q)}\right)^{1/3}\right).$$

Table 4. Complexity of exTNFS variants.

Algorithm	C	Conditions
exTNFS-JLSV$_2$	$(64/9)^{\frac{1}{3}}$	$\kappa = o\left(\left(\frac{\log Q}{\log\log Q}\right)^{\frac{1}{3}}\right)$
exTNFS-GJL	$(64/9)^{\frac{1}{3}}$	$\kappa \le \left(\frac{8}{3}\right)^{-\frac{1}{3}}\left(\frac{\log Q}{\log\log Q}\right)^{\frac{1}{3}}$
exTNFS-Conj	$(48/9)^{\frac{1}{3}}$	$\ell_p < 2/3$ or $\ell_p = 2/3$ and $c_p < 12^{\frac{1}{3}}$
		$\kappa = 12^{-\frac{1}{3}}\left(\frac{\log Q}{\log\log Q}\right)^{\frac{1}{3}}$
SexTNFS	$(32/9)^{\frac{1}{3}}$	$\kappa = o\left(\left(\frac{\log Q}{\log\log Q}\right)^{\frac{1}{3}}\right)$
		p is d-SNFS with $d = \frac{(2/3)^{\frac{1}{3}}+o(1)}{\kappa}\left(\frac{\log Q}{\log\log Q}\right)^{\frac{1}{3}}$
MexTNFS-JLSV$_2$	$\left(\frac{92+26\sqrt{13}}{27}\right)^{\frac{1}{3}}$	$\kappa = o\left(\left(\frac{\log Q}{\log\log Q}\right)^{\frac{1}{3}}\right)$
MexTNFS-GJL	$\left(\frac{92+26\sqrt{13}}{27}\right)^{\frac{1}{3}}$	$\kappa \le \left(\frac{7+2\sqrt{13}}{6}\right)^{-1/3}\left(\frac{\log Q}{\log\log Q}\right)^{\frac{1}{3}}$
MexTNFS-Conj	$\frac{3+\sqrt{3(11+4\sqrt{6})}}{\left(18(7+3\sqrt{6})\right)^{1/3}}$	$\ell_p < 2/3$ or $\ell_p = 2/3$ and $c_p < \left(\frac{56+24\sqrt{6}}{12}\right)^{1/3}$
		$\kappa = \left(\left(\frac{56+24\sqrt{6}}{12}\right)^{-1/3}+o(1)\right)\left(\frac{\log Q}{\log\log Q}\right)^{\frac{1}{3}}$

Table 5. Comparison of norm sizes. $\tau = \deg r(x)$ while D and K are integer parameters subject to the conditions in the last column.

Method	Norms product	Conditions and parameters
NFS-JLSV$_1$	$E^{\frac{4n}{\tau}}Q^{\frac{\tau-1}{n}}$	
NFS-JLSV$_2$	$E^{\frac{2(n+D)}{\tau}}Q^{\frac{\tau-1}{D+1}}$	$D = \deg(g) \ge n$
NFS-GJL	$E^{\frac{2(2D+1)}{\tau}}Q^{\frac{\tau-1}{D+1}}$	$D \ge n$
NFS-Conj	$E^{\frac{6n}{\tau}}Q^{\frac{\tau-1}{2n}}$	
NFS-SS	$E^{\frac{2\eta(2K+1)}{\tau}}Q^{\frac{\tau-1}{\eta(K+1)}}$	$n = \eta\kappa, K \ge \kappa, \deg(g) = \eta K$
TNFS	$E^{\frac{2(d+1)}{\tau}}Q^{\frac{2(\tau-1)}{d+1}}$	n small, $d = \deg(f)$
exTNFS-JLSV$_1$	$E^{\frac{4\kappa}{\tau}}Q^{\frac{\tau-1}{\kappa}}$	$n = \eta\kappa, \gcd(\eta,\kappa) = 1, \eta$ small
exTNFS-JLSV$_2$	$E^{\frac{2(\kappa+D)}{\tau}}Q^{\frac{\tau-1}{D+1}}$	$n = \eta\kappa, \gcd(\eta,\kappa) = 1, \eta$ small, $D \ge \kappa$
exTNFS-GJL	$E^{\frac{2(2D+1)}{\tau}}Q^{\frac{\tau-1}{D+1}}$	$n = \eta\kappa, \gcd(\eta,\kappa) = 1, \eta$ small, $D \ge \kappa$
exTNFS-Conj	$E^{\frac{6\kappa}{\tau}}Q^{\frac{(\tau-1)}{2\kappa}}$	$n = \eta\kappa, \gcd(\eta,\kappa) = 1, \eta$ small
exTNFS-SS	$E^{\frac{2\kappa_0(2K+1)}{\tau}}Q^{\frac{\tau-1}{\kappa_0(K+1)}}$	$n = \eta\kappa_0\kappa_1, \gcd(\eta,\kappa_1) = 1,$
		η small, $K \ge \kappa_1, \deg(g) = \kappa_0 K$

Let us introduce $\|h\|_\infty = O(1)$ and the values of $\|f\|_\infty, \|g\|_\infty \approx p^{\kappa/(D+1)}$ coming from the JLSV$_2$ method (Sect. 2.2) in Eq. (2). Then we get

$$|N_f(a,b)| < \left(A^{\eta\kappa}(p^{\frac{\kappa}{D+1}})^\eta\right)^{1+o(1)} = \left(E^\kappa P^{\frac{\kappa}{D+1}}\right)^{1+o(1)}, \tag{4}$$

$$|N_g(a,b)| < \left(A^{\eta D}(p^{\frac{\kappa}{D+1}})^\eta\right)^{1+o(1)} = \left(E^D P^{\frac{\kappa}{D+1}}\right)^{1+o(1)}, \tag{5}$$

where we set $E := A^\eta$ and $P := |R/pR| = p^\eta$.

One recognizes the expressions for the norms in the large prime case [22, Appendix A.3.], where $P = p$ and $\kappa = n$. We conclude that we have the same complexity:

$$\text{complexity(exTNFS with JLSV}_2) = L_Q(1/3, \sqrt[3]{64/9}).$$

3.2 exTNFS-GJL

We relax a bit the condition from the previous section: we assume that n has a factor κ such that

$$\kappa \leq (8/3)^{-\frac{1}{3}} \left(\frac{\log(Q)}{\log\log(Q)} \right)^{1/3}.$$

Recall the characteristics of our polynomials: $\|h\|_\infty = O(1)$ and $\deg h = \eta$; $\|f\|_\infty = O(1)$ and $\deg f = D + 1$ for a parameter $D \geq \kappa$; $\|g\|_\infty \approx p^{\kappa/(D+1)}$ and $\deg g = D$. We inject these values in Eq. (2) and we get

$$|N_f(a,b)| < E^{D+1} L_Q(2/3, o(1)), \tag{6}$$

$$|N_g(a,b)| < E^D Q^{1/(D+1)} L_Q(2/3, o(1)), \tag{7}$$

where we set $E := A^\eta$ and $P := |R/pR| = p^\eta$. We recognize the expression in the first equation of Sect. 4.2 in [5], so

$$\text{complexity(exTNFS with GJL)} = L_Q(1/3, \sqrt[3]{64/9}).$$

3.3 exTNFS-Conj

We propose here a variant of NFS which combines exTNFS with the Conjugation method of polynomial selection.

Let us consider the case when $n = \eta\kappa$ with

$$\kappa = \left(\frac{1}{12^{1/3}} + o(1) \right) \left(\frac{\log(Q)}{\log\log(Q)} \right)^{1/3}.$$

Note that this implies $\ell_p \leq 2/3$ so that we are in the medium characteristic or boundary case.

As   before,   evaluating the values coming from the Conjugation method (Sect. 2.2) in Eq. (2), we have

$$|N_f(a,b)| < E^{2\kappa} L_Q(2/3, o(1)), \tag{8}$$

$$|N_g(a,b)| < E^\kappa (p^{\kappa\eta})^{1/(2\kappa)} L_Q(2/3, o(1)). \tag{9}$$

When we combine Eqs. (8) and (9) we obtain

$$|N_f(a,b)| \cdot |N_g(a,b)| < E^{3\kappa} Q^{(1+o(1))/(2\kappa)}.$$

But this is Eq. (5) in [5] when $\tau = 2$ (the parameter τ is written as t in [5], the number of coefficients of the sieving polynomial r). The rest of the computations are identical as in point 3. of Theorem 1 in [5], so

$$\text{complexity(exTNFS-Conj)} = L_Q(1/3, (48/9)^{1/3}).$$

4   Variants

4.1   The Case When p has a Special Form (SexTNFS)

In some pairing-based constructions p has a special form, e.g. in the Barreto-Naehrig curves [9] $p = 36u^4 + 36u^3 + 24u^2 + 6u + 1$ of embedding degree 12 and in the Freeman pairing-friendly constructions of embedding degree 10 [18, Sect. 5.3] $p = 25u^4 + 25u^3 + 25u^2 + 10u + 3$. For a given integer d, an integer p is d-SNFS if there exists an integer u and a polynomial $\Pi(x)$ with integer coefficients so that

$$p = \Pi(u),$$

$\deg \Pi = d$ and $\|\Pi\|_\infty$ is bounded by an absolute constant.

We consider the case when $n = \eta\kappa$, $\gcd(\eta, \kappa) = 1$ with $\kappa = o\left(\left(\frac{\log Q}{\log \log Q}\right)^{1/3}\right)$ and p is d-SNFS. In this case exTNFS is unchanged: we select h, f and g three polynomials with integer coefficients so that

- h is irreducible modulo p, $\deg h = \eta$ and $\|h\|_\infty = O(1)$;
- f and g have a common factor $k(x)$ modulo p which is irreducible of degree κ.

Choice of f and g Using the Method of Joux and Pierrot (as in SNFS-JP). Find a polynomial S of degree $\kappa - 1$ with coefficients in $\{-1, 0, 1\}$ so that $k(x) = x^\kappa + S(x) - u$ is irreducible modulo p. Since the proportion of irreducible polynomials in $\mathbb{F}_p$ of degree κ is $1/\kappa$ and there are 3^κ choices we expect this step to succeed. Then we set

$$\begin{cases} g = x^\kappa + S(x) - u \\ f = \Pi(x^\kappa + S(x)). \end{cases}$$

If f is not irreducible over $\mathbb{Z}[x]$, which happens with small probability, start over. Note that g is irreducible modulo p and that f is a multiple of g modulo p. Precisely, as in [23], we choose $S(x)$ so that it is of degree $O(\log \kappa / \log 3)$. Since $3^{O(\log \kappa / \log 3)} > \kappa$, we still have enough chance to have irreducible g. By construction we have:

- $\deg(g) = \kappa$ and $\|g\|_\infty = u = p^{1/d}$;
- $\deg(f) = \kappa d$ and $\|f\|_\infty = O\big((\log \kappa)^d\big)$.

Let us compute the analysis of this particular case of exTNFS. We inject these values in Eq. (2) and obtain

$$|N_f(a, b)| \leq E^{\kappa d} L_Q(2/3, o(1))$$
$$|N_g(a, b)| \leq E^\kappa P^{1/d} L_Q(2/3, o(1)),$$

where $E := A^\eta$ and $P := |R/pR| = p^\eta$. We recognize the size of the norms in the analysis by Joux and Pierrot [23, Sect. 6.3], so we obtain the same complexity as in their paper:

$$\text{complexity(SexTNFS)} = L_Q(1/3, (32/9)^{1/3}).$$

4.2   The Multiple Polynomial Variants (MexTNFS)

Virtually every variant of NFS can be accelerated using multiple polynomials and exTNFS makes no exception. The multiple variant of exTNFS is as follows: choose f and g which have a common factor $k(x)$ modulo p which is irreducible of degree κ using any of the methods given in Sect. 2.2. Next we set $f_1 = f$ and $f_2 = g$ and select other $V - 2$ irreducible polynomials $f_i := \mu_i f_1 + \nu_i f_2$ where $\mu_i = \sum_{j=0}^{\eta-1} \mu_{i,j} \iota^j$ and $\nu_i = \sum_{j=0}^{\eta-1} \nu_{i,j} \iota^j$ are elements of $R = \mathbb{Z}[t]/h\mathbb{Z}[t]$ such that $\|\mu_i\|_\infty, \|\nu_i\|_\infty \leq V^{\frac{1}{2\eta}}$ where $V = L_Q(1/3, c_v)$ is a parameter which will be selected later. Denote α_i a root of f_i for $i = 1, 2, \ldots, V$.

Once again the complexity depends on the manner in which the polynomials f and g are selected.

MexTNFS-JLSV$_2$. Barbulescu and Pierrot [8, Sect. 5.3] analysed the complexity of MNFS with JLSV$_2$, so we only need to check that the size of the norm is the same for NFS and exTNFS for each polynomial f_i with $1 \leq i \leq V$. By construction we have:

- $\deg(f_1) = \kappa$ and $\|f_1\|_\infty = p^{\frac{\kappa}{D+1}}$;
- $\deg(f_i) = D \geq \kappa$ and $\|f_i\|_\infty = V^{\frac{1}{2\eta}} p^{\frac{\kappa}{D+1}}$ for $2 \leq i \leq V$.

As before, we inject these values in Eq. (2) and obtain

$$|N_{f_1}(a,b)| < E^\kappa (p^{\kappa\eta})^{\frac{1}{D+1}} L_Q(2/3, o(1))$$
$$|N_{f_i}(a,b)| < E^D (p^{\kappa\eta})^{\frac{1}{D+1}} L_Q(2/3, o(1)) \text{ for } 2 \leq i \leq V.$$

We emphasize that $(V^{\frac{1}{2\eta}})^\eta = V^{\frac{1}{2}} = L_Q(1/3, c_v/2) = L_Q(2/3, o(1))$ which is true without any condition on η. Hence we obtain

$$\text{complexity(MexTNFS-JLSV}_2) = L_Q\left(1/3, \left(\frac{92 + 26\sqrt{13}}{27}\right)^{1/3}\right).$$

MexTNFS-Conj and GJL. Pierrot [30] studied the multiple polynomial variant of NFS when the Conjugation method or GJL are used. To show that we obtain the same complexities we need to show that the norm with respect to each polynomial is the same as in the classical NFS, except for a factor $L_Q(2/3, o(1))$, which boils down to testing again that $(V^{\frac{1}{2\eta}})\eta = L_Q(2/3, o(1))$ which is always true. When $P = p^\eta = L_Q(2/3, c_P)$ such that $c_P > (\frac{7+2\sqrt{13}}{6})^{1/3}$ and τ is the number of coefficients of the enumerated polynomials r, then the complexity obtained is $L_Q(1/3, C(\tau, c_P))$ where

$$C(\tau, c_P) = \frac{2}{c_P \tau} + \sqrt{\frac{20}{9(c_P \tau)^2} + \frac{2}{3} c_P(\tau - 1)}.$$

The best case is when $c_P = \left(\frac{56+24\sqrt{6}}{12}\right)^{1/3}$ and $\tau = 2$ (linear polynomials):

$$\text{complexity(best case of MexTNFS-Conj)} = L_Q\left(1/3, \frac{3 + \sqrt{3(11 + 4\sqrt{6})}}{\left(18(7 + 3\sqrt{6})\right)^{1/3}}\right),$$

where the second constant being approximated by 1.71.

5   Comparison and Examples

NFS, TNFS and exTNFS have the same main lines:

– we compute a large number of integer numbers;
– we factor these numbers to test if they are B-smooth for some parameter B;
– we solve a linear system depending on the previous steps.

If we reduce the size of the integers computed in the algorithm we reduce the work needed to find a subset of integers which are B-smooth, which further allows us to adapt the other parameters so that the linear algebra is also cheap. A precise analysis is complex because in some variants one tests smoothness using ECM while in others one can sieve (which is faster). Nevertheless, as a first comparison we use the criterion in which one must minimize the bitsize of the product of the norms.

5.1   Precise Comparison When p is Arbitrary

Each method of polynomial selection has a different expression of the norm bitsize, which depends on the number τ of coefficients of the polynomials $r(x)$ that are enumerated during the relation collection. Let us reproduce Table 2 in [31], which we extend with TNFS and exTNFS:

Note that the method of Sarkar and Singh requires that n is composite. The settings based on TNFS (TNFS, exTNFS-GJL etc.) have an overhead due to the combinatorial factor which is not written in this table, so we add the condition that the degree of the intermediate number field must be small. Finally, exTNFS requires the additional condition that κ and η are relatively prime.

Extrapolation of E. The parameter E depends on the implementation of NFS and might be different for one variant to another. Let us take for example three computations with NFS which tackle various problems of the same bitsize:

– Danilov and Popovyan [16] factored a 180-digit RSA modulus using $\log_2 E \approx 30$ (although the size of the pairs (a, b) in theirs computations is not written explicitly, one can compute E using the range of special-q's and the default cardinality of the sieving space per special-q, which is 2^{30});
– Bouvier et al. [12] computed discrete logarithms in a 180-digit field $\mathbb{F}_p$ using $\log_2 E \approx 30$ (computed from other parameters).

- Barbulescu et al. [5] computed discrete logarithms in a 180-digit field $\mathbb{F}_{p^2}$ using $\log_2 E \approx 29$.

We see that in the first approximation E depends only on the bitsize of the field that we target and has the same value as in the factoring variant of NFS. Let us extrapolate E from the pair $(\log_2 Q = 600, \log_2 E = 30)$ using the formula

$$E = cL_Q(1/3, (8/9)^{1/3}).$$

Since exTNFS requires that $\gcd(\eta, \kappa) = 1$, the first case to study is $n = 6$. *The case of fields.* $\mathbb{F}_{p^6}$ When $n = 6$ we can use the general methods

- NFS-JLSV$_1$ (bitsize $E^{\frac{24}{\tau}} Q^{\frac{\tau-1}{6}}$, best values of τ are 3 and 2)
- NFS-GJL with D equal to its optimal value, 6 (bitsize $E^{\frac{26}{\tau}} Q^{\frac{\tau-1}{7}}$, best values of τ are 3 and 2)
- TNFS with $\deg f = 5$, its optimal value for this range of fields (bitsize $E^{\frac{12}{\tau}} Q^{\frac{\tau-1}{3}}$, best value of τ is 2)

as well as the methods which exploit the fact that n is composite

- Sarkar-Singh (NFS-SS) with $\eta = 2$ and $K = 3$, best value so that $K \geq n/\eta$ for this range of fields, $(E^{\frac{28}{\tau}} Q^{\frac{\tau-1}{8}})$ respectively $\eta = 3$ and $K = 2$, best value so that $K \geq n/\eta$ for this range of fields, (bitsize $E^{\frac{30}{\tau}} Q^{\frac{\tau-1}{9}}$, best τ are 4 and 3)
- exTNFS with $\eta = 2$ or $\eta = 3$ and one of two methods for selecting f and g
  - exTNFS-GJL with $\eta = 3$, $D = 2$ its best value so that $D \geq n/\eta$, (bitsize $E^{\frac{10}{\tau}} Q^{\frac{\tau-1}{3}}$, best value of τ is 2)
  - exTNFS-GJL with $\eta = 2$, $D = 3$ its best value so that $D \geq n/\eta$, ($E^{\frac{14}{\tau}} Q^{\frac{\tau-1}{4}}$, best values of τ are 3 and 2)
  - exTNFS-Conj with $\eta = 2$ (bitsize $E^{\frac{18}{\tau}} Q^{\frac{\tau-1}{6}}$, best values of τ is 2).
  - exTNFS-Conj with $\eta = 3$ (bitsize $E^{\frac{12}{\tau}} Q^{\frac{\tau-1}{4}}$, best values of τ are 3 and 2).

We plot the values of the norms product in Fig. 2. Note that exTNFS with the Conjugation method seems to be the best choice for fields between 300 and 1000 bits.

For even more insight we enter into details on a specific field.

Example 1: Let us consider the field $\mathbb{F}_{p^6}$ when

$$p = 314159265358979323846264338\overline{3589}.$$

The bitsize of $Q = p^6$ is 608 and its number of decimal digits is 182. Since the parameter E can only be chosen after an effective computation we are bound to make the hypothesis that it will have a similar value as in a series of record computations with NFS having the same input size:

In the following $\log_2 E = 30$. Let us make a list with the norm sizes obtained with each version of NFS:

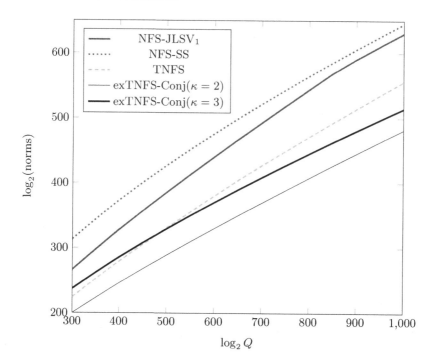

Fig. 2. Plot of the norms bitsize for several variants of NFS. Horizontal axis indicates the bitsize of p^n while the vertical axis the bitsize of the norms product. (Color figure online)

1. NFS-JLSV$_1$. We take for example $f = x^6 - 1772453850905518$ and $g = 1772453850905514x^6 + 96769484157337$. The sieving space contains polynomials of degree two $r(x) = a + xb + cx^2 \in \mathbb{Z}[x]$, i.e. $\tau = 3$, and the absolute value of the coefficients is bounded by $E^{2/3}$. The upper bound on the norms' product is

$$\text{norms bitsize(NFS-JLSV}_1) = 8\log_2 E + \frac{1}{3}\log_2 Q \approx 440.$$

2. NFS-Conj. We take $f = x^{12} + 3$ and $g = 1016344366092854x^6 - 206700367981621$. We sieve polynomials $r \in \mathbb{Z}[x]$ of degree 4, i.e. $\tau = 5$, and the absolute value of the coefficients is bounded by $E^{2/5}$. Then we obtain

$$\text{norms bitsize(NFS-Conj)} = \frac{36}{5}\log_2 E + \frac{1}{3}\log_2 Q \approx 418.$$

3. TNFS. We take $f = x^5 + 727139x^3 + 538962x^2 + 513716x + 691133$, $g = x - 1257274$ and $h = t^6 + t^4 + t + 1$. Here, h is chosen so that $\mathbb{F}_{p^6} = (\mathbb{Z}[t]/h(t))/p(\mathbb{Z}[t]/h(t))$. The sieving polynomials are of the form $r(x) = a(\iota) - b(\iota)x$, i.e. $\tau = 2$. Here, $a = \sum_{i=0}^{5} a_i \iota^i$ and $b = \sum_{i=0}^{5} a_i \iota^i$ are elements in $\mathbb{Z}(\iota) = \mathbb{Z}[t]/h(t)$ with the coefficients whose absolute values

bounded by $A = E^{1/\deg(h)} = E^{1/6}$. Note that the parameter $d = \deg f$ is equal to 5, so that we have

$$\text{norms bitsize(TNFS)} = 6 \log_2 E + \frac{1}{3} \log_2 Q \approx 380.$$

4. exTNFS-Conj with $\eta = 2$ and $\kappa = 3$. We take $f = x^6 - 3$, $g = 309331385734750x^3 - 1851661516636217$ and $h = t^2 + 2$. We sieve polynomials of the form $a(\iota) - b(\iota)x$, i.e. $\tau = 2$, where a and b are linear in ι with their coefficients bounded by $A = E^{1/2}$. Hence we obtain

$$\text{norms bitsize(exTNFS}\eta = 2) = 9 \log_2 E + \frac{1}{6} \log_2 Q \approx 370.$$

5. exTNFS-Conj with $\eta = 3$ and $\kappa = 2$. We take $f = x^4 - 2x^3 + x^2 - 3$, $g = 1542330130901467x^2 - 1542330130901467x - 923667359431967$ and $h = t^3 + t + 1$. Again we sieve polynomials of the form $a(\iota) - b(\iota)x$, i.e. $\tau = 2$, where a and b are quadratic in ι with coefficients bounded by $A = E^{1/3}$. This leads to

$$\text{norms bitsize(exTNFS } \kappa = 2) = 6 \log_2 E + \frac{1}{4} \log_2 Q \approx 330.$$

We conclude that in this example the best choice is exTNFS with $\kappa = 2$.

The condition $\gcd(\eta, \kappa) = 1$ is also satisfied by $n = 10, 12, 14, 18, 20, 24$ etc., but we do not discuss these cases in detail.

5.2 Precise Comparison When p is SNFS

To compare precise norm sizes when p is a d-SNFS prime, let us consider Table 6.

Note that SexTNFS encompass SNFS-JP when $\eta = 1$, and STNFS when $\eta = n$, so we only call it SexTNFS when $2 \leq \eta < n$.

As in the case when p is arbitrary, we do not have precise estimations of E, especially in the large range of fields $\log_2 Q \in [1000, 10000]$. We are going to extrapolate from the pair $(\log_2 Q = 1039, \log_2 E = 30.38)$, due to the record of [1], using the formula

$$E = cL_Q(1/3, (4/9)^{\frac{1}{3}}).$$

Table 6. Comparison of norm sizes when p is d-SNFS prime.

Method	Norms product	Conditions
STNFS	$E^{\frac{2(d+1)}{\tau}} Q^{\frac{\tau-1}{d}}$	
SNFS-JP	$E^{\frac{2n(d+1)}{\tau}} Q^{\frac{\tau-1}{nd}}$	
SexTNFS	$E^{\frac{2\kappa(d+1)}{\tau}} Q^{\frac{\tau-1}{\kappa d}}$	$n = \eta\kappa$
		$\gcd(\kappa, \eta) = 1$
		$2 \leq \eta < n$

Let us introduce a notation for the bitsize of SexTNFS, for any integers $\kappa \geq 1$ and $\tau \geq 2$:

$$C_{norm}(\tau, \kappa) = \frac{2\kappa(d+1)}{\tau} \log E + \frac{\tau - 1}{\kappa d} \log Q.$$

For each κ, $C_{norm}(\tau, \kappa)$ has a minimum at the integer $\tau \geq 2$ which best approximates $\left(\frac{2\kappa^2 d(d+1) \log E}{\log Q} \right)^{1/2}$.

The Case of 4-SNFS Primes. To fix ideas, we restrict at the case $d = 4$. When $\kappa = 1$, i.e. STNFS, the norm size has its minimum at $\tau = 2$ as soon as $\frac{\log Q}{\log E} \geq 40/2^2 = 10$. In our range of interest $(300 \leq \log_2 Q \leq 10000)$, the ratio $\log Q / \log E$ is always larger than 19. So, we only take care of sieving linear polynomials in the case of STNFS with $d = 4$. Similarly, it suffices to consider sieving linear polynomials in the case of SexTNFS with $\kappa = 2$ (resp. $\kappa = 3$) whenever $\log Q / \log E \geq 40$ (resp. $\log Q / \log E \geq 90$). It is satisfied when Q is of at least 1450 bits (resp. 6300 bits).

Let us compare the norm sizes of STNFS and SexTNFS when we sieve only linear polynomials $(\tau = 2)$ in both cases. The value $C_{norm}(2, \kappa)$ has a minimum at $\kappa = \left(\frac{\log Q}{d(d+1) \log E} \right)^{1/2}$. In the case of $d = 4$, this value has minimum at $\kappa = 2$ or $\kappa = 3$ whenever $20 \leq \log Q / \log E \leq 180 = 20 \cdot 3^2$. Thus, in fields with large size, SexTNFS with $\kappa = 2$ or $\kappa = 3$ is better than STNFS.

In Fig. 3 we plot the norm sizes of SNFS-JP, STNFS, and SexTNFS for $n = 12$ and $d = 4$ for Q is of from 300 bits to 5000 bits. We also compare these values with the best choice for general prime cases (exTNFS with Conjugation when $\kappa = 3$). From the plots we remark that STNFS could be a best choice for small Q otherwise SexTNFS with small κ becomes an important challenger against any other methods as the size of Q grows.

To get a better intuition, let us see in detail a specific field.

Example 2: We consider the prime $p = P_4(u_4)$ where

$$P_4(x) = 36x^4 + 36x^3 + 24x^2 + 6x + 1 \text{ and } u_4 = 2^{158} - 2^{128} - 2^{68} + 1$$

(Sect. 6 in [2]), and note that p is 4-SNFS. The bitsize of p^{12} is 7647 for which we predict by extrapolation that $\log_2 E = 76.15$.

Let us make a list with the norm sizes obtained with each version of NFS:

1. STNFS. The size of the norms is $E^{2(d+1)/\tau} Q^{(\tau-1)/d}$ and has its minimum for $\tau = 2$. Take for example $h = x^{12} + x^{10} + x^9 - x^6 - 1$, $f = P_4$ and $g = x - u_4$.

$$\text{norms bitsize(STNFS)} = 5 \log_2 E + \frac{1}{4} \log_2 Q \approx 2292.$$

2. SNFS-JP. The size of the norms is $E^{2n(d+1)/\tau} Q^{(\tau-1)/(nd)}$ and has its minimum when $\tau = 8$. Take for example $f = P_4(x^{12} + x^6 + x^3 + 1)$ and $g = (x^{12} + x^6 + x^3 + 1) - u_4$.

$$\text{norms bitsize(SNFS-JP)} = \frac{120}{7} \log_2 E + \frac{1}{8} \log_2 Q \approx 2257.$$

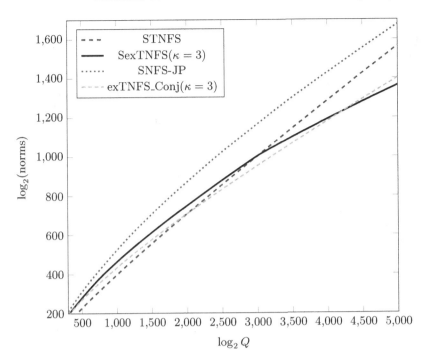

Fig. 3. Comparison when $n = 12$ and $d = 4$ for $300 \leq \log_2 Q \leq 5000$. Horizontal axis indicates the bitsize of p^n while the vertical axis the bitsize of the norms product. (Color figure online)

3. SexTNFS-JP $\eta = 4$. In this case the norm size is $E^{2\kappa(d+1)/\tau} Q^{\frac{(\tau-1)}{\kappa d}}$ and has its minimum when $\tau = 2$. Take for example $h = x^4 - x - 1$, $f = P_4(x^3 - x^2)$ and $g = x^3 - x^2 - u_4$.

$$\text{norms bitsize(SexTNFS)} = 15 \log_2 E + \frac{1}{12} \log_2 Q \approx 1779.$$

One can do a similar analysis in the cases $d = 5$, $d = 6$ etc., but we do not present the details here.

6   On the Necessity to Update Key Sizes

Pairings are not included in the 2012 report of NIST [28] but they are included in the 2013 report of ENISA [17, Table 3.6] where pairings and RSA have the same recommended key sizes. This is in accordance with a general belief stated for example by Lenstra [25, Sect. 5.1]:

> 'An RSA modulus n and a finite field $\mathbb{F}_{p^k}$ therefore offer about the same level of security if n and p^k are of the same order of magnitude.'

Freeman et al. [19] compiled key size recommendations from different sources in Table 1.1, all of which make or are coherent with the above supposition.

The currently recommended key sizes are derived from the complexity $L[c] := L_{p^n}(1/3, (c/9)^{1/3})$ with $c = 64$, which corresponds to NFS over fields whose characteristic is large and doesn't have a special form. This complexity has been a safe choice until recently because the constant $c = 64$ has been the smallest among the variants of NFS over fields of non-small characteristic.

The Case of Primes of General Form. However, exTNFS has a constant $c = 48$ for a vast range of fields, so the safe choice becomes to derive key sizes using $L[48]$. A more precise evaluation would require to determine what embedding degree is large enough to be in the medium prime case, i.e. $c = 48$, and what degree is smaller so that we use $c = 64$. This seems to be hard to tell, especially after the record computation presented in [5, Sect. 7] showed that the attack in $\mathbb{F}_{p^2}$ was 260 times faster than the attack in $\mathbb{F}_{p'}$ where p and p' are primes so that $2\log_2(p) \approx \log_2(p')$.

A crude and naive estimation, when a constant c_{old} is replaced by c_{new}, is to write

$$L_{Q_{new}}(1/3, c_{new}) = L_{Q_{old}}(1/3, c_{old})$$

which is equivalent to

$$\frac{\log Q_{new}}{\log Q_{old}} = \frac{c_{old}}{c_{new}} + o(1). \tag{10}$$

Overall, we might say that the key size should be increased by $64/48 \approx 1.33$ in an asymptotic sense (simply ignoring the factor $o(1)$), which allows to comprehend what means a change in the second constant of NFS. We avoid to derive a table of key sizes using the methods in [29, Appendix H] and [25] not because the formulae are difficult but because we lack the experience with record computations needed to validate the formulae.

The Special Prime Case. When the characteristic has a special form the constant c changed twice in three years and there are some subtle points to understand about how the key sizes were computed. Before the algorithm of Joux and Pierrot there was no variant of NFS for $\mathbb{F}_{p^n}$ with $n > 1$ and p of special form. Hence, the recommended values correspond to $c = 64$. Their SNFS algorithm updated the constants to 32 in large characteristic and 64 in the middle prime case. A pessimistic choice would have been to update the key sizes using $c = 32$. Nevertheless, the very important example of Barreto-Naehrig pairings has an embedding degree $n = 12$ which seems to be considered as medium sized (the difference between large and medium characteristic is asymptotic and is hard to translate in practice). Due to SexTNFS the constant is now $c = 32$ for all fields of non-small characteristic, so we don't need a precise examination anymore, as long as n has a factor ≥ 2. We conclude that the key sizes of pairings where p has a special form, in a polynomial of degree ≥ 3, should increase roughly by a factor $c_{old}/c_{new} = 2$.

7   Cryptologic Consequences

Our work comes in a context of recent progress on the DLP in finite fields p^n of degree $n \geq 2$. The case $n = 2$ has been the object of precise estimations and real-life computations and is now known to be weaker than the case of prime fields. On the contrary, the cases $n = 6$ and $n = 12$ remained difficult according to precise practical estimations.

In this paper we proposed the exTNFS which allowed us to apply the polynomials constructed in the case $n = 2$, which have good properties, to the highly important case $n = 6$, where the polynomials had less good properties. A precise estimation showed that this invalidates the key sizes currently used and we recommend that they should be updated (see Sect. 6). When p is of special form, as in the Barreto-Naehrig construction, one needs to update the key sizes for large characteristic because of the algorithm proposed by Joux and Pierrot in 2013 but it is not clear if the keys of the Barreto-Naehrig keys had to be updated. Due to exTNFS the key sizes of all pairings of SNFS characteristic need to be updated.

It is interesting to remark that the new variants of NFS exploit those properties of some pairings which made them fast:

- **Special form characteristic.** The advantage of using special form characteristic is that it eliminates the cost of modular reductions (see for example [10, Algorithm 4]). It is the same special form of p which allows to use the fastest variant of exTNFS, i.e. SexTNFS, rather than the general case algorithm.
- **Composite embedding degree.** In this case the pairings computations are done using tower extension field arithmetic, as explained for example in [10, Sect. 3.1]. The same structure of tower extension field is a main ingredient of exTNFS, as explained in Remark 1.

A large number of pairings have either special form characteristic or an embedding degree divisible by 2 or 3, for example the Barreto-Naehrig curves have both properties. In a recent preprint Chatterjee et al. [13] discussed the pairing constructions which are not affected by our algorithms, in particular the pairings of embedding degree one which are as secure as DSA and RSA. This shows that, regardless on the progress on DLP in $\mathbb{F}_{p^n}$ with $n > 1$, pairings are a secure tool for cryptography. Nevertheless, safe pairings might be very slow and determine cryptographers to use alternatives, as Chillotti et al. did in [14] for an e-voting protocol. We conclude with the question asked by our referee: "Is this the beginning of the end for pairing-based based cryptography?"

A   Non-linear Polynomials

In all the variants of exTNFS that we have discussed, one puts linear polynomials $r(x) \in R[x]$ in the diagram of Fig. 1. This is justified by the fact that exTNFS is a way of copying the setting from large characteristic to the medium prime case. Since in the large characteristic, the best choice is to take linear polynomials in all the variants, NFS, MNFS, SNFS, we have done the same thing in exTNFS, MexTNFS and SexTNFS.

The estimation of the norms sizes given in Lemma 1 is central in the analysis of exTNFS. For completion reasons we generalize this result to arbitrary degrees.

Lemma 2. *Let h be an irreducible polynomial over $\mathbb{Z}$ of degree η and f be an irreducible polynomial over $\mathbb{Z}[\iota]$ of degree κ. Let ι (resp. α) be a root of h (resp. f) in its number field and set $K_f := \mathbb{Q}(\iota, \alpha)$. Let $A > 0$ be a real number and τ an integer such that $2 \le \tau \le \kappa$. For each $i = 0, \ldots, \kappa - 1$, let $a_i(t) \in \mathbb{Z}[t]$ be polynomials of degree $\le \eta - 1$ with $\|a_i\|_\infty \le A$. Put $r(t, x) = \sum_{i=0}^{\tau-1} a_i(t)x^i$. Then we have*

$$\left| N_{K_f/\mathbb{Q}}\big(r(\iota, \alpha)\big) \right| < A^{\eta\kappa} \|f\|_\infty^{(\tau-1)\eta} \|h\|_\infty^{(\tau+\kappa-1)(\eta-1)} D(\eta, \kappa),$$

where $D(\eta, \kappa) = \big((2\kappa - 1)(\eta - 1) + 1\big)^{\eta/2}(\eta + 1)^{(2\kappa-1)(\eta-1)/2}\big((2\kappa - 1)!\eta^{2\kappa}\big)^\eta$. The above formula remains the same when we restrict the coefficients of f to be integers.

Proof. By abusing the notation, we write $f(t, x) := \sum_i f_i(t)x^i$ with $\deg_t(f_i) \le \kappa - 1$ for $f(x) = \sum_i f_i(\iota)x^i \in \mathbb{Z}[\iota][x]$. We write $\mathcal{R}(t) := \mathrm{Res}_x\big(A(t, x), f(t, x)\big)$ and have

$$N_{K_f/\mathbb{Q}(\iota)}(r(\iota, \alpha)) = \mathcal{R}(\iota).$$

By Theorems 8 and 10 in [11], the degree of $\mathcal{R}(t)$ is given by $(\kappa + \tau - 1)(\eta - 1)$ and

$$\|\mathcal{R}(t)\|_\infty \le (\tau + \kappa - 1)!\eta^{\tau+\kappa-2}A^\kappa\|f\|_\infty^{\tau-1}.$$

Then by Theorem 7 in the same article, we have

$$|N_{\mathbb{Q}(\iota)/\mathbb{Q}}(\mathcal{R}(\iota))| \le (\deg \mathcal{R} + 1)^{\deg h/2}(\deg h + 1)^{\deg \mathcal{R}/2}\|\mathcal{R}\|_\infty^{\deg h}\|h\|_\infty^{\deg \mathcal{R}}.$$

Combining all together, we obtain the desired result.   □

This result allows to analyze MexTNFS-SS when $\kappa = \frac{1}{c_p}\big(\frac{\log Q}{\log \log Q}\big)^3$ and $c_p < (\sqrt{78}/9 + 29/36)^{\frac{1}{3}} \approx 1.21$. Indeed, in this case one puts non-linear polynomials in the diagram, as indicated in Table 4 of [31].

Once again we check when $D(\eta, \kappa) = L_Q(2/3, o(1))$ and obtain the condition $\eta\kappa = o\big((\frac{\log Q}{\log \log Q})^{\frac{2}{3}}\big)$. The factor $\|h\|_\infty^{(T+\kappa-1)(\eta-1)}$ is also negligible under the same condition. Hence the overhead is negligible for all range $\ell_p > 1/3$.

B   Individual Logarithm

Let $s \in \mathbb{F}_{p^n}^* = \mathbb{F}_{p^{\eta\kappa}}^*$ be an element for which we want to compute the discrete logarithm. In general, the discrete logarithm of s can be found by following two steps: smoothing step and special-q descent.

In the smoothing step, the value s is randomized by $z := s^e$ for random value e and B_1-smoothness of z (for pre-determined value $B_1 > B$) is tested.

Then, for each prime ideal $\mathfrak{D}$ which is not in the factor base, one finds a linear relation involving $\mathfrak{D}$ and other smaller ideals. This step is called special-q descent. We recursively produce the special-q descent tree, and finally deduce the desired discrete logarithm.

The complexity of the individual logarithm step differs by polynomial selection methods. In the following, to fix ideas, we consider only the JLSV$_2$ and Conjugation methods (exTNFS-JLSV$_2$ and exTNFS-Conj), but similar argument directly applies to any other polynomial selection method.

Smoothing. For each $z \in \mathbb{F}_{p^n}$ we compute an element $\bar{z} \in K_f = \mathbb{Q}(\iota, \alpha_f)$ which is sent to z when ι is mapped to a root of h in $\mathbb{F}_{p^\eta}$ and α_f in a root of f in $\mathbb{F}_{p^{\eta\kappa}}$. Then we test if $N_{K_f/\mathbb{Q}}(\bar{z})$ is B_1-smooth and squarefree. Let us discuss how to compute and what is the size of its norm.

JLSV$_2$ As before, we consider the target field $\mathbb{F}_{p^n}$ as an extension field $\mathbb{F}_{p^{\eta\kappa}} = \mathbb{F}_{p^\eta}(m) = \mathbb{F}_{p^\eta}[x]/k(x)$ over $\mathbb{F}_{p^\eta} = \mathbb{F}_p(\iota) = \mathbb{F}_p[t]/h(t)$. For a given z in $\mathbb{F}_{p^n}^*$, we write $z = \sum_i z_i(\iota)m^i$, where the coefficients of z_i are non-negative integers bounded by p. We set

$$\bar{z} = \sum_{i=0}^{\kappa-1} z_i(\iota)\alpha_f^i$$

and, by Lemma 2 for $T = \kappa$, we obtain

$$|N_{K_f/\mathbb{Q}}(\bar{z})| \leq \left(p^n(p^{\kappa/(D+1)})^{n-\eta}\right)^{1+o(1)} \leq Q^{2-2/(\kappa+1)+o(1)},$$

where, in the last inequality, we used the condition that $D \geq \kappa$.

Conjugation. In this case, a direct lift would make that $\bar{z}$ has degree κ instead of $2\kappa = \deg K_f$, and the coefficients $z_i(t)$ have norm bounded by p. In order to "spread" the coefficients, i.e. compute another polynomial with the same image in $\mathbb{F}_{p^n}$ of degree 2κ and coefficients of norm $p^{1/2}$, we need to use the LLL algorithm. With no extra cost we can obtain a further improvement: use the Waterloo improvement which consists in replacing the smoothness condition of integers of a given size X by the smoothness condition of two integers of size $X^{1/2}$.

The Waterloo improvement for exTNFS-Conj is as follows: we find two bivariate polynomials $u(t,x) = \sum_{i=0}^{2\kappa-1} u_i(t)x^i$ and $v(t,x) = \sum_{i=0}^{2\kappa-1} v_i(t)x^i \in \mathbb{Z}[t,x]$ such that z is the image in $\mathbb{F}_{p^n}$ of

$$\bar{z} := \frac{u(\iota, \alpha_f)}{v(\iota, \alpha_f)}$$

where $\|u_i\|_\infty, \|v_j\|_\infty \leq 2^n p^{1/4}$. For this we LLL-reduce the lattice of dimension $4n$ defined by the lines of the matrix

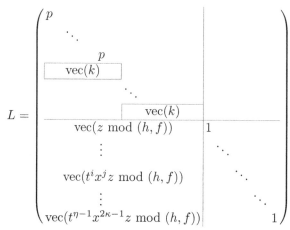

the first n rows contain only the diagonal coefficient equal to p and where, for all bivariate polynomial $w(t, x) = \sum_{i=0}^{2\kappa-1} w_i(t)x^i$ with $w_i(t) = \sum_{j=0}^{\eta-1} w_{i,j}t^{\eta-1-j}$, $\text{vec}(w) = (w_{0,0}, \ldots, w_{0,\eta-1}, \ldots, w_{2\kappa-1,0}, \ldots, w_{2\kappa-1,\eta-1})$ of dimension $2n$. In particular, $k \in \mathbb{F}_{p^n}[x]$ has been seen as a two-variate polynomial.

By dividing if necessary by the leading coefficient, we can assume that $k(x)$ is monic, hence the right-most coordinate of $\text{vec}(k)$ is 1. Then $\det L = p^n$ and we have u, v with $\|u_i\|_\infty, \|v_j\|_\infty \leq 2^{(4n-1)/4}Q^{\frac{1}{4n}} \leq 2^n Q^{\frac{1}{4n}}$. By Lemma 2 we obtain that

$$|N_{K_f/\mathbb{Q}}(u(\iota, \alpha_f))N_{K_f/\mathbb{Q}}(v(\iota, \alpha_f))| \leq 2^{n^2}Q\left(\|f\|_\infty^{(2\kappa-1)\eta}\|h\|_\infty^{(3\kappa-1)}(\eta-1)D(\eta, 2\kappa)\right)^2.$$

The term in the later bracket is $L_Q(2/3, o(1))$ and 2^{n^2} is negligible compared to Q if and only if $\ell_p > 1/2$. We conclude that when $\ell_p > 1/2$

$$|N_{K_f/\mathbb{Q}}(u(\iota, \alpha_f))N_{K_f/\mathbb{Q}}(v(\iota, \alpha_f))| = Q^{1+o(1)}.$$

Once the lift $\bar{z}$ has been computed, the smoothing step is carried out as usual: one tests that the norm of $\bar{z}$ (or u and v) is squarefree and B_1-smooth where $B_1 = L_Q(2/3, \beta_1)$ for some constant $\beta_1 > 0$. We recognize the complexity analysis done in [15] in the case of prime fields: the complexity of the smoothing step is $L_Q(1/3, c_{\text{smooth}})$ with

- $c_{\text{smooth}} = 6^{\frac{1}{3}}$ for exTNFS-JLSV$_2$;
- $c_{\text{smooth}} = 3^{\frac{1}{3}}$ for exTNFS-Conj.

Descent by Special-q. Recall how the special-q descent is done in the large characteristic case of NFS (for example NFS-JLSV$_2$). Due to the condition that $N_{K_f/\mathbb{Q}}(\bar{z})$ is squarefree the ideal generated by $\bar{z}$ factors only into prime ideals of degree 1. For a prime ideal $\mathfrak{q}$ of degree 1 in K_f that appears in the factorization of the principal ideal $(\bar{z})$, we write the logarithm of $\mathfrak{q}$ as a formal sum of virtual logarithms of ideals in K_f and K_g of norm less than $N(\mathfrak{q})^c$ for a constant $c < 1$. For this, we enumerate pairs $(a, b) \in \mathbb{Z} \times \mathbb{Z}$ such that $\mathfrak{q}$ divides $(a - b\alpha_f)$ to find one pair such that

- $(a - b\alpha_f)/\mathfrak{q}$ factors into prime ideals of norm less than $N(\mathfrak{q})^c$, and
- the ideal $(a - b\alpha_g)$ factors into prime ideals of norm less than $N(\mathfrak{q})^c$.

To do this we find two pairs $(a^{(1)}, b^{(1)})$ and $(a^{(2)}, b^{(2)})$ of euclidean norm less than a constant times $N(\mathfrak{q})^{\frac{1}{2}}$, using LLL. Then we enumerate the pairs $i_1 + i_2$ for all rational integers with $|i_1|, |i_2| \leq E'$. The complexity of the descent is mainly determined by the size of the norms:

$$|N_{K_f/\mathbb{Q}}(a - b\alpha_f)| \leq \left((E')^\kappa N(\mathfrak{D})^{\kappa/2} Q^{1/(D+1)}\right)^{1+o(1)},$$

$$|N_{K_g/\mathbb{Q}}(a - b\alpha_g)| \leq \left((E')^D N(\mathfrak{D})^{D/2} Q^{1/(D+1)}\right)^{1+o(1)}.$$

In our two cases, exTNFS-JLSV$_2$ and exTNFS-Conj, we enumerate $a(\iota), b(\iota) \in R \subset \mathbb{Q}(\iota)$ where $a(t), b(t) \in \mathbb{Z}[x]$ of degree $\leq \eta - 1$ and $\|a\|_\infty, \|b\|_\infty \leq (E')^{\frac{1}{\eta}}$ so that $a(\iota) - b(\iota)\alpha_f \equiv 0 \bmod \mathfrak{q}$. This can be done in the following manner (cf Appendix 7.1 in [7]). First, we construct the lattice

$$L(\mathfrak{q}) := \{(a, b) = (a_0, \ldots, a_{\eta-1}, b_0, \ldots, b_{\eta-1}) \in \mathbb{Z}^{2\eta} : a(\iota) - b(\iota)\alpha_f \equiv 0 \bmod \mathfrak{q}\},$$

which has determinant $N(\mathfrak{q})$. Let $(a^{(k)}, b^{(k)})$, $k = 1, 2, \ldots, 2\eta$, be the LLL-reduced basis of this lattice. Then we test the above smoothness conditions for pairs $(a, b) = \sum_{k=1}^{2\eta} i_k(a^{(k)}, b^{(k)})$, where i_k are rational integers with absolute value less than $I := (E')^{\frac{1}{\eta}}$. By Lemma 1, in the case of exTNFS-JLSV$_2$ the size of the norms is

$$|N_{K_f/\mathbb{Q}}(a - b\alpha_f)| \leq \left((E')^\kappa N(\mathfrak{q})^{\kappa/2} Q^{1/(D+1)}\right)^{1+o(1)},$$

$$|N_{K_g/\mathbb{Q}}(a - b\alpha_g)| \leq \left((E')^D N(\mathfrak{q})^{D/2} Q^{1/(D+1)}\right)^{1+o(1)}.$$

Then, the rest of the analysis is similar to that of Chap. 7.3 in [3] and we conclude that in exTNFS-JLSV$_2$ the special-q descent is negligible compared to the smoothing step.

In the case of exTNFS-Conj, we use again Lemma 1 and obtain:

$$|N_{K_f/\mathbb{Q}}(a - b\alpha_f)| \leq \left((E')^{2\kappa} N(\mathfrak{q})^\kappa\right)^{1+o(1)},$$

$$|N_{K_g/\mathbb{Q}}(a - b\alpha_g)| \leq \left((E')^\kappa N(\mathfrak{q})^{\kappa/2} Q^{1/(2\kappa)}\right)^{1+o(1)}.$$

We make an usual heuristic argument that a number x is y-smooth with the probability of $\rho(\log x / \log y)$ for Dickman function ρ. So, the probability of the pair (a, b) to be descended is given by

$$\text{Prob}[(a, b) \text{ descends}] \geq \rho\left(\frac{3\kappa \log E' + (3\kappa/2) \log \nu + (1/(2\kappa)) \log Q}{c \log \nu}\right)^{1+o(1)},$$

$$(11)$$

where $\nu := N(\mathfrak{q})$.

In the case when ν is large, i.e. $\nu = L_Q(2/3, \beta_1)$, where β_1 is imposed by the smoothing step described above, the inverse of the probability can be approximated by

$$\rho\left(\frac{3\kappa}{2c}\right)^{-1+o(1)} = L_Q\left(\frac{1}{3}, \frac{c_\kappa}{2c}\right)^{1+o(1)},$$

where $c_\kappa = \kappa/(\frac{\log Q}{\log\log Q})^{\frac{1}{3}} = 12^{-\frac{1}{3}}$. Multiplying this by the time for ν^c-smoothness test the total cost becomes

$$L_Q\left(1/3, \frac{c_\kappa}{2c} + 2\sqrt{\frac{c\beta_1}{3}}\right)^{1+o(1)}.$$

This value is minimized by $L_Q(1/3, (9\beta_1 c_\kappa/2)^{1/3})$ when $c = \left(\frac{3c_\kappa^2}{4\beta_1}\right)^{1/3}$. When we use that $\beta_1 = (1/3)^{1/3}$ and $c_\kappa = 12^{-\frac{1}{3}}$, we deduce the complexity

$$L_Q\left(1/3, (81/32)^{\frac{1}{9}}\right)$$

that is less than the complexity of the smoothing step.

In the case of small ν, i.e. $\nu = L_Q(1/3)$, the hardest descent step corresponds to the case when $\nu^c = B$ (the smoothness bound for the factor base). In this case, again by Eq. (11), we have the probability of the descent,

$$L_Q\left(1/3, \frac{c_\kappa}{2c} + \frac{c_\kappa\epsilon}{\beta} + \frac{1}{6c_\kappa\beta}\right)^{-1+o(1)}.$$

The complexity is minimized when the size of sieving space equals to the inverse of the above probability. This translates to

$$2\epsilon = \frac{c_\kappa}{2c} + \frac{c_\kappa\epsilon}{\beta} + \frac{1}{6c_\kappa\beta}.$$

This shows that the optimal value for c can be any value close but not equal to 1, e.g. $c = 0.999$, and the optimal complexity of descent step for small ν is $L_Q(1/3, 2\epsilon)$ where

$$\epsilon = \left(\frac{c_\kappa}{2} + \frac{1}{6\beta c_\kappa}\right)/\left(2 - \frac{c_\kappa}{\beta}\right) = 12^{-1/3} \approx 0.44,$$

where we used $\beta = (2/3)^{1/3}$ and $c_\kappa = 12^{-1/3}$. This complexity is negligible to the smoothing step.

For medium ν, i.e. $\nu = L_Q(\ell)$ with $1/3 < \ell < 2/3$, it is obviously faster than the case of large ν. So, we omit detailed analysis for this case and refer to Chap. 7.3 in [3].

We conclude this section of the Appendix with a summary of our results in Table 7.

Table 7. Complexity of individual logarithm

Algorithm	Rels collection + lin. algebra	Smoothing	Special-q descent	Extra conditions
exTNFS-JLSV$_2$	$(64/9)^{\frac{1}{3}}$	$(54/9)^{\frac{1}{3}}$	negligible	-
exTNFS-Conj	$(48/9)^{\frac{1}{3}}$	$(27/9)^{\frac{1}{3}}$	negligible	$\ell_p > 1/2$

References

1. Aoki, K., Franke, J., Kleinjung, T., Lenstra, A.K., Osvik, D.A.: A kilobit special number field sieve factorization. In: Kurosawa, K. (ed.) ASIACRYPT 2007. LNCS, vol. 4833, pp. 1–12. Springer, Heidelberg (2007)
2. Aranha, D.F., Fuentes-Castañeda, L., Knapp, E., Menezes, A., Rodríguez-Henríquez, F.: Implementing pairings at the 192-bit security level. In: Abdalla, M., Lange, T. (eds.) Pairing 2012. LNCS, vol. 7708, pp. 177–195. Springer, Heidelberg (2013)
3. Barbulescu, R.: Algorithms of discrete logarithm in finite fields. Ph.D. thesis, Université de Lorraine, December 2013
4. Barbulescu, R.: An appendix for a recent paper of Kim. IACR Cryptology ePrint Archive 2015:1076 (2015)
5. Barbulescu, R., Gaudry, P., Guillevic, A., Morain, F.: Improving NFS for the discrete logarithm problem in non-prime finite fields. In: Oswald, E., Fischlin, M. (eds.) EUROCRYPT 2015. LNCS, vol. 9056, pp. 129–155. Springer, Heidelberg (2015)
6. Barbulescu, R., Gaudry, P., Joux, A., Thomé, E.: A heuristic quasi-polynomial algorithm for discrete logarithm in finite fields of small characteristic. In: Nguyen, P.Q., Oswald, E. (eds.) EUROCRYPT 2014. LNCS, vol. 8441, pp. 1–16. Springer, Heidelberg (2014)
7. Barbulescu, R., Gaudry, P., Kleinjung, T.: The towed number field sieve. In: Iwata, T., Cheon, J.H. (eds.) ASIACRYPT 2015. LNCS, vol. 9453, pp. 31–55. Springer, Heidelberg (2015). doi:10.1007/978-3-662-48800-3_2
8. Barbulescu, R., Pierrot, C.: The multiple number field sieve for medium- and high-characteristic finite fields. LMS J. Comput. Math. **17**, 230–246 (2014). The published version contains an error which is corrected in https://hal.inria.fr/hal-00952610
9. Barreto, P.S.L.M., Naehrig, M.: Pairing-friendly elliptic curves of prime order. In: Preneel, B., Tavares, S. (eds.) SAC 2005. LNCS, vol. 3897, pp. 319–331. Springer, Heidelberg (2006)
10. Beuchat, J.-L., González-Díaz, J.E., Mitsunari, S., Okamoto, E., Rodríguez-Henríquez, F., Teruya, T.: High-speed software implementation of the optimal ate pairing over Barreto–Naehrig curves. In: Joye, M., Miyaji, A., Otsuka, A. (eds.) Pairing 2010. LNCS, vol. 6487, pp. 21–39. Springer, Heidelberg (2010)
11. Bistritz, Y., Lifshitz, A.: Bounds for resultants of univariate, bivariate polynomials. Linear Algebra Appl. **432**(8), 1995–2005 (2010). Special Issue Devoted to the 15th ILAS Conference at Cancun, Mexico, 16–20 June 2008
12. Bouvier, C., Gaudry, P., Imbert, L., Jeljeli, H., Thom, E.: Discrete logarithms in GF(p) – 180 digits. Announcement available at the NMBRTHRY Archives, item 004703 (2014)

13. Chatterjee, S., Menezes, A., Rodriguez-Henriquez, F.: On implementing pairing-based protocols with elliptic curves of embedding degree one. Cryptology ePrint Archive, Report 2016/403 (2016). http://eprint.iacr.org/2016/403

14. Chillotti, I., Gama, N., Georgieva, M., Izabachène, M.: A homomorphic LWE based E-voting scheme. In: Takagi, T., et al. (eds.) PQCrypto 2016. LNCS, vol. 9606, pp. 245–265. Springer, Heidelberg (2016). doi:10.1007/978-3-319-29360-8_16

15. Commeine, A., Semaev, I.A.: An algorithm to solve the discrete logarithm problem with the number field sieve. In: Yung, M., Dodis, Y., Kiayias, A., Malkin, T. (eds.) PKC 2006. LNCS, vol. 3958, pp. 174–190. Springer, Heidelberg (2006)

16. Danilov, S., Popovyan, I.: Factorization of RSA-180 (2010). http://eprint.iacr.org/2010/270

17. European Union Agency of Network and Information Security (ENISA): Algorithms, key sizes and parameters report, 2013 recommandations, version 1.0, October 2013. Publucation https://www.enisa.europa.eu/publications/algorithms-key-sizes-and-parameters-report

18. Freeman, D.: Constructing pairing-friendly elliptic curves with embedding degree 10. In: Hess, F., Pauli, S., Pohst, M. (eds.) ANTS 2006. LNCS, vol. 4076, pp. 452–465. Springer, Heidelberg (2006)

19. Freeman, D., Scott, M., Teske, E.: A taxonomy of pairing-friendly elliptic curves. J. Cryptol. **23**(2), 224–280 (2010)

20. Gordon, D.M.: Discrete logarithms in $GF(p)$ using the number field sieve. SIAM J. Discret. Math. **6**(1), 124–138 (1993)

21. Granger, R., Kleinjung, T., Zumbrägel, J.: On the powers of 2. Cryptology ePrint Archive, Report 2014/300 (2014). http://eprint.iacr.org/2014/300

22. Joux, A., Lercier, R., Smart, N.P., Vercauteren, F.: The number field sieve in the medium prime case. In: Dwork, C. (ed.) CRYPTO 2006. LNCS, vol. 4117, pp. 326–344. Springer, Heidelberg (2006)

23. Joux, A., Pierrot, C.: The special number field sieve in $\mathbb{F}_{p^n}$. In: Cao, Z., Zhang, F. (eds.) Pairing 2013. LNCS, vol. 8365, pp. 45–61. Springer, Heidelberg (2014)

24. Kim, T.: Extended tower number field sieve: a new complexity for medium prime case. IACR Cryptology ePrint Archive 2015:1027 (2015)

25. Lenstra, A.K.: Unbelievable security. In: Boyd, C. (ed.) ASIACRYPT 2001. LNCS, vol. 2248, p. 67. Springer, Heidelberg (2001)

26. Lenstra Jr., H.W.: Factoring integers with elliptic curves. Ann. Math. **126**, 649–673 (1987)

27. Matyukhin, D.V.: Effective version of the number field sieve for discrete logarithm in a field $GF(p^k)$. Trudy po Diskretnoi Matematike **9**, 121–151 (2006)

28. National Institute of Standards and Technology (NIST): NIST Special Publication 800–57 Part 1 (Revised): Recommendation for Key Management, Part 1: General (Revised), July 2012. Publication http://csrc.nist.gov/publications/nistpubs/800-57/sp800-57-Part1-revised2_Mar08-2007.pdf

29. Odlyzko, A.M.: The future of integer factorization. CryptoBytes (Tech. Newsl. RSA Lab.) **1**(2), 5–12 (1995)

30. Pierrot, C.: The multiple number field sieve with conjugation and generalized Joux-Lercier methods. In: Oswald, E., Fischlin, M. (eds.) EUROCRYPT 2015. LNCS, vol. 9056, pp. 156–170. Springer, Heidelberg (2015)

31. Sarkar, P., Singh, S.: New complexity trade-offs for the (multiple) number field sieve algorithm in non-prime fields. In: Fischlin, M., Coron, J.-S. (eds.) EUROCRYPT 2016. LNCS, vol. 9665, pp. 429–458. Springer, Heidelberg (2016). doi:10.1007/978-3-662-49890-3_17

32. Schirokauer, O.: Discrete logarithms and local units. Philos. Trans. Roy. Soc. Lond. A: Math. Phys. Eng. Sci. **345**(1676), 409–423 (1993)
33. Schirokauer, O.: Using number fields to compute logarithms in finite fields. Math. Comput. **69**(231), 1267–1283 (2000)
34. Wiedemann, D.: Solving sparse linear equations over finite fields. IEEE Trans. Inf. Theor. **32**(1), 54–62 (1986)

Efficient Algorithms for Supersingular Isogeny Diffie-Hellman

Craig Costello$^{(\boxtimes)}$, Patrick Longa, and Michael Naehrig

Microsoft Research, Redmond, USA
{craigco,plonga,mnaehrig}@microsoft.com

Abstract. We propose a new suite of algorithms that significantly improve the performance of supersingular isogeny Diffie-Hellman (SIDH) key exchange. Subsequently, we present a full-fledged implementation of SIDH that is geared towards the 128-bit quantum and 192-bit classical security levels. Our library is the first constant-time SIDH implementation and is up to 2.9 times faster than the previous best (non-constant-time) SIDH software. The high speeds in this paper are driven by compact, inversion-free point and isogeny arithmetic and fast SIDH-tailored field arithmetic: on an Intel Haswell processor, generating ephemeral public keys takes 46 million cycles for Alice and 52 million cycles for Bob, while computing the shared secret takes 44 million and 50 million cycles, respectively. The size of public keys is only 564 bytes, which is significantly smaller than most of the popular post-quantum key exchange alternatives. Ultimately, the size and speed of our software illustrates the strong potential of SIDH as a post-quantum key exchange candidate and we hope that these results encourage a wider cryptanalytic effort.

Keywords: Post-quantum cryptography · Diffie-Hellman key exchange · Supersingular elliptic curves · Isogenies · SIDH

1 Introduction

Post-quantum Cryptography. The prospect of a large scale quantum computer that is capable of implementing Shor's algorithm [43] has given rise to the field of post-quantum cryptography (PQC). Its goal is to develop and ultimately deploy cryptographic primitives that resist cryptanalysis by both classical *and* quantum computers. Recent developments in quantum computing (see, e.g., [16,23,34]) have helped catalyze government and corporate action in this arena. For example, in April 2015, the National Institute of Standards and Technology (NIST) held a "Workshop on Cybersecurity in a Post-Quantum World", reaching out to academia and industry to discuss potential future standardization of PQC. Later, in August 2015, the National Security Agency (NSA) released a major policy statement that announced plans to "transition to quantum resistant algorithms in the not too distant future" [35]. In February 2016, NIST published a draft "Report on Post-Quantum Cryptography" [11], which

© International Association for Cryptologic Research 2016
M. Robshaw and J. Katz (Eds.): CRYPTO 2016, Part I, LNCS 9814, pp. 572–601, 2016.
DOI: 10.1007/978-3-662-53018-4_21

emphasizes the need to start working towards the deployment of post-quantum cryptography in our information security systems, and outlines NIST's plans to "initiate a standardization effort in post-quantum cryptography".

In terms of public-key PQC, there are four well-known and commonly cited classes of cryptographic primitives that are believed to remain secure in the presence of a quantum computer: code-based cryptography, lattice-based cryptography, hash-based cryptography, and multivariate cryptography. Specific examples for each of these are McEliece's code-based encryption scheme [29]; Hoffstein, Pipher and Silverman's lattice-based encryption scheme "NTRU" [21]; Merkle's hash-tree signatures [30]; and Patarin's "HFE^{v-}" signature scheme [38]. A positive trait shared by all of these examples is a resistance to decades of attempted classical and quantum cryptanalysis which has inspired widespread confidence in their suitability as a post-quantum primitive. However, most of these examples also share the trait of having enormous public key and/or signature sizes, particularly when compared to traditional primitives based on the hardness of integer factorization or (elliptic curve) discrete logarithm computation.

Supersingular Isogeny Diffie-Hellman. In this paper, we study a different primitive that does not fall into any of the above classes, but is currently believed to offer post-quantum resistance: the supersingular isogeny Diffie-Hellman (SIDH) key exchange protocol proposed by Jao and De Feo in 2011 [22]. The SIDH key exchange protocol is more than a decade younger than all of the above schemes, so its security is yet to withstand the tests of time and of a wide cryptanalytic effort. Nevertheless, the current picture of its security properties looks promising. The best known classical and quantum attacks against the underlying problem are both exponential in the size of the underlying finite field, and their complexities make current SIDH key sizes significantly smaller than their post-quantum key exchange and/or encryption counterparts[1].

Our Contributions. We present a full-fledged, high-speed implementation of (unauthenticated) ephemeral SIDH that currently provides 128 bits of quantum security and 192 bits of classical security. This implementation uses 48-byte private keys to produce 564-byte ephemeral Diffie-Hellman public keys, is written in C and includes an optimized version of the field arithmetic written in assembly. To our knowledge, our library (see [14]) presents the first SIDH software that runs in *constant-time*, i.e., that is designed to resist timing [26] and cache-timing [37] attacks. On x64 platforms, our implementation runs up to 2.9 times faster than the (previously fastest) implementation of SIDH by Azarderakhsh et al. [2]. Note that this performance comparison does not take into account the fact that the implementation from [2] is not protected against timing attacks. The main technical contributions that lead to these improvements are:

---

[1] An exception here is NTRUEncrypt [21], which has comparable public key sizes – see https://github.com/NTRUOpenSourceProject/ntru-crypto.

Projective Curve Coefficients. A widely-deployed technique in traditional ECC involves avoiding inversions by working with elliptic curve points in projective space. Following Jao and De Feo [22], we also employ this technique to work efficiently with points in $\mathbb{P}^1$ by making use of the fast arithmetic associated with the Kummer varieties of Montgomery curves. A crucial difference in this work, however, is that we also work projectively with the curve coefficients; unlike traditional ECC where the curve is fixed, every SIDH key exchange requires computations on many different isogenous curves. In Sect. 3 we show that the Montgomery model also allows all of the necessary isogeny arithmetic to be performed efficiently in $\mathbb{P}^1$. This gives rise to more compact algorithms, significantly simplifies the overall computation, and means that key generation and shared secret computations only require one and two field inversions, respectively.

Prime Selection and Tailor-Made Montgomery Multiplication. We select a prime with form $p = \ell_A^{e_A} \ell_B^{e_B} f - 1$, where $\ell_A = 2$, $\ell_B = 3$, and the bit lengths of 2^{e_A} and 3^{e_B} are slightly smaller than a multiple of 64. This supports efficient arithmetic on a wide range of platforms and allows access to a large variety of optimizations such as the efficient use of vector instructions, Karatsuba multiplication, and lazy reduction. Moreover, it is well-known that primes of a special form can lead to faster algorithms for computing modular arithmetic in comparison with general-purpose algorithms. In this work, we note the special shape of these *SIDH-friendly* primes and modify the popular Montgomery multiplication algorithm to speed up modular arithmetic.

Ground Field Scalar Multiplications for Key Generation. Secure key generation in the SIDH protocol requires the definition of two independent cyclic subgroups of a fixed order (see Sect. 2). Jao and De Feo [22, Sect. 4.1] propose that generators of these two groups can be computed by multiplying random curve points by an appropriate cofactor, and that their linear independence can be checked via the Weil pairing. In Sect. 4 we employ a well-known technique from the pairing literature [42, Sect. 5] to work with two advantageous choices of torsion subgroups: the *base-field* and *trace-zero* subgroups. These choices allow the initial scalar multiplications that are required during key generation to be performed entirely over the base field. While these scalar multiplications only constitute a small fraction of the overall key generation time, and therefore the overall speedup from this technique is only moderate, a more visible benefit is the significant decrease in the size of the public parameters – see Sect. 6. We discuss possible security implications of this choice in Sect. 4.

Several of the above choices not only aid efficiency, but also the overall simplicity and compactness of the SIDH scheme. Choosing to unify points with their inverses and to unify Montgomery curves with their quadratic twists (see Sect. 3) effectively compresses the elements that are sent over the wire, i.e., the public keys, by a factor of two. Moreover, our software never requires the computation of square roots.

The timings we present in Sect. 7 reveal that high-security SIDH key exchange is more efficient than it was previously known to be. Our constant-time software shows that, if confidence in the security of SIDH warrants real-world deployment in the future, the same level of side-channel protection can be achieved in the SIDH setting as in traditional number-theoretic schemes. We therefore hope that this paper encourages a wider cryptanalytic effort on the problems underlying the security of SIDH (see Sect. 2). Moreover, even if cryptanalytic improvements are made in the future, the huge difference between current SIDH key sizes and those of other PQC primitives suggest that the problem could remain of interest to practitioners. So long as the best known attacks remain exponential with a reasonable exponent (see the discussion below), it is reasonable to suggest that elliptic curves could offer the same benefit in post-quantum cryptography that they did in classical cryptography.

Beyond the efficiency improvements above, we present several techniques that help to bridge the gap between the theoretical SIDH scheme in [22] and its real-world deployment. Of particular importance are the contributions discussed in the following two paragraphs.

A Strong ECDH + SIDH Hybrid. Given the uncertainty surrounding the arrival date of large-scale quantum computers (as well as the time it takes for new primitives to be thoroughly cryptanalyzed, standardized and deployed), many real-world cryptographers are hastily pushing for deployment of post-quantum primitives sooner rather than later. Subsequently, a proposal that is gaining popularity in the PQC community is the deployment of *hybrid* schemes, i.e., schemes where a long-standing classically-secure primitive $\mathcal{P}$ is partnered alongside a newer post-quantum candidate $\mathcal{Q}$ (cf. [5]). The simple reasoning here is that, even if further cryptanalysis weakens $\mathcal{Q}$'s resistance to classical computers, the hybrid scheme $\mathcal{P} + \mathcal{Q}$ is likely to remain classically secure; conversely, $\mathcal{P}$'s presumed weakness against a quantum computer does not affect the post-quantum security of $\mathcal{P} + \mathcal{Q}$. Taking such a prudent measure in the case of SIDH, which is much newer than other post-quantum primitives, seems especially wise. In Sect. 8 we present a possibility to partner SIDH public keys alongside traditional elliptic curve Diffie-Hellman (ECDH) public keys that are extremely strong. In particular, while our proposed SIDH parameters respectively offer around 128 and 192 bits of security against the best known quantum and classical attacks, the proposed hybrid offers around 384 bits of classical security based on the elliptic curve discrete logarithm problem (ECDLP). While this might seem like overkill, we show that this partnering is a very natural choice and comes at a relatively small cost: compared to a standalone SIDH, the size of the public keys and the overall runtime in our SIDH + ECDH hybrid increase by no more than 17 % and 13 %, respectively, and there is almost no additional code required to include ECDH in the scheme.

Public Key Validation. The security of unauthenticated ephemeral key exchange is modeled using passive adversaries, in which case we can assume that both parties' public keys are honestly generated. As was pointed out in April 2015 by a group at the NSA [24], in static key exchange when private keys are reused, validating public keys in the case of isogeny-based cryptography becomes both necessary and non-trivial. The suggested indirect public key validation procedure described in [24] is costly and requires one party to reveal their secret key, such that only the other party can reuse theirs. In Sect. 9 we detail a form of direct validation for the public keys used in our scheme, and show how to achieve this validation efficiently in our compact framework.

SIDH History and Security. Beginning with an unpublished preprint with Rostovtsev in early 2006 [40], and then in a series of Russian papers that culminated in his thesis [45], Stolbunov proposed a Diffie-Hellman-like cryptosystem based on the difficulty of computing isogenies between ordinary (i.e., non-supersingular) elliptic curves. The best algorithm to solve this problem on a classical computer runs in exponential time and is due to Galbraith and Stolbunov [18].

In late 2010, however, Childs et al. [12] gave a quantum algorithm that computes isogenies between ordinary curves in subexponential time, assuming the Generalized Riemann Hypothesis (GRH). Subsequently, in late 2011, Jao and De Feo [22] put forward SIDH, which is instead based on the difficulty of computing isogenies between supersingular elliptic curves. This problem is immune to the quantum attack in [12], since this attack crucially relies on the endomorphism ring being commutative, which is not the case for a supersingular curve whose endomorphism ring is isomorphic to an order in a quaternion algebra [44, Sect. V.3.1].

Given two isogenous supersingular elliptic curves defined over a field of characteristic p, the *general* supersingular isogeny problem is to construct an isogeny between them. The best known classical algorithm for this problem is due to Delfs and Galbraith [15] and requires $\tilde{O}(p^{1/2})$ bit operations, while the best known quantum algorithm is due to Biasse et al. [6] and requires $\tilde{O}(p^{1/4})$ bit operations. The problems underlying SIDH (see Sect. 2) are not general in that the degree of the isogeny, which is smooth and in $O(\sqrt{p})$, is known and public. As is discussed by De Feo et al. [17, Sect. 5.1][2], this specialized problem can be viewed as an instance of the *claw problem*, and the optimal asymptotic classical and quantum complexities for the claw problem are known to be $O(p^{1/4})$ and $O(p^{1/6})$, respectively [47,52]. Currently, this approach yields the best known classical and quantum attacks against SIDH.

Organization . In Sect. 2 we recall the key concepts from [17] that are needed in SIDH. In Sect. 3 we show that all isogeny and point computations can be performed in $\mathbb{P}^1$; here we derive all of the lower-level functions that are called during the key generation and shared secret operations. In Sect. 4 we fix the underlying isogeny class used in our software, describe the high-level key exchange operations, and discuss other implementation choices. In Sect. 5 we detail the special field arithmetic that is tailored towards our chosen prime (as well as many other well-chosen SIDH-friendly primes).

We give a summary of the scheme in Sect. 6 and present performance results of our implementation in Sect. 7. In Sect. 8 we describe our proposal for a strong

---

[2] This is an extended version of the original SIDH paper by Jao and De Feo [22].

hybrid key exchange scheme that combines classical ECDH with post-quantum SIDH, and in Sect. 9 we show how to efficiently validate SIDH public keys in static key exchange settings. We conclude the paper in Sect. 10.

To promote future implementations of SIDH, we have endeavored to make this paper as self-contained as possible. Essentially, all functions that are needed to implement SIDH are described in Sect. 3. High level functions can be found in the appendix of the full version [13]. All other details can be found in the released code [14].

2 Diffie-Hellman Key Exchange from Supersingular Elliptic Curve Isogenies

This section sets the stage by introducing notation, giving some basic properties of torsion subgroups and isogenies, and recalling the supersingular isogeny Diffie-Hellman key exchange protocol. This is all described in a similar fashion by De Feo et al. in [17, Sect. 2].

Smooth Order Supersingular Elliptic Curves. SIDH uses isogeny classes of supersingular elliptic curves with smooth orders so that rational isogenies of exponentially large (but smooth) degree can be computed efficiently as a composition of low degree isogenies. Fix two small prime numbers ℓ_A and ℓ_B, an integer cofactor f, and let p be a prime of the form $p = \ell_A^{e_A} \ell_B^{e_B} f \pm 1$. It is then easy to construct a supersingular elliptic curve E defined over $\mathbb{F}_{p^2}$ of order $(\ell_A^{e_A} \ell_B^{e_B} f)^2$ [9].

For $\ell \in \{\ell_A, \ell_B\}$ and $e \in \{e_A, e_B\}$ the corresponding exponent, we have that the full ℓ^e-torsion group on E is defined over $\mathbb{F}_{p^2}$, i.e. $E[\ell^e] \subseteq E(\mathbb{F}_{p^2})$. Since ℓ is coprime to p, $E[\ell^e] \cong (\mathbb{Z}/\ell^e\mathbb{Z}) \times (\mathbb{Z}/\ell^e\mathbb{Z})$ [44, III. 6.4]. Let $P, Q \in E[\ell^e]$ be two points that generate $E[\ell^e]$ such that the above isomorphism is given by $(\mathbb{Z}/\ell^e\mathbb{Z}) \times (\mathbb{Z}/\ell^e\mathbb{Z}) \to E[\ell^e]$, $(m, n) \mapsto [m]P + [n]Q$. Roughly speaking, the SIDH secret keys are degree ℓ^e isogenies of the base curve E, which are in one-to-one correspondence with the cyclic subgroups of order ℓ^e that form their kernels. A point $[m]P + [n]Q$ has full order ℓ^e if and only if at least either m or n are not divisible by ℓ. There are $\ell^{2e-2}(\ell^2 - 1)$ such points. Since distinct cyclic subgroups only intersect in points of order less than ℓ^e and all full-order points in a single subgroup are coprime multiples of one such point, it follows that there are $\ell^{e-1}(\ell + 1)$ distinct cyclic subgroups of order ℓ^e.

Computing Large Degree Isogenies. Given a cyclic subgroup $\langle R \rangle \subseteq E[\ell^e]$ of order ℓ^e, there is a unique isogeny ϕ_R of degree ℓ^e, defined over $\mathbb{F}_{p^2}$ with kernel $\langle R \rangle$ [44, III. 4.12], mapping E to an isogenous elliptic curve $E/\langle R \rangle$. The isogeny ϕ_R can be computed as the composition of e isogenies of degree ℓ which in turn can be computed by using Vélu's formulas [49]. As described in [17, Sect. 4.2.2], we can start with $E_0 := E$ and $R_0 := R$ and then iteratively compute $E_{i+1} = E_i/\langle [\ell^{e-i-1}]R_i \rangle$ for $0 \leq i < e$ as follows. Each iteration computes the degree-ℓ isogeny $\phi_i : E_i \to E_{i+1}$ whose kernel is the cyclic group $\langle [\ell^{e-i-1}]R_i \rangle$ of order ℓ, before applying the isogeny to compute $R_{i+1} = \phi_i(R_i)$. The point R_i is an (ℓ^{e-i})-torsion point and so $[\ell^{e-i-1}]R_i$ has order ℓ. Thus, the composition $\phi_R = \phi_{e-1} \circ \cdots \circ \phi_0$ has degree ℓ^e, which together with $(\phi_{e-1} \circ \cdots \circ \phi_0)(R) = R_e = \mathcal{O}$ shows that $\ker(\phi_R) = \langle R \rangle$, and therefore that $\phi = \phi_{e-1} \circ \cdots \circ \phi_0$.

There are two obvious ways of computing ϕ using the above decomposition. One of them follows directly from the description above: in each iteration, one first computes the scalar multiplication $[\ell^{e-i-1}]R_i$ to obtain a point of order ℓ, then uses Vélu's formulas to compute ϕ_i, and evaluates it at R_i to obtain the next point R_{i+1}. Jao and De Feo [22, Fig. 2] call this the multiplication-based strategy because it is dominated by the number of scalar multiplications by ℓ that are needed to obtain the ℓ-torsion points. The second obvious approach is called the isogeny-based method [22, Fig. 2] because it is dominated by the number of isogeny evaluations. It requires only one loop of scalar-multiplications that stores all ℓ-multiples of R, i.e., all intermediate results $Q_i = [\ell^i]R$ for $0 \le i < e$. The point Q_{e-1} has order ℓ and can be used to obtain the isogeny ϕ_0 as above. One then replaces all Q_i for $0 \le i \le (e-2)$ by $\phi_0(Q_i)$. At this point Q_{e-2} has order ℓ and is used to obtain ϕ_1. This is repeated until one obtains ϕ_{e-1} and hence the composition ϕ.

De Feo et al. [17, Sect. 4.2.2] demonstrate that both of these methods are rather wasteful and that there is a much more efficient way to schedule the multiplications-by-ℓ and ℓ-isogeny evaluations. We briefly touch on this in Sect. 4, and defer the finer details to the full version [13].

SIDH Key Exchange. This paragraph recalls the SIDH key exchange protocol from [17, Sect. 3.2]. The public parameters are the supersingular curve $E_0/\mathbb{F}_{p^2}$ whose group order is $(\ell_A^{e_A} \ell_B^{e_B} f)^2$, two independent points P_A and Q_A that generate $E_0[\ell_A^{e_A}]$, and two independent points P_B and Q_B that generate $E_0[\ell_B^{e_B}]$. To compute her public key, Alice chooses two secret integers $m_A, n_A \in \mathbb{Z}/\ell_A^{e_A}\mathbb{Z}$, not both divisible by ℓ_A, such that $R_A = [m_A]P_A + [n_A]Q_A$ has order $\ell_A^{e_A}$. Her secret key is computed as the degree $\ell_A^{e_A}$ isogeny $\phi_A : E_0 \to E_A$ whose kernel is R_A, and her public key is the isogenous curve E_A together with the image points $\phi_A(P_B)$ and $\phi_A(Q_B)$. Similarly, Bob chooses two secret integers $m_B, n_B \in \mathbb{Z}/\ell_B^{e_B}\mathbb{Z}$, not both divisible by ℓ_B, such that $R_B = [m_B]P_B + [n_B]Q_B$ has order $\ell_B^{e_B}$. He then computes his secret key as the degree $\ell_B^{e_B}$ isogeny $\phi_B : E_0 \to E_B$ whose kernel is R_B, and his public key is E_B together with $\phi_B(P_A)$ and $\phi_B(Q_A)$. To compute the shared secret, Alice uses her secret integers and Bob's public key to compute the degree $\ell_A^{e_A}$ isogeny $\phi_A' : E_B \to E_{BA}$ whose kernel is the point $[m_A]\phi_B(P_A) + [n_A]\phi_B(Q_A) = \phi_B([m_A]P_A + [n_A]Q_A) = \phi_B(R_A)$. Similarly, Bob uses his secret integers and Alice's public key to compute the degree $\ell_B^{e_B}$ isogeny $\phi_B' : E_B \to E_{AB}$ whose kernel is the point $[m_B]\phi_A(P_B) + [n_B]\phi_A(Q_B) = \phi_A(R_B)$. It follows that E_{BA} and E_{AB} are isomorphic, so Alice and Bob can compute a shared secret as the common j-invariant $j(E_{BA}) = j(E_{AB})$.

Security Under SSDDH. In [17, Sect. 5], De Feo et al. give a number of computational problems related to SIDH and discuss their complexity. In [17, Sect. 6], they prove that SIDH is *session-key secure* in the authenticated-links adversarial model of Canneti and Krawczyk [10] under the Supersingular Decision Diffie-Hellman (SSDDH) problem, which we recall as follows. With the public parameters as above, one is given a tuple sampled with probability $1/2$ from either $(E_A, E_B, \phi_A(P_B), \phi_A(Q_B), \phi_B(P_A), \phi_B(Q_A), E_{AB})$ or from $(E_A, E_B, \phi_A(P_B), \phi_A(Q_B), \phi_B(P_A), \phi_B(Q_A), E_C)$, where $E_{AB} \cong E_0/\langle [m_A]P_A + [n_A]Q_A, [m_B]P_B + [n_B]Q_B \rangle$, $E_C \cong E_0/\langle [m_A']P_A + [n_A']Q_A, [m_B']P_B + [n_B']Q_B \rangle$, and the values m_A', n_A', m_B' and n_B' are chosen randomly from the same respective distributions as m_A, n_A, m_B and n_B. The SSDDH problem is to determine from which distribution the tuple is sampled.

3 Projective Points and Projective Curve Coefficients

In this section we present one of our main technical contributions by showing that, just as the Montgomery form allows point arithmetic to be carried out efficiently in $\mathbb{P}^1$, in the context of SIDH it also allows isogeny arithmetic to be carried out in $\mathbb{P}^1$. This gives rise to fast, inversion-free point-and-isogeny operations that significantly boost the performance of SIDH. In comparison to the software[3] accompanying [17] that computes at least one inversion per isogeny computation, and therefore $O(\ell)$ inversions per round of the protocol, our software only requires one inversion during key generation and two inversions during the computation of the shared secret.

Montgomery Curves. Over a field K, a Montgomery curve [33] is defined by the two constants $(a, b) \in \mathbb{A}^2(K)$ as $E_{(a,b)} \colon by^2 = x^3 + ax^2 + x$. Unlike traditional ECC, in this work the defining curve does not stay fixed, but changes as we *move around* an isogeny class. As we discuss further below, it is therefore convenient to work projectively both with points on curves and with the curve coefficients themselves. Let $(A \colon B \colon C) \in \mathbb{P}^2(K)$ with $C \in \bar{K}^\times$ be such that $a = A/C$ and $b = B/C$. Then $E_{(a,b)}$ can alternatively be written as $E_{(A \colon B \colon C)} \colon By^2 = Cx^3 + Ax^2 + Cx$. The K-rational points on $E_{(a,b)}$ or $E_{(A \colon B \colon C)}$ are contained in $\mathbb{P}^2(K)$, so as usual we use the notation $(X \colon Y \colon Z) \in \mathbb{P}^2(K)$ with $Z \neq 0$ to represent all points $(x, y) = (X/Z, Y/Z)$ in $\mathbb{A}^2(K)$, and the point at infinity is $\mathcal{O} = (0 \colon 1 \colon 0)$. The j-invariants of the curves given by these models are $j(E_{a,b}) = \frac{256(a^2 - 3)^3}{a^2 - 4}$ and $j(E_{(A \colon B \colon C)}) = \frac{256(A^2 - 3C^2)^3}{C^4(A^2 - 4C^2)}$.

Kummer Varieties and Points in $\mathbb{P}^1$. Following [33], viewing the x-line $\mathbb{P}^1$ as the Kummer variety of $E_{(a,b)}$ allows for particularly efficient arithmetic in $E_{(a,b)}/\langle \pm 1 \rangle \cong \mathbb{P}^1$. Let $x \colon E_{(a,b)} \setminus \{\mathcal{O}\} \to \mathbb{P}^1$, $(X \colon Y \colon Z) \mapsto (X \colon Z)$. For the points $P, Q \in E_{(a,b)} \setminus \{\mathcal{O}\}$ and $m \in \mathbb{Z}$, Montgomery [33] gave efficient formulas

---
[3] See https://github.com/defeo/ss-isogeny-software/.

for computing the doubling function xDBL: $(x(P), a) \mapsto x([2]P)$, the function xADD: $(x(P), x(Q), x(Q-P)) \mapsto x(Q+P)$ for *differential additions*, and the function xDBLADD: $(x(P), x(Q), x(Q - P), a) \mapsto (x([2]P), x(Q - P))$ for the merging of the two. These are all ingredients in the Montgomery ladder function to compute the $\mathbb{Z}$-action on $E_{(a,b)}/\langle \pm 1 \rangle \cong \mathbb{P}^1$, i.e., LADDER: $(x(P), a, m) \mapsto x([m]P)$. We also make use of the Montgomery tripling function xTPL: $(x(P), a) \mapsto x([3]P)$ on $E_{(a,b)}/\{\pm 1\}$, which is taken from [17].

We note that the xADD function works identically for $E_{(a,b)}$ and $E_{(A:B:C)}$, while the other functions on $E_{(a,b)}$ that involve a can be trivially modified to work on $E_{(A:B:C)}$ by substituting $a = A/C$ and avoiding the inversion by carrying the denominator C through to the projective output. All of these functions are summarized in Table 1. Conveniently, all of these subroutines are only needed to work entirely in only one of $E_{(A:B:C)}$ and $E_{(a,b)}$.

During the computations of shared secrets, we found it advantageous to employ the function LADDER_3_pt: $(x(P), x(Q), x(Q - P), a, m) \mapsto x(P + [m]Q)$, which is precisely the "three point ladder" given by De Feo et al. [17, Algorithm 1].

Minimizing the Number of Inversions via Curves in $\mathbb{P}^1$. Observe that all of the functions mentioned above on $E_{(a,b)}/\{\pm 1\}$ (resp. $E_{(A:B:C)}/\{\pm 1\}$) depend entirely on a (resp. A and C) and are independent of b (resp. B). This is because, for a fixed $a = A/C$ and up to isomorphism, there are only two curves found by varying b (resp. B) over K: the curve E and its non-trivial quadratic twist. Indeed, an elliptic curve and its twist are unified under the quotient by $\{\pm 1\}$, i.e., have the same Kummer variety, so it is no surprise that the Kummer arithmetic is independent of the Montgomery b (resp. B) coefficient. Moreover, we see above that the j-invariant is also independent of b (resp. B).

Our implementation profits significantly from these observations, and the choice of Montgomery form provides two advantages in parallel. The first is the well-known Montgomery-style *point* arithmetic that unifies points and their inverses by ignoring the Y coordinate to work with $(X: Z) \in \mathbb{P}^1$; the second is new *isogeny* arithmetic that unifies curves and their quadratic twists by ignoring the B coefficient to instead work only with $(A: C) \in \mathbb{P}^1$. In this way all point operations and isogeny computations are performed in $\mathbb{P}^1$, meaning that only one inversion is required (at the very end) when generating public keys or computing shared secrets. In the latter case, the inversion is computed during the j-invariant function j_inv: $(A, C) \mapsto j(E_{(A:B:C)})$, while in the former case we use a 3-way simultaneous inversion [33] to normalize all of the components of the public key prior to transmission; see Table 1 for more details on these functions.

Projective Three Isogenies. Let $x(P) = (X_3: Z_3) \in \mathbb{P}^1$ be such that P has order 3 in $E_{(A:C)}$. Let $E'_{(A':C')} = E_{(A:C)}/\langle P \rangle$, $\phi: E_{(A:C)} \to E'_{(A':C')}$, $Q \in E_a \setminus \ker(\phi)$, and write $x(Q) = (X: Z) \in \mathbb{P}^1$ with $x(\phi(Q)) = (X': Z') \in \mathbb{P}^1$. Our goal is to derive two sets of explicit formulas: the first set computes the isogenous curve $E_{(A':C')}$ from $(X_3: Z_3)$ and $E_{(A:C)}$, while the second set is used to evaluate the corresponding isogeny by computing $(X': Z')$ from the additional input $(X\colon Z)$. The projective version of [17, Eq. (17)] gives $(A'\colon C') = ((AX_3Z_3 + 6(Z_3^2 - X_3^2))X_3\colon CZ_3^3)$, which can be computed in $6\mathbf{M} + 2\mathbf{S} + 5\mathbf{a}^4$. However, it is possible to do much better by using $Z_3 \neq 0$ and the fact that X_3/Z_3 is a root of the 3-division polynomial $\psi_3(x) = 3x^4 + 4(A/C)x^3 + 6x^2 - 1$ on $E_{(A\colon C)}$. This yields the alternative expression $(A'\colon C') = (Z_3^4 + 18X_3^2Z_3^2 - 27X_3^4\colon 4X_3Z_3^3)$, which is independent of the coefficients of $E_{(A\colon C)}$ and can be computed in $3\mathbf{M} + 3\mathbf{S} + 8\mathbf{a}$; see the function get_3_isog in Table 1. For the evaluation of the isogeny, we modify the map in [17, Eq. (17)] to give $(X'\colon Z') = (X(X_3X - Z_3Z)^2\colon Z(Z_3X - X_3Z)^2)$. This costs $6\mathbf{M} + 2\mathbf{S} + 2\mathbf{a}$; see the function eval_3_isog in Table 1.

Projective Four Isogenies. We now let $x(P) = (X_4\colon Z_4) \in \mathbb{P}^1$ be such that P has exact order 4 in $E_{(A\colon C)}$, and leave all other notation and definitions as above. As is discussed in [17, Sect. 4.3.2], there are some minor complications in the derivation of 2- and 4-isogenies, either because a direct application of Vélu's formulas [49] for a 2-isogeny do not preserve the Montgomery form, or because repeated application of the 4-isogeny resulting from Vélu's formulas is essentially degenerate. For our purposes, i.e., in the case of 4-isogenies (overall, we found using 4-isogenies to be significantly faster than using 2-isogenies), the latter problem is remedied by application of the simple isomorphism in [17, Eq. (15)]. When building the 4e isogenies as a composition of 4-isogenies, this isomorphism is needed in every 4-isogeny computation except for the very first one, and we derive explicit formulas for both of these cases.

Note that for the very first 4-isogeny $\phi_0\colon E_{(A\colon C)} \to E_{(A'\colon C')}$ computed in the public key generation phase, the curve $E_{(A\colon C)}$ is that which is specified in the system parameters; and, for the first 4-isogeny in the shared secret computation, $E_{(A\colon C)}$ is the curve that is received as part of a public key sent over the wire. In both cases the curve is normalized so that $A = a$ and $C = 1$. In this case we use [17, Eq. (20)] directly, which gives $(A'\colon C') = (2(a + 6)\colon a - 2)$, and projectivize the composition of [17, Eqs. (19) and (21)] to give $(X'\colon Z') = ((X + Z)^2(aXZ + X^2 + Z^2)\colon (2 - a)XZ(X - Z)^2)$. This costs $4\mathbf{M} + 2\mathbf{S} + 9\mathbf{a}$; see the function first_4_isog in Table 1.

For the general 4-isogeny, we projectivized the composition of the above isogeny with the isomorphism in [17, Eq. (15)], making some modifications as follows. We made use of the xDBL function to parameterize the point of order 2 in [17, Eq. (15)] in terms of the point $(X_4\colon Z_4)$ of order 4. For the isogeny evaluation function, we again found it advantageous to simplify under the applicable component of the 4-division polynomial $\psi_4(x, y) = 4y(x - 1)(x + 1)\hat{\psi}_4(x)$, which is $\hat{\psi}_4(x) = x^4 + 2(A/C)x^3 + 6x^2 + 2(A/C)x + 1$ and which vanishes at X_4/Z_4. For the computation of the isogenous curve, we get $(A'\colon C') = (2(2X_4^4 - Z_4^4)\colon Z_4^4)$, and for the evaluation of the isogeny, we get the image

---

[4] As usual, $\mathbf{M}$, $\mathbf{S}$ and $\mathbf{a}$ represent the costs of field multiplications, squarings, and additions, respectively. We always count multiplications by curve coefficients as full multiplications, since these coefficients change within an isogeny class and thus we cannot expect any savings by treating them differently to generic elements.

point $(X': Z')$ where $X' = X \left(2X_4 Z_4 Z - X(X_4^2 + Z_4^2)\right)(X_4 X - Z_4 Z)^2$ and $Z' = Z \left(2X_4 Z_4 X - Z(X_4^2 + Z_4^2)\right)(Z_4 X - X_4 Z)^2$. Since each 4-isogeny is evaluated at multiple points, during the above computation of the isogenous curve, we also compute and store five values that can be (re)used in the evaluation: $\mathbf{c} = [X_4^2 + Z_4^2, X_4^2 - Z_4^2, 2X_4 Z_4, X_4^4, Z_4^4]$.

The computation of the isogenous curve and of the five values in $\mathbf{c}$ above costs $5\mathbf{S} + 7\mathbf{a}$, and on input of $\mathbf{c}$ and $Q = (X : Z)$, the isogeny evaluation costs $9\mathbf{M} + 1\mathbf{S} + 6\mathbf{a}$; see the functions get_4_isog and eval_4_isog in Table 1.

Summary of Subroutines. All of the point and isogeny operations are summarized in Table 1. We note that the input $\mathbf{c} \in K^5$ into the eval_4_isog function is the same tuple of constants output from get_4_isog, as described above.

Table 1. Summary of the subroutines used in our SIDH implementation. Here the points P and Q are on the curve $E_{(a,b)} = E_{(A:B:C)}$, and $E' = E_{(A':B':C')}$ is used to denote the isogenous curve. We use $n = \log_2 m - 1$ to count operations in loops. For a more detailed table, see the full version [13].

Function	Input (s)	Output (s)	M	S	a	I
j_inv	(A, C)	$j(E)$	3	4	8	1
xDBLADD	$\left(x(P), x(Q), x(Q-P), \frac{a+2}{4}\right)$	$(x([2]P), x(Q+P))$	6	4	8	-
xADD	$(x(P), x(Q), x(Q-P))$	$x(Q+P)$	3	2	6	-
xDBL	$(x(P), A+2C, 4C)$	$x([2]P)$	4	2	4	-
xDBLe	$(x(P), A, C, e)$	$x([2^e]P)$	4e	2e	4e	-
LADDER	$(x(P), a, m)$	$x([m]P)$	5n	4n	9n	-
LADDER_3_pt	$(x(P), x(Q), x(Q-P), a, m)$	$x(P + [m]Q)$	9n	6n	14n	-
xTPL	$(x(P), A+2C, 4C)$	$x([3]P)$	8	4	8	-
xTPLe	$(x(P), A, C, e)$	$x([3^e]P)$	8e	4e	8e	-
get_3_isog	$x(P)$	(A', C')	3	3	8	-
eval_3_isog	$(x(P), x(Q))$	$x(\phi(Q))$	6	2	2	-
first_4_isog	$(x(Q), a)$	$(x(\phi_0(Q)), A', C')$	4	2	9	-
get_4_isog	$x(P)$	$(A', C', \mathbf{c})$	-	5	7	-
eval_4_isog	$(\mathbf{c}, x(Q))$	$x(\phi(Q))$	9	1	6	-
secret_pt	$(P, Q = \tau(P), m)$	$x(P + [m]Q)$	5n	4n	9n	-
distort_and_diff	x_P	$x(\tau(P) - P)$	-	1	2	-
get_A	(x_P, x_Q, x_{Q-P})	A	4	1	7	1
inv_3_way	(z_1, z_2, z_3)	$(z_1^{-1}, z_2^{-1}, z_3^{-1})$	6	-	-	1

4   Parameters and Implementation Choices

Prime Field and Isogeny Class. From here on, the field K is fixed as $K = \mathbb{F}_{p^2}$, where $p := 2^{372} \cdot 3^{239} - 1$, and $\mathbb{F}_{p^2} = \mathbb{F}_p(i)$ for $i^2 = -1$. In terms of the notation from Sect. 2, this means that $\ell_A = 2$, $\ell_B = 3$, $e_A = 372$, $e_B = 239$ and $f = 1$. We searched for primes of the form $2^{e_A} 3^{e_B} f - 1$ with a bit length close to (but no larger than) 768, aiming to strike a balance $\ell_A^{e_A} \approx \ell_B^{e_B}$ to ensure that one side of the key exchange is not appreciably easier to attack than the other (more on this below), and to balance the computational costs for Alice and Bob. We originally searched with no restriction on the cofactor f, but did not find an example of another prime that would perform as fast as ours and where the overall security was increased enough to warrant $f \neq 1$. Given the best known classical and quantum attack complexities (see Sect. 1), choosing a prime close to 768 bits aims to reach a claim of 192 bits of classical security and 128 bits of quantum security. The arithmetic advantages of this prime choice are detailed in Sect. 5.

Our implementation works in the isogeny class of elliptic curves over $\mathbb{F}_{p^2}$ that contains the supersingular Montgomery curve $E_0/\mathbb{F}_{p^2} : y^2 = x^3 + x$. Every curve in this isogeny class has $(p + 1)^2 = (2^{372} \cdot 3^{239})^2$ points and is also supersingular [44, Exercise 5.4 and 5.10(a)]. The curve E_0 is the public parameter that is the starting point for the key exchange protocol.

The Base-Field and Trace-Zero Torsion Subgroups. A valuable technique that was introduced by Verheul [50] and that has played a key role in the implementation of symmetric pairings on supersingular elliptic curves [42], is that of using a distortion map. Verheul showed that every supersingular elliptic curve has a distortion map [50]. For a prime power $\ell^e \mid \#E_0(\mathbb{F}_p)$, such a map connects the cyclic torsion subgroup $E_0(\mathbb{F}_p)[\ell^e]$ defined over the base field $\mathbb{F}_p$ with the trace-zero subgroup of $E_0(\mathbb{F}_{p^2})[\ell^e]$. The distortion map we use for E_0 is given by the endomorphism $\tau : E_0(\mathbb{F}_{p^2}) \to E_0(\mathbb{F}_{p^2})$, $(x, y) \mapsto (-x, iy)$.

An ℓ^e torsion point $P \in E_0(\mathbb{F}_p)$ is mapped to an ℓ^e-torsion point $\tau(P) \in E_0(\mathbb{F}_{p^2})$ and the Weil pairing $e_{\ell^e}(P, \tau(P)) \neq 1$ is non-trivial. It is easy to see that the trace of the image point is zero, namely $\mathrm{Tr}(\tau(P)) = \tau(P) + \pi_p(\tau(P)) = \mathcal{O}$, where π_p is the p-power Frobenius endomorphism on E_0. An advantage of using the trace-zero subgroup is that its points can be represented by two $\mathbb{F}_p$-elements only and are therefore half the size of a general curve point defined over $\mathbb{F}_{p^2}$.

Choosing Generator Points for Torsion Subgroups. We apply a similar idea in that we fix the public $\ell_A^{e_A}$-torsion points P_A, Q_A and $\ell_B^{e_B}$-torsion points P_B, Q_B as generators of the (respective) base field and trace-zero subgroups, chosen as follows. Let $P_A \in E_0(\mathbb{F}_p)[2^{372}]$ be the point given as $[3^{239}](z, \sqrt{z^3 + z})$, where z is the smallest positive integer such that $\sqrt{z^3 + z} \in \mathbb{F}_p$ and P_A has order 2^{372}. The point P_B is selected in the same way with order and cofactor swapped. We then take $Q_A = \tau(P_A)$ and $Q_B = \tau(P_B)$, which produces the following generators: $P_A = [3^{239}](11, \sqrt{11^3 + 11})$, $Q_A = \tau(P_A)$, $P_B = [2^{372}](6, \sqrt{6^3 + 6})$, and $Q_B = \tau(P_B)$.

In addition to the base field representations mentioned above, the simple relationship between the coordinates of Q_A and P_A and the coordinates of Q_B and P_B helps to further compactify the public parameters; see Sect. 6. However, choosing $\{P_A, Q_A\}$ and $\{P_B, Q_B\}$ as the bases for generating isogeny kernels from the base-field and trace-zero torsion subgroups can have caveats. For example, in the case $\ell = \ell_A = 2$, one obtains the following lemma (the proof of which is in the full version [13]).

Lemma 1. *Let $E : y^2 = x^3 + x$ be a supersingular elliptic curve defined over $\mathbb{F}_p$, $p > 3$, $p \equiv 3 \pmod{4}$, such that $\#E(\mathbb{F}_p) = 2^e \cdot N$ with N odd. Let $\mathbb{F}_{p^2} = \mathbb{F}_p(i)$, $i^2 = -1$, and let $E[\ell^e] \subseteq E(\mathbb{F}_{p^2})$. Let $P \in E(\mathbb{F}_p)[2^e]$ be any point of order 2^e and let $Q \in E(\mathbb{F}_{p^2})[2^e]$ be any point of order 2^e with $\mathrm{Tr}(Q) = Q + \pi_p(Q) = \mathcal{O}$. Then the order of $P + Q$ equals 2^{e-1}.*

In particular, Lemma 1 proves that any point of the form $P + [m]Q$ for odd m has order less than 2^e. Also note that if m is even, then the order of $P + [m]Q$ is 2^e because $[2^{e-1}](P + [m]Q) = [2^{e-1}]P \neq \mathcal{O}$. Furthermore, this means that the points P and Q do not generate the full 2^e-torsion subgroup, and strictly speaking, the two points are not independent[5].

In the following two paragraphs we show how Alice and Bob can choose their secret scalars to guarantee that the degrees of their isogenies are maximal, i.e., $\ell_A^{e_A}$ and $\ell_B^{e_B}$ respectively.

Sampling Full Order 2-Torsion Points. To sample a 2-torsion point R_A of full order, we sample a uniform random integer $m' \in \{1, 2, \ldots, 2^{e_A-1} - 1 = 2^{371} - 1\}$ and set $R_A = P_A + [2m']Q_A$; R_A is guaranteed to have order 2^{e_A} by the above discussion. Because two distinct choices for m' lead to two distinct cyclic subgroups generated by the corresponding R_A, one can reach $2^{e_A-1} - 1 = 2^{371} - 1$ distinct subgroups and thus isogenies with this sampling procedure. We have seen in Sect. 2 that there are $3 \cdot 2^{e_A-1}$ distinct full order subgroups in $E_0[2^{e_A}]$, and thus our sampling procedure only reaches about one third of those.

Sampling Full Order 3-Torsion Points. To sample a 3-torsion point R_B of full order, we sample a uniform random integer $m' \in \{1, 2, \ldots, 3^{e_B-1} - 1 = 3^{238} - 1\}$ and set $R_B = P_B + [3m']Q_B$. Since $[3^{e_B-1}]R_B = [3^{e_B-1}]P_B \neq \mathcal{O}$, R_B is guaranteed to have order 3^{e_B}. In this way, we reach $3^{238} - 1$ of the possible subgroups and corresponding isogenies. Since there are $4 \cdot 3^{e_B-1}$ such subgroups in $E_0[3^{e_B}]$, we sample from about one quarter of those.

Strategies for Isogeny Computation and Evaluation. For computing and evaluating $\ell_A^{e_A}$- and $\ell_B^{e_B}$-isogenies, we closely follow the methodology described in [17, Sect. 4.2]. As already described in Sect. 2, such isogenies are composed of e_A isogenies of degree ℓ_A and e_B isogenies of degree ℓ_B, respectively. Figure 2 in [17] illustrates this computation with the help of a directed acyclic graph. In order to be able to evaluate the desired isogeny, one needs to compute all points that are

---

[5] Whenever we use the term independent for the points P and Q in what follows, we mean that the Weil pairing evaluated at P and Q is non-trivial.

represented by the final vertices, i.e., the leaves in the graph. As described earlier in Sect. 2, using the multiplication-based or isogeny-based methods to traverse this graph yields a simple but costly algorithm. De Feo et al. [17, Sect. 4.2.2] provide a discussion of how to obtain an optimal algorithm. They formally define the notion of a *strategy* for evaluating ϕ along a directed acyclic graph and show how to find an optimal strategy depending on the relative costs of scalar multiplication-by-ℓ and ℓ-isogeny evaluation. For the details on the optimal strategies for our chosen parameters, we refer to the full version [13].

5   Field Arithmetic

In this section, we describe the advantages of the chosen prime and optimizations to speed up the modular reduction inside SIDH, which were inspired by similar work on so-called Montgomery-friendly primes (e.g., see [19,27]). We remark that similar ideas can be easily applied to selecting primes and implementing their modular arithmetic at different security levels.

In our case, arithmetic is performed modulo the prime $p = 2^{372} \cdot 3^{239} - 1$. As described in Sect. 4, choosing an SIDH prime such that $\ell_A^{e_A} \approx \ell_B^{e_B}$ ensures a certain security strength across the whole key exchange scheme. Additionally, some implementations benefit from having a prime with a bit length slightly smaller than a multiple of a word size. Since 768 is the next multiple of 32 and 64 above the bit length of our prime, and $\log_2 p = 751 = 768 - 17$, the *extra room* available at the word boundaries enables the efficient use of other optimization techniques such as carry-handling elimination, and eases the efficient use of vector instructions. Working on a field of size slightly smaller than 2^{768} enables us to, e.g., use 12×64-bit limbs to represent field elements, whereas a prime slightly larger than 2^{768}, such as $p_{768} = 2^{387} \cdot 3^{242} - 1$ from [2], requires 13×64-bit limbs; the latter choice brings a relatively small increase in security at the expense of a significant increase in the cost of the modular arithmetic.

Since we work over $\mathbb{F}_{p^2}$, where $\mathbb{F}_{p^2} = \mathbb{F}_p(i)$ for $i^2 = -1$, we can leverage the extensive research done on the efficient implementation of such quadratic extension fields. In the context of pairings, high-speed implementations have exploited the combination of Karatsuba multiplication, lazy reduction, and carry-handling elimination; e.g., these techniques have been combined in optimized implementations on the curve BN254 [1]. Here we can follow a similar strategy since our field definition and underlying prime share several common traits with BN254, e.g., our prime being slightly smaller than a multiple of the word size enables the computation of several additions without carry-outs in the most significant word.

Efficient Modular Reduction. The cost of modular arithmetic (and, in particular, of modular multiplication) dominates the cost of the isogeny-based key exchange, so its efficient implementation is crucial for achieving high performance. At first glance, it would seem that SIDH primes prompt the use of generic Montgomery [32] or Barrett [3] reduction algorithms, which are relatively expensive in comparison with the efficient reduction of certain primes with special form (e.g., pseudo-Mersenne primes). For example, Azarderakhsh et al. [2] use a generic Barrett reduction for computing the modular multiplication in their SIDH implementation. However, we note that primes of this form *do* have a special shape that is amenable to faster modular reduction. Consider the case of the well-known Montgomery reduction [32]: letting $R = 2^{768}$ and $p' = -p^{-1} \bmod R$, then one can compute the Montgomery residue $c = aR^{-1} \bmod p$ for an input $a < pR$, by using $c = (a + (ap' \bmod 2^{768}) \cdot p)/2^{768}$, which costs approximately $s^2 + s$ multiplications for a $2s$-limb value a. For $p = 2^{372} \cdot 3^{239} - 1$, however, this computation simplifies to $c = (a + (ap' \bmod 2^{768}) \cdot 2^{372} \cdot 3^{239})/2^{768}$.

Moreover, $p' = -p^{-1} \bmod 2^{768}$ also exhibits a special form which reduces the cost of computing $ap' \bmod 2^{768}$ (e.g., $p' - 1$ contains five 64-bit limbs or eleven 32-bit limbs of value 0). In total, the cost of computing c in this case is $s(s - \lfloor 372/w \rfloor)$ multiplications for a word-size w. For example, if $w = 64$ (i.e., $s = 12$), the theoretical speedup for the simplified modular reduction is about 1.85x when applying these optimizations.

It is straightforward to extend the above optimizations to the different Montgomery reduction variants that exist in the literature. For our implementation, we adapted the Comba-based Montgomery reduction algorithm from [41]. Although merged multiplication/reduction algorithms, such as the coarsely integrated operand scanning (CIOS) Montgomery multiplication [25], offer performance advantages in certain scenarios, we prefer an implementation variant that consists of separate routines for integer multiplication and modular reduction. This approach enables the use of lazy reduction for the $\mathbb{F}_{p^2}$ arithmetic and allows easy-to-implement improvements in the integer multiplication, e.g., by using Karatsuba.

Algorithm 1 is based on the Montgomery reduction algorithm in product scanning form (a.k.a. Comba) presented in [41]. It has been especially tailored for efficient computation modulo the prime $p = 2^{372} \cdot 3^{239} - 1$ following the optimizations discussed above. As usual, given a radix-2^r field element representation using s limbs, the algorithm receives as input an operand $a < 2^{rs}p$ (e.g., the integer product of two Montgomery residues) and outputs the Montgomery residue $c = a \cdot 2^{-rs} \bmod p$. Here c is typically computed as $(a + (ap' \bmod 2^r) \cdot p)/2^r$ (s times) in a Comba-like fashion, where $p' = -p^{-1} \bmod 2^r$. However, as mentioned above, this expression simplifies to $(a + (a \bmod 2^r) \cdot \hat{p})/2^r$ where $\hat{p} = p + 1 = 2^{372} \cdot 3^{239}$, since $p' = 1$ for our prime. In addition, Algorithm 1 eliminates several multiplications due to the fact that the $\lfloor e_A/r \rfloor$ least significant limbs in $\hat{p}$ have value 0.

Since our scheme forces the availability of extra room in the radix-2^r representation (which is made possible by having the additional condition that $p < 2^{rs-2}$), there is no overflow in the most significant word during the computation of c in Algorithm 1 (i.e., its intermediate value can be held on exactly s r-bit registers). Moreover, if field elements are represented as elements in $[0, 2p - 1]$ (instead of the typical range $[0, p - 1]$), the output of Algorithm 1 remains bounded without the need of the conditional subtraction in Steps 19–20 [51].

Algorithm 1. Optimized Comba-based Montgomery reduction for the prime $p = 2^{372} \cdot 3^{239} - 1$.

Input: The prime $p = 2^{e_A} \cdot 3^{e_B} - 1$; the value $\hat{p} = p + 1$ containing $z = \lfloor e_A/r \rfloor$ 0-value terms in its r-bit representation, where $e_A = 372$, $e_B = 239$ and 2^r is the radix; the Montgomery constant 2^{rs} such that $2^{r(s-1)} \leq p < 2^{rs-1}$; and, the operand $a = (a_{2s-1}, ..., a_1, a_0)$ with $a < 2^{rs}p$ and $s = \lceil \log_2 p/r \rceil$.
Output: The Montgomery residue $c = a \cdot 2^{-rs} \bmod p$.

1: $(t, u, v) = 0$
2: **for** $i = 0$ **to** $s - 1$ **do**
3:    **for** $j = 0$ **to** $i - 1$ **do**
4:       **if** $j < i - z + 1$ **then**
5:          $(t, u, v) = c_j \times \hat{p}_{i-j} + (t, u, v)$
6:       $(t, u, v) = (t, u, v) + a_i$
7:    $c_i = v$
8:    $v = u, \ u = t, \ t = 0$
9: **for** $i = s$ **to** $2s - 2$ **do**
10:    **if** $z > 0$ **then**
11:       $z = z - 1$
12:    **for** $j = i - s + 1$ **to** $s - 1$ **do**
13:       **if** $j < s - z$ **then**
14:          $(t, u, v) = c_j \times \hat{p}_{i-j} + (t, u, v)$
15:       $(t, u, v) = (t, u, v) + a_i$
16:    $c_{i-s} = v$
17:    $v = u, \ u = t, \ t = 0$
18: $c_{s-1} = v + a_{2s-1}$
19: **if** $c \geq p$ **then**
20:    $c = c - p$
21: **return** c

Although typical values for r would be $w = 32$ or 64 to match w-bit architectures, some redundant representations might benefit from the use of $r < w$ in order to avoid additions with carries or to facilitate the efficient use of vector instructions. To this end, the chosen prime is very flexible and supports different efficient alternatives; for example, it supports the use of a 58-bit representation with $s = 13$ limbs when using 64-bit multipliers or the use of a 26-bit representation with $s = 29$ limbs when using 32-bit multipliers.

In our 64-bit implementation, we opted for a generic radix-2^{64} representation using $s = 12$ limbs, in which case the Montgomery constant is $2^{rs} = 2^{768}$. In this case, given that the initial and final loop iterations can be simplified in an unrolled implementation of Algorithm 1, the cost of the modular reduction is 83 multiplication instructions. This result almost halves the number of multiplication instructions compared to a naïve Montgomery reduction, which requires $12^2 + 12 = 156$ multiplication instructions (per reduction).

Inversions. Our SIDH implementation requires one modular inversion during key generation, and two modular inversions during the computation of the shared secret. These inversions can be implemented using Montgomery inversion based on, e.g., the binary GCD algorithm. However, this method does not run in constant time by default, and therefore requires additional countermeasures to protect it against timing attacks (e.g., the application of input randomization). Since inversion is used scarcely in our software, we instead opted for the use of Fermat's little theorem, which inverts the field element a via the exponentiation a^{p-2} mod p that uses a fixed addition chain. Our experiments showed that the cost of this exponentiation is around 9 times slower than (an average run of) the GCD-based method, however even the more expensive inversion only contributes to less than 1% of the overall latency of each round of the protocol. Thus, our choice to compute each isolated inversion via a fixed exponentiation protects the implementation without impacting the performance in any meaningful way, and avoids the need for any additional randomness.

6   SIDH Implementation Summary

In this section we pull together all of the main ingredients from Sects. 2–5 to give a brief overview of the scheme and its implementation. For high-level Magma code that illustrates the entire SIDH protocol, see SIDH.mag in [14].

Public Parameters. Together with the curve $E_0 : y^2 = x^3 + x$ and the prime $p = 2^{372}3^{239} - 1$, the public parameters are $P_A = [3^{239}](11, \sqrt{11^3 + 11})$, $Q_A = \tau(P_A)$, $P_B = [2^{372}](6, \sqrt{6^3 + 6})$, and $Q_B = \tau(P_B)$. Given that all these square roots are in $\mathbb{F}_p$ (we choose the "odd" ones), and that Q_A and Q_B require no storage, this means that only 4 $\mathbb{F}_p$-elements (or 3004 bits) are required to fully specify the public generators. If we were to instead randomly choose extension field torsion generators without use of the distortion map, as is suggested in [17], then 16 $\mathbb{F}_p$ elements (or 12016 bits) would be required to specify the public generators.

Key Generation. On input of the public parameters above, and the secret key m_A chosen as in Sect. 4, Alice proceeds as in [13, Algorithm 3] (see [13, Algorithm 2] for the simple, but slower *multiplication-based* main loop). She calls the secret_pt function, which computes $P_A + [m_A]Q_A$ by calling LADDER to compute $x([m_A]Q_A)$, before recovering the corresponding y-coordinate using the Okeya-Sakurai strategy [36]; this allows the addition of P_A and $[m_A]Q_A$. All of these operations are performed over the ground field and we proceed by taking only $x(P_A + [m_A]Q_A)$ through the main loop.

We note that our implementation requires that Alice's secret isogeny is evaluated at both of the public parameters x_{P_B} and x_{Q_B}, as well as at the x-coordinate of the difference, $x_{Q_B - P_B}$; this allows Bob to kickstart the three_pt_ladder function (from [17, Algorithm 1]) during his shared secret phase. Conversely, Bob must also evaluate his secret isogeny at $x_{Q_A - P_A}$. In both cases, rather than setting x_{Q-P} as a public parameter, it can be computed on-the-fly from x_P, since in this special instance, $x_{Q-P} = x_{\tau(P)-P} = i \cdot (x_P^2 + 1)/(2x_P)$. This is fed directly into our projective isogeny evaluation function, so we do not need $x_{Q-P} \in \mathbb{A}$, but can instead compute $x(Q - P) = (i(x_P^2 + 1) : 2x_P) \in \mathbb{P}^1$, which costs just one squaring and two additions in $\mathbb{F}_p$; this operation is performed with the distort_and_diff function.

At the conclusion of [13, Algorithm 3], Alice outputs her public key as $\mathrm{PK}_{\mathrm{Alice}} = [x_{\phi_A(P_B)}, x_{\phi_A(Q_B)}, x_{\phi_A(Q_B-P_B)}] \in \mathbb{F}_{p^2}^3$. Bob proceeds similarly, as shown in [13, Algorithm 5] (again, see [13, Algorithm 4] for a simpler, but slower multiplication-based approach), and outputs his public key as $\mathrm{PK}_{\mathrm{Bob}} = [x_{\phi_B(P_A)}, x_{\phi_B(Q_A)}, x_{\phi_B(Q_A-P_A)}] \in \mathbb{F}_{p^2}^3$.

Alice's fast key generation via [13, Algorithm 3], using the strategies for computing the isogeny trees as given in Sect. 4, requires 638 multiplications-by-4 and the evaluation of 1330 4-isogenies; calling the simpler [13, Algorithm 2] requires 17020 multiplications-by-4 an 744 4-isogeny evaluations. On Bob's side, the optimal strategy (i.e., fast key generation) requires 811 multiplications-by-3 and the evaluation of 1841 3-isogenies; the simpler version requires 28441 multiplications-by-3 and 956 3-isogeny evaluations. See Sect. 7 for the benchmarks and further discussion.

Remark 1. Observe that the public keys above only contain x-coordinates of points, and do not contain the Montgomery coefficient, a, that defines the isogenous curve E_a. This is because a can be recovered (on the other side) by exploiting the relation $a = \frac{(1-x_P x_Q - x_P x_{Q-P} - x_Q x_{Q-P})^2}{4 x_P x_Q x_{Q-P}} - x_P - x_Q - x_{Q-P}$, which holds if x_P, x_Q and x_{Q-P} are the respective x-coordinates of three points P, Q and $Q - P$ on the Montgomery curve with coefficient a [19, Sect. A.2]. Here public key compression (i.e., dropping the a coefficient) is free, and decompression via the above equation amounts to $4\mathbf{M} + 1\mathbf{S} + 7\mathbf{a} + 1\mathbf{I}$; see the function get_A in [14]. Compared to the overall shared secret computation, this decompression comes at a minor cost. In an earlier draft of this paper, we provided an option for a compression that instead transmitted the a coefficient, together with x_P, x_Q, and a *sign bit* that was used to choose the correct square root (during the recovery of x_{Q-P}). The above compression has the obvious advantage of saving the sign bit, and, more importantly, means that decompression only requires an inversion (instead of a square root). Since our software already required inversions, but did not use square roots anywhere else, the amount of additional code required to include this compression is minimal. We thank Luca De Feo and Ben Smith for pointing out this simpler compression.

Shared Secret. On input of $\mathrm{PK}_{\mathrm{Bob}} = [x_{\phi_B(P_A)}, x_{\phi_B(Q_A)}, x_{\phi_B(Q_A-P_A)}]$ and her secret key m_A, Alice first computes $a_B = \text{get\_A}(x_{\phi_B(P_A)}, x_{\phi_B(Q_A)}, x_{\phi_B(Q_A-P_A)})$, then calls [13, Algorithm 7] (again, see [13, Algorithm 6] for a more compact, but significantly slower main loop) to generate her shared secret. This starts by calling the three_pt_ladder function (from [17, Algorithm 1]) to compute $x(\phi_B(P_A) + [m_A]\phi_B(Q_A))$, which is used to generate the kernel of the isogeny that is computed in the main loop. Finally, Alice uses the j_inv function to compute her shared secret. For Bob's analogous shared key generation, see [13, Algorithms 8–9].

Alice's fast key generation via [13, Algorithm 7], again using the strategies in Sect. 4, requires 638 multiplications-by-4 and the evaluation of 772 4-isogenies; calling the simpler [13, Algorithm 6] requires 17020 multiplications-by-4 and 186 4-isogeny evaluations. On Bob's side, the optimal strategy (i.e., fast key generation) requires 811 multiplications-by-3 and the evaluation of 1124 3-isogenies; the simpler version requires 28441 multiplications-by-3 and 239 3-isogeny evaluations. See Sect. 7 for the benchmarks and further discussion.

7   SIDH Performance

To evaluate the performance of the proposed supersingular isogeny system and the different optimizations, we wrote a software library supporting ephemeral SIDH key-exchange. The software is mostly written in the C language and has been designed to facilitate the addition of specialized code for different platforms and applications. The first release of the library comes with a fully portable C implementation supporting 32- and 64-bit platforms and two optional x64 implementations of the field arithmetic: one implementation based on intrinsics (which is, e.g., supported on Windows OS by Visual Studio) and one implementation written in x64 assembly (which is, e.g., supported on Linux OS using GNU GCC and clang compilers). The latter two optional modules are intended for high-performance applications. All of the software is publicly available in [14].

In Table 2, we present the performance of our software using the x64 assembly implementation in comparison with the implementation proposed by [2]. Results for the implementation in [2] were obtained by benchmarking their software[6] on the same Intel Sandy Bridge and Haswell machines, running Ubuntu 14.04 LTS. Note that the results in Table 2 differ from what was presented in Table 3 in [2]. The differences might be due to the use of overclocking (i.e., TurboBoost technology). For our comparisons, we disabled TurboBoost for a more precise and fair comparison.

Table 2 shows that the total cost of computing one Diffie-Hellman shared key (adding Alice's and Bob's individual costs together) using our software is, on both platforms, over 2.8 times faster than the software from [2]. These results are due to the different optimizations discussed throughout this work, the most prominent two being (i) the elimination of inversions during isogeny computations by working with projective curve coefficients, and (ii) the faster modular arithmetic triggered by the selected prime and the tailor-made Montgomery reduction for SIDH primes. It is important to note that, in particular, the advantage over [2] is not even larger because the numerous inversions used during the isogeny computations in [2] are not computed in constant time. Making such inversions constant-time would significantly degrade their performance (see the related paragraph in Sect. 5).

---

[6] See http://djao.math.uwaterloo.ca/thesis-code.tar.bz2.

Table 2. Performance results (expressed in millions of clock cycles) of the proposed SIDH implementation in comparison with the implementation by Azarderakhsh et al. [2] on x64 platforms. Benchmark tests were taken with Intel's TurboBoost disabled and the results were rounded to the nearest 10^6 clock cycles. Benchmarks were done on a 3.4 GHz Intel Core i7-2600 Sandy Bridge and a 3.4 GHz Intel Core i7-4770 Haswell processor running Ubuntu 14.04 LTS.

Operation	This work		Prior work [2]	
	Sandy Bridge	Haswell	Sandy Bridge	Haswell
Alice's keygen	50	46	165	149
Bob's keygen	57	52	172	152
Alice's shared key	47	44	133	118
Bob's shared key	55	50	137	122
Total	207	192	608	540

Remark 2. In Sect. 4 we discussed several specialized choices that were made for reasons unrelated to performance, e.g., in the name of simplicity and/or compactness. We stress that, should future cryptanalysis reveal that these choices introduce a security vulnerability, the performance of SIDH and the performance improvements in Sects. 3 and 5 are unlikely to be affected (in any meaningful way) by reverting back to the more general case(s). In particular, if it turns out that sampling from a fraction of the possible 2- and 3-torsion subgroups gives an attacker some appreciable advantage, then modifying the code to sample from the full set of torsion subgroups is merely an exercise, and the subsequent performance difference would be unnoticeable. Similarly, if any of (i) starting on a subfield curve (see [13, Remark 2]), (ii) using of the base-field and trace-zero subgroups, or (iii) using the distortion map, turns out to degrade SIDH security, then the main upshot of reverting to randomized public generators or starting on a curve minimally defined over $\mathbb{F}_{p^2}$ would be the inflated public parameters (see Sect. 6); the slowdown during key generation would be minor and the shared secret computations would be unchanged.

8 BigMont: A Strong ECDH + SIDH Hybrid

We now return to the discussion (from Sect. 1) of a hybrid scheme. Put simply, and in regards to both security and suitability, at present there is not enough confidence and consensus within the PQC community to warrant the standalone deployment of one particular post-quantum key exchange primitive. Subsequently, there is interest (cf. [5]) in deploying classical primitives alongside post-quantum primitives in order to hedge one's bets until a confidence-inspiring PQC key exchange standard arrives. This is particularly interesting in the case of SIDH, whose security has (because of its relatively short lifespan) received less cryptanalytic scrutiny than its post-quantum counterparts.

In this section we discuss how traditional ECDH key exchange can be included alongside SIDH key exchange at the price of a very small overhead. The main benefit of our approach is its simplicity; while SIDH could be partnered with ECDH on any of the standardized elliptic curves, this would mean that a lot more code needs to be written and/or maintained. In particular, it is often the case that the bulk of the code in high-speed ECC implementations relates to the underlying field arithmetic. Given that none of the fields underlying the standardized curves are SIDH-friendly[7], such a partnership would require either a generic implementation that would be much less efficient, or two unrelated implementations of field arithmetic. Our proposal avoids this additional complexity by performing ECDH on an elliptic curve defined over the same ground field as the one used for SIDH.

For $p = 2^{372}3^{239} - 1$, recall that our SIDH software works with isogenous curves $E_a/\mathbb{F}_{p^2}: y^2 = x^3 + ax^2 + x$ whose group orders are of the form $\#E_a = 2^i \cdot 3^j$, meaning that elliptic curve discrete logarithms are easy on all such curves by the Pohlig-Hellman algorithm [39]. However, there are also (exponentially many) ordinary curves of the form $E_a/\mathbb{F}_{p^2}$ that *are* cryptographically secure. In particular, over the base field $\mathbb{F}_p$, we can hope to find $a \in \mathbb{F}_p$ such that $E_a/\mathbb{F}_p$ and its quadratic twist $E'_a/\mathbb{F}_p$ are cryptographically strong, i.e., such that $E_a/\mathbb{F}_p$ is *twist-secure* [4].

Since $p \equiv 3 \bmod 4$, we searched for such a curve in exactly the same way as, e.g., Hamburg's Goldilocks curve [20] was found. Namely, since the value $(a+2)/4$ is the constant that appears in Montgomery's ladder computation [33], we searched for the value of a that gave rise to the smallest absolute value of $(a + 2)/4$ (when represented as an integer in $[0, p)$), and such that $\#E_a$ and $\#E'_a$ are both 4 times a large prime. For p as above, the first such value is $a = 624450$; to make a clear distinction between curves in the supersingular isogeny class and the strong curve used to perform ECDH, we (re)label this curve as $M_a/\mathbb{F}_p: y^2 = x^3 + ax^2 + x$ with $a = 624450$. The trace t_{M_a} of the Frobenius endomorphism on M_a (see [13]) gives $\#M_a = p + 1 - t_{M_a} = 4r_a$ and $\#M'_a = p + 1 + t_{M_a} = 4r'_a$, where r_a and r'_a are both 749-bit primes.

Following [4], every element in $\mathbb{F}_p$ corresponds to the x-coordinate of a point on either M_a or on M'_a. Together with the fact that Montgomery's LADDER function correctly computes underlying scalar multiplications independently of the quadratic twist, M_a being twist-secure allows us to treat all $\mathbb{F}_p$ elements as valid public keys and to perform secure ECDH without the need for any point validation.

The ECDH secret keys are integers in $[0, r_a)$. To ensure an easy constant-time LADDER function, we search for the smallest $\alpha \in \mathbb{N}$ such that αr_a and $(\alpha + 1)r_a - 1$ are the same bit length, which is $\alpha = 3$; accordingly, secret keys are parsed into $[3r_a, 4r_a)$ prior to the execution of scalar multiplications via LADDER. Subsequently, for $m \in [0, r_a)$ and $x(P) \in \mathbb{P}^1(\mathbb{F}_p)$, computing $x([m]P) = $ LADDER$(x(P), m, a)$ requires 1 call to xDBL and 750 calls to xDBLADD (see Table 1 for the operation counts of these functions, but note that here we can take

---

[7] Nor are any of the fields large enough to support highly (quantum-)secure SIDH.

Table 3. Comparison of standalone SIDH versus hybrid SIDH + ECDH. Timing benchmarks were taken on a 3.4 GHz Intel Core i7-4770 Haswell processor running Ubuntu 14.04 LTS with TurboBoost disabled and results rounded to the nearest 10^6 clock cycles. For simplicity, the bit-security of the primitives was taken to be the target security level and is not intended to be precise.

Comparison		Standalone SIDH	Hybrid SIDH + ECDH
≈ bit-security (hard problem)	Classical	192 (SSDDH)	384 (ECDHP)
	PQ	128 (SSDDH)	128 (SSDDH)
Public key size		564	658
Speed (cc $\times 10^6$)	Alice's keygen	46	52
	Bob's keygen	52	58
	Alice's shared key	44	50
	Bob's shared key	50	57

advantage of the fixed, small constant a). As all of these computations take place over the ground field, the total time taken to compute ECDH public keys and shared secrets is only a small fraction of the total time taken to compute the analogous SIDH keys – see Table 3.

From an implementation perspective, partnering SIDH with ECDH as above is highly advantageous because the functions required to compute $x([m]P) =$ LADDER$(x(P), m, a)$ are already available from our Montgomery SIDH framework. In particular, the key generation (see Sect. 6) already has a tailored Montgomery LADDER function that works entirely over the base field, i.e., on the starting curve E_0, so computing ECDH keys is as simple as calling pre-existing functions on input of a different constant.

Though the speed overhead incurred by adding ECDH to SIDH in this way is small (see Table 3), choosing to use such a large elliptic curve group makes concatenated keys larger than they would be if a smaller elliptic curve was used for ECDH. For example, suppose we were to instead use the curve currently recommended in Suite B [35], Curve P-384, and (noting that uncompressed Curve P-384 points are larger than our proposed ECDH public keys) were to compress ECDH public keys as an x-coordinate and a sign bit. The total public key size with SIDH-compressed keys would then be 612 bytes, instead of the 658 bytes reported in Table 3. Though this difference is noticeable, it must be weighed up against the cost of the extensive additional code required to support Curve P-384, which would almost certainly share nothing in common with the existing SIDH code. Moreover, the simplicity of adding ECDH to SIDH as we propose is not the only reason to justify slightly larger public keys; the colossal 384-bit security achieved by M_{624450} also puts it in a position to tolerate the possibility of significant future advancements in ECDLP attacks. Due to the complexity of the ECDLP on M_{624450} in comparison with all of the elliptic curves in the standards, we dub this curve "BigMont".

In Table 3 we compare hybrid SIDH + ECDH versus standalone SIDH. The take-away message is that for a less than 1.17x increase in public key sizes and less than 1.13x increase in the overall computing cost, we can increase the classical security of the key exchange from 192 bits (based on the relatively new SSDDH problem) to 384 bits (based on the long-standing ECDLP).

9   Validating Public Keys

Recall from Sect. 2 that De Feo et al. [17] prove that SIDH is *session-key secure* (under SSDDH) in the authenticated-links adversarial model [10]. This model assumes perfectly authenticated links which effectively forces adversaries to be passive eavesdroppers; in particular, it assumes that public keys are correctly generated by honest users. While this model can be suitable for key exchange protocols that are instantiated in a truly ephemeral way, in real-world scenarios it is often the case that (static) private keys are reused. This can incentivize malicious users to create faulty public keys that allow them to learn information about the other user's static private key, and in such scenarios validating public keys becomes a mandatory practical requirement.

In traditional elliptic curve Diffie-Hellman (ECDH), validating public keys essentially amounts to checking that points are on the correct and cryptographically secure curve [7]. Such *point validation* is considered trivial in ECDH, since checking that a point satisfies a curve equation requires only a handful of field multiplications and additions, and this is negligible compared to the overall cost (e.g., of a scalar multiplication).

In contexts where SIDH private keys are reused, public key validation is equally as important but is no longer as trivial. In April 2015, a group from the NSA [24] pointed out that "direct public key validation is not always possible for [...] isogeny based schemes" before describing more complicated options that validate public keys *indirectly*. In this section we describe ways to directly validate various properties of our public keys that, in particular, work entirely in our compact framework, i.e., without the need of y-coordinates or of the Montgomery b coefficient that fixes the quadratic twist.

Recall from Sect. 6 that an honest user generates public keys of the form $\text{PK} = [x_P, x_Q, x_{Q-P}] \in \mathbb{F}_{p^2}^3$, where $P = (x_P, y_P)$ and $Q = (x_Q, y_Q)$ are of the same order ℓ^e on a Montgomery curve E_a that is $\mathbb{F}_{p^2}$-isogenous to E_0, and are such that $Q \neq [\lambda]P$ for any $\lambda \in \mathbb{Z}$; the algorithms we describe below will only deem a purported public key as valid if this is indeed the case. Recall from Remark 1 that the three x-coordinates in the public key are immediately used to recover the Montgomery a coefficient that was dropped during compression; this coefficient must also be considered as part of the public key during validation.

Public key validation must check that the (underlying) points P and Q are of the full order ℓ^e. If not, then an SIDH-like analogue of the Lim-Lee [28] small subgroup attack becomes a threat; e.g., an attacker could send x_Q where Q has small order q and guess the shared secret (i.e., the kernel $\langle P + [m]Q \rangle$) to learn $m \bmod q$. In addition, the procedure must also assert that $Q \neq [\lambda]P$ (or equivalently, that $P \neq [\lambda]Q$) for some $\lambda \in \mathbb{Z}$; if this assertion is not made, then a malicious user can simply send a public key where $Q = [\lambda]P$, which ultimately forces the shared secret to be independent of the honest party's private key. Such capabilities could be catastrophic if the authentication mechanism does not detect them.

The validation procedure we describe below guards against all of these attacks by asserting that P and Q both have order ℓ^e, and that the Weil pairing $e_{\ell^e}(P,Q)$ has the maximum possible order, namely the same order as the Weil pairing of the corresponding public parameters[8]; this means that the points P and Q generate as much of the ℓ^e torsion as is possible (according to the definition of the public parameters). This second assertion can be made in a very simple way, thanks to an observation by Ben Smith, who pointed out the following (using [31, Lemma 16.2]). If the points P and Q are in $E[mn]$, then the n-th power of the Weil pairing $e_{mn}(P,Q)$ can be computed as $e_{mn}(P,Q)^n = e_m([n]P,[n]Q)$, which allows us to efficiently check that the order of the Weil pairing is as it should be[9].

The application of the above validation procedure (to the three x-coordinates in a public key) is different for Alice and Bob, so we now describe these cases separately. We then discuss how both parties validate that the curve E_a corresponds to a supersingular curve in the correct isogeny class, and conclude the section with performance benchmarks for the validation process. All of the procedures described below can be found in the file `Validate.mag` [14].

Alice's Validation of Bob's Public Key. Alice must determine whether Bob's transmission $[x_P, x_Q, x_R] \in \mathbb{F}_{p^2}^3$ passes the tests described above. Recall from Sect. 4 that a consequence of Lemma 1 is that if the public parameters P_A and Q_A are chosen from the base field and trace-zero subgroups, then they do not form a basis for the full $\ell_A^{e_A}$-torsion. In particular, the order of the Weil pairing $e_{\ell_A^{e_A}}(P_A, Q_A)$ in our case is $\ell_A^{e_A - 1} = 2^{371}$; although this order is less than $\ell_A^{e_A}$, it is as large as is possible when the two basis elements are chosen from these particular torsion subgroups.

If Bob's public key is honestly generated, then x_P and x_Q correspond to points P and Q whose Weil pairing also has order $\ell_A^{e_A - 1}$; indeed, checking that this is the case ensures that we maximize the number of torsion subgroups that are spanned by $P + [2m']Q$. Let a be computed from x_P, x_Q and x_R as in Remark 1, and let $m = 4$ and $n = 2^{370}$ so that $mn = \ell_A^{e_A} = 2^{372}$. We assert that the exact order of $e_{\ell_A^{e_A}}(P,Q)$ is $\ell_A^{e_A - 1}$ by showing that $e_{\ell_A^{e_A}}(P,Q)^{\ell_A^{e_A - 2}}$ is non-trivial, making use of the identity above which gives $e_{\ell_A^{e_A}}(P,Q)^{\ell_A^{e_A - 2}} = e_{mn}(P,Q)^n, = e_m([n]P,[n]Q) = e_4([2^{370}]P,[2^{370}]Q)$. Together with the assertion that P and Q both have exact order 2^{372}, the assertion that the Weil pairing $e_4([2^{370}]P,[2^{370}]Q)$ is non-trivial completes the validation of x_P and x_Q. If indeed P and Q have order 2^{372}, the points $P' = [2^{370}]P$ and $Q' = [2^{370}]Q$

---

[8] We thank Steven Galbraith and David Jao, who independently pointed out that the Pohlig-Hellman algorithm [39] can also be used to efficiently check whether P and Q are dependent.

[9] A prior version of this paper made a weaker assertion using a more elaborate computation.

have exact order 4. In that case, $e_4(P', Q') \neq 1$ if, and only if, $x(P') \neq x(Q')$. This can be seen by an elementary proof using [8, Theorem IX.10(5.)] and [8, Corollary IX.11] together with the fact that $Q' \in \langle P' \rangle$ implies $x(P') = x(Q')$. All of these checks can be performed entirely with x-coordinates as follows. We compute $x(P') = x([2^{370}]P) = \mathtt{xDBLe}(x(P), a, 370)$ and $x(Q') = x([2^{370}]Q) = \mathtt{xDBLe}(x(Q), a, 370)$. Next, we assert that $x(P') \neq x(Q')$, which is done projectively via a cross-multiplication. To check that P has full order 2^{372}, we then use two more calls to $\mathtt{xDBL}$ to assert that $(X : Z) = x([2]P')$ has $Z \neq 0$ and that $(\tilde{X} : \tilde{Z}) = x([4]P')$ has $\tilde{Z} = 0$; we do exactly the same for Q. If any of these checks fail, the public key is deemed invalid and rejected.

The assertion that x_R is the correct *difference* x_{Q-P} on E_a is implicit from the computation of a during decompression, and from the combined validation of x_P, x_Q and a. Validating that a indeed corresponds to a supersingular curve in the correct isogeny class is performed in the same way for Alice and Bob, so we postpone it until after describing Bob's validation.

Bob's Validation of Alice's Public Key. Bob must determine whether Alice's transmission $[x_P, x_Q, x_R] \in \mathbb{F}_{p^2}^3$ passes the tests described above. In this case our choice of the base field and trace-zero subgroups does not impede the possibility of the Weil pairing having full order; indeed, the public generators P_B and Q_B are such that the order of $e(P_B, Q_B)$ is $\ell_B^{e_B}$. Thus, honest public keys also give rise to the Weil pairing $e_{\ell_B^{e_B}}(P, Q)$ having order $\ell_B^{e_B}$. To make use of the identity above, we set $m = 3$ and $n = 3^{238}$ so that $mn = \ell_B^{e_B} = 3^{239}$, which gives $e_{\ell_B^{e_B}}(P, Q)^{\ell_B^{e_B-1}} = e_{mn}(P, Q)^n = e_m([n]P, [n]Q) = e_3([3^{238}]P, [3^{238}]Q)$. Together with the assertion that P and Q both have exact order 3^{239}, the assertion that the Weil pairing $e_3([3^{238}]P, [3^{238}]Q)$ is non-trivial completes the validation of x_P and x_Q. If $P' = [3^{238}]P$ and $Q' = [3^{238}]Q$ have order 3, then $e_3(P', Q') \neq 1$ if, and only if, $x(P') \neq x(Q')$. This follows directly from [8, Corollary IX.11]. Again, we perform all of these checks using only x-coordinates as follows. We compute $x(P') = x([3^{238}]P) = \mathtt{xTPLe}(x(P), a, 238)$ and $x(Q') = x([2^{238}]Q) = \mathtt{xTPLe}(x(Q), a, 238)$ and assert that $x(P') \neq x(Q')$, which is again done projectively via a cross-multiplication. To check that P has full order 3^{239}, we assert that $(X : Z) = x(P')$ has $Z \neq 0$, and use one more call to $\mathtt{xTPL}$ to assert that $(\tilde{X} : \tilde{Z}) = x([3]P')$ has $\tilde{Z} = 0$; again, we do the same for Q. If any of these checks fail, the public key is deemed invalid and rejected.

Validating the Curve. We now show how to validate that a (i.e., the curve coefficient that is computed during the decompression of Alice or Bob's public key) corresponds to a Montgomery curve E_a that is a member of the correct supersingular isogeny class. The validation has two steps: we firstly assert that $j(E_a) \notin \mathbb{F}_p$ so that E_a is not a subfield curve, then we assert that E_a is in the correct supersingular isogeny class.

The first step is easy and totals a handful of multiplications in $\mathbb{F}_p$ (see the full version [13]); the less trivial step is to validate that E_a is supersingular.

To do this, we make use of Sutherland's probabilistic algorithm [46, Algorithm 1], which (for our purposes) says to pick a random point $P \in E_a(\mathbb{F}_{p^2})$, and to check whether $[p-1]P = \mathcal{O}$ or $[p+1]P = \mathcal{O}$. If this is the case, then E_a is supersingular with overwhelming probability: the probability that this test would pass if E_a was actually an ordinary curve is at most $8p/(p-1)^2 < 1/2^{747}$ [46, Proposition 1].

We now point out that E_a being supersingular is equivalent to either E_a or its quadratic twist, E'_a, belonging to the correct isogeny class. Namely, by [44, V.5.10(a)], E_a is supersingular if and only if its trace, t_{E_a}, satisfies $t_{E_a} \equiv 0$ mod p. Together with [48, Theorem 1], and recalling that $-2p \leq t_{E_a} \leq 2p$ [44, V.1.1], this means that there are (at most) 5 possible isógeny classes of supersingular elliptic curves, those which are described by $t_{E_a} \in \{-2p, -p, 0, p, 2p\}$. Since $p \equiv 3$ mod 4, there are only two possibilities for t_{E_a} that correspond to a Montgomery curve, i.e., two possible t_{E_a} such that $4 \mid \#E_a$ [33], namely $t_{E_a} = -2p$ and $t_{E_a} = 2p$. These traces respectively correspond to curves with $\#E_a = (p+1)^2$ that are in the correct isogeny class, and to curves with $\#E'_a = (p-1)^2$ that are in the isogeny class containing all of their non-trivial quadratic twists.

In our case we are trying to validate that a corresponds to a curve with $\#E_a = (p+1)^2$, so at first glance it would seem that the best route is to pick a random point $P \in E_a(\mathbb{F}_{p^2})$ and to assert that $[p+1]P = \mathcal{O}$. However, generating such a random point requires a square-root computation, and it turns out that we can (again) avoid the need for a square root altogether. For a given a, recall from Sect. 8 (or, in turn, from [4]) that elements in $\mathbb{F}_{p^2}$ are either the x-coordinate of a point on $E_a/\mathbb{F}_{p^2}$ or the x-coordinate of a point on $E'_a/\mathbb{F}_{p^2}$. This means that if E_a is supersingular, every element in $\mathbb{F}_{p^2}$ is the x-coordinate of a point whose order divides either $p-1$ or $p+1$. This gives us a way to quickly assert (with overwhelming probability) that a corresponds to a supersingular Montgomery curve in the correct isogeny class. With the Montgomery LADDER function as described in Sect. 3, we simply take a random element r in $\mathbb{F}_{p^2}$, compute $(X : Z) = \text{LADDER}((r : 1), a, p+1)$ and $(X' : Z') = \text{LADDER}((r : 1), a, p-1)$, and ensure that $Z \cdot Z' = 0$; otherwise, we reject the public key as invalid. We can compute a condition equivalent to $Z \cdot Z' = 0$ using only one call to the LADDER function as follows. The condition $\mathcal{O} \in \{[p-1]P, [p+1]P\}$ is equivalent to the condition $x(P) = x([p]P)$, which can be checked by computing $(X : Z) = \text{LADDER}(x(P), a, p)$ with $x(P) = (x_P : 1)$ and checking that $Z \cdot x_P = X$. However, calling LADDER to compute $x([p]P)$ directly is undesirable; given that $p+1 = 2^{\ell_A} 3^{\ell_B}$, it is instead preferable to write a tailored ladder (consisting only of xDBL and xTPL operations) that computes a scalar multiplication by $p+1$. We do this by noting that the condition $x(P) = x([p]P)$ is equivalent to the condition that either $x([p+1]P) = x([2]P)$ or $[p+1]P = \mathcal{O}$ is satisfied.

The Price of Our Public Key Validation Procedure. On our target platforms, i.e., a 3.4 GHz Intel Core i7-2600 Sandy Bridge and a 3.4 GHz Intel Core i7-4770 Haswell processor running Ubuntu 14.04 LTS, the validation of Alice's public key costs (according to the above procedure) around 23 million and 21 million clock cycles, respectively. Similarly, the validation of Bob's public key costs around 20 million and 18 million clock cycles, respectively. Referring back to Table 2, this means that both Alice and Bob's validation procedures cost between 0.39 and 0.43 times their key generation and shared secret computations.

Unlike public key validation in some other contexts, e.g., point validation in ECC, the compute time of the above SIDH public key validation is non-negligible compared to the compute time of each round of the key exchange. Nevertheless, in scenarios where static keys are desirable, the above overhead might be preferred over changes in the protocol description, e.g., the *indirect* validation proposed in [24].

10   Conclusion

We presented several new algorithms that have given rise to more efficient SIDH key exchange. We built a software library around a supersingular isogeny class determined by a fixed base curve that was chosen to target 128 bits of quantum security, and showed that these techniques give rise to a factor speedup of up to 2.9x over the previous fastest SIDH software. To our knowledge, our SIDH key exchange software is the first such implementation to run in constant time, and offers a range of additional benefits, such as compactness. In addition, we introduced two new techniques that bridge the gap between theoretical and real-world deployment of SIDH key exchange: the ECDH+SIDH hybrid and efficient algorithms for validating properties of public keys. The speed of our software (and the size of the public keys it generates) highlights the potential that SIDH currently offers as a candidate for post-quantum key exchange.

Acknowledgements. This paper has been significantly improved due to the feedback we received on a previous version. We are especially thankful to Ben Smith who pointed out a much simpler and faster method of our public key validation (see Sect. 9). We thank Luca De Feo and Ben Smith for pointing out a simplified compression of public keys (see Sect. 6). We thank Luca De Feo, Steven Galbraith and David Jao for their useful feedback, and the anonymous reviewers for their comments.

References

1. Aranha, D.F., Karabina, K., Longa, P., Gebotys, C.H., López, J.: Faster explicit formulas for computing pairings over ordinary curves. In: Paterson, K.G. (ed.) EUROCRYPT 2011. LNCS, vol. 6632, pp. 48–68. Springer, Heidelberg (2011)
2. Azarderakhsh, R., Fishbein, D., Jao, D.: Efficient implementations of a quantum-resistant key-exchange protocol on embedded systems. Technical report (2014). http://cacr.uwaterloo.ca/techreports/2014/cacr2014-20.pdf
3. Barrett, P.: Implementing the Rivest Shamir and Adleman public key encryption algorithm on a standard digital signal processor. In: Odlyzko, A.M. (ed.) CRYPTO 1986. LNCS, vol. 263, pp. 311–323. Springer, Heidelberg (1987)

4. Bernstein, D.J.: Curve25519: new Diffie-Hellman speed records. In: Yung, M., Dodis, Y., Kiayias, A., Malkin, T. (eds.) PKC 2006. LNCS, vol. 3958, pp. 207–228. Springer, Heidelberg (2006)
5. Bernstein, D.J.: The post-quantum internet. Invited talk at PQCrypto 2016, February 2016. https://cr.yp.to/talks/2016.02.24/slides-djb-20160224-a4.pdf
6. Biasse, J., Jao, D., Sankar, A.: A quantum algorithm for computing isogenies between supersingular elliptic curves. In: Meier, W., Mukhopadhyay, D. (eds.) INDOCRYPT 2014. LNCS, vol. 8885, pp. 428–442. Springer, Berlin (2014)
7. Biehl, I., Meyer, B., Müller, V.: Differential fault attacks on elliptic curve cryptosystems. In: Bellare, M. (ed.) CRYPTO 2000. LNCS, vol. 1880, p. 131. Springer, Heidelberg (2000)
8. Blake, I.F., Seroussi, G., Smart, N.P. (eds.): Advances in Elliptic Curve Cryptography. London Mathematical Society Lecture Notes Series, vol. 317. Cambridge University Press, Cambridge (2004)
9. Bröker, R.: Constructing supersingular elliptic curves. J. Comb. Number Theory 1(3), 269–273 (2009)
10. Canetti, R., Krawczyk, H.: Analysis of key-exchange protocols and their use for building secure channels. In: Pfitzmann, B. (ed.) EUROCRYPT 2001. LNCS, vol. 2045, pp. 453–474. Springer, Heidelberg (2001)
11. Chen, L., Jordan, S., Liu, Y.-K., Moody, D., Peralta, R., Perlner, R., Smith-Tone, D.: Report on post-quantum cryptography. NISTIR 8105, DRAFT (2016). http://csrc.nist.gov/publications/drafts/nistir-8105/nistir_8105_draft.pdf
12. Childs, A.M., Jao, D., Soukharev, V.: Constructing elliptic curve isogenies in quantum subexponential time. J. Math. Cryptology 8(1), 1–29 (2014)
13. Costello, C., Longa, P., Naehrig, M.: Efficient algorithms for supersingular isogeny Diffie-Hellman (full version). Cryptology ePrint Archive, Report 2016/413 (2016). http://eprint.iacr.org/
14. Costello, C., Longa, P., Naehrig, M.: SIDH Library (2016). https://www.microsoft.com/en-us/research/project/sidh-library/
15. Delfs, C., Galbraith, S.D.: Computing isogenies between supersingular elliptic curves over $\mathbb{F}_p$. Des. Codes Crypt. 78(2), 425–440 (2016)
16. Devoret, M.H., Schoelkopf, R.J.: Superconducting circuits for quantum information: an outlook. Science 339(6124), 1169–1174 (2013)
17. De Feo, L., Jao, D., Plût, J.: Towards quantum-resistant cryptosystems from supersingular elliptic curve isogenies. J. Math. Cryptology 8, 209–247 (2014)
18. Galbraith, S.D., Stolbunov, A.: Improved algorithm for the isogeny problem for ordinary elliptic curves. Appl. Algebra Eng. Commun. Comput. 24(2), 107–131 (2013)
19. Hamburg, M.: Fast and compact elliptic-curve cryptography. IACR CryptologyePrint Archive, 2012:309 (2012)
20. Hamburg, M.: Ed448-Goldilocks, a new elliptic curve. Cryptology ePrint Archive, Report 2015/625 (2015). http://eprint.iacr.org/
21. Hoffstein, J., Pipher, J., Silverman, J.H.: NTRU: a ring-based public key cryptosystem. In: Buhler, J.P. (ed.) ANTS 1998. LNCS, vol. 1423, pp. 267–288. Springer, Heidelberg (1998)
22. Jao, D., De Feo, L.: Towards quantum-resistant cryptosystems from supersingular elliptic curve isogenies. In: Yang, B.-Y. (ed.) PQCrypto 2011. LNCS, vol. 7071, pp. 19–34. Springer, Heidelberg (2011)

23. Kelly, J., Barends, R., Fowler, A.G., Megrant, A., Jeffrey, E., White, T.C., Sank, D., Mutus, J.Y., Campbell, B., Chen, Y., Chen, Z., Chiaro, B., Dunsworth, A., Hoi, I.-C., Neill, C., O'Malley, P.J.J., Quintana, C., Roushan, P., Vainsencher, A., Wenner, J., Cleland, A.N., Martinis, J.M.: State preservation by repetitive error detection in a superconducting quantum circuit. Nature **519**, 66–69 (2015)
24. Kirkwood, D., Lackey, B.C., McVey, J., Motley, M., Solinas, J.A., Tuller, D.: Failure is not an option: standardization issues for post-quantum key agreement. Talk at NIST Workshop on Cybersecurity in a Post-Quantum World, April 2015. http://www.nist.gov/itl/csd/ct/post-quantum-crypto-workshop-2015.cfm
25. Koc, C.K., Acar, T., Kaliski, B.S.: Analyzing and comparing Montgomery multiplication algorithms. IEEE Micro **16**(3), 26–33 (1996)
26. Kocher, P.C.: Timing attacks on implementations of Diffie-Hellman, RSA, DSS, and other systems. In: Koblitz, N. (ed.) CRYPTO 1996. LNCS, vol. 1109, pp. 104–113. Springer, Heidelberg (1996)
27. Lenstra, A.K.: Generating RSA moduli with a predetermined portion. In: Ohta, K., Pei, D. (eds.) ASIACRYPT 1998. LNCS, vol. 1514, pp. 1–10. Springer, Heidelberg (1998)
28. Lim, C.H., Lee, P.J.: A key recovery attack on discrete log-based schemes using a prime order subgroup. In: Kaliski Jr., B.S. (ed.) CRYPTO 1997. LNCS, vol. 1294, pp. 249–263. Springer, Heidelberg (1997)
29. McEliece, R.J.: A public-key cryptosystem based on algebraic coding theory. Coding Thv **4244**, 114–116 (1978)
30. Merkle, R.C.: Secrecy, authentication, and public key systems. Ph.D. thesis, Stanford University (1979)
31. Milne, J.S.: Abelian Varieties. In: Cornell, G., Silverman, J.H. (eds.) Arithmetic Geometry, pp. 103–150. Springer, New York (1986)
32. Montgomery, P.L.: Modular multiplication without trial division. Math. Comput. **44**(170), 519–521 (1985)
33. Montgomery, P.L.: Speeding the Pollard and elliptic curve methods of factorization. Math. Comput. **48**(177), 243–264 (1987)
34. Mosca, M.: Cybersecurity in an era with quantum computers: will we be ready? Cryptology ePrint Archive, Report 2015/1075 (2015). http://eprint.iacr.org/
35. National Security Agency (NSA): Cryptography today, August 2015. https://www.nsa.gov/ia/programs/suiteb_cryptography/
36. Okeya, K., Sakurai, K.: Efficient elliptic curve cryptosystems from a scalar multiplication algorithm with recovery of the y-coordinate on a Montgomery-form elliptic curve. In: Koç, Ç.K., Naccache, D., Paar, C. (eds.) CHES 2001. LNCS, vol. 2162, pp. 126–141. Springer, Heidelberg (2001)
37. Page, D.: Theoretical use of cache memory as a cryptanalytic side-channel. Technical report CSTR-02-003, Department of Computer Science, University of Bristol (2002). http://www.cs.bris.ac.uk/Publications/Papers/1000625.pdf
38. Patarin, J.: Hidden fields equations (HFE) and isomorphisms of polynomials (IP): two new families of asymmetric algorithms. In: Maurer, U.M. (ed.) EUROCRYPT 1996. LNCS, vol. 1070, pp. 33–48. Springer, Heidelberg (1996)
39. Pohlig, S.C., Hellman, M.E.: An improved algorithm for computing logarithms over GF(p) and its cryptographic significance. IEEE Trans. Inf. Theory **24**(1), 106–110 (1978)
40. Rostovtsev, A., Stolbunov, A.: Public-key cryptosystem based on isogenies. Cryptology ePrint Archive, Report 2006/145 (2006). http://eprint.iacr.org/
41. Scott, M.: Fast machine code for modular multiplication (1995). Manuscript, available for download at ftp://ftp.computing.dcu.ie/pub/crypto/fastmodmult2.ps

42. Scott, M.: Computing the Tate pairing. In: Menezes, A. (ed.) CT-RSA 2005. LNCS, vol. 3376, pp. 293–304. Springer, Heidelberg (2005)
43. Shor, P.W.: Algorithms for quantum computation: discrete logarithms and factoring. In: 35th Annual Symposium on Foundations of Computer Science, Proceedings, pp. 124–134. IEEE (1994)
44. Silverman, J.H.: The Arithmetic of Elliptic Curves. Graduate Texts in Mathematics, 2nd edn. Springer, New York (2009)
45. Stolbunov, A.: Cryptographic schemes based on isogenies. Ph.D. thesis, Norwegian University of Science and Technology (2012). http://www.item.ntnu.no/_media/people/personalpages/phd/anton/stolbunov-crypthographic_schemes_based_on_isogenies-phd_thesis_2012.pdf
46. Sutherland, A.V.: Identifying supersingular elliptic curves. LMS J. Comput. Math. **15**, 317–325 (2012)
47. Tani, S.: Claw finding algorithms using quantum walk. Theor. Comput. Sci. **410**(50), 5285–5297 (2009)
48. Tate, J.: Endomorphisms of abelian varieties over finite fields. Inventiones Math. **2**(2), 134–144 (1966)
49. Vélu, J.: Isogénies entre courbes elliptiques. CR Acad. Sci. Paris Sér. AB **273**, A238–A241 (1971)
50. Verheul, E.R.: Evidence that XTR is more secure than supersingular elliptic curve cryptosystems. J. Cryptology **17**(4), 277–296 (2004)
51. Walter, C.D.: Montgomery exponentiation needs no final subtractions. Electron. Lett. **35**(21), 1831–1832 (1999)
52. Zhang, S.: Promised and distributed quantum search. In: Wang, L. (ed.) COCOON 2005. LNCS, vol. 3595, pp. 430–439. Springer, Heidelberg (2005)

Symmetric Primitives

New Insights on AES-Like SPN Ciphers

Bing Sun[1,2,3]([⊠]), Meicheng Liu[3,4], Jian Guo[3],
Longjiang Qu[1]([⊠]), and Vincent Rijmen[5]

[1] College of Science, National University of Defense Technology,
Changsha 410073, Hunan, People's Republic of China
happy_come@163.com, ljqu_happy@hotmail.com
[2] State Key Laboratory of Cryptology,
P.O. Box 5159, Beijing 100878, People's Republic of China
[3] Nanyang Technological University, Central Area, Singapore
meicheng.liu@gmail.com, ntu.guo@gmail.com
[4] State Key Laboratory of Information Security,
Institute of Information Engineering, Chinese Academy of Sciences,
Beijing 100093, People's Republic of China
[5] Department of Electrical Engineering (ESAT),
KU Leuven and iMinds, Leuven, Belgium
vincent.rijmen@esat.kuleuven.be

Abstract. It has been proved in Eurocrypt 2016 by Sun *et al.* that if
the details of the S-boxes are not exploited, an impossible differential and
a zero-correlation linear hull can extend over at most 4 rounds of the
AES. This paper concentrates on distinguishing properties of AES-like
SPN ciphers by investigating the details of both the underlying S-boxes
and the MDS matrices, and illustrates some new insights on the security of these schemes. Firstly, we construct several types of 5-round zero-
correlation linear hulls for AES-like ciphers that adopt *identical S-boxes*
to construct the round function and that have *two identical elements in a
column of the inverse of their MDS matrices*. We then use these linear hulls
to construct 5-round integrals provided that the difference of two sub-key
bytes is known. Furthermore, we prove that we can always distinguish 5
rounds of such ciphers from random permutations even when the difference of the sub-keys is unknown. Secondly, the constraints for the S-boxes
and special property of the MDS matrices can be removed if the cipher is
used as a building block of the Miyaguchi-Preneel hash function. As an
example, we construct two types of 5-round distinguishers for the hash
function Whirlpool. Finally, we show that, in the chosen-ciphertext mode,
there exist some nontrivial distinguishers for 5-round AES. To the best
of our knowledge, this is the longest distinguisher for the round-reduced
AES in the secret-key setting. Since the 5-round distinguisher for the
AES can only be constructed in the chosen-ciphertext mode, *the security*

The work in this paper is supported by the National Basic Research Program of
China (973 Program) (2013CB338002), the National Natural Science Foundation of
China (No: 61272484, 61379139, 61402515, 61572016, 11526215), the Program for
New Century Excellent Talents in University (NCET) and the Strategic Priority
Research Program of the Chinese Academy of Science, Grant No. XDA06010701).

M. Robshaw and J. Katz (Eds.): CRYPTO 2016, Part I, LNCS 9814, pp. 605–624, 2016.
DOI: 10.1007/978-3-662-53018-4_22

margin for the round-reduced AES under the chosen-plaintext attack may be different from that under the chosen-ciphertext attack.

Keywords: Distinguishinger · AES · Whirlpool · Zero correlation linear · Integral

1   Introduction

Block ciphers are among the most important primitives in constructing symmetric cryptographic schemes such as encryption algorithms, hash functions, authentication schemes and pseudo-random number generators. The Advanced Encryption Standard (AES) [12] is currently the most interesting candidate to build different schemes. For example, in the on-going Competition for Authenticated Encryption: Security, Applicability, and Robustness (CAESAR) [10], among many others, the permutation of PRIMATEs [1] is designed based on an AES-like SPN structure, AEGIS [40] uses 4 AES round-functions in the state update functions, ELmD [13] recommends to use some round-reduced AES including the 5-round AES to partially encrypt the data, and 4-round AES is adopted by Marble [21], and used to build the AESQ permutation in PAEQ [5]. Although the security of these candidates does not completely depend on the underlying primitives, we believe that security of the round-reduced AES could give some new insights to both the design and cryptanalysis of the authenticated encryption algorithms.

1.1   Distinguishers

The distinguishing properties refer to those properties of a cipher that random permutations do not have thus we can distinguish a cipher from random permutations. For example, in *differential cryptanalysis* [4], one always finds an r-round differential characteristic with high probability while for random permutations such a differential characteristic does not exist.

In [11], Daemen *et al.* proposed a new method that can break more rounds of SQUARE than differential and linear cryptanalysis, which is named the SQUARE attack consequently. Some similar ideas such as the saturation attack [30], the multi-set attack [6], and the higher-order differential attack [23,27] have also been proposed. In [26], Knudsen and Wagner proposed the integral cryptanalysis as a generalized case of these attacks. In an integral attack, with some special inputs, one checks whether the sum of the corresponding ciphertexts is zero or not. Integral attacks on the round-reduced AES are based on the following distinguisher:

Property 1 [12,17]. Let 15 bytes of the input be constants and the remaining byte take all possible values from $\mathbb{F}_{2^8}$. Such a set is called a Λ-set. Then, the sum of each byte of the output of the third round is 0. Furthermore, let the 4 bytes in the diagonal of the state take all possible values from $\mathbb{F}_{2^8}^4$ and the other 12 bytes be constants, then the output of 1-round AES can be divided into 2^{24} Λ-sets. Therefore, the sum of each byte of the output of the fourth round is 0.

Gilbert and Minier showed that the set of functions mapping one active byte to one byte after 3 rounds depends on 9-byte parameters [20]. Therefore, the whole set can be described by using a table of 2^{72} entries of 256-byte sequences. This idea was later generalized by Demirci and Selçuk in [14] using meet-in-the-middle techniques. They showed that on 4 rounds, the value of each byte of the ciphertext can be described by a function of the active byte parameterized by 25 in [14] and 24 8-bit parameters in [15].

Property 2 [15]. The set of functions mapping one active byte to one byte after 4 rounds AES depends on 24 one-byte parameters.

Knudsen [24] and Biham *et al.* [3] independently proposed impossible-differential cryptanalysis. The main idea of impossible-differential cryptanalysis is to use differentials that hold with probability zero to discard the wrong keys that lead to the impossible differential. Now, it is one of the most effective methods towards many different ciphers. One of the 4-round impossible differentials is shown as follows:

Property 3 [31,32,34]. The differential, where there is only one nonzero (active) byte of the input difference and output difference, respectively, is a 4-round impossible differential of the AES.

Zero-correlation linear cryptanalysis was proposed by Bogdanov and Rijmen in [9]. They try to construct some linear hulls with correlation exactly zero. The 4-round zero correlation linear hull of the AES is shown as follows:

Property 4 [9]. If there is only one nonzero (active) byte of the input mask and output mask, respectively, then the correlation of 4-round AES is 0.

In summary, although there exist some 5-round distinguishers for AES-192 and AES-256 [16], the known distinguishers for all version of the AES only cover at most 4 rounds.

All the above distinguishers are in the secret-key setting, which were used in key recovery attacks. At Asiacrypt 2007, Knudsen and Rijmen proposed the *known-key distinguisher* for block ciphers [25]. In the setting that the key is public to the attacker, one can construct 7-round known-key distinguisher for the AES, which was improved to 8-round and 10-round in [19]. Allowing even more degrees of freedom to attackers so that they can even choose keys, distinguishers of 9-round AES were proposed [18] in the *chosen-key setting*. In this paper, we restrict ourselves to the secret-key setting, and the distinguishers to be presented are natural extensions of those used in key recovery attacks.

1.2   Key-Recovery Attacks

The aim of a key-recovery attack is to recover some round keys of a cipher. Usually, the attack is applied once some distinguishing property of the reduced-round block cipher has been found. Up to date, the biclique attack can recover some subkeys of the full round AES with slightly less than exhaustive complexity [7]. We briefly list some results of the key-recovery attacks against round-reduced AES as in Table 1, together with the number of rounds of the underlying distinguishers used.

Table 1. Some key-recovery attacks against AES-128

Rounds	Technique	Data	Memory	Time	Reference	Rounds of distinguisher
6	Integral	6×2^{32}	2^8	2^{72}	[17]	4
7	Integral	$2^{127.997}$	2^{64}	2^{120}	[17]	4
7	Impossible differential	$2^{112.2}$	$2^{112.2}$	$2^{117.2}$	[31]	4
7	Impossible differential	$2^{106.2}$	$2^{90.2}$	$2^{110.2}$	[32]	4
7	Meet-in-the-middle	2^{105}	2^{90}	2^{99}	[16]	4
7	Meet-in-the-middle	2^{97}	2^{98}	2^{99}	[16]	4

1.3 Details of the Components of a Cipher

If we choose the parameters carefully, the dedicated cipher based on the AES-like structure can be resilient to both differential [4] and linear cryptanalysis [33]. For example, based on the fact that the branch number of the MixColumns is 5, it is proved in [12] that the number of active S-boxes of 4-round AES is at least 25. Since the maximal differential probability of the S-box is 2^{-6}, there does not exist any differential characteristic of 4-round AES with probability larger than $2^{-6 \times 25} = 2^{-150}$.

In most cases, especially in the cryptanalysis of AES, one does not have the necessity to investigate the details of the S-boxes. Thus, the corresponding results are independent of the non-linear components. In other words, if some other S-boxes with similar differential/linear properties are chosen in a cipher, the corresponding cryptanalytic results remain almost the same. To characterize what "being independent of the choice of the S-boxes" means, in [37], Sun *et al.* proposed the concept of *Structure* of a block cipher. By structural evaluation, we mean the domain of cryptography that analyzes a cryptosystem in terms of generic constructions which keep the linear parts of the cipher and omit the details of the non-linear components.

The influence of the choices of S-boxes in constructing integral distinguishers has been studied in [22,29,35]. For example, if ARIA adopts only one S-box, more balanced bytes could be determined and if the order of different S-boxes is changed (There are 4 different S-boxes in ARIA), one will get different integral distinguishers from the one constructed in [29]. In [35], the authors pointed out that in some cases, the key-recovery attacks based on the integral distinguisher may fail. Very recently, Todo proposed the division property [38] by which one could build longer integral distinguishers provided the algebraic degree of the

S-boxes is known. For example, a 6-round integral for MISTY1 was built in [39] based on which the first cryptanalysis result against the full MISTY1 was found.

Although there are already 4-round impossible differentials and zero-correlation linear hulls for the AES, the effort to find new impossible differentials and zero-correlation linear hulls that could cover more rounds has never been stopped. In Eurocrypt 2016, Sun *et al.* proved that, unless the details of the S-boxes are exploited, one cannot find any impossible differential or zero-correlation linear hull of the AES that covers 5 or more rounds:

Property 5 [36]. There does not exist any impossible differential or zero-correlation linear hull of $\mathcal{E}^{AES}$ which covers $r \geq 5$ rounds. Or equivalently, there does not exist any 5-round impossible differential or zero-correlation linear hull of the AES unless the details of the S-boxes are considered.

To increase the performance of a block cipher, one usually uses an MDS (Maximal Distance Seperatable) matrix whose elements are restricted to low hamming weights in order to reduce the workload of the multiplications over finite fields. Furthermore, it is noticed that not only the MDS matrices are always circulant, but also there are identical elements in each row. For example, in AES, the first row of the MDS matrix is $(02, 03, 01, 01)$. However, most known techniques have not made use of these observations and there is little literature concentrating on the choices of these matrices in constructing distinguishers of round-reduced AES. Since known impossible differentials and zero-correlation linear hulls of round-reduced AES are constructed based on the fact that the branch number of the MixColumns is 5, these two types of distinguishers still hold even if a different 4×4 MDS matrix over $\mathbb{F}_{2^8}$ is used. Furthermore, since the inverse of an MDS matrix also has the MDS property, these distinguishers hold not only in the *chosen-plaintext* setting, but also in the *chosen-ciphertext* setting.

1.4   Our Contributions

This paper concentrates on the details of both the S-boxes and MDS matrices that are used in AES-like SPN structures. Denote by M_{MC} the MDS matrix used in a cipher. If there are two identical elements in a row of $(M_{MC}^{-1})^{T}$ and if the cipher adopts identical S-boxes, then we can construct a 5-round distinguisher. This implies that applied to AES, our distinguisher covers the most number of rounds up till now.

(1) If the difference of two sub-key bytes is known, we can construct several types of 5-round zero-correlation linear hulls for such ciphers *without* MixColumns operation in the last round which could be turned into 5-round integrals both *with* and *without* MixColumns operations in the last round. Furthermore, we not only prove that 5 rounds of such ciphers *with* MixColumns operation in the last round can be distinguished from a random permutation, but also that some sub-keys can be recovered from the distinguisher directly.

(2) In a hash function setting, where an AES-like SPN structure is used as a building block and the chaining value acts as the key, there always exist 5-round distinguishers. As a proof of concept, we give two types of 5-round distinguishers for the hash function Whirlpool.

For the AES, every row of $(M_{\mathsf{MC}}^{-1})^{\mathrm{T}}$ contains 4 different elements. Thus we cannot apply the results to the AES directly. However, for the decryption of the AES, every row of $(M_{\mathsf{MC}}^{-1})^{\mathrm{T}}$ contains twice the same element 01, therefore we can construct a 5-round distinguisher for the AES in a chosen-ciphertext mode:

(3) For 5-round AES, divide the whole space of plaintext-ciphertext pairs into the following 2^8 subsets:

$$A_\Delta = \{(p,c)|c_{0,0} \oplus c_{1,3} = \Delta\}.$$

Then, there always exists a Δ such that $\sum_{(p,c)\in A_\Delta} p = 0$, while for random permutations, this happens with probability $1 - (1 - 2^{-128})^{2^8} \approx 2^{-120}$. Furthermore, we can deduce $k_{0,0} \oplus k_{1,3} = \Delta$ from the distinguisher.

Since this property only applies in the chosen-ciphertext setting, we conclude that the security margin of the AES under the chosen-plaintext setting may be different from the one under the chosen-ciphertext setting. Furthermore, since we have proved that 5-round AES can be distinguished from a random permutation, more attention should be paid when round-reduced AES is used as a building block in some new cryptographic schemes.

Though we have already found some 5-round distinguisher, we leave as an open problem whether we could mount more efficient key-recovery attack against round-reduced AES or other AES-based schemes.

2 Preliminaries

Before proceeding to our results, we first introduce some notations here on both boolean functions and the ciphers we are analyzing.

2.1 Boolean Functions

Given a boolean function $G : \mathbb{F}_2^n \to \mathbb{F}_2$, the *correlation* of G is defined by

$$c(G(x)) \triangleq \frac{1}{2^n} \sum_{x \in \mathbb{F}_2^n} (-1)^{G(x)}.$$

Given a vectorial function $H : \mathbb{F}_2^n \to \mathbb{F}_2^k$, the *correlation* of the linear approximation for a k-bit output mask b and an n-bit input mask a is defined by

$$c(a \cdot x \oplus b \cdot H(x)) \triangleq \frac{1}{2^n} \sum_{x \in \mathbb{F}_2^n} (-1)^{a \cdot x \oplus b \cdot H(x)},$$

where "$\cdot$" is the inner product of two elements. If $c(a \cdot x \oplus b \cdot H(x)) = 0$, then $a \rightarrow b$ is called a *zero-correlation linear hull* of H, following the same definition in [9]. Let $A \subseteq \mathbb{F}_2^n$, $B \subseteq \mathbb{F}_2^k$, if for all $a \in A$, $b \in B$, $c(a \cdot x \oplus b \cdot H(x)) = 0$, then $A \rightarrow B$ is called a *zero-correlation linear hull* of H.

In this paper, we denote by $\mathrm{circ}(a_0, a_1, \ldots, a_{n-1})$ a *circulant matrix* defined as follows:

$$\mathrm{circ}(a_0, a_1, \ldots, a_{n-1}) = \begin{pmatrix} a_0 & a_1 & \ldots & a_{n-1} \\ a_{n-1} & a_0 & \ldots & a_{n-2} \\ \vdots & \vdots & \vdots & \vdots \\ a_1 & a_2 & \cdots & a_0 \end{pmatrix}.$$

For any vector $v = (v_0, v_1, \ldots, v_{n-1}) \in \mathbb{F}_{2^b}^n$, the *Hamming Weight* of v is defined as the number of non-zero components of v:

$$\mathrm{wt}(v) = \#\{i | v_i \neq 0, i = 0, 1, \ldots, n-1\}.$$

Let $P \in \mathbb{F}_{2^b}^{n \times n}$, then the *branch number* of P is defined as

$$\mathcal{B}(P) = \min_{0 \neq x \in \mathbb{F}_{2^b}^n} \{\mathrm{wt}(x) + \mathrm{wt}(Px)\}.$$

Obviously, for any $x \in \mathbb{F}_{2^b}^n$, we always have $\mathrm{wt}(Px) \leq n$. Therefore, we can choose x such that $\mathrm{wt}(x) = 1$ which indicates that $\mathcal{B}(P) \leq n + 1$. A matrix $P \in \mathbb{F}_{2^b}^{n \times n}$ is called *Maximum Distance Separable* (MDS) matrix if and only if $\mathcal{B}(P) = n + 1$. In the proof of the security of a cipher against differential and linear cryptanalysis, one can make use of the branch number to bound the number of active S-boxes. Since a larger branch number usually gives more active S-boxes, MDS matrices are widely used in modern block ciphers including AES.

2.2   SPN and AES-Like SPN Ciphers

To keep our results as general as possible, we are going to give a generic description of the Substitution-Permutation Network (SPN) ciphers and AES-like ciphers, respectively. We assume that the input can be viewed as an $n \times n$ square matrix over $\mathbb{F}_{2^b}$, which implies that both the input (plaintext) and output (ciphertext) of the block ciphers count $n^2 b$ bits. The cipher successively applies R round functions, and we denote respectively by $s^{(r)}$ and $k^{(r)}$ the input and sub-key states of the r-th round. The state $s^{(0)}$ is initialized with the input plaintext. One round function is composed of the following layers: a key addition layer (KA) where an $n^2 b$-bit roundkey $k^{(r-1)}$ is xored to $s^{(r-1)}$, a block cipher permutation layer BC that updates the $n^2 b$-bit current state of the block cipher after addition of the subkey, i.e. $s^{(r)} = \mathsf{BC}(s^{(r-1)} \oplus k^{(r-1)})$. For an SPN cipher, the permutation BC is composed of SubBytes (SB) which applies non-linear transformations to the n^2 b-bit bytes in parallel, and then a layer P which is linear over $\mathbb{F}_2^{n^2 b}$, i.e. $\mathsf{BC} = \mathsf{P} \circ \mathsf{SB}$. The final ciphertext is then defined as $s^{(r)} \oplus k^{(r)}$.

In the following, we will simply use $\mathcal{E}(n, b, r)$ to denote an r-round AES-like SPN cipher which operates on $n \times n$ b-bit bytes.

In the case of AES-like ciphers, the internal state of BC can be viewed as a square matrix of b-bit cells with n rows and n columns. A cell of $s^{(r)}$ is denoted by $s_{i,j}^{(r)}$, where i is its row position and j its column position in the square matrix, starting counting from 0. Then, the linear layer itself is composed of the ShiftRows transformation (SR), which can be defined as a permutation $\pi_{SR} = (l_0, l_1, \ldots, l_{n-1})$ on $\mathbb{Z}_n = \{0, 1, \ldots, n-1\}$ that moves cell $s_{i,j}^{(r)}$ by l_i positions to the left in its own row, and the MixColumns transformation (MC), which linearly mixes all the columns of the matrix. Overall, for AES-like ciphers, we always have $\mathsf{BC} = P \circ S = \mathsf{MC} \circ \mathsf{SR} \circ \mathsf{SB}$.

The AES Block Cipher. AES only uses a single S-box which is based on the inverse function over $\mathbb{F}_{2^8}$ to construct the round function. The SR and the MC of AES are defined as follows:

$$\pi_{SR} = (0, 1, 2, 3),$$

$$M_{MC} = \begin{pmatrix} 02 & 03 & 01 & 01 \\ 01 & 02 & 03 & 01 \\ 01 & 01 & 02 & 03 \\ 03 & 01 & 01 & 02 \end{pmatrix} = \mathrm{circ}(02, 03, 01, 01).$$

Since we do not investigate the key-recovery attacks, please refer to [12] for the details of the key schedule.

3 Zero-Correlation Linear Cryptanalysis of AES-Like SPN Ciphers

3.1 Zero-Correlation Linear Hull of 4-round AES-Like Ciphers

In zero-correlation linear cryptanalysis, we construct some linear hulls with correlation exactly zero. One of the most efficient methods to construct zero correlation linear hulls is based on the miss-in-the-middle technique, i.e., we start from the beginning and the end of the cipher, partially encrypt the plaintext and decrypt the ciphertext, respectively. Then some contradiction could be found in the middle round of the cipher with probability 1. For example, the 4-round zero-correlation linear hull of the AES is built as follows [9] (see Fig. 1): if only the first byte of the input mask is active, then after 1 round, all the 4 bytes in the first column of the output mask are active. Thus in each column of the input mask to the second MixColumns, the number of active bytes is 1. Using the same technique, we find that if there is only 1 active byte in the output mask of the forth round, in each column of the output mask to the second MixColumns round, the number of active bytes is 1. Since the branch number of MixColumns is 5, we find a contradiction which indicates that the correlation of such a linear hull is 0.

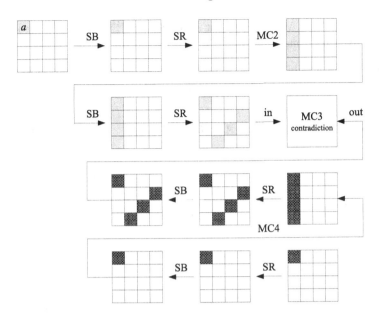

Fig. 1. 4-round zero-correlation linear hull of the AES

To enhance the performance of a cipher, designers usually use identical S-boxes and a diffusion layer whose elements often have relatively low hamming weights, which not necessarily but often cause some weakness as shown in the following.

3.2 New Cryptanalysis of 5-round AES-Like Ciphers

Though it has been proven that the longest zero-correlation linear hull of the AES only covers 4 rounds if we do not investigate the details of the S-box, we can improve this result exactly by exploiting these details.

In this section, we are going to use the miss-in-the-middle technique to construct some novel distinguishers of AES-like SPN ciphers, provided that the difference of two sub-keys bytes is known. Firstly, we recall the following propositions for the propagation of input-output masks/differentials of linear functions:

Proposition 1. *Let $\mathcal{L}$ be a linear transformation defined on $\mathbb{F}_2^T$, and $L \in \mathbb{F}_2^{t \times t}$ be the matrix representation of $\mathcal{L}$. Then,*

(1) For any input-output mask $\Gamma_I \to \Gamma_O$, if the correlation is nonzero, we always have $\Gamma_O = (L^{-1})^T \Gamma_I$.

(2) For any input-output difference $\Delta_I \to \Delta_O$, if the differential probability is nonzero, we always have $\Delta_O = L\Delta_I$.

Since ShiftRows in the first round does not influence the results, in this section, we omit SR in the first round. Denote by $(M_{\mathsf{MC}}^{-1})^T = (m_{i,j}^*)$ the transpose of the inverse of M_{MC}. We assume that an AES-like SPN cipher $\mathcal{E}(n, b, r)$ satisfies the following conditions:

(1) There exists a triplet (i, j_0, j_1) such that $m_{i,j_0}^* = m_{i,j_1}^*$ where $j_0 \neq j_1$;
(2) Without loss of generality, the S-boxes used at positions $(j_0, 0)$ and $(j_1, 0)$ are identical.

Lemma 1. *Let $\mathcal{E}(n, b, r)$ be an AES-like SPN cipher satisfying conditions (1) and (2). Define*

$$V = \{(s_{i,j}) \in \mathbb{F}_{2^b}^{n \times n} | s_{j_0,0} \oplus s_{j_1,0} = k_{j_0,0} \oplus k_{j_1,0}\}.$$

For any $0 \neq a \in \mathbb{F}_{2^b}$, let the input mask be

$$\Gamma_I = (\alpha_{i,j})_{0 \leq i,j \leq n-1}, \quad \alpha_{i,j} = \begin{cases} a & (i,j) = (j_0, 0), (j_1, 0), \\ 0 & otherwise, \end{cases}$$

and the output mask be $\Gamma_O = (\beta_{i,j}) \in \mathbb{F}_{2^b}^{n \times n}$. Then, if the correlation $\Gamma_I \to \Gamma_O$ of $\mathcal{E}(n, b, 1)$ on V is non-zero, we have

$$wt(\beta_{0,0}, \beta_{1,0}, \ldots, \beta_{n-1,0}) = n - 1,$$

$\beta_{i,j} = 0$ for $j \geq 1$, and the absolute value of the correlation is 1.

Proof. Let the output mask of the SB layer be

$$\Gamma_{\mathsf{SB}} = (\gamma_{i,j}) \in \mathbb{F}_{2^b}^{n \times n}.$$

To make the correlation non-zero, $\gamma_{i,j} = 0$ should hold if $\alpha_{i,j} = 0$. Next, we will show $\gamma_{j_0,0} = \gamma_{j_1,0}$. Since $s_{j_0,0} \oplus s_{j_1,0} = k_{j_0,0} \oplus k_{j_1,0}$, denote by

$$x = s_{j_0,0} \oplus k_{j_0,0} = s_{j_1,0} \oplus k_{j_1,0},$$

then

$$\Gamma_I \cdot X \oplus \Gamma_{\mathsf{SB}} \cdot S(X) = a \cdot x \oplus a \cdot x \oplus \gamma_{j_0,0} \cdot S(x) \oplus \gamma_{j_1,0} \cdot S(x)$$
$$= (\gamma_{j_0,0} \oplus \gamma_{j_1,0}) \cdot S(x),$$

Since $S(x)$ is a permutation on $\mathbb{F}_{2^b}$, if $\gamma_{j_0,0} \oplus \gamma_{j_1,0} \neq 0$, the correlation of $(\gamma_{j_0,0} \oplus \gamma_{j_1,0}) \cdot S(x)$ is always 0. On the other hand, if $\gamma_{j_0,0} \oplus \gamma_{j_1,0} = 0$, the correlation is always 1.

Therefore, to make the correlation non-zero, according to Proposition 1, the output mask of $\mathcal{E}(n, b, 1)$ should be

$$\Gamma_O = (M_{\mathsf{MC}}^{-1})^{\mathsf{T}} \Gamma_{\mathsf{SB}}.$$

Taking this into consideration, the absolute value of the correlation is always 1 which ends our proof. $\qquad\square$

Lemma 2. *Let $\mathcal{E}(n, b, r)$ be an AES-like SPN cipher satisfying conditions (1) and (2). Let $\Delta = k_{j_0,0}^{(0)} \oplus k_{j_1,0}^{(0)}$, and define*

$$V_\Delta = \{(s_{i,j}^{(0)}) \in \mathbb{F}_{2^b}^{n \times n} | s_{j_0,0}^{(0)} \oplus s_{j_1,0}^{(0)} = \Delta\}.$$

For any $0 \neq a \in \mathbb{F}_{2^b}$, let the input mask be

$$\Gamma_I = (\alpha_{i,j})_{0 \leq i,j \leq n-1}, \quad \alpha_{i,j} = \begin{cases} a & (i,j) = (j_0, 0), (j_1, 0), \\ 0 & otherwise, \end{cases}$$

and for any $0 \neq d \in \mathbb{F}_{2^b}$, $(u, v) \in \mathbb{Z}_n \times \mathbb{Z}_n$, let the output mask be

$$\Gamma_O^{(u,v)} = (\beta_{i,j})_{0 \leq i,j \leq n-1}, \quad \beta_{i,j} = \begin{cases} d & (i,j) = (u, v), \\ 0 & otherwise. \end{cases}$$

Then for $\mathcal{E}(n, b, 5)$ without MixColumns in the last round, the correlation for $\Gamma_I \to \Gamma_O^{(u,v)}$ on V_Δ is always 0.

Proof. The proof is divided into 2 halves (Fig. 2 gives the procedure of the proof for the case $n = 4$ and $\pi_{\mathsf{SR}} = (0, 3, 2, 1)$):

Firstly, from the encryption direction, let the input mask be Γ_I as defined above. According to Lemma 1, the output mask of the first round has the following properties: there are $n - 1$ non-zero elements in the first column and all of the elements in other columns are zero.

Then, in the second round, the output mask of the SB layer keeps the pattern of the input mask and SR shifts the $n - 1$ non-zero elements to $n - 1$ different columns. Since MC has the MDS property, we can conclude that the output mask of the second round has the following properties: there exists 1 column such that all elements in this columns are 0's, and all elements in the other columns are non-zero.

In the third round, the output mask of the SB layer keeps the pattern of the input mask and SR shifts the n zero elements to n different columns, i.e., there are $n - 1$ non-zero elements in each column of the input mask of MC in the third round.

Using the same technique, we can find that from the decryption direction, there is only 1 non-zero element in each column of the output mask of MC in the third round.

Since the MC has the MDS property, i.e., the sum of number of non-zero elements from both the input and output mask of MC is at least $n + 1$, the correlation of $\Gamma_I \to \Gamma_O^{(u,v)}$ is 0. □

4   Integrals for the AES-Like SPN Ciphers

Links between integrals and zero correlation linear hulls were first studied by Bogdanov *et al.* at Asiacrypt 2012 [8], and then refined at CRYPTO 2015 [37].

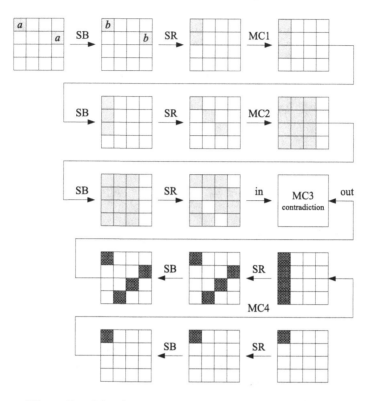

Fig. 2. Proof for the zero correlation linear hull of $\mathcal{E}(n, b, 5)$

In [37], Sun *et al.* proved that a zero correlation linear hull of a block cipher always implies the existence of an integral distinguisher which gives a novel way to construct integrals of a cipher. For example, the 4-round zero-correlation linear hull of the AES implies the following distinguisher: Let 15 bytes of the input take all possible values from $\mathbb{F}_{2^8}^{15}$ and the other 1 byte be constant, then each byte of the output before the MixColumns operation in the forth round takes each value from $\mathbb{F}_{2^8}$ exactly 2^{112} times.

This section mainly discusses the integral properties of the AES-like ciphers based on the links between zero correlation linear hulls and integrals. It was pointed out at CRYPTO 2015 [37] that a zero-correlation linear hull always implies the existence of an integral, based on which we can get the following results.

Corollary 1. *Let $\mathcal{E}(n, b, r)$ be an AES-like SPN cipher satisfying conditions (1) and (2). Let $\Delta = k_{j_0,0}^{(0)} \oplus k_{j_1,0}^{(0)}$ and the input set be*

$$V_\Delta = \{(s_{i,j}^{(0)})_{0 \le i,j \le n-1} \in \mathbb{F}_{2^b}^{n \times n} | s_{j_0,0}^{(0)} \oplus s_{j_1,0}^{(0)} = \Delta\}.$$

Then for each output byte of $\mathcal{E}(n, b, 5)$ without MixColumns, every value of $\mathbb{F}_{2^b}$ appears exactly $2^{(n^2-2)b}$ times, and the sum of every output byte of $\mathcal{E}(n, b, 5)$ with MixColumns is 0.

Since there exists exactly one value in $\{0, 1, \cdots, 2^b - 1\}$ which is equal to $\delta = k_{j_0,0}^{(0)} \oplus k_{j_1,0}^{(0)}$, we have:

Theorem 1. *Denote by $\mathcal{E}(n, b, r)$ an r-round AES-like SPN cipher with Mix-Columns in the last round, where b and n are the sizes of the S-boxes and the MDS matrix, respectively. Let $(M_{MC}^{-1})^T = (m_{i,j}^*) \in \mathbb{F}_{2^b}^{n \times n}$ be the transpose of the inverse of M_{MC}. Assume that there exists a triplet (i, j_0, j_1) such that $m_{i,j_0}^* = m_{i,j_1}^*$. Then $\mathcal{E}(n, b, 5)$ can be distinguished from a random permutation $\mathcal{R}$ as follows: for $F \in \{\mathcal{E}(n, b, 5), \mathcal{R}\}$ and $\Delta = 0, 1, \ldots, 2^b - 1$, divide the whole input-output space into the following 2^b subsets:*

$$A_\Delta^F = \{(p, c) | c = F(p), p_{j_0,a_0} \oplus p_{j_1,a_1} = \Delta\},$$

where SR moves p_{j_0,a_0} and p_{j_1,a_1} to the same column, and let

$$T_\Delta^F = \sum_{(p,c) \in A_\Delta^F} c.$$

If the S-boxes applied to p_{j_0,a_0} and p_{j_1,a_1} are identical, there always exists a Δ such that $T_\Delta^{\mathcal{E}(n,b,5)} = 0$, while for random permutations, this happens with probability $1 - (1 - 2^{-n^2 b})^{2^b} \approx 2^{-(n^2-1)b}$. Furthermore, we can deduce that the value of $k_{j_0,a_0} \oplus k_{j_1,a_1}$ is Δ.

This theorem can be clearly deduced from Corollary 1 above. We can further give a direct proof as follows.

Proof. Without loss of generality, let $(M_{MC}^{-1})^T = (m_{i,j}^*)$ and $m_{0,0}^* = m_{0,1}^* = 01$. Let the input and output of the MixColumns operation be $(x_0, x_0, x_1, \ldots, x_{n-2})^T$ and $(y_0, y_1, \ldots, y_{n-1})^T$, respectively. Then we have

$$\begin{pmatrix} x_0 \\ x_0 \\ x_1 \\ \vdots \\ x_{n-2} \end{pmatrix} = \begin{pmatrix} 01 & * & \cdots & * & * \\ 01 & * & \cdots & * & * \\ * & * & \cdots & * & * \\ & & \cdots & & \\ * & * & \cdots & * & * \end{pmatrix} \begin{pmatrix} y_0 \\ y_1 \\ y_2 \\ \vdots \\ y_{n-1} \end{pmatrix},$$

which implies

$$x_0 = y_0 \oplus l_1(y_1, \ldots, y_{n-1}) = y_0 \oplus l_2(y_1, \ldots, y_{n-1}),$$

where l_1 and l_2 are different linear functions on $(y_1, \ldots, y_{n-1})$. Accordingly, we always have

$$(l_1 \oplus l_2)(y_1, \ldots, y_{n-1}) = 0.$$

Since the dimension of the input is $n - 1$, we conclude that y_0 is independent of $y_1, \ldots, y_{n-1}$, i.e., the number of possible values for $(y_1, \ldots, y_{n-1})$ is $2^{(n-2)b}$. Thus the output of the first round can be divided into the following $2^{(n-2)b}$ subsets: the last $n - 1$ bytes of the first columns are fixed to $(y_1, \ldots, y_{n-1})$ and the other $n^2 - n + 1$ bytes take all possible value from $\mathbb{F}_{2^b}^{n^2-n+1}$. Taking the 4-round integral distinguisher into consideration, we conclude that the sum of the output of the fifth round with MixColumns is 0.   □

Since a lot of AES-based ciphers adopt circulant MDS matrices, now we will list a result when a cipher uses a circulant MDS matrix:

Corollary 2. *Let $\mathcal{E}(n, b, r)$ be an AES-like SPN cipher which uses a circulant MDS matrix $M_{MC} = circ(m_0, m_1, \ldots, m_{n-1}) \in \mathbb{F}_{2^b}^{n \times n}$. Denote by $(M_{MC}^{-1})^T = circ(m_0^*, m_1^*, \ldots, m_{n-1}^*)$ the transpose of the inverse of M_{MC}. If there exists a (j_0, j_1) where $j_0 \neq j_1$ such that $m_{j_0}^* = m_{j_1}^*$, then the plaintext-ciphertext space of $\mathcal{E}(n, b, 5)$ can be divided into 2^{nb} subsets A_Δ and $|A_\Delta| = 2^{(n^2-n)b}$, and there exists a Δ such that the sum of ciphertexts in A_Δ is 0. Moreover, some sub-keys can also be deduced from the partition.*

5   Application to Hashing Schemes

To apply these results to block ciphers directly, we need to know the difference of the corresponding sub-key bytes which is impossible in most cases. However, if the cipher is used as a building block of a hash function and the chain value acts as the key, we can always get a new distinguisher of the hash function based on these new observations. We use Whirlpool [2] as an example in this section.

5-Round Distinguisher for Whirlpool. Whirlpool [2] is a hash function proposed by Barreto and Rijmen as a candidate for the NESSIE project. It iterates the Miyaguchi-Preneel hashing scheme over t padded message blocks m_i, $0 \leq i \leq t - 1$, using the dedicated 512-bit block cipher W:

$$H_i = W_{H_{i-1}}(m_{i-1}) \oplus H_{i-1} \oplus m_{i-1}, \quad i = 1, 2, \ldots, t.$$

The W block cipher only employs one S-box, and the SR and the MC are defined as follows:

$$\pi_{SR} = (0, 1, 2, 3, 4, 5, 6, 7),$$
$$M_{MC} = circ(01, 01, 04, 01, 08, 05, 02, 09).$$

Notice that the SR of Whirlpool applies to columns and MC applies to rows, respectively (Fig. 3).

Noting that the matrix

$$(M_{MC}^{-1})^T = circ(04, 3E, CB, C2, C2, A4, 0E, AE),$$

has two identical elements in each row, according to Theorem 1, we have the following distinguishing property for Whirlpool:

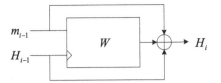

Fig. 3. The structure of Whirlpool Hash Function.

Corollary 3. *Let* $V_1 = \{(p_{i,j}) \in \mathbb{F}_{2^8}^{8\times8} | p_{0,3} \oplus p_{0,4} = h_{0,3}^{(0)} \oplus h_{0,4}^{(0)}\}$. *Then for Whirlpool reduced to* 5 *rounds, the sum of all the outputs over* V_1 *is* 0.

Although this distinguisher covers less rounds than the rebound attack [28], our result shows some new features of Whirlpool that could be exploited in the future. From the direct proof of Theorem 1, the key point is that the outputs of the first round could be divided into some known structures which lead to 4-round integrals. Therefore we have the following property:

Corollary 4. *Let* $V_2 = \{(p_{i,j}) \in \mathbb{F}_{2^8}^{8\times8} | \mathsf{AE} \cdot S(p_{0,0} \oplus h_{0,0}^{(0)}) = \mathsf{04} \cdot S(p_{1,1} \oplus h_{1,1}^{(0)})\}$. *Then for Whirlpool reduced to* 5 *rounds, the sum of all the outputs over* V_2 *is* 0.

Proof. Let the input of the first column to the first MixColumns be $X = (x_0, \ldots, x_7)^{\mathrm{T}}$ and $Y = (y_0, \ldots, y_7)^{\mathrm{T}}$ be the corresponding output. Then $x_0 = S(p_{0,0} \oplus h_{0,0}^{(0)})$, $x_1 = S(p_{1,1} \oplus h_{1,1}^{(0)})$ and we have $\mathsf{AE} \cdot x_0 = \mathsf{04} \cdot x_1$. Since $X = M_{\mathsf{MC}}^{-1} Y$, therefore,

$$\begin{cases} x_0 = \mathsf{04} \cdot y_0 \oplus \mathsf{3E} \cdot y_1 \oplus \mathsf{CB} \cdot y_2 \oplus \mathsf{C2} \cdot y_3 \oplus \mathsf{C2} \cdot y_4 \oplus \mathsf{A4} \cdot y_5 \oplus \mathsf{0E} \cdot y_6 \oplus \mathsf{AE} \cdot y_7 \\ x_1 = \mathsf{AE} \cdot y_0 \oplus \mathsf{04} \cdot y_1 \oplus \mathsf{3E} \cdot y_2 \oplus \mathsf{CB} \cdot y_3 \oplus \mathsf{C2} \cdot y_4 \oplus \mathsf{C2} \cdot y_5 \oplus \mathsf{A4} \cdot y_6 \oplus \mathsf{0E} \cdot y_7. \end{cases}$$

Consequently,

$$\mathsf{AE}(\mathsf{3E} \cdot y_1 \oplus \mathsf{CB} \cdot y_2 \oplus \mathsf{C2} \cdot y_3 \oplus \mathsf{C2} \cdot y_4 \oplus \mathsf{A4} \cdot y_5 \oplus \mathsf{0E} \cdot y_6 \oplus \mathsf{AE} \cdot y_7)$$
$$= \mathsf{04}(\mathsf{04} \cdot y_1 \oplus \mathsf{3E} \cdot y_2 \oplus \mathsf{CB} \cdot y_3 \oplus \mathsf{C2} \cdot y_4 \oplus \mathsf{C2} \cdot y_5 \oplus \mathsf{A4} \cdot y_6 \oplus \mathsf{0E} \cdot y_7),$$

which implies that there exists a linear function l such that

$$y_4 = l(y_1, y_2, y_3, y_5, y_6, y_7).$$

Since the dimension of the input is $n - 1$, we know that y_0 is independent of $y_1, \ldots, y_7$. As in constructing the 4-round integral distinguisher of the AES based on the 3-round distinguisher, place this property in front of the known 4-round integral distinguisher for Whirlpool and we conclude that the sum of the outputs is 0.   □

Furthermore, we can extend the results to the structures with different S-boxes and no constraints on the elements of $(M_{\mathsf{MC}}^{-1})^{\mathrm{T}}$.

Theorem 2. *In a Miyaguchi-Preneel hashing mode, if the block cipher adopts a 5-round AES-like structure, there always exists a subset V such that when the input takes all possible value in V, the sum of output is 0.*

Let the first two elements in the first column of the inverse MDS matrix be a_0 and a_1, and the input to these two positions be $S_0(p_{0,0} \oplus h_{0,0})$ and $S_1(p_{1,1} \oplus h_{1,1})$. For any $p_{0,0}$, we can always choose $p_{1,1}$ such that

$$a_1 S_0(p_{0,0} \oplus h_{0,0}) = a_0 S_1(p_{1,1} \oplus h_{1,1}).$$

Then the conclusion follows from the proof of Corollary 4.

6   Application to AES

AES is one of the most widely used block ciphers since 2000, and many cryptographic primitives adopt round-reduced AES as a building block. The first known integral distinguisher for the AES covers 3 rounds [12] which was later improved to a 4-round higher-order integral [17]. However, the technique that improved the 3-round integral to a 4-round one cannot be directly used to improve the integral from 4 rounds to 5 rounds. In the following, we will show that the improvement is possible provided the difference of some sub-key bytes is known.

Since for M_{MC} adopted in the AES, we have

$$(M_{\mathsf{MC}}^{-1})^{\mathrm{T}} = \begin{pmatrix} \texttt{0E 09 0D 0B} \\ \texttt{0B 0E 09 0D} \\ \texttt{0D 0B 0E 09} \\ \texttt{09 0D 0B 0E} \end{pmatrix} = \mathrm{circ}(\texttt{0E}, \texttt{09}, \texttt{0D}, \texttt{0B}),$$

i.e., the elements in each row are different from each other, it seems that we cannot construct such distinguishers for 5-round AES. However, since there are two 1's in each columns of $M_{\mathsf{MC}} = \mathrm{circ}(\texttt{02}, \texttt{03}, \texttt{01}, \texttt{01})$, we can construct a distinguisher for AES^{-1}, i.e., we can turn the chosen-plaintext distinguishers shown in Theorem 1 into a chosen-ciphertext one.

Lemma 3. *Let $V = \{(x_{i,j}) \in \mathbb{F}_{2^8}^{4 \times 4} | x_{0,0} \oplus x_{1,3} = k_{0,0} \oplus k_{1,3}\}$ be the input set. Then for each output byte of 5-round AES^{-1} without MixColumns operation in the last round, every value of $\mathbb{F}_{2^8}$ appears 2^{112} times and the sum of every output byte of the 5-round AES^{-1} with MixColumns operation in the last round is 0.*

Theorem 3. *5-round AES with MixColumns in the last round can be distinguished from a random permutations as follows. Divide the whole input-output space into the following 2^8 subsets:*

$$A_\Delta = \{(p, c) | c_{0,0} \oplus c_{1,3} = \Delta\},$$

and let

$$T_\Delta = \sum_{(p,c) \in A_\Delta} p.$$

Then there always exists a Δ such that $k_{0,0} \oplus k_{1,3} = \Delta$ and $T_\Delta = 0$. For random permutations, this happens with probability $1 - (1 - 2^{-128})^{2^8} \approx 2^{-120}$.

To the best of our knowledge, Theorem 3 gives the best distinguisher[1] of the AES with respect to the rounds it covers. Since the AES adopts a circulant MDS matrix, we can get many other different variants of this property by dividing the whole set into different subsets. For example,

Corollary 5. *5-round AES with MixColumns in the last round can be distinguished from a random permutation as follows. Divide the whole input-output space into the following 2^{32} subsets:*

$$A_{\alpha,\beta,\gamma,\phi} = \{(p,c) | c_{0,0} \oplus c_{1,3} = \alpha, c_{0,1} \oplus c_{3,2} = \beta, c_{1,2} \oplus c_{2,1} = \gamma, c_{2,0} \oplus c_{3,3} = \phi\},$$

and let

$$T_{\alpha,\beta,\gamma,\phi} = \sum_{(p,c) \in A_{\alpha,\beta,\gamma,\phi}} p.$$

Then there always exists an $(\alpha, \beta, \gamma, \phi) \in \mathbb{F}_{2^8}^4$ such that $T_{\alpha,\beta,\gamma,\phi} = 0$. For random permutations, this happens with probability $1 - (1 - 2^{-128})^{2^{32}} \approx 2^{-96}$.

7   Conclusion

Distinguishers on AES-like SPN structures are covered extensively in the literature. For example, we already have 4-round zero-correlation linear hulls for AES-like structures without MixColumns in the last round and 4-round integral distinguishers for AES-like structures with MixColumns in the last round. Note that these distinguishers do not depend on which S-box and MDS matrix are used in the cipher. This paper gives some new insights on such ciphers especially with detailed S-boxes and MDS matrices.

Firstly, we observe that if there are two identical elements in a row of the transpose of the inverse matrix of the MixColumns operation, and the S-boxes used in these two positions are identical, then we can construct some 5-round zero-correlation linear hull for a 5-round AES-like SPN structure provided some differences of the sub-key bytes are known. Then, under the same setting, and based on the link between zero-correlation linear hulls and integrals, we construct 5-round integrals for such AES-like SPN structures both with and without the MixColumns operation in the last round. These results show that such 5-round AES-like SPN structures can be theoretically distinguished from random permutations.

Secondly, in a hashing scheme where the chaining value serves as the secret key in block ciphers, we can further remove the constraint on the matrices and S-boxes. We apply the new results to the Whirlpool hash function and construct 5-round integral-like distinguishers.

Furthermore, since these results do not apply to the AES directly, we find that although we cannot build a distinguisher in a chosen-plaintext mode, we can construct a 5-round distinguisher for the AES in the chosen-ciphertext mode which is the best distinguisher for the AES with respect to the number of rounds it covers.

---

[1] This property could be used for instance, when the codebook is provided, to determine whether it is AES when both the block cipher and the keys are unknown.

Our results show that despite the key schedule, there may be some difference between the security margins of round-reduced AES under chosen-plaintext attacks and that under chosen-ciphertext attacks. Since we can distinguish 5-round AES from random permutations, some dedicated cryptographic schemes should be carefully investigated to guarantee the security claims. Furthermore, when we design an AES-like cipher, it is better to choose those MDS matrices M_{MC} such that both M_{MC} and M_{MC}^{-1} do not have identical elements in the same columns.

Now that we get some new features of 5-round AES, we leave as an open problem whether one could mount better key-recovery attack against round-reduced AES or some other schemes based on the AES-like SPN structure.

Acknowledgment. The authors would like to thank the anonymous reviewers for their useful comments, and Ruilin Li, Shaojing Fu, Wentao Zhang and Ming Duan for fruitful discussions.

References

1. Andreeva, E., Bilgin, B., Bogdanov, A., Luykx, A., Mendel, F., Mennink, B., Mouha, N., Wang, Q., Yasuda, K.: PRIMATEs v1.02 Submission to the CAESAR Competition. http://competitions.cr.yp.to/round2/primatesv102.pdf
2. Barreto, P., Rijmen, V.: NESSIE proposal: Whirlpool (2000). https://www.cosic.esat.kuleuven.be/nessie/
3. Biham, E., Biryukov, A., Shamir, A.: Cryptanalysis of skipjack reduced to 31 rounds using impossible differentials. In: Stern, J. (ed.) EUROCRYPT 1999. LNCS, vol. 1592, pp. 12–23. Springer, Heidelberg (1999)
4. Biham, E., Shamir, A.: Differential Cryptanalysis of the Data Encryption Standard. Springer, New York (1993)
5. Biryukov, A., Khovratovich, D.: PAEQ v1. http://competitions.cr.yp.to/round1/paeqv1.pdf
6. Biryukov, A., Shamir, A.: Structural cryptanalysis of SASAS. In: Pfitzmann, B. (ed.) EUROCRYPT 2001. LNCS, vol. 2045, pp. 394–405. Springer, Heidelberg (2001)
7. Bogdanov, A., Khovratovich, D., Rechberger, C.: Biclique cryptanalysis of the full AES. In: Lee, D.H., Wang, X. (eds.) ASIACRYPT 2011. LNCS, vol. 7073, pp. 344–371. Springer, Heidelberg (2011)
8. Bogdanov, A., Leander, G., Nyberg, K., Wang, M.: Integral and multidimensional linear distinguishers with correlation zero. In: Wang, X., Sako, K. (eds.) ASIACRYPT 2012. LNCS, vol. 7658, pp. 244–261. Springer, Heidelberg (2012)
9. Bogdanov, A., Rijmen, V.: Linear hulls with correlation zero and linear cryptanalysis of block ciphers. Des. Codes Crypt. **70**(3), 369–383 (2014)
10. CAESAR: Competition for Authenticated Encryption: Security, Applicability, and Robustness. http://competitions.cr.yp.to/caesar.html
11. Daemen, J., Knudsen, L.R., Rijmen, V.: The block cipher SQUARE. In: Biham, E. (ed.) FSE 1997. LNCS, vol. 1267, pp. 149–165. Springer, Heidelberg (1997)
12. Daemen, J., Rijmen, V.: The Design of Rijndael: AES-the Advanced Encryption Standard. Springer, Heidelberg (2002)

13. Datta, N., Nandi, M.: ELmD v2.0. http://competitions.cr.yp.to/round2/elmdv20. pdf
14. Demirci, H., Selçuk, A.A.: A meet-in-the-middle attack on 8-round AES. In: Nyberg, K. (ed.) FSE 2008. LNCS, vol. 5086, pp. 116–126. Springer, Heidelberg (2008)
15. Demirci, H., Taşkın, I., Çoban, M., Baysal, A.: Improved meet-in-the-middle attacks on AES. In: Roy, B., Sendrier, N. (eds.) INDOCRYPT 2009. LNCS, vol. 5922, pp. 144–156. Springer, Heidelberg (2009)
16. Derbez, P., Fouque, P.-A., Jean, J.: Improved key recovery attacks on reduced-round AES in the single-key setting. In: Johansson, T., Nguyen, P.Q. (eds.) EUROCRYPT 2013. LNCS, vol. 7881, pp. 371–387. Springer, Heidelberg (2013)
17. Ferguson, N., Kelsey, J., Lucks, S., Schneier, B., Stay, M., Wagner, D., Whiting, D.L.: Improved cryptanalysis of Rijndael. In: Schneier, B. (ed.) FSE 2000. LNCS, vol. 1978, pp. 213–230. Springer, Heidelberg (2001)
18. Fouque, P.-A., Jean, J., Peyrin, T.: Structural evaluation of AES and chosen-key distinguisher of 9-round AES-128. In: Canetti, R., Garay, J.A. (eds.) CRYPTO 2013, Part I. LNCS, vol. 8042, pp. 183–203. Springer, Heidelberg (2013)
19. Gilbert, H.: A simplified representation of AES. In: Sarkar, P., Iwata, T. (eds.) ASIACRYPT 2014. LNCS, vol. 8873, pp. 200–222. Springer, Heidelberg (2014)
20. Gilbert, H., Minier, M.: A collision attack on 7 rounds of Rijndael. In: AES Candidate Conference, pp. 230–241 (2000)
21. Guo, J.: Marble Version 1.1. https://competitions.cr.yp.to/round1/marblev11.pdf
22. Hatano, Y., Sekine, H., Kaneko, T.: Higher order differential attack of camellia(II). In: Nyberg, K., Heys, H.M. (eds.) SAC 2002. LNCS, vol. 2595, pp. 129–146. Springer, Heidelberg (2003)
23. Knudsen, L.R.: Truncated and higher order differentials. In: Preneel, B. (ed.) FSE 1994. LNCS, vol. 1008, pp. 196–211. Springer, Heidelberg (1995)
24. Knudsen, L.R.: DEAL – a 128-bit block cipher. Technical report, Department of Informatics, University of Bergen, Norway (1998)
25. Knudsen, L.R., Rijmen, V.: Known-key distinguishers for some block ciphers. In: Kurosawa, K. (ed.) ASIACRYPT 2007. LNCS, vol. 4833, pp. 315–324. Springer, Heidelberg (2007)
26. Knudsen, L.R., Wagner, D.: Integral cryptanalysis. In: Daemen, J., Rijmen, V. (eds.) FSE 2002. LNCS, vol. 2365, pp. 112–127. Springer, Heidelberg (2002)
27. Lai, X.: Higher order derivatives and differential cryptanalysis. In: Blahut, R.E., Costello Jr., D.J., Maurer, U., Mittelholzer, T. (eds.) Communications and Cryptography: Two Sides of One Tapestry. The Springer International Series in Engineering and Computer Science, vol. 276, pp. 227–233. Springer, New York (1994)
28. Lamberger, M., Mendel, F., Schläffer, M., Rechberger, C., Rijmen, V.: The rebound attack and subspace distinguishers: application to whirlpool. J. Cryptol. (JOC) 28(2), 257–296 (2015)
29. Li, P., Sun, B., Li, C.: Integral cryptanalysis of ARIA. In: Bao, F., Yung, M., Lin, D., Jing, J. (eds.) Inscrypt 2009. LNCS, vol. 6151, pp. 1–14. Springer, Heidelberg (2010)
30. Lucks, S.: The saturation attack - a bait for twofish. In: Matsui, M. (ed.) FSE 2001. LNCS, vol. 2355, pp. 1–15. Springer, Heidelberg (2002)
31. Lu, J., Dunkelman, O., Keller, N., Kim, J.-S.: New impossible differential attacks on AES. In: Chowdhury, D.R., Rijmen, V., Das, A. (eds.) INDOCRYPT 2008. LNCS, vol. 5365, pp. 279–293. Springer, Heidelberg (2008)

32. Mala, H., Dakhilalian, M., Rijmen, V., Modarres-Hashemi, M.: Improved impossible differential cryptanalysis of 7-round AES-128. In: Gong, G., Gupta, K.C. (eds.) INDOCRYPT 2010. LNCS, vol. 6498, pp. 282–291. Springer, Heidelberg (2010)
33. Matsui, M.: Linear cryptanalysis method for DES cipher. In: Helleseth, T. (ed.) EUROCRYPT 1993. LNCS, vol. 765, pp. 386–397. Springer, Heidelberg (1994)
34. Phan, R.: Impossible differential cryptanalysis of 7-round Advanced Encryption Standard (AES). Inf. Process. Lett. **91**(1), 33–38 (2004)
35. Sun, B., Li, R., Qu, L., Li, C.: SQUARE attack on block ciphers with low algebraic degree. Sci. China Inf. Sci. **53**(10), 1988–1995 (2010)
36. Sun, B., Liu, M., Guo, J., Rijmen, V., Li, R.: Provable security evaluation of structures against impossible differential and zero correlation linear cryptanalysis. In: Fischlin, M., Coron, J.-S. (eds.) EUROCRYPT 2016. LNCS, vol. 9665, pp. 196–213. Springer, Heidelberg (2016). doi:10.1007/978-3-662-49890-3_8
37. Sun, B., Liu, Z., Rijmen, V., Li, R., Cheng, L., Wang, Q., Alkhzaimi, H., Li, C.: Links among impossible differential, integral and zero correlation linear cryptanalysis. In: Gennaro, R., Robshaw, M. (eds.) CRYPTO 2015. LNCS, vol. 9215, pp. 95–115. Springer, Heidelberg (2015)
38. Todo, Y.: Structural evaluation by generalized integral property. In: Oswald, E., Fischlin, M. (eds.) EUROCRYPT 2015. LNCS, vol. 9056, pp. 287–314. Springer, Heidelberg (2015)
39. Todo, Y.: Integral cryptanalysis on full MISTY1. In: Gennaro, R., Robshaw, M. (eds.) CRYPTO 2015. LNCS, vol. 9215, pp. 413–432. Springer, Heidelberg (2015)
40. Wu, H., Preneel, B.: A fast authenticated encryption algorithm. http://competitions.cr.yp.to/round1/aegisv1.pdf

Lightweight Multiplication in $GF(2^n)$ with Applications to MDS Matrices

Christof Beierle[✉], Thorsten Kranz, and Gregor Leander

Horst Görtz Institute for IT Security, Ruhr-Universität Bochum, Bochum, Germany
{christof.beierle,thorsten.kranz,gregor.leander}@rub.de

Abstract. In this paper we consider the fundamental question of optimizing finite field multiplications with one fixed element. Surprisingly, this question did not receive much attention previously. We investigate which field representation, that is which choice of basis, allows for an optimal implementation. Here, the efficiency of the multiplication is measured in terms of the number of XOR operations needed to implement the multiplication. While our results are potentially of larger interest, we focus on a particular application in the second part of our paper. Here we construct new MDS matrices which outperform or are on par with all previous results when focusing on a round-based hardware implementation.

Keywords: Finite fields · Multiplication · XOR-count · Lightweight cryptography · MDS matrices · Block cipher

1 Introduction

Many cryptographic schemes build on finite fields as their underlying mathematic structure. In almost all cases, the schemes can be designed without having to specify a concrete representation of the finite field in advance. However, when finally being implemented in practice, one necessarily has to choose a particular representation of the finite field, basically as bit strings. In general, this choice does not influence the security of the scheme, but might well influence the performance of the resulting implementation. In this work we focus on this choice of field representations and derive theoretical results on how to choose an optimal field representation with respect to multiplication with fixed field elements. Before going into details, we elaborate on this setup in the special case of symmetric cryptography.

Symmetric Cryptographic Primitives. Build the back-bone of virtually any secure communication today. Block ciphers and hash functions can be seen as the workhorses in cryptography, used for encrypting and authenticating the largest part of the workload.

Today, we are in the comfortable situation of having at hand a choice of strong block ciphers and hash functions that seem secure against even the strongest

© International Association for Cryptologic Research 2016
M. Robshaw and J. Katz (Eds.): CRYPTO 2016, Part I, LNCS 9814, pp. 625–653, 2016.
DOI: 10.1007/978-3-662-53018-4_23

adversaries with practically unlimited computational resources. Moreover, those primitives are based on rather well-understood design principles that allow to construct efficient, simple and easy to analyze ciphers. Especially in the case of substitution-permutation (SP) networks, following the seminal ideas of AES [9] and its predecessor SQUARE [8], arguing the security of ciphers against the two most powerful generic attacks, that is differential- and linear attacks [6,23], became significantly easier. In an SP-network the cipher (or the cryptographic permutation) consists of a number of almost identical rounds, each of which consists of a layer of S-boxes and an $\mathbb{F}_2$-linear layer to mix those parts.

One of the most important design strategies for those primitives is the so-called wide-trail strategy, initiated in [7], that aims at lower bounding the number of active S-boxes. Here, for a given linear or differential trail, an S-box is called active if its input-mask (resp. input-difference) is non-zero. The main observations of the wide-trail strategy is that it is actually the linear layer that is to a large extent responsible for the security of the primitive against linear and differential attacks. Moreover, the wide-trail strategy allows a natural decoupling of the design choice for a linear layer and an S-box.

Interestingly, for the linear layer not many general constructions are known. Two basic approaches can be identified. On the one hand, an ad-hoc approach, where lower bounding the number of active S-boxes requires computer-aided tools that search (sometimes heuristically) for optimal trails. This approach is used e.g. for Serpent [5] or Keccak [4]. On the other hand, a code-based approach, where the linear layers are chosen in such a way that they correspond to good (often locally optimal) linear codes. This is, most prominently, the case for AES where a Maximum Distance Separable (MDS) code is implemented via the MixColumns operation.

Even in the theoretically better circumstantiated code-based approach many fundamental questions are left open. Here, when using an MDS matrix for (parts of) the linear layer, the main challenge is to choose an MDS matrix that is most suitable for an efficient implementation. As those MDS matrices are usually defined over a finite field with characteristic two, i.e. $\mathbb{F}_{2^n}$, one important and so far almost unstudied question is the choice of an $\mathbb{F}_2$-basis of $\mathbb{F}_{2^n}$ and its impact on the implementation efficiency.

From a design point of view, one has to choose a linear layer given as a mapping on $\mathbb{F}_{2^n}^b$ and an $\mathbb{F}_2$-basis of $\mathbb{F}_{2^n}$ to concretely specify the primitive. This is actually a very natural separation of the design of the cipher and its specification (and thus implementation) on bit level. As nicely explained in [10] by introducing RIJNDAEL-GF this separation is probably most obvious for AES itself, but in principle possible for any cipher. Following [10], the choice of basis is to a large extent independent of the design and the security of the cipher. However, the choice of basis might have a significant impact on the efficiency of the cipher on certain platforms.

For software implementations, depending on the details, the choice of basis is either irrelevant (in e.g. a table-based implementation) or hard to capture (in e.g. a bit-sliced implementation) as the efficiency might depend on the exact instructions offered by a given platform. For hardware implementations, one has to distinguish between a serial implementation or a round-based implementation. As the round-based implementation seems most relevant in practice (cf. [27]), we mainly focus on this use-case here. Surprisingly, compared to a serial hardware implementation, the case of a round-based hardware implementation has attracted less attention so far.

For a round-based hardware implementation, the impact of the choice of basis already becomes apparent when focusing on how to implement the multiplication with one given element α in $\mathbb{F}_{2^n}$. For different choices of bases, the efficiency of implementations of the resulting $\mathbb{F}_2$-linear mappings differs significantly. Thus, the very fundamental task we study in the first part of the paper is:

For a given element $\alpha \in \mathbb{F}_{2^n}$ find a basis such that multiplication by α can be implemented most efficiently.[1]

It is worth pointing out that the related question of how to efficiently multiply *two* arbitrary field elements has been studied extensively in the past.

While the above question is of independent interest, with potentially very different applications, we use our results for designing efficient linear layers. Thus, in the second part, we will give several constructions of MDS matrices. Echoing the above, the construction of our MDS matrices are independent of the choice of the basis – actually to a large extent independent of the field size as well.

The combination of the first part, i.e. how to choose a basis that allows for an optimal implementation, and the second part, i.e. the construction of MDS matrices, finally results in implementations of MDS matrices that are more efficient for a large variety of parameters than the best matrices discussed so far in literature.

Thus, this application serves as a nice example were an improved understanding on how to choose the field representation immediately leads to improved results. This is even more interesting as the construction of efficient MDS matrices has been an active field of research recently.

1.1 Related Work

In particular the construction of efficient serial MDS matrices is a well-studied subject. Considering serial implementations of MDS matrices is based on the initial idea of Guo, Peyrin, and Poschmann used in the design of PHOTON [13] and later in the block cipher LED [14]. In a nutshell the idea is not to implement an MDS matrix directly, but rather implement a matrix A such that A^k is MDS for some small k. When considering a hardware implementation, it reduces the chip area if implementing A is significantly cheaper than A^k. The circuit implementing A is then iterated k times, which does not increase its size significantly. This basic idea has been further generalized and improved in a series of subsequent papers. In [24,30] the authors focus on even more efficient

---

[1] Note that the choice of basis is of course not restricted to choosing different irreducible polynomials to represent the finite field.

choices for A by considering additive, i.e. $\mathbb{F}_2$-linear MDS codes. Their approach uses symbolic computations in order to derive general conditions on how to choose the matrix entries independent of the dimension.

In [31] Xu et al. furthermore took into account the cost of implementing the inverse matrix. At FSE 2014, in [2] Augot and Finiasz improved significantly upon the efficiency of the search algorithm of [24], allowing them to search for MDS matrices of much larger dimension than previously possible.

For a round-based implementation, less work has been done so far. The authors of [27] focus on MDS matrices that have an efficient implementation (in terms of the XOR-count) and put special emphasis on involutory MDS matrices, i.e. MDS matrices that are their own inverse. They derive several constructions and rather efficient search methods for MDS matrices meeting their goals. Very recently, Liu and Sim [21] improved upon some of those results by characterizing equivalences in circulant (and circulant-like) MDS matrices and thus further reduced the search space. In both works, in order to improve the efficiency for a given MDS matrix defined over a finite field, the authors considered different representations of the underlying finite fields by running through all possible irreducible polynomials of the given degree. However, in view of the question of how to choose an optimal basis, this corresponds to investigating only a small subset of all possible bases. Work on investigating the XOR-count distribution for other than the polynomial bases has been done very recently in [25].

Also recently, Li and Wang constructed circulant involutory $\mathbb{F}_2$-linear MDS matrices [19]. While it was already known that circulant MDS matrices over a finite field cannot be involutory [15], they have shown their existence in the additive case. Independently, the authors of [21] have shown the existence of left-circulant involutory MDS matrices over finite fields.

1.2   Our Contribution

After fixing our notation and recalling basic facts in Sect. 2, in the first part of the paper we focus on the question on how to find an optimal implementation of the multiplication by a given field element α (cf. Sect. 3). Here efficiency is measured in terms of the number of XOR operations needed to implement the corresponding binary matrix. Note that this metric differs from the XOR-count used in [27]. In [27] the XOR-count of an $n \times n$ matrix M was defined as the number of ones in M minus n. However, the number of (additional) ones in a matrix does not necessarily correspond to the number of XOR operations needed for implementation. Thus, while the number of ones in M is certainly an easier to handle metric, in our opinion it is more appropriate to consider the actual number of XOR operations as the efficiency metric. Note that this improved notion was also discussed in [16]. For technical reasons, we focus on the number of XOR operations without temporary registers, i.e. in-place XOR operations. One of our main results in this first part of the paper is, that for a non-trivial element α one can find a basis such that the resulting matrix can be implemented with one XOR operation if and only if the characteristic polynomial of α is an irreducible trinomial. Note that an XOR-count equal to one in our notion coincides with the definition of the XOR-count in [27]. The interesting part here is that the condition on the characteristic polynomial is not only sufficient but also necessary. As an immediate consequence, one cannot hope to implement the multiplication by any element $\alpha \neq 1$ in $\mathbb{F}_{2^8}^*$ with one XOR only. This follows by the above and the well-known fact that there are no irreducible trinomials of degree 8 [28].

We furthermore show that, for any given basis, there are at most two (non-trivial) elements α and β such that the multiplication with those elements can be implemented with one XOR operation. In fact, β is necessarily the multiplicative inverse of α.

While the weight of the (irreducible) characteristic polynomial of an element α clearly gives an upper bound of the number of XOR operations needed to implement the corresponding multiplication, we show that this bound is in general not tight in the case were the characteristic polynomial is of weight larger than three.

In particular, for all elements $\alpha \in \mathbb{F}_{2^n}^*$ with $n \leq 8$ we present an optimal representation such that the multiplication with α can be implemented with a minimal number of XOR operations. For all those elements α, that are not contained in a proper subfield of $\mathbb{F}_{2^n}$, the multiplication can be implemented with at most 3 XOR operations (and often with two only). Those results are given in Tables 3, 4, 5, 6 and 7 and cover the cases which are most relevant for symmetric cryptography. Interestingly, and maybe counter-intuitive, multiplication with non trivial elements in a proper subfield turns out to be among the most expensive in all the cases explored here.

Moreover, for all $n \leq 2048$ for which no irreducible trinomial of degree n exists, we present one element $\alpha \in \mathbb{F}_{2^n}$ such that multiplication by α requires two XOR operations, cf. Table 8. Those results are proven optimal by the above mentioned necessary and sufficient condition.

In the second part of the paper (cf. Sect. 4) we present several (circulant) matrices. Entries in those matrices are represented as powers of a generic field element α. By symbolically computing all minors, i.e. the determinants of all square submatrices, we derive a list of polynomials in $\mathbb{F}_2[\alpha]$. Now, whenever α is chosen such that it is not a root of any of those polynomials, the matrix is MDS. One nice consequence of this approach is that, as the degree of those polynomials is limited, our matrices are MDS for almost all elements in $\mathbb{F}_{2^n}$ as soon as n is large enough, i.e. larger than the maximal degree of those polynomials.

Finally, the first and second part are combined in Sect. 4.2 to result in the most efficient MDS matrices in terms of the XOR-count known so far. A summary of our results and comparison with previous work is given in Tables 1 and 2, respectively. The main observation here is that if multiplication by α can be implemented with t XOR operations, then multiplication by $\alpha^{\pm i}$ for $i \geq 0$ can be implemented with at most $t \cdot i$ XOR operations.[2] Thus, by simply mini-

---

[2] It is exactly this part where considering only in-place XOR operations becomes very helpful, as otherwise multiplication by α and by α^{-1} might differ in their XOR-count.

mizing the sum of the (absolute) exponents for our circulant MDS matrices, we immediately reduce the XOR-count.

As an interesting side result, we like to point out that the *XOR-count per bit actually decreases with increasing field size*.[3] For example, our 4×4 MDS matrices have a per bit XOR-count of $3 + \frac{3}{n}$, or $3 + \frac{6}{n}$ in the case that no irreducible trinomial of degree n exists.

Thus, even so reducing the number of XOR operations has already received considerable attention recently, this part nicely shows that our improved understanding of how to choose an optimal basis allows us to easily improve upon known constructions. Note that such improvements are possible independent from which XOR-count definition is used, that is, we were able to improve existing results also in the old XOR-count definition by changing the basis. For example, we found an element in $\mathbb{F}_{2^8}$ with only 2 additional non-zero entries which directly improves the results of [27].

Finally, in Sect. 5 we give a perspective on non-linear, additive MDS matrices. In particular, we point out that while there exists no $\alpha \in \mathbb{F}_{2^8}$ (resp. $\mathbb{F}_{2^{13}}$, $\mathbb{F}_{2^{16}}$) which can be implemented with only one XOR operation, there does exist an 8×8 (resp. 13×13, 16×16) binary matrix, that can be used in place for the multiplication by α in the above mentioned 4×4 matrix to result in an additive MDS matrix with reduced cost.[4] Again, the idea of considering the entries of the matrix as powers of a single field element is beneficial as the conditions for the matrix to be MDS remain basically unchanged.

We conclude the paper by pointing to some interesting questions for future investigations.

2   Preliminaries

If p is a prime, we denote the *finite field with p* elements by $\mathbb{F}_p$ and the *extension field with p^n elements* by $\mathbb{F}_{p^n}$, respectively. In this work, we consider binary fields, thus $p = 2$. Although there exists up to isomorphism only one finite field for every possible order, we are interested in the specific representation. For instance, if $q \in \mathbb{F}_2[x]$ is an irreducible polynomial of degree n, then $\mathbb{F}_{2^n} \cong \mathbb{F}_2[x]/(q)$ where (q) denotes the ideal generated by q. The multiplicative group of some field K is denoted by K^*. By the term *matrix*, we refer to matrices with entries in $\mathbb{F}_2$. In general, the ring of $n \times n$ matrices over a field K will be denoted by $\mathrm{Mat}_n(K)$. The symbol $\mathbf{0}_n$ will denote the *zero matrix* and I_n will be the *identity matrix*. As a third important type of matrix in $\mathrm{Mat}_n(\mathbb{F}_2)$, we introduce $E_{i,j}$ which consist of all zeros except in the i-th row of the j-th column for $i, j \in \{1, \ldots, n\}$. We denote a block diagonal matrix consisting of d matrix blocks A_k as $\bigoplus_{k=1}^{d} A_k$. By $\mathrm{wt}(A)$, we denote the number of non-zero entries of a matrix A. Analogously, $\mathrm{wt}(q)$ denotes the number of non-zero coefficients of a polynomial q.

---

[3] This is also true for the constructions given in [30], but does not hold for the subfield (or code-interleaving) construction.

[4] Note that the authors of [19] recently constructed a similar 32×32 $\mathbb{F}_2$-linear MDS matrix.

2.1   Some Basic Facts About Linear Transformations

We next recall some basics about finite fields and matrix representations. For more background the reader is referred to e.g. [20, Sect. 2.5] and [29]. Let $V \cong K^n$ be a finite-dimensional vector space over the field K. Every linear mapping $f : V \to V$ can be described as $v \mapsto A_B v$ by a left-multiplication with a matrix $A_B \in \mathrm{Mat}_n(K)$. This representation is dependent on the choice of the basis B for V. For instance, if $B = \{b_1, \dots b_n\}$, the j-th column of A_B consists of the coefficients $a_{1,j}, \dots, a_{n,j}$ of $f(b_j) = \sum_{i=1}^{n} a_{i,j} b_i$. Thus, changing the basis from B to B' results in a different matrix representation of f. This transformation is called the *change of basis* transformation, which is simply a conjugation of A_B. Thus, $A_{B'} = T A_B T^{-1}$ using an invertible matrix T. In this case, A_B and $A_{B'}$ are are called *similar* (resp. *permutation-similar* if T is a permutation matrix).

There is a natural way of representing the elements in a finite field with characteristic p as vectors with coefficients in $\mathbb{F}_p$. In the following, we consider the representation of the multiplication by α by a matrix as described in the following diagram.

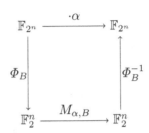

The bijection Φ_B maps elements $\alpha \in \mathbb{F}_{2^n}$ to its vectorial representation over $\mathbb{F}_2$ with regard to a basis B (and Φ_B^{-1} vice versa). $M_{\alpha,B}$ denotes the $n \times n$ matrix representing (left-) multiplication by the element α. For different bases B and B', one can obtain $M_{\alpha,B'}$ from $M_{\alpha,B}$ by the change of basis transformation, in particular $M_{\alpha,B'} = T M_{\alpha,B} T^{-1}$ for an invertible T. We denote similarity of matrices with the relation symbol $\sim$, (resp. $\sim_\pi$ for permutation-similarity). The *characteristic polynomial* of a matrix A is defined as $\chi_A := \det(\lambda I - A) \in \mathbb{F}_2[\lambda]$ and the *minimal polynomial* is denoted by m_A. Recall that the minimal polynomial is the (monic) polynomial p of least degree, such that $p(A) = \mathbf{0}_n$. It is a well-known fact that the minimal polynomial divides the characteristic polynomial, thus $\chi_A(A) = \mathbf{0}_n$. As the minimal polynomial and the characteristic polynomial are actually properties of the underlying linear mapping, similar matrices have the same characteristic and the same minimal polynomial.

A special type of matrix, that will play an important role in the following is the companion matrix of a polynomial. For a polynomial

$$q = x^n + q_{n-1} x^{n-1} + \cdots + q_1 x + q_0 \in \mathbb{F}_2[x]$$

of degree n, the *companion matrix* of q is defined as

$$
C_q = \begin{pmatrix}
0 & & & & q_0 \\
1 & 0 & & & q_1 \\
& \ddots & \ddots & & \vdots \\
& & 1 & 0 & q_{n-2} \\
& & & 1 & q_{n-1}
\end{pmatrix}.
$$

It is known from linear algebra that the characteristic polynomial and the minimal polynomial of C_q are equal to q itself, i.e. $\chi_{C_q} = m_{C_q} = q$. In addition, any matrix A is similar to a companion matrix if and only if its characteristic polynomial coincides with its minimal polynomial. In particular, C_q is exactly the *rational canonical form* [11, Sect. 12.2] of A in this case.

2.2   The XOR-Count and the Cycle Normal Form

The *XOR-count* of a field element was already studied in [17,27]. In the formal definition in [27], an invertible n-dimensional matrix A has an *XOR-count* of t if and only if A can be written as a permutation matrix with t additional non-zero entries. Formally, $A = P + \sum_{k=1}^{t} E_{i_k,j_k}$ and $\mathrm{wt}(A) = n + t$. Although all matrices of that structure can be implemented with at most t XOR operations (not necessarily without temporary registers), the construction does not contain all possible matrices which are realizable with at most t XOR operations. For instance, there are matrices with three additional non-zero entries such that the result of their defining linear function can be computed with just two additions. As an example, consider

$$
\begin{pmatrix}
1 & 0 & 1 \\
1 & 1 & 1 \\
0 & 0 & 1
\end{pmatrix}
\begin{pmatrix}
v_1 \\
v_2 \\
v_3
\end{pmatrix}
=
\begin{pmatrix}
v_1 + v_3 \\
(v_1 + v_3) + v_2 \\
v_3
\end{pmatrix}.
$$

In the following, we provide an alternative definition which includes the cases described above.

Definition 1. *An invertible matrix A has an XOR-count of t, denoted $\mathrm{wt}_\oplus(A) = t$, if t is the minimal number such that A can be written as*

$$
A = P \prod_{k=1}^{t}(I + E_{i_k,j_k})
$$

with $i_k \neq j_k$ for all k.

Note that if a matrix can be represented in the form $P \prod_{k=1}^{t}(I + E_{i_k,j_k})$, the number of factors $(I + E_{i_k,j_k})$ clearly gives an upper bound on the actual XOR-count. It is worth pointing out that the definition above just counts the number of XOR operations without using temporary registers. Those are technically somewhat easier to handle. However, this restriction does not make a difference for matrices with XOR-count less or equal to 2, which we are most concerned about in in the following. In general, allowing temporary registers might well reduce the number of XOR operations needed for an implementation.

Our definition coincides with the one from [27] for the case that $t = 1$, that is, for matrices of XOR-count 1. For other cases, the number of additional non-zero entries can increase. We will often consider $t = 2$ within this work. By evaluating the product, it follows that any A with $\mathrm{wt}_\oplus(A) = 2$ is of the form

$$A = \begin{cases} P + P(E_{i_1,j_1} + E_{i_2,j_2}) & \text{iff } i_2 \neq j_1 \\ P + P(E_{i_1,j_1} + E_{i_2,j_2} + E_{i_1,j_2}) & \text{iff } i_2 = j_1. \end{cases}$$

The XOR-count is invariant under permutation-similarity. Moreover, naturally in the setting not allowing temporary registers, the XOR-count is invariant under taking the inverse. This is summarized and formally proven in the following Lemma and Corollary.

Lemma 1. *If $A \sim_\pi A'$, then $\mathrm{wt}_\oplus(A) = \mathrm{wt}_\oplus(A')$.*

Proof. Let $A' = QAQ^{-1}$ where Q is the permutation matrix representing the permutation $\sigma \in S_n$. Let $I + E_{i_k,j_k}$ be a factor in the XOR-count representation of $A = P \prod_{k=1}^t (I + E_{i_k,j_k})$ where $t = \mathrm{wt}_\oplus(A)$. Then the following identity holds:

$$(I + E_{i_k,j_k})Q^{-1} = Q^{-1} + E_{i_k,\sigma^{-1}(j_k)} = Q^{-1}(I + E_{\sigma(i_k),\sigma^{-1}(j_k)}).$$

One is able to commute Q^{-1} to the front before the first factor by proceeding for all of the t factors and finally obtain

$$A' = QPQ^{-1} \prod_{k=1}^t (I + E_{\sigma(i_k),\sigma^{-1}(j_k)}).$$

It follows that $\mathrm{wt}_\oplus(A') \leq \mathrm{wt}_\oplus(A)$. By reverting the above steps we obtain $\mathrm{wt}_\oplus(A) \leq \mathrm{wt}_\oplus(A')$.   □

Corollary 1. *If $\mathrm{wt}_\oplus(A) = t$, then also $\mathrm{wt}_\oplus(A^{-1}) = t$.*

Proof. We show that A^{-1} is permutation-similar to a matrix with an XOR-count of t.

$$\left(P \prod_{k=1}^t (I + E_{i_k,j_k}) \right)^{-1} = \prod_{k=t}^1 (I + E_{i_k,j_k}) P^{-1} \sim_\pi P^{-1} \prod_{k=t}^1 (I + E_{i_k,j_k})$$

□

Later, we would like to be able to exhaustively search over all matrices with low XOR-count for a given dimension n. Since the number of permutation matrices (which is $n!$) rapidly increases with n, an exhaustive search will quickly become infeasible if we do not restrict the structure of P. By a well-known fact from combinatorics, one is able to assume P to be in a specific form.

Lemma 2. *For any permutation matrix P of dimension n, it is*

$$P \sim_\pi \bigoplus_{k=1}^{d} C_{x^{m_k}+1}$$

for some m_k with $\sum_{k=1}^{d} m_k = n$ and $m_1 \geq \cdots \geq m_d \geq 1$.

Proof. It is well-known that two permutations with the same cycle type are conjugate [11, Chapter 4.3, Proposition 11]. That is, given the permutations $\sigma, \tau \in S_n$ as

$$\sigma = (s_1, s_2, \ldots, s_{d_1})(s_{d_1+1}, \ldots, s_{d_2}) \ldots (s_{d_{m-1}+1}, \ldots, s_{d_m})$$
$$\tau = (t_1, t_2, \ldots, t_{d_1})(t_{d_1+1}, \ldots, t_{d_2}) \ldots (t_{d_{m-1}+1}, \ldots, t_{d_m})$$

in cycle notation, one can find some $\pi \in S_n$ such that $\pi \sigma \pi^{-1} = \tau$. This π operates as a relabeling of indices.

Let σ in the form above be the permutation defined by P. Now, there exits a permutation π such that $\pi \sigma \pi^{-1} = (d_1, 1, 2, \ldots, d_1 - 1)(d_2, d_1 + 1, d_1 + 2, \ldots, d_2 - 1) \ldots (d_m, d_{m-1} + 1, d_{m-1} + 2, \ldots, d_m - 1)$. If Q denotes the permutation matrix defined by π, one obtains QPQ^{-1} in the desired form.   □

We say that any permutation matrix of this structure is in *cycle normal form*. The cycle normal form of P is denoted by $C(P)$. Up to permutation-similarity, we can always assume that the permutation matrix P of a given matrix with XOR-count t is in cycle normal form, as stated in the following corollary.

Corollary 2.

$$P \prod_{k=1}^{t} (I + E_{i_k, j_k}) \sim_\pi C(P) \prod_{k=1}^{t} (I + E_{\sigma(i_k), \sigma^{-1}(j_k)})$$

for some permutation $\sigma \in S_n$.

3   Efficient Multiplication in Finite Fields

In this section, we first present some theoretic results towards understanding the structure of matrices $M_{\alpha,B}$ representing (left-) multiplication by some finite field element $\alpha \in \mathbb{F}_{2^n}^*$. The parameter B indicates a basis of $\mathbb{F}_{2^n}$ considered as an n-dimensional vector space over $\mathbb{F}_2$. The XOR-count of $M_{\alpha,B}$ is indeed depending on the choice of the basis B. As described in Corollary 2, we can assume a certain normal form for matrices with an XOR-count of t.

Not every (invertible) matrix is a representation of a field multiplication. For example, an obvious condition for that, is that the multiplicative order of the matrix divides $2^n - 1$. In order to understand exactly which matrices indeed represent multiplication with some field element α, Theorem 1 below gives a characterization that allows to efficiently decide when a given matrix corresponds to multiplication by a field element. The crucial part is the minimal polynomial of α. It is a property of the linear mapping

$$f_\alpha : \mathbb{F}_{2^n} \to \mathbb{F}_{2^n}, \beta \mapsto \alpha\beta$$

and is invariant under changing the specific representation of f_α to $\beta \mapsto M_{\alpha,B}\beta$.

Theorem 1. *Let $A \in \mathrm{Mat}_n(\mathbb{F}_2) \setminus \{0_n\}$. Then $A = M_{\alpha,B}$ for some element $\alpha \in \mathbb{F}_{2^n}^*$ with respect to some basis B if and only if m_A is irreducible.*

Proof. As described in [29], the ring generated by some matrix A defines a field of order 2^n if and only if the characteristic polynomial χ_A is irreducible. This is the case since $\chi_A(A) = 0$ and thus A is the root of an irreducible polynomial of degree n. One can see that $\mathbb{F}_2(A) = \{\sum_{i=0}^{n-1} \alpha_i A^i \mid \alpha_i \in \mathbb{F}_2\}$ since it must contain all sums of powers of A. However, for $\mathbb{F}_2(A)$ being a field it is not necessary that A has an irreducible characteristic polynomial. It can be possible that A generates a subfield $\mathbb{F}_{2^m}$ of $\mathbb{F}_{2^n}$. As we show now, this is the case if and only if the minimal polynomial of α is irreducible and has degree m.

If m_A is not irreducible, $\mathbb{F}_2(A)$ is not a field and thus A cannot represent a field multiplication. Let now m_A be irreducible. The characteristic polynomial χ_A is necessarily a power of m_A, since both of these polynomials share the same irreducible factors. So, $\chi_A = (m_A)^d$ for some positive integer d. Both d and $\deg(m_A)$ divide n. Because of the irreducibility of m_A, the rational canonical form of A consists of d blocks of C_{m_A}. Thus, we obtain the similarity

$$A \sim \bigoplus_{k=1}^{d} C_{m_A}.$$

Since $\chi_{C_{m_A}} = m_A$, the matrix A defines a multiplication with some element in a subfield of $\mathbb{F}_{2^n}$. $\square$

Note that, any field element α is, up to its conjugates $\alpha, \alpha^2, \alpha^{2^2}, \ldots, \alpha^{2^{n-1}}$, uniquely identified by its minimal polynomial. For every field element α, the minimal polynomial m_α is exactly the minimal polynomial m_A of a matrix A representing multiplication with α. Furthermore, two matrices $A, A' \in \mathrm{Mat}_n(\mathbb{F}_2)$ with the same irreducible minimal polynomial are similar. Thus, given a matrix A, identifying the element α such that $A = M_{\alpha,B}$ is equivalent to computing the (irreducible) minimal polynomial of A.

The main question is which field elements can be implemented with a minimal number of XOR operations, or in particular, what is the minimal XOR-count for a given (non-trivial) field element $\alpha \in \mathbb{F}_{2^n}^*$. Trivially, multiplication with $\alpha = 1$ can be implemented with zero additions since $M_{1,B} = I_n$ for all bases B. On the other hand, if the XOR-count is 0, the element is equal to 1. In a first place, we thus aim for an XOR-count of 1 whenever possible. By a simple observation, this optimal result can be realized if the minimal polynomial of α is a trinomial of degree n.

Example 1. Let the field with 2^n elements be represented as $\mathbb{F}_{2^n} = \mathbb{F}_2[x]/(q)$ for an irreducible q of degree n. For the (left-) multiplication with x in the canonical basis $B = \{1, x, x^2, \ldots, x^{n-1}\}$, it is $M_{x,B} = C_q$. Thus, $\mathrm{wt}_\oplus(M_{x,B}) = \mathrm{wt}(q) - 2$ and the XOR-count of $M_{x,B}$ equals 1 if q is a trinomial.

Since our approach is about finding any (non-trivial) element $\alpha \in \mathbb{F}_{2^n}^*$ such that multiplication with α can be implemented with minimal additions, this fact implies that we cannot hope to improve upon the implementation costs if there exists an irreducible trinomial of degree n. However, for several n, including the interesting case where n is a multiple of 8, there does not exist such a trinomial [28]. The question is what happens for these cases. As one of our main results, we show that the condition on the minimal polynomial is not only sufficient but also necessary.

3.1   Characterizing Elements with Optimal XOR-Count

In this section, we prove the converse of the fact described in Example 1, namely the necessary condition on the minimal (resp. characteristic) polynomial of α resulting in an XOR-count of 1.

Theorem 2. *Let $\alpha \in \mathbb{F}_{2^n}$. Then there exists a matrix A with $\mathrm{wt}_\oplus(A) = 1$ such that $A = M_{\alpha,B}$ for some basis B if and only if m_α is a trinomial of degree n.*

Proof. Let $M_{\alpha,B}$ represent multiplication by some element $\alpha \in \mathbb{F}_{2^n}$ with respect to the basis $B = \{b_1, \ldots, b_n\}$ and let further $\mathrm{wt}_\oplus(M_{\alpha,B}) = 1$. We show that the characteristic polynomial $\chi_{M_{\alpha,B}}$ is a trinomial and coincides with m_α. Since the XOR-count is 1, we can assume w.l.o.g. that $M_{\alpha,B} = P + E_{i,j}$ such that $P = \bigoplus_{k=1}^l C_{x^{m_k}+1}$ is in cycle normal form. We first show that $l = 1$. Suppose $l > 1$, then, depending on $E_{i,j}$, the matrix $M_{\alpha,B}$ is either in upper or lower triangular form consisting of at least two diagonal blocks. Since one of them must be of the form C_{x^m+1}, the polynomial $x^m + 1$ must divide the characteristic polynomial $\chi_{M_{\alpha,B}}$. Since further $(x+1) \mid (x^m + 1)$, the minimal polynomial of α is necessarily a multiple of $x + 1$. This is a contradiction since $\alpha \neq 1$ and m_α must be irreducible. Hence, $M_{\alpha,B}$ is permutation-similar to $C_{x^n+1} + E_{i,j}$. It is further $i \neq j + 1 \mod n$ since otherwise $M_{\alpha,B}$ would be singular.

We now investigate how α operates on the basis elements $b_k \in B$. Considering the structure of $M_{\alpha,B}$, we obtain the following list of equations.

$$\alpha b_1 = b_2$$

$$\vdots$$

$$\alpha b_{j-1} = b_j$$
$$\alpha b_j = b_{j+1} + b_i$$
$$\alpha b_{j+1} = b_{j+2}$$

$$\vdots$$

$$\alpha b_n = b_1.$$

By defining $\gamma := b_{j+1}$, one can express every basis element b_k as a power of α multiplied by γ. In particular,

$$b_{j+k \mod n} = \alpha^{k-1}\gamma \tag{1}$$

for $k \in \{1,\ldots,n\}$. Combining this observation with the identity $\alpha b_j = b_{j+1} + b_i$, one obtains

$$\alpha^n \gamma = \gamma + \alpha^t \gamma \tag{2}$$

for some exponent $t \neq 0$. Since $\gamma \neq 0$, the field element α is a root of the trinomial $p = x^n + x^t + 1$. It is left to show that p is exactly the minimal polynomial of α. Suppose that $m_\alpha = x^m + \sum_{k=0}^{m-1} c_k x^k$ with constants $c_k \in \{0,1\}$ and $m < n$. By multiplying $m_\alpha(\alpha)$ with γ, one obtains

$$\alpha^m \gamma = \sum_{k=0}^{m-1} c_k \alpha^k \gamma$$

and thus $b_{t_m} = \sum_{k=0}^{m-1} c_k b_{t_k}$ for some basis elements b_{t_k}. We are now able to express one basis element b_{t_k} as a sum of other elements from B which is contradictory to the linear independence of the basis. Hence, $\deg(m_\alpha) = n$ and thus $m_\alpha = p$ which finally proves the theorem. $\square$

Note that the polynomial p is exactly the characteristic polynomial of $M_{\alpha,B}$ since it must be a monic multiple of m_α having degree n. An alternative way of proving that the characteristic polynomial of a matrix $C_{x^n+1} + E_{i,j}$ is a trinomial is given in Appendix A. As a simple corollary one obtains that any $\alpha \in \mathbb{F}_{2^n}^*$ with an XOR-count of 1 cannot be contained in a proper subfield.

Corollary 3. *Let $\alpha \in \mathbb{F}_{2^n}^* \setminus \{1\}$ and let further $\deg(m_\alpha) < n$, indicating that α lies in a proper subfield of $\mathbb{F}_{2^n}$. Then, any matrix $M_{\alpha,B}$ representing multiplication by a field element α with respect to some basis B has $\mathrm{wt}_\oplus(M_{\alpha,B}) > 1$.*

This result implies that building MDS layers using a block interleaving construction [1], also called subfield construction in [17], almost always results in suboptimal implementation costs. Note that specific instances of this construction are also implicitly used in the AES, LS-Designs [12] and the hash function Whirlwind [3].

Now let α be an element with XOR-count 1. From Corollary 1 we know that α^{-1} has the same XOR-count. Next, we show that there do not exist any further elements with an XOR-count equal to 1.

Theorem 3. *For any given basis B of $\mathbb{F}_{2^n}$, there exist at most two field elements α and α^{-1} with $\mathrm{wt}_\oplus(M_{\alpha,B}) = \mathrm{wt}_\oplus(M_{\alpha^{-1},B}) = 1$.*

Proof. Let $\alpha \in \mathbb{F}_{2^n}^*$ with $\mathrm{wt}_\oplus(M_{\alpha,B}) = 1$ for the basis $B = \{b_1,\ldots,b_n\}$. We show that for any $\beta \in \mathbb{F}_{2^n}$ with $\mathrm{wt}_\oplus(M_{\beta,B}) = 1$ it holds that $\beta = \alpha^{\pm 1}$.

Since w.l.o.g. $M_{\alpha,B}$ can be assumed to be of the form $C_{x^n+1} + E_{i,j}$, we know that (1) and (2) hold. We further know that $M_{\beta,B}$ is of the form $P + E_{i',j'}$ and thus there exist $l, m \in \{1, \ldots, n\}$ with $l \neq m$ and $\beta b_{j+l \mod n} = b_{j+m \mod n}$. Using Eq. (1), we can write $\beta = \alpha^{m-l} =: \alpha^s$ where $s \in \{-(n-1), \ldots, n-1\}$. We directly see that $s \neq 0$. It remains to show that $-1 \leq s \leq 1$.

Assume $s \geq 2$. We use Eqs. (1) and (2) to obtain

$$\beta b_{j+(n-s+1) \mod n} = \alpha^n \gamma = \gamma + \alpha^t \gamma = b_{j+1 \mod n} + b_{j+t+1 \mod n}.$$

Since $0 < t < n$, it holds that $b_{j+1 \mod n} \neq b_{j+t+1 \mod n}$ and thus the according column contains an additional 1. For the next column, we have

$$\beta b_{j+(n-s+2) \mod n} = \alpha^{n+1}\gamma = \alpha\gamma + \alpha^{t+1}\gamma$$

$$= \begin{cases} b_{j+2 \mod n} + b_{j+t+2 \mod n}, & \text{for } t < n-1 \\ b_{j+2 \mod n} + b_{j+1 \mod n} + b_{j \mod n}, & \text{for } t = n-1 \end{cases}$$

Hence, this column also contains at least one additional 1 which is contradictory to the XOR-count of 1.

For $-s \geq 2$ we can construct the same contradiction by considering β^{-1}. $\square$

We now understand the structure of field elements α that can be implemented with a single addition. One might think that also for the other cases, the weight of the minimal polynomial of α strictly lower-bounds XOR-count as $\text{wt}(m_\alpha) - 2$. As we will see next, this is not the case.

3.2 Experimental Search for Optimal XOR-Counts

Surprisingly, we often can improve the XOR-count, compared to using the companion matrix for multiplication, if the weight of the minimal polynomial is greater than 3. For instance, if m_α is an irreducible pentanomial, that is of weight 5, of degree n there often exists a basis B such that $\text{wt}_\oplus(M_{\alpha,B}) = 2$. Indeed, for all $n \leq 2048$ for which no irreducible trinomial of degree n exists, we found some element $\alpha \in \mathbb{F}_{2^n}^*$ with an XOR-count of 2 for some basis B. For every such dimension, we present an example of such a matrix in Table 8. Thus, for all practically relevant fields, we are able to identify an element such that multiplication can be implemented with one or two XOR operations. By Theorem 2, these results are proven to be optimal.

Moreover, as fields of small size are most interesting for SP-networks, we investigated those in full detail. For the fields $\mathbb{F}_{2^4}$, $\mathbb{F}_{2^5}$, $\mathbb{F}_{2^6}$, $\mathbb{F}_{2^7}$ and $\mathbb{F}_{2^8}$ we present the optimal XOR-count for each non-trivial element α in Tables 3, 4, 5, 6 and 7, respectively. The main observation is that each element which is not contained in a proper subfield can be implemented with at most 3 additions. Furthermore, whenever an XOR-count of 2 is possible, the minimal polynomial of α is a pentanomial in all those cases. However, a more thorough characterization of elements with non-optimal XOR-count is left as an open problem (see Sect. 6 for more details).

Those results are based on a search. Since we are only interested in matrices up to similarity (due to the change of basis), we just need to consider all matrices in the normal form described in Corollary 2. This will exhaust all possibilities of similarity classes for a given XOR-count t. In particular, the search space is reduced from $n!(n(n-1))^t$ to only $p(n)(n(n-1))^t$ where $p(n)$ denotes the number of partitions of n, which is exactly the number of possible cycle normal forms of dimension n. This allows us to exhaustively search over all similarity classes up to $t = 3$ XOR operations for the fields of small size. The key-point here is that, instead of searching for an optimal basis for a given field element, we generated all matrices with small XOR-count and used Theorem 1 in order to check which field element (if any) the given matrix corresponds to.

In order to identify a single lightweight element for larger field sizes, we identified conditions in which cases the characteristic polynomial of a matrix with XOR-count 2 has weight 5, cf. Theorem 4 below. During the search, one only has to check for irreducibility. This allows to compute the results presented in Table 8 extremely fast, that is within a couple of minutes on a standard PC. The proof of Theorem 4 is given in Appendix A.

Theorem 4. *Let $M = C_{x^n+1} + E_{i_1,j_1} + E_{i_2,j_2}$ such that the following relations hold:*

$$i_1 < j_1 \neq n, \quad i_2 > j_2 + 1, \quad i_1 \leq j_2, \quad i_2 \leq j_1, \quad j_1 - (i_1 - 1) \neq n, \quad n - (j_1 - i_1) \neq i_2 - j_2$$

The characteristic polynomial of M is a pentanomial of degree n. In particular

$$\chi_M = \lambda^n + \lambda^{n+i_1-j_1+i_2-j_2-2} + \lambda^{n+i_1-j_1-1} + \lambda^{i_2-j_2-1} + 1.$$

4   Constructing Lightweight MDS Matrices

Our goal is now to construct lightweight MDS matrices. We use the results obtained in the previous sections and restrict our search to circulant matrices and entries with low XOR-count. This simplifies checking the MDS property and computing an upper bound of the XOR-count of the whole matrix. The complexity of our algorithm enables us to easily search for MDS matrices up to dimension 8. Our construction is generic and works for all finite fields $\mathbb{F}_{2^m}$ with $m > b$ for a given bound b.

More precisely, we construct circulant matrices with entries of the form $\alpha^{\pm i}$ where α is an element in $\mathbb{F}_{2^m}$. Choosing entries of this form enables us to easily upper-bound the XOR-count of the elements since

$$\mathrm{wt}_\oplus(x^{\pm k}) \leq k\,\mathrm{wt}_\oplus(x).$$

This can be easily seen by using Corollary 1 and the fact that α^k can be implemented by k times implementing α. We want to keep the size of the finite field over which the matrix is defined generic. Thus, we choose the matrix entries from a subgroup of the *field of fractions* of the polynomial ring $\mathbb{F}_2[x]$, denoted $\mathrm{Quot}(\mathbb{F}_2[x])$. That is, every element is of the form

$$\frac{x^s + a_{s-1}x^{s-1} + \cdots + a_1 x + a_0}{x^t + b_{t-1}x^{t-1} + \cdots + b_1 x + b_0}.$$

More precisely, and as mentioned above, we restrict our search to elements from $\langle x \rangle$ which is the multiplicative subgroup of $\text{Quot}(\mathbb{F}_2[x])$ generated by x. Our search works by constructing MDS conditions for an $n \times n$ matrix M with entries in $\langle x \rangle$. This approach later allows us to substitute the indeterminate x by any $\alpha \in \mathbb{F}_{2^m}$ that fulfills all of the conditions given below. In this context, we let $M(\alpha) \in \text{Mat}_n(\mathbb{F}_{2^m})$ denote the matrix obtained by substituting x with $\alpha \in \mathbb{F}_{2^m}$.

We define the *weight* of some circulant matrix with entries in $\langle x \rangle$ as the sum of the absolute values of the exponents in its first row, that is, the number of times α has to be applied *per row*. Then, for a given dimension, we are interested in finding the lightest matrix M which can be made MDS for as many finite fields as possible. Note that the higher priority here was to find a lightweight matrix. Thus, there might exist matrices which can be made MDS for even more fields, but with a probably higher cost.

MDS Conditions. Note that a matrix is MDS, if and only if all its square submatrices are invertible [22, page 321, Theorem 8]. Thus, given a matrix $M \in \text{Mat}_n(\text{Quot}(\mathbb{F}_2[x]))$, we compute the determinants of all square submatrices (called *minors*) of M in order to check the MDS property. This way one obtains a list of conditions (polynomials in $\mathbb{F}_2$) for a matrix to be MDS. Since the determinant of a matrix with elements from a field is an element of the field itself, all of these determinants can be represented as the fraction of two polynomials. Thus, M is MDS if and only if the numerator of all minors is non-zero. One can decompose the numerators into their irreducible factors and collect all of them in a set T. This set now defines the MDS conditions. In particular, $M(\alpha)$ is MDS if and only if α is not a root of any of these irreducible polynomials in T, that is, iff $m_\alpha \notin T$. This trivially holds for $m > \max_{p \in T}\{\deg(p)\}$ and any $\alpha \in \mathbb{F}_{2^m}$ which is not contained in a proper subfield. In general, if α is not contained in a proper subfield, the necessary and sufficient condition for the existence of an MDS matrix $M(\alpha)$ is that not all irreducible polynomials of degree m are contained in T. We note that there exists a value b which lower bounds the field size for which M can always be made MDS. That is, for all $t > b$, there exists an irreducible polynomial of degree t which is not in T.

4.1 Generic Lightweight MDS Matrices

We now present some results obtained by the approach described above. Given the restrictions, these matrices achieve the smallest weight, i.e. the smallest sum of (absolute) exponents of x. Later, we will use these generic matrices to build concrete instantiations of $n \times n$ MDS matrices $M(\alpha)$ for $n \in \{2, 3, \ldots, 8\}$ over a finite field $\mathbb{F}_{2^m}$ with $m > b$. We note that the given results are not necessarily the only possible constructions with the smallest weight.

We also present the conditions for the matrix to be MDS, that is, the irreducible polynomials that must not be equal to m_α. However, since the number of conditions rapidly increases with the dimension of the matrix, we refrain from presenting a complete list for dimensions 6 to 8. Instead, we give the SageMath

Listing 1.1. Sage code for computing the set T.

```
P.<x> = GF(2)[]
K = FractionField(P)

def mds_equations(M):
    R = [P(x)]
    for i in range(len(M.rows())+1)[1:]:
        L = M.minors(i)
        for l in L:
            if (l != 0):
                F = list(l.numerator().factor())
                for f in F:
                    R.append(f[0])
            else:
                return
    return list(set(R))
```

source code that was used to compute the set T of irreducible polynomials in Listing 1.1.

2×2 **and** 3×3 **matrices.** The matrices

$$\mathrm{circ}(1, \alpha) = \begin{pmatrix} 1 & \alpha \\ \alpha & 1 \end{pmatrix}$$

and

$$\mathrm{circ}(1, 1, \alpha) = \begin{pmatrix} 1 & 1 & \alpha \\ \alpha & 1 & 1 \\ 1 & \alpha & 1 \end{pmatrix}$$

are MDS for all $\alpha \neq 0, 1$.

4×4 **matrices.** For $m > 3$, there exists an $\alpha \in \mathbb{F}_{2^m}$ such that the matrix $\mathrm{circ}(1, 1, \alpha, \alpha^{-2})$ is MDS. More precisely, the matrix is MDS iff α is not a root of any of the following polynomials:

$$x$$
$$x + 1$$
$$x^2 + x + 1$$
$$x^3 + x + 1$$
$$x^3 + x^2 + 1$$
$$x^4 + x^3 + x^2 + x + 1$$
$$x^5 + x^2 + 1$$

5×5 **matrices.** For $m > 3$, there exists an $\alpha \in \mathbb{F}_{2^m}$ such that the matrix $\mathrm{circ}(1, 1, \alpha, \alpha^{-2}, \alpha)$ is MDS. More precisely, the matrix is MDS iff α is not a root of any of the following polynomials:

$$x$$
$$x + 1$$
$$x^2 + x + 1$$
$$x^3 + x + 1$$
$$x^3 + x^2 + 1$$
$$x^4 + x + 1$$
$$x^4 + x^3 + 1$$

6×6 **matrices.** For $m > 5$, there exists an $\alpha \in \mathbb{F}_{2^m}$ such that the matrix $\mathrm{circ}(1, \alpha, \alpha^{-1}, \alpha^{-2}, 1, \alpha^3)$ is MDS.

7×7 **matrices.** For $m > 5$, there exists an $\alpha \in \mathbb{F}_{2^m}$ such that the matrix $\mathrm{circ}(1, 1, \alpha^{-2}, \alpha, \alpha^2, \alpha, \alpha^{-2})$ is MDS.

8×8 **matrices.** For $m > 7$, there exists an $\alpha \in \mathbb{F}_{2^m}$ such that the matrix $\mathrm{circ}(1, 1, \alpha^{-1}, \alpha, \alpha^{-1}, \alpha^3, \alpha^4, \alpha^{-3})$ is MDS.

4.2   Instantiating Lightweight MDS Matrices

We now combine the efficient multiplication in finite fields from Sect. 3 with our construction of MDS matrices. That is, the presented generic MDS matrices are instantiated with elements α with low XOR-count.

In a matrix multiplication every element is computed as the sum over multiplications. The according XOR-count was already discussed in [17,27]. For our matrices, the total number of XOR operations needed *per row* is upper bounded by

$$(n - 1)m + w \cdot \mathrm{wt}_{\oplus}(\alpha).$$

Here, $(n - 1)m$ XORs are the static part which comes from summing over the multiplication results and w is the weight as defined above. The *overhead* of $w \cdot \mathrm{wt}_{\oplus}(\alpha)$ XORs is needed for multiplying with the single elements. The static part cannot be changed by fast multiplication. Therefore, this overhead is the part that has to be minimized.

The cost per bit for the whole matrix is given by

$$\frac{n((n - 1)m + w\,\mathrm{wt}_{\oplus}(\alpha))}{nm} = n - 1 + \frac{w\,\mathrm{wt}_{\oplus}(\alpha)}{m}.$$

One can notice that it decreases for larger field sizes.

For each of the matrices M described in Sect. 4.1, Table 1 presents choices for α such that $M(\alpha)$ is MDS. Note that concrete instantiations are only given up to the field size $m = 13$. The reason is that for larger m, all possible C_p with

Table 1. Optimal instantiations of the generic MDS matrices for $2 \leq n \leq 8$. In each cell, the first entry describes the minimal polynomial of $\alpha \in \mathbb{F}_2^m$ and the second entry describes the overhead of the instantiated $n \times n$ matrix $M(\alpha)$. The trinomial $x^m + x^a + 1$ is denoted by (a) and the pentanomial $x^m + x^a + x^b + x^c + 1$ is denoted by (a, b, c).

n \ m	2	3	4	5	6	7	8	9	10	11	12	13
2	(1), 1	(1), 1	(1), 1	(2), 1	(1), 1	(1), 1	(6,5,1), 2	(1), 1	(3), 1	(2), 1	(3), 1	(10,9,1), 2
3	(1), 1	(1), 1	(1), 1	(2), 1	(1), 1	(1), 1	(6,5,1), 2	(1), 1	(3), 1	(2), 1	(3), 1	(10,9,1), 2
4	-	-	(1), 3	(3), 3	(1), 3	(1), 3	(6,5,1), 6	(1), 3	(3), 3	(2), 3	(3), 3	(10,9,1), 6
5	-	-	(3,2,1), 8	(2), 4	(1), 4	(1), 4	(6,5,1), 8	(1), 4	(3), 4	(2), 4	(3), 4	(10,9,1), 8
6	-	-	-	-	(1), 7	(1), 7	(6,5,1), 14	(1), 7	(3), 7	(2), 7	(3), 7	(10,9,1), 14
7	-	-	-	-	(1), 8	(1), 8	(6,5,1), 16	(1), 8	(3), 8	(2), 8	(3), 8	(10,9,1), 16
8	-	-	-	-	-	-	(6,5,2), 26	(8), 13	(3), 13	(2), 13	(3), 13	(10,9,1), 26

p as an irreducible degree-m polynomial of weight 3 are valid choices. If no such trinomial exists, one can choose $M_{\alpha,B}$ as in Table 8.

Table 2 compares the results presented in this section to the best constructions known so far. It turned out that our construction of the 4×4 MDS matrix in $\mathbb{F}_{2^4}$ is identical to the $\mathbb{F}_2$-linear matrix constructed in [19,21]. We stress that our construction leads to the lightest MDS matrices known, improving the results described in [21,27] for 8×8 MDS matrices in $\mathbb{F}_{2^4}$ and $\mathbb{F}_{2^8}$ respectively. This is also the case when considering an unrolled implementation of the serial implementations in [30]. Unrolled variants of their implementations have an XOR-count that is slightly larger than ours. Moreover, and more importantly, the circuit depth is considerably increased due to the optimization with respect to a serial implementation.

Table 2. Comparison of our results with the (non-involutory) $\mathbb{F}_{2^m}$-linear MDS matrices from [27, Sect. 6.2], [19,21] by overhead. a: In these constructions, the XOR-count is measured by counting the number of additional 1's in the corresponding matrix.

(n,m)	Our construction	Construction in [27][a]	Construction in [21][a]	Construction in [19][a]
(4,4)	3	5	3	3
(4,8)	6	10	8	
(8,8)	26	40	30	

Note that our results in Table 2 are measured by the XOR-count from Definition 1 while the results from [19,21,27] use the old XOR-count definition. Additionally to these results, our understanding of how to choose an optimal basis can also be used to improve existing results in the old XOR-count definition. For example, we can represent the 8×8 MDS matrix in $\mathbb{F}_{2^8}$ from [21] with 28 additional ones instead of 30 by change of basis.

5 Generalizing the MDS Property

Here, following e.g. [30], we consider a generalization to additive MDS codes in order to improve efficiency.

There are some dimensions for which no field element with an XOR-count of 1 exists, for instance $m = 8$. However, especially this dimension is very important since lots of block cipher designs are byte oriented. One would wish to have some element α with $\mathrm{wt}_{\oplus}(\alpha) = 1$. A way of solving this problem is to not restrict to field elements. Instead, α can be chosen to be some other matrix in the ring $R = \mathrm{Mat}_m(\mathbb{F}_2)$. Given an $n \times n$ matrix M with elements in $\mathrm{Quot}(\mathbb{F}_2[x])$, the substitution $M(\alpha)$ now consists of elements in a commutative ring with unity, which is the subring of R generated by α. In general, given a commutative ring with unity R, one can define the determinant $\det_R : \mathrm{Mat}_n(R) \to R$ in a similar way than for matrices over fields. As described in [18, pp. 212–215], any $A \in \mathrm{Mat}_n(R)$ is invertible if and only if $\det_R(A)$ is a unit in R. We now define the MDS property for matrices over a commutative ring.

Definition 2. *Let R be a commutative ring with unity. A matrix $M \in \mathrm{Mat}_n(R)$ is MDS if and only if for every $1 \le s \le n$, any $s \times s$ submatrix of M is invertible.*

For checking the MDS property in our case, we use a well-known fact about block matrices.

Theorem 5 (Theorem 1 in [26]). *Let K be a field and let R be a commutative subring of $\mathrm{Mat}_m(K)$ for some integer m. For any matrix $M \in \mathrm{Mat}_d(R)$, it is*

$$\det(M) = \det(\det_R(M)),$$

where $\det(M)$ is the determinant of M considered as $M \in \mathrm{Mat}_{dm}(K)$.

As an implication, $M(\alpha)$ is MDS if and only if $p(\alpha)$ is invertible for all $p \in T$, if and only if $\det(p(\alpha)) \ne 0$ for all $p \in T$.

2×2 and 3×3 matrices. Given $M = \mathrm{circ}(1, x)$ (resp. $M = \mathrm{circ}(1, 1, x)$), one has to make sure that both x and $x + 1$ are invertible for M to be MDS. This is the case if x is substituted by the companion matrix $C_{x^m + x + 1}$ for $m \ge 2$. Thus, $M(C_{x^m + x + 1})$ is MDS and each entry has an XOR-count of 1.

4×4 matrices. The MDS conditions are more complex than above. So, we only present some improvements for $m \in \{8, 13, 16\}$. The matrix $M = \mathrm{circ}(1, 1, \alpha, \alpha^{-2})$ is MDS for

$$\alpha \in \{C_{x^8 + x^2 + 1}, C_{x^{13} + x + 1}, C_{x^{16} + x + 1}\}.$$

Note that a similar matrix for $m = 8$ was recently constructed in [19].

6 Conclusion and Open Problems

We presented a study of optimal multiplication bases with respect to the XOR-count. When applied to MDS matrices those lead to very efficient round-based implementations. We expect our results to be applied in other domains as well.

Our investigations leave many possibilities for future research. While we have been able to characterize exactly which field elements can be implemented with one XOR operation only, the general case is still open. For small fields of dimension smaller or equal to eight, we were able to compute the optimal bases with the help of an exhaustive computer search. However, for larger dimensions, this approach turns quickly inefficient and more insight would be needed. As a first step, we conjecture the following statement.

Conjecture 1. If $\mathrm{wt}_\oplus(M_{\alpha,B}) = 2$, then m_α is of weight smaller or equal to 5.

Note that the converse of the conjectured statement is (unlike the case of trinomials) wrong. As can be seen in Table 7, there exist a pentanomial of degree 8 which cannot be implemented with two XOR operations only. Beyond that, our intuition is that the larger the weight of the minimal polynomial, the larger the gap between the most efficient multiplication and the efficiency of multiplying by means of the companion matrix. Quantifying and demonstrating such a statement is an interesting and challenging open problem. Another interesting question is to get an improved understanding of how to most efficiently multiply with elements in proper subfields. More specifically, as a generalization of Corollary 3, one may ask the following question.

Question 1. Is the most efficient way to multiply with a subfield element given by multiplying in the subfield d times, where d is the extension degree of the field when viewed as an extension of the subfield. More precisely, given an $\alpha \in \mathbb{F}_{2^m}^* \subset \mathbb{F}_{2^n}^*$ in a proper subfield of dimension $m = \frac{n}{d}$ and let $M_{\alpha \in \mathbb{F}_{2^m},B'}$ be the multiplication matrix in $\mathbb{F}_{2^m}$ with an optimal XOR-count. Is $M_{\alpha \in \mathbb{F}_{2^n},B} = \bigoplus_{k=1}^{d} M_{\alpha \in \mathbb{F}_{2^m},B'}$ a matrix with the lowest possible XOR-count for multiplication with $\alpha \in \mathbb{F}_{2^n}$? In particular, is $\mathrm{wt}_\oplus(M_{\alpha \in \mathbb{F}_{2^n},B}) = d\,\mathrm{wt}_\oplus(M_{\alpha \in \mathbb{F}_{2^m},B'})$?

Finally, for MDS matrices, it should be noted that we *locally* achieve the optimal solution. What would be needed to finally settle the search for lightweight matrices is a global optimal solution. That is for a given dimension, find an MDS matrix that can be implemented with the minimal number of XOR operations.

Finally, when optimizing for software, similar questions can be phrased and investigating solutions that are valid for more than one specific platform is a challenging research topic.

Acknowledgements. We would like to thank Thomas Peyrin for some valuable discussions on the notion of the XOR-count. We would also like to thank Gottfried Herold. This work was partly supported by the DFG Research Training Group GRK 1817 Ubicrypt and by the BMBF Project UNIKOPS (01BY1040).

A   Proofs

In the following, we present an alternative way of proving the fact that the characteristic polynomial of some matrix $M = C_{x^n+1} + E_{i,j}$ with $\mathrm{wt}_\oplus(M) = 1$ is a trinomial of degree n. This is true in general, even if M does not represent a multiplication with a field element.

Lemma 3. *For $M = C_{x^n+1} + E_{i,j}$ with $\mathrm{wt}(M) = n + 1$, the characteristic polynomial χ_M of M is a trinomial of degree n.*

Proof. It is to compute $\chi_M = \det(\lambda I_n - M) = \det(\lambda I_n + C_{x^n+1} + E_{i,j})$. If $j = n$, then $M = C_{x^n+x^{i-1}+1}$ and $\chi_M = \lambda^n + \lambda^{i-1} + 1$ is a trinomial of degree n. Thus, w.l.o.g. one can assume $j < n$. To compute the determinant we use Laplace's formula by expanding along the n-th column. One obtains

$$\chi_M = \det\left(\begin{pmatrix} 1 & \lambda & & & \\ & 1 & \lambda & & \\ & & \ddots & \ddots & \\ & & & 1 & \lambda \\ & & & & 1 \end{pmatrix} + E_{i-1,j}\right) + \lambda \det\left(\begin{pmatrix} \lambda & & & \\ 1 & \lambda & & \\ & \ddots & \ddots & \\ & & 1 & \lambda \\ & & & 1 & \lambda \end{pmatrix} + E_{i,j}\right),$$

where $E_{0,j} := \mathbf{0}$ and $E_{n,j} := \mathbf{0}$. Both of these remaining matrices are of dimension $(n-1) \times (n-1)$. We now distinguish three cases:

(i) $i < j$: The additional 1 lies in the upper triangle of M. Now, χ_M reduces to $\chi_M = 1 + \lambda \det(\lambda I_{n-1} + C_{x^{n-1}} + E_{i,j})$. In order to compute the remaining determinant, we keep on expanding along the last column for $n-1-j$ times until the additional 1 is located in the rightmost column. We now obtain the determinant of a companion matrix. Thus,

$$\chi_M = 1 + \lambda^{n-j} \det(\lambda I_j + C_{x^j + x^{i-1}})$$
$$= 1 + \lambda^{n-j}(\lambda^j + \lambda^{i-1}) = \lambda^n + \lambda^{n-j+i-1} + 1.$$

(ii) $i = j$: In this case, the additional 1 lies on the main diagonal of M and

$$\chi_M = 1 + \lambda(\lambda^{n-2}(\lambda + 1)) = \lambda^n + \lambda^{n-1} + 1.$$

(iii) $i > j$: The additional 1 lies in the lower triangle of M. Because of the structure of M, it is further $i > (j+1)$. Defining the $m \times m$ matrix S_m^λ as

$$S_m^\lambda := \begin{pmatrix} 1 & \lambda & & & \\ & 1 & \lambda & & \\ & & \ddots & \ddots & \\ & & & 1 & \lambda \\ & & & & 1 \end{pmatrix},$$

the characteristic polynomial of M reduces to $\chi_M = \det(S^\lambda_{n-1}+E_{i-1,j})+\lambda^n$. We expand along the last row of $S^\lambda_{n-1} + E_{i-1,j}$ for $n - i$ times and get $\chi_M = \det(S^\lambda_{i-1} + E_{i-1,j}) + \lambda^n$.

Now, the additional 1 lies in the last row of the remaining $(i - 1) \times (i - 1)$-dimensional matrix. The goal is now to shift this 1 to the first column. This is done by expanding $j - 1$ times along the first column. We now obtain $\chi_M = \det(S^\lambda_{i-j}+E_{i-j,1})+\lambda^n$ and the additional 1 is in the lower left corner of the matrix. As a last step, we expand along the first column for one more time and finally get

$$\chi_M = \lambda^n + \det(S^\lambda_{i-j} + E_{i-j,1}) = \lambda^n + \det(\lambda I_{i-j-1} + C_{x^i{}_j{}_1}) + 1$$
$$= \lambda^n + \lambda^{i-j-1} + 1.$$

We now present the proof of Theorem 4 which makes use of Lemma 3. $\square$

Theorem 4. Let $M = C_{x^n+1} + E_{i_1,j_1} + E_{i_2,j_2}$ such that the following relations hold:

$$i_1 < j_1 \neq n, \quad i_2 > j_2 + 1, \quad i_1 \leq j_2, \quad i_2 \leq j_1, \quad j_1 - (i_1 - 1) \neq n, \quad n - (j_1 - i_1) \neq i_2 - j_2$$

The characteristic polynomial of M is a pentanomial of degree n. In particular

$$\chi_M = \lambda^n + \lambda^{n+i_1-j_1+i_2-j_2-2} + \lambda^{n+i_1-j_1-1} + \lambda^{i_2-j_2-1} + 1.$$

Proof. The first two conditions ensure that M has exactly one additional non-zero entry in the upper and one in the lower triangle (not on the main diagonal). Since $j_1, j_2, i_2 \neq n$, we can expand along the last column and obtain

$$\chi_M = \det(S^\lambda_{n-1} + E_{i_1-1,j_1} + E_{i_2-1,j_2}) + \lambda \det(\lambda I_{n-1} + C_{x^{n-1}} + E_{i_1,j_1} + E_{i_2,j_2}).$$

For simplicity, we define $A := S^\lambda_{n-1} + E_{i_1-1,j_1} + E_{i_2-2,j_2}$ and $B := \lambda I_{n-1} + C_{x^{n-1}} + E_{i_1,j_1} + E_{i_2,j_2}$. In order to compute the latter part, we "push" the additional non-zero entry from the upper triangle to the top-right corner by first expanding $n - 1 - j_1$ times along the last column and then expanding $i_1 - 1$ times along the first row. The condition $i_2 \leq j_1$ ensures that E_{i_2,j_2} will not be eliminated from expanding along the last column and the condition $i_1 \leq j_2$ ensures that E_{i_2,j_2} will not be eliminated from expanding along the first row. Using Lemma 3, one obtains

$$\lambda \det(B) = \lambda\lambda^{n-1-j_1}\lambda^{i_1-1} \det(\lambda I_{j_1-i_1+1} + C_{x^{j_1-i_1+1}+1} + E_{i_2-i_1+1,j_2-i_1+1})$$
$$= \lambda^{n-1-j_1+i_1}(\lambda^{j_1-i_1+1} + \lambda^{i_2-i_1+1-j_2+i_1-1-1} + 1)$$
$$= \lambda^n + \lambda^{n+i_1-j_1+i_2-j_2-2} + \lambda^{n+i_1-j_1-1}.$$

For $\det(A)$, we proceed similar to case (iii) in Lemma 3. We first expand $j_2 - 1$ times along the first column in order to get the additional non-zero value from the lower triangle to the leftmost column. Because of the condition $i_1 \leq j_2$, this elimintates E_{i_1-1,j_1}. Now, one can expand $n - j_2 - (i_2 - j_2)$ times along the last row, until the remaining additional non-zero entry lies in the lower left corner of the remaining matrix. We finally expand along the first column one more time and obtain

$$\det(A) = \det(S^\lambda_{n-j_2} + E_{i_2-j_2,1}) = \det(S^\lambda_{i_2-j_2} + E_{i_2-j_2,1}) = \lambda^{i_2-j_2-1} + 1.$$

The last two assumptions make sure that all of the five coefficients of $\det(A) + \lambda \det(B)$ are distinct such that χ_M is indeed a pentanomial.   □

B   Minimal XOR-Counts in $\mathbb{F}_{2^n}$

Table 3. Minimal XOR-counts for all elements in $\mathbb{F}^*_{2^4}$.

Minimal polynomial m_α	Min $\mathrm{wt}_\oplus(\alpha)$	Matrix
$x + 1$	0	I
$x^2 + x + 1$	2	$C_{m_\alpha} \oplus C_{m_\alpha}$
$x^4 + x + 1$	1	C_{m_α}
$x^4 + x^3 + 1$	1	C_{m_α}
$x^4 + x^3 + x^2 + x + 1$	2	$C_{x^4+1} + E_{2,2} + E_{3,4}$

Table 4. Minimal XOR-counts for all elements in $\mathbb{F}^*_{2^5}$.

Minimal polynomial m_α	Min $\mathrm{wt}_\oplus(\alpha)$	Matrix
$x + 1$	0	I
$x^5 + x^2 + 1$	1	C_{m_α}
$x^5 + x^3 + 1$	1	C_{m_α}
$x^5 + x^3 + x^2 + x + 1$	2	$C_{x^5+1} + E_{2,4} + E_{4,2}$
$x^5 + x^4 + x^2 + x + 1$	2	$C_{x^5+1} + E_{2,2} + E_{3,5}$
$x^5 + x^4 + x^3 + x + 1$	2	$C_{x^5+1} + E_{2,3} + E_{3,1} + E_{3,3}$
$x^5 + x^4 + x^3 + x^2 + 1$	2	$C_{x^5+1} + E_{2,2} + E_{3,4}$

Table 5. Minimal XOR-counts for all elements in $\mathbb{F}_{2^6}^*$.

Minimal polynomial m_α	Min $\mathrm{wt}_\oplus(\alpha)$	Matrix
$x + 1$	0	I
$x^2 + x + 1$	3	$C_{m_\alpha} \oplus C_{m_\alpha} \oplus C_{m_\alpha}$
$x^3 + x + 1$	2	$C_{m_\alpha} \oplus C_{m_\alpha}$
$x^3 + x^2 + 1$	2	$C_{m_\alpha} \oplus C_{m_\alpha}$
$x^6 + x + 1$	1	C_{m_α}
$x^6 + x^3 + 1$	1	C_{m_α}
$x^6 + x^4 + x^2 + x + 1$	2	$(C_{x^4+1} \oplus C_{x^2+1})(I + E_{1,5} + E_{5,4})$
$x^6 + x^4 + x^3 + x + 1$	2	$C_{x^6+1} + E_{2,3} + E_{4,6}$
$x^6 + x^5 + 1$	1	C_{m_α}
$x^6 + x^5 + x^2 + x + 1$	2	$C_{x^6+1} + E_{2,2} + E_{3,6}$
$x^6 + x^5 + x^3 + x^2 + 1$	2	$C_{x^6+1} + E_{2,2} + E_{3,5}$
$x^6 + x^5 + x^4 + x + 1$	2	$C_{x^6+1} + E_{2,3} + E_{3,1} + E_{3,3}$
$x^6 + x^5 + x^4 + x^2 + 1$	2	$(C_{x^4+1} \oplus C_{x^2+1})(I + E_{1,5} + E_{6,1} + E_{6,5})$

Table 6. Minimal XOR-counts for all elements in $\mathbb{F}_{2^7}^*$.

Minimal polynomial m_α	Min $\mathrm{wt}_\oplus(\alpha)$	Matrix
$x + 1$	0	I
$x^7 + x + 1$	1	C_{m_α}
$x^7 + x^3 + 1$	1	C_{m_α}
$x^7 + x^3 + x^2 + x + 1$	2	$C_{x^7+1} + E_{2,6} + E_{4,2}$
$x^7 + x^4 + 1$	1	C_{m_α}
$x^7 + x^4 + x^3 + x^2 + 1$	2	$(C_{x^4+1} \oplus C_{x^3+1})(I + E_{1,5} + E_{5,3})$
$x^7 + x^5 + x^2 + x + 1$	2	$(C_{x^5+1} \oplus C_{x^2+1})(I + E_{1,6} + E_{6,5})$
$x^7 + x^5 + x^3 + x + 1$	2	$C_{x^7+1} + E_{2,3} + E_{4,7}$
$x^7 + x^5 + x^4 + x^3 + 1$	2	$(C_{x^4+1} \oplus C_{x^3+1})(I + E_{1,5} + E_{7,2})$
$x^7 + x^5 + x^4 + x^3 + x^2 + x + 1$	3	$C_{x^7+1} + E_{2,3} + E_{4,6} + E_{4,7}$
$x^7 + x^6 + 1$	1	C_{m_α}
$x^7 + x^6 + x^3 + x + 1$	2	$(C_{x^6+1} \oplus C_{x^1+1})(I + E_{1,7} + E_{7,4})$
$x^7 + x^6 + x^4 + x + 1$	2	$(C_{x^6+1} \oplus C_{x^1+1})(I + E_{1,7} + E_{7,3})$
$x^7 + x^6 + x^4 + x^2 + 1$	2	$C_{x^7+1} + E_{2,4} + E_{4,1} + E_{4,4}$
$x^7 + x^6 + x^5 + x^2 + 1$	2	$(C_{x^5+1} \oplus C_{x^2+1})(I + E_{1,6} + E_{7,1} + E_{7,6})$
$x^7 + x^6 + x^5 + x^3 + x^2 + x + 1$	3	$C_{x^7+1} + E_{2,2} + E_{2,3} + E_{4,7}$
$x^7 + x^6 + x^5 + x^4 + 1$	2	$C_{x^7+1} + E_{2,2} + E_{3,4}$
$x^7 + x^6 + x^5 + x^4 + x^2 + x + 1$	3	$C_{x^7+1} + E_{2,2} + E_{3,4} + E_{3,7}$
$x^7 + x^6 + x^5 + x^4 + x^3 + x^2 + 1$	3	$C_{x^7+1} + E_{2,2} + E_{2,3} + E_{4,6}$

Table 7. Minimal XOR-counts for all elements in $\mathbb{F}_{2^8}^*$.

Minimal polynomial m_α	Min $\mathrm{wt}_\oplus(\alpha)$	Matrix
$x+1$	0	I
x^2+x+1	4	$\bigoplus_{k=1}^4 C_{m_\alpha}$
x^4+x+1	2	$C_{m_\alpha} \oplus C_{m_\alpha}$
x^4+x^3+1	2	$C_{m_\alpha} \oplus C_{m_\alpha}$
$x^4+x^3+x^2+x+1$	4	$\bigoplus_{k=1}^2 (C_{x^4+1} + E_{2,2} + E_{3,4})$
$x^8+x^4+x^3+x+1$	2	$C_{x^8+1} + E_{2,6} + E_{4,2}$
$x^8+x^4+x^3+x^2+1$	3	C_{m_α}
$x^8+x^5+x^3+x+1$	2	$(C_{x^5+1} \oplus C_{x^3+1})(I + E_{1,6} + E_{6,5})$
$x^8+x^5+x^3+x^2+1$	2	$C_{x^8+1} + E_{2,6} + E_{5,2}$
$x^8+x^5+x^4+x^3+1$	2	$(C_{x^5+1} \oplus C_{x^3+1})(I + E_{1,6} + E_{6,2})$
$x^8+x^5+x^4+x^3+x^2+x+1$	3	$C_{x^8+1} + E_{2,5} + E_{2,7} + E_{4,2}$
$x^8+x^6+x^3+x^2+1$	2	$(C_{x^6+1} \oplus C_{x^2+1})(I + E_{1,7} + E_{8,5})$
$x^8+x^6+x^4+x^3+x^2+x+1$	3	$C_{x^8+1} + E_{2,3} + E_{4,7} + E_{4,8}$
$x^8+x^6+x^5+x+1$	2	$C_{x^8+1} + E_{2,4} + E_{4,2}$
$x^8+x^6+x^5+x^2+1$	2	$(C_{x^6+1} \oplus C_{x^2+1})(I + E_{1,7} + E_{7,2})$
$x^8+x^6+x^5+x^3+1$	2	$C_{x^8+1} + E_{2,3} + E_{4,6}$
$x^8+x^6+x^5+x^4+1$	3	C_{m_α}
$x^8+x^6+x^5+x^4+x^2+x+1$	3	$C_{x^8+1} + E_{2,3} + E_{2,4} + E_{5,8}$
$x^8+x^6+x^5+x^4+x^3+x+1$	3	$C_{x^8+1} + E_{2,3} + E_{2,5} + E_{6,8}$
$x^8+x^7+x^2+x+1$	2	$C_{x^8+1} + E_{2,2} + E_{3,8}$
$x^8+x^7+x^3+x+1$	2	$(C_{x^7+1} \oplus C_{x+1})(I + E_{1,8} + E_{8,5})$
$x^8+x^7+x^3+x^2+1$	2	$C_{x^8+1} + E_{2,2} + E_{3,7}$
$x^8+x^7+x^4+x^3+x^2+x+1$	3	$C_{x^8+1} + E_{2,2} + E_{3,6} + E_{3,8}$
$x^8+x^7+x^5+x+1$	2	$(C_{x^7+1} \oplus C_{x+1})(I + E_{1,8} + E_{8,3})$
$x^8+x^7+x^5+x^3+1$	2	$(C_{x^5+1} \oplus C_{x^3+1})(I + E_{1,6} + E_{8,1} + E_{8,6})$
$x^8+x^7+x^5+x^4+1$	2	$C_{x^8+1} + E_{2,2} + E_{3,5}$
$x^8+x^7+x^5+x^4+x^3+x^2+1$	3	$C_{x^8+1} + E_{2,2} + E_{3,5} + E_{3,7}$
$x^8+x^7+x^6+x+1$	2	$C_{x^8+1} + E_{2,3} + E_{3,1} + E_{3,3}$
$x^8+x^7+x^6+x^3+x^2+x+1$	3	$C_{x^8+1} + E_{2,2} + E_{2,3} + E_{4,8}$
$x^8+x^7+x^6+x^4+x^2+x+1$	3	$(C_{x^6+1} \oplus C_{x^2+1})(I + E_{1,7} + E_{7,3} + E_{7,8})$
$x^8+x^7+x^6+x^4+x^3+x^2+1$	3	$C_{x^8+1} + E_{2,2} + E_{2,3} + E_{4,7}$
$x^8+x^7+x^6+x^5+x^2+x+1$	3	$C_{x^8+1} + E_{2,2} + E_{3,4} + E_{3,8}$
$x^8+x^7+x^6+x^5+x^4+x+1$	3	$C_{x^8+1} + E_{2,3} + E_{3,1} + E_{3,3} + E_{8,3}$
$x^8+x^7+x^6+x^5+x^4+x^2+1$	3	$C_{x^8+1} + E_{2,2} + E_{2,5} + E_{6,7}$
$x^8+x^7+x^6+x^5+x^4+x^3+1$	3	$C_{x^8+1} + E_{2,2} + E_{2,3} + E_{4,6}$

Table 8. For each $n \leq 2048$ for which no irreducible trinomial of degree n exists, this table presents a matrix of the form $C_{x^n+1}+E_{i_1,j_1}+E_{i_2,j_2}$ with irreducible characteristic pentanomial. Such a matrix is represented as a 4-tuple (i_1, j_1, i_2, j_2). In all cases, the characteristic polynomial is equal to $\lambda^n+\lambda^{n+i_1-j_1+i_2-j_2-2}+\lambda^{n+i_1-j_1-1}+\lambda^{i_2-j_2-1}+1$.

n		n		n		n		n		n		n		n		n		n	
8	(1,3,3,1)	237	(1,168,3,1)	451	(1,104,3,1)	659	(1,250,3,1)	869	(1,128,3,1)	1067	(1,960,5,1)	1274	(1,1176,3,1)	1480	(1,413,3,1)	1680	(1,645,3,1)	1867	(1,670,3,1)
13	(1,4,3,1)	240	(1,121,4,1)	452	(1,90,3,1)	661	(1,224,3,1)	872	(1,405,3,1)	1068	(1,54,3,1)	1275	(1,1265,3,1)	1483	(1,412,3,1)	1682	(1,5,3,1)	1868	(1,420,3,1)
16	(1,9,4,1)	243	(1,38,3,1)	453	(1,302,3,1)	664	(1,149,3,1)	874	(1,83,3,1)	1069	(1,338,3,1)	1277	(1,230,3,1)	1484	(1,41,3,1)	1683	(1,1278,3,1)	1869	(1,384,3,1)
19	(1,8,3,1)	245	(1,38,3,1)	454	(1,314,3,1)	666	(1,117,3,1)	875	(1,386,3,1)	1070	(1,228,3,1)	1280	(1,81,3,1)	1485	(1,1086,3,1)	1684	(1,730,3,1)	1872	(1,183,3,1)
24	(1,5,3,1)	246	(1,71,3,1)	456	(1,129,3,1)	667	(1,38,4,1)	877	(1,248,5,1)	1072	(1,789,4,1)	1283	(1,344,3,1)	1488	(1,1017,3,1)	1685	(1,816,3,1)	1874	(1,35,4,1)
26	(1,11,3,1)	248	(1,29,3,1)	459	(1,270,3,1)	669	(1,48,3,1)	878	(1,3,3,1)	1073	(1,362,3,1)	1285	(1,1174,3,1)	1491	(1,666,3,1)	1686	(1,72,3,1)	1875	(1,1386,3,1)
27	(1,3,3,1)	251	(1,24,3,1)	461	(1,170,3,1)	672	(1,567,3,1)	880	(1,11,3,1)	1074	(1,801,3,1)	1288	(1,379,3,1)	1493	(1,1002,3,1)	1688	(1,255,3,1)	1876	(1,1204,3,1)
32	(1,3,3,1)	254	(1,19,3,1)	464	(1,55,3,1)	674	(1,307,4,1)	883	(1,589,3,1)	1075	(1,142,3,1)	1290	(1,149,3,1)	1494	(1,620,3,1)	1690	(1,733,3,1)	1877	(1,628,3,1)
37	(1,16,4,1)	256	(1,157,3,1)	466	(1,451,3,1)	675	(1,225,3,1)	885	(1,512,3,1)	1076	(1,49,3,1)	1291	(1,302,3,1)	1496	(1,21,3,1)	1691	(1,26,3,1)	1880	(1,207,3,1)
38	(1,4,3,1)	259	(1,20,3,1)	467	(1,72,3,1)	677	(1,647,3,1)	886	(1,10,3,1)	1077	(1,706,4,1)	1292	(1,473,3,1)	1498	(1,223,3,1)	1693	(1,394,3,1)	1882	(1,399,4,1)
40	(1,14,3,1)	261	(1,12,3,1)	469	(1,188,3,1)	678	(1,312,3,1)	888	(1,501,3,1)	1080	(1,75,3,1)	1293	(1,212,3,1)	1499	(1,3,3,1)	1696	(1,19,3,1)	1883	(1,680,3,1)
43	(1,8,3,1)	262	(1,13,3,1)	472	(1,385,3,1)	680	(1,21,3,1)	891	(1,12,3,1)	1083	(1,92,3,1)	1296	(1,257,3,1)	1501	(1,1222,3,1)	1699	(1,404,3,1)	1885	(1,352,3,1)
45	(1,6,3,1)	264	(1,63,3,1)	475	(1,94,3,1)	681	(1,51,3,1)	893	(1,827,4,1)	1088	(1,3,3,1)	1299	(1,144,3,1)	1502	(1,5,3,1)	1701	(1,540,3,1)	1888	(1,905,3,1)
48	(1,21,3,1)	267	(1,182,3,1)	477	(1,286,4,1)	683	(1,104,3,1)	896	(1,87,3,1)	1091	(1,1026,3,1)	1301	(1,160,3,1)	1504	(1,559,3,1)	1702	(1,262,3,1)	1891	(1,280,3,1)
50	(1,7,3,1)	269	(1,64,3,1)	480	(1,273,5,1)	685	(1,172,3,1)	899	(1,64,3,1)	1093	(1,310,3,1)	1303	(1,380,3,1)	1506	(1,215,3,1)	1704	(1,1617,4,1)	1892	(1,440,3,1)
51	(1,12,3,1)	272	(1,165,3,1)	482	(1,115,3,1)	688	(1,149,3,1)	901	(1,504,4,1)	1096	(1,947,3,1)	1304	(1,391,3,1)	1507	(1,200,3,1)	1706	(1,843,3,1)	1893	(1,344,3,1)
53	(1,4,3,1)	275	(1,20,3,1)	483	(1,26,3,1)	691	(1,606,7,1)	904	(1,241,5,1)	1099	(1,644,3,1)	1307	(1,1200,3,1)	1509	(1,128,5,1)	1707	(1,150,3,1)	1894	(1,391,3,1)
56	(1,13,3,1)	277	(1,208,3,1)	485	(1,158,3,1)	693	(1,278,3,1)	907	(1,142,3,1)	1101	(1,474,3,1)	1309	(1,26,3,1)	1512	(1,381,3,1)	1709	(1,688,3,1)	1896	(1,1053,4,1)
59	(1,14,3,1)	280	(1,73,3,1)	488	(1,359,3,1)	696	(1,77,3,1)	909	(1,480,3,1)	1104	(1,515,3,1)	1312	(1,901,3,1)	1515	(1,14,3,1)	1712	(1,95,3,1)	1897	(1,80,3,1)
61	(1,4,3,1)	283	(1,154,3,1)	491	(1,477,3,1)	699	(1,360,3,1)	910	(1,8,3,1)	1107	(1,936,3,1)	1315	(1,508,3,1)	1517	(1,698,3,1)	1714	(1,1021,3,1)	1898	(1,241,3,1)
64	(1,61,3,1)	285	(1,158,3,1)	493	(1,20,3,1)	701	(1,238,3,1)	912	(1,627,3,1)	1109	(1,278,3,1)	1316	(1,204,3,1)	1520	(1,131,3,1)	1715	(1,250,3,1)	1899	(1,986,3,1)
67	(1,58,3,1)	288	(1,206,3,1)	496	(1,149,3,1)	703	(1,19,3,1)	914	(1,81,3,1)	1112	(1,35,3,1)	1317	(1,820,4,1)	1522	(1,985,3,1)	1717	(1,142,3,1)	1901	(1,230,3,1)
69	(1,42,4,1)	290	(1,96,3,1)	499	(1,40,3,1)	704	(1,195,5,1)	915	(1,320,3,1)	1114	(1,143,3,1)	1318	(1,109,3,1)	1523	(1,906,3,1)	1718	(1,242,3,1)	1904	(1,535,3,1)
70	(1,19,3,1)	291	(1,200,3,1)	501	(1,144,4,1)	706	(1,503,3,1)	917	(1,572,3,1)	1115	(1,328,3,1)	1320	(1,167,3,1)	1525	(1,602,3,1)	1720	(1,133,3,1)	1907	(1,780,3,1)
72	(1,15,5,1)	293	(1,16,3,1)	502	(1,245,3,1)	707	(1,376,3,1)	920	(1,535,3,1)	1117	(1,220,3,1)	1322	(1,405,3,1)	1528	(1,79,3,1)	1723	(1,322,3,1)	1909	(1,46,3,1)
75	(1,36,3,1)	296	(1,109,3,1)	504	(1,141,3,1)	709	(1,230,3,1)	922	(1,299,4,1)	1118	(1,168,3,1)	1323	(1,272,3,1)	1531	(1,910,3,1)	1725	(1,90,3,1)	1910	(1,266,3,1)
77	(1,47,3,1)	298	(1,55,3,1)	507	(1,429,6,1)	710	(1,17,3,1)	923	(1,52,4,1)	1120	(1,1043,3,1)	1325	(1,22,3,1)	1532	(1,66,3,1)	1727	(1,207,3,1)	1912	(1,157,3,1)
78	(1,23,3,1)	299	(1,116,3,1)	509	(1,10,3,1)	712	(1,455,3,1)	925	(1,859,3,1)	1123	(1,410,3,1)	1328	(1,917,3,1)	1533	(1,1160,3,1)	1728	(1,1227,3,1)	1914	(1,639,4,1)
80	(1,5,3,1)	301	(1,154,3,1)	512	(1,425,3,1)	715	(1,110,3,1)	928	(1,537,4,1)	1124	(1,424,3,1)	1330	(1,187,3,1)	1536	(1,39,3,1)	1730	(1,443,3,1)	1915	(1,40,3,1)
82	(1,47,5,1)	304	(1,109,3,1)	515	(1,236,3,1)	717	(1,448,4,1)	929	(1,51,3,1)	1125	(1,254,3,1)	1331	(1,978,3,1)	1538	(1,509,3,1)	1731	(1,338,3,1)	1916	(1,663,3,1)
83	(1,4,3,1)	306	(1,81,3,1)	517	(1,172,3,1)	720	(1,369,4,1)	931	(1,544,3,1)	1128	(1,21,3,1)	1333	(1,530,3,1)	1539	(1,30,3,1)	1732	(1,1250,3,1)	1917	(1,1026,3,1)
85	(1,16,3,1)	307	(1,192,4,1)	520	(1,331,3,1)	723	(1,414,3,1)	933	(1,264,3,1)	1131	(1,552,3,1)	1336	(1,97,3,1)	1541	(1,698,3,1)	1733	(1,128,3,1)	1920	(1,39,3,1)
88	(1,11,3,1)	309	(1,155,3,1)	523	(1,140,3,1)	725	(1,68,3,1)	934	(1,424,3,1)	1132	(1,52,3,1)	1339	(1,820,3,1)	1544	(1,411,3,1)	1736	(1,493,3,1)	1922	(1,1497,3,1)
91	(1,8,3,1)	311	(1,25,3,1)	525	(1,328,4,1)	728	(1,393,3,1)	936	(1,23,3,1)	1133	(1,238,3,1)	1341	(1,732,3,1)	1546	(1,351,4,1)	1739	(1,420,3,1)	1923	(1,326,3,1)
96	(1,9,3,1)	312	(1,7,5,1)	528	(1,35,3,1)	731	(1,80,3,1)	939	(1,288,3,1)	1136	(1,957,3,1)	1342	(1,610,3,1)	1547	(1,62,3,1)	1741	(1,476,3,1)	1925	(1,210,3,1)
99	(1,78,3,1)	315	(1,50,3,1)	530	(1,17,3,1)	733	(1,310,3,1)	940	(1,8,3,1)	1139	(1,246,3,1)	1344	(1,185,3,1)	1549	(1,1220,3,1)	1744	(1,1183,3,1)	1928	(1,83,3,1)
101	(1,20,3,1)	317	(1,90,3,1)	531	(1,44,3,1)	734	(1,56,3,1)	941	(1,382,3,1)	1141	(1,196,3,1)	1346	(1,57,3,1)	1552	(1,325,3,1)	1747	(1,1402,3,1)	1930	(1,569,3,1)
104	(1,94,3,1)	320	(1,53,3,1)	533	(1,56,3,1)	736	(1,271,5,1)	944	(1,191,3,1)	1143	(1,191,3,1)	1347	(1,18,3,1)	1555	(1,404,3,1)	1749	(1,56,3,1)	1931	(1,3,3,1)
107	(1,44,4,1)	323	(1,120,3,1)	535	(1,104,3,1)	739	(1,286,3,1)	946	(1,469,3,1)	1144	(1,335,3,1)	1349	(1,138,5,1)	1557	(1,822,3,1)	1752	(1,1585,4,1)	1933	(1,940,3,1)
109	(1,20,3,1)	325	(1,244,3,1)	536	(1,117,3,1)	741	(1,18,3,1)	947	(1,16,3,1)	1147	(1,598,3,1)	1352	(1,595,3,1)	1560	(1,1027,5,1)	1754	(1,103,3,1)	1936	(1,1657,3,1)
112	(1,91,3,1)	326	(1,27,3,1)	539	(1,3,3,1)	744	(1,635,3,1)	949	(1,550,4,1)	1149	(1,902,5,1)	1355	(1,216,3,1)	1563	(1,198,3,1)	1755	(1,308,3,1)	1939	(1,110,3,1)
114	(1,33,3,1)	328	(1,237,4,1)	541	(1,364,5,1)	747	(1,548,3,1)	950	(1,154,3,1)	1150	(1,130,3,1)	1357	(1,302,5,1)	1565	(1,66,3,1)	1757	(1,988,3,1)	1941	(1,1262,3,1)
115	(1,32,3,1)	331	(1,8,3,1)	542	(1,50,3,1)	749	(1,316,3,1)	952	(1,573,4,1)	1152	(1,11,3,1)	1360	(1,485,3,1)	1568	(1,231,3,1)	1758	(1,213,3,1)	1942	(1,76,3,1)
116	(1,20,3,1)	334	(1,64,3,1)	544	(1,65,5,1)	752	(1,117,3,1)	955	(1,80,3,1)	1155	(1,8,3,1)	1363	(1,1018,3,1)	1571	(1,430,3,1)	1760	(1,273,3,1)	1944	(1,429,3,1)
117	(1,99,4,1)	335	(1,20,3,1)	546	(1,213,3,1)	755	(1,230,3,1)	957	(1,372,3,1)	1157	(1,328,3,1)	1365	(1,488,3,1)	1573	(1,112,3,1)	1761	(1,186,3,1)	1947	(1,210,3,1)
120	(1,15,3,1)	336	(1,35,3,1)	547	(1,302,4,1)	757	(1,751,3,1)	958	(1,80,3,1)	1160	(1,87,3,1)	1368	(1,533,3,1)	1574	(1,284,3,1)	1762	(1,277,3,1)	1949	(1,686,3,1)
122	(1,60,3,1)	338	(1,15,3,1)	548	(1,18,3,1)	760	(1,187,3,1)	960	(1,765,3,1)	1162	(1,451,3,1)	1370	(1,35,3,1)	1576	(1,131,3,1)	1763	(1,266,3,1)	1952	(1,205,3,1)
125	(1,18,3,1)	339	(1,20,3,1)	549	(1,304,4,1)	763	(1,334,3,1)	962	(1,213,3,1)	1163	(1,162,3,1)	1371	(1,528,3,1)	1579	(1,706,3,1)	1765	(1,458,3,1)	1954	(1,71,4,1)
128	(1,61,3,1)	341	(1,278,3,1)	552	(1,195,3,1)	764	(1,40,3,1)	963	(1,38,3,1)	1165	(1,32,3,1)	1373	(1,154,3,1)	1581	(1,1446,3,1)	1766	(1,453,3,1)	1955	(1,1256,4,1)
131	(1,36,3,1)	344	(1,85,3,1)	554	(1,191,3,1)	765	(1,216,3,1)	965	(1,88,3,1)	1168	(1,629,3,1)	1376	(1,151,3,1)	1584	(1,621,3,1)	1768	(1,157,3,1)	1957	(1,1486,3,1)
133	(1,28,3,1)	347	(1,10,3,1)	555	(1,10,4,1)	766	(1,161,3,1)	968	(1,319,3,1)	1171	(1,466,3,1)	1378	(1,611,4,1)	1587	(1,250,7,1)	1771	(1,518,3,1)	1960	(1,31,3,1)
136	(1,11,3,1)	349	(1,58,3,1)	557	(1,318,3,1)	768	(1,117,3,1)	970	(1,415,3,1)	1172	(1,288,3,1)	1379	(1,306,3,1)	1589	(1,128,7,1)	1773	(1,1352,3,1)	1963	(1,1504,3,1)
138	(1,53,3,1)	352	(1,125,3,1)	560	(1,125,3,1)	770	(1,547,3,1)	971	(1,346,3,1)	1173	(1,108,3,1)	1381	(1,1276,3,1)	1592	(1,1135,3,1)	1776	(1,669,3,1)	1965	(1,732,3,1)
139	(1,8,5,1)	355	(1,164,3,1)	562	(1,25,3,1)	771	(1,138,3,1)	973	(1,56,3,1)	1176	(1,57,3,1)	1382	(1,270,3,1)	1594	(1,821,3,1)	1779	(1,1046,3,1)	1968	(1,9,3,1)
141	(1,47,3,1)	356	(1,10,3,1)	563	(1,60,3,1)	773	(1,138,3,1)	974	(1,30,3,1)	1179	(1,56,3,1)	1384	(1,221,3,1)	1595	(1,192,3,1)	1781	(1,36,3,1)	1970	(1,649,3,1)
143	(1,19,3,1)	357	(1,266,3,1)	565	(1,178,3,1)	776	(1,569,3,1)	976	(1,715,3,1)	1181	(1,650,4,1)	1387	(1,140,3,1)	1597	(1,1150,3,1)	1784	(1,1481,3,1)	1971	(1,6,3,1)
144	(1,39,3,1)	360	(1,5,3,1)	568	(1,373,3,1)	779	(1,338,3,1)	978	(1,311,3,1)	1184	(1,701,3,1)	1389	(1,204,3,1)	1598	(1,289,3,1)	1786	(1,1429,3,1)	1972	(1,820,3,1)
149	(1,40,3,1)	361	(1,25,3,1)	571	(1,10,4,1)	781	(1,278,3,1)	980	(1,369,3,1)	1187	(1,530,3,1)	1392	(1,1283,3,1)	1600	(1,529,3,1)	1787	(1,112,4,1)	1973	(1,1918,3,1)
152	(1,3,3,1)	363	(1,258,3,1)	572	(1,35,3,1)	784	(1,169,3,1)	981	(1,42,3,1)	1189	(1,136,3,1)	1394	(1,371,3,1)	1603	(1,134,3,1)	1789	(1,578,3,1)	1976	(1,361,3,1)
157	(1,50,3,1)	365	(1,294,3,1)	573	(1,6,3,1)	786	(1,387,3,1)	984	(1,917,5,1)	1192	(1,397,3,1)	1395	(1,380,3,1)	1605	(1,1328,5,1)	1792	(1,1001,4,1)	1978	(1,289,3,1)
158	(1,12,3,1)	368	(1,25,3,1)	576	(1,41,3,1)	787	(1,52,3,1)	987	(1,308,3,1)	1194	(1,153,3,1)	1397	(1,128,3,1)	1608	(1,273,3,1)	1794	(1,27,4,1)	1979	(1,1190,4,1)
160	(1,91,3,1)	371	(1,20,3,1)	578	(1,507,3,1)	788	(1,26,3,1)	989	(1,138,3,1)	1195	(1,962,3,1)	1400	(1,489,3,1)	1610	(1,561,3,1)	1795	(1,130,3,1)	1981	(1,1930,3,1)
163	(1,104,3,1)	373	(1,34,3,1)	579	(1,114,3,1)	789	(1,294,3,1)	992	(1,33,3,1)	1197	(1,12,3,1)	1403	(1,3,3,1)	1611	(1,506,3,1)	1796	(1,48,3,1)	1982	(1,260,3,1)
164	(1,18,3,1)	374	(1,38,3,1)	581	(1,318,3,1)	790	(1,269,3,1)	995	(1,558,5,1)	1200	(1,179,3,1)	1405	(1,658,3,1)	1613	(1,892,4,1)	1797	(1,8,3,1)	1984	(1,739,3,1)
165	(1,135,3,1)	376	(1,113,3,1)	584	(1,63,3,1)	792	(1,131,3,1)	997	(1,76,3,1)	1203	(1,788,3,1)	1406	(1,97,3,1)	1614	(1,138,3,1)	1800	(1,819,3,1)	1987	(1,1196,3,1)
168	(1,21,3,1)	379	(1,8,3,1)	586	(1,469,3,1)	795	(1,450,3,1)	1000	(1,101,3,1)	1205	(1,590,5,1)	1408	(1,817,3,1)	1616	(1,801,3,1)	1803	(1,1158,3,1)	1989	(1,232,5,1)
171	(1,60,3,1)	381	(1,198,4,1)	587	(1,256,3,1)	796	(1,26,3,1)	1002	(1,225,3,1)	1208	(1,319,3,1)	1411	(1,634,3,1)	1619	(1,578,3,1)	1805	(1,1312,3,1)	1992	(1,177,3,1)
173	(1,32,3,1)	384	(1,195,3,1)	589	(1,70,3,1)	797	(1,336,3,1)	1003	(1,716,3,1)	1211	(1,686,3,1)	1413	(1,880,4,1)	1621	(1,852,4,1)	1808	(1,545,3,1)	1995	(1,710,3,1)
176	(1,19,3,1)	387	(1,102,3,1)	591	(1,95,3,1)	800	(1,523,3,1)	1004	(1,106,3,1)	1213	(1,664,3,1)	1416	(1,461,3,1)	1622	(1,158,3,1)	1811	(1,618,3,1)	1997	(1,366,3,1)
179	(1,10,3,1)	389	(1,38,3,1)	592	(1,241,3,1)	802	(1,47,4,1)	1005	(1,692,3,1)	1216	(1,709,3,1)	1418	(1,149,3,1)	1624	(1,1375,3,1)	1812	(1,702,3,1)	1998	(1,155,3,1)
181	(1,175,3,1)	392	(1,47,3,1)	595	(1,364,3,1)	803	(1,60,3,1)	1006	(1,40,3,1)	1219	(1,676,4,1)	1419	(1,324,3,1)	1626	(1,573,3,1)	1813	(1,14,3,1)	2000	(1,333,3,1)
184	(1,83,3,1)	395	(1,126,3,1)	597	(1,540,3,1)	805	(1,248,3,1)	1008	(1,521,3,1)	1221	(1,200,3,1)	1421	(1,496,3,1)	1627	(1,130,3,1)	1816	(1,83,3,1)	2002	(1,5,3,1)
187	(1,22,3,1)	397	(1,208,3,1)	598	(1,248,3,1)	808	(1,379,3,1)	1011	(1,330,3,1)	1222	(1,181,3,1)	1424	(1,323,3,1)	1629	(1,500,3,1)	1819	(1,926,3,1)	2003	(1,710,3,1)
188	(1,21,3,1)	398	(1,104,3,1)	600	(1,87,3,1)	811	(1,394,3,1)	1013	(1,296,3,1)	1224	(1,545,3,1)	1427	(1,1110,3,1)	1632	(1,791,3,1)	1821	(1,668,3,1)	2005	(1,1074,4,1)
189	(1,63,3,1)	400	(1,283,3,1)	603	(1,453,3,1)	813	(1,98,3,1)	1016	(1,131,3,1)	1227	(1,212,3,1)	1429	(1,400,3,1)	1635	(1,545,3,1)	1822	(1,856,3,1)	2008	(1,445,3,1)
190	(1,16,3,1)	403	(1,124,3,1)	605	(1,248,3,1)	816	(1,285,3,1)	1017	(1,261,3,1)	1229	(1,40,3,1)	1432	(1,1241,3,1)	1637	(1,90,3,1)	1824	(1,1125,4,1)	2011	(1,922,3,1)
192	(1,155,3,1)	405	(1,68,5,1)	608	(1,73,3,1)	819	(1,496,4,1)	1018	(1,937,3,1)	1232	(1,477,3,1)	1435	(1,484,3,1)	1640	(1,401,3,1)	1826	(1,335,3,1)	2012	(1,40,3,1)
195	(1,158,3,1)	408	(1,27,3,1)	611	(1,334,3,1)	821	(1,424,3,1)	1019	(1,3,3,1)	1235	(1,50,3,1)	1437	(1,380,3,1)	1643	(1,3,3,1)	1827	(1,1308,3,1)	2013	(1,108,3,1)
197	(1,126,3,1)	410	(1,117,3,1)	613	(1,368,4,1)	824	(1,233,3,1)	1021	(1,832,3,1)	1237	(1,380,3,1)	1439	(1,300,3,1)	1644	(1,221,3,1)	1829	(1,760,3,1)	2014	(1,170,3,1)
200	(1,53,3,1)	411	(1,108,7,1)	616	(1,597,4,1)	827	(1,680,3,1)	1024	(1,643,3,1)	1240	(1,826,3,1)	1440	(1,901,4,1)	1645	(1,176,3,1)	1832	(1,295,3,1)	2016	(1,645,3,1)
203	(1,3,3,1)	413	(1,36,3,1)	619	(1,118,3,1)	829	(1,560,3,1)	1027	(1,830,3,1)	1243	(1,34,3,1)	1443	(1,348,3,1)	1646	(1,68,3,1)	1834	(1,1820,3,1)	2018	(1,173,3,1)
205	(1,176,3,1)	416	(1,87,3,1)	621	(1,344,3,1)	830	(1,60,3,1)	1032	(1,541,4,1)	1244	(1,46,3,1)	1445	(1,68,3,1)	1648	(1,1249,3,1)	1835	(1,1352,3,1)	2019	(1,1506,3,1)
206	(1,7,3,1)	419	(1,40,5,1)	624	(1,609,3,1)	832	(1,685,3,1)	1035	(1,208,5,1)	1245	(1,360,3,1)	1448	(1,93,3,1)	1651	(1,490,3,1)	1837	(1,1162,4,1)	2021	(1,102,3,1)
208	(1,197,4,1)	421	(1,20,7,1)	627	(1,377,3,1)	835	(1,92,3,1)	1037	(1,758,3,1)	1248	(1,327,3,1)	1450	(1,1183,3,1)	1653	(1,834,3,1)	1840	(1,409,3,1)	2024	(1,1949,4,1)
211	(1,112,3,1)	424	(1,359,3,1)	629	(1,268,3,1)	836	(1,8,3,1)	1038	(1,708,3,1)	1250	(1,313,3,1)	1451	(1,22,3,1)	1654	(1,1000,3,1)	1842	(1,303,3,1)	2027	(1,1980,3,1)
213	(1,24,3,1)	427	(1,22,3,1)	630	(1,80,3,1)	837	(1,542,3,1)	1040	(1,311,3,1)	1251	(1,38,3,1)	1453	(1,146,3,1)	1656	(1,179,3,1)	1843	(1,904,3,1)	2029	(1,4,3,1)
216	(1,11,3,1)	429	(1,18,3,1)	632	(1,117,3,1)	840	(1,173,3,1)	1043	(1,304,3,1)	1253	(1,318,3,1)	1456	(1,29,3,1)	1658	(1,801,3,1)	1845	(1,786,3,1)	2030	(1,337,3,1)
219	(1,201,3,1)	430	(1,49,3,1)	635	(1,60,3,1)	843	(1,200,3,1)	1045	(1,298,3,1)	1254	(1,56,3,1)	1459	(1,496,3,1)	1659	(1,446,3,1)	1848	(1,1077,3,1)	2032	(1,1479,5,1)
221	(1,80,3,1)	432	(1,117,5,1)	637	(1,38,6,1)	848	(1,243,3,1)	1046	(1,38,3,1)	1256	(1,605,3,1)	1461	(1,122,3,1)	1661	(1,156,3,1)	1850	(1,1015,4,1)	2035	(1,442,3,1)
222	(1,71,3,1)	434	(1,119,3,1)	638	(1,3,3,1)	851	(1,58,3,1)	1048	(1,539,3,1)	1258	(1,689,3,1)	1462	(1,913,3,1)	1662	(1,285,3,1)	1851	(1,720,3,1)	2037	(1,210,3,1)
224	(1,83,3,1)	435	(1,122,3,1)	640	(1,35,3,1)	853	(1,142,3,1)	1051	(1,12,4,1)	1259	(1,168,3,1)	1464	(1,1155,3,1)	1664	(1,135,3,1)	1852	(1,46,3,1)	2038	(1,1498,3,1)
226	(1,169,3,1)	437	(1,318,3,1)	643	(1,412,4,1)	854	(1,138,3,1)	1053	(1,1044,3,1)	1261	(1,640,3,1)	1467	(1,782,3,1)	1667	(1,1146,3,1)	1853	(1,906,3,1)	2040	(1,77,3,1)
227	(1,182,3,1)	440	(1,297,3,1)	644	(1,236,3,1)	856	(1,691,3,1)	1056	(1,891,3,1)	1262	(1,285,3,1)	1469	(1,164,4,1)	1669	(1,106,3,1)	1856	(1,861,3,1)	2042	(1,417,3,1)
229	(1,80,3,1)	442	(1,187,4,1)	645	(1,50,3,1)	859	(1,178,3,1)	1059	(1,260,3,1)	1264	(1,283,3,1)	1472	(1,83,3,1)	1670	(1,472,3,1)	1858	(1,1099,3,1)	2043	(1,398,3,1)
230	(1,3,3,1)	443	(1,48,3,1)	648	(1,626,3,1)	863	(1,92,3,1)	1061	(1,268,3,1)	1267	(1,800,4,1)	1474	(1,829,3,1)	1672	(1,881,4,1)	1859	(1,4,3,1)	2045	(1,1554,3,1)
232	(1,133,3,1)	445	(1,88,3,1)	653	(1,478,3,1)	864	(1,39,3,1)	1064	(1,351,3,1)	1269	(1,907,4,1)	1475	(1,68,3,1)	1675	(1,970,4,1)	1861	(1,398,3,1)	2046	(1,240,3,1)
235	(1,26,3,1)	448	(1,119,3,1)	656	(1,295,3,1)	867	(1,48,3,1)	1066	(1,401,3,1)	1272	(1,383,3,1)	1477	(1,140,3,1)	1677	(1,852,3,1)	1864	(1,61,3,1)	2048	(1,83,3,1)

References

1. Albrecht, M.R., Driessen, B., Kavun, E.B., Leander, G., Paar, C., Yalçın, T.: Block ciphers – focus on the linear layer (feat. PRIDE). In: Garay, J.A., Gennaro, R. (eds.) CRYPTO 2014, Part I. LNCS, vol. 8616, pp. 57–76. Springer, Heidelberg (2014)

2. Augot, D., Finiasz, M.: Direct construction of recursive MDS diffusion layers using shortened BCH codes. In: Cid, C., Rechberger, C. (eds.) FSE 2014. LNCS, vol. 8540, pp. 3–17. Springer, Heidelberg (2015)

3. Barreto, P., Nikov, V., Nikova, S., Rijmen, V., Tischhauser, E.: Whirlwind: a new cryptographic hash function. Des. Codes Crypt. 56(2–3), 141–162 (2010)

4. Bertoni, G., Daemen, J., Peeters, M., Assche, G.: The Keccak reference. Submission to NIST (Round 3) (2011)

5. Biham, E., Anderson, R., Knudsen, L.R.: Serpent: a new block cipher proposal. In: Vaudenay, S. (ed.) FSE 1998. LNCS, vol. 1372, p. 222. Springer, Heidelberg (1998)

6. Biham, E., Shamir, A.: Differential cryptanalysis of DES-like cryptosystems. In: Menezes, A., Vanstone, S.A. (eds.) CRYPTO 1990. LNCS, vol. 537, pp. 2–21. Springer, Heidelberg (1991)

7. Daemen, J.: Cipher and hash function design strategies based on linear and differential cryptanalysis. Ph.D. thesis, Doctoral Dissertation, KU Leuven, March 1995

8. Daemen, J., Knudsen, L.R., Rijmen, V.: The block cipher SQUARE. In: Biham, E. (ed.) FSE 1997. LNCS, vol. 1267, pp. 149–165. Springer, Heidelberg (1997)

9. Daemen, J., Rijmen, V.: AES Proposal: Rijndael (1998). http://csrc.nist.gov/archive/aes/rijndael/Rijndael-ammended.pdf

10. Daemen, J., Rijmen, V.: Correlation analysis in $GF(2^n)$. In: Advanced Linear Cryptanalysis of Block and Stream Ciphers. Cryptology and Information Security, pp. 115–131 (2011)

11. Dummit, D.S., Foote, R.M.: Abstract Algebra. Wiley, Hoboken (2004)

12. Grosso, V., Leurent, G., Standaert, F.-X., Varici, K.: LS-designs: bitslice encryption for efficient masked software implementations. In: Cid, C., Rechberger, C. (eds.) FSE 2014. LNCS, vol. 8540, pp. 18–37. Springer, Heidelberg (2015)

13. Guo, J., Peyrin, T., Poschmann, A.: The PHOTON family of lightweight hash functions. In: Rogaway, P. (ed.) CRYPTO 2011. LNCS, vol. 6841, pp. 222–239. Springer, Heidelberg (2011)

14. Guo, J., Peyrin, T., Poschmann, A., Robshaw, M.: The LED block cipher. In: Preneel, B., Takagi, T. (eds.) CHES 2011. LNCS, vol. 6917, pp. 326–341. Springer, Heidelberg (2011)

15. Gupta, K.C., Ray, I.G.: Cryptographically significant MDS matrices based on circulant and circulant-like matrices for lightweight applications. Crypt. Commun. 7(2), 257–287 (2015)

16. Jean, J., Peyrin, T., Sim, S.M.: Minimal implementations of linear and non-linear lightweight building blocks. Personal communication (2015)

17. Khoo, K., Peyrin, T., Poschmann, A.Y., Yap, H.: FOAM: searching for hardware-optimal SPN structures and components with a fair comparison. In: Batina, L., Robshaw, M. (eds.) CHES 2014. LNCS, vol. 8731, pp. 433–450. Springer, Heidelberg (2014)

18. Knapp, A.W.: Basic Algebra. Birkhäuser, Boston (2006)

19. Li, Y., Wang, M.: On the construction of lightweight circulant involutory MDS matrices. In: Fast Software Encryption (FSE), LNCS. Springer, Heidelberg (2016, to appear)

20. Lidl, R., Niederreiter, H.: Introduction to Finite Fields and Their Applications. Cambridge University Press, Cambridge (1994)
21. Liu, M., Sim, S.M.: Lightweight MDS generalized circulant matrices. In: Fast Software Encryption (FSE). LNCS. Springer, Heidelberg (2016, to appear)
22. MacWilliams, F.J., Sloane, N.J.A.: The Theory of Error-Correcting Codes. North-Holland Publishing Company, Amsterdam (1977)
23. Matsui, M.: Linear cryptanalysis method for DES cipher. In: Helleseth, T. (ed.) EUROCRYPT 1993. LNCS, vol. 765, pp. 386–397. Springer, Heidelberg (1994)
24. Sajadieh, M., Dakhilalian, M., Mala, H., Sepehrdad, P.: Recursive diffusion layers for block ciphers and hash functions. In: Canteaut, A. (ed.) FSE 2012. LNCS, vol. 7549, pp. 385–401. Springer, Heidelberg (2012)
25. Sarkar, S., Sim, S.M.: A deeper understanding of the XOR count distribution in the context of lightweight cryptography. In: Pointcheval, D., et al. (eds.) AFRICACRYPT 2016. LNCS, vol. 9646, pp. 167–182. Springer, Heidelberg (2016). doi:10.1007/978-3-319-31517-1_9
26. Silvester, J.R.: Determinants of block matrices. Math. Gaz. **84**(501), 460–467 (2000)
27. Sim, S.M., Khoo, K., Oggier, F., Peyrin, T.: Lightweight MDS involution matrices. In: Leander, G. (ed.) FSE 2015. LNCS, vol. 9054, pp. 471–493. Springer, Heidelberg (2015)
28. Swan, R.G.: Factorization of polynomials over finite fields. Pacific J. Math. **12**(3), 1099–1106 (1962)
29. Wardlaw, W.P.: Matrix representation of finite fields. Math. Mag. **67**(4), 289–293 (1994)
30. Wu, S., Wang, M., Wu, W.: Recursive diffusion layers for (lightweight) block ciphers and hash functions. In: Knudsen, L.R., Wu, H. (eds.) SAC 2012. LNCS, vol. 7707, pp. 355–371. Springer, Heidelberg (2013)
31. Xu, H., Zheng, Y., Lai, X.: Construction of perfect diffusion layers from linear feedback shift registers. IET Inf. Secur. **9**(2), 127–135 (2015)

Another View of the Division Property

Christina Boura[1]([⊠]) and Anne Canteaut[2]

[1] University of Versailles, Versailles, France
Christina.Boura@uvsq.fr
[2] Inria, Paris, France
Anne.Canteaut@inria.fr

Abstract. A new distinguishing property against block ciphers, called the division property, was introduced by Todo at Eurocrypt 2015. Our work gives a new approach to it by the introduction of the notion of parity sets. First of all, this new notion permits us to formulate and characterize in a simple way the division property of any order. At a second step, we are interested in the way of building distinguishers on a block cipher by considering some further properties of parity sets, generalising the division property. We detail in particular this approach for substitution-permutation networks. To illustrate our method, we provide low-data distinguishers against reduced-round PRESENT. These distinguishers reach a much higher number of rounds than generic distinguishers based on the division property and demonstrate, amongst others, how the distinguishers can be improved when the properties of the linear and the Sbox layers are taken into account. At last, this work provides an analysis of the resistance of Sboxes against this type of attacks, demonstrates links with the algebraic normal form of an Sbox as well as its inverse Sbox and exhibit design criteria for Sboxes to resist such attacks.

Keywords: Division property · Integral attacks · Sboxes · PRESENT

1 Introduction

A new distinguishing property against block ciphers, called the division property, was recently introduced by Todo [25]. This property, that can be seen as a generalization of integral [9,16] and higher-order differential [15,17] distinguishers, was used to present new generic distinguishers against both the SPN and the Feistel constructions. Later, this attack was used by the same author to present the first cryptanalysis of the full block cipher MISTY [24].

If $u = (u_1, \ldots, u_n)$ is a vector of $\mathbf{F}_2^n$, we denote by x^u the coordinate product $x = (x_1, \ldots, x_n) \mapsto \prod_{i=1}^n x_i^{u_i}$. The division property, as introduced by Todo [25], is interested in the sum of this quantity taken over all vectors of X. More precisely, we say that a set $X \subseteq \mathbf{F}_2^n$ has the division property $\mathcal{D}_k^n$, for some

---

Partially supported by the French Agence Nationale de la Recherche through the BRUTUS project under Contract ANR-14-CE28-0015.

M. Robshaw and J. Katz (Eds.): CRYPTO 2016, Part I, LNCS 9814, pp. 654–682, 2016.
DOI: 10.1007/978-3-662-53018-4_24

$1 \leq k \leq n$, if the sum over all vectors x in X of the product x^u equals 0, for all vectors u that have a Hamming weight strictly less than k, i.e.

$$\bigoplus_{x \in X} x^u = 0 \text{ for all } u \in \mathbf{F}_2^n \text{ such that } wt(u) < k.$$

The division property then generalizes integral attacks in the sense that $\mathcal{D}_2^n$ means that the set X is balanced, while $\mathcal{D}_n^n$ means that it is saturated. But the novelty is that it introduces intermediate properties, $\mathcal{D}_k^n$ for $3 \leq k \leq n-1$, which do not appear in classical integral attacks. Even if these intermediate properties do not have a simple interpretation like $\mathcal{D}_2^n$ and $\mathcal{D}_n^n$, they allow to easily propagate the property through the successive rounds of a cipher by capturing some information resulting from the algebraic degree of the round function. In a nutshell, the distinguishers described by Todo in [24,25] are classical higher-order differential distinguishers, but they are exhibited by exploiting the classical properties used in integral attacks together with some algebraic properties related to the degree of several iterations of a nonlinear function like in [6,8].

Our Contribution. This work aims at providing new insights into the division property, presenting a new approach to it. This new approach enables us to provide a simpler formulation and interpretation of the division property of any order. It also improves the strength of the distinguishers that exploit this type of properties. For this, we introduce a new notion, that we call the *parity set*. The parity set of a set $X \subseteq \mathbf{F}_2^n$ is nothing more than the set of all exponents $u \in \mathbf{F}_2^n$ such that $\bigoplus_{x \in X} x^u = 1$. The main advantage of this new notion is that it completely characterises a set X in the sense that there is a one-to-one correspondence between sets and their parity sets. It also provides a very simple formulation of the division property. In particular, we show that the division property of any order can be expressed in an elegant way by using the theory of Reed-Muller codes. One of the first questions we investigate in this work is what does it mean for a set X to have the division property $\mathcal{D}_k^n$ for some special values of k. As previously explained, this question was treated for $k \in \{1, 2, n\}$ in previous works [14,22,25]. However, our approach, and especially the link with the Reed-Muller codes, permits us to recover in a much simpler way these previous results and to characterize the property for some other values of k.

We investigate next the question of how to build distinguishers for keyed permutations by means of parity sets. For this, we start by analyzing the distinguishers built by Todo in [25] and formulate them in terms of parity sets. These distinguishers were of a generic nature as they only exploited the classical integral properties and the propagation of the algebraic degree through the successive non-linear layers. It is thus natural to believe that these distinguishers can be improved if additional information besides the degree is taken into account, in the same spirit as in cube distinguishers [1,12]. We investigate this issue here and provide a way to exploit these more precise properties. We then further show how to find distinguishers on iterated block ciphers, especially on substitution-permutation networks, by propagating some information on the parity set of the output set through the successive rounds of the cipher. For this, we provide a detailed analysis of the evolution of the parity set through the basic operations of the round function of an SPN cipher.

We illustrate the above technique by constructing low-data distinguishers on the PRESENT block cipher. With this example we aim at particularly showing how the generic distinguishers can be improved when the properties of the linear and the Sbox layer are taken into account. We manage to provide a distinguisher on 6 rounds of PRESENT with data complexity 2^{12}, while the generic distinguisher from [25], reaches only 3 rounds for the same quantity of data. Finally, we analyze the resistance of Sboxes against this type of attack, show a link with the algebraic normal forms of the Sbox and of its inverse, and we give a criterion that an Sbox must satisfy in order to resist this kind of attacks.

Organization of the Paper. The rest of the paper is organized as follows. Section 2 introduces the notion of the parity set of a set and shows how it is related to Reed-Muller codes. Section 3 presents the link between this new notion and the division property, and it characterizes the division property of any order. It also focuses on the division property of low and high orders. Section 4 explains how to build distinguishers by means of parity sets and Sect. 5 analyzes the special case of SPN ciphers. In Sect. 6, low-data distinguishers for the block cipher PRESENT are presented. Finally, Sect. 7 discusses the properties that an Sbox must exhibit in order to resist the above attacks.

2   Parity Set of a Set

2.1   Preliminaries

A Boolean function of n variables can be alternatively represented as a multivariate polynomial, named the Algebraic Normal Form (aka ANF) of the function, or as a 2^n-bit vector (named the value vector) corresponding to all $f(x), x \in \mathbf{F}_2^n$.

Polynomial Representation. We use the following notation for the monomials of n variables where u is an element of $\mathbf{F}_2^n$:

$$x^u = \prod_{i=1}^{n} x_i^{u_i}.$$

The following well-known lemma will be extensively used for evaluating a monomial at a given point.

Lemma 1. *Let x and u be two n-bit words. Then $x^u = 1$ if and only if $u \preceq x$, i.e., $u_i \leq x_i$ for all $1 \leq i \leq n$.*

The previous relation between n-bit words is a partial order. It equivalently means that the support of u is included in the support of x. In the whole paper, we will use the following notation for the set of all words less than (resp. greater than) a given word with respect to this partial order.

Notation 1. *Let $u \in \mathbf{F}_2^n$. Then, we define*

$$\mathsf{Prec}(u) = \{x \in \mathbf{F}_2^n : x \preceq u\}$$
$$\mathsf{Succ}(u) = \{x \in \mathbf{F}_2^n : u \preceq x\}.$$

It is worth noticing that $\mathsf{Prec}(u)$ is a linear subspace of dimension $wt(u)$, while $\mathsf{Succ}(u)$ is an affine subspace of dimension $(n - wt(u))$.

Value Vector. When a Boolean function is represented by its value vector, it is often convenient to use the terminology coming from coding theory since Boolean functions have been widely studied for error-correction. In this context, the value vector of a function is seen as a codeword from a Reed-Muller code [20,21].

Definition 1 (Reed-Muller Codes). *Let n be a positive integer and r an integer such $0 \leq r \leq n$. The r-th order binary Reed-Muller code of length 2^n, denoted by $\mathcal{R}(r, n)$, is the set formed by the value vectors of all Boolean functions of n variables with degree at most r:*

$$\mathcal{R}(r, n) = \{(f(x), x \in \mathbf{F}_2^n), \ f : \mathbf{F}_2^n \to \mathbf{F}_2 \ with \ \deg f \leq r\}.$$

2.2   Parity Set of a Set

We now define a new notion named *parity set*. We will show that any set is characterized by its parity set. Any property of a set can then also be expressed in terms of its parity set, and we will show that the division property has a very simple expression by means of parity sets.

Definition 2. *Let X be a set of elements in $\mathbf{F}_2^n$. The parity set of X, denoted by $\mathcal{U}(X)$, is the subset of $\mathbf{F}_2^n$ defined by*

$$\mathcal{U}(X) = \{u \in \mathbf{F}_2^n : \bigoplus_{x \in X} x^u = 1\}.$$

The parity set provides a complete characterization of a set, as shown by the following results.

Lemma 2. *Let G be the $2^n \times 2^n$ binary matrix whose entries are indexed by n-bit vectors and defined by*

$$G_{u,a} = a^u, \ a, u \in \mathbf{F}_2^n.$$

For any subset X of $\mathbf{F}_2^n$, the incidence vector of $\mathcal{U}(X)$ is equal to the product of G by the incidence vector of X.

Proof. The incidence vector of a set X, v_X, is the 2^n-bit vector having a one at position $x \in \mathbf{F}_2^n$ if and only if $x \in X$. Then, Gv_X is equal to the sum of all columns of G indexed by the elements in the support of v_X, i.e., indexed by the elements in X:

$$(Gv_X)_u = \bigoplus_{x \in X} x^u.$$

By definition, the support of $v_{\mathcal{U}(X)}$ is then the set of all positions u such that $(Gv_X)_u = 1$. $\square$

We can now deduce that there is a one-to-one correspondence between sets and their parity sets.

Theorem 1. *Let G be the $2^n \times 2^n$ binary matrix defined by*

$$G_{u,a} = a^u, \ a, u \in \mathbf{F}_2^n.$$

Then, G is non-singular and $G^{-1} = G$. Therefore, for any subset U of $\mathbf{F}_2^n$, there exists a unique set $X \subset \mathbf{F}_2^n$ such that $\mathcal{U}(X) = U$.

Proof. The fact that G is non-singular can be deduced by using that it is a generator matrix of the Reed-Muller code of length 2^n and order n. This code has dimension 2^n [19, p. 376], i.e., G is invertible. Its inverse is equal to G itself. Indeed, for any $u, w \in \mathbf{F}_2^n$, we have

$$
\begin{aligned}
(G \times G)_{u,w} &= \bigoplus_{v \in \mathbf{F}_2^n} G_{u,v} G_{v,w} = \bigoplus_{v \in \mathbf{F}_2^n} v^u w^v \\
&= |\{v \in \mathbf{F}_2^n : u \preceq v \text{ and } v \preceq w\}| \bmod 2 \\
&= \begin{cases} 2^{wt(w)-wt(u)} \bmod 2 \text{ if } u \preceq w \\ 0 \text{ otherwise.} \end{cases}
\end{aligned}
$$

We then deduce that $(G \times G)_{u,w} = 1$ if and only if $u = w$, i.e., $G \times G = \mathsf{Id}$. As a direct consequence, we get that the mapping $v_X \mapsto v_{\mathcal{U}(X)}$ is an isomorphism of the set of 2^n-bit vectors. □

The fact that G is involutive provides a simple way to find the set X corresponding to a given parity set U. Indeed, X corresponds to the parity set of U. Some useful examples are described in the following corollary.

Corollary 1. *Let X be a subset of $\mathbf{F}_2^n$. Then,*

- $\mathcal{U}(X)$ *is empty if and only if X is empty.*
- $\mathcal{U}(X) = \mathsf{Prec}(x)$ *if and only if $X = \{x\}$.*
- $\mathcal{U}(X) = \{u\}$ *if and only if $X = \mathsf{Prec}(u)$.*
- $\mathcal{U}(X) = \{\underline{1}\}$ *if and only if $X = \mathbf{F}_2^n$,*

where $\underline{1}$ denotes the all-one vector in $\mathbf{F}_2^n$.

3   New Insights into the Division Property

3.1   The Division Property by Means of Parity Sets

The division property introduced by Todo in [25] is a distinguishing property of the set $E_k(X)$ for a given choice of the input set X, where E_k is (typically) a keyed permutation. This property must be independent from the choice of the secret key. We now reformulate the division property of order k, $\mathcal{D}_k^n$, on a set X by a simple property of $\mathcal{U}(X)$. Indeed, $\mathcal{D}_k^n$ corresponds to a lower bound on the weights of all elements in $\mathcal{U}(X)$.

Definition 3. *A set X of elements in $\mathbf{F}_2^n$ is said to fulfill the division property of order k, $\mathcal{D}_k^n$, if all elements in $\mathcal{U}(X)$ have weight at least k, i.e.,*

$$\mathcal{U}(X) \subseteq \{u \in \mathbf{F}_2^n : wt(u) \geq k\}.$$

It is worth noticing that in [25], the division property is defined for a multiset, i.e., the elements in X may appear with some multiplicity. However, the original division property for a multiset X equivalently corresponds to the division property for the set composed of all elements in X having an odd multiplicity. Therefore, we will only focus on sets, instead of multisets.

As a direct consequence of the matrix relationship exhibited in Lemma 2, we deduce the following two characterizations of the incidence vectors of the sets satisfying the division property of order k.

Proposition 1. *Let X be a set of elements in $\mathbf{F}_2^n$ and k be an integer $1 \leq k \leq n$. Then, the following assertions are equivalent:*

(i) *X fulfills the division property of order k, $\mathcal{D}_k^n$.*
(ii) *The incidence vector of X belongs to the Reed-Muller code of length 2^n and order $(n - k)$.*
(iii) *The incidence vector of X belongs to the dual of the Reed-Muller code of length 2^n and order $(k - 1)$.*

Proof. Assertion (ii) equivalently means that the incidence vector of $\mathcal{U}(X)$ vanishes at all positions u with $wt(u) \leq k - 1$. This means that, if G' denotes the restriction of G to the rows of index u with $wt(u) < k$, then $G'v_X$ is the all-zero vector. But G' is a generator matrix of the Reed-Muller code of length 2^n and order $(k - 1)$. The set of all v_X such that $G'v_X = 0$ is therefore the dual (i.e., the orthogonal) of $\mathcal{R}(k - 1, n)$. It is well-known (see e.g. [19, p. 375]) that, for any r, the dual of $\mathcal{R}(r, n)$ is the Reed-Muller code $\mathcal{R}(n - r - 1, n)$. We deduce that $G'v_X = 0$ if and only if $v_X \in \mathcal{R}(n - k, n)$. $\square$

The first one of the previous characterization, (ii), has been independently exhibited by Khovratovich [14], while the equivalent formulation (iii) is new.

Using the minimum distance of the Reed-Muller codes, we recover very easily a result from [22] on the minimal size of a set satisfying $\mathcal{D}_k^n$. More importantly, we are able to characterize the sets of minimal size satisfying $\mathcal{D}_k^n$.

Proposition 2. *Let X be a non-empty set of elements in $\mathbf{F}_2^n$ satisfying $\mathcal{D}_k^n$. Then*

$$|X| \geq 2^k.$$

Moreover, a set X of size 2^k satisfies $\mathcal{D}_k^n$ if and only if X is an affine subspace[1] of dimension k.

---

[1] In the whole paper, the terminology *affine subspace* includes any linear subspace or any coset of a linear subspace.

Proof. We here use that X satisfies $\mathcal{D}_k^n$ if and only if v_X belongs to $\mathcal{R}(n-k,n)$. It is well-known that the minimum distance of $\mathcal{R}(n-k,n)$ is 2^k [19, p. 375]. Using that $|X| = wt(v_X)$, we deduce that $|X| \geq 2^k$.

Moreover, it is known that the minimum-weight codewords in $\mathcal{R}(n-k,n)$ are the incidence vectors of the affine subspaces of dimension k [19, p. 380]. It follows that a set of size 2^k satisfies $\mathcal{D}_k^n$ if and only if it is an affine subspace of dimension k. □

3.2 Division Property of Low Order

Since the codewords of Reed-Muller codes of low order have a very simple form, a simple characterization of the division properties of low order directly follows. The Reed-Muller code $\mathcal{R}(0,n)$ consists of the all-zero and all-one words, $\mathcal{R}(0,n) = \{\underline{0}, \underline{1}\}$, and $\mathcal{R}(1,n) \setminus \mathcal{R}(0,n)$ is composed of all incidence vectors of affine hyperplanes. Then, we easily recover the characterization of the division properties of order 1 and 2 exhibited in [25]:

- X fulfills $\mathcal{D}_1^n$ if and only if its cardinality is even.
- X fulfills $\mathcal{D}_2^n$ if and only if its cardinality is even and it has the *Balance property* [16], i.e., $\bigoplus_{x \in X} x = 0$.

Then, the division property of order $k > 2$ generalizes the balance property used in integral attacks [16] in the following sense.

Proposition 3. *Let X be a set of elements in $\mathbf{F}_2^n$. Then, the following assertions are equivalent:*

(i) *X fulfills the division property of order k, $\mathcal{D}_k^n$.*
(ii) *For any set of coordinates $\{i_1, \ldots, i_t\} \subseteq \{1, \ldots, n\}$ of size $t < k$ and any constant $\alpha \in \mathbf{F}_2^t$, the number of elements in X such that $x_{i_j} = \alpha_j$ for all $1 \leq j \leq t$ is even.*
(iii) *For any set of coordinates $\{i_1, \ldots, i_t\} \subseteq \{1, \ldots, n\}$ of size $t < k$, the number of elements in X such that $x_{i_j} = 0$ for all $1 \leq j \leq t$ is even.*

Proof. **(i)** $\Rightarrow$ **(ii)** Let $I = \{i_1, \ldots, i_t\}$ be a set of coordinates of size $t < k$, and u be the vector in $\mathbf{F}_2^n$ having support I. Then,

$$\{x \in X : x_{i_j} = \alpha_j, 1 \leq j \leq t\} = \{x \in X : (x \oplus \beta) \succeq u\}$$

where β is the n-bit vector such that $\beta_{i_j} = \alpha_j \oplus 1$ for $1 \leq j \leq t$, and $\beta_i = 0$ if $i \notin I$. It follows that

$$|\{x \in X : x_{i_j} = \alpha_j, 1 \leq j \leq t\}| \bmod 2 = \bigoplus_{x \in X} (x \oplus \beta)^u = \bigoplus_{x \in X} \bigoplus_{v \preceq u} x^v \beta^{u \oplus v}$$

$$= \bigoplus_{v \preceq u} \beta^{u \oplus v} \left(\bigoplus_{x \in X} x^v \right) = 0$$

since the division property of order k implies that all $\bigoplus_{x \in X} x^v$ vanish when $wt(v) \leq wt(u) < k$.

(ii) ⇒ (iii) Trivial.

(iii) ⇒ (i) Let $u \in \mathbf{F}_2^n$ with $wt(u) < k$. We have

$$\bigoplus_{x \in X} x^u = \bigoplus_{x \in X} ((x \oplus u) \oplus u)^u = \bigoplus_{v \preceq u} \bigoplus_{x \in X} (x \oplus u)^v u^{u \oplus v}$$

$$= \bigoplus_{v \preceq u} \bigoplus_{x \in X} (x \oplus u)^v = \bigoplus_{v \preceq u} |\{x \in X : x_i = 0, \forall i \in \mathsf{Supp}(v)\}| \bmod 2.$$

From (iii), all sets involved in the previous sum have even size because $wt(v) < k$. We then deduce that $\bigoplus_{x \in X} x^u = 0$, i.e. X fulfills the division property of order k.

□

It is worth noticing that, more generally, the division property $\mathcal{D}_k^n$ implies that the number of elements in $X \cap A$ is even for any affine subspace A of dimension $t > n - k$ (see e.g. [7, Proposition III.1]).

As a direct corollary, we can for instance characterize the division property of order 3.

Corollary 2. *Let X be a set of elements in $\mathbf{F}_2^n$. Then, X fulfills the division property of order 3, $\mathcal{D}_3^n$, if and only if X and all the n subsets*

$$\{x \in X \text{ with } x_i = 0\}, \ 1 \le i \le n,$$

satisfy the balance property.

Example 1. The following set $X \in \mathbf{F}_2^5$ composed of 12 elements satisfy the division property of order 3:

x_1	0	0	0	0	1	1	1	1	1	1	1	1
x_2	0	1	0	1	0	1	0	1	1	0	1	0
x_3	1	0	0	1	1	0	0	1	1	0	1	0
x_4	0	1	0	1	0	1	0	0	0	1	1	1
x_5	1	1	1	1	1	1	0	1	0	0	0	1

3.3   Division Property of High Order

The division property of maximal order, i.e., $\mathcal{D}_n^n$, obviously corresponds to the fact that X is either empty, or equal to the whole set $\mathbf{F}_2^n$ (see the last item in Corollary 1). But, we are also able to characterize all sets satisfying $\mathcal{D}_{n-1}^n$.

Proposition 4. *Let X be a set of elements in $\mathbf{F}_2^n$. Then X fulfills $\mathcal{D}_{n-1}^n$ and not $\mathcal{D}_n^n$ if and only if X is an (affine) hyperplane of $\mathbf{F}_2^n$.*

Proof. Let v denote the incidence vector of X. From the previous proposition, we have X satisfies $\mathcal{D}_{n-1}^n$ and not $\mathcal{D}_n^n$ if and only if $v \in \mathcal{R}(1,n) \setminus \mathcal{R}(0,n)$. This set consists of the incidence vectors of all (affine) hyperplanes of $\mathbf{F}_2$. Then, this equivalently means that X is an (affine) hyperplane.   □

For instance, it can be easily checked that the multiset of elements of $\mathbf{F}_2^4$ defined in [25, p. 293]

$$\{\mathtt{0x0}, \mathtt{0x3}, \mathtt{0x3}, \mathtt{0x3}, \mathtt{0x5}, \mathtt{0x6}, \mathtt{0x8}, \mathtt{0xb}, \mathtt{0xd}, \mathtt{0xe}\},$$

satisfies $\mathcal{D}_3^4$ because the corresponding set composed of all elements with an odd multiplicity

$$\{\mathtt{0x0}, \mathtt{0x3}, \mathtt{0x5}, \mathtt{0x6}, \mathtt{0x8}, \mathtt{0xb}, \mathtt{0xd}, \mathtt{0xe}\}$$

is a linear subspace of dimension 3 spanned by $\{\mathtt{0x3}, \mathtt{0x5}, \mathtt{0x8}\}$.

4   Distinguishers Based on Parity Sets

We now investigate how we can build some distinguishers for a given keyed permutation E_K by means of parity sets. The basic idea consists in choosing an input set X such that the parity set of the corresponding output set $E_K(X)$ has some specific property for any choice of the key.

Since the size of X determines the data complexity of the distinguisher, it has to be as small as possible. For this reason, X is always chosen to be an affine subspace since subspaces are the smallest possible sets satisfying the division property of a given order (see Proposition 2).

4.1   Todo's Distinguishers

The strategy proposed by Todo to build a distinguisher is to exhibit an affine subspace $a + V$ such that the corresponding output set $E_K(a + V)$ satisfies the division property of order 2, i.e., such that $E_K(a+V)$ is balanced. This property can be easily interpreted in terms of higher-order derivatives in the sense of the following definition.

Definition 4. [17] *Let F be a function from $\mathbf{F}_2^n$ into $\mathbf{F}_2^m$. Let $a \in \mathbf{F}_2^n$. The derivative of F with respect to a is the function from $\mathbf{F}_2^n$ into $\mathbf{F}_2^m$ defined by*

$$D_a F(x) = F(x \oplus a) \oplus F(x).$$

For any k-dimensional subspace V of $\mathbf{F}_2^n$ and for any basis of V, $\{a_1, \ldots, a_k\}$, the k-th order derivative of F with respect to V is the function defined by

$$D_V F(x) = D_{a_1} D_{a_2} \ldots D_{a_k} F(x) = \bigoplus_{v \in V} F(x + v)$$

We introduce now the following notation.

Notation 2. *Let P be a permutation of $\mathbf{F}_2^n$ and $P_1, P_2, \ldots, P_n$ be the n coordinates of P. If $x = (x_1, \ldots, x_n)$ and $u = (u_1, \ldots, u_n)$ are vectors of $\mathbf{F}_2^n$, we denote by $P^u(x)$ the coordinate product $\prod_{i=1}^{n} P_i(x)^{u_i}$.*

Proposition 5. *Let P be a permutation of $\mathbf{F}_2^n$. Let V be the linear subspace of $\mathbf{F}_2^n$ and $a \in \mathbf{F}_2^n$. Then, an element u belongs to $\mathcal{U}(P(a + V))$ if and only if the derivative of P^u with respect to V satisfies*

$$D_V P^u(a) = 1.$$

In the particular case where $V = \mathsf{Prec}(v)$ for some $v \in \mathbf{F}_2^n$, the following formulations are equivalent:

(i) *For all $a \in \mathbf{F}_2^n$, $u \notin \mathcal{U}(P(a + V))$*
(ii) *The algebraic normal form of the Boolean function $x \mapsto P^u(x)$ contains no monomial multiple of x^v*
(iii) *The superpoly of v in $x \mapsto P^u(x)$ vanishes.*

Proof. The fact that $D_V P^u(a) = 1$ if and only if $u \in \mathcal{U}(P(a + V))$ is directly deduced from the definition of the derivative with respect to V using that

$$D_V P^u(a) = \bigoplus_{x \in a + V} P^u(x).$$

The superpoly of v in P^u is defined [12] as the Boolean function p_v such that

$$P^u(x) = x^v p_v(x) + q(x)$$

where q does not contain any monomial multiple of x^v. Moreover, it has been proved in [12, Theorem 1] that

$$p_v(a) = \bigoplus_{x \preceq v} P^u(a \oplus x) = D_V P^u(a).$$

It then follows that, for $V = \mathsf{Prec}(v)$, $u \notin \mathcal{U}(P(a + V))$ for all a if and only if the superpoly of v in P^u vanishes, which equivalently means that P^u does not contain any monomial multiple of x^v. □

4.2   Improving Todo's Distinguishers

The distinguishers presented in [25] correspond to the existence of a word v such that $E_K(a + \mathsf{Prec}(v))$ satisfies the division property of order 2 for all a, which equivalently means that the monomial x^v does not appear in any coordinate of E_K. Since these distinguishers are constructed by propagating some information on the smallest Hamming weight of the elements in the parity set, they are based on the fact that the weight of v exceeds the degree of E_K, where the degree of the coordinates of the cipher after several iterations can be upper bounded by exploiting the techniques introduced in [5,6]. However, it clearly appears that this type of distinguishers can be improved in the following two directions:

– it may happen that a given monomial x^u does not appear in the coordinates of E_K even if $wt(u) \leq \deg P$. This type of property, derived from the sparsity of some coordinates of the cipher, has been extensively used in cube attacks, e.g. [1,11,13].
– it may happen that a given monomial x^u appears in one coordinate of E_K but not in all functions $x \mapsto E_K^v(x)$. Then, $E_K(a + \mathsf{Prec}(u))$ does not fulfill the division property of order 2; instead we obtain a weaker distinguisher based on the fact that a given v does not belong to the parity set of $E_K(a + \mathsf{Prec}(u))$.

These two ways of exploiting some additional information besides the degree of the function are illustrated in the following toy example. In the rest of the paper, binary words are represented in hexadecimal notation where the least significant bit corresponds to the rightmost bit in the binary word.

Example 2. Let us consider the 4-bit Sbox S used in PRESENT [2]. This Sbox has degree 3 which is the maximal degree for a permutation of $\mathbf{F}_2^4$. The rows of Table 1 describes all sets

$$V_s(u) = \{v \in \mathbf{F}_2^n : S^v(x) \text{ contains } x^u\}.$$

In others words, the entry at Row u and Column v in this table is an x if and only if x^u appears in the ANF of $x \mapsto S^v(x)$. Equivalently, the column of index v in this table corresponds the list of all monomials in the ANF of $x \mapsto S^v(x)$.

Clearly, we cannot exhibit any u such that $S(a + \mathsf{Prec}(u))$ fulfills the division property of order 2 for all a (i.e. such that no coordinate of S contains x^u) if we exploit the degree of the Sbox only. Indeed, using that $\deg(S) = 3$, we deduce that the all-one vector $u = \mathtt{0xf}$ is the only value satisfying this property. This does not provide any distinguisher since this holds for all permutations.

However, distinguishers can be found by using that not all $u \in \mathbf{F}_2^4$ of weight less than or equal to 3 appear in the ANF of the coordinates of S. Indeed $u = \mathtt{0xe}$ does not appear in any of the coordinates of S, i.e., for $u = \mathtt{0xe}$, $S(a + \mathsf{Prec}(u))$ fulfills the division property of order 2 for all a. It is worth noticing that this is the only value of u which provides such a distinguisher.

Instead of searching for u such that the parity set of $S(a + \mathsf{Prec}(u))$ does not contain any word of weight 1, we can exhibit a few values v which belong to none of the sets $S(a + \mathsf{Prec}(u))$, $a \in \mathbf{F}_2^4$. For instance, we can observe that neither S^1 nor $S^{\mathtt{e}}$ contains a multiple of $x^3 = x_1 x_2$. This means that $v = \mathtt{0x1}$ and $v = \mathtt{0xe}$ do not belong to a set $S(a + \mathsf{Prec}(\mathtt{0x3}))$. Similarly, the sets $S(a + \mathsf{Prec}(\mathtt{0x9}))$ do not contain $v = \mathtt{0x1}$ and $v = \mathtt{0xb}$. The sets $S(a + \mathsf{Prec}(\mathtt{0xc}))$ do not contain $v = \mathtt{0x1}$ and $v = \mathtt{0x6}$.

Clearly, these two ideas allow us to decrease the dimension of the input subspace $X = a + \mathsf{Prec}(u)$ involved in the distinguisher. But the distinguisher based on the division property of order 2 is obviously stronger than the second one. A precise evaluation of the advantages of these distinguishers is provided in Appendix.

Table 1. Sets $V_S(u)$ for all $u \in \mathbf{F}_2^4$ for the PRESENT Sbox. All 4-bit words are represented in hexadecimal notation, and the rightmost bit of the word corresponds to the least significant bit.

	$V_S(u)$															
	0	1	2	4	8	3	5	9	6	a	c	7	b	d	e	f
0	x			x	x						x					
1		x		x		x					x					
2			x	x			x				x					
4		x	x				x				x					
8		x	x	x	x	x					x					
3			x			x	x	x	x	x	x		x			
5							x	x			x					
9			x			x	x		x	x					x	
6		x		x			x	x	x	x						
a		x	x			x	x		x			x	x	x	x	x
c		x		x		x					x					
7		x		x	x			x	x					x	x	
b		x	x	x	x			x	x	x		x		x		x
d		x	x	x			x		x			x			x	
e					x							x	x	x	x	x
f																x

5   Exhibiting Distinguishers on SPN by Means of Parity Sets

We now show how to find some distinguishers on iterated block ciphers, especially on substitution-permutation networks, by propagating some information on the parity set of the output set through the successive rounds of the cipher. As previously explained, we choose as input set an affine subspace.

5.1   Propagation Through Key Addition

One of the difficulties for finding a distinguisher for a block cipher is that the distinguishing property must hold for any value of the secret key. For this reason, we need to exploit a property which can be easily propagated through the operation inserting the round key, which is usually an XOR. This is the case of differential properties, or of the algebraic degree. We can show that the parity set can also be easily propagated.

Proposition 6. *Let X be a subset of $\mathbf{F}_2^n$ with parity set $\mathcal{U}(X)$. Then, for any $k \in \mathbf{F}_2^n$, the parity set of $(k + X)$ satisfies*

$$\mathcal{U}(k + X) \subseteq \bigcup_{u \in \mathcal{U}(X)} \mathsf{Succ}(u).$$

Proof. We use that

$$(x \oplus k)^v = \bigoplus_{u \preceq v} x^v k^{v \oplus u}.$$

It follows that

$$\bigoplus_{x \in X} (x \oplus k)^v = \bigoplus_{x \in X} \bigoplus_{u \preceq v} x^u k^{v \oplus u} = \bigoplus_{u \preceq v} k^{v \oplus u} \left(\bigoplus_{x \in X} x^u \right).$$

Then, this sum equals zero for all $k \in \mathbf{F}_2^n$ if $\bigoplus_{x \in X} x^u = 0$ for all u such that $u \preceq v$. In other words, if the sum equals one, then there exists u in $\mathcal{U}(X)$ such that $u \preceq v$, i.e., v satisfies $v \succeq u$ for at least one $u \in \mathcal{U}(X)$. $\square$

It is worth noticing that there is no general improvement of the previous result which holds without any further assumption on k or on X. Indeed, it is easy to check that, when $X = \{u\}$, we have that any $v \in \mathsf{Succ}(u)$ belongs to $\mathcal{U}(k + X)$ for some k (e.g. $k = (u + v)$ satisfies this property). Thus, $\mathsf{Succ}(u)$ is the smallest set which contains the parity sets of all cosets $(k + X)$ in this case.

5.2   Propagation Through an Sbox

We now investigate how a parity set propagates through a permutation, for instance through an Sbox or through a linear permutation.

Proposition 7. *Let S be a permutation of $\mathbf{F}_2^n$. For any $v \in \mathbf{F}_2^n$, we define*

$$V_S(u). = \{v \in \mathbf{F}_2^n : S^v(x) \text{ contains } x^u\}$$

Then, for any set X of elements of $\mathbf{F}_2^n$,

$$\mathcal{U}(S(X)) \subseteq \bigcup_{u \in \mathcal{U}(X)} V_S(u).$$

Proof. By definition, the vectors v which may be in $\mathcal{U}(S(X))$ are those such that $S(x)^v$ contains a monomial x^u with $u \in \mathcal{U}(X)$. Otherwise, we have that $\bigoplus_{x \in X} S(x)^v = 0$. The result then directly follows. $\square$

We will discuss in more details in Sect. 7 the properties of an Sbox which make it resistant or not to this attack. It is worth noticing that the previous proposition applies to any permutation S, including the case where S corresponds to the linear layer of the cipher.

Another case of interest is the case where the Sbox can be seen as the concatenation of several independent Sboxes, like in a typical Sbox layer.

Proposition 8. *Let X be a set of elements in $\mathbf{F}_2^{mt}$ and let S be an Sbox over $\mathbf{F}_2^{mt}$ which consists of the parallel application of t Sboxes $S_1, \ldots, S_t$ over $\mathbf{F}_2^m$: $S(x_1, \ldots, x_t) = (S_1(x_1), \ldots, S_t(x_t))$. Then,*

$$\mathcal{U}(S(X)) \subseteq \bigcup_{(u_1, \ldots, u_t) \in \mathcal{U}(X)} V_{S_1}(u_1) \times \ldots \times V_{S_t}(u_t)$$

where $V_{S_i}(u) = \{v \in \mathbf{F}_2^m : S_i^v(x) \text{ contains } x^u\}$.

Proof. From Proposition 7, we know that

$$\mathcal{U}(S(X)) \subseteq \bigcup_{(u_1, \ldots, u_t) \in \mathcal{U}(X)} V_S(u).$$

We then have to determine all $v = (v_1, \ldots, v_t) \in (\mathbf{F}_2^m)^t$ such that $S^v(x)$ contains $u = (u_1, \ldots, u_t)$. We use that

$$S^v(x) = S_1^{v_1}(x_1) S_2^{v_2}(x_2) \ldots S_t^{v_t}(x_t).$$

Since only $S_i^{v_i}(x_i)$ may contain $x_i^{u_i}$, we deduce that $v \in V_S(u)$ if and only if $v_i \in V_{S_i}(u_i)$ for each $1 \leq i \leq n$. Therefore, $V_S(u)$ is the Cartesian product of all $V_{S_i}(u_i)$. □

5.3   Propagation Through One Round

We now consider an SPN where the round key is inserted by addition at the end of the round. This implies that each Sbox layer comes after a round-key addition. Thus, if $\mathcal{U}(X)$ denotes the parity set of the input set X before the key addition, then the parity set after the key addition is included in a union of sets of the form $\mathsf{Succ}(u)$, for some $u \in \mathbf{F}_2^n$. It follows that the parity set after the Sbox layer satisfies

$$\mathcal{U}(S(X+k)) \subseteq \bigcup_{u \in \mathcal{U}(X)} \left(\bigcup_{v \in \mathsf{Succ}(u)} V_S(v) \right).$$

Therefore, propagating the information from $\mathcal{U}(X)$ to $\mathcal{U}(S(X+k))$ involves the sets

$$\mathcal{V}_S(u) = \bigcup_{v \in \mathsf{Succ}(u)} V_S(v)$$

which depend on the Sbox only.

These sets $\mathcal{V}_S(u)$ are then the relevant quantities involved in the propagation through the Sbox, instead of the sets $V_S(u), u \in \mathbf{F}_2^n$. For instance, Table 2 provides all sets $\mathcal{V}_S(u)$ for the PRESENT Sbox.

The table representing all $\mathcal{V}_S(u)$ has a few generic properties which hold for any bijective Sbox. The first obvious remark is that the all-zero vector does not belong to any $\mathcal{V}_S(u)$ except when $u = \underline{0}$. Indeed S^0 is the all-one function and then does not contain any monomial except x^0. The following property is much more interesting.

Table 2. Sets $\mathcal{V}_S(u)$ for all $u \in \mathbf{F}_2^4$ for the PRESENT Sbox. All 4-bit words are represented in hexadecimal notation, and the rightmost bit of the word corresponds to the least significant bit.

		$\mathcal{V}_S(u)$														
	0	1	2	4	8	3	5	9	6	a	c	7	b	d	e	f
0	x	x	x	x	x	x	x	x	x	x	x	x	x	x	x	x
1		x	x	x	x	x	x	x	x	x	x	x	x	x	x	x
2		x	x	x	x	x	x	x	x	x	x	x	x	x	x	x
4		x	x	x	x	x	x	x	x	x	x	x	x	x	x	x
8		x	x	x	x	x	x	x	x	x	x	x	x	x	x	x
3			x	x	x	x	x	x	x	x	x	x	x	x		x
5			x	x	x	x	x	x	x	x	x	x	x	x	x	x
9			x	x	x	x	x	x	x	x	x		x	x		x
6	x	x		x	x	x	x	x	x	x	x	x	x	x	x	x
a			x	x	x	x	x	x	x	x	x	x	x	x	x	x
c			x	x	x	x	x	x				x	x	x	x	x
7		x		x	x		x	x					x	x		x
b		x	x	x	x			x	x	x	x	x				x
d		x	x	x		x		x			x			x	x	x
e						x						x	x	x	x	x
f																x

Proposition 9. *Let S be any permutation of $\mathbf{F}_2^n$. Then,*

$$\mathcal{V}_S(\underline{1}) = \{\underline{1}\},$$

where $\underline{1}$ denotes the all-one vector in $\mathbf{F}_2^n$.

Proof. Since $\mathsf{Succ}(\underline{1}) = \underline{1}$, we have

$$\mathcal{V}_S(\underline{1}) = V_S(\underline{1}) = \{v \in \mathbf{F}_2^n : S^v(x) \text{ contains } x^{\underline{1}}\}$$

or equivalently $\mathcal{V}_S(\underline{1})$ is the set of all v such that $x \mapsto S^v(x)$ has degree n. It is known [6, Proposition 1] that $\deg(S^v) = n$ if and only if $v = \underline{1}$.   □

Some further properties of Table 2, specific to the PRESENT Sbox will be studied in Sect. 7.

6   Low-Data Distinguishers on a Few Rounds of PRESENT

6.1   Distinguisher on 3 Rounds

In [25] Todo presents generic distinguishers for ciphers based on the SPN construction. In particular, it is shown in Table 4 of this same paper that 3 rounds of an SPN construction whose nonlinear layer is composed of 16 Sboxes over $\mathbf{F}_2^4$ of degree 3 can be distinguished from a random permutation with data complexity 2^{12}. These results are therefore valid for PRESENT. This distinguisher improves upon the distinguishers exploiting the algebraic degree since the best upper bound on the degree of three rounds of such ciphers is $3^3 = 27$, leading to a distinguisher with data complexity 2^{28}.

Todo's distinguisher can be easily explained in the following way (see also [14]). Suppose that the input space X is composed of vectors taking all possible 2^4 values on the first three nibbles and where the last 13 nibbles are fixed to a same constant value for all vectors. In this case $|X| = 2^{12}$ and X can be seen now as a coset of V, i.e. $X = a + V$, where

$$V = \mathsf{Prec}(\mathtt{0x0000000000000fff}).$$

After the application of the round-key addition and the Sbox layer to X, the output space Y satisfies the same integral property as X, i.e. $Y = b + V$. Denote now by $F = \mathcal{R}^2 \circ P$, where $\mathcal{R}$ stands for the round function and where P denotes the linear layer. One can easily see that as F contains two non-linear layers, of degree 3 each, $\deg(F) \leq 9$. Therefore, as $\dim Y = 12 > 9$, we have that

$$\bigoplus_{x \in a+V} E_K(X) = \bigoplus_{y \in b+V} F(Y) = \bigoplus_{y \in V} F(b \oplus y) = D_V F(b) = 0.$$

Equivalently, this generic distinguisher on 3 rounds uses the feature that none of the coordinates of E_K contains a multiple of x^u for $u = \mathtt{0x0000000000000fff}$ (see Proposition 5). This distinguisher can therefore be very easily explained in terms of parity sets. Since $X = a + V$, where $V = \mathsf{Prec}(\mathtt{0x0000000000000fff})$, we have that $\mathcal{U}(V) = \{\mathtt{0x0000000000000fff}\}$ (from Corollary 1), implying that

$$\mathcal{U}(X) \subseteq \mathsf{Succ}(\mathtt{0x0000000000000fff}).$$

Since the Sbox S is a permutation, for each Sbox $\mathcal{V}_S(\mathtt{0xf}) = \{\mathtt{0xf}\}$ (Proposition 9) meaning that after the first Sbox layer the parity set U of the resulting set is again included in $\mathsf{Succ}(\mathtt{0x0000000000000fff})$. By defining the function F as before, we have that

$$\mathcal{U}(E_K(X)) \subseteq \bigcup_{u \in U} V_F(u).$$

But, $V_F(u) = \{v : F^v(x) \text{ contains } x^u\}$ contains no vector v with $wt(v) \leq 1$ when $wt(u) \geq 12$ since $\deg(F) \leq 9$. Therefore,

$$\mathcal{U}(E_K(X)) \subseteq \{v : wt(v) \geq 2\},$$

meaning that the output of the cipher restricted to 3 rounds has the balanced property (i.e., satisfies the division property of order 2).

6.2   Distinguisher on 4 Rounds

As explained in [25] in the case of AES-like ciphers, such generic distinguishers can be improved by exploiting the structure of the linear layer. In the particular case of PRESENT, the linear layer (see Fig. 1) is a bit permutation. Moreover, because of its structure, two rounds of PRESENT (without the last permutation layer) can be seen as the concatenation of four independent Superboxes operating on $\mathbf{F}_2^{16}$. With this structure, it is clear that any coordinate of the output at round $(r+1)$ of the cipher only contains monomials involving inputs from the same Superbox at round r.

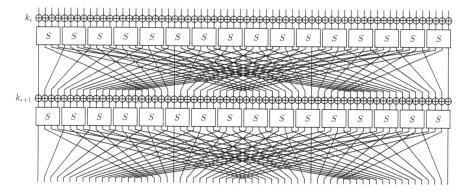

Fig. 1. Round function of the block cipher PRESENT

Therefore, by exploiting the linear layer of PRESENT one can extend the previous distinguisher to one more round as follows. Suppose now that the input set X has the form $X = a + V$, with

$$V = \mathsf{Prec}(\mathtt{0x000000000000fff0}).$$

In this case $\mathcal{U}(X) \subseteq \mathsf{Succ}(\mathtt{0x000000000000fff0})$. The parity set remains unchanged after the application of the first nonlinear layer. By applying now the permutation layer we see that the parity set of the resulting set is included in

$$\mathsf{Succ}(\mathtt{0x000e000e000e000e}),$$

leading to four active Superboxes. As previously explained, after the application of the Superboxes, so after the non-linear layer of the third round, any output coordinate only contains monomials coming from the same Superbox. Therefore, the resulting parity set is included in

$$\{u : wt(u) \geq 4\}.$$

By applying now the linear layer we have that the parity set is included in

$$\{u \text{ with } \geq 2 \text{ active nibbles}\} \cup \{\mathtt{0x00\ldots0f}, \ldots, \mathtt{0xf0\ldots0}\}.$$

This parity set is invariant under the application of the fourth nonlinear layer, meaning that
$$\mathcal{U}(E_K(X)) \subseteq \{v : wt(v) \geq 2\}.$$

We see therefore that the output set has the balanced property after 4 rounds.

However, it can be shown, that by only exploiting the properties of the linear layer of PRESENT, this distinguisher cannot always be extended to five rounds. The following table shows a possible propagation of values in the parity sets, where some output coordinate of the 12th Sbox after 5 rounds may contain the monomial x^u, with $u = $ 0x0000000000000fff0. By looking at rows 2 and 3 of this table, we can see that this propagation can be realised, among others, if the Sbox makes the propagations 0xe → 0x2 and 0xe → 0x1 possible. All the elements of this table should be interpreted as hexadecimal values.

input	0	0	0	0	0	0	0	0	0	0	0	0	f	f	f	0
output S-layer 1st round	0	0	0	0	0	0	0	0	0	0	0	0	f	f	f	0
output P-layer 1st round	0	0	0	e	0	0	0	e	0	0	0	e	0	0	0	e
output S-layer 2nd round	0	0	0	2	0	0	0	1	0	0	0	1	0	0	0	1
output P-layer 2nd round	0	0	0	0	0	0	0	0	1	0	0	0	0	1	1	1
output S-layer 3rd round	0	0	0	0	0	0	0	0	1	0	0	0	0	1	1	1
output P-layer 3rd round	0	0	0	0	0	0	0	0	0	0	0	0	0	0	8	7
output S-layer 4th round	0	0	0	0	0	0	0	0	0	0	0	0	0	0	2	8
output P-layer 4th round	0	0	0	3	0	0	0	0	0	0	0	0	0	0	0	0
output S-layer 5th round	0	0	0	1	0	0	0	0	0	0	0	0	0	0	0	0

However, by looking closer at Table 1, one can see that these two propagations are not possible for the PRESENT Sbox, as 0xe can only be propagated to values of Hamming weight of at least 2.

6.3 Distinguishers on 5 and 6 Rounds

The previous 4-round distinguisher on PRESENT exploited the algebraic degree of the Sbox together with the structure of the linear layer. We show here, that by further taking into account the particular form of the PRESENT Sbox, the previous distinguisher can be extended to one more round. For doing so, we consider the same input set X as for the 4-round distinguisher, i.e. $X = a + V$, with $V = \mathsf{Prec}(\text{0x000000000000fff0})$. As previously shown (see also the second row of the table above), after the application of the first round transformations, the parity set is included in $\mathsf{Succ}(\text{0x000e000e000e000e})$. However, the value 0xe presents, in the case of PRESENT, a particular interest. Indeed, notice that the row indexed by 0xe in Table 2 contains only a single element of weight 2 and no element of weight 1. This means that compared to other values u, $\mathcal{V}_S(u)$ for $u = $ 0xe contains exceptionally few elements, making more than half of the transitions impossible. In particular, the transitions, 0xe → 0x2 and 0xe → 0x1 in the above 5-round path are not possible.

We checked by computer programming that in this setting, there is no u of weight 1 in $\mathcal{U}(E_k(X))$ if E_k is 5-round PRESENT. Therefore, the output set after 5 rounds has the balanced property and can then be distinguished from a random permutation.

However, it is not possible to extend the distinguisher in its actual form to 6 rounds. Indeed, we checked that after 6 rounds, many elements of weight 1 can be found in $\mathcal{U}(E_k(X))$. Nevertheless, it is still possible to exhibit a weaker distinguisher for 6 rounds of PRESENT, by exploiting the fact that the column corresponding to the element 0x1 in Table 2 is very sparse, meaning that more than half of the transitions u → 0x1 are not possible. We were able to check that in fact, among all the elements of weight 1 present in $\mathcal{U}(E_k(X))$, only the nibble values 0x2, 0x4 and 0x8 were possible. Therefore, we have exhibited 16 values which do not belong to $\mathcal{U}(E_k(X))$ after six rounds of PRESENT, leading to a distinguisher with data complexity 2^{12} and advantage $(1 - 2^{-16})$ (see Appendix for the evaluation of the advantage).

6.4   Distinguishers Using More Data

We provide in Table 3 a summary of all the distinguishers obtained for reduced-round versions of PRESENT by using different sizes of the input space. These distinguishers were obtained by implementing the propagation of the parity set of the input subspace in a compact way. We mention here only strong distinguishers in the sense that we give only the results where the output set has the balanced property.

Table 3. Input sets leading to the division property of order 2 for reduced-round PRESENT.

Input set	$\log_2(\#\text{texts})$	Rounds
0x000000000000000f	4	4
0x000000000000fff0	12	5
0x00000000ffffffff	32	6
0xfffffffffffff000	52	7
0xfffffffffffffffe	63	8

6.5   Changing the Sbox

We discuss in this section the strength of the above distinguishers in the case where the Sbox of PRESENT is replaced by some other permutation of degree 3. For this, we consider exactly the same cipher but we change the nonlinear permutation. For instance, we consider the Sbox used in the block cipher PRINCE [3] and we study the propagation of the parity set of the resulting cipher. Table 4 provides the sets $\mathcal{V}_S(u)$ for all $u \in \mathbf{F}_2^4$ for this Sbox. However, it has to be noted that we do not propose in any case to replace the Sbox of PRESENT for obtaining a more robust design in general. This change is done here only for being able to run the same experiments by keeping exactly the same other parameters. Applying for example the same attack to the PRINCE cipher and compare the results would not make any sense as the results could be different because of the different linear layers for example, making the impact of the Sbox unclear. This paragraph aims only at demonstrating that PRESENT with some other Sbox can better resist against this type of distinguishers, but we do not argue that this would necessarily be the case for other type of attacks.

The following table is the equivalent of Table 2 for PRESENT. As one can see, this table, is much less sparse than Table 2. In particular, even the rows the more sparse (rows corresponding to 0xb and 0xd), make at least half of the transitions possible. One can further notice that all rows and columns contain elements of weight 1 and 2. All the above indicate that whatever the input u of a particular Sbox, the set $\mathcal{V}_S(u)$ contains a high number of values, making only very few transitions impossible.

Table 4. Sets $\mathcal{V}_S(u)$ for all $u \in \mathbf{F}_2^4$ for the PRINCE Sbox. All 4-bit words are represented in hexadecimal notation.

	0	1	2	4	8	3	5	9	6	a	c	7	b	d	e	f
0	x	x	x	x	x	x	x	x	x	x	x	x	x	x	x	x
1		x	x	x	x	x	x	x	x	x	x	x		x	x	x
2		x	x	x	x	x	x	x	x	x	x	x	x	x	x	x
4		x	x	x	x	x	x	x	x	x	x	x	x	x	x	x
8		x	x	x	x	x	x	x	x	x			x	x	x	x
3		x	x	x	x	x	x	x	x	x	x	x		x	x	x
5		x	x		x	x	x	x	x	x	x	x		x	x	x
9		x		x	x	x	x		x		x			x	x	x
6		x	x	x	x	x	x	x	x	x		x	x	x	x	x
a			x	x	x		x	x	x	x	x		x	x	x	x
c		x	x	x	x	x	x	x		x	x		x	x	x	x
7		x	x		x	x	x	x	x	x		x		x	x	x
b			x	x				x		x			x	x	x	x
d				x	x	x				x			x		x	
e		x	x			x	x		x			x	x	x	x	
f															x	

$\mathcal{V}_S(u)$

Table 5. Example of a 5-round path that is satisfied when the PRINCE Sbox is plugged into the PRESENT block cipher. All entries should be interpreted as hexadecimal values.

input	0	0	0	0	0	0	0	0	0	0	0	0	f	f	f	0
output S-layer 1st round	0	0	0	0	0	0	0	0	0	0	0	0	f	f	f	0
output P-layer 1st round	0	0	0	e	0	0	0	e	0	0	0	e	0	0	0	e
output S-layer 2nd round	0	0	0	4	0	0	0	2	0	0	0	2	0	0	0	2
output P-layer 2nd round	0	0	0	0	1	0	0	0	0	1	1	1	0	0	0	0
output S-layer 3rd round	0	0	0	0	1	0	0	0	0	1	1	1	0	0	0	0
output P-layer 3rd round	0	0	0	0	0	0	0	0	0	0	0	0	0	8	7	0
output S-layer 4th round	0	0	0	0	0	0	0	0	0	0	0	0	0	1	1	0
output P-layer 4th round	0	0	0	0	0	0	0	0	0	0	0	0	0	0	0	6
output S-layer 5th round	0	0	0	0	0	0	0	0	0	0	0	0	0	0	0	1

We were able to verify, that if we start with the same input space as before, the output set after five rounds does not satisfy the division property of order 2, while this was the case for the PRESENT Sbox. Indeed, Table 5 provides a path that is satisfied when this Sbox is used. As one can see, we end up with a vector of weight 1 in $\mathcal{U}(E_k)$. This was not the case with the original PRESENT Sbox, proving that the PRINCE Sbox is more resistant against this kind of property. A more detailed study of why this is so is provided in the following section.

7   Related Security Criterion for the Sbox

As illustrated by the previous attack, one of the main ingredients of the distinguisher is the particular form of the sets $\mathcal{V}_S(u)$ for the PRESENT Sbox, i.e., the particular form of Table 2. Two properties of this table are exploited:

1. Column of index 0x1 is very sparse and in particular it does not contain any element of weight 3 and it contains a single element of weight 2. This property is exploited at the end of the attack on 6 rounds, where we use that 0x1 only belongs to a very few sets $\mathcal{V}_S(u)$, i.e., only a few transitions $u \rightarrow$ 0x1 are possible.
2. Row of index 0xe is very sparse and in particular it does not contain any element of weight 1 and it contains a single element of weight 2. This property is exploited when we use that $\mathcal{V}_S(0xe)$ contains a few elements only, implying that many transitions 0xe $\rightarrow v$ are impossible.

We now show where these unsuitable properties of the Sbox come from, and how they can be avoided. The first property has an obvious algebraic interpretation since each column in the table is derived from the ANF of a Boolean function $x \mapsto S^v(x)$. In particular, the columns having an index of weight 1 are derived from the ANF of the coordinates of the Sbox. The ANF of the PRESENT Sbox is

$$S_1(x_1, x_2, x_3, x_4) = x_1 + x_3 + x_4 + x_2 x_3$$
$$S_2(x_1, x_2, x_3, x_4) = x_2 + x_4 + x_2 x_4 + x_3 x_4 + x_1 x_2 x_3 + x_1 x_2 x_4 + x_1 x_3 x_4$$
$$S_3(x_1, x_2, x_3, x_4) = 1 + x_3 + x_4 + x_1 x_2 + x_1 x_4 + x_2 x_4 + x_1 x_2 x_4 + x_1 x_3 x_4$$
$$S_4(x_1, x_2, x_3, x_4) = 1 + x_1 + x_2 + x_4 + x_2 x_3 + x_1 x_2 x_3 + x_1 x_2 x_4 + x_1 x_3 x_4.$$

The first weakness of the PRESENT Sbox then comes from the fact that its first coordinate has degree 2 only and contains a single quadratic term.

The second weakness is related to the fact that the monomial $x^e = x_2 x_3 x_4$ appears neither in the coordinates of S, nor in most functions $S_i S_j$. This second property can be deduced in a much simpler way from the ANF of the inverse Sbox, as shown by the following proposition.

Lemma 3. *Let S be a permutation of $\mathbf{F}_2^n$. Then, for any $u, v \in \mathbf{F}_2^n$, the ANF of $x \mapsto S(x)^v$ contains x^u if and only if the ANF of $x \mapsto S^*(x)^{\overline{u}}$ contains $x^{\overline{v}}$, where $\overline{u}$ denotes the vector $u \oplus \mathbf{1}$ and S^* is the permutation $x \mapsto \overline{S^{-1}(\overline{x})}$.*

Proof. Let a_u denote the coefficient of x^u in the ANF of $x \mapsto S(x)^v$. Then

$$a_u = \bigoplus_{x \preceq u} S(x)^v$$
$$= |\{x \in \mathbf{F}_2^n : x_i = 0, i \in \mathsf{Supp}(\overline{u}) \text{ and } S(x)_j = 1, j \in \mathsf{Supp}(v)\}| \bmod 2$$
$$= |\{y \in \mathbf{F}_2^n : S^{-1}(y)_i = 0, i \in \mathsf{Supp}(\overline{u}) \text{ and } y_j = 1, j \in \mathsf{Supp}(v)\}| \bmod 2$$

where the last equality comes from the fact that S is a permutation, implying that there is a one-to-one correspondence between x and $y = S(x)$. We now replace y by $z = \overline{y}$ and use that $S^{-1}(y) = \overline{S^*(\overline{y})}$. Then,

$$a_u = |\{z \in \mathbf{F}_2^n : S^{-1}(\overline{z})_i = 0, i \in \mathsf{Supp}(\overline{u}) \text{ and } z_j = 0, j \in \mathsf{Supp}(v)\}| \bmod 2$$
$$= |\{z \in \mathbf{F}_2^n : S^*(z)_i = 1, i \in \mathsf{Supp}(\overline{u}) \text{ and } z_j = 0, j \in \mathsf{Supp}(v)\}| \bmod 2$$
$$= \bigoplus_{z \preceq \overline{v}} (S^*(z))^{\overline{u}}$$

which means that a_u is the coefficient of $x^{\overline{v}}$ in the ANF of $x \mapsto S^*(x)^{\overline{u}}$.   □

The function S^* corresponding to the PRESENT Sbox has the following ANF:

$$S_1^*(x_1, x_2, x_3, x_4) = 1 + x_1 + x_2 + x_3 + x_4 + x_2 x_4$$
$$S_2^*(x_1, x_2, x_3, x_4) = x_1 + x_4 + x_1 x_3 + x_2 x_3 + x_1 x_2 x_3 + x_1 x_2 x_4 + x_1 x_3 x_4$$
$$S_3^*(x_1, x_2, x_3, x_4) = 1 + x_2 + x_4 + x_1 x_2 + x_1 x_3 + x_1 x_4 + x_3 x_4 + x_1 x_2 x_3$$
$$+ x_1 x_2 x_4 + x_1 x_3 x_4$$
$$S_4^*(x_1, x_2, x_3, x_4) = x_2 + x_3 + x_2 x_3 + x_1 x_4 + x_3 x_4 + x_1 x_2 x_3 + x_1 x_3 x_4$$

These ANF correspond to the rows with an index of weight 3 in Table 2. In particular, the form of the row of index 0xe (defining $\mathcal{V}_S(\texttt{0xe})$) comes from the fact that the first coordinate of S^* has degree 2 only, and that its ANF contains a single monomial of degree 2.

Conversely, we can easily guarantee that, when an n-bit bijective Sbox is considered, all sets $\mathcal{V}_S(u)$ with $wt(u) < n$ contain an element of weight 1.

Proposition 10. *Let S be a permutation of $\mathbf{F}_2^n$. Then, all coordinates of S^{-1} have degree $(n-1)$ if and only if all sets $\mathcal{V}_S(u)$, with $u \neq \underline{1}$, contain at least one element of weight 1.*

Proof. Since the sets $\mathcal{V}_S(u)$ include all sets $\mathcal{V}_S(u')$ for $u' \succeq u$, we only have to prove the result for the sets $\mathcal{V}_S(u)$ with $wt(u) = n - 1$. When $wt(u) = n - 1$, we have

$$\mathcal{V}_S(u) = V_S(u) \cup V_S(\underline{1}) = V_S(u) \cup \{\underline{1}\}$$

from Proposition 9. Therefore, all sets $\mathcal{V}_S(u)$ with $wt(u) < n$ contain an element of weight 1 if and only if all sets $V_S(u)$ with $wt(u) = n-1$ contain an element of weight 1. This equivalently means that all monomials of degree $(n-1)$ appear in the ANF of the coordinates of S, or from Lemma 3 that all coordinates of S^* have degree $(n-1)$. But, this last condition equivalently means that all coordinates of S^{-1} have degree $(n-1)$. Indeed, the monomials of highest degree in the ANF of a Boolean function f and in the ANF of $f^* : x \mapsto f(\overline{x}) + 1$ are the same. Then the i-th coordinates of S^* and S^{-1} have the same degree.   □

It is worth noticing that the fact that all coordinates of S^{-1} have maximal degree is not equivalent to the fact that the same property holds for S, as shown in the following example.

Example 3. We consider the permutation of $\mathbf{F}_2^4$ corresponding to the inverse of the function G_{10} defined in the classification in [18, Table 6]. This Sbox has optimal cryptographic properties in the sense that it has the smallest possible differential uniformity and highest nonlinearity. Its coordinates are given by

$$S_1(x_1, x_2, x_3, x_4) = x_1 + x_1 x_2 + x_2 x_3 + x_3 x_4 + x_1 x_3 x_4 + x_2 x_3 x_4$$
$$S_2(x_1, x_2, x_3, x_4) = x_2 + x_1 x_3 + x_2 x_4 + x_3 x_4 + x_1 x_2 x_4 + x_1 x_3 x_4 + x_2 x_3 x_4$$
$$S_3(x_1, x_2, x_3, x_4) = x_3 + x_1 x_2 + x_1 x_3 + x_1 x_4 + x_2 x_4 + x_1 x_2 x_4 + x_1 x_3 x_4$$
$$S_4(x_1, x_2, x_3, x_4) = x_4 + x_1 x_2 + x_1 x_3 + x_1 x_2 x_4.$$

It can then be checked that all coordinates of S have degree 3 while the monomial $x_1 x_2 x_3$ appears in none of the coordinates, implying that $\mathcal{V}_S(\mathtt{0x7})$ contains no element of weight 1. Indeed,

$$\mathcal{V}_S(\mathtt{0x7}) = \{\mathtt{0x9}, \mathtt{0xc}, \mathtt{0x7}, \mathtt{0xb}, \mathtt{0xe}, \mathtt{0xf}\}.$$

Actually, S does not satisfy the hypotheses of Proposition 10 since the first coordinate of S^{-1} has degree 2 only.

It is also worth noticing that the condition on the degree of the coordinates of S^{-1} is not invariant under composition (to the right or left) by an affine permutation.

A simple way to guarantee that all coordinates of the inverse Sbox have maximal degree consists in choosing for S an Sbox such that any linear combination of its coordinates (i.e. any of its components) has maximal degree. In this case, we obtain the following stronger result on the number of elements of weight 1 in $\mathcal{V}_S(u)$.

Corollary 3. *Let S be a permutation of $\mathbf{F}_2^n$ such that the Boolean functions $x \mapsto \lambda \cdot S(x)$ have degree $(n-1)$ for all nonzero $\lambda \in \mathbf{F}_2^n$. Then, for any $u \in \mathbf{F}_2^n$, $\mathcal{V}_S(u)$ contains at least $(n - wt(u))$ elements of weight 1.*

Proof. We will first prove the result for all u with $wt(u) = n - 1$ by showing that all coordinates of S^{-1} have degree $(n-1)$. Let A denote the $n \times n$ binary matrix such that $a_{i,j}$ is the coefficient of the monomial of degree $(n-1)$ $x^{\overline{e_j}}$ in the ANF of the i-th coordinate of S where e_j denotes the n-bit word having a single one at position j. A component of S, $x \mapsto \lambda \cdot S(x)$, $\lambda \neq 0$, has degree less than $(n-1)$ if and only if the corresponding linear combination of the rows of A vanishes, *i.e.*, $\lambda A = 0$. It follows that the number of non-trivial components of S with degree $(n-1)$ is equal to $2^{\mathrm{rk}(A)} - 1$. From Lemma 3, the coefficients of the monomials of degree $(n-1)$ in the coordinates of S^{-1} are defined by the transpose of A. Then, the number of non-trivial components of S^{-1} with degree $(n-1)$ is equal to $2^{\mathrm{rk}(A^T)} - 1$. We deduce from Proposition 10 that, if all components of S have degree $(n-1)$ (i.e., if $\mathrm{rk}(A) = \mathrm{rk}(A^T) = n$), all $\mathcal{V}_S(u)$ for $u \neq \mathbf{1}$ contain at least one element of weight 1.

Let us now consider any $u \in \mathbf{F}_2^n$. By definition, $\mathcal{V}_S(u)$ contains all sets $\mathcal{V}_S(\overline{e_i})$ with $i \notin \mathsf{Supp}(u)$ since the $k = (n - wt(u))$ words $\overline{e_i}$ of weight $(n-1)$ belong to $\mathsf{Succ}(u)$. As Matrix A has full rank, the k columns of A corresponding to the monomials $x^{\overline{e_i}}$ with $i \notin \mathsf{Supp}(u)$ have rank k, implying that this $k \times n$-submatrix has at least k nonzero rows. These rows correspond to the words of weight 1 which belong to $\mathcal{V}_S(u)$, implying that this set contains at least k vectors of weight 1.   □

Example 4. The 4-bit Sbox used in PRINCE [3] (as well as all Sboxes with similar properties recommended for any cipher within the PRINCE-family [4, Appendix B]) has been chosen in such a way that all its nontrivial components have degree 3. Then, we can guarantee that any set $\mathcal{V}_S(u)$ contains at least $(4 - wt(u))$ elements of weight 1. This can be checked on Table 4.

It is worth noticing that the condition exhibited in Corollary 3 offers a similar guarantee for the inverse Sbox. Indeed, making the decryption function also immune to this type of attacks may be relevant, even if mounting the attack on the decryption function is much more difficult in practice because it requires the knowledge of plaintext-ciphertext pairs corresponding to chosen ciphertexts.

Application to the MISTY *Sboxes.* So far, the most important application of the division property is the cryptanalysis of the full MISTY1 [24]. It is then relevant to study the MISTY Sboxes in the light of the previous criteria. MISTY1 is an unbalanced Feistel network. It then uses two different Sboxes, which are both

Table 6. Minimum Hamming weight of $\mathcal{V}_{S_7}(u)$ depending on the Hamming weight of u for the 7-bit Sbox in MISTY1.

$wt(u)$	1	2	3	4	5	6	7
$\min\{wt(v) : v \in \mathcal{V}_{S_7}(u)\}$	1	1	1 or 2	2	2	4	7

Table 7. Minimum Hamming weight of $\mathcal{V}_{S_9}(u)$ depending on the Hamming weight of u for the 9-bit Sbox in MISTY1.

$wt(u)$	1	2	3	4	5	6	7	8	9
$\min\{wt(v) : v \in \mathcal{V}_{S_9}(u)\}$	1	1	2	2	3	3	4	4	9

linearly equivalent to a power permutation. More precisely, S_7 is a permutation of degree 3 of $\mathbf{F}_2^7$ and S_9 is a permutation of degree 2 of $\mathbf{F}_2^9$.

We have then computed all sets $\mathcal{V}_S(u)$ for each of these Sboxes. Most notably, Tables 6 and 7 give the values of the minimal Hamming weight of an element in $\mathcal{V}_S(u)$ depending on the Hamming weight of u for the 7-bit Sbox S_7 and for the 9-bit Sbox S_9 respectively. These tables then recover the results on the propagation of the division property described in [24, p. 420]. The fact that, for both Sboxes, many sets $\mathcal{V}_S(u)$ have a large minimal weight show that the two MISTY Sboxes are weak regarding the division property. An interesting new observation is that, for S_7, for some vectors u of weight 3, for instance $u = \texttt{0x0b}$, $\mathcal{V}_{S_7}(u)$ does not contain any vector of weight 1. This equivalently means that some monomials of degree 3 do not appear in any of the coordinates of S_7. This property of S_7, which is more precise that the propagation of the division property studied in [24], may then be exploited in an attack.

8   Conclusions

In some contexts, the notion of parity set provides a powerful tool for representing subsets. We have shown for instance that the division property has a simple formulation in terms of parity sets, which allows to easily deduce some properties of the sets satisfying the division property of a given order. Also, focusing on the parity set, and not only on the minimal weight of its elements, enables the attacker to capture some algebraic properties of the nonlinear functions used in the cipher, besides the algebraic degree. This general view also brings to light the properties of the Sbox which avoid this type of attacks. The counterpart is that computing the whole parity set after many rounds of a cipher is obviously more expensive than considering its minimal weight only, as this is done in the division property. However, a promising technique consists in combining both approaches where the first and last rounds are analysed with the whole parity set, while the propagation through the middle rounds only exploits the degree of the function. Another direction could be to use parity sets for identifying some sets of weak keys. We have focused on distinguishers which hold for all keys.

But, the addition of an unknown key increases the size of the parity set, since all words greater than or equal to the words in the input parity set must be considered. A different approach then consists in considering only round keys of a particular form. Then, the resulting parity set after key addition may simplify a lot, and these particular round keys may then be easily detected.

Acknowledgements. We thank the anonymous reviewers for their helpful comments. We also thank Willi Meier and Qingju Wang for insightful discussions.

A   Advantages of the Distinguishers Based on Parity Sets

In order to estimate the advantages of the distinguishers exhibited in this paper, we need to evaluate the probability that, given an input set X, a randomly chosen permutation π is such that $\pi(X)$ does not satisfy the division property of order 2. For the weaker distinguisher, we similarly need to evaluate the probability that a given u does not belong to $\mathcal{U}(\pi(X))$. Clearly, the probability that $\pi(X)$ satisfies the division property of order 2 (i.e., is balanced) is close to 2^{-n}, while the probability that a given u does not belong $\mathcal{U}(\pi(X))$ is close to $1/2$. However, these probabilities may vary with the size of X: for instance, if u is the all-zero word, we have that $u \in \mathcal{U}(\pi(X))$ if and only if $|X|$ is odd. Also, if $|X|$ is odd, $\pi(X)$ cannot satisfy the division property of order 2. A more careful analysis seems therefore needed. The exact values of these probabilities confirming these estimates are then given by the following propositions.

Proposition 11. *Let $u \in \mathbf{F}_2^n$. The probability over all sets $X \subseteq \mathbf{F}_2^n$ of size k that u does not belong to $\mathcal{U}(X)$ is equal to*

$$\frac{1}{2}\left(1 + \frac{P_k(2^{n-wt(u)})}{\binom{2^n}{k}}\right),$$

where $P_k(w)$ is the Krawtchouk polynomial

$$P_k(x) = \sum_{i=0}^{k}(-1)^i\binom{x}{i}\binom{2^n - x}{k - i}.$$

In particular, if $wt(u) = 1$, this probability equals

$$\begin{cases} \frac{1}{2} & \text{if } k \text{ is odd} \\ \frac{1}{2}\left(1 + (-1)^{k/2}\frac{\binom{2^{n-1}}{k/2}}{\binom{2^n}{k}}\right) \simeq \frac{1}{2} & \text{if } k \text{ is even.} \end{cases}$$

Proof. From Lemma 2, $u \notin \mathcal{U}(X)$ if and only if the product between the row G_u of index u in Matrix G and the incidence vector of X vanishes. The row G_u is a word of length 2^n and weight $2^{n-wt(u)}$ since $G_{u,v} = 1$ if and only if $u \preceq v$. Then we need to count the number of vectors v_X of weight k such that the scalar product $G_u \cdot v_X = 0$. It is known [10, Theorem 4.1] that, for any vector g of length $N = 2^n$,

$$\sum_{v \in \mathbf{F}_2^N, wt(v)=k} (-1)^{g \cdot v} = P_k(wt(g)).$$

Then,

$$\Pr_{X, |X|=k} [u \notin \mathcal{U}(X)] = \frac{1}{2} \left(1 + \frac{P_k(2^{n-wt(u)})}{\binom{2^n}{k}} \right).$$

In the special case where $wt(u) = 1$, we need to estimate the value of $P_k(2^{n-1})$. The generating function of the Krawtchouk polynomials is [23]

$$(1+z)^{N-i}(1-z)^i = \sum_{\ell=0}^{N} P_\ell(i) z^\ell.$$

We deduce that, for $i = N/2$, this generating function is $(1-z^2)^{N/2}$. It contains monomials of even degree only, and then

$$P_{2\ell}(N/2) = (-1)^\ell \binom{N/2}{\ell},$$

implying that

$$P_k(2^{n-1}) = \begin{cases} 0 & \text{if } k \text{ is odd} \\ (-1)^{k/2} \binom{2^{n-1}}{k/2} & \text{if } k \text{ is even.} \end{cases}$$

The probability that u does not belong to $\mathcal{U}(X)$ is then is very close to $1/2$ in all cases. Indeed, the ratio of the two binomial coefficients satisfies

$$\frac{\binom{2^{n-1}}{k/2}}{\binom{2^n}{k}} = \Theta \left(2^{-2^{n-1} H_2 \left(\frac{k}{2^n} \right)} \right)$$

where H_2 is the binary entropy, $H_2(x) = -x \log_2(x) - (1-x) \log_2(1-x)$.   □

Similarly, the probability that a set X of given size k satisfies the division property of order d is determined by the number of codewords of weight k in the Reed-Muller code of length 2^n and order $(n-d)$. There is no known formula for this number in general, but it can be computed when $d = 2$, which is the case corresponding to Todo's distinguishers.

Proposition 12. *The probability that a set $X \subseteq \mathbf{F}_2^n$ of size k satisfies the division property of order 2 is 0 if k is odd and*

$$2^{-n} + (-1)^{k/2} (1 - 2^{-n}) \frac{\binom{2^{n-1}}{k/2}}{\binom{2^n}{k}} \simeq 2^{-n},$$

if k is even.

Proof. The result comes from the weight distribution of the Reed-Muller code $\mathcal{R}(n-2, n)$. Since this code is the dual of $\mathcal{R}(1, n)$, its weight distribution can be derived from the weight distribution of the dual code by the MacWilliams transform [19, p. 129]: the number of codewords of weight k in $\mathcal{R}(n-2, n)$ is

$$A_k = 2^{-(n+1)} \left(P_k(0) + (2^{n+1} - 2)P_k(2^{n-1}) + P_k(2^n) \right),$$

where the $P_k(i)$ are the previously defined Krawtchouk polynomials. From the generating function, we get that $P_k(0) = \binom{2^n}{k}$ and $P_k(2^n) = (-1)^k \binom{2^n}{k}$, and the value of $P_k(2^{n-1})$ has been computed in the proof of the previous proposition. The result then directly follows. □

References

1. Aumasson, J.-P., Dinur, I., Meier, W., Shamir, A.: Cube testers and key recovery attacks on reduced-round MD6 and trivium. In: Dunkelman, O. (ed.) FSE 2009. LNCS, vol. 5665, pp. 1–22. Springer, Heidelberg (2009)
2. Bogdanov, A.A., Knudsen, L.R., Leander, G., Paar, C., Poschmann, A., Robshaw, M., Seurin, Y., Vikkelsoe, C.: PRESENT: an ultra-lightweight block cipher. In: Paillier, P., Verbauwhede, I. (eds.) CHES 2007. LNCS, vol. 4727, pp. 450–466. Springer, Heidelberg (2007)
3. Borghoff, J., et al.: PRINCE – a low-latency block cipher for pervasive computing applications. In: Wang, X., Sako, K. (eds.) ASIACRYPT 2012. LNCS, vol. 7658, pp. 208–225. Springer, Heidelberg (2012)
4. Borghoff, J., Canteaut, A., Güneysu, T., Kavun, E.B., Knežević, M., Knudsen, L.R., Leander, G., Nikov, V., Paar, C., Rechberger, C., Rombouts, P., Thomsen, S.S., Yalçin, T.: PRINCE - a low-latency block cipher for pervasive computing applications (full version). Cryptology ePrint Archive, Report 2012/529 (2012). http://eprint.iacr.org/2012/529
5. Boura, C., Canteaut, A.: On the Influence of the algebraic degree of F^{-1} on the algebraic degree of $G \circ F$. IEEE Trans. Inf. Theor. **59**(1), 691–702 (2013). http://hal.inria.fr/hal-00738398
6. Boura, C., Canteaut, A., De Cannière, C.: Higher-order differential properties of KECCAK and *Luffa*. In: Joux, A. (ed.) FSE 2011. LNCS, vol. 6733, pp. 252–269. Springer, Heidelberg (2011)
7. Canteaut, A., Carlet, C., Charpin, P., Fontaine, C.: On cryptographic properties of the cosets of $R(1, m)$. IEEE Trans. Inf. Theor. **47**(4), 1494–1513 (2001)
8. Canteaut, A., Videau, M.: Degree of composition of highly nonlinear functions and applications to higher order differential cryptanalysis. In: Knudsen, L.R. (ed.) EUROCRYPT 2002. LNCS, vol. 2332, pp. 518–533. Springer, Heidelberg (2002)
9. Daemen, J., Knudsen, L.R., Rijmen, V.: The block cipher SQUARE. In: Biham, E. (ed.) FSE 1997. LNCS, vol. 1267, pp. 149–165. Springer, Heidelberg (1997)
10. Delsarte, P.: An algebraic approach to the association schemes of coding theory. Ph.D. thesis, Université catholique de Louvain, Belgium (1973)
11. Dinur, I., Liu, Y., Meier, W., Wang, Q.: Optimized interpolation attacks on LowMC. Cryptology ePrint Archive, Report 2015/418 (2015). http://eprint.iacr.org/2015/418

12. Dinur, I., Shamir, A.: Cube attacks on tweakable black box polynomials. In: Joux, A. (ed.) EUROCRYPT 2009. LNCS, vol. 5479, pp. 278–299. Springer, Heidelberg (2009)
13. Dinur, I., Shamir, A.: Breaking Grain-128 with dynamic cube attacks. In: Joux, A. (ed.) FSE 2011. LNCS, vol. 6733, pp. 167–187. Springer, Heidelberg (2011)
14. Khovratovich, D.: Private Communication (2016)
15. Knudsen, L.R.: Truncated and higher order differentials. In: Preneel, B. (ed.) FSE 1994. LNCS, vol. 1008, pp. 196–211. Springer, Heidelberg (1995)
16. Knudsen, L.R., Wagner, D.: Integral cryptanalysis. In: Daemen, J., Rijmen, V. (eds.) FSE 2002. LNCS, vol. 2365, pp. 112–127. Springer, Heidelberg (2002)
17. Lai, X.: Higher order derivatives and differential cryptanalysis. In: Proceedings of Symposium on Communication, Coding and Cryptography, in Honor of Massey, J.L. on the Occasion of his 60'th Birthday. Kluwer Academic Publishers (1994)
18. Leander, G., Poschmann, A.: On the classification of 4 bit S-Boxes. In: Carlet, C., Sunar, B. (eds.) WAIFI 2007. LNCS, vol. 4547, pp. 159–176. Springer, Heidelberg (2007)
19. MacWilliams, F.J., Sloane, N.J.: The Theory of Error-Correcting Codes. North-Holland, Secaucus (1977)
20. Muller, D.E.: Application of Boolean algebra to switching circuit design and to error detection. IEEE Trans. Comput. **3**, 6–12 (1954)
21. Reed, I.S.: A class of multiple-error-correcting codes and the decoding scheme. IEEE Trans. Inf. Theor. **4**, 38–49 (1954)
22. Sun, B., Hai, X., Zhang, W., Cheng, L., Yang, Z.: New observation on division property. Cryptology ePrint Archive, Report 2015/459 (2015). http://eprint.iacr.org/2015/459
23. Szegö, G.: Orthogonal Polynomials. American Mathematical Society Colloquium Publications, New York (1959)
24. Todo, Y.: Integral cryptanalysis on full MISTY1. In: Gennaro, R., Robshaw, M.J.B. (eds.) CRYPTO 2015, Part I. LNCS, vol. 9215, pp. 413–432. Springer, Heidelberg (2015)
25. Todo, Y.: Structural evaluation by generalized integral property. In: Oswald, E., Fischlin, M. (eds.) EUROCRYPT 2015. LNCS, vol. 9056, pp. 287–314. Springer, Heidelberg (2015)

Author Index

Printed in the United States
By Bookmasters